Physics for Scientists and Engineers: Foundations and Connections

with Modern Physics
Volume 2

To Zak and Jeff
with love.

Physics for Scientists and Engineers: Foundations and Connections
with Modern Physics
Volume 2

Debora M. Katz

Australia • Brazil • Mexico • Singapore • United Kingdom • United States

Physics for Scientists and Engineers: Foundations and Connections with Modern Physics, Volume 2
Debora M. Katz

Product Director: Mary Finch
Product Manager: Rebecca Berardy Schwartz
Managing Developer: Peter McGahey
Senior Content Developer: Susan Dust Pashos
Content Developer: Ed Dodd
Product Assistant: Margaret O'Neill
Senior Marketing Manager: Janet del Mundo
Senior Content Project Manager: Alison Eigel Zade
Art Director and Executive Director of Design: Bruce Bond
Manufacturing Planner: Beverly Breslin
IP Project Manager: Farah J. Fard
IP Analyst: Christine Myaskovsky
Production Service and Compositor:
 Cenveo® Publisher Services
Photo Researchers: Carly Bergey (Chapters 23–39) and
 Pandisathya Paul (Chapters 40–43)
Cover Designer: Bruce Bond
Cover Image: United Launch Alliance

© 2017 Cengage Learning
WCN: 01-100-101

ALL RIGHTS RESERVED. No part of this work covered by the copyright herein may be reproduced, transmitted, stored, or used in any form or by any means graphic, electronic, or mechanical, including but not limited to photocopying, recording, scanning, digitizing, taping, Web distribution, information networks, or information storage and retrieval systems, except as permitted under Section 107 or 108 of the 1976 United States Copyright Act, without the prior written permission of the publisher.

> For product information and technology assistance, contact us at
> **Cengage Learning Customer & Sales Support, 1-800-354-9706.**
> For permission to use material from this text or product, submit all
> requests online at **www.cengage.com/permissions**.
> Further permissions questions can be e-mailed to
> **permissionrequest@cengage.com**.

Library of Congress Control Number: 2015949144

Package ISBN: 978-1-305-95597-4

Book-only ISBN: 978-1-305-95608-7

Cengage Learning
20 Channel Center Street
Boston, MA 02210
USA

Cengage Learning is a leading provider of customized learning solutions with employees residing in nearly 40 different countries and sales in more than 125 countries around the world. Find your local representative at **www.cengage.com.**

Cengage Learning products are represented in Canada by Nelson Education, Ltd.

To learn more about Cengage Learning Solutions, visit **www.cengage.com.**

Purchase any of our products at your local college store or at our preferred online store **www.cengagebrain.com.**

Printed in the United States of America
Print Number: 01   Print Year: 2015

About the Author

Debora Katz is on the physics faculty at the United States Naval Academy, where she teaches calculus-based introductory physics. Soon after beginning her job at the Academy, she co-authored a book, *The Physics Toolbox*, meant to help students struggling in introductory physics. She was born in Chicago, Illinois and grew up in a suburb of Chicago as well as in Hawaii. She went to Brandeis University in Waltham, Massachusetts, where she earned a Bachelor of Arts degree in physics. She then went to the University of California in Irvine, California, where she earned a Master of Science degree in physics. Finally, she attended the University of Minnesota, earning a doctorate in astrophysics. After her PhD, she immediately began teaching in the physics department at the United States Naval Academy. She lives in Annapolis with her husband Jeff, her son Zak, and their dog Simon.

Contents

About the Author v
Preface for the Instructor xii
Acknowledgments xx
Preface for the Students xxiii

Part III Electricity

Chapter 23 Electric Forces 683
23-1 Another Fundamental Force 684
23-2 Models of Electrical Phenomena 685
   Charge 687
   Sketching Charge 687
23-3 A Qualitative Look at the Electrostatic Force 688
23-4 Insulators and Conductors 690
   Microscopic Model 691
   Charging by Direct Contact 691
   Charging by Induction 692
   Polarization and Induced Dipoles 693
   Sparks 693
23-5 Coulomb's Law 696
   Coulomb's Experiments 696
   Inverse-Square Laws 697
   Vector Form of Coulomb's Law 697
23-6 Applications of Coulomb's Law 699

Chapter 24 Electric Fields 713
24-1 What Are Fields? 714
24-2 Special Case: Electric Field of a Charged Sphere 716
24-3 Electric Field Lines 718
24-4 Electric Field of a Collection of Charged Particles 720
   Electric Dipole 723
   Far from an Electric Dipole: The Dipole Moment 725
24-5 Electric Field of a Continuous Charge Distribution 726
   Charge Density 726
24-6 Special Cases of Continuous Distributions 728
24-7 Case Study: The Shape of Lightning Rods 733
   Lightning Rods on Buckingham Palace 736
24-8 Charged Particle in an Electric Field 737
24-9 Special Case: Dipole in an Electric Field 741
   Potential Energy of a Dipole in an Electric Field 743

Chapter 25 Gauss's Law 753
25-1 Qualitative Look at Gauss's Law 754
   What's in the Box? 754
   Symmetry 756
25-2 Flux 757
   Area Vectors and Electric Flux 758
   A More General Expression for Electric Flux 759
25-3 Gauss's Law 761
25-4 Special Case: Linear Symmetry 765
25-5 Special Case: Spherical Symmetry 767
25-6 Special Case: Planar Symmetry 771
25-7 Special Case: Conductors 774

Chapter 26 Electric Potential 787
26-1 Scalars Versus Vectors 788
26-2 Gravity Analogy 788
26-3 Electric Potential Energy U_E 791
   Special Case: Electric Potential Energy Involving a Charged Spherical Source 791
   Special Case: Electric Potential Energy Stored in a Collection of Charged Particles 793
26-4 Electric Potential V 796
   Special Case: Electric Potential (Voltage) Due to a Charged Particle 798
   Visualizing Electric Potential 799
26-5 Special Case: Electric Potential Due to a Collection of Charged Particles 801
   Special Case: Electric Potential Due to a Dipole 801
26-6 Electric Potential Due to a Continuous Distribution 804
26-7 Connection Between Electric Field $\vec{E}$ and Electric Potential V 807
   The Relationship Between Equipotential Surfaces and Electric Field Lines 809
26-8 Finding $\vec{E}$ from V 813
   Case Study: $\vec{E}$ and V in the Human Heart 814
26-9 Graphing E and V 817
   Special Case: Particle Source 818
   Special Case: Two Infinite Charged Sheets 818
   Special Case: Isolated Charged Conductor 819

Chapter 27 Capacitors and Batteries 827
27-1 The Leyden Jar 828
27-2 Capacitors 829
   General Expression for Energy Stored by a Capacitor 830
27-3 Batteries 833
   Charging a Capacitor 835
27-4 Capacitors in Parallel and Series 837
   Series Capacitors 838
   Parallel Capacitors 840
27-5 Capacitance: Special Cases 843
27-6 Dielectrics 847
27-7 Energy Stored by a Capacitor with a Dielectric 851
   Energy Density 853
27-8 Gauss's Law in a Dielectric 855

Chapter 28 Current and Resistance 864
28-1 Microscopic Model of Charge Flow 865
   Basic Model 865
28-2 Current 866
28-3 Current Density 870
28-4 Resistivity and Conductivity 874
   Gravitational Analogy 875
   Resistivity 876
   Temperature Dependence 877
28-5 Resistance and Resistors 878
28-6 Ohm's Law 881
28-7 Power in a Circuit 884
   Gravitational Analogy Revisited 884
   Calculating Power 885
   Kilowatt-Hours 887

Chapter 29 Direct Current (DC) Circuits 895
29-1 Measuring Potential Differences Between Two Points 896
   Using a Voltmeter 897
   Expected Voltages 897
   Real Versus Ideal Emf Devices 899
   Grounded Circuits 899
29-2 Kirchhoff's Loop Rule 903
29-3 Resistors in Series 906
29-4 Kirchhoff's Junction Rule 908
29-5 Resistors in Parallel 909
29-6 Circuit Analysis 912
29-7 DC Multimeters 917
   Ammeter Design 918
   Voltmeter Design 919
   Ohmmeter Design 919
29-8 RC Circuits 919
   Charging a Capacitor 919
   Time Constant 922
   Discharging a Capacitor 922

Part IV Magnetism

Chapter 30 Magnetic Fields and Forces 934
30-1 Another Fundamental Force 935
30-2 Revealing Magnetic Fields 936
   Magnetic Monopoles and Dipoles 937
   Internal and External Magnetic Field Lines 938
   The Earth's Magnetic Field 938
30-3 Ørsted's Discovery 939
30-4 The Biot-Savart Law 940
   The Magnetic Field Due to a Moving Charged Particle 940
   The Magnetic Field Due to a Current 941
30-5 Using the Biot-Savart Law 942
   A Very Long, Straight Wire 944
30-6 The Magnetic Dipole Moment and Modeling Atoms 948
   Modeling Atoms 949
   The Magnetic Moment of Electrons and Atoms 949
30-7 Ferromagnetic Materials 951
30-8 Magnetic Force on a Charged Particle 952
   The Lorentz Force 953
30-9 Motion of Charged Particles in a Magnetic Field 954
30-10 Case Study: The Hall Effect 957
30-11 Magnetic Force on a Current-Carrying Wire 960
30-12 Force Between Two Long, Straight, Parallel Wires 963
30-13 Current Loop in a Uniform Magnetic Field 964
   Magnetic Potential Energy 966
   Galvanometers 967
   DC Motors 968

Chapter 31 Gauss's Law for Magnetism and Ampère's Law 978
31-1 Measuring the Magnetic Field 979
31-2 Gauss's Law for Magnetism 980
   Quick Review of Gauss's Law for Electricity 980
   Magnetic Monopoles Revisited 981
31-3 Ampère's Law 984
   Finding the Current Through a Loop 985
31-4 Special Case: Linear Symmetry 987
31-5 Special Case: Solenoids 992
31-6 Special Case: Toroids 996
31-7 General Form of Ampère's Law 999

Chapter 32 Faraday's Law of Induction 1010
32-1 Another Kind of Emf 1011
32-2 Faraday's Law 1013
32-3 Lenz's Law 1016
   Observed Current Directions 1016
   Another Right-Hand Rule 1016
   Combining Faraday and Lenz 1018
32-4 Lenz's Law and Conservation of Energy 1019
32-5 Case Study: Slide Generator 1020
   Another Look at the Motional Emf 1023
   Eddy Currents 1025

32-6 Case Study: AC Generators 1026
    Features and Operation of an AC Generator 1026
    The AC Motor 1030
32-7 Case Study: Faraday's Generator and Other DC Generators 1030
    Time-Varying DC Emf and Current 1031
32-8 Case Study: Power Transmission and Transformers 1033
    Root Mean Square Emf, Current, and Power 1033
    Power Transmission in Wires 1034
    Transformers 1035
    Who Won the War? 1037

Chapter 33 Inductors and AC Circuits 1045
33-1 Inductors and Inductance 1046
33-2 Back Emf 1049
    Transformers Revisited 1051
33-3 Special Case: Resistor–Inductor (*RL*) Circuit 1052
33-4 Energy Stored in a Magnetic Field 1055
33-5 Special Case: Inductor–Capacitor (*LC*) Circuit 1057
    Graphs of U_E and U_B for the LC Circuit 1060
33-6 Special Case: AC Circuit with Resistance 1062
    Phasor Diagrams 1063
33-7 Special Case: AC Circuit with Capacitance 1064
    Special Case: RC Filters 1067
33-8 Special Case: AC Circuit with Inductance 1068
    Special Case: RL Filters 1070
33-9 Special Case: AC Circuit with Resistance, Inductance, and Capacitance 1072
    Current and Voltage in an RLC Circuit 1072
    Impedance 1074
    Resonance in an RLC Circuit 1075

Chapter 34 Maxwell's Equations and Electromagnetic Waves 1085
34-1 Light: One Last Classical Topic 1086
34-2 Generalized Form of Faraday's Law 1087
34-3 Five Equations of Electromagnetism 1092
    Gauss's Law for Electricity 1092
    Gauss's Law for Magnetism 1092
    Faraday's Law 1093
    Ampère–Maxwell's Law 1093
    The Fifth Equation 1093
34-4 Electromagnetic Waves 1093
    Hertz's Experiments 1094
    Electromagnetic Wave Transmission 1094
    The Electromagnetic Wave Equation 1095
    Properties of Electromagnetic Waves 1098
    The Connection Between $\vec{E}$ and $\vec{B}$ 1099
34-5 The Electromagnetic Spectrum 1101
34-6 Energy and Intensity 1103
34-7 Momentum and Radiation Pressure 1107
    Reflected Versus Absorbed Waves 1108
    Radiation Pressure 1108
34-8 Polarization 1112
    Polarized Versus Unpolarized Radiation 1112
    Polarizers 1113

Part V Light

Chapter 35 Diffraction and Interference 1123
35-1 Light Is a Wave 1124
    Representing Light as a Wave 1125
    Propagation of Light 1125
35-2 Sound Wave Interference Revisited 1126
    Young's Double-Slit Experiment 1127
35-3 Young's Experiment: Position of the Fringes 1128
35-4 Single-Slit Diffraction 1132
    Effect of Slit Width on the Diffraction Pattern 1134
35-5 Young's Experiment: Intensity 1137
    Finding the Maxima and Minima from the Intensity 1138
    Plotting Intensity 1139
35-6 Single-Slit Diffraction Intensity 1140
    Where the Intensity Expression Comes From 1141
    Dark Fringes 1142
    Visualizing Intensity 1143
35-7 Double-Slit Diffraction 1144

Chapter 36 Applications of the Wave Model 1154
36-1 Implications of the Wave Model 1155
36-2 Circular Aperture Diffraction 1155
    Why Is There a Factor of 1.22? 1156
36-3 Thin-Film Interference 1158
    Speed of Light 1158
    Reflection and Phase Changes 1159
    Thin Films: Conditions for Constructive and Destructive Interference 1160
36-4 Diffraction Gratings 1166
    The Hydrogen Spectrum 1166
    Width of the Maxima in a Diffraction Grating 1168
36-5 Dispersion and Resolving Power of Gratings 1170
    Dispersion 1170
    Resolving Power 1171
    X-ray Diffraction 1173
36-6 Case Study: Michelson's Interferometer 1174
    Swimming Race Analogy 1174
    The Michelson–Morley Experiment 1176

Chapter 37 Reflection and Images Formed by Reflection 1184
37-1 Geometric Optics 1185
    Ray Model 1185
    Camera Obscura 1185
37-2 Law of Reflection 1187

37-3 Images Formed by Plane Mirrors 1190
 Ray Diagrams 1191
 Virtual and Real Images 1191
 Finding the Image Formed by a Plane Mirror 1191
 Left Hand or Right Hand? 1192
37-4 Spherical Mirrors 1195
 Finding the Focal Point 1196
 Primary Rays 1197
 Sign Conventions 1199
37-5 Images Formed by Convex Mirrors 1199
37-6 Images Formed by Concave Mirrors 1204
 Concave Mirror: Object Closer than the Focal Point 1205
 Concave Mirror: Object Farther than the Focal Point 1206
 Concave Mirror: Object at the Focal Point 1207
37-7 Spherical Aberration 1210

Chapter 38 Refraction and Images Formed by Refraction 1218
38-1 Law of Refraction 1219
38-2 Total Internal Reflection 1221
38-3 Dispersion 1223
 Rainbows 1224
38-4 Refraction at Spherical Surfaces 1226
 Paraxial Assumption 1226
 Sign Conventions for Spherical Refractors 1228
38-5 Thin Lenses 1232
 Sign Conventions for Thin Lenses 1235
 Converging Versus Diverging Lenses 1236
 Ray Diagrams for Thin Lenses 1237
 Primary Rays for Thin Lenses 1238
38-6 Images Formed by Diverging Lenses 1238
38-7 Images Formed by Converging Lenses 1240
 Converging Lens: Object Farther than Focal Point 1240
 Converging Lens: Object Closer than Focal Point 1241
 Converging Lens: Object at Focal Point 1242
38-8 The Human Eye 1243
 Anatomy of the Human Eye 1243
 Modeling the Eye 1244
 How the Human Eye Focuses 1244
 Corrective Lenses 1245
38-9 One-Lens Systems 1248
 The Camera 1248
 Case Study: The Magnifying Glass 1250
38-10 Multiple-Lens Systems 1252
 Case Study: The Compound Microscope 1252
 Case Study: The Refracting Telescope 1254

Part VI 20th Century Physics

Chapter 39 Relativity 1267
39-1 It's in the Eye of the Observer 1268
 Experimenting in a Noninertial Reference Frame 1268
 Looking for the Inertial Reference Frame 1268
 The Aberration of Starlight 1268
39-2 Special Case: Galilean Relativity 1271
 Position and Distance 1272
 Displacement and Velocity 1272
 Acceleration 1273
39-3 Postulates of Special Relativity 1275
39-4 Lorentz Transformations 1276
 The Correspondence Principle 1277
 Lorentz Transformations for x, y, and z 1277
 Lorentz Transformation for t 1278
 Simultaneity Is Relative 1278
39-5 Length Contraction 1279
39-6 Time Dilation 1281
39-7 The Relativistic Doppler Effect 1285
39-8 Velocity Transformation 1288
39-9 Mass and Momentum Transformation 1290
 The Ultimate Speed Limit 1292
39-10 Newton's Second Law and Energy 1293
 The Famous Equation $E = mc^2$ 1294
39-11 General Relativity 1297
 The Principle of Equivalence 1297
 Gravity Is an Illusion 1297
 Time Dilation and the Gravitational Doppler Shift 1298
39-12 Gravitational Lenses and Black Holes 1300
 Using an Eclipse to Test General Relativity 1301
 Gravitational Lenses 1301
 Black Holes 1301

Chapter 40 The Origin of Quantum Physics 1308
40-1 Another Modern Idea 1309
40-2 Black-Body Radiation and the Ultraviolet Catastrophe 1309
 Modeling Black-Body Curves 1311
 Planck's Solution 1312
40-3 The Photoelectric Effect 1316
40-4 The Compton Effect 1320
40-5 Wave-Particle Duality 1326
40-6 The Wave Properties of Matter 1328

Chapter 41 Schrödinger's Equation 1338
41-1 The New Quantum Theory 1339
41-2 A Trapped Particle 1340
 Energy Levels 1342
41-3 The Double-Slit Experiment Revisited: Probability Waves 1343

41-4 Schrödinger's Equation **1346**
   A Plausible Equation **1346**
   Finding Solutions **1347**
41-5 Special Case: A Particle in an Infinite Square Well **1349**
   Normalizing the Wave Function **1350**
   Interpreting the Wave Function **1350**
41-6 Special Case: A Particle in a Finite Square Well **1352**
41-7 Barrier Tunneling **1355**
   Scanning Tunneling Microscope **1357**
41-8 Special Case: Quantum Simple Harmonic Oscillator **1359**
41-9 Heisenberg's Uncertainty Principle **1362**
   A Free Particle **1362**
   Uncertainty Is Inherent **1363**

Chapter 42 Atoms **1372**
42-1 Early Atomic Models **1373**
42-2 Rutherford's Model of the Atom **1375**
42-3 Bohr's Model and Atomic Spectra **1377**
   The Hydrogen Spectrum **1378**
   Bohr's Atomic Model **1378**
   Energy Levels in Hydrogen **1381**
   Excited Atoms and Ions **1382**
42-4 De Broglie's Theory and Atoms **1386**
42-5 Schrödinger's Equation Applied to Hydrogen **1388**
   Hydrogen's Quantum Numbers **1388**
   Hydrogen's ground state **1389**
   Hydrogen in an Excited State **1390**
42-6 Magnetic Dipole Moments and Spin **1392**
   Stern–Gerlach Experiment **1393**
   Spin **1394**
42-7 Other Atoms **1395**
   The Pauli Exclusion Principle **1395**
42-8 Organizing Atoms **1397**
   Moseley's Experiment **1397**
   Shielding **1397**
   Atomic Size and Ionization Energy **1397**
   The Periodic Table **1399**
42-9 The Zeeman Effect **1401**
42-10 Practical Devices **1405**
   The Cesium Clock **1405**
   Fluorescent Bulbs **1405**

Chapter 43 Nuclear and Particle Physics **1413**
43-1 Describing the Nucleus **1414**
   Nuclear Radius **1415**
43-2 The Strong Force **1416**
43-3 Models of Nuclei **1417**
   The Shell Model **1417**
   Other Models **1419**
43-4 Radioactive Decay **1420**
   Gamma Rays **1420**
   Alpha Particles **1421**
   Decay Rate **1421**
   Radioactive Dating **1422**
43-5 The Weak Force **1424**
   Neutrinos and the Weak Force **1424**
   Beta Decay, Inverse Beta Decay, and Electron Capture **1425**
   Uranium: A Naturally Occurring Radioactive Element **1426**
43-6 Binding Energy **1428**
   Mass Deficit and Binding Energy **1428**
   Convenient Units of Mass and Binding Energy **1430**
   The Binding Energy Curve **1430**
   Energy Released by Reactions **1431**
43-7 Fission Reactions **1433**
   Nuclear Power Plants **1434**
   Nuclear Fission Bombs **1435**
43-8 Fusion Reactions **1436**
   Fusion in the Sun's Core **1437**
   Fusion on Earth **1438**
43-9 Human Exposure to Radiation **1441**
   Dosimetry **1443**
43-10 The Standard Model **1446**
   Electricity, Magnetism, and Relativity **1446**
   Quantum Electrodynamics (QED) **1447**
   A Zoo of Particles **1447**
   Forces in the Standard Model **1449**
   The Higgs Boson **1449**
   A Final Word **1450**

Appendix A Mathematics **App-1**
A-1 Algebra and Geometry **App-1**
A-2 Trigonometry **App-2**
A-3 Calculus **App-3**
   Derivatives **App-3**
   Integrals **App-4**
A-4 Propagation of Uncertainty **App-5**
   Sums and Differences **App-5**
   Products, Quotients, and Powers **App-6**
   Multiplication by an Exact Number **App-6**

Appendix B Reference Tables **App-7**
B-1 Symbols and Units **App-7**
B-2 Conversion Factors **App-9**
B-3 Some Astronomical Data **App-10**
B-4 Rough Magnitudes and Scales **App-11**

Periodic Table of the Elements **App-14**

Answers to Concept Exercises and Odd-Numbered Problems **Ans-1**

Index **I-1**

Preface for the Instructor

When I decided to pursue physics and astrophysics, I didn't think about teaching. I just knew I was interested in the subjects. As a graduate student at the University of Minnesota, however, I won a fellowship that included pedagogical training by well-known figures in physics education research (PER). I learned to take teaching seriously, and when it was time to find a job, I knew I wanted to work at an institution committed to undergraduate education.

Fortunately, I found that job at the United States Naval Academy. I have had the opportunity to teach our year-long calculus-based course not just to physics majors, but also to a varied audience with a diverse set of abilities, interests, learning styles, and preconceptions. In order to engage my students in the process of learning physics, I use many of the techniques that have come out of PER, and over the past two decades my students have taught me how to teach them. I have tried my best to integrate both PER results and the lessons learned from my students into the pages of this text.

One of the major results of PER is that lectures don't work. Students who sit passively in a classroom, listening to a brilliant lecture, retain very little. Fortunately, PER offers many ways to approach teaching. I have used a great number of these approaches: *peer instruction, active learning groups, interactive lectures,* and, most recently, a *flipped classroom.* All of these approaches require students to take an active part in their education, while I act as a coach, urging them to take the necessary steps to learn physics.

Although there is a wide range of physics pedagogy, including lectures in introductory physics, one thing is always true: *students do better if they read their textbook.* So my first goal in writing this textbook was to write a book that students would actually read and value. I know students don't generally like to read a science textbook, and if they do try to read it, they don't get much for their effort. So I looked into what makes other reading material, such as a novel or a magazine article, so effective. One answer is that the human brain enjoys stories about people. Generally, physics textbooks omit stories; science is presented as a series of results, without the process that led to these results or the impact these results have on our lives. Students cannot connect to this traditional, dogmatic approach; they don't see how physics fits into their lives. So, wherever possible, I included the stories of our field. Some of these are stories of discovery, whereas others are stories that show the impact of physics results on the human experience.

Another major result of PER is the insight that students are not blank slates. They come to our classrooms with *preconceptions* about physics, which they have developed over years of observations made during the course of their everyday lives. One of the important jobs of physics education is to help students tie their preconceptions to the appropriate formalisms of physics. For example, students know that seat belts secure passengers in cars, but they are not usually aware of the connection between this fact from their everyday lives and Newton's first law. To make and reinforce these connections, I use dialogues between fictional students (named Avi, Cameron, and Shannon) to highlight and clarify commonly held preconceptions. Our real students are then tasked with critiquing the fictional dialogue, leading them to connect their preconceptions to the correct physics concepts.

As university instructors, our goal is to help our students work their way up to the top of Bloom's taxonomy (see the figure on page xiii), where they create and answer their own questions using physics. Of course, students must start at the bottom and work their way up to the top. For example, at first we want our students to remember and understand Newton's second law. We spend most of our time working in the middle of the pyramid, where we may expect our students to apply Newton's second law to analyze problems. Occasionally, we expect our students to work higher up in Bloom's taxonomy. Often we rely on a laboratory component, which requires students to make judgments, or we find some problems that call for such evaluation. However, there are very few tools at our disposal for helping students to reach the top of Bloom's taxonomy.

But only when students reach the top of Bloom's taxonomy can they see the value in learning physics because it is at the top of the pyramid that students use physics to answer their own questions. So to help students reach the top and to make physics engaging, I use case studies in my classroom and in this book. Case studies relate interesting topics to the concepts, principles, and tools of physics. In my class, students write their own case studies, using physics concepts to understand situations drawn from such subjects as sports, movies, and history. In my textbook, a case study is woven into each chapter, to make

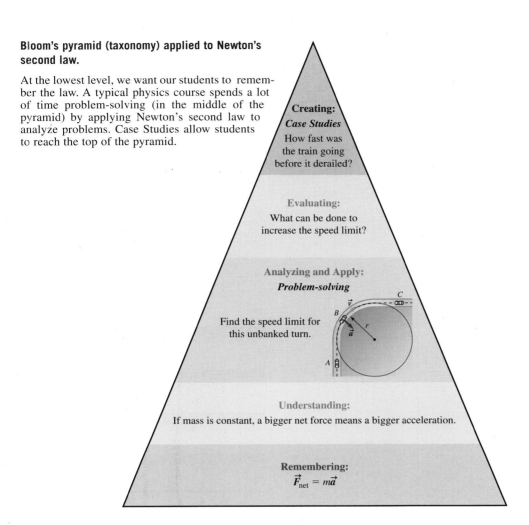

Bloom's pyramid (taxonomy) applied to Newton's second law.

At the lowest level, we want our students to remember the law. A typical physics course spends a lot of time problem-solving (in the middle of the pyramid) by applying Newton's second law to analyze problems. Case Studies allow students to reach the top of the pyramid.

physics more relevant and engaging—for example, Chapter 5 uses a newsworthy train collision to teach Newton's laws of motion.

Case studies fulfill other goals as well. First, because case studies are based on human experience, they also help students realize how the preconceptions they have developed over decades of observation tie into the formalism of physics. Second, case studies are stories—so they are fun to read and memorable. I know you might find it hard to believe, but both my own students and those at other institutions report that they actually *enjoy* the case studies. When was the last time a student said he or she enjoyed reading a science textbook? I hope you'll find this textbook program to be a better learning tool for your students than what has previously been available.

If you have any comments or questions, feel free to contact me at deborakatz@yahoo.com.

How do you engage your students to go beyond the quantitative?

The main goal of *Physics for Scientists and Engineers: Foundations and Connections* is to offer a calculus-based introductory physics textbook designed to assist you in taking your students "beyond the quantitative." Physics Education Research (PER) best practices and the author's extensive classroom experience are leveraged to motivate readers and address the areas where students struggle the most—bridging the gap between abstract language and application, overcoming common preconceptions, and connecting mathematical formalism and physics concepts.

Case studies to motivate students and make abstract concepts concrete.

This text uses case studies to draw readers into the story of physics. Case studies are introduced and revisited throughout the chapters in pedagogy such as the concept exercises, examples, and end-of-chapter problems. Some case studies are based on students' contemporary "real-world" experiences, including events they may read about in the news. These case studies make abstract physics concepts understandable and help bridge the gaps between the key concepts, the formal language, and the mathematics of physics. For example, the case study in Chapter 5 illustrates Newton's laws of motion by examining an actual train collision described in news articles. The author introduces free-body diagrams and Newton's second law, asking students to determine why backward-facing passengers experienced less bodily harm than did the forward-facing passengers. Students apply physics "tools"—Newton's laws and free-body diagrams—to get to the bottom of this real-world example.

FIGURE 5.1 Aerial view of emergency workers helping the injured from a train crash near Los Angeles (April 23, 2002).

> **CASE STUDY** Train Collision
>
> On April 23, 2002, a passenger train about 35 miles outside of Los Angeles was hit by a freight train (Fig. 5.1). The accident killed two people and injured more than 260, with all the injured being on the passenger train. Witnesses reported that those people who were seated facing backward suffered little or no injury. News reports said that the passenger train came to a quick stop before the collision and that the impact with the freight train pushed the passenger train 370 ft backward (Fig. 5.2).
>
> One of the most controversial parts of the early reports was how fast the freight train was going at the moment of impact. In Chapter 11, we will reconstruct the accident and estimate the speed of the freight train upon impact. In this chapter, we are concerned only with the following questions:
>
> 1. Why did passengers seated facing backward fare better than those who were either standing or seated facing forward?
> 2. The passenger train was at rest before the collision and was pushed backward. What do those facts tell us about how hard the freight train pushed on the passenger train? Did the passenger train push on the freight train? If so, how hard did it push?

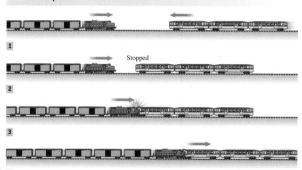

FIGURE 5.2 **1** A freight train and a passenger train move toward each other. **2** The passenger train stops. **3** The freight train continues and collides with the passenger train. **4** The passenger train is shoved backward.

Student dialogues to address preconceptions.

Students often come to the classroom with preconceptions (what some call misconceptions). By acknowledging and addressing these preconceptions, we can transform them into building blocks toward proper understanding. Preconceptions are primarily addressed by dialogues between fictional students that allow readers to discover their preconceptions without a sense of failure. Dialogues may be incorporated into case studies, concept exercises, examples, and end-of-chapter problems.

> **CONCEPT EXERCISE 5.2**
>
> **CASE STUDY** Train Collision and Newton's First Law
>
> A group of college students discusses the train collision case study. Use Newton's first law to decide which underlined statements are correct and which are false. Explain your answers.
>
>
>
> **Shannon:** This newspaper says that the people who got really hurt were either standing up or sitting in a forward-facing seat. Those people got thrown forward when the train stopped.
>
> **Avi:** That's why there are seat belts in cars. <u>If you get into a crash, the force can throw you through the windshield.</u>
>
> **Cameron:** There is no force that throws you through the windshield. <u>You fly through the windshield because you are already moving and it would take a force to stop you from going forward.</u> That's why there's a seat belt.
>
> **Avi:** That doesn't make sense. Because then <u>you would need a force to stop you from flying through the windshield even when you just stop slowly at a red light.</u>
>
> **Cameron:** That's right, but <u>when you slow down slowly, you don't need such a big force and the car seat can take care of it.</u>
>
> **Shannon:** The seat? <u>I don't think a seat can exert a force. It can't move on its own or hold you. That's why the people who were sitting forward on the train were hurt. The people who were sitting backward had the back of the seat to block them.</u>

Two-column format for examples and derivations to connect mathematical formalism and physics concepts.

Research shows that students struggle to make these connections. By presenting many of the **examples and derivations in two columns (what an expert problem-solver thinks on the left and what that expert would write on the board on the right)**, students can make the connections between the concept being taught and the mathematical steps to follow. It is like having an instructor within the text: in one column he or she explains the concept, and in the other he or she shows the mathematical steps to follow, just as an instructor would verbally explain a problem in class while simultaneously solving the problem on the board.

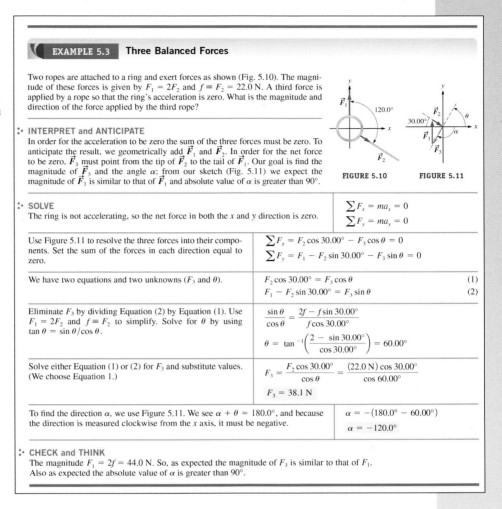

EXAMPLE 5.3 Three Balanced Forces

Two ropes are attached to a ring and exert forces as shown (Fig. 5.10). The magnitude of these forces is given by $F_1 = 2F_2$ and $f \equiv F_2 = 22.0$ N. A third force is applied by a rope so that the ring's acceleration is zero. What is the magnitude and direction of the force applied by the third rope?

: INTERPRET and ANTICIPATE
In order for the acceleration to be zero the sum of the three forces must be zero. To anticipate the result, we geometrically add $\vec{F}_1$ and $\vec{F}_2$. In order for the net force to be zero, $\vec{F}_3$ must point from the tip of $\vec{F}_2$ to the tail of $\vec{F}_1$. Our goal is find the magnitude of $\vec{F}_3$ and the angle α; from our sketch (Fig. 5.11) we expect the magnitude of $\vec{F}_3$ is similar to that of $\vec{F}_1$ and absolute value of α is greater than 90°.

FIGURE 5.10   FIGURE 5.11

: SOLVE
The ring is not accelerating, so the net force in both the x and y direction is zero.

$$\sum F_x = ma_x = 0$$
$$\sum F_y = ma_y = 0$$

Use Figure 5.11 to resolve the three forces into their components. Set the sum of the forces in each direction equal to zero.	$\sum F_x = F_2 \cos 30.00° - F_3 \cos \theta = 0$ $\sum F_y = F_1 - F_2 \sin 30.00° - F_3 \sin \theta = 0$
We have two equations and two unknowns (F_3 and θ).	$F_2 \cos 30.00° = F_3 \cos \theta$ (1) $F_1 - F_2 \sin 30.00° = F_3 \sin \theta$ (2)
Eliminate F_3 by dividing Equation (2) by Equation (1). Use $F_1 = 2F_2$ and $f \equiv F_2$ to simplify. Solve for θ by using $\tan \theta = \sin \theta / \cos \theta$.	$\dfrac{\sin \theta}{\cos \theta} = \dfrac{2f - f \sin 30.00°}{f \cos 30.00°}$ $\theta = \tan^{-1}\left(\dfrac{2 - \sin 30.00°}{\cos 30.00°}\right) = 60.00°$
Solve either Equation (1) or (2) for F_3 and substitute values. (We choose Equation 1.)	$F_3 = \dfrac{F_2 \cos 30.00°}{\cos \theta} = \dfrac{(22.0 \text{ N}) \cos 30.00°}{\cos 60.00°}$ $F_3 = 38.1$ N
To find the direction α, we use Figure 5.11. We see $\alpha + \theta = 180.0°$, and because the direction is measured clockwise from the x axis, it must be negative.	$\alpha = -(180.0° - 60.00°)$ $\alpha = -120.0°$

: CHECK and THINK
The magnitude $F_1 = 2f = 44.0$ N. So, as expected the magnitude of F_3 is similar to that of F_1. Also as expected the absolute value of α is greater than 90°.

Problem-Solving Strategy

Physics is not a spectator sport. Physics students are expected to *do* physics. So, problem-solving is a major component to learning physics. In keeping with the sports analogy, a novice player is given detailed instructions on how to position and move his or her body, but a professional athlete often forgets these details and seems to play the sport naturally. Likewise, physics students need a detailed problem-solving strategy when they first start off. The strategy in this book is streamlined and designed to mimic expert methods. All worked examples in the text have been solved through a three-procedure approach: 1) *Interpret and Anticipate*, 2) *Solve*, and 3) *Check and Think*. In addition, further problem-solving strategies are provided for specific types of problems. As shown on page xvi, these specific strategies flesh out one or more of the three procedures.

PROBLEM-SOLVING STRATEGY

Applying Newton's Second Law

∴ INTERPRET and ANTICIPATE
Identify the system (often, a single object) that is subject to external force(s). Draw a free-body diagram for that system, making sure that the diagram has all four elements given above. Once the free-body diagram is complete, there are three steps that help in the **SOLVE** procedure when an algebraic or numerical result is required.

∴ SOLVE
Step 1 Apply Newton's second law in component form. Use the coordinate system on the free-body diagram to apply Equation 5.2. You will have one equation for each direction in which there is at least one force.

$$\sum F_x = ma_x \qquad \sum F_y = ma_y \qquad \sum F_z = ma_z$$

Step 2 Write down any other equations that are relevant to the forces involved. Table 5.1 lists magnitudes of the gravitational force (weight), the spring force, and kinetic friction. If the situation involves one or more of these forces, write down the appropriate equation.

Step 3 Do algebra before substitution. Review your equations. Which parameters are known? Which are unknown? Which do you need to solve for? You may have more equations than you need. Find an algebraic expression for the parameter you need before you substitute any numerical values. This practice makes it easier to find a mistake if you make one and makes it easier for another person to understand your work.

Conceptual Framework

You probably wouldn't be surprised to learn that studies have enumerated the ways physics instructors differ from their students. For example, physicists tend to be *linear* learners, and physics students tend to be *global* learners. This means that physics students need to see the big picture before they can settle into learning the details. To give students a global perspective, each chapter begins with an overview of the material covered in the chapter.

Another difference between physics instructors and their students is that instructors often see physics topics arranged in a conceptual hierarchy, but students have trouble organizing topics. So to help students develop an organization, each chapter begins with a list of the topics arranged into categories: *Underlying Principles*, *Major Concepts*, *Special Cases*, and *Tools*; an example of this structure is shown below:

Throughout each chapter, notes in the margins refer to the topics using these categories (see examples below).

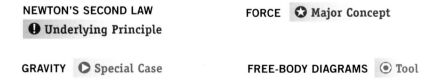

Finally, each chapter has a **Summary** at the end, before the problems and questions. In a traditional textbook, the summary is arranged according to the order in which the topics appear in the chapter. By contrast, the summary in this textbook (see page xvii for an example) is arranged using the same hierarchy found at the beginning of the chapter. This reinforces the conceptual organization by revisiting the concepts found on the chapter's opening page with more complete descriptions. In this way, students can use this information for review before attempting to solve the problems assigned for homework.

SUMMARY

❶ Underlying Principles: Newton's three laws of motion

1. Close translation of **Newton's first law**:

   Every body continues in its state of rest, or of uniform motion in a straight line, unless it is compelled to change that state by forces impressed on it.

   More contemporary statement of Newton's first law:

   If no force acts on an object, then the object cannot accelerate.

   Newton's second law contains his first law:

   The net force on an object is zero if and only if the acceleration of the object is also zero.

2. **Newton's second law**: The acceleration of an object is proportional to the total force acting on it:
   $$\vec{F}_{tot} \equiv \sum \vec{F} = m\vec{a} \quad (5.1)$$

   Newton's second law is often applied in component form:
   $$\sum F_x = ma_x \quad \sum F_y = ma_y \quad \sum F_z = ma_z \quad (5.2)$$

3. **Newton's third law**: If two objects A and B interact, the force exerted by A on B is equal in magnitude to the force exerted by B on A but in the opposite direction. Mathematically,
   $$\vec{F}_{[B \text{ on } A]} = -\vec{F}_{[A \text{ on } B]} \quad (5.10)$$

❷ Major Concepts

1. **Dynamics**: a branch of mechanics that focuses on the causes of motion.
2. **Force**: a push or pull; required to make an object accelerate. A force has three properties:
   a. Force does not exist in isolation; it must act on a "subject."
   b. If a force is exerted on a subject, there must be some "source" responsible for that force.
   c. Force is a vector quantity having both magnitude and direction.
3. **System**: any collection of objects.
4. **Inertia**: the tendency of an object to maintain its constant velocity. **Mass**, also known as **inertial mass**, is the intrinsic property of an object that determines its inertia.
5. **Inertial reference frame**: one in which Newton's first law is valid. Inertial reference frames may move with constant velocity, but they do not accelerate. Accelerating frames are known as **noninertial** reference frames, and Newton's first law is not valid in them.

❸ Special Cases

Table 5.1 on page 134 summarizes the five specific forces examined in this chapter.

❹ Tools

Four elements of a **free-body diagram**:

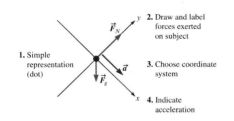

1. Simple representation (dot)
2. Draw and label forces exerted on subject
3. Choose coordinate system
4. Indicate acceleration

PROBLEM-SOLVING STRATEGY

Applying Newton's Second Law,
$$\vec{F}_{tot} \equiv \sum \vec{F} = m\vec{a}$$

:• INTERPRET and ANTICIPATE

Draw a free-body diagram.

:• SOLVE

1. Apply Newton's second law in component form. Write one equation for each direction that has at least one force.
2. Write any other relevant force equations.
3. Do algebra before substituting values.

Problems Set

Over 3,500 problems have been written for this edition. Each part of every problem has been classified as either *Algebraic, Conceptual, Estimation, Graphical,* or *Numerical* in nature. In most traditional textbooks, problems are generally either numerical or algebraic, and in recent years conceptual questions have been added to the end-of-chapter exercises in their own separate section. These books have few, if any, estimation or graphical problems. However, one goal of physics instruction is to teach students to think like physicists. By organizing the questions and problems by topic rather than by type, and by introducing graphical and estimation problems, this textbook teaches students to think more like practicing physicists, who approach a topic by solving a variety of types of problems — *Algebraic, Conceptual, Estimation, Graphical,* or *Numerical*—at once.

Engagement Solutions that Fit Your Teaching Goals and Your Students' Learning Needs

Whether you offer a more traditional lecture-based course or are interested in flipping the classroom, *Physics for Scientists and Engineers: Foundations and Connections* offers print and digital solutions for an active learning environment. The learning resources and analytics in Enhanced WebAssign will provide you with tools that will easily help you check student comprehension before class; create in-class activities and discussions; identify students who need help based on homework results; and compare homework scores with test results. Your students will gain confidence and improve their test scores with ongoing assessments and engagement activities throughout the entire course.

Pre-Lecture
Enhanced WebAssign for *Physics for Scientists and Engineers: Foundations and Connections*. Exclusively from Cengage Learning, Enhanced WebAssign offers an extensive online program for physics to encourage exploration and practice that's so critical for concept mastery. With Enhanced WebAssign you can assign content before the beginning of the class and use powerful analytics to monitor your students' comprehension and engagement. Options include:

- The **Cengage YouBook.** Students should read to succeed. WebAssign has a customizable and interactive eBook, the Cengage YouBook, that lets you tailor the textbook to fit your course and connect with your students. You can remove and rearrange chapters in the table of contents and tailor assigned readings that match your syllabus exactly. Powerful editing tools let you change as much as you'd like—or leave it just like it is.

- **Reading Check Questions.** These questions give your students ample opportunity to test their conceptual understanding before class.

- **PreLecture Explorations (PLE). This option uses** HTML5 interactive simulations enabling students to make predictions, change parameters, and observe results. Each PreLecture Exploration presents an engaging simulation based on a relevant scenario and then asks conceptual and analytic questions, guiding students to a deeper understanding and helping promote a robust physical intuition. This is the perfect resource for a flipped classroom or for professors looking for new ways to increase student engagement and interest in the material prior to lecture.

In-Class Group Discussions and Active Learning
Assign case study problems from the chapter in class or use the text's Student Engagement slides to encourage peer learning and group discussions. The Student Engagement slides, when used with Lecture Tools or any clicker device, will give your students a chance to think critically in the classroom and collaborate in an engaging environment.

Homework
Assessments throughout the semester give you ample opportunity to tailor your lecture and provide targeted help and feedback. Your Enhanced WebAssign course includes:

- **All of the quantitative end-of-chapter problems.**
- **Problem-Solving strategies for when your students need help the most:**
  - **Master It tutorials** help students work through the problems one step at a time.
  - **Watch It solution videos** explain fundamental problem-solving strategies, helping students step through the problem. In addition, instructors can choose to include video hints of problem-solving strategies.

- **Integrated Tutorials (IT)**, written by the author, strengthen students' skills by guiding them through the problem-solving steps identified in their textbook. Tutorials take students through the process, asking questions they will learn to ask themselves when faced with new problems. An example of an Integrated Tutorial in Enhanced WebAssign is shown below.

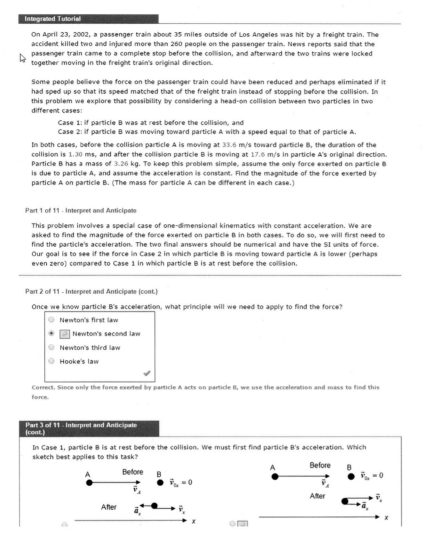

After Class

Using Enhanced WebAssign, you can administer high-stakes assessments with security measures such as IP restriction, password protection, and LockDown Browser. But before the exam you should make sure your students are ready by assigning a Personalized Study Plan.

- **Personalized Study Plan.** The Personalized Study Plan in Enhanced WebAssign provides chapter and section assessments that show students what material they know and what areas require more work. For items that they answer incorrectly, students can click on links to related study resources such as videos, tutorials, or reading materials. Color-coded progress indicators let them see how well they are doing on different topics. You decide what chapters and sections to include—and whether to include the plan as part of the final grade or as a study guide with no scoring involved.

Please visit **http://www.webassign.net/features/textbooks/katzpse1/details.html** to view an interactive demonstration of Enhanced WebAssign.

Supporting Materials

Please visit **Cengage.com** for information about student and instructor resources for this text, including custom versions and laboratory manuals.

Acknowledgments

Before writing this textbook, I thought books were written by their authors. Period. Of course, they are *written* by their authors, but writing is a small part of *creating* a book. Books are written, edited, reviewed, revised, and tested. Figures are scribbled by the author, marked up and, in many cases, redrawn by the editors before they are sent to professional graphic artists, and then they are further reviewed, tested, and revised. Photos are imagined by the author, then sometimes taken by professional photographers, but often by an editor, or the author's spouse, or a friend or a colleague. Of course, I really want this book to touch the lives of students and instructors, and it is simply impossible to imagine how I could have ever gotten this book into your hands without help. So this book exists and is being read by you in large part due to the tremendous effort of my publishing team, my contributors, my reviewers, my colleagues, my friends, and most of all my family. I cannot possibly thank each individual who touched this book, but I will do my best to mention many here.

First, I want to thank the people at Cengage Learning: Mary Finch, Rebecca Berardy Schwartz, Nicole Hurst, Peter McGahey, Janet del Mundo, Alison Eigel Zade, Cathy Brooks, Bruce Bond, Cate Barr, Brandi Kirksey, Jonathan McDonald, John Wimer, Brendan Killion, Ed Dodd, and especially Susan Dust Pashos. I would also like to thank Irene Nunes, who worked with me in the early stages of development and Chris Hall, who showed great confidence in my abilities.

Second, as I am sure you are aware, textbooks have many components such as the end-of-chapter problems and questions, the online supplements, the student solutions manual, and the instructor's solutions manual. I express my sincere gratitude to the contributors listed below who made these components possible. Most of all, I want to express my deep-felt thanks to Eric S. Mandell for his tremendous fortitude, insight, and creativity.

Contributors of End-of-Chapter Problems

David Bannon, Oregon State University
Andrew Boudreaux, Western Washington University
Juan Cabenela, Minnesota State University, Moorhead
Mark James, Northern Arizona University
Ronald Jodoin, Rochester Institute of Technology
Tom Krause, Towson University

Kingshuk Majumdar, Grand Valley State University
Eric Mandell, Bowling Green State University
Vahe Peroomian, University of California, Los Angeles
Yuri Sikorski, Kettering University
John Stamm, University of Evansville
Brian Utter, James Madison University
Shannon Willoughby, Montana State University

Third, I love to learn, especially about physics and physics education. Writing this book has given me a great opportunity to talk with physics instructors from all around North America. I have learned so much from all of you. I thank all of you, including those who formally reviewed drafts of this book.

Reviewer Consultants

The following dedicated individuals provided valuable long-term guidance for one part of the book (one group of related chapters) through multiple stages of development.

Yildirim Aktas, University of North Carolina at Charlotte
Jason Brown, Clemson University
Andrew Cornelius, University of Nevada, Las Vegas
John DiTusa, Louisiana State University
James Dove, Metropolitan State University of Denver
Scott Dwyer, Rensselaer Polytechnic University
Thomas Hemmick, Stony Brook University
Robert Johnson, University of Pennsylvania
Sally Koutsoliotas, Bucknell University
Jorge Lopez, University of Texas at El Paso
Rafael Lopez-Mobilia, University of Texas at San Antonio

Matthew Mackie, Temple University
James Olsen, Princeton University
Amy Pope, Clemson University
Ramon Ravelo, University of Texas at El Paso
Michael Richmond, Rochester Institute of Technology
Joseph Rothberg, University of Washington
Joseph Scanio, University of Cincinnati
Douglas Sherman, San Jose State University
Chuck Stone, Colorado School of Mines
Catalin Teodorescu, Montgomery College
Jeff Winger, Mississippi State University
Michael Ziegler, Ohio State University

Board of Advisors

These instructors provided guidance during early stages of development, especially while the two-column design for worked examples and derivations was formulated.

Mirela Fetea, University of Richmond
Thomas Keil, late of Worcester Polytechnic Institute
Eric Mandell, Bowling Green State University
Amy Pope, Clemson University
Michael Richmond, Rochester Institute of Technology
Jeff Winger, Mississippi State University

Manuscript Reviewers

Anthony Aguirre, University of California, Santa Cruz
Robert Balogh-Robinson, Marist College
David Bannon, Oregon State University
Arun Bansil, Northeastern University
Peter Barnes, Clemson University
Linda Barton, Rochester Institute of Technology
Dave Besson, University of Kansas
Ken Bolland, Ohio State University
Andrew Boudreaux, Western Washington University
Ryan Case, Elizabethtown Community and Technical College
Amit Chakrabarti, Kansas State University
Raymond Chastain, Louisiana State University
Liao Chen, University of Texas at San Antonio
Krishna Chowdary, Evergreen State College
David Cole, Northern Arizona University
Doug Copely, Sacramento City College
Andrew Cornelius, University of Nevada, Las Vegas
Stephane Coutu, Pennsylvania State University
Grant Denn, Metropolitan State University of Denver
Stephanie Diemel, Shoreline Community College
Susan DiFranzo, Hudson Valley Community College
John DiTusa, Louisiana State University
James Dunne, Mississippi State University
Dipangkar Dutta, Mississippi State University
Mirela Fetea, University of Richmond
Christopher Fischer, University of Kansas
Elena Flitsiyan, University of Central Florida
Rica Sirbaugh French, MiraCosta College
Alejandro Garcia, University of Washington
J. D. Garcia, University of Arizona
Delena Gatch, Georgia Southern University
Christopher Gould, University of Southern California
Benjamin Grinstein, University of California, San Diego
D. Eitan Gross, University of Arkansas
Kevin Haglin, St. Cloud State University
Nathan Harshman, American University
Thomas Hemmick, Stony Brook University
David Heskett, University of Rhode Island
Wendell Hill, University of Maryland, College Park
Jo Hopp, University of Wisconsin—Stout
Yung Huh, South Dakota State University
Anthony Hyder, University of Notre Dame
Robert Jacobsen, University of California, Berkeley
Mark James, Northern Arizona University
John Jaszczak, Michigan Technological University
Ronald Jodoin, Rochester Institute of Technology
Charlie Johnson, University of Pennsylvania
Darrin Johnson, University of Minnesota, Duluth
Philip Johnson, American University
Robert Johnson, University of Pennsylvania
Linda Jones, College of Charleston
Neda Katz, University of Southern California
Thomas Keil, late of Worcester Polytechnic Institute
Yangsoo Kim, Virginia Tech
David Klassen, Rowan University
Igor Kogoutiouk, Minnesota State University, Mankato
Zlatko Koinov, University of Texas at San Antonio
Ichishiro Konno, University of Texas at San Antonio
Dorina Kosztin, University of Missouri, Columbia
Michael Kotlarchyk, Rochester Institute of Technology
Theo Koupelis, University of Wisconsin—Marathon
Sally Koutsoliotas, Bucknell University
Monika Kress, San Jose State University
James LaBelle, Dartmouth College
David Lamp, Texas Tech University
Allen Landers, Auburn University
Pedram Leilabady, University of North Carolina at Charlotte
David Lind, Florida State University
Susannah Lomant, Georgia Perimeter College
Jorge Lopez, University of Texas at El Paso
Rafael Lopez-Mobilia, University of Texas at San Antonio
Daniel Ludwigsen, Kettering University
Kristina Lynch, Dartmouth College
Matthew Mackie, Temple University
Kingshuk Majumdar, Grand Valley State University
Eric Mandell, Bowling Green State University
Andrea Markelz, University at Buffalo
Peter Markowitz, Florida International University
Mark Mattson, James Madison University
Will McElgin, Louisiana State University
David McIntyre, Oregon State University
Marina Milner-Bolotin, University of British Columbia
Drew Milsom, University of Arizona
Rabindra Mohapatra, University of Maryland, College Park
Teruo Morishige, University of Central Oklahoma
Gerhard Muller, University of Rhode Island
Eric Murray, Georgia Institute of Technology
Taha Mzoughi, Kennesaw State University
Craig Ogilvie, Iowa State University
James Olsen, Princeton University
Halina Opyrchal, New Jersey Institute of Technology

Peter Persans, Rensselaer Polytechnic Institute
Doug Petkie, Wright State University
Amy Pope, Clemson University
Richard Quimby, Worcester Polytechnic University
Corneliu Rablau, Kettering University
Gloria Ramos, Citrus College
Ramon Ravelo, University of Texas at El Paso
Michael Richmond, Rochester Institute of Technology
Andreas Riemann, Western Washington University
John Rollino, Rutgers University—Newark
Joseph Rothberg, University of Washington
Baharam Roughani, Kettering University
Dubravka Rupnik, Louisiana State University
Mehmet Sahiner, Seton Hall University
Mahdi Sanati, Texas Tech University
Vladimir Savinov, University of Pittsburgh
Joseph Scanio, University of Cincinnati
Ann Schmiedekamp, Pennsylvania State University—Abington
Ben Shaevitz, Slippery Rock University
Kim Sharp, University of Pennsylvania
Douglas Sherman, San Jose State University
Ethan Siegel, University of Portland
Chandralekha Singh, University of Pittsburgh
Henry Smith, Northeastern University
Bryndol Sones, United States Military Academy
Phillip Sprunger, Louisiana State University
Gay Stewart, University of Arkansas
Chuck Stone, Colorado School of Mines
Jay Strieb, Villanova University
Tad Thurston, Oklahoma City Community College
Ionel Tifrea, California State University, Fullerton
Somdev Tyagi, Drexel University
Brian Utter, James Madison University
Ravi Vadapalli, Texas Tech University
Trina Van Ausdal, Salt Lake Community College
Joan Vogtman, Potomac State College of West Virginia
Keith Warren, North Carolina State University
Laura Weinkauf, Jacksonville State University
Edward Whittaker, Stevens Institute of Technology
Shannon Willoughby, Montana State University
Jeff Winger, Mississippi State University
David Young, Louisiana State University
Michael Ziegler, Ohio State University
Bernard Zygelman, University of Nevada, Las Vegas

Fourth, I thank my friends and colleagues at the United States Naval Academy. You are an inspiration. At the risk of leaving someone out, I must at least mention a few colleagues who helped me with some tricky physics: C. Elise Albert, Peter G. Brereton, Charles A. Edmonson, John P. Ertel, Irene Engle, Jeffrey A. Larsen, Paul T. Mikulski, and Carl E. Mungan. I also thank colleagues who are my role models: Francis David Correll, Robert Shelby, and Donald Treacy.

Fifth, I thank my friends who helped keep me going both emotionally and intellectually: Sallie Gentry, J. Allie Hajian, Milena Higgins, Max and Tess Light, Heidi Manning, Stephen Winchell, and Hui Yang, and the way too many people to name from Ridgley Retreat. Adam Geremia, you've been a great friend, and your photo saved one of my favorite case studies from the chopping block; thanks. I also thank Lawrence Rudnick—my Ph.D. thesis adviser—who taught me how to teach, how to learn, how to write, and how to think. Finally, thanks to Flo and Mary for providing me with a writer's paradise in Northampton.

Most importantly, thanks to my family for support and for allowing me to take the enormous amount of time I needed to work on this book. How can I ever express how grateful I am to Jeff and Zak for allowing me to pursue this dream? Special thanks to Jeff—the best husband on the planet—and also my ghost editor. To the rest of my family: I am not mentioning you by name here because if you read my book carefully, you will find your names throughout the book as my thank-you to you. I love you all!

Sincerely,
Debora

Preface for the Students

If you have taken the time to read this, I am truly proud of and impressed by you! You are off to a great start. Physics is a difficult subject for most students, but *reading your textbook will help you to succeed.* Many textbooks were originally written for previous generations of students. (Now these students are old folks like me.) Although these textbooks have been updated for content, today's students don't find them to be readable. My students easily read 100 pages of history but find it difficult to get through five pages of one of these old physics textbooks. My primary goal in writing this book was to write something that *you* would actually enjoy reading—something that you could learn from.

I may not have met your physics instructor, but I can tell you one thing about him or her. He or she *loves* physics because physics is fun. Physics allows you to answer so many interesting questions. Your instructor wants you to find the joy in applying physics to questions that you care about.

Our job—mine as the author of this book, your instructor's as your mentor, and *yours* as the only person who can teach you anything—is to get you to *think like a physicist*. In other words, we want you to learn how to ask and answer questions using physics. So you will spend a lot of your time in this course working on assigned questions and problems. You must *actively* answer and solve these problems. Physics is not learned passively. But don't worry: you are not alone. Your instructor will find good problems for you to work on, and you will have other people to help you when you're stuck. And you have this textbook. Recent editions of traditional textbooks have begun to include problem-solving strategies. However, many steps in their example problems are still not well explained, and often seem mysterious and opaque to students. I have presented my example problems and major derivations using a two-column format. The column on the right contains the formalism you would typically find in a traditional textbook, or what an instructor might display in a classroom lecture. The left column includes my thoughts. *I have done my best to let you see inside my head while I solve example problems*. Use these two-column examples to see how a physicist thinks.

I have included other features to help you learn physics. For example, it turns out that most physicists (for example, your instructor and me) are *linear learners*. We are perfectly happy learning one detail after another without first seeing the big picture. Most physics students are *global learners*. You probably want the big picture before you are willing to get into the details. So the first page of every chapter begins with an outline and a list of the topics to be covered in the chapter.

Another difference between physics instructors and their students is that instructors see the physics topics arranged in a hierarchy. To them, some topics are more general, and therefore more important, than others. But students have trouble organizing physics topics in this way. So the list of the topics at the beginning of each chapter is arranged into categories. Then throughout each chapter, you will find notes in the margins that refer to the topics using these categories. Finally, each chapter has a summary at the end, before the problems and questions. The summary is arranged using the same hierarchy.

Throughout each chapter are *concept exercises*. These are short exercises you should do to break up the reading. They will help you make sure that you have understood what you have just read. The answers to these *concept exercises* appear near the end of the book, where you will also find the answers to a number of selected end-of-chapter problems and questions.

Students are often surprised to find that physics requires much of the mathematics they have learned in earlier classes. If you have forgotten some of this math, check Appendix A ("Mathematics"). I have included a short review of mathematical topics and formulas. You will find that Appendix B ("Reference Tables") is full of handy lists, such as a list of Greek letters (yes, we use them a lot in physics), and lists of conversion factors and data.

Some of you will find physics to be an easy subject. That is fantastic. Perhaps you will go on to get a Ph.D. in physics, and to solve some very cool problems. Others will find physics a tough subject. I cannot change how you feel about physics, but I hope I have written a book that will help you learn physics—and that you come to find pleasure in studying a subject that has brought me much joy and happiness.

All my best to you!

Debora

PART THREE
Electricity

Electric Forces

23

❶ Underlying Principles

1. Conservation of charge
2. Electrostatic force
3. Coulomb's law

⭐ Major Concepts

1. Charge
2. Quantization of charge
3. Conductors and insulators
4. Ground

⊙ Tools

Hybrid free-body diagram

Key Questions

What are two ways to give objects a net charge?

How do we find the magnitude and direction of the electrostatic force exerted by one charged object on another?

Why do charged objects attract neutral ones?

23-1	Another fundamental force	684
23-2	Models of electrical phenomena	685
23-3	A qualitative look at the electrostatic force	688
23-4	Insulators and conductors	690
23-5	Coulomb's law	696
23-6	Applications of Coulomb's law	699

You stick a decal on the rear window of your car. The sticker can easily be peeled off and put back on again. You decorate for a party by rubbing balloons on your head, and then they stay on the wall (Fig. 23.1). You slide out of the car and find that your skirt is clinging to your legs. After running the dryer without a dryer sheet, you find that your socks are stuck to your clothes. These are all examples of what is commonly called "static cling." Static cling is a result of the *electric force* acting between objects. In physics terms, we say that an electrostatic force attracts the decal to the car's window, the balloons to the wall, your skirt to your legs, and the socks to your clothes. The electric force is one of the fundamental forces in physics and forms the basis of what we study throughout Part III.

FIGURE 23.1 Visible evidence of the electric force—static cling makes the air-filled balloons stick to the wall.

23-1 Another Fundamental Force

Until the middle of the 19th century, scientists believed that electricity and magnetism were two separate forces. Then the Scottish mathematician and theoretical physicist James Clerk Maxwell discovered that the two forces were *unified*. In other words, the electric force and the magnetic force are parts of a more fundamental force known today as the **electromagnetic force**.

Today, the electromagnetic force is considered one of the four fundamental forces (Section 5-10), and Maxwell's work has inspired physicists to look for theories and experimental evidence showing that all the fundamental forces are unified into a broader, more general force (Fig. 23.2). The electromagnetic and weak nuclear forces are also combined in what is called the **electroweak force**. There are theories—known as grand unification theories (GUTs)—that support the idea that the electroweak force and the strong force may be unified. However, as yet no experimental evidence supports these GUTs. Finally, theories of how gravity may be unified with the other three forces are still being worked out. Physicists hope that such a theory will be developed and supported so that there will be a single *theory of everything*.

The theory of everything and GUTs are beyond the scope of this book. In fact, the strong and weak forces are only briefly mentioned. However, Figure 23.2 gives us perspective. In this part of the book, we study electricity—one of the fundamental branches of physics. In the next part, we tackle magnetism. By the end of Part IV, we learn how Maxwell discovered the unification of these two forces. Part V (optics) is devoted to phenomena that are governed by their unification.

Of all the forces shown in Figure 23.2, the only force we have encountered by name in earlier chapters is gravity. So you might wonder how all the other forces we have studied—friction, the normal force, tension, drag, the spring force—fit into the framework of only four fundamental forces. All of these forces (other than gravity) are macroscopic manifestations of the electromagnetic force acting on the microscopic level between atoms and molecules. In fact, much of what we study in the rest of this book involves the forces between particles on the microscopic level, which have important consequences on the macroscopic level, as illustrated by the next case study.

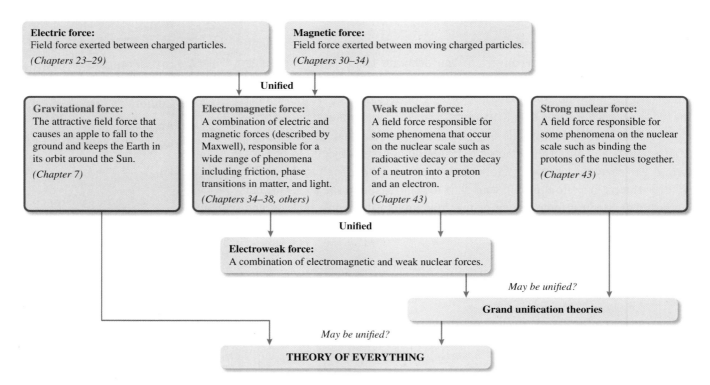

FIGURE 23.2 A concept map showing the relationships of the four fundamental forces (outlined in red) and their known or proposed unification.

CASE STUDY Fires at the Gas Pump

On the evening of December 12th, 2004, a young woman drove up to a gas pump. She removed the gas cap and put the pump's nozzle in her SUV's tank. (Fig. 23.3 shows a similar situation.). Then she went back inside her SUV and slid across the seat. When she got back out of the car and reached for the nozzle, flames erupted from the nozzle. Fortunately the fire was relatively small, and no one was harmed.

Such accidents can be more serious. Vehicles have been totaled; people have been seriously burned, and some have even died. From September 1999 through January 2000, 36 fires during refueling were reported to Robert N. Renkes, Executive VP and General Counsel for the Petroleum Equipment Institute (PEI). As a result, PEI asked readers of its newsletter and website to report incidents of refueling fires. Here is what PEI found:

1. Most people (94%) who reported on the type of footwear they were wearing at the time of the fire said they were wearing rubber-soled shoes. We'll assume the woman in the case study was also wearing rubber-soled shoes.
2. Many of the people reported that they had gone back inside the vehicle after they put the nozzle in their tank.
3. One person reported that she had been petting her small dog, which was in her arms. She put the dog down, and when she reached for the nozzle, a fire started.
4. Many people reported hearing a popping sound or feeling a shock just before the fire began.
5. No one reported using a cell phone while fueling. In other studies, researchers tried to ignite gasoline vapor with cell phones and were unable to do so.

Warning signs are posted at gas pumps. Typically these signs indicate that smoking and using a cell phone or other electronic devices are forbidden. What advice would you post on warning signs in order to prevent the sort of refueling fire started by the woman in our case study?

FIGURE 23.3 A woman pumps gas into a vehicle.

23-2 Models of Electrical Phenomena

Let's begin by turning our attention back 250 years. In the mid 1700's, people performed electrostatic demonstrations for entertainment in the evenings. Figure 23.4 shows a gathering in 1746. The boy suspended in the center of the group has been *charged*. The woman on the right is about to be shocked as she touches the boy's nose.

Benjamin Franklin (1706–1790) hosted and attended such parties, and he developed many of the concepts we study in this chapter. Franklin—one of the Founding Fathers of the United States and a signer of the Declaration of Independence—was first a scientist and inventor. As we'll discuss, Franklin made major contributions to the field of electricity.

On a dry day, such as in the middle of winter, you might find that after you comb your hair, it is attracted to the comb. Such phenomena were the basis of evening entertainment in the 18th century. People rubbed objects together such as wool against amber or silk against glass (Fig. 23.5A). They found that after rubbing, the objects were attracted to each other; a silk cloth was attracted to a glass rod that it was rubbed against (Fig. 23.5B)—just as your comb is attracted to your hair. They also found that rubbed objects could be used to attract other objects. For instance, small bits of paper were attracted to a comb that had been rubbed through hair (Fig. 23.5C).

Franklin's explanation of such phenomena is at the heart of our contemporary understanding of electricity. According to Franklin, all objects are full of an *electric fluid*. When you bring two objects close together, it is possible to transfer some electric fluid from one object to the other. Afterward, one object has a surplus of

FIGURE 23.4 During Benjamin Franklin's time, people entertained one another with static electricity demonstrations. Here Jean-Antoine Nollet is using a rubbed rod to charge a boy suspended by a silk rope. The charged boy makes bits of paper fly upward. A woman is about to discharge the boy and create a spark by touching his nose.

FIGURE 23.5 A. Silk is rubbed against glass, and afterward **B.** the silk is attracted to the glass. **C.** After running a comb through hair on a dry day, the comb attracts paper.

electric fluid and the other has a deficit of electric fluid. Franklin said an object with a surplus of electric fluid is *positive* or *plus*, and an object with a deficit of electric fluid is *negative* or *minus*. According to Franklin, both a glass rod and a silk cloth have some amount of electric fluid initially, and after they are rubbed together, some of the electric fluid is transferred from the silk to the glass (Fig. 23.6A). The silk then has a deficit of electric fluid, so Franklin said the silk is *negative*. The glass has a surplus of electric fluid, so he said it is *positive*. Franklin made an arbitrary choice in this model; he assumed the electric fluid was transferred from the silk to the glass. There was no way for Franklin to know whether that was true. Because there was no experimental evidence to determine which way the electric fluid flowed, Franklin could have imagined that electric fluid was transferred from the glass to the silk. Subsequent scientists have kept his arbitrary choice, however, and that has important implications for our contemporary model of electric charge.

Today we have experimental evidence supporting the model that all objects are made up of atoms, and atoms are made up of three types of particles—neutrons, protons, and electrons. The neutrons and protons are tightly packed into the central region of the atom known as the nucleus. The electrons move rapidly outside the nucleus, forming a cloud. According to this contemporary model, when two objects are brought near each other, it is possible for electrons that are loosely bound to the atoms in one object to move to the other object. The protons and neutrons are tightly bound in the nucleus, and so they are not transferred. Therefore, when glass is rubbed against silk, electrons are transferred from the glass to the silk. Afterward, the glass has a surplus of protons and the silk has a deficit of protons (Fig. 23.6B).

Our sign convention is based on the relative number of electrons and protons in a given object. An object that has an equal number of electrons and protons is said to be **neutral**. An object that has a deficit of electrons has a surplus *of protons*; such an object is said to be **positive**. An object that has a surplus of electrons has a *deficit of protons*; such an object is said to be **negative**.

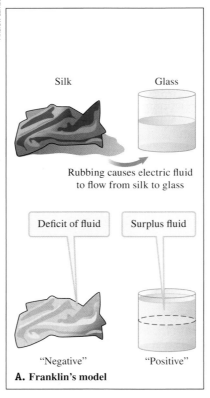
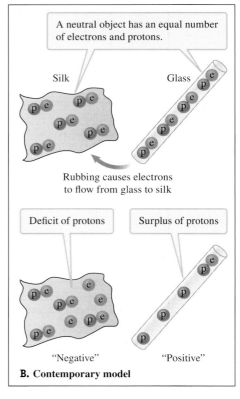

FIGURE 23.6 Comparison of Ben Franklin's model and our contemporary model of electric charge.

Charge

Subatomic particles have an intrinsic property called **charge** that determines the particle's role in electrical phenomena. The charge (symbolized by q or Q) of any object or particle may be positive, negative, or zero.

The SI unit of charge is the **coulomb**, abbreviated C. Just as the mass of subatomic particles is an intrinsic property determined by experimentation, so is charge. Experiments have shown that a neutron has no charge, a proton's charge is positive, and an electron's charge is negative. Experiments have also shown that the magnitudes of the proton and electron charges are equal. This magnitude is known as the **elementary charge** e. The charge of a proton to four significant figures is

$$q_{\text{proton}} = +e = 1.602 \times 10^{-19} \text{ C}$$

and so the charge of an electron is

$$q_{\text{electron}} = -e = -1.602 \times 10^{-19} \text{ C}$$

An object is said to be **neutral** or **uncharged** if it contains an equal number of electrons and protons. An object is **positively charged** if the number of protons is greater than the number of electrons, and the object's charge q is

$$q = +Ne \tag{23.1}$$

where N is the number of excess protons. An object is **negatively charged** if the number of electrons is greater than the number of protons, and the object's charge q is

$$q = -Ne \tag{23.2}$$

where N is the number of excess electrons. Generally, the number N of excess protons or electrons is small compared to the total number of protons or electrons in the object.

Because the charge of any object is an integer N times the elementary charge e, we say that charge is **quantized**, which means that charge is delivered in small packets or multiples of the elementary charge. So the charge of an object may be

$$q = \pm 1e, \pm 2e, \pm 3e, \ldots \pm Ne$$

An object cannot have a fractional number of excess electrons or protons—an object cannot have a charge of $\pm 1.75e$, for example.

When we study macroscopic objects such as glass rods and plastic combs, we are unaware of the quantization of charge because we cannot detect a charge as small as e in ordinary observations. For example, suppose you rub a glass rod with silk just until you can tell that the rod is charged (due to its behavior). Afterward, suppose you measure the charge of this glass rod, and you find it is $q = (53.4 \pm 0.1)$ nC (a pretty accurate measurement). The error in your measurements is $q = 0.1$ nC, which amounts to

$$\delta N = \frac{\delta q}{e} = \frac{1 \times 10^{-10} \text{ C}}{1.602 \times 10^{-19} \text{ C}} = 6 \times 10^{8} \text{ protons}$$

You cannot come close to detecting single units of the elementary charge by conducting such an experiment. So, on the macroscopic scale, it seems like charge is continuous because each packet of elementary charge is so small.

In Franklin's model, electric fluid is transferred from one object to another without being created, lost, or destroyed in the process. **Conservation of charge** is also an underlying principle in our contemporary model of physics. According to this principle, the net charge in the universe is constant. If you rub glass with silk and find that the glass rod's charge has increased by, say, 15 nC, then the silk's charge must have decreased by the same amount, or –15 nC.

Sketching Charge

By now you know that a good sketch is important in solving problems and understanding physics in general. We cannot possibly draw all the atoms in an object or even the number of excess charged particles, which can be in the millions.

CHARGE ✪ Major Concept

QUANTIZATION OF CHARGE ✪ Major Concept

It has been discovered that protons are not fundamental particles. Instead, they are made up of quarks. Quarks have a fractional charge, but because an object cannot have a single "excess" quark, this will not affect our discussions.

CONSERVATION OF CHARGE ❗ Underlying Principle

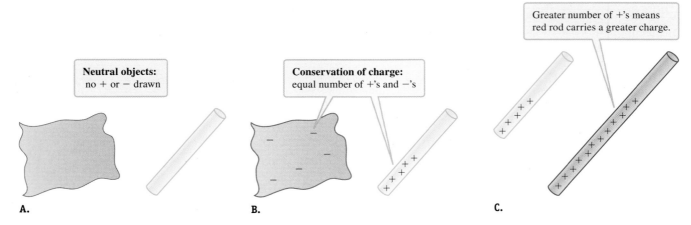

FIGURE 23.7 When you are sketching, indicate only the excess charged particles. **A.** Neutral objects are shown without any + or − signs. **B.** The rod and cloth are charged by rubbing. When charge is transferred from one neutral object to another, both become charged. The number of excess positive particles on one object must equal the number of excess negative particles on the other object. **C.** The two rods were charged independently. (The cloths used to charge these rods are not shown in the sketch.)

Instead, we represent the excess charge with plus or minus signs (Fig. 23.7). The actual number of signs is somewhat arbitrary, but usually between 1 and 12 will work. A sketch, however, should illustrate the fact that charge is conserved. So, if charge is transferred from one neutral object to another (Fig. 23.7A), the number of pluses on one object must equal the number of minuses on the other object (Fig. 23.7B). Charge conservation does not mean that the number of pluses and minuses must be the same on all sketches. For example, if one object has more positive charge than another object, you should draw more plus signs on that object (Fig. 23.7C).

CONCEPT EXERCISE 23.1

Initially a glass rod and a piece of silk are neutral. After you rub the silk against the rod, the glass rod has a surplus of 3.33×10^{11} protons. What is the charge q of the silk?

23-3 A Qualitative Look at the Electrostatic Force

Franklin learned that some rubbed objects attract one another and some repel one another. Suppose you try some of Franklin's experiments with two glass rods and two silk handkerchiefs. Initially, you find that there is no attraction or repulsion between any of these objects. Then you rub one glass rod with one of the silk handkerchiefs and the other rod with the other handkerchief (Fig. 23.8A). You find that (1) both glass rods attract the silk handkerchiefs (Fig. 23.8B); (2) the silk handkerchiefs repel each other (Fig. 23.8C); and (3) the glass rods repel each other (Fig. 23.8D). This simple experiment illustrates important properties of the **electrostatic force**, the force exerted by one charged object on another charged object. In particular, the electrostatic force depends on the charges of the objects involved:

ELECTROSTATIC FORCE
❶ Underlying Principle

1. *Neutral objects do not attract or repel each other.* The silk and glass were initially uncharged, and there was no attraction or repulsion between them.
2. *Oppositely charged objects are attracted to each other.* The negatively charged silk is attracted to the positively charged glass (Fig. 23.8B).
3. Two negatively charged objects repel each other, such as the silk handkerchiefs (Fig. 23.8C). Two positively charged objects repel each other, such as the two glass rods (Fig. 23.8D). To summarize: Objects that have charges of the same sign are mutually repelled.

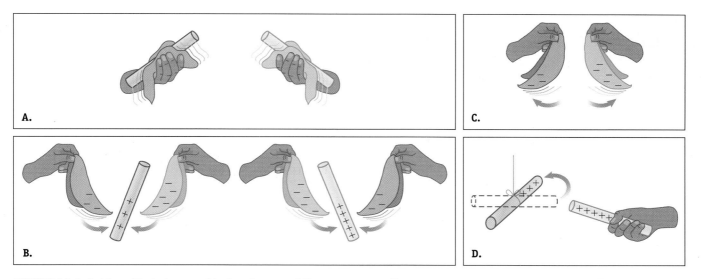

FIGURE 23.8 **A.** Two silk cloths are rubbed against two different glass rods. **B.** Each silk cloth is attracted to each rod. **C.** The two silk cloths repel each other. **D.** The two glass rods repel each other.

You may find it helpful to use the phrase *opposites attract and likes repel* to remember that the electrostatic force between objects that have charges of the opposite sign is attractive, and objects that have charges of the same sign are mutually repelled.

You may wonder why the electrostatic force isn't called simply the *electric* force. The *static* part is included because we are looking at the force between static (nonmoving) charged particles. Of course, electrons must move from one object to another in order to charge the objects, but, for now, we are not concerned about the forces exerted during that transfer. We are concerned with the force one charged object exerts on the other after the movement has stopped.

When charged particles move, we must also think about magnetism. Until Chapter 30, we will assume that magnetism is negligible even if charged objects move.

CONCEPT EXERCISE 23.2

a. In Figure 23.8, why are there three plus signs on the red rod and three minus signs on the red cloth?
b. Which object in Figure 23.8 has the greatest positive charge? How do you know?

CONCEPT EXERCISE 23.3

When wool is rubbed against amber, the wool becomes positively charged and the amber becomes negatively charged. If you rub a glass rod with a silk cloth and an amber rod with a wool cloth, are the rods attracted to or repelled by each other? Are the cloths attracted to or repelled by each other? Is the silk cloth attracted to the amber rod or repelled by the amber rod? Is the wool cloth attracted to the glass rod or repelled by the glass rod? Explain your answers.

CONCEPT EXERCISE 23.4

Three objects A, B, and C are charged. Suppose object A is repelled by object B and attracted to object C.

a. Is object B attracted to object C? Explain.
b. Is it possible to find three charged objects A, B, and C such that A and B attract each other, and A and B are both attracted to C? Explain.

FIGURE 23.9 Why does this technician wrap rubber over these electrical lines before repairing them?

INSULATOR ⭐ Major Concept

23-4 Insulators and Conductors

Before beginning repair work, a technician covers electrical lines with rubber sheets (Fig. 23.9). He may also wear a rubber suit or rubber sheets over his arms. Why does all this rubber protect him from getting shocked?

The electrons in some materials—such as rubber—do not move freely. Such materials are known as **insulators**. When a surplus of charged particles (positive or negative) builds up on some part of an insulator, the excess remains there. So, if you hold one end of a rubber rod while the far end is being rubbed with silk (Fig. 23.10A), the far end of the rod will acquire a surplus of electrons and those electrons will remain at that end, never flowing into your hand (Fig. 23.10B).

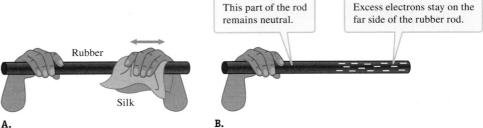

FIGURE 23.10 A. When a rubber rod is rubbed with a silk cloth, **B.** the excess charge stays on the part of the rod that was in contact with the cloth.

CONDUCTOR ⭐ Major Concept

If you try to charge a copper rod in the same way, you will find that you cannot build up charge on the copper. Copper is an example of a **conductor**. A conductor is a material in which the charged particles (usually electrons) can flow freely. When you hold one end of the copper rod and rub the other end with silk, electrons are transferred from the silk to the copper rod, and those excess electrons are free to flow. Because *likes repel*, the electrons move away from one another, which means they travel through the rod into your hand (Fig. 23.11). The human body is also a conductor, so charged particles move freely through your body toward the Earth. If there are no insulators between you and the ground—such as when you are barefoot—the charge will continue to flow into the Earth. As a result, you and the copper rod remain neutral despite the rod being rubbed with silk.

GROUND ⭐ Major Concept

The Earth often serves as a charge reservoir known as a **ground**. A ground can accept or provide electrons freely, and it is so large that the addition or subtraction of electrons has a negligible effect on it. So, the ground remains essentially neutral at all times. The copper rod in Figure 23.11 is connected to the ground through your body. When something is connected to the ground by a conductor, we say that it is *grounded*. Every building (with a contemporary electrical system) has a wire connected between the electrical system and a copper pipe. The copper pipe is

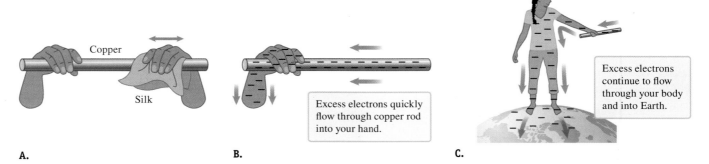

FIGURE 23.11 A. When you rub a copper rod with a cloth, **B.** the excess charge moves all over the rod and **C.** through your body into the Earth.

connected through other copper pipes to the Earth. This ensures that the third connection on electrical sockets is *grounded* (Fig. 23.12). The connection to the ground keeps electrical appliances from building up excess charge.

Microscopic Model

Metals, tap water (which contains impurities), and the human body (which is similar to tap water) are all good conductors. Nonmetals such as rubber, plastic, glass, silk, wool, dry air, and pure water are good insulators. What makes some materials conductors and others insulators?

To answer this question, let's take a microscopic look at a metal conductor such as copper. Each atom of copper has 29 protons in the nucleus and 29 electrons swarming around that nucleus in a cloud. Much as the swarm of asteroids in the asteroid belt is held in place by gravity, the cloud of electrons is bound to the copper nucleus by electrical attraction. The outermost electrons, however, are more weakly bound to the nucleus than are the inner electrons. When many copper atoms are grouped together as they are in a piece of metal, the binding of each atom's outermost electron is further weakened by the presence of the other atomic nuclei. In fact, the outermost electron is no longer bound to a particular atom. Instead, the outermost electrons from all the atoms are free to move around the entire piece of metal (Fig. 23.13A). Typically, a metal has one free electron per atom, but some metals may have more than one free electron per atom.

In an insulator, the electrons are more tightly bound to their particular nucleus. When many insulator atoms are grouped together in a substance, the electrons of each particular atom stay bound to that atom (Fig. 23.13B).

Charging by Direct Contact

From Figure 23.11, we see that if we wish to build up charge on a conductor, we must insulate it from any large objects that might serve to ground the conductor. Figure 23.14A shows one possible solution—a spherical conductor is placed on a pedestal made of an insulator. As always, electrons may flow throughout the conductor, but now they do not have a conducting pathway to a ground. You can charge the conductor by touching it with a charged object such as a negatively charged plastic rod (Fig. 23.14B). When the rod is in contact with the sphere, some of the electrons are transferred from the rod to the sphere. Remember that charge cannot flow through the rod because it is an insulator, so you may need to roll the rod around the surface of the sphere in order to transfer charge from many parts of the rod. Because electrons move freely throughout the conductor and because *likes repel*, the electrons quickly redistribute themselves in the conductor, moving as far apart as possible. For a spherical conductor, "as far apart as possible" means that the electrons are uniformly distributed on the outside surface of the sphere. If you try to charge a conductor of another shape, the charge is again distributed on the outside surface, although for nonspherical shapes the charge distribution is not uniform.

FIGURE 23.12 The third prong of a three-prong electrical plug is connected to the ground to prevent an appliance from building up dangerous excess charge.

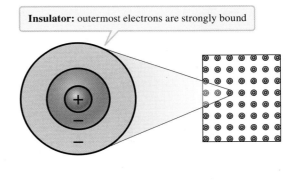

FIGURE 23.13 A. The outermost electrons move freely throughout a conductor. **B.** In an insulator, the electrons of each particular atom stay bound to that atom.

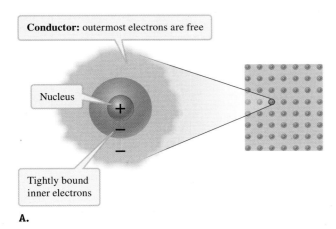

A.   B.

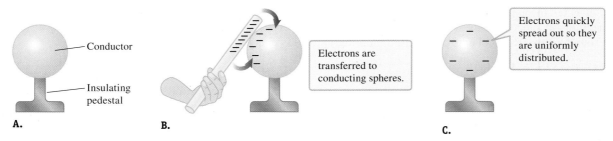

FIGURE 23.14 A. A conductor rests on top of an insulating pedestal. **B.** When the conductor is touched by a charged rod, some charge is transferred from the rod to the conductor. **C.** The excess charge quickly distributes over the surface of the conductor. The insulating pedestal prevents the excess charge from flowing into the Earth.

Charging by Induction

There is another way to build up charge on a conductor. Suppose you use the same equipment as in Figure 23.14. This time, however, you hold the negatively charged rod near but *not* touching the sphere (Fig. 23.15). Because the electrons in the sphere are free to move, they flow to the side opposite the rod. If you ground the side of the conducting sphere where there is a surplus of electrons, those electrons will flow to the ground. When you remove the connection to the ground, the net charge of the sphere is positive and the ground remains essentially neutral. No electrons are lost from the rod; it has the same charge throughout the process. In fact, you could reuse the rod to charge another conductor by induction without having to recharge the rod itself.

In this process—known as **induction**—the charged rod never touches the sphere. The charge that is induced on the sphere has the opposite sign as the charged rod. This is true whenever you charge an object by induction. Compare this to Figure 23.14, in which the sphere is in direct contact with the rod and the sphere ends up being negatively charged—the same as the rod. When you charge an object through direct contact with another charged object, both objects have charges of the same sign as a result of their contact.

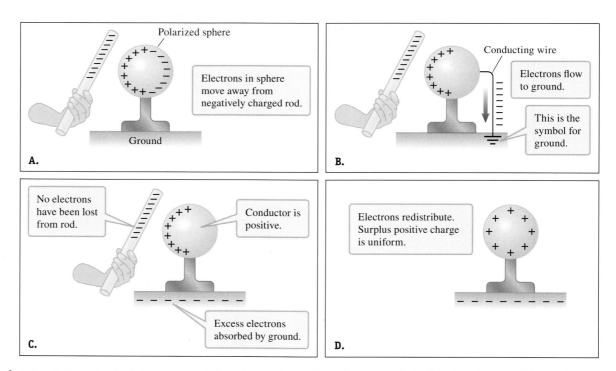

FIGURE 23.15 A. When a charged rod is held near a neutral conductor, the conductor becomes polarized. **B.** The far side of the conductor is connected to the ground, and some electrons flow out of the conductor into the Earth. **C.** The connection is removed. **D.** When the rod is removed, the electrons redistribute, leaving the surplus positive charge uniformly distributed over the surface of the conductor.

Polarization and Induced Dipoles

The conducting sphere in Figure 23.15A has been *polarized* by the charged rod. The term **polarized** describes an object that is neutral but has a net separation between its positive and negative charged particles. If you remove the rod, the electrons will move back and the sphere will not be polarized.

If you wish to entertain your friends at a Franklin-style party, here are instructions for a great trick. You need a glass swizzle stick, an empty metal can, and a silk handkerchief. Place the empty metal can on its side on a flat, horizontal surface. Charge the glass stick by rubbing it with your silk handkerchief. The glass stick will then be positively charged. Hold the charged glass stick near the can. It is best to align the long axis of the stick with the long axis of the can (Fig. 23.16A). The can is neutral, but because it is a conductor, the presence of the positively charged glass stick polarizes the can when electrons in the can move closer to the stick. The negatively charged particles are closer to the stick than are the positively charged particles, so the net effect is that the can is attracted to the stick and begins to roll along the table (Fig. 23.16B). With practice, you can get the can to roll quickly. Then by holding the stick on the other side of the can, you can make it slow down, stop, and reverse direction.

This party trick demonstrates why neutral insulators and charged objects are attracted to one another. For example, rubber balloons rubbed on hair are negatively charged and attracted to the ceiling, a neutral insulator. Likewise, a comb rubbed through hair is positively charged and can be used to pick up paper, another neutral insulator. The charged particles in an insulator are not free, so an insulator cannot be polarized as a whole. However, the atoms and molecules that make up an insulator can be polarized. When this happens, there is a slight separation of the positive from the negative charge within the atom or molecule (Fig. 23.17). We say that the atoms (or molecules) have become *dipoles*. A **dipole** is a neutral object in which the charged particles are distributed so that there is a net positive side and a net negative side. So, if a positively charged object (a comb) is brought near an insulator (paper), the atoms (or molecules) in the paper become dipoles such that their individual electron clouds are shifted slightly toward the charged object. The net effect of this slight shift is that the insulator is attracted by the charged object.

Sparks

Any insulator will allow conduction under extreme conditions. Under normal conditions, dry air is a good insulator. (Moist air, such as on a humid day or in a rainstorm, does not insulate as well as dry air.) When two oppositely charged objects are separated by air, charge is not usually transferred through air from one object to the other. However, if the charged objects are extremely large or the objects are very close together, it is possible for the air to transfer the charge momentarily. When this happens, we often see and hear a dramatic spark—as in the case of lightning.

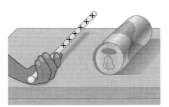

A.

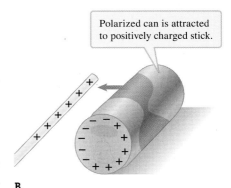

B.

FIGURE 23.16 A. A charged rod is held near a neutral metal can. **B.** The can becomes polarized.

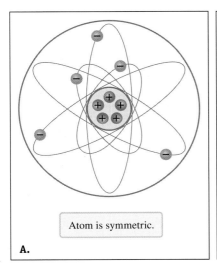

Atom is symmetric.

A.

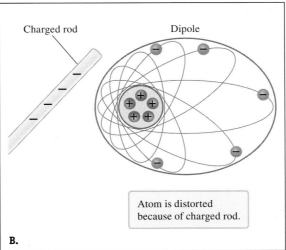

Atom is distorted because of charged rod.

B.

FIGURE 23.17 A. A neutral symmetrical atom. **B.** When a charged rod is held near the atom, the atom is distorted so that it is best modeled as a dipole.

You have probably experienced somewhat less dramatic cases of sparks. For example, if you pull apart clothes stuck together by static cling, you may hear a crackling sound due to charged particles moving through the air between the clothes. On a dry winter day, if you rub your socks on the carpet as you walk and then reach for the doorknob, you may see, hear, and feel a spark just before you touch the knob. Sparks were most entertaining for people in the 18th century. The woman in Figure 23.4 is about to draw a spark by touching the charged boy's nose. Some people found it amusing to charge up a woman so that when a man kissed her, he felt a shock.

CONCEPT EXERCISE 23.5

The following scenarios involve a metal ball and a charged glass rod. No other objects are involved.

a. A small metal ball is attracted to a negatively charged glass rod. Is the metal ball necessarily positively charged? Explain.
b. A small metal ball is repelled by a negatively charged glass rod. Is the metal ball necessarily negatively charged? Explain.

CONCEPT EXERCISE 23.6

a. Why is the charged boy in Figure 23.4 suspended by a silk rope? Could the rope be replaced by a metal chain?
b. Why is the boy able to attract paper?
c. Why will the woman draw a spark when she touches the boy's nose?

EXAMPLE 23.1 "Improvise!" or "You Don't Have to Dress Like James Bond to be Cool"

Suppose you wish to perform the metal can trick (Fig. 23.16), but you cannot find a glass stick and you don't wear a silk handkerchief in your lapel pocket. You can find a plastic swizzle stick, and you are wearing a wool sweater. If you rub the plastic stick on your wool sweater, the plastic stick will become negatively charged—the opposite of the glass stick rubbed with silk. Will the trick still work? Explain.

:• INTERPRET and ANTICIPATE
We know from Figure 23.16 that a positively charged stick attracts a neutral conductor because electrons in the conductor that are free to move are attracted to the stick. The conductor becomes polarized, so the negative electrons are closer to the positive stick and there is a net attraction. The question is, "Will a negatively charged object attract a neutral conductor?"

:• SOLVE
When the negatively charged stick is held near the metal can, the electrons in the can are repelled. Those electrons that are free to move go to the far side of the can (Fig. 23.18). That leaves a surplus of protons on the side near the stick. The net effect is that the can is attracted to the stick. So the can rolls toward the stick just as if you had used a positively charged glass stick.

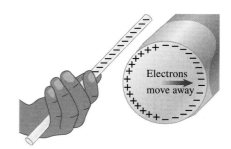

FIGURE 23.18

:• CHECK and THINK
We can make a more general statement about conductors: A neutral conductor is always attracted to a charged object, independent of the sign of the object's charge. (A similar statement can be made about neutral insulators.)

EXAMPLE 23.2 — CASE STUDY: Safety Precautions at the Gas Pump

In our case study, a woman returned to her car after putting the pump nozzle in her tank. She slid across the seat. When she got out of the car and reached for the nozzle, a fire erupted.

A Why did a fire erupt?

SOLVE
This is much like the situation in Figure 23.4. When the 18th-century woman reaches for the nose of the charged boy, there is a spark as charged particles are transferred between them (Concept Exercise 23.6). When the woman at the gas pump reached for the nozzle, charge was transferred and there was a spark. This time the spark passed through a mixture of gasoline vapor and air. Just as in the engine of your car, when the spark plug ignites the gasoline vapor-air mixture, the spark caused the gas fumes around the nozzle to ignite.

B Why is it significant that the woman slid across her seat in the car? Does the type of clothing matter?

SOLVE
In order for there to be a spark between two objects, one or both of the objects must have a net charge. In Figure 23.4, the boy has a net charge. The woman at the gas pump built up a net charge on herself by sliding across the seat. The seat was probably leather or vinyl, and her clothes were probably wool, cotton, or some synthetic fabric. Just as rubbing a plastic rod with a wool cloth transfers electrons and results in charging both objects, the woman's clothes and car seat were charged by her sliding motion. Without knowing the type of fabric in her clothes or the type of material on her car seat, we cannot be sure whether her clothes were positive or negative after sliding. In order to be specific, let's assume her clothes were negatively charged. This assumption won't affect the major points of the case study, however.

C Is it significant that the woman was wearing rubber-soled shoes?

SOLVE
The human body is a good conductor, and the woman's charged clothing either transferred charged particles to her or caused her to be polarized like the can in Figure 23.16. If she had been grounded before she came close to the nozzle, then she and her clothes would have been discharged as electrons flowed from her into the Earth. However, rubber is a good insulator. So, like the pedestal in Figure 23.14, her rubber shoes prevented the woman from discharging; thus, she had a net charge when she touched the nozzle.

D What advice should be posted at the gas pump?

SOLVE
1. Don't get back into your car while your gas pump nozzle is in the tank and there is likely to be gas vapor near the nozzle. Most of the time, you need to slide across the seat to get into or out of your car. When you do this, you are probably building up a net charge.
2. If you must get back into your car, be sure to ground yourself far from the nozzle. You may want to touch the ground or the metal supports of the gas station awning with your bare hand far from your open gas tank. Notice that these warnings agree with the precautions posted at the gas pump.

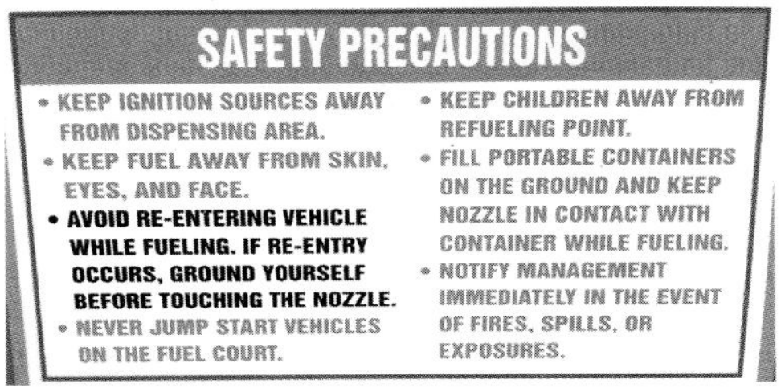

FIGURE 23.19

According to classical mechanics, photons (massless particles of light) are not subject to gravity, but according to general relativity, photons are affected by the presence of massive objects (Chapter 39).

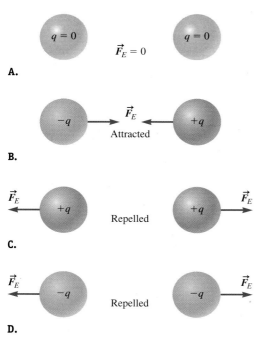

FIGURE 23.20 A. There is no electrostatic force between neutral objects. **B.** The electrostatic force between two oppositely charged objects is attractive. **C.** The electrostatic force between two positively charged objects is repulsive. **D.** The electrostatic force between two negatively charged objects is repulsive.

23-5 Coulomb's Law

Having explored some qualitative aspects of the electrostatic force, we are ready to develop some quantitative relationships by building on what we already know about another fundamental force, gravity. Both are field forces—as opposed to contact forces (Section 5-3). So, objects do not need to be in direct contact in order to exert an electrostatic or gravitational force on one another. Remember the Earth is not in direct contact with the Moon, yet they experience a mutual gravitational force.

According to the law of universal gravity $F_G = Gm_1m_2/r^2$ (Eq. 7.4), there is no gravitational force between massless particles; similarly, there is no electrostatic force between two neutral particles (Fig. 23.20A). Between two particles, one with a positive charge and one with a negative charge, however, the electrostatic force is attractive (Fig. 23.20B). The major difference between the gravitational force and the electrostatic force is that the gravitational force is always attractive, whereas the electrostatic force may be repulsive, as when particles have charges of the same sign (Fig. 23.20C and D). The gravitational force between two particles depends on an intrinsic property of the particles—their mass, and there is only one type of mass. Similarly, the electrostatic force between two particles depends on an intrinsic property of the particles—their charge. However, unlike mass, there are two different types of charge, positive and negative.

Coulomb's Experiments

The French engineer and scientist Charles Augustin de Coulomb (1736–1806) invented a torsion pendulum (Section 16-8) and adapted it as shown in Figure 23.21 to study electrostatic force. One small metal sphere is charged and fixed in place. A pair of small metal spheres is attached to a lightweight insulated rod. The rod is suspended from a torsion spring of known torsion spring constant.

Imagine that the pair of spheres is initially uncharged. The rod is twisted, bringing one of those spheres in contact with the charged fixed sphere and thus transferring charge. If the spheres are identical, half the charge on the fixed sphere will be transferred to the movable sphere. So these spheres repel each other, causing the rod to twist. The amount of twist is measured. Because Coulomb knew the torsion spring constant, he could calculate the force exerted between the spheres. In many trials, Coulomb varied the amount of charge on the spheres as well as the torsion spring constant. From these trials, he found:

The electrostatic force between two charged spheres is directly proportional to the product of their charges and inversely proportional to the square of the distance between them.

FIGURE 23.21 Torsion balance designed by Coulomb. Two (blue) spheres are attached to a rod suspended from a torsion spring. The fixed (red) sphere is positively charged, and then one of the movable spheres is brought in contact with the charged, fixed one. Because they are both conductors, charge is shared between the spheres, and the movable (blue) sphere becomes positive. Both spheres have charge of the same sign, so they are repelled, which causes the rod to rotate. The rod's rotation is measured using a scale (shown in yellow).

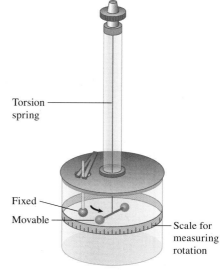

Today we call this **Coulomb's law**. It is best expressed mathematically:

$$F_E = k \frac{|q_1 q_2|}{r^2} \quad (23.3)$$

where F_E is the magnitude of the electrostatic force between two spherical objects that have charges q_1 and q_2 and whose center-to-center distance is r. The constant of proportionality k is called **Coulomb's constant**. It has a value of

$$k = 8.99 \times 10^9 \, \text{N} \cdot \text{m}^2/\text{C}^2$$

in SI units. Coulomb's constant k is easy to remember if you think of its value to one significant figure as *nine times ten to the ninth*.

Coulomb's experiments involved solid spherical metal balls, but his law holds for any two objects that have spherical symmetry. Coulomb's law can be applied to spheres, spherical shells, pointlike objects, particles, and any combination of such objects (Fig. 23.22). In the next chapter, we will learn how to find the electrostatic force between charged objects that do not have spherical shapes. Equation 23.3 cannot be used in such cases. For the rest of this chapter, we consider only objects that may be described by Equation 23.3, and for convenience we refer to these objects as "particles," although they may be any objects with spherical symmetry.

Inverse-Square Laws

Compare $F_G = G m_1 m_2 / r^2$ (Eq. 7.4) for universal gravity with $F_E = k|q_1 q_2|/r^2$ (Eq. 23.3) for the electrostatic force. Both forces follow an inverse-square law, meaning that the force is inversely proportional to the square of the separation of interacting objects. For gravity and the electrostatic force, the farther apart the particles are, the weaker the force. The magnitude of the gravitational force between two particles is directly proportional to the product of their masses. Likewise, the magnitude of the electrostatic force between two particles is directly proportional to the product of their charges.

Vector Form of Coulomb's Law

All forces are vectors, which means we must describe the direction as well as the magnitude of the electrostatic force. Because $F_G = G m_1 m_2 / r^2$ and $F_E = k|q_1 q_2|/r^2$ have mathematically similar forms, we can use our study of universal gravity to guide us in writing a vector equation. To write such an equation for universal gravity, we place one of the two particles at the origin of a polar coordinate system (Fig. 7.11, page 191). If we place particle 1 at the origin, then a unit vector $\hat{r}$ points from particle 1 toward particle 2 and $\vec{r} = r\hat{r}$. Because gravity is always an attractive force, the force exerted *by* particle 2 *on* particle 1 is in the positive r direction, and we write as in Section 7-3:

$$\vec{F}_{[2 \text{ on } 1]} = G \frac{m_1 m_2}{r^2} \hat{r}$$

Of course, by Newton's third law, the force exerted *by* particle 1 *on* particle 2 has the same magnitude but is in the opposite direction:

$$\vec{F}_{[1 \text{ on } 2]} = -G \frac{m_1 m_2}{r^2} \hat{r}$$

Let's write vector equations for the electrostatic force that acts on two charged particles by imitating our procedure for gravity. As in the case of gravity, we place particle 1 at the origin of a polar coordinate system (Fig. 23.23). A unit vector $\hat{r}$ points from particle 1 toward particle 2. If q_1 and q_2 are both positive (or both negative), they repel each other and the electrostatic force exerted *by* particle 2 *on* particle 1 is in the *negative* r direction. So Coulomb's law is written in vector form as

$$\vec{F}_{[2 \text{ on } 1]} = -k \frac{q_1 q_2}{r^2} \hat{r} \quad (23.4)$$

where $\vec{r}$ extends from particle 1 to particle 2 and $\vec{r} = r\hat{r}$. Compare Equation 23.4 with $\vec{F}_{[2 \text{ on } 1]} = (G m_1 m_2 / r^2)\hat{r}$. The gravitational force is always attractive, so the

COULOMB'S LAW
⚠️ **Underlying Principle**

We have used the letter k to symbolize other quantities, including the spring constant and Boltzmann's constant. One of your jobs is to keep these multiple uses of symbols straight.

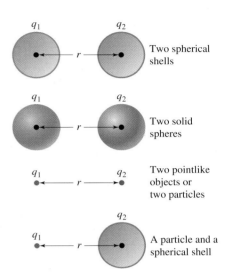

FIGURE 23.22 Coulomb's law may be applied directly to pairs or collections of objects, each with spherical symmetry.

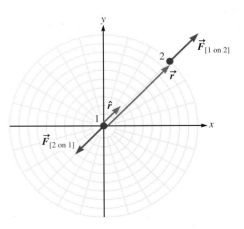

FIGURE 23.23 A coordinate system used to write Coulomb's law in vector form. Particle 1 with charge q_1 is at the origin. The unit vector $\hat{r}$ points from particle 1 toward particle 2.

gravitational force exerted by particle 2 on particle 1 is in the *positive r* direction. When particles have charges of the same sign, the electrostatic force is repulsive, so the electrostatic force exerted by particle 2 on particle 1 is in the *negative r* direction, and we must insert a minus sign in Equation 23.4, as we have done. If q_1 and q_2 have opposite signs, then the product $q_1 q_2$ is negative, which cancels out the negative sign in Equation 23.4. In this case, the force exerted by particle 2 on particle 1 is in the *positive r* direction. So Equation 23.4 shows that oppositely charged particles are attracted as expected.

Finally, by Newton's third law, the force exerted by particle 1 on particle 2 has the same magnitude but is in the opposite direction to that exerted by particle 2 on particle 1:

$$\vec{F}_{[1 \text{ on } 2]} = k \frac{q_1 q_2}{r^2} \hat{r} \quad (23.5)$$

It is left as an exercise (Problem 24) to show that Equation 23.5 is consistent with the rules *like charges repel* and *opposite charges attract*.

EXAMPLE 23.3 Electricity Wins!

A typical nucleus has a diameter on the order of 10^{-15} m and is surrounded by a cloud of electrons that extend out to distances of order 10^{-10} m from the center of the nucleus. Compare the electrostatic force exerted by an electron on a proton when they are separated by $r = 1.00 \times 10^{-10}$ m to the gravitational force exerted by that same electron on the proton at that separation.

:• INTERPRET and ANTICIPATE
Let's label the proton as particle 1 and the electron as particle 2. Then the unit vector $\hat{r}$ points from particle 1 toward particle 2 (Fig. 23.24).

FIGURE 23.24

:• SOLVE
Use Equation 23.4 to find the electrostatic force exerted by the electron on the proton. This is an attractive force (in the positive $\hat{r}$ direction). The charges on the proton and electron are $+e$ and $-e$, respectively.

$$\vec{F}_{[2 \text{ on } 1]} = -k \frac{q_1 q_2}{r^2} \hat{r} \quad (23.4)$$

$$\vec{F}_E = -k \frac{(e)(-e)}{r^2} \hat{r} = k \frac{e^2}{r^2} \hat{r} = k \left(\frac{e}{r}\right)^2 \hat{r}$$

$$\vec{F}_E = (8.99 \times 10^9 \, \text{N} \cdot \text{m}^2/\text{C}^2) \left(\frac{1.60 \times 10^{-19} \, \text{C}}{1.00 \times 10^{-10} \, \text{m}}\right)^2 \hat{r}$$

$$\vec{F}_E = 2.30 \times 10^{-8} \, \hat{r} \, \text{N}$$

Similarly, use the first equation on page 192 to find the gravitational force exerted by the electron on the proton. This is also an attractive force (in the positive $\hat{r}$ direction). The masses of the proton and electron are listed in the inside front cover.

$$\vec{F}_{[2 \text{ on } 1]} = G \frac{m_1 m_2}{r^2} \hat{r}$$

$$\vec{F}_G = G \frac{(m_p)(m_e)}{r^2} \hat{r}$$

$$\vec{F}_G = (6.67 \times 10^{-11} \, \text{N} \cdot \text{m}^2/\text{kg}^2) \frac{(1.67 \times 10^{-27} \, \text{kg})(9.11 \times 10^{-31} \, \text{kg})}{(1.00 \times 10^{-10} \, \text{m})^2} \hat{r}$$

$$\vec{F}_G = 1.01 \times 10^{-47} \, \hat{r} \, \text{N}$$

:• CHECK and THINK
Both forces are attractive. The gravitational force is much weaker (39 orders of magnitude smaller) because the particle masses are very small and because the gravitational constant G is very small. When we think about charged particles, it is usually safe to ignore the gravitational attraction between them because it is so much weaker than the electrostatic force. (It is hard to comprehend 39 orders of magnitude. It might help to know that the diameter of a proton is "only" about 36 orders of magnitude smaller than the diameter of the Milky Way galaxy; see Appendix B-4.)

EXAMPLE 23.4 Gravity Wins!

We assume the Earth and the Moon are electrically neutral, so there is no electrostatic force between them. However, if each possessed a net positive charge, it would be possible for the gravitational force to be balanced by electrostatic repulsion. If each object had the same charge q, what charge would be needed to balance ("cancel") the gravitational force between the Earth and the Moon?

INTERPRET and ANTICIPATE
If the gravitational force is balanced by the electrostatic force, we can set their magnitudes equal and solve for q.

SOLVE
Both charges equal q in Equation 23.3.

$$F_E = k\frac{|q_1 q_2|}{r^2} = k\frac{q^2}{r^2} \qquad (23.3)$$

Set the magnitude of the electrostatic force equal to the magnitude of the gravitational force. The Earth's mass is $M_\oplus$, and the Moon's mass is M_{Moon}. The center-to-center distance r cancels, and we solve for q.

$$F_E = F_G \qquad k\frac{q^2}{r^2} = G\frac{M_\oplus M_{\text{Moon}}}{r^2}$$

$$kq^2 = GM_\oplus M_{\text{Moon}} \qquad q = \sqrt{\frac{GM_\oplus M_{\text{Moon}}}{k}}$$

The masses of the Earth and the Moon are listed on the inside front cover.

$$q = \sqrt{\frac{(6.67 \times 10^{-11} \text{ N}\cdot\text{m}^2/\text{kg}^2)(5.98 \times 10^{24} \text{ kg})(7.35 \times 10^{22} \text{ kg})}{8.99 \times 10^9 \text{ N}\cdot\text{m}^2/\text{C}^2}}$$

$$q = 5.71 \times 10^{13} \text{ C}$$

CHECK and THINK
In order for gravity to be balanced by the electrostatic force, the Moon and the Earth would each need nearly 4×10^{32} surplus protons. The Moon has about 10^{48} atoms, and the Earth has about 10^{50} atoms. That means the Moon would need to lose only one electron for every 10^{16} atoms, and the Earth would need to lose one electron for every 10^{18} atoms. It is surprising how relatively few atoms would need to lose just a single electron in order for the electrostatic force to cancel the gravitational force between the Earth and the Moon.

23-6 Applications of Coulomb's Law

In this section, we practice applying Coulomb's law. We begin with a problem-solving strategy.

PROBLEM-SOLVING STRATEGY

Applying Coulomb's Law

Problems that involve Coulomb's law are no different from other problems that require us to apply Newton's second law, so the strategy developed in Section 5-8 works here. However, there are specific tips that you may find helpful.

:• INTERPRET and ANTICIPATE
As in many problem-solving strategies, a sketch is often a good start. Here we draw a *"hybrid"* free-body diagram, similar to the traditional free-body diagram used when applying Newton's second law, but also showing the *interaction* between charged particles, which is governed by Newton's third law.

1. Because Coulomb's law involves the interaction between at least two charged objects, it is often best to **represent each object**.
2. Coulomb's law applies to particles and objects that can be modeled as particles, so on hybrid free-body diagrams, **represent the objects as dots**. The distance r that appears in Coulomb's law is the center-to-center distance. When a spherical object is represented as a dot, **place the dot at the sphere's center**.
3. **Label information** about the charged particles on the diagram. For example, if object 1 has a charge of

Problem-Solving Strategy continues on page 700 ▶

−15 nC, write "$q_1 = -15$ nC" near its dot. If no information is given, it is best to assume the particle's charge is positive.

4. Usually you are interested in finding the total electrostatic force exerted on one of the particles, called the *subject*. **Circle the subject**. Then use your knowledge of the electrostatic force to **draw vectors representing the force exerted by all the other objects on the subject**. Place the tail of each vector on the subject. Although the magnitude of the forces is not critical at this stage, it is important to indicate the direction. Draw the vector toward a particular object if that object has the opposite sign as the subject (because the force is attractive). Draw the vector away from the object if that object has the same sign as the subject (because the force is repulsive).

5. **Indicate the subject's acceleration.**
6. **Choose a coordinate system.**

:• SOLVE

The drawing of the subject—with force vectors extending from it—is treated as a free-body diagram. Then Newton's second law is applied. Often that means:

Step 1. Writing the electrostatic force (Eq. 23.3) in component form, and then

Step 2. Algebraically combining the components.

Step 3. Be careful about signs. You already took the sign of the charge into account when you determined the direction of each vector. When you substitute in values for the charge, you (probably) need only the **absolute value of the charge**.

EXAMPLE 23.5 Static Equilibrium

Two small spherical insulators are separated by 2.5 cm, which is much greater than either of their diameters. Both carry positive charge, one $+60.0\ \mu C$ and the other $+6.66\ \mu C$. A third charged sphere remains at rest between the two spheres and along the line joining them. What is the position of this third charged sphere? In the **CHECK and THINK** step, answer this question: What can you say about the sign and magnitude of this charge?

:• INTERPRET and ANTICIPATE

Start with a hybrid free-body diagram (Fig. 23.25).

1. and 2. Each sphere is represented by a dot at its center.
3. The charges on the first two spherical insulators are labeled q_1 and q_2. Because we are given numerical information about the charges, that information is included. The charge on the third sphere is (arbitrarily) labeled q_0, and no numerical information is given about it.
4. We have circled the subject, q_0. Only two electrostatic forces are exerted on the subject: $\vec{F}_1$, the force exerted by q_1, and $\vec{F}_2$, the force exerted by q_2. We have assumed that q_0 is positive, so both objects repel the subject. Therefore, $\vec{F}_1$ and $\vec{F}_2$ are drawn *away* from their respective objects.
5. The acceleration is zero as indicated.
6. A coordinate system has been chosen.

The forces acting on the third sphere are balanced, so we expect the third charged sphere to be closer to the sphere that has the smaller charge.

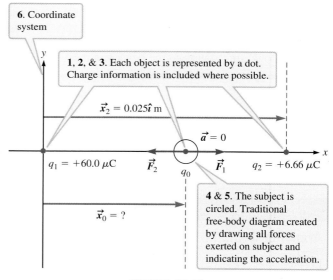

FIGURE 23.25

:• SOLVE

The region in the center of Figure 23.25 is like any traditional free-body diagram. We apply Newton's second law as usual, having chosen a coordinate system. Because the acceleration is zero, the sum of the forces is zero.	$\sum \vec{F} = \vec{F}_1 + \vec{F}_2 = 0$
These forces must have equal magnitude and point in opposite directions.	$\vec{F}_1 = -\vec{F}_2$ $F_1 = F_2$

The magnitude of the electrostatic force is given by Coulomb's law, Equation 23.3. Coulomb's constant k cancels, as does the subject's charge q_0.	$k\dfrac{\|q_1 q_0\|}{r_1^2} = k\dfrac{\|q_2 q_0\|}{r_2^2}$
	$\dfrac{\|q_1\|}{r_1^2} = \dfrac{\|q_2\|}{r_2^2}$ (1)
The center-to-center distances r_1 and r_2 come from Figure 23.25.	$r_1 = x_0$
	$r_2 = x_2 - x_0$
Substitute r_1 and r_2 into Equation (1), and solve for x_0. We choose the positive square root; the negative square root gives a value for x_0 that is not between the original two spheres.	$\dfrac{\|q_1\|}{(x_0)^2} = \dfrac{\|q_2\|}{(x_2 - x_0)^2}$
	$\left(\dfrac{x_2 - x_0}{x_0}\right) = \pm\sqrt{\dfrac{\|q_2\|}{\|q_1\|}}$
	$x_0 = x_2\left(\sqrt{\dfrac{\|q_2\|}{\|q_1\|}} + 1\right)^{-1} = (0.025 \text{ m})\left(\sqrt{\dfrac{6.66\ \mu\text{C}}{60.0\ \mu\text{C}}} + 1\right)^{-1}$
	$\boxed{x_0 = 1.9 \times 10^{-2}\text{ m} = 1.9 \text{ cm}}$

CHECK and THINK

According to our results, the subject is nearly 2 cm from sphere 1 and only about half a centimeter from sphere 2. So, as we expected, the subject is closer to the sphere that has the smaller charge. Because q_0 cancels, neither its magnitude nor its sign matters. In other words, we cannot tell whether q_0 is positive or negative, and we cannot determine its magnitude.

EXAMPLE 23.6 Right Triangle

Three small charged spheres lie on the vertices of a right isosceles triangle as shown in Figure 23.26. The sides of equal length are 0.654 m long. The right angle is at the origin of the coordinate system. The sphere at the origin has a charge of −15 mC. A sphere with a charge of +40 mC is on the x axis, and a sphere with a charge of +20 mC is on the y axis. Find the net electrostatic force exerted on the sphere at the origin. Give your answer in component form and in terms of magnitude and direction.

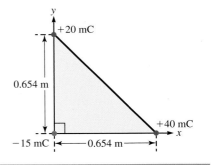

FIGURE 23.26

INTERPRET and ANTICIPATE

Figure 23.27 is the hybrid free-body diagram.

1. and 2. Each sphere is represented by a dot at its center.
3. The charge on each sphere is given in the figure. The sphere at the origin is the subject because we are asked to find the net electrostatic force exerted on it.
4. The subject is circled. Sphere 1 exerts an attractive force along the x axis, so $\vec{F}_1$ points along the x axis toward sphere 1. Sphere 2 also exerts an attractive force, so $\vec{F}_2$ points along the y axis toward sphere 2. Both spheres 1 and 2 are a distance ℓ away from sphere 0. However, sphere 1 has twice as much charge as sphere 2, so we have drawn $\vec{F}_1$ twice as long as $\vec{F}_2$. To anticipate our result, we use geometric addition to find $\vec{F}_{\text{tot}} = \vec{F}_1 + \vec{F}_2$.
5. We don't have any information about the acceleration, so it is not indicated on the diagram.

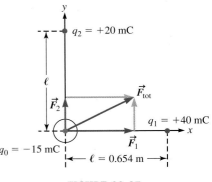

FIGURE 23.27

Example continues on page 702 ▶

SOLVE

$\vec{F}_1$ has only an x component and $\vec{F}_2$ has only a y component. Adding these components gives a general expression for the net electrostatic force $\vec{F}_{tot}$ in component form.

$$\vec{F}_{tot} = \vec{F}_1 + \vec{F}_2$$
$$\vec{F}_{tot} = F_1 \hat{\imath} + F_2 \hat{\jmath} \quad (1)$$

The magnitudes F_1 and F_2 come from Equation 23.3.

$$F_1 = k\frac{|q_1 q_0|}{\ell^2} = (8.99 \times 10^9 \text{ N} \cdot \text{m}^2/\text{C}^2)\frac{|(40 \times 10^{-3}\text{C})(-15 \times 10^{-3}\text{C})|}{(0.654 \text{ m})^2}$$

$$F_1 = 1.26 \times 10^7 \text{ N}$$

$$F_2 = k\frac{|q_2 q_0|}{\ell^2} = (8.99 \times 10^9 \text{ N} \cdot \text{m}^2/\text{C}^2)\frac{|(20 \times 10^{-3}\text{C})(-15 \times 10^{-3}\text{C})|}{(0.654 \text{ m})^2}$$

$$F_2 = 6.31 \times 10^6 \text{ N}$$

To write the net electrostatic force in component form, substitute F_1 and F_2 into Equation (1).

$$\vec{F}_{tot} = (1.26 \times 10^7 \hat{\imath} + 6.31 \times 10^6 \hat{\jmath}) \text{ N}$$

The magnitude and direction come from $A = \sqrt{A_x^2 + A_y^2}$ and $\theta_A = \tan^{-1}(A_y/A_x)$ (Eqs. 3.12 and 3.14, respectively).

$$F_{tot} = \sqrt{(F_{tot})_x^2 + (F_{tot})_y^2}$$
$$F_{tot} = \sqrt{(1.26 \times 10^7 \text{ N})^2 + (6.31 \times 10^6 \text{ N})^2}$$
$$F_{tot} = 1.41 \times 10^7 \text{ N}$$

$$\theta = \tan^{-1}\frac{(F_{tot})_y}{(F_{tot})_x} = \tan^{-1}\left(\frac{6.31 \times 10^6 \text{ N}}{1.26 \times 10^7 \text{ N}}\right)$$
$$\theta = 26.6°$$

CHECK and THINK

Our results fit the vector we anticipated in Figure 23.27. Both the anticipated vector and the vector we found are in quadrant I and have a magnitude greater than that of either $\vec{F}_1$ or $\vec{F}_2$ alone.

EXAMPLE 23.7 Square

Suppose a fourth small charged sphere is added to the three in Example 23.6 so that now the four spheres are at the vertices of a square. If the fourth sphere has a charge of −15 mC, what is the net electrostatic force on the sphere at the origin? Use the same coordinate system. Give your answer in component form and in terms of magnitude and direction.

INTERPRET and ANTICIPATE

In this problem, we can update Figure 23.27 in order to draw our hybrid free-body diagram (Fig. 23.28). The fourth charge is represented by a dot and called sphere 3. Its charge is written on the diagram. Because sphere 3 has a charge of the same sign as sphere 0, sphere 3 repels sphere 0. Therefore, force $\vec{F}_3$ is drawn pointing away from sphere 3. The other forces are unchanged. Sphere 3 has the smallest magnitude of charge and because it is the farthest from sphere 0, we have drawn it shorter than the other forces. We anticipate that our result will be similar to that for Example 23.6.

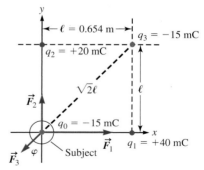

FIGURE 23.28

SOLVE

Since $\vec{F}_1$ and $\vec{F}_2$ are unchanged, we only need to find an expression for $\vec{F}_3$ in component form. $\vec{F}_3$ is along the same line as the diagonal of the square, so φ is 45° ($\sin 45° = \cos 45° = \sqrt{2}/2$).

$$\vec{F}_3 = F_{3x}\hat{\imath} + F_{3y}\hat{\jmath}$$
$$\vec{F}_3 = (-F_3 \sin \varphi)\hat{\imath} + (-F_3 \cos \varphi)\hat{\jmath}$$
$$\vec{F}_3 = \left(-\tfrac{\sqrt{2}}{2}F_3\right)\hat{\imath} + \left(-\tfrac{\sqrt{2}}{2}F_3\right)\hat{\jmath}$$

The magnitude of F_3 comes from Equation 23.3. Remember that $q_0 = q_3 = -15$ mC.	$F_3 = k\dfrac{|q_3 q_0|}{(\sqrt{2}\ell)^2} = (8.99 \times 10^9 \text{ N·m}^2/\text{C}^2)\dfrac{(-15 \times 10^{-3}\text{C})^2}{2(0.654 \text{ m})^2}$ $F_3 = 2.36 \times 10^6$ N
Write $\vec{F}_3$ in component form.	$\vec{F}_3 = (-\tfrac{\sqrt{2}}{2} 2.36 \times 10^6 \hat{\imath} - \tfrac{\sqrt{2}}{2} 2.36 \times 10^6 \hat{\jmath})$ N $\vec{F}_3 = (-1.67 \times 10^6 \hat{\imath} - 1.67 \times 10^6 \hat{\jmath})$ N
Add $\vec{F}_3$ to $\vec{F}_1 + \vec{F}_2$ from Example 23.6 to find the new $\vec{F}_{tot}$.	$\vec{F}_{tot} = (F_1 + F_{3x})\hat{\imath} + (F_2 + F_{3y})\hat{\jmath}$ $\vec{F}_{tot} = (1.26 \times 10^7 \text{ N} - 1.67 \times 10^6 \text{ N})\hat{\imath} + (6.31 \times 10^6 \text{ N} - 1.67 \times 10^6 \text{ N})\hat{\jmath}$ $\vec{F}_{tot} = (1.09 \times 10^7 \hat{\imath} + 4.64 \times 10^6 \hat{\jmath})$ N
Find the magnitude and direction as in Example 23.6.	$F_{tot} = \sqrt{(F_{tot})_x^2 + (F_{tot})_y^2} = \sqrt{(1.09 \times 10^7 \text{ N})^2 + (4.64 \times 10^6 \text{ N})^2} = 1.18 \times 10^7$ N $\theta = \tan^{-1}\dfrac{(F_{tot})_y}{(F_{tot})_x} = \tan^{-1}\left(\dfrac{4.64 \times 10^6 \text{ N}}{1.09 \times 10^7 \text{ N}}\right) = 23.1°$

:• **CHECK and THINK**
As anticipated, our results here are similar to the results from Example 23.6.

EXAMPLE 23.8 Equilateral Triangle

Three particles with charges q_0, q_1, and q_2 are at the vertices of an equilateral triangle of side ℓ (Fig. 23.29). Particle 1 and particle 2 lie on the x axis at $x = -\ell/2$ and $x = +\ell/2$, respectively. Particle 0 is on the y axis. Show that the net electrostatic force exerted on particle 0 is given by

$$\vec{F}_{tot} = \dfrac{kq_0}{2\ell^2}\left[(q_1 - q_2)\hat{\imath} + \sqrt{3}(q_1 + q_2)\hat{\jmath}\right]$$

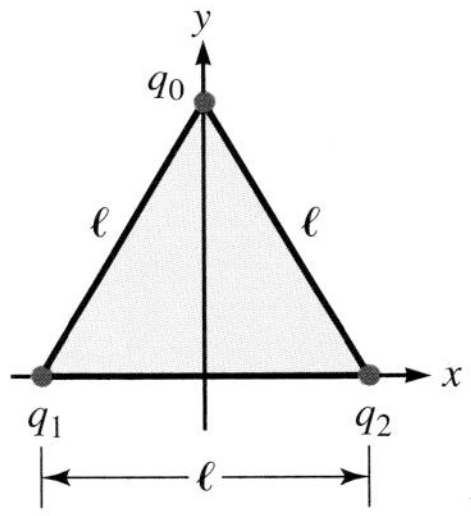

FIGURE 23.29

:• **INTERPRET and ANTICIPATE**
The three particles and the coordinate system are shown in Figure 23.30. No information is given about the charge of each particle, so only the symbols q_0, q_1, and q_2 are included on the diagram. We'll assume these charges are positive, but that does not affect the general form of the vector equation for force that we must develop. The subject is charge q_0. Two repulsive forces are exerted on the subject, but we do not know their magnitudes. (To keep the figure clean, we haven't indicated the acceleration.)

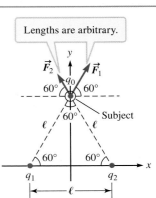

FIGURE 23.30

:• **SOLVE** Find a general expression for $\vec{F}_{tot}$ in component form.	$\vec{F}_{tot} = \vec{F}_1 + \vec{F}_2 = (F_{1x} + F_{2x})\hat{\imath} + (F_{1y} + F_{2y})\hat{\jmath}$ $\vec{F}_{tot} = (F_1 \cos 60° - F_2 \cos 60°)\hat{\imath} + (F_1 \sin 60° + F_2 \sin 60°)\hat{\jmath}$ (1)
Use Equation 23.3 to find F_1 and F_2.	$F_1 = k\dfrac{q_1 q_0}{\ell^2}$ $F_2 = k\dfrac{q_2 q_0}{\ell^2}$

Example continues on page 704 ▶

Substitute F_1 and F_2 into Equation (1), and use $\cos 60° = \frac{1}{2}$ and $\sin 60° = \frac{\sqrt{3}}{2}$ to reduce the expression for $\vec{F}_{tot}$.

$$\vec{F}_{tot} = \left(k\frac{q_1 q_0}{\ell^2}\cos 60° - k\frac{q_2 q_0}{\ell^2}\cos 60°\right)\hat{\imath} + \left(k\frac{q_1 q_0}{\ell^2}\sin 60° + k\frac{q_2 q_0}{\ell^2}\sin 60°\right)\hat{\jmath}$$

$$\vec{F}_{tot} = \left(k\frac{q_1 q_0}{\ell^2}\frac{1}{2} - k\frac{q_2 q_0}{\ell^2}\frac{1}{2}\right)\hat{\imath} + \left(k\frac{q_1 q_0}{\ell^2}\frac{\sqrt{3}}{2} + k\frac{q_2 q_0}{\ell^2}\frac{\sqrt{3}}{2}\right)\hat{\jmath}$$

$$\vec{F}_{tot} = \frac{kq_0}{2\ell^2}[(q_1 - q_2)\hat{\imath} + \sqrt{3}(q_1 + q_2)\hat{\jmath}] \quad \checkmark \qquad (2)$$

:• **CHECK and THINK**
This is exactly what we were asked to find. Notice that if q_0 were negative, there would be an overall negative sign, and the total force would be in the opposite direction. What if $q_1 = q_2$? Then we expect that (1) $F_1 = F_2$; (2) the x components of $\vec{F}_1$ and $\vec{F}_2$ cancel; and therefore (3) the total force points straight up along the y axis. That is exactly what we find with Equation (2); the first term is zero if $q_1 = q_2$, and the second term combines to give $\vec{F}_{tot} = (\sqrt{3}kq_0 q_1/\ell^2)\hat{\jmath}$.

EXAMPLE 23.9 **A Show Stopper, or There Has to Be a Typo**

In a physics demonstration, two identical conducting spheres are attached on top of two insulating pedestals. Each sphere, including its pedestal, has a mass of 0.58 kg. Their center-to-center distance is 2.50 m (Fig. 23.31). Initially, sphere 1 has a charge of $+10.0$ C and sphere 2 is neutral. A conducting wire is momentarily placed between the two spheres. The spheres remain at rest after the wire is removed.

A What is the minimum coefficient of static friction μ_s between the pedestal and the tabletop?

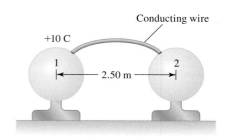

FIGURE 23.31

:• **INTERPRET and ANTICIPATE**
Electrons in sphere 2 are attracted to sphere 1 because it is positively charged. When the spheres are joined by the conducting wire, electrons are free to flow from sphere 2 to sphere 1. Because the spheres are identical, each ends up with half of the initial charge. When the wire is removed, each sphere has a charge of $+5.0$ C. Because they have charges of the same sign, the spheres repel each other.

Let's arbitrarily choose sphere 1 as the subject and place it at the origin of the coordinate system (Fig. 23.32). Two forces are exerted on the subject—the repulsive electrostatic force exerted by sphere 2 and static friction due to the tabletop.

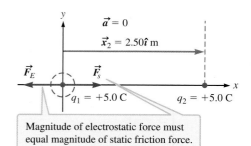

Magnitude of electrostatic force must equal magnitude of static friction force.

FIGURE 23.32

:• **SOLVE**
The spheres remain at rest, so the electrostatic force exerted on sphere 1 must be balanced by static friction.

$F_E = F_s$

The magnitude of the electrostatic force is very large in this case.

$F_E = k\frac{|q_1 q_2|}{r^2}$

$F_E = (8.99 \times 10^9 \text{ N} \cdot \text{m}^2/\text{C}^2)\frac{(5.0\,\text{C})^2}{(2.50\,\text{m})^2}$

$F_E = 3.6 \times 10^{10}$ N

Because the electrostatic force is so great, let's imagine that static friction is at its maximum. This gives us the minimum value of μ_s using $F_{s,\text{max}} = \mu_s F_N$ (Eq. 6.1) for $F_{s,\text{max}}$. The tabletop is level, so the magnitude of the normal force equals the sphere's weight.

$$F_E = F_{s,\text{max}} = \mu_s F_N = \mu_s mg$$

$$\mu_s = \frac{F_E}{mg} = \frac{3.6 \times 10^{10}\text{ N}}{(0.58\text{ kg})(9.81\text{ m/s}^2)}$$

$$\mu_s = 6.3 \times 10^9$$

:• **CHECK and THINK**
This answer seems **highly improbable**. According to Table 6.1, our coefficient of static friction is more than nine orders of magnitude greater than the typical value (and our answer is the *minimum* possible value). We can resolve this by recognizing that 5.0 C is a great amount of charge. If you see a result like this, you might guess that a prefix such as μ was missing so that the charge was $+5.0$ μC instead of $+5.0$ C. If that were the case, the electrostatic force would be 0.036 N. This is 12 orders of magnitude smaller than what we found, and the minimum value for μ_s would have been 0.006—a reasonable value.

B Now let's consider a more reasonable physics demonstration. After the conducting wire is removed, each conducting sphere has a charge of $+7.5$ μC. The coefficient of static friction between the pedestal and the tabletop is $\mu_s = 0.006$. What is magnitude of either sphere's initial acceleration?

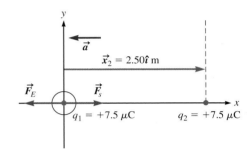

FIGURE 23.33

:• **INTERPRET and ANTICIPATE**
As in part A, we choose sphere 1 as the subject and place it at the origin of the coordinate system (Fig. 23.33). Now the acceleration may be nonzero, and we'll assume it is in the same direction as the electrostatic force.

:• **SOLVE**
Now, when we apply Newton's second law, the sum of the force is not zero.

$$\sum F_x = F_s - F_E = -ma$$

The magnitude of the electrostatic force is not nearly as great as it was in part A.

$$F_E = k\frac{|q_1 q_2|}{r^2}$$

$$F_E = (8.99 \times 10^9\text{ N}\cdot\text{m}^2/\text{C}^2)\frac{(7.5 \times 10^{-6}\text{ C})^2}{(2.50\text{ m})^2}$$

$$F_E = 8.1 \times 10^{-2}\text{ N}$$

To know whether the sphere moves, we need to compare F_E to the maximum value of static friction $F_{s,\text{max}} = \mu_s F_N$ (Eq. 6.1).

$$F_{s,\text{max}} = \mu_s F_N = \mu_s mg$$

$$F_{s,\text{max}} = (0.006)(0.58\text{ kg})(9.81\text{ m/s}^2)$$

$$F_{s,\text{max}} = 3.4 \times 10^{-2}\text{ N}$$

$$F_{s,\text{max}} < F_E$$

Because $F_{s,\text{max}} < F_E$, the sphere does accelerate in the direction of the electrostatic force.

$$a = \frac{F_{s,\text{max}} - F_E}{-m} = \frac{3.4 \times 10^{-2}\text{ N} - 8.1 \times 10^{-2}\text{ N}}{-0.58\text{ kg}}$$

$$a = 8.1 \times 10^{-2}\text{ m/s}^2$$

:• **CHECK and THINK**
In this demonstration, we see that the spheres accelerate slowly away from each other. Keep in mind that this is a very low coefficient of static friction. The spheres may need to be on an air table to achieve such low friction.

EXAMPLE 23.10 The Strong Force

The helium nucleus has two neutrons and two protons. If the distance between the protons is about 2×10^{-15} m, find the electrostatic force between the protons.

:• INTERPRET and ANTICIPATE
You may find this situation easy to visualize without a diagram. (It is very similar to Examples 23.3 and 23.4.) However, we include a diagram (Fig. 23.34) for clarity and practice. The protons are dots labeled q_1 and q_2. Both carry the same amount of charge $+e$. We have chosen proton 1 as the subject, and we use a coordinate system so that unit vector $\hat{r}$ points from proton 1 toward proton 2. The protons are both positive, so the force on each of them is repulsive.

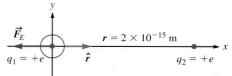

FIGURE 23.34

:• SOLVE
Use Equation 23.4 to find the electrostatic force on proton 1. The magnitude of the force on proton 2 is the same, but the two forces are in opposite directions.

$$\vec{F}_{[2\text{ on }1]} = -k\frac{q_1 q_2}{r^2}\hat{r} \qquad (23.4)$$

$$\vec{F}_E = -k\frac{e^2}{r^2}\hat{r}$$

$$\vec{F}_E = -(8.99 \times 10^9 \text{ N}\cdot\text{m}^2/\text{C}^2)\frac{(1.60 \times 10^{-19}\text{ C})^2}{(2 \times 10^{-15}\text{ m})^2}\hat{r}$$

$$\vec{F}_E = -60\,\hat{r} \text{ N}$$

:• CHECK and THINK
This is an enormous repulsive force acting on a particle of very low mass. If this were the only force exerted on a proton, its acceleration would be $a = F_E/m_p = (60 \text{ N})/(2 \times 10^{-27} \text{ kg}) = 3 \times 10^{28} \text{ m/s}^2$. Protons subjected to such a net force would go flying out of the nucleus. Protons stay in the nucleus because another force—known as the strong nuclear force—is attractive and holds them inside.

In this chapter, we began to study a new branch of physics—electricity. When an object has a net charge, it exerts an electrostatic force on other charged objects or on neutral objects. The electrostatic force between particles is similar to gravity because both forces are inversely proportional to the square of the distance between the particles. However, unlike gravity, the electrostatic force can be either attractive or repulsive.

SUMMARY

❶ Underlying Principles

1. **Conservation of charge.** The net charge in the Universe is constant. Charge may transfer from one object to another, but it is not created, lost, or destroyed.
2. **Electrostatic force** between charged objects is (part of) one of the fundamental forces in physics. If two objects are oppositely charged, they are mutually attracted. If two objects have charges of the same sign, they are mutually repelled. (*Opposites attract; likes repel.*)
3. **Coulomb's law.** For two charged objects that can be modeled as particles with charges q_1 and q_2, the magnitude of the electrostatic force is given by

$$F_E = k\frac{|q_1 q_2|}{r^2} \qquad (23.3)$$

where r is the distance between the particles. If polar coordinates are used so that the unit vector $\hat{r}$ points from q_1 toward q_2, Coulomb's law is written in vector form as

$$\vec{F}_{[2\text{ on }1]} = -k\frac{q_1 q_2}{r^2}\hat{r} \qquad (23.4)$$

✪ Major Concepts

1. **Charge** q or Q is an intrinsic property that determines an object's role in electrical phenomena. A proton carries charge $+e$ and an electron carries charge $-e$, where e is the **elementary charge**:

   $e = (1.602\,176\,565 \times 10^{-19} \pm 000\,000\,035 \times 10^{-19})\text{C}$

2. Charge is **quantized**.
   a. An object is **positively charged** if the number of protons is greater than the number of electrons:
   $$q = +Ne \quad (23.1)$$
   where N is the number of excess protons.
   b. An object is **negatively charged** if the number of electrons is greater than the number of protons:
   $$q = -Ne \quad (23.2)$$
   where N is the number of excess electrons.

3. a. A material in which electrons do not move freely is an **insulator**. In the microscopic structure of an insulator, the electrons of each particular atom stay bound to that atom.
   b. A **conductor** is a material in which the charged particles (usually electrons) can freely flow. In a metal atom, the outermost electrons are weakly bound to the nucleus. Within a sample of metal, the outermost electrons from all the atoms are free to move around in the entire sample. Typically in a metal there is one free electron per atom.

4. A **ground** is a large reservoir of charge that can accept or provide electrons freely while remaining neutral at all times. The Earth often serves as a ground. When something is connected to ground by a conductor, we say that it is **grounded**.

PROBLEM-SOLVING STRATEGY | Applying Coulomb's Law

∶• INTERPRET and ANTICIPATE

Draw a "*hybrid*" free-body diagram:
1. Represent each object.
2. Represent the objects as dots.
3. Label information about the charges. If no information is given, it is best to assume the particle's charge is positive.
4. Circle the subject, and draw vectors representing the forces exerted on the subject.
5. Indicate the subject's acceleration.
6. Choose a coordinate system.

∶• SOLVE

The subject—with force vectors extending from it—is treated as a free-body diagram. Then Newton's second law is applied:

1. Write the electrostatic force in component form, and then
2. Algebraically combine the components.
3. When you substitute in values for the charge, you probably need only the **absolute value of the charge**.

PROBLEMS AND QUESTIONS

A = algebraic   C = conceptual   E = estimation   G = graphical   N = numerical

23-1 Another Fundamental Force

1. **C** What is the difference between a contact force and a field force? List all the forces presented in Chapters 1 through 22. Which are field forces? Which are contact forces? Which forces are macroscopic manifestations of the electromagnetic force? What does that tell you about contact forces? Explain your answers.

23-2 Models of Electrical Phenomena

2. **C** Many textbooks claim Franklin decided that moving charged particles are positive. How would you correct this claim? Think about these questions in developing your answer: Did Franklin's model include particles? What did the terms *positive* and *negative* mean to Franklin?

3. **N** An object has a charge of 35 nC. How many excess protons does it have?

4. **E** As part of a demonstration, a physics professor rubs wool against a plastic disk about the size and mass of a small dinner plate. Afterward, the disk has a charge of about -75 μC. Estimate the fractional increase in the number of electrons.

5. **N** A single coulomb represents a large amount of charge. A sphere has a net charge of -1.00 C. How many excess electrons does the sphere have?

6. **N** A sphere has a net charge of 8.05 nC, and a negatively charged rod has a charge of -6.03 nC. The sphere and rod undergo a process such that 5.00×10^9 electrons are transferred from the rod to the sphere. What are the charges of the sphere and the rod after this process?

Problems 7 and 8 are paired.

7. A glass rod is initially neutral. After it is rubbed with silk, its charge is 45.7 μC.
   a. **C** Has the rod gained or lost mass? Explain.
   b. **N** How much mass has the rod gained or lost?

8. **E** After an initially neutral glass rod is rubbed with an initially neutral silk scarf, the rod has a charge of 45.7 μC. Estimate the fractional increase or decrease in the scarf's mass. Would this change in mass be easily noticed?

9. **N** A 50.0-g piece of aluminum has a net charge of +4.20 μC. Aluminum has an atomic mass of 27.0 g/mol, and its atomic number is 13.
   a. Calculate the number of electrons that were removed from the initially neutral aluminum to produce this charge.
   b. Determine what fraction of the original number of electrons this removed number represents.
   c. By how much did the original mass of the aluminum decrease after charging?

23-3 A Qualitative Look at the Electrostatic Force

10. **C** You walk around a carpeted floor in your socks and get a shock when you reach for a doorknob.
    a. Use Franklin's model to explain why you feel the shock.
    b. Use the contemporary model to explain why you feel the shock. Compare your answers.

11. **C** A silk scarf is rubbed against glass, and a wool scarf is rubbed against plastic. (Initially, all four objects were neutral.) Afterward, it is found that the glass is attracted to the plastic. Will the silk be attracted to the wool? Explain.

23-4 Insulators and Conductors

12. **C CASE STUDY** A person in Franklin's time may have been able to provide the same advice we came up with in Example 23.2. However, an 18th-century person may have had a different reason for his or her advice. What part of that person's explanation would be the same as ours? What would be different?

13. **C** Why does the technician in Figure 23.9 cover the electrical lines with rubber and perhaps wear a rubber suit?

14. **C** Here is another party trick for you. Take a wooden stick, such as a chopstick. It is best if the stick has a uniform circular cross section. Suspend the stick from a string, or place it on an insulated pivot. Charge an object such as a comb or plastic stir stick by rubbing it on some material such as a sweater or your hair. Now bring the charged object near the wooden stick. What happens to the wooden stick? Explain your answer.

15. **N** A charge of −36.3 nC is transferred to a neutral copper ball of radius 4.35 cm. The ball is not grounded. The excess electrons spread uniformly on the surface of the ball. What is the number density (number of electrons per unit surface area) of excess electrons on the surface of the ball?

Problems 16 and 17 are paired.

16. **N** Two identical conductors are brought into contact. Initially, one conductor has a charge of +30.0 μC. What is the charge of each conductor afterward? Does it matter how the contact is made?

17. **N** Two identical insulators are brought into contact. Initially, one insulator has a charge of +30.0 μC. What is the charge of each insulator afterward? Does it matter how the contact is made?

18. **C** An **electrophorus** is a device developed more than 200 years ago for the purpose of charging objects. The insulator on top of a pedestal is rubbed with a cloth, such as wool (Fig. P23.18A). A conductor is placed on top of the insulator, and the conductor is connected to ground by a conducting wire (Fig. P23.18B). (The conductor has an insulating handle, so charge cannot be transferred between the person and the conductor.) The conductor is then removed (Fig. P23.18C). The conductor may then be used to transfer charge to other objects. If the insulator's charge after being rubbed with the wool is negative, what is the charge of the conductor when it is removed?

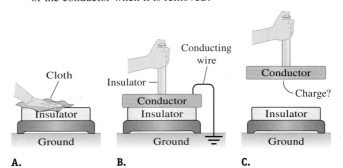

FIGURE P23.18 Problems 18 and 19.

19. **C** Consider the electrophorus from Problem 18. Suppose only the left half of the insulator is rubbed before the conductor is placed on top and then grounded. Then the conductor is removed. How is the conductor's charge distributed?

20. **C** An **electroscope** is a device used to measure the (relative) charge on an object (Fig. P23.20). The electroscope consists of two metal rods held in an insulated stand. The bent rod is fixed, and the straight rod is attached to the bent rod by a pivot. The straight rod is free to rotate. When a positively charged object is brought close to the electroscope, the straight movable rod rotates. Explain your answers to these questions:
    a. Why does the rod rotate in Figure P23.20?
    b. If the positively charged object is removed, what happens to the electroscope?
    c. If a negatively charged object replaces the positively charged object in Figure P23.20, what happens to the electroscope?
    d. If a charged object touches the top of the fixed conducting rod and is then removed, what happens to the electroscope?

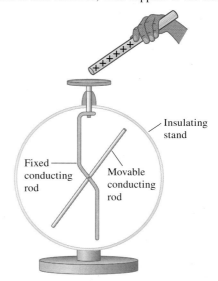

FIGURE P23.20

23-5 Coulomb's Law

21. Two particles with charges of +5.50 nC and −8.95 nC are separated by 3.00 m.
    a. **N** What is the magnitude of the electrostatic force between the particles?
    b. **C** Is this force attractive or repulsive?

22. **N** Particle A has a charge of 34.5 nC, and particle B has a charge of -54.3 nC. The attractive force between them has a magnitude of 2.70×10^{-4} N. How far apart are the particles?
23. **N** Two coins are placed on a horizontal insulating surface a distance of 1.5 m apart and given equal charges. They experience a repulsive force of 2.3 N. Calculate the magnitude of the charge on each coin.
24. **A** Show that Equation 23.5 is consistent with the rules *like charges repel* and *opposite charges attract*.
25. **A** Particle A has charge q_A and particle B has charge q_B. When they are separated by a distance r_i, they experience an attractive force F_i. The particles are moved without altering their charges. Now they experience an attractive force with a magnitude of $4F_i$. Find an expression for their new separation.
26. Two charged particles are placed along the y axis. The first particle, at the origin, has a charge of -25.0 μC, and the second particle, at $y = 55.0$ cm, has a charge of 15.0 μC.
    a. **C** Is the electric force between the two particles attractive or repulsive? Explain.
    b. **N** What is the magnitude of the electric force between the two particles?

Problems 27 and 28 are paired.

27. **N** A 1.75-nC charged particle located at the origin is separated by a distance of 0.0825 m from a 2.88-nC charged particle located farther along the positive x axis. Both particles are held at their locations by an external agent.
    a. What is the electrostatic force on the 2.88-nC particle?
    b. What is the electrostatic force on the 1.75-nC particle?
28. **N** A 1.75-nC charged particle located at the origin is separated by a distance of 0.0825 m from a 2.88-nC charged particle located farther along the positive x axis. If the 1.75-nC particle is kept fixed at the origin, where along the positive x axis should the 2.88-nC particle be located so that the magnitude of the electrostatic force it experiences is twice as great as it was in Problem 27?
29. **A** Two particles with charges q_1 and q_2 are separated by a distance d, and each exerts an electric force on the other with magnitude F_E.
    a. In terms of these quantities, what separation distance would cause the magnitude of the electric force to be halved?
    b. In terms of these quantities, what separation distance would cause the magnitude of the electric force to be doubled?
30. **A** An electron with charge $-e$ and mass m moves in a circular orbit of radius r around a nucleus of charge Ze, where Z is the atomic number of the nucleus. Ignore the gravitational force between the electron and the nucleus. Find an expression in terms of these quantities for the speed of the electron in this orbit.
31. **N** Two electrons in adjacent atomic shells are separated by a distance of 5.00×10^{-11} m.
    a. What is the magnitude of the electrostatic force between the electrons?
    b. What is the ratio of the electrostatic force to the gravitational force between the electrons?
32. Two small, identical metal balls with charges 5.0 μC and 15.0 μC are held in place 1.0 m apart. In an experiment, they are connected for a short time by a conducting wire.
    a. **C** What will be the charge on each ball after this experiment?
    b. **N** By what factor will the magnitude of the electrostatic force on either ball change after this experiment is performed?
33. **N** Two identical spheres each have a mass of 5.0 g and they are 1.0 m apart. What should be the identical charges on each of the spheres so that the magnitude of their electrostatic repulsion equals the magnitude of the gravitational force between them?
34. **N** One end of a light spring with force constant $k = 125$ N/m is attached to a wall, and the other end to a metal block with charge $q_A = 2.00$ μC on a horizontal, frictionless table (Fig. P23.34). A second block with charge $q_B = -3.60$ μC is brought close to the first block. The spring stretches as the blocks attract each other so that at equilibrium, the blocks are separated by a distance $d = 12.0$ cm. What is the displacement x of the spring?

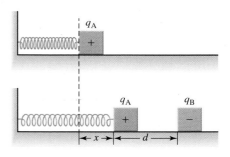

FIGURE P23.34

35. **N** Two 25.0-g copper spheres are placed 75.0 cm apart. Each copper atom has 29 electrons, and the molar mass of copper is 63.5463 g/mol. What fraction of the electrons from the first sphere must be transferred to the second sphere for the net electrostatic force between the spheres to equal 100 kN?

23-6 Applications of Coulomb's Law

36. **C** Three charged particles lie along a single line. Is it possible for one of the charges at either end to have zero net electrostatic force exerted on it? Explain.
37. **N** Given the arrangement of charged particles shown in Figure P23.37, find the net electrostatic force on the 5.00-nC charged particle located at the origin.

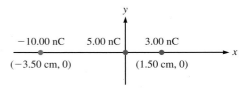

FIGURE P23.37

Problems 38 and 39 are paired.

38. **N** Given the arrangement of charged particles in Figure P23.38, find the net electrostatic force on the 5.65-μC charged particle.

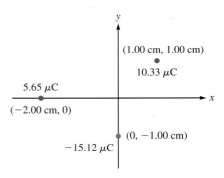

FIGURE P23.38 Problems 38 and 39.

39. **N** Given the arrangement of charged particles in Figure P23.38, find the net electrostatic force on the 10.33-μC charged particle.

40. **N** Three charged metal spheres are arrayed in the xy plane so that they form an equilateral triangle (Fig. P23.40). What is the net electrostatic force on the sphere at the origin?

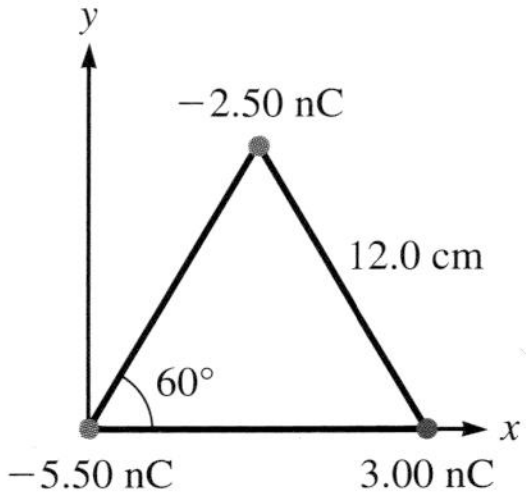

41. **N** Charges A, B, and C are arrayed along the y axis, with $q_A = -4.00$ nC, $q_B = 2.40$ nC, and $q_C = 5.30$ nC (Fig. P23.41). The distance between charges A and B is $y_1 = 1.00$ m, and the distance between charges B and C is $y_2 = 2.50$ m.
   a. What is the net electrostatic force on charge A?
   b. What is the net electrostatic force on charge B?
   c. What is the net electrostatic force on charge C?

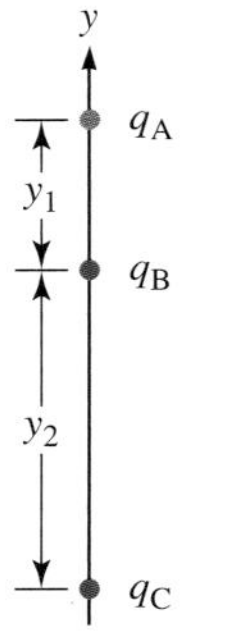

42. Three identical conducting spheres are fixed along a single line. The middle sphere is equidistant from the other two so that the center-to-center distance between the middle sphere and either of the other two is 0.125 m. Initially, only the middle sphere is charged, with $q_{middle} = +35.6$ nC. The middle sphere is later connected by a conducting wire to the sphere on the left. The wire is removed and then used to connect the middle sphere to the sphere on the right. The wire is again removed.
   a. **C** What is the charge on each sphere?
   b. **C** Which sphere experiences the greatest electrostatic force?
   c. **N** What is the magnitude of that force?

43. **N** Charges A, B, and C are arranged in the xy plane with $q_A = -5.60$ μC, $q_B = 4.00$ μC, and $q_C = 2.30$ μC (Fig. P23.43). What are the magnitude and direction of the electrostatic force on charge B?

44. **N** In an early attempt to understand atomic structure, Niels Bohr modeled the hydrogen atom as an electron in uniform circular motion about a proton with the centripetal force caused by Coulomb attraction. He predicted the radius of the electron's orbit to be 5.29×10^{-11} m. Calculate the speed of the electron and the frequency of its circular motion.

45. **A** A particle with charge q is located at the origin, and a particle with charge $2q$ is on the positive x axis, a distance d from the origin. The particles are not free to move. In terms of q and d, at what coordinates should a third particle with charge q be placed so that it experiences no net electrostatic force?

46. **N** Figure P23.46 shows four identical conducting spheres with charge $q = +6.50$ nC placed at the corners of a rectangle of width 20.0 cm and height 80.0 cm. What are the magnitude and direction

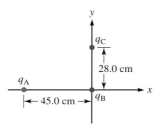

FIGURE P23.40

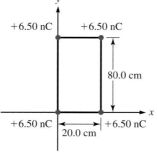

FIGURE P23.41

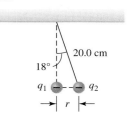

FIGURE P23.43

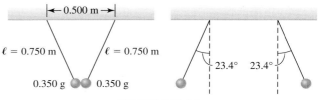

FIGURE P23.46

of the net electrostatic force on the charge on the lower right-hand corner?

47. **N** A sphere of mass 5.00 g carries a positive charge of 30.0 nC and remains stationary when placed 5.00 cm directly above a second charged sphere that is fixed to a tabletop. What must be the charge of the second sphere?

48. **N** Two metal spheres of identical mass $m = 4.00$ g are suspended by light strings 0.500 m in length. The left-hand sphere carries a charge of 0.800 μC, and the right-hand sphere carries a charge of 1.50 μC. What is the equilibrium separation between the centers of the two spheres?

49. **N** Figure P23.49 shows two identical small, charged spheres. One of mass 4.0 g is hanging by an insulating thread of length 20.0 cm. The other is held in place and has charge $q_1 = -3.6$ μC. The thread makes an angle of 18° with the vertical, resulting in the spheres being aligned horizontally, a distance r apart. Determine the charge q_2 on the hanging sphere.

FIGURE P23.49

50. Two small spherical conductors are suspended from lightweight vertical insulating threads. The conductors are brought into contact (Fig. P23.50, left) and released. Afterward, the conductors and threads stand apart as shown at right.
   a. **C** What can you say about the charge of each sphere?
   b. **N** Use the data given in Figure P23.50 to find the tension in each thread.
   c. **N** Find the magnitude of the charge on each sphere.

FIGURE P23.50

Problems 51 and 52 are paired.

51. **A** Four equally charged particles with charge q are placed at the corners of a square with side length L, as shown in Figure P23.51. A fifth charged particle with charge Q is placed at the center of the square so that the entire system of charges is in static equilibrium. What are the magnitude and sign of the charge Q?

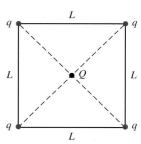

FIGURE P23.51

52. **C** Four charged particles q, $-q$, q, and $-q$ are fixed at the corners of a square with side length L as shown in Figure P23.52. If another charged particle of magnitude Q is placed at the center of the square, will it be in static equilibrium? Does the sign of the charge Q matter? Explain.

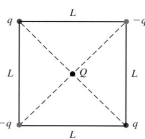

FIGURE P23.52

53. **N** A metal sphere with charge $+8.00$ nC is attached to the left-hand end of a nonconducting rod of length $L = 2.00$ m. A second sphere with charge $+2.00$ nC is fixed

to the right-hand end of the rod (Fig. P23.53). At what position d along the rod can a charged bead be placed for the bead to be in equilibrium?

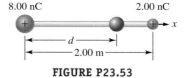

FIGURE P23.53

54. **A** A particle with charge q is located at the origin, and a particle with charge $-2q$ is on the positive x axis, a distance d away from the origin. The particles are not free to move. In terms of q and d, at what coordinate should a third particle with charge q be placed so that it experiences no net electrostatic force?

55. **A** Three small metallic spheres with identical mass m and identical charge $+q$ are suspended by light strings from the same point (Fig. P23.55). The left-hand and right-hand strings have length L and make an angle θ with the vertical. What is the value of q in terms of k, g, m, L, and θ?

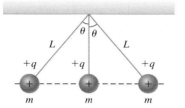

FIGURE P23.55

General Problems

56. **C** How does a negatively charged rubber balloon stick to a neutral wooden ceiling?

57. **N** How many electrons are in a 1.00-g electrically neutral steel paper clip? The molar mass of steel is approximately that of iron, or 55.845 g/mol, and a neutral iron atom has 26 electrons.

58. The Ni^{2+} and S^{2-} ions in nickel sulfide are separated by a distance of 0.680 nm.
    a. **N** What is the magnitude of the electrostatic force between the ions in nickel sulfide?
    b. **C** How would the magnitude of the electrostatic force found in part (a) change if the nickel ion was replaced by an iron (Fe^{2+}) ion?

59. **N** A small metal sphere with a charge of 2.25 μC is placed at the origin, and a second sphere with unknown charge q is placed at $y = 80.0$ cm. A negatively charged sphere is found to be in equilibrium when placed at $y = 62.0$ cm. What is the charge q on the second sphere?

60. **N** Two otherwise identical, small conducting spheres have charges $+5.0$ μC and -2.0 μC. When placed a distance r apart, each experiences an attractive force of 3.0 N. The spheres are then touched together and moved back to a distance r apart. Find the magnitude of the new force on each sphere.

61. **N** Three charged particles are arranged in the xy plane as shown in Figure P23.61, with $q_A = 6.40$ μC, $q_B = -2.30$ μC, and $q_C = 3.80$ μC. What is the net electrostatic force on the particle with charge q_A?

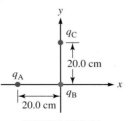

FIGURE P23.61

62. **A** We saw in Figure 23.16 that a neutral metal can was attracted to a positively charged glass rod because the rod polarized the can. Let's model this situation in a very rough sense. Suppose a charged sphere C is held near two other charged spheres A and B as shown in Figure P23.62. Let's call the magnitude of the electrostatic force exerted

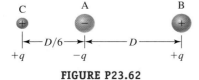

FIGURE P23.62

by sphere C on A "F_A" and that on B "F_B." Derive an expression for the ratio F_A/F_B. Comment on why a neutral conductor is attracted to a charged object.

63. **N** A $+4.0$-μC charged particle is located at the origin of a coordinate system, and a -1.0-μC charged particle is located at $x = 2.0$ cm.
    a. Calculate the net electrostatic force on a third particle with charge $+2.0$ μC located at $x = 5.0$ cm.
    b. Compare your answer to part (a) with the force due to a $+3.0$-μC charged particle located at the origin on a $+2.0$-μC charged particle at $x = 5.0$ cm.

64. **E** In Figure P23.64, a boy's hair is attracted to his comb after the comb has been run through his hair. Estimate the amount of charge transferred between his raised hair and the comb.

FIGURE P23.64

65. **A** Figure P23.65 shows two identical conducting spheres, each with charge q, suspended from light strings of length L. If the equilibrium angle the strings make with the vertical is θ, what is the mass m of the spheres?

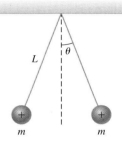

FIGURE P23.65

Problems 66 and 67 are paired.

66. **N** Two helium-filled, spherical balloons, each with charge q, are tied to a 5.00-g mass with strings of negligible mass, and the system floats in equilibrium as shown in Figure P23.66. The distance between the balloons is 60.0 cm, and the strings are 100.0 cm long. Ignore the weight of the balloon material, and assume that the density of air is 1.29 kg/m³ and the density of helium inside the balloons is 0.200 kg/m³.
    a. Find the magnitude of the charge q on each balloon. Assume that the charge on each balloon acts as if it were concentrated at its center.
    b. Find the volume of each balloon.

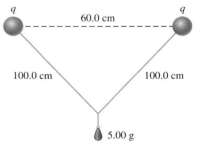

FIGURE P23.66

67. **N** Two small metallic spheres, each with a mass of 2.00 g, are suspended from a common point by two strings of negligible mass and of length 10.0 cm. When the spheres have an equal amount of charge, the two strings make an

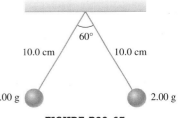

FIGURE P23.67

angle of 60° with each other as shown in Figure P23.67. Calculate the magnitude of the charge on each sphere.

68. **A** Two positively charged spheres with charges $4e$ and e are separated by a distance L and held motionless. A third charged sphere with charge Q is set between the two spheres and along the line joining them. The third sphere is in static equilibrium. What is the distance between the third charged sphere and the sphere that has charge $4e$?

69. **N** The two ends of a light spring with force constant $k = 145$ N/m are connected to identical metal blocks that are at rest on a horizontal, frictionless table. The equilibrium length of the spring is $x_0 = 34.0$ cm. Electrons are slowly stripped from both blocks, giving each an identical charge $+q$. The repulsive electric force between the blocks stretches the spring by 14.0 cm. Assuming the blocks can be modeled as particles, what is the charge q on each block?

70. **N** Three charged spheres are at rest in a plane as shown in Figure P23.70. Spheres A and B are fixed, but sphere C is attached to the ceiling by a lightweight thread. The tension in the string is 0.240 N. Spheres A and B have charges $q_A = 28.0$ nC and $q_B = -28.0$ nC. What charge is carried by sphere C?

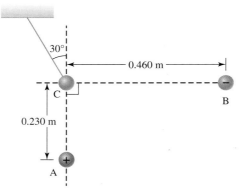

FIGURE P23.70

71. **N** A small sphere with charge $Q = 2.0$ μC is held in place directly above another similar sphere with total charge $q = -1.0$ μC. The lower sphere has a mass of 40 g and is suspended in place by the electrostatic force of attraction with the upper sphere. Calculate the distance r between the centers of the spheres.

72. **N** Three particles with charges of 1.0 μC, -1.0 μC, and 0.50 μC are placed at the corners A, B, and C of an equilateral triangle with side length 0.10 m as shown in Figure P23.72. Find the net force on the charge at point C.

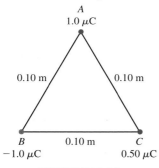

FIGURE P23.72

Problems 73 and 74 are paired.

73. **A** Two positively charged particles, each with charge Q, are held at positions $(-a, 0)$ and $(a, 0)$ as shown in Figure P23.73. A third positively charged particle with charge q is placed at $(0, h)$.
   a. Find an expression for the net electric force on the third particle with charge q.

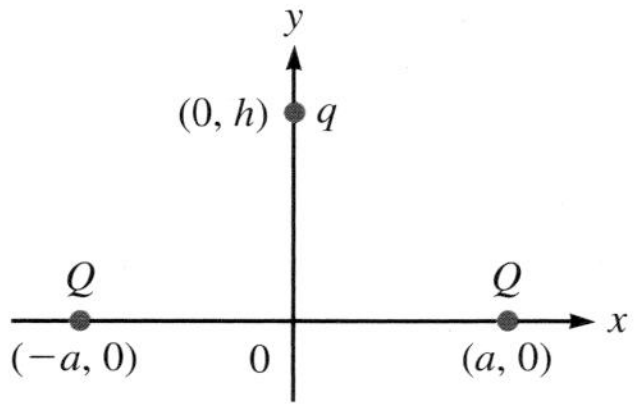

FIGURE P23.73 Problems 73 and 74.

b. Show that the two charges Q behave like a single charge $2Q$ located at the origin when the distance h is much greater than a.

74. Consider the arrangement of charges in Problem 73.
   a. **A** Determine the location on the y axis where the charge q will experience the greatest force.
   b. **G** Sketch the magnitude of the force on charge q as a function of y for positive values of y.

75. **N** Eight small conducting spheres with identical charge $q = -2.00$ μC are placed at the corners of a cube of side $d = 0.500$ m (Fig. P23.75). What is the total force on the sphere at the origin (sphere A) due to the other seven spheres?

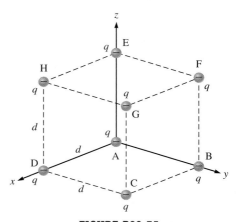

FIGURE P23.75

Problems 76, 77, and 78 are grouped.

76. **C** Two identical spheres carry identical positive charge. Static friction between the spheres and the level surface they sit on just barely keeps them at rest. Then they are placed closer together on a different level surface, and they are barely at rest due to static friction. Is the coefficient of static friction between a sphere and the second surface higher or lower than on the first? Explain.

77. **A** Two identical spheres of mass m carry identical positive charge q. Static friction between the spheres and the level surface they sit on just barely keeps them at rest. Their center-to-center separation is r. Find an expression for the coefficient of static friction between a sphere and the surface.

78. **N** Two identical spheres of mass $m = 7.5 \times 10^{-5}$ kg carry an identical positive charge $q = 89$ nC. Static friction between the spheres and the level surface they sit on just barely keeps them at rest. Their center-to-center separation is $r = 0.50$ m. Find the coefficient of static friction between a sphere and the surface.

79. **A** Two positive charges, each with charge Q, are held at positions $x = -a$ and $x = +a$ along the x axis of a coordinate system. A small sphere of mass m and positive charge q is placed at the origin, but it can move freely back and forth on the x axis. Show that if the sphere is displaced by an amount $x \ll a$ and released, it will undergo simple harmonic motion about the origin with an angular frequency given by $\omega = \sqrt{4kQq/ma^3}$.

Electric Fields

24

❶ Underlying Principles
Electrostatic field

❷ Major Concepts
1. Source and subject
2. Electric dipole
3. Electric dipole moment
4. Electrostatic force exerted by electrostatic field

❸ Special Cases
1. Electric field sources
   a. Particle (or spherical object)
   b. Dipole
   c. Charged rod
   d. Charged ring
   e. Charged disk
   f. Near a charged surface
2. Dipole in an electric field

❹ Tools
Electric field lines

Key Questions

What is an electric field?

How can we find expressions for the electric field due to sources of various geometries?

How do we find the electrostatic force exerted on subjects in an electric field?

- 24-1 What are fields? 714
- 24-2 Special case: Electric field of a charged sphere 716
- 24-3 Electric field lines 718
- 24-4 Electric field of a collection of charged particles 720
- 24-5 Electric field of a continuous charge distribution 726
- 24-6 Special cases of continuous distributions 728
- 24-7 Case study: The shape of lightning rods 733
- 24-8 Charged particle in an electric field 737
- 24-9 Special case: Dipole in an electric field 741

An electric precipitator cleans the air on a submarine; a drop of ink in your printer is accelerated toward the paper; a potato cooks in a microwave oven. These objects are accelerated by electrostatic forces. The charged objects used inside many common devices are not spherical, so we cannot use Coulomb's law or the methods from Chapter 23 to calculate the forces involved. In this chapter, we learn to account for the nonspherical shapes that often play a significant role in electrostatics. To do this, we'll introduce the concept of an electrostatic field, which is analogous to the gravitational field.

24-1 What Are Fields?

The best way to study nonspherical shapes is by calculating the *field* they produce. To understand what a field is, we start by reviewing the more familiar gravitational field. Suppose you were thinking about placing several spacecraft with different masses into orbit at some distance r from the center of the Earth (Section 7-4). You would need to find the gravitational force exerted by the Earth on each spacecraft using Newton's law of universal gravity. Following the form of Equation 7.4, we write

$$\vec{F}_G(r) = -G\frac{M_\oplus m_{SC}}{r^2}\hat{r} \tag{24.1}$$

where the unit vector $\hat{r}$ points outward from the Earth toward a spacecraft, and m_{SC} is the mass of a particular spacecraft.

Because the gravitational force exerted on each spacecraft depends only on its mass m_{SC}, it is helpful to rewrite Equation 24.1 as

$$\vec{F}_G(r) = m_{SC}\left(-G\frac{M_\oplus}{r^2}\hat{r}\right) = m_{SC}\vec{g}(r)$$

where $\vec{g}(r)$ is known as the **gravitational field** of the Earth for points outside the Earth or on its surface, $r \geq R_\oplus$, as given by Equation 7.12 (page 197):

$$\vec{g}(r) \equiv -\frac{GM_\oplus}{r^2}\hat{r} \tag{7.12}$$

The dimensions of the gravitational field are force per mass.

The influence exerted by some physical source over a region of space is called a **field**. In this case, the field's region is the environment extending from the Earth's surface outward and the source is the Earth. Every position in a field is assigned a particular value of some quantity. For example, a temperature field is a *scalar field* and every position is associated with a temperature. If the field is a **vector field**, each position in the field is associated with a particular magnitude and direction. The Earth's gravitational field is a vector field, and Equation 7.12 assigns a magnitude and direction to every position in that field.

Although the terms *force* and *field* may sound alike, they have different meanings. Let's use this case of spacecraft near the Earth (Fig. 24.1) to distinguish between the two:

1. The field has a **source** that influences its surroundings. We are interested in knowing the gravitational force on a variety of spacecraft, so we choose the Earth as the source of the gravitational field. At each position r, the magnitude and direction of the Earth's gravitational field $\vec{g}(r)$ **depend only on properties of the source**—in this case, the Earth's mass $M_\oplus$. In fact, even if all the spacecraft vanished, $\vec{g}(r)$ would still have the same magnitude and direction given by Equation 7.12.

The word *vector* is often omitted when we refer to specific vector fields, so we use *gravitational field* rather than *gravitational vector field*. The term *field* also refers to the mathematical description of the source's influence.

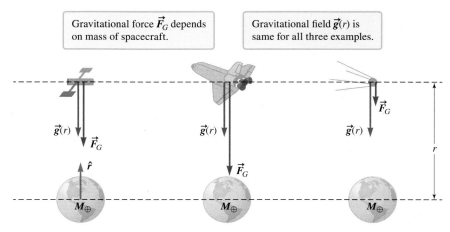

FIGURE 24.1 The gravitational field of the Earth and the gravitational force exerted on three different spacecraft. The gravitational field is the same in all three cases, but $\vec{F}_G$ is strongest for the middle spacecraft, because it has the greatest mass, and weakest for the rightmost spacecraft, because it has the smallest mass.

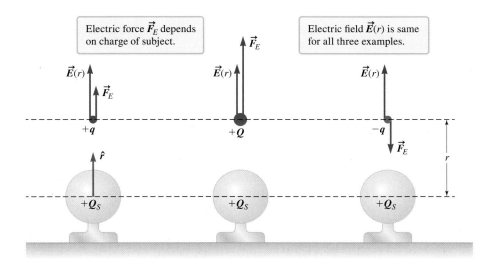

FIGURE 24.2 The electrostatic field of a charged sphere and the electrostatic force exerted on various charged particles. The electrostatic field is the same in all three cases, but the electrostatic force varies, depending on the charged particle.

2. **Subjects** are objects that may be placed in the source's field. The **force** exerted on a subject **depends on properties of the field and the subject**. In this case, the gravitational force exerted on the spacecraft depends on the mass m_{SC} of the spacecraft and the Earth's gravitational field $\vec{g}(r)$ at the location of the spacecraft.

SOURCE AND SUBJECT

✪ **Major Concept**

Like gravity, the electrostatic force is also a field force. We apply the concept of a field to the electrostatic force by analogy with gravity. Figure 24.2 shows a large sphere resting on an insulated pedestal. The sphere has positive charge $+Q_S$. We consider the sphere to be the source of the field. Three subjects—a particle with charge $+q$, a sphere with a greater charge $+Q \ (> q)$, and a particle with charge $-q$—are each placed in turn a distance r from the center of the source.

The electrostatic force exerted by the source on each subject depends on both the source's charge $+Q_S$ and the subject's charge q_{subject}. The electrostatic force $\vec{F}_E$ is strongest for the middle subject because it has the greatest charge. Also notice that $\vec{F}_E$ on the rightmost subject points toward the source because that subject has a negative charge, whereas for the other two subjects $\vec{F}_E$ points away from the source because both source and subjects are positively charged.

The **electrostatic** *field* (also known as the **electric** *field*) $\vec{E}(r)$ does not depend on the subject at all. The electrostatic field $\vec{E}(r)$—like the gravitational field—is a vector field that depends only on properties of the source and position r from the source.

The concept of a field is very convenient when we want to study the influence of one particular source on a number of different subjects. The field concept is also helpful when we consider sources of different shapes. In our study of gravity and in electrostatics so far, we have considered only sources with spherical symmetry (planets, moons, spheres, shells, and particles). However, in electrostatics, we are often interested in sources that are *not* spherical. For example, in the next case study, we explore the 18th-century controversy surrounding the best shape to use for lightning rods.

CASE STUDY The Shape of Lightning Rods

In the middle of the 18th century, people loved to play with electricity (Fig. 23.4, page 685), but most people thought lightning was another phenomenon altogether—explosions of atmospheric gas, something like the explosions of gunpowder.

Benjamin Franklin thought otherwise. He believed that lightning was a colossal electrical spark just like the small sparks people found so amusing. To support his theory, he suggested that a metal rod be placed on top of a tall structure to capture "electric fluid"—what we call charged particles. While Franklin was waiting for

FIGURE 24.3 Franklin's kite experiment showed that lightning is a giant electrical spark. His son William was outside, assisting him in flying the kite.

the completion of Christ Church in Philadelphia (there were no other tall structures in Philadelphia at the time), he came up with another way to do his experiment. He used a kite with a metal wire attached (Fig. 24.3). The wire was connected to a string, which when wet would act as a conductor. At the end of the string was a metal key connected to a Leyden jar—a device used to store charge. (Leyden jars were among the props normally used in demonstrations of electricity at that time.) Franklin held a piece of dry silk, which insulated him, and then proceeded to collect charge from his flying kite. Franklin showed that a Leyden jar charged by clouds produced all the same effects as Leyden jars charged in the home. So he concluded that lightning is an electrical phenomenon, like a giant spark.

Don't try this experiment yourself. Franklin took a number of precautions—for example, he didn't fly the kite during a storm but only as the storm was approaching. He kept himself and the silk string dry. He may have just been lucky, however. The Swedish professor George William Richman was killed performing a similar experiment when a spark about a foot long jumped from a metal rod to Richman's head.

Franklin—like other people at the time—knew about the dangers posed by lightning. As people started building taller structures, those structures became more likely targets of lightning strikes that could cause fires, destroying property and lives. In 1769, lightning struck a British gunpowder magazine, killing more than a thousand people and destroying a whole town. Once Franklin understood that lightning was a giant spark, he invented a way to protect against lightning strikes—the lightning rod.

Franklin published the following recommendations for lightning rods in *Poor Richard* (1753):

1. Just outside each building an iron rod **should be planted 3 feet to 4 feet into the moist ground**.
2. The rod should extend 6 feet to 8 feet **above the tallest part of the structure**.
3. On top of the rod should be a foot of brass wire **sharpened to a fine point**.

24-2 Special Case: Electric Field of a Charged Sphere

We wish to find the electric fields generated by sources of various shapes. We start by finding the electric fields outside sources that have spherical symmetry (solid spheres, spherical shells, and particles).

Compare Figures 24.1 and 24.2. Because the charged, spherical source has the same spherical shape as the Earth, we can rely on our work on the Earth's gravitational field. The electrostatic force exerted by the source on each subject is found by using Coulomb's law:

$$\vec{F}_E = k\frac{Q_S q_{\text{subject}}}{r^2}\hat{r} \tag{23.5}$$

where the unit vector $\hat{r}$ points outward from the source (at the origin) toward the subject, and q_{subject} is the charge of a particular subject.

The electrostatic force $\vec{F}_E$ exerted on the three possible subjects in Figure 24.2 depends on each subject's charge q_{subject}. So—just as we rewrote Newton's law of gravity—we can rewrite Coulomb's law as

$$\vec{F}_E(r) = q_{\text{subject}}\left(k\frac{Q_S}{r^2}\hat{r}\right)$$

$$\vec{F}_E(r) = q_{\text{subject}}\vec{E}(r) \tag{24.2}$$

where $\vec{E}(r)$ is the electric field generated outside of any spherical source and

$$\vec{E}(r) = k\frac{Q_S}{r^2}\hat{r} \tag{24.3}$$

ELECTRIC FIELD DUE TO A CHARGED PARTICLE ⓞ Special Case

Like $\vec{g}(r)$, $\vec{E}(r)$ depends only on the source and the distance r from that source. In fact, if the three subjects in Figure 24.2 vanished, $\vec{E}(r)$ would be unchanged; *the electric field $\vec{E}(r)$ never depends on the properties of the subject.*

Compare the electric field (Eq. 24.3) with the Earth's gravitational field (Eq. 7.12). Mathematically, these two fields are similar:

1. The magnitude of each field drops off with the distance squared.
2. The direction is radial (along $\hat{r}$). In the case of the gravitational field, the direction is given by $-\hat{r}$ (toward the source), whereas for the electrostatic field, the direction is $+\hat{r}$ (outward from the source toward the subject) if the source has a positive charge.
3. Both fields have a constant of proportionality (G in the case of gravity and k in the case of the electrostatic field).
4. Each field depends on only one property of the source (the mass of the Earth in the case of gravity and the charge of the source in the case of electrostatics).

There is one very important difference between the electrostatic field and the gravitational field. The gravitational field always points toward the center of the source because there is only one kind of mass. However, there are two kinds of charge—positive and negative. If the source is positive, the electrostatic field points away from the source (Fig. 24.4). If source is negative, the electrostatic field points toward the source.

Figure 24.4 and Equation 24.2 can be used to develop a general prescription for finding the electric field due to a source of any shape. Imagine a test particle with a small amount of positive charge $+q_{test}$. You move this particle to various locations in the field of the source (Fig. 24.4). For each position, you measure or calculate the electrostatic force exerted on the test particle. Because the test particle is positive, the electrostatic force points in the same direction as the electrostatic field. The magnitude of the electric field is found from Equation 24.2:

$$E(r) = \frac{F_E(r)}{q_{test}} \qquad (24.4)$$

where the test particle is the subject and its charge is always positive: $q_{test} > 0$. So the magnitude of the electric field vectors is directly proportional to the magnitude of the electrostatic force vectors. We see from Figure 24.4C that outside a positively

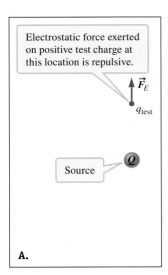
A.

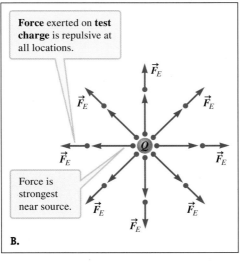
B.

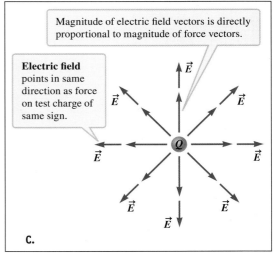
C.

FIGURE 24.4 A. The electric force exerted by a positively charged sphere on a positively charged test particle. **B.** The electric force exerted on a test particle is repulsive at 16 different locations. **C.** The electric field due to the sphere at all 16 locations points outward. The magnitude of the electric field vectors is directly proportional to the magnitude of the force vectors, so the field is strongest near the source. Compare to Figure 7.19 (page 197) for the gravitational field.

charged spherical source, the electrostatic field points radially away from the source and is strongest near the source.

We see from Equation 24.4 that the dimensions of the electric field are force per charge, so the SI units of electric field are N/C. Some values of the electric fields associated with various sources are given in Table 24.1.

TABLE 24.1 Rough magnitude of the electric field near various sources.

Source	Magnitude of E (N/C)
Near surface of uranium nucleus	3×10^{21}
At orbit of ground-state electron in hydrogen atom	6×10^{11}
Near surface of charged comb or balloon rubbed on hair	10^{3}
Earth's lower atmosphere (fair weather)	10^{2}

CONCEPT EXERCISE 24.1

In a few sentences, explain how you know that $\vec{E}(r) = (kQ_s/r^2)\hat{r}$ (Eq. 24.3) is consistent with Figure 24.4C.

CONCEPT EXERCISE 24.2

What is the magnitude of the electric field due to a charged particle at its exact location ($r = 0$)?

24-3 Electric Field Lines

As with all vectors, the length of an arrow represents the vector's magnitude. In Figure 24.4C, longer vectors show that the electric field is stronger near the source, and shorter vectors show that the electric field farther from the source is weaker. Drawing vectors for each situation takes considerable time and some skill, because many must be carefully drawn at a number of locations. A much easier way to represent a vector field is by drawing *field lines* as in Figure 24.5. **Electric field lines** provide a way to visualize the magnitude and direction of the electric field. *Electric field lines must either originate at a positively charged object or terminate on a*

ELECTRIC FIELD LINES ⊙ Tool

We draw the arrowhead in the middle of the line, not at the end. This helps us avoid confusing field lines with vectors.

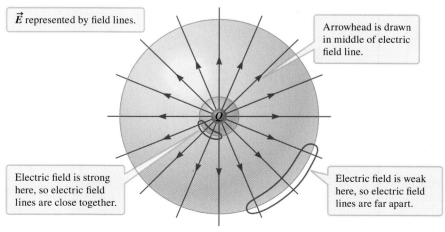

FIGURE 24.5 The electric field of a spherical positive source represented by field lines. The surfaces of the two concentric spheres are perpendicular to the field lines. Compare this to Figure 24.4C, which is the arrow representation.

negatively charged object (or both). For example, in Figure 24.5, each electric field line originates at the positively charged source.

The magnitude of the electric field is related to the density of the electric field lines. To find the density of the electric field lines, imagine a set of surfaces perpendicular to the field lines; the more lines per unit area passing through one of these surfaces, the stronger the electric field. In Figure 24.5, the electric field lines are farther apart (less dense) in the region far from the source, showing that the electric field is weaker there.

The direction of the electric field at any position is tangent to the electric field line and in the direction indicated by the arrowhead on the field line. Figure 24.6 shows electric field lines along with vectors representing the electric fields at four locations. Notice that the electric field lines cannot cross because then there would be two possible directions for the electric field at their point of intersection.

When representing electric fields visually, we are often forced to draw two-dimensional representations of three-dimensional situations. The electric field in Figure 24.5 is three-dimensional; we must imagine the third dimension because we cannot draw it on a two-dimensional surface. Likewise, you must imagine that the vector arrows in Figure 24.4C emerge from the paper (or screen) both toward and away from you so that they point radially outward in all directions. Electric field lines from a spherical positive charge extend outward in all directions like the quills of a balled-up porcupine.

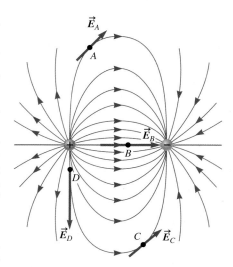

FIGURE 24.6 The electric field at any point such as A, B, C, or D is tangent to an electric field line. The direction of the field is determined by the arrow on the field line. The magnitude of the electric field is represented by the field line density. Here $E_D > E_B > E_A = E_C$.

CONCEPT EXERCISE 24.3

Which lines in Figure 24.7 *cannot* represent an electric field? Explain.

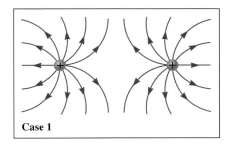
Case 1

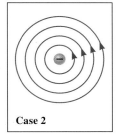
Case 2

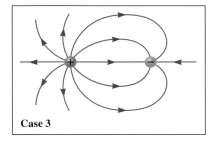
Case 3

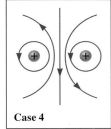
Case 4

FIGURE 24.7

EXAMPLE 24.1   A Negative Spherical Source

Consider a negatively charged spherical source, and use $\vec{E}(r) = (kQ_S/r^2)\hat{r}$ (Eq. 24.3) for the electric field. Draw a diagram of the sphere's electric field using field lines, similar to Figure 24.5.

:• INTERPRET and ANTICIPATE

Equation 24.3 works regardless of the sign of the source charge. When Q_S has a negative value, the magnitude of the electric field is still higher near the sphere and lower farther away. The main difference from Figure 24.5 is that the negative charge changes the direction of the electric field, so now the electric field points inward—toward the source.

Example continues on page 720 ▶

SOLVE
Draw electric field lines that reflect this description (Fig. 24.8). The electric field lines are closer together near the sphere than they are farther away, and the arrowheads point inward.

CHECK and THINK
Imagine placing a *positive* test particle at several locations near a negatively charged sphere. The positive test charge is attracted to the negative sphere. This is consistent with Figure 24.8 because the electrostatic force on a test charge points in the same direction as the electric field. Both point inward in this case.

FIGURE 24.8

24-4 Electric Field of a Collection of Charged Particles

In this section, we focus on sources that are collections of charged particles. When we consider the electric field due to an unfamiliar source, it is helpful to visualize it by sketching electric field lines. For example, Figure 24.9 shows two identical positively charged particles. Here are a few tips for drawing electric field lines when the source is a collection of particles (or charged spherical objects):

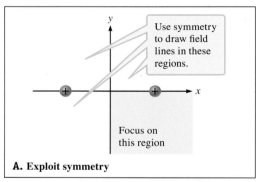

A. Exploit symmetry

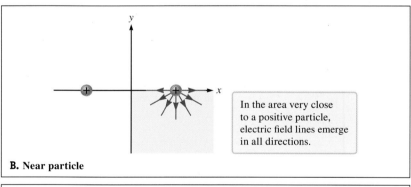

B. Near particle

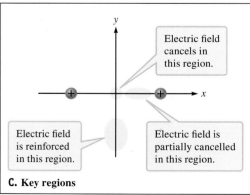

C. Key regions

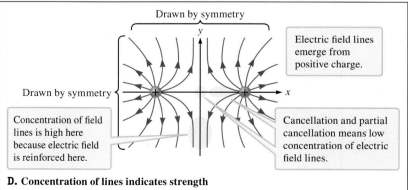

D. Concentration of lines indicates strength

FIGURE 24.9 This source consists of two positively charged particles. Four tips for drawing electric field lines: **A.** Exploit symmetry and focus on one region. **B.** Draw field lines emerging from a positive source (or terminating on a negative source). **C.** Determine how the presence of other charged particles affects the electric field in the various regions. **D.** Make sure the concentration of field lines represents the relative strength of the field in various regions.

1. **Exploit symmetry (Fig. 24.9A).** Often the collection of particles is symmetrically arranged. You can draw the electric field lines in one region, and then exploit the symmetry to draw the electric field lines in the remaining regions. So start by focusing your attention on one region.
2. **Draw field lines near one or more particles.** Electric field lines must originate at a positive particle, terminate on a negative particle, or both (Fig. 24.9B). Because we are considering a collection of particles, the electric field lines *near* each particle must look something like the electric field lines in Figure 24.5 or 24.8.
3. **Determine the relative strength of the electric field in a few key regions.** Regions that are far from a particular particle may be strongly influenced by the presence of the other particles (Fig. 24.9C). The electric fields due to all charged particles must be added vectorially. In a region, it is possible for the electric fields due to various particles to cancel, partially cancel, or reinforce one another.
4. Sketch the field lines so that the **concentration of lines corresponds to the relative strength of the electric field** in the various regions you considered (Fig. 24.9D). Be sure the concentration of electric field lines is higher near the particles and in regions where the electric field is reinforced by the presence of other charged particles. Also be sure the concentration of electric field lines is low where the electric field is partially cancelled.

CONCEPT EXERCISE 24.4

Figure 24.10 shows a source that consists of two charged particles.

a. What is the sign of the charge on each particle?
b. In which region (A, B, or C) is the electric field the weakest?
c. In which region (A, B, or C) is the electric field the strongest?

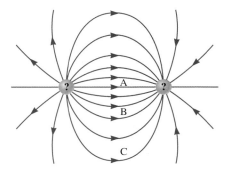

FIGURE 24.10

PROBLEM-SOLVING STRATEGY — Electric Field due to a Collection of Charged Particles

The electric field due to a single particle is given by $\vec{E}(r) = (kQ_S/r^2)\hat{r}$ (Eq. 24.3). How do we find the electric field when a source consists of more than one charged particle? To find a mathematical expression or a numerical value of the electric field at some particular position, we must add the electric fields due to each particle, taking into account the magnitude and direction of the vector fields.

:• INTERPRET and ANTICIPATE

Draw a diagram showing *all the electric fields* present at some particular position. There are four elements to this diagram:

1. **Labeled source particles and test particle.** Draw and label the particles that make up the source. Also draw an imaginary test particle at the location where you wish to find the electric field. The **test particle is always assumed to have a positive charge**.
2. The **electric field vectors** at the location of the test charge. These field vectors are found from the electric force exerted on the test charge by each particle in the source. The direction of the force is along the line joining the test particle to the particular source particle. To get the length of the vector, estimate the magnitude of the force by taking into account the magnitude of the source particle's charge and the distance to the test particle. Because the electrostatic force is parallel and proportional to the electric field, **label these vectors as electric fields**, using a subscript to indicate the source particle that produced that electric field vector.
3. The **net electric field vector** at the location of interest. Find this graphically in order to anticipate the result. Draw the resultant vector and label it $\vec{E}$.
4. A **coordinate system** and other **geometric details**. Your diagram is similar to a free-body diagram for the test charge, and your experience with free-body diagrams will help you choose a good coordinate system. Be sure to include details such as distances and angles.

:• SOLVE

Once the diagram is complete, find a mathematical expression for the electric field at the point of interest by adding the electric field vectors algebraically as you would any vectors. (We are *not* applying Newton's second law, so acceleration is not involved.)

EXAMPLE 24.2 Two Positively Charged Particles

Find an expression for the electric field at a point equidistant between two identical particles, each with charge $+Q$. The separation between the particles is $2a$, and the distance between each particle and the point is $\sqrt{2}a$.

INTERPRET and ANTICIPATE

The two charged particles exert a repulsive force on each other, but we can assume they are fixed in place, and we are not interested in calculating the force they exert on each other. We are interested in the electric field created by the combination of these two charged particles.

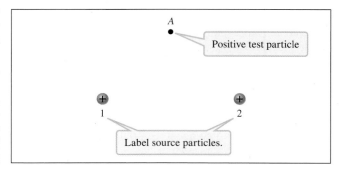

FIGURE 24.11

We draw a diagram with all four elements. **Label the source particles** 1 and 2. The **test particle** is drawn at A—a point equidistant from the two source particles as stated in the problem (Fig. 24.11).

There are two **electric field vectors** at A, the location of the (imaginary) test particle—one from particle 1 and another from particle 2. We estimate the magnitude and direction of these vectors from the force the two source particles exert on the (positively charged) test particle. The source particles carry the same charge and are equidistant from A, so they exert forces of equal magnitude on the test particle. These vectors are **labeled as electric fields** (Fig. 24.12).

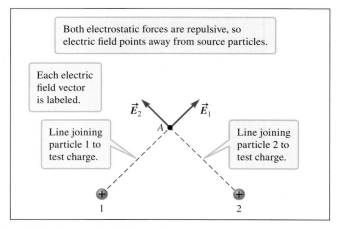

FIGURE 24.12

Graphically, add the individual electric field vectors to show the **net electric field vector** at A (Fig. 24.13).

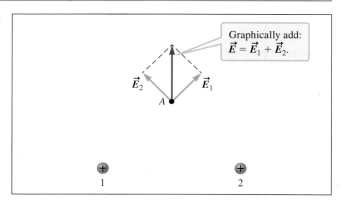

FIGURE 24.13

Add a **coordinate system** and other **geometric details**. In this case, we see that point A and the two source particles form a triangle, which we label in terms of the parameters given. From Figure 24.14, we expect that the x components of the field cancel and the y components add, so that the resulting electric field points in the positive y direction.

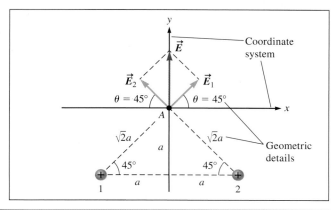

FIGURE 24.14

SOLVE Write $\vec{E}_1$ and $\vec{E}_2$ from Figure 24.14 in component form.	$\vec{E}_1 = (E_1 \cos\theta)\hat{\imath} + (E_1 \sin\theta)\hat{\jmath}$ $\vec{E}_2 = (-E_2 \cos\theta)\hat{\imath} + (E_2 \sin\theta)\hat{\jmath}$
Since $\theta = 45°$, simplify the expressions using $\sin 45° = \cos 45° = \sqrt{2}/2$.	$\vec{E}_1 = \dfrac{\sqrt{2}}{2} E_1 (\hat{\imath} + \hat{\jmath})$ $\vec{E}_2 = \dfrac{\sqrt{2}}{2} E_2 (-\hat{\imath} + \hat{\jmath})$
Add these vectors.	$\vec{E} = \vec{E}_1 + \vec{E}_2 = \dfrac{\sqrt{2}}{2} E_1 (\hat{\imath} + \hat{\jmath}) + \dfrac{\sqrt{2}}{2} E_2 (-\hat{\imath} + \hat{\jmath})$ $\vec{E} = \dfrac{\sqrt{2}}{2} [(E_1 - E_2)\hat{\imath} + (E_1 + E_2)\hat{\jmath}]$ (1)
The magnitudes E_1 and E_2 can be found from $E(r) = kQ_S/r^2$ (Eq. 24.3). Both source particles have positive charge $+Q$ and both are $\sqrt{2}a$ away from point A. We find that the magnitudes E_1 and E_2 are equal.	$E_1 = k\dfrac{Q}{(\sqrt{2}a)^2} = \dfrac{1}{2}\dfrac{kQ}{a^2}$ $E_2 = k\dfrac{Q}{(\sqrt{2}a)^2} = \dfrac{1}{2}\dfrac{kQ}{a^2} = E_1$
Substitute the magnitudes of E_1 and E_2 in the expression for $\vec{E}$ (Eq. 1).	$\vec{E} = \dfrac{\sqrt{2}}{2} [(E_1 - E_1)\hat{\imath} + (E_1 + E_1)\hat{\jmath}]$ $\vec{E} = \sqrt{2} E_1 \hat{\jmath} = \dfrac{\sqrt{2}}{2}\dfrac{kQ}{a^2}\hat{\jmath}$

:• **CHECK and THINK**
As expected, the x components cancel and the y components add; the net electric field is in the positive y direction.

Electric Dipole

An **electric dipole** consists of two charged particles of magnitude Q but with opposite signs (Fig. 24.15). The particles are attracted to each other, but their separation is maintained so that the distance between them is d. The electric dipole is an important special case because in many circumstances, atoms and molecules can be modeled as dipoles. For example, in Section 23-4, we saw that the atoms or molecules in an insulator can be modeled as dipoles when the insulator is near a charged object. Other molecules are natural dipoles; for example, a water molecule is best modeled as a dipole even when it is not near any charged objects.

ELECTRIC DIPOLE ✪ **Major Concept**

FIGURE 24.15 An electric dipole consists of two charged particles, one with charge $+Q$ and the other with charge $-Q$. The charged particles are attracted to each other, but maintain a separation d.

EXAMPLE 24.3 Special Case: Electric Field due to a Dipole

Find an expression for the electric field at a point a distance x to the right of the center of the dipole shown in Figure 24.15. (Other positions are given as homework; see Problem 18.)

:• **INTERPRET and ANTICIPATE**
Include all four elements in your sketch.

1. Label the source particles and draw the (positive) **test charge** (Fig. 24.16).

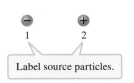
Label source particles.

Positive test particle

FIGURE 24.16

Example continues on page 724 ▶

2. **Sketch the electric field vectors for each source particle** (Fig. 24.17). The test charge is positive and particle 1 is negative, so particle 1 exerts an attractive force on the test charge, whereas particle 2 is positive and exerts a repulsive force on the test charge. (Remember the electric field points in the same direction as the electric force for a positive test charge.) Because particle 1 is farther away than particle 2, E_1 , E_2.

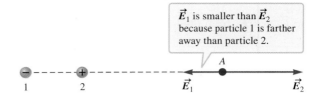

FIGURE 24.17

3. **Add the electric field vectors graphically** (Fig. 24.18). The sum of the electric fields at A is $\vec{E} = \vec{E}_1 + \vec{E}_2$.

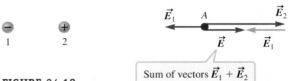

FIGURE 24.18

4. **Choose a coordinate system and add geometric details.** We have chosen the origin of coordinates to be midway between the source charges. From Figure 24.19, we expect that the electric field vector $\vec{E}$ points in the positive x direction.

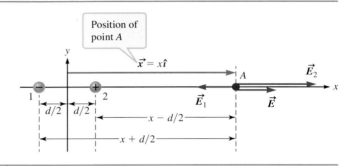

FIGURE 24.19

SOLVE	
Vectors $\vec{E}_1$ and $\vec{E}_2$ have only x components; $\vec{E}_1$ is in the negative x direction and $\vec{E}_2$ is in the positive x direction.	$\vec{E}_1 = -E_1 \hat{\imath} \qquad \vec{E}_2 = E_2 \hat{\imath}$
Add these vectors.	$\vec{E} = \vec{E}_1 + \vec{E}_2 = -E_1 \hat{\imath} + E_2 \hat{\imath} = (E_2 - E_1) \hat{\imath}$ (1)
The magnitudes of E_1 and E_2 can be found from Equation 24.3. Each particle has a charge of magnitude Q. The center-to-center distances r between the test charge and each of the two sources are shown in Figure 24.19. For particle 1 this distance is $x + d/2$, and for particle 2 the distance is $x - d/2$.	$E(r) = k\dfrac{\lvert Q_s \rvert}{r^2}$ (24.3) $E_1 = k\dfrac{\lvert -Q \rvert}{(x + d/2)^2} = k\dfrac{Q}{(x + d/2)^2}$ $E_2 = k\dfrac{Q}{(x - d/2)^2}$
Substitute E_1 and E_2 into Equation (1).	$\vec{E} = kQ\left(\dfrac{1}{(x - d/2)^2} - \dfrac{1}{(x + d/2)^2}\right)\hat{\imath}$
Find a common denominator and simplify this expression.	$\vec{E} = kQ\left(\dfrac{(x + d/2)^2 - (x - d/2)^2}{(x - d/2)^2(x + d/2)^2}\right)\hat{\imath}$ $\vec{E} = kQ\dfrac{2xd}{(x^2 - d^2/4)^2}\hat{\imath}$ (2)
CHECK and THINK If the denominator in Equation (2) is zero, then E is infinite. Let's see under what circumstance the denominator will be zero.	$x^2 - d^2/4 = 0$ $x = \pm\dfrac{d}{2}$

The electric field is infinite if point A is on either of the charged particles. This makes sense because the electric field of any charged particle goes to infinity as r goes to zero (Concept Exercise 24.2). Finally, as expected, for $x > d/2$ as in Figure 24.19, the electric field $\vec{E}$ points in the positive x direction.

Far from an Electric Dipole: The Dipole Moment

The electric field found in Example 24.3 depends on our choice of coordinates. If another person decided to use an x axis pointing in a different direction, that person would have to modify our expression to make it fit that choice of coordinate system. Because dipoles are particularly important special cases, we can avoid such complications by writing the electric field for a dipole in terms of an **electric dipole moment**, $\vec{p}$. The electric dipole moment—also called the *dipole moment*—is a vector that points from the negatively charged particle toward the positively charged particle in a dipole. So, in Figure 24.19, the dipole moment points in the positive x direction. The magnitude of the dipole moment is Qd. The result in Example 24.3 is written as

$$\vec{E} = \frac{2kx\vec{p}}{(x^2 - d^2/4)^2} \quad (24.5)$$

Equation 24.5 does not include a unit vector and so it does not depend on a particular coordinate system; it is good for finding the electric field due to a dipole at a distance x (from the midpoint of the dipole) along the line that passes through both particles.

Often we are interested in the electric field of a dipole at a position that is far compared to the separation between the two particles. We find an expression for the electric field in that case by making the approximation $x \gg d/2$. So the denominator of Equation 24.5 reduces to x^4:

$$\vec{E} \approx \frac{2kx\vec{p}}{(x^2 - 0)^2} \approx \frac{2kx\vec{p}}{x^4}$$

and the electric field is approximately

$$\vec{E} \approx \frac{2k\vec{p}}{x^3} \text{ when } x \gg d/2 \quad (24.6)$$

Equation 24.6 is a good approximation for any position far from an electric dipole along the line that passes through both charges. In fact, the magnitude of Equation 24.6 is good for *any* position r in *any* direction far from an electric dipole:

$$E \approx \frac{2kp}{r^3} \quad (24.7)$$

where $r \gg d$ and d is the separation between the charges.

If a point is *very* far from the dipole—in other words, if $r \to \infty$—then the two charged particles that make up the dipole appear to cancel each other. We expect that the electric field should be zero for a position so far away, and that is exactly what we find if we take the limit of Equation 24.7:

$$\lim_{r \to \infty} E = \lim_{r \to \infty} \frac{2kp}{r^3} \to 0$$

If you model a water molecule as a dipole (Fig. 24.20), you can use Equation 24.5 to find the electric field near the molecule along the line that passes through its midpoint (the x axis shown in Fig. 24.20). You can use Equation 24.7 to find the magnitude of the electric field for any position far ($r \gg d$) from the water molecule. Finally, the electric field at a point very far from the water molecule is zero because the charged particles are at essentially the same distance from that point and so the electric field due to each particle cancels out.

ELECTRIC DIPOLE MOMENT
⭐ **Major Concept**

MAGNITUDE OF THE ELECTRIC FIELD DUE TO A DIPOLE FOR $r \gg d$ ⏵ **Special Case**

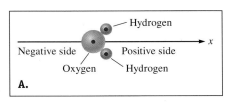

A.

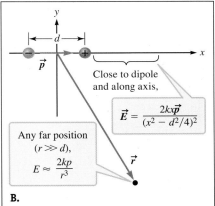

B.

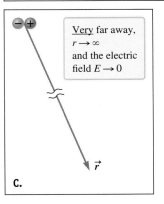

C.

FIGURE 24.20 A. A water molecule consists of two hydrogen atoms and one oxygen atom. **B.** The electric field of a water molecule may be modeled as a dipole. The electric field on the x axis is given by Equation 24.5. The magnitude of the electric field for $r \gg d$ is given by Equation 24.7. **C.** The electric field very far from a water molecule is zero.

CONCEPT EXERCISE 24.5

A water molecule is made up of two hydrogen atoms and one oxygen atom, with a total of 10 electrons and 10 protons. The molecule is modeled as a dipole with an effective separation $d = 3.9 \times 10^{-12}$ m between its positive and negative charges. What is the magnitude of the water molecule's dipole moment?

24-5 Electric Field of a Continuous Charge Distribution

When you rub a plastic rod with wool or shuffle your socks across the carpet, charged particles are transferred between the objects. In fact, the charged particles are so numerous that they essentially form a continuous or smooth charge distribution. Because of this great number of particles, adding up the electric field due to each particle means solving an integral. Before we present a problem-solving strategy for finding the electric field of a continuous charge distribution, we need to discuss charge density.

Charge Density

We use the same symbol for both volume charge density and mass density (Eq. 1.1). Both represent an intrinsic property (charge or mass) per unit volume. Be careful not to confuse these two uses of the same symbol.

When applying calculus to a continuous distribution of charged particles, we use the source's *charge density* and integrate over the appropriate region of space to account for the total charge. There are three types of charge density:

1. **Volume charge density** ρ (lowercase Greek letter rho) is the amount of charge per unit volume (Fig. 24.21A). For a uniform charge distribution,

$$\rho \equiv \frac{Q}{V} \qquad (24.8)$$

Volume charge density has the dimensions charge per volume, with SI units C/m³. The amount of charge dq in a small volume dV can be written in terms of the volume charge density:

$$\begin{pmatrix} \text{charge contained} \\ \text{in small volume} \end{pmatrix} = \begin{pmatrix} \text{charge per} \\ \text{unit volume} \end{pmatrix} \times (\text{volume})$$

$$dq = \rho dV \qquad (24.9)$$

2. **Surface charge density** σ (lowercase Greek letter sigma) is the amount of charge per unit area (Fig. 24.21B). For a uniform charge distribution,

$$\sigma \equiv \frac{Q}{A} \qquad (24.10)$$

The dimensions of σ are charge per area, with SI units C/m². The amount of charge dq in a small area dA can be written in terms of the surface charge density:

$$\begin{pmatrix} \text{charge contained} \\ \text{in small area} \end{pmatrix} = \begin{pmatrix} \text{charge per} \\ \text{unit area} \end{pmatrix} \times (\text{area})$$

$$dq = \sigma dA \qquad (24.11)$$

3. **Linear charge density** λ (lowercase Greek letter lambda) is the amount of charge per unit length (Fig 24.21C). For a uniform charge distribution,

$$\lambda \equiv \frac{Q}{L} \qquad (24.12)$$

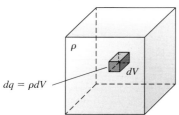

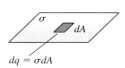

FIGURE 24.21 A. Volume charge density. **B.** Surface charge density. **C.** Linear charge density.

The dimensions of λ are charge per length, with SI units C/m. The amount of charge dq in a small length element dL can be written in terms of the linear charge density:

$$\begin{pmatrix} \text{charge contained in} \\ \text{small length element} \end{pmatrix} = \begin{pmatrix} \text{charge per} \\ \text{unit length} \end{pmatrix} \times (\text{length})$$

$$dq = \lambda dL \qquad (24.13)$$

CONCEPT EXERCISE 24.6

a. Figure 24.22A shows a rod of length L and radius R with excess positive charge Q. The excess charge is uniformly distributed over the entire outside surface of the rod. Write an expression for the surface charge density σ. Write an expression in terms of σ for the amount of charge dq contained in a small segment of the rod of length dx.

b. Figure 24.22B shows a very narrow rod of length L with excess positive charge Q. The rod is so narrow compared to its length that its radius is negligible and the rod is essentially one-dimensional. The excess charge is uniformly distributed over the length of the rod. Write an expression for the linear charge density λ. Write an expression in terms of λ for the amount of charge dq contained in a small segment of the rod of length dx. Compare your answers with those for part (a). Explain the similarities and differences.

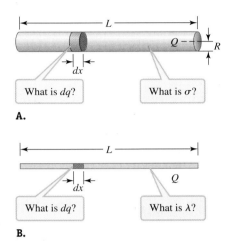

FIGURE 24.22

PROBLEM-SOLVING STRATEGY

Electric Field due to a Continuous Distribution of Charged Particles

The overall plan for finding the electric field due to a continuous distribution of charged particles is based on the procedure we followed for a collection of charged particles. The first step is to imagine slicing up the distribution into tiny pieces, with each piece so small that it can be modeled as a single charged particle. We consider just a small number of pieces—usually one or two—to get an expression for the electric field produced by those pieces. Then we integrate that expression to find the electric field produced by the entire continuous distribution.

:• INTERPRET and ANTICIPATE

As for a collection of charged particles, draw a diagram showing the electric field at some particular position. There are four elements to this diagram:

1. **Labeled pieces** of the source and **an imaginary test particle.** Imagine slicing up the distribution into small pieces. It is usually helpful to indicate two of these pieces on your sketch. In choosing which two pieces, exploit the symmetry of the problem. Also draw an imaginary test particle (assumed positive) at the location where you wish to find the electric field. If you are not told whether the charge distribution is positive or negative, assume it is positive.
2. The **electric field vectors** at the location of the test charge, as found from the electric force exerted on the test charge by each piece of the source you've indicated. The direction of the force is along the line joining the test particle to the particular piece of the source. To get the length of the vector, estimate the magnitude of the force by taking into account the distance to the test particle. As long as the charge distribution is uniform, each piece has the same small amount of charge dq, which makes your estimation task easy. **Label these vectors as electric fields** $d\vec{E}_{\text{sub}}$, using a subscript to indicate the piece of the source that produced each electric field vector.
3. The **net electric field vector** at the location of interest. Find this graphically in order to anticipate the result. Draw the resultant vector and label it $d\vec{E}$.
4. A **coordinate system** and other **geometric details**.

:• SOLVE

There are two major steps:

Step 1 Find a **mathematical expression** for the infinitesimal electric field $d\vec{E}$ in terms of an infinitesimal spatial variable (dx, dy, dz, dr, ds, or $d\theta$). The magnitude of the electric field due to an infinitesimal amount of charge dq comes from modifying $\vec{E}(r) = (kQ_S/r^2)\hat{r}$ (Eq. 24.3) by replacing Q_S with dq:

$$dE(r) = k\frac{dq}{r^2} \qquad (24.14)$$

The direction will come from your diagram. In this step, you are likely to write dq in terms of the charge density, so Equations 24.8 through 24.13 are very useful.

Problem-Solving Strategy continues on page 728 ▶

728 CHAPTER 24 Electric Fields

Step 2 **Integrate** this expression over the entire charge distribution.

:• CHECK and THINK
Here are two things to check once you have found an expression for the electric field $\vec{E}$:
1. Make sure the **dimensions** of the electric field are force per charge.

2. If the charge distribution is finite, the **electric field** for a point far from the distribution compared to its size **should approach the electric field of a charged particle**, $\vec{E}(r) = (kQ_s/r^2)\hat{r}$.

24-6 Special Cases of Continuous Distributions

ELECTRIC FIELD DUE TO A CHARGED ROD ▶ Special Case

In this section we practice finding the electric field due to three different continuous charge distributions using the problem-solving strategy from Section 24-5. All three of these are special cases, but they are important ones that you are likely to refer to many times. In the first two cases, we start by following all the steps outlined in Section 24-5. In the third case, we use a shortcut.

EXAMPLE 24.4 Special Case: Charged Rod

In Chapter 23, we considered rods of glass or plastic that have an excess charge. Now find an expression for the electric field at point A directly above the midpoint of a very thin rod of length 2ℓ (Fig. 24.23). The excess charge Q is uniformly distributed along the length of the rod.

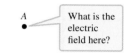

FIGURE 24.23

:• INTERPRET and ANTICIPATE
If you need to calculate the electric field due to any continuous charge distribution such as the charged rod in this problem, you cannot use Equation 24.3 directly because that equation is only for a single charged particle or sphere. We must derive an expression for the charged rod, using the strategy in Section 24-5. Here we start with a diagram that has all four elements. We imagine a positive **test particle** at point A and slice the rod into thin pieces (Fig. 24.24). Two pieces are **labeled** R and L for right and left. We have exploited the symmetry of the problem by choosing two pieces equidistant from point A.

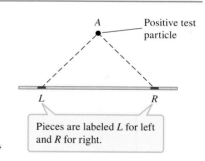

FIGURE 24.24

Draw the **electric field vectors** at the location of the test charge (Fig. 24.25). Model each piece L and R as a charged particle, and draw each electric field vector. Each piece has a positive charge dq.

Add the electric field vectors of the left and right pieces graphically, and sketch the **net electric field vector** at point A. The horizontal components cancel and the vertical components add, so we need to find only an expression for the vertical component (Fig. 24.26).

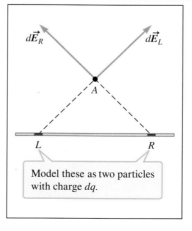

FIGURE 24.25

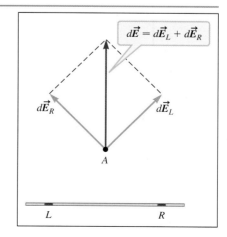

FIGURE 24.26

Add a **coordinate system** and other **geometric details.** Now compare Figure 24.27 with Figure 24.14; they are very similar. In both figures, two vectors are added such that their x components cancel. So we expect that $\vec{E}$ points in the positive y direction.

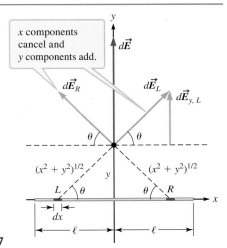

FIGURE 24.27

SOLVE

Step 1 Find a **mathematical expression** for the infinitesimal electric field $d\vec{E}$. From Figure 24.27, we reason that $d\vec{E}$ equals two times the y component of either $d\vec{E}_R$ or $d\vec{E}_L$. Let's arbitrarily work with $d\vec{E}_L$.

Write $d\vec{E}_L$ in component form.	$d\vec{E}_L = (dE_L \cos\theta)\hat{\imath} + (dE_L \sin\theta)\hat{\jmath}$
$d\vec{E}$ equals two times the y component of $d\vec{E}_L$.	$d\vec{E} = 2\, dE_L \sin\theta\, \hat{\jmath}$ (1)
The magnitude dE_L is found from Equation 24.14. The distance from piece L to point A is $r = \sqrt{x^2 + y^2}$ (Fig. 24.27).	$dE(r) = k\dfrac{dq}{r^2}$ (24.14) $dE_L = k\dfrac{dq}{(\sqrt{x^2+y^2})^2} = \dfrac{k\,dq}{x^2+y^2}$
Substitute the magnitude dE_L into the expression for $d\vec{E}$ (Eq. 1).	$d\vec{E} = \dfrac{2k\,dq}{x^2+y^2}\sin\theta\,\hat{\jmath}$
The length of either piece L or R is dx. So we modify Equation 24.13 to write $dq = \lambda\,dx$, and replace dq.	$d\vec{E} = \dfrac{2k\lambda\,dx}{x^2+y^2}\sin\theta\,\hat{\jmath}$ (2)
Step 2 Integrate. Equation (2) gives only the y component of the electric field produced by two small pieces of the charged rod as shown in Figure 24.27. We must now integrate this expression over the entire rod. We can integrate Equation (2) over only one variable—x. Because $\sin\theta$ depends on x, we must rewrite it in terms of x. Use trigonometry and the right triangle shown in Figure 24.27 to come up with an expression, and eliminate $\sin\theta$.	$\sin\theta = \dfrac{y}{(x^2+y^2)^{1/2}}$ $d\vec{E} = \dfrac{2k\lambda\,dx}{x^2+y^2}\dfrac{y}{(x^2+y^2)^{1/2}}\hat{\jmath} = \dfrac{2k\lambda y\,dx}{(x^2+y^2)^{3/2}}\hat{\jmath}$ (3)
Now we are ready to integrate over the left half of the rod from $x = -\ell$ to 0. The integral on the left of the equal sign is what we want, the electric field at point A. Pull the constants out of the integral on the right, treating y as a constant for a given point A.	$\displaystyle\int d\vec{E} = \int_{-\ell}^{0}\dfrac{2k\lambda y\,dx}{(x^2+y^2)^{3/2}}\hat{\jmath}$ $\vec{E} = 2k\lambda y\,\hat{\jmath}\displaystyle\int_{-\ell}^{0}\dfrac{dx}{(x^2+y^2)^{3/2}}$
This integral can be found in Appendix A.	$\vec{E} = 2k\lambda y\,\hat{\jmath}\left[\dfrac{1}{y^2}\dfrac{x}{\sqrt{x^2+y^2}}\right]_{-\ell}^{0}$
Substitute the limits of integration and simplify the expression.	$\vec{E} = \dfrac{2k\lambda}{y}\dfrac{\ell}{\sqrt{\ell^2+y^2}}\hat{\jmath}$
Finally we must write our expression in terms of the parameters given in the problem—in this case, the total charge Q of the source. The rod's length is 2ℓ and its linear charge density is λ, so $Q = 2\ell\lambda$.	$\vec{E} = \dfrac{kQ}{y}\dfrac{1}{\sqrt{\ell^2+y^2}}\hat{\jmath}$ (24.15)

Example continues on page 730 ▶

730 CHAPTER 24 Electric Fields

: CHECK and THINK There are two things for us to check. First, are the **dimensions** force per charge? Our dimensional analysis looks good.	$[E] = \dfrac{(F \cdot L^2/C^2)C}{L} \dfrac{1}{\sqrt{L^2 + L^2}}$ $[E] = \dfrac{(F \cdot L^2/C)}{L^2} = \dfrac{F}{C}$
Second, make sure the electric field very far away ($y \gg \ell$) **approaches that of a charged particle** (Eq. 24.3). Indeed, we arrive at the form we expect for a charged particle at a distance y.	$\vec{E} = \dfrac{kQ}{y} \dfrac{1}{\sqrt{\ell^2 + y^2}} \hat{j} \approx \dfrac{kQ}{y} \dfrac{1}{\sqrt{y^2}} \hat{j}$ if $y \gg \ell$ $\vec{E} \approx \dfrac{kQ}{y^2} \hat{j}$ ✓

EXAMPLE 24.5 **Special Case: Charged Ring**

ELECTRIC FIELD DUE TO A CHARGED RING ▶ Special Case

A person wearing wool clothes uses a plastic hula hoop of radius $R = 1.20$ m. The hula hoop has negative charge $Q = -75$ nC spread uniformly on it. First find an expression for the electric field at point A, and then find a numerical result for a point 0.60 m above the center of the hoop (Fig. 24.28).

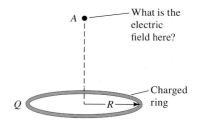

FIGURE 24.28

: INTERPRET and ANTICIPATE
We must derive an expression for a charged ring (or hoop). We draw a diagram with all four elements.

Figure 24.29 shows the first two elements. We have chosen and **labeled two pieces**—L and R—positioned symmetrically on either side of point A, and drawn a positive **test charge** at point A. We model each piece as a charged particle and draw each **electric field vector**. Let's assume for now that each piece has a positive charge dq so we can come up with a general expression that is good for any charged ring. In a later substitution, we will take into account the hoop's negative charge.

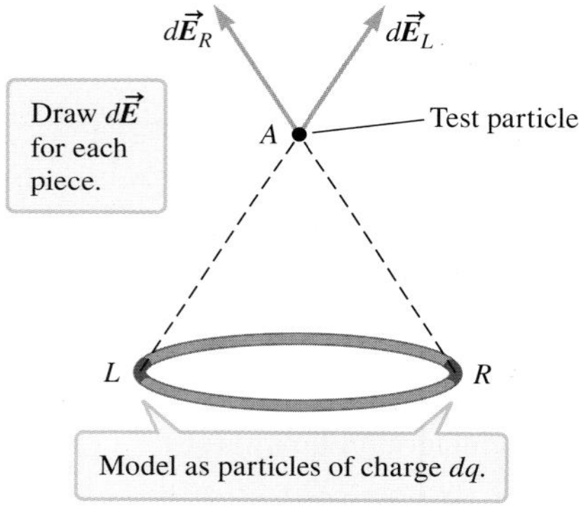

FIGURE 24.29

Figure 24.30 includes the next two elements. The electric field vectors are added to find the **net electric field vector**. The horizontal components cancel and the vertical components add, so we need to find only an expression for the vertical component. Our figure includes a **coordinate system** (y axis only) and **geometric details**.

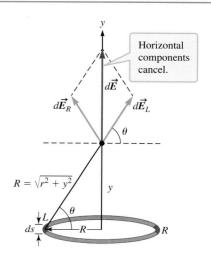

FIGURE 24.30

SOLVE

Step 1 Find an **expression** for $d\vec{E}$ in terms of a spatial variable—in this case, a small length ds of the ring (Fig. 24.30). As in the case of the charged rod, we see that $d\vec{E}$ equals two times the y component of either $d\vec{E}_R$ or $d\vec{E}_L$. Let's arbitrarily work with $d\vec{E}_L$.

$$d\vec{E} = 2dE_L \sin\theta \hat{j} \quad (1)$$

The magnitude dE_L is found from Equation 24.14 for an infinitesimal amount of charge. The distance from the piece to the point of interest is given by $r = \sqrt{R^2 + y^2}$ (Fig. 24.30).

$$dE(r) = k\frac{dq}{r^2} \quad (24.14)$$

$$dE_L = k\frac{dq}{(R^2 + y^2)}$$

Substitute dE_L into Equation (1).

$$d\vec{E}_y = 2k\frac{dq}{(R^2 + y^2)} \sin\theta \,\hat{j}$$

Replace dq by modifying Equation 24.13, this time referring to a small length of the ring as ds.

$$dq = \lambda ds$$

$$d\vec{E} = 2k\frac{\lambda ds}{(R^2 + y^2)} \sin\theta \,\hat{j} \quad (2)$$

Step 2 As we **integrate** around the left half the ring from $s = 0$ to $s = \pi R$, all the other parameters (R, y, and θ) in Equation (2) remain constant. Therefore, it is not necessary to replace $\sin\theta$ before integration. The left side is what we want—the electric field at point A.

$$\int d\vec{E} = \int_0^{\pi R} 2k\frac{\lambda ds}{(R^2 + y^2)} \sin\theta \,\hat{j}$$

$$\vec{E} = 2k\frac{\lambda}{(R^2 + y^2)} \sin\theta \,\hat{j} \int_0^{\pi R} ds$$

$$\vec{E} = 2k\frac{\lambda}{(R^2 + y^2)} \sin\theta \,\hat{j} [s]\Big|_0^{\pi R}$$

$$\vec{E} = \frac{k 2\pi R \lambda}{(R^2 + y^2)} \sin\theta \,\hat{j} \quad (3)$$

We need to find the electric field in terms of the parameters given. The problem statement gave us Q, R, and y but not λ or θ, so now we eliminate these unknown parameters. To eliminate λ, use Equation 24.12.

$$\lambda = \frac{Q}{L} \quad (24.12)$$

$$Q = L\lambda = 2\pi R \lambda \quad (4)$$

To eliminate θ, use trigonometry and the right triangle shown in Figure 24.30.

$$\sin\theta = \frac{y}{r} = \frac{y}{(R^2 + y^2)^{1/2}} \quad (5)$$

Finally, substitute Equations (4) and (5) into Equation (3).

$$\vec{E} = \frac{kQy}{(R^2 + y^2)^{3/2}} \hat{j} \quad (24.16)$$

CHECK and THINK

Before substituting values, let's check our expression. First, the **dimensions** are force per charge as expected (Problem 25). Second, if point A is far from the ring ($y \gg R$), the electric field **approaches that of a point charge**.

$$\lim_{y \gg R} \vec{E} = \frac{kQy}{(R^2 + y^2)^{3/2}} \hat{j} \to \frac{kQy}{(y^2)^{3/2}} \hat{j}$$

$$\vec{E} \approx \frac{kQy}{y^3} \hat{j} \approx \frac{kQ}{y^2} \hat{j} \checkmark$$

SOLVE

Substitute numerical values.

$$\vec{E} = \frac{(8.99 \times 10^9 \,\text{N} \cdot \text{m}^2/\text{C}^2)(-75 \times 10^{-9} \,\text{C})(0.60 \,\text{m})}{((1.20 \,\text{m})^2 + (0.60 \,\text{m})^2)^{3/2}} \hat{j}$$

$$\vec{E} = -1.7 \times 10^2 \hat{j} \,\text{N/C}$$

CHECK and THINK

Because the hoop has a negative charge, the electric field points downward in the negative y direction. The magnitude of the electric field seems reasonable because it is about an order of magnitude less than the electric field near a charged comb (Table 24.1).

EXAMPLE 24.6 Special Case: Charged Disk

Use the algebraic result of Example 24.5:

$$\vec{E} = \frac{kQy}{(R^2 + y^2)^{3/2}}\hat{j} \quad (24.16)$$

to derive an expression for the electric field at a point directly above the center of a positively charged disk as shown in Figure 24.31.

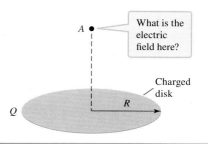

FIGURE 24.31

INTERPRET and ANTICIPATE

Using our result from Example 24.5—the electric field due to a charged ring—makes our task simpler here. Figure 24.32 combines elements 1 through 3. There is a positive **test charge** at point A. We have divided and **labeled** the disk into ring-shaped **pieces** 1 and 2. From Example 24.5, the **electric field** at point A due to each ring is straight up. **Add the electric fields** produced by these two thin rings to find $d\vec{E}$.

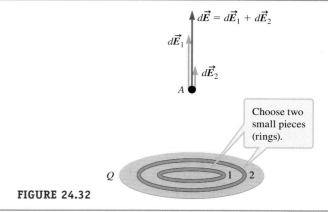

FIGURE 24.32

The electric field due to each ring adds without any component cancelling, so there is no need to consider more than one ring. Figure 24.33 shows a simplified sketch with a **coordinate system** and **geometric details**. Only one thin ring is shown, with radius r and thickness dr. If you imagine cutting the ring and uncurling it, it would be a rectangle of length $2\pi r$ and width dr, so the area of the ring is $dA = 2\pi r\, dr$.

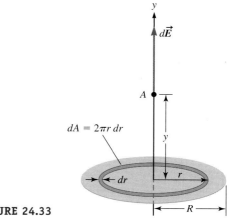

FIGURE 24.33

SOLVE
Step 1 Find an expression for $d\vec{E}$, the electric field due to that one ring of radius r. In this case we can use Equation 24.16, replacing Q with dq and R with lowercase r.

$$d\vec{E} = \frac{k(dq)y}{(r^2 + y^2)^{3/2}}\hat{j} \quad (1)$$

Step 2 Substitute $dq = \sigma(2\pi r)dr$ into Equation (1) and **integrate** over the entire disk from $r = 0$ to $r = R$ (Problem 81).

$$d\vec{E} = \frac{k\sigma 2\pi r y\, dr}{(r^2 + y^2)^{3/2}}\hat{j} \qquad \int d\vec{E} = \int_0^R \frac{k\sigma 2\pi r y\, dr}{(r^2 + y^2)^{3/2}}\hat{j}$$

$$\vec{E} = k\sigma 2\pi y\,\hat{j} \int_0^R \frac{r\, dr}{(r^2 + y^2)^{3/2}}$$

$$\boxed{\vec{E} = 2\pi k\sigma \left[1 - \frac{y}{\sqrt{R^2 + y^2}}\right]\hat{j}} \quad (24.17)$$

ELECTRIC FIELD DUE TO A CHARGED DISK ▶ **Special Case**

:• **CHECK and THINK**
The **dimensions** are correct—force per charge (Problem 26). To check that the electric field **approaches that of a charged particle** if point A is far away ($y \gg R$), start by rearranging Equation 24.17.

$$\vec{E} = 2\pi k\sigma \left[1 - \frac{y}{y\sqrt{(R/y)^2 + 1}} \right] \hat{j}$$

$$\vec{E} = 2\pi k\sigma \left[1 - (1 + (R/y)^2)^{-1/2} \right] \hat{j}$$

Then approximate $[1 + (R/y)^2]^{-1/2}$ for the case when R/y is small (Appendix A) and simplify.

$$\vec{E} \approx 2\pi k\sigma \left[1 - \left(1 - \frac{1}{2}\frac{R^2}{y^2} \right) \right] \hat{j}$$

$$\vec{E} \approx k\frac{\sigma \pi R^2}{y^2} \hat{j}$$

Use $\sigma = Q/A$ (Eq. 24.10) to write this in terms of the total charge on the disk $Q = \sigma A = \sigma \pi R^2$. This is the electric field of a charged particle at a point y as required.

$$\vec{E} \approx k\frac{Q}{y^2} \hat{j} \checkmark$$

24-7 Case Study: The Shape of Lightning Rods

After conducting his kite-flying experiments that supported his theory of lightning as a giant spark, Franklin came up with a plan to avoid lightning strikes. He knew that objects could be charged by rubbing. He also knew that you cannot charge a metal object—a conductor—by rubbing it if you connect the conductor to the ground. Today we know why: The Earth acts as a source and sink of electrons. If you build up a few excess electrons on a grounded conductor, these electrons quickly travel through the conductor and along the pathway to ground (Fig. 24.34A).

Franklin reasoned that during a storm, the atmosphere builds up excess charge much as a glass rod builds up excess charge when it is rubbed with silk. He knew from his experience with charged objects that a spark can make its way through the air. His indoor experiments demonstrated that if he used a pointed object (like a knitting needle) to draw charge from an object through the air, only a small spark occurred compared to the larger spark drawn to a blunt object (like his thumb). Franklin reasoned that when a small charge builds up in the atmosphere, a pointed lightning rod will draw the small charge through the air and into the Earth continuously (Fig. 24.34B). If only a small charge travels through the air, it will not be visible and there will be no giant spark of lightning.

Without a lightning rod, charge still builds up in the atmosphere. Because buildings are connected to the Earth, they attract charge of the opposite sign (Fig. 24.34C). When sufficient charge builds up, the air acts as a conductor and a large charge is transferred in a giant lightning spark. Such a violent spark causes great damage to buildings and can be very dangerous.

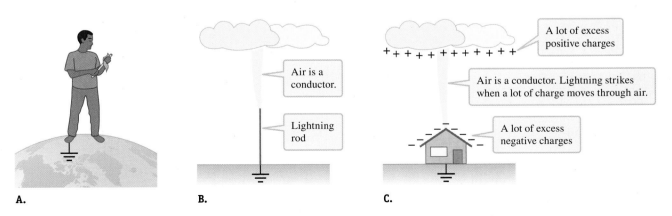

FIGURE 24.34 A. Franklin knew that a grounded person cannot charge a conducting rod by rubbing. **B.** According to Franklin, lightning strikes could be prevented if a town put up a lot of lightning rods. Then the atmosphere would be connected to ground, so it could not build up charge. **C.** Without a lightning rod, Franklin reasoned, the atmosphere discharges in a violent strike.

Benjamin Wilson—a contemporary of Franklin's—believed that lightning rods should be blunt. He argued that a pointed rod would draw down lightning that might have just passed harmlessly overhead. Wilson argued that pointed lightning rods were more dangerous than having no rods at all.

In this section, we explore both types of rods (Fig. 24.35). In order for air to break down and become a conductor, the electric field in the air must be 3×10^6 N/C. Let's assume that in order for a lightning rod to work, the electric field at its surface must equal that breakdown electric field. We will calculate the amount of charge on the surface of each conductor. The one with the least amount of charge is the better design because a smaller amount of charge on the surface of the conductor means a smaller amount of charge travels through the air. Calculating the charge on each rod gives us a way to compare their effectiveness. A rod that does not require much charge to have a strong electric field on its surface will not generate a big spark to discharge the atmosphere. Because both designs use conductors, we will assume the excess charge is uniformly spread over the surface of the lightning rod.

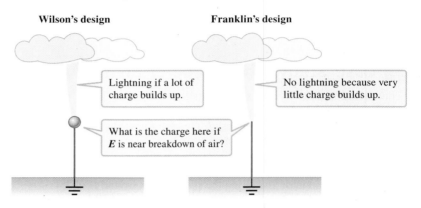

FIGURE 24.35 If there is a ball at the end of the lightning rod, a lot of charge builds up before the atmosphere discharges. If there is no ball at the end of the lightning rod, little charge builds up because the atmosphere continually discharges through the rod.

EXAMPLE 24.7 | CASE STUDY Wilson's Blunt Lightning Rod

At the end of Wilson's rod was a cannonball (Fig. 24.35). Cannonballs in the middle of the 18th century varied in size, but let's assume a moderate-sized cannonball with a radius of 4 inches, or $R = 0.1$ m. Because the ball is so much larger than the thickness of the supporting rod, we will ignore that rod and model Wilson's device as a suspended sphere connected to ground. If the electric field on the surface of the ball is $E = 3 \times 10^6$ N/C, what is the charge on the ball?

INTERPRET and ANTICIPATE
The ball is a sphere, so we can use the magnitude of $\vec{E}(r) = (kQ/r^2)\hat{r}$ (Eq. 24.3) to find the charge on the ball.

SOLVE
Solve for the charge Q on the surface of the sphere, where $r = R$.

$$E(R) = k\frac{Q}{R^2}$$

$$Q = \frac{ER^2}{k} = \frac{(3 \times 10^6 \text{ N/C})(0.1 \text{ m})^2}{(8.99 \times 10^9 \text{ N} \cdot \text{m}^2/\text{C}^2)}$$

$$Q = 3.3 \times 10^{-6} \text{ C} = 3.3 \mu\text{C}$$

CHECK and THINK
This is not very much charge, especially compared to the amount of charge involved in a lightning strike, which is on the order of hundreds of coulombs. However, our job is not done. We need to see how Franklin's pointed rod would do under the same conditions.

24-7 Case Study: The Shape of Lightning Rods

EXAMPLE 24.8 | CASE STUDY | Franklin's Pointed Lightning Rod

Franklin's lightning rod is very thin. We don't have an expression for the electric field near the end of a very thin rod. (Equation 24.15 gives the field at a distance y from the midpoint of a charged rod but is not valid near the tip.)

A Find an expression for the electric field at the tip of Franklin's pointed lightning rod. Evaluate Franklin's design by considering a point very close to the end of the rod.

• INTERPRET and ANTICIPATE
Figure 24.36 shows all four elements of our usual sketch. A positive **test charge** is at point A just above the end of the rod. The rod is divided into many pieces. **One piece** is shown as well as the **electric field** $d\vec{E}$ produced by that piece. Any piece will produce an upward-pointing electric field, so there is no cancellation. The figure includes a **coordinate system** and **geometric details**.

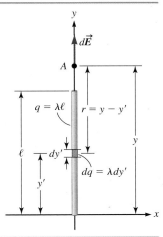

FIGURE 24.36

• SOLVE
Step 1 Use Equation 24.14 to derive an **expression** for $d\vec{E}$.

$$d\vec{E} = k\frac{dq}{r^2}\hat{j}$$

Use the coordinate system in Figure 24.36 to express the distance from the piece of charge to the point of interest as $r = y - y'$. The infinitesimal charge dq comes from $dq = \lambda dy'$ (Eq. 24.13).

$$d\vec{E} = k\frac{\lambda dy'}{(y-y')^2}\hat{j}$$

Step 2 Integrate over the entire length of the rod from $y = 0$ to $y = \ell$. Complete the integration and evaluate between the limits (Problem 82).

$$\int d\vec{E} = \int_0^\ell k\frac{\lambda dy'}{(y-y')^2}\hat{j}$$

$$\vec{E} = k\lambda \hat{j}\int_0^\ell \frac{dy'}{(y-y')^2} = \frac{k\lambda\ell}{y(y-\ell)}\hat{j}$$

The total charge is $q = \lambda\ell$. Substitute this expression for q.

$$\vec{E} = \frac{kq}{y(y-\ell)}\hat{j} \quad (1)$$

• CHECK and THINK
Before testing Franklin's design, first check Equation (1) for the **dimensions** of force per charge (Problem 40). Also, if point A is far from the rod ($y \gg \ell$), the electric field **approaches the electric field due to a charged particle**.

$$\lim_{y\gg\ell} \vec{E} = \frac{kq}{y(y-\ell)}\hat{j} \to \frac{kq}{y(y)}\hat{j}$$

$$\vec{E} \approx \frac{kq}{y^2}\hat{j} \checkmark$$

• SOLVE
Now use Equation (1) to evaluate Franklin's design. As point A gets closer to the end of the rod, the electric field increases. At the tip of the rod, the electric field approaches infinity. So, in principle, even a very tiny charge on the rod would lead to a very high electric field at its end. Franklin's rod would continually draw a small amount of charge from the atmosphere and prevent a large lightning strike.

$$\lim_{y\to\ell} E = \lim_{y\to\ell}\left[\frac{kq}{y(y-\ell)}\right]$$

$$\lim_{y\to\ell} E \to \frac{1}{0} \to \infty$$

Example continues on page 736 ▶

B It seems that an ideal Franklin rod (one that is infinitely thin) would easily win over the blunt Wilson rod. As one final check, let's model the *end* of a Franklin rod as a tiny ball of radius $R = 2$ mm—something like the end of a knitting needle. If the electric field on the surface of the ball is $E = 3 \times 10^6$ N/C, what is the charge on the ball? (We'll consider a thin-rod model in Problem 42.)

SOLVE
Repeat the calculation from Example 24.7 for the Wilson rod, this time for a much smaller ball. Solve for the charge q on the surface of the sphere, where $r = R$. The charge on Franklin's rod is about 2500 times lower than on Wilson's.

$$E(R) = k\frac{q}{R^2}$$

$$q = \frac{ER^2}{k} = \frac{(3 \times 10^6 \text{ N/C})(0.002 \text{ m})^2}{(8.99 \times 10^9 \text{ N} \cdot \text{m}^2/\text{C}^2)}$$

$$q = 1.3 \times 10^{-9} \text{ C} = 1.3 \text{ nC}$$

CHECK and THINK
Even if we model the end of Franklin's rod as a tiny ball, we find that it requires far less charge than Wilson's rod ($q \ll Q$) in order to produce a very large electric field at its tip. According to Franklin, this small amount of charge would travel across the air and keep the atmosphere discharged. Franklin actually suggested that if lightning rods were used all over a town, it would be possible to prevent the local atmosphere from building up any charge at all.

Lightning Rods on Buckingham Palace

In the 18th century, no one could carry out the calculations in Examples 24.7 and 24.8 showing that less charge travels through the air in the case of pointed rods. Nevertheless, the practical question of what shape of rods should be placed on a building needed to be answered. In 1771, the British Parliament needed to know how the arsenal at Purfleet on the Thames should be protected from lightning. The Royal Society formed a committee to investigate and make a recommendation. Both Franklin and Wilson served on the committee. The committee considered the path the lightning took when it struck an unprotected building, and it considered the effectiveness of lightning rods that had been in use for a few decades. Although Wilson objected, the committee recommended that pointed rods be installed on the arsenal. King George III must have been impressed by the committee's recommendation because he ordered that his house—now called *Buckingham Palace*—be protected by pointed lightning rods as well.

The story may have ended there, but two things happened. First, Franklin fell out of favor with British royalty when he supported America's war of independence. Second, lightning struck a building on the grounds of the Purfleet arsenal, although the protected gunpowder magazine was not hit. With Franklin out of the country, the king gave Wilson permission to test the two types of lightning rods in the London Pantheon. Wilson charged an enormous cylinder that was 16 inches in diameter and 155 feet long. The cylinder was hung from the ceiling by silk to simulate a charged-up atmosphere. Under the cylinder was a model of the Purfleet arsenal with both blunt and pointed lightning rods. The model was on a trolley so that it could be rolled under the charged cylinder, simulating a storm rolling in over a town. Both rods were found to draw sparks, but the pointed rods drew sparks from a greater distance than the blunt rods. Wilson argued that this made the pointed rods more dangerous. Franklin's supporters argued they were safer because the "damaging power" of the strikes was weaker.

In fact, the difference between a blunt and a pointed rod is irrelevant in the case of a full-scale lightning storm. The charged particles in the atmosphere are at an altitude of several kilometers. Imagine looking at the ends of a blunt rod and a pointed rod from a distance of a few kilometers. You could not distinguish one from the other, and the exact shape of the rod makes little difference. So neither Franklin nor Wilson was completely correct. Wilson was wrong to think that pointed rods were more dangerous than no rods at all, and Franklin's rod was too far from the charged particles to quietly discharge the atmosphere.

So, lightning still strikes structures equipped with lightning rods. Lightning rods are important, however, because they provide a pathway to ground that prevents damage to structures. Although some contemporary lightning protection systems are fairly complicated, many are relatively simple rods based on Franklin's original pointed design. We will continue to explore lightning protection in Chapter 25.

24-8 Charged Particle in an Electric Field

The payoff of writing an expression for the electric field produced by some source is that once you have done that, it is relatively easy to find the force exerted on some subject. In this section, we calculate the force exerted on a single charged particle in an electric field.

No matter how complicated the electric field due to some source, the force exerted by that source on a single particle with charge q is

$$\vec{F}_E = q\vec{E} \qquad (24.18)$$

ELECTROSTATIC FORCE EXERTED BY ELECTROSTATIC FIELD

✪ Major Concept

Keep in mind that the source—a charged particle, a collection of particles, or a continuous distribution of charged particles—sets up the electric field $\vec{E}$. A subject—in this case, a single particle with charge q—is in the source's electric field, and the force on the subject is given by Equation 24.18. If the subject is positively charged, the force exerted on it by the source is in the same direction as the electric field. If the subject is negatively charged, the force is in the opposite direction to the electric field. After you have calculated the force exerted on the subject, you can use all the tools of Part I to calculate the kinematics and dynamics of the particle.

EXAMPLE 24.9 Millikan's Oil Drop Experiment

Figure 24.37 shows the apparatus used by Robert Millikan (1868–1953) in the early 20th century to measure the elementary charge e. Although his measurement was off, he won the Nobel Prize as a result of his experiment. Today, students use similar equipment to reproduce Millikan's measurement. The apparatus is a cylinder divided into two chambers. Oil drops are sprayed into the upper chamber, and some of them become positively or negatively charged in the process. A few drops make it through a small hole in the disk that separates the two chambers. An experimenter can view these drops with the aid of a microscope. The disks that make up the ceiling and floor of the lower chamber are connected to a battery. When the experimenter switches on the battery, the upper disk is positively charged and the lower disk is grounded.

FIGURE 24.37 A. Millikan measured the elementary charge e using this apparatus. **B.** Diagram showing the major components of Millikan's experiment.

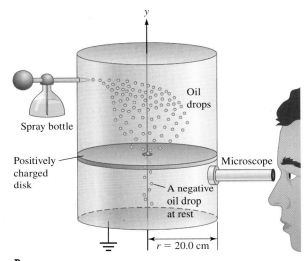

A.

B.

Example continues on page 738 ▶

In this example, we consider a single charged oil drop of mass $m = 7.26 \times 10^{-16}$ kg. The oil drop has one excess electron and is at rest in front of the microscope.

A The oil drop is relatively close to the charged disk, corresponding to $y \ll R$ in Figure 24.33. Use the result of Example 24.6 to show that the electric field near a charged surface is

$$\vec{E} \approx 2\pi k\sigma \hat{j} \quad (24.19)$$

ELECTRIC FIELD NEAR A CHARGED SURFACE ⊙ Special Case

INTERPRET and ANTICIPATE
Begin with Equation 24.17, the result of Example 24.6, to find an approximate expression for the electric field due to a charged disk at point A (Fig. 24.33) close to the disk.

SOLVE
If point A is close to the disk compared to the disk's radius ($y \ll R$), then the second term in Equation 24.17 is essentially zero. The electric field is uniform—a constant value—for all points near the disk.

$$\vec{E} = 2\pi k\sigma \left[1 - \frac{y}{\sqrt{R^2 + y^2}}\right]\hat{j} \quad (24.17)$$

$$\vec{E} \approx 2\pi k\sigma \hat{j} \checkmark \quad (24.19)$$

CHECK and THINK
Equation 24.19 has the right dimensions—force per charge. What seems odd is that the electric field is uniform. We cannot check this expression in the limit of a very distant point because it is good only for points close to the disk. However, Equation 24.19 is a good approximation for any point near a charged plane; the shape of the plane does not matter (disk or rectangle) as long as the point of interest is close to the plane and not near its edges. This is an important equation, so it appears in the summary of this chapter.

B What are the magnitude and direction of the electric field produced by the charged disk at the position of the oil drop?

INTERPRET and ANTICIPATE
We can approach this part of the problem as we would any problem involving Newton's second law. A free-body diagram for the charged oil drop (Fig. 24.38) shows two forces: the downward force of gravity and the upward force exerted by the positively charged disk above the drop.

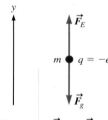

FIGURE 24.38 $\quad \vec{a} = 0 \Rightarrow \vec{F}_E = -\vec{F}_g$

SOLVE
The oil drop is at rest, so its acceleration is zero. According to Newton's second law, the electrostatic force exerted on the drop must be balanced by the gravitational force.

$$\sum F_y = F_E - F_g = 0$$

$$F_E = F_g$$

The magnitude of the electric force can be found from $\vec{F}_E = q\vec{E}$ (Eq. 24.18). As usual, the magnitude of gravitational force is the oil drop's weight.

$$|q|E = mg$$

Solve for the magnitude of the electric field. The mass of the oil drop is given, and $|q|$ is the elementary charge.

$$E = \frac{mg}{|q|} = \frac{(7.26 \times 10^{-16} \text{ kg})(9.81 \text{ m/s}^2)}{|-1.60 \times 10^{-19} \text{ C}|}$$

$$E = 4.45 \times 10^4 \text{ N/C}$$

The oil drop is negatively charged, so the force exerted on it by the electric field is in the *opposite* direction to the electric field. Because the electrostatic force must be upward in the positive y direction, the electric field must be downward in the negative y direction.

$$\vec{E} = -4.45 \times 10^4 \,\hat{\jmath}\, \text{N/C}$$

C If the radius of the cylinder is $R = 20.0$ cm, estimate the amount of excess charge on the disk.

:• **INTERPRET and ANTICIPATE**
Because we know the magnitude of the electric field from part B, we can use Equation 24.19 (part A) to find the surface charge density on the disk. Then we find the area of the disk from its radius, enabling us to find its excess charge.

:• **SOLVE**
Solve for the charge density σ.

$$E \approx 2\pi k\sigma$$

$$\sigma \approx \frac{E}{2\pi k} \approx \frac{4.45 \times 10^4 \text{ N/C}}{2\pi (8.99 \times 10^9 \text{ N}\cdot\text{m}^2/\text{C}^2)}$$

$$\sigma \approx 7.88 \times 10^{-7} \text{ C/m}^2$$

Use $\sigma = Q/A$ (Eq. 24.10) to find the total charge on the disk.

$$Q = \sigma A = \sigma(\pi R^2) \approx (7.88 \times 10^{-7} \text{ C/m}^2)\pi(0.200 \text{ m})^2$$

$$Q \approx 9.90 \times 10^{-8} \text{ C} \approx 99.0 \text{ nC}$$

:• **CHECK and THINK**
The excess charge must be positive in order for the electric field to point downward in the lower chamber.

EXAMPLE 24.10 Jumping Through the Hoop

A particle with charge $q = -30.0$ nC and mass $m = 50.0\ \mu g$ is placed 2.00 m directly above the middle of a charged ring (Fig. 24.39). The ring has positive charge $Q = +60.0$ nC and radius $R = 0.500$ m.

A Consider the ring to be the source, and plot the (scalar component of) electric field produced by the ring from $y = -2.00$ m to $y = 2.00$ m. According to your graph, where is the magnitude of the electric field the greatest?

:• **INTERPRET and ANTICIPATE**
Our job amounts to plotting Equation 24.16—the result of Example 24.5—for the values given in the problem statement.

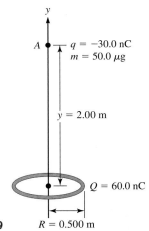

FIGURE 24.39

:• **SOLVE**
Start with the scalar component. Then substitute the numerical values of all constants and reduce the function to its simplest form before plotting.

$$E_y(y) = \frac{kQy}{(R^2 + y^2)^{3/2}}$$

$$E_y(y) = \frac{(8.99 \times 10^9 \text{ N}\cdot\text{m}^2/\text{C}^2)(60.0 \times 10^{-9} \text{ C})y}{[(0.500 \text{ m})^2 + y^2]^{3/2}}$$

$$E_y(y) = \frac{(5.39 \times 10^2 \text{ N}\cdot\text{m}^2/\text{C})y}{[(0.250 \text{ m}^2) + y^2]^{3/2}} \qquad (1)$$

Example continues on page 740 ▶

Using a spreadsheet program or working by hand, calculate E_y for values of y between $y = -2.00$ m and $y = 2.00$ m. The results are shown in Figure 24.40.

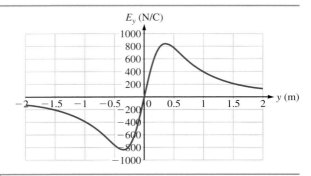

FIGURE 24.40

Inspect the graph (Fig. 24.40) to find the greatest magnitude of the electric field and read off the positions. There are two points where the electric field reaches its greatest magnitude. | The magnitude of the electric field is greatest at $y \approx \pm 0.4$ m.

∴ CHECK and THINK
Because of the symmetry, it is not surprising that the electric field has a maximum at points equidistant from the ring.

B The charged particle is the subject in the ring's electric field. Plot the (scalar component of) force exerted by the ring on the particle from $y = -2.00$ m to $y = 2.00$ m. According to your graph, where is the magnitude of the force the greatest?

∴ INTERPRET and ANTICIPATE
According to $\vec{F}_E = q\vec{E}$ (Eq. 24.18), to find an expression for the force, we multiply the expression for the electric field by the charge of the subject.

∴ SOLVE
Multiply Equation 24.16 by q, the charge of the subject.

$$F_y = qE_y = q\frac{kQy}{(R^2 + y^2)^{3/2}} = \frac{kQqy}{(R^2 + y^2)^{3/2}}$$

As in part A, substitute numerical values for the constants before plotting them. We already did some of this work in Equation (1).

$$F_y(y) = \frac{(-30.0 \times 10^{-9}\,\text{C})(5.39 \times 10^2\,\text{N} \cdot \text{m}^2/\text{C})y}{[(0.250\,\text{m}^2) + y^2]^{3/2}}$$

$$F_y(y) = -\frac{(1.62 \times 10^{-5}\,\text{N} \cdot \text{m}^2)y}{[(0.250\,\text{m}^2) + y^2]^{3/2}}$$

Calculate F_y for values of y between $y = -2.00$ m and $y = 2.00$ m. The results are shown in Figure 24.41.

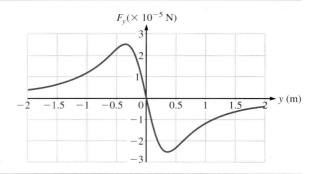

FIGURE 24.41

Inspect the graph (Fig. 24.41) to find the greatest magnitude of the force and read off the positions. There are two points where the force reaches its greatest magnitude. | The magnitude of the force is greatest at $y \approx \pm 0.4$ m.

∴ CHECK and THINK
Our results here are consistent with part A. First, because force is proportional to the electric field, you should expect the strongest force to occur where the electric field is greatest. Second, the subject has a negative charge, so the force exerted on it is in the opposite direction to the electric field. Figure 24.41 is an inverted (upside down) version of Figure 24.40.

C Assume that the only force acting on the particle is the electrostatic force. Use your graph from part B to find the magnitude of the maximum force exerted on the particle. Compare your answer to the particle's weight. What is the magnitude of the particle's maximum acceleration?

INTERPRET and ANTICIPATE
The strongest force is found from reading the value of the peak in Figure 24.41. Once we know that force, we can find the acceleration from Newton's second law.

SOLVE

Read the value of the peak from Figure 24.41.	$F_{max} = 2.5 \times 10^{-5} \text{ N}$
Find the weight of the particle from its mass.	$F_g = mg$ $F_g = (50.0 \times 10^{-9} \text{ kg})(9.81 \text{ m/s}^2)$ $F_g = 4.90 \times 10^{-7} \text{ N}$
Compare the maximum force exerted by the ring to the weight of the particle by forming a ratio. The maximum electrostatic force is about 50 times greater than the particle's weight, so we expect the particle's acceleration to be about 50 times greater than the free-fall acceleration.	$\dfrac{F_{max}}{F_g} = \dfrac{2.5 \times 10^{-5} \text{ N}}{4.90 \times 10^{-7} \text{ N}} = 51$ $F_{max} = 51 F_g$
Use Newton's second law to find the particle's maximum acceleration, assuming the only force exerted on the particle is the electrostatic force exerted by the charged disk.	$F_{max} = m a_{max}$ $a_{max} = \dfrac{F_{max}}{m} = \dfrac{2.5 \times 10^{-5} \text{ N}}{50.0 \times 10^{-9} \text{ kg}} = \boxed{500 \text{ m/s}^2}$

CHECK and THINK
As expected, the particle's acceleration is nearly 50 times greater than free-fall acceleration. This means that if we were to take gravity into account, rather than assuming that the only force is the electrostatic force, we would find that the particle's acceleration is only slightly different from our value.

24-9 Special Case: Dipole in an Electric Field

When you make popcorn in a microwave oven, the oven creates an oscillating electric field. (The electric field inside the oven points first one way, then the opposite way, and so on.) Food contains water molecules, and the oscillating electric field causes the water molecules to rotate back and forth. So, the electric field increases the thermal energy of the water molecules. Those water molecules transfer thermal energy to other molecules, and the food cooks. How does the electric field cause water molecules to rotate? Remember that water is modeled as a dipole (Fig 24.20). Therefore, to answer the question, we need to consider a dipole in an electric field.

Figure 24.42 shows a dipole in a uniform electric field. (It might help to imagine that the dipole is near the surface of a charged disk.) The electric field lines are evenly spaced, which indicates an electric field with the same value everywhere. The force on the positive particle $\vec{F}_{(+)}$ points in the positive x direction, and the force on the negative particle $\vec{F}_{(-)}$ points in the negative x direction. Because the electric field has the same strength at both locations and the two charges have the same magnitude, these two forces have equal magnitudes. The result is that the net force on the electric dipole in a uniform electric field is zero, so the dipole will have no translational acceleration.

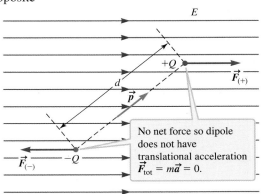

FIGURE 24.42 No net force is exerted on an electric dipole in a uniform electric field. The magnitude of the dipole moment is $p = Qd$.

FIGURE 24.43 **A.** The net force is zero, but the net torque is not zero. **B.** The net torque may be written in terms of the dipole moment and the electric field.

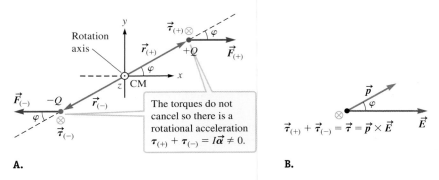

A.

B.

However, the two forces create a net torque (Fig. 24.43A). You may imagine the dipole being like a dumbbell; if you press on the ends of the dumbbell with antiparallel forces, the dumbbell will rotate. The same thing happens here: The electric field causes the dipole to rotate around its center of mass. We can use $\tau = rF \sin\varphi$ (Eq. 12.16) and the right-hand rule to calculate the torques around the center of mass. For the torque on the positive charge,

$$\vec{\tau}_{(+)} = -r_{(+)} F_{(+)} \sin\varphi \hat{k}$$

The torque on the negative charge is

$$\vec{\tau}_{(-)} = -r_{(-)} F_{(-)} \sin\varphi \hat{k}$$

The magnitudes of the forces are equal; let's use the symbol F_E for both:

$$F_E \equiv F_{(+)} = F_{(-)}$$

The magnitudes of the position vectors both equal $d/2$:

$$r_{(+)} = r_{(-)} = \frac{d}{2}$$

Therefore, the torque exerted on the positive charge equals the torque exerted on the negative charge:

$$\vec{\tau}_{(+)} = \vec{\tau}_{(-)} = -\frac{d}{2} F_E \sin\varphi \hat{k}$$

The total torque exerted by the electric field on the dipole is the sum of the torques on both charges:

$$\vec{\tau} = \vec{\tau}_{(+)} + \vec{\tau}_{(-)}$$

$$\vec{\tau} = -\frac{d}{2} F_E \sin\varphi \hat{k} - \frac{d}{2} F_E \sin\varphi \hat{k}$$

$$\vec{\tau} = -d F_E \sin\varphi \hat{k} \tag{24.20}$$

The magnitude of the force comes from $\vec{F}_E = q\vec{E}$ (Eq. 24.18), so the torque on the dipole is

$$\vec{\tau} = -d(QE) \sin\varphi \hat{k}$$

The dipole moment is $p = Qd$, so the torque is

$$\vec{\tau} = -pE \sin\varphi \hat{k} \tag{24.21}$$

Equation 24.21 is the cross product of $\vec{p}$ and $\vec{E}$ (Fig. 24.43B):

$$\vec{\tau} = \vec{p} \times \vec{E} \tag{24.22}$$

TORQUE ON DIPOLE IN AN ELECTRIC FIELD ○ Special Case

When a water molecule is in a microwave oven, the electric field reverses direction rapidly. Imagine the electric field in Figure 24.43 pointing to the right, then to the left, and so on. The torque on the water molecule switches direction, pointing in the negative z direction, then in the positive z direction, and so on. The water molecule rotates clockwise, then counterclockwise, and so on. All this rotation increases the thermal energy of the water and cooks your food. (This is something like the paddle wheels Joule used in his container of water. See Fig. 21.2, page 612.)

Potential Energy of a Dipole in an Electric Field

To derive an expression for potential energy, let's consider the dipole and the source of the electric field to be part of a system. With the source inside the system, potential energy can be stored by the electric field just as when we consider the Earth to be part of a system and then potential energy is stored by the Earth's gravitational field. Imagine an external agent exerts a torque in the positive z direction, so that the dipole in Figure 24.43 rotates counterclockwise at constant angular velocity with no change in kinetic energy or thermal energy. Then the work done by the external agent equals the change in the system's potential energy:

$$K_i + U_i + W = K_f + U_f + \Delta E_{th}$$
$$W = U_f - U_i = \Delta U$$

The work done by the external agent is given by

$$\Delta U = W = \int_{\varphi_i}^{\varphi_f} \tau \, d\varphi \quad (13.18)$$

As always, only changes in potential energy are significant, so potential energy is usually measured from some reference configuration. For a dipole in an electric field, the reference configuration is shown in Figure 24.44A, where the dipole moment is perpendicular to the electric field, $\varphi = 90°$. Then a counterclockwise rotation (Fig. 24.44B) to $\varphi > 90°$ represents an increase in potential energy, and a clockwise rotation (Fig. 24.44C) to $\varphi < 90°$ represents a decrease in potential energy. With this convention for the reference configuration, Equation 13.18 becomes

$$U = W = \int_{90°}^{\varphi} \tau \, d\varphi \quad (24.23)$$

If the dipole rotates at a constant angular velocity, the torque exerted by the external agent must be balanced by the torque exerted by the electric field. Therefore, the magnitude of the torque exerted by the external agent is $\tau = pE \sin \varphi$ (Eq. 24.21). The external agent does positive work in rotating the dipole from $\varphi_i = 90°$ to $\varphi_f = \varphi$, and Equation 24.23 becomes

$$U = W = \int_{90°}^{\varphi} pE \sin \varphi \, d\varphi \quad (24.24)$$

We integrate (Appendix A) and drop the W:

$$U = -pE(\cos \varphi - \cos 90°)$$
$$U = -pE \cos \varphi \quad (24.25)$$

Equation 24.25 is the (negative) dot product of $\vec{p}$ and $\vec{E}$:

$$U = -\vec{p} \cdot \vec{E} \quad (24.26)$$

with the reference configuration set to $\varphi_i = 90°$.

When there is no external agent, as in the case of a water molecule in a microwave oven, rotation of the dipole results in a change in the potential energy and a change in the dipole's kinetic energy, thermal energy, or both.

POTENTIAL ENERGY OF DIPOLE IN AN ELECTRIC FIELD ▶ Special Case

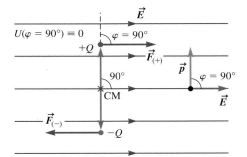
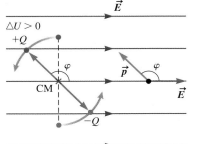
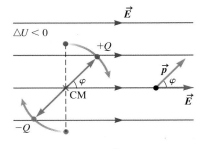

A. Reference configuration B. Potential energy increases. C. Potential energy decreases.

FIGURE 24.44 A. Choose the reference configuration for the angular position φ. **B.** When $\varphi > 90°$, the potential energy increases. **C.** When $\varphi < 90°$, the potential energy decreases.

EXAMPLE 24.11 Rotating Water Molecules

In Concept Exercise 24.5, we found that a water molecule is made up of two hydrogen atoms and one oxygen atom, for a total of 10 electrons and 10 protons. The molecule is modeled as a dipole with an effective separation $d = 3.9 \times 10^{-12}$ m between its positive and negative charges. The water molecule's dipole moment is $p = 6.2 \times 10^{-30}$ C·m. Consider water molecules in the lower atmosphere where the electric field has a magnitude of about 325 N/C. Assume the field is uniform.

A What is the maximum torque exerted on a water molecule?

INTERPRET and ANTICIPATE
The magnitude of the torque depends on the sine of the angle between $\vec{p}$ and $\vec{E}$. The sine function is maximum when $\varphi = 90°$, so we need to find the torque when $\varphi = 90°$.

SOLVE
Substitute numerical values for the magnitude of $\vec{\tau} = -pE\sin\varphi \hat{k}$ (Eq. 24.21).

$\tau = pE\sin\varphi = (6.2 \times 10^{-30} \text{C} \cdot \text{m})(325 \text{ N/C})\sin 90°$

$\tau = 2.0 \times 10^{-27}$ N·m

CHECK and THINK
This torque may seem tiny, but the rotational inertia for an object as small and light as a water molecule is also very small, so the molecule's angular acceleration is still significant.

B What is the change in potential energy if a water molecule rotates from $\varphi = 170°$ to $180°$? From $\varphi = 90°$ to $100°$? From $\varphi = 10°$ to $0°$?

INTERPRET and ANTICIPATE
In this part of the problem we use $U = -\vec{p} \cdot \vec{E}$ (Eq. 24.26) to calculate the change in potential energy due to the rotation of a water molecule. From Figure 24.44, we expect an increase in potential energy when the molecule rotates counterclockwise and a decrease when it rotates clockwise.

SOLVE
For $\varphi = 170°$ to $180°$, substitute numerical values to find ΔU. Remember that 1 N·m = 1 J.

$\Delta U = U(180°) - U(170°)$

$\Delta U = (-pE\cos 180°) - (-pE\cos 170°)$

$\Delta U = -(6.2 \times 10^{-30} \text{ C}\cdot\text{m})(325 \text{ N/C})(\cos 180° - \cos 170°)$

$\Delta U = 3.1 \times 10^{-29}$ N·m $= 3.1 \times 10^{-29}$ J

CHECK and THINK
As expected, this counterclockwise rotation results in an increase in potential energy.

SOLVE
For $\varphi = 90°$ to $100°$, substitute numerical values to find ΔU. Remember that at the reference configuration ($\varphi = 90°$), the potential energy is zero.

$\Delta U = U(100°) - U(90°) = U(100°) - 0 = -pE\cos 100°$

$\Delta U = -(6.2 \times 10^{-30} \text{ C}\cdot\text{m})(325 \text{ N/C})\cos 100°$

$\Delta U = 3.5 \times 10^{-28}$ J

CHECK and THINK
As expected, this counterclockwise rotation also results in an increase in potential energy.

SOLVE
For $\varphi = 10°$ to $0°$, the molecule rotates clockwise.

$\Delta U = U(0) - U(10°) = (-pE\cos 0) - (-pE\cos 10°)$

$\Delta U = -(6.2 \times 10^{-30} \text{ C}\cdot\text{m})(325 \text{ N/C})(\cos 0 - \cos 10°)$

$\Delta U = -3.1 \times 10^{-29}$ J

:• CHECK and THINK

As expected, this clockwise rotation results in a decrease in potential energy.

Notice that all the rotations go through 10°, but the change in potential energy is not the same in all three cases. To understand this, note that the torque exerted by the electric field works to align the dipole moment with that field. The system has the lowest potential energy ($-pE$) when $\vec{p}$ is parallel to $\vec{E}$. The system has the highest potential energy when $\vec{p}$ is antiparallel to $\vec{E}$. Because of the cosine function in $|U| = pE \cos \varphi$, the change in potential energy is greatest when the dipole moves between two configurations nearly perpendicular to the field ($\varphi \approx 90°$). In this example, the rotation of the molecule when it was nearly perpendicular to the field ($\varphi = 90°$ to $100°$) has an order-of-magnitude greater change in potential energy than in the other two cases.

SUMMARY

❶ Underlying Principles

Electrostatic field or **electric field** $\vec{E}(r)$ is a vector field resulting from a charged source.

✪ Major Concepts

1. When two objects interact through a field force, one object is the **source** and the other is the **subject**. The source produces the electric field, and the subject is influenced by that field. The electrostatic field depends only on properties of the source, whereas the electrostatic force depends on both the source and the subject.

2. An **electric dipole** consists of two charged particles of magnitude Q but with opposite signs (Fig. 24.15). The particle separation is maintained at a distance d.

3. The **electric dipole moment** $\vec{p}$ is a vector pointing from the negative charge toward the positive charge. The magnitude of the dipole moment is $p = Qd$.

4. The **electrostatic force exerted by an electric field** on a single particle with charge q is:

$$\vec{F}_E = q\vec{E} \quad (24.18)$$

▶ Special Cases

1. Electric field sources
   a. The electric field due to a **particle** or outside any **charged spherical object**:

   $$\vec{E}(r) = k\frac{Q_S}{r^2}\hat{r} \quad (24.3)$$

   b. The magnitude of the electric field at a position r far away in any direction from an electric **dipole** ($r \gg d$):

   $$E \approx \frac{2kp}{r^3} \quad (24.7)$$

   where p is the magnitude of the electric dipole moment.

   c. The electric field at a point on the perpendicular bisector (y axis) of a **charged rod** of length 2ℓ and net charge Q (Fig. 24.23):

   $$\vec{E} = \frac{kQ}{y}\frac{1}{\sqrt{\ell^2 + y^2}}\hat{j} \quad (24.15)$$

   d. The electric field at a point on the y axis passing through the center of a **charged ring** of radius R and net charge Q (Fig. 24.28):

   $$\vec{E} = \frac{kQy}{(R^2 + y^2)^{3/2}}\hat{j} \quad (24.16)$$

   e. The electric field at a point on the y axis that passes through the center of a **charged disk** of surface charge density σ and radius R (Fig. 24.31):

   $$\vec{E} = 2\pi k\sigma \left[1 - \frac{y}{\sqrt{R^2 + y^2}}\right]\hat{j} \quad (24.17)$$

   f. For **points near a charged surface** (such as $y \ll R$ in Fig. 24.33), the electric field is a constant:

   $$\vec{E} \approx 2\pi k\sigma \hat{j} \quad (24.19)$$

2. Dipole in an electric field

A dipole with dipole moment $\vec{p}$ in electric field $\vec{E}$ experiences a torque

$$\vec{\tau} = \vec{p} \times \vec{E} \quad (24.22)$$

and the potential energy of the dipole-source system is

$$U = -\vec{p} \cdot \vec{E} \quad (24.26)$$

with the reference configuration set to $\varphi_i = 90°$.

Tools

Electric field lines provide a way to visualize the magnitude and direction of an electric field. Electric field lines must originate or terminate (or both) on a charged object (Figs. 24.5 and 24.6). The magnitude of the electric field is shown by the density of the electric field lines: A high density of field lines indicates a region with a relatively strong electric field.

PROBLEM-SOLVING STRATEGY

Electric Field due to a Collection of Charged Particles or due to a Continuous Distribution of Charged Particles

The strategies for finding the electric field due to a collection of particles or due to a continuous distribution of particles are similar. The main difference is that for a continuous distribution you must first imagine slicing up the distribution into tiny pieces, with each piece so small that it can be modeled as a single charged particle.

:• INTERPRET and ANTICIPATE

Make a diagram showing *all the electric fields* at a particular location. There are four elements to this diagram:
1. **Labeled source particles (or pieces)** and the **test particle**
2. The **electric field vectors** at the location of the test charge
3. The **net electric field vector**
4. A **coordinate system** and other **geometric details**

:• SOLVE

For a **collection of particles**, find a mathematical expression for the electric field at point A by adding the electric field vectors.

For a **continuous distribution**, there are two steps:
1. Find a **mathematical expression** for the infinitesimal electric field $d\vec{E}$ in terms of an infinitesimal spatial variable. The magnitude of the electric field due to an infinitesimal amount of charge dq is

$$dE(r) = k\frac{dq}{r^2} \quad (24.14)$$

The direction will come from your diagram.
2. **Integrate** this expression over the entire charge distribution.

:• CHECK and THINK

1. Make sure the **dimensions** of the electric field are force per charge.
2. If the charge distribution is finite, the **electric field for a point far from the distribution compared to its size should approach the electric field of a charged particle,** $\vec{E}(r) = (kQ_S/r^2)\hat{r}$.

PROBLEMS AND QUESTIONS

A = algebraic   C = conceptual   E = estimation   G = graphical   N = numerical

24-1 What Are Fields?

1. **C** The terms electrostatic *force* and electrostatic *field* may sound alike. To help keep them straight, identify and write down the standard symbol and SI units for each one. Which one requires a source and a subject? Which requires only a source?
2. **C** Imagine you are planning to send a probe to orbit Jupiter, a planet that has many moons. Explain how you would use the concept of a gravitational field. What would you consider to be part of the source? What would be the subject?

24-2 Special Case: Electric Field of a Charged Sphere

3. **N** A sphere has a charge of −89.5 nC and a radius of 4.65 cm. What is the magnitude of its electric field 3.15 cm from its surface?
4. **G** Plot E_r (scalar component) versus r for a positively charged sphere and then for a negatively charged sphere using $\vec{E}(r) = (kQ_S/r^2)\hat{r}$ (Eq. 24.3). Compare your graphs and check for consistency.

5. **N** A sphere with a charge of -3.50 nC and a radius of 1.00 cm is located at the origin of a coordinate system.
   a. What is the electric field 1.75 cm away from the center of the sphere along the positive y axis?
   b. If a particle with a charge of 5.39 nC were placed at that location, what would be the electrostatic force on this charge?
6. **N** Is it possible for a conducting sphere of radius 0.10 m to hold a charge of 4.0 μC in air? The minimum field required to break down air and turn it into a conductor is 3.0×10^6 N/C.
7. **N** What is the net charge of the Earth if the magnitude of its electric field near the terrestrial surface is 1.30×10^2 N/C? Assume the Earth is a sphere of radius 6.40×10^6 m.

24-3 Electric Field Lines

8. **G** For each sketch of electric field lines in Figure P24.8, compare the magnitude of the electric field in region A to the magnitude of the electric field in region B.

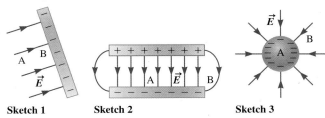

FIGURE P24.8

9. **G** Sketch the electric field lines for a source that consists of two particles carrying opposite charges. The negative particle has more excess charge than the positive one.
10. **C,G** Two large neutral metal plates, fitted tightly against each other, are placed between two particles with charges of equal magnitude but opposite sign, such that the plates are perpendicular to the line connecting the charges (Fig. P24.10). What will happen to each plate when they are released and allowed to move freely? Draw the electric field lines for the particles-plates system.

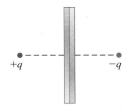

FIGURE P24.10

24-4 Electric Field of a Collection of Charged Particles

11. **N** Given the two charged particles shown in Figure P24.11, find the electric field at the origin.
12. **N** A particle with charge $q_1 = +5.0$ μC is located at $x = 0$, and a second particle with charge $q_2 = -3.0$ μC is located at $x = 15$ cm. Determine the location of a third particle with charge $q_3 = +4.0$ μC such that the net electric field at $x = 25$ cm is zero.

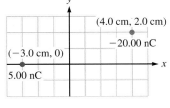

FIGURE P24.11

13. **N** A small sphere with charge $q_A = +5.20$ μC is at the origin, and a second sphere with charge $q_B = -3.70$ μC is located at $x = 2.00$ m. At what finite distance from the origin is the electric field equal to zero?

Problems 14 and 15 are paired.

14. **A** A particle with charge q on the negative x axis and a second particle with charge $2q$ on the positive x axis are each a distance d from the origin. Where should a third particle with charge $3q$ be placed so that the magnitude of the electric field at the origin is zero?
15. **N** A particle with charge 2.57 μC on the negative x axis and a second particle with charge 5.14 μC on the positive x axis are each a distance 0.0569 m from the origin. Where should a third particle with charge 7.71 μC be placed so that the magnitude of the electric field at the origin is zero?
16. **N** Figure P24.16 shows three charged particles arranged in the xy plane at the coordinates shown, with $q_A = q_B = -3.30$ nC and $q_C = 4.70$ nC. What is the electric field due to these particles at the origin?

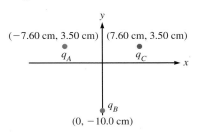

FIGURE P24.16

Problems 17, 18, and 19 are grouped.

17. **N** Figure P24.17 shows a dipole. If the positive particle has a charge of 35.7 mC and the particles are 2.56 mm apart, what is the electric field at point A located 2.00 mm above the dipole's midpoint?
18. **A** Find an expression for the electric field at point A for the dipole source shown in Figure P24.17. Show that when $y \gg d$, the electric field is given by $\vec{E} \approx -k\vec{p}/y^3$.
19. **N** Figure P24.17 shows a dipole (not drawn to scale). If the positive particle has a charge of 35.7 mC and the particles are 2.56 mm apart, what is the (approximate) electric field at point A located 2.00 m above the dipole's midpoint?

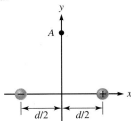

FIGURE P24.17
Problems 17, 18, and 19.

20. **N** Figure P24.20 shows three charged spheres arranged along the y axis.
    a. What is the electric field at $x = 0$, $y = 3.00$ m?
    b. What is the electric field at $x = 3.00$ m, $y = 0$?

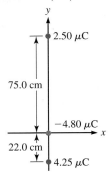

FIGURE P24.20

24-5 Electric Field of a Continuous Charge Distribution

21. **N** Often we have distributions of charge for which integrating to find the electric field may not be possible in practice. In such cases, we may be able to get a good approximate solution by dividing the distribution into small but finite particles and taking the vector sum of the contributions of each. To see how this might work, consider a very thin rod of length $L = 16$ cm with

uniform linear charge density $\lambda = 50.0$ nC/m. Estimate the magnitude of the electric field at a point P a distance $d = 8.0$ cm from the end of the rod by dividing it into n segments of equal length as illustrated in Figure P24.21 for $n = 4$. Treat each segment as a particle whose distance from point P is measured from its center. Find estimates of E_P for $n = 1, 2, 4,$ and 8 segments.

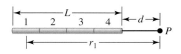

FIGURE P24.21

22. **G** Plot the magnitude of the electric field of a charged rod for points along its perpendicular bisector where Equation 24.15 holds:

$$\vec{E} = \frac{kQ}{y} \frac{1}{\sqrt{\ell^2 + y^2}} \hat{j}$$

How does your graph compare to that for a charged sphere (Problem 4)?

Problems 23 and 24 are paired.

23. **A** A positively charged rod with linear charge density λ lies along the x axis (Fig. P24.23). Find an expression for the magnitude of the electric field at the position P a distance x away from the origin, where $x > L$.

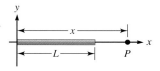

FIGURE P24.23 Problems 23 and 24.

24. **N** A positively charged rod of length $L = 0.250$ m with linear charge density $\lambda = 2.33$ mC/m lies along the x axis (Fig. P24.23). Find the electric field at the position P a distance 0.375 m away from the origin.

24-6 Special Cases of Continuous Distributions

25. **A** Check the dimensions of

$$\vec{E} = \frac{kQy}{(R^2 + y^2)^{3/2}} \hat{j}$$

from Example 24.5 (Special Case: Charged Ring, page 730).

26. **A** Check the dimensions of

$$\vec{E} = 2\pi k\sigma \left[1 - \frac{y}{\sqrt{R^2 + y^2}}\right] \hat{j}$$

from Example 24.6 (Special Case: Charged Disk, page 732).

27. **A** Find an expression for the position y (along the positive axis perpendicular to the ring and passing through its center) where the electric field due to a charged ring is a maximum. Also find an expression for the electric field at that point.

28. The electric field at a point on the perpendicular bisector of a charged rod was calculated as the first example of a continuous charge distribution, resulting in Equation 24.15:

$$\vec{E} = \frac{kQ}{y} \frac{1}{\sqrt{\ell^2 + y^2}} \hat{j}$$

   a. **A** Find an expression for the electric field when the rod is infinitely long.
   b. **C** An infinitely long rod with uniform linear charge density λ also contains an infinite amount of charge. Explain why this still produces an electric field near the rod that is finite.

29. **N** In Example 24.6 (page 732), we found that the electric field of a charged disk approaches that of a charged particle for distances y that are large compared to R, the radius of the disk. To see a numerical instance of this, calculate the magnitude of the electric field a distance $y = 3.0$ m from a disk of radius $R = 3.0$ cm that has a total charge of 6.0 μC using the exact formula, Equation 24.17:

$$\vec{E} = 2\pi k\sigma \left[1 - \frac{y}{\sqrt{R^2 + y^2}}\right] \hat{j}$$

Then calculate the magnitude of the electric field at $y = 3.0$ m, assuming that the disk is a particle of the same total charge at the origin, and compare your answers.

Problems 30 and 31 are paired.

30. **A** Find an expression for the magnitude of the electric field at point A midway between the two rings of radius R shown in Figure P24.30. The ring on the left has a uniform charge q_1 and the ring on the right has a uniform charge q_2. The rings are separated by distance d. Assume the positive x axis points to the right, through the center of the rings.

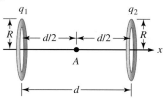

FIGURE P24.30 Problems 30 and 31.

31. **N** What is the electric field at point A in Figure P24.30 if $d = 1.40$ m, $R = 0.500$ m, $q_1 = 15.0$ nC, and $q_2 = -25.0$ nC? Assume the positive x axis points to the right, through the center of the rings.

Problems 32 and 33 are paired.

32. **A** A charged rod is curved so that it is part of a circle of radius R (Fig. P24.32). The excess positive charge Q is uniformly distributed on the rod. Find an expression for the electric field at point A in the plane of the curved rod in terms of the parameters given in the figure.

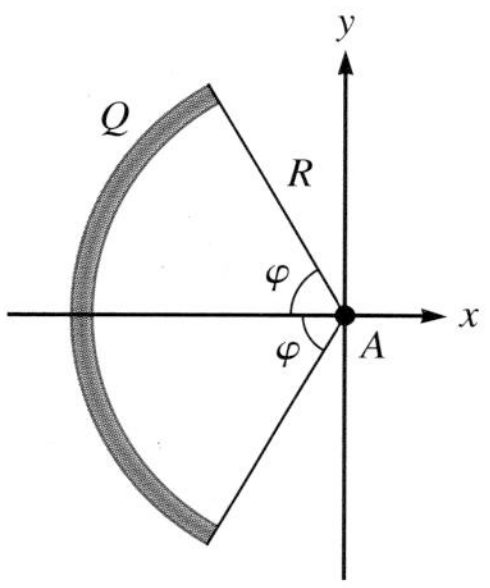

33. **N** If the curved rod in Figure P24.32 has a uniformly distributed charge $Q = 35.5$ nC, radius $R = 0.785$ m, and $\varphi = 60.0°$, what is the magnitude of the electric field at point A?

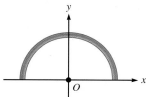

FIGURE P24.32 Problems 32 and 33.

34. **N** A plastic rod of length $\ell = 24.0$ cm is uniformly charged with a total charge of $+12.0$ μC. The rod is formed into a semicircle with its center at the origin of the xy plane (Fig. P24.34). What are the magnitude and direction of the electric field at the origin?

FIGURE P24.34

Problems 35 and 36 are paired.

35. **N** A positively charged disk of radius $R = 0.0366$ m and total charge 56.8 μC lies in the xz plane, centered on the y axis (Fig. P24.35). Also centered on the y axis is a charged ring with the same radius as the disk and a total charge of -34.1 μC. The ring is a distance $d = 0.0050$ m above the disk. Determine the electric field at the point P on the y axis, where P is $y = 0.0100$ m above the origin.

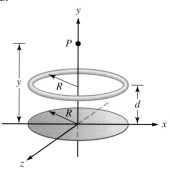

FIGURE P24.35 Problems 35 and 36.

36. **A** A positively charged disk of radius R and total charge Q_{disk} lies in the xz plane, centered on the y axis (Fig. P24.35). Also centered on the y axis is a charged ring with the same

radius as the disk and total charge Q_{ring}. The ring is a distance d above the disk. Determine the electric field at the point P on the y axis, where P is above the ring a distance y from the origin.

37. **N** A uniformly charged conducting rod of length $\ell = 30.0$ cm and charge per unit length $\lambda = -3.00 \times 10^{-5}$ C/m is placed horizontally at the origin (Fig. P24.37). What is the electric field at point A with coordinates (0, 0.400 m)?

38. **G** Plot the scalar component of the electric field due to a charged disk for points along its axis as given by Equation 24.17:

$$\vec{E} = 2\pi k\sigma \left[1 - \frac{y}{\sqrt{R^2 + y^2}}\right]\hat{j}$$

How does your graph compare to that for a charged ring (Fig. 24.40, page 740)?

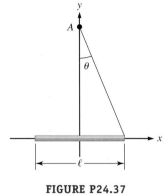

FIGURE P24.37

24-7 Case Study: The Shape of Lightning Rods

39. **N** Assume the electric field on Franklin's pointed lightning rod is the breakdown electric field of 3×10^6 N/C, and estimate the charge on the tip. How does that compare to the charge on Wilson's rod (Example 24.7, page 734)?

40. **A** Check the dimensions of

$$\vec{E} = \frac{kq}{y(y - \ell)}\hat{j}$$

from Example 24.8 (page 735).

41. **N** In Example 24.7 (page 734), we analyzed Wilson's lightning rod. The large spherical conductor was later shown to be less effective than Franklin's pointed rod. In fact, large spherical conductors are often used in high-charge applications to prevent breakdown of the surrounding air (or another insulator). Suppose a spherical conductor must be able to hold a charge of 50.0 μC in air. What must be its minimum diameter?

42. **A** In Example 24.8, Franklin's thin lightning rod has a uniform linear charge density. Consider a different situation in which the thin rod is an insulator with a nonuniform distribution of charge. Referring to the geometry in Figure 24.36 (page 735), suppose the charge density of the rod is given by $\lambda(y) = \lambda_0 y/\ell$. This represents a charge per unit length that linearly increases in magnitude from zero at the bottom to λ_0 at the top of the insulating rod. Find an expression for the electric field at point A, at a position $y > \ell$ above the end of the rod.

24-8 Charged Particle in an Electric Field

43. **N** What are the magnitude and direction of a uniform electric field perpendicular to the ground that is able to suspend a particle of mass $m = 2.00$ g carrying a charge of $+6.00$ μC in midair, assuming gravity and the electrostatic force are the only forces exerted on the particle?

44. **N** An electron is in a uniform upward-pointing electric field.
    a. If the electron experiences a downward acceleration of 9.81 m/s², what is the magnitude of the electric field? (Ignore gravity.)
    b. What is the gravitational force on this electron? Is it okay to ignore gravity? Explain.

45. **N** A proton is at rest and in equilibrium, floating above a positively charged disk. What is the surface charge density of the disk? Assume the proton is very near the surface of the charged disk.

46. **N** A uniform electric field of 725 N/C is produced within a particle accelerator. Starting from rest, what are
    a. the speed of an electron placed in this field after 23.0 ns have elapsed, and
    b. the speed of a proton placed in this field after 23.0 ns have elapsed?

47. **N** A very large disk lies horizontally and has surface charge density $\sigma = -2.3$ nC/m². An electron is released at the surface. (It begins from rest and moves vertically upward.) Ignoring gravity, find the speed of the electron when it is 1.0 mm above the disk.

48. **N** An electron is released from rest in a uniform electric field of 465 N/C near a particle detector. The electron arrives at the detector with a speed of 2.40×10^6 m/s.
    a. What was the uniform acceleration of the electron?
    b. How long did the electron take to reach the detector?
    c. What distance was traveled by the electron?
    d. What is the kinetic energy of the electron when it reaches the detector?

49. **N** In Figure P24.49, a charged particle of mass $m = 4.00$ g and charge $q = 0.250$ μC is suspended in static equilibrium at the end of an insulating thread that hangs from a very long, charged, thin rod. The thread is 12.0 cm long and makes an angle of 35.0° with the vertical. Determine the linear charge density of the rod.

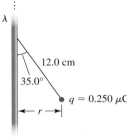

FIGURE P24.49

50. **N** Three charged spheres are suspended by nonconducting light rods of length $L = 0.625$ m from the point O. The rods are fixed in position so that the middle rod is vertical and the rods on the left and right make an angle of 30.0° with the vertical (Fig. P24.50).
    a. What is the total electric field due to the charged spheres at point O?
    b. What is the electric force on a particle with charge $q = +5.00$ μC placed at point O?

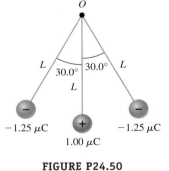

FIGURE P24.50

51. **N** Figure P24.51 shows four small charged spheres arranged at the corners of a square with side $d = 25.0$ cm.
    a. What is the electric field at the location of the sphere with charge $+2.00$ nC?
    b. What is the total electric force exerted on the sphere with charge $+2.00$ nC by the other three spheres?

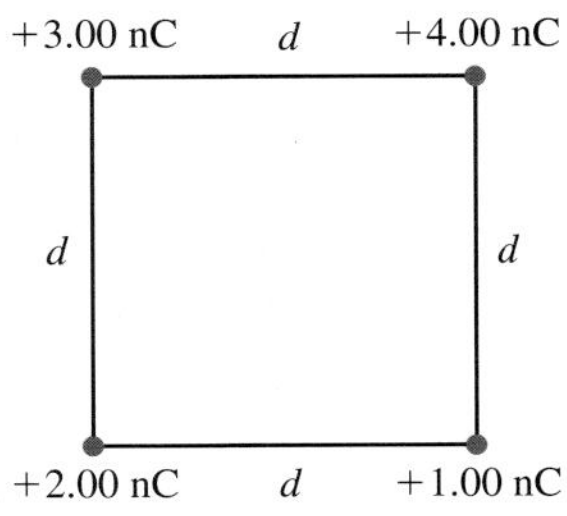

FIGURE P24.51

52. **N** A beam of protons in a particle accelerator is observed to be moving to the right with a kinetic energy per proton of 3.00×10^{-15} J. What are the magnitude and direction of the electric field that will stop the protons in this beam in 1.00 m?

53. **N** A uniform electric field given by $\vec{E} = (2.65\hat{i} - 5.35\hat{j}) \times 10^5$ N/C permeates a region of space in which a small negatively charged sphere of mass 1.30 g is suspended by a light cord (Fig. P24.53). The sphere is found to be in equilibrium when the string makes an angle $\theta = 23.0°$.
    a. What is the charge on the sphere?
    b. What is the magnitude of the tension in the cord?

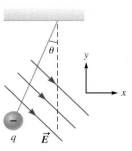

FIGURE P24.53

54. A uniformly charged ring of radius $R = 25.0$ cm carrying a total charge of $-15.0\ \mu C$ is placed at the origin and oriented in the yz plane (Fig. P24.54). A 2.00-g particle with charge $q = 1.25\ \mu C$, initially at the origin, is nudged a small distance x along the x axis and released from rest. The particle is confined to move only in the x direction.
    a. **A** Show that the particle executes simple harmonic motion about the origin.
    b. **N** What is the frequency of oscillation for the particle?

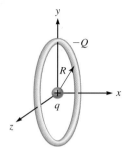

FIGURE P24.54

24-9 Special Case: Dipole in an Electric Field

55. **C** Avery runs a comb through his hair, which causes the comb to acquire a net charge. When he holds the comb near a lightly flowing stream of water coming out of a faucet, the water is attracted toward the comb (Fig. P24.55). Given that a water molecule can be modeled as an electric dipole, explain why the stream of water is deflected by the comb.

56. **C** In Chapter 23, we saw that neutral insulating objects, like bits of paper, can be attracted to charged objects because dipoles are induced in the bits of paper. Suppose instead of a comb (Fig. 23.5C, page 686), we put the bits of paper near a charged object that produces a uniform electric field, like a large, uniformly charged flat sheet. Would the paper be attracted to the sheet? Explain.

FIGURE P24.55

57. **N** A potassium chloride molecule (KCl) has a dipole moment of 8.9×10^{-30} C·m. Assume the KCl molecule is in a uniform electric field of 325 N/C. What is the change in the system's potential energy when the molecule rotates
    a. from $\varphi = 170°$ to $180°$,   b. from $\varphi = 90°$ to $100°$, and
    c. from $\varphi = 10°$ to $0°$?

58. **N** Review Find the magnitude of the angular acceleration α for the water molecule in Example 24.11 (page 744).

General Problems

59. **N** Two metallic spheres of the same size with charges $+9.0\ \mu C$ and $-4.0\ \mu C$ are kept 0.20 m apart. Find the location where the electric field is zero if the $+9.0\text{-}\mu C$ charge is at the origin and the $-4.0\text{-}\mu C$ charge is at $x = 0.20$ m.

60. **C** Which equations in the chapter are worth memorizing? Be sure to state if the equation has certain limitations such as "good only for spherical source."

61. A total charge Q is distributed uniformly on a metal ring of radius R.
    a. **N** What is the magnitude of the electric field in the center of the ring at point O (Fig. P24.61)?
    b. **A** What is the magnitude of the electric field at the point A lying on the axis of the ring a distance R from the center O (same length as the radius of the ring)?

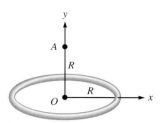

FIGURE P24.61

62. **A** A simple pendulum has a small sphere at its end with mass m and charge q. The pendulum's rod has length L and its weight is negligible. The pendulum is placed in a uniform electric field of strength E directed vertically upward. What is the period of oscillation of the sphere if the electric force is less than the gravitational force on the sphere? Assume the oscillations are small.

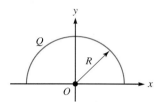

FIGURE P24.63

63. **A** A thin, semicircular wire of radius R is uniformly charged with total positive charge Q (Fig. P24.63). Determine the electric field at the midpoint O of the diameter.

Problems 64 and 65 are paired.

64. **A** An infinite number of positively charged particles, each with charge q, are placed along the x axis at $x = 1$ m, $x = 2$ m, $x = 4$ m, $x = 8$ m, …, as shown in Figure P24.64. Determine the electric field at $x = 0$ due to this set of charges. *Hint*: Use the formula for a geometric series: $1 + r + r^2 + r^3 + r^4 + \cdots = 1/(1-r)$.

FIGURE P24.64

65. **A** An infinite number of charged particles, each with charge $\pm q$, are placed along the x axis at $x = 1$ m, $x = 2$ m, $x = 4$ m, $x = 8$ m, and so on. What is the electric field at $x = 0$ if consecutive charges have opposite sign as shown in Figure P24.65? *Hint*: Use the formula for a geometric series: $1 + r + r^2 + r^3 + r^4 + \cdots = 1/(1-r)$.

FIGURE P24.65

66. **N** Three identical cylinders made of solid plastic are 15.0 cm in length and 4.50 cm in radius. What is the total charge on
    a. the first cylinder, which has a uniform charge density of $235\ \mu C/m^3$ throughout the cylinder;
    b. the second cylinder, which has a uniform charge density of $12.5\ \mu C/m^2$ on its entire surface including its two ends; and

c. the third cylinder, which has a uniform charge density of 12.5 μC/m^2 only on its curved lateral surface, not including its two ends?

67. **N** The total charge on a uniformly charged ring with diameter 25.0 cm is -54.0 μC. What is the magnitude of the electric field along the ring's axis at a distance of
a. 2.50 cm, b. 12.5 cm, c. 25.0 cm, and
d. 2.00 m from its center?

68. **N** The charge density of a uniformly charged disk 0.500 m in diameter is 2.45×10^{-2} C/m^2. What is the magnitude of the electric field along the disk's axis at a distance of
a. 2.50 cm, b. 25.0 cm, c. 50.0 cm, and
d. 5.00 m from its center?

69. **N** A thin wire with linear charge density

$$\lambda = \lambda_0 y_0 \left(\frac{1}{4} + \frac{1}{y} \right)$$

extends from $y_0 = 1.00$ m to infinity. If $\lambda_0 = 1.45 \times 10^{-5}$ C/m, what is the magnitude of the electric field due to this wire at the origin (y is measured in meters)?

Problems 70, 71, and 72 are grouped.

70. **A** Two positively charged spheres are shown in Figure P24.70. Sphere 1 has twice as much charge as sphere 2. Find an expression for the electrostatic field at point A in terms of the parameters shown in the figure.

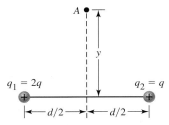

FIGURE P24.70 Problems 70, 71, and 72.

71. **N** Two positively charged spheres are shown in Figure P24.70. Sphere 1 has twice as much charge as sphere 2. If $q = 6.55$ nC, $d = 0.250$ m, and $y = 1.25$ m, what is the electric field at point A?

72. **N** Two positively charged spheres are shown in Figure P24.70. Sphere 1 has twice as much charge as sphere 2; $q = 6.55$ nC, $d = 0.250$ m, and $y = 1.25$ m. Suppose an electron is placed at point A and released.
a. What is the electron's acceleration?
b. If the electron were replaced by a proton, what is the proton's acceleration?

73. **N** The total charge on a uniformly charged, horizontal rod of length $\ell = 26.0$ cm is $+43.0$ μC. What are the magnitude and direction of the electric field produced by the rod at a distance of 20.0 cm from the left end of the rod along its axis?

74. **N** Two parallel plates are placed 5.60 cm apart, creating a uniform electric field of magnitude 525 N/C between the plates. At the same instant in time, a proton is released from rest from the positive plate and an electron is released from rest from the negative plate. If we ignore the electrical attraction between the two particles, at what distance from the positive plate do the particles pass each other?

75. **N** A conducting rod carrying a total charge of $+9.00$ μC is bent into a semicircle of radius $R = 33.0$ cm, with its center of curvature at the origin (Fig.P24.75). The charge density along the rod is given by $\lambda = \lambda_0 \sin \theta$, where θ is measured clockwise from the $+x$ axis. What is the magnitude of the electric force on a 1.00-μC charged particle placed at the origin?

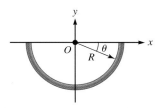

FIGURE P24.75

76. **A** A uniformly charged ring of radius d and total charge Q is oriented in the xz plane and placed at $y = 0$. Show that the maximum electric field due to this ring occurs at $y = d/\sqrt{2}$ and has magnitude $E_{max} = 2kQ/(3\sqrt{3}d^2)$.

Problems 77, 78, 79, and 80 are grouped.

77. **A** When we find the electric field due to a continuous charge distribution, we imagine slicing that source up into small pieces, finding the electric field produced by the pieces, and then integrating to find the electric field. Let's see what happens if we break a finite rod up into a small number of finite particles. Figure P24.77 shows a rod of length 2ℓ carrying a uniform charge Q modeled as two particles of charge $Q/2$. The particles are at the ends of the rod. Find an expression for the electric field at point A located a distance ℓ above the midpoint of the rod using each of two methods:
a. modeling the rod with just two particles and
b. using the exact expression

$$\vec{E} = \frac{kQ}{y} \frac{1}{\sqrt{\ell^2 + y^2}} \hat{j}$$

c. Compare your results to the exact expression for the rod by finding the ratio of the approximate expression to the exact expression.

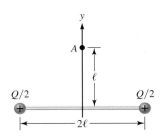

FIGURE P24.77 Problems 77 and 78.

78. **N** Consider the model discussed in Problem 77. Figure P24.77 shows a rod of length 1.00 m carrying a uniform charge $Q = 0.42$ nC modeled as two particles of charge $Q/2$. Find the electric field at point A located a distance 0.50 m above the midpoint of the rod using each of two methods:
a. modeling the rod with just two particles and
b. using the exact expression

$$\vec{E} = \frac{kQ}{y} \frac{1}{\sqrt{\ell^2 + y^2}} \hat{j}$$

c. What is the ratio of your approximate result to the electric field calculated exactly?

79. **A** Consider the model discussed in Problem 77. Figure P24.79 shows a rod of length 2ℓ carrying a uniform charge Q modeled as five particles of charge $Q/5$. Two particles are at the ends of the rod, and the rest are evenly spaced along the rod. Find an expression for the electric field at point A located a distance ℓ above the midpoint of the rod using each of two methods

(to make your work easier, you can calculate any square root to three significant figures):
a. modeling the rod with just five particles and
b. using the exact expression

$$\vec{E} = \frac{kQ}{y}\frac{1}{\sqrt{\ell^2 + y^2}}\hat{\jmath}$$

c. Compare your approximate result to your exact expression for the rod by finding the ratio of the approximate expression to the exact expression.

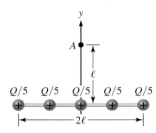

FIGURE P24.79 Problems 79 and 80.

80. **N** Consider the model discussed in Problem 77. Figure P24.79 shows a rod of length 1.00 m carrying a uniform charge $Q = 0.42$ nC modeled as five particles of charge $Q/5$. Two particles are at the ends of the rod, and the rest are evenly spaced along the rod. Find the electric field at point A located a distance 0.50 m above the midpoint of the rod using each of two methods (to make your work easier, you can calculate any square root to three significant figures):
a. modeling the rod with five particles and
b. using the exact expression

$$\vec{E} = \frac{kQ}{y}\frac{1}{\sqrt{\ell^2 + y^2}}\hat{\jmath}$$

c. What is the ratio of your approximate result to the electric field calculated exactly?

81. **A** Return to Example 24.6 (page 732) for the charged disk. By integrating, verify Equation 24.17:

$$\vec{E} = k\sigma 2\pi y\hat{\jmath}\int_0^R \frac{r\,dr}{(r^2 + y^2)^{3/2}} = 2\pi k\sigma\left[1 - \frac{y}{\sqrt{R^2 + y^2}}\right]\hat{\jmath}$$

82. **A** CASE STUDY Return to Example 24.8 (page 735) about Franklin's lightning rod. By integrating, show that

$$\vec{E} = k\lambda\hat{\jmath}\int_0^\ell \frac{dy'}{(y - y')^2} = \frac{k\lambda\ell}{y(y - \ell)}\hat{\jmath}$$

Gauss's Law

25

❶ Underlying Principles
Gauss's law

✪ Major Concepts
1. Gaussian surface
2. Electric flux

▶ Special Cases
1. Electric field due to sources with
   a. Linear symmetry
   b. Spherical symmetry (outside, shell theorem, and inside solid sphere)
   c. Planar symmetry
2. Electric field inside and outside a charged conductor

Key Questions

What is electric flux?
What is Gauss's law?
How do we use Gauss's law to find the electric field due to charged sources with various symmetries?

- 25-1 **Qualitative look at Gauss's law** 754
- 25-2 **Flux** 757
- 25-3 **Gauss's law** 761
- 25-4 **Special case: Linear symmetry** 765
- 25-5 **Special case: Spherical symmetry** 767
- 25-6 **Special case: Planar symmetry** 771
- 25-7 **Special case: Conductors** 774

On your birthday, you find a gift-wrapped box with your name on it hidden in the closet. You are excited because the box is the right size to hold that special gift you are hoping for. Still, you cannot just rip off the paper to see what it is. Instead, you pick it up to estimate its weight. You tap it lightly. Does it sound hollow? You might even shake it to see whether it contains loose pieces.

Now, instead of a birthday present, your physics professor hands you a sealed box. She tells you that there is a plastic object inside, and your job is to figure out what the object's net charge is—positive, negative, or zero. You ask yourself, "What would Ben Franklin do?" Then you suspend a lightweight ball from an insulating thread and give the test ball a net positive charge. You hold the test ball near the box, looking for a response on all sides. If the ball is repelled, there must be a net positive charge in the box. If the ball is attracted, the box must contain a net negative charge. If the ball does not respond at all, the box must have no net charge. In this chapter, we study Gauss's law, which relates the amount of net charge inside the box to the electric field passing through the box.

25-1 Qualitative Look at Gauss's Law

Another title for this chapter could have been "Another Look at the Electric Field" because we will learn how to find the electric fields due to sources of various shapes. In Chapter 24, we found the electric field by considering the electrostatic force the source exerts on a small positive test charge. This approach is important: It illustrates the close connection between the electric field produced by a source and the electrostatic force exerted by that source on some charged particle. In this chapter, we use another method—based on *Gauss's law*—to find the electric field.

Let's compare the method we learned in Chapter 24 to the method in this chapter (Fig. 25.1). A source—a collection of charged particles or a charged object—produces an electric field. In Chapter 24, we used Coulomb's law to come up with an expression for the electric field produced by a single charged particle (or outside a charged spherical object). With this as our only tool, we used "brute force" to calculate the electric field produced by sources with other shapes by adding up (or integrating) the electric fields produced by the many tiny charged pieces that make up the source.

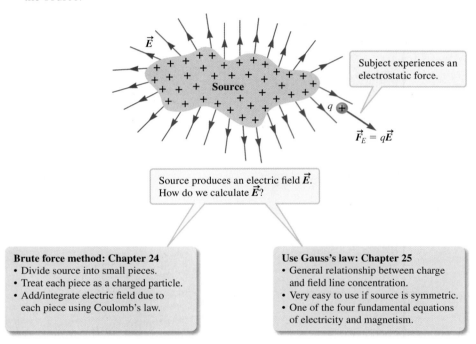

FIGURE 25.1 Concept map showing two approaches to calculating a source's electric field.

The method we learn in this chapter—based on Gauss's law—is more abstract. Instead of considering test charges and the electrostatic force, we find the electric field from the source's charge by calculating the number of electric field lines that pass through an area. Gauss's law is more general than Coulomb's law because it is good for charge distributions of any shape, whereas Coulomb's law is good only for spherically shaped charge distributions. In practice, we apply Gauss's law to sources that are symmetrical, and in those cases, Gauss's law is much easier to use than the brute force method based on Coulomb's law.

What's in the Box?

Gauss's law is expressed mathematically. To get an overview, let's start with a qualitative understanding. Gauss's law makes use of an imaginary, closed, three-dimensional surface known as a **Gaussian surface**. Figure 25.2 shows three Gaussian surfaces—boxes, in this case—each with a set of electric field lines. Like a gift box, these boxes do not interfere with their contents in any way; the boxes only prevent us from seeing directly inside. Although you cannot see inside, the electric field lines pass through the walls of the boxes without being altered. In each case, we can figure

GAUSSIAN SURFACE

⭐ Major Concept

out something about the contents of the box from the electric field lines that pass through it. In Figure 25.2A, the electric field lines emerge from the box, so we conclude that whatever is in the box must have a net positive charge. The electric field lines enter the box in Figure 25.2B, so we conclude that the box must contain a net negative charge. Finally, the electric field lines pass through (enter and exit) the box in Figure 25.2C, so we conclude that there is no net charge inside the box. It must either be empty or contain a neutral object.

In Figure 25.2, we imagined that each Gaussian surface was opaque, like a cardboard box you might use to hold a present. Gauss's law does not require that such a box actually exist. Instead, a Gaussian surface is always created in our imagination, so we are free to imagine a surface of any shape or size that will simplify our job of determining the electric field due to a charged source.

For example, to find the magnitude of the electric field in Fig. 25.2A, it is best to use a *spherical* Gaussian surface. You can imagine a hollow ball with a positively charged source inside (Fig. 25.3). Because Gaussian spheres exist only in our imagination, they never affect the charged source inside them or the electric field due to that source. The source inside the Gaussian sphere and the resulting electric field in Figure 25.3 are identical to the charge inside the box and the electric field in Figure 25.2A.

We can learn about Gauss's law by considering two concentric Gaussian spheres (Fig. 25.3). So far, we don't know the shape of the source inside, but we know that the source has a net positive charge. Let's assume that it is a positively charged particle (charge Q) at the common center of the two Gaussian spheres.

The electric field in Figure 25.3 is arbitrarily represented by eight lines. You might have thought that more or fewer lines would make the drawing better. But no matter how many electric field lines are drawn, we find that the same number of lines penetrate each sphere. Therefore, we suspect some physical quantity is constant for both spheres. That physical quantity is called *electric flux*, or simply *flux*, a major part of Gauss's law. Before we define flux precisely in the next section, let's continue to develop a conceptual understanding of Gauss's law by coming up with a physical interpretation for the constant number of electric field lines that penetrate each Gaussian sphere in Figure 25.3. We'll explore three different physical quantities: (1) the electric field E, (2) the surface area A, and (3) the product of the electric field and the surface area EA. One of these three quantities is constant for the two spheres.

We know from $E = kQ/r^2$ (Eq. 24.3) that the electric field decreases with increasing distance from the source. So, the electric field at positions on the larger spherical surface is weaker than the electric field at positions on the smaller spherical surface. The electric field is not constant for the two spheres.

The surface area of a sphere depends on the sphere's radius, $A(r) = 4\pi r^2$, so the surface area of the small sphere is less than the surface area of the large sphere. Area is obviously not constant between spheres of different radii.

Because the electric field is weaker on the sphere with the larger area, we can find our constant quantity by multiplying the electric field strength on the surface of one sphere by the surface area of that sphere. For the small sphere,

$$E(r)A(r) = \frac{kQ}{r^2}4\pi r^2 = 4\pi kQ \quad (25.1)$$

where $E(r) = kQ/r^2$ is the electric field on the surface of the small sphere and $A(r) = 4\pi r^2$ is its surface area. For the large sphere,

$$E(R)A(R) = \frac{kQ}{R^2}4\pi R^2 = 4\pi kQ \quad (25.2)$$

where $E(R) = kQ/R^2$ is the electric field on the surface of the large sphere and $A(R) = 4\pi R^2$ is its surface area.

From Equations 25.1 and 25.2, we see that when we multiply the electric field on the surface of a sphere by its surface area, we get the constant $EA = 4\pi kQ$. The physical interpretation of the constant number of field lines that penetrate the two Gaussian surfaces in Figure 25.3 is that the quantity EA is a constant. As we'll soon define, EA is the electric flux through either Gaussian sphere. Furthermore, from Equations 25.1 and 25.2, we see that the flux EA is a constant that depends only on

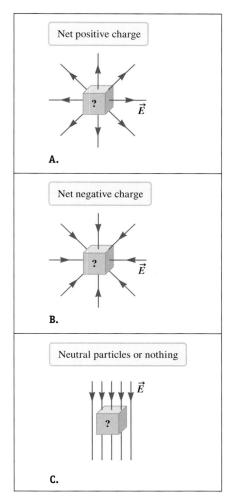

FIGURE 25.2 What is inside each box?

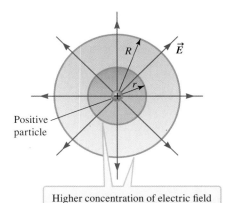

FIGURE 25.3 Gaussian spheres surround a positively charged particle. The number of electric field lines (eight in this case) that pass through each sphere is a constant proportional to EA.

the amount of charge Q inside the Gaussian sphere. Conceptually, Gauss's law relates the flux to the amount of charge enclosed by that Gaussian surface.

Symmetry

Why did we decide to use spherical Gaussian surfaces in Figure 25.3 instead of boxes as in Figure 25.2A? The reason is that spheres *exploit the symmetry* of the source. Whenever you come up with a Gaussian surface, you should try to exploit such symmetry.

"Exploiting the symmetry" means choosing a Gaussian surface that has the same symmetry as the source. There are many ways to test for symmetry, but we need to think about only one test—rotation. Consider the uppercase letter **A**. In Figure 25.4A, we have chosen a coordinate system such that **A** lies in the xy plane. Because a written letter must lie on the paper (in this case, the xy plane), we imagine rotating **A** by 180° around the x axis and then around the y axis. Both of these rotations return the letter to the xy plane. If we rotate the letter **A** around the x axis by 180°, the letter is upside down. If, instead, we rotate **A** around the y axis by 180°, the letter remains unchanged. We say that **A** is symmetrical with respect to a 180° rotation around the y axis, but not the x axis. How about the z axis? The letter lies in the xy plane, so we can test rotation around the z axis by imagining that we spin the letter around the z axis. For example, in the fourth panel of Figure 25.4A, the letter **A** has been rotated by 90° around the z axis, and the letter looks sideways. In fact, except for a 360° rotation, when we rotate **A** around the z axis, it does not look like an upright **A** again. So, **A** is not symmetrical with respect to rotations around the z axis.

Now suppose you want to find another letter that has the same symmetry as **A**. You might choose **U**, the only other vowel that has the same symmetry. To test it, imagine rotating **U** by 180° around the x axis, and you see that—just like **A**—it is upside down. Then imagine rotating **U** by 180° around the y axis; it remains unchanged. Finally, try rotating **U** by any amount (except 360°) around the z axis. You find that **U**—like **A**—is not symmetrical with respect to rotations around the z axis.

Not all letters have the same symmetry as **A**. For example, look at **Z** in Figure 25.4B. If you rotate **Z** around the x axis or the y axis by 180°, the letter looks upside down. However, if you rotate **Z** around the z axis by 180°, the letter is right side up. So, **Z** is symmetrical with respect to 180° rotations around the z axis.

The letters **A** and **Z** are equally symmetrical; both are symmetrical with respect to a 180° rotation around a single axis. Some letters have much more symmetry, however. For example, consider the letter **O**, which is unchanged by a 180° rotation around either the x axis or the y axis. You can also rotate **O** by any amount around the z axis, and it will look the same. So, **O** is symmetrical with respect to a 180° rotation around either the x axis or the y axis, and with respect to any rotation around the z axis.

In this chapter, we study three types of symmetrical charge distributions: linear, spherical, and planar. We'll use rotation to test for these types of symmetry.

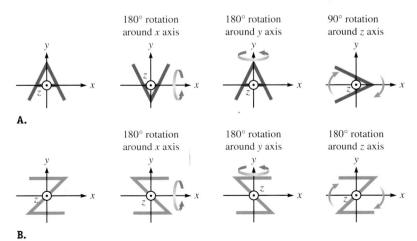

FIGURE 25.4 A. A symmetry. Rotation by 180° around the y axis leaves the letter unchanged. **B. Z** symmetry. Rotation by 180° around the z axis leaves the letter unchanged.

CONCEPT EXERCISE 25.1

a. List all the uppercase letters that have the same symmetry as **A**.
b. List all the uppercase letters that have the same symmetry as **Z**.
c. List all the uppercase letters that have the same symmetry as **O**.

CASE STUDY   Stuck in the Rain

Avi, Cameron, and Shannon are driving along a rural road when a powerful thunderstorm rolls in. They see lightning strikes nearby.

Avi: We should pull over and find shelter.

Shannon: There's nothing around here. Just fields. Not even a farmhouse in sight.

Cameron: The safest place is in the car. Just pull over so we don't slide off the road. We'll wait it out in the car.

Avi: No way. Didn't you see how close that lightning is? We're right in the middle of this, and I don't want to get struck by lightning. If lightning hits the car, it will blow up the gas tank, and I am not interested in going up in smoke. Remember how a small spark can cause gasoline to explode (page 685)?

Cameron: Look. The gas is in a sealed metal tank. And we are in a sealed metal car. The lightning can't get to the gas or us. We are much safer in here.

Avi: I think it is safer out there. There are some trees in the fields. The trees will act like lightning rods. We'll get wet out there, but at least we won't get hurt.

Cameron: I totally disagree. Let's let Shannon decide.

What would you decide? Your answer will help explain why cell phones don't work in closed elevators and how antistatic bags protect sensitive electronics (Fig. 25.5).

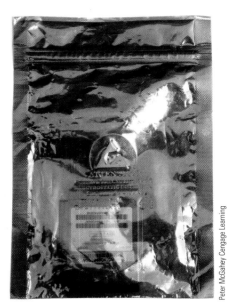

FIGURE 25.5 Antistatic bags protect sensitive electronics.

25-2 Flux

Gauss's law involves finding the *electric flux*. We can use a flowing fluid as a mechanical analogy. Figure 25.6A shows streamlines (Section 15-6) for water flowing through a pipe that narrows. Streamlines are like electric field lines. In both cases, the absolute number of field lines or streamlines is chosen arbitrarily,

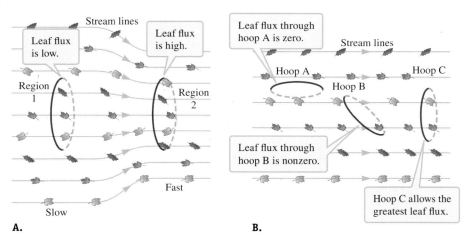

FIGURE 25.6 A. Streamlines for water in a pipe that narrows. The streamlines are more concentrated in region 2, indicating that the water's speed is higher in region 2. **B.** Three hoops in a stream. Hoop A is parallel to the flow, allowing no water to flow through it. Hoop B is tilted at an angle φ to the streamlines, allowing some water to flow through it. Hoop C is perpendicular to the water's flow, allowing for the maximum flow of water through it.

but once the choice has been made, the relative density of the lines in the drawing is important. The density of electric field lines indicates the relative magnitude of the electric field. The density of streamlines indicates the fluid's relative speed (Fig. 25.6A). Imagine that leaves are scattered on top of the water, making it easier to observe the fluid flow. Suppose you place a hoop of area A in the stream so that the hoop is perpendicular to the flow. If you place the hoop in region 1, relatively few leaves penetrate the hoop per second. If you place the hoop in region 2, many more leaves penetrate the hoop per second. The number of leaves that flow through the hoop (per unit time) can be defined as the "leaf flux." The word *flux* is Latin for "flow," so leaf flux is the number of leaves that flow through the hoop (per unit time).

We can learn about flux from this analogy. First, the leaf flux depends on the area of the hoop: A larger area allows more leaves to flow through. Second, the angle between the hoop and the streamlines affects the flux (Fig. 25.6B).

In Gauss's law, the magnitude of the electric field on a Gaussian surface is represented by the number of field lines that penetrate the surface and is related to the *electric flux*. The word *flux* is used here for illustration: Nothing is flowing when we study an electrostatic field. However, the electric field lines drawn on paper look like streamlines, so the analogy with fluid flow helps us visualize the calculation of electric flux. In Figure 25.7, a uniform electric field is indicated by evenly spaced field lines, with three identical circular disks in the region covered by the field lines. (These disks are not closed surfaces, so they are not Gaussian surfaces.) The number of electric field lines penetrating each disk represents the electric flux through that disk. Disk A is parallel to the field lines, so no electric field can penetrate it; the electric flux through disk A is zero. Disk B is tilted at an angle φ to the field lines, so those lines can penetrate it; the electric flux is nonzero. However, the electric flux through disk B is not as great as the electric flux through disk C. Disk C is perpendicular to the electric field lines, so it allows the greatest electric flux.

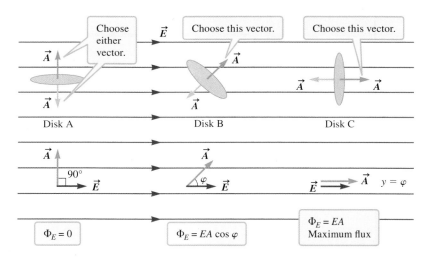

FIGURE 25.7 Electric field lines look like streamlines, but nothing is really flowing. To find the electric flux through the three disks, you must choose the direction of $\vec{A}$. Then you must maintain that direction throughout all calculations. The electric flux in each case is found from $\Phi_E = \vec{E} \cdot \vec{A} = EA\cos\varphi$. For disk A, $\varphi = 90°$, so the electric flux is zero. For disk C, $\varphi = 0$, so the electric flux is EA, which is the maximum flux. The electric flux through disk B has a value between zero and EA.

Area Vectors and Electric Flux

A mathematical description of electric flux requires us to define an **area vector** $\vec{A}$. The magnitude of the area vector is the area of the surface, with the dimensions of length squared. The direction is defined to be perpendicular to the surface. Of course, for a disk or other "open" surface that is not part of a three-dimensional volume, such as the lid taken off a tub of yogurt, there are two possible directions perpendicular to the surface. Choose the direction of $\vec{A}$ so that it makes the smallest possible angle with respect to the electric field. For example, for disk B in Figure 25.7, the area vector may point either up and to the right, or down and to the left. Choose up and to the right because that gives the smallest angle between $\vec{A}$ and $\vec{E}$. If the surface is closed (part of a surface surrounding a three-dimensional volume, such as the lid attached to a yogurt tub), choose $\vec{A}$ so that it points outward.

The **electric flux** Φ_E is given by the dot product of the electric field and the area vector:

ELECTRIC FLUX ⭐ **Major Concept**

$$\Phi_E = \vec{E} \cdot \vec{A} \quad (25.3)$$

The electric flux has the dimensions electric field times area, so its SI units are $N \cdot m^2/C$. Like any dot product, the electric flux can be written in terms of the magnitude of each vector (E and A) and the angle φ between them:

Don't confuse the uppercase Greek letter phi Φ, which stands for flux, with the lowercase phi φ, which is often used to represent an angle.

$$\Phi_E = EA \cos \varphi \quad (25.4)$$

When we apply either Equation 25.3 or Equation 25.4, it is often best to sketch $\vec{E}$ and $\vec{A}$ with their tails touching and φ between the two vectors (Fig. 25.7).

Here's another way to think about Equations 25.3 and 25.4: The electric flux Φ_E is the magnitude of the electric field $|\vec{E}|$ times the magnitude of the component of the area vector $\vec{A}_{\parallel}$ that is *parallel* to the electric field (Fig. 25.8):

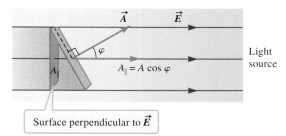

$$\Phi_E = EA_{\parallel} \quad (25.5)$$

where $A_{\parallel} = A \cos \varphi$. If you were to shine a light parallel to the electric field, the shadow cast by a surface of area A would have an area equal to the magnitude of $\vec{A}_{\parallel}$ (Fig. 25.8).

FIGURE 25.8 You can find $A_{\parallel}$—the component of $\vec{A}$ that is parallel to the electric field—from the shadow cast by $\vec{A}$. Then the electric flux is $\Phi_E = EA_{\parallel}$.

A More General Expression for Electric Flux

Equations 25.3 and 25.4 work well when the angle between $\vec{A}$ and $\vec{E}$ is a constant over the entire surface, as it is for all three disks in Figure 25.7, but that is not always the case. The surface in Figure 25.9 is a hemisphere—like what you might get by blowing soap film through a ring. The electric field is uniform, but the angle between the electric field and the area is not uniform over the whole surface. We must imagine dividing the surface into many small pieces, each with an area vector $d\vec{A}$. Each piece is small enough that the angle between $d\vec{A}$ and $\vec{E}$ is constant. The flux through one small piece is $d\Phi_E = \vec{E} \cdot d\vec{A}$, and the flux through the entire surface comes from adding up—that is, integrating—the fluxes through each small piece:

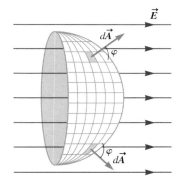

$$\Phi_E = \int \vec{E} \cdot d\vec{A} \quad (25.6)$$

Equation 25.6 is a more general equation for the electric flux than Equations 25.3 and 25.4. In fact, Equation 25.6 is valid when the electric field is not uniform. In practice, calculating the integral in Equation 25.6 may be difficult or even impossible if the angle φ between $d\vec{A}$ and $\vec{E}$ is not constant, so we will choose a Gaussian surface such that the angle is a constant usually equal to 0, 90°, or 180°.

FIGURE 25.9 When we need to calculate the flux through a curved surface, we imagine breaking that surface up into many small, flat surfaces.

CONCEPT EXERCISE 25.2

The terms *electric force*, *electric field*, and *electric flux* sound similar and are easily confused. Write a short description of each term, including the SI units.

CONCEPT EXERCISE 25.3

An electric field is described by $\vec{E} = (15\hat{\imath} + 25\hat{\jmath})$ N/C. Find the electric flux through a surface whose area vector is $\vec{A} = (0.65\hat{\imath} + 0.35\hat{\jmath})$ m².

EXAMPLE 25.1 Three Surfaces in a Uniform Electric Field

Figure 25.10 shows a uniform electric field of magnitude 15.0 N/C that has three surfaces. Surface A is a disk perpendicular to the electric field so that $\vec{A}$ is parallel to $\vec{E}$ and $\varphi = 0$. The radius of the disk is $r = 0.37$ m. Surface B is tilted with respect to the electric field so that $\varphi = 25°$ but is otherwise identical to disk A. Surface C is an open hemisphere of radius $r = 0.37$ m. The hemisphere is oriented so that its cross-sectional area—a disk—is perpendicular to the electric field. Find the electric flux through each surface.

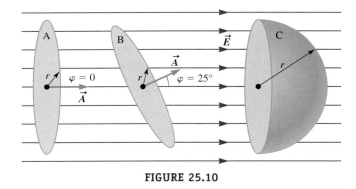

FIGURE 25.10

INTERPRET and ANTICIPATE
When calculating numerical values for the electric flux, we expect our answers to have the units $N \cdot m^2/C$. Imagine shining a light parallel to the electric field. We would find that surface B has the smallest shadow, so we expect surface B to have the smallest electric flux.

SOLVE

Disk A. Because the electric field and angle φ are both uniform over the entire surface A, we can use either Equation 25.3 or Equation 25.4 to find the flux through the disk.	$\Phi_A = \vec{E} \cdot \vec{A}$ (25.3) $\Phi_A = EA \cos \varphi$ (25.4)
Because $\varphi = 0$, the electric flux is EA.	$\Phi_A = EA \cos \varphi = EA \cos 0 = EA$
The area of the disk is $A = \pi r^2$.	$\Phi_A = EA = E(\pi r^2)$ $\Phi_A = (15.0 \text{ N/C})(\pi)(0.37 \text{ m})^2$ $\Phi_A = 6.5 \text{ N} \cdot m^2/C$
Disk B. The process is similar to that for disk A, but in this case $\varphi = 25°$.	$\Phi_B = EA \cos \varphi$ (25.4) $\Phi_B = E(\pi r^2) \cos \varphi$ $\Phi_B = (15.0 \text{ N/C})(\pi)(0.37 \text{ m})^2 \cos 25°$ $\Phi_B = 5.8 \text{ N} \cdot m^2/C$
Hemisphere C. The hemisphere seems like a much more complicated calculation; you may be tempted to break up the surface into many small pieces and integrate using Equation 25.6. This would work, but there is a much simpler way. Imagine shining a light parallel to the electric field so that the shadow cast by the hemisphere is a disk of radius $r = 0.37$ m. According to Equation 25.5, we can find the electric flux from the area of the shadow, $A_\parallel$. In this case the area of the shadow is identical to the area of disk A, so $\Phi_C = \Phi_A$.	$\Phi_C = EA_\parallel$ (25.5) $\Phi_C = E(\pi r^2)$ $\Phi_C = \Phi_A = 6.5 \text{ N} \cdot m^2/C$

CHECK and THINK
Our answers have the units we expected, and the electric flux is smallest through surface B as expected. For the hemisphere, we did not have to integrate because the dot product tells us that we need to use only the component of the area vector that is parallel to the electric field when we calculate electric flux.

25-3 Gauss's Law

According to **Gauss's law**, the net electric flux Φ_E through a closed (Gaussian) surface is proportional to the net charge q_{in} inside the surface. Mathematically, we write Gauss's law as

$$\Phi_E = \frac{q_{in}}{\varepsilon_0} \tag{25.7}$$

where $1/\varepsilon_0$ is the constant of proportionality. The quantity ε_0 is called the **permittivity of free space** or the **permittivity constant** and has the value

$$\varepsilon_0 = 8.85 \times 10^{-12} \frac{C^2}{N \cdot m^2} \tag{25.8}$$

The permittivity constant ε_0 is related to Coulomb's constant k by

$$k = \frac{1}{4\pi\varepsilon_0} = 8.99 \times 10^9 \, N \cdot m^2/C^2 \tag{25.9}$$

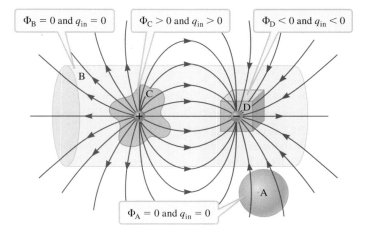

FIGURE 25.11 Dipole with four Gaussian surfaces. Surface A is an empty sphere, and all electric field lines pass through it. Surface B is a cylinder that contains equal amounts of positive and negative charge, so the net charge enclosed is zero. There are as many field lines pointing outward from surface B as pointing into it. Surface C is irregularly shaped like a potato skin. It contains a positive charge, and field lines point outward from it. Surface D is a cube that contains negative charge, and field lines point into it.

Figure 25.11 illustrates Gauss's law for a dipole electric source and four Gaussian surfaces. Just by looking at the field lines that penetrate each Gaussian surface, you can determine whether the net flux is positive, negative, or zero: If the net electric field lines point *outward*, the electric flux is *positive*; if they point *inward*, the electric flux is *negative*; if there are as many lines pointing inward as outward, the electric flux is *zero*. According to Gauss's law, the net flux tells the sign of the charge enclosed by the Gaussian surface. For example, you can see that the net electric flux Φ_E through surfaces A and B is zero, so according to Gauss's law, there is no net charge q_{in} inside either of these surfaces. You can see that the net flux through C is positive, so according to Gauss's law, the charge inside is positive. Finally, the flux through D is negative and the charge inside is negative.

To find the electric field using Gauss's law, we calculate the net electric flux through a Gaussian surface. We use the general equation for electric flux (Eq. 25.6) to write an expression for the net flux through a Gaussian surface:

$$\Phi_E = \oint \vec{E} \cdot d\vec{A} \tag{25.10}$$

The circle on the integral symbol indicates that the integral is taken over a *closed* surface. Combining Equations 25.7 and 25.10 is a convenient way to write Gauss's law:

$$\Phi_E = \oint \vec{E} \cdot d\vec{A} = \frac{q_{in}}{\varepsilon_0} \tag{25.11}$$

GAUSS'S LAW
❶ Underlying Principle

CONCEPT EXERCISE 25.4

Which of the following expressions are correct forms of Gauss's law?

a. $\Phi_E = \oint \vec{E} \cdot d\vec{A}$ **b.** $q_{in} = \varepsilon_0 \oint \vec{E} \cdot d\vec{A}$ **c.** $\oint \vec{E} \cdot d\vec{A} = 4\pi k q_{in}$

CONCEPT EXERCISE 25.5

Find the electric flux through the three Gaussian surfaces in Figure 25.12.

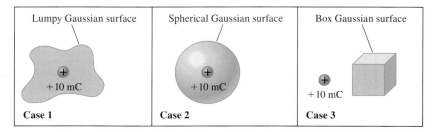

FIGURE 25.12

EXAMPLE 25.2 Comparing Gaussian Surfaces

Figure 25.13 shows two Gaussian surfaces in uniform electric fields. In part A the Gaussian surface is a square box, and in part B the Gaussian surface is a closed cylinder. In each case, integrate $\Phi_E = \oint \vec{E} \cdot d\vec{A} = q_{in}/\varepsilon_0$ (Eq. 25.11) to find the amount of charge enclosed in each Gaussian surface.

:• **INTERPRET and ANTICIPATE**
We can easily determine whether the charge is positive, negative, or zero by looking at the electric field lines (Fig. 25.13). Because the electric field lines pass completely through the Gaussian surfaces, the net charge inside each surface is zero.

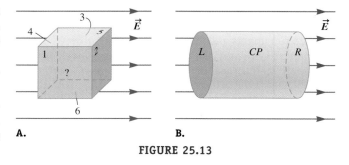

FIGURE 25.13

:• **SOLVE**
Although we already know the results, we are asked to use integration in order to practice using Gauss's law with different Gaussian surfaces.

| A **closed box** is made up of six sides. According to Gauss's law, we must integrate over the entire closed box, so we break up the integral into six pieces, one for each side of the box. The six subscripts on the integrals correspond to the six sides of the box (Fig. 25.13A). | $\Phi_E = \oint \vec{E} \cdot d\vec{A}$

 $\Phi_E = \int_1 \vec{E} \cdot d\vec{A} + \int_2 \vec{E} \cdot d\vec{A} + \int_3 \vec{E} \cdot d\vec{A} + \int_4 \vec{E} \cdot d\vec{A} + \int_5 \vec{E} \cdot d\vec{A} + \int_6 \vec{E} \cdot d\vec{A}$ |

To calculate the six dot products, we need to know the angle φ between $\vec{E}$ and $d\vec{A}$ for each side. A separate drawing of each side is helpful (Fig. 25.14). The area vector $d\vec{A}$ for each side is perpendicular to that side and points outward.

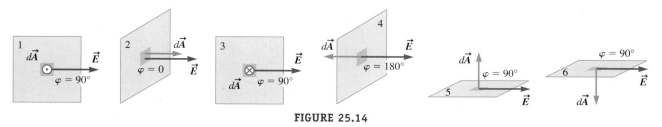

FIGURE 25.14

The dot product is zero whenever the angle φ between $\vec{E}$ and $d\vec{A}$ is 90°, so the dot product is zero for sides 1, 3, 5, and 6. That means these four integrals are zero, and we are left with just two integrals.	$\Phi_E = \oint \vec{E} \cdot d\vec{A}$ $\Phi_E = 0 + \int_2 \vec{E} \cdot d\vec{A} + 0 + \int_4 \vec{E} \cdot d\vec{A} + 0 + 0$ $\Phi_E = \int_2 \vec{E} \cdot d\vec{A} + \int_4 \vec{E} \cdot d\vec{A}$
Write each dot product in terms of the magnitude of the vectors and the angle between them.	$\Phi_E = \int_2 E\, dA \cos\varphi + \int_4 E\, dA \cos\varphi$ $\Phi_E = \int_2 E\, dA \cos 0 + \int_4 E\, dA \cos 180°$ $\Phi_E = \int_2 E\, dA + \int_4 -E\, dA$
The electric field is uniform, so as we integrate over dA, E is a constant that we can pull outside the integrals.	$\Phi_E = E \int_2 dA - E \int_4 dA$
The integrals are identical: Each equals the area A of one square side.	$\Phi_E = EA - EA = 0$
The net electric flux is zero, and according to Gauss's law, that means the charge enclosed by the Gaussian surface is zero.	$\Phi_E = \dfrac{q_{in}}{\varepsilon_0} = 0$ $q_{in} = 0$

:• **CHECK and THINK**
This is exactly what we predicted because the number of electric field lines entering the box is equal to the number of electric field lines leaving the box.

:• **SOLVE**

The **closed cylinder** is made up of three surfaces—the left cap, the right cap, and the curved part. These are labeled L, R, and CP (Fig. 25.13B). Break the integral up into three pieces, one for each surface in Figure 25.13B.	$\Phi_E = \oint \vec{E} \cdot d\vec{A}$ $\Phi_E = \int_L \vec{E} \cdot d\vec{A} + \int_{CP} \vec{E} \cdot d\vec{A} + \int_R \vec{E} \cdot d\vec{A}$

Figure 25.15 shows the angle φ between $\vec{E}$ and $d\vec{A}$ for each surface.

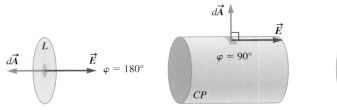

FIGURE 25.15

The area vector points outward for all small pieces of the curved part, while the electric field points to the right. Therefore, the angle $\varphi = 90°$ for the curved part and the corresponding dot product and integral are zero, leaving two integrals.	$\Phi_E = \int_L \vec{E} \cdot d\vec{A} + \int_R \vec{E} \cdot d\vec{A}$

Example continues on page 764 ▶

Write each dot product in terms of the magnitude of the vectors and the angle between the vectors.	$\Phi_E = \int_L E\,dA\cos 180° + \int_R E\,dA\cos 0$
	$\Phi_E = -\int_L E\,dA + \int_R E\,dA$ (25.12)
As for the box, the electric field is constant and the integrals are identical. This time, each is the area of the end cap.	$\Phi_E = -EA + EA = 0$
The electric flux is zero, so the charge inside the Gaussian surface is zero.	$\Phi_E = \dfrac{q_{in}}{\varepsilon_0} = 0 \quad q_{in} = 0$

:• **CHECK and THINK**
We find the same result whether we use a Gaussian cylinder or a box. The process is similar, but there is slightly less work when we use the cylinder because the Gaussian cylinder is made up of just three surfaces instead of the six surfaces of a box. Because we are free to choose a convenient Gaussian surface, it may be helpful to choose a closed cylinder instead of a box when that is possible.

EXAMPLE 25.3 Nonuniform Electric Field

The Gaussian surface in Figure 25.16 is a closed cylinder. An electric field points to the right throughout, made up of two uniform components with magnitude $E_L = 40.9$ N/C on the left side of the cylinder and $E_R = 81.8$ N/C on the right side. Integrate $\Phi = \oint \vec{E}\cdot d\vec{A} = q_{in}/\varepsilon_0$ (Eq. 25.11) to find the amount of charge enclosed in the Gaussian surface. Express your result in terms of the surface charge density (charge per unit area) σ. (Assume σ is uniform.)

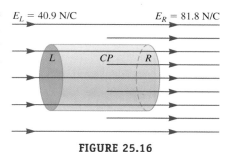

FIGURE 25.16

:• **INTERPRET and ANTICIPATE**
This is similar to Example 25.2 (Fig. 25.13B), except the electric field is not uniform. Because more electric field lines emerge from the Gaussian surface than enter it, we expect the net charge inside to be positive.

:• **SOLVE**

Much of the work we need to do was already done in Example 25.2. We can start with Equation 25.12.	$\Phi_E = -\int_L E\,dA + \int_R E\,dA$ (25.12)
The electric field is constant over each end cap, so it can be pulled outside the integrals. The magnitude of the electric field depends on the position, and the subscripts refer to the magnitudes in Figure 25.16.	$\Phi_E = -E_L\int_L dA + E_R\int_R dA$
The integrals are identical, equaling the area A of each end cap.	$\Phi_E = -E_L A + E_R A = A(E_R - E_L)$
According to Gauss's law, the electric flux is proportional to the charge inside.	$\Phi_E = A(E_R - E_L) = \dfrac{q_{in}}{\varepsilon_0}$
The surface charge density σ is charge per unit area (Eq. 24.10).	$\sigma = \dfrac{q_{in}}{A} = \varepsilon_0 (E_R - E_L)$ (24.10) $\sigma = (8.85\times 10^{-12}\,\text{C}^2/\text{N}\cdot\text{m}^2)(81.8\,\text{N/C} - 40.9\,\text{N/C})$ $\sigma = 3.62\times 10^{-10}\,\text{C/m}^2$

:• **CHECK and THINK**
As expected, the charge inside the Gaussian surface is positive.

PROBLEM-SOLVING STRATEGY: Finding the Electric Field Using Gauss's Law

Gauss's law says that the net electric flux Φ_E through a closed (Gaussian) surface is proportional to the net charge q_{in} inside the surface, and in the preceding examples we used Gauss's law to find that net enclosed charge. In the rest of this chapter, we focus on using Gauss's law to find the electric field for different special cases. So now we present a four-step problem-solving strategy.

:• INTERPRET and ANTICIPATE
Step 1 Sketch the electric field lines (Section 24-3). You may need to draw more than one perspective. Your sketch will help you choose the best Gaussian surface and anticipate your results.

Step 2 Choose a Gaussian surface. You should choose a closed surface that exploits the symmetry of the situation. In other words, you must pick a Gaussian surface that has the same symmetry as the source and its electric field—just as the letter **A** has the same symmetry as the letter **U**. In practice, you should try to pick a Gaussian surface so that each piece of the surface has an angle φ between $\vec{E}$ and $d\vec{A}$ equal to 0, 90°, or 180°. The most commonly used Gaussian surfaces are spherical shells, closed cylinders, and closed boxes. Add the Gaussian surface to your sketch.

:• SOLVE
Step 3 Integrate to find the flux in $\Phi_E = \oint \vec{E} \cdot d\vec{A} = q_{in}/\varepsilon_0$ (Eq. 25.11). Because your Gaussian surface exploits the symmetry of the situation, the electric field is often constant over some portion of the integral and the integral is usually very simple. It may help to add a few area vectors $d\vec{A}$ to your sketch, perpendicular to the surface and pointing outward so that you can easily see the angle between $\vec{E}$ and $d\vec{A}$.

Step 4 Determine the amount of charge inside the Gaussian surface. There may be charged objects inside or outside the Gaussian surface, or a portion of a continuous distribution may be inside the Gaussian surface and the rest outside. Only the net charge q_{in} inside the Gaussian surface is part of Gauss's law. This net charge may be positive, negative, or zero.

:• SOLVE
As in Chapter 24, there are two things to check once you have found an expression for the electric field $\vec{E}$:
1. Make sure the **dimensions** of the electric field are force per charge.
2. If the charge distribution is finite, the electric field for a point far from the distribution compared to its size should approach the electric field of a charged particle $\vec{E}(r) = (kQ_S/r^2)\hat{r}$. We can also **compare to another known electric field**. For example, after we find the electric field produced by an infinitely long charged rod, we can compare that result to the electric field of a finite rod found in Chapter 24. (See Concept Exercise 25.6.)

25-4 Special Case: Linear Symmetry

Imagine a rod such as a pencil. If you look down the end of the pencil and rotate it by any amount, the rod is symmetrical with respect to such a rotation (Fig. 25.17A). If you rotate that rod by 180° around a perpendicular axis that passes through its center, the rod is also symmetrical with respect to such a rotation (Fig. 25.17B). Now imagine an infinitely long rod. Like the finite rod (Fig. 25.17A), the infinite rod is symmetrical with respect to any rotation around its long axis. However, an infinitely long rod is more symmetrical than a finite rod because an infinitely long rod is symmetrical with respect to a 180° rotation around *any* perpendicular axis (Fig. 25.17C). An infinitely long rod is said to possess *linear symmetry*. Of course, no real rod is infinitely long, but a real rod can be modeled as infinitely long for positions close to the rod but not near its ends.

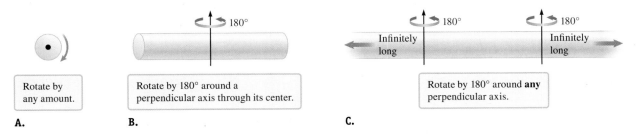

FIGURE 25.17 Special case of linear symmetry. **A.** Rotation by any amount around the long axis. **B.** Rotation by 180° around a perpendicular axis passing through the center. **C.** Rotation by 180° around any axis perpendicular to the long axis.

EXAMPLE 25.4 $\vec{E}$ for a Charge Distribution with Linear Symmetry

It is relatively easy to find the electric field due to a source with linear symmetry using Gauss's law and the four steps listed in the problem-solving strategy (page 765). You can imagine an infinitely long rod with excess positive charge spread uniformly. Because the rod is infinitely long, the total charge is infinite, so it is not very practical to use the total charge. Instead, we work with the linear charge density λ (Eq. 24.12). Show that the electric field due to this source is

$$\vec{E} = \frac{1}{2\pi\varepsilon_0} \frac{\lambda}{r} \hat{r} \qquad (25.13)$$

where r is the perpendicular distance from the infinitely long charged rod.

ELECTRIC FIELD FOR A SOURCE WITH LINEAR SYMMETRY
▶ Special Case

:• INTERPRET and ANTICIPATE
Step 1 Sketch the electric field lines, which point outward and are perpendicular to the rod (Fig. 25.18). We use two views to show the electric field clearly.

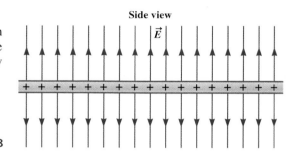
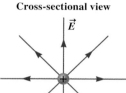

FIGURE 25.18

Step 2 Choose a Gaussian surface that exploits symmetry. A closed cylinder with radius r has the symmetry we need. We add the Gaussian surface to our sketch (Fig. 25.19). The Gaussian cylinder does not need to be infinitely long. Instead, the linear symmetry of the charge distribution means that the length of the cylinder and its position are arbitrary. To be specific, we use a cylinder of length ℓ shown in the middle of the side-view sketch.

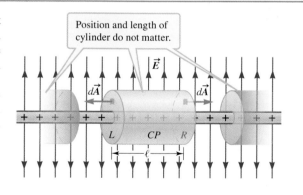
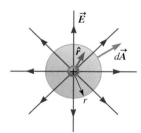

FIGURE 25.19

:• SOLVE
Step 3 Integrate. As we found in Example 25.2, the closed cylinder is made up of three surfaces—the left cap, the right cap, and the curved part—labeled in Figure 25.19 as L, R, and CP.

$$\Phi_E = \oint \vec{E} \cdot d\vec{A}$$

$$\Phi_E = \int_L \vec{E} \cdot d\vec{A} + \int_{CP} \vec{E} \cdot d\vec{A} + \int_R \vec{E} \cdot d\vec{A}$$

The area vector $d\vec{A}$ for surface L points to the left, while that for R points to the right. Because the electric field points outward, the angle φ between $\vec{E}$ and $d\vec{A}$ is 90° for both of these surfaces. Therefore, the dot products are zero and the integrals are zero for surfaces L and R.

$$\Phi_E = 0 + \int_{CP} \vec{E} \cdot d\vec{A} + 0$$

$$\Phi_E = \int_{CP} \vec{E} \cdot d\vec{A}$$

Only one integral remains. The area vector $d\vec{A}$ for surface CP points outward just like the electric field itself, so the angle φ is zero for this entire surface.

$$\Phi_E = \int_{CP} E\, dA \cos 0 = \int_{CP} E\, dA$$

The electric field is not uniform, but the density of the field lines is the same all over the curved surface, so $\vec{E}$ has the same magnitude over that surface and we can pull E outside the integral. The integral is then just the area of the curved part A_{CP}.	$\Phi_E = E \int_{CP} dA = EA_{CP}$
The area of the curved part is the circumference of the circle times the length of the cylinder.	$A_{CP} = 2\pi r \ell$ $\Phi_E = E 2\pi r \ell$ (1)
Step 4 Find the **charge inside** the Gaussian cylinder. Because the charge per unit length is λ, multiply λ by the length of the cylinder to find the charge inside.	$q_{in} = \lambda \ell$ (2)
Finally, we substitute Equations (1) and (2) into Gauss's law (Eq. 25.7) and solve for E. As expected, the length of the cylinder does not matter (ℓ cancels out).	$\Phi_E = \dfrac{q_{in}}{\varepsilon_0}$ (25.7) $E 2\pi r \ell = \dfrac{\lambda \ell}{\varepsilon_0}$ $E = \dfrac{1}{2\pi \varepsilon_0} \dfrac{\lambda}{r}$
From Figure 25.19, we see that the electric field points outward from the line of charge in the positive $\hat{r}$ direction, so we write the electric field vector in component form.	$\vec{E} = \dfrac{1}{2\pi \varepsilon_0} \dfrac{\lambda}{r} \hat{r}$ ✓ (25.13)
∴ CHECK and THINK At a minimum, we need to check that our expression has the dimensions of electric field—force per charge. In Concept Exercise 25.6, you will perform a second check.	$[E] = \left[\dfrac{1}{\varepsilon_0}\right]\left[\dfrac{[\lambda]}{[r]}\right] = \dfrac{F \cdot L^2}{C^2} \dfrac{C}{L} \dfrac{1}{L}$ $[E] = \dfrac{F}{C}$

This example gave us practice using Gauss's law to find the electric field produced by a source. The result is used to model many real charge distributions, so you can expect to see it often.

CONCEPT EXERCISE 25.6

Check Equation 25.13 by showing that it is consistent with

$$\vec{E} = \dfrac{kQ}{y} \dfrac{1}{\sqrt{\ell^2 + y^2}} \hat{j} \qquad (24.15)$$

for a finite rod of length 2ℓ.

25-5 Special Case: Spherical Symmetry

A spherical object is highly symmetrical. Imagine rotating a sphere by any amount through any axis that passes through its center. The sphere remains unchanged. In this section, we find the electric field due to a source with spherical symmetry, such as a charged particle, solid sphere, or spherical shell. The charge is uniformly distributed throughout the solid sphere and over the shell. Both a solid sphere and a spherical shell have interiors, but a charged particle does not (because a particle has no spatial extent). We will use Gauss's law to find the electric field both inside and outside these spherical sources. Coulomb's law is valid outside these charge distributions, so we expect our answer will be the same as what we found using Coulomb's law directly (Eq. 24.3).

EXAMPLE 25.5 $\vec{E}$ Outside a Charge Distribution with Spherical Symmetry

For a solid sphere or spherical shell, show that the electric field outside the source (at $r \geq R$, where R is the radius of the sphere or shell) is

$$\vec{E} = \frac{1}{4\pi\varepsilon_0} \frac{q}{r^2} \hat{r} \quad (25.14)$$

ELECTRIC FIELD OUTSIDE A SOURCE WITH SPHERICAL SYMMETRY
▶ Special Case

INTERPRET and ANTICIPATE
Step 1 Sketch the electric field lines. When the source (a particle, solid sphere, or shell) is positively charged, the electric field lines point outward in all directions, like the quills of a porcupine (Fig. 25.20).

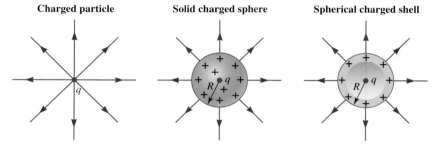

FIGURE 25.20

Step 2 Choose a Gaussian surface. A spherical Gaussian surface has the same symmetry as a spherically symmetrical source (Fig. 25.21). Because we are interested in the electric field outside the source, the radius r of the Gaussian sphere must be greater than or equal to the radius R of the source: $r \geq R$.

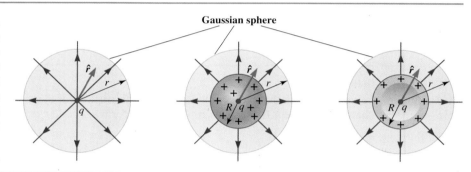

FIGURE 25.21

SOLVE
Step 3 Integrate. A sphere has only one surface, so there is no need to break up the integral into multiple parts. The area vectors point outward like the electric field vectors, and the angle φ is zero over the entire sphere.

$$\Phi_E = \oint \vec{E} \cdot d\vec{A}$$

$$\Phi_E = \oint E \, dA \cos 0 = \oint E \, dA$$

The electric field is not uniform, but it has constant magnitude over the surface of the sphere. We pull E out of the integral, which then becomes the surface area of the Gaussian sphere, $A = 4\pi r^2$.

$$\Phi_E = E \oint dA = EA$$

$$\Phi_E = E(4\pi r^2) \quad (1)$$

Step 4 Find the charge inside. The charge inside the Gaussian sphere in all three cases is the charge q of the source.

$$q_{in} = q \quad (2)$$

Substitute Equations (1) and (2) into Gauss's law (Eq. 25.7), and solve for E.

$$\Phi_E = \frac{q_{in}}{\varepsilon_0} \quad (25.7)$$

$$E(4\pi r^2) = \frac{q}{\varepsilon_0}$$

$$E = \frac{1}{4\pi\varepsilon_0} \frac{q}{r^2}$$

Write the electric field using the unit vector $\hat{r}$ (Fig. 25.21).

$$\vec{E} = \frac{1}{4\pi\varepsilon_0} \frac{q}{r^2} \hat{r} \checkmark \quad (25.14)$$

CHECK and THINK
With the substitution of Coulomb's constant (Eq. 25.9), we find that Equation 25.14 is the same as Equation 24.3, as expected outside a source with spherical symmetry.

$$k = \frac{1}{4\pi\varepsilon_0} \quad (25.9)$$

$$\vec{E}(r) = k\frac{q}{r^2}\hat{r} \quad (24.3)$$

According to Equation 25.14, the electric field outside a charged sphere or shell ($r \geq R$) is the same as the electric field due to a charged particle located at the center of the distribution. But Equation 25.14 tells us nothing about the electric field inside the sphere or shell ($r < R$). We can use Gauss's law to find the electric field inside these sources. The case of a charged shell is a little easier, so let's start with it.

EXAMPLE 25.6 $\vec{E}$ Inside a Charged Spherical Shell

According to the **shell theorem**, the electric field inside any charged shell is zero and the electric field outside is given by Equation 25.14. Show that the electric field inside a charged spherical shell is $\vec{E} = 0$.

SHELL THEOREM ▶ Special Case

INTERPRET and ANTICIPATE
Finding the electric field inside the shell does not require all four steps. Figure 25.22 shows a **Gaussian sphere** inside the charged spherical shell, but no field lines are drawn. You might guess that you could draw inward-pointing field lines, but what will happen when they get to the center? They cannot terminate there because that would require a negatively charged particle in the center. They cannot cross the center and terminate on the opposite wall of the shell because the shell would have to be negative.

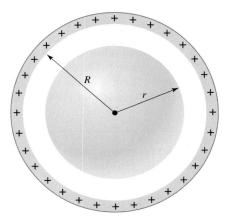

FIGURE 25.22

SOLVE
Step 4 We skip to finding the **charge inside**. There is no charge inside the Gaussian sphere.

$q_{in} = 0$

Because there is no charge inside the Gaussian sphere, the flux through it is zero.

$\Phi_E = \dfrac{q_{in}}{\varepsilon_0} = 0$

So, the integral in Gauss's law is zero.

$\Phi_E = \oint \vec{E} \cdot d\vec{A} = 0$

Because the area of the Gaussian surface cannot be zero and the angle φ is not necessarily zero, the electric field must be zero.

$\vec{E} = 0$ ✓
inside a charged shell

CHECK and THINK
We had trouble even imagining how we could draw the electric field inside the spherical shell, so it makes sense that the electric field inside the shell is zero. In a sense, the electric field lines do point inward from every segment of the wall, but they cancel in the shell so there is no net electric field inside.

EXAMPLE 25.7 — $\vec{E}$ Inside a Charged Solid Sphere

Show that the electric field inside a uniformly charged solid sphere is

$$\vec{E} = \frac{1}{4\pi\varepsilon_0} \frac{q}{R^3} r\hat{r} \qquad (25.15)$$

ELECTRIC FIELD INSIDE A SOLID SPHERE
▶ Special Case

INTERPRET and ANTICIPATE
We can combine the first two steps of the problem-solving strategy. The **electric field** points outward in all directions (Fig. 25.23). (You might be tempted to think the electric field points inward, but remember it must point outward beyond the sphere. So, unless the net electric field reverses direction at the surface, it must point outward everywhere.) Also, a **Gaussian sphere** with $r < R$ exploits the symmetry.

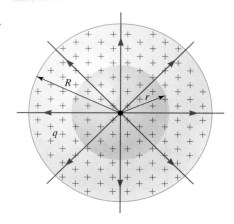

FIGURE 25.23

SOLVE
Step 3 Integrate in much the same way as in Example 25.5. The angle φ is zero over the entire sphere. The electric field is not uniform, but it has constant magnitude over the surface of the sphere. We pull E out of the integral, which then becomes the surface area of the Gaussian sphere, $A = 4\pi r^2$.

$$\Phi_E = \oint \vec{E} \cdot d\vec{A}$$

$$\Phi_E = \oint E\, dA \cos 0 = \oint E\, dA$$

$$\Phi_E = E \oint dA = EA = E(4\pi r^2) \qquad (1)$$

Step 4 Find the charge inside. The charged sphere has radius R and total charge q. The Gaussian surface is smaller than the charged sphere ($r < R$), so the Gaussian surface encloses some fraction (less than 1) of the sphere's charge. To find that fraction, first write an expression for the volume charge density ρ (Eq. 24.8) of the entire charged sphere.

$$\rho = \frac{q}{V} \qquad (24.8)$$

The volume V of the charged sphere can be found from its radius R.

$$V = \frac{4}{3}\pi R^3$$

Write an expression for the volume charge density.

$$\rho = \frac{3q}{4\pi R^3} \qquad (2)$$

The amount of charge inside the Gaussian surface is the volume of the Gaussian surface V_G times the volume charge density.

$$q_{in} = V_G \rho \qquad (3)$$

Find the volume enclosed by the Gaussian surface V_G from its radius r.

$$V_G = \frac{4}{3}\pi r^3 \qquad (4)$$

Substitute Equations (2) and (4) into Equation (3) to find the charge inside the Gaussian surface.

$$q_{in} = \left(\frac{4}{3}\pi r^3\right)\left(\frac{3q}{4\pi R^3}\right) = \left(\frac{r}{R}\right)^3 q \qquad (5)$$

CHECK and THINK
As expected, $r < R$, so the fraction $(r/R)^3 < 1$ and the charge inside the Gaussian surface is smaller than the total charge.

SOLVE
To find the electric field inside the sphere, substitute Equation (5) into Equation 25.7 (Gauss's law) and use Equation (1) for the electric flux.

$$\Phi_E = \frac{q_{in}}{\varepsilon_0} \qquad (25.7)$$

$$E(4\pi r^2) = \frac{1}{\varepsilon_0}\left(\frac{r}{R}\right)^3 q \qquad E = \frac{1}{4\pi\varepsilon_0}\frac{q}{R^3} r$$

Use the unit vector $\hat{r}$ to write the electric field vector.

$$\vec{E} = \frac{1}{4\pi\varepsilon_0} \frac{q}{R^3} r\hat{r} \quad ✓ \quad (25.15)$$

:• CHECK and THINK

Equation 25.15 shows that the electric field gets stronger with increasing distance from the center of the sphere. It is left as a homework exercise (Problem 38) to show that at the surface of the sphere ($r = R$), Equations 25.15 and 25.14 are identical.

25-6 Special Case: Planar Symmetry

You decide to show your school pride and put a static decal on the back window of your car (Fig. 25.24). Because it has excess charge, the decal clings to the window. We can model this source as a charged, very thin sheet.

Think about the symmetry of a sheet such as a piece of printer paper. It is symmetrical with respect to 180° rotations around an axis perpendicular to the sheet and passing through its center. Now imagine an infinite sheet. An infinite sheet is symmetrical with respect to any rotation around any axis that is perpendicular to its surface. Such a sheet is said to possess **planar symmetry**. It is easier to apply Gauss's law to a source with planar symmetry than to a finite sheet. Of course, no real sheet is infinite: A real sheet may be modeled as an infinite sheet for points near its surface but not near its edges.

FIGURE 25.24 A static decal can be modeled as a charged thin sheet.

EXAMPLE 25.8 $\vec{E}$ for a Charge Distribution with Planar Symmetry

Figure 25.25 shows a continuous distribution of positively charged particles uniformly distributed to form a very thin sheet. Its surface charge density is σ. Use Gauss's law to show that the magnitude of the electric field due to such a source is

$$E = \frac{\sigma}{2\varepsilon_0} \quad (25.16)$$

ELECTRIC FIELD FOR A SOURCE WITH PLANAR SYMMETRY ▶ Special Case

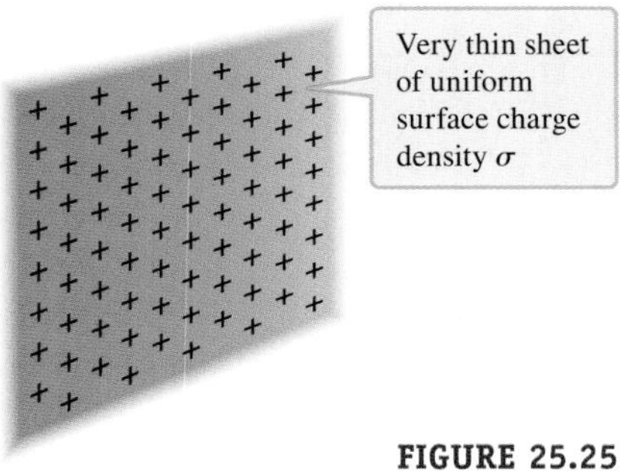

Very thin sheet of uniform surface charge density σ

FIGURE 25.25

:• INTERPRET and ANTICIPATE

Step 1 Sketch the electric field lines. Because the sheet is positively charged, the electric field points outward. Because of the planar symmetry, the electric field is uniform and perpendicular to the surface, seen edge-on in Figure 25.26.

Step 2 Choose a Gaussian surface. A closed cylinder perpendicular to and intersecting the sheet has the symmetry we need (Fig. 25.27). Because of the planar symmetry, the cylinder may have any size and be located anywhere on the sheet. Because we chose the size of the Gaussian cylinder arbitrarily, we don't expect its length or area to be in our final expression.

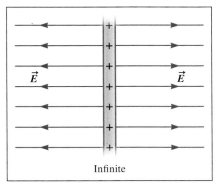

FIGURE 25.26

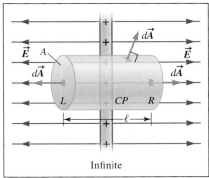

FIGURE 25.27

Example continues on page 772 ▶

:• **SOLVE**
Step 3 Integrate. The Gaussian cylinder is made up of three surfaces: the right cap R, the curved part CP, and the left cap L. The area vectors $d\vec{A}$ on the curved part are perpendicular to the electric field, so the integral over surface CP is zero.

$$\Phi_E = \oint \vec{E} \cdot d\vec{A}$$

$$\Phi_E = \int_L \vec{E} \cdot d\vec{A} + \int_{CP} \vec{E} \cdot d\vec{A} + \int_R \vec{E} \cdot d\vec{A}$$

$$\Phi_E = \int_L \vec{E} \cdot d\vec{A} + \int_R \vec{E} \cdot d\vec{A}$$

The area vectors on surface R point to the right, as does the electric field on that side of the sheet. The angle φ is thus zero for surface R. The area vectors on surface L point to the left, as does the electric field on that side of the sheet. The angle φ is thus also zero for surface L.

$$\Phi_E = \int_L E\, dA \cos 0 + \int_R E\, dA \cos 0$$

$$\Phi_E = \int_L E\, dA + \int_R E\, dA$$

The electric field is uniform, so E can be pulled outside both integrals. These integrals both equal the area A of an end cap.

$$\Phi_E = E \int_L dA + E \int_R dA$$

$$\Phi_E = EA + EA = 2EA \qquad (1)$$

Step 4 Find the **charge inside** the Gaussian surface. The sheet has a uniform surface charge density σ. To find the amount of charge enclosed by the Gaussian surface, multiply the surface charge density by the area A of an end cap.

$$q_{in} = \sigma A \qquad (2)$$

Substitute Equations (1) and (2) into Gauss's law (Eq. 25.7), and solve for E.

$$\Phi_E = \frac{q_{in}}{\varepsilon_0} \qquad (25.7)$$

$$2EA = \frac{\sigma A}{\varepsilon_0}$$

$$E = \frac{\sigma}{2\varepsilon_0} \checkmark \qquad (25.16)$$

:• **CHECK and THINK**
As expected, the area A of the end caps cancels out, so the electric field does not depend on the size of the Gaussian surface.

The infinite sheet of charge is used to model some real charge distributions. You will see Equation 25.16 often, so here are some things to know. First, $E = \sigma/2\varepsilon_0$ gives the magnitude of the electric field due to an infinite charged sheet. The direction of the field depends on the sign of the charge and is reversed when going from one side of the sheet to the other. If the sheet is positively charged, the field direction is away from the sheet. If the sheet is negatively charged, the field direction is toward the sheet.

Second, according to $E = \sigma/2\varepsilon_0$, the electric field due to an infinite charged sheet is uniform, with the same value at all positions. Of course, this is true for the ideal source—an infinite sheet. In practice, it is true for positions that are close to the surface of a finite sheet and not near an edge.

EXAMPLE 25.9 Two Infinite Sheets

Figure 25.28 shows a side view of two infinite sheets. The sheet on the left has excess positive charge uniformly distributed so that its surface charge density is $\sigma_+ = 48.0\,\mu\text{C}/\text{m}^2$. The sheet on the right has excess negative charge uniformly distributed so that its surface charge density is $\sigma_- = -24.0\,\mu\text{C}/\text{m}^2$. Find the net electric field at points A, B, and C.

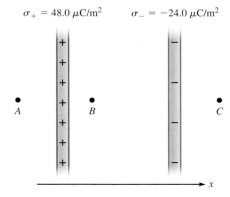

FIGURE 25.28

:• INTERPRET and ANTICIPATE

Because we already have found an expression for the electric field produced by an infinite sheet, there is no need to start with Gauss's law. The electric field at each point is a result of the electric field due to both sheets. For an infinite sheet, the magnitude of the electric field depends only on the magnitude of the surface charge density (Eq. 25.16). Because the positive sheet has a higher surface charge density, it produces a stronger electric field. Sketch the electric field vectors due to the positive sheet ($\vec{E}_+$) and the negative sheet ($\vec{E}_-$) at each point A, B, and C (Fig. 25.29). $\vec{E}_+$ points away from the positive sheet, and $\vec{E}_-$ points toward the negative sheet. Add these vectors graphically in order to anticipate the results.

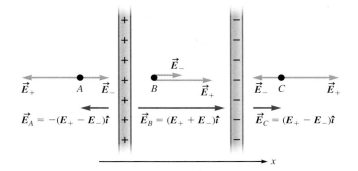

FIGURE 25.29

:• SOLVE

The magnitude of the electric field produced by each sheet is a constant given by $E = \sigma/2\varepsilon_0$ (Eq. 25.16).

$$E_+ = \frac{\sigma}{2\varepsilon_0} = \frac{48.0 \times 10^{-6}\,\text{C}/\text{m}^2}{2(8.85 \times 10^{-12}\,\text{C}^2/\text{N}\cdot\text{m}^2)}$$

$$E_+ = 2.71 \times 10^6\,\text{N}/\text{C}$$

$$E_- = \frac{\sigma}{2\varepsilon_0} = \frac{24.0 \times 10^{-6}\,\text{C}/\text{m}^2}{2(8.85 \times 10^{-12}\,\text{C}^2/\text{N}\cdot\text{m}^2)}$$

$$E_- = 1.36 \times 10^6\,\text{N}/\text{C}$$

Write the electric field at each point in component form, and then add the electric fields due to each sheet.

Point A: $\vec{E}_+$ due to the positive sheet points in the negative x direction, and $\vec{E}_-$ due to the negative sheet points in the positive x direction.

$$\vec{E}_A = E_-\hat{\imath} - E_+\hat{\imath} = (E_- - E_+)\hat{\imath}$$

$$\vec{E}_A = [(1.36 - 2.71) \times 10^6]\hat{\imath}\,\text{N}/\text{C}$$

$$\boxed{\vec{E}_A = -1.35 \times 10^6\,\hat{\imath}\,\text{N}/\text{C}}$$

Point B: Both $\vec{E}_+$ and $\vec{E}_-$ point in the positive x direction.

$$\vec{E}_B = E_+\hat{\imath} + E_-\hat{\imath} = (E_+ + E_-)\hat{\imath}$$

$$\vec{E}_B = [(2.71 + 1.36) \times 10^6]\hat{\imath}\,\text{N}/\text{C} = \boxed{4.07 \times 10^6\,\hat{\imath}\,\text{N}/\text{C}}$$

Example continues on page 774 ▶

Point C: $\vec{E}_+$ points in the positive x direction, and $\vec{E}_-$ points in the negative x direction.

$$\vec{E}_C = -E_-\hat{i} + E_+\hat{i} = (E_+ - E_-)\hat{i}$$

$$\vec{E}_C = [(2.71 - 1.36) \times 10^6]\hat{i} \text{ N/C} = 1.35 \times 10^6 \hat{i} \text{ N/C}$$

CHECK and THINK

Our results match the electric field vectors we found graphically (Fig. 25.29). At point A, the electric field points to the left (negative x direction), and at points B and C, the electric field points to the right (positive x direction). The magnitude of the electric field is greatest at point B, and the magnitude at point A equals the magnitude at point C.

As a final thought, imagine that the sheets were identical except that one was positive and the other negative. The magnitudes of the surface charge densities on the sheets would be equal, so $E_+ = E_-$ and the electric field outside the two sheets would cancel ($E_A = E_C = 0$). The electric field between the sheets would be reinforced ($E_B = 2E_+$ or $E_B = 2E_-$).

25-7 Special Case: Conductors

In this section, we consider the special case of a charged conductor. Imagine charging a solid spherical conductor such as a cannonball on an insulating pedestal (Fig. 25.30A), perhaps by touching a positively charged rod to the surface of the ball. Very quickly the excess charge on the cannonball spreads out because of the repulsive force between like charges. After the charges stop moving, the conductor is in **electrostatic equilibrium**. When a conductor is in electrostatic equilibrium, the excess charge is found only on its surface; there is no excess charge in the body of the conductor as shown in the cross-sectional view of the charged cannonball in Figure 25.30B.

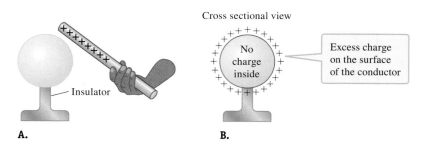

FIGURE 25.30 A. Charging an isolated conductor. **B.** The excess charge rests on the conductor's outer surface.

ELECTRICAL FIELD INSIDE A CONDUCTOR
▶ Special Case

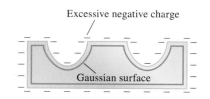

FIGURE 25.31 The excess charge is found only on the surface of this solid conductor, and the electric field inside is zero. We choose a Gaussian surface just inside the conductor.

According to Gauss's law, the electric field inside a charged conductor in electrostatic equilibrium is zero. If we choose a Gaussian spherical surface that lies just beneath the surface of the conductor, there is no charge inside it and the electric flux must be zero:

$$\Phi_E = \oint \vec{E} \cdot d\vec{A} = \frac{q_{\text{in}}}{\varepsilon_0} = 0$$

Because the area of the Gaussian sphere is *not* zero and the angle φ is not necessarily zero, the electric field inside the conductor must be zero:

$$\vec{E} = 0 \text{ inside a charged conductor}$$

The shape of the conductor and the sign of the excess charge don't change the fact that the electric field inside a conductor is zero. Figure 25.31 shows a solid isolated conductor. If you give the conductor excess negative charge, the excess charge will rest on the surface. Again, we can choose a Gaussian surface that lies just beneath the surface of the conductor. There is no charge inside that surface, so the electric field inside the conductor is zero.

Where is the excess charge if the conductor has a cavity, like a hollow locket (Fig. 25.32A)? Figure 25.32B shows a cross-sectional view of such a cavity within a conductor. In this case, the conductor has excess negative charge. To figure out how much charge is on the inner surface of the cavity, imagine embedding a Gaussian surface in the conductor just outside the cavity. There is no electric field in the body of the conductor, so the electric flux through the Gaussian surface must be zero:

$$\Phi_E = \oint \vec{E} \cdot d\vec{A} = \oint 0 \, d\vec{A} = 0$$

Because the electric flux is zero, there is no charge inside the Gaussian surface:

$$\Phi_E = \frac{q_{in}}{\varepsilon_0} = 0$$

So, there is no charge inside the cavity of a charged conductor and no electric field in the cavity either.

When you charge a conductor, the excess charge quickly distributes on the conducting surface. If the conductor is spherical or an infinitely large sheet, the surface charge density σ is uniform. However, if the conductor is some other shape, such as the conductor in Figure 25.31 or the heart-shaped locket in Figure 25.32, the surface charge density is greater at the more pointed parts of the conductor (Fig. 25.33). We expect that the electric field outside those pointed places is stronger. In fact, the electric field just outside a conductor is proportional to the local surface charge density.

A.

B.

FIGURE 25.32 A. An open locket. When the locket is closed, it has an empty cavity. Any excess charge is found on the outside surface of the closed locket, and the electric field inside the locket is zero. **B.** Gaussian surface just outside the cavity of a heart-shaped locket.

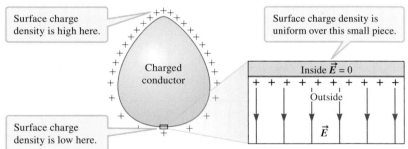

FIGURE 25.33 Excess charge is concentrated on the pointed parts of a conductor. *Inset*: Close to a charged conductor, the electric field is uniform.

EXAMPLE 25.10 $\vec{E}$ Just Outside a Charged Conductor

Use Gauss's law to show that the magnitude of the electric field just outside any charged conductor is

$$E = \frac{\sigma}{\varepsilon_0} \qquad (25.17)$$

ELECTRIC FIELD OUTSIDE A CONDUCTOR
  ▶ Special Case

Hint: Near the conducting surface, the surface appears to be a flat (Fig. 25.33, close-up), infinite plane, just as the surface of the ocean seems to be flat from our perspective standing on a beach. So, the surface charge density σ over a small part of the conductor is uniform.

Example continues on page 776 ▶

INTERPRET and ANTICIPATE

Step 1 Sketch the electric field lines. Figure 25.34 shows a close-up of a small portion of a charged conductor. Because the surface appears flat on this scale, the electric field must be perpendicular to the surface and directed outward. Inside the body of the conductor, the electric field is zero.

Step 2 Choose a Gaussian surface. Near the conductor, the source has planar symmetry. So, as in the case of an infinite sheet, we can choose any size Gaussian cylinder and place it anywhere. Only one such cylinder is shown here.

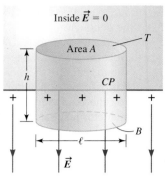

FIGURE 25.34

SOLVE

Step 3 Integrate. The Gaussian cylinder is made up of three surfaces—the top cap T, the curved part CP, and the bottom cap B.

$$\Phi_E = \oint \vec{E} \cdot d\vec{A}$$

$$\Phi_E = \int_T \vec{E} \cdot d\vec{A} + \int_{CP} \vec{E} \cdot d\vec{A} + \int_B \vec{E} \cdot d\vec{A}$$

The electric field over surface T is zero because the electric field is zero inside the conductor. The area vectors for the surface CP are perpendicular to the electric field outside the conductor, and the electric field inside the conductor is zero, so the integral over the surface CP is zero. This leaves only the integral over surface B.

$$\Phi_E = 0 + 0 + \int_B \vec{E} \cdot d\vec{A}$$

$$\Phi_E = \int_B \vec{E} \cdot d\vec{A}$$

The area vectors for surface B point downward in the same direction as $\vec{E}$, so $\varphi = 0$.

$$\Phi_E = \int_B E\, dA \cos\varphi = \int_B E\, dA \cos 0$$

The electric field is uniform over surface B, so E can be pulled outside the integral. The integral is the area A of the end cap.

$$\Phi_E = \int_B E\, dA = E \int_B dA$$

$$\Phi_E = EA \qquad (1)$$

Step 4 Find the charge inside the Gaussian surface. As in the case of the infinite sheet, the charge inside the Gaussian surface is the surface charge density times the area of the end cap.

$$q_{in} = \sigma A \qquad (2)$$

Substitute Equations (1) and (2) into Gauss's law (Eq. 25.7), and solve for E.

$$\Phi_E = \frac{q_{in}}{\varepsilon_0} \qquad (25.7)$$

$$EA = \frac{\sigma A}{\varepsilon_0}$$

$$E = \frac{\sigma}{\varepsilon_0} \checkmark \qquad (25.17)$$

CHECK and THINK

Equation 25.17 is valid for an infinite conductor with uniform charge density σ. It can also be used *just outside* a conductor of any shape, where σ is the local surface charge density. The electric field is perpendicular to the surface, pointing outward if the conductor is positively charged and inward if it is negatively charged.

A *capacitor* is a common device consisting of two large, oppositely charged conducting parallel plates. Imagine that initially the large plates are far apart and each plate has the same surface charge density σ_i (Fig. 25.35). Each plate has two surfaces, and when the plates are far apart, the excess charge is uniformly distributed on both surfaces of each plate as shown. The electric field due to the positive plate is perpendicular to the surfaces and outward, and the electric field due to the

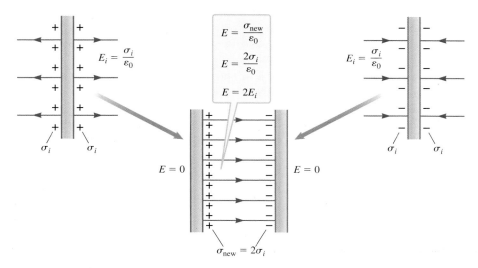

FIGURE 25.35 Two oppositely charged conducting plates, initially far apart, are brought close together (but not touching). The resulting device is a type of *capacitor*. We'll encounter these devices in Chapters 26 through 34.

negative plate is similarly perpendicular but inward. Because the plates have the same surface charge density, the magnitude of each one's electric field is the same, $E_i = \sigma_i/\varepsilon_0$ (Eq. 25.17).

Now imagine bringing these two charged plates closer together, keeping them parallel as shown in the lower middle portion of Figure 25.35. Because the plates have charges of opposite sign, the excess charge on each plate moves to the inner surfaces. This doubles the surface charge density on these inner surfaces: $\sigma_{new} = 2\sigma_i$. The electric field between the plates is still given by $E = \sigma/\varepsilon_0$ (Eq. 25.17), but the surface charge density is now σ_{new} on either plate, so the electric field has doubled:

$$E_{new} = \frac{\sigma_{new}}{\varepsilon_0} = \frac{2\sigma_i}{\varepsilon_0} \qquad (25.18)$$

$$E_{new} = 2E_i$$

Because no excess charge is left on either of the outer faces, the surface charge density there is zero. So, according to Equation 25.17, the electric field outside the two plates is zero (Fig. 25.35).

CONCEPT EXERCISE 25.7

Is it possible for the charged solid sphere in Figure 25.23 to be a conductor in electrostatic equilibrium? Explain.

EXAMPLE 25.11 Hanging Inside a Conductor

Figure 25.36 shows a cross-sectional view of a thick spherical conductor. The conductor is neutral, and a small charged sphere ($q = +29.5\ \mu C$) hangs from an insulating thread. The sphere is not in the center of the conductor; instead, it is closer to the left side as shown.

A Find the charge q_{wall} on the wall of the cavity and the charge q_{out} on the outer surface of the conductor. Start by sketching the electric field for all regions—inside the cavity, inside the body of the conductor, and outside the conductor. In **CHECK and THINK**, discuss this sketch.

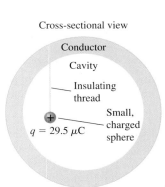

FIGURE 25.36

Example continues on page 778 ▶

INTERPRET and ANTICIPATE

The positive charge on the small sphere attracts electrons in the conductor. These electrons move close to the walls of the cavity. If the positively charged sphere were in the center of the cavity, the electrons would be uniformly distributed on the cavity wall. However, the electrons are more concentrated on the left side of the cavity because the positive sphere is closer to the left side (Fig. 25.37). The electric field in the body of any conductor in electrostatic equilibrium is zero. By choosing a Gaussian sphere that is concentric with the conductor and embedded in it, we can determine the amount of charge on the walls of the cavity.

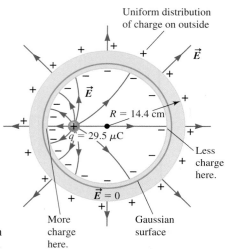

FIGURE 25.37 The Gaussian surface chosen for this problem is inside the conducting shell.

SOLVE

The electric flux through the Gaussian surface (Fig. 25.37) is zero because the electric field in the body of the conductor is zero. Therefore, the net charge inside the Gaussian sphere is zero, so the total charge on the inside wall of the cavity is negative, equal in magnitude to the charge on the small sphere inside.	$\Phi_E = \oint \vec{E} \cdot d\vec{A} = 0 = \dfrac{q_{in}}{\varepsilon_0}$ $q_{in} = 0 = q + q_{wall}$ $q_{wall} = -q = \boxed{-29.5\,\mu C}$
Because the conductor is neutral, the charge on its surface must be positive and equal to the charge on the cavity wall.	$q_{out} = -q_{wall}$ $q_{out} = -(-29.5\,\mu C) = \boxed{+29.5\,\mu C}$

The positive charge q_{out} on the outer surface of the conductor is uniformly distributed, as excess charge always is on the surface of a spherical conductor. To see why this is so, imagine that free electrons move toward the cavity wall, leaving positively charged ions in place.

CHECK and THINK

Figure 25.37 shows the electric field in all regions. Inside the cavity, the electric field is stronger on the left side where there is a higher concentration of charge. Outside the spherical conductor, the electric field looks like that of a charged particle located at the center of the entire collection of objects (and replacing them). This last sentence may seem confusing. Think of it this way. Suppose the positive sphere inside the conductor were at the center; then by symmetry the surface charge density on the outside of the conductor would be uniform. Because the electric field is zero inside the body of the conductor, moving the sphere off the center doesn't change the uniform charge distribution on the conductor's outer surface. (The particles on the outer surface having no way of "knowing" that the sphere moved.)

B If the radius of the conductor is $R = 14.4$ cm, what is the magnitude of the electric field just outside the conductor?

INTERPRET and ANTICIPATE

The electric field just outside any conductor depends only on the surface charge density, so we need to find the surface charge density so that we can find the electric field.

SOLVE

The charge is uniformly distributed on a sphere of radius R. We divide the charge q_{out} by the surface area of the sphere.	$\sigma = \dfrac{q_{out}}{A} = \dfrac{q_{out}}{4\pi R^2} = \dfrac{29.5 \times 10^{-6}\,C}{4\pi(14.4 \times 10^{-2}\,m)^2}$ $\sigma = 1.13 \times 10^{-4}\,C/m^2$

The magnitude of the electric field just outside a conductor is given by Equation 25.17.	$E = \dfrac{\sigma}{\varepsilon_0}$ (25.17)
	$E = \dfrac{1.13 \times 10^{-4}\ \text{C/m}^2}{8.85 \times 10^{-12}\ \text{C}^2/\text{N} \cdot \text{m}^2}$
	$= 1.28 \times 10^7\ \text{N/C}$
CHECK and THINK From Figure 25.37, we see that the electric field outside the conductor is equivalent to the electric field produced by a charged particle located at the center of the objects. We can check our result by using the relationship for electric field derived from Coulomb's law (Eq. 24.3) to calculate the field at a distance of 14.4 cm from such a fictitious particle with charge $+29.5\ \mu\text{C}$.	$E = \dfrac{kq}{r^2}$ (24.3) $E = \dfrac{(8.99 \times 10^9\ \text{N} \cdot \text{m}^2/\text{C}^2)(29.5 \times 10^{-6}\ \text{C})}{(14.4 \times 10^{-2}\ \text{m})^2}$ $E = 1.28 \times 10^7\ \text{N/C}$

EXAMPLE 25.12 CASE STUDY What Should Shannon Do?

Avi, Cameron, and Shannon are caught in a thunderstorm. Cameron thinks they should wait out the storm in the car. Avi is worried that if lightning hits the car, the gas tank may explode. Shannon is expected to make a decision. What would you decide?

SOLVE
Cameron is right. The car is made of metal, a conductor. It is true that the car has windows made of glass (an insulator), but the car can still be modeled as a conductor with a cavity. The inside of the car—where passengers sit—is like the cavity inside the locket (Fig. 25.32). When a car is struck by lightning, the charge stays on the outside surface of the car and the electric field inside is zero. The cavity inside a conductor is shielded from the external electric field (Fig. 25.38; such a device is called a *Faraday cage*). In addition, the gasoline is in a metal tank that acts as a Faraday cage.

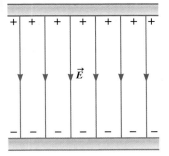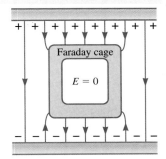

FIGURE 25.38 A Faraday cage is a hollow conductor that shields its interior from external electric fields.

We can use these ideas to explain how antistatic bags work (Fig. 25.5) and why your cell phone doesn't work in a closed elevator. The antistatic bag is coated with a conductor, and sensitive equipment is sealed inside. If the bag comes in contact with a charged object, the outside of the bag becomes charged but the inside remains uncharged, and the electric field inside the bag is zero. The antistatic bag is a Faraday cage. In order for your cell phone to work, it must be able to receive signals that consist of changing electric and magnetic fields. If your cell phone is inside a closed elevator, the elevator acts as a Faraday cage. The electric field inside the elevator is zero, so your cell phone cannot receive signals.

Avi is correct that the trees will attract lightning, and because trees are taller than people, they are more likely to get hit (Section 24-7). When a tree is struck by lightning, however, excess charge travels through the surface of the Earth. If you are in contact with the Earth nearby, considerable excess charge is likely to travel through you.
  Trees and people have been known to survive a direct lightning strike. If a tree is very wet, the water (which is not pure) acts as a Faraday cage. The excess charge stays on the outside coating of water. However, the large amount of energy delivered by lightning can cause the water to vaporize and burn the tree.

SUMMARY

❶ Underlying Principles

Gauss's law relates the electric flux that penetrates a closed surface to the charge enclosed by that surface:

$$\Phi_E = \oint \vec{E} \cdot d\vec{A} = \frac{q_{\text{in}}}{\varepsilon_0} \tag{25.11}$$

where Φ_E is the electric flux through a Gaussian surface.

✪ Major Concepts

1. A **Gaussian surface** is an imaginary, closed, three-dimensional surface.
2. **Electric flux** is represented by the number of field lines penetrating a surface and is given by

$$\Phi_E = \int \vec{E} \cdot d\vec{A} \tag{25.6}$$

For the net flux through a Gaussian surface,

$$\Phi_E = \oint \vec{E} \cdot d\vec{A} \tag{25.10}$$

where the symbol $\oint$ indicates that the integral is taken over a closed surface.

▷ Special Cases

1. Electric field due to sources with
   a. **Linear symmetry** (an infinitely long charged rod):

   $$\vec{E} = \frac{1}{2\pi\varepsilon_0} \frac{\lambda}{r} \hat{r} \tag{25.13}$$

   b. **Spherical symmetry** (a particle, sphere, or spherical shell)
   For a particle outside a sphere or shell,

   $$\vec{E} = \frac{1}{4\pi\varepsilon_0} \frac{q}{r^2} \hat{r} \tag{25.14}$$

   Inside a charged spherical shell, $\vec{E} = 0$ (the **shell theorem**).
   Inside ($r < R$) a solid sphere of radius R with uniform volume charge density,

   $$\vec{E} = \frac{1}{4\pi\varepsilon_0} \frac{q}{R^3} r\hat{r} \tag{25.15}$$

   c. **Planar symmetry** (infinite charged sheet):

   $$E = \frac{\sigma}{2\varepsilon_0} \tag{25.16}$$

2. Electric field inside and outside a charged conductor
   The **electric field inside a conductor** in electrostatic equilibrium is zero, $\vec{E} = 0$.

   Just **outside a charged conductor**,

   $$E = \frac{\sigma}{\varepsilon_0} \tag{25.17}$$

   where σ is the local surface charge density of a conductor of any shape, or the uniform surface charge density on an infinite conducting sheet.

PROBLEM-SOLVING STRATEGY Finding the Electric Field Using Gauss's Law

⋮ INTERPRET and ANTICIPATE
1. Sketch the electric field lines.
2. Choose a Gaussian surface.

⋮ SOLVE
3. Integrate $\Phi_E = \oint \vec{E} \cdot d\vec{A}$.
4. Determine the amount of **charge inside** the Gaussian surface.

⋮ CHECK and THINK
There are two checks:
1. Make sure the **dimensions** of the electric field are force per charge.
2. Compare to another known electric field.

PROBLEMS AND QUESTIONS

A = algebraic   C = conceptual   E = estimation   G = graphical   N = numerical

25-1 Qualitative Look at Gauss's Law

1. **C** Which word or name has the same symmetry as the letters in the name **ZAK**? (Explain your answer.)
   a. NUT
   b. SUE
   c. CAL
   d. BIG
2. **C** Describe the symmetry of the number **8**.

Problems 3, 4, and 5 are grouped.

3. **C** A nonconducting source is hidden inside a box. Your goal is to find the sign of the source's net charge. You suspend a small, lightweight, positively charged ball just outside each of the six faces of the box. The ball is attracted to all the faces. What do you conclude about the source's net charge?
4. **C** A nonconducting source is hidden inside a box. Your goal is to find the sign of the source's net charge. You suspend a small, lightweight, positively charged ball just outside each of the six faces of the box. The ball shows no deflection. What do you conclude about the source's net charge?
5. **C** A nonconducting source is hidden inside a box. Your goal is to find the sign of the source's net charge. You suspend a small, lightweight, positively charged ball just outside each of the six faces of the box. The ball is attracted to one face and repelled by the face on the opposite side. What do you conclude about the source's net charge?
6. **C** Discuss the case study on page 757 with your friends, roommates, or family. What do people think is the right decision—wait out the storm in the car or wait outside? What makes sense to you, and why?
7. **C** A positively charged sphere and a negatively charged sphere are in a sealed container. The only way the charged spheres can be examined is by observing the electric field outside the container.
   a. Given the depiction of the electric fields in Figure P25.7A, is the net electric flux through the container zero, positive, or negative? Explain your answer.
   b. Two different spheres are placed inside a container. Given the depiction of the electric fields in Figure P25.7B, is the net electric flux through the container zero, positive, or negative? Explain your answer.

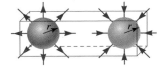
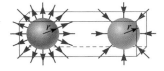

A.   B.
FIGURE P25.7

25-2 Flux

8. **N** A circular hoop of radius 0.50 m is immersed in a uniform electric field of 12.0 N/C. The electric field is at an angle of 30.0° to the plane of the hoop. Determine the electric flux through the hoop.
9. **N** A rectangular loop with dimensions 5.50 cm by 10.0 cm is placed in a uniform electric field and rotated to find the orientation that produces a maximum electric flux through the loop of 7.65×10^4 N·m²/C. What is the magnitude of this electric field?
10. **N** If the hemisphere (surface C) in Figure 25.10 (page 760) is tilted so that its disk-shaped cross section makes a 25° angle with the electric field, what is the electric flux through the hemisphere? Use Example 25.1 to check your result.
11. **N** A Ping-Pong paddle with surface area 3.80×10^{-2} m² is placed in a uniform electric field of magnitude 1.10×10^6 N/C.
    a. What is the magnitude of the electric flux through the paddle when the electric field is parallel to the paddle's surface?
    b. What is the magnitude of the electric flux through the paddle when the electric field is perpendicular to the paddle's surface?
12. The electric field in some region is $\vec{E} = (35\hat{\imath} + 70\hat{\jmath})$ N/C.
    a. **G** Sketch an electric field vector.
    b. **N, G** What is the maximum electric flux through a disk of radius 15 cm? Write an expression for $\vec{A}$, and sketch the disk along with the electric field vector.
    c. **N, G** Write an expression for $\vec{A}$ that produces no electric flux through the disk, and sketch the disk along with the electric field vector.

Problems 13, 14, and 15 are grouped.

13. **N** A pyramid has a square base with an area of 4.00 m² and a height of 3.5 m. Its walls are four isosceles triangles. The pyramid is in a uniform electric field of 655 N/C pointing downward (Fig. P25.13). What is the electric flux through the square base?
14. **N** For the pyramid in Problem 13, what is the electric flux through all four walls combined?
15. **N** For the pyramid in Problem 13, what is the electric flux through one of the four walls?

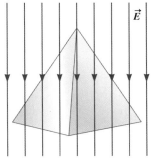

FIGURE P25.13 Problems 13, 14, and 15.

Problems 16 and 17 are paired.

16. **A** A circular loop with radius r is rotating with constant angular velocity ω in a uniform electric field with magnitude E. The axis of rotation is perpendicular to the electric field direction and is along the diameter of the loop. Initially, the electric flux through the loop is at its maximum value. Write an equation for the electric flux through the loop as a function of time in terms of r, E, and ω.
17. **A** A circular loop with radius r is rotating with constant angular velocity ω in a uniform electric field with magnitude E. The axis of rotation is perpendicular to the electric field direction and is along the diameter of the loop. Initially, the electric flux through the loop is zero. Write an equation for the electric flux through the loop as a function of time in terms of r, E, and ω.

25-3 Gauss's Law

18. **N** The net electric flux through a Gaussian surface is -456 N·m²/C. What is the net charge of the source inside the surface?

19. **N** What is the net electric flux through each of the four surfaces shown in Figure P25.19?

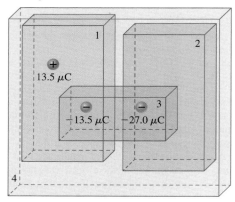

FIGURE P25.19

20. An isolated system consists of a single particle with charge 56.0 μC placed at the center of a cube of side 1.25 m.
    a. **N** What is the flux through each of the faces of the cube?
    b. **C** Would the answer to part (a) change if the particle was moved away from the center of the cube?

21. **N** The colored regions in Figure P25.21 represent four three-dimensional Gaussian surfaces A through D. The regions may also contain three charged particles, with $q_A = +5.00$ nC, $q_B = -5.00$ nC, and $q_C = +8.00$ nC, that are nearby as shown. What is the electric flux through each of the four surfaces?

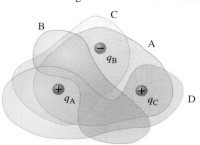

FIGURE P25.21

22. **N** A particle with a charge of 3.0 μC is placed at the center of a cubical Gaussian surface with side length 5.0 cm. Find the electric flux through one face of the cube.

23. **A** A particle with charge q is placed at a corner of a cube with side length a. Determine the net electric flux through the cube.

24. **N** Three particles and three Gaussian surfaces are shown in Figure P25.24. All the surfaces are three-dimensional. Use the net electric flux through each surface indicated on the figure to find the charge of each particle.

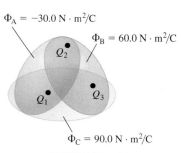

$\Phi_A = -30.0$ N·m²/C
$\Phi_B = 60.0$ N·m²/C
$\Phi_C = 90.0$ N·m²/C

FIGURE P25.24

Problems 25 and 26 are paired.

25. **A** Using Gauss's law, find the electric flux through each of the closed Gaussian surfaces A, B, C, and D shown in Figure P25.25.

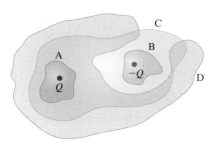

FIGURE P25.25

26. **N** Three point charges $q_1 = 2.0$ nC, $q_2 = -4.0$ nC, and $q_3 = -3.0$ nC are placed as shown in Figure P25.26. Find the electric flux through each of the closed Gaussian surfaces C_1, C_2, C_3, and C_4.

27. **N** A spherical shell of radius 1.00 m contains a single charged particle with $q = 78.0$ nC at its center.

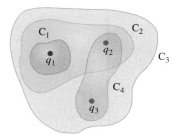

FIGURE P25.26

   a. What is the total electric flux through the surface of the shell?
   b. What is the total electric flux through any hemispherical portion of the shell's surface?

25-4 Special Case: Linear Symmetry

28. **N** A very long, thin wire fixed along the x axis has a linear charge density of 3.2 μC/m.
    a. Determine the electric field at point P a distance of 0.50 m from the wire.
    b. If there is a test charge $q_0 = +2.0$ μC at point P, what is the magnitude of the net force on this charge? In which direction will the test charge accelerate?

29. **A** Figure P25.29 shows a very long tube of inner radius a and outer radius b that has uniform volume charge density ρ. Find an expression for the electric field between the walls of the tube—that is, for $a < r < b$.

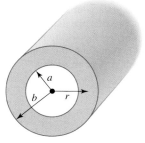

FIGURE P25.29

30. **G** Two very long, thin, charged rods lie in the same plane (Fig. P25.30). One rod is positively charged with charge per unit length $+\lambda$, and the other is negatively charged with charge per unit length $-\lambda$. The perpendicular distance between the rods is R. Using the coordinate system shown in the figure, sketch the electric field as a function of r from $-R$ to $+2R$.

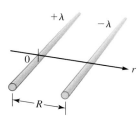

FIGURE P25.30

31. **N** A long, straight copper rod has charge per unit length $\lambda = 0.200\ \mu C/m$ and a diameter of 8.00 cm. What is the electric field at each of the following radial distances from the axis of the rod?
   a. 2.00 cm
   b. 8.00 cm
   c. 2.00 m

32. **N** Two long, thin rods each have linear charge density $\lambda = 6.0\ \mu C/m$ and lie parallel to each other, separated by 20.0 cm as shown in Figure P25.32. Determine the magnitude and direction of the net electric field at point P, a distance of 15.0 cm directly above the right rod.

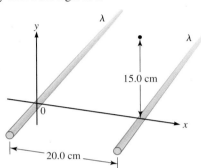

FIGURE P25.32

33. **A** Figure P25.33 shows a very long, thick rod with radius R, uniformly charged throughout. Find an expression for the electric field inside the rod ($r < R$). Use Equation 25.13,

$$\vec{E} = \frac{1}{2\pi\varepsilon_0}\frac{\lambda}{r}\hat{r}$$

to check your solution at the surface, where $r = R$.

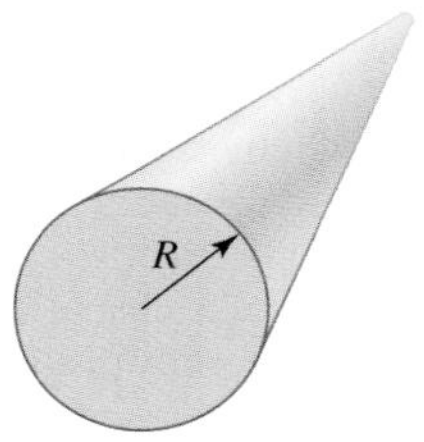

FIGURE P25.33

34. **A** A very long line of charge with a linear charge density, λ, is parallel to another very long line of charge with a linear charge density, -2λ. Both lines are parallel to the y-axis, and are the same distance r from the y-axis, where the first wire is to the left of the origin and the second is to the right. Use Gauss's law and the principle of superposition to find an expression for the magnitude of the electric field at the origin.

35. **N** Two infinitely long, parallel lines of charge with linear charge densities $3.2\ \mu C/m$ and $-3.2\ \mu C/m$ are separated by a distance of 0.50 m. What is the net electric field at points A, B, and C as shown in Figure P25.35?

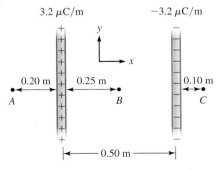

FIGURE P25.35

36. **N** An infinitely long wire with uniform linear charge density $\lambda = 2.80\ \mu C/m$ is surrounded by an insulating shell with inner radius $R_1 = 4.00$ cm and outer radius $R_2 = 6.00$ cm and uniform volume charge density $\rho = 1.05$ mC/m³. What is the electric field at radial distances of
   a. $r = 3.00$ cm,
   b. $r = 5.00$ cm, and
   c. $r = 10.0$ cm?

25-5 Special Case: Spherical Symmetry

37. **N** A particle with a charge of 55.0 μC is at the center of a thin spherical shell of radius $R = 12.0$ cm that has uniform surface charge density σ. Determine the value of σ such that the net electric field outside the shell is zero.

38. Excess charged particles are uniformly distributed throughout a sphere as shown in Figure 25.23 (page 770).
   a. **A** Show that at the surface of a sphere ($r = R$),

   $$\vec{E} = \frac{1}{4\pi\varepsilon_0}\frac{q}{r^2}\hat{r}\ \text{and}\ \vec{E} = \frac{1}{4\pi\varepsilon_0}\frac{q}{R^3}r\hat{r}$$

   (Eqs. 25.14 and 25.15) are identical.
   b. **G** Use Equations 25.14 and 25.15 to plot the electric field for a spherical charge distribution for all values of r.

Problems 39 and 40 are paired.

39. **N** A uniform spherical charge distribution (as shown in Fig. 25.23, page 770) has a total charge of 45.3 mC and radius $R = 15.2$ cm. Find the magnitude of the electric fields at $r = 0$, 7.60 cm, 15.2 cm, and 22.8 cm.

40. **N** For the uniform spherical charge distribution in Problem 39, find the electric force exerted on an electron placed at $r = 0$, 7.60 cm, 15.2 cm, and 22.8 cm.

Problems 41 and 42 are paired.

41. **N** Two uniform spherical charge distributions (Fig. P25.41) each have a total charge of 45.3 mC and radius $R = 15.2$ cm. Their center-to-center distance is 37.50 cm. Find the magnitude of the electric field at point A midway between the two spheres.

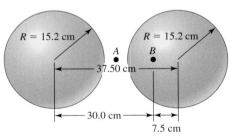

FIGURE P25.41 Problems 41 and 42.

42. **N** Two uniform spherical charge distributions (Fig. P25.41) each have a total charge of 45.3 mC and radius $R = 15.2$ cm. Their center-to-center distance is 37.50 cm. Find the magnitude of the electric field at point B, 7.50 cm from the center of one sphere and 30.0 cm from the center of the other sphere.

43. **A** The nonuniform charge density of a solid insulating sphere of radius R is given by $\rho = cr^2$ ($r < R$), where c is a positive constant and r is the radial distance from the center of the sphere. For a spherical shell of radius r and thickness dr, the volume element $dV = 4\pi r^2 dr$.
   a. What is the magnitude of the electric field outside the sphere ($r > R$)?
   b. What is the magnitude of the electric field inside the sphere ($r < R$)?

44. **N** A sphere with a radius of 0.230 m has a uniform charge density and a total charge of 70.9 mC. What is the magnitude of the electric field at each of the following locations:
   a. a distance of 0.100 m from the center,
   b. a distance of 0.230 m from the center, and
   c. a distance of 0.500 m from the center?

25-6 Special Case: Planar Symmetry

45. **N** What is the magnitude of the electric field just above the middle of a large, flat, horizontal sheet carrying a charge density of 98.0 nC/m²?
46. **N** What is the force on a proton placed at point A in Figure 25.28 (page 773)?
47. **N** The infinite sheets in Figure P25.47 are both positively charged. The sheet on the left has a uniform surface charge density of 48.0 μC/m², and the one on the right has a uniform surface charge density of 24.0 μC/m².
   a. What are the magnitude and direction of the net electric field at points A, B, and C?
   b. What is the force exerted on an electron placed at points A, B, and C?

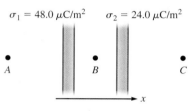

FIGURE P25.47

48. A large vertical sheet of nonconducting material has a uniform charge density of 47.0 μC/m².
   a. **N** What is the electric field a horizontal distance of 3.00 cm from the sheet?
   b. **C** How would the result in part (a) change if the distance from the sheet was increased by a factor of 10 or more?
49. **N** A charge-neutral, square aluminum plate with edge $d = 33.0$ cm is placed horizontally in a uniform electric field of 4.50×10^4 N/C directed vertically, or in the $+y$ direction.
   a. What is the charge density on each face of the aluminum plate?
   b. What is the total charge on each face of the aluminum plate?
50. **G** Two large sheets are perpendicular to each other (Fig. P25.50). One sheet has an excess charge per unit area of $-\sigma$, and the other has an excess charge per unit area of $+\sigma$. Sketch the electric field in the region bordered by the two sheets.

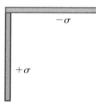

FIGURE P25.50

Problems 51 and 52 are paired.
51. A very large, flat slab has uniform volume charge density ρ and thickness $2t$. A side view of the cross section is shown in Figure P25.51.
   a. **A** Find an expression for the magnitude of the electric field inside the slab at a distance x from the center.
   b. **N** If $\rho = 2.00$ μC/m³ and $2t = 8.00$ cm, calculate the magnitude of the electric field at $x = 3.00$ cm.
52. **N** Find the surface charge density σ of a sheet of charge that would produce the same electric field as that of a very large flat slab of uniform charge density $\rho = 2.00$ μC/m³ and thickness $2t = 5.00$ cm (Fig. P25.51).

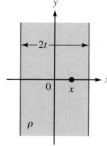

FIGURE P25.51 Problems 51 and 52.

25-7 Special Case: Conductors

53. **N** The electric field at a point 1.0 m from the center of a charged spherical conductor of radius 0.20 m is -12 N/C. Determine the number of excess electrons in the conductor.
54. **G** Model a metal can such as an empty oil barrel as having linear symmetry. If positive charge is uniformly distributed on the walls of the barrel, sketch the electric field both inside ($r < R$) and outside ($r \geq R$) the barrel as a function of r.
55. **N** If the magnitude of the surface charge density of the plates in Figure P25.55 is $\sigma = 99.5$ nC/m², what is the magnitude of the electric field between the plates? If an electron is placed between the plates, what is the magnitude of the electric force on it?

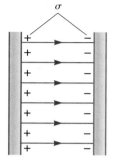

FIGURE P25.55

56. **N** A spherical conducting shell with a radius of 0.200 m has a very small charged sphere suspended in its center. The sphere has a charge of 24.6 mC, and the conducting shell has a charge of -24.6 mC. What is the magnitude of the electric field at distances of
   a. 0.100 m and
   b. 0.300 m from the center of the shell?
57. **N** A charged rod is placed in the center along the axis of a neutral metal cylinder (Fig. P25.57). The rod has a total charge of 38.3 μC uniformly distributed. What are the charges on the inner and outer surfaces of the metal cylinder? (Ignore the ends.)

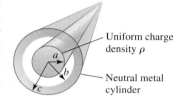

FIGURE P25.57 Problems 57 and 58.

58. A charged rod is placed in the center along the axis of a neutral metal cylinder (Fig. P25.57). The rod has positive charge uniformly distributed. (Ignore the ends.)
   a. **A** Find expressions for the electric fields in all regions: $r < a$, $a < r < b$, $b < r < c$, and $r > c$.
   b. **G** Plot your expressions on one graph. Is the electric field continuous or discontinuous? Explain.

Problems 59 and 60 are paired.
59. **N** A thick spherical conducting shell with an inner radius of 0.200 m and an outer radius of 0.250 m has a very small charged sphere suspended in its center. The sphere has a charge of 24.6 mC, and the conducting shell has a net charge of -24.6 mC. What is the net amount of charge that resides on
   a. the inner and
   b. the outer surface of the shell?
   What is the charge density on
   c. the inner and
   d. the outer surface of the shell?
60. **N** A thick spherical conducting shell with an inner radius of 0.200 m and an outer radius of 0.250 m has a very small charged sphere suspended in its center. The sphere has a charge of 24.6 mC, and the conducting shell has a net charge of -24.6 mC. What is the magnitude of the electric field at distances of
   a. 0.100 m,
   b. 0.225 m, and
   c. 0.500 m from the center of the shell?

General Problems

61. **N** A rectangular plate with sides 0.60 m and 0.40 m long is lying in the xy plane. The plate is placed in a uniform electric field $\vec{E} = (-4.0\hat{i} + 5.0\hat{j} + 3.0\hat{k})$ N/C. Calculate the flux through the plate.

62. **N** A circular loop of radius 15.0 cm is placed in a uniform electric field given by $\vec{E} = (5.65 \times 10^5 \text{ N/C})\hat{k}$. What is the electric flux through the loop if it is oriented
    a. parallel to the xy plane and
    b. parallel to the xz plane?
    c. What is the electric flux through the loop if it is oriented at a 45.0° angle to the xy plane?

63. **N** A total charge of 78.0 μC is uniformly distributed on the surface of a thin spherical shell of radius 22.0 cm. What is the magnitude of the electric field at distances of
    a. 5.00 cm and
    b. 44.0 cm from the center of the spherical shell?

64. **N** A uniform spherical charge distribution has a total charge of 45.3 mC and a radius of 15.2 cm. It is surrounded by a thin spherical shell with a uniform charge distribution. The uniform shell's net charge is 35.5 mC. The shell's radius is 20.2 cm, and it is concentric with the solid sphere. Find the electric field at points A and B located as shown on Figure P25.64.

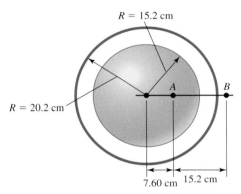

FIGURE P25.64

65. **N** A rectangular surface extends from $x = 0$ to $x = 50.0$ cm and from $y = 0$ to $y = 24.0$ cm in the xy plane in a region of space permeated by a nonuniform electric field given by $\vec{E} = (3.00z\hat{i} + 2.00y\hat{j} - 4.00x\hat{k})$ N/C. What is the electric flux through the surface?

66. **N** A uniform electric field $\vec{E} = 1.57 \times 10^4 \hat{i}$ N/C passes through a closed surface with a slanted top as shown in Figure P25.66.
    a. Given the dimensions and orientation of the closed surface, what is the electric flux through the slanted top of the surface?
    b. What is the net electric flux through the entire closed surface?

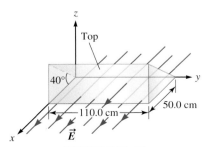

FIGURE P25.66

67. **N** A solid plastic sphere of radius $R_1 = 8.00$ cm is concentric with an aluminum spherical shell with inner radius $R_2 = 14.0$ cm and outer radius $R_3 = 17.0$ cm (Fig. P25.67). Electric field measurements are made at two points: At a radial distance of 34.0 cm from the center, the electric field has magnitude 1.70×10^3 N/C and is directed radially outward, and at a radial distance of 12.0 cm from the center, the electric field has magnitude 9.10×10^4 N/C and is directed radially inward. What are the net charges on
    a. the plastic sphere and
    b. the aluminum spherical shell?
    c. What are the charges on the inner and outer surfaces of the aluminum spherical shell?

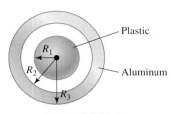

FIGURE P25.67

68. **C** Examine the summary on page 780. Why are conductors and charged sources with linear symmetry, spherical symmetry, and planar symmetry categorized as special cases rather than major concepts or underlying principles?

69. **N** A total charge of -33.0 μC is uniformly distributed throughout the volume of a solid sphere 60.0 cm in diameter.
    a. What is the magnitude of the electric field at the center of the sphere?
    What is the magnitude of the electric field
    b. 15.0 cm,
    c. 30.0 cm, and
    d. 90.0 cm from the center of the sphere?

Problems 70 and 71 are paired.

70. **N** A cube with sides of 20.0 cm is placed in a nonuniform electric field $\vec{E}(x,y,z) = -4.0x\hat{i} + 5.0y\hat{j} + 3.0z\hat{k}$, where x, y, and z are measured in meters and E is in N/C. Determine the electric flux through each of the six faces of the cube, assuming the cube has a corner at the origin and sides along the x, y, and z axes.

71. **N** A cube with side length 20.0 cm is placed in a uniform electric field $\vec{E}$, directed along the z axis. The cube is oriented so that the field is perpendicular to two of its sides. What is the net flux through the cube?

72. **A** A coaxial cable is formed by a long, straight wire and a hollow conducting cylinder with axes that coincide. The wire has charge per unit length $\lambda = 2\lambda_0$, and the hollow cylinder has net charge per unit length $\lambda = 3\lambda_0$. Use Gauss's law to answer these questions: What are the charges per unit length on
    a. the inner surface and
    b. the outer surface of the hollow cylinder?
    c. What is the electric field a radial distance d from the axis of the coaxial cable?

73. **N** The electric field due to a very long wire at a point 0.50 m from the wire is 12 N/C. Determine the amount of charge in a 15-cm length of the wire.

74. **N** Charge is distributed uniformly on the curved surface of a horizontal cylindrical shell 10.0 cm in diameter and 1.50 m in length. The midpoint of the axis of the cylinder is at the origin. The magnitude of the electric field at $y = 25.0$ cm is measured to be 2.35×10^4 N/C.
    a. What is the total charge on the cylindrical shell?
    b. What is the magnitude of the electric field at $y = 5.00$ cm?

Problems 75 and 76 are paired.

75. **A** A solid sphere of radius R has a spherically symmetrical, nonuniform volume charge density given by $\rho(r) = A/r$, where

r is the radial distance from the center of the sphere in meters, and A is a constant such that the density has dimensions M/L^3.
a. Calculate the total charge in the sphere.
b. Using the answer to part (a), write an expression for the magnitude of the electric field outside the sphere—that is, for some distance $r > R$.
c. Find an expression for the magnitude of the electric field inside the sphere at position $r < R$.

76. **G** A solid sphere of radius R has a spherically symmetrical, nonuniform volume charge density given by $\rho(r) = A/r$, where r is the radial distance from the center of the sphere in meters, and A is a constant such that the density has dimensions of M/L^3. Sketch a graph of the magnitude of the electric field as a function of distance for $0 < r < 3R$.

77. **N** A very large, horizontal conducting square plate with sides of length 2.0 m and thickness 3.0 mm is given a total charge of 40.0 μC. Calculate the magnitude of the electric field near the center of the plate and just above its surface.

78. **A** The charge density within a spherically symmetrical charge distribution is given by $\rho = c/r^2$, where r is the radial distance from the center of the distribution and c is a positive constant. What is the electric field as a function of radial distance within this charge distribution?

79. **N** A particle with charge $q = 7.20$ μC is surrounded by a spherical shell of radius $R = 1.50$ m. What is the electric flux through the spherical cap with half angle $\theta = 30.0°$ (Fig. P25.79)?

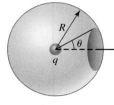

FIGURE P25.79

80. **A** A sphere with radius R has a charge density given by $\rho = cr^3$. Use Gauss's law to find an expression for the magnitude of the electric field at a distance r from the center of the sphere, where **a.** $r < R$ and **b.** $r > R$.

Electric Potential

26

❶ Underlying Principles

No new principles are introduced in this chapter.

✪ Major Concepts

1. Electric potential energy
2. Electric potential
3. Finding the electric potential V from the electric field $\vec{E}$
4. Finding the electric field $\vec{E}$ from the electric potential V

▷ Special Cases

1. Electric potential energy for a system of two charged particles
2. Electric potential due to
   a. Source with spherical symmetry
   b. Dipole
   c. Ring
   d. Disk
   e. Source with planar symmetry

⊙ Tools

Contour map

Key Questions

What are electric potential and electric potential energy?

How are electric potential and electric potential energy related to the electric force and the electric field?

How do electric potential and electric potential energy make our study of electricity easier?

26-1	Scalars versus vectors 788
26-2	Gravity analogy 788
26-3	Electric potential energy U_E 791
26-4	Electric potential V 796
26-5	Special case: Electric potential due to a collection of charged particles 801
26-6	Electric potential due to a continuous distribution 804
26-7	Connection between electric field $\vec{E}$ and electric potential V 807
26-8	Finding $\vec{E}$ from V 813
26-9	Graphing $\vec{E}$ and V 817

You have heard of voltage (electric potential). A car battery is 12 volts (V). The D batteries in your flashlight and the AAA batteries in your calculator are 1.5 V. In Europe the wall outlet is 240 V, and in the United States it is 120 V. On average, there is about 1 mV across your chest and 70 mV across your nerve cells. What do these numbers mean? In this chapter, we explain how these numbers are related to electric fields and forces.

This chapter is about *electric potential* (also called *voltage*), a scalar quantity. We know from our experience with forces (vectors) and energy (scalars) that using scalars allows us to analyze complicated situations. The concept of electric potential enables us to study complicated systems such as the circuits in common devices like flashlights and cameras as well as the human body.

26-1 Scalars Versus Vectors

In mechanics, we focused on forces, drew free-body diagrams, and used Newton's second law (Chapters 5 and 6). Then, in Chapter 8, we introduced energy, a scalar. For a conservative force such as gravity, we associated potential energy with the configuration of a system. The benefit of this new scalar was that we could use the conservation of energy principle to study complicated situations.

Our study of electricity will follow a similar path. We have learned to manipulate two vector quantities—electrostatic force $\vec{F}_E$ and electric field $\vec{E}$—with tools similar to those we used in mechanics. Like gravity, the electrostatic force is conservative, so now we can apply the conservation of energy principle to electricity, which allows us to study complicated and practical situations. (In Problem 62, you will show that the electrostatic force passes the tests for conservative forces.) We now introduce two scalar quantities: (1) *potential energy*, which is familiar from mechanics, and (2) *electric potential*, also known as *voltage*—an entirely new idea. Because gravity is similar to the electrostatic force both conceptually and mathematically, we use gravity throughout this chapter as an analogy.

CASE STUDY Electric Potential in the Human Body

As in a car, much of the mechanics of the human body is run by a complex electrical system. The first evidence that electricity plays a role in muscle action was discovered in 1786 by the Italian anatomist Luigi Galvani. Galvani connected the legs of a dead frog to pieces of two different metals (forming a crude battery) and found that the legs contracted. A dispute between Galvani and the Italian physicist Alessandro Volta led Volta to invent the battery.

The role that electricity plays in the human body goes beyond muscle contraction. Electricity controls organs and nerves. The brain—and therefore human thought—is controlled by electricity. The senses take in information and transmit that information via electrical signals through nerves. The brain sends electrical signals through nerves to the rest of the body. As in mechanics, a complex system like the human body is often best studied in terms of scalar quantities. In this chapter's case study, we consider parts of the body's electrical system in terms of its electric potential.

26-2 Gravity Analogy

Imagine a system consisting of the Earth and a spacecraft at some distance r from the Earth's center, with the Earth as the source and the spacecraft as the subject. We associate four quantities with this situation: (1) gravitational force $\vec{F}_G$, (2) gravitational field $\vec{g}$, (3) gravitational potential energy U_G, and (4) gravitational potential V_G. These four quantities are arranged in a pattern in Figure 26.1. The first row is for vectors: gravitational force $\vec{F}_G$ and gravitational field $\vec{g}$. The second row is for scalars: gravitational potential energy U_G and gravitational potential V_G. The first column is for quantities that depend on both the source and the subject: gravitational force $\vec{F}_G$ and gravitational potential energy U_G. The second column is for quantities that depend only on the source: gravitational field $\vec{g}$ and gravitational potential V_G.

All but the last quantity—gravitational potential V_G—are familiar from earlier chapters. The **gravitational potential** V_G, like the gravitational potential energy, is a scalar quantity. The most important difference between the gravitational potential energy and the gravitational potential is that *the gravitational potential depends only on the source*. The gravitational potential is like the gravitational field in that both depend on the mass of the source, but neither depends on the mass of the subject.

The terms *gravitational potential energy* and *gravitational potential* are very similar; you must pay close attention when reading so you do not confuse them. One way to make sure you are reading carefully is to highlight the word *energy* every time you see it.

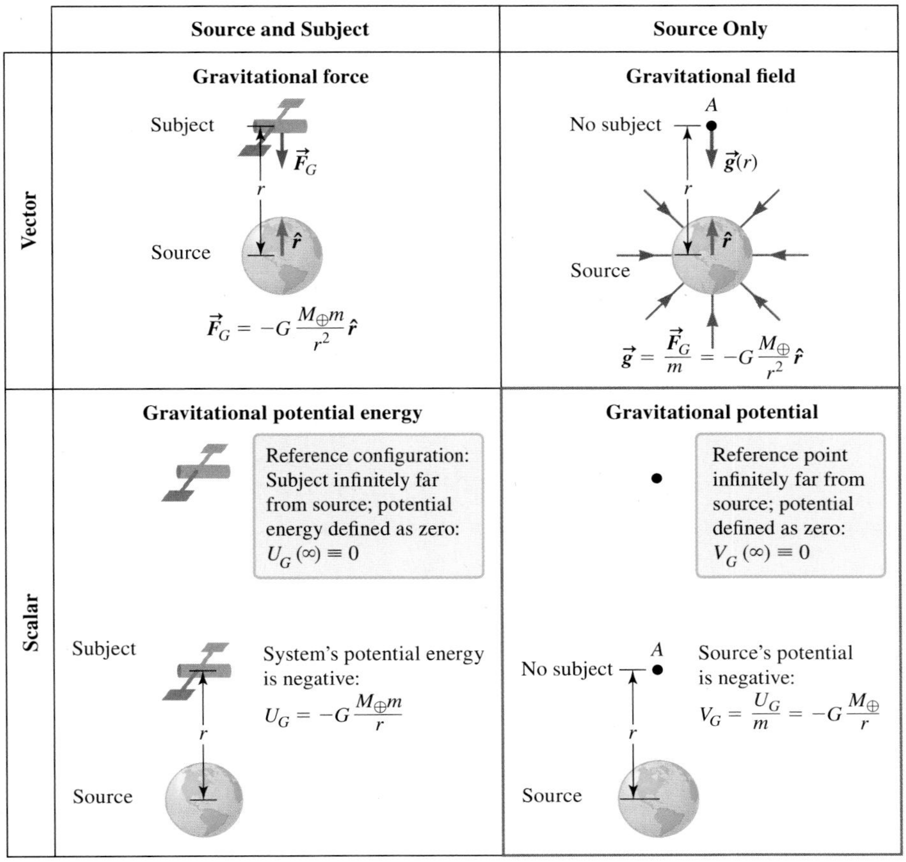

26-2 Gravity Analogy

	Source and Subject	Source Only
Vector	**Gravitational force** Subject — $\vec{F}_G$ Source — $\hat{r}$ $$\vec{F}_G = -G\frac{M_\oplus m}{r^2}\hat{r}$$	**Gravitational field** No subject — A, $\vec{g}(r)$ Source — $\hat{r}$ $$\vec{g} = \frac{\vec{F}_G}{m} = -G\frac{M_\oplus}{r^2}\hat{r}$$
Scalar	**Gravitational potential energy** Reference configuration: Subject infinitely far from source; potential energy defined as zero: $U_G(\infty) \equiv 0$ System's potential energy is negative: $$U_G = -G\frac{M_\oplus m}{r}$$	**Gravitational potential** Reference point infinitely far from source; potential defined as zero: $V_G(\infty) \equiv 0$ Source's potential is negative: $$V_G = \frac{U_G}{m} = -G\frac{M_\oplus}{r}$$

FIGURE 26.1 The gravitational influence of the Earth can be described in terms of two vector quantities and two scalar quantities.

Just as we can find the gravitational field by dividing the gravitational force by the mass m of a test particle, we can find the gravitational potential by dividing the gravitational potential energy by the mass of a test subject:

$$V_G = \frac{U_G}{m} \tag{26.1}$$

So, dividing $U_G = -GM_\oplus m/r$ (Eq. 8.7) by m, we find that the gravitational potential is given by

$$V_G = -G\frac{M_\oplus}{r} \tag{26.2}$$

The gravitational potential has the dimensions energy per mass, so it has the SI units J/kg.

Only *differences* in the potential are physically meaningful, just like only differences in potential energy are physically meaningful. When working with potential, it is usually convenient to set a reference point at which the potential is defined to be zero. In the case of universal gravity, the most convenient reference point is at infinity: $V_G(\infty) \equiv 0$. At all other points, the gravitational potential is negative and given by Equation 26.2.

CONCEPT EXERCISE 26.1

Complete the analogies by filling in the blanks, and explain your answers. *Hint*: Consult Figure 26.1.

a. Gravitational force is to gravitational field as gravitational potential energy is to _____.

b. Gravitational force is to gravitational potential energy as gravitational field is to _____.

EXAMPLE 26.1 Gravitational Potential and Gravitational Potential Energy Near the Earth

A Find the gravitational potential at the surface of the Earth ($r = R_\oplus$). Compare it to the gravitational potential at $r = 2R_\oplus$ and at $r = 3R_\oplus$.

INTERPRET and ANTICIPATE
Use Equation 26.2 three times to find the gravitational potential at three places near the source (the Earth).

SOLVE
Substitute values to find V_1, the gravitational potential at the surface of the Earth ($r = R_\oplus$).

$$V_G = -G\frac{M_\oplus}{r} \qquad (26.2)$$

$$V_1 = -G\frac{M_\oplus}{R_\oplus} = -\frac{(6.67 \times 10^{-11}\,\text{N}\cdot\text{m}^2/\text{kg}^2)(5.97 \times 10^{24}\,\text{kg})}{(6.38 \times 10^6\,\text{m})}$$

$$V_1 = -6.26 \times 10^7\,\text{J/kg} \text{ at the Earth's surface}$$

The term in parentheses at right equals V_1. Write V_2, the gravitational potential at $r = 2R_\oplus$, in terms of V_1 as found above.

$$V_2 = -G\frac{M_\oplus}{2R_\oplus} = \frac{1}{2}\left(-\frac{GM_\oplus}{R_\oplus}\right) = \frac{1}{2}V_1$$

$$V_2 = -3.13 \times 10^7\,\text{J/kg at } 2R_\oplus$$

Similarly, write V_3, the gravitational potential at $r = 3R_\oplus$, in terms of V_1 and substitute.

$$V_3 = -G\frac{M_\oplus}{3R_\oplus} = \frac{1}{3}\left(-\frac{GM_\oplus}{R_\oplus}\right) = \frac{1}{3}V_1$$

$$V_3 = -2.09 \times 10^7\,\text{J/kg at } 3R_\oplus$$

CHECK and THINK
Because of the convention that the gravitational potential is zero at infinity and negative at all other points, the gravitational potential is lowest at the surface of the Earth and greater (less negative) at points that are farther away.

B Now imagine a system that consists of the Earth and subject 1 with mass $m_1 = 15.0$ kg. Find the gravitational potential energy of the system when subject 1 is at the surface of the Earth ($r = R_\oplus$). Compare your result to the gravitational potential energy when subject 2 with mass $m_2 = 30.0$ kg is at $r = 2R_\oplus$ and subject 3 with mass $m_3 = 45.0$ kg is at $r = 3R_\oplus$. As part of the **CHECK and THINK** step, answer this question: If the mass of the subject were the same in each case—for instance, $m_1 = m_2 = m_3 = 15.0$ kg—what would change and what would stay the same?

INTERPRET and ANTICIPATE
Use the gravitational potential found in part A to find the gravitational potential energy in the three cases. According to Equation 26.1, multiplying the gravitational potential by the mass of the subject gives the gravitational potential energy.

SOLVE
To find the gravitational potential energy U_1 when subject 1 is at the surface, multiply V_1 by the mass m_1 of the first subject.

$$V_G = \frac{U_G}{m} \qquad (26.1)$$

$$U_G = mV_G$$

$$U_1 = m_1 V_1 = (15.0\,\text{kg})(-6.26 \times 10^7\,\text{J/kg}) = -9.39 \times 10^8\,\text{J}$$

Find the gravitational potential energy U_2 when subject 2 is at $r = 2R_\oplus$ by multiplying the mass of subject 2 by the gravitational potential at this position. Write the result in terms of U_1.

$$U_2 = m_2 V_2 = (30.0\,\text{kg})(-3.13 \times 10^7\,\text{J/kg}) = -9.39 \times 10^8\,\text{J}$$

$$U_2 = U_1$$

Find the gravitational potential energy U_3 when subject 3 is at $r = 3R_\oplus$ by multiplying the mass of subject 3 by the gravitational potential at this position. Write the result in terms of U_1. (If we ignore the slight rounding error, $U_3 = U_1$.)

$U_3 = m_3 V_3 = (45.0 \text{ kg})(-2.09 \times 10^7 \text{ J/kg}) = -9.40 \times 10^8 \text{ J}$

$U_3 = U_1$

CHECK and THINK

The gravitational potential depends only on the Earth, and V_G increases (becomes less negative) for points that are farther from the surface of the Earth. The gravitational potential *energy* U_G, however, is the same in all three cases because that energy depends on both the gravitational potential and the mass of the subject. The masses of the three subjects were carefully chosen to compensate for the variation in the gravitational potential. If all three masses were equal ($m_1 = m_2 = m_3 = 15.0$ kg), the gravitational potential energy of the system at position 1 ($r = R_\oplus$) would be unchanged. However, the gravitational potential energy at the other positions would be $U_2 = U_1/2$ and $U_3 = U_1/3$, mirroring the relationships we found in part A for the gravitational potential.

26-3 Electric Potential Energy U_E

Now we use gravitational potential energy (lower left panel in Fig. 26.1) as an analogy for *electric potential energy*. Like gravity, the electrostatic force is conservative, so we can associate the potential energy with it. The **electric potential energy** U_E—like the gravitational potential energy U_G—is a scalar that depends on both the source and the subject. In the case of gravity, the important property of each object is its mass. In the case of electricity, the important property is the charge. Electric potential energy depends on the charge of both the source and the subject. Also, like gravity, electric potential energy depends on the configuration of the system.

In this section, we find the electric potential energy of a system that consists of two objects: a spherical source with charge Q and a particle with charge q. Although this is a special case, in subsequent sections we will find the electric potential energy associated with systems that have sources of other shapes.

ELECTRIC POTENTIAL ENERGY
✪ Major Concept

The SI unit for electric potential energy is the same as the SI unit for any type of energy—the joule (J).

Special Case: Electric Potential Energy Involving a Charged Spherical Source

We can find an expression for the electric potential energy of the charged particle-sphere system in two ways. First, by comparing the gravitational force exerted on a spacecraft $\vec{F}_G = -(GM_\oplus m/r^2)\hat{r}$ (Eq. 7.4) to the electrostatic force exerted on a positively charged particle $\vec{F}_E = (kQq/r^2)\hat{r}$ (Eq. 23.5), we can come up with a translation between the variables:

$$-G \to k$$
$$M_\oplus \to Q$$
$$m \to q$$

We can use this translation of variables in Equation 8.7 to find an expression for the electric potential energy of the charged particle-sphere system:

$$U_G = (-G)\frac{M_\oplus m}{r} \qquad (8.7)$$

$$\downarrow \quad \downarrow \quad \downarrow$$

$$U_E = k\frac{Qq}{r} \qquad (26.3)$$

DERIVATION: Electric Potential Energy for a Particle-Sphere System

Our second way of arriving at Equation 26.3 is more formal than the first. Using the same procedure we used in Section 8-4 to find the gravitational potential energy of a spacecraft-Earth system, we show that the electric potential energy for the particle-sphere system is:

$$U_E(r) = \frac{kQq}{r} \quad (26.3)$$

ELECTRIC POTENTIAL ENERGY FOR A SYSTEM OF TWO CHARGED PARTICLES ▶ Special Case

The change in potential energy is found by taking the path integral of any conservative force from an initial position r_i to a final position r_f (Eq. 8.3).	$\Delta U = -\int_{r_i}^{r_f} F_r dr \quad (8.3)$
Substitute Coulomb's law $F_E = kQq/r^2$ (Eq. 23.5) for F_r.	$\Delta U_E = -\int_{r_i}^{r_f} F_E dr \quad (26.4)$ $\Delta U_E = -\int_{r_i}^{r_f}\left(k\frac{Qq}{r^2}\right) dr = -kQq\int_{r_i}^{r_f}\left(\frac{1}{r^2}\right) dr$
Integrate, substitute limits, and reduce.	$\Delta U_E = -kQq\left[-\frac{1}{r}\right]\Big|_{r_i}^{r_f} = kQq\left[\frac{1}{r}\right]\Big|_{r_i}^{r_f}$ $\Delta U_E = \frac{kQq}{r_f} - \frac{kQq}{r_i}$
Using the usual convention that the electric potential energy is zero when the particle and sphere are infinitely far apart, we imagine that the particle started at infinity ($r_i = \infty$) and its final position is $r_f = r$.	$\Delta U_E = U_E(r) - U_E(\infty) = U_E(r) - 0$ $\Delta U_E = U_E(r)$ $U_E(r) = \frac{kQq}{r}$ ✓ $\quad (26.3)$

∴ COMMENTS

Although we assumed that the source is a sphere with charge Q and the subject is a particle with charge q, Equation 26.3 gives the **electric potential energy** for any system that consists of two charged, spherically symmetrical objects. For convenience, we'll refer to these objects as particles.

The electric potential energy of a particle-sphere system (Eq. 26.3) is mathematically similar to the gravitational potential energy of the spacecraft-Earth system: $U_G = -GM_\oplus m/r$ (Eq. 8.7). Both the gravitational potential energy and the electric potential energy are zero when the particles are infinitely far apart. However, there is one major difference when $r < \infty$. In that case, the gravitational potential energy is negative, whereas the electric potential energy may be negative or positive depending on the signs of the charges. If Q and q have the same sign, the system's potential energy is positive, but if Q and q have opposite signs, the system's potential energy is negative. This difference between the gravitational potential energy and the electric potential energy arises because gravity is always an attractive force, whereas the electric force is attractive only when the source and the subject have charges of opposite sign.

CONCEPT EXERCISE 26.2

A system consisting of a charged sphere and a proton has positive electric potential energy $U_E = +15.0 \times 10^{-20}$ J (with $U_E(\infty) = 0$).
  a. What is the sign of the sphere's charge?
  b. If the proton were replaced by an electron, what would be the new system's electric potential energy?

Special Case: Electric Potential Energy Stored in a Collection of Charged Particles

If a system consists of more than two charged particles, you can find the system's electric potential energy by applying $U_E(r) = kQq/r$ (Eq. 26.3) to each pair and adding the results. Because electric potential energy (like any type of energy) is a scalar, there are no directions to take into account.

One way to understand the electric potential energy stored in a collection of charged particles is to imagine assembling the collection from particles that are infinitely far apart. Starting with the first particle in place and the other particles infinitely far away, imagine moving each particle into place one at a time, keeping a running tally of the system's total potential energy as shown in the next example.

EXAMPLE 26.2 System of Four Charged Particles

Calculate the electric potential energy of the system shown in Figure 26.2. Model all four objects as charged particles.

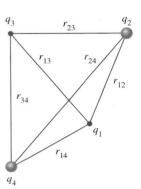

FIGURE 26.2

:• **INTERPRET and ANTICIPATE**
We expect to find a sum of terms, each with the form of Equation 26.3.

:• **SOLVE**

Start with particle 1 (Fig. 26.3). The system's potential energy is initially zero because the particles are infinitely far apart: $U_i = U(\infty) \equiv 0$.	FIGURE 26.3 $\quad q_1 \bullet \quad U_{tot} = 0$
Next, we imagine that particle 2 moves into place so that it is at a distance r_{12} from particle 1 (Fig. 26.4).	 FIGURE 26.4
The system's electric potential energy U_{12} after these two particles are in place is given by Equation 26.3.	$U_{12} = k\dfrac{q_1 q_2}{r_{12}}$
Initially, the system's electric potential energy was zero because the particles were infinitely far apart.	$U_i = U(\infty) \equiv 0$
The system's change in potential energy is equal to the final potential energy. So far, the system's total electric potential energy is U_{12} because the other two particles (q_3 and q_4) are infinitely far away.	$\Delta U_{\text{2nd particle}} = U_f - U_i$ $\Delta U_{\text{2nd particle}} = U_{12} - U(\infty) = U_{12}$

Example continues on page 794 ▶

Next, imagine that particle 3 comes into place (Fig. 26.5)

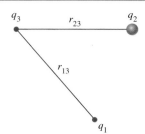

FIGURE 26.5

Once particle 3 is in position, we find the change in the system's potential energy by adding the potential energies of each pair that includes particle 3.	$\Delta U_{\text{3rd particle}} = U_{13} + U_{23}$
Before particle 3 moved into place, the system already had electric potential energy U_{12} due to the first two particles. So, the total electric potential energy of the system (so far) comes from adding the energy the system had when just q_1 and q_2 were present to the change the system undergoes when q_3 is added.	$U_{\text{tot}}(3 \text{ particles}) = \Delta U_{\text{3rd particle}} + U_{12}$ $U_{\text{tot}}(3 \text{ particles}) = U_{13} + U_{23} + U_{12}$

Finally, imagine that particle 4 comes into place (Fig. 26.6).

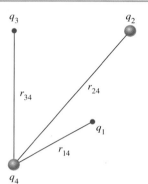

FIGURE 26.6

Once particle 4 is in position, we find the change in the system's potential energy by adding the potential energies of each pair involving particle 4.	$\Delta U_{\text{4th particle}} = U_{14} + U_{24} + U_{34}$ (26.5)
Before particle 4 moved into place, the system had the electric potential energy due to the first three particles. So, the total electric potential energy of the system comes from adding the energy the system had with only q_1, q_2, and q_3 present to the change the system undergoes when q_4 is added.	$U_{\text{tot}}(4 \text{ particles}) = \Delta U_{\text{4th particle}} + (U_{13} + U_{23} + U_{12})$ $U_{\text{tot}}(4 \text{ particles}) = (U_{14} + U_{24} + U_{34}) + (U_{13} + U_{23} + U_{12})$ $U_{\text{tot}}(4 \text{ particles}) = U_{12} + U_{13} + U_{14} + U_{23} + U_{24} + U_{34}$
Each term is given by $U_E(r) = kQq/r$ (Eq. 26.3), and the electric potential energy of the system shown in Figure 26.6 is given by this expression for U_{tot}.	$U_{\text{tot}} = k\dfrac{q_1 q_2}{r_{12}} + k\dfrac{q_1 q_3}{r_{13}} + k\dfrac{q_1 q_4}{r_{14}} + k\dfrac{q_2 q_3}{r_{23}} + k\dfrac{q_2 q_4}{r_{24}} + k\dfrac{q_3 q_4}{r_{34}}$

:• CHECK and THINK

Equation 26.3 is used once for each pair in the system, and the total electric potential energy is the sum of all such terms. The total electric potential energy may be positive, negative, or zero depending on the signs and magnitudes of all the charges in the collection.

Another way to think about the electric potential energy of a collection of charged particles is to imagine the amount of work W_{tot} necessary for an external force to assemble the collection. As in Example 26.2, consider a system of four particles that are initially infinitely far apart. An external force—perhaps a person—brings the four particles together one at a time. The system's kinetic energy does not change in any of the steps—the particles are initially at rest when they are infinitely far apart, and they are at rest when they are in their places in the system. Therefore, the work done by the external force equals the change in the system's potential energy, which is U_{tot} because initially the potential energy is zero, $U(\infty) = 0$:

$$W_{tot} = U_f - U_i = U_{tot} - U(\infty)$$

$$W_{tot} = U_{tot}$$

So, for the four-particle system in Figure 26.6, the work done *by an external force* is

$$W_{tot} = k\frac{q_1 q_2}{r_{12}} + k\frac{q_1 q_3}{r_{13}} + k\frac{q_1 q_4}{r_{14}} + k\frac{q_2 q_3}{r_{23}} + k\frac{q_2 q_4}{r_{24}} + k\frac{q_3 q_4}{r_{34}}$$

Remember that this is the work done by an *external* force, not the internal electric force each particle exerts on the other particles in the system.

The work done by the external force may be positive, negative, or zero depending on the signs and magnitudes of all the charges. If the work done is positive, energy is transferred from the environment to the system, as is the case when the charges have the same sign. Imagine bringing two positively (or two negatively) charged particles together. The particles exert a repulsive force on each other, and the external force must do positive work to bring them together. In other words, the system's electric potential energy increases as a result of bringing two like charges together.

If the work done is negative, energy is transferred out of the system and into the environment, as is the case when the particles have opposite signs. A negatively and a positively charged particle exert an attractive force on each other just as two massive objects exert an attractive gravitational force on each other. The work done by the external force in this case is negative; in other words, the system's electric potential energy decreases as a result of bringing the particles together. This is exactly like when you lower a brick off the back of a truck to the ground. If the system consists of the brick and the Earth, you do negative work in lowering the brick, and the system loses gravitational potential energy.

CONCEPT EXERCISE 26.3

A water molecule is made up of two hydrogen atoms and one oxygen atom, with a total of 10 electrons and 10 protons. The molecule is modeled as a dipole with an effective separation $d = 3.9 \times 10^{-12}$ m between its positive and negative particles. What is the electric potential energy stored in the dipole? What does the sign of your answer mean?

EXAMPLE 26.3 Particles in a Triangle

A Three negatively charged particles are held together at the vertices of an equilateral triangle with sides $r = 0.09$ m (Fig. 26.7). The magnitudes of the charges are q, $2q$, and $3q$ where $q = -360$ nC. Find the electric potential energy of the system shown in Figure 26.7.

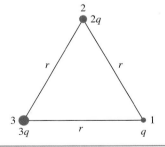

FIGURE 26.7

INTERPRET and ANTICIPATE
Conceptually, we can think of bringing the three particles together from infinity. Because all the charges have the same sign, an external force would do positive work on the system in order to put this configuration together. Because the system's electric potential energy would increase in the assembly process, we expect $U_{tot} > 0$.

Example continues on page 796 ▶

SOLVE
Apply $U_E(r) = kQq/r$ (Eq. 26.3) to each pair of charges, and add.

$$U_{tot} = k\frac{q_1 q_2}{r_{12}} + k\frac{q_1 q_3}{r_{13}} + k\frac{q_2 q_3}{r_{23}}$$

Substitute the charge of each particle and the distance between the particles.

$$U_{tot} = k\frac{q(2q)}{r} + k\frac{q(3q)}{r} + k\frac{(2q)(3q)}{r}$$

$$U_{tot} = 2\frac{kq^2}{r} + 3\frac{kq^2}{r} + 6\frac{kq^2}{r} = 11\frac{kq^2}{r}$$

Substitute values.

$$U_{tot} = 11\frac{(8.99 \times 10^9 \text{ N} \cdot \text{m}^2/\text{C}^2)(-360 \times 10^{-9} \text{ C})^2}{(0.09 \text{ m})}$$

$$U_{tot} = 0.14 \text{ J}$$

CHECK and THINK
As expected, the system has positive electric potential energy. Because there is a repulsive force between all the negatively charged particles that make up the system, an external force would do positive work in bringing them together.

B If the particles are released, what is the system's kinetic energy when the particles are infinitely far apart?

INTERPRET and ANTICIPATE
An energy bar chart (Fig. 26.8) is helpful when applying the conservation of energy principle. Initially, the particles are held in place, so their kinetic energy is zero. In part A, we found that the system's electric potential energy is positive. The particles repel one another, so when they are released they will fly apart, reducing their electric potential energy and increasing their kinetic energy. When the particles are infinitely far apart, their electric potential energy is zero. There are no external forces acting on the system and no change in the system's internal (thermal) energy.

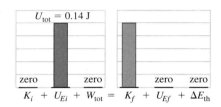

FIGURE 26.8

SOLVE
Apply the work-energy theorem (Eq. 9.31).

$$K_i + U_{Ei} + W_{tot} = K_f + U_{Ef} + \Delta E_{th} \quad (9.31)$$

$$0 + U_{tot} + 0 = K_f + U(\infty) + 0$$

$$U_{tot} = K_f$$

The kinetic energy of the system when the particles are very far apart is equal to the electric potential energy of the system in its triangle configuration.

$$K_f = U_{tot} = 0.14 \text{ J}$$

CHECK and THINK
Our result matches the bar chart (Fig. 26.8), which shows the final kinetic energy bar at the same height as the initial potential energy bar.

26-4 Electric Potential V

Electric potential is analogous to gravitational potential (lower right panel of Fig. 26.1). The **electric potential** V_E is a scalar that depends on the source's charge. Although the terms *electric potential* and *electric potential energy* are similar, it is important to distinguish between them. Any type of potential *energy* is associated with a system, so it must involve properties of both the source and the subject. The term *potential* refers to a property of the source alone. In our Earth-spacecraft analogy (Fig. 26.1), the gravitational potential depends only on the mass of the Earth, whereas the gravitational potential energy depends on the masses of both the Earth and the spacecraft. A less formal term for electric potential is **voltage**.

To emphasize the distinction between electric potential and electric potential energy in the next few paragraphs, we add the term *voltage* in parentheses whenever we refer to electric potential.

In Figure 26.1, we found the gravitational potential by dividing the gravitational potential energy by the mass of a test subject. The electric potential (voltage) V_E can be found from the electric potential energy in a similar way: Divide the electric potential energy U_E by the charge q of a test subject:

$$V_E = \frac{U_E}{q} \qquad (26.6)$$

The dimensions of electric potential (voltage) are energy per charge, so the SI units are joules per coulomb, or volts (V):

$$1 \text{ J/C} = 1 \text{ V}$$

Just as only *changes* in potential energy are physically important, only *differences* in electric potential (voltage) are physically important. The electric potential (voltage) at any particular point is assigned a convenient value by convention or by definition; then you measure the potential at some other point *relative* to this assigned potential (Fig. 26.9).

Suppose a source sets up an electric potential (voltage) difference between points A and B and a subject moves from point A to point B. The source–subject system's electric potential *energy* changes as a result of the subject's motion from A to B. The change in the system's electric potential energy ΔU_E is found by multiplying the subject's charge q by the difference in the electric potential (voltage) ΔV_E:

$$\Delta U_E = q\Delta V_E \qquad (26.7)$$

The SI units for ΔU_E are joules, as you would expect for energy. However, the joule is often inconvenient to use when the subject has a small number of excess protons (or electrons). In such cases, the change in energy is very small. Consider a proton as the subject, with charge $q = e$. If the proton moves from point A to point B (Fig. 26.9), the change in the system's electric potential energy is

$$\Delta U_E = q\Delta V_E = e(V_B - V_A)$$
$$\Delta U_E = (1.60 \times 10^{-19} \text{C})(4.0 \text{V})$$
$$\Delta U_E = 6.40 \times 10^{-19} \text{J}$$

To avoid such a small number, another energy unit known as the **electron volt** (eV) may be used. It is related to SI units by

$$1 \text{ eV} = 1.60 \times 10^{-19} \text{J}$$

It is convenient to think of the electron volt this way: If a single proton (or electron) moves through an electric potential difference (voltage) of 1 V, the (magnitude of the) change in the system's electric potential energy is 1 eV.

ELECTRIC POTENTIAL (VOLTAGE) ✪ **Major Concept**

The symbol for the unit—volts—is an uppercase roman (not italic) V, and the symbol for electric potential (voltage) is an uppercase italic *V*. When you see these symbols handwritten, you must determine their meaning from the context.

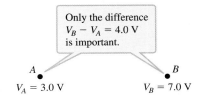

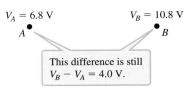

FIGURE 26.9 You are free to choose the zero point of the electric potential, but electric potential difference is independent of your choice. Here the difference in the electric potential (voltage) between points A and B is $V_B - V_A = 4.0$ V. If $V_A = 3.0$ V, then $V_B = 7.0$ V. However, if $V_B = 10.8$ V, then $V_A = 6.8$ V.

EXAMPLE 26.4 CASE STUDY Synapses

When your senses detect an event, an electrical signal travels to your brain along long, thin nerve cells known as *neurons*. Gaps between neurons (called synapses, Fig. 26.10) play an important role in regulating the signals. In order for a signal to reach the brain, it must pass through many synapses. One way for a signal to get through the synapse is by sending a chemical compound known as a neurotransmitter through the gap. The first neurotransmitter to be discovered was acetylcholine (ACh). In this problem, we model the transmission of one ACh molecule through a synapse. This first model is crude, but as we continue to study electricity in the next chapters, we will develop better models.

Consider a single ACh molecule traveling from point i to point f through a synapse. The potential difference (voltage) is $V_f - V_i = -55$ mV. The ACh molecule has one excess proton and mass $m = 2.4 \times 10^{-25}$ kg. The system consists of the ACh molecule and the two nerve cells on either side of the synapse.

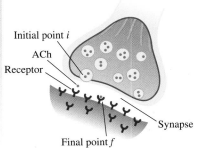

FIGURE 26.10 A synapse is a gap between nerve cells.

Example continues on page 798 ▶

A Find the change in the system's electric potential energy in electron volts and in joules.

B Assume the ACh molecule starts from rest, and find its speed when it reaches point f.

:• INTERPRET and ANTICIPATE

The source of the electric potential (voltage) is the nerve cells. The subject is the ACh molecule. The electric potential difference (voltage) is negative, $V_f - V_i < 0$, and the subject (ACh) is positively charged, so the system's electric potential energy will decrease: $U_f - U_i < 0$.

As shown in the bar chart (Fig 26.11), we have chosen a reference configuration so that $U_f = 0$. We have assumed there are no external forces doing work on the system ($W = 0$) and no dissipative forces ($\Delta E_{th} = 0$). These assumptions allow us to solve for the ACh molecule's final speed by setting $U_i = K_f$.

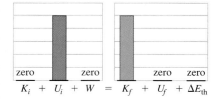

FIGURE 26.11

:• SOLVE

A The change in the system's electric potential energy comes from Equation 26.7. The ACh has one excess proton, so its charge is $q = e$. Working in electron volts is convenient because we do not need to substitute a value for e. As usual, the "m" in front of the unit stands for "milli."

$$\Delta U_E = q\Delta V_E \quad (26.7)$$

$$\Delta U_E = e(-55\,\text{mV}) = -55\,\text{meV}$$

Convert from electron volts to joules using $1\,\text{eV} = 1.60 \times 10^{-19}\,\text{J}$.

$$\Delta U_E = (-55 \times 10^{-3}\,\text{eV})\left(\frac{1.6 \times 10^{-19}\,\text{J}}{1\,\text{eV}}\right)$$

$$\Delta U_E = -8.8 \times 10^{-21}\,\text{J}$$

:• SOLVE

B Use $U_i = K_f$ to find the ACh molecule's final speed when it arrives at the other side of the synapse.

$$\Delta U_E = U_f - U_i = 0 - U_i$$

$$U_i = -\Delta U_E = K_f$$

$$-\Delta U_E = \frac{1}{2}mv_f^2$$

$$v_f = \sqrt{\frac{2(-\Delta U_E)}{m}} = \sqrt{\frac{2(8.8 \times 10^{-21}\,\text{J})}{(2.4 \times 10^{-25}\,\text{kg})}}$$

$$v_f = 2.7 \times 10^2\,\text{m/s}$$

:• CHECK and THINK

Our results have the correct dimensions and signs. The speed of the signal in human nerve cells is about 100 m/s. The speed of the ACh molecule we found is probably too high. Contrary to our assumptions, there are probably dissipative forces, so not all the potential energy goes into the molecule's kinetic energy. In Chapter 28, we learn to deal with such dissipative forces, which create resistance to the flow of charged particles. The speed of nerve signal transmission is important for good health. Diseases such as multiple sclerosis result when the transmission speed is atypical.

Special Case: Electric Potential (Voltage) Due to a Charged Particle

The source in Figure 26.12 has spherical symmetry, such as a charged sphere, spherical shell, or particle. For convenience, we refer to this source as a particle with charge Q.

According to $V_E = U_E/q$ (Eq. 26.6), we can find the electric potential by dividing the electric potential energy by the charge q of a test subject. Dividing $U_E(r) = kQq/r$ (Eq. 26.3) by q, we find the electric potential due to a charged particle:

$$V_E = k\frac{Q}{r} \quad (26.8)$$

ELECTRIC POTENTIAL (VOLTAGE) OUTSIDE A SPHERICAL SOURCE
▶ Special Case

Equation 26.8 is valid outside any source with spherical symmetry that possesses excess charge Q.

As we know from our experience with gravitational potential (Fig. 26.1), it is usually convenient to set a reference point at which the potential is defined to be zero. In the case of universal gravity, the most convenient reference point is at infinity. The same is true in the case of a charged particle; the most convenient reference point is at infinity where the electric potential is defined to be zero: $V_E(\infty) \equiv 0$. At all other points, the electric potential is given by Equation 26.8.

Although the gravitational potential is zero at infinity and negative at all other points (Fig. 26.1), the electric potential due to a charged source particle is zero at infinity and may be positive or negative at other points depending on the sign of the source. If the source is positive, the electric potential is positive at points other than infinity. If the source is negative, the electric potential is negative (like the gravitational potential) at points other than infinity. Figure 26.12 summarizes the electric force, electric field, electric potential energy, and electric potential for a charged particle. Compare Figures 26.1 and 26.12, and you see that equations and concepts concerning electricity are similar to those concerning gravity.

Visualizing Electric Potential

There are two main tools for visualizing electric potential. One is a graph of electric potential as a function of position. Figure 26.13A is a graph of V_E versus r, where r is the distance from a charged particle (Eq. 26.8). The graph shows how quickly the

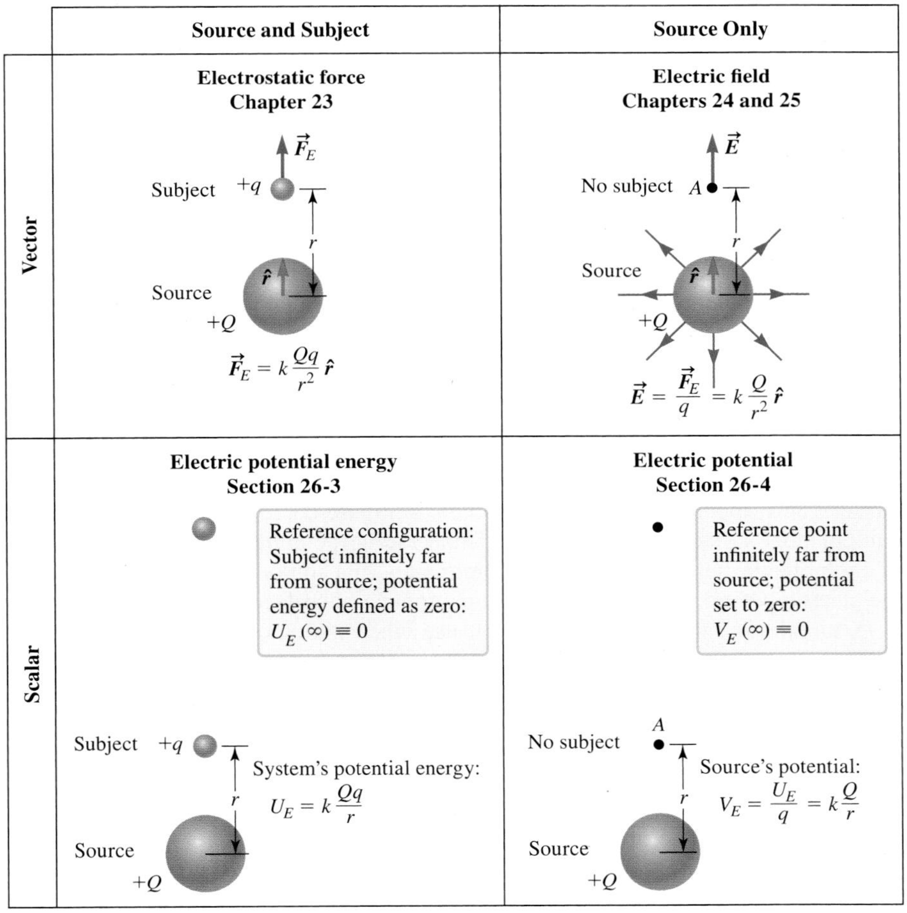

	Source and Subject	Source Only
Vector	**Electrostatic force** Chapter 23 Subject $+q$, Source $+Q$ $\vec{F}_E = k\dfrac{Qq}{r^2}\hat{r}$	**Electric field** Chapters 24 and 25 No subject A, Source $+Q$ $\vec{E} = \dfrac{\vec{F}_E}{q} = k\dfrac{Q}{r^2}\hat{r}$
Scalar	**Electric potential energy** Section 26-3 Reference configuration: Subject infinitely far from source; potential energy defined as zero: $U_E(\infty) \equiv 0$ Subject $+q$, Source $+Q$ System's potential energy: $U_E = k\dfrac{Qq}{r}$	**Electric potential** Section 26-4 Reference point infinitely far from source; potential set to zero: $V_E(\infty) \equiv 0$ No subject A, Source $+Q$ Source's potential: $V_E = \dfrac{U_E}{q} = k\dfrac{Q}{r}$

FIGURE 26.12 Compare this figure to Figure 26.1. The quantities electric force, electric field, electric potential energy, and electric potential are all analogous to the quantities for gravity. Use this analogy to help build your understanding of electricity.

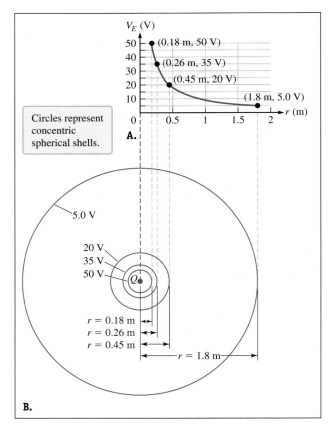

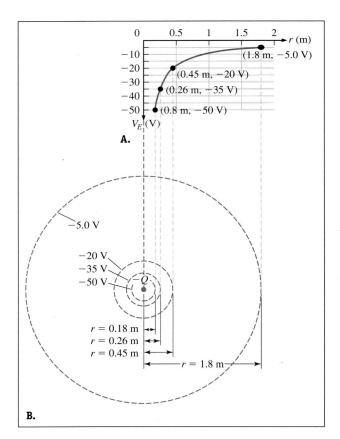

FIGURE 26.13 A. Graph of the electric potential near a positively charged source particle. **B.** A contour map corresponding to the graph in part A. A contour map consists of a number of equally spaced equipotential surfaces. In this case, there is 15 V between adjacent surfaces.

FIGURE 26.14 A. Graph of the electric potential near a negatively charged source particle. **B.** Corresponding contour map. Notice that we use dashed lines for negative potentials.

CONTOUR MAP ⊙ Tool

In contour maps, we represent positive equipotential surfaces by solid curves and negative equipotential surfaces by dashed curves.

electric potential drops off with distance from the source. For this particle, the net charge is $Q = 1 \times 10^{-9}$ C, and the electric potential decreases from 50 V at 0.18 m from the source to 5 V at 1.8 m away, dropping off sharply near the source as can be seen by the steep slope for positions closer than about 1 m.

All points that are at the same distance r from a given charged particle have the same electric potential (Fig. 26.13). Imagine a series of concentric spherical shells centered on the particle; the electric potential is the same for all points on one shell. For example, all points on a shell of radius $r = 0.18$ m are at an electric potential of 50 V. A surface on which all the points are at the same electric potential is called an **equipotential surface**. In addition to a graph, a series of equipotential surfaces provides another way to visualize the electric potential.

Figure 26.13B shows four equipotential surfaces for a positively charged source particle, corresponding to the labeled points in Figure 26.13A. The number on each equipotential surface is the electric potential for all the points on that surface. A set of equipotential surfaces is called a **contour map**. In a contour map, the difference in the electric potentials between adjacent equipotential surfaces is constant.

A topographical map that you might use when hiking is another example of a contour map. In a topographical map, the contours represent elevations, and the spacing of the contours corresponds to changes in the terrain. For example, closely spaced contours indicate steep terrain. Likewise, on a contour map of equipotential surfaces, regions where the contours are close together indicate a sharp change in the electric potential. The surfaces in Figure 26.13B are close together near the source because the electric potential drops off quickly there, as also shown by the steep slope in Figure 26.13A at distances less than about 1 m.

Figure 26.14A is a graph of V_E versus r for a negatively charged particle. In this case, the electric potential is zero at infinity and negative at all other points. Figure 26.14A looks like an upside-down version of Figure 26.13A. Again, we use four labeled points to make a contour map (Fig. 26.14B). Compare the contour maps (Figs. 26.13B and 26.14B): The equipotential surfaces are close together near the source in both cases, indicating that the electric potential changes sharply near the source. If these were topographical hiking maps, Figure 26.13B would represent a hill that gets very steep as you climb toward the summit, whereas Figure 26.14B would represent a valley that gets very steep the closer you get to the bottom.

CONCEPT EXERCISE 26.4

Match the topographical maps in Figure 26.15 with the corresponding landscapes.

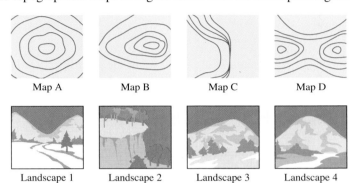

FIGURE 26.15

26-5 Special Case: Electric Potential Due to a Collection of Charged Particles

In this section, we drop the subscript E from the symbol V_E, so that the symbol for electric potential is simply V. We drop the word *electric* from the phrase *electric potential difference*.

If the source of an electric field is a collection of charged particles, the electric potential at a particular point in space is found by adding the electric potentials due to each particle. Because the potential is a scalar, there is no direction to take into account when we do this addition, although for each particle we must (1) use the same convention for the zero of electric potential, $V(\infty) = 0$, and (2) take into account the signs of the electric potentials due to each charge. The electric potential at some point in space due to a collection of n charged particles is given by

$$V = k \sum_{i=1}^{n} \frac{q_i}{r_i} \quad (26.9)$$

where q_i is the charge of the ith particle and r_i is its distance to the point at which we calculate the electric potential.

Special Case: Electric Potential Due to a Dipole

Let's find the electric potential at point A due to the dipole shown in Figure 26.16A. Each particle in a dipole has a charge whose absolute value is q. The negative particle is at a distance r_- and the positive particle is at a distance r_+ from point A. According to Equation 26.9, the electric potential at A is the sum of the electric potentials due to each charged particle:

$$V = V_+ + V_-$$

$$V = k\frac{q}{r_+} + k\frac{(-q)}{r_-}$$

Finding a common denominator and adding these terms, we find that the electric potential is

$$V = kq\left(\frac{r_- - r_+}{r_+ r_-}\right) \quad (26.10)$$

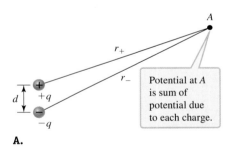

A.

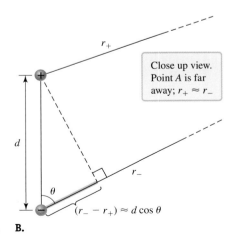

B.

FIGURE 26.16 A. Finding the electric potential at point A near a dipole. **B.** Use trigonometry to find the extra distance from the negative particle to point A.

CHAPTER 26 Electric Potential

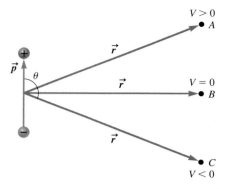

FIGURE 26.17 The dipole moment points from the negative to the positive charge. The electric potential is positive at point A, zero at point B, and negative at point C.

We are often interested in regions of space that are relatively far from the dipole compared to the charge separation d. In such cases, the distances r_- and r_+ are nearly equal, and we write

$$r \equiv r_+ \approx r_-$$

so that the denominator in Equation 26.10 becomes $r_+ r_- \approx r^2$. The small difference in the numerator can be written in terms of the separation d and the angle θ (Fig. 26.16B):

$$r_- - r_+ \approx d \cos\theta$$

Substituting these approximations into Equation 26.10, we find the electric potential at a distant point from a dipole is given by

$$V \approx k \frac{qd \cos\theta}{r^2}$$

The dipole moment $\vec{p}$ (Section 24-4) points from the negative charge toward the positive charge and has magnitude qd. Write the electric potential in terms of the dipole moment p,

ELECTRIC POTENTIAL DUE TO A DIPOLE ▶ Special Case

$$V \approx k \frac{p \cos\theta}{r^2} \qquad (26.11)$$

where angle θ is measured from the dipole moment $\vec{p}$ to the position vector $\vec{r}$ that extends from the center of the dipole to the point of interest. The sign of the electric potential depends on the angle θ (Fig. 26.17).

Figure 26.18 shows a contour map for the dipole in Figure 26.16. Most of the equipotential surfaces look like flattened spherical shells. Solid lines indicate positive electric potentials, and dashed lines indicate negative electric potentials. The solid black line in the center of the figure represents the zero-potential surface, a flat plane. For a dipole, the electric potential is zero both at infinity and at any point on the plane shown in black. Notice that the sharpest change in the potential is between the two charges, near the plane.

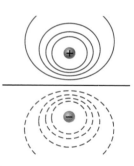

FIGURE 26.18 Contour map for an electric dipole.

CONCEPT EXERCISE 26.5

Which term or phrase is a synonym for *electric potential*?

a. gravitational potential
b. potential energy
c. voltage
d. electric field

EXAMPLE 26.5 Electric Potential Due to Three Particles

Figure 26.19 shows a source that consists of three charged objects. Objects 1 and 3 are particles with charges q_1 and q_3, respectively. Object 2 is a sphere with charge q_2. Point A is in the same plane as the source; its distance to each object is given in the figure.

A Use the numerical values in Figure 26.19 to find the electric potential at point A. (All values provided have two significant figures.)

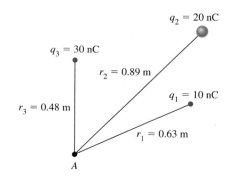

FIGURE 26.19

26-5 Special Case: Electric Potential Due to a Collection of Charged Particles

:• INTERPRET and ANTICIPATE
All objects that make up the source are positively charged, so we expect the electric potential at point A to be positive (using the usual convention that the potential at infinity is zero).

:• SOLVE	
The electric potential outside a spherical charge distribution is the same as the electric potential due to a charged particle, so all three objects in the source may be modeled as particles. The electric potential at point A is the sum of the electric potentials due to three charged particles (Eq. 26.9).	$$V = k\sum_{i=1}^{n} \frac{q_i}{r_i} \quad (26.9)$$ $$V = k\left(\frac{q_1}{r_1} + \frac{q_2}{r_2} + \frac{q_3}{r_3}\right) \quad (1)$$
Substitute the numerical values of charges and distances given in Figure 26.19. Remember that $1\text{ J/C} = 1\text{ V}$. (We'll need to keep an extra significant figure for part B: $V = 9.07 \times 10^2$ V.)	$V = (8.99 \times 10^9\text{ N}\cdot\text{m}^2/\text{C}^2) \times$ $\left[\dfrac{10 \times 10^{-9}\text{C}}{0.63\text{ m}} + \dfrac{20 \times 10^{-9}\text{C}}{0.89\text{ m}} + \dfrac{30 \times 10^{-9}\text{C}}{0.48\text{ m}}\right]$ $V = 9.1 \times 10^2\text{ V}$

:• CHECK and THINK
As expected, the electric potential is positive.

B How much work is required (by an external force) to move a 40-nC charge from very far away (from infinity) to point A?

:• INTERPRET and ANTICIPATE
There are at least two ways we can do this calculation. We'll use one to anticipate our results and the other to solve the problem. Of course, we expect the two answers to agree. The first method is based on an idea from Section 26-3: The work done by an external force is equal to the change in the system's electric potential energy. So, we need to calculate ΔU when the fourth charge is moved from infinitely far away to point A.

The required task is essentially what we did in Example 26.2 to find Equation 26.5 when we added the fourth particle to the system.	$$W = \Delta U_{\text{4th particle}} = U_{14} + U_{24} + U_{34} \quad (26.5)$$
The electric potential energy stored by each pair of particles is given by $U_E(r) = kQq/r$ (Eq. 26.3).	$W = k\dfrac{q_1 q_4}{r_1} + k\dfrac{q_2 q_4}{r_2} + k\dfrac{q_3 q_4}{r_3} = kq_4\left(\dfrac{q_1}{r_1} + \dfrac{q_2}{r_2} + \dfrac{q_3}{r_3}\right) \quad (2)$ $W = (8.99 \times 10^9\text{ N}\cdot\text{m}^2/\text{C}^2)(40 \times 10^{-9}\text{C}) \times$ $\left(\dfrac{10 \times 10^{-9}\text{C}}{0.63\text{ m}} + \dfrac{20 \times 10^{-9}\text{C}}{0.89\text{ m}} + \dfrac{30 \times 10^{-9}\text{C}}{0.48\text{ m}}\right)$ $W = 3.6 \times 10^{-5}\text{ J}$ (anticipated result)

:• SOLVE	
We can also find the change in the electric potential energy and therefore the work done by an external force by multiplying the potential difference between two points (A and ∞) by the amount of charge that moves between those two points (Eq. 26.7). The electric potential at infinity is zero, so the potential difference is equal to our results in part A.	$W = \Delta U = q\Delta V \quad (26.7)$ $W = q_4[V - V(\infty)] = q_4 V$ $W = (40 \times 10^{-9}\text{ C})(9.1 \times 10^2\text{ V})$ $W = 3.6 \times 10^{-5}\text{ J} \checkmark$

:• CHECK and THINK
We found the same result using both methods. In fact, the two methods really aren't different. Compare Equation (2) for work W to Equation (1) for electric potential V. Equation (2) can be found by multiplying q_4 by V in Equation (1).

26-6 Electric Potential Due to a Continuous Distribution

In this section, we show how to find the electric potential V due to a continuous distribution, such as a charged rod or disk, by modeling the distribution as a large number of charged particles. The process is much like finding the electric field $\vec{E}$ for the distribution (Section 24-5), but finding the electric potential (a scalar) is somewhat easier. In Section 26-7, we show how to find the electric potential from the electric field.

PROBLEM-SOLVING STRATEGY

Electric Potential for a Continuous Charge Distribution

: INTERPRET and ANTICIPATE
As in many problem-solving strategies, draw a **sketch** of the source and choose a **coordinate system**.

: SOLVE
Step 1 Divide the source into small **pieces**, and write an expression for dV for each piece. Each small piece has charge dq and can be modeled as a charged particle. The electric potential dV for each small piece is given by a modified form of Equation 26.8:

$$dV = k\frac{dq}{r} \qquad (26.12)$$

where r is the distance between a particular piece and the point where you would like to find the electric potential. Write your expression in terms of your coordinate system, and express dq in terms of the charge density (Section 24-5).
Step 2 Integrate over the entire source to find V.

: CHECK and THINK
In addition to checking the dimensions, there are two other checks for a finite source:
1. Verify that at **a point far away** compared to the size of the source, the electric potential has the same form as that of a **charged particle** ($V = kQ/r$).
2. Verify that at a point **infinitely far away, the electric potential is zero**.

EXAMPLE 26.6 Electric Potential due to a Charged Ring (Special Case)

Show that the electric potential at point A due to a uniformly charged ring of radius R and charge Q is

$$V = \frac{kQ}{\sqrt{R^2 + x^2}} = \frac{1}{4\pi\varepsilon_0}\frac{Q}{\sqrt{R^2 + x^2}} \qquad (26.13)$$

Point A is a distance x from the center of the ring (Fig. 26.20).

ELECTRIC POTENTIAL DUE TO A RING ▶ Special Case

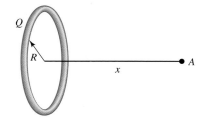

FIGURE 26.20

: INTERPRET and ANTICIPATE
Sketch the source and choose a **coordinate system**. The origin of our coordinate system is at the center of the ring, and the x axis is aligned with the ring's axis so that point A is a distance x from the center of the ring (Fig. 26.21). The y axis points upward, and the z axis is coming out of the page.

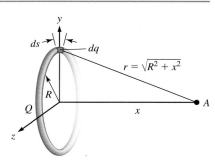

FIGURE 26.21

SOLVE

Step 1 Divide the source into **pieces**, and write an **expression for dV** for each piece. Figure 26.21 shows one piece at the top of the ring. The electric potential due to that one piece is given by Equation 26.12.

$$dV = k\frac{dq}{r} \quad (26.12)$$

The distance r to point A is the hypotenuse of a right triangle with sides x and R. Write r in terms of x and R.

$$r = \sqrt{R^2 + x^2} \quad (1)$$

The piece has charge dq and length ds. Write dq in terms of ds and the ring's linear charge density λ (Eq. 24.13).

$$dq = \lambda ds \quad (24.13)$$

The electric potential at point A due to one piece comes from substituting for r (Eq. 1) and dq (Eq. 24.13).

$$dV = k\frac{\lambda ds}{\sqrt{R^2 + x^2}}$$

Step 2 Integrate over the source. In this case, we must integrate from $s = 0$ to $s = 2\pi R$ (the circumference of the ring). The integral on the left is the electric potential at point A due to the entire ring. The radius R of the ring and the distance x do not depend on s, so they are pulled outside the integral along with the constants k and λ.

$$\int dV = \int_0^{2\pi R} k\frac{\lambda ds}{\sqrt{R^2 + x^2}}$$

$$V = \frac{k\lambda}{\sqrt{R^2 + x^2}} \int_0^{2\pi R} ds$$

Complete the integral and evaluate between the limits.

$$V = \frac{k\lambda}{\sqrt{R^2 + x^2}} s \Big|_0^{2\pi R} = \frac{k\lambda}{\sqrt{R^2 + x^2}}(2\pi R)$$

The ring's charge Q is the linear charge density times the ring's circumference, $Q = (\lambda)(2\pi R)$.

$$V = \frac{kQ}{\sqrt{R^2 + x^2}} \quad \checkmark$$

It is common practice to replace Coulomb's constant k with $k = 1/(4\pi\varepsilon_0)$ (Eq. 25.9).

$$V = \frac{1}{4\pi\varepsilon_0}\frac{Q}{\sqrt{R^2 + x^2}} \quad \checkmark \quad (26.13)$$

CHECK and THINK

To check our result, we imagine point A moving away from the ring ($x \gg R$). First, we see that the electric potential has the same form as that from a charged particle. Then, as we continue to imagine point A moving toward infinity ($x \to \infty$), we find that the electric potential drops off to zero.

$$\lim_{x \gg R} V = \lim_{x \gg R} \frac{kQ}{\sqrt{R^2 + x^2}} \to \frac{kQ}{\sqrt{x^2}} = \frac{kQ}{x} \quad \checkmark$$

$$\lim_{x \to \infty} V = \frac{kQ}{x} \to 0 \quad \checkmark$$

EXAMPLE 26.7 Electric Potential due to a Charged Disk (Special Case)

Show that the electric potential at point A due to a uniformly charged disk of radius R and excess surface charge density σ is

$$V = 2\pi k\sigma[(\sqrt{R^2 + x^2}) - x] = \frac{\sigma}{2\varepsilon_0}[(\sqrt{R^2 + x^2}) - x] \quad (26.14)$$

Point A is a distance x from the center of the disk (Fig. 26.22).

ELECTRIC POTENTIAL DUE TO A CHARGED DISK ▶ Special Case

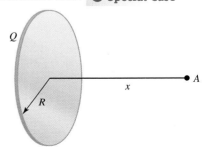

FIGURE 26.22

Example continues on page 806 ▶

CHAPTER 26 Electric Potential

:• **INTERPRET and ANTICIPATE**
Sketch the source and choose a **coordinate system.** This source is very similar to the charged ring, which we can use as a shortcut. The origin of our coordinate system is at the center of the disk, and the x axis is aligned with the disk's axis so that point A is a distance x from the center of the disk (Fig. 26.23). The y axis points upward, and the z axis points out of the page.

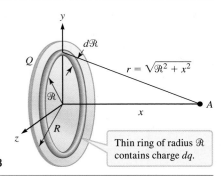

FIGURE 26.23

:• **SOLVE**
Step 1 Divide the source into **pieces,** and write an **expression for dV** for each piece. As a shortcut, we divide the disk into concentric rings. One such ring with charge dq and radius $\mathcal{R}$ is shown in Figure 26.23. (Notice that $\mathcal{R}$ is the radius of a ring, and R is the radius of the disk.) The potential dV due to a ring is given by Equation 26.13. (We could also start with Equation 26.12 and substitute $r = \sqrt{\mathcal{R}^2 + x^2}$).

$$dV = \frac{k\,dq}{\sqrt{\mathcal{R}^2 + x^2}} \quad (26.13)$$

The area of the ring is its circumference $2\pi\mathcal{R}$ times its width $d\mathcal{R}$. Write dq in terms of the area of the ring and the surface charge density σ (Eq. 24.11).

$$dq = \sigma\,dA \quad (24.11)$$
$$dq = \sigma(2\pi\mathcal{R})d\mathcal{R}$$

The electric potential at point A due to one ring comes from substituting for dq.

$$dV = k\frac{\sigma(2\pi\mathcal{R})d\mathcal{R}}{\sqrt{\mathcal{R}^2 + x^2}}$$

Step 2 Integrate over the source. In this case, we must integrate over the surface of the disk from a ring with $\mathcal{R} = 0$ to a ring with $\mathcal{R} = R$. The integral on the left is the electric potential at point A due to the entire disk. This time we cannot pull the term involving $\mathcal{R}$ outside the integral.

$$\int dV = \int_0^R k\frac{\sigma(2\pi\mathcal{R})d\mathcal{R}}{\sqrt{\mathcal{R}^2 + x^2}}$$
$$V = k2\pi\sigma\int_0^R \frac{\mathcal{R}\cdot d\mathcal{R}}{\sqrt{\mathcal{R}^2 + x^2}}$$

Complete the integral (Problem 81), and evaluate between the limits.

$$\int_0^R \frac{\mathcal{R}\cdot d\mathcal{R}}{\sqrt{\mathcal{R}^2 + x^2}} = (\sqrt{R^2 + x^2}) - x$$

Use Equation 25.9 to rewrite Coulomb's constant k in terms of the permittivity constant ε_0.

$$V = 2\pi k\sigma[(\sqrt{R^2 + x^2}) - x] \quad \checkmark$$
$$V = \frac{\sigma}{2\varepsilon_0}[(\sqrt{R^2 + x^2}) - x] \quad \checkmark \quad (26.14)$$

:• **CHECK and THINK**
To check the result, take the limit as the point A moves away from the disk. First, replace the surface charge density by $\sigma = Q/\pi R^2$.

$$V = 2\pi k \frac{Q}{\pi R^2}[(\sqrt{R^2 + x^2}) - x]$$

Then factor x out of the square root term.

$$V = 2k\frac{Q}{R^2}\left[(x\sqrt{(R/x)^2 + 1}) - x\right]$$

So, as A moves away from the disk ($x \gg R$), the electric potential takes on the same form as that due to a charged particle. As A continues toward infinity ($x \to \infty$), the electric potential drops off to zero.

$$\lim_{x \gg R} V \to 2k\frac{Q}{R^2}\left[x\left(\frac{1}{2}\frac{R^2}{x^2} + 1\right) - x\right]$$
$$\lim_{x \gg R} V \to 2k\frac{Q}{R^2}\left(\frac{1}{2}\frac{R^2}{x}\right) \to k\frac{Q}{x} \quad \checkmark$$
$$\lim_{x \to \infty} V = \lim_{x \to \infty}\left(k\frac{Q}{x}\right) \to 0 \text{ as } x \to \infty \quad \checkmark$$

26-7 Connection Between Electric Field $\vec{E}$ and Electric Potential V

In the preceding sections, we found an expression for the electric potential that results from a complicated charged source by breaking up the source into small pieces and adding (or integrating) the electric potentials associated with each small piece. In this section, we show another approach for finding the electric potential V based on the source's electric field $\vec{E}$.

We seek to relate the electric potential (a scalar) to the electric field (a vector); both quantities depend on only the source and not the subject. The relationship between the electric potential V and the electric field $\vec{E}$ is analogous to the relationship between the potential energy U and the force $\vec{F}$ (Fig. 26.12). To find the change in the potential energy, we take a path integral of the force:

$$\Delta U = -\int_{r_i}^{r_f} \vec{F} \cdot d\vec{r} \qquad (9.24)$$

By analogy, the electric potential difference is found by taking the path integral of the electric field:

$$\Delta V = -\int_{r_i}^{r_f} \vec{E} \cdot d\vec{r} \qquad (26.15)$$

FINDING THE ELECTRIC POTENTIAL V FROM THE ELECTRIC FIELD $\vec{E}$
⭐ **Major Concept**

We can derive Equation 26.15 formally from $\Delta U = -\int_{r_i}^{r_f} \vec{F} \cdot d\vec{r}$ (Eq. 9.24). Substitute $\vec{F}_E = q\vec{E}$ (Eq. 24.2) for the electric force:

$$\Delta U = -\int_{r_i}^{r_f} q\vec{E} \cdot d\vec{r} = -q\int_{r_i}^{r_f} \vec{E} \cdot d\vec{r}$$

where q is the subject's charge. According to $\Delta V = \Delta U/q$ (Eq. 26.7), we simply divide ΔU by q to find the electric potential difference:

$$\Delta V = -\int_{r_i}^{r_f} \vec{E} \cdot d\vec{r} \quad \checkmark \qquad (26.15)$$

EXAMPLE 26.8 | Electric Potential due to a Charged Particle

From the electric field $\vec{E}(r) = (kQ_S/r^2)\hat{r}$ (Eq. 24.3) produced by a positively charged particle, find the (usual) expression for its electric potential.

:• INTERPRET and ANTICIPATE
Sketch the electric field and include a path from r_i to r_f (Fig. 26.24). We expect our result to be the same as $V = kQ/r$ (Eq. 26.8).

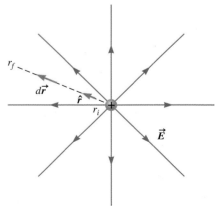

FIGURE 26.24

:• SOLVE
For a particle, the electric field $\vec{E}$ and $d\vec{r}$ (a small portion of the path from r_i to r_f) both point in the $\hat{r}$ direction, so their dot product is $E\,dr$. Substitute the magnitude of the electric field E for a charged particle.

$$\Delta V = -\int_{r_i}^{r_f} \vec{E} \cdot d\vec{r} \qquad (26.15)$$

$$\Delta V = -\int_{r_i}^{r_f} E\,dr = -\int_{r_i}^{r_f} \left(k\frac{Q}{r^2}\right)dr$$

Example continues on page 808 ▶

808 CHAPTER 26 Electric Potential

Pull the constants outside the integral and integrate.	$\Delta V = -kQ \int_{r_i}^{r_f} \left(\frac{1}{r^2}\right) dr$ $\Delta V = -kQ\left[-\frac{1}{r}\right]\Big|_{r_i}^{r_f} = kQ\left[\frac{1}{r}\right]\Big|_{r_i}^{r_f}$
Evaluate between limits and reduce.	$\Delta V = \frac{kQ}{r_f} - \frac{kQ}{r_i}$
Using the usual convention that the electrostatic potential is zero at infinity, we find an expression for the electric potential at a distance r from the source, in agreement with Equation 26.8.	$\Delta V = V(r) - V(\infty) = V(r) - 0 = V$ $V = \frac{kQ}{r}$ ✓ $\qquad\qquad$ (26.8)

:• CHECK and THINK
We started with the particle's electric field and found its electric potential as expected.

EXAMPLE 26.9 Electric Potential due to a Charged Sheet

Sources with planar symmetry (Section 25-6), such as a very large charged sheet (Fig. 26.25, edge view), are common and important. The magnitude of the electric field for such a source is uniform and given by $E = \sigma/2\varepsilon_0$ (Eq. 25.16).

A Use path 1 shown in Figure 26.25 to find the potential difference ΔV_{AB} between points A and B. (We'll consider path 2 in the next part, and path 3 in a later derivation.)

ELECTRIC POTENTIAL DUE TO A SOURCE WITH PLANAR SYMMETRY ▶ **Special Case**

:• INTERPRET and ANTICIPATE
To find the potential difference between a pair of points, we must calculate the path integral $\Delta V = -\int_{r_i}^{r_f} \vec{E} \cdot d\vec{r}$. Path 1 is a vertical line between points A and B.

FIGURE 26.25

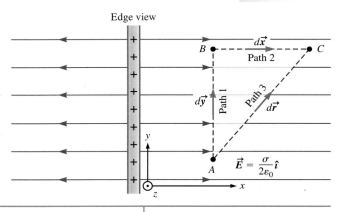

:• SOLVE Using the coordinate system in the figure, write the electric field in component form.	$\vec{E} = \frac{\sigma}{2\varepsilon_0}\hat{\imath}$ $\qquad\qquad$ (1)
In general, the path element $d\vec{r}$ may be written in terms of its vector components.	$d\vec{r} = d\vec{x} + d\vec{y} + d\vec{z}$
Because point B is directly above point A, path 1 is purely in the y direction. The x and z components of $d\vec{r}_1$ are zero.	$d\vec{r}_1 = 0 + d\vec{y} + 0 = d\vec{y}$ $d\vec{r}_1 = dy\,\hat{\jmath}$ $\qquad\qquad$ (2)
We set up the path integral (Eq. 26.15) from A to B by substituting Equation (1) for $\vec{E}$ and Equation (2) for $d\vec{r}_1$.	$\Delta V_{AB} = -\int_{r_i}^{r_f} \vec{E} \cdot d\vec{r}_1$ $\qquad\qquad$ (26.15) $\Delta V_{AB} = -\int_A^B \left(\frac{\sigma}{2\varepsilon_0}\hat{\imath}\right) \cdot d\vec{y} = -\int_A^B \left(\frac{\sigma}{2\varepsilon_0}dy\right)\hat{\imath}\cdot\hat{\jmath}$
The dot product of the perpendicular vectors $\hat{\imath} \cdot \hat{\jmath} = 0$, so the potential difference between points A and B is zero.	$\Delta V_{AB} = 0$

CHECK and THINK
Our result shows that $V_A = V_B$, so points A and B are on the same equipotential surface. If we choose any two other points along the same vertical line, we find that they are also at the same electric potential. Furthermore, from the symmetry of the situation, if we choose a point directly in front of or behind point A in the z direction perpendicular to the page, that point is also at the same electric potential as point A. So, the equipotential surfaces are sheets parallel to the charged source sheet (Fig. 26.25). In fact, the charged sheet itself is an equipotential surface.

B Use path 2 to show that the potential difference ΔV_{BC} between points B and C is given by

$$\Delta V = -\frac{\sigma}{2\varepsilon_0}\Delta x \qquad (26.16)$$

where $\Delta x = x_C - x_B$ is their separation.

INTERPRET and ANTICIPATE
This is similar to part A. The main difference here is that path 2 is horizontal.

SOLVE	
Path 2 from B to C is in the x direction, so path element $d\vec{r}_2$ has only an x component.	$d\vec{r} = d\vec{x} + d\vec{y} + d\vec{z} = d\vec{x} + 0 + 0$ $d\vec{r}_2 = d\vec{x} = dx\hat{\imath}$ (3)
The electric field is still given by Equation (1), so we substitute Equations (1) and (3) into Equation 26.15 (the path integral).	$\Delta V_{BC} = -\int_{r_i}^{r_f} \vec{E} \cdot d\vec{r}_2$ (26.15) $\Delta V_{BC} = -\int_B^C \left(\frac{\sigma}{2\varepsilon_0}\hat{\imath}\right) \cdot d\vec{x}$ $\Delta V_{BC} = -\int_B^C \left(\frac{\sigma}{2\varepsilon_0}dx\right)\hat{\imath} \cdot \hat{\imath}$
The dot product $\hat{\imath} \cdot \hat{\imath} = 1$.	$\Delta V_{BC} = -\int_B^C \frac{\sigma}{2\varepsilon_0} dx$
Pull the constants out and integrate.	$\Delta V_{BC} = -\frac{\sigma}{2\varepsilon_0}\int_B^C dx = -\frac{\sigma}{2\varepsilon_0}x\Big|_B^C$ $V_C - V_B = -\frac{\sigma}{2\varepsilon_0}(x_C - x_B)$ ✓ (26.16)

CHECK and THINK
The negative sign in Equation 26.16 means that the electric potential at point B is higher than the electric potential at point C. If point B were closer to the charged sheet, it would be at an even higher electric potential. In fact, the highest electric potential is found on the charged sheet. Writing Equation 26.16 in terms of the electric field ($\Delta V = -E\Delta x$) allows you to find the potential difference due to any source that has planar symmetry. It depends only on the horizontal separation Δx between the points. There is no potential difference between points that are vertically separated.

The Relationship Between Equipotential Surfaces and Electric Field Lines

Let's continue to think about the infinite sheet in Example 26.9. We argued that the equipotential surfaces are parallel sheets and the charged sheet is also an equipotential surface. Figure 26.26 is a contour map for the infinite sheet, including its electric field lines. This figure shows two general relationships between the equipotential surfaces and the electric field lines: (1) *the electric field lines are perpendicular to the equipotential surfaces*, and (2) *the electric field lines point from high electric potential to low electric potential*.

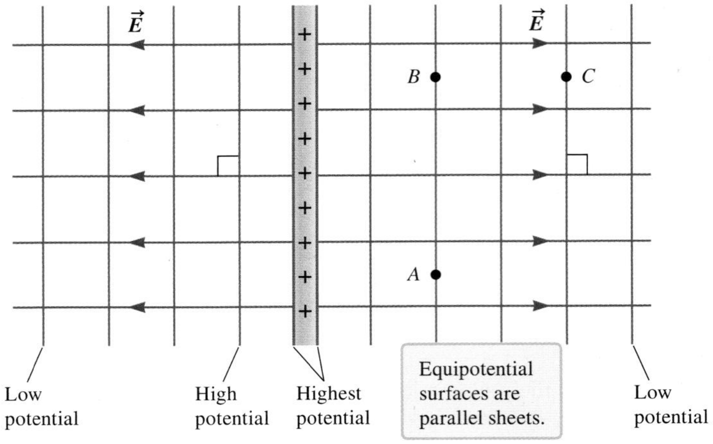

810 CHAPTER 26 Electric Potential

FIGURE 26.26 The equipotential surfaces for an infinite sheet. The electric potential of these parallel sheets depends on their distance from the source. The closer sheets (on both sides of the source) are at higher electric potential than the more distant sheets.

Furthermore, often the surface of the source defines an equipotential surface. (This is always the case for a charged conductor in electrostatic equilibrium.) If the source is positively charged (and an equipotential surface), it is at the highest electric potential. If the source is negatively charged (and an equipotential surface), it is at the lowest electric potential.

Figure 26.27A shows the electric field lines and equipotential surfaces outside a charged spherical source. The equipotential surfaces are concentric spheres perpendicular to the electric field lines. The field lines point outward in the direction of decreasing electric potential. The surface of the positively charged sphere has the highest electric potential. If the source were a particle instead of a sphere, it would still be at the highest electric potential, but it would consist only of a point, not a surface.

Figure 26.27B shows a dipole source that consists of two charged spheres. The equipotential surfaces are perpendicular to the electric field lines. The surfaces look like squashed spheres except for the surface directly between the two charged spheres, which is a flat sheet. The electric field lines point from high electric potential to low electric potential. The surface of the positive sphere is at the highest (positive) electric potential, and the surface of the negative sphere is at the lowest (negative) electric potential. The flat equipotential surface is at zero electric potential.

CONCEPT EXERCISE 26.6

If the contours in Figure 26.26 represent the topography of some landscape, describe that landscape in words and in a sketch.

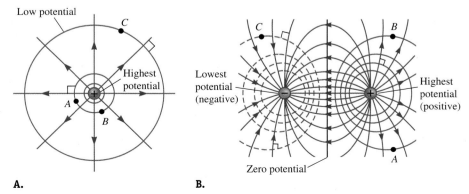

FIGURE 26.27 A. The equipotential surfaces and electric field lines for a charged sphere. **B.** The equipotential surfaces and electric field lines for a dipole. In both cases, the electric field lines are perpendicular to the surfaces and point from high electric potential to low electric potential.

DERIVATION   Electric Potential Difference Is Path-Independent

For all three sources shown in Figures 26.26 and 26.27, points A and B are at the same electric potential and point C is at a lower electric potential. When we calculate the potential difference, the particular path we use does not matter; only the endpoints are important. To illustrate this, we return to the infinite charged sheet in Figure 26.25 and calculate the potential difference between points A and C using two different paths. The first path is L-shaped, made up of path 1 from A to B and path 2 from B to C. The other path is the diagonal line labeled path 3 in Figure 26.25. We will calculate ΔV_{AC} using the L-shaped path and path 3, and show that the two results are the same.

Using the L-shaped path is fairly easy because we have already calculated the potential difference along paths 1 and 2. All we need to do is add our results.	$\Delta V_{AC} = \Delta V_{AB} + \Delta V_{BC}$
We found that the potential difference between points A and B is zero ($\Delta V_{AB} = 0$), so the potential difference between A and C equals the potential difference between B and C.	$\Delta V_{AC} = 0 + \Delta V_{BC}$ $\Delta V_{AC} = \Delta V_{BC}$ (1)
Now we use path 3 from A to C. The path is in the xy plane, so path element $d\vec{r}_3$ has no z component.	$d\vec{r} = d\vec{x} + d\vec{y} + d\vec{z}$ $d\vec{r}_3 = d\vec{x} + d\vec{y} + 0 = dx\hat{\imath} + dy\hat{\jmath}$
As before, the electric potential difference is found by integrating Equation 26.15. The electric field is in the positive x direction, and its magnitude comes from $E = \sigma/2\varepsilon_0$ (Eq. 25.16).	$\Delta V_{AC} = -\int_{r_i}^{r_f} \vec{E} \cdot d\vec{r}_3$ (26.15) $\Delta V_{AC} = -\int_A^C \left(\frac{\sigma}{2\varepsilon_0}\hat{\imath}\right) \cdot (d\vec{x} + d\vec{y})$ $\Delta V_{AC} = -\int_A^C \left[\left(\frac{\sigma}{2\varepsilon_0}dx\right)\hat{\imath}\cdot\hat{\imath} + \left(\frac{\sigma}{2\varepsilon_0}dy\right)\hat{\imath}\cdot\hat{\jmath}\right]$
The dot products $\hat{\imath}\cdot\hat{\imath} = 1$ and $\hat{\imath}\cdot\hat{\jmath} = 0$.	$\Delta V_{AC} = -\int_B^C \frac{\sigma}{2\varepsilon_0}dx$
This is exactly the same integral we solved in going along path 2 from B to C. Because point A is directly below point B, they have the same x coordinate, $x_A = x_B$. We just found Equation (1) using path 3. Thus, the potential difference between points A and C is path-independent.	$\Delta V_{AC} = -\frac{\sigma}{2\varepsilon_0}\int_A^C dx = -\frac{\sigma}{2\varepsilon_0}x\Big|_A^C$ $V_C - V_A = -\frac{\sigma}{2\varepsilon_0}(x_C - x_A)$ $V_C - V_A = -\frac{\sigma}{2\varepsilon_0}(x_C - x_B)$ $V_C - V_A = V_C - V_B$ $\Delta V_{AC} = \Delta V_{BC}$ ✓

:• COMMENTS

We showed path independence for this particular case, but our result is generally true. The *electric potential difference between two points is path-independent*—only the location of the endpoints matters. We have seen this sort of path independence before: In Section 8-3, we showed that the change in the potential energy depends only on the final and initial configurations of the system, no matter what configurations the system takes in between.

EXAMPLE 26.10 A Sheet of Charge

Figure 26.28 shows a large charged sheet and four equipotential surfaces (A–D) above it. The equipotential surfaces are equally spaced so that the distance between adjacent surfaces, including the charged sheet, is 0.25 m. The reference point has been chosen such that the charged sheet is at +16.0 V. The electric potentials at the other surfaces are shown in the figure.

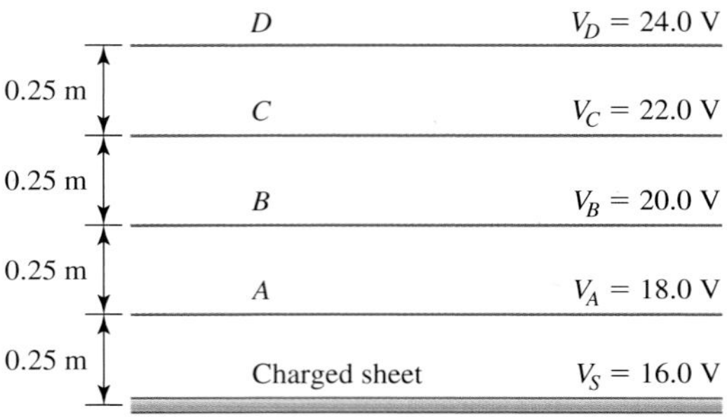

A Add electric field lines to the contour map in Figure 26.28, and then find the sign of the excess charge and the charge density of the sheet.

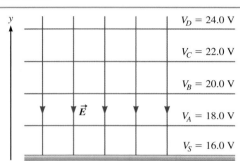

FIGURE 26.28

• INTERPRET and ANTICIPATE
Electric field lines must be perpendicular to equipotential surfaces, so the field lines must be vertical. Electric field lines point from high electric potential to low electric potential. In this case, the equipotential surface at the top of the figure is at a high electric potential and the charged sheet is at the lowest electric potential, so the electric field lines must point downward (Fig. 26.29). Electric field lines terminate on negative charges, so we expect that the sheet must have an excess negative charge. In the figure, we have included an upward-pointing y axis.

FIGURE 26.29 Electric field lines for the sheet in Figure 26.28.

• SOLVE
The surface charge density is found from the potential difference between any two equipotential surfaces (Eq. 26.16). Let's work this twice, first using surfaces A and B.

$$V_B - V_A = -\frac{\sigma}{2\varepsilon_0}(y_B - y_A) \quad (26.16)$$

$$\sigma = -2\varepsilon_0 \frac{V_B - V_A}{y_B - y_A}$$

$$\sigma = -2(8.85 \times 10^{-12}\,\text{C}^2/\text{N}\cdot\text{m}^2)\frac{(20.0\,\text{V} - 18.0\,\text{V})}{(0.50\,\text{m} - 0.25\,\text{m})}$$

$$\sigma = -1.42 \times 10^{-10}\,\text{C/m}^2 = -0.142\,\text{nC/m}^2$$

• CHECK and THINK
As expected, the excess charge is negative. Now let's work the problem again using the charged sheet and the equipotential surface D to check our result. We find the same answer no matter which surfaces we choose to work with.

$$\sigma = -2\varepsilon_0 \frac{V_D - V_S}{y_D - y_S}$$

$$\sigma = -2(8.85 \times 10^{-12}\,\text{C}^2/\text{N}\cdot\text{m}^2)\frac{(24.0\,\text{V} - 16.0\,\text{V})}{(1.00\,\text{m} - 0\,\text{m})}$$

$$\sigma = -1.42 \times 10^{-10}\,\text{C/m}^2$$

$$\sigma = -0.142\,\text{nC/m}^2 \checkmark$$

B Imagine a subject with excess charge $q = +0.50$ C moving from equipotential surface A to surface C (Fig. 26.28). If the system consists of the charged sheet and the subject, what is the change in the system's electric potential energy?

• INTERPRET and ANTICIPATE
The positively charged subject is attracted to the negatively charged source, so an external force must do work to lift the subject. The electric potential energy of the subject–sheet system must increase. The situation is analogous to an object near the Earth's surface being lifted against gravity; in that case, the system's gravitational potential energy increases.

:• **SOLVE**
The change in the system's electric potential energy is found by multiplying the subject's charge by the difference in the potential between the two points (Eq. 26.7).

$$\Delta U_E = q\Delta V = q(V_C - V_A) \tag{26.7}$$

$$\Delta U_E = (0.50\,\text{C})(22.0\,\text{V} - 18.0\,\text{V}) = 2.0\,\text{J}$$

:• **CHECK and THINK**
As expected, the system gains potential energy (+2.0 J) as the positively charged subject is moved away from the negatively charged sheet.

26-8 Finding $\vec{E}$ from V

In the preceding section, we integrated the electric field $\vec{E}$ of a source to find the electric potential V. In this section, we reverse that process by starting with the electric *potential* and finding the electric *field*. We expect this reverse process to involve derivatives, but before we start taking derivatives, let's consider a very simple case that doesn't require calculus.

Figure 26.30A shows a contour map that consists of equally spaced planes (Δx is the same for each pair of surfaces). The potential difference ΔV between the contours is uniform. Because the geometry of this situation is simple, it is easy to find the electric field. First, we draw the electric field lines (Fig. 26.30B) using the two rules on page 809 (1) electric field lines are perpendicular to equipotential surfaces, and (2) electric field lines point from high electric potential to low electric potential. We are interested in finding the electric field $\vec{E}$, so we combine $E = \sigma/2\varepsilon_0$ and $\Delta V = -(\sigma/2\varepsilon_0)\Delta x$ (Eqs. 25.16 and 26.16) to write E in terms of ΔV:

$$\Delta V = -E\Delta x \tag{26.17}$$

$$E = -\frac{\Delta V}{\Delta x} \tag{26.18}$$

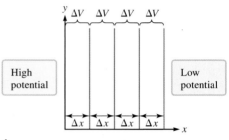

A.

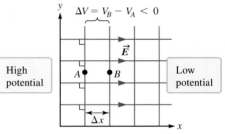

B.

FIGURE 26.30 A. The equipotential surfaces are uniform. **B.** Electric field lines are perpendicular to equipotential surfaces.

The electric field is a vector. In this case, $\vec{E}$ points in the x direction, so we write it as

$$\vec{E} = -\frac{\Delta V}{\Delta x}\hat{\imath} \tag{26.19}$$

Why does Equation 26.19 have a negative sign when the electric field points in the positive x direction? Let's take a closer look. The potential difference between the equipotential surfaces is uniform, so let's arbitrarily consider contours through points A and B (Fig. 26.30B), for which the potential difference is

$$\Delta V = V_B - V_A$$

Contour A is at a higher electric potential than B ($V_A > V_B$), so their potential difference is negative:

$$\Delta V = V_B - V_A < 0$$

Because the potential difference is negative, the negative sign in Equation 26.19 ensures that $\vec{E}$ points in the correct direction—in this case, the positive x direction.

Equation 26.19 gives us another set of SI units for the electric field. Potential difference is measured in volts and distance is measured in meters, so the electric field may be measured in either volts per meter or newtons per coulomb:

$$1\,\text{N/C} = 1\,\text{V/m}$$

CASE STUDY: $\vec{E}$ and V in the Human Heart

In the case of a uniform electric field—the sort of field you expect from a very large charged sheet—we can use Equation 26.19 to find the electric field from the electric potential. What do we do when the source produces a more complicated electric field, such as in the human chest?

Your heart's rhythm is controlled by electrical signals that can be monitored with an electrocardiogram (EKG), a recording of the electric potential on the chest. For an instant during each heartbeat, the equipotential surfaces look something like Figure 26.31A. We can use the electric potential surfaces on the chest to illustrate how to find the electric field.

Consider a small region near the center of the pattern (Fig. 26.31, inset) where the contours look equally spaced and parallel—like a rotated version of Figure 26.30B. We can always focus on a small enough region so that the equipotential surfaces are well modeled as parallel sheets, and we can draw the electric field lines in this tiny region so that they are (1) perpendicular to the equipotential surfaces and (2) pointing from high electric potential to low electric potential (Fig. 26.31B).

We have chosen a coordinate system such that the x axis is aligned with the electric field. The separation dx between the equipotential sheets is uniform and infinitesimal. The difference in the potential dV between adjacent contours is uniform and small. Equation 26.19 in this case becomes a derivative:

$$\vec{E} = -\frac{dV}{dx}\hat{\imath} \qquad (26.20)$$

Equation 26.20 is all you need to find the electric field if the electric potential changes along only one dimension. However, Figure 26.31A shows that the electric potential over the whole chest region is more complicated. Let's use this contour map of the whole chest to see how we can extend Equation 26.20 to an expression that applies even in a more complicated situation. This time, we model the heart as a dipole (Fig. 26.32) in order to better take into account the complicated contours. Consider a region on the side near the person's left arm. The inset in Figure 26.32 shows a close-up of this region along with the same coordinate system we used in Figure 26.31B. The electric field in this region is nearly straight up, so $\vec{E}$ has both x and y components: $\vec{E} = E_x\hat{\imath} + E_y\hat{\jmath}$. The field components are found by taking the derivative of the electric potential in each direction separately:

$$E_x = -\frac{\partial V}{\partial x}, \quad E_y = -\frac{\partial V}{\partial y}, \quad E_z = -\frac{\partial V}{\partial z} \qquad (26.21)$$

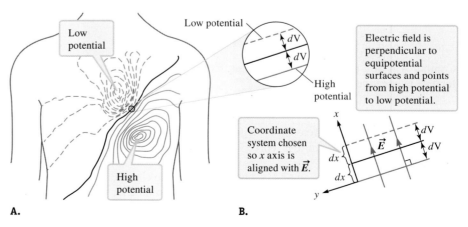

FIGURE 26.31 A. Equipotential surfaces on the human chest. *Inset*: Equipotential surfaces for a small region on the chest. **B.** Electric field lines for the small region in the inset.

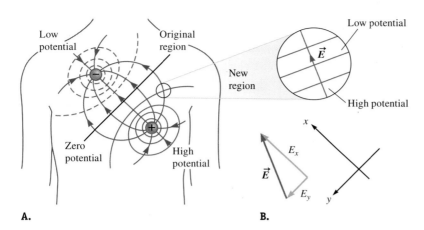

FIGURE 26.32 Modeling the human heart as a dipole electric source.

We have included the z component for completeness, although it is not shown in Figure 26.32.

The symbol ∂ indicates a **partial derivative** (Section 17-10, page 514). In a partial derivative, all variables are held fixed except the one you are differentiating with respect to. For example, if V is a function of x and y as in $V(x, y) = (3x - 6y)$ V, the electric field found by taking the partial derivatives is

$$E_x = -\frac{\partial V}{\partial x} = -\frac{\partial(3x - 6y)}{\partial x} = -3 \text{ V/m}$$

$$E_y = -\frac{\partial V}{\partial y} = -\frac{\partial(3x - 6y)}{\partial y} = 6 \text{ V/m}$$

$$\vec{E} = (-3\hat{\imath} + 6\hat{\jmath}) \text{ V/m}$$

Equations 26.21 are valid for Cartesian coordinates. When the source has spherical or linear symmetry, you may wish to use polar coordinates so that the radial component of the electric field is given by

$$E_r = -\frac{\partial V}{\partial r} \qquad (26.22)$$

(We don't need to worry about the azimuthal θ component in this book.)

FINDING THE ELECTRIC FIELD $\vec{E}$ FROM THE ELECTRIC POTENTIAL V
⭐ **Major Concept**

EXAMPLE 26.11 CASE STUDY The Electric Field Inside You

Measurements of the electric potential at several sites on the skin allow doctors to monitor important internal activities. For example, electroencephalograms (EEGs) involve recordings from 10 to 20 electrodes placed on a person's head and can aid in diagnosing brain diseases such as epilepsy (Fig. 26.33). Use the data in Table 26.1 to estimate the magnitude of the electric field in a few body regions.

TABLE 26.1

Location and activity	Approximate potential difference	Approximate separation between electrodes
Head; resting	50 μV	10 cm
Forearm; stimulation	0.7 mV	3 cm
Center of chest; light activity	1 mV	4 cm

FIGURE 26.33 Electrodes placed on a person's head measure the electric potential on the scalp at each location.

Example continues on page 816 ▶

INTERPRET and ANTICIPATE
Because we are given only the potential difference between two electrodes, we cannot calculate the derivative, but we can use the data to estimate the average electric field between the electrodes in each case.

SOLVE
In this case, the absolute value of Equation 26.18 gives the average electric field.

$$E_{av} = \left| -\frac{\Delta V}{\Delta x} \right| \quad (26.18)$$

Head:
$$E_{av} \approx \frac{50 \times 10^{-6} \text{ V}}{0.1 \text{ m}} \approx 5 \times 10^{-4} \text{ V/m}$$
$$E_{av} \approx 0.5 \text{ mV/m} \quad (\text{head})$$

Forearm:
$$E_{av} \approx \frac{0.7 \times 10^{-3} \text{ V}}{0.03 \text{ m}} \approx 2 \times 10^{-2} \text{ V/m}$$
$$E_{av} \approx 20 \text{ mV/m} \quad (\text{forearm})$$

Chest:
$$E_{av} \approx \frac{1 \times 10^{-3} \text{ V}}{0.04 \text{ m}} \approx 2.5 \times 10^{-2} \text{ V/m}$$
$$E_{av} \approx 30 \text{ mV/m} \quad (\text{chest})$$

CHECK and THINK
These electric fields are weak compared to those due to many other common sources. For example, near a charged comb, the electric field is about 100,000 times stronger. Of course, the values we calculated are for measurements made on the skin. Inside the body, the electric field is considerably stronger.

Although the electric field on the skin is very weak, it could possibly be used for networking! In spring 2005, a Japanese company developed a technology that can send data over the surface of the skin at a typical broadband connection speed.[1] The hope is to use this technology to connect headphones in your ears to a phone in your pocket with no wire or infrared connection.

EXAMPLE 26.12 Linear Symmetry

Figure 26.34 shows a positively charged, infinitely long rod of radius R and linear charge density λ. The equipotential surfaces are infinitely long, concentric cylinders. The rod itself is an equipotential surface and is chosen as the reference so that its potential is zero. The electric potential on all the other surfaces is then negative and given by

$$V(r) = \frac{\lambda}{2\pi\varepsilon_0} \ln \frac{R}{r} \quad (26.23)$$

where $r > R$. Using Equation 26.23, find an expression for the electric field due to this source.

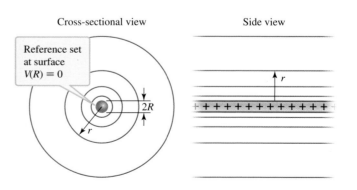

FIGURE 26.34 Two views of a long, positively charged rod.

---

[1] See *The Guardian*, March 20, 2005, p. 12.

:• **INTERPRET and ANTICIPATE**
We add electric field lines to Figure 26.34, perpendicular to the equipotential surfaces and pointing from high electric potential to low electric potential. In this case, the electric field lines point outward (Fig. 26.35). This situation has linear symmetry, and in Example 25.4 (page 766) we used Gauss's law to find the electric field due to such a source. So, we expect to find that the electric field is given by Equation 25.13, $\vec{E} = (\lambda/2\pi\varepsilon_0 r)\hat{r}$.

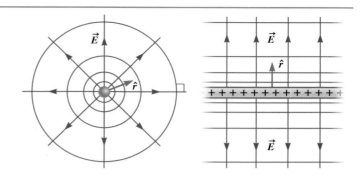

FIGURE 26.35 Electric field lines and unit vectors for the rod.

:• **SOLVE**

Because of the symmetry in this problem, we use a polar coordinate system (Fig. 26.35) with the unit vector $\hat{r}$ pointing outward from the center of the rod. Use Equation 26.22 to find the electric field, substituting Equation 26.23 for V.	$E_r = -\dfrac{\partial V}{\partial r}$	(26.22)
	$E_r = -\dfrac{\partial}{\partial r}\left(\dfrac{\lambda}{2\pi\varepsilon_0}\ln\dfrac{R}{r}\right) = -\dfrac{\lambda}{2\pi\varepsilon_0}\dfrac{\partial}{\partial r}\left(\ln\dfrac{R}{r}\right)$	
Write the natural logarithm of a fraction as two terms.	$E_r = -\dfrac{\lambda}{2\pi\varepsilon_0}\dfrac{\partial}{\partial r}(\ln R - \ln r)$	
	$E_r = -\dfrac{\lambda}{2\pi\varepsilon_0}\left(\dfrac{\partial}{\partial r}\ln R - \dfrac{\partial}{\partial r}\ln r\right)$	
The radius R of the rod is a constant, so the first derivative is zero.	$E_r = -\dfrac{\lambda}{2\pi\varepsilon_0}\left(0 - \dfrac{\partial}{\partial r}\ln r\right) = \dfrac{\lambda}{2\pi\varepsilon_0}\dfrac{\partial}{\partial r}\ln r$	
The derivative of the natural logarithm can be found in Appendix A.	$E_r = \dfrac{\lambda}{2\pi\varepsilon_0}\dfrac{1}{r}$	
Finally, write the electric field in component form.	$\vec{E} = \dfrac{1}{2\pi\varepsilon_0}\dfrac{\lambda}{r}\hat{r}$	

:• **CHECK and THINK**
As expected, this is exactly what we found using Gauss's law. Equation 26.23 is the electric potential due to a source with linear symmetry. Nerves and muscles in the body are often modeled as such sources.

26-9 Graphing *E* and *V*

We have studied sources of various geometries and compiled formulas for the electric field $\vec{E}$ and the electric potential V in numerous special cases. One way to organize all this information is by sketching graphs.

Because the electric potential is found by taking a path integral of the electric field, it is helpful to make a pair of graphs for each source to display the connection between V and E. The two graphs are V versus position and E versus position (usually r or x). Because the electric field is found by taking the (negative) spatial derivative of the electric potential, each point on the E-versus-position graph is the negative of the slope of the tangent to the V-versus-position graph. In this section, we will show such paired graphs for three charged sources: a particle, two infinite sheets, and an isolated conductor.

Special Case: Particle Source

Figure 26.36A shows a positively charged particle as the source. As usual for a particle, the reference point has been chosen at infinity where the electric potential is zero: $V(\infty) = 0$. The electric potential is high (positive) near the particle. The electric field points outward in all directions. Each point on the E-versus-r graph is -1 times the slope of the tangent line at the corresponding point on the V-versus-r plot (Fig. 26.36B), and the electric field follows an inverse-square law $(1/r^2)$ (Fig. 26.36C).

Special Case: Two Infinite Charged Sheets

Figure 26.37A shows a source that consists of two infinite parallel sheets. We have chosen a coordinate system with the x axis pointing to the right and the origin on the left sheet. The sheet on the left has excess negative charge, and the sheet on the right has excess positive charge. The charge on each sheet is uniformly distributed, and the surface charge densities are equal in magnitude. In Figure 25.35 (page 777), we found that the electric field is zero outside the sheets and uniform between them, pointing from the positive sheet to the negative sheet.

The electric field points from high to low electric potential, so the negative sheet is at a lower electric potential than the positive sheet. It is convenient to choose our reference for the electric potential such that the surface of the negative sheet is at $V = 0$. The electric potential V increases uniformly from the negative sheet to the positive sheet (Fig. 26.37B). For all positions to the left of the sheets, the electric potential is zero, and for all positions to the right of the sheets, the electric potential is a constant equal to the electric potential of the positive sheet.

Each point on the graph of E_x versus x is the negative of the slope of the V-versus-x graph (Fig. 26.37C). For the regions outside the sheets, $E = 0$ whenever V is constant, whether $V = 0$ or some other constant.

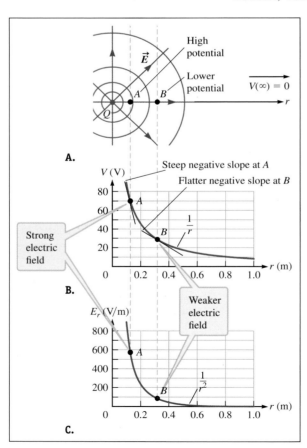

FIGURE 26.36 Special case of point particle source. **A.** Its electric field lines and equipotential surfaces. **B.** V versus r and **C.** E versus r for this source found from $E_r = -\partial V / \partial r$.

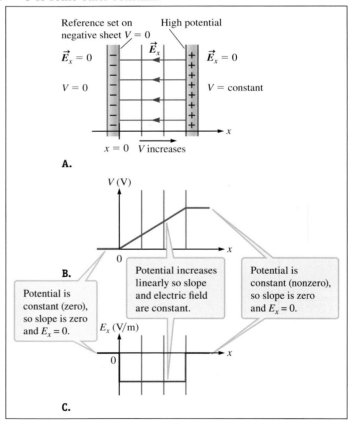

FIGURE 26.37 Special case of two infinite sheets. **A.** The negative sheet is on the left and the positive sheet is on the right, so the electric field points to the left. The electric potential on the right is higher than that on the left. **B.** V versus x and **C.** E versus x for this source. E is negative because it points in the negative x direction.

Special Case: Isolated Charged Conductor

Conductors play important practical roles in our use of electricity. So our last special case about an arbitrarily shaped charged conductor will help us in the next three chapters when we study the practical uses of electricity.

For now, we are concerned only with a conductor in electrostatic equilibrium. Imagine charging a conductor by touching it with a charged rod. At first, the conductor is not in equilibrium. The charged particles are free to travel throughout the conductor, and they do until (1) the excess charge rests only on the surface, (2) the electric field inside the body of the conductor is zero, and (3) the electric potential is uniform over the surface. Because the electric field inside the conductor is zero, the electric potential inside the body of the conductor is constant, although not necessarily zero. In fact, the electric potential in the body of the conductor is equal to the electric potential V_S at its surface. Think of it this way: If the electric field was not zero inside the conductor, or if the electric potential V_S was not the same all over the conductor, charged particles would move and the conductor would no longer be in electrostatic equilibrium.

Figure 26.38A shows an irregularly shaped, positively charged conductor with an r axis that runs from the interior of the conductor to the right. The electric field is zero in the body of the conductor. Just outside the conductor, the electric field is uniform ($E = \sigma/\varepsilon_0$, Eq. 25.17). Far from the conductor, the electric field looks like the electric field due to a positively charged particle ($E = kQ/r^2$).

Because the electric field far from a charged conductor of any shape looks like the electric field due to a charged particle, we typically set the reference at infinity where the electric potential is zero. With this convention, the electric potential V_S at the surface of the conductor is positive if the conductor is positively charged and negative if the conductor is negatively charged.

Now look at the electric potential as we work our way from $r = 0$ outward (Fig. 26.38B). Inside the body of the conductor, the electric potential is a constant, V_S. Outside the conductor but close to its surface, the electric potential drops off linearly. Finally, far from the conductor, the electric potential drops off as $1/r$ as it does when the source is a particle.

As usual, the graph of E_r versus r is the negative of the slope of the V-versus-r graph. Note that this graph (Fig. 26.38C) is consistent with the sketch (Fig. 26.38A).

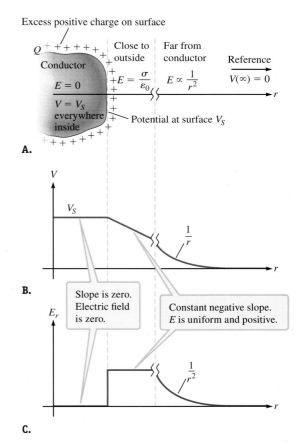

FIGURE 26.38 Special case of an isolated conductor. **A.** The conductor is positively charged and irregularly shaped. **B.** V versus r and **C.** E versus r for this source.

CONCEPT EXERCISE 26.7

CASE STUDY EKG

One way for doctors to monitor a patient's heart is with an EKG. A potential difference is measured across two (or more) points on the patient's chest. The EKG is a graph of electric potential versus time (Fig 26.39). Can you use an EKG to plot electric field versus position? If so, explain how to find that graph. If not, explain why not.

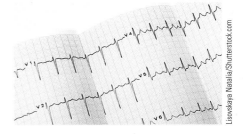

FIGURE 26.39 In an EKG, electric potential as a function of time is recorded at several different points.

SUMMARY

❶ Underlying Principles

No new principles are introduced in this chapter.

✪ Major Concepts

1. **Electric potential energy U_E** is a scalar that depends on both the source's charge and the subject's charge. Only changes in U_E are physically important.
2. **Electric potential V_E** is a scalar that depends only on the source's charge. Find V_E by dividing the electric potential energy U_E by the charge q of a test subject:

$$V_E = \frac{U_E}{q} \quad (26.6)$$

Only differences in V_E are physically important. Electric potential differences depend only on the endpoints, not on the particular path between them.

3. **Finding the electric potential from the electric field.** The electric potential difference ΔV between two points can be found by taking the path integral of the electric field $\vec{E}$ between those points:

$$\Delta V = -\int_{r_i}^{r_f} \vec{E} \cdot d\vec{r} \quad (26.15)$$

4. **Finding the electric field from the electric potential.** The electric field is found by taking the partial derivative of V. In Cartesian coordinates,

$$E_x = -\frac{\partial V}{\partial x}, \quad E_y = -\frac{\partial V}{\partial y}, \quad E_z = -\frac{\partial V}{\partial z} \quad (26.21)$$

In polar coordinates,

$$E_r = -\frac{\partial V}{\partial r} \quad (26.22)$$

Each point on an E-versus-position graph is the negative of the slope of the tangent on the corresponding V-versus-position graph.

▶ Special Cases

1. **Electric potential energy U_E in systems with a spherically symmetrical source** (a particle, outside a sphere or spherical shell where r is greater than the radius of the sphere or spherical shell):

$$U_E(r) = \frac{kQq}{r} \quad (26.3)$$

By the usual convention, $U_E = 0$ when the source and subject are infinitely far apart.

If a system has more than two charged particles, the system's electric potential energy is found by applying Equation 26.3 to each pair and then adding the results.

2. **Electric potential V due to:**
   a. A source with **spherical symmetry**. For a particle source with excess charge Q,

$$V = k\frac{Q}{r} \quad (26.8)$$

The electric potential due to a collection of n charged particles is

$$V = k\sum_{i=1}^{n} \frac{q_i}{r_i} \quad (26.9)$$

where q_i is the charge of the ith particle and r_i is its distance to the point at which the electric potential is calculated.

   b. **Dipole** source: $V \approx k\dfrac{p\cos\theta}{r^2} \quad (26.11)$

   c. **Ring** source:

$$V = \frac{kQ}{\sqrt{R^2 + x^2}} = \frac{1}{4\pi\varepsilon_0}\frac{Q}{\sqrt{R^2 + x^2}} \quad (26.13)$$

   d. **Disk** source:

$$V = 2\pi k\sigma[(\sqrt{R^2 + x^2}) - x] = \frac{\sigma}{2\varepsilon_0}[(\sqrt{R^2 + x^2}) - x] \quad (26.14)$$

   e. A source with **planar symmetry**:

$$\Delta V = -E\Delta x \quad (26.17)$$

◉ Tools

A surface on which all the points are at the same electric potential (voltage) is called an **equipotential surface**. A set of equipotential surfaces separated by equal steps in electric potential is known as a **contour map**.

For any source:
1. Electric field lines are perpendicular to equipotential surfaces.
2. Electric field lines point from high electric potential to low electric potential.
3. If the source is a continuous distribution, its surface defines an equipotential surface; if the source is positively charged, the electric potential at the source is highest; if the source is negatively charged, the source is at the lowest electric potential.

PROBLEM-SOLVING STRATEGY: Electric Potential for a Continuous Charge Distribution

INTERPRET and ANTICIPATE
Draw a **sketch** of the source and choose a **coordinate system**.

SOLVE
1. Divide the source into small **pieces** and write an expression for **dV** for each piece.
2. **Integrate** over the entire source to find V.

CHECK and THINK
For a finite source:
1. Verify that **at a point far away** compared to the size of the source, the electric potential has the same form as that of a **charged particle**.
2. Verify that **at a point infinitely far away, the electric potential is zero**.

PROBLEMS AND QUESTIONS

A = algebraic   C = conceptual   E = estimation   G = graphical   N = numerical

26-1 Scalars Versus Vectors

1. **C** What does it mean when a force is negative? What does it mean when the potential energy is negative?
2. **C Review** Return to Chapter 8 and the potential energy associated with both gravity near the surface of the Earth and universal gravity.
   a. What does the term *reference configuration* mean in the context of gravity?
   b. Suppose a system consists of an apple and the Earth, with the reference configuration set so that the system's gravitational potential energy is zero when the apple is on your desk. If the apple is above the desk, is the potential energy positive, negative, or zero?
   c. What is the conventional reference configuration for universal gravity? Why is that a convenient reference configuration? If a system consists of an apple and the Earth, does the system's gravitational potential energy increase, decrease, or stay the same as you raise the apple upward?

26-2 Gravity Analogy

3. **C Review** A system consists of a planet and a star, with the planet in an elliptical orbit. As the planet orbits, which of these quantities change because they depend on the mass of the planet? Explain your answers.
   a. Gravitational potential energy
   b. Gravitational potential
   c. Kinetic energy
   d. Gravitational force
   e. Gravitational field
4. **C** Try to complete Table P26.4 from memory. If you must look back in this chapter or other chapters for information, note the page number, figure number, or equation number that helped you.

TABLE P26.4

Term	Mathematical symbol	Vector or scalar?	Source and subject or source only?	SI units
Gravitational force				
Gravitational field				
Gravitational potential energy				
Gravitational potential				

5. **C** Try to complete Table P26.5 from memory. If you must look back in the chapter for information, note the page number, figure number, or equation number that helped you.

TABLE P26.5

Term	Mathematical symbol	Vector or scalar?	Source and subject or source only?	SI units
Electrostatic force				
Electric field				
Electric potential energy				
Electric potential				

26-3 Electric Potential Energy U_E

6. **C** Can you associate electric potential energy with an isolated charged particle? Explain.

Problems 7 and 28 are paired.

7. **N** Consider the final arrangement of charged particles shown in Figure P26.7. What is the work necessary to build such an arrangement of particles, assuming they were originally very far from one another?

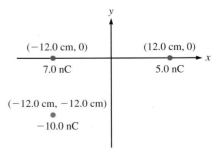

FIGURE P26.7 Problems 7 and 28.

Problems 8 and 9 are paired.

8. **C** Using the usual convention that the electric potential energy is zero when charged particles are infinitely far apart, rank the electric potential energy from least to greatest for the systems shown in Figure P26.8. Explain your answers.

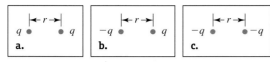

FIGURE P26.8

9. **A** Find an expression for the electric potential energy associated with each system in Figure P26.8 in terms of the quantities provided on the figure.

10. A hydrogen atom consists of an electron and a proton. Model the hydrogen atom as a dipole with separation $d = 10^{-10}$ m.
    a. **E** Estimate the electric potential energy of the hydrogen atom.
    b. **N** How much work does an external force do in liberating the electron from the atom?
    c. **C** If the external force does more than the work you found in part (b), what can you say about the electron's motion when it is very far from the proton?

11. **N** What is the work that a generator must do to move 1.80×10^{10} protons from a location with electric potential 12.4 V to a location with electric potential 43.0 V?

12. **N** How far should a $+3.0$-μC charged particle be from a -5.5-μC charged particle so that the electric potential energy of the pair of particles is -0.90 J?

13. **N** A proton is fired from very far away directly at a fixed particle with charge $q = 1.28 \times 10^{-18}$ C. If the initial speed of the proton is 2.4×10^5 m/s, what is its distance of closest approach to the fixed particle? The mass of a proton is 1.67×10^{-27} kg.

Problems 14, 15, and 16 are grouped.

14. Four charged particles are at rest at the corners of a square (Fig. P26.14). The net charges are $q_1 = q_2 = 2.65$ μC and $q_3 = q_4 = 5.15$ μC. The distance between particle 1 and particle 3 is $r_{13} = 1.75$ cm.
    a. **N** What is the electric potential energy of the four-particle system?
    b. **C** If the particles are released from rest, what will happen to the system? In particular, what will happen to the system's kinetic energy as their separations become infinite?

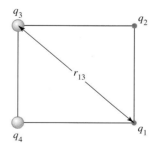

FIGURE P26.14 Problems 14, 15, and 16.

15. Four charged particles are at rest at the corners of a square (Fig. P26.14). The net charges are $q_1 = q_2 = -2.65$ μC and $q_3 = q_4 = -5.15$ μC. The distance between particle 1 and particle 3 is $r_{13} = 1.75$ cm.
    a. **N** What is the electric potential energy of the four-particle system?
    b. **C** If the particles are released from rest, what will happen to the system? In particular, what will happen to the system's kinetic energy as their separations become infinite?

16. Four charged particles are at rest at the corners of a square (Fig. P26.14). The net charges are $q_1 = q_2 = +2.65$ μC and $q_3 = q_4 = -5.15$ μC. The distance between particle 1 and particle 3 is $r_{13} = 1.75$ cm.
    a. **N** What is the electric potential energy of the four-particle system?
    b. **C** If the particles are released from rest, what will happen to the system? In particular, what will happen to the system's kinetic energy?

17. **N** Eight identical charged particles with $q = 1.00$ nC are to be arrayed at the vertices of a cube of side $d = 1.00$ cm. What is the work required to assemble the particles into this arrangement?

26-4 Electric Potential V

18. **N** A conducting sphere with a radius of 0.25 m has a total charge of 6.00 mC. A particle with a charge of -2.00 mC is initially 0.35 m from the sphere's center and is moved to a final position 0.50 m from the sphere's center.
    a. What is the difference in electric potential between the particle's final and initial positions, $\Delta V = V_f - V_i$?
    b. What is the change in the system's electric potential energy?

19. **N** The speed of an electron moving along the y axis increases from 4.40×10^6 m/s at $y = 10.0$ cm to 7.00×10^6 m/s at $y = 2.00$ cm.
    a. What is the electric potential difference between these two points?
    b. Which of the two points is at a higher electric potential?

20. **G** Figure P26.20 is a topographic map.
    a. Rank A, B, and C by elevation from the lowest point to the highest point.
    b. Rank A, B, and C by slope from the steepest slope to the flattest slope.

21. **N** At a point in space, the electric potential due to a charged particle is measured to be 2.55×10^3 V, and the magnitude of the electric field is measured to be 875 V/m.
    a. How far is the charged particle from this point?
    b. What is the magnitude of the charge on the particle?

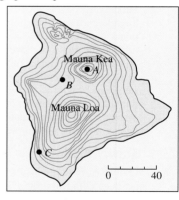

FIGURE P26.20

22. **C** Explain the difference between $U_E(r) = kQq/r$ and $V_E = kQ/r$ (Eqs. 26.3 and 26.8). Which depends on both the source and the subject? Describe the source in each case. When do you use each equation? Can you use either equation if the source has linear symmetry? Explain.

23. **N** Suppose a single electron moves through an electric potential difference of 1.5 V (the potential difference between the terminals of an AAA battery). What is the change in the system's electric potential energy? Give your answer in eV and in J.

24. Two point charges, $q_1 = -2.0$ μC and $q_2 = 2.0$ μC, are placed on the x axis at $x = 1.0$ m and $x = -1.0$ m, respectively (Fig. P26.24).
    a. **N** What are the electric potentials at the points P (0, 1.0 m) and R (2.0 m, 0)?
    b. **N** Find the work done in moving a 1.0-μC charge from P to R along a straight line joining the two points.
    c. **C** Is there any path along which the work done in moving the charge from P to R is less than the value from part (b)? Explain.

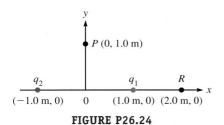

FIGURE P26.24

25. **N** Separating the electron from the proton in a hydrogen atom takes 2.21×10^{-18} J of work. Through what electric potential difference does the electron move?

26. **C** Can a contour map help you visualize the electric potential energy? Explain.

26-5 Special Case: Electric Potential Due to a Collection of Charged Particles

27. **N** A particle with charge $q_A = -6.75$ μC is located at (0, 3.25 cm), and a second particle with charge $q_B = 3.20$ μC is located at (0, −2.75 cm). What is the electric potential due to the two charges
   a. at the origin and
   b. at (3.00 cm, 0)?

28. **N** Find the electric potential at the origin given the arrangement of charged particles shown in Figure P26.7.

29. **N** Twenty-seven identical spherical drops of mercury are charged simultaneously to the same electric potential of 10.0 V. What will the electric potential be if all the charged drops are combined to form one large spherical drop?

30. Figure P26.30 shows a source that consists of two negatively charged particles, one with charge $-q$ and the other with charge $-2q$.
   a. **C** Assuming the usual convention, where is the electric potential zero?
   b. **C** Is the electric potential higher at point A or at point B? Explain.
   c. **N** Find the electric potential at point A and at point B.

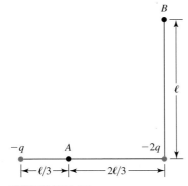

FIGURE P26.30

31. **N** A spherical conductor of radius 15.0 cm has net charge $Q = 0.72$ μC. Calculate the electric potential at the following distances from its center, assuming the electric potential goes to zero at an infinite distance from the sphere:
   a. 30.0 cm,
   b. 15.0 cm, and
   c. 4.0 cm.

Problems 32 and 33 are paired.

32. **A** A source consists of three charged particles located at the vertices of a square (Fig. P26.32). Find an expression for the electric potential at point A located at the fourth vertex.

33. **N** A source consists of three charged particles located at the vertices of a square (Fig. P26.32), where the square has sides of length 0.243 m. The charges are $q_1 = 35.0$ nC, $q_2 = -65.0$ nC, and $q_3 = 56.5$ nC. Find the electric potential at point A located at the fourth vertex.

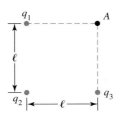

FIGURE P26.32 Problems 32 and 33.

34. **N** Two identical metal balls of radii 2.50 cm are at a center-to-center distance of 1.00 m from each other (Fig. P26.34). Each ball is charged so that a point at the surface of the first ball has an electric potential of $+1.20 \times 10^3$ V and a point at the surface of the other ball has an electric potential of -1.20×10^3 V. What is the total charge on each ball?

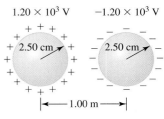

FIGURE P26.34

35. **N** Figure P26.35 shows four particles with identical charges of $+5.75$ μC arrayed at the vertices of a rectangle of width 25.0 cm and height 55.0 cm. What is the change in the electric potential energy of this system if particles A, B, and C are held in place and particle D is brought from infinity to the position shown in the figure?

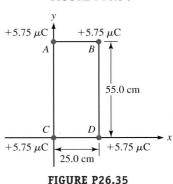

FIGURE P26.35

36. **N** Two charged particles with $q_A = 9.75$ μC and $q_B = -12.60$ μC are at the vertices of an equilateral triangle of side $a = 3.50$ cm (Fig. P26.36).
   a. What is the electric potential due to these charges at point A, halfway between the charges?
   b. What is the electric potential due to these charges at point P, the apex of the triangle?

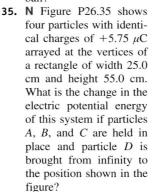

FIGURE P26.36

37. **N** Two charged particles with $q_1 = 5.00$ μC and $q_2 = -3.00$ μC are placed at two vertices of an equilateral tetrahedron whose edges all have length $s = 4.20$ m (Fig. P26.37). Determine what charge q_3 should be placed at the third vertex so that the total electric potential at the fourth vertex is 2.00 kV.

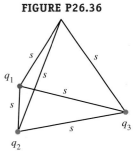

FIGURE P26.37

26-6 Electric Potential Due to a Continuous Distribution

38. Consider the charged ring in Figure 26.20 (page 804).
   a. **G** Sketch V versus x.
   b. **C** Where is the electric potential greatest?
   c. **C** Compare a sketch of V versus x for a small ring to that for a large ring. What differences are there in the graphs? Does the point at which the electric potential maximum is found depend on the size of the ring? Explain.

39. **N** A ring of radius 0.75 m has an excess charge of -892 μC, uniformly distributed. Find the electric potential at the center of the ring.

40. **N** A uniformly charged ring with total charge $q = 3.00\ \mu C$ and radius $R = 10.0$ cm is placed with its center at the origin and oriented in the xy plane. What is the difference between the electric potential at the origin and the electric potential at the point $(0, 0, 30.0$ cm$)$?

Problems 41 and 42 are paired.

41. **A** A line of charge with uniform charge density λ lies along the x axis from $x = -a$ to $x = a$.
 a. What is the magnitude of the electric potential at $(0, y)$?
 b. How much work is necessary to move a particle with charge q from very far away to $(0, y)$?

42. **N** A line of charge with uniform charge density $\lambda = 2.00 \times 10^{-3}$ C/m lies along the x axis from $x = -0.250$ m to $x = 0.250$ m.
 a. What is the magnitude of the electric potential at $(0, 1.000$ m$)$?
 b. How much work is necessary to move a particle with a charge of -5.00 nC from very far away to $(0, 1.000$ m$)$?

Problems 43 and 54 are paired.

43. **A** Consider a thin rod of total charge Q and length L (Fig. P26.43). Show that the electric potential at point P, a distance x from the end of the rod, is given by
$$V(x) = \frac{kQ}{L}\ln\left(\frac{x + L}{x}\right)$$

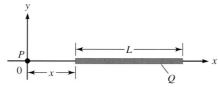

FIGURE P26.43 Problems 43 and 54.

44. **N** Figure P26.44 shows a rod of length $\ell = 1.00$ m aligned with the y axis and oriented so that its lower end is at the origin. The charge density on the rod is given by $\lambda = a + by$, with $a = 2.00\ \mu C/m$ and $b = -1.00\ \mu C/m^2$. What is the electric potential at point P with coordinates $(0, 25.0$ cm$)$? A table of integrals will aid you in solving this problem.

45. **N** The charge density on a disk of radius $R = 12.0$ cm is given by $\sigma = ar$, with $a = 1.40\ \mu C/m^3$ and r measured radially outward from the origin (Fig. P26.45). What is the electric potential at point A, a distance of 40.0 cm above the disk? *Hint:* You will need to integrate the nonuniform charge density to find the electric potential. You will find a table of integrals helpful for performing the integration.

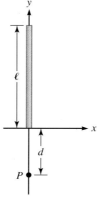

FIGURE P26.44

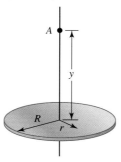

FIGURE P26.45

26-7 Connection Between Electric Field $\vec{E}$ and Electric Potential V

46. **A** Start with the expression for the electric field due to an infinitely long rod of radius R and linear charge density λ at a perpendicular distance r away from the rod: $\vec{E} = (\lambda/2\pi\varepsilon_0 r)\hat{r}$ (Eq. 25.13). Find an expression for the electric potential at a point r outside the rod. *Hint:* You already know the answer from Example 26.12 (page 816), and now you must show how V is derived from $\vec{E}$.

47. **N** In some region of space, the electric field is given by $\vec{E} = Ax\hat{i} + By^2\hat{j}$. Find the electric potential difference between points whose positions are $(x_i, y_i) = (a, 0)$ and $(x_f, y_f) = (0, b)$. The constants A, B, a, and b have the appropriate SI units.

48. **N** A particle with charge 1.60×10^{-19} C enters midway between two charged plates, one positive and the other negative. The initial velocity of the particle is parallel to the plates and along the midline between them (Fig. P26.48). A potential difference of 300.0 V is maintained between the two charged plates. If the lengths of the plates are 10.0 cm and they are separated by 2.00 cm, find the greatest initial velocity for which the particle will not be able to exit the region between the plates. The mass of the particle is 12.0×10^{-24} kg.

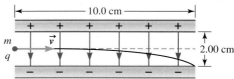

FIGURE P26.48

49. A conducting sphere with radius R has total charge Q.
 a. **A** Find the relationship between the magnitude of the electric field and the electric potential on the surface of the conducting sphere.
 b. **N** For a sphere of radius 16 cm, calculate the maximum surface electric potential at which the surrounding air begins to break down. Take the dielectric strength of (maximum sustainable electric field in) air to be 3.0×10^6 V/m.

50. **A** Show that the interior of a conductor in electrostatic equilibrium is always an equipotential volume.

26-8 Finding $\vec{E}$ from V

51. **A** The electric potential for a system is given by $V(r) = V_0 e^{-ar}/r$, where V_0 and a ($a > 0$) are constants. Determine the magnitude of the electric field.

52. **E** **CASE STUDY** The resting electric potential across the surface or membrane (perpendicular to the long axis) of a nerve cell is -70 mV (Fig. P26.52). The thickness of the membrane that separates the inside from the outside of the nerve cell is about 6×10^{-9} m. Estimate the magnitude of the electric field. Which way does the electric field point?

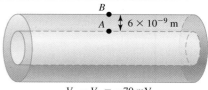

FIGURE P26.52 $V_B - V_A = -70$ mV

53. **N** The function $V = b + cy$ describes the electric potential in the region between $y = -4.00$ m and $y = 4.00$ m, with $b = 12.5$ V and $c = 4.50$ V/m.
 a. What is the electric potential at $y = -4.00$ m, at $y = 0$, and at $y = 4.00$ m?
 b. What are the magnitude and direction of the electric field at $y = -4.00$ m, at $y = 0$, and at $y = 4.00$ m?

54. **A** According to Problem 43, the electric potential at point P, a distance x from the end of the rod, is given by

$$V(x) = \frac{kQ}{L} \ln\left(\frac{x+L}{x}\right)$$

Use this equation to find the electric field at a distance x from the end of the charged rod.

55. The electric potential is given by $V = 4x^2z + 2xy^2 - 8yz^2$ in a region of space, with x, y, and z in meters and V in volts.
  a. **A** What are the x, y, and z components of the electric field in this region?
  b. **N** What is the magnitude of the electric field at the coordinates (2.00 m, $-$2.00 m, 1.00 m)?

56. **N** The electric potential $V(x, y, z)$ in a region of space is given by $V(x, y, z) = V_0(2x^2 - 3y^2 - z^2)$, where $V_0 = 12.0$ V and x, y, and z are measured in meters. Find the electric field at the point (1.00 m, 1.00 m, 0).

26-9 Graphing E and V

57. Two thin, rectangular plates have charges of equal magnitude and opposite sign as shown in Figure P26.57. The electric field between the plates is constant and equal to E.
  a. **G** Sketch the electric field lines and the equipotential lines between the plates.
  b. **N** Determine the work required to move a particle with charge q in the closed rectangular loop $ABCD$ shown in the figure.

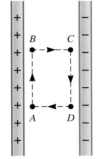

FIGURE P26.57

58. In three regions of space, the electric potential is given by

$$V(r) = 0 \text{ for } r < R$$
$$V(r) = \frac{V_0}{4R^2}r^2 \text{ for } R \leq r < 2R$$
$$V(r) = V_0 \text{ for } r \geq 2R$$

  a. **G** Plot V as a function of r.
  b. **A** Find expressions for the electric field in all three regions.
  c. **G** Plot E versus r in all three regions.

General Problems

59. **N** An aluminum sphere 22.0 cm in radius carries a total charge of 845 nC. What are the magnitudes of the electric field and the electric potential at these radial distances from the center of the sphere:
  a. 5.00 cm,
  b. 22.0 cm, and
  c. 50.0 cm?

60. **N** A particle with charge $q = 2.7$ nC and mass $m = 0.57$ g is in a circular orbit of radius $r = 1.3$ mm around a fixed particle with charge $Q = -8.6$ μC.
  a. Find the speed of the orbiting particle.
  b. Calculate the electric potential energy of the system.
  c. What is the total energy of the system?

61. **N** The distance between two small charged spheres with charges $q_A = -8.35$ μC and $q_B = +4.90$ μC is 48.0 cm.
  a. What is the electric potential energy due to the two spheres?
  b. What is the electric potential halfway between the two spheres along the line connecting them?

62. **A** From Section 9-5, there are two ways to test a force to see whether it is path-independent—a necessary property of a conservative force. The first test involves calculating the work done by the force on a subject along several different paths. If the work done is not the same along all the paths, the force is nonconservative. The alternative test involves calculating the work done by the force on a particle along a closed path. If the work done along a closed path does not equal zero, the force is nonconservative. Show that the electrostatic force does not fail the two tests for a conservative force. *Hint*: Use Examples 9.5 and 9.6 (pages 257–259) as a guide.

63. **N** A glass sphere with radius 4.00 mm, mass 85.0 g, and total charge 4.00 μC is separated by 150.0 cm from a second glass sphere 2.00 mm in radius, with mass 300.0 g and total charge -5.00 μC. The charge distribution on both spheres is uniform. If the spheres are released from rest, what is the speed of each sphere the instant before they collide?

64. **N** A charged particle with $q_A = 1.00$ nC is located at the origin, and a second particle with charge $q_B = -3.00$ nC is located at $y = 1.50$.
  a. What are the finite values of y for which the electric field is zero?
  b. What are the finite values of y for which the electric potential is zero?

Problems 65, 66, and 67 are grouped.

65. Two 5.00-nC charged particles are in a uniform electric field with a magnitude of 625 N/C. Each of the particles is moved from point A to point B along two different paths, labeled in Figure P26.65.
  a. **N** Given the dimensions in the figure, what is the change in the electric potential experienced by the particle that is moved along path 1 (black)?
  b. **N** What is the change in the electric potential experienced by the particle that is moved along path 2 (red)?
  c. **C** Is there a path between the points A and B for which the change in the electric potential is different from your answers to parts (a) and (b)? Explain.

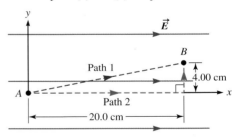

FIGURE P26.65 Problems 65, 66, and 67.

66. **N** A 5.00-nC charged particle is at point B in a uniform electric field with a magnitude of 625 N/C (Fig. P26.65). What is the change in electric potential experienced by the charge if it is moved from B to A along
  a. path 1 and
  b. path 2?

67. **C** A charged particle is moved in a uniform electric field between two points, A and B, as depicted in Figure P26.65. Does the change in the electric potential or the change in the electric potential energy of the particle depend on the sign of the charged particle? Consider the movement of the particle from A to B, and vice versa, and determine the signs of the electric potential and the electric potential energy in each possible scenario.

68. **N** Figure P26.68 shows three small spheres with identical charges of -3.00 nC placed at the vertices of an equilateral triangle with side $d = 2.50$ cm.

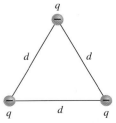

FIGURE P26.68

a. Is the electric potential due to the three spheres zero anywhere in the plane that contains the triangle, other than at infinity?
b. What is the electric potential at the location of each sphere due to the other two spheres?

69. N What is the work required to charge a spherical shell of radius $r = 43.0$ cm to a total charge of 67.0 μC with charges that are brought to the shell from infinity?

70. G For a system consisting of two identical negatively charged particles, sketch U_E versus r and V_E versus r. Comment on your graphs.

71. N Figure P26.71 shows three charged particles arranged at the vertices of an isosceles triangle with base $b = 1.00$ m. What is the electric potential due to the particles at point P, which is at the midpoint of the base?

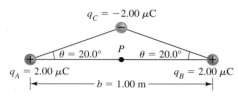

FIGURE P26.71

Problems 72, 73, and 74 are grouped.

72. A Figure P26.72 shows a source consisting of two identical parallel disks of radius R. The x axis runs through the center of each disk. Each disk carries an excess charge uniformly distributed on its surface. The disk on the left has a total positive charge Q, and the disk on the right has a total negative charge $-Q$. The distance between the disks is $3R$, and point A is $2R$ from the positively charged disk. Find an expression for the electric potential at point A between the disks on the x axis. Approximate any square roots to three significant figures.

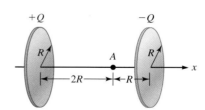

FIGURE P26.72 Problems 72, 73, and 74.

73. A Start with $V = 2\pi k\sigma\left[(\sqrt{R^2 + x^2}) - x\right]$ for the electric potential of a disk of radius R and excess surface charge density σ at a position x from the center of a disk on its axis, and derive an expression for the electric field at this position. *Hint*: See Example 24.6 (page 732) to check your answer.

74. A **Review** Consider the charged disks in Problem 72 (Fig. P26.72). Find an expression for the electric field at point A between the disks on the x axis. Approximate any square roots to three significant figures.

75. A long thin wire is used in laser printers to charge the photoreceptor before exposure to light. This is done by applying a large potential difference between the wire and the photoreceptor.
a. A Use Equation 26.23,
$$V(r) = \frac{\lambda}{2\pi\varepsilon_0} \ln\frac{R}{r}$$
to determine a relationship between the electric potential V and the magnitude of the electric field E at a distance r from the center of the wire of radius R ($r > R$).
b. N Determine the electric potential at a distance of 2.0 mm from the surface of a wire of radius $R = 0.80$ mm that will produce an electric field of 1.8×10^6 V/m at that point.

76. An electric potential exists in a region of space such that $V = 8x^4 - 2y^2 + 9z^3$ and V is in units of volts, when x, y, and z are in meters.
a. A Find an expression for the electric field as a function of position.
b. N What is the electric field at $(2.0$ m, -4.5 m, -2.0 m$)$?

77. A A disk with a nonuniform charge density $\sigma = ar^2$ has total charge Q and radius R. Derive an expression for the electric potential at a point along the axis of the disk a distance x from the center of the disk. A table of integrals will aid you in solving this problem.

Problems 78 and 79 are paired.

78. N An infinite number of charges with $q = 2.0$ μC are placed along the x axis at $x = 1.0$ m, $x = 2.0$ m, $x = 4.0$ m, $x = 8.0$ m, and so on, as shown in Figure P26.78. Determine the electric potential at the point $x = 0$ due to this set of charges. *Hint*: Use the mathematical formula for a geometric series,
$$1 + r + r^2 + r^3 + r^4 + \cdots = \frac{1}{1-r}$$

![q q q q along x axis at 0, 1.0 m, 2.0 m, 4.0 m, 8.0 m]

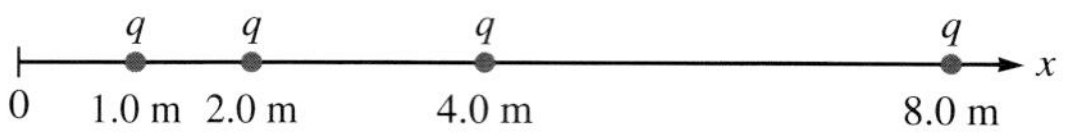

FIGURE P26.78

79. N An infinite number of charges with $|q| = 2.0$ μC are placed along the x axis at $x = 1.0$ m, $x = 2.0$ m, $x = 4.0$ m, $x = 8.0$ m, and so on, as shown in Figure P26.79. What will be the electric potential at $x = 0$ if the consecutive charges have alternating signs as shown in Figure P26.79? *Hint*: Use the mathematical formula for a geometric series,
$$1 + r + r^2 + r^3 + r^4 + \cdots = \frac{1}{1-r}$$

![q -q q -q along x axis at 0, 1.0 m, 2.0 m, 4.0 m, 8.0 m]

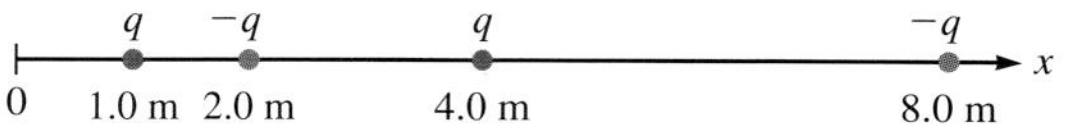

FIGURE P26.79

80. N Figure P26.80 shows a wire with uniform charge per unit length $\lambda = 2.25$ nC/m comprised of two straight sections of length $d = 75.0$ cm and a semicircle with radius $r = 25.0$ cm. What is the electric potential at point P, the center of the semicircular portion of the wire?

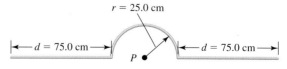

FIGURE P26.80

81. A Integrate to show
$$\int_0^R \frac{\mathcal{R} \cdot d\mathcal{R}}{\sqrt{\mathcal{R}^2 + x^2}} = \left(\sqrt{R^2 + x^2}\right) - x$$
which we needed in Example 26.7 (page 805).

Capacitors and Batteries

27

❶ Underlying Principles

General form of Gauss's law

✪ Major Concepts

1. Capacitor
2. Capacitance
3. Energy stored by a capacitor
4. Ideal battery
5. Equivalent capacitance of capacitors in series
6. Equivalent capacitance of capacitors in parallel
7. Dielectric; capacitance in the presence of a dielectric
8. Energy density stored in an electric field

▶ Special Cases

1. Parallel-plate capacitor
2. Cylindrical capacitor

⊙ Tools

Schematic diagrams (circuit diagrams)

Key Questions

How do we apply the theoretical principles and concepts of electricity to practical devices?

Can an electric field be used to store energy?

How do we make Gauss's law more general?

27-1 **The Leyden jar** 828
27-2 **Capacitors** 829
27-3 **Batteries** 833
27-4 **Capacitors in parallel and series** 837
27-5 **Capacitance: Special cases** 843
27-6 **Dielectrics** 847
27-7 **Energy stored by a capacitor with a dielectric** 851
27-8 **Gauss's law in a dielectric** 855

Ben Franklin, like most scientists, believed the most important goal of science is to understand nature. Scientists try to answer fundamental questions about nature, such as "How does the Moon orbit the Earth?" and "What is lightning?"

Like many inventors, Franklin also believed scientific discoveries should be used to help people. Franklin invented many useful things, including bifocals, the Franklin stove, and lightning rods. This chapter is about practical ways to create an electric field and to store energy as electric potential energy. We will study two types of devices: (1) those that maintain a potential difference, such as batteries and hand-operated electric generators, and (2) those that store electric potential energy. The original device for storing electric potential energy—the Leyden jar—comes from Franklin's time.

FIGURE 27.1 In the late 1700s, generators enabled experimenters to build up charge efficiently. By turning the crank, the experimenter rubbed two materials together, causing the transfer of charge from one material to the other.

27-1 The Leyden Jar

In order to perform his experiments, Franklin needed a way to generate and store charged particles. He could charge a glass rod by rubbing it with a piece of silk, but this soon grew tiring and was not efficient for building up a large amount of charge. So, 18th-century experimenters used generators like the ones shown in Figure 27.1. A large glass sphere replaces the glass rod. A pad covered in a piece of cloth such as silk is held against the sphere, which is attached to a crank. When the experimenter turns the crank, the sphere rotates and charge is transferred between the sphere and the cloth.

Using our contemporary understanding of work and energy, consider the silk cloth and the glass sphere to be the system. The experimenter does work on the system by turning the crank, which transfers energy from the environment (the experimenter) to the silk-glass system. Although some energy may be dissipated by friction, much of the energy goes into transferring charged particles. Because the glass sphere ends up positively charged and the silk is negatively charged as a result of the work done, electric potential energy is stored in the system. Furthermore, if we consider the glass sphere as the source, we can say that the work done by the experimenter sets up the sphere's electric potential.

Experimenters in the 18th century found that after turning the crank for a few minutes, they were unable to build up any more charge on the sphere. After it acquires some positive charge, the sphere attracts negatively charged particles, so it becomes difficult to transfer more electrons to the silk. Experimenters needed a way around this limitation. In the 18th century, people thought of electricity as a fluid, so they stored the "fluid" in a **Leyden jar,** which consisted of two conductors separated by an insulator. To store even more charge, they used several Leyden jars.

Franklin built his own Leyden jars and described their construction in a 1758 letter. He lined the inside and outside of a glass jar with tinfoil. You can easily make your own Leyden jar; the key is to have two conductors (the foil) separated by an insulator (glass). The jar is charged by means of a metal conductor in contact with the glass sphere of a generator. Charged particles move from the rotating sphere to the conductor and through a thin wire to the tinfoil on the inside of the Leyden jar (Fig. 27.2). If the experimenter holds the tinfoil on the outside of the jar and is in

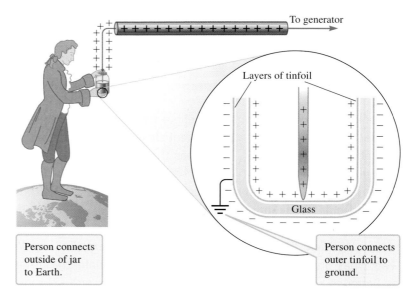

FIGURE 27.2 Positive particles are transferred to the inner surface of the Leyden jar. The outer surface is connected to ground through the researcher's body, so it picks up negative particles.

contact with the ground, an equal magnitude of charge of opposite sign builds up on the outside of the jar (Fig. 27.2, close-up).

With our contemporary understanding of work and energy, we know that the work done by the experimenter in turning the crank is stored as electric potential energy in the Leyden jar, so the 18th-century experimenter had a way to store energy. The experimenter could later use that energy as needed by connecting the inside of the jar to the outside with a conductor (Fig. 27.3). A Leyden jar is like a simple spring-loaded toy dart gun. Compressing the spring does work on the dart-spring system, storing potential energy. Energy can be released from a loaded dart gun or a charged Leyden jar as in this chapter's case study.

FIGURE 27.3 Discharging a Leyden jar.

CASE STUDY Thompson Coil

If you search on the Internet for *Thompson coils*, you are likely to find many sites describing how enthusiastic experimenters build such a device. Potential energy stored in an electric field may be used to launch a metal ring (Fig. 27.4). No springs are used, only electromagnetism. In Chapter 33, we will study the mechanism for converting the stored electric potential energy into kinetic energy. In this chapter, we will explore claims made on some websites in terms of the conservation of energy principle. You may see a demonstration of a Thompson coil in your class. Typically, a ring reaches a height of a few feet (1 m to 5 m); however, our case study is about one group of experimenters who claimed to have launched a 90-g aluminum ring about 120 m straight up!

FIGURE 27.4 A Thompson coil in use.

CONCEPT EXERCISE 27.1

CASE STUDY How Big a Spring?

Imagine the ring in the case study is launched using a spring-loaded gun instead of a Thompson coil. If the gun has a stiff spring with a spring constant of 1000 N/m, how much does the spring need to be compressed in order to propel a 90-g object 120 m up? Comment on the size of the spring. (Ignore air resistance.)

27-2 Capacitors

A Leyden jar is an example of a **capacitor**, a device that consists of two conductors known as **plates** separated by an insulator or vacuum (Fig. 27.5). The plates don't have to be flat. For Leyden jars, the plates are the metal foil lining the inside and outside of the jar, and the insulator is the glass. The purpose of the capacitor is to store electric potential energy. When the capacitor is charged, one of the plates is positive and the other has an equal magnitude of negative charge.

CAPACITOR ⭐ Major Concept

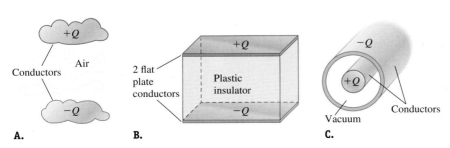

FIGURE 27.5 A capacitor consists of two conducting plates separated by an insulator such as **A.** air, **B.** plastic, or **C.** a vacuum.

The phrases "the capacitor has a charge Q" and "charge Q is stored on the capacitor" mean that the charge on the positive plate is $+Q$ and the charge on the negative plate is $-Q$.

The net charge of the entire capacitor is zero (positive charge $+Q$ on one plate and negative charge $-Q$ on the other). However, we usually refer to the charge on the positive plate rather than to the net charge of the entire capacitor. If a capacitor is charged, there is a potential difference between the two plates. Because the potential is uniform on the surface of any conductor, no matter how oddly shaped it is, the potential difference between any point on one plate and any point on the other plate is the same. Now, imagine that each plate is neutral, so the potential difference between the plates is zero. If one plate acquires a small amount of positive charge and the other a small amount of negative charge, you would expect a small potential difference between the plates. As the amount of charge on each plate increases, so does the potential difference between the plates. The potential difference ΔV between the plates is directly proportional to the magnitude of charge Q on each plate:

$$Q \propto \Delta V$$

CAPACITANCE ✪ **Major Concept**

The constant of proportionality is called the **capacitance** C:

$$Q = C\Delta V \qquad (27.1)$$

The SI unit for capacitance is the **farad** (F). Using Equation 27.1, we can express farads in terms of coulombs and volts:

$$1\,\mathrm{F} = 1\,\mathrm{C/V}$$

You must distinguish between the symbol for capacitance (uppercase italic C) and the abbreviation for the SI unit coulomb (uppercase nonitalic C).

The capacitance depends only on geometric factors and the type of insulator between the plates. For two capacitors with the same potential difference between their plates, the one with greater capacitance has a greater charge on its plates and stores more electric potential energy.

General Expression for Energy Stored by a Capacitor

Let's use gravity as an analogy to find a general expression for the electric potential energy stored by a capacitor. Figure 27.6 shows a tray full of marbles on a table. Above the marbles is another tray, which is initially empty. The system consists of the Earth and the marbles, with the reference configuration chosen so that the gravitational potential energy is zero when all the marbles are in the lower tray. In Figure 27.6A, Rochelle (outside the system) raises the marbles one at a time to the upper tray at height h above the lower tray. If each marble has mass m, the work w she does to raise one marble to the upper tray is

$$w = mgh$$

If Rochelle lifts all N marbles, the total work W she does on the system is the sum of the work she does in lifting each marble:

$$W = \sum_{i=1}^{N} m_i g h = Nmgh = Mgh$$

where M is the total mass of the marbles. Because there is no change in the system's kinetic energy ($\Delta K = 0$) and no dissipative forces are present ($\Delta E_{\mathrm{th}} = 0$), according

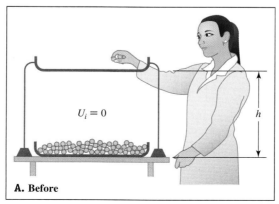
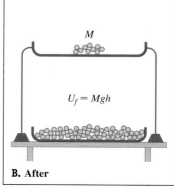

FIGURE 27.6 A person does work in lifting marbles from a lower tray to an upper tray.

A. Before **B. After**

to the work-energy theorem ($W = \Delta K + \Delta U + \Delta E_{th}$), the total work done by Rochelle equals the change in the system's potential energy: $W = \Delta U$. The system started with zero potential energy, so the system's potential energy after Rochelle lifted the marbles is equal to the work she did: $U_f = Mgh$ (Fig. 26.7B). In the next derivation, we use similar reasoning to find the potential energy stored by a capacitor.

DERIVATION Potential Energy Stored by a Capacitor

Figure 27.7A shows two neutral conductors separated by a vacuum. Each plate has an equal number of positive and negative particles, but for simplicity we have omitted the charges on the upper plate. As in Figure 27.6, we imagine an external force moves positive particles from the lower plate to the upper plate. (This is equivalent to moving negative particles in the opposite direction.) We will show that the potential energy stored by a capacitor (Fig. 27.7B) is given by

$$U_E = \frac{1}{2}\frac{Q^2}{C} \quad (27.2)$$

and

$$U_E = \frac{1}{2}C(\Delta V)^2 \quad (27.3)$$

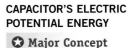

CAPACITOR'S ELECTRIC POTENTIAL ENERGY

✪ Major Concept

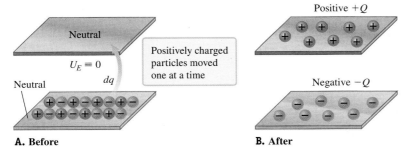

A. Before **B. After**

FIGURE 27.7

The system consists of the two plates. When the plates are neutral, the system is in the reference configuration with zero electric potential energy (Fig. 27.7A). The work done by the external force equals the change in the system's potential energy: $W = \Delta U_E$.

Each particle has a small amount of positive charge dq. Equation 26.7 gives the change in the system's electric potential energy dU when a single charged particle is moved to the upper plate.	$dU_E = (dq)\Delta V \quad (26.7)$
After N particles are moved, the total charge on the upper plate is $+Q$ and the total charge on the lower plate is $-Q$ (Fig. 27.7B). The total change in the system's potential energy is found by adding (or integrating) the changes in potential energy due to the relocation of all the charged particles.	$\int_0^{U_E} dU_E = \int_0^Q (dq)\Delta V$
The potential difference ΔV depends on the amount of charge already on the plates. Substitute $\Delta V = q/C$ (Eq. 27.1) before integrating.	$\int_0^{U_E} dU_E = \int_0^Q (dq)\left(\frac{q}{C}\right) = \frac{1}{C}\int_0^Q q\,dq$ $U_E = \frac{1}{C}\frac{q^2}{2}\bigg|_0^Q$
Evaluate the integral between limits to arrive at the energy stored by a capacitor with charge Q.	$U_E = \frac{1}{2}\frac{Q^2}{C} \checkmark \quad (27.2)$

Derivation continues on page 832 ▶

It is sometimes convenient to write the potential energy stored by a capacitor in terms of the potential difference ΔV between its plates instead of its charge. Substitute $Q = C\Delta V$ (Eq. 27.1) for Q in Equation 27.2.

$$U_E = \frac{1}{2}C(\Delta V)^2 \quad \checkmark \qquad (27.3)$$

COMMENTS

The amount of energy stored by a capacitor depends on three factors. According to $U_E = Q^2/2C$ (Eq. 27.2), more stored charge Q means more stored energy U_E, and according to $U_E = \frac{1}{2}C(\Delta V)^2$ (Eq. 27.3), a greater potential difference means more energy stored. However, the third factor is the capacitance, which is a property of the geometry and the insulator between the plates. The energy stored is inversely proportional to C in $U_E = Q^2/2C$ and directly proportional to C in $U_E = \frac{1}{2}C(\Delta V)^2$. You will explore this apparent contradiction further in Problem 6.

CONCEPT EXERCISE 27.2

Consider two different capacitors, A and B. Figure 27.8 shows a graph of the potential difference ΔV between the two plates of each capacitor versus the charge Q on the plates. Use these graphs to find the capacitance of each capacitor.

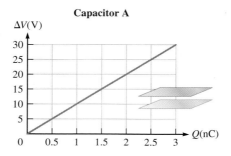
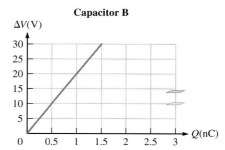

FIGURE 27.8

CONCEPT EXERCISE 27.3

a. If capacitor B in Figure 27.8 has a charge of 0.5 nC, what is the potential difference between its plates and how much potential energy does it store?
b. If capacitor A in Figure 27.8 stores the same amount of potential energy found in part (a), what is the magnitude of the excess charge on its plates? What is the potential difference between its plates? Use the capacitance C_A found in Concept Exercise 27.2.

EXAMPLE 27.1 — CASE STUDY: Careful Around Those Plates!

The Thompson coil experimenters in the case study on page 829 said they launched the ring by charging a bank of capacitors something like a collection of Leyden jars. We can model this bank of capacitors as a single device known as an **equivalent** capacitor (Section 27-4). The experimenters reported a potential difference of 1500 V between the plates of an equivalent capacitor. Model their collection of capacitors as a single capacitor, and find its (equivalent) capacitance. How much charge is stored by this capacitor when its potential difference is 1500 V? Ignore air resistance, and assume that the ring undergoes only translational motion. Recall that the mass of the ring is 90 g and its maximum height is 120 m. (Report your answers to two significant figures.)

:• **INTERPRET and ANTICIPATE**
Consider the capacitor, the ring, and the Earth to be the system. The ring begins at rest, and it is momentarily at rest at the top of its flight. No external force does work on the system, and there is no change in the thermal energy. A bar chart (Fig. 27.9) shows that the electric potential energy stored by the bank of capacitors is equal to the gravitational potential energy of the system when the ring is at its maximum height. Because we can find the gravitational potential energy, we can also find the electric potential energy. We expect the capacitance and stored charge to be fairly large.

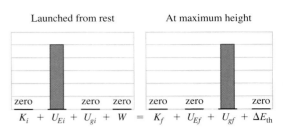

FIGURE 27.9

:• **SOLVE**
Set the potential energy stored by the capacitor equal to the maximum gravitational potential energy. U_E is related to the capacitance by $U_E = \frac{1}{2}C(\Delta V)^2$ (Eq. 27.3).

$$U_{Ei} = U_{gf}$$

$$\frac{1}{2}C(\Delta V)^2 = mgy_{max}$$

$$C = \frac{2mgy_{max}}{(\Delta V)^2} = \frac{2(0.09\,\text{kg})(9.81\,\text{m/s}^2)(120\,\text{m})}{(1500\,\text{V})^2}$$

$$C = 9.4 \times 10^{-5}\,\text{F} = \boxed{94\,\mu\text{F}}$$

Use Equation 27.1 to find the charge stored by the capacitor.

$$Q = C\Delta V \quad (27.1)$$

$$Q = (9.4 \times 10^{-5}\,\text{F})(1500\,\text{V}) = \boxed{0.14\,\text{C}}$$

:• **CHECK and THINK**
The capacitance of a typical off-the-shelf capacitor ranges from a few picofarads to a few microfarads, and as expected, the capacitance here is rather large. We'll continue to explore how the experimenters were able to get such a large capacitance. Also, notice that the charge stored by the capacitor is very large. (The charge that builds up when two objects are rubbed together is on the order of a few microcoulombs.) As discussed on their website, the experimenters had to be careful not to touch both plates of the capacitor at the same time because if the capacitor had discharged through a person, the result would have probably been fatal.

27-3 Batteries

The invention of the battery was the byproduct of a heated disagreement between 18th-century scientists. The Italian anatomy professor Luigi Galvani was dissecting a frog near an electric generator. When he touched the scalpel to the frog's nerves, its legs twitched and a spark was drawn from the generator. Galvani discovered that the frog's legs also twitched when he touched the animal with two different metals (for instance, copper on the frog's spine and iron on its feet). Galvani was convinced that the frog's body produced electricity ("animal electricity").

Another Italian scientist, Alessandro Giuseppe Antonio Volta, successfully reproduced Galvani's experiment using two different metals in contact with a frog. Eventually, Volta found that if he used the same metal at both contact places, the frog's legs would *not* twitch. Volta reasoned that connecting dissimilar metals somehow produced electricity and there was no "animal electricity." Many scientists joined the argument, which went on for decades beyond the lifetimes of both Galvani and Volta. Although Galvani was correct that electricity plays a key role in muscle function, Volta's skepticism led him to invent the battery by "piling" (stacking) dissimilar types of metal.

The SI unit of electric potential—the volt—is named in Volta's honor.

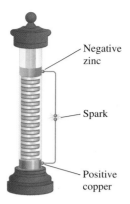

FIGURE 27.10 Volta's pile: a metal rod attached to the zinc plate at the top and another rod attached to the copper plate at the bottom. At the other end of each rod is a small metal ball with a short gap between the balls. Volta could demonstrate the electric nature of his pile by narrowing the gap between the balls so that a spark would jump that gap.

Figure 27.10 shows one of Volta's piles consisting of many cells. Each cell is made up of three layers: two types of metal (here, copper and zinc) separated by a layer of moist paper or cloth. There is a potential difference between the top and bottom of each cell. Volta achieved a greater potential difference by stacking many cells on top of one another. A pile of such cells is called a **battery**. Today, we use the term *battery* to mean any device—cell or collection of cells—that maintains a potential difference through chemical reactions.

Many students think batteries are like capacitors, but batteries and capacitors have different purposes. A capacitor's purpose is to store charge and therefore electric potential energy, whereas a battery's purpose is to maintain a potential difference through chemical reactions. A battery is more like an 18th-century electric generator (Fig. 27.1). In both a generator and a battery, work is done in order to separate charged particles, creating an electric potential difference. In a generator, this work comes from a person rotating the crank. In a battery, the energy needed to separate charges comes from chemical reactions.

Volta found that he could also build a cell by placing strips of different metals in a cup of liquid such as salt water, forming a **wet cell**. Figure 27.11 shows a wet cell with a strip of zinc and a strip of copper in a sulfuric acid (H_2SO_4) solution. The metal strips are called **electrodes**, and the liquid solution is called an **electrolyte**. The parts of the electrodes that are above the solution are called the **terminals**.

Chemical reactions generate a potential difference between the zinc and copper terminals. Sulfuric acid (H_2SO_4) molecules break up when they are in solution into two positively charged H^+ ions and one negatively charged SO_4^{2-} ion. The SO_4^{2-} ion reacts with zinc, pulling positively charged zinc ions off that electrode. The zinc electrode begins to dissolve and becomes negatively charged. At the same time, positively charged H^+ ions are attracted to copper and remove electrons from the copper electrode. This reaction causes the copper to become positively charged. The result is that the zinc terminal becomes negative and the copper terminal becomes positive. So, there is a potential difference between the terminals; this is called the **terminal potential** or **terminal voltage**.

When the electrodes in Figure 27.11 are first put into the sulfuric acid (H_2SO_4) solution, they are neutral and the terminal potential is zero. As the chemical reactions proceed, the zinc electrode becomes negative and the copper becomes positive, so the terminal potential increases. After a period of time, the terminal potential reaches a maximum because as the zinc electrode becomes more negative, positive ions cannot easily escape; they are attracted to that electrode and are quickly pulled back. The same sort of argument can be made about electrons pulled off the positive copper electrode. The maximum terminal potential is reached when no more positive ions can successfully escape from the zinc and no more electrons can successfully escape from the copper. The terminal potential depends on the specific chemical reaction, so different electrodes in different electrolytes produce different terminal potentials. Many batteries that are commonly used in portable devices such as cell phones, portable media players, and flashlights are known as **dry cells** because the electrolyte is a paste instead of a messy liquid. Table 27.1 lists some commonly used batteries by their electrodes and gives their typical terminal potentials.

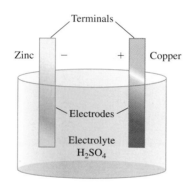

FIGURE 27.11 A wet cell consists of electrodes in a liquid electrolyte. The part of an electrode outside the electrolyte is known as a terminal. The cell maintains a potential difference between its terminals through chemical reactions.

TABLE 27.1 Common types of batteries.

Electrodes	Terminal potential (V)	Uses
Zinc and carbon	1.5	AA, AAA, C, and D dry cells used in portable devices
Lead and lead dioxide	12 (6 cells)	Cars
Nickel-hydroxide and cadmium	1.2	Rechargeable
Lithium and carbon	3.7	High-end rechargeable batteries used in computers and cell phones
Lithium-iodide and lead-iodide	2.2	Cameras
Zinc and air	1.4	Hearing aids

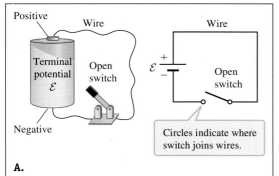
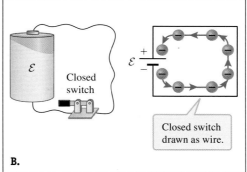

FIGURE 27.12 A. The switch is open, so charged particles cannot move through the wire. **B.** The switch is closed, and electrons move away from the negative terminal through the wire to the positive terminal. The battery maintains the potential difference between its terminals.

Figure 27.12 shows a typical battery, some wire, and a switch connected to form an **electrical network**. The wire and switch are conductors. If the switch is open, air in the gap prevents electrons from moving through the network (Fig. 27.12A). When the switch is closed, electrons can flow from one part of the electrical network to the other (Fig. 27.12B).

Simple sketches used to illustrate an electrical network are called **schematic diagrams**. The schematic diagrams on the right of each part of Figure 27.12 represent the electrical networks shown on the left. A wire is represented by lines that may be bent or straight as shown. A switch is represented by a short, straight line with a circle on the end. Another circle is drawn on the part of the wire across the gap. When the switch is closed, it is not necessary to draw these circles (Fig. 27.12B). A battery is represented by two parallel lines: a long line representing the positive terminal and a short line representing the negative terminal. The terminal potential $\mathcal{E}$, which looks like a curly Greek letter epsilon, is written next to the battery's symbol.

Suppose the terminals of a zinc-copper battery are connected by a wire (Fig. 27.12B). Electrons flow from the negative electrode (zinc) through the external circuit to the positive electrode (copper). As the electrons leave the zinc electrode, the zinc can once again react with the electrolyte, and more positive zinc ions move from the electrode into the electrolyte. As a result, the zinc is able to maintain its negative charge. The same sort of process maintains the copper's positive charge. Because each electrode maintains its net charge, the terminal potential remains constant. Throughout this book, we will use the term **ideal battery** to describe a device that maintains its terminal potential through chemical reactions. The symbol $\mathcal{E}$ represents the terminal potential of an ideal battery, which is actually the potential *difference* between its terminals. All the batteries in this chapter may be modeled as ideal.

SCHEMATIC DIAGRAMS ⦿ Tool

IDEAL BATTERY ✪ Major Concept

Charging a Capacitor

In the 18th century, experimenters charged Leyden jars with electric generators. Today, capacitors are charged with batteries. Figure 27.13A shows an electrical network that may be used to charge a capacitor that consists of two parallel plates separated by air. The positive terminal of the battery is connected to the top plate of the capacitor, and the negative terminal is connected through a switch to the bottom plate. Figure 27.13B shows a schematic diagram for this network. All capacitors,

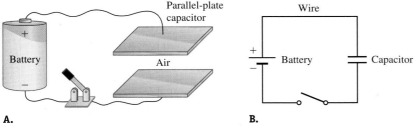

FIGURE 27.13 A. A battery, parallel-plate capacitor, and switch form a network. **B.** Schematic diagram for the network shown in part A. The symbol for a capacitor is similar to the symbol for a battery. However, for a capacitor, the parallel lines are the same length, and often the capacitance C is written next to the capacitor symbol.

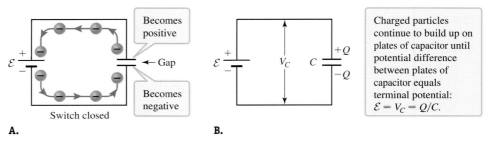

FIGURE 27.14 A. When the switch is closed, negative charge flows from the negative terminal of the battery toward the positive terminal. Charge cannot pass through the capacitor, so it builds up on the plates. **B.** The fully charged capacitor has the same potential difference across its plates as the battery's terminal potential.

regardless of shape, are represented in circuit diagrams by two short parallel lines of equal length.

When the switch is closed, electrons from the negative terminal begin to flow toward the positive terminal (Fig. 27.14A). This is similar to Figure 27.12B, except in that case there is no capacitor, so the particles are free to flow through the wire from the negative terminal to the positive terminal. The particles in Figure 27.14A cannot flow through the gap between the plates of the capacitor; instead, electrons build up on the lower plate. The lower plate becomes negatively charged, and then the electrons on the upper plate are repelled by those on the lower plate. Electrons flow away from the upper plate toward the battery's positive terminal. As a result, the upper plate acquires an excess positive charge exactly equal in magnitude to the excess negative charge on the lower plate.

For a while, electrons continue to build up on the lower plate and the upper plate continues to lose an equal number of electrons, which increases the potential difference between the plates of the capacitor. When that potential difference equals the battery's terminal potential, electrons stop flowing and the capacitor is said to be *fully charged* (Fig. 27.14B). It is common practice to omit the symbol Δ and represent the potential difference between the plates of a capacitor as V_C. The capacitor is fully charged when

$$V_C = \mathcal{E} \tag{27.4}$$

The amount of charge on a fully charged capacitor is found by substituting Equation 27.4 into $Q = C\Delta V$ (Eq. 27.1):

$$Q = C\mathcal{E} \tag{27.5}$$

CONCEPT EXERCISE 27.4

Explain why electrons stop flowing when the potential difference between the plates of a capacitor equals the battery's terminal potential.

CONCEPT EXERCISE 27.5

A large parallel-plate capacitor is attached to a battery that has terminal potential $\mathcal{E}$ (Fig. 27.15A). After a period of time, the capacitor stores charge Q so that its top plate is positive and its bottom plate is negative, and the potential difference between the plates is $V_C = \mathcal{E}$. An **I**-shaped neutral conductor consisting of two parallel plates connected by a wire is slipped between the plates of the capacitor so that all four plates are parallel (Fig. 27.15B). What are the charges q_1 and q_2 on the plates of the **I**-shaped conductor? What is the potential difference V_C between the top and bottom plates of the capacitor?

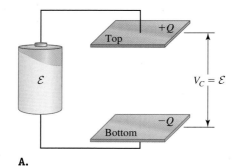

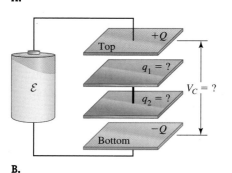

FIGURE 27.15

EXAMPLE 27.2 Comparing Capacitors

A battery with a terminal potential of 11.7 V fully charges two different capacitors in turn. First, a capacitor with capacitance $C_1 = 31.9$ μF is connected to the battery. It is then removed and a second capacitor with double the capacitance is connected to the battery. Find and compare the charges on each capacitor and the energy they store after they have been fully charged. In the **CHECK and THINK** step, answer this question: What will happen if the second capacitor is fully charged by a battery whose terminal potential is $\frac{1}{2}(11.7)$ V $= 5.85$ V?

INTERPRET and ANTICIPATE
The same battery is used to charge both capacitors, so the terminal potential $\mathcal{E}$ is the same for both capacitors. Therefore, the capacitor with the greater capacitance will store a greater charge and more electric potential energy.

SOLVE
Substitute values into Equation 27.5 for the first capacitor. Remember that $1\,\text{F} = 1\,\text{C/V}$.

$$Q_1 = C_1 \mathcal{E} \qquad (27.5)$$
$$Q_1 = (31.9 \times 10^{-6}\,\text{F})(11.7\,\text{V})$$
$$Q_1 = 3.73 \times 10^{-4}\,\text{F} \cdot \text{V} = \boxed{373\,\mu\text{C}}$$

Now find the charge on the second capacitor, where $C_2 = 2C_1$.

$$Q_2 = C_2 \mathcal{E} = 2C_1 \mathcal{E}$$
$$Q_2 = 2(31.9 \times 10^{-6}\,\text{F})(11.7\,\text{V})$$
$$Q_2 = 746\,\mu\text{C}$$

The second capacitor stores twice as much charge as the first capacitor, as expected.

$$Q_2 = 2Q_1$$

We can use either $U_E = Q^2/2C$ or $U_E = \frac{1}{2}C(\Delta V)^2$ (Eqs. 27.2 or 27.3) to find the electric potential energy stored by each capacitor. Let's arbitrarily use Equation 27.2.

Substitute values for the first capacitor.

$$U_1 = \frac{1}{2}\frac{Q_1^2}{C_1} = \frac{1}{2}\frac{(3.73 \times 10^{-4}\,\text{C})^2}{(31.9 \times 10^{-6}\,\text{F})}$$
$$U_1 = 2.18\,\text{mJ}$$

Use $Q_2 = 2Q_1$ and $C_2 = 2C_1$, and then substitute values for the second capacitor.

$$U_2 = \frac{1}{2}\frac{Q_2^2}{C_2} = \frac{1}{2}\frac{(2Q_1)^2}{2C_1} = \frac{1}{2}\frac{4Q_1^2}{2C_1}$$
$$U_2 = \frac{Q_1^2}{C_1} = \frac{(3.73 \times 10^{-4}\,\text{C})^2}{(31.9 \times 10^{-6}\,\text{F})} = 4.36\,\text{mJ}$$

As expected, the second capacitor stores twice as much energy as the first capacitor.

$$U_2 = 2U_1$$

CHECK and THINK
If the second capacitor were charged with a battery that has a terminal potential of 5.85 V, it would store the same amount of charge $Q_1 = Q_2$ and *half* as much energy as the first capacitor: $U_2 = \frac{1}{2}U_1$.

27-4 Capacitors in Parallel and Series

Imagine the task of designing and building an electronic network for a particular device such as a camera or travel alarm clock. The device should be powered by batteries that may be easily purchased, so your design must work with one of a small number of terminal potentials (Table 27.1). In addition, commercially available capacitors are manufactured with a limited variety of capacitances. In your design, you are

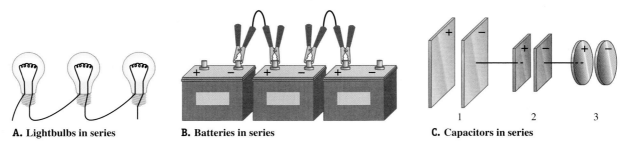

A. Lightbulbs in series   **B. Batteries in series**   **C. Capacitors in series**

FIGURE 27.16 When elements are connected in series, one wire connects each element to the next.

likely to find you need a capacitance that cannot be purchased from any manufacturer. The way around this problem is to combine available capacitors to come up with the equivalent capacitance that you need. In this section, we learn how to find the equivalent capacitance of a combination of capacitors.

Series Capacitors

In this book, we study two ways to combine elements such as lightbulbs, batteries, or capacitors to form a network. Let's first consider a **series** network, in which one element is connected to the next with a single wire between elements. The elements in each network in Figure 27.16 are connected in series.

You have probably loaded several batteries into a device. These batteries are connected in series to achieve a greater potential difference. For example, if you put two AA batteries into a flashlight, the potential difference from the free positive terminal of one battery to the free negative terminal of the other battery is 1.5 V + 1.5 V = 3.0 V. So, from your experience, you know that the potential difference across the two batteries is the sum of the terminal potentials. Likewise, the potential difference across elements connected in series is the sum of their individual potential differences. For example, the potential difference ΔV from the positive plate of capacitor 1 to the negative plate of capacitor 3 in Figure 27.17 is

$$\Delta V = V_1 + V_2 + V_3 \quad (27.6)$$

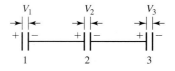

FIGURE 27.17 A schematic diagram of three capacitors connected in series.

Now that we know how to find the potential difference across a series of capacitors, we can derive their equivalent capacitance.

DERIVATION   Capacitors in Series

Suppose you design a network that uses a battery with terminal potential $\mathcal{E}$ to charge a capacitor, which for convenience we call the **equivalent capacitor** with capacitance C_{eq} (Fig. 27.18A). Because this equivalent capacitor is not available, you connect in series a number of capacitors that are equivalent to this one capacitor (Fig. 27.18B). We will show that the equivalent capacitance of N capacitors in series is

$$\frac{1}{C_{eq}} = \sum_{i=1}^{N} \frac{1}{C_i} = \frac{1}{C_1} + \frac{1}{C_2} + \cdots + \frac{1}{C_N} \quad (27.7)$$

To make the derivation more manageable, we'll start with $N = 3$ (Fig. 27.18) and use ideas from Concept Exercise 27.5.

All the capacitors in Figure 27.18 are fully charged, and in order for the two networks to be equivalent, the batteries must be identical.

CAPACITORS IN SERIES ○ **Major Concept**

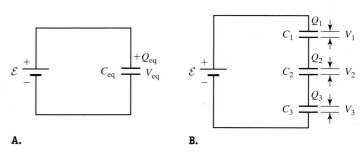

FIGURE 27.18 The capacitor in part A can be replaced by a number of capacitors connected in series as in part B.

The equivalent capacitor (Fig. 27.18A) is fully charged when the potential difference V_{eq} across its plates equals the battery's terminal potential (Eq. 27.4).	$V_{eq} = \mathcal{E}$ (1)
Similarly, when a series of capacitors is connected to the battery (Fig. 27.18B), the electrons stop flowing when the potential difference ΔV from the positive plate of capacitor 1 to the negative plate of capacitor 3 equals the battery's terminal potential $\mathcal{E}$.	$\Delta V = \mathcal{E}$ (2)
Substitute Equations 27.6 and (1) into Equation (2). Equation (3) shows that in order for the two networks to be equivalent, the voltage across the equivalent capacitor must equal the voltage across the three capacitors in series.	$\Delta V = V_1 + V_2 + V_3$ (27.6) $V_1 + V_2 + V_3 = V_{eq} = \mathcal{E}$ (3)
Next, we find the charge stored by the two networks. The amount of charge stored on the equivalent (single) capacitor when it is fully charged is $Q_{eq} = C_{eq} V_{eq}$ (Eq. 27.1). Then solve for V_{eq}.	For the single equivalent capacitor, $V_{eq} = \mathcal{E} = \dfrac{Q_{eq}}{C_{eq}}$ (4)

To find the charge stored by the series of capacitors (Fig. 27.18B), imagine the process of charging them. Electrons flow from the battery's negative terminal and collect on the lower plate of capacitor 3, building an excess negative charge $-Q$ (Fig. 27.19). The highlighted portion in the close-up of capacitors 2 and 3 looks like an **I**-shaped conductor. Because the lower plate of capacitor 3 has a negative charge, electrons in the bottom of the **I**-shaped conductor move upward. The result is that the bottom part of the **I**-shaped conductor acquires an excess positive charge $+Q$ and the upper part acquires an excess negative charge $-Q$. Thus, capacitor 3's upper plate develops charge $+Q$ and capacitor 2's lower plate develops charge $-Q$.

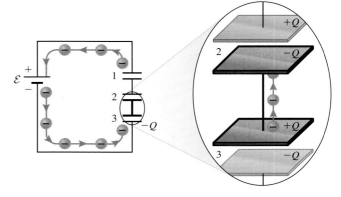

FIGURE 27.19

A similar situation occurs for capacitors 2 and 1, with each plate acquiring excess charge of either $+Q$ or $-Q$. The result is that each capacitor stores the same amount of charge.	$Q_1 = Q_2 = Q_3 \equiv Q$
In order for the series of capacitors (Fig. 27.18B) to function just like the equivalent capacitor in Fig. 27.18A, the charge stored on the equivalent capacitor must equal the charged stored by each individual capacitor in the series.	$Q_1 = Q_2 = Q_3 = Q = Q_{eq}$ (5)
The potential difference across each individual capacitor V_i depends on its capacitance and the amount of charge Q_{eq} it stores. So, we can rewrite Equation (3) in terms of charge and capacitance by substituting $V = Q/C$ (Eq. 27.1) for each individual capacitor.	$V_1 + V_2 + V_3 = \mathcal{E}$ $\dfrac{Q_1}{C_1} + \dfrac{Q_2}{C_2} + \dfrac{Q_3}{C_3} = \mathcal{E}$
Use Equation (5) to write this in terms of Q.	$\dfrac{Q_{eq}}{C_1} + \dfrac{Q_{eq}}{C_2} + \dfrac{Q_{eq}}{C_3} = \mathcal{E}$ For a series of three capacitors, $Q\left(\dfrac{1}{C_1} + \dfrac{1}{C_2} + \dfrac{1}{C_3}\right) = \mathcal{E}$ (6)
We find the relationship between the capacitances of the equivalent capacitor and the series capacitors by comparing Equation (4) for the equivalent capacitor to Equation (6) for the series of three capacitors.	$\dfrac{1}{C_{eq}} = \dfrac{1}{C_1} + \dfrac{1}{C_2} + \dfrac{1}{C_3}$
We considered a series of three capacitors, but it could have been N capacitors, each with capacitance C_i. The charge stored by each capacitor equals the charge Q	$\dfrac{1}{C_{eq}} = \displaystyle\sum_{i=1}^{N} \dfrac{1}{C_i}$ ✓ (27.7)

Derivation continues on page 840 ▶

stored by the equivalent capacitor, and the potential differences between the plates of each capacitor add up to the potential difference between the plates of the equivalent capacitor: $V_{eq} = \sum V_i$. If we follow the same procedure in this more general case, we arrive at the general expression for the capacitance of a series of capacitors.

COMMENTS

According to Equation 27.7, the equivalent capacitance C_{eq} is smaller than the capacitance of the individual capacitors: $C_{eq} < C_1$, $C_{eq} < C_2$, ..., $C_{eq} < C_N$.

Often the quantity you need to solve for is in the denominator of Equation 27.7, so be sure to invert both sides of the equation. For example, let's find an expression for the equivalent capacitance in the case of a network consisting of two capacitors in series.

$$\frac{1}{C_{eq}} = \sum_{i=1}^{2} \frac{1}{C_i} = \frac{1}{C_1} + \frac{1}{C_2}$$

$$C_{eq} = \left(\frac{C_2 + C_1}{C_1 C_2}\right)^{-1} = \frac{C_1 C_2}{C_2 + C_1}$$

Parallel Capacitors

When you connect capacitors in series, their equivalent capacitance is *smaller* than that of the individual capacitors. When your design requires a *greater* capacitance than is available from individual capacitors, you must connect the capacitors in **parallel**. In order to connect elements in parallel, you need two wires joining each element to the next. The circuit elements in Figure 27.20 are connected in parallel.

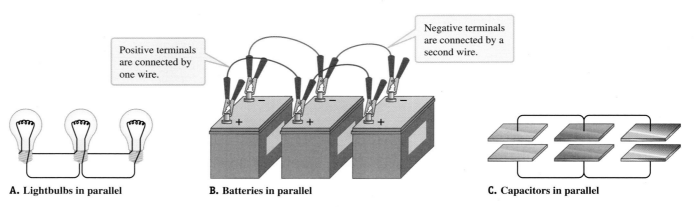

A. Lightbulbs in parallel **B. Batteries in parallel** **C. Capacitors in parallel**

FIGURE 27.20 When circuit elements are connected in parallel, two wires connect each element to the next.

DERIVATION Capacitors in Parallel

We will show that N parallel capacitors, each with capacitance C_i, have an equivalent capacitance given by

$$C_{eq} = \sum_{i=1}^{N} C_i \qquad (27.8)$$

CAPACITORS IN PARALLEL

⭐ Major Concept

To make this derivation more manageable, we start by deriving an expression in the case of just $N = 2$ capacitors in parallel.

To develop a conceptual understanding, imagine that the two capacitors are constructed from one large one as follows: Start with a capacitor consisting of two large parallel plates separated by air. The plates have been charged so that the top plate has a net positive charge $+Q$ and the bottom plate has a net negative charge $-Q$ (Fig. 27.21A). The potential difference between the plates is V_C. Two wires have been used to connect a point on the left side of each plate to a point on the right side. The capacitance of this original capacitor is $C_{eq} = Q/V_C$.

Now imagine cutting each plate into two unequal pieces so that one plate is 1/3 the size of the original plate and the other is 2/3 (Fig. 27.21B). The smaller plate has 1/3 of the original charge, and the larger plate has 2/3 of the original charge. So, the charges on the top plates are $+Q/3$ and $+2Q/3$, while the charges on the bottom plates are $-Q/3$ and $-2Q/3$. The potential is uniform over the surface of a conductor, so the top two plates plus the connecting wire are at the same potential and the bottom two plates plus the connecting wire are at a different, lower potential. Cutting the plates does not change the potential difference between the top plates and the bottom plates; the potential difference across those plates still equals V_C.

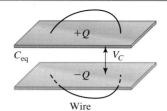

A.

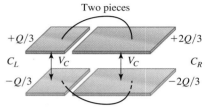

B.

FIGURE 27.21 Constructing capacitors connected in parallel by cutting capacitor plates that are connected by thin wires.

Cutting the plates makes two new parallel-plate capacitors. Because capacitance depends on geometry, these two newly made capacitors do not have the same capacitance as the original device. Label them C_L and C_R for *left* and *right* capacitors, and use $C = q/V_C$, where q is the charge stored by each.	$C_L = \dfrac{Q/3}{V_C} = \dfrac{1}{3}\dfrac{Q}{V_C}$ (1) $C_R = \dfrac{2Q/3}{V_C} = \dfrac{2}{3}\dfrac{Q}{V_C}$ (2)
Substitute $C_{eq} = Q/V_C$ into Equations (1) and (2).	$C_L = \dfrac{1}{3} C_{eq}$ (3) $C_R = \dfrac{2}{3} C_{eq}$ (4)
The sum of C_L and C_R is equal to the capacitance of the original large capacitor.	$C_L + C_R = \dfrac{1}{3}C_{eq} + \dfrac{2}{3}C_{eq} = C_{eq}$
We imagined cutting the original capacitor into two pieces, but we could have cut it into any number of pieces N. The potential difference between the plates of each capacitor would still be the same as the potential difference V_C between the plates of the equivalent capacitor, and the total charge stored on the capacitors in parallel would equal the charge stored on the equivalent capacitor: $Q = \sum Q_i$. When there are N capacitors in parallel, the equivalent capacitance C_{eq} is found by adding the individual capacitances.	$C_{eq} = \sum_{i=1}^{N} C_i$ ✓ (27.8)

⁖ COMMENTS

Equations (3) and (4) show that the smallest capacitor (the left one) has the smallest capacitance.

CONCEPT EXERCISE 27.6

CASE STUDY **Capacitors for a Thompson coil**

The Thompson coil experimenters (page 829) claimed they used six identical capacitors connected in parallel. In Example 27.1, we found that the equivalent capacitance is 94 μF. What is the capacitance of each capacitor?

EXAMPLE 27.3 Building a Network of Capacitors

You are working in a laboratory to build a network from a design (Fig. 27.22). You have a dozen batteries with terminal potential $\mathcal{E}_0 = 1.5$ V and a dozen capacitors with capacitance $C_0 = 9.0$ μF. Draw a schematic diagram showing how you would build this network from the components in the laboratory.

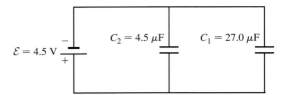

FIGURE 27.22 How can we build this network from 1.5-V batteries and 9.0-μF capacitors?

:• INTERPRET and ANTICIPATE

The design requires a greater terminal potential than that of an individual battery. When batteries are connected in series, their terminal potentials add, so we will use several batteries in series. The design requires C_1 to be greater than the capacitance of the laboratory's capacitors. When capacitors are connected in parallel, the resulting capacitance is the sum of the individual capacitances. So, to get C_1, we need to connect some of the laboratory's capacitors in parallel. The design also requires C_2 to be smaller than the capacitance of the laboratory's capacitors. When capacitors are connected in series, the resulting capacitance is smaller than the individual capacitances. So, to get C_2, we need to connect some of the laboratory's capacitors in series.

:• SOLVE

All the laboratory's batteries have the same terminal potential $\mathcal{E}_0 = 1.5$ V, so connecting $n = 3$ of these in series gives the required $\mathcal{E} = 4.5$ V.

$$\mathcal{E} = n\mathcal{E}_0$$
$$n = \frac{\mathcal{E}}{\mathcal{E}_0} = \frac{4.5 \text{ V}}{1.5 \text{ V}} = 3$$

All the laboratory capacitors have the same capacitance $C_0 = 9.0$ μF, so connecting $n_1 = 3$ of these in parallel gives the required C_1 (Eq. 27.8).

$$C_{eq} = \sum_{i=1}^{N} C_i \qquad (27.8)$$
$$C_1 = n_1 C_0$$
$$n_1 = \frac{C_1}{C_0} = \frac{27.0 \, \mu\text{F}}{9.0 \, \mu\text{F}} = 3$$

We need to connect $n_2 = 2$ of the 9.0-μF capacitors in series to achieve the required C_2 (Eq. 27.7).

$$\frac{1}{C_{eq}} = \sum_{i=1}^{N} \frac{1}{C_i} \qquad (27.7)$$
$$\frac{1}{C_2} = n_2 \frac{1}{C_0}$$
$$n_2 = \frac{C_0}{C_2} = \frac{9.0 \, \mu\text{F}}{4.5 \, \mu\text{F}} = 2$$

As shown in Figure 27.23, we need three batteries connected in series, three capacitors connected in parallel, and two more capacitors connected in series.

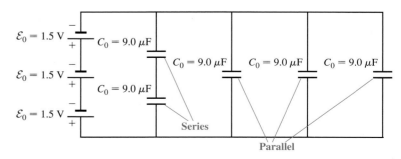

FIGURE 27.23

:• **CHECK and THINK**
One way to check our network is to think about other equivalent networks. For example, the original design (Fig. 27.22) shows two capacitors in parallel. We can find the equivalent capacitance of these two capacitors and compare it to the equivalent capacitance of the five capacitors in Figure 27.23. First, use Equation 27.8 to find the equivalent capacitance of C_1 and C_2.

For the original design (Fig. 27.22),
$$C_{eq} = C_1 + C_2 = 27.0\,\mu\text{F} + 4.5\,\mu\text{F}$$
$$C_{eq} = 31.5\,\mu\text{F}$$

For a combination of capacitors in series and parallel, as in Figure 27.23, we must use both Equations 27.7 (series) and 27.8 (parallel). As required, the equivalent capacitances are equal.

For the actual network (Fig. 27.23),
$$C_{eq} = C_{parallel} + C_{series}$$
$$C_{eq} = 3C_0 + \left(\frac{2}{C_0}\right)^{-1} = 3C_0 + \frac{C_0}{2}$$
$$C_{eq} = \frac{7}{2}C_0 = \frac{7}{2}(9\,\mu\text{F}) = 31.5\,\mu\text{F}$$

27-5 Capacitance: Special Cases

The capacitance of a capacitor depends only on its geometry and on the insulator between the plates. In this section, we develop a problem-solving strategy for finding the capacitances for different geometries, assuming a vacuum between the plates. For most practical purposes, air between the plates can be modeled as a vacuum. We apply this strategy to find the capacitances for two special cases: a parallel-plate capacitor and a cylindrical capacitor. The special case of a spherical capacitor is left as homework (Problem 32).

PROBLEM-SOLVING STRATEGY

Finding Capacitance

:• **SOLVE**
Step 1 Assume the capacitor plates are charged so that one has positive charge $+q$ and the other has negative charge $-q$. Then use Gauss's law to find an expression for the **electric field** between the plates. Refer to the problem-solving strategy "Finding the Electric Field Using Gauss's Law" on page 765.
Step 2 After the electric field is found, you need to complete a **path integral** to find the electric potential difference V_C between the plates. Choose a path from the positive plate to the negative plate, so the limits of integration are $r_i = r_+$ and $r_f = r_-$. With this path, the integral you need to solve is

$$V_C = \int_{r_+}^{r_-} E\, dr \qquad (27.9)$$

Because the positive plate is at a higher potential than the negative plate, this potential difference is positive ($V_C > 0$).
Step 3 When V_C is written in terms of Q (the charge stored by the capacitor), use $C = Q/V_C$ (Eq. 27.1) to find the capacitance.

:• **CHECK and THINK**
In addition to checking the **dimensions**, you should make sure the expression you find for capacitance **depends only on geometric properties**.

EXAMPLE 27.4 Special Case: Parallel-plate Capacitor

Two large conducting plates make up a parallel-plate capacitor (Fig. 27.24). The left plate has positive charge $+Q$, and the right plate has negative charge $-Q$. The area of each plate is A, and the plates are separated by a distance d. Show that the capacitance is

$$C = \frac{\varepsilon_0 A}{d} \quad (27.10)$$

CAPACITANCE OF PARALLEL-PLATE CAPACITOR ▶ Special Case

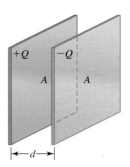

FIGURE 27.24 A parallel-plate capacitor.

INTERPRET and ANTICIPATE

The area of the plates is large compared to their separation, and because we are interested in the region between the plates, we can model each one as an infinitely large plane. Thus, we can ignore the slight bend in the electric field near the edge of the plates and model them as sources with planar symmetry (an important step for using Gauss's law). Use the three problem-solving steps to find the capacitance, assuming either a vacuum or air between the plates.

SOLVE

Step 1 Use Gauss's law to find $\vec{E}$. Finding the electric field between the plates of a parallel-plate capacitor is similar to finding the electric field just outside any charged conductor (Section 25-7). The electric field points from the positive plate to the negative plate (Fig. 27.25). Due to planar symmetry, the electric field is uniform and the electric field lines are evenly spaced and parallel. A Gaussian cylinder exploits this symmetry.

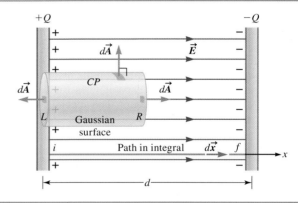

FIGURE 27.25

Integrate to find the flux through the closed Gaussian surface. Divide the integral into three parts: one for the left cap L, one for the curved part CP, and one for the right cap R. The integral over L is zero because that cap is inside the conductor, and the electric field is zero within the body of a conductor. The integral over CP is zero because $\vec{E}$ is perpendicular to $d\vec{A}$ for the curved part. The integral over R is positive because $\vec{E}$ is parallel to $d\vec{A}$. Because E is constant, pull it outside the integral, and then the integral is just the area of either cap, A_{cap}.	$\Phi_E = \oint \vec{E} \cdot d\vec{A}$ $\Phi_E = \int_L \vec{E} \cdot d\vec{A} + \int_{CP} \vec{E} \cdot d\vec{A} + \int_R \vec{E} \cdot d\vec{A}$ $\Phi_E = 0 + 0 + \int_R E\, dA = E A_{\text{cap}}$
The surface charge density σ is the total charge Q on the plate's surface divided by the area A of the plate (Eq. 24.10). The charge enclosed by the Gaussian surface is the surface charge density multiplied by the area A_{cap} of either cap.	$\sigma = \dfrac{Q}{A} \quad (24.10)$ $q_{\text{in}} = \sigma A_{\text{cap}} = Q A_{\text{cap}} / A$
Find the electric field by substituting the flux Φ_E and charge q_{in} into Gauss's law (Eq. 25.7).	$\Phi_E = \dfrac{q_{\text{in}}}{\varepsilon_0} \quad (25.7)$ $E A_{\text{cap}} = \dfrac{1}{\varepsilon_0} \dfrac{Q}{A} A_{\text{cap}}$ $E = \dfrac{1}{\varepsilon_0} \dfrac{Q}{A} \quad (27.11)$

Step 2 Complete the **path integral** to find V_C. When using Equation 27.9, choose a path parallel to an electric field line and pointing from the positive plate to the negative plate. In this case, the path is in the positive x direction (Fig. 27.25). Both $d\vec{x}$ and $\vec{E}$ are in the positive x direction. Because we already took these directions into account, all we need are the magnitudes.	$V_C = \int_{r_+}^{r_-} E\,dr$ (27.9) $V_C = \int_{x_+}^{x_-} \left(\frac{1}{\varepsilon_0}\frac{Q}{A}\,dx\right)$ $V_C = \left(\frac{1}{\varepsilon_0}\frac{Q}{A}\right)\int_{x_+}^{x_-} dx$
The integral is the length of the path from the positive to the negative plate. We have labeled the separation between the plates d (Fig. 27.25).	$V_C = \left(\frac{1}{\varepsilon_0}\frac{Q}{A}\right)(x_- - x_+)$ $V_C = \frac{1}{\varepsilon_0}\frac{Q}{A}d$ (27.12)
Step 3 Use $C = Q/V_C$ (Eq. 27.1) to find the capacitance. Solve Equation 27.12 for Q/V_C.	$C = \frac{\varepsilon_0 A}{d}$ ✓ (27.10)
∴ CHECK and THINK 1. The permittivity constant ε_0 has the dimensions $C^2/(F \cdot L^2)$, where C is charge (Eq. 25.8). So, the **dimensions** of Equation 27.10 are charge per voltage as expected. 2. Because ε_0 is a constant, the capacitance **depends only on geometric properties** (A and d) as expected. The capacitance does not depend on the charge stored or the potential difference between the plates.	$[C] = \left[\frac{\varepsilon_0 A}{d}\right]$ $[C] = \frac{C^2}{F \cdot L^2} \cdot \frac{L^2}{L} = \frac{C^2}{F \cdot L} = C \cdot \frac{C}{F \cdot L}$ $[C] = \frac{C}{V}$

Before we consider the next special case, let's express the permittivity constant $\varepsilon_0 = 8.85 \times 10^{-12}\,C^2/(N \cdot m^2)$ (Eq. 25.8) in more convenient SI units. According to Equation 27.10, the dimensions of the permittivity constant are capacitance (CAP) per length (L): $[\varepsilon_0] = \text{CAP}/L$. So, we can use $1\,F = 1\,C/V$ to write $\varepsilon_0 = 8.85 \times 10^{-12}\,F/m = 8.85\,pF/m$. These units for ε_0 are convenient when we work with capacitance.

EXAMPLE 27.5 Special Case: Cylindrical Capacitor

A cylindrical capacitor of length L is made up of an inner plate that is a long solid cylinder with radius r_{in} and an outer plate that is a long hollow cylinder with inner radius r_{out} (Fig. 27.26). The inner plate has positive charge $+Q$, and the outer plate has negative charge $-Q$. Show that the capacitance is

$$C = \frac{2\pi\varepsilon_0 L}{\ln(r_{\text{out}}/r_{\text{in}})} \quad (27.13)$$

CAPACITANCE OF CYLINDRICAL CAPACITOR
▶ Special Case

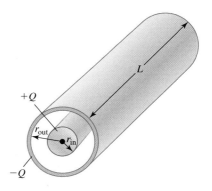

FIGURE 27.26 A cylindrical capacitor.

∴ INTERPRET and ANTICIPATE
The plates are very long compared to the radius r_{out}, so we can ignore the slight bend in the electric field near the edge of the plates and model them as sources with linear symmetry. We find the capacitance of a cylindrical capacitor assuming either a vacuum or air between the plates.

Example continues on page 846 ▶

SOLVE

Step 1 Use Gauss's law to find $\vec{E}$. Finding the **electric field** between the plates of a cylindrical capacitor is similar to finding the electric field due to a charged rod (Section 25-4). The electric field points from the positive plate to the negative plate (Fig. 27.27). A Gaussian cylinder of radius r and length ℓ aligned with the axes of the cylindrical plates exploits this symmetry.

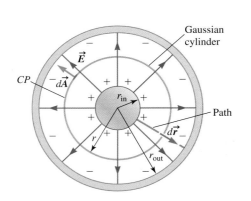
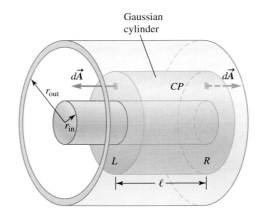

FIGURE 27.27 Gaussian cylinder aligned with the capacitor's axis.

Integrate to find the flux through the closed Gaussian surface. Divide the integral into three parts: one for the left cap L, one for the curved part CP, and one for the right cap R. The integrals over L and R are zero because $\vec{E}$ is perpendicular to $d\vec{A}$ for both caps. The integral over CP is positive because $\vec{E}$ is parallel to $d\vec{A}$ for every point on the curved part. Bring E outside the integral because it is constant. Write the area A_{CP} of the curved part in terms of the circumference $2\pi r$ of the Gaussian cylinder and its length ℓ.	$\Phi_E = \oint \vec{E} \cdot d\vec{A}$ $\Phi_E = \int_L \vec{E} \cdot d\vec{A} + \int_{CP} \vec{E} \cdot d\vec{A} + \int_R \vec{E} \cdot d\vec{A}$ $\Phi_E = 0 + \int_{CP} E\, dA + 0$ $\Phi_E = E A_{CP} = E 2\pi r \ell$
The linear charge density λ is the total charge Q on the plate's surface divided by the length L of the plate (Eq. 24.12). The charge enclosed by the Gaussian surface is the linear charge density multiplied by the length ℓ of the Gaussian cylinder.	$\lambda = \dfrac{Q}{L}$ (24.12) $q_{in} = \lambda \ell = \dfrac{Q}{L}\ell$
Find the electric field by substituting the flux Φ_E and charge q_{in} into Gauss's law (Eq. 25.7).	$\Phi_E = \dfrac{q_{in}}{\varepsilon_0}$ (25.7) $E 2\pi r \ell = \dfrac{1}{\varepsilon_0}\dfrac{Q}{L}\ell$ $E = \dfrac{1}{2\pi\varepsilon_0}\dfrac{Q}{L}\dfrac{1}{r}$
Step 2 Complete the **path integral** to find V_C. As shown in Figure 27.27, the path points along an electric field line in the positive r direction. Both $d\vec{r}$ and $\vec{E}$ are in the positive r direction, so only their magnitudes are needed. The antiderivative is the natural logarithm (Appendix A).	$V_C = \displaystyle\int_{r_{in}}^{r_{out}} E\, dr$ (27.9) $V_C = \displaystyle\int_{r_{in}}^{r_{out}} \left(\dfrac{1}{2\pi\varepsilon_0}\dfrac{Q}{L}\dfrac{1}{r}\right) dr$ $V_C = \dfrac{1}{2\pi\varepsilon_0}\dfrac{Q}{L}(\ln r_{out} - \ln r_{in})$ $V_C = \dfrac{1}{2\pi\varepsilon_0}\dfrac{Q}{L}\ln\dfrac{r_{out}}{r_{in}}$ (1)

Step 3 Use $C = Q/V_C$ (Eq. 27.1) to find the capacitance. Solve Equation (1) for Q/V_C.

$$C = \frac{2\pi\varepsilon_0 L}{\ln(r_{out}/r_{in})} \quad (27.13)$$

CHECK and THINK
1. The denominator has no **dimensions**, so the dimensions come from the numerator. We find the dimensions are capacitance.
2. As expected, the capacitance **depends only on geometric properties**—the radii of the cylinders and their length.

$$[C] = \left[\frac{2\pi\varepsilon_0 L}{\ln(r_{out}/r_{in})}\right] = [\varepsilon_0] \cdot [L]$$

$$[C] = \frac{CAP}{L} \cdot L = CAP$$

EXAMPLE 27.6 | Making a 1-F Capacitor

The capacitance of most commonly used capacitors is small ($\sim 1\,\text{pF} < C < \sim 1\,\mu\text{F}$). A capacitance of 1 F is very large. In this problem, we estimate the size of a 1-F capacitor using cylindrical geometry. (In Problem 41, you will consider parallel-plate geometry.) If the radius of the outer plate of a 1-F cylindrical capacitor is $r_{out} = 1.5$ mm (about the size of sewing needle) and the radius of the inner plate is $r_{in} = 50\,\mu\text{m}$, how long must the capacitor be?

INTERPRET and ANTICIPATE
The capacitance of a cylindrical capacitor is directly proportional to the length of its plates, so we expect the capacitor to be very long.

SOLVE
Solve Equation 27.13 for L.

$$L = \frac{C\ln(r_{out}/r_{in})}{2\pi\varepsilon_0} = \frac{(1\,\text{F})\ln[(1.5\times 10^{-3}\,\text{m})/(50\times 10^{-6}\,\text{m})]}{2\pi(8.85\times 10^{-12}\,\text{F/m})}$$

$$L \approx 6 \times 10^{10}\,\text{m}$$

CHECK and THINK
As expected, this is a very long capacitor. In fact, it would stretch to about 40% of the distance to the Sun! Figure 27.28 shows an inexpensive 1-F capacitor that is about the size of a piece of candy. How is it possible to build such a small capacitor with such a large capacitance? The answer is that the space between the plates must be filled with an insulator (other than air).

FIGURE 27.28 A 1-F capacitor, just a little larger than a piece of candy.

27-6 Dielectrics

So far, we have been considering capacitors with either a vacuum or air between the plates. Capacitance can be increased by filling the space between the plates with an insulator such as plastic, paper, or pure water. When the capacitor is charged, an electric field is induced within the insulator. The term **dielectric** describes an insulator that can support an induced electric field. In this section, we modify the problem-solving strategy on page 843 to find capacitance in the presence of a dielectric. (Later, we'll develop a shortcut for finding the capacitance.)

DIELECTRIC ⊗ Major Concept

With a dielectric between the plates, the **electric field** found in **Step 1** must be modified to include the field induced in the dielectric. Figure 27.29 shows how this electric field is induced for the special case of a charged parallel-plate capacitor. Start by imagining the familiar vacuum between the plates; then the capacitor's

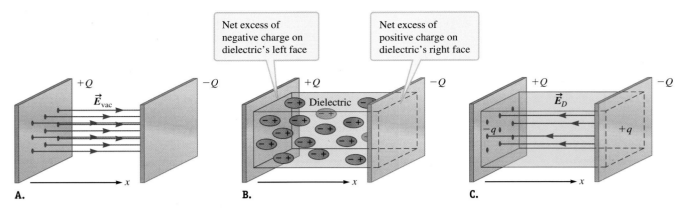

FIGURE 27.29 A. The electric field produced by the plates of the capacitor points in the positive x direction. **B.** The dielectric's molecules are polarized. **C.** The dielectric produces an electric field in the negative x direction.

electric field $\vec{E}_{\text{vac}}$ points to the right in the positive x direction (Fig. 27.29A). This electric field polarizes the molecules inside the dielectric so that each molecule's electrons are shifted slightly to the left (Fig. 27.29B). (If the dielectric is made up of dipole molecules such as water molecules, the capacitor's electric field causes the dipole molecules to line up.) The result is that the dielectric creates a second electric field $\vec{E}_D$ pointing to the left in the negative x direction (Fig. 27.29C). The net electric field between the capacitor plates is

$$\vec{E}_C = \vec{E}_{\text{vac}} + \vec{E}_D = (E_{\text{vac}} - E_D)\hat{\imath} \qquad (27.14)$$

As represented by the relative number of field lines in Figure 27.29, the electric field produced by the dielectric is weaker than the vacuum-filled capacitor's electric field: $E_D < E_{\text{vac}}$. So, the net electric field points in the positive x direction (the same direction as the vacuum-filled capacitor's electric field) and is weaker than the empty capacitor's electric field. Mathematically, we write the total electric field in terms of the vacuum-filled capacitor's electric field:

$$\vec{E}_C = \frac{1}{\kappa}\vec{E}_{\text{vac}} \qquad (27.15)$$

where κ is called the **dielectric constant**. So, in Step 1, we modify the electric field we found for a vacuum-filled capacitor by dividing it by the dielectric constant.

The dielectric constant is a positive number that depends only on the type of material; κ is unitless and its value is always greater than or equal to 1. If no dielectric is present (if there is a vacuum between the plates), then $\kappa = 1$. Table 27.2 lists the dielectric constants for a number of materials. Because the dielectric constant of air is very close to 1, we are able to model an air-filled capacitor as a vacuum-filled capacitor.

In **Step 2**, we need to replace the vacuum-filled electric field with E_C (Eq. 27.15). Then the potential difference between the plates is given by a modified **path integral**:

$$V_C = \int_{r_+}^{r_-} E_C \, dr = \int_{r_+}^{r_-} \frac{E_{\text{vac}}}{\kappa} dr = \frac{1}{\kappa} \int_{r_+}^{r_-} E_{\text{vac}} \, dr$$

where E_{vac} is the electric field magnitude if there were a vacuum between the plates. So, we can write the potential difference between the plates filled with a dielectric in terms of the vacuum potential difference:

$$V_C = \frac{1}{\kappa} V_{\text{vac}} \qquad (27.16)$$

Because the dielectric constant $\kappa \geq 1$, the potential difference when a dielectric is present is *less* than when there is a vacuum.

TABLE 27.2 Dielectrics at room temperature.

Material	Dielectric constant κ	Dielectric strength ($\times 10^6$ V/m)
Air (at 1 atm)	1.00054	3
Bakelite	4.9	24
Mica	7	150
Mylar	3.2	7
Neoprene rubber	6.7	12
Nylon	3.4	14
Paper	3.5	16
Paraffin	2.2	10
Polycarbonate	2.8	30
Polyester	3.3	60
Polystyrene	2.6	24
Polyvinyl chloride	3.4	40
Porcelain	6.5	12
Pyrex	4.7	14
Quartz	4.3	8
Silicone oil	2.5	15
Strontium titanate	310	8
Teflon	2.1	60
Titania ceramic	130	
Vacuum	1	n/a
Water (liquid)	80	

Finally, in **Step 3**, we divide the charge Q stored by the potential difference V_C between the plates (Eq. 27.16):

$$C = \frac{Q}{V_C} = \frac{Q}{V_{vac}/\kappa} = \kappa \frac{Q}{V_{vac}}$$

$$C = \kappa C_{vac} \tag{27.17}$$

CAPACITANCE IN THE PRESENCE OF A DIELECTRIC ⭐ **Major Concept**

where C_{vac} is the capacitance the capacitor would have if there were a vacuum between the plates. In practice, once you have an expression for the capacitance of a vacuum-filled capacitor, you only need to multiply by the dielectric constant to find the capacitance when a dielectric is present. (You may not need the entire problem-solving strategy.) Because the dielectric constant $\kappa \geq 1$, the capacitance when a dielectric is present is *greater* than when there is a vacuum.

Besides increasing the capacitance, a dielectric has another practical function. It provides a practical way to bring the plates of a capacitor close together without touching, which also increases the capacitance. For example, the capacitance of a parallel-plate capacitor with a dielectric is

$$C = \kappa C_{vac} = \kappa \frac{\varepsilon_0 A}{d} \tag{27.18}$$

where C_{vac} is given by Equation 27.10. If the dielectric is sandwiched between the parallel plates, the separation between the plates d equals the thickness of the dielectric. Thus, a very thin dielectric increases the capacitance.

However, there are practical limits to how thin a dielectric can be and how much charge can be stored on a capacitor's plates. As the charge increases, the electric field inside the dielectric becomes stronger. At some point, the electric field is strong enough that the dielectric *breaks down* and behaves as a conductor, allowing charge to flow through. This is not a new idea to this chapter; we considered lightning in Chapter 24. Air is normally a good insulator, but when a lot of charge builds up in a storm, air breaks down and becomes a conductor, and lightning results. The **dielectric strength** of an insulator is the maximum electric field it can tolerate without breaking down and becoming a conductor. The dielectric strengths of many materials are listed in Table 27.2. The electric field and dielectric strength both depend on the potential difference between the plates and the plate separation. In practice, the plate separation is fixed, and off-the-shelf capacitors are rated in terms of their maximum operating potential. In addition, dielectric strength depends on temperature, so capacitors are rated in terms of their operating temperatures also.

EXAMPLE 27.7 | **CASE STUDY** Size of the Plates

We found in Concept Exercise 27.6 that each of the six capacitors used by the experimenters in the case study had a fairly large capacitance ($C = 16 \ \mu F$). Such capacitors may be made by layering the plates of the capacitor with two dielectric sheets (Fig. 27.30). The layers are then rolled together so that the capacitor fits inside a cylindrical case something like a soda can. Imagine uncoiling one of these capacitors so that it resembles a large parallel-plate capacitor with a polyvinyl chloride dielectric.

FIGURE 27.30 Layers of dielectric and metal are rolled together to form a compact capacitor with a large capacitance.

A The experimenters charged the capacitors until the potential difference between the plates was 1500 V. If this produces an electric field that is about half the dielectric strength, what is the thickness of the polyvinyl chloride dielectric?

Example continues on page 850 ▶

INTERPRET and ANTICIPATE
We must first relate the electric field E_C produced by a parallel-plate capacitor to the potential difference and the distance between the plates. Then we set E_C equal to half the dielectric strength and solve for the thickness of the polyvinyl chloride.

SOLVE
Divide $V_C = V_{vac}/\kappa$ by $E_C = E_{vac}/\kappa$ and the dielectric constant cancels out.

$$\frac{V_C}{E_C} = \frac{V_{vac}}{E_{vac}} \quad (1)$$

Substitute Equations 27.11 and 27.12 for the vacuum-filled parallel-plate capacitor's electric field and the potential difference into Equation (1). Solve for the plate separation d. [Equation 2 is used frequently; also notice that it is similar to Equation 26.16, $\Delta V = -(\sigma/2\varepsilon_0)\Delta x$.]

$$\frac{V_C}{E_C} = \left(\frac{1}{\varepsilon_0}\frac{Q}{A}d\right) \Big/ \left(\frac{1}{\varepsilon_0}\frac{Q}{A}\right)$$

$$d = \frac{V_C}{E_C} \quad (2)$$

The potential between the plates is 1500 V. For the electric field, use half the dielectric strength of polyvinyl chloride found in Table 27.2.

$$d = \frac{1500 \text{ V}}{\frac{1}{2}(40 \times 10^6 \text{ V/m})} = 7.5 \times 10^{-5} \text{ m}$$

CHECK and THINK
This is a little thinner than a lightweight sheet of paper.

B Assuming there is no space between the plates and the dielectric, find the area of the plates. If the plates and the dielectric are rolled into a cylinder 0.18 m tall, what is the length of one plate?

INTERPRET and ANTICIPATE
We estimated the thickness of the dielectric in part A. We can use this thickness to find the area of the plates because we already know the capacitance.

SOLVE
The capacitance of a dielectric-filled parallel-plate capacitor is given by Equation 27.18. Use Table 27.2 to find κ for polyvinyl chloride.

$$C = \kappa \frac{\varepsilon_0 A}{d} \quad (27.18)$$

$$A = \frac{Cd}{\kappa \varepsilon_0} = \frac{(1.6 \times 10^{-5} \text{ F})(7.5 \times 10^{-5} \text{ m})}{(3.4)(8.85 \times 10^{-12} \text{ F/m})}$$

$$A = 39.9 \text{ m}^2 \approx 40 \text{ m}^2$$

The plates are long, skinny rectangles of height $h = 0.18$ m.

$$A = h\ell$$

$$\ell = \frac{A}{h} = \frac{40 \text{ m}^2}{0.18 \text{ m}} = 2.2 \times 10^2 \text{ m}$$

$$\ell = 220 \text{ m}$$

CHECK and THINK
The plates are longer than two football fields. It may be difficult to imagine rolling plates this long into a device of a reasonable size. However, the paper on a single roll of toilet paper is about the length of one football field.

C Review the facts presented by the experimenters. Does their Thompson coil apparatus seem feasible, or do you think their description is a hoax?

The apparatus seems feasible. The energy required to launch the ring could come from a large potential difference (1500 V) across six capacitors, each with a large capacitance (16 μF). Such capacitors are possible to construct and available from many manufacturers, although high-voltage capacitors can be expensive. The experimenters claim that they worked hard to find affordable capacitors. Other experimenters claim to have built their own high-voltage capacitors in order to save money, which increases already serious safety concerns.

EXAMPLE 27.8 Cylindrical Nerves

The nerve cells in the human body and in other animals are modeled as very long cylindrical capacitors. Portions of some nerves are covered with a layer of fat, known as myelin, functioning as the dielectric ($\kappa = 7$) between two plates in the cylindrical capacitor model. The immune system in some people attacks this layer of fat, changing their nerve cells' capacitance and resulting in multiple sclerosis (MS). Normally, the inner radius of the myelin layer is 2×10^{-6} m and its thickness is 5×10^{-6} m. Find the capacitance per unit length for a nerve cell. (Give your answer to two significant figures.)

:• **INTERPRET and ANTICIPATE**
The capacitance of a long, cylindrical capacitor with a vacuum between the plates (Section 27-5) is proportional to the length of the cylinder. In this problem, we need just the capacitance per unit length.

:• **SOLVE**

The expression for the capacitance of a long, cylindrical, vacuum-filled capacitor (Eq. 27.13) must be multiplied by κ because in our model of a nerve cell, the plates are filled with a dielectric.	$C = \dfrac{2\pi\varepsilon_0 L}{\ln(r_{out}/r_{in})}$ (27.13) $\dfrac{C}{L} = \kappa\left[\dfrac{2\pi\varepsilon_0}{\ln(r_{out}/r_{in})}\right]$ (1)
The inner radius r_{in} of the myelin is given, and the outer radius r_{out} must be found from the thickness.	$r_{out} - r_{in} = 5 \times 10^{-6}$ m $r_{out} = r_{in} + (5 \times 10^{-6}$ m$) = (2 \times 10^{-6}$ m$) + (5 \times 10^{-6}$ m$)$ $r_{out} = 7 \times 10^{-6}$ m
Substitute values into Equation (1).	$\dfrac{C}{L} = \dfrac{(7)2\pi(8.85 \times 10^{-12}\text{ F/m})}{\ln\left[(7 \times 10^{-6}\text{ m})/(2 \times 10^{-6}\text{ m})\right]}$ $\dfrac{C}{L} = 3.1 \times 10^{-10}\text{ F/m} = 0.31\text{ nF/m}$

:• **CHECK and THINK**
Our result has the expected dimensions.

27-7 Energy Stored by a Capacitor with a Dielectric

Does the presence of a dielectric increase or decrease the amount of energy stored by a capacitor? The short answer is that *it depends* on whether or not the charged capacitor is connected to a battery when the dielectric is inserted. In this section, we consider a few situations in order to see which factors are important.

First, imagine there are two identical batteries, each connected to a capacitor. The capacitors are identical except that one is vacuum-filled and has capacitance C_{vac}, and the other contains a dielectric (dielectric constant κ) and has capacitance κC_{vac}. Each battery has terminal potential $\mathcal{E}$, so when either capacitor is fully charged, the potential difference across the plates is $V_C = \mathcal{E}$. The energy stored by the vacuum-filled capacitor is $U_{vac} = \frac{1}{2} C_{vac} \mathcal{E}^2$ (Eq. 27.3), and the energy stored by the dielectric-filled capacitor is $U = \frac{1}{2} \kappa C_{vac} \mathcal{E}^2$. So, the dielectric-filled capacitor stores more energy than the vacuum-filled capacitor:

$$U = \kappa U_{vac} \qquad (27.19)$$

Why didn't we use $U = Q^2/2C$ (Eq. 27.2) to find the energy stored by each capacitor? Because we would first need to find the charge stored by each capacitor, and each stores a different amount. The charge stored by the vacuum-filled capacitor is $Q_{vac} = C_{vac}\mathcal{E}$ (Eq. 27.1), and the charge stored in the dielectric-filled capacitor is $Q = \kappa C_{vac}\mathcal{E}$. So, the dielectric-filled capacitor stores more charge:

$$Q = \kappa Q_{vac} \qquad (27.20)$$

Now when we use $U = Q^2/2C$, we find the same result as when we used $U = \frac{1}{2}CV_C^2$ (Eq. 27.3):

$$U = \frac{1}{2}\frac{Q^2}{C} = \frac{1}{2}\frac{(\kappa Q_{vac})^2}{\kappa C_{vac}} = \kappa \frac{1}{2}\frac{Q_{vac}^2}{C_{vac}}$$

$$U = \kappa U_{vac} \qquad (27.19)$$

So, we see it is better to use $U = \frac{1}{2}CV_C^2$ (Eq. 27.3) when the potential V_C is constant and $U = Q^2/2C$ (Eq. 27.2) when the charge Q is constant.

To further make this point, consider a parallel-plate capacitor is connected to a battery (Fig. 27.31A). The battery maintains a constant potential ($V_C = \mathcal{E}$) across the capacitor's plates as a dielectric is inserted between them. Inserting the dielectric *increases* (1) the capacitance $C = \kappa C_{vac}$ (Eq. 27.17), (2) the energy stored by the capacitor $U = \kappa U_{vac}$ (Eq. 27.19), and (3) the amount of charge stored by the capacitor $Q = \kappa Q_{vac}$ (Eq. 27.20).

The situation in Figure 27.31B is somewhat different; a parallel-plate capacitor is connected to a battery. The battery is then disconnected, so the potential V_C is no longer held constant. The charge Q stored by the capacitor is constant because neither plate is connected to a conductor. A dielectric is then inserted between the plates. As before, inserting the dielectric increases the capacitance, $C = \kappa C_{vac}$ (Eq. 27.17).

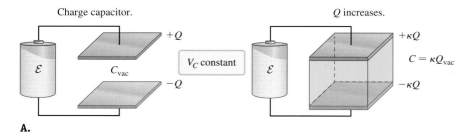

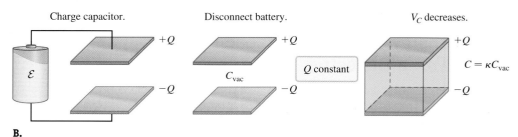

FIGURE 27.31 A. The battery connected to the capacitor ensures that the potential is held constant when a dielectric is slipped between the plates. As a result, the charge on the plates and the energy stored by the capacitor both increase. **B.** If the capacitor is charged and the battery is disconnected, when the dielectric is slipped between the plates, the potential decreases while the charge remains constant, and the energy stored by the capacitor decreases.

However, because the charge is constant, we find using Equation 27.2 that the energy stored by the capacitor decreases:

$$U = \frac{1}{2}\frac{Q^2}{\kappa C_{vac}} = \frac{1}{\kappa}U_{vac}$$

and the potential between the plates also decreases (Eq. 27.1):

$$V_C = \frac{Q}{\kappa C_{vac}} = \frac{1}{\kappa}\mathcal{E}$$

The difference between the two situations in Figure 27.31 can be explained as follows: Inserting a dielectric partially cancels the electric field between the plates because the electric field $\vec{E}_D$ created by the dielectric is in the opposite direction as the vacuum-filled capacitor's electric field $\vec{E}_{vac}$ (Fig. 27.29C). When no battery is connected to the capacitor, this reduced electric field means a reduced potential difference V_C and less energy stored by the capacitor. However, when the capacitor is connected to a battery and a dielectric is inserted, the battery supplies more charge to each plate in order to maintain a constant potential difference. Because the capacitor now stores more charge, it also stores more energy.

Energy Density

The electric potential energy stored by a capacitor can also be considered energy stored in the electric field between the capacitor plates. In fact, any electric field stores energy, no matter what the source of that field. We can derive an expression for the energy stored per unit volume by an electric field by thinking about the special case of a parallel-plate capacitor.

Consider a dielectric-filled parallel-plate capacitor (Fig. 27.29). The net electric field between the plates is $\vec{E}_C$, and the potential between the plates is

$$V_C = E_C d \qquad (27.21)$$

The capacitance is given by $C = \kappa \varepsilon_0 A/d$ (Eq. 27.18), so the energy stored by the electric field $\vec{E}_C$ is (Eq. 27.3):

$$U_E = \frac{1}{2}CV_C^2 = \frac{1}{2}\frac{\kappa \varepsilon_0 A}{d}(E_C d)^2$$

$$U_E = \frac{1}{2}\kappa \varepsilon_0 E_C^2 (Ad)$$

The dielectric is a rectangular box with volume Ad, so the energy density (energy per unit volume) of the field is

$$u_E = \frac{U_E}{Ad} = \frac{1}{2}\kappa \varepsilon_0 E^2 \qquad (27.22)$$

ENERGY DENSITY STORED IN AN ELECTRIC FIELD ✪ Major Concept

where we have dropped the subscript from the electric field, and the symbol for energy density is a lowercase u.

Equation 27.22 was derived for the special case of a dielectric-filled parallel-plate capacitor, but it is a general expression for the energy density stored in any electric field. If the field is in a vacuum, then $\kappa = 1$. Because the dielectric constant of any insulator is greater than 1, more energy is stored in a region that contains an insulator than in a vacuum, for electric fields of equal magnitude.

CONCEPT EXERCISE 27.7

An X-ray tube at a dentist's office produces X-rays by striking a metal plate with a beam of electrons in an evacuated tube. The electric field generated in an X-ray tube is approximately 5×10^6 V/m. What is the energy density stored in this electric field?

EXAMPLE 27.9 Removing a Dielectric

A large parallel-plate capacitor consists of two metal disks of radius 9.45 cm. A Teflon dielectric has the same radius and is 0.55 cm thick. The dielectric just barely fits between the capacitor's plates, and the capacitor is connected to a battery that has a terminal potential of 3.2 V. The capacitor is fully charged, and then the battery is disconnected. Next, a person carefully removes the dielectric without touching the plates. How much work must the person do to remove the dielectric? (Assume the capacitor is horizontal so the person removes the dielectric horizontally and there is no change in the microscopic internal energy.) In the **CHECK and THINK** step, compare your result to the amount of work needed to lift the dielectric by a distance equal to its diameter. The density of Teflon is 2200 kg/m³.

:• INTERPRET and ANTICIPATE
Consider a system consisting of the capacitor and the dielectric. The bar chart in Figure 27.32 shows that the initial electrostatic potential energy plus the work done by the person equals the final potential energy. We expect that the dielectric is attracted to the plates of the capacitor, so the person must exert positive work to remove it. This might seem odd because removing the dielectric increases the energy stored in the capacitor in this case.

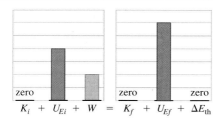

FIGURE 27.32

:• SOLVE

The work done by the person is the final potential energy minus the initial potential energy.	$W = U_{Ef} - U_{Ei}$
Initially, the capacitor contains a dielectric and is fully charged, so the potential difference across its plates is $\mathcal{E}$. The energy stored by the capacitor is given by Equation 27.3. Substitute the capacitance given by $C = \kappa C_{vac}$ (Eq. 27.17).	$U_{Ei} = \frac{1}{2} C V_C^2$ (27.3) $U_{Ei} = \frac{1}{2} \kappa C_{vac} \mathcal{E}^2$ (1)
It is convenient (though not necessary) to find a value for the initial potential energy at this stage. The capacitance C_{vac} of a parallel-plate capacitor is given by Equation 27.10. The area of a disk is πr^2. The dielectric constant comes from Table 27.2.	$U_{Ei} = \frac{1}{2} \kappa \left(\frac{\varepsilon_0 A}{d}\right) \mathcal{E}^2 = \frac{1}{2} \kappa \left(\frac{\varepsilon_0 \pi r^2}{d}\right) \mathcal{E}^2$ $U_{Ei} = \frac{1}{2}(2.1)\left(\frac{(8.85 \times 10^{-12}\,\text{F/m})\pi(9.45 \times 10^{-2}\,\text{m})^2}{0.55 \times 10^{-2}\,\text{m}}\right)(3.2\,\text{V})^2$ $U_{Ei} = 4.9 \times 10^{-10}\,\text{J}$
The battery is disconnected, so the charge on the plates is constant and the potential difference changes when the dielectric is removed. The charge is found from $Q = CV_C$ (Eq. 27.1).	$Q = CV_C = \kappa C_{vac} \mathcal{E}$ (2)
Because the charge is constant, it is best to find the final potential energy using Equation 27.2. The charge comes from Equation (2), and the capacitance is C_{vac} because the dielectric has been removed.	$U_{Ef} = \frac{1}{2}\frac{Q^2}{C}$ (27.2) $U_{Ef} = \frac{1}{2}\frac{(\kappa C_{vac}\mathcal{E})^2}{C_{vac}} = \frac{1}{2}\kappa^2 C_{vac}\mathcal{E}^2$
The final potential energy may be written in terms of the initial potential energy (Eq. 1).	$U_{Ef} = \kappa U_{Ei}$
The work done by the person is the final potential energy minus the initial potential energy.	$W = \kappa U_{Ei} - U_{Ei} = (\kappa - 1)U_{Ei} = (2.1 - 1)(4.9 \times 10^{-10}\,\text{J})$ $W = 5.4 \times 10^{-10}\,\text{J}$

:• **CHECK and THINK**
As expected, the person does positive work to remove the dielectric. This amount of work is tiny compared to the work needed to lift the dielectric up by $2r$. The mass of the dielectric can be found from its volume and density; then use $W = mg(2r)$.

$$m = \rho V = \rho \pi r^2 d$$
$$m = (2200 \text{ kg/m}^3)\pi(9.45 \times 10^{-2} \text{ m})^2(0.55 \times 10^{-2} \text{ m})$$
$$m = 0.34 \text{ kg}$$
$$W = mg(2r) = (0.34 \text{ kg})(9.81 \text{ m/s}^2)2(9.45 \times 10^{-2} \text{ m})$$
$$W = 0.63 \text{ J} \gg 5.3 \times 10^{-10} \text{ J}$$

27-8 Gauss's Law in a Dielectric

When we studied Gauss's law in Chapter 25, we assumed the electric field existed in a vacuum. In that case, the electric flux through a Gaussian surface is proportional to the charge inside that surface:

$$\Phi = \oint \vec{E} \cdot d\vec{A} = \frac{q_{\text{in}}}{\varepsilon_0} \tag{25.11}$$

How can we modify Gauss's law to include the presence of an insulator such as a dielectric between the plates of a capacitor?

To answer that question, let's consider a parallel-plate capacitor. First, remember that we have used Gauss's law to find the electric field between the plates of a vacuum-filled parallel-plate capacitor: $E_{\text{vac}} = Q_{\text{free}}/\varepsilon_0 A$ (Eq. 27.11); the reason for adding the subscript "free" is explained below. Next, imagine a dielectric consisting of dipole molecules between the plates of an uncharged capacitor (Fig. 27.33A). Because there is no electric field, the dipole molecules are randomly oriented. Any Gaussian surface you could imagine would contain no net charge because every molecule is neutral. (Even if the Gaussian surface happened to pass right through the middle of a few molecules, it would include positive charge from some molecules and negative charge from other molecules, with no net charge overall.)

Now, imagine charging the plates of the capacitor so that there is an electric field inside the dielectric. The electric field causes the dipole molecules to line up such that the negative side of each molecule is closer to the capacitor's positive plate (Fig. 27.33B). Now our Gaussian surface encloses the excess positive charge Q_{free} on the capacitor's plate. Also, our Gaussian surface no doubt passes through the middle of some molecules, preferentially enclosing an excess negative charge because the negative sides of the molecules all face the same way. So, the total charge q_{in} inside the Gaussian cylinder is the charge $+Q_{\text{free}}$ on the surface of the capacitor's plate plus the excess negative charge $-q_D$ from the dielectric:

$$q_{\text{in}} = Q_{\text{free}} - q_D \tag{27.23}$$

We substitute Equation 27.23 into Gauss's law and solve for the electric field inside the dielectric:

$$E_{\text{tot}} = \frac{Q_{\text{free}} - q_D}{\varepsilon_0 A} \tag{27.24}$$

This is the *total* electric field that results from the charged capacitor and the dielectric itself. The dielectric's electric field is in the opposite direction as the electric field produced by the vacuum-filled capacitor $\vec{E}_{\text{vac}}$ (Fig. 27.29), so the total electric field is weaker by a factor of κ (Eq. 27.15):

$$E_{\text{tot}} = \frac{E_{\text{vac}}}{\kappa} \tag{27.25}$$

To find an expression for the charge q_{in} inside the Gaussian surface, substitute Equations 27.24 for E_{tot} and 27.11 for E_{vac} into Equation 27.25:

$$\frac{Q_{\text{free}} - q_D}{\varepsilon_0 A} = \frac{1}{\kappa \varepsilon_0} \frac{Q_{\text{free}}}{A}$$

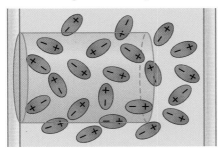

Capacitor is uncharged.

A.

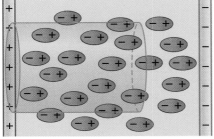

Capacitor is charged.

B.

FIGURE 27.33 A. The dipole molecules of a dielectric are randomly oriented in the absence of an electric field. The net charge enclosed by the Gaussian surface is zero. **B.** When an electric field is present, the molecules of the dielectric line up as shown. The net charge enclosed by the Gaussian surface is $q_{\text{in}} = Q_{\text{free}} - q_D$.

$$q_{in} = Q_{free} - q_D = \frac{Q_{free}}{\kappa} \tag{27.26}$$

So we modify Gauss's law (Eq. 25.11) by replacing q_{in} with Equation 27.26:

$$\oint \kappa \vec{E} \cdot d\vec{A} = \frac{Q_{free}}{\varepsilon_0} \tag{27.27}$$

GENERAL FORM OF GAUSS'S LAW
❶ Underlying Principle

Equation 27.27 was derived for a dielectric-filled capacitor, but it is valid in any situation. It is a more general expression of Gauss's law than Equation 25.11. (If the charged source is in a vacuum, then $\kappa = 1$ and the two expressions are identical.)

There are two major differences between the general expression for Gauss's law (Eq. 27.27) and the vacuum expression (Eq. 25.11). First, the flux integral in the general expression involves the dielectric constant κ, and κ is kept inside the integral because the dielectric constant may not have the same value throughout the entire Gaussian surface. Second, the right side of the general expression for Gauss's law involves Q_{free} instead of q_{in}. The subscript "free" is a reminder that the charged particles on the plates of the capacitor are free to move throughout the conductor and would move if—for example—we changed the potential across the plates. The free charge Q_{free} (on the plates) is not the same as the charge enclosed by the Gaussian surface because it does not include the excess charge q_D embedded in the dielectric. We already took q_D into account when we introduced κ in Gauss's law.

Let's find a way to summarize how quantities are affected by the presence of a dielectric. For example, compare the equations for the capacitance of a vacuum-filled parallel-plate capacitor $C_{vac} = \varepsilon_0 A/d$ (Eq. 27.10) to that of dielectric-filled one $C = \kappa\varepsilon_0 A/d$ (27.18). Mathematically we can alter the equation for a vacuum-filled capacitor by replacing ε_0 with $\kappa\varepsilon_0$ to arrive at the equation for a dielectric-filled capacitor. We extend this comparison to come up with a general rule: *the electrostatic equations involving the permittivity constant ε_0 are altered by replacing ε_0 with $\kappa\varepsilon_0$ to take into account the effects of the presence of a dielectric material with dielectric constant κ.*

The general form of Gauss's law does not play a major role in this textbook. In practice, you may not need to use this general form, but it is the underlying principle. All the principles of electricity and magnetism can be expressed in four equations, and Gauss's law is one of them.

SUMMARY

❶ Underlying Principle

General form of Gauss's law:

$$\oint \kappa \vec{E} \cdot d\vec{A} = \frac{Q_{free}}{\varepsilon_0} \tag{27.27}$$

where κ is the dielectric constant and Q_{free} is the free charge enclosed in the Gaussian surface. In a vacuum, $\kappa = 1$, $q_{in} = Q_{free}$, and Equation 27.27 is the same as Equation 25.11.

✪ Major Concepts

1. A **capacitor** is a device that consists of two conductors (**plates**) separated by an insulator or vacuum. When charged, the plates carry charges of equal magnitude and opposite sign. A capacitor stores electric potential energy.

2. **Capacitance** C is the constant of proportionality in
$$Q = CV_C \tag{27.1}$$
where $V_C = \Delta V$, the potential difference across the capacitor's plates. The SI unit for capacitance is the **farad**: $1\,F = 1\,C/V$.

Capacitance depends only on geometry and the type of insulator between the plates.

3. The **energy stored by a capacitor** with charge Q is
$$U_E = \frac{1}{2}\frac{Q^2}{C} \quad (27.2)$$
Another convenient expression for this potential energy is
$$U_E = \frac{1}{2}CV_C^2 \quad (27.3)$$

4. An **ideal battery** is a device that maintains its terminal potential $\mathcal{E}$ through chemical reactions.

5. The **equivalent capacitance of N capacitors in series** is
$$\frac{1}{C_{eq}} = \sum_{i=1}^{N}\frac{1}{C_i} \quad (27.7)$$

6. The **equivalent capacitance of N capacitors in parallel** is
$$C_{eq} = \sum_{i=1}^{N}C_i \quad (27.8)$$

7. A **dielectric** is an insulator that can support an induced electric field. A dielectric placed between the plates of a capacitor increases its capacitance by a factor κ, the **dielectric constant**:
$$C = \kappa C_{vac} \quad (27.17)$$

8. The **energy density** u_E (energy per unit volume) stored in an electric field $\vec{E}$ is:
$$u_E = \frac{1}{2}\kappa\varepsilon_0 E^2 \quad (27.22)$$

▶ Special Cases

1. **Parallel-plate capacitor** of plate area A and separation d:
$$C = \frac{\varepsilon_0 A}{d} \quad (27.10)$$

2. **Cylindrical capacitor** of inner radius r_{in} and outer radius r_{out}:
$$C = \frac{2\pi\varepsilon_0 L}{\ln(r_{out}/r_{in})} \quad (27.13)$$

PROBLEM-SOLVING STRATEGY — Finding Capacitance

∴ SOLVE

1. Use Gauss's law to find an expression for the **electric field** between the plates. Refer to the problem-solving strategy "Finding the Electric Field Using Gauss's Law" on page 765.

2. Complete a **path integral** to find the electric potential difference V_C:
$$V_C = \int_{r_+}^{r_-} E\, dr \quad (27.9)$$

3. Use $C = Q/V_C$ (Eq. 27.1) to find the capacitance.

∴ CHECK and THINK

In addition to checking the **dimensions**, you should make sure that the expression you found for capacitance depends only on geometric properties.

PROBLEMS AND QUESTIONS

A = algebraic   C = conceptual   E = estimation   G = graphical   N = numerical

27-1 The Leyden Jar

1. **C CASE STUDY** In Concept Exercise 27.1 (page 829), we ignored air resistance acting on the ring launched by the experimenters in the case study. How would including air resistance change your answer to the question about the size of the spring? Does air resistance require that more, less, or the same amount of electric potential energy be stored in the spring? Explain your answers.

2. **C** Imagine turning the crank on a generator like the one shown in Figure 27.1 (page 828). Describe the experience of turning the crank for a long time. Would it get easier or harder to turn? Explain your reasoning.

3. **C** In Franklin's time, a device for storing electric potential energy was called a **Leyden jar**. Today, we call that device a **capacitor**. Another term that is sometimes used is **condenser**. What ideas do these three terms bring to mind? What are the advantages and disadvantages of each?

27-2 Capacitors

4. **E** The first Leyden jar was probably discovered by a German clerk named E. Georg von Kleist. Because von Kleist was not a scientist and did not keep good records, the credit for the discovery of the Leyden jar usually goes to physicist Pieter Musschenbroek from Leyden, Holland. Musschenbroek accidentally discovered the Leyden jar when he tried to charge a jar of water and shocked himself by touching the wire on the inside of the jar while holding the jar on the outside. He said that the shock was no ordinary shock and his body shook violently as though he had been hit by lightning. The energy from the jar that passed through his body was probably around 1 J, and his jar probably had a capacitance of about 1 nF.
   a. Estimate the charge that passed through Musschenbroek's body.
   b. What was the potential difference between the inside and outside of the Leyden jar before Musschenbroek discharged it?

5. **N** A parallel-plate capacitor has a capacitance of 6.0 μF. If the potential difference across the plates of the capacitor is 48 V, what must be the charge on the plates of the capacitor?

6. **C** According to $U_E = \frac{1}{2}C(\Delta V)^2$ (Eq. 27.3), a greater capacitance means more energy is stored by the capacitor, but according to $U_E = Q^2/2C$ (Eq. 27.2), a greater capacitance means less energy is stored. How can both of these equations be correct?

7. **N** In Figure P27.7, capacitor 1 ($C_1 = 20.0$ μF) initially has a potential difference of 50.0 V and capacitor 2 ($C_2 = 5.00$ μF) has none. The switches are then closed simultaneously. a. Find the final charge on each capacitor after a long time has passed. b. Calculate the percentage of the initial stored energy that was lost when the switches were closed.

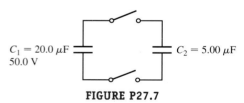

FIGURE P27.7

8. **N** A capacitor has a charge of 3.587 nC and a potential difference of 5.00 V. a. How much energy is stored in the capacitor? b. What is its capacitance?

9. **N** A 4.50-μF capacitor is connected to a battery for a long time. a. If the voltage of the battery is 9.00 V, how much energy is stored in the capacitor? b. If the voltage of the battery is increased to 24.0 V, how much energy is now stored in the capacitor?

10. **G** **CASE STUDY** Students working in a physics lab use a Thompson coil to launch a light ring ($m = 25.0$ g) in much the same way as the experimenters in the case study. The students vary the potential difference across their capacitor bank and measure the maximum height of the ring. Their data are shown in the table to the right. Plot these data to find the capacitance of their capacitor bank.

V (V)	h (cm)
0	0.00
10	0.32
20	2.7
30	5.8
40	9.9
50	16.2
60	21.2
70	29.3
80	39.1
90	49.6
100	61.1

Problems 11 and 12 are paired.

11. **N** A capacitor stores 37.5 mJ when the potential difference between its plates is 16.0 V. a. What is the charge stored by this capacitor? b. What is its capacitance?

12. **N** A capacitor stores 37.5 mJ when the potential difference between its plates is 16.0 V. Then the potential difference is decreased to 8.00 V. a. After the potential difference is reduced, what is the charge stored by the capacitor? b. What is its final capacitance?

27-3 Batteries

13. **N** A 1.50-V battery is used to charge a capacitor. When the capacitor is fully charged, it stores 37.5 mJ. If the capacitor were fully charged by a 3.00-V battery instead, how much energy would it store?

14. **C** When a Leyden jar is charged by a hand generator (Fig. 27.1, page 828), the work done by the person turning the crank is stored as electric potential energy in the jar. When a capacitor is charged by a battery, where does the electric potential energy come from?

15. **N** A battery is used to fully charge a capacitor of capacitance 85.25 μF. When the capacitor is fully charged, it stores 3.454 mJ. What is the terminal potential of the battery?

16. **N** A 6.50-μF capacitor is connected to a battery. What is the charge on each plate of the capacitor if the voltage of the battery is a. 10.0 V and b. 2.00 V?

27-4 Capacitors in Parallel and Series

17. **N** A pair of capacitors with capacitances $C_A = 3.70$ μF and $C_B = 6.40$ μF are connected in a network. What is the equivalent capacitance of the pair of capacitors if they are connected a. in parallel and b. in series?

18. **C** Two 1.5-V batteries are required in a flashlight. a. If the batteries are connected as shown in configuration 1 in Figure P27.18, what is the potential difference between points A and B? b. If, instead, the batteries are connected as shown in configuration 2, what is the potential difference between points A and B? c. Use your answers to figure out why a flashlight with two good batteries may not light up.

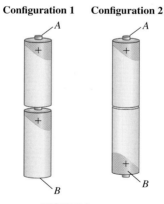

FIGURE P27.18

19. **N** Two capacitors have capacitances of 6.0 μF and 3.0 μF. Initially, the first one has a potential difference of 18.0 V and the second one 9.0 V. If the capacitors are now connected in parallel, what is their common potential difference?

20. **C** When 18th-century experimenters such as Ben Franklin and Abbé Nollet needed a large amount of electric potential energy (or a lot of charged particles), they used several Leyden jars. Did they connect the jars in series or in parallel? Draw a sketch and explain your answer.

21. **N** Calculate the equivalent capacitance between points *a* and *b* for each of the two networks shown in Figure P27.21. Each capacitor has a capacitance of 1.00 μF.

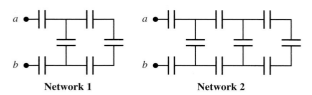

FIGURE P27.21

22. **N** A 12.0-V battery is connected to a pair of capacitors in parallel with capacitances $C_A = 9.00\ \mu\text{F}$ and $C_B = 14.0\ \mu\text{F}$. **a.** What is the equivalent capacitance of the pair of capacitors? **b.** What charge is stored by each of the capacitors? **c.** What is the potential difference across each of the capacitors?

Problems 23 and 24 are paired.

23. **A** Given the arrangement of capacitors in Figure P27.23, find an expression for the equivalent capacitance between points a and b.

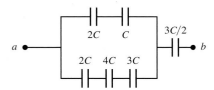

FIGURE P27.23 Problems 23 and 24.

24. **N** An arrangement of capacitors is shown in Figure P27.23. **a.** If $C = 9.70 \times 10^{-5}$ F, what is the equivalent capacitance between points a and b? **b.** A battery with a potential difference of 12.00 V is connected to a capacitor with the equivalent capacitance. What is the energy stored by this capacitor?

25. **N** You find three capacitors that have capacitances $20.00\ \mu\text{F}$, $33.00\ \mu\text{F}$, and $47.00\ \mu\text{F}$. What are the smallest and greatest equivalent capacitances that you can make with these capacitors?

Problems 26 and 27 are paired.

26. **A** Find the equivalent capacitance for the network shown in Figure P27.26.

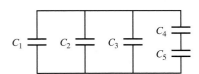

FIGURE P27.26 Problems 26 and 27.

27. **N** Find the equivalent capacitance for the network shown in Figure P27.26 if $C_1 = 1.00\ \mu\text{F}$, $C_2 = 2.00\ \mu\text{F}$, $C_3 = 3.00\ \mu\text{F}$, $C_4 = 4.00\ \mu\text{F}$, and $C_5 = 5.00\ \mu\text{F}$.

28. **N** Figure P27.28 shows three capacitors with capacitances $C_A = 1.00\ \mu\text{F}$, $C_B = 2.00\ \mu\text{F}$, and $C_C = 4.00\ \mu\text{F}$ connected to a 9.00-V battery. **a.** What is the equivalent capacitance of the three capacitors? **b.** What charge is stored in each of the capacitors? **c.** What is the potential difference across each of the capacitors?

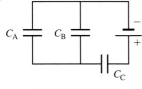

FIGURE P27.28

29. **N** The capacitances of three capacitors are in the ratio 1:2:3. Their equivalent capacitance when all three are in parallel is 120.0 pF greater than when all three are in series. Determine the capacitance of each capacitor.

30. **N** For the four capacitors in the circuit shown in Figure P27.30, $C_A = 1.00\ \mu\text{F}$, $C_B = 4.00\ \mu\text{F}$, $C_C = 2.00\ \mu\text{F}$, and $C_D = 3.00\ \mu\text{F}$. What is the equivalent capacitance between points a and b?

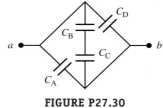

FIGURE P27.30

27-5 Capacitance: Special Cases

31. **N** The separation between the 4.40-cm² plates of an air-filled parallel-plate capacitor is 0.230 cm. **a.** What is the capacitance of this capacitor? If the capacitor is connected to a 9.00-V battery, find **b.** the charge stored by the capacitor and **c.** the magnitude of the electric field between its plates.

Problems 32, 33, and 34 are grouped.

32. **A** A spherical capacitor is made up of two concentric spherical conductors. The inner sphere has positive charge $+Q$, and the outer sphere has negative charge $-Q$. The radius of the inner sphere is r_{in}, and the radius of the outer sphere is r_{out}. Show that its capacitance is $C = 4\pi\varepsilon_0[r_{\text{in}}r_{\text{out}}/(r_{\text{out}} - r_{\text{in}})]$.

33. **A** Derive an expression for the capacitance of an isolated sphere of radius r by taking the limit of $C = 4\pi\varepsilon_0[r_{\text{in}}r_{\text{out}}/(r_{\text{out}} - r_{\text{in}})]$ (see Problem 32) as r_{out} approaches infinity.

34. **E** Use the expression you derived in Problem 33 to estimate the capacitance of **a.** the Earth and **b.** a cannonball placed at the top of a lightning rod (Fig. 24.35, page 734).

35. **N** The plates of an air-filled parallel-plate capacitor carry a surface charge density of $56.0\ \mu\text{C/m}^2$ when the potential difference across the plates is 225 V. What is the separation between the plates of this capacitor?

36. **N** A variable capacitor like the one shown in Figure P27.36 is often used as a tuning element in a high-frequency radio circuit. Suppose the air gap between a pair of plates is 0.300 mm and the pair behaves like a parallel-plate capacitor. **a.** Determine the effective area of the plates when they are rotated to be fully meshed (overlapped), where the capacitance of the pair has a maximum value of 100.0 pF. **b.** Determine the effective area of the plates when they are rotated to the point where the pair of plates has the minimum capacitance of 2.00 pF.

FIGURE P27.36

Problems 37 and 38 are paired.

37. **C** A charged parallel-plate capacitor is connected to a battery. The plates of the capacitor are pulled apart so that their separation doubles. What happens to **a.** the charge and **b.** the amount of energy stored by the capacitor?

38. **C** A charged parallel-plate capacitor is isolated (not connected to a battery). The plates of the capacitor are pulled apart so that their separation doubles. What happens to **a.** the charge and **b.** the amount of energy stored by the capacitor?

39. **N Review** One of the plates of a parallel-plate capacitor is suspended from the beam of a balance as shown in Figure P27.39. The distance d between the capacitor plates is 5.00 mm, and the cross-sectional area of the plates is 625 cm². Determine the potential difference between the capacitor plates if a mass of 4.00 g is placed on the other pan of the balance to obtain static equilibrium.

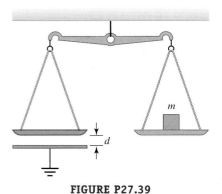

FIGURE P27.39

40. **A** Two fully charged cylindrical capacitors are connected to two identical batteries. The capacitors are identical except that the radius of the outer plate of capacitor A is twice the radius of the outer plate of capacitor B. Both inner plates have a radius that is half the outer radius of capacitor B. **a.** What is the relative capacitance of each capacitor? Express your answer as a ratio C_A/C_B. **b.** What is the relative energy stored by each capacitor? Express your answer as a ratio U_A/U_B. **c.** What is the relative charge stored by each capacitor? Express your answer as a ratio Q_A/Q_B.

41. **N** The capacitance of most commonly used capacitors is small (~ 1 pF $< C < \sim 1 \mu$F). A capacitance of 1 F is very large. In this problem, we estimate the size of a 1-F capacitor using parallel-plate geometry. If the plates of a 1.0-F, air-filled, parallel-plate capacitor are separated by 50.0 μm (about the diameter of a human hair), what is the area of each plate?

Problems 42 and 43 are paired.

42. **N** A 56.90-pF cylindrical capacitor carries a charge of 1.540 μC. The capacitor has a length of 1.000×10^{-3} m. **a.** What is the potential difference across the capacitor? **b.** If the radial separation between the two cylinders is 6.520×10^{-4} m, what are the inner and outer radii of the cylindrical conductors?

43. **N** A 5.69-pF spherical capacitor carries a charge of 1.54 μC. **a.** What is the potential difference across the capacitor? **b.** If the radial separation between the two spherical shells is 6.52×10^{-3} m, what are the inner and outer radii of the spherical conductors? *Hint*: See Problem 32.

27-6 Dielectrics

44. **C** If a parallel-plate capacitor is to have its capacitance doubled, describe three possible ways this can be accomplished.

45. **N** How much charge is stored by a parallel-plate capacitor with plate area 2.25×10^{-2} m² and plate separation 5.00×10^{-3} m that is filled with a dielectric with $\kappa = 5$ and connected to a battery with a potential difference of 12.0 V?

46. **E** Model a Leyden jar as a cylindrical capacitor with a Pyrex dielectric. Assume the jar is about the size of a large coffee cup or large soup can. **a.** Estimate the capacitance of this model Leyden jar. **b.** If you charged the jar so that the electric field was just less than the dielectric strength of Pyrex, how much energy could you store? Check your answer by using the information in Problem 4.

47. **N** The plates of an air-filled parallel-plate capacitor with a plate area of 16.0 cm² and a separation of 9.00 mm are charged to a 145-V potential difference. After the plates are disconnected from the source, a porcelain dielectric with $\kappa = 6.5$ is inserted between the plates of the capacitor. **a.** What is the charge on the capacitor before and after the dielectric is inserted? **b.** What is the capacitance of the capacitor after the dielectric is inserted? **c.** What is the potential difference between the plates of the capacitor after the dielectric is inserted? **d.** What is the magnitude of the change in the energy stored in the capacitor after the dielectric is inserted?

48. **E** You can build a capacitor very cheaply by using two rolls of aluminum foil and an equal amount of paper towels. Imagine layering the paper and foil together and rolling the layers as in Figure 27.30 (page 849). **a.** Estimate the capacitance of this homemade capacitor. Explain your work, including any assumptions you make. **b.** What is the maximum potential you can have across this capacitor without dielectric breakdown? **c.** What steps could you take to increase the capacitance without purchasing more equipment?

49. **N** The robot Johnny-Five is undergoing some repairs and needs to replace one of his parallel-plate capacitors. The original capacitor consisted of plates with an area of 0.0500 m² separated by 0.0300 m, and there was a vacuum between the plates. He finds a replacement that has the same plate separation, but the plate area is only 0.00143 m². In order to achieve the same capacitance as the original part, he can insert a dielectric between the plates of the new capacitor. What dielectric constant does he require?

50. **N** A vacuum-filled, cylindrical capacitor connected to a battery stores 6.75 J when it is fully charged. Paraffin is inserted between the plates without removing the capacitor from the battery. How much energy is stored in the paraffin-filled capacitor?

51. **N** A parallel-plate capacitor with a Pyrex dielectric ($\kappa = 4.70$) and a capacitance of 3.40 μF is charged to a potential difference of 230.0 V and disconnected from the source. **a.** How much work is required to withdraw the Pyrex dielectric from the capacitor? (Ignore any change in the gravitational potential energy.) **b.** What is the change in the potential difference of the capacitor after the Pyrex is removed?

27-7 Energy Stored by a Capacitor with a Dielectric

52. **C** Imagine charging two capacitors with a hand generator (Fig. 27.1). The capacitors are identical except that one has air between its plates and the other has a dielectric. **a.** If you wished to store the same amount of charge on each capacitor, which one would be more difficult to charge? **b.** If you wished to have the same potential difference between the plates of each capacitor, which one would be more difficult to charge? Explain your answers in terms of the amount of work needed in each case.

Problems 53 and 54 are paired.

53. **A** A parallel-plate capacitor with an air gap has capacitance C_0. It is connected to a battery with potential V_0 that gives it charge Q_0 and stored energy U_0. While the capacitor is still connected to the battery, a dielectric with constant $\kappa = 3$ is inserted into the air gap, completely filling it. In terms of the initial values, find the new capacitance C, charge Q, potential V, and stored energy U.

54. **A** A parallel-plate capacitor with an air gap has capacitance C_0. It is connected to a battery with potential V_0 that gives it charge Q_0 and stored energy U_0. After the capacitor is disconnected from the battery, a dielectric with constant $\kappa = 3$ is inserted into the air gap, completely filling it. In terms of the initial values, find the new capacitance C, charge Q, potential V, and stored energy U.

55. **N** A parallel-plate capacitor with plates of area $A = 0.100$ m^2 separated by distance $d = 2.25 \times 10^{-3}$ m is connected to a battery with a potential difference of 9.00 V for a very long time. Initially, a vacuum exists between the plates. **a.** How much energy is stored in the capacitor? **b.** What is the magnitude of the electric field between the plates? **c.** The capacitor is disconnected from the battery and a dielectric with $\kappa = 4.23$ is inserted, filling the space between the plates. What is the magnitude of the electric field in the region between the plates? **d.** What is the magnitude of the electric field produced by the dielectric? **e.** How much energy is stored in the capacitor after the dielectric has been inserted?

56. Model a charged rubber balloon as a spherical source with charge Q and radius R.
    a. **A** Find an expression for the energy density of the electric field just outside the balloon. If the balloon is deflated without losing charge, does the energy density of the electric field increase, decrease, or stay the same? Explain your answer.
    b. **C** How does your answer to part (a) change if the balloon is made of Mylar instead of rubber?

57. **N** Five hundred 8.00-μF capacitors are connected in parallel and then charged to a potential of 25.0 kV. For how long will the stored energy light a 100.0-W bulb until no energy remains in the capacitors?

58. **E** Another simple way to make a homemade capacitor requires only a soft-lead pencil and paper. You can make two conducting parallel plates by covering the paper with graphite on both sides. The finish should be dark and glossy when you are done. It may be helpful to use thick, strong paper so that you don't puncture it when you are coloring. If you use a standard letter-sized sheet of paper, estimate the maximum amount of electric potential energy you could store with such a capacitor. Explain your work, including any assumptions you make.

59. **N** A parallel-plate capacitor with a gap of 0.200 mm is filled with a dielectric with constant $\kappa = 4.60$. **a.** When the capacitor's potential difference is 50.0 V, what is the energy density? **b.** Find the average energy density in an AA battery. A typical new alkaline AA battery stores 9.00×10^3 J of electrical energy and has a volume of 8.30 cm^3.

27-8 Gauss's Law in a Dielectric

60. A rubber balloon with a radius of 18.4 cm has a charge of 30 nC spread uniformly over its surface.
    a. **N** Suppose the balloon is underwater. What is the electric field at a distance of 36.8 cm from the center of the balloon?
    b. **C** How does your answer to part (a) change if the balloon is surrounded by oil instead of water?
    c. **C** How does your answer to part (a) change if the balloon is made of Mylar instead of rubber?

61. **A** Find an expression for the electric field between the two conducting disks in Figure P27.61. Make sure your expression is general enough to include the possibility of a dielectric between the disks. Check your answer using the information given in Section 27-8.

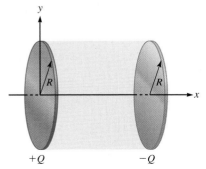

FIGURE P27.61

62. **A** An air-filled parallel-plate capacitor is charged to a certain potential difference. A dielectric is then inserted in the capacitor to completely fill the space between the plates. Then the charge on the plates is increased by a factor of three to restore the original potential difference. Determine the dielectric constant.

63. **N** Two Leyden jars are similar in size and shape, but one has glass as the dielectric and the other ebonite. The glass jar is charged, but when the charge is shared between the two jars (connected in parallel), the electric potential drops by 40% of its initial value. If the dielectric constant of glass is 3.0, find the dielectric constant of ebonite.

Problems 64, 65, and 66 are grouped.

64. **A** Nerve cells in the human body and in other animals are modeled as very long cylindrical capacitors. Portions of some nerves are covered in a layer of fat known as myelin, which functions as the dielectric between two plates in the cylindrical capacitor model. Find an expression for the electric field in the myelin as a function of the distance r from the cell's central axis (assuming a cylindrical cell with $r = 0$ at the axis).

65. **N** Nerve cells in the human body and in other animals are modeled as very long cylindrical capacitors. Portions of some nerves are covered with a layer of fat known as myelin, which functions as the dielectric ($\kappa = 7$) between two plates in the cylindrical capacitor model. The potential difference between the inner and outer walls of myelin in resting nerve cells is roughly $V_{\text{inner}} - V_{\text{outer}} = -70$ mV. Find the linear charge density on the inner (positive) plate. *Hint*: Use the result of Example 27.8.

66. **G** Nerve cells in the human body and in other animals are modeled as very long cylindrical capacitors. Portions of some nerves are covered with a layer of fat known as myelin, which functions as the dielectric ($\kappa = 7$) between two plates in the cylindrical capacitor model. The potential difference between the inner and outer walls of myelin in resting nerve cells is roughly $V_{\text{inner}} - V_{\text{outer}} = -70$ mV. Normally, the inner radius of the myelin layer is 2×10^{-6} m and its thickness is 5×10^{-6} m. Plot the magnitude of the electric field as a function of position r for a nerve cell.

General Problems

67. **N** A 1.50-nF capacitor has a charge of 6.00 nC. **a.** How much energy is stored by the capacitor? **b.** What is the potential difference across the capacitor?

68. **C** If you carefully open a 9-V battery, you discover it is made up of six small batteries (Fig. P27.68). How are these batteries connected to one another, and what is the terminal potential across each of them?

FIGURE P27.68

69. **N** A 9.00-V battery is connected across two capacitors, $C_A = 12.0\ \mu F$ and $C_B = 3.0\ \mu F$, connected in series. **a.** What is the equivalent capacitance of the two capacitors? **b.** How much energy is stored by such an equivalent capacitor? **c.** What is the charge on each of the capacitors? **d.** How much energy is stored in each of the capacitors in this circuit?

70. **N** Three capacitors with capacitances $2.00 \times 10^{-3}\ \mu F$, $4.00 \times 10^{-3}\ \mu F$, and $6.00 \times 10^{-3}\ \mu F$ are connected in series. Is it possible to apply a potential difference of $11.00 \times 10^3\ V$ across the set if the breakdown voltage of each capacitor is $4.00 \times 10^3\ V$?

71. **N** What is the maximum charge that can be stored on the 8.00-cm^2 plates of an air-filled parallel-plate capacitor before breakdown occurs? The dielectric strength of air is $3.00\ MV/m$.

Problems 72 and 73 are paired.

72. **N, C** In a laboratory, you find a 9.00-V battery and a 12.0-V battery. You also find a 30.0-μF capacitor and a 45.0-μF capacitor. Your challenge is to use only one battery and one capacitor to store the maximum possible energy. Which battery and which capacitor do you choose, and how much energy will the capacitor store?

73. **N, C** In a laboratory, you find a 9.00-V battery and a 12.0-V battery. You also find a 30.0-μF capacitor and a 45.0-μF capacitor. Your challenge is to store the maximum possible energy. You may use as much of this equipment as you wish. Describe your solution, and draw a schematic diagram of your network. How much energy is stored by the capacitor(s)?

74. **N** A spherical capacitor has an inner-shell radius of 4.00 cm and an outer-shell radius of 8.00 cm. (See Problem 32.) **a.** What is the capacitance of this capacitor? **b.** When connected to a battery, the capacitor carries a charge of 7.45 μC. What is the voltage of the battery?

75. **N** Figure P27.75 shows four capacitors with $C_A = 4.00\ \mu F$, $C_B = 8.00\ \mu F$, $C_C = 6.00\ \mu F$, and $C_D = 5.00\ \mu F$ connected across points a and b, which have potential difference $\Delta V_{ab} = 12.0\ V$. **a.** What is the equivalent capacitance of the four capacitors? **b.** What is the charge on each of the four capacitors?

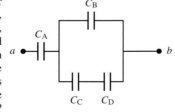

FIGURE P27.75

76. **N, C** As in Example 27.3 (page 842), you are working in a laboratory to build a network from a design (Fig. P27.76). You have a dozen batteries with terminal potential $\mathcal{E}_0 = 1.5\ V$ and a dozen capacitors with capacitance $C_0 = 9.0\ \mu F$. Draw a schematic diagram showing how you can build this network from the components in the laboratory.

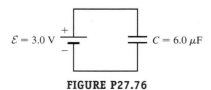

FIGURE P27.76

77. **N** The plates of an air-filled parallel-plate capacitor each have an area of $5.75\ cm^2$ and are separated by 0.145 cm. A 12.0-V battery is connected to the capacitor. **a.** What is the capacitance of the capacitor? **b.** How much charge is stored by the capacitor? **c.** What is the magnitude of the electric field between the plates of the capacitor? **d.** What is the surface charge density on the positive plate of the capacitor?

78. **A** A parallel-plate capacitor with plates of area A and spacing d is filled with a dielectric with constant κ. The dielectric is then pulled halfway out of the gap. Find an expression for the capacitance of the resulting configuration.

79. **N** When connected in series, two capacitors have an equivalent capacitance of 3.00 μF. The same two capacitors have an equivalent capacitance of 13.0 μF when connected in parallel. What is the capacitance of each of the capacitors?

80. **A** Show that $V_C = \int_{r_+}^{r_-} E\,dr$ (Eq. 27.9) is derived from Equation 26.15, $\Delta V = V_f - V_i = -\int_{r_i}^{r_f} \vec{E} \cdot d\vec{r}$ when a path is chosen that is parallel to an electric field line and $\Delta V = V_f - V_i = V_- - V_+ = -V_C$.

81. **N** A 90.0-V battery is connected to a capacitor with capacitance C_A. The capacitor is charged and then disconnected from the battery. Capacitor C_A is next connected to a second, uncharged capacitor with capacitance $C_B = 22.0\ \mu F$. If the voltage across the capacitors in parallel is measured to be 55.0 V, what is the capacitance C_A?

82. **N** Consider an infinitely long network with identical capacitors arranged as shown in Figure P27.82. Determine the equivalent capacitance of such a network. Each capacitor has a capacitance of 1.00 μF.

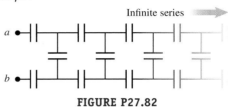

FIGURE P27.82

83. **A** A capacitor C_1 has potential difference V and stores energy U_1. It is then connected without loss of charge in parallel with a second capacitor C_2 that was initially uncharged. **a.** Find an expression for the energy U_2 stored in C_2 in terms of C_1, C_2, and U_1. **b.** Show that U_2 is a maximum when $C_1 = C_2$. *Hint*: Consider the derivative of U_2 as a function of the ratio $x = C_1/C_2$.

84. **N** What is the equivalent capacitance of the five capacitors shown in Figure P27.84? *Hint*: Note the symmetry of the circuit.

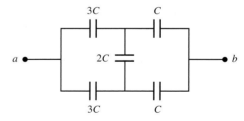

FIGURE P27.84

85. **N** The circuit in Figure P27.85 shows four capacitors connected to a battery. The switch S is initially open, and all capacitors have reached their final charge. The capacitances are $C_1 = 6.00\ \mu F$, $C_2 = 12.00\ \mu F$, $C_3 = 8.00\ \mu F$, and $C_4 = 4.00\ \mu F$. **a.** Find the potential difference across each capacitor and the charge stored in each. **b.** The switch is now closed. What is the new final potential difference across each capacitor and the new charge stored in each?

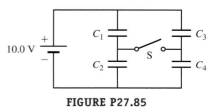

FIGURE P27.85

86. **A** The 1.00-cm gap between the horizontal plates of an air-filled, 44.0-μF, parallel-plate capacitor is slowly filled with peanut oil ($\kappa = 3.0$). How does the capacitance of the capacitor change as a function of the height h of the oil that fills the gap between its plates?

87. **A** Pairs of parallel wires or coaxial cables are two conductors separated by an insulator, so they have a capacitance. For a given cable, the capacitance is independent of the length if the cable is very long. A typical circuit model of a cable is shown in Figure P27.87. It is called a lumped-parameter model and represents how a unit length of the cable behaves. Find the equivalent capacitance of **a.** one unit length (Fig. P27.87A), **b.** two unit lengths (Fig. P27.87B), and **c.** an infinite number of unit lengths (Fig. P27.87C). *Hint*: For the infinite number of units, adding one more unit at the beginning does not change the equivalent capacitance.

88. **N** A parallel-plate capacitor has square plates of side $s = 2.50$ cm and plate separation $d = 2.50$ mm. The capacitor is charged by a battery to a charge $Q = 4.00\ \mu$C, after which the battery is disconnected. A porcelain dielectric ($\kappa = 6.5$) is then inserted a distance $y = 1.00$ cm into the capacitor (Fig. P27.88). *Hint*: Consider the system as two capacitors connected in parallel. **a.** What is the effective capacitance of this capacitor? **b.** How much energy is stored in the capacitor? **c.** What are the magnitude and direction of the force exerted on the dielectric by the plates of the capacitor?

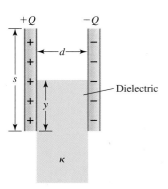

FIGURE P27.88

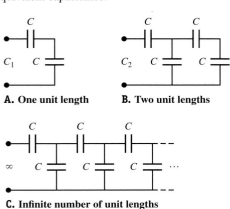

A. One unit length **B. Two unit lengths**

C. Infinite number of unit lengths

FIGURE P27.87

28 Current and Resistance

Key Questions

How does the microscopic model for the motion of electrons in a conductor relate to our macroscopic observations of current and electric potential?

What does a battery supply in a circuit?

28-1 **Microscopic model of charge flow** 865

28-2 **Current** 866

28-3 **Current density** 870

28-4 **Resistivity and conductivity** 874

28-5 **Resistance and resistors** 878

28-6 **Ohm's law** 881

28-7 **Power in a circuit** 884

❶ Underlying Principles

No new principles are introduced in this chapter.

✪ Major Concepts

1. Circuits and circuit elements
2. Current
3. Current density
4. Conductivity, resistivity, and their temperature dependence
5. Resistance and resistors
6. Ohm's law
7. Power supplied to or used by a circuit element

You dry your hair with an electric blow dryer. Your car's headlights illuminate the road. You send e-mail messages to your professor and call your friends and family on your cell phone. These are examples of the ways in which your daily routine depends on the motion of charged particles.

The past five chapters dealt with *electrostatics*. Although an electric field causes charged particles to rearrange themselves, we ignored their motion and imagined waiting until they reached equilibrium. Now we take a close look at the motion of charged particles. In metallic conductors, those charged particles are electrons whose motion is described in terms of a *current*. Electrons that carry a current do not flow smoothly like toothpaste out of a tube. Moving electrons collide with ion cores in the conducting metal and zigzag through the conductor, so that the conductor *resists* the current. In this chapter, we will connect a microscopic model of the motion of electrons to macroscopic observations of current and resistance in conductors. What do quantities we can measure (macroscopic properties) tell us about the motion of charged particles inside a conductor (the microscopic model)?

28-1 Microscopic Model of Charge Flow

This is the second of three chapters devoted to practical applications of the principles of electricity. Chapter 27 focused on capacitors (devices for storing electric potential energy) and batteries (devices that use chemical reactions to maintain a potential difference). This chapter is about the motion of charged particles within conductors. We develop a microscopic model and then connect that model to macroscopic observations. The microscopic model for the motion of charged particles underlies many practical applications, as in this chapter's case study.

CASE STUDY Dead Phone[1]

Read the following student discussion. As always, some parts of the explanations are incorrect, and other parts are on the right track. For now, think about what makes sense to you. We'll return to this case study after we learn more about current and resistance.

Shannon: Avi, you look upset. What's going on?

Avi: I went to Electronics Hut to get a car adapter for my new smartphone, and the guy wrecked my phone. He grabbed some adapter off the rack, and when he plugged it into my phone, the words kind of melted off the screen. I think I even smelled smoke. We unplugged it, but it wouldn't go back on. We tried everything—recharging the battery off the wall socket, rebooting. Nothing worked. Then I had to get the manager to say they would buy me a new phone. I have to go back later to deal with that.

Shannon: That's awful. I'm so sorry.

Cameron: Sounds like the guy fried your phone. I bet he used a high-voltage adapter. It's like the phone's equivalent of touching a high-voltage power line.

Shannon: I think you mean the current was too high. The adapter is like a battery. It's supposed to maintain a constant voltage. Besides, you normally charge your phone off the wall socket, which is about 120 V, and you would use that adapter in the car, where the battery is only 12 V. There's no way the adapter voltage was higher than 120 V.

Cameron: Look, my friend made his own adapter for his MP3 player, and he had to make sure the output voltage of the adapter was around 5 V.

Avi: The thing I really don't get is how the metal in my phone could melt. It's not like there was a big spark like lightning hitting a tree. Why did my phone go up in smoke?

Basic Model

In electronic devices such as Avi's phone, charged particles move through conductors. To understand how these devices work, we must model the motion of charged particles in a conductor. Consider copper. Each neutral copper atom has 29 protons in its nucleus surrounded by 29 swarming electrons. Electrical attraction keeps the electrons bonded to the nucleus, but the outer electrons are only weakly attracted. When many copper atoms are close together, as they are in a metal wire, the outer electrons are also attracted to the nuclei of other nearby atoms. In fact, the outermost electron is no longer bound to any one particular atom and is free to move throughout the entire conductor. These free electrons are known as **conduction electrons**. Typically, a metal conductor has one free electron per atom. We think of the metal as having two components: (1) conduction electrons free to move throughout the conductor (Fig. 28.1) and (2) ions essentially fixed in place in a lattice.

Perhaps we can model the motion of the conduction electrons in the same way we modeled the motion of gas molecules in a container. In fact, our conduction model is based on kinetic theory (Chapter 20), and the five assumptions we made about the

Conduction electrons are free to move throughout conductor; positive ions are fixed in place.

FIGURE 28.1 A model for the motion of conduction electrons in a conductor.

---
[1] This case study is based on the real experience of the textbook author.

molecules in an ideal gas are a good place to start. Table 28.1 lists the assumptions we make about the conduction electrons in a metal and, for comparison, the assumptions we made about molecules in an ideal gas.

TABLE 28.1

Ideal Gas	Conduction Electrons
1. A gas consists of a large number of molecules or atoms that can be modeled as particles.	1. A metal contains a large number of conduction electrons that can be modeled as particles. (Conduction electrons are small compared to the average distance between fixed ions.)
2. Particles in an ideal gas do not interact with one another.	2. Conduction electrons do not interact with one another. Electrons repel one another, but because there are many free electrons, we assume the net force exerted on any one electron by all the other electrons is zero.
3. Particles make elastic collisions with the walls. The duration of each collision is short.	3. Conduction electrons collide with the positively charged ions, which are assumed to be (essentially) at rest. The duration of a collision is short.
4. Particles are free to move in any direction at any speed.	4. Between collisions, conduction electrons are free to move in any direction at relatively high speeds.
5. The gas is made up of identical particles.	5. Conduction electrons are identical. Each has the same mass m_e and charge $-e$.

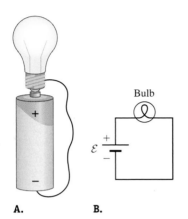

FIGURE 28.2 A. A lightbulb's side is connected to the negative terminal of a battery, and the bulb's base is connected to the battery's positive terminal.
B. Schematic diagram for the circuit in part A.

CIRCUITS AND CIRCUIT ELEMENTS
⭐ **Major Concepts**

Throughout this chapter, we will apply this microscopic model to a simple practical device—a battery with terminal potential $\mathcal{E}$ connected to a lightbulb (Fig. 28.2A). There are several symbols for a lightbulb; in this textbook, we use the symbol shown in the schematic diagram in Figure 28.2B. The smooth lines drawn between the battery and the bulb represent wires. The network in Figure 28.2 is a closed pathway in which charged particles can flow. Such a closed network is called a **circuit**. The various devices that may be in the circuit—such as the battery and the bulb—are called the **circuit elements**. In this chapter, we consider simple circuits with only one pathway for charge to flow through. In the next chapter, we will consider more complicated circuits that may contain many branching pathways.

CONCEPT EXERCISE 28.1

One of the assumptions we make about the conduction electrons in a metal is that the net force exerted on any one electron by all the other conduction electrons is zero (Table 28.1). Justify that assumption.

28-2 Current

Picture the sea of activity inside a copper penny in your pocket, where presumably there is no electric field. Conduction electrons in the penny whiz around, colliding with ions. Imagine watching just one conduction electron for a few moments (Fig. 28.3A). Because the electron's starting position at A is directly above its ending position at G, its displacement along x is zero during this time. The electron's displacement depends on the time interval during which it is observed. For example, if we had observed the electron for a slightly shorter time, it would have been at position F, and its displacement would have had a component in the negative x direction. However, if we observe a large number of conduction electrons over any length of time, we find there is no net motion of the electrons in any particular direction. (This is also true for ideal gas particles inside a container: Their motions are also random, so there is no net displacement of the gas particles.)

What happens if you put your penny in an electric field—for example, by placing it between the plates of a charged capacitor? The electric field causes the charged

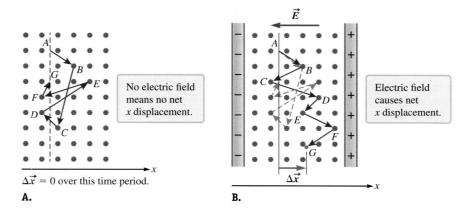

FIGURE 28.3 A. Path of a conduction electron in a metal with $\vec{E} = 0$. **B.** Path of a conduction electron in a metal with $\vec{E} = E_y \hat{\jmath}$. In each a case, the electron starts at point A and ends at point G. In between, it collides with five different ions at points B–F, changing direction at each of these points.

particles in the penny to rearrange themselves to reach equilibrium with no electric field inside the penny. During the short time interval *before* the particles reach equilibrium, however, the electric field inside the penny is not zero. Suppose the electric field is in the negative x direction, $\vec{E} = -E_x \hat{\imath}$, and consider the path of a single conduction electron (Fig. 28.3B). The electric field exerts a force on that electron in the positive x direction:

$$\vec{F}_E = q\vec{E} = (-e)(-E_x \hat{\imath})$$
$$\vec{F}_E = eE_x \hat{\imath}$$

When the electron is moving between collisions in this field, it is pulled in the positive x direction. As a result, the electron drifts to the right. If we observe the same conduction electron over the same period of time as in Figure 28.3A, we find the ending position at G is to the right of the starting position at A, so the electron's displacement has a positive x component (Fig. 28.3B).

Figure 28.3B shows the path of the electron both with and without an electric field in the penny for comparison. The electric field does not reduce the average number of collisions the conduction electron makes with the ions. Instead, the electron's path is shifted to the right. If we looked at all the conduction electrons in the penny, we would find (1) they undergo many collisions with the ions; (2) their motion between collisions is very fast, with speeds around 10^6 m/s (Problem 4); and (3) they tend to move toward the right at a slow speed known as the **drift velocity** in the direction opposite to the electric field. The typical drift speed (magnitude of drift velocity) v_{drift} is between 10^{-5} m/s and 10^{-4} m/s.

Let's see how this microscopic model of electron flow is connected to our macroscopic observations. You might use a voltmeter to measure the potential difference ΔV between two points R and L on the penny, where $\Delta V = V_R - V_L$. In Figure 28.4A, there is no electric field, so there is no net displacement of conduction electrons and no excess charge anywhere in the penny. The result is that the voltmeter's reading is $\Delta V = V_R - V_L = 0$, which indicates no potential difference across the penny.

In Figure 28.4B, the penny is placed in a leftward-pointing electric field so that the conduction electrons drift toward the right. If we wait until the charged particles in the penny reach equilibrium, the penny is still neutral overall but the right side of the penny has excess negative charge and the left side has excess positive charge. The voltmeter's reading is $\Delta V = V_R - V_L < 0$, indicating that the right side of the penny is now at a lower potential than the left side.

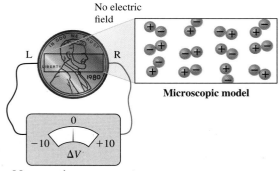

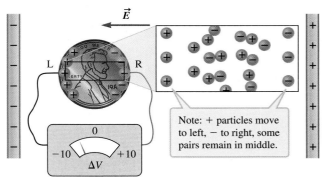

FIGURE 28.4 A. When $\vec{E} = 0$, $V_R = V_L$. **B.** When the penny is in an electric field $\vec{E} = -E_x \hat{\imath}$, the conduction electrons drift so that the left side of the penny becomes positive and the right side becomes negative, and $V_R - V_L < 0$.

We cannot directly observe the motion of conduction electrons in a metal, and a voltmeter is no help in determining whether the free negatively charged electrons really move or perhaps the free positively charged particles move. Without better information, 18th- and 19th-century experimenters assumed the positive particles were mobile. Suppose the *positive* particles move toward the left as a result of the leftward-pointing electric field. The bound negative particles would remain in place, and the net result is that the left side would have excess positive charge and the right side excess negative charge—exactly the way the penny appears (Fig. 28.4B). The voltmeter measures the potential difference $\Delta V = V_R - V_L < 0$ exactly as before. The case study in Chapter 30 is about an experiment showing that in metals, conduction electrons move and positive ions stay relatively motionless, but today we still find it convenient to think about positively charged particles moving in the opposite direction as electrons.

Current is the apparent motion of positively charged particles. In a metal conductor, the current is a result of the conduction electrons' motion in the opposite direction. So, in a metal conductor, the *current* is in the same direction as the electric field. Current is not always the result of electrons flowing. In ionic solutions or in semiconductors, current may be the result of the motion of positive particles.

If a penny is placed between the plates of a charged capacitor, conduction electrons flow in the opposite direction as the electric field, and very soon the particles are in equilibrium again. Such a current does not last very long. In order to study current, we must set up a steady current in some conductor, such as in a lightbulb connected to a battery. An incandescent bulb consists of a conducting wire known as a *filament* encased in a glass globe. The filament may be made of tungsten and may be coiled (Fig. 28.5). One end of the filament is connected to the metal screw threads at the base of the globe. The other end of the filament is connected to the metal foot in the center of the base. Between this foot and the metal threads is an insulator. If you want to light the bulb with a battery, you connect one end of the battery to the metal foot and the other end to the metal threads (Fig. 28.2A). Chemical reactions inside the battery maintain its terminal potential, so electrons continue to flow away from the negative terminal toward the positive terminal. In terms of current, we say that the current is directed away from the positive terminal through the filament and toward the negative terminal.

The amount of current in the filament depends on temperature. The bulb is cool when you first turn it on, and it can get very hot after it has been on for a while. Assume the bulb has been on for a few moments and its temperature is in equilibrium so that the current is steady. Figure 28.6A shows a close-up of a small portion of the filament, small enough to look like a simple cylinder. Because in this small portion of the filament the electric field points from left to right, the conduction electrons drift from right to left. This flow of electrons is modeled as a flow of imaginary positive particles from left to right (Fig. 28.6B).

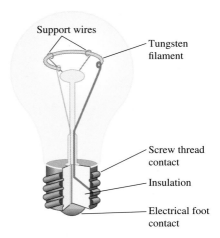

FIGURE 28.5 An incandescent lightbulb. Although compact fluorescent bulbs are in wide use, incandescent bulbs are still used in some applications. Compact fluorescent bulbs operate on a different principle and are not discussed in this chapter.

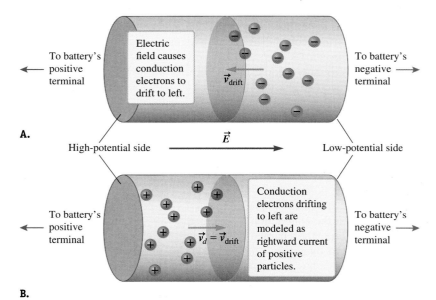

FIGURE 28.6 A. A small portion of a filament connected to a battery. Because the electric field points from high potential to low potential, the field inside the filament points from left to right. So, the conduction electrons in the filament drift to the left. **B.** We model each conduction electron as an imaginary positive particle traveling in the opposite direction.

Current is defined as the rate at which these imaginary positive particles pass through a cross section of the filament:

$$I = \frac{dq}{dt} \quad (28.1)$$

The SI unit for current is the **ampere**, abbreviated with an upper case A. An ampere is one coulomb per second:

$$1\,\text{A} = 1\,\text{C/s}$$

The ampere is one of the seven fundamental SI units. In fact, a coulomb is defined in terms of the ampere: *If there is a steady current of one ampere (1 A), then one coulomb (1 C) is the amount of charge that passes a particular cross section in one second (1 s).* The amount of charge q that passes a particular cross section in some amount of time t is found by integration:

$$q = \int dq = \int_0^t I\,dt \quad (28.2)$$

Current has magnitude given by Equation 28.1 and is directed from high potential to low potential. However, current is a *scalar*. The current follows the geometry of the wire. In the filament in Figure 28.5, the current follows the curling wire from high potential toward low potential. We often draw an arrow near a wire to indicate the direction of the current, as in Figure 28.7, but these arrows are not vectors. Because current is a scalar, vector algebra is not needed when we deal with current combinations. For example, in Figure 28.7, a current-carrying conductor is split into two pieces. The relationship between the currents in each branch is found by simple addition: $I_0 = I_1 + I_2$.

CURRENT ✪ **Major Concept**

The term *amp* is used informally for ampere, and *amperage* is used for current.

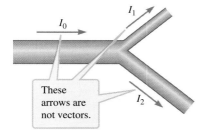

These arrows are not vectors.

FIGURE 28.7 Current is a scalar, and no vector algebra is needed. Here, $I_0 = I_1 + I_2$.

CONCEPT EXERCISE 28.2

Typical currents in three common devices are listed in the table. Assume the currents are constant. Find the amount of charge that passes through a cross section of wire carrying these currents in 1 s.

Device	Current	Charge in 1 s?
a. Ordinary flashlight	0.5 A	
b. Wires in car starter motor	200 A	
c. Laptop computer circuit	1 pA	

EXAMPLE 28.1   CASE STUDY   The Battery's Job

In an effort to understand what happened to Avi's phone, the three students discuss a very simple circuit made up of just one battery whose terminals are connected by a simple wire (Fig. 28.8). After reading their discussion, decide whom you agree with and why. You may agree with more than one student.

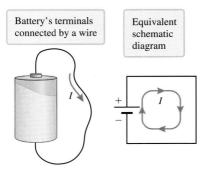

Battery's terminals connected by a wire

Equivalent schematic diagram

FIGURE 28.8 A wire is connected to the terminals of a simple battery.

Example continues on page 870 ▶

Avi: The battery is a source of heat. If you touch the wire, it is going to get very hot, just like a lightbulb.

Shannon: No. The battery is a source of current. The bulb will not light up unless charged particles flow through the filament. The chemical reactions in the battery generate charges, like there is a fountain inside the battery. Positive charges bubble out of the positive end. Most of the particles are converted to light by the bulb. Any particles that are left over get sucked into the negative end of the battery and reused.

Cameron: OK, current is from the positive side of the battery toward the negative side, but the battery's job is to maintain a potential difference between its terminals. It does this through chemical reactions. That's why you buy a battery by its voltage, not by its amperage.

Shannon: If the battery didn't make the current, there would be no current in the wire. It would be like a dry riverbed. It has the potential for current, but it is empty.

Cameron: The wire is not like a dry riverbed! It is more like a flat pipe full of water. The water just sits there until you pick up one end so that the water can flow down. The battery is more like the thing that picks up the end of the pipe. The battery makes sure there is a high-potential side of the wire and a low-potential side of the wire. Then the current just goes from the high side to the low side.

Shannon: If you were right, batteries would last forever. Batteries die because they cannot make any more charged particles. They just run out of juice.

Avi: I think you are both missing the bigger picture. Heat is what killed my phone. Just as the wire or a lightbulb gets hot when you connect it to a battery, my phone got *very* hot. The battery has to be a source of heat. The chemical reactions give off heat.

SOLVE

Some parts of Avi's and Cameron's statements are correct, but Shannon is simply wrong. The battery is not a source of current. Chemical reactions in the battery maintain a potential difference between its terminals. When a wire is connected to those terminals, there is a current in the wire because conduction electrons drift toward the positive end of the battery, carrying energy throughout the wire and causing the wire to get hot. Thus, the battery must be a source of *energy*. Avi is misusing the word *heat*. **Heat** is energy transferred from the environment to a system or from a system to the environment due to their temperature difference (Chapter 21).

28-3 Current Density

If all we need is a macroscopic description of a circuit, current as a scalar parameter is usually sufficient. However, our goal is to connect the macroscopic description to a microscopic model, so we need a vector quantity—**current density**. The magnitude of the current density is the current I per unit cross-sectional area A,

$$J \equiv \frac{I}{A} \tag{28.3}$$

where I is uniform over the area A. Current density is a vector that points in the same direction as the electric field,

$$\vec{J} \propto \vec{E} \tag{28.4}$$

and is a macroscopic property.

To relate current density to the microscopic motion inside a wire, consider Figure 28.6. When a battery's terminals are connected by a wire, the potential difference causes conduction electrons in the wire to drift from the negative terminal toward the positive terminal (Fig. 28.6A), equivalent to the motion of positive particles in the opposite direction. The "moving" positive particles are not the metal ions, but rather imaginary particles that replace actual conduction electrons. Each imaginary particle has the mass of an electron m_e and carries a positive charge $+e$. We imagine that (1) these positive particles undergo many collisions with fixed ions,

(2) between collisions their speed is very high, and (3) their direction of motion after each collision is random. If there is an electric field in the wire, the positive particles drift in the direction of the electric field (Fig. 28.6B). Their drift velocity $\vec{v}_d$ has the same magnitude as the drift velocity $\vec{v}_{\text{drift}}$ of the conduction electrons, but it is in the opposite direction: $\vec{v}_d = -\vec{v}_{\text{drift}}$.

DERIVATION Current Density in a Wire

We show that the current density $\vec{J}$ in a wire is directly proportional to the drift velocity $\vec{v}_d$ of the imaginary positive particles:

$$\vec{J} = ne\vec{v}_d \quad (28.5)$$

where n is the number density of atoms in the conductor.

Figure 28.9 shows a straight wire of cross-sectional area A. An electric field in the positive x direction causes conduction electrons to drift in the negative x direction, modeled as positive particles drifting in the positive x direction. These equivalent positive particles are shown at two instants: just as they are about to pass through the cross section labeled 1 and just as they are about to pass through another section labeled 2.

CURRENT DENSITY
★ Major Concept

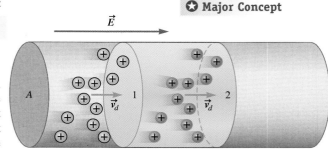

FIGURE 28.9

The positive particles travel with velocity $\vec{v}_d$, so in a time interval of dt, their displacement is $d\vec{x}$.	$d\vec{x} = \vec{v}_d\, dt$	(1)
The amount of charge dq passing through 1 in the time interval dt depends on the number of particles N passing through and the amount of charge e that each one carries.	$dq = Ne$	(2)
We find N from the number density of the actual conduction electrons. Most conductors have one conduction electron per atom, so the number density of the imaginary particles is equal to the number density n of atoms in the metal, and N is the number density of the particles n times the volume dV of that portion of the wire.	$N = n\, dV$	(3)
The volume of the orange portion of the wire is its cross-sectional area A times the length of that portion dx. Substitute Equation (1) for dx.	$dV = A\, dx = Av_d\, dt$	(4)
Find the number of particles that pass through cross section 1 in the time interval dt by substituting Equation (4) into Equation (3).	$N = nAv_d\, dt$	(5)
Find the amount of charge dq that passes through cross section 1 in the time interval dt by substituting Equation (5) into Equation (2).	$dq = neAv_d\, dt$	(6)
Substitute Equation (6) into $I = dq/dt$ (Eq. 28.1). The time interval dt cancels out.	$I = \dfrac{dq}{dt} = nev_d A$	(28.6)
Find the magnitude of the current density by dividing by the cross-sectional area (Eq. 28.3).	$J = \dfrac{I}{A} = nev_d$	
The current density points in the same direction as the electric field. From Figure 28.9, the electric field is in the same direction as the drift velocity of the imaginary positive particles. So, the current density is in the same direction as the drift velocity of the imaginary positive particles.	$\vec{J} = ne\vec{v}_d$ ✓	(28.5)

Derivation continues on page 872 ▶

COMMENTS

Equation 28.5 achieves our goal of finding a connection between the macroscopic observation of current density and the microscopic model of particle motion. On the left side of the equation is the current density—a quantity we can observe on the macroscopic level. The right side involves properties (charge and drift velocity) of the microscopic particles. The current density is also proportional to the number density n. Table 28.2 lists values of n for various metals.

TABLE 28.2 Number density, conductivity, resistivity, and temperature coefficient of resistivity for various materials near room temperature (20°C).

Substance	Number density n ($\times 10^{28}$ m^{-3})	Conductivity σ ($\Omega^{-1} \cdot$ m^{-1})	Resistivity ρ ($\Omega \cdot$ m)	Temperature coefficient of resistivity α (°C^{-1})
Conductors				
Aluminum	6.03	3.767×10^7	2.655×10^{-8}	0.00429
Constantan		2.0×10^6	4.9×10^{-7}	0.00001
Copper	8.42	5.959×10^7	1.678×10^{-8}	0.00393
Gold	5.90	4.46×10^7	2.24×10^{-8}	0.0083
Lead	3.31	4.843×10^6	2.065×10^{-7}	0.00336
Manganin		2.3×10^6	4.4×10^{-7}	0.00001
Mercury	4.07	1.02×10^6	9.84×10^{-7}	0.00089
Nichrome		1.00×10^6	1.00×10^{-6}	0.0004
Silver	5.86	6.302×10^7	1.586×10^{-8}	0.0061
Tungsten	6.22	1.77×10^7	5.65×10^{-8}	0.0045
Semiconductors[a]				
Carbon (graphite)		1.7 to 29×10^3	3.5 to 60×10^{-5}	-0.0005
Germanium		0.001 to 2	1 to 500×10^{-3}	-0.05
Silicon		0.02 to 10	0.1 to 60	-0.07
Insulators				
Amber		2×10^{-15}	5×10^{14}	
Glass[b]		2×10^{-4}	5×10^3	
Rubber (hard)[b]		2×10^{-15}	5×10^{14}	
Quartz		1.3×10^{-18}	7.5×10^{17}	

[a] The conductivity and resistivity of a semiconductor depend on the presence of impurities in the material, and the ranges here reflect ranges in the amount of impurities present.
[b] There is a wide range of values for glass and hard rubber. The values here are in the midrange.

CONCEPT EXERCISE 28.3

We found $\vec{J} = ne\vec{v}_d$ (Eq. 28.5) by considering the motion of the imaginary positive particles (Fig. 28.9). If we consider the actual conduction electrons moving in the negative x direction and carrying charge $-e$, will the current density still point in the positive x direction?

EXAMPLE 28.2 That's Slow!

Assume the wire between the battery and the bulb in Figure 28.2 is made of copper and has a diameter of 1.022 mm. If the current in the wire is 1.33 A, find the magnitude of the current density and the drift speed in the wire.

INTERPRET and ANTICIPATE
Because the current is given, we only need to divide by the cross-sectional area of the wire to find the magnitude of the current density in A/m^2. Once we know the current density, we use the number density n for copper from Table 28.2 to find the drift speed. We expect to find a speed between 10^{-5} m/s and 10^{-4} m/s as stated on page 867.

SOLVE

Find the cross-sectional area of the wire, assuming it is circular.	$A = \pi r^2 = \pi \left(\dfrac{d}{2}\right)^2$ $A = \pi \left(\dfrac{1.022 \times 10^{-3}\,\text{m}}{2}\right)^2$ $A = 8.203 \times 10^{-7}\,\text{m}^2$
Divide the current by the cross-sectional area (Eq. 28.3).	$J = \dfrac{I}{A} = \dfrac{1.33\,\text{A}}{8.203 \times 10^{-7}\,\text{m}^2}$ $J = 1.62 \times 10^6\,\text{A}/\text{m}^2$
To calculate the drift speed, use $\vec{J} = ne\vec{v}_d$ (Eq. 28.5) and look up the number density for copper in Table 28.2. Use the definition 1 A = 1 C/s.	$J = nev_d$ $v_d = \dfrac{J}{ne}$ $v_d = \dfrac{1.62 \times 10^6\,\text{A}/\text{m}^2}{(8.42 \times 10^{28}\,\text{m}^{-3})(1.60 \times 10^{-19}\,\text{C})}$ $v_d = 1.20 \times 10^{-4}\,\text{m/s}$

CHECK and THINK
The current density has the correct units, and the drift speed is in the range we expected (very slow). At this speed, it would take an electron more than 2 hours to travel through a wire the length of your arm. The current density is a very large number because there are many charge carriers per unit volume (more than 10^{28} per cubic meter) spread throughout the wire. If you leave a lightbulb on for a short time, many of the charge carriers from the wire will never make it into the bulb. Instead, the bulb is lit by the motion of charge carriers that were already in the filament.

EXAMPLE 28.3 That's Slow Too!

The filament in an incandescent bulb is often very thin compared to the wires that connect the bulb to the battery. Suppose the filament in Figure 28.5 is made of tungsten and has a diameter of 0.045 mm. Use the information in Example 28.2 to find the current density and drift speed in the filament.

Example continues on page 874 ▶

INTERPRET and ANTICIPATE

This problem is essentially like Example 28.2. The same current must exist in both the filament and the wire. If not, charged particles would "pile up." Because the current is the same in the filament but its cross-sectional area is smaller, we expect the current density to be higher in the filament than in the wire. If the filament has the same number density of charge carriers as the copper wire, we expect the increase in the current density to mean an increase in the drift speed. However, the number density of tungsten is a little lower than that of copper (Table 28.2).

SOLVE

Find the cross-sectional area of the filament, assuming it is circular.

$$A = \pi r^2 = \pi \left(\frac{d}{2}\right)^2 = \pi \left(\frac{0.045 \times 10^{-3}\,\text{m}}{2}\right)^2$$

$$A = 1.59 \times 10^{-9}\,\text{m}^2$$

Divide the current by the cross-sectional area (Eq. 28.3).

$$J = \frac{I}{A} = \frac{1.33\,\text{A}}{1.59 \times 10^{-9}\,\text{m}^2}$$

$$J = 8.36 \times 10^8\,\text{A/m}^2$$

To calculate the drift speed, use $\vec{J} = ne\vec{v}_d$ (Eq. 28.5) and look up the number density for tungsten in Table 28.2.

$$v_d = \frac{J}{ne}$$

$$v_d = \frac{8.36 \times 10^8\,\text{A/m}^2}{(6.22 \times 10^{28}\,\text{m}^{-3})(1.60 \times 10^{-19}\,\text{C})}$$

$$v_d = 8.40 \times 10^{-2}\,\text{m/s}$$

CHECK and THINK

As predicted, the current density is higher in the filament than in the wire that connects the bulb to the battery. Because the number density of tungsten is not much lower than that of copper, the drift speed in the filament is about 700 times higher than the drift speed in the wire.

How can the current I be the same in both the wire and the filament, but the drift speed v_d is higher in the filament? It might help to think of an analogy. Imagine beads placed four abreast on a conveyor belt leading into a paint box (Fig. 28.10A). The speed v_A of this conveyor belt is set so that four beads pass into the paint box per second. Now imagine a narrower conveyor belt on which the beads must be placed single file (Fig. 28.10B). If these beads are to pass into the paint box at the same rate (four beads per second), this narrow conveyor belt must move at a speed four times higher: $v_B = 4v_A$. In this analogy, the rate at which beads enter the paint box is like the current, four beads per second for both conveyor belts. The width of the conveyor belt is like the cross-sectional area of each conductor (wide for the wire, narrow for the filament). The conveyor belt's speed is analogous to the drift speed. The narrow conveyor belt must have a higher speed in order for the same number of beads to enter the paint box per second, just like charge carriers in the filament must have a higher drift speed in order to maintain the same current as in the thicker wire.

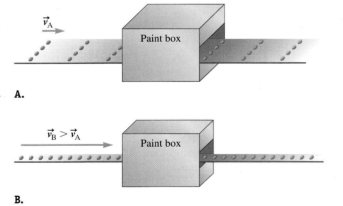

FIGURE 28.10 Paint box analogy for thick versus thin conductors.

28-4 Resistivity and Conductivity

We have modeled a conductor as a material in which charged particles (usually electrons) are free to flow, but Figure 28.3 shows that this model is not exactly correct. Because moving conduction electrons in a metal collide frequently with positive ions, a better description is: *Conduction electrons are free to walk randomly (or bump) around the entire conductor.* The frequency of the collisions determines how well the material conducts electricity. If the frequency of collisions is low, the material is a good conductor. **Electrical conductivity** (or simply **conductivity**) is a measure of a material's ability to conduct current, so it is a measure of how freely charged particles are able to flow in a given material.

Gravitational Analogy

To develop an expression for the conductivity of a material, we use a gravitational analogy. Imagine an array of pegs sandwiched between two vertical boards and organized into an array (Fig. 28.11A). The vertical boards (not shown) are parallel to the page, with one board behind the page and the other in front. Now a small rubber ball is released with some initial velocity between the vertical pegboards. The Earth's gravity accelerates the ball downward. As the ball falls, it is likely to collide with a peg. It bounces off the peg and travels in a parabolic path, just as all projectiles do, and then collides with another peg. Between collisions, the ball is a projectile accelerated downward by gravity and traveling on a parabolic path. Of course, the ball eventually makes it to the bottom.

We have chosen an upward-pointing y axis, so the acceleration of the ball between collisions is

$$\vec{a} = -g\hat{j} \quad (28.7)$$

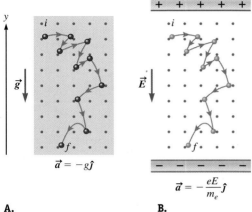

FIGURE 28.11 A. A rubber ball falls through a pegboard. **B.** An imaginary positively charged particle falls through fixed ions. In both cases, the path of the ball or particle between collisions is a parabola.

Because the acceleration is constant, the y component of the ball's velocity between two collisions is given by Equation 2.9:

$$v_y = v_{0y} + a_y t$$

$$v_y = v_{0y} - gt \quad (28.8)$$

The initial time in this case is the instant the ball bounces off of some peg.

We need to think about a large number of balls falling through the pegs because a large number of conduction electrons move through a conductor. If you took a snapshot of the balls at some arbitrary time, you would find the balls that had just collided with a peg moving in random directions. Thus, the average y component of the balls' velocities just after colliding with a peg is zero:

$$(v_{0y})_{av} = 0 \quad (28.9)$$

This is the average initial speed along y.

The average y component of all the balls' velocities at any arbitrary time depends on the average time between collisions. The average time between collisions is called the *mean free time*, t_{mf} (Section 20-5). To find the average y component of velocity, replace t in Equation 28.8 with the mean free time t_{mf} and use Equation 28.9 for the initial speed:

$$v_{av,y} = (v_{0y})_{av} - gt_{mf} = 0 - gt_{mf}$$

$$v_{av,y} = -gt_{mf} \quad (28.10)$$

In Figure 28.11B, a conductor is between the plates of a charged capacitor, with a downward-pointing electric field:

$$\vec{E} = -E\hat{j} \quad (28.11)$$

The ions are shown in a lattice similar to the pegs, and the downward-pointing electric field is analogous to the gravitational field in Figure 28.11A and causes conduction electrons to move upward along a zigzag path. However, following the usual convention, we show an imaginary positive particle of mass m_e and charge e moving in the opposite direction—heading downward along a zigzag path. Like the ball in Figure 28.11A, the particle makes many collisions and its path between collisions is a parabola. Between collisions, each particle has an acceleration given by

$$\vec{a} = -\frac{eE}{m_e}\hat{j} \quad (28.12)$$

The average downward velocity of many such particles is their drift velocity $\vec{v}_d$, which can be found by analogy with Equation 28.10. The quantity $v_{av,y}$ in Equation 28.10 is analogous to the drift speed v_d in a conductor. Replace the acceleration due to gravity with the acceleration due to the electric field (Eq. 28.12):

$$\vec{v}_d = -\left(\frac{eE}{m_e}\right)t_{mf}\hat{\jmath} \qquad (28.13)$$

According to Equation 28.13, the drift velocity is higher when there is a longer average time between collisions. A higher drift velocity (longer average time between collisions) means a higher current density. Substitute Equation 28.13 into $\vec{J} = ne\vec{v}_d$ (Eq. 28.5):

$$\vec{J} = ne\left[-\left(\frac{eE}{m_e}\right)t_{mf}\hat{\jmath}\right] = \frac{ne^2}{m_e}t_{mf}(-E\hat{\jmath})$$

Then substitute $\vec{E} = -E\hat{\jmath}$ (Eq. 28.11):

$$\vec{J} = \left(\frac{ne^2}{m_e}t_{mf}\right)\vec{E} \qquad (28.14)$$

The quantity $(ne^2 t_{mf}/m_e)$ is the constant of proportionality missing from $\vec{J} \propto \vec{E}$ (Eq. 28.4).

Now, imagine putting two different conducting materials between the plates of a capacitor. The electric field in each conductor is the same, but the current density in each conductor depends on the proportionality constant $(ne^2 t_{mf}/m_e)$. The conductor with the longer average time between collisions and the greater density of conduction electrons will be the better conductor and have the higher current density. This constant of proportionality is the **conductivity** σ of the material:

CONDUCTIVITY ⊕ **Major Concept**

$$\sigma \equiv \frac{ne^2}{m_e}t_{mf} \qquad (28.15)$$

Rewriting Equation 28.14 in terms of conductivity, we have

$$\vec{J} = \sigma\vec{E} \qquad (28.16)$$

The symbol for conductivity is σ (lowercase Greek letter sigma), the same symbol used for surface charge density and the Stefan–Boltzmann constant. You must use context to interpret this symbol.

Conductivity is a scalar, so Equation 28.16 shows that the current density points in the same direction as the electric field. Conductivity depends only on the type of material, with values for various substances given in Table 28.2. The SI units for conductivity can be found from Equation 28.16:

$$[\sigma] = \frac{A/m^2}{V/m} = \frac{A/V}{m}$$

In the SI system, the combination of volts per ampere (V/A) is called an **ohm**:

$$1\,V/A = 1\,\Omega \qquad (28.17)$$

where Ω is the uppercase Greek letter omega. The SI units for conductivity are usually given in ohms and meters ($\Omega^{-1} \cdot m^{-1}$).

Resistivity

In many practical applications, it is more convenient to work with the reciprocal of the conductivity, known as the resistivity. **Resistivity** ρ is a measure of a material's ability to resist conducting electricity:

RESISTIVITY ⊕ **Major Concept**

$$\rho \equiv \frac{1}{\sigma} = \frac{m_e}{ne^2 t_{mf}} \qquad (28.18)$$

The SI units for resistivity are $\Omega \cdot m$. The symbol for resistivity is ρ (lowercase Greek letter rho), the same symbol used for volume charge density and mass density. You must use context to interpret this symbol.

Resistivity is a scalar that depends only on the type of substance, with values for various substances given in Table 28.2.

Conductivity in Equation 28.16 may be replaced by resistivity: $\vec{J} = (1/\rho)\vec{E}$, and then

$$\vec{E} = \rho\vec{J} \qquad (28.19)$$

In Table 28.2, the substances categorized as *conductors* have high conductivities and low resistivities. The substances categorized as *insulators* have low conductivities and high resistivities. The substances between conductors and insulators are known as **semiconductors**; their conductivities and resistivities depend on the amount of impurities in a given sample. Semiconductors play an important role in devices such as computers.

Temperature Dependence

The conductivity and therefore the resistivity of a metal depend on (1) the lattice structure of the ions that make up the metal, (2) impurities in the metal, and (3) temperature. Circuit elements such as a lightbulb can get very hot when there is current in the circuit, so it is important to know how temperature affects resistivity.

When a conductor is hot, the ions in the lattice vibrate vigorously. Think back to the analogy of a ball dropping through an array of pegs (Fig. 28.11A), and imagine each peg is vibrating. Each ball will then make more collisions with the pegs because effectively the pegs take up more space. Thus, the mean free path for the balls is reduced.

In a conductor, a higher temperature means more vigorously vibrating ions and more collisions for conduction electrons. An increase in the collision frequency means a decrease in the mean free time between collisions. So, as the temperature of a conductor increases, its conductivity goes down and its resistivity goes up. Mathematically, we express the resistivity at some temperature T as

$$\rho(T) = \rho_0[1 + \alpha(T - T_0)] \qquad (28.20)$$

where ρ_0 is the resistivity at temperature T_0. Usually T_0 is set to room temperature. The resistivities listed in Table 28.2 are for substances at $T_0 = 20°C$. The **temperature coefficient of resistivity** α depends on the type of material. It has the dimensions of 1/temperature; for convenience, the values listed in Table 28.2 are given in terms of the Celsius scale (non-SI units). For most conductors, $\alpha > 0$, so resistivity increases at higher temperatures (Eq. 28.20).

The semiconductors in Table 28.2 have midrange conductivities. A semiconductor is similar to an insulator in that the electrons are bound to particular atoms even when many atoms are pressed together as they are in a solid. However, under the right conditions, it is possible to free an outer electron in each atom. One way to do this is to increase the temperature of the semiconductor. The increase in thermal energy causes the atoms to vibrate more vigorously, and the outermost electrons may shake free. Thus, in semiconductors, resistivity decreases as temperature increases. Mathematically, this means semiconductors have a negative temperature coefficient of resistivity: $\alpha < 0$.

This is a classical physics description; a better description involves quantum physics.

EXAMPLE 28.4 Conductors, Semiconductors, and Superconductors

In this example, we compare the temperature dependence of resistivity in a conductor and a semiconductor.

A Make a graph of resistivity as a function of temperature $\rho(T) = \rho_0[1 + \alpha(T - T_0)]$ (Eq. 28.20) for tungsten between 0 and 40°C.

B Make a similar graph for carbon in the same temperature range. For carbon, assume $\rho_0 = 3.5 \times 10^{-5}\ \Omega \cdot m$.

• INTERPRET and ANTICIPATE
Equation 28.20 is the equation of a line. The sign of the slope (positive or negative) is the same as the sign of the temperature coefficient α. Because α for tungsten is positive (Table 28.2), we expect the graph for part A will be a line tilted up toward the right. Because α for carbon is negative, we expect the graph for part B will be a line tilted down toward the right.

Example continues on page 878 ▶

878 CHAPTER 28 Current and Resistance

:• **SOLVE**
Substitute values of ρ_0 and α from Table 28.2 for each material into Equation 28.20, with $T_0 = 20°C$.

For tungsten,
$$\rho(T) = [5.65 \times 10^{-8} \, \Omega \cdot m][1 + 0.0045°C^{-1}(T - 20°C)]$$

For carbon,
$$\rho(T) = [3.5 \times 10^{-5} \, \Omega \cdot m][1 - 0.0005°C^{-1}(T - 20°C)]$$

Calculate $\rho(T)$ for T between 0 and 40°C. Plot these values for each material (Fig. 28.12).

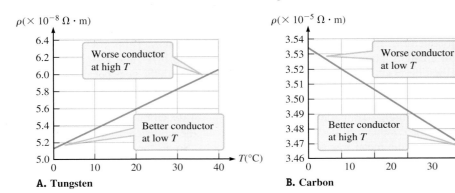

FIGURE 28.12   **A.** Tungsten   **B.** Carbon

:• **CHECK and THINK**
The graphs are as we expected. Figure 28.12A shows that when tungsten is below room temperature, its resistivity is lower and it is a better conductor. The graph for carbon in Figure 28.12B shows just the opposite. When carbon is below room temperature, its resistivity is higher and it is a worse conductor.

The temperature dependence of resistivity is dramatic in certain materials known as **superconductors**. In 1911, Dutch physicist Heike Kamerlingh discovered that at very low temperatures (about 4 K), mercury's resistivity drops to zero. Liquid helium is used to cool the mercury to such a low temperature, but liquid helium is expensive and impractical to use. Since Kamerlingh's discovery, other materials have been found to be superconductors at somewhat higher temperatures that can be reached with liquid nitrogen, a much cheaper coolant. Superconductors are used in some magnetic-levitation (Maglev) trains.

28-5 Resistance and Resistors

In the two preceding sections, we used a microscopic model to find the current density $\vec{J}$ when there is an electric field $\vec{E}$. The macroscopic vector quantities $\vec{J}$ and $\vec{E}$ are not very useful when we are building circuits. Instead, it is more convenient to know two scalar quantities: the electric potential difference ΔV and current I.

In order to write $\vec{E} = \rho \vec{J}$ (Eq. 28.19) in terms of electric potential ΔV and current I, consider a small portion of the filament in a lightbulb as shown in Figure 28.13. The left side of the filament is at a higher potential than the right side. Let's call the slight potential difference in that small portion of the filament δV and the length of this small portion $\delta \ell$. The magnitude of the electric field E is uniform throughout the filament, so the potential difference is proportional to the electric field (Eq. 26.17):

$$\delta V = E \delta \ell \qquad (28.21)$$

(We are concerned only with the magnitude of the electric field, so we don't need the negative sign in Equation 26.17.) Because the magnitude of the electric field is uniform, we can add up δV for each segment to arrive at the potential difference ΔV between the ends of the filament:

$$\Delta V = E \ell \qquad (28.22)$$

where ℓ is the length of the entire filament.

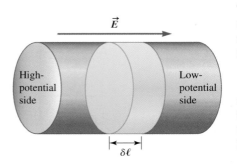

FIGURE 28.13 A small portion of a wire such as a lightbulb filament.

Working with magnitudes only, we substitute $E = \Delta V/\ell$ and $J \equiv I/A$ (Eq. 28.3) into $E = \rho J$ (Eq. 28.19):

$$\frac{\Delta V}{\ell} = \rho \frac{I}{A}$$

Solve for the potential difference ΔV between the ends of the filament:

$$\Delta V = I\left(\rho \frac{\ell}{A}\right) \quad (28.23)$$

The term in parentheses depends only on properties of the filament: (1) its resistivity ρ, (2) its length ℓ, and (3) its cross-sectional area A. The combination of these properties is called the **resistance** of the filament:

RESISTANCE ○ Major Concept

$$R = \rho \frac{\ell}{A} \quad (28.24)$$

Equation 28.23 is usually expressed in terms of resistance:

$$\Delta V = IR \quad (28.25)$$

Resistance R is another measure of an object's ability to resist an electric current. It is different from resistivity ρ, which depends on the type of material and temperature. Resistance depends on the object's resistivity, so resistance also depends on the type of material and temperature. However, resistance R also depends on the geometry of the object. If two conductors are made from the same material and have the same temperature, they have the same resistivity ρ. Now imagine that one of the conductors is a long, thin wire like the filament in a lightbulb, and the other is a short, thick one. According to $R = \rho\ell/A$ (Eq. 28.24), the resistance R of the long, thin wire is higher than the resistance of the short, thick one.

RESISTORS ○ Major Concept

The difference between resistivity ρ and resistance R can be explained using the analogy of balls falling through pegs (Fig. 28.14). Resistivity is represented by the spacing of the pegs. Current is represented by the number of balls that emerge from the bottom per unit time. If the arrangement of the pegs is long and narrow, very few balls emerge from the bottom per unit time, representing a small current in a long, thin wire.

The SI units of resistance are ohms (Ω). Usually the resistance of the wires in a circuit is fairly low. For example, 100 m of wire in a typical household circuit has a resistance of about 1 Ω. The resistance of an incandescent lightbulb's filament when it is operating is about 150 Ω.

A **resistor** is a circuit element designed to have a particular constant resistance (Fig. 28.15). According to $\Delta V = IR$ (Eq. 28.25), for a given potential difference ΔV, the current in a conductor depends on the conductor's resistance. Resistors are made with resistances ranging from 10^{-2} Ω to 10^7 Ω and are used in circuits to control the amount of current. The colored bands on a resistor indicate its nominal resistance. As shown in Table 28.3, the first two bands give the first two digits of the resistance, and the third band is a power-of-10 multiplier. If a resistor's first three bands are green, blue, and brown, its resistance is $56 \times 10^1 = 560$ Ω. The fourth band, if present, is the precision of the resistance. If no band is present, the precision is $\pm 20\%$.

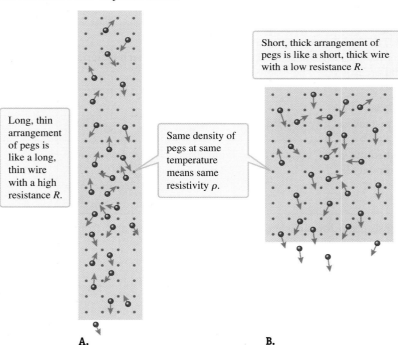

FIGURE 28.14 A. A gravitational analogy for a long, narrow resistor. When the arrangement of the pegs is long and narrow, very few balls emerge from the bottom per unit time. **B.** A model for a short, thick resistor. When the arrangement of the pegs is short and wide, many balls emerge per unit time. In both cases, balls falling through the pegs are analogous to the motion of charged particles in a resistor. So, the current in a long, thin wire is less than in a short, thick wire of the same resistivity.

880 CHAPTER 28 Current and Resistance

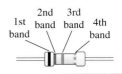

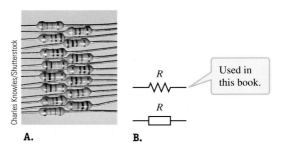

FIGURE 28.15 A. Some resistors. **B.** Schematic symbols for a resistor. The top symbol is used in North America and Japan; the bottom one is used in Europe.

Color	1st band (1st figure)	2nd band (2nd figure)	3rd band (multiplier)	4th band (tolerance)
Black	0	0	10^0	
Brown	1	1	10^1	
Red	2	2	10^2	±2%
Orange	3	3	10^3	
Yellow	4	4	10^4	
Green	5	5	10^5	
Blue	6	6	10^6	
Violet	7	7	10^7	
Gray	8	8	10^8	
White	9	9	10^9	
Gold			10^{-1}	±5%
Silver			10^{-2}	±10%

TABLE 28.3 Color codes for resistance 4-band resistors.

EXAMPLE 28.5 Wires and Filaments

Suppose the total length of copper wire used to connect a lightbulb to a battery (Fig. 28.2) is 35 cm and the length of the tungsten filament is 9.0 cm. Use the cross-sectional areas found in Examples 28.2 and 28.3: 8.203×10^{-7} m² for the wire and 1.59×10^{-9} m² for the filament.

A Calculate the resistances of the copper wire and the tungsten filament at room temperature.

:• INTERPRET and ANTICIPATE
To find the resistances, look up the resistivity of each material in Table 28.2, and use the lengths and areas given. The results should be in Ω.

:• SOLVE
Resistance is given by Equation 28.24.

$$R = \rho \frac{\ell}{A} \tag{28.24}$$

Substitute ρ for copper from Table 28.2 and the area from Example 28.2.

$$R_{\text{wire}} = (1.678 \times 10^{-8}\, \Omega\cdot\text{m})\left(\frac{0.35\,\text{m}}{8.203 \times 10^{-7}\,\text{m}^2}\right)$$

$$R_{\text{wire}} = 7.2 \times 10^{-3}\, \Omega$$

Substitute ρ for tungsten from Table 28.2 and the area from Example 28.3.

$$R_{\text{filament}} = (5.65 \times 10^{-8}\, \Omega\cdot\text{m})\left(\frac{0.09\,\text{m}}{1.59 \times 10^{-9}\,\text{m}^2}\right)$$

$$R_{\text{filament}} = 3.2\, \Omega$$

:• CHECK and THINK
The results have the expected SI units. The tungsten filament has a much higher resistance than the wires. In most circuits, the wires are designed to have a very low resistance compared to the circuit elements. Usually we can ignore the resistance of the wires.

B Calculate the potential difference between the ends of the wire (one end adjacent to the battery and the other end adjacent to the bulb) if the current is 1.33 A (as in Example 28.2). Then calculate the potential difference between the ends of the filament.

:• INTERPRET and ANTICIPATE
The currents in the wire and in the filament are the same but the resistance of the filament is much higher, so we expect the potential difference across the ends of the filament to be much greater than that across the wire.

:• SOLVE Use the resistance found in part A for the wire. Use $1\,\text{V/A} = 1\,\Omega$ (Eq. 28.17) to write the potential difference in volts.	$\Delta V_{\text{wire}} = IR_{\text{wire}}$ $\Delta V_{\text{wire}} = (1.33\,\text{A})(7.2 \times 10^{-3}\,\Omega)$ $\Delta V_{\text{wire}} = 9.6 \times 10^{-3}\,\text{V}$	(28.25)
Repeat this calculation for the filament.	$\Delta V_{\text{filament}} = IR_{\text{filament}} = (1.33\,\text{A})(3.2\,\Omega)$ $\Delta V_{\text{filament}} = 4.3\,\text{V}$	

:• CHECK and THINK

As expected, the potential difference between the ends of the wire is much less than the potential difference across the filament. Again, this is part of the design used in most circuits. There is essentially no potential difference along the wire, and we usually assume $\Delta V = 0$ for any wire. In the simple circuit in Figure 28.2, the potential difference across the bulb is equal to the terminal potential of the battery. If the battery were built from the typical 1.5-V D-cell batteries in a flashlight, there would be three D cells in this circuit. (The actual terminal potential doesn't always equal the nominal value given by the manufacturer.)

CONCEPT EXERCISE 28.4

When a lightbulb burns out, its filament breaks so that there is a gap between the two sides of the filament. What happens to the current in and the resistance of the lightbulb when it burns out? Explain your answers.

28-6 Ohm's Law

In the decades that followed Franklin's work on electricity, scientists continued to experiment and develop a microscopic model of electrical conduction. At that time, it was difficult to make precise macroscopic observations in order to test models. A major breakthrough was made by German mathematics and physics teacher Georg S. Ohm, who published an article entitled *Mathematical Theory of the Galvanic Circuit* in 1827. Although many details of his microscopic model differ from our current model, his article is the basis of what we call *Ohm's law*. The ideas in his article were initially so scorned that Ohm was forced to resign from his teaching position at the Jesuit school in Cologne. Later, Ohm's law was accepted and he won the Copley Prize in 1841.

The best way to state Ohm's law is still somewhat controversial today. For example, Ohm's law is often stated mathematically as $\Delta V = IR$ (Eq. 28.25), but this statement is misleading. In this book, we will present three statements of Ohm's law and show why Equation 28.25 alone is not sufficient.

Ohm's law is not a *law* in the same sense as Newton's laws of motion. Ohm's law is more like Hooke's law for springs. Both Hooke and Ohm came up with their "laws" by empirically fitting their data to a mathematical function. Hooke hung objects of various masses from a spring or rod and measured the amount the spring or rod stretched (Chapter 14). And, just as Hooke's law does not hold for all stretched objects, Ohm's law does not hold for all conductors. Hooke's law and Ohm's law are valid in many practical situations, however, and both apply successfully in a wide variety of circumstances.

Hooke's law and Ohm's law are also mathematically similar. Hooke found that the magnitude of the force applied to a spring is directly proportional to the amount it stretches (Eq. 5.8). Ohm found that *the current I in a conductor is directly*

1ST STATEMENT OF OHM'S LAW
✪ Major Concept

882 CHAPTER 28 Current and Resistance

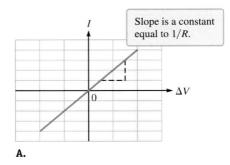

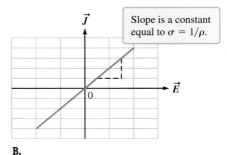

FIGURE 28.16 Alternative expressions for Ohm's law: **A.** $I \propto \Delta V$, **B.** $\vec{J} \propto \vec{E}$.

2ND AND 3RD STATEMENTS OF OHM'S LAW ✪ Major Concept

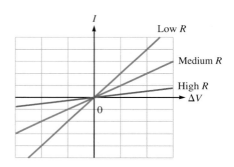

FIGURE 28.17 Current versus potential difference for three ohmic devices. The slope of each line is the conductance $1/R$ of the corresponding circuit element.

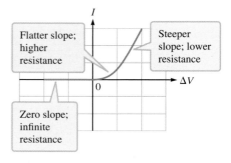

FIGURE 28.18 Current as a function of potential difference for a nonohmic device. The resistance is *not* constant.

proportional to the potential difference applied across it. This is our first statement of **Ohm's law**, usually expressed mathematically as

$$\Delta V \propto I \qquad (28.26)$$

This expression of Ohm's law is shown graphically in Figure 28.16A.

If you compare this proportionality relationship $\Delta V \propto I$ (Eq. 28.26) with $\Delta V = IR$ (Eq. 28.25), you can see why Equation 28.25 is called Ohm's law. If the resistance of the conductor is constant—independent of both ΔV and I—then $\Delta V = IR$ is a statement of Ohm's law, where R is the constant of proportionality. Because we have plotted I on the vertical axis and ΔV on the horizontal axis, the slope of the line in Figure 28.16A is $1/R$. For a circuit element that obeys Ohm's law, that slope is constant. The quantity $1/R$ is called the **conductance**, so the slope of the curve on a graph of I versus ΔV is the conductance.

A second statement of Ohm's law is in terms of the electric field within a conductor and the current density: *The electric field $\vec{E}$ in a conductor is directly proportional to the current density $\vec{J}$ in that conductor*:

$$\vec{E} \propto \vec{J} \qquad (28.27)$$

This statement of Ohm's law is shown graphically in Figure 28.16B. According to $\vec{J} = \sigma\vec{E}$ (Eq. 28.16), the slope of the line is the conductivity σ. Because the conductivity is the reciprocal of the resistivity ($\sigma = 1/\rho$), the slope is also $1/\rho$. For a device that obeys Ohm's law, the slope in Figure 28.16B is constant, so the conductivity and resistivity are constants.

A third statement of Ohm's law is that the resistance R, conductance $1/R$, resistivity ρ, and conductivity σ are all constants. If so, the resistance and conductance of a circuit element do not depend on the potential difference across the element or the current in it, and the resistivity and conductivity do not depend on the electric field or current density within the element. No actual circuit element obeys Ohm's law under all conditions, but a circuit that obeys Ohm's law over a wide range of potential differences across it is described as **ohmic**. Resistors are examples of ohmic circuit elements.

Figure 28.17 shows graphs of I versus ΔV for three different ohmic circuit elements (such as three different resistors). The curve for each circuit element is a straight line with constant slope equal to the conductance $1/R$. The line with the steepest slope represents the circuit element with the lowest resistance, and the line with the flattest slope represents the element with the highest resistance.

A circuit element that does not obey Ohm's law over any significant range of ΔV is described as **nonohmic**. Figure 28.18 is a graph of I versus ΔV for a diode, a nonohmic device. The curve for the diode is a flat line for $\Delta V < 0$; for $\Delta V > 0$, the curve is approximately given by $I \propto \Delta V^{3/2}$.

Although $\Delta V = IR$ (Eq. 28.25) holds for a nonohmic device, the resistance is not constant. Let's use the diode as an example. Although a diode does not obey Ohm's law, any point on the curve in Figure 28.18 is described by $\Delta V = IR$. When $\Delta V < 0$, the current is very nearly zero and we'll assume $I = 0$. According to $\Delta V = IR$, the diode has (nearly) infinite resistance ($R \to \infty$) when $\Delta V < 0$. The resistance is finite but not constant when $\Delta V > 0$. Because $I \propto \Delta V^{3/2}$, the curve is flatter (has lower slope) for small values of ΔV than for large values, so the resistance is higher for small values of ΔV than for large values and is given by

$$R = \frac{\Delta V}{I} = \frac{\Delta V}{\Delta V^{3/2}} = \Delta V^{-1/2}$$

CONCEPT EXERCISE 28.5

A battery with terminal potential $\mathcal{E}$ is connected to a lightbulb (Fig. 28.2). After a long time, the bulb stops working because the filament snaps. After the filament breaks, what is the potential difference across the filament? Explain your answer.

EXAMPLE 28.6 Tungsten and Carbon Filaments: Which Is Which?

Students in a physics laboratory are given two incandescent lightbulbs, one with a carbon filament and the other with a tungsten filament. Their assignment is to figure out which is which based on electrical measurements. The students connect each bulb to a variable power supply, which is like a battery except that the terminal voltage may be adjusted by simply turning a dial. The students vary the terminal potential from roughly 10 V to 100 V and measure the current through each filament. As usual, the resistance in the wires is very low compared to the resistance in the bulb, so the terminal potential of the power supply is about equal to the potential difference between the ends of the filament. The current and voltage for each bulb are given in Table 28.4. Calculate the resistance of each bulb for each current-voltage measurement, and then plot resistance as a function of potential difference to determine which bulb is carbon and which is tungsten.

TABLE 28.4 Data taken to determine the type of filament used in each bulb. The ending zero in each case is a significant figure.

Bulb A		Bulb B	
ΔV (V)	I (A)	ΔV (V)	I (A)
10	0.10	10	0.21
21	0.27	21	0.30
30	0.41	30	0.38
41	0.59	41	0.46
50	0.75	50	0.51
60	0.97	60	0.56
70	1.17	70	0.63
80	1.36	80	0.67
90	1.55	91	0.71
		101	0.77
		110	0.79

INTERPRET and ANTICIPATE

As the students increase the terminal potential, the bulbs get brighter and hotter. We expect each bulb's temperature to increase as the students increase the terminal potential of the power supply. Carbon is a semiconductor with a negative temperature coefficient of resistivity α, whereas tungsten is a conductor with a positive α (Table 28.2). So, by plotting R as a function of V, the students can determine whether the filament is carbon or tungsten. Carbon's resistance should decrease as the terminal potential increases, and tungsten's resistance should increase.

SOLVE

Use $\Delta V = IR$ (Eq. 28.25) to find the resistance of each bulb for each measurement. Although lightbulbs are nonohmic devices, Equation 28.25 still holds (but R is not constant).

TABLE 28.5 Tabulation of R (shaded columns) for each measurement, where $R = (\Delta V)/I$. Values of ΔV and I are from Table 28.4.

Bulb A			Bulb B		
Δ (V)	I (A)	R (Ω)	Δ (V)	I (A)	R (Ω)
10	0.10	100	10	0.21	48
21	0.27	78	21	0.30	70
30	0.41	73	30	0.38	79
41	0.59	69	41	0.46	89
50	0.75	67	50	0.51	98
60	0.97	62	60	0.56	107
70	1.17	60	70	0.63	111
80	1.36	59	80	0.67	119
90	1.55	58	91	0.71	128
			101	0.77	131
			110	0.79	139

Plot R on the vertical axis and ΔV on the horizontal axis for each bulb (Fig. 28.19). Bulb A must be carbon because its resistance decreases as the terminal potential increases. Bulb B must be tungsten because its resistance increases as the terminal potential increases.

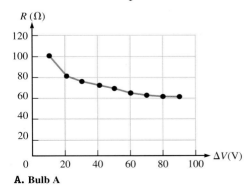
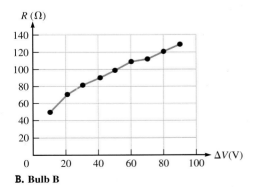

FIGURE 28.19   A. Bulb A   B. Bulb B

Example continues on page 884 ▶

CHECK and THINK

We could have plotted R as a function of I because an increase in current also means an increase in the bulb's temperature. In the next section, we will show that the power lost (through heat) by a lightbulb is proportional to both $(\Delta V)^2$ and I^2.

28-7 Power in a Circuit

In Example 28.1, Shannon argues that a battery is a source of charged particles and that devices such as lightbulbs use up those charged particles. Shannon is mistaken. The function of the battery is to maintain a specific potential difference between its terminals through chemical reactions (Section 27-3). When a device such as a lightbulb is connected to the terminals of a battery (Fig. 28.2), a current is set up in the filament because charged particles are forced to drift due to the potential difference between the filament's ends. Those drifting charged particles were already in the filament; the battery did not supply them. Shannon's intuition is only slightly misguided, however. The battery does supply something that is converted to light by the lightbulb. That "something" is energy.

Recall the basic physics of a wet cell (battery) as shown in Figure 27.11 (page 834). Initially a neutral zinc terminal and a neutral copper terminal are placed in a solution of sulfuric acid. After a period of time, the zinc terminal is negatively charged and the copper terminal is positively charged. Because positively charged particles are attracted to negatively charged particles, we conclude that the chemical reactions must have supplied energy to separate the charged particles. Once charged particles have been separated, the battery has stored electric potential energy that we can tap by connecting a device such as a lightbulb to the battery terminals.

Gravitational Analogy Revisited

The analogy with rubber balls falling through a pegboard (Fig. 28.11) can help us apply the conservation of energy principle to circuits. In order for the balls to drop through the pegboard, a person must first raise the balls to the top, doing positive work and increasing the system's gravitational potential energy. The person is analogous to a battery. The battery does work in separating the charged particles, depositing positive particles on the positive terminal and negative particles on the negative terminal. We can carry this analogy one step further. In order for a person to do the work necessary in lifting the balls, she must eat. As she digests food, chemical reactions in her body give her the energy to do the work. Chemical reactions in the battery enable the battery to do work.

For the moment, suppose the person drops the balls through free space so they do not collide with any pegs. The gravitational potential energy of the system is converted to kinetic energy as the balls fall back to the ground. If the balls fall through a vacuum so that there is no drag force, all the gravitational potential energy is converted to kinetic energy (Fig. 28.20).

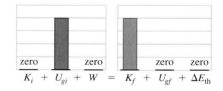

FIGURE 28.20 A bar chart for balls falling through free space. In the absence of drag forces, all the initial gravitational potential energy of the system is converted to kinetic energy.

Now suppose the balls are dropped through the array of pegs (Fig. 28.11A). Although the gravitational potential energy of the system decreases as before, not all of this potential energy goes into kinetic energy of the balls. A peg is at rest when a ball collides with it. Much of the ball's kinetic energy is transferred to the peg as a result of the collision, and the peg begins to vibrate. Let's include the pegs, the balls, and the Earth in the system so that we can take the kinetic energy associated with peg vibration into account as a change in internal energy. Consider this vibrational kinetic energy as an increase in thermal energy ΔE_{th}. (**Thermal energy** is associated with the kinetic energy of the particles that make up the objects in a system; Chapter 19.) Figure 28.21A is a bar chart for this system.

Finally, imagine the pegboard is very long and each ball undergoes many collisions. Under these conditions, the balls emerge from the bottom of the pegboard at the rate the person drops them into the top. Effectively, the balls drift downward at a constant drift speed, which they reach soon after they are dropped. The initial time

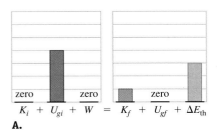
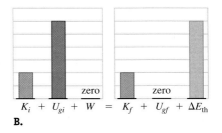

FIGURE 28.21 Bar charts for balls falling through pegboard. **A.** The balls are initially at rest, and the gravitational potential energy is converted to the kinetic energy of the balls plus the thermal energy associated with the pegs' vibration. **B.** The balls are moving at the drift speed, and their kinetic energy is unchanged.

is shortly after the balls have been dropped into the pegboard, so that the balls have already reached their constant drift speed. In this case, there is no change in kinetic energy and all the potential energy is converted to thermal energy (Fig. 28.21B).

This gravitational analogy helps explain what happens on the microscopic level when a conductor such as a lightbulb filament is connected to the terminals of a battery. The charge carriers collide with "fixed" ions just as the balls collide with pegs, causing the ions to increase their vibration so that the system's thermal energy increases.

As in the case where the balls undergo many collisions, soon after the filament is connected to the battery, the average kinetic energy of the charge carriers reaches equilibrium because collisions with the fixed ions keep the charge carriers drifting at a constant average drift speed v_d. The bar chart in Figure 28.22 is similar to Figure 28.21B, showing that the electric potential energy stored by the battery's electric field is converted to thermal energy of the ions in the conductor lattice.

Collisions between balls and pegs in the pegboard system cause the system's internal energy to increase. This energy may leave the system in many ways. If you were in a room with such an apparatus, you would probably find it very noisy. The pegs vibrations cause air molecules to vibrate, so energy leaves the system as sound waves.

The same sort of thing happens with current in a conductor. The increase in the conductor's thermal energy leaves the system through heat by conduction, convection, and radiation. (**Heat** describes the energy transferred between a system and its environment due to their temperature difference; Chapter 21.) In the case of a lightbulb's filament, radiation takes the form of light. So, there was something right about Shannon's intuition. The battery does work on the charge carriers, increasing the system's potential energy, and the potential energy is converted to thermal energy, some of which is radiated away in the form of light.

FIGURE 28.22 The bar chart for charged carriers moving in a conductor is similar to the bar chart for balls falling through a pegboard (Fig. 28.21B). The kinetic energy of the charged particles does not change. The electric potential energy stored in the battery goes into the thermal energy of the filament.

Calculating Power

In practice, the total amount of energy supplied by a battery is not as important as the rate at which the energy is delivered. The rate of energy transfer is known as **power** (Section 9-9). To find the power delivered by a battery, consider a battery whose terminals are connected by a wire (Fig. 28.8). Let's simplify our microscopic conduction model somewhat by imagining that small, positively charged bundles move through the wire from the positive terminal to the negative terminal at the drift speed v_d. In Figure 28.23, each bundle of charge looks like a bead on a string moving clockwise.

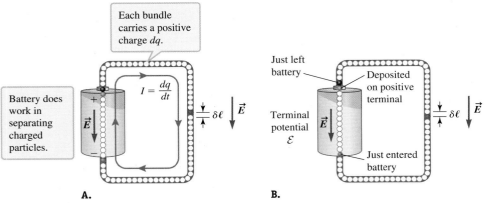

FIGURE 28.23 A. A bundle of charge dq shown in orange is about to be deposited on the positive terminal, and at the same time a bundle shown in blue is about to enter the battery at the negative terminal. **B.** A short interval of time dt later, the orange bundle has been deposited and the blue bundle has entered the battery. And, in this short time interval, each bundle (such as the brown ones) between the blue bundle and the orange one has advanced one step clockwise toward the positive terminal.

Our system includes the bundles of positive charge and the electric field produced by the battery. We'll consider the battery itself to be outside the system. As a result of chemical reactions, the battery moves each positive bundle inside it from the negative terminal to the positive terminal. Because there is an attraction between like charges, the battery must put energy into the system through *positive* work, so the electric potential energy of the bundles must increase. In Figure 28.23, each bundle of charge dq moves one step clockwise in a short time interval dt. The net result of the motion of all these bundles in the battery is equivalent to one bundle having moved from the negative terminal to the positive terminal through a potential difference equal to the terminal potential $\mathcal{E}$ of the battery. The movement of one bundle of charge dq from the negative to the positive terminal of the battery increases the system's electric potential energy by a small amount dU (Eq. 26.6):

$$dU = \mathcal{E} dq \tag{28.28}$$

The rate at which bundles of charge are deposited on the positive terminal is equal to the rate at which bundles of charge move around the circuit, which is the current $I = dq/dt$:

$$dq = I dt \tag{28.29}$$

Substitute Equation 28.29 into Equation 28.28:

$$dU = \mathcal{E} I dt$$

and divide by dt:

$$\frac{dU}{dt} = I\mathcal{E} \tag{28.30}$$

Equation 28.30 is the rate at which the potential energy of the system increases. Because the battery is responsible for the increase in potential energy, Equation 28.30 is the power supplied by the battery:

POWER SUPPLIED BY A BATTERY
✪ Major Concept

$$P_{\text{bat}} = I\mathcal{E} \tag{28.31}$$

The SI unit for power is the watt, and from Equation 28.31,

$$1 \text{ W} = 1 \text{ A} \cdot \text{V}$$

Now let's find an expression for the rate at which energy leaves the system. The battery moves positively charged bundles inside it from the negative terminal toward the positive terminal in a direction opposite to the electric field. The battery must do work to make that happen, increasing the system's potential energy. On the other hand, a charge bundle in the wire, such as the brown one in Figure 28.23, moves from the positive terminal toward the negative terminal in the same direction as the electric field. The motion of bundles in the wire decreases the system's potential energy. The brown bundle moves a distance $\delta\ell$ in the short time interval dt. The electric field in the wire is uniform, and the brown bundle moves through a small potential difference given by $\delta V = E\delta\ell$ (Eq. 28.21). The bundle carries a positive charge dq, so the system's potential energy decreases by a small amount δU:

$$\delta U = dq E \delta\ell$$

This is the decrease in potential energy that results from the motion of just one bundle. However, every bundle in the wire must move a distance $\delta\ell$ in the time interval dt.

Therefore, the total decrease dU in potential energy in the time interval dt is found by adding up the decreases in potential energy due to each bundle:

$$dU = \sum \delta U = \sum dq E \delta\ell$$

Each bundle carries the same amount of charge and the electric field is uniform throughout the wire, so we can pull those constants outside the summation:

$$dU = dq E \sum \delta\ell = dq E \ell$$

where ℓ is the entire length of the wire. Substitute $dq = Idt$ (Eq. 28.29) for dq and $\Delta V = E\ell$ (Eq. 28.22) for $E\ell$ to find $dU = Idt\Delta V$. Then divide by dt:

$$\frac{dU}{dt} = I\Delta V \tag{28.32}$$

Equation 28.32 is the rate at which the system's potential energy decreases as a result of the current in a wire. It is a general expression for the power P used by any circuit element that carries current I and has potential difference ΔV between its ends:

$$P = I\Delta V \tag{28.33}$$

Other convenient expressions for the power used by a circuit element with resistance R can be found by substituting $\Delta V = IR$ (Eq. 28.25) into Equation 28.33, eliminating either I or ΔV:

$$P = I^2 R = \frac{(\Delta V)^2}{R} \tag{28.34}$$

For a wire connected to a battery (Fig. 28.8), the potential difference between the ends of the wire equals the terminal potential $\mathcal{E}$ of the battery. In this case, Equation 28.33 becomes

$$P = I\mathcal{E} = P_{bat}$$

so the power supplied by the battery is used by the wire.

What does the wire do with that power? In the case of wire or a resistor, the power goes into vibrating the ions in the lattice (due to collisions between conduction electrons and ions); thus, the power goes into the system's thermal energy. The wire or resistor becomes hotter than its environment (the circuit board on which it is mounted and the surrounding air). As a result, energy leaves the system through heat. You have experienced this energy transfer when you hold your phone, computer, or other electronic device and feel it warm up your hand. In a poorly designed or dysfunctional device, the heat transfer rate may not be sufficient to cool the circuit. When this happens, the device may smoke, melt, or even catch fire (Fig. 28.24).

POWER USED BY A CIRCUIT ELEMENT
✪ Major Concept

FIGURE 28.24 This computer's circuit was unable to transfer energy to its environment fast enough to avoid catching on fire.

Kilowatt-Hours

Of course, electric circuits would not be particularly useful if the only thing they could do is get warmer. The power that goes into a circuit can do many useful things, such as move the speakers in your car stereo system, rotate the tub in your washing machine, and keep your food cold. In your home or dormitory, power is supplied by a utility company. Typically, households and businesses pay for the total amount of energy they consume in a month.

The SI unit for power is the watt, where

$$1 \text{ W} = 1 \text{ J/s}$$

You might expect that the utility company's bill would indicate your energy consumption in joules, but most companies use **kilowatt-hours** (kW · h) to measure energy. One kilowatt-hour is the amount of energy consumed by a 1000-W device such as a small blow dryer in 1 hour.

CONCEPT EXERCISE 28.6

A battery of terminal potential $\mathcal{E}$ is connected to a lightbulb (Fig. 28.2). After a long time, the bulb stops working because the filament snaps. After the filament breaks, how much power is dissipated by the filament? Explain your answer.

EXAMPLE 28.7 Record-Breaking Lightbulb

One of the earliest household devices to use electricity was the lightbulb. In 1901, Dennis Bernal donated a carbon-filament lightbulb to the fire department in Livermore, California. The bulb has been lit nearly continuously since that year. It went without power for very short time periods when the fire department moved. (It was also off for about 10 hours in 2013 when the power to the bulb was cut.) The book *Guinness World Records* has recognized this "Centennial Bulb" as the longest-burning in the world (Fig. 28.25). The bulb's power output is 4 W.

A Estimate the total amount of energy W consumed by the lightbulb since it was first turned on. Give your answer in kilowatt-hours and in joules.

B Today we pay about 10¢ per kW·h. How much does the fire station have to pay each month to keep the bulb lit?

C How much money would it cost to keep this bulb lit at this price for 100 years? (Assume the power radiated has been constant during this period.)

FIGURE 28.25 The Centennial Bulb holds the world record for the longest-illuminated lightbulb.

INTERPRET and ANTICIPATE

We can find the total energy radiated by the bulb during the past 100 years from its power output. The power radiated must equal the power supplied. Because this is an estimate, we will assume the bulb has been lit continuously for 100 years.

SOLVE

A The total energy consumed in 100 years is the power times 100 years. (One year is about $\pi \times 10^7$ s—good enough for an estimate.)	$100 \text{ yr} \left(\dfrac{\pi \times 10^7 \text{ s}}{1 \text{ yr}} \right) = \pi \times 10^9 \text{ s} \approx 3 \times 10^9 \text{ s}$ $W \approx P\Delta t \approx (4 \text{ W})(3 \times 10^9 \text{ s}) \approx \boxed{1.2 \times 10^{10} \text{ J}}$
Convert to kilowatt-hours.	$W = (1.2 \times 10^{10} \text{ J})\left(\dfrac{1 \text{ kW} \cdot \text{h}}{3.6 \times 10^6 \text{ J}} \right)$ $W \approx \boxed{3.3 \times 10^3 \text{ kW} \cdot \text{h}}$
B The energy used in 1 month is found in a similar way, but we need the number of hours in a month.	$\Delta t = (30 \text{ days})\left(\dfrac{24 \text{ h}}{1 \text{ day}} \right) = 720 \text{ h}$ $W = P\Delta t = (4 \text{ W})(720 \text{ h}) = 2880 \text{ W} \cdot \text{h}$ $W \approx 2.9 \text{ kW} \cdot \text{h}$
To find the cost, multiply by the price per kilowatt-hour.	$(2.9 \text{ kW} \cdot \text{h})\left(\dfrac{\$0.10}{\text{kW} \cdot \text{h}} \right) = \boxed{\$0.29 \text{ per month}}$
C There are 1200 months in 100 years, so multiply by 1200 to find the cost for 100 years at this rate.	$\$0.29 \times 1200 = \boxed{\$348.00}$

CHECK and THINK

Because this is a low-power bulb, it hasn't consumed (or given off) much energy during 100 years. The typical American home uses about 10,000 kW·h per year at an annual cost of about $1000. During the 100+ years that the Centennial Bulb has been glowing, it has used about one-third the energy a typical American household consumes in 1 year.

EXAMPLE 28.8 CASE STUDY Where There's Smoke...

Answer Avi's question, "Why did my phone go up in smoke?"

SOLVE
The thermal energy of the phone increased at a rate faster than its circuits could remove energy through heat. Electronic devices usually have some mechanism to promote heat flow from the device. For example, computers often have fans that remove thermal energy from the circuitry by forced convection. Some electrical components are equipped with fins that increase their surface area and allow them to lose energy to the surroundings more efficiently.

Avi's phone was designed to lose thermal energy at some particular rate, which was sufficient before the adapter was connected to it. The adapter probably supplied the phone with too much power. (The adapter provided a voltage V that was too high, and because $P = IV$, the power was too high.) The phone was unable to radiate fast enough, so thermal energy built up and destroyed the phone.

SUMMARY

❶ Underlying Principles

No new principles are introduced in this chapter.

✪ Major Concepts

1. A **circuit** is a closed network of circuit elements, including metal wires, that form a pathway in which charged particles can flow. A **circuit element** is a component or device found in a circuit, such as a battery, capacitor, resistor, or lightbulb.

2. **Current** I is the (apparent) motion of positively charged particles:
$$I = \frac{dq}{dt} \quad (28.1)$$
In a metal conductor, current is a result of the motion of conduction electrons. For a particular location in the conductor, current is in the direction opposite to the electrons' drift velocity and is in the same direction as the electric field.

3. **Current density** $\vec{J}$ is a vector whose magnitude is the current I per unit cross-sectional area A:
$$J \equiv \frac{I}{A} \quad (28.3)$$
and whose direction is the same as that of the electric field, $\vec{J} \propto \vec{E}$. Current density is a macroscopic observation connected to the microscopic motion of charged particles in the conductor:
$$\vec{J} = ne\vec{v}_d \quad (28.5)$$
where $\vec{v}_d$ is the drift velocity of (imaginary) positive charge carriers and n is the number density of charge carriers in the conductor.

4. **Conductivity, resistivity, and their temperature dependence**
   a. **Conductivity** σ is a measure of a material's ability to conduct current:
   $$\sigma \equiv \frac{ne^2}{m_e} t_{mf} \quad (28.15)$$
   where e is the electron's charge, m_e is the electron's mass, and t_{mf} is the mean free time between collisions of conduction electrons and lattice ions.
   The current density in a conductor is determined by the electric field in that conductor and its conductivity:
   $$\vec{J} = \sigma \vec{E} \quad (28.16)$$
   b. **Resistivity** ρ is a measure of a material's ability to resist conducting electricity. Resistivity is the reciprocal of conductivity:
   $$\rho \equiv \frac{1}{\sigma} = \frac{m_e}{ne^2 t_{mf}} \quad (28.18)$$
   and
   $$\vec{E} = \rho \vec{J} \quad (28.19)$$
   c. Conductivity and resistivity depend on the material's temperature. The **temperature dependence** of resistivity is given by
   $$\rho(T) = \rho_0[1 + \alpha(T - T_0)] \quad (28.20)$$
   where α is the temperature coefficient of resistivity and ρ_0 is the resistivity at temperature T_0, usually 20°C.

Major Concepts cont'd

5. **Resistance and resistors**
   a. **Resistance** R is a measure of an object's ability to resist an electric current. Resistance depends on the object's resistivity ρ and the geometry of the object:
   $$R = \rho \frac{\ell}{A} \quad (28.24)$$
   where ℓ is the object's length and A is its cross-sectional area.

   The current in a conductor depends on the potential difference between the ends of the conductor and its resistance:
   $$\Delta V = IR \quad (28.25)$$
   b. A **resistor** is a circuit element designed to have a particular constant resistance.

6. **Ohm's law** is an empirically derived rule that holds for some circuit elements such as resistors. Ohm's law may be stated in three ways:
   a. The potential difference ΔV between the ends of a conductor is directly proportional to the current I in that conductor:
   $$\Delta V \propto I \quad (28.26)$$
   b. The electric field $\vec{E}$ in a conductor is directly proportional to the current density $\vec{J}$ in that conductor:
   $$\vec{E} \propto \vec{J} \quad (28.27)$$
   c. The resistance R, conductance $1/R$, resistivity ρ, and conductivity σ are all constants.

   A circuit element that obeys Ohm's law over a wide range of potential differences across it is described as **ohmic**.

7. The **power supplied to or used by a circuit element** depends on the current in the circuit element and the potential difference across it:
   $$P = I\Delta V \quad (28.33)$$
   The power supplied by a battery with terminal potential $\mathcal{E}$ is
   $$P_{\text{bat}} = I\mathcal{E} \quad (28.31)$$
   The power used by a circuit element with resistance R is
   $$P = I^2 R = \frac{(\Delta V)^2}{R} \quad (28.34)$$

PROBLEMS AND QUESTIONS

A = algebraic   C = conceptual   E = estimation   G = graphical   N = numerical

28-1 Microscopic Model of Charge Flow

1. **C** You have two different metal conductors. Each has the same number of conduction electrons per unit volume, but the conduction electrons in metal 1 undergo fewer collisions per unit time than those in metal 2. Which is the better conductor: metal 1 or metal 2? Explain.

2. **C** In most metals, there is one conduction electron per atom. Suppose you find a metal that on average has more than one conduction electron per atom. Do you expect it to be a better conductor than most other metals? Explain.

Problems 3, 4, and 5 are grouped.

3. **C Review** You have two antique copper pennies. One is much cooler than the other, but both have the same density and both have one conduction electron per atom. Which penny is a better conductor? Explain.

4. **E Review** Suppose you have an antique copper penny in your pocket. Modeling the conduction electrons in the penny as an ideal gas, use results from Chapter 20 to estimate their kinetic energy and root-mean-square (rms) velocity.

5. **E Review** Suppose that a heated gas comprised of electrons glows bright orange when it is in use. Estimate the average speed of the electrons in the gas. Model this gas as an ideal gas.

28-2 Current

6. **N** In a low-energy transmission electron microscope (TEM), the typical average beam current is about $-1.00\ \mu\text{A}$. If a material sample is exposed to this electron current for 10.0 min, how many electrons impact the material during that time?

7. **N** Imagine a horizontal wire in which a steady stream of 3.0×10^{18} electrons per second flow to the right. What is the current in the wire? What is the direction of the current?

8. **N** In Concept Exercise 28.2 (page 869), we found that the current in a computer circuit is about 1 pA and the current in a car starter motor is about 200 A. How long do you need to wait to have the same amount of charge pass through the wires of a computer circuit that passes through a starter motor in 1 s?

Problems 9, 10, and 11 are grouped.

9. **A** Positively charged ions move along a single axis, forming a beam. The amount of charge that passes through a cross section of the beam is given by $q = q_0 \cos \omega t$, where q_0 is the amount of charge that passes through at $t = 0$ and ω is the angular frequency. Find an expression for the current as a function of time.

10. **C, A** As described in Problem 9, positively charged ions move along a single axis to form a beam. **a.** Describe what is happening to the total charge that has passed by as time goes on. **b.** Are there moments in time when the total charge that has passed by is zero? If so, find an expression for when this occurs and explain your result.

11. **C, A** As described in Problem 9, positively charged ions move along a single axis to form a beam. **a.** Describe what is happening to the current as time goes on. **b.** Are there moments in time when the current is zero? If so, find an expression for when this occurs and explain your result.

12. **C** CASE STUDY Figure P28.12 shows a series of six identical lightbulbs connected to a battery. The bulbs could be a short strand of old-style holiday lights. Assume all the bulbs are operational. Reread the discussion among Avi, Shannon, and Cameron in the case study (Section 28-1 and Example 28.1). Then decide how Shannon and Cameron might answer these questions: Which bulb has the greatest current? Which bulb is the brightest? Which bulb is the hottest? **a.** How do you think Shannon might answer the questions? Explain. **b.** How do you think Cameron might answer the questions? Explain. **c.** What are your answers to the questions? Explain.

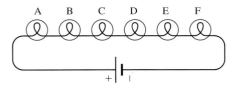

FIGURE P28.12

13. **N** The current in a metal wire of cross-sectional area 1.0×10^{-6} m^2 and length 1.0 m is 16 A. If the metal contains 6.0×10^{28} free electrons per cubic meter, find the drift speed in the wire.

14. **E, C** Consider the length of wire that connects the batteries to the lightbulb in a typical flashlight. **a.** Estimate the time it would take an imaginary positive charge to make its way through this length of wire. Explain how you found any values you use. **b.** How does your answer compare to the amount of time it takes before the light appears? (Use your experience from switching on a flashlight.) Comment on your results.

15. **N** The current in a wire varies with time (measured in seconds) as $I = 24 \text{ A} - (0.12 \text{ A/s}^2)t^2$. Determine the amount of charge that flows through a cross-sectional area of the wire between $t = 0$ and $t = 12$ s.

28-3 Current Density

16. **C Review** Many people say that the electrons in a metal wire travel at nearly the speed of light. Think of how you would respond to that statement. How does the drift speed compare to the speed of light? How about the rms speed? *Hint*: Consider the results from Example 28.2 and Problem 4.

17. **N** The amount of charge that flows through a copper wire 1.00 cm in radius as a function of time is given by $q = 2t^3 + 8t^2 + 2$, where q is in coulombs and t is in seconds. **a.** What is the current through the copper wire at $t = 2.00$ s? **b.** What is the current density in the copper wire at $t = 2.00$ s?

Problems 18 and 19 are paired.

18. **N** A copper wire that has a cross-sectional diameter of 3.500 mm has a current of 1.241 A. **a.** What is the current density? **b.** What is the total charge that passes by a certain location along the wire in 2.000 s?

19. **N** Consider the current described in Problem 18. How long would it take for an electron to travel a 1.00-m length of the wire?

20. **C** Imagine that you can change the thickness of a wire while the electric field in the wire remains constant. Suppose you double the cross-sectional area. What happens to **a.** the current density, **b.** the current, **c.** the number density of charge carriers, and **d.** the drift speed in the wire?

Problems 21 and 22 are paired.

21. **N** A current-carrying conductor made of aluminum gradually narrows as shown in Figure P28.21. The cross-sectional area of region 1 is twice that of region 2. In region 1, the current is 6.25 mA and the current density is 2.00 kA/m^2. Find the current and current density in region 2, assuming a steady-state current has already been achieved.

FIGURE P28.21 Problems 21 and 22.

22. **N** Consider the situation described in Problem 21. Find the drift speed in both regions of the conductor.

23. **N** A copper wire that is 2.00 mm in radius with density 8.94 g/cm^3 has a current of 8.00 A. The molar mass of copper is 63.5463, and each copper atom contributes one free electron. What is the drift speed of the electrons in the copper wire?

24. **C** Many people believe that when you turn on the lamp in your room, electrons from the wall outlet are immediately delivered to the lightbulb filament and then flow back to the wall via the cord running between the wall and the lamp. Explain why this could not be the case given your knowledge of the behavior of the electrons in the wire. Why, then, does the lightbulb light immediately?

28-4 Resistivity and Conductivity

Problems 25 and 26 are paired.

25. **N** The resistivity of a metallic, single-walled carbon nanotube is 3.40×10^{-8} Ω · m. What is the conductivity of this nanotube?

26. **N** Consider the nanotube described in Problem 25. The electron number density is 6.57×10^{28} m^{-3}. What is the mean free time for the electrons flowing in a current along the carbon nanotube?

27. **N** What is the electric field in an aluminum wire if the drift speed of the free electrons in the wire is measured to be 4.50×10^{-5} m/s? The resistivity of aluminum is 2.655×10^{-8} Ω · m.

28. **E** Fifty pennies are stacked together to make one roll. From 1793 to 1857, the American penny was made of copper. After 1857, the penny was made of copper alloys, in which copper was the predominant but not the only metal. Today's penny is copper-plated zinc. **a.** Estimate the resistivity and conductivity of a roll of pure copper pennies. **b.** Estimate the resistance of a roll of pure copper pennies. **c.** In the mid-1970s, the price of copper became so high that the U.S. Mint struck a batch of pennies from aluminum as part of a search for a replacement metal. Those pennies were never released into circulation. One of them was donated to the Smithsonian Institution, where it is now part of the coin collection in the National Museum of American History. A few aluminum pennies may be in the hands of collectors, but such pennies are illegal. Estimate the resistance of a roll of aluminum pennies.

Problems 29 and 30 are paired.

29. **C** You increase the temperature of a metal conductor by a few degrees. What happens to the **a.** mean free time t_{mf}, **b.** frequency of collisions between the charge carriers and lattice ions, **c.** density of the ions, **d.** density of the conduction electrons, and **e.** conductivity? Explain your answers.

30. **C** You increase the temperature of a semiconductor by a few degrees. What happens to the **a.** density of the ions, **b.** density of the conduction electrons, and **c.** conductivity? Explain your answers. If appropriate, compare your answers to those in Problem 29.

31. **N** Two long concentric cylinders of radii 4.0 cm and 8.0 cm are separated by aluminum. The inner cylinder has a charge per unit length of λ at any time. When the two cylinders are maintained at a constant potential difference of 2.0 V via an external source, calculate the current from one cylinder to the other if the cylinders are 100.0 cm long.

32. **G** Plot resistivity versus temperature for the data in Table P28.32. Use your graph to find the temperature coefficient of resistivity. Then use Table 28.2 to identify the material.

TABLE P28.32

T (°C)	ρ ($\Omega \cdot$ m)
0	1.87×10^{-8}
10	2.05×10^{-8}
20	2.24×10^{-8}
30	2.42×10^{-8}
40	2.61×10^{-8}
50	2.80×10^{-8}
60	2.99×10^{-8}
70	3.17×10^{-8}
80	3.36×10^{-8}
90	3.54×10^{-8}
100	3.27×10^{-8}

TABLE P28.38

I (A)	V (V)	T (K)
0.268	7.603	2209
0.269	7.661	2216
0.266	7.682	2241
0.271	7.744	2222
0.272	7.815	2231
0.273	7.894	2243
0.274	7.947	2249
0.277	8.075	2258
0.279	8.213	2276
0.281	8.347	2292
0.284	8.457	2297
0.287	8.656	2321
0.291	8.842	2335
0.293	8.954	2346
0.295	9.059	2355
0.297	9.163	2364
0.298	9.227	2371
0.300	9.335	2380
0.302	9.445	2385
0.301	9.449	2390

33. **N** Two concentric, metal spherical shells of radii $a = 4.0$ cm and $b = 8.0$ cm are separated by aluminum as shown in Figure P28.33. The inner sphere has a total charge Q at any time. If the two spheres are maintained at a potential difference of 2.0 V via an external source, calculate the current from one sphere to the other.

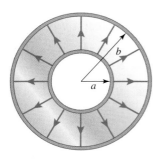

FIGURE P28.33

34. **C** A battery is connected across a conductor that has an irregular cross-sectional area as shown in Figure P28.34.
   a. Is the rate of flow of electrons per unit area at point A less than, equal to, or greater than the rate of flow of electrons per unit area at point B?
   b. Is the magnitude of the electric field at A less than, equal to, or greater than that at B?

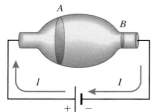

FIGURE P28.34

28-5 Resistance and Resistors

35. **N** When 110.0 V of potential difference is applied across a lightbulb, its filament develops a resistance of 195 Ω. What is the current through the bulb?

36. **A** Show that the dependence of resistance on temperature is given by $R(T) = R_0[1 + \alpha(T - T_0)]$, where α is the temperature coefficient of resistivity and R_0 is the resistance at temperature T_0.

37. **N** A copper wire has a diameter of 0.400 mm. What length of the wire has a resistance of 2.00 Ω?

Problems 38, 50, and 52 are grouped.

38. A lightbulb is connected to a variable power supply. As the potential across the bulb is varied, the resulting current and the filament's temperature are measured. The data are listed in Table P28.38.
   a. **G** Find R for each entry in Table P28.38, and then plot R as a function of T.
   b. **N** Assume that room temperature is at 293 K. Find R_0 (resistance at room temperature). Comment on your result.

39. **N** A 10.0-g sample of copper with density 8.94×10^3 kg/m^3 is made into a wire 1.00×10^3 m long. If the resistivity of copper is 1.678×10^{-8} $\Omega \cdot$ m, what is the resistance of this wire?

40. **C** For each of the following quantities, determine its fundamental dimensions (length, mass, time, current, and temperature): **a.** conductivity, **b.** resistivity, **c.** conductance, and **d.** resistance. Use your results to write the SI unit ohm (Ω) in terms of the fundamental units (meter, kilogram, second, ampere, and kelvin).

41. **C** Which of the following quantities depends only on the microscopic properties of a conductor? Which depend on both microscopic and macroscopic properties? Explain. **a.** conductivity **b.** resistivity **c.** conductance **d.** resistance

42. **N** A long, thin cylindrical conductor is made of silver. Its resistance is 78.5 Ω, and its mass is 0.0250 kg. Find the length and radius of the conductor.

43. **N** An aluminum wire is heated from 20.0°C to 90.0°C. By what percentage does the resistance of the wire change? The temperature coefficient for aluminum is $\alpha = 4.29 \times 10^{-3}$ °C^{-1}.

Problems 44 and 45 are paired.

44. **A** Two wires with different resistivities, ρ_1 and ρ_2, are supposed to have the same resistance. If the radii of the wires are r_1 and r_2, find a function for the length of the second wire in terms of the length of the first wire.

45. **N** A copper and a gold wire are supposed to have the same resistance. The copper wire has a radius of 2.50 mm, and the gold wire has a radius of 3.25 mm. If the copper wire is 1.20 m long, what must be the length of the gold wire?

46. **N** Gold bricks are formed with the dimensions $7 \times 3\frac{5}{8} \times 1\frac{3}{4}$ inches. What is the maximum resistance of a gold brick? What is the minimum resistance? Include sketches to explain how each of these resistances is achieved.

47. **N** A 75.0-cm length of a cylindrical silver wire with a radius of 0.150 mm is extended horizontally between two leads. The potential at the left end of the wire is 3.20 V, and the potential at the right end is zero. The resistivity of silver is 1.586×10^{-8} $\Omega \cdot$ m.

a. What are the magnitude and direction of the electric field in the wire? **b.** What is the resistance of the wire? **c.** What are the magnitude and direction of the current in the wire? **d.** What is the current density in the wire?

28-6 Ohm's Law

48. **G** In 2001, a group of scientists measured the conductivity of a single molecule. Such work is key to developing molecular-based electronics, the next step in further reducing the size of electronic devices and possibly leading to nanoscale computers. Imagine a computer no larger than a single bacterium. Computers that small could be programmed to enter the body and cure diseases on the cellular level. Figure P28.48 is based on the paper published by X. D. Cui and collaborators in 2001. It shows current (in nanoamperes) as a function of potential difference across a single molecule. **a.** Does the molecule obey Ohm's law over the entire range from -1 V to 1 V? **b.** Use the inset in Figure P28.48 to find the conductance and resistance of the molecule over the range from -0.1 V to 0.1 V.

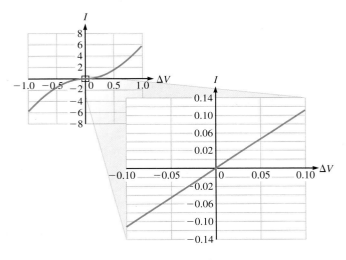

FIGURE P28.48

49. **C** Juanita is hard at work in her lab measuring the resistance of a new material. She observes a current of 0.10 A when applying a potential difference of 9.00 V and a current of 0.20 A when applying a potential difference of 12.00 V. Is the material ohmic? Explain your answer.

50. **G** A conductor is connected to a variable power supply. As the potential across the conductor is varied, the resulting current is measured. The data are listed in Table P28.38. (Ignore the temperature data.) Plot current versus potential and determine whether this conductor obeys Ohm's law. Explain your reasoning.

51. **C** The word *law* in physics can be misleading. In the text, we said that Ohm's law and Hooke's law are not laws in the same sense as Newton's three laws of motion. Perhaps they should be renamed "Ohm's rule" and "Hooke's rule." Explain why these may be better terms by comparing Ohm's law and Hooke's law to Newton's laws of motion.

28-7 Power in a Circuit

52. **Review** A lightbulb is connected to a variable power supply. As the potential across the bulb is varied, the resulting current and the filament's temperature are measured. The data are listed in Table P28.38.

a. **N** Find the power for each entry in Table P28.38.
b. **N** Find T^4 for each entry.
c. **G** Plot P as a function of T^4. Comment on your graph.

53. **N** A 100.0-W lightbulb is connected to a 220.0-V source. **a.** What is the current through the bulb? **b.** What is the internal resistance of the bulb?

54. **C** A current I exists in a loop of superconducting wire. How much power is dissipated by the wire?

55. **N** A two-slice bread toaster consumes 850.0 W of power when plugged into a 120.0-V source. **a.** What is the current in the toaster? **b.** What is the resistance of the coils in the toaster?

56. **N, E CASE STUDY** According to information on the back of Avi's phone, it draws a current of 3.0 A. Assume it uses a lithium–ion battery that has a terminal potential of 3.7 V. **a.** What is the phone's power requirement? **b.** Avi's phone was destroyed in seconds after it was plugged into the wrong adapter. Estimate how much energy was delivered by the battery to the phone in that amount of time. **c.** If the battery were replaced by a power supply with a terminal voltage of 120 V, estimate how much energy would be delivered by the power supply to Avi's phone in the amount of time it took to destroy the phone.

57. **N** Household water heaters use a 240.0-V rather than a 120.0-V source. What is the resistance of a water heater's heating element if it heats 40.0 gallons (151 kg) of water from 15.0°C to 60.0°C in 15.0 min?

58. **Review** Assume the filaments in the bulbs in Example 28.6 (page 883) have the same dimensions, 0.045 mm in diameter and 580 mm in length. Assume all the power supplied is radiated away by the bulb.
   a. **N** Find the power delivered by the power supply for each measurement in Table 28.4.
   b. **N** Assuming the emissivity of the filaments is 1, find the temperature of each filament for each entry in the table.
   c. **G** Plot P as a function of T for each bulb on the same graph.

59. **A** A resistor is connected to a variable power supply. Initially, the power supply's terminal potential is V_0 and the current through the resistor is I_0. The terminal potential is then doubled. Find an expression for the new **a.** current through the resistor, **b.** power delivered by the power supply, and **c.** power used by the resistor in terms of the initial parameters.

60. **C** Two separate resistors, R_1 and R_2, are each connected across identical batteries with terminal potential $\mathcal{E}$. In each case, the power used by each resistor is the same. Explain how this is possible. What must be true about the resistors?

61. **N Review** The current through a 1.50-L electric kettle operating at 120.0 V is 12.5 A. If the energy delivered to the heating element of the kettle is absorbed entirely by the water, how much time does it take to get 1.50 L of water initially at 20.0°C to start boiling?

62. **N** High-voltage transmission lines carrying electricity from a generating station to a local switching station 125 km away carry a current of 850 A. How much power is lost because of resistance in each wire during transmission from the generating station to the local station if the wire's resistance is 0.240 Ω/km?

General Problems

63. **N** The USB charger for a smartphone delivers 500.0 mA of current at 5.00 V while charging the phone. What is the power consumption of the smartphone while it is charging?

64. You measure the amount of charge that passes a certain location as a function of time for several seconds in three separate trials. Your measurements in each of three trials are represented

by the functions (i) $q_1 = 3t^2 + 6t$, (ii) $q_2 = 4t^2 + 2t$, and (iii) $q_3 = 2t^2 + 10t$.
  a. **A** Find the function that represents the current in each of the three trials.
  b. **G** Plot each of the currents you found on the same graph for $t = 0$ to $t = 5$ s (the first 5 s of each trial).
  c. **C** In which case is the current greatest after 10 s? In which case does the most charge pass by the location in 10 s?

65. **N** The current through a copper wire as a function of time is given by $I(t) = 65.0 \cos(33\pi t)$, with I in amperes and t in seconds. How much charge, in coulombs, flows through the wire during the first 0.0050 s after the current is switched on?

66. **N** A beam of electrons incident on a target carries -0.325 mA of current. How many electrons strike the target each minute?

67. **N** A variable resistance is connected to a battery that has a potential difference of 24.0 V. What is the resistance if a current of **a.** 1.00 A and **b.** 1.00×10^{-2} A is observed?

Problems 68 and 69 are paired.

68. **G** The resistivity ρ of a cylindrical conductor with uniform cross-sectional area A increases linearly from the left end to the right end as $\rho = \rho_0 + \alpha x$, where ρ_0 and α are constants. If the current I is constant through it, obtain an expression for the electric field as a function of the distance x from the left end, and plot the electric field as a function of the distance.

69. **A** If the electric field in a uniform conductor is E and v_d is the corresponding drift velocity of the free electrons in the conductor, obtain an expression for the drift speed as a function of the electric field E. Assume the cross-sectional area of the conductor is A, the number density of electrons is n, and ρ is the resistivity of the conductor.

70. **N** A 60.0-V battery is connected across a 175-Ω resistor. What is the **a.** power delivered by the battery and **b.** power used by the resistor?

71. **N, C** Two separate 60.0-V batteries are each connected across separate resistors. One is connected across a 250.0-Ω resistor, and the other is connected across a 350.0-Ω resistor.
  a. How much power is delivered by the battery in each case?
  b. Which resistor consumes 1000 J of energy in the shortest time?

72. **C** Terrance is hard at work fixing his robot, Victor, but finds that a resistor made of a coiled-up gold wire has gone bad. In looking for a replacement part, Terrance finds he has several lead bullets he can melt down and form into a new wire. How should Terrance choose the dimensions of the new lead wire? Describe the reasons for your choices and how each dimension compares to the original gold wire's dimensions.

73. **N** What is the ratio of the cross-sectional areas of a silver wire and a gold wire of equal length that have an equal amount of resistance? The resistivity of silver is $1.586 \times 10^{-8}\ \Omega \cdot \text{m}$, and the resistivity of gold is $2.24 \times 10^{-8}\ \Omega \cdot \text{m}$.

74. **N** The resistance of a copper wire is measured to be 3.50 Ω at 22.0°C. What is the resistance of this wire at 43.0°C? The temperature coefficient for copper is $\alpha = 3.9 \times 10^{-3}$ C^{-1}.

75. **N Review** When a metal rod is heated, its resistance changes both because of a change in resistivity and because of a change in the length of the rod. If a silver rod has a resistance of 2.00 Ω at 22.0°C, what is its resistance when it is heated to 200.0°C? The temperature coefficient for silver is $\alpha = 6.1 \times 10^{-3}$ °C^{-1}, and its coefficient of linear expansion is 18×10^{-6} C^{-1}. Assume that the rod expands in all three dimensions.

Problems 76, 77, and 78 are grouped.

76. **A** A 1.50-m-long wire of aluminum is constructed such that it has inner and outer radii as shown in Figure P28.76.

A potential difference is maintained on either end of the wire so that current I is from one end of the wire to the other. Find an expression for the current density as a function of I and the two radii r_a and r_b.

FIGURE P28.76 Problems 76, 77, and 78.

77. **A** Consider the situation described in Problem 76. Find an expression for the drift velocity of the electrons flowing in the wire as a function of I, the number density n, and the two radii r_a and r_b.

78. **C** Consider the situation described in Problem 76. Describe what happens to the current density and drift velocity as the radius r_a is decreased while the radius r_b remains constant. Why, in principle, do these quantities change or remain constant as the inner radius is decreased?

79. **N** Assuming that its coefficient of resistivity remains constant at 3.93×10^{-3} °C^{-1}, at what temperature is the resistance of a copper wire 50% higher than its resistance at 22.0°C?

80. **A** Two long concentric cylinders with radii a and b are separated by a material with conductivity σ that varies as $\sigma(r) = k/r$ ($r > a$), where r is the radial distance from the axis of the two cylinders and k is a constant. A steady current I flows radially between the cylinders. Find the resistance between the cylinders. Assume the length of the cylinders is L.

81. **N** A generating station and a switching station 140.0 km away are connected by a high-voltage aluminum transmission line that has a current of 850.0 A. Each wire of the transmission line has a cross-sectional area of 750.0 mm^2, and the density of free electrons in aluminum is 6.022×10^{28} electrons/m^3. How much time does it take for one electron to travel from the generating station to the switching station in this wire?

82. **A** A conducting material with resistivity ρ is shaped into a wire of length ℓ with a tapered cross section that decreases from radius r_1 on the left end to r_2 on its right end (Fig. P28.82). If the current density in the wire is a constant as a function of the horizontal distance x along the wire, what is the resistance of this tapered wire?

FIGURE P28.82

Direct Current (DC) Circuits

29

❶ Underlying Principles

No new principles are introduced in this chapter.

✪ Major Concepts

1. Kirchhoff's loop rule
2. Kirchhoff's junction rule
3. Resistors in series
4. Resistors in parallel

▷ Special Cases

1. Emf devices (ideal and real)
2. Voltage rules in a circuit
   a. Wire rule
   b. Resistor rule
   c. Switch rule
   d. Emf rule (ideal and real)
3. RC circuits
   a. Charging
   b. Discharging

Key Questions

How do the concepts of electricity apply to practical devices?

How do we analyze existing circuits and design new ones?

29-1 Measuring potential differences between two points 896
29-2 Kirchhoff's loop rule 903
29-3 Resistors in series 906
29-4 Kirchhoff's junction rule 908
29-5 Resistors in parallel 909
29-6 Circuit analysis 912
29-7 DC multimeters 917
29-8 RC circuits 919

An astronomer finds that her telescope won't move into position. Is the telescope's motor broken? The astronomer calls in an electrician, who uses a multimeter to measure the electric potential difference, current, and resistance in the circuit that contains the motor.

Scientists are not the only people who rely on electrical devices. Your home and car are full of devices such as lights, hot water heaters, refrigerators, clothes washers, and radios. When a device like a telescope motor fails, the fault may be mechanical or electrical. To find out whether something is wrong with an electric circuit, you must first know what to expect. What is the potential difference supposed to be across each circuit element? What is the expected current? What is the nominal resistance of each circuit element? Calculating these values is known as *circuit analysis*, the focus of this chapter. Once the expected values are known, a multimeter can be used to find any discrepancies and determine the reason for the failure.

CHAPTER 29 Direct Current (DC) Circuits

29-1 Measuring Potential Differences Between Two Points

This chapter continues our practical application of the principles of electricity. Because this chapter focuses on practical applications, we often use the terminology heard in an electronics laboratory (Table 29.1). You need to be comfortable using both the formal and informal terms. Our goal is to analyze circuits, which means calculating the potential difference across, the current in, and the resistance of the various circuit elements. Circuit analysis can be used to design a new device or troubleshoot a malfunctioning one.

TABLE 29.1 Electrical terminology.

Informal term(s)	Formal term(s)	Definition	Usage
Voltage, voltage across, or voltage drop	Electric potential difference or potential difference (ΔV)	Difference in electric potential energy per unit charge (Section 26-4)	"The voltage drop across the resistor is 6.0 V." "The electric potential difference between the two ends of the resistor is -6.0 V."
Amperage or juice	Current (I)	*Apparent* motion of positively charged particles per unit time	"The amperage or current in those speakers should be 1.2 A."
DC power supply	DC emf device	Device that does work in separating charge, creating a constant electric potential difference between its terminals	"Connect a DC power supply or DC emf device to your lightbulb."
Input voltage	Terminal potential (of emf device)	Electric potential difference between the terminals of an emf device	"The input voltage must be 5.0 V." "The terminal potential must be 5.0 V."

EMF DEVICE ⊙ Special Case

Of course, when an electrical device doesn't work at all, one of the first questions we ask is about the power source: "Does my phone need a new battery?" "Is the TV plugged in?" So far, the only electrical power source we have focused on is a chemical battery. There are many other types of power supplies, however, such as the electric generators cranked by a person in Ben Franklin's time or contemporary generators, which run on fossil fuels, nuclear material, wind, falling water, or sunlight. Another term for a power supply is an **emf device**. The initials "emf" come from the outdated term *electromotive force*. That term is misleading because it implies that the job of an emf device is to exert a force on the charged particles and that we can measure that force in newtons. Instead, an emf device is meant to maintain a potential difference between its terminals, which we measure in volts. So, it is common practice to refer to a power supply as an emf device without using the outdated term *electromotive force*.

Emf devices come in two major types—direct current (DC) and alternating current (AC). Again, these terms are somewhat misleading; they imply that the function of the emf device is to supply current, which is not correct. Charged particles are not made inside the emf device. Instead, the current in a circuit depends on the potential difference of the emf device and the resistance in the circuit. In a **direct current (DC)** circuit, the direction of the current does not change. In an **alternating current (AC)** circuit, the direction of the current switches back and forth, first pointing in one direction and then in the opposite direction. A DC power supply, such as a battery, maintains a constant potential difference between its terminals. In this chapter, we study only DC circuits.

If a circuit has a working power supply but is still malfunctioning, one of the next things to check is the potential difference between various points in the circuit. Let's illustrate the procedure with a simple circuit consisting of a lightbulb, an emf device, and a switch (Fig. 29.1). The schematic diagram (Fig. 29.1B) is easier to draw and makes it easier to see how the circuit elements are connected. So, from now on, we'll mostly use schematic diagrams. Learn to draw the symbols in Table 29.2 for the circuit elements.

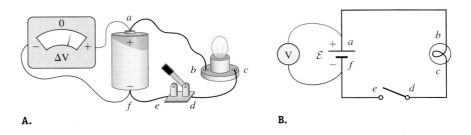

FIGURE 29.1 A. A simple circuit consisting of a lightbulb, a switch, wires, and an emf device. The voltmeter measures the potential difference across the battery. **B.** Schematic diagram for the circuit shown in part A.

TABLE 29.2 Circuit elements and their symbols.

Name	Symbol	Picture	Name	Symbol	Picture
DC emf device (battery, DC power supply)	$\mathcal{E} \dashv\vdash$		Switches		
Resistor	$\mathrm{-\!\!\!\!\!/\!}$ R				
Capacitor	$\dashv\vdash$ C		Ground		
Wire			Voltmeter	$-\!\!(V)\!\!-$	
Lightbulb			Ammeter	$-\!\!(A)\!\!-$	
			Ohmmeter	$-\!\!(\Omega)\!\!-$	

Using a Voltmeter

A **multimeter** is a device that has various settings for measuring voltage, current, and resistance (the last row in Table 29.2). We'll refer to a multimeter as a **voltmeter**, **ammeter**, or **ohmmeter** when it is set to measure potential difference, current, or resistance, respectively.

The voltmeter is connected *in parallel* to a particular circuit element using two wires known as leads. Traditionally, one lead is red and the other is black. The voltmeter reading is the potential difference between the two leads: $\Delta V = V_{red} - V_{black}$. Although there is nothing special about these colors, for convenience throughout this book, we will assume that the voltmeter has two colored leads—a red one and a black one—and that these leads are connected to the voltmeter so that it always gives the difference $\Delta V = V_{red} - V_{black}$. For example, if you wish to measure the potential difference $\Delta V = V_a - V_f$ across the battery in Figure 29.1, connect the voltmeter's red lead to point a and the black lead to point f ($V_a - V_f = V_{red} - V_{black}$).

Expected Voltages

To diagnose a circuit using a voltmeter, you first need to know what the potential difference should be across each circuit element. We use $\Delta V = IR$ (Eq. 28.25) to come up with five rules for finding the expected voltage across the circuit elements in Figure 29.1:

Battery: pryzmat/Shutterstock.com
Resistor: TheSnake19/Shutterstock.com
Capacitor: Stu49/Shutterstock.com
Wire: Ingvar Bjork/Shutterstock.com
Lightbulb: ThomasLenne/Shutterstock.com
Two-way switch: Hemera Technologies/Getty Images
Ground connection: Ntdanai/Shutterstock.com
Multimeter: StefanoT/Shutterstock.com

FIGURE 29.2 The voltmeter measures the potential difference between its leads: $V_{red} - V_{black}$.

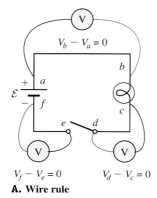

A. Wire rule

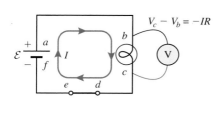

B. Resistor rule

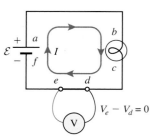

C. Switch rule (closed)

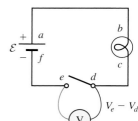

D. Switch rule (open)

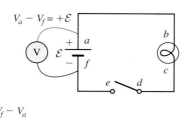

E. Ideal emf rule

1. **Wire rule.** Wires are designed to have very low resistance compared to other circuit elements. We will assume that the resistance of a wire is zero. So, the voltage across a wire is zero whether there is current in the wire or not (Fig. 29.2A):

WIRE RULE ▶ Special Case

$$\Delta V_{wire} = 0$$

2. **Resistor rule.** The current in a resistor or lightbulb is from high potential to low potential. In Figure 29.2B, the potential at point b is higher than the potential at point c, and the current through the bulb is from b to c when the switch is closed. If you connect a voltmeter in parallel across a resistor or lightbulb with the *red lead on the low-potential side* and the *black lead on the high-potential side* (Fig. 29.2B), the potential difference is negative:

$$V_{red} - V_{black} < 0$$

So, the reading on the voltmeter is negative. Informally, we say that there is a "voltage drop" across the resistor or bulb. The magnitude of that drop is given by $\Delta V = IR$ (Eq. 28.25).

Of course, if you reverse the leads so that the red lead is at point b and the black at c, the voltage reading is positive. So, to come up with a rule for the voltage across a resistor, we must first state how the voltmeter is connected. It is most convenient to write the rule in the case shown in Figure 29.2B. Then the resistor rule is: *When the current is from the black to the red lead, the voltage measured across a resistor is negative and given by*

RESISTOR RULE ▶ Special Case

$$\Delta V_R = V_{red} - V_{black} = -IR \quad (29.1)$$

(If the current is from the red to the black lead, then $\Delta V_R = +IR$.)

A resistor obeys Ohm's law (Section 28-6), so its resistance is constant and thus the current is directly proportional to the voltage drop across the resistor. Circuits that contain lightbulbs are more complicated because lightbulbs do not obey Ohm's law. When there is current in an incandescent bulb, the filament gets hot, increasing its resistance. Equation 29.1 still holds, but R is not constant.

3. **Switch rule.** If a switch is closed, it acts like a wire. A closed switch has very low resistance, and the voltage drop across it should be very small (Fig. 29.2C).

We will assume that the resistance of a closed switch and the voltage drop across it are zero:

$$\Delta V_{\text{closed switch}} = 0$$

SWITCH RULE ▶ Special Case

If the switch is open, there is air between the two sides of the switch. Because air is normally a good insulator, the resistance of an open switch is very high. There is no current through an open switch, but there may be a potential difference across it. For example, in Figure 29.2D, the voltage across the open switch equals the terminal voltage of the power supply.

4. **(Ideal) emf rule.** In DC power supplies, the high-potential terminal and the low-potential terminal of the emf device remain fixed and do not alternate. The plus and minus signs next to the emf device in schematic diagrams (Table 29.2) indicate the positive and negative terminals. So, the emf rule is: *If the red lead is connected to the positive terminal and the black lead to the negative terminal, the voltage measured is positive and equals the terminal potential* (Fig. 29.2E):

$$\Delta V = V_{\text{red}} - V_{\text{black}} = \Delta V_{\text{terminal}} = \mathcal{E} \qquad (29.2)$$

IDEAL EMF RULE ▶ Special Case

5. **Circuit elements in series.** When you connect a voltmeter so that there are two or more circuit elements between its leads, the voltage reading is the sum of the voltages across each element.

Real Versus Ideal Emf Devices

So far, we have considered only *ideal* emf devices. An **ideal DC emf device**, such as an ideal battery, maintains a constant terminal potential $\mathcal{E}$ whether there is current in the emf device or not. No real emf device can maintain its terminal potential when there is current in the device. If you close the switch while measuring the voltage across the emf device (Fig. 29.3), the terminal voltage decreases slightly. If you open the switch again, the terminal voltage returns to its earlier higher value. We can explain this observation: When you close the switch, there is current in the entire circuit, including the emf device. The current in a real emf device—like the current anywhere else in the circuit—is hindered by resistance, even inside the emf itself.

The **real emf device** in Figure 29.3 is modeled as two circuit elements: (1) an ideal emf device with no internal resistance and (2) a resistor with resistance r. Of course, we cannot put our voltmeter inside the emf device. Instead, we connect our voltmeter to the emf terminals labeled H and L (Fig. 29.3) and measure the potential difference between them. In Figure 29.3A, the emf is connected to an open switch, so there is no current in the leads or the battery, and therefore no voltage drop across the battery's internal resistor: $\Delta V_r = -Ir = 0$. The terminal potential measured by the voltmeter is

$$\Delta V_{\text{terminal}} = V_{\text{red}} - V_{\text{black}} = \mathcal{E}$$

Thus, when there is no current in a real emf device, its terminal potential is the same as that of an ideal emf device.

When the switch is closed (Fig 29.3B), current I is present in the circuit and therefore in the battery. Now the voltage drop across the internal resistor is given by $\Delta V_r = -Ir$. The terminal potential is the sum of the voltage across the ideal emf device and the voltage drop across the internal resistor. Therefore, the emf rule for a real emf device is

$$\Delta V_{\text{real emf}} = \Delta V_{\text{terminal}} = \mathcal{E} - Ir \qquad (29.3)$$

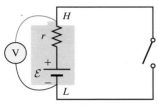

A.

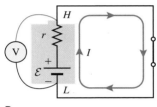

B.

FIGURE 29.3 A real emf device is modeled as an ideal emf in series with a resistor. **A.** The terminal voltage across a real emf device is higher when there is no current in the device. **B.** The terminal voltage is lower across a real emf device when a current is present.

REAL EMF RULE ▶ Special Case

Unless otherwise specified, all emf devices in this textbook are assumed to be *ideal*, obeying $\Delta V_{\text{terminal}} = \mathcal{E}$ (Eq. 29.2).

Grounded Circuits

Only *differences* in potential are physically meaningful (Section 26-2). The rules given above for finding expected voltages are used to find the potential *difference* across a circuit element. Some circuits are **grounded**, meaning that a wire is connected from some point in the circuit to the Earth (or another large object such as the frame of a car). A point in a circuit connected to ground is at zero potential. The electric potential of all other points in the circuit is then measured with respect to the grounded point. Table 29.2 includes the ground symbol.

Because only potential differences are physically important, a grounded circuit behaves in the same way as an ungrounded circuit. Circuits are grounded for safety reasons. The circuit in most household appliances is enclosed in some sort of insulating case. Under normal conditions, the case remains neutral, but if the circuit malfunctions, it is possible for the case of an ungrounded appliance to become charged. Then, if a person touches the charged case, he or she provides the excess charge with a pathway to the ground and the person could be shocked or even electrocuted. If the circuit is grounded, the case remains neutral.

CONCEPT EXERCISE 29.1

What are the SI units of $\mathcal{E}$?

CONCEPT EXERCISE 29.2

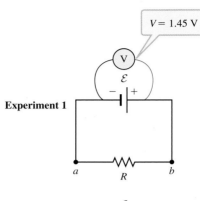

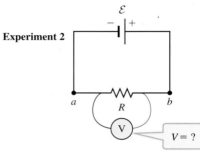

FIGURE 29.4

An emf device is connected to a resistor as shown in Figure 29.4. In experiment 1, a student measures the terminal potential of the emf device and finds that it is 1.45 V. In experiment 2, the student moves the voltmeter to points a and b. What will the voltmeter read? Will the reading be positive or negative? Explain.

CASE STUDY Helping Around the House

Avi, Cameron, and Shannon are staying at Cameron's house for spring break. Cameron's mom (Carol) wants them to help fix a couple of things: (1) The hot water heater is not working properly. The hot water isn't very warm, and the supply quickly runs out when anyone tries to take a shower or bath. (2) Carol recently replaced a headlight in her antique car. Now both headlights work, but the old headlight is dimmer than the new one.

EXAMPLE 29.1 CASE STUDY What's Wrong with the Water Heater?

Carol's water heater has two large resistors. When the circuit is operating correctly, the thermal energy of these resistors heats the water primarily by conduction. The emf device is the wall outlet (an AC device), but we can treat it as a DC emf device with a terminal potential of roughly 240 V. In trying to diagnose the water heater's problems, Avi, Cameron, and Shannon discuss a simpler circuit—a single resistor connected to a 240-V battery.

Avi: If you measure the voltage across the battery, it's 240 V. If the resistor is working, you should find that the voltage drop across it is -240 V.

Cameron: Yes, but I think that's also what you would find if the resistor is cracked. So, measuring the voltage across the resistor won't help us find out whether it's dead.

Shannon: If the resistor is cracked, there's no current in it. No current means no voltage.

Decide which student's statements are true and which are false. Explain your answer.

SOLVE

Avi is correct. If the resistor is intact, the voltage drop across the resistor should equal the (negative) terminal potential of the emf device.

Cameron is correct. A crack in the resistor is like an open switch. In this case, the voltage drop across the broken resistor is 240 V. (The absolute value of the voltage is the same whether the resistor is intact or broken.)

Shannon is incorrect. No current does not necessarily mean no voltage. Consider voltage across the open switch in Figure 29.2D. As long as the emf is working, the voltage across the switch is nonzero.

EXAMPLE 29.2 Ideal and Real Batteries

A An ideal 12.0-V battery is connected to a 135-Ω resistor (Fig. 29.5). Find the current in the circuit, the power delivered by the battery, and the power consumed by the resistor.

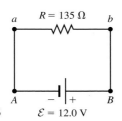

FIGURE 29.5 $\mathcal{E} = 12.0$ V

INTERPRET and ANTICIPATE

We expect to find that the power supplied by the battery equals the power consumed by the resistor (Section 28-7). Always indicate the direction of the current on the schematic diagram; the current in this circuit is counter-clockwise (from the emf's positive terminal through the circuit to its negative terminal; Fig. 29.6).

To find the required quantities, we first need to know the voltage across the resistor and across the emf device. So, imagine connecting the red and black leads of the voltmeter so that the voltmeter measures $\Delta V = V_{red} - V_{black}$.

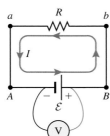

FIGURE 29.6

SOLVE

Only a wire connects point a to A, and another wire connects point b to B. Because the voltage across a wire is zero, point a is at the same potential as point A, and point b is at the same potential as point B. If you connect the red lead of the voltmeter to point B and the black lead to point A, you get the same reading as if you had connected the red lead of the voltmeter to point b and the black lead to point a.

The terminal voltage is equal to the voltage across the resistor.	$\Delta V_{terminal} = \Delta V_R$	(1)
Because this is an ideal battery, the terminal voltage is equal to the emf $\mathcal{E}$.	$\Delta V_{terminal} = \mathcal{E}$	(2)
The current is from the red lead to the black lead. So, according to the resistor rule, the voltage across the resistor is $+IR$.	$\Delta V_R = IR$	(3)
Substitute Equations (2) and (3) into Equation (1), and solve for I.	$\mathcal{E} = IR$ $I = \dfrac{\mathcal{E}}{R} = \dfrac{12.0 \text{ V}}{135 \text{ Ω}} = 8.89 \times 10^{-2}$ A = $\boxed{88.9 \text{ mA}}$	
The power delivered by the battery is given by Equation 28.31.	$P_{bat} = I\mathcal{E}$ $P_{bat} = (8.89 \times 10^{-2} \text{ A})(12.0 \text{ V}) = \boxed{1.07 \text{ W}}$	(28.31)
The power used by the resistor is given by Equation 28.34.	$P_{res} = I^2 R$ $P_{res} = (8.89 \times 10^{-2} \text{ A})^2 (135 \text{ Ω}) = \boxed{1.07 \text{ W}}$	(28.34)

CHECK and THINK

As expected, the power supplied by the battery is equal to the power consumed by the resistor. In fact, this is a result of the conservation of energy. When a particle passes through the battery from the negative to the positive terminal, the electric potential energy of the system increases. When the particle moves through the resistor, it collides with the molecules, increasing the system's thermal energy. The increase in electric potential energy is equal to the increase in thermal energy. In the next section, we present a convenient rule—known as *Kirchhoff's loop rule*—to express the conservation of energy principle for circuits.

Example continues on page 902

B Now, a real battery is connected to a 135-Ω resistor (Fig. 29.7). The real battery can be modeled as an ideal emf device with $\mathcal{E} = 12.0$ V in series with a resistor with $r = 1.0$ Ω. Find the current in the circuit, the terminal potential of the real battery, the power consumed by the 135-Ω resistor, and the power consumed by the internal resistance of the battery.

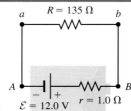

FIGURE 29.7

INTERPRET and ANTICIPATE
Follow a procedure similar to that in part A. Because this circuit has a higher resistance than the ideal emf circuit (given the added internal resistance of the emf), we expect that the current is smaller and the terminal potential is lower than 12.0 V. We also expect that the power consumed by the 135-Ω resistor plus the power consumed by the battery's internal resistance equals the power supplied by the ideal battery in part A. Figure 29.8 shows the current's direction.

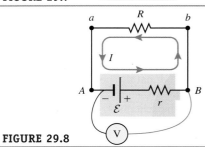

FIGURE 29.8

SOLVE
As in part A, the red lead is attached to point b or B, and the black lead is attached to point a or A. The potential difference across the resistor is $+IR$ because the current points from the red lead to the black lead.

$$\Delta V_R = IR \tag{4}$$

To find the terminal potential of the real emf, add the potential $\mathcal{E}$ across the ideal emf to the voltage drop across its internal resistance r. According to the resistor rule, the voltage across the internal resistor is negative because here the current is from the black lead to the red lead. (Notice that our result is the same as Eq. 29.3.)

$$\Delta V_{\text{terminal}} = \mathcal{E} - Ir \tag{29.3}$$

As in part A, the terminal voltage equals the voltage across the 135-Ω resistor. Substitute Equations (4) and 29.3 into $\Delta V_{\text{terminal}} = \Delta V_R$ (Eq. 1).

$$\mathcal{E} - Ir = IR$$
$$\mathcal{E} = I(R + r)$$
$$I = \frac{\mathcal{E}}{R + r} = \frac{12.0 \text{ V}}{135 \text{ Ω} + 1.0 \text{ Ω}} = 8.82 \times 10^{-2} \text{ A}$$
$$I = 88.2 \text{ mA}$$

CHECK and THINK
As expected, the current is smaller in the circuit that has a real battery than in the earlier circuit with an ideal battery.

SOLVE
To find the terminal potential, substitute values into Equation 29.3.

$$\Delta V_{\text{terminal}} = \mathcal{E} - Ir$$
$$\Delta V_{\text{terminal}} = 12.0 \text{ V} - (8.82 \times 10^{-2} \text{ A})(1.0 \text{ Ω})$$
$$\Delta V_{\text{terminal}} = 11.9 \text{ V}$$

CHECK and THINK
The terminal potential is only slightly lower than that of an ideal emf device. For this reason, we usually treat all emf devices as ideal.

SOLVE
The power consumed by the 135-Ω resistor and by the internal resistance is given by Equation 28.34.

For the 135-Ω resistor,
$$P_{\text{res}} = I^2 R \tag{28.34}$$
$$P_{\text{res}} = (8.82 \times 10^{-2} \text{ A})^2 (135 \text{ Ω}) = 1.05 \text{ W}$$

For the internal resistance r,
$$P_{\text{int}} = I^2 r \tag{28.34}$$
$$P_{\text{int}} = (8.82 \times 10^{-2} \text{ A})^2 (1.0 \text{ Ω})$$
$$P_{\text{int}} = 7.8 \times 10^{-3} \text{ W} = 7.8 \text{ mW}$$

:• **CHECK and THINK**
As expected, the power consumed by the 135-Ω resistor plus the power consumed by the emf device's internal resistance (≈ 1.06 W) equals the power supplied by the ideal battery in part A: 1.07 W. (The slight discrepancy is due to rounding error.) Another reason we usually model a real emf device as ideal is that the power consumed by the internal resistance of the emf device is much lower than the power consumed by the external 135-Ω resistor.

29-2 Kirchhoff's Loop Rule

KIRCHHOFF'S LOOP RULE
✪ **Major Concept**

When the German physicist Gustav Robert Kirchhoff was still a student, he discovered two rules that are essential for circuit analysis. The first is **Kirchhoff's loop rule**: *The total change in potential around any closed loop in a circuit is always zero.* Kirchhoff's loop rule is based on the principle of conservation of energy; the rule is expressed in terms of potential energy per charge (electric potential), and so it is convenient to use in circuit analysis.

Kirchhoff's loop rule may not hold when a changing magnetic field is present. (See Section 34-4.)

PROBLEM-SOLVING STRATEGY

Using Kirchhoff's Loop Rule

The loop rule is used to write an equation involving the potential difference across each circuit element in a closed loop.

:• **INTERPRET and ANTICIPATE**
Start with a schematic diagram to which you add two elements: (1) the **current** and (2) the **starting point**.
Element 1. Indicate the **current** as a curve or line drawn near the circuit elements, with an arrow indicating the direction.
  a. If there is an open switch in the circuit, there will be no current.
  b. If there is a current, choose a direction for it. If there is one emf device in the circuit, the current is away from the positive terminal through the loop and toward the negative terminal. The current continues through the emf from the negative terminal toward the positive one. If there is more than one emf device, the current is in the direction of the stronger emf.
  c. Sometimes you will not be able to determine the direction of the current. In those cases, make your best guess. If you are wrong, when you solve for current, you will get a negative value.

Element 2. Choose a convenient **starting point** and indicate it with a dot. The key to using the loop rule is to imagine measuring the voltage across each circuit element in the loop. By choosing and labeling a starting point, you know when you have considered the complete loop.

:• **SOLVE**
Step 1 Write expressions for the expected **voltage across each circuit element**. Imagine using a voltmeter to measure the voltage across each circuit element, always keeping the voltmeter's leads oriented the same way as you imagine moving it around the loop. The resistor rule and the emf rule in Section 29-1 can be summarized: (1) Measuring in the direction of the current results in a voltage drop across the resistor $-IR$. (2) Measuring from the negative to the positive terminal of the emf results in a positive voltage $+\mathcal{E}$. (Of course, measurements made in the opposite direction result in the opposite signs.)
Step 2 Set the sum of the potential differences to zero. Suppose the first voltage measurement is taken with the black lead at the starting point. When the red lead is at the starting point, you have measured the potential differences around the closed loop, and according to Kirchhoff's loop rule, the sum of the measurements must be zero.

EXAMPLE 29.3 Two Emf Devices, a Resistor, and a Bulb

The circuit shown in Figure 29.9 consists of two emf devices, a resistor, and a bulb. Use Kirchhoff's loop rule to show that

$$\mathcal{E}_1 - IR_1 - \mathcal{E}_2 - IR_2 = 0$$

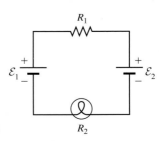

FIGURE 29.9

Example continues on page 904 ▶

:• INTERPRET and ANTICIPATE
Start with a schematic diagram (Fig. 29.10). Because there are two emf devices and we don't know their relative strengths, we don't know the direction of the **current**. We arbitrarily choose clockwise for the current and the bottom left corner as the **starting point**.

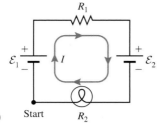

FIGURE 29.10

:• SOLVE
Step 1 Write expressions for the **voltage across each circuit element**. Imagine that a voltmeter is moved clockwise around the loop, with the red lead in front of the black lead at all times. (In practice, you do not need to redraw the circuit diagram as we have done here. Instead, use your finger or a pencil to represent the voltmeter as you imagine measuring the voltage around the closed loop.)

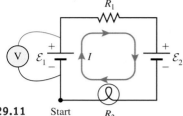

FIGURE 29.11

The emf rule across emf device 1 gives a positive potential difference because the red lead is on the positive terminal and the black lead is on the negative terminal (Fig. 29.11). We are measuring from the negative to the positive terminal, and we find

$$\Delta V_{\mathcal{E}1} = V_{red} - V_{black} = +\mathcal{E}_1 \qquad (1)$$

The resistor rule for R_1 gives a negative potential difference because the current is from the black to the red lead (Fig. 29.12). Measuring in the direction of the current, we have

$$\Delta V_{R1} = V_{red} - V_{black} = -IR_1 \qquad (2)$$

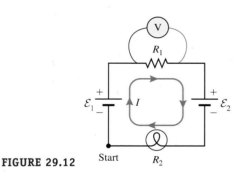

FIGURE 29.12

The voltage across emf device 2 is negative because now we are measuring from the positive to the negative terminal (Fig. 29.13):

$$\Delta V_{\mathcal{E}2} = V_{red} - V_{black} = -\mathcal{E}_2 \qquad (3)$$

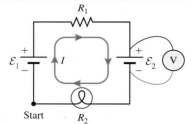

FIGURE 29.13

Finally, the resistor rule gives a negative voltage across lightbulb R_2 because we are measuring in the direction of the current (Fig. 29.14):

$$\Delta V_{R2} = V_{red} - V_{black} = -IR_2 \qquad (4)$$

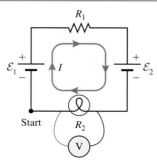

FIGURE 29.14

Step 2 Sum of potential differences is zero. According to Kirchhoff's loop rule, the sum of the four measurements (Eqs. 1–4) must be zero.

$$\Delta V_{\mathcal{E}1} + \Delta V_{R1} + \Delta V_{\mathcal{E}2} + \Delta V_{R2} = 0$$

$$\mathcal{E}_1 - IR_1 - \mathcal{E}_2 - IR_2 = 0 \quad \checkmark$$

:• CHECK and THINK
This is exactly what we expected to find.

CONCEPT EXERCISE 29.3

You may wonder why anyone would ever build the circuit in Figure 29.9 with two emfs in opposite directions. One possibility is that the circuit is designed to recharge a battery. Suppose emf device 2 is a rechargeable battery and emf device 1 is another battery such as your car's battery. Resistor R_1 is in the circuit to regulate the amount of current. So, why is there a lightbulb? Lightbulbs are often used in circuits as indicators. (Does your phone have an indicator light to tell you whether it is plugged in or powered by its internal battery?) If a bulb is off, there is no current. A dim bulb indicates a small current; a bright bulb indicates a large current.

a. Suppose battery 2 starts off with a very low terminal potential and then its terminal potential gradually increases. Describe the brightness of the bulb during this process.
b. The term *recharge* is misleading. Why? What is a better term?

EXAMPLE 29.4   CASE STUDY   Headlights

When Carol replaces a burnt-out headlight in her antique car, the remaining old headlight is dimmer than the new one. Avi and Shannon have learned that the old headlight has a higher resistance than the new headlight. According to their research, the old headlight has operating resistance $R_{old} = 3.2\ \Omega$, and the new headlight has operating resistance $R_{new} = 1.6\ \Omega$. Avi believes both bulbs are connected in series with the car's 12.0-V battery (Fig. 29.15).

A Find the current in each bulb and the power consumed by each bulb when the switch is closed.

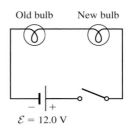

FIGURE 29.15 Avi's proposed headlight circuit.

INTERPRET and ANTICIPATE
The circuit consists of only one loop, so we expect the current to be the same through each circuit element. The power consumed by each bulb depends on the current and its resistance. We expect that the bulb with the higher resistance (the old bulb) will consume more power. To find the current, we will use Kirchhoff's loop rule.

Sketch the circuit when the switch is closed (Fig. 29.16). There is only one emf device, so it is easy to find the **current's** direction—counterclockwise as indicated. The bottom left corner is our (arbitrary) **starting point**.

SOLVE
Step 1 Write expressions for the **voltage across each circuit element.** We arbitrarily decided to keep the red lead behind the black one, so the current is from black to red. For the three circuit elements, we imagine making three voltage measurements as indicated in Figure 29.16. (Drawing the voltmeter is not necessary.)

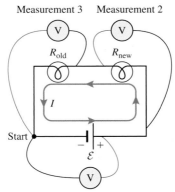

FIGURE 29.16

Measurement 1. The voltage across the emf is positive.	$\Delta V_{\mathcal{E}} = +\mathcal{E}$
Measurements 2 and 3. The voltage across each bulb is negative because the measurement is in the direction of the current (from black to red).	$\Delta V_{new} = -IR_{new}$ $\Delta V_{old} = -IR_{old}$

Example continues on page 906 ▶

Step 2 **Sum of potential differences is zero.** In the third measurement, the red lead is at the starting point, so we have measured all the potential differences. Applying Kirchhoff's loop rule, we find the sum of the potential differences is zero.	$\Delta V_\mathcal{E} + \Delta V_\text{new} + \Delta V_\text{old} = 0$ $\mathcal{E} - IR_\text{new} - IR_\text{old} = 0$ (1)
Solve Equation (1) for current.	$\mathcal{E} = IR_\text{new} + IR_\text{old}$ $I = \dfrac{\mathcal{E}}{R_\text{new} + R_\text{old}} = \dfrac{12.0\,\text{V}}{1.6\,\Omega + 3.2\,\Omega}$ $I = 2.5\,\text{A}$
The power consumed by each bulb is given by Equation 28.34.	For the new bulb, $P_\text{new} = I^2 R_\text{new}$ (28.34) $P_\text{new} = (2.5\,\text{A})^2(1.6\,\Omega) = \boxed{10\,\text{W}}$ For the old bulb, $P_\text{old} = I^2 R_\text{old}$ (28.34) $P_\text{old} = (2.5\,\text{A})^2(3.2\,\Omega) = \boxed{20\,\text{W}}$

CHECK and THINK

As expected, the current is the same in all the circuit elements and the power consumed by the old bulb is greater than the power consumed by the new bulb because the old bulb has a higher resistance. In fact, the old bulb consumes about twice as much power as the new bulb. That doesn't fit Carol's observation that the old bulb is dimmer than the new bulb. Because the old bulb consumes more power, we would expect it to give off more light than the new bulb. There is likely a problem with Avi's circuit diagram. In the next part, we will evaluate this diagram.

B If Avi is correct about the bulbs being connected in series, what would happen to the good bulb when one bulb burns out?

SOLVE

When we say "a bulb burns out," what we really mean is that the filament breaks. A broken filament is like an open switch; there can be no current through either. We would expect both bulbs to be off; neither headlight would light up when one of them burns out. Of course, one headlight often burns out on a car while the other bulb remains lit. Therefore, the bulbs must not be connected to the battery in series as Avi thought.

29-3 Resistors in Series

The function of a resistor in a circuit is to regulate the amount of current. Manufacturers make a wide range of resistances, but often you must combine a number of standard resistors to come up with the resistance required. As we know from our work with capacitors, we may need to connect a number of resistors in series or parallel to come up with the desired resistance. For example, the three resistors in series in Figure 29.17A are equivalent to the one resistor in Figure 29.17B, and these two circuits are said to be *equivalent* because when the same emf device is used, both have the same current.

We use Kirchhoff's loop rule (and the problem-solving strategy on page 903) for each circuit in Figure 29.17 to come up with an expression for the equivalent resistance R_eq in terms of the three resistances R_1, R_2, and R_3. In both circuits, there is only one emf device, and the current is clockwise in both. Start in the bottom left corner of each circuit, and imagine measuring the potential across each circuit element clockwise around the loop. Apply the emf rule and the resistor rule.

After completing the loop, set the sum of the potential differences to zero. For the circuit in Figure 29.17A,

$$\mathcal{E} - IR_1 - IR_2 - IR_3 = 0$$
$$\mathcal{E} = I(R_1 + R_2 + R_3) \tag{29.4}$$

For the circuit in Figure 29.17B,

$$\mathcal{E} - IR_{eq} = 0$$
$$\mathcal{E} = IR_{eq} \tag{29.5}$$

Comparing Equation 29.5 to Equation 29.4, we find that

$$R_{eq} = R_1 + R_2 + R_3$$

We generalize our results for a circuit with any number of resistors in series. The equivalent resistance R_{eq} of N resistors connected in series is

$$R_{eq} = R_1 + R_2 + R_3 + \cdots + R_N = \sum_{i=1}^{N} R_i \tag{29.6}$$

RESISTORS IN SERIES
✪ **Major Concept**

Compare Equation 29.6 to $C_{eq} = \sum_{i=1}^{N} C_i$ (Eq. 27.8) for capacitors in parallel; they are mathematically the same. *Resistors in series* combine like *capacitors in parallel*.

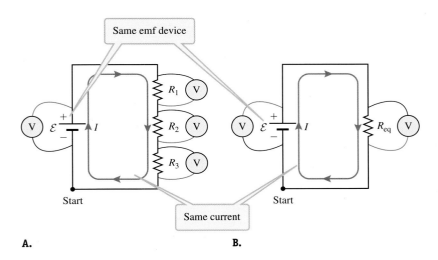

FIGURE 29.17 A. Three resistors connected in series. **B.** An equivalent circuit.

CONCEPT EXERCISE 29.4

Use the gravitational analogy from Section 28-4 (page 875) to explain why Equation 29.6 makes sense.

EXAMPLE 29.5   CASE STUDY   Another Look at Avi's Circuit

When you analyze a circuit that involves resistors in series, it may help if you imagine the equivalent circuit. Find the equivalent resistance in Avi's circuit diagram (Fig. 29.15) for the car's headlights, and then find the current. In the **CHECK and THINK** step, compare your result to the current found in Example 29.4.

:• **INTERPRET and ANTICIPATE**
The total resistance should be higher than the resistance of either bulb, and the current we find here should equal the current we found earlier.

Example continues on page 908 ▶

:• SOLVE	
Find the equivalent resistance by adding the resistance of each bulb (Eq. 29.6).	$R_{eq} = \sum_{i=1}^{N} R_i$ (29.6) $R_{eq} = R_{old} + R_{new} = 3.2\ \Omega + 1.6\ \Omega$ $R_{eq} = 4.8\ \Omega$
The current is given by Equation 29.5.	$\mathcal{E} = IR_{eq}$ (29.5) $I = \dfrac{\mathcal{E}}{R_{eq}} = \dfrac{12.0\ \text{V}}{4.8\ \Omega} = 2.5\ \text{A}$

:• **CHECK and THINK**

As expected, the equivalent resistance is higher than that of either bulb, and the current is exactly what we found in Example 29.4. When a circuit involves a large number of resistors in series, it may simplify the problem to draw a schematic diagram for the equivalent circuit and then apply Kirchhoff's loop rule to the equivalent circuit.

29-4 Kirchhoff's Junction Rule

Consider a charged particle moving in a circuit that consists of a single loop (Fig. 29.17). This charged particle must pass through each circuit element because there is nowhere else for it to go. So, the number of charged particles that pass through each circuit element per unit time is constant, and the same current exists in each circuit element.

Now consider the circuit in Figure 29.18 made up of three branches and two junctions. A **junction** is a place where three or more wires meet. A **branch** is a part of a circuit in between junctions. Starting at point a, a charged particle moves through "branch 0" to point b, which is a junction. Just as when a driver encounters a junction in a roadway, the particle can go into either branch 1 or branch 2.

In this circuit, the current is different in each branch. Because charged particles are not created or destroyed by any circuit element, and because no particle leaks out of the circuit or flows into it, all the current going *into* a junction must equal all the current going *out*. This is a loose statement of **Kirchhoff's junction rule**: *At any junction, the sum of the all the currents entering the junction equals the sum of all the currents exiting the junction.*

KIRCHHOFF'S JUNCTION RULE
✪ Major Concept

Consider the junction at point b (Fig. 29.18). The current I_0 from branch 0 enters the junction. The sum of the currents exiting that junction is the current I_1 in branch 1 plus the current I_2 in branch 2. According to Kirchhoff's junction rule,

$$I_0 = I_1 + I_2$$

If we consider the junction at a instead of the junction at b, we find the same equation. The sum of the currents entering the junction at a is the current I_1 in branch 1 plus the current I_2 in branch 2, and the current leaving the junction is the current in branch 0:

$$I_1 + I_2 = I_0$$

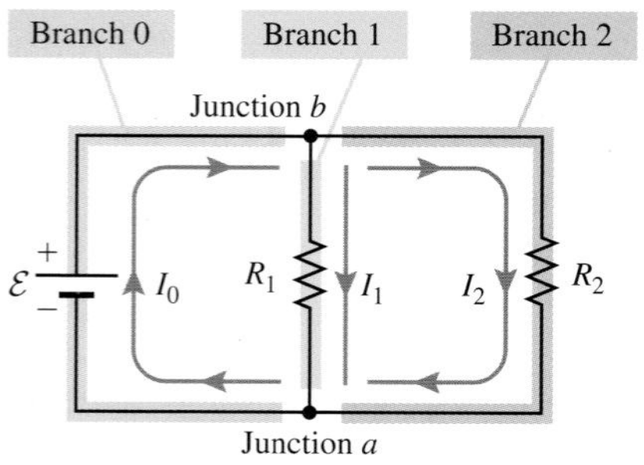

FIGURE 29.18 Kirchhoff's junction rule gives the same result when applied to the junction at a or the junction at b.

CONCEPT EXERCISE 29.5

Imagine connecting a voltmeter to points a and b in Figure 29.18 so that the voltmeter reads the potential difference $\Delta V = V_b - V_a$. *Hint*: The red lead is at point b.

a. Use the emf rule to write an expression for ΔV in terms of the emf $\mathcal{E}$.
b. Use the resistor rule to write an expression for ΔV in terms of the resistances R_1 and R_2.

CONCEPT EXERCISE 29.6

If $I_0 = 3.0$ A and $R_1 = R_2$ in Figure 29.18, what are I_1 and I_2? *Hint*: Use your answer to Concept Exercise 29.5, part (b).

29-5 Resistors in Parallel

If you try to drink a milkshake with only one straw, you find it difficult to get a satisfying amount into your mouth with each sip. The best way to drink a milkshake is with two straws. Think of it this way: Each straw resists the motion of the milkshake, but two straws used in parallel decrease the overall resistance to the milkshake flow. What if you use the two straws in series—by forcing the end of one straw into the end of the other? You would find it even more difficult to drink the milkshake with two straws in series than if you used only one straw, because resistances in series add (Eq. 29.6).

Drinking a milkshake with several straws is analogous to charged particles flowing through several resistors. If the resistors are in series (Fig. 29.17), the equivalent resistance R_{eq} is higher than the resistance of any one resistor. If the resistors are in parallel (Fig. 29.19), the equivalent resistance R_{eq} is lower than the resistance of any one of the resistors.

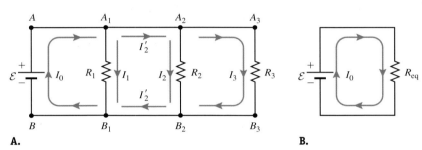

FIGURE 29.19 A. Three resistors connected in parallel. **B.** An equivalent circuit.

DERIVATION Resistors In Parallel

We will show that the equivalent resistance for resistors in parallel is

$$\frac{1}{R_{eq}} = \sum_{i=1}^{N} \frac{1}{R_i} = \frac{1}{R_1} + \frac{1}{R_2} + \frac{1}{R_3} + \cdots + \frac{1}{R_N} \quad (29.7)$$

RESISTORS IN PARALLEL
✪ **Major Concept**

To make this derivation concrete, consider three parallel resistors and find an expression for the equivalent resistor (Fig. 29.19). Both circuits have the same emf $\mathcal{E}$ and the same amount of current I_0 in the emf device.

The circuit in Figure 29.19A is made up of three loops that involve the emf device. Let's use the loop rule for each.

For the clockwise loop from B to A to A_1 to B_1 and back to B:	$\mathcal{E} - I_1 R_1 = 0$
	$\mathcal{E} = I_1 R_1 \quad (1)$
For the clockwise loop from B to A to A_2 to B_2 and back to B:	$\mathcal{E} - I_2 R_2 = 0$
	$\mathcal{E} = I_2 R_2 \quad (2)$
For the clockwise loop from B to A to A_3 to B_3 and back to B:	$\mathcal{E} - I_3 R_3 = 0$
	$\mathcal{E} = I_3 R_3 \quad (3)$

Derivation continues on page 910 ▶

The current splits twice—once at junction A_1 and once at junction A_2. Let's apply the junction rule at these two points.

The current entering junction A_1 is I_0. The current leaving this junction is I_1 plus I'_2. (We get the same result if we apply the junction rule at junction B_1.)	$I_0 = I_1 + I'_2$	(4)
The current entering junction A_2 is I'_2. The current leaving this junction is I_2 plus I_3. (We get the same result if we apply the junction rule at junction B_2.)	$I'_2 = I_2 + I_3$	(5)
Combine these two applications of the junction rule (Eqs. 4 and 5).	$I_0 = I_1 + I_2 + I_3$	(6)
Substitute Equations (1) through (3) into Equation (6).	$I_0 = \dfrac{\mathcal{E}}{R_1} + \dfrac{\mathcal{E}}{R_2} + \dfrac{\mathcal{E}}{R_3}$ $I_0 = \mathcal{E}\left(\dfrac{1}{R_1} + \dfrac{1}{R_2} + \dfrac{1}{R_3}\right)$	(7)
Now apply the loop rule to the equivalent circuit in Figure 29.19B.	$\mathcal{E} - I_0 R_{eq} = 0$ $\mathcal{E} = I_0 R_{eq}$	
Solve for I_0.	$I_0 = \dfrac{\mathcal{E}}{R_{eq}}$	(8)
Compare Equations (8) and (7) to find an expression for the equivalent resistance R_{eq} in terms of the individual resistances R_1, R_2, and R_3.	$\dfrac{1}{R_{eq}} = \dfrac{1}{R_1} + \dfrac{1}{R_2} + \dfrac{1}{R_3}$	(9)
We derived Equation (9) using three resistors in parallel, and we can extend that equation for any number N of resistors in parallel.	$\dfrac{1}{R_{eq}} = \sum_{i=1}^{N} \dfrac{1}{R_i}$ ✓	(29.7)

:• **COMMENTS**
As expected from our analogy with drinking straws, Equation 29.7 shows that the equivalent resistance R_{eq} is lower than the resistance of the individual resistors: $R_{eq} < R_1$ and $R_{eq} < R_2$ and ... and $R_{eq} < R_N$. Compare Equation 29.7 for resistors in parallel to

$$\frac{1}{C_{eq}} = \sum_{i=1}^{N} \frac{1}{C_i} \qquad (27.7)$$

for capacitors in series; these equations are mathematically identical. *Resistors in parallel combine like capacitors in series.*

CONCEPT EXERCISE 29.7

When you solve for R_{eq} using Equation 29.7, the quantity you need is in the denominator, so you must be sure to invert both sides of the equation. It is common to forget this step. To help avoid this mistake, show that the equivalent resistance of a network consisting of two resistors in parallel is

$$R_{eq} = \frac{R_1 R_2}{R_1 + R_2} \qquad (29.8)$$

EXAMPLE 29.6 — CASE STUDY: Current Through a Car's Battery

Avi, Cameron, and Shannon have decided that a car's headlights must be connected in parallel (Fig. 29.20).

A Find the equivalent resistance. From Example 29.4, $R_{old} = 3.2\ \Omega$ and $R_{new} = 1.6\ \Omega$.

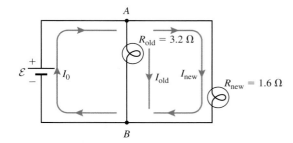

FIGURE 29.20 Parallel circuit for car headlights.

INTERPRET and ANTICIPATE
Sketch the equivalent circuit (Fig. 29.21). Because the bulbs are in parallel, we expect the equivalent resistance to be lower than the resistance of either bulb: $R_{eq} < R_{old}$ and $R_{eq} < R_{new}$.

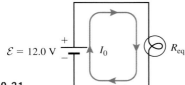

FIGURE 29.21

SOLVE
Because there are only two bulbs, apply Equation 29.8 to find the equivalent resistance.

$$R_{eq} = \frac{R_{old} R_{new}}{R_{old} + R_{new}} \quad (29.8)$$

$$R_{eq} = \frac{(3.2\ \Omega)(1.6\ \Omega)}{3.2\ \Omega + 1.6\ \Omega} = 1.07\ \Omega$$

$$\boxed{R_{eq} = 1.1\ \Omega}$$

CHECK and THINK
As expected, the equivalent resistance is lower than that of either bulb.

B Use your result from part A to find the current I_0 through the car's 12-V battery. (In the **CHECK and THINK** step, consider this fact: In a car's headlight circuit, there is a 15-A fuse. If the current reaches 15 A, the fuse breaks. Just like an open switch, current cannot pass through a broken fuse.)

INTERPRET and ANTICIPATE
Now that we have the equivalent resistance, we can consider the simple equivalent circuit (Fig. 29.21).

SOLVE
Use the loop rule.

$$\mathcal{E} - I_0 R_{eq} = 0 \qquad I_0 = \frac{\mathcal{E}}{R_{eq}}$$

Substitute values.

$$I_0 = \frac{12.0\ \text{V}}{1.07\ \Omega} = 11.2\ \text{A} \approx \boxed{11\ \text{A}}$$

CHECK and THINK
This seems like a reasonable value for the current because it is somewhat smaller than the maximum current (15 A) that can pass through the fuse. Fuses protect circuit elements from overheating. Recall from Section 28-7 that when there is current in a circuit element with resistance, that circuit element gets hot. For this reason, circuit designers include mechanisms for transferring heat to the environment. For example, your computer has a fan to help transfer heat through forced convection. Many circuits include a fuse to further protect the circuit. Fuses are rated by the maximum current they can endure. If the current reaches that value, the fuse breaks and acts as an open switch, cutting off current.

29-6 Circuit Analysis

At the beginning of this chapter, we imagined using a multimeter to troubleshoot a failed electric circuit. The multimeter can measure potential difference, current, resistance, and capacitance. Circuit analysis is the process of determining the expected values that you can measure with a multimeter. The first five sections of this chapter as well as Section 27-4 introduced rules for circuit analysis. The rules fall into three broad categories:

1. **Voltage rules** for finding the potential difference across various circuit elements (Section 29-1)
2. **Kirchhoff's rules** for writing equations (Sections 29-2 and 29-4)
3. **Capacitance and resistance combination rules** for simplifying a circuit that contains more than one capacitor, resistor, or lightbulb:

$$\frac{1}{C_{eq}} = \sum_{i=1}^{N} \frac{1}{C_i} \text{ (Eq. 27.7) and } R_{eq} = \sum_{i=1}^{N} R_i \text{ (Eq. 29.6) for \textbf{series}}$$

$$C_{eq} = \sum_{i=1}^{N} C_i \text{ (Eq. 27.8) and } \frac{1}{R_{eq}} = \sum_{i=1}^{N} \frac{1}{R_i} \text{ (Eq. 29.7) for \textbf{parallel}}$$

Now we bring these rules together in a problem-solving strategy for circuit analysis in the case of complicated circuits involving multiple branches.

PROBLEM-SOLVING STRATEGY

Circuit Analysis

:• INTERPRET and ANTICIPATE
As in the problem-solving strategy "Using Kirchhoff's Loop Rule" (page 903), start with a schematic diagram (Fig. 29.22) that includes these elements: (1) arrows and labels for the **currents**, (2) labels for the **junctions** and the **circuit elements** (such as resistors, capacitors, and emf devices), and (3) a dot for the **starting point** (if you plan to use Kirchhoff's loop rule).

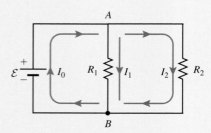

FIGURE 29.22 This schematic diagram has all three elements: (1) The three current directions are indicated and labeled I_0, I_1, and I_2. (2) The two junctions are labeled A and B, the two resistors are labeled R_1 and R_2, and the emf device is labeled $\mathcal{E}$. (3) The starting point is indicated by a dot at junction B.

:• SOLVE
There are three possible tactics for you to try depending on the situation. In many cases, you will not use all three tactics.

Tactic 1. Find an **equivalent circuit** and draw its schematic diagram. When a circuit involves more than one type of circuit element, finding an equivalent circuit is sometimes helpful and is particularly useful when you need to find the current in one branch of the circuit. One of the hardest tasks in finding the equivalent circuit is determining which elements are in series and which are in parallel. Use these facts to make your decision.

Elements in series (Fig. 27.16, page 838)
1. are connected by a single wire and
2. have the same current in each element.

Elements in parallel (Fig. 27.20, page 840)
1. are connected by two wires and
2. have the same voltage across each element.

Tactic 2. Apply **Kirchhoff's junction rule.** If the circuit has more than one loop, the junction rule may be used to relate the current in the various branches. Often you need to apply the junction rule to only half the junctions. For example, applying the junction rule to either junction A or B in Figure 29.22 gives the same relationship.

Tactic 3. Apply **Kirchhoff's loop rule** using the problem-solving strategy on page 903. You do not need to apply Kirchhoff's loop rule to every closed loop because the equations are mathematically dependent.

29-6 Circuit Analysis

EXAMPLE 29.7 Resistors in Series and in Parallel

When there is current in the circuit in Figure 29.23, all the resistors have $R = 144\ \Omega$ and the two emf devices have the same terminal potential $\mathcal{E} = 1.48$ V. Find the current in either emf device.

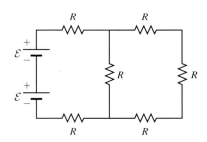

FIGURE 29.23

• INTERPRET and ANTICIPATE
Draw a schematic diagram (Fig. 29.24), indicating the **currents** and labeling the **junctions** and **circuit elements**. At this time, there is no need to indicate a **starting point** because we don't plan to apply Kirchhoff's loop rule (yet).

The emf devices are in series with the two resistors adjacent to them, so the current is the same (I_0) in all four of those circuit elements. At junction A, the current splits into two branches.

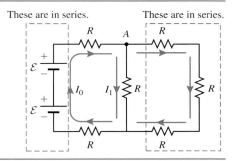

FIGURE 29.24

• SOLVE
Tactic 1. Find an **equivalent circuit** and draw its schematic diagram (Fig. 29.25). It is difficult to find the equivalent circuit in just one step because of the large number of resistors and emf devices. In such a case, consider a small number of resistors at a time and find several intermediate equivalent circuits.

Because the emf devices are in series (Fig. 29.24), we can replace them with an equivalent emf device that has an emf of $2\mathcal{E}$. The three resistors in the branch farthest from the emf devices are connected by a single wire between each pair and have the same current (Fig. 29.24). Therefore, these resistors are in series and have an equivalent resistance of $3R$ (Eq. 29.6).

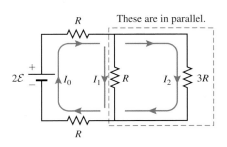

FIGURE 29.25

The two resistors in Figure 29.25 that are farthest from the equivalent emf device are connected by two wires and have the same potential difference across them. Therefore, these resistors are in parallel. Use Equation 29.8 to find their equivalent resistance, $3R/4$, and draw the new equivalent circuit (Fig. 29.26).

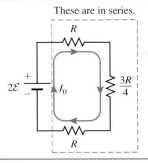

FIGURE 29.26

• CHECK and THINK
Check the dimensions at intermediate steps in a complicated circuit analysis, especially if the circuit involves resistors in parallel. The equivalent resistance we found is a number times R (correct). Forgetting to invert the result in the earlier step leaves R in the denominator (wrong dimensions).

• SOLVE
The three resistors in Figure 29.26 are connected by a single wire between each pair, and they have the same current. Therefore, these three resistors are in series and have an equivalent resistance of $11R/4$ (Eq. 29.6). The circuit in Figure 29.27 is equivalent to the one in Figure 29.23.

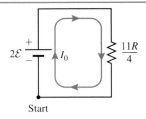

FIGURE 29.27

Example continues on page 914 ▶

Tactic 3. Apply **Kirchhoff's loop rule** to the equivalent circuit. Use the starting point indicated (Fig. 29.27), and imagine taking voltage measurements in the direction of the current.	Loop rule: $$2\mathcal{E} - I_0\left(\frac{11R}{4}\right) = 0$$
Solve for I_0.	$$I_0 = 2\mathcal{E}\left(\frac{4}{11R}\right)$$ $$I_0 = 2(1.48\,\text{V})\left[\frac{4}{11(144\,\Omega)}\right]$$ $$I_0 = 7.47 \times 10^{-3}\,\text{A}$$

CHECK and THINK
As one final dimensional check, the SI units are correct because $1\,\text{V}/\Omega = 1\,\text{A}$.

EXAMPLE 29.8 Can't Be Simplified

Four ideal emf devices and three resistors are connected as in Figure 29.28. The emf devices are identical (terminal potential $\mathcal{E} = 3.22\,\text{V}$). The resistors have resistances $R_1 = R$, $R_2 = 2R$, and $R_3 = 3R$, where $R = 57.4\,\Omega$. Find the current through each resistor.

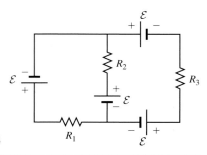

FIGURE 29.28

INTERPRET and ANTICIPATE
None of the resistors in this circuit are in series or in parallel, so we can't use Tactic 1 (page 912). We must apply Kirchhoff's rules and the voltage rules to find the current in each resistor. We'll be able to check at the end that the current in each resistor is different and the voltage across each resistor is different.

Draw a schematic diagram (Fig. 29.29), indicating the **currents** and labeling the **junction** and **circuit elements**. At this time, there is no need to indicate a **starting point**. With this many emf devices, the direction of the current in each branch is not obvious. Make arbitrary choices; if a choice is incorrect, you will end up with a negative value when you solve for that current. (There is no need to label both junctions because we only need to apply the junction rule to one.)

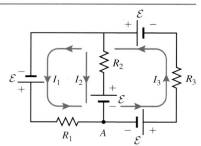

FIGURE 29.29

SOLVE
Tactic 2. Apply **Kirchhoff's junction rule** to junction A.

$$I_1 + I_2 = I_3 \quad (1)$$

Tactic 3. Apply **Kirchhoff's loop rule**. The circuit has three loops (left, right, and outer). Apply the loop rule to any two of the three loops. (If you apply the loop rule to all three loops, you will not gain extra information.) We arbitrarily choose the left and right loops.

Left loop. Start in the bottom left corner and measure counterclockwise around the left loop (Fig. 29.30).

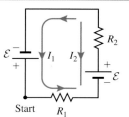

FIGURE 29.30

The voltage across R_1 is negative (a voltage drop), and the voltage across R_2 is positive. The voltage across each emf device is positive.	$-I_1R_1 + \mathcal{E} + I_2R_2 + \mathcal{E} = 0$ (2)

Right loop. Start in the bottom right corner and measure clockwise around the right loop, applying the voltage rules (Fig. 29.31).

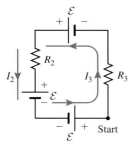

FIGURE 29.31

The voltage across both resistors is positive. The voltage across one emf device is positive and across the other two is negative.	$-\mathcal{E} + \mathcal{E} + I_2R_2 - \mathcal{E} + I_3R_3 = 0$ (3)

Algebra is all that is left. There are three unknowns—the currents I_1, I_2, and I_3—and three independent equations (1, 2, and 3). There are several ways to solve three equations simultaneously. We'll use the substitution method, showing only a few key steps.

Use Equation (1) to eliminate I_1 from Equation (2), and simplify using $R_1 = R$ and $R_2 = 2R$.	$-(I_3 - I_2)R_1 + I_2R_2 = -2\mathcal{E}$ $3I_2R - I_3R = -2\mathcal{E}$ (4)
Solve Equation (3) for I_3 and simplify using $R_2 = 2R$ and $R_3 = 3R$.	$I_2R_2 + I_3R_3 = \mathcal{E}$ $I_3 = \dfrac{\mathcal{E} - I_2R_2}{R_3} = \dfrac{\mathcal{E} - 2I_2R}{3R}$ (5)
Substitute Equation (5) for I_3 into Equation (4), solving for I_2.	$3I_2R - \left(\dfrac{\mathcal{E} - 2I_2R}{3R}\right)R = -2\mathcal{E}$ $I_2\left(\dfrac{11R}{3}\right) = -\dfrac{5\mathcal{E}}{3}$ $I_2 = -\dfrac{5\mathcal{E}}{11R} = -\dfrac{5(3.22\text{ V})}{11(57.4\text{ }\Omega)} = -2.54 \times 10^{-2}\text{ A}$ $I_2 = -25.4\text{ mA}$
Now that we have a value for I_2, substitute it into Equation (5) to find I_3.	$I_3 = \dfrac{3.22\text{ V} - (-2.54 \times 10^{-2}\text{A})(2)(57.4\text{ }\Omega)}{(3)(57.4\text{ }\Omega)}$ $I_3 = 3.56 \times 10^{-2}\text{A} = 35.6\text{ mA}$
To find I_1, substitute I_2 and I_3 into Equation (1).	$I_1 = I_3 - I_2 = 35.6\text{ mA} - (-25.4\text{ mA})$ $I_1 = 61.0\text{ mA}$

:• **CHECK and THINK**
The values found for I_1 and I_3 are both positive, so the directions we chose for those currents are correct. However, the value for I_2 is negative, so the direction we picked for that current is incorrect. Instead of pointing downward, I_2 points upward. The currents I_1, I_2, and I_3 through the three resistors are different, which confirms that these resistors are *not* in series.

Now let's check that the resistors are not in parallel by finding the potential difference across each one. For the moment, we are not concerned about the sign of the voltage, so we will just use $\Delta V = |IR|$ for each resistor.

Potential difference across R_1:	$\Delta V_1 = \|I_1 R_1\| = (61.0 \times 10^{-3}\,\text{A})(57.4\,\Omega)$ $\Delta V_1 = 3.50\,\text{V}$
Potential difference across R_2:	$\Delta V_2 = \|I_2 R_2\| = \|I_2(2R)\|$ $\Delta V_2 = \|(-25.4 \times 10^{-3}\,\text{A})(2)(57.4\,\Omega)\|$ $\Delta V_2 = 2.92\,\text{V}$
Potential difference across R_3:	$\Delta V_3 = \|I_3 R_3\| = \|I_3(3R)\|$ $\Delta V_3 = (35.6 \times 10^{-3}\,\text{A})(3)(57.4\,\Omega) = 6.31\,\text{V}$

The potential differences across the resistors are not equal, so they cannot be in parallel.

EXAMPLE 29.9 CASE STUDY Why Is the Old Headlight So Dim?

Avi, Cameron, and Shannon now know that car headlights are connected in parallel. In Example 29.6, we found the equivalent circuit (Fig. 29.21) and the current in the car's battery ($I_0 = 11.2$ A). In the original circuit (Fig. 29.20), find the current in each bulb and the power each consumes.

INTERPRET and ANTICIPATE
There are two unknown currents, I_{old} and I_{new}, so we need two independent equations to solve for them. More than one solution method exists; here we use Kirchhoff's junction rule once and Kirchhoff's loop rule once. Because the new headlight is brighter, we expect it to consume more power than the old bulb. We can use Figure 29.20 as our complete schematic diagram.

SOLVE
Tactic 2. Apply **Kirchhoff's junction rule**. There are two junctions (A and B) in this circuit. Apply the junction rule to either.

$$I_0 = I_{\text{old}} + I_{\text{new}} \tag{1}$$

Tactic 3. Apply **Kirchhoff's loop rule**. Start at B and measure clockwise around the right loop to come up with one equation that involves both currents I_{old} and I_{new}.

$$I_{\text{old}} R_{\text{old}} - I_{\text{new}} R_{\text{new}} = 0 \tag{2}$$

CHECK and THINK
Rewrite Equation (2). According to Equation (3), the voltage drop across the old bulb equals the voltage drop across the new bulb, as should be true for two bulbs in parallel.

$$I_{\text{old}} R_{\text{old}} = I_{\text{new}} R_{\text{new}} \tag{3}$$

SOLVE
Solve Equation (3) for I_{old} and substitute into Equation (1).

$$I_{\text{old}} = \frac{I_{\text{new}} R_{\text{new}}}{R_{\text{old}}}$$

$$I_0 = \frac{I_{\text{new}} R_{\text{new}}}{R_{\text{old}}} + I_{\text{new}} = I_{\text{new}}\left(\frac{R_{\text{new}} + R_{\text{old}}}{R_{\text{old}}}\right)$$

Solve for I_{new}.

$$I_{\text{new}} = I_0\left(\frac{R_{\text{old}}}{R_{\text{new}} + R_{\text{old}}}\right) \tag{4}$$

$$I_{\text{new}} = (11.2\,\text{A})\left(\frac{3.2\,\Omega}{1.6\,\Omega + 3.2\,\Omega}\right) = 7.47\,\text{A}$$

Find I_{old} from Equation (1).	$I_{old} = I_0 - I_{new} = 11.2\text{ A} - 7.47\text{ A} = \boxed{3.73\text{ A}}$
Find the power consumed by each bulb from Equation 28.34.	For the new bulb, $$P_{new} = I_{new}^2 R_{new} = (7.47\text{ A})^2(1.6\ \Omega) = \boxed{90\text{ W}}$$ For the old bulb, $$P_{old} = I_{old}^2 R_{old} = (3.73\text{ A})^2(3.2\ \Omega) = \boxed{45\text{ W}}$$

:• CHECK and THINK

This fits Carol's observation. The old bulb appears dimmer because it consumes less power than the new bulb. To fix the problem, use two bulbs with the same operating resistance.

We can apply our experience with this circuit to similar circuits. This circuit (Fig. 29.20) is sometimes called a current splitter because at junction A the current splits into two branches. If the two bulbs have the same resistance, the current through each bulb is half the current passing through the battery: $I_{new} = I_{old} = I_0/2$. If the bulbs have unequal resistances, as they do in this case, the branch with the lower resistance has greater current. I_{new} is twice as much as I_{old} because the resistance of the new bulb is half that of the old bulb.

29-7 DC Multimeters

This chapter is about the practical application of the principles of electricity. In Section 29-6, we studied circuit analysis—a way to predict the measurements made by a DC multimeter. In this section, we learn how a multimeter is used and designed to measure current, voltage, and resistance. Learning about multimeters not only will help you build your laboratory skills; it will also help you better understand circuits. Many laboratories have a digital multimeter like the one in Figure 29.32A. We will describe an analog multimeter (Fig. 29.32B), however, because it is partially mechanical and so its principles of operation are simpler.

The essential element of an analog meter is a **galvanometer**, which consists of an armature (a coil of wire wrapped around an iron core) that is free to rotate between the poles of a permanent magnet (Fig. 29.33). When there is current in the coil of wire, the coil becomes an electromagnet (Chapter 30). The electromagnet inside the galvanometer acts like a compass needle, a small magnet that aligns itself with the Earth's magnetic field so that the needle points toward the North Pole. The electromagnet produced by the coil in the galvanometer aligns itself with the galvanometer's permanent magnet, causing the needle attached to the armature to be deflected. The greater the current in the coil, the stronger the electromagnet and the greater the

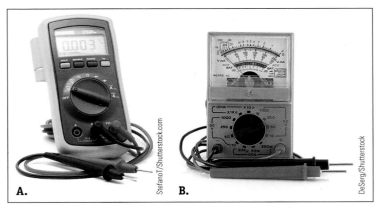

FIGURE 29.32 A multimeter is an ammeter, voltmeter, ohmmeter, and capacitance meter all in one. **A.** Digital multimeters are commonly found in student laboratories. **B.** Analog multimeters are easier to understand.

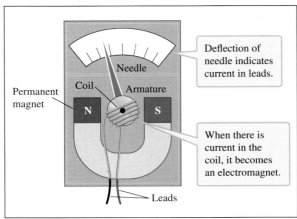

FIGURE 29.33 Current in the coil of the galvanometer causes it to become magnetic, and the coil rotates to align with the permanent magnet.

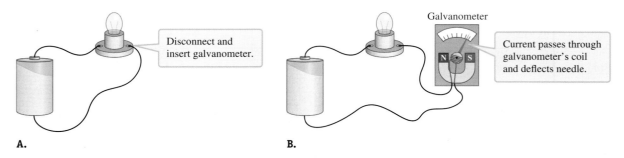

FIGURE 29.34 **A.** A circuit with a battery and a lightbulb. **B.** The circuit is broken so that a galvanometer can be inserted.

needle's deflection. The direction of deflection depends on the direction of the current in the coil. Thus, the deflection of the needle indicates the direction and amount of current in the coil. A spring attached to the armature (not shown in the figure) restores the needle to its original position when there is no current.

When you want to measure current in a circuit, you must break the circuit and insert the galvanometer in series (Fig. 29.34). In this textbook, the symbol for an ideal galvanometer is a circle with a **G** inside (Fig. 29.35A). An ideal galvanometer has no resistance, so it does not change the amount of current in a circuit when it is inserted. The coil of a real galvanometer contains many turns of wire, so it has an internal resistance r. We represent a real galvanometer by an ideal galvanometer in series with a resistor of resistance r (Fig. 29.35B).

In practice, a galvanometer can measure small currents up to several microamperes. If you try to use the galvanometer to measure a somewhat greater current, the needle will be deflected to its maximum position. Suppose the galvanometer you are using is designed to operate at currents up to 50 μA and you are trying to measure a current of 60 μA. The needle will deflect so that the meter reads 50 μA. You then know that the current is at least 50 μA, but it could be greater.

Ammeter Design

An **ammeter** is designed to measure a range of currents by placing a galvanometer in parallel with a **shunt resistor**. (*Shunt* is another word for "parallel.") The shunt resistor is inside the case of the multimeter, so you cannot easily see it. In Figure 29.36A, an ammeter is shown in a circuit that consists of a lightbulb and a battery. Figure 29.36B shows the same circuit with schematic detail of the inside of the ammeter. The shunt resistor has resistance R_{sh}. The galvanometer's coil has resistance r and is shown as a resistor in series with an ideal galvanometer. The shunt resistance is generally low compared to the internal resistance of the galvanometer, so only a small current I_G passes through the galvanometer and the needle is not deflected to its maximum position. The current through the shunt resistor is I_{sh}, and the current through the lightbulb and battery is $I = I_{sh} + I_G$. The designer knows the internal galvanometer resistance, and the user chooses the shunt resistance, thus determining what fraction of the total current I goes through the galvanometer so that the appropriate scale can be displayed on the meter's face.

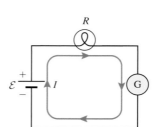

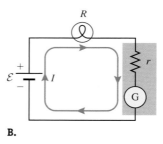

FIGURE 29.35 **A.** An ideal galvanometer measures current without altering the circuit. **B.** A real galvanometer has some internal resistance, so its presence alters the current. Figure 29.35B is a schematic diagram for the circuit shown in Figure 29.34B.

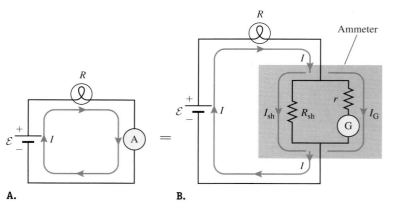

FIGURE 29.36 **A.** An ammeter is connected in series in a circuit. **B.** An ammeter is a galvanometer in parallel with a shunt resistor. The shunt resistor divides the current. When you use an ammeter, you select the value of the shunt resistor, which determines the range of measurable current.

Voltmeter Design

A galvanometer is also the basis for an analog **voltmeter**. Measuring voltage is often more practical than measuring current because using a voltmeter does not require you to break the circuit, as you must when using an ammeter. A voltmeter is placed in parallel with a circuit element in order to measure the potential difference across it (Fig. 29.37A). Because you do not want to disturb the circuit by deflecting much current through the voltmeter, the branch that contains the voltmeter should have a high resistance. A voltmeter is constructed by placing a galvanometer in series with a large resistor R_{ser} (Fig. 29.37B). The galvanometer measures the current I through R_{ser} and r. The voltage drop across them, $\Delta V = I(R_{ser} + r)$, equals the voltage across the circuit element—in this case, the lightbulb (and emf). The scale on the face of the meter is calibrated in volts, with that calculation "built in."

Ohmmeter Design

An **ohmmeter** measures the resistance of a circuit element. To measure the resistance of an element such as the lightbulb in Figure 29.38A, connect the ohmmeter across the element (which should not be in a working circuit). The ohmmeter is made up of an emf device and an ammeter (Fig. 29.38B). The emf device inside the ohmmeter ensures that there is current through the circuit element. The ammeter measures that current. The current is inversely proportional to the resistance. The scale on the ohmmeter is calibrated in ohms, with that calculation "built in."

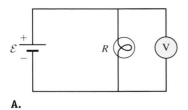
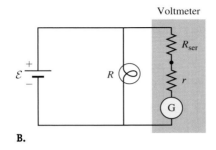

FIGURE 29.37 A. A voltmeter is connected in parallel with a circuit element. **B.** A voltmeter is a galvanometer in series with a large resistor.

29-8 RC Circuits

So far, our focus has been on DC circuits in which the current does not change over time. There are many circumstances where a time-varying current is required, however. A camera's flashbulb requires a short burst of current. A car's turn signal, a patient's electronic pacemaker, and a car's intermittent windshield wipers require a current that repeatedly increases and decreases.

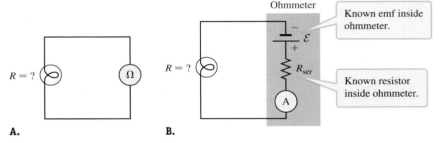

FIGURE 29.38 A. A lightbulb of unknown resistance is connected to an ohmmeter. **B.** An ohmmeter is an ammeter in series with an emf device and a resistor.

A circuit consisting of a resistor, a capacitor, and an emf (Fig. 29.39) can produce such time-varying currents and is called an **RC circuit**, where R stands for resistance and C for capacitance. The switch S has three possible positions; in Figure 29.39 it is shown open. When the switch is closed at point A, an emf device with terminal potential $\mathcal{E}$ is in the circuit. In this case, the capacitor is *charging*. When the switch is closed at point B, the emf device is not in the circuit and the capacitor is *discharging*.

We will use circuit analysis to find expressions for the current $I(t)$ in the circuit and the charge $q(t)$ stored by the capacitor, both of which are functions of time. We can imagine monitoring the current in the circuit by placing a voltmeter across the resistor. The current through the resistor is then

$$I(t) = \frac{\Delta V_R(t)}{R} \quad (28.25)$$

where $\Delta V_R(t)$ is the potential difference across the resistor as a function of time. We can monitor the charge stored by the capacitor with a second voltmeter placed across the capacitor. The charge stored by the capacitor is given by

$$q(t) = C\Delta V_C(t) \quad (27.1)$$

where $\Delta V_C(t)$ is the potential difference across the capacitor as a function of time.

Charging a Capacitor

Initially, the capacitor is uncharged and the switch is open (Fig. 29.39). At time $t = 0$, the switch is closed at point A so that the circuit consists of a resistor, a capacitor, and the emf device (Fig. 29.40; the extraneous parts of the network are not

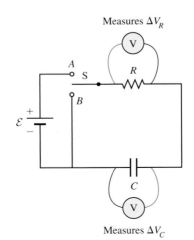

FIGURE 29.39 An *RC* circuit.

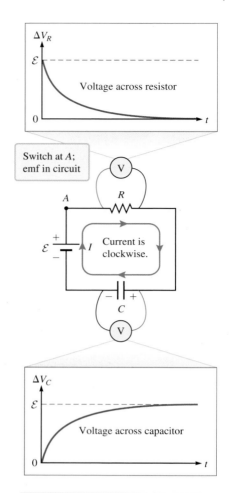

shown). The current in the circuit is initially clockwise as shown. Positive charge builds up on the right plate of the capacitor, and negative charge builds up on the left plate. Charge continues to flow until the voltage across the capacitor $\Delta V_C(t)$ equals the terminal potential $\mathcal{E}$ across the emf device.

A graph of the capacitor's voltage $\Delta V_C(t)$ versus t (lower portion of Fig. 29.40) shows: (1) Initially, the capacitor is uncharged, so the voltage is zero: $\Delta V_C(0) = 0$. (2) As the charge builds up, the measured potential difference across the capacitor is positive and increases. (3) As time approaches infinity, the capacitor reaches its maximum potential difference, the terminal potential $\mathcal{E}$ of the emf device. In practice, this maximum potential difference is reached after a long time.

Now consider the graph of the resistor's voltage $\Delta V_R(t)$ versus t (upper portion of Fig. 29.40). The voltage across the emf device must equal the sum of the voltage across the resistor and the voltage across the capacitor:

$$\mathcal{E} = \Delta V_R(t) + \Delta V_C(t) \tag{29.9}$$

The initial capacitor voltage $\Delta V_C(0) = 0$, so the voltage across the resistor equals the voltage across the emf device: $\Delta V_R(0) = \mathcal{E}$. As the voltage across the capacitor increases, the voltage across the resistor must decrease. When the voltage across the capacitor equals the voltage of the emf device ($\mathcal{E}$), the voltage across the resistor must be zero.

The graphs in Figure 29.40 give us a qualitative understanding of what happens when the capacitor is charging. The voltage across the capacitor increases as it stores more charge, and the voltage across the resistor drops as the current drops. Our goal now is to derive expressions for the charge stored by the capacitor and for the current in the circuit as functions of time.

FIGURE 29.40 A charging RC circuit. With the voltmeter's red lead connected to the capacitor's positive plate and the black lead connected to the negative plate, the measured potential difference across the capacitor is positive. A graph of the voltage across the capacitor as a function of time rises from zero to $\mathcal{E}$. The current is from the red lead toward the black lead, so the potential difference across the resistor is positive. The voltage across the resistor as a function of time drops from $\mathcal{E}$ to zero.

DERIVATION $q(t)$ for a Charging Capacitor

We show that the charge $q(t)$ stored by the charging capacitor (Fig. 29.40) as a function of time is given by

$$q(t) = C\mathcal{E}(1 - e^{-t/RC}) \tag{29.10}$$

CHARGE ON A CHARGING CAPACITOR
▶ Special Case

Because the voltage across the capacitor is proportional to the charge it stores, we expect a graph of $q(t)$ versus t to look similar to the graph of $\Delta V_C(t)$ versus t.

We begin when the switch has been closed at point A for some time but the capacitor has not yet reached its maximum voltage. Use Kirchhoff's loop rule, starting in the lower left corner of the circuit and measuring clockwise in the direction of the current (Fig. 29.41). The voltmeter leads here are in the opposite orientation as those in Figure 29.40.

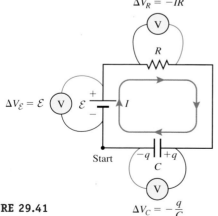

FIGURE 29.41

Use the emf rule and the resistor rule. The voltage across the capacitor is given by $q = C\Delta V_C$ (Eq. 27.1), and with the red lead connected to the negative plate of the capacitor, the measured potential difference is negative.	$\Delta V_\mathcal{E} + \Delta V_R + \Delta V_C = 0$ $\mathcal{E} - IR - \dfrac{q}{C} = 0 \tag{29.11}$

After substituting $I = dq/dt$ (Eq. 28.1) for the current in Equation 29.11, we are left with a differential equation.	$\mathcal{E} - \dfrac{dq}{dt}R - \dfrac{q}{C} = 0$ (2) $\dfrac{C\mathcal{E} - q}{RC} = \dfrac{dq}{dt}$ (3)
Isolate q on one side of the equation.	$\dfrac{dt}{RC} = \dfrac{dq}{C\mathcal{E} - q}$
Integrate the left side from $t = 0$ (when the switch was closed at point A) to some arbitrary time t later. Integrate the right side from $q = 0$ (because the capacitor is initially uncharged) to q (the charge stored at some arbitrary time t).	$\displaystyle\int_0^t \dfrac{dt}{RC} = \int_0^q \dfrac{dq}{C\mathcal{E} - q}$ $\dfrac{1}{RC}\displaystyle\int_0^t dt = \int_0^q \dfrac{dq}{C\mathcal{E} - q}$
The integral on the left is straightforward. The integral on the right leads to a natural logarithm (Problem 82).	$-\dfrac{t}{RC} = \ln\left(\dfrac{C\mathcal{E} - q}{C\mathcal{E}}\right)$
To eliminate the natural logarithm, take the exponential of both sides.	$e^{-t/RC} = \left(\dfrac{C\mathcal{E} - q}{C\mathcal{E}}\right)$
Solve for q.	$q(t) = C\mathcal{E}(1 - e^{-t/RC})$ ✓ (29.10)

: COMMENTS

Figure 29.42 is a graph of q as a function of t. This graph looks similar to the capacitor's voltage graph in Figure 29.40, as expected. When the switch is closed at $t = 0$, the charge is zero:

$$q(0) = C\mathcal{E}(1 - e^{-0/RC}) = 0$$

as it should be. After a long time, $t \to \infty$ and the charge stored by the capacitor is at its maximum: $q_{max} = C\mathcal{E}(1 - e^{-\infty}) = C\mathcal{E}$, also as it should be.

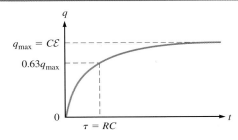

FIGURE 29.42 Charge versus time for a charging capacitor.

DERIVATION $I(t)$ for a Charging Capacitor

We show that the current $I(t)$ in the circuit (Fig. 29.40) as a function of time is

$$I(t) = \left(\dfrac{\mathcal{E}}{R}\right)e^{-t/RC} \qquad (29.12)$$

CURRENT IN A CHARGING RC CIRCUIT ⊙ Special Case

when the capacitor is charging.

Because the current through the resistor is proportional to the voltage across it, we expect a graph of $I(t)$ versus t to look similar to the graph of $\Delta V_R(t)$ versus t.

Take the time derivative of Equation 29.10.	$I(t) = \dfrac{dq}{dt} = \dfrac{d}{dt}[C\mathcal{E}(1 - e^{-t/RC})]$ $I(t) = C\mathcal{E}\left[0 - \left(-\dfrac{1}{RC}\right)e^{-t/RC}\right]$ $I(t) = C\mathcal{E}\left[\left(\dfrac{1}{RC}\right)e^{-t/RC}\right]$ $I(t) = \left(\dfrac{\mathcal{E}}{R}\right)e^{-t/RC}$ ✓ (29.12)

Derivation continues on page 922 ▶

COMMENTS

As expected, a graph of I as a function of t (Fig. 29.43) looks similar to the resistor's voltage graph in Figure 29.40. When the switch is first closed at $t = 0$, the current is initially

$$I_0 = \left(\frac{\mathcal{E}}{R}\right)e^{-0/RC} = \frac{\mathcal{E}}{R}$$

as expected. After the switch has been closed for a long time, $t \to \infty$ and the current goes to zero:

$$I(t \to \infty) = \left(\frac{\mathcal{E}}{R}\right)e^{-\infty/RC} = 0$$

as we expect for a fully charged capacitor.

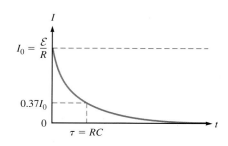

FIGURE 29.43 Current versus time in a charging RC circuit.

Time Constant

Equations 29.10 for $q(t)$ and 29.12 for $I(t)$ both depend on the quantity RC, which has dimensions of time (see Problem 83). This quantity is called the **time constant** τ:

$$\tau \equiv RC \qquad (29.13)$$

The time constant is completely determined by the resistance and capacitance in the circuit. The emf does not affect the time constant.

We make our expressions for the charge (Eq. 29.10) and current (Eq. 29.12) more intuitive and convenient to use when we rewrite them in terms of the time constant τ. The charge is

$$q(t) = C\mathcal{E}(1 - e^{-t/\tau}) = q_{max}(1 - e^{-t/\tau}) \qquad (29.14)$$

where $q_{max} = C\mathcal{E}$, and the current is

$$I(t) = \left(\frac{\mathcal{E}}{R}\right)e^{-t/\tau} = I_0 e^{-t/\tau} \qquad (29.15)$$

where $I_0 = \mathcal{E}/R$.

The time constant is a measure of how quickly the capacitor charges up. By substituting $t = \tau$ into Equation 29.14, we have

$$q(\tau) = q_{max}(1 - e^{-\tau/\tau}) = q_{max}(1 - e^{-1}) \approx 0.63 q_{max}$$

We find that the time constant τ is the time it takes the capacitor to reach about 63% of its maximum charge (Fig. 29.42). The time constant is also a measure of how quickly the current in the circuit drops. By substituting $t = \tau$ into Equation 29.15, we have

$$I(\tau) = I_0 e^{-\tau/\tau} = I_0 e^{-1} \approx 0.37 I_0$$

We see that the time constant τ is the time the current takes to drop to 37% of its initial value (Fig. 29.43).

Discharging a Capacitor

When the switch in Figure 29.39 has been closed at point A for a long time (at least several time constants), the capacitor is fully charged, which means $q = q_{max}$. There is no current in the circuit. To discharge the capacitor, the switch is thrown to point B so that there is no emf device in the circuit (Fig. 29.44). For convenience, we reset the time to $t = 0$ at the moment the switch is thrown to point B. (This is like resetting a stopwatch.)

Just before the switch is thrown to B, the capacitor is fully charged and its voltage must equal the terminal potential of the emf device, so now the initial voltage across the capacitor must be $\Delta V_C(0) = q_{max}/C = \mathcal{E}$. After the switch is thrown to B, the capacitor discharges through a counterclockwise current in the circuit. When the capacitor is completely discharged, the voltage across it must be zero. So, the voltage across the capacitor goes from $\mathcal{E}$ to zero (Fig. 29.44). Similarly, we expect the charge stored by the capacitor to decay from its maximum value.

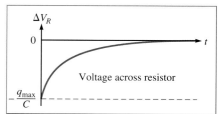

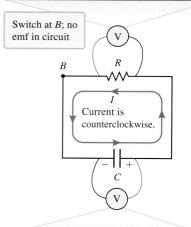

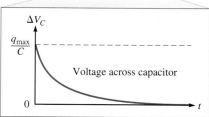

FIGURE 29.44 A discharging RC circuit. With the voltmeter's red lead connected to the capacitor's positive plate and the black lead connected to the negative plate, the measured potential difference across the capacitor is positive. A graph of the voltage across the capacitor as a function of time falls to zero. The current is from the black lead toward the red lead, so the potential difference across the resistor is negative. The voltage across the resistor as a function of time goes to zero.

In Problem 93, you will show that for a discharging RC circuit, the charge $q(t)$ on the capacitor as a function of time is

$$q(t) = q_{max} e^{-t/\tau} \quad (29.16)$$

CHARGE IN A DISCHARGING RC CIRCUIT ▶ Special Case

As expected, a graph of q as a function of t (Fig. 29.45) looks similar to the capacitor's voltage graph (Fig. 29.44). When the switch is closed at point B, $t = 0$ and

$$q(0) = q_{max} e^{-0/\tau} = q_{max}$$

so the charge is at its maximum. After a long time, $t \to \infty$ and $q(t \to \infty) = q_{max}(e^{-\infty}) \to 0$, so the charge stored by the capacitor falls to zero. Finally, the time constant is the time it takes the capacitor's stored charge to drop to about 37% of its maximum: $q(\tau) = q_{max}(e^{-1}) \approx 0.37 q_{max}$.

For the charging capacitor, the current is clockwise, which we set to positive (Eq. 29.15). In the discharging RC circuit, the current is counterclockwise (Fig. 29.44), so, as you will find in Problem 95, the current is negative and given by

$$I(t) = -I_{max} e^{-t/\tau} \quad (29.17)$$

CURRENT IN A DISCHARGING RC CIRCUIT ▶ Special Case

A graph of I as a function of t (Fig. 29.46) looks similar to the resistor's voltage graph in Figure 29.44. (If the capacitor was fully charged, then $I_{max} = I_0 = \mathcal{E}/R$.) After the switch has been closed for a long time, $t \to \infty$ and the current goes to zero:

$$I(t \to \infty) = -I_{max} e^{-\infty/\tau} \to 0$$

We see that the time constant is also the time it takes the current in a discharging RC circuit to drop to 37% of its initial value: $I(\tau) \approx 0.37 I_0$.

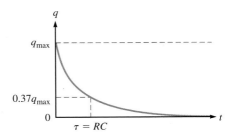

FIGURE 29.45 Charge versus time for a discharging RC circuit.

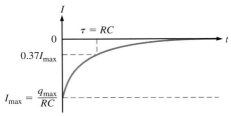

FIGURE 29.46 Current versus time for a discharging RC circuit.

In practice, "a long time" is about 3 to 5 times the time constant.

EXAMPLE 29.10 Photography in a Cave

Your camera probably has a flash in the form of a discharge tube like a small version of the fluorescent tubes that light up your classroom. You may have seen an old-style flashbulb (Fig. 29.47) used by a newspaper photographer in a movie set in the 1950s or a disposable flashbulb from the 1970s. These old flashbulbs were similar to incandescent lightbulbs. When the photographer pressed a button, current passed through the flashbulb's filament. The flashbulb gave off considerable energy (light and thermal energy) in a short amount of time, breaking the filament and sometimes the glass housing. The photographer waited a moment for the flashbulb to cool, then tossed it out and replaced it with a new bulb. You might think flashbulbs are a thing of the past, but many photographers still use large flashbulbs to take photos in very dark conditions, such as in a cave.

The flashbulb is part of an RC circuit that contains two parallel switches. When switch A is closed (Fig. 29.48A), the capacitor is charging. When switch B is closed (Fig. 29.48B), the capacitor discharges through the flash. The charging circuit includes an indicator light. Assume most of the resistance in the charging circuit is due to the resistance of the indicator light, $R = 4.0$ kΩ. The terminal potential of the battery is 22.5 V, and the capacitance $C = 150$ μF.

FIGURE 29.47 An old-fashioned camera with a flashbulb.

Example continues on page 924 ▶

A Find the time constant of the charging circuit. In the **CHECK and THINK** step, answer this question: What happens to the indicator light when the capacitor is fully charged?

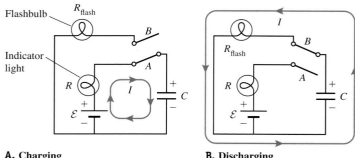

A. Charging   **B.** Discharging

FIGURE 29.48 A. Charging flashbulb RC circuit. **B.** Discharging circuit.

INTERPRET and ANTICIPATE
The time constant depends only on the resistance and the capacitance, both of which are given. From our experience with modern flash tubes or from watching old movies, we know it takes a second or so before the flash is ready. So we expect the time constant to be roughly 1 second.

SOLVE
Substitute values into Equation 29.13.

$$\tau = RC \quad (29.13)$$
$$\tau = (4.0 \times 10^3 \, \Omega)(150 \times 10^{-6} \, \text{F}) = 0.60 \, \text{s}$$

CHECK and THINK
The time constant is just a little less than 1 second, which makes sense because τ is the time it takes the capacitor to reach about 63% of its maximum charge. We should wait somewhat longer than one time constant to have a fully charged capacitor.

The indicator light will be illuminated when there is current in it. When switch A is first closed, the current in the circuit is at its maximum ($I_0 = \mathcal{E}/R$) and the indicator light is bright. As the capacitor charges up, the current decreases (Fig. 29.43), and the indicator light grows dimmer. When the capacitor is fully charged, there is no current in the circuit, so the indicator light goes off.

B Flashbulbs are used to illuminate a scene from 4 ms to 2 s. Photographers typically carry several different flashbulbs depending on the amount of time they need to illuminate the scene. Estimate the range in the resistance of these flashbulbs.

INTERPRET and ANTICIPATE
The time the flashbulb illuminates the scene is somewhat longer than the time constant of the discharging capacitor. Let's assume the illumination time is about three times as long as the time constant. We expect the range in the resistance to correspond to the range in time.

SOLVE
Find the range in time constants using the assumption that the illumination time is 3τ.

$$\frac{4 \text{ ms}}{3} \leq \tau \leq \frac{2 \text{ s}}{3}$$
$$1.3 \times 10^{-3} \text{ s} \leq \tau \leq 0.67 \text{ s}$$

Use Equation 29.13 to find a range of resistances. The capacitance is the same as in the charging circuit.

$$\tau = RC \quad (29.13)$$
$$R = \frac{\tau}{C}$$
$$\frac{1.3 \times 10^{-3} \text{ s}}{150 \times 10^{-6} \text{ F}} \leq R \leq \frac{0.67 \text{ s}}{150 \times 10^{-6} \text{ F}}$$
$$8.7 \, \Omega \leq R \leq 4.47 \text{ k}\Omega$$
$$9 \, \Omega \leq R \leq 4 \text{ k}\Omega$$

CHECK and THINK

As expected, the range in the resistance corresponds to the range in illumination time (and is about a factor of 500 in each case). If you compare flashbulbs with low resistance to those with high resistance, you will see that the ones with high resistance have many more coils of filament inside.

The terminal potential of the battery was irrelevant information for this problem. Why?

SUMMARY

❶ Underlying Principles

No new principles are introduced in this chapter.

✪ Major Concepts

1. **Kirchhoff's loop rule:** The total change in electric potential around any closed loop in a circuit is always zero.
2. **Kirchhoff's junction rule:** At any junction point in a circuit, the sum of all the currents entering the junction equals the sum of all the currents exiting the junction.
3. The equivalent resistance R_{eq} of N **resistors connected in series** is

$$R_{eq} = \sum_{i=1}^{N} R_i \qquad (29.6)$$

4. The equivalent resistance R_{eq} of N **resistors connected in parallel** is

$$\frac{1}{R_{eq}} = \sum_{i=1}^{N} \frac{1}{R_i} \qquad (29.7)$$

▶ Special Cases

1. An **emf device** is another term for a power supply. The function of an emf device is to maintain a potential difference $\mathcal{E}$ between its terminals.
   a. An **ideal DC emf device**, such as an ideal battery, maintains a constant terminal potential $\mathcal{E}$ whether there is current in the emf device or not.
   b. A **real emf device** is modeled as an ideal emf device in series with an internal resistor of resistance r.
2. Potential differences across circuit elements follow these **voltage rules**:
   a. **Wire rule:** The voltage across a wire is zero whether there is current in the wire or not:
   $$\Delta V_{wire} = 0$$
   b. **Resistor rule:** The voltage measured across a resistor when the current is from the black lead to the red lead is
   $$\Delta V_R = V_{red} - V_{black} = -IR \qquad (29.1)$$
   c. **Switch rule:** The voltage drop across a closed switch is zero:
   $$\Delta V_{closed\ switch} = 0$$
   If the switch is open, it has a very high resistance. There is no current through an open switch, but there may be a potential difference across it.
   d. **Emf rules:** (The red lead is assumed to be connected to the positive terminal in either case.) For an ideal emf device, the emf rule is
   $$\Delta V_{ideal\ emf} = V_{red} - V_{black} = \mathcal{E}.$$
   For a real emf device, the emf rule is
   $$\Delta V_{real\ emf} = V_{red} - V_{black} = \mathcal{E} - Ir.$$
3. An **RC circuit** consists of an emf device, a resistor, and a capacitor. The time constant τ is defined as
   $$\tau \equiv RC \qquad (29.13)$$
   a. For a **charging RC circuit**, the charge stored by the capacitor depends on time:
   $$q(t) = C\mathcal{E}(1 - e^{-t/\tau}) = q_{max}(1 - e^{-t/\tau}) \quad (29.14)$$
   The current in the resistor is also time-dependent and given by
   $$I(t) = \left(\frac{\mathcal{E}}{R}\right)e^{-t/\tau} = I_0 e^{-t/\tau} \qquad (29.15)$$
   b. For a **discharging RC circuit**, the charge stored by the capacitor depends on time:
   $$q(t) = q_{max} e^{-t/\tau} \qquad (29.16)$$
   The time-dependent current in the resistor is given by
   $$I(t) = -I_0 e^{-t/\tau} \qquad (29.17)$$

PROBLEM-SOLVING STRATEGY: Using Kirchhoff's Loop Rule

INTERPRET and ANTICIPATE
Start with a schematic diagram to which you add two elements: (1) an arrow for the **current** and (2) a dot for the **starting point**.

SOLVE
1. Write expressions for the expected **voltage across each circuit element**.
2. Set the sum of the expected voltages to zero.

PROBLEM-SOLVING STRATEGY: Circuit Analysis

INTERPRET and ANTICIPATE
As in the problem-solving strategy "Using Kirchhoff's Loop Rule," start with a schematic diagram that includes these elements: (1) arrows and labels for the **currents**, (2) labels for the **junctions** and the **circuit elements**, and (3) a dot for the **starting point** (if you plan to use Kirchhoff's loop rule).

SOLVE
There are three possible tactics for you to try depending on the situation.
Tactic 1. Find an **equivalent circuit** and draw its schematic diagram.
Tactic 2. Apply **Kirchhoff's junction rule**.
Tactic 3. Apply **Kirchhoff's loop rule**.

PROBLEMS AND QUESTIONS

A = algebraic   C = conceptual   E = estimation   G = graphical   N = numerical

29-1 Measuring Potential Differences Between Two Points

1. **C** Study the symbols in Table 29.2. Then, without looking at the table, draw the symbols for these circuit elements:
   a. a wire,
   b. a switch,
   c. a resistor,
   d. an emf device, and
   e. a lightbulb.
2. **C** Study the terms in Table 29.1. Then, without looking at the table, provide formal synonyms for these terms:
   a. voltage,
   b. amperage,
   c. DC power supply, and
   D. input voltage.
3. **C** *True or false*: The terminal potential of a DC power supply is *not* required to be constant. Explain your answer.

Problems 4, 5, and 6 are grouped.

4. **C** Suppose you need to measure the potential difference between the points in Figure P29.4. Assume the voltmeter reading is the potential difference between the two leads: $\Delta V = V_{red} - V_{black}$. For each of the following measurements, determine at which point you would connect the red lead and at which point you would connect the black lead:
   a. $V_b - V_a$,
   b. $V_c - V_b$,
   c. $V_d - V_c$,
   d. $V_a - V_d$.

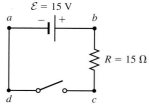

FIGURE P29.4 Problems 4, 5, and 6.

5. **N** Figure P29.4 shows a circuit with an open switch, an emf device, and a resistor. If we assume the switch remains open, use the values given in the figure to find the potential difference between the points
   a. $V_b - V_a$,
   b. $V_c - V_b$,
   c. $V_d - V_c$, and
   d. $V_a - V_d$.
6. **N** If the switch in Figure P29.4 is closed, use the values given in the figure to find the potential difference between the points
   a. $V_b - V_a$,
   b. $V_c - V_b$,
   c. $V_d - V_c$, and
   d. $V_a - V_d$.
   If you worked Problem 5, compare your answers in each case.
7. **C** A real battery (modeled as an ideal emf device in series with an internal resistor) is connected to a lightbulb. As the battery ages, its internal resistance increases while the internal ideal emf device remains unchanged. As the battery ages, what happens to
   a. the terminal potential of the real battery,
   b. the current in the circuit, and
   c. the light emitted?

Problems 8 and 9 are paired.

8. **A** Two circuits made up of identical ideal emf devices and resistors are shown in Figure P29.8. What is the potential difference $V_b - V_a$
   a. for circuit 1 and
   b. for circuit 2?
   c. Find expressions for both the current in the resistor in circuit 1 and the current in the resistor in circuit 2, and compare them.

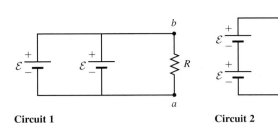

FIGURE P29.8 Problems 8 and 9.

9. **N** Two circuits made up of identical ideal emf devices ($\mathcal{E} = 1.67$ V) and resistors ($R = 35.9\ \Omega$) are shown in Figure P29.8. What is the potential difference $V_b - V_a$
   a. for circuit 1 and
   b. for circuit 2?
   What is the current in the resistor
   c. in circuit 1 and
   d. in circuit 2?

10. A lightbulb is connected to an ideal emf device that has an emf of 1.5 V.
    a. **N** What is the magnitude of the potential difference across the lightbulb?
    b. **C** If the bulb burns out, what is the potential difference across the lightbulb? What is the current in the circuit?
    c. **C** If the bulb is unscrewed from its socket, what is the potential difference across the empty socket?

11. **N** The terminal voltage of a real battery that delivers 15.0 W of power to a load resistor is 13.4 V, and its emf is 16.0 V. What is the resistance of
    a. the load resistor in this circuit and
    b. the internal resistor of the battery?

12. **C** Two circuit elements are connected directly end to end. Can the potential difference across the first circuit element ever be different from the potential difference across the second circuit element? Explain your answer.

13. **N** Eight real batteries, each with an emf of 5.00 V and an internal resistance of 0.200 Ω, are connected end to end in a loop as in Figure P29.13. What is the terminal voltage across one of the batteries between points a and b?

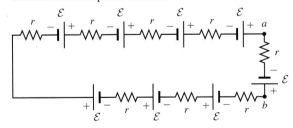

FIGURE P29.13

14. **C** Two circuit elements are connected directly end to end. Can the current through the first circuit element ever be different from the current through the second circuit element? Explain your answer.

29-2 Kirchhoff's Loop Rule
Problems 15 and 16 are paired.

15. **N** In Figure P29.15, three resistors are connected to an ideal emf device, where $\mathcal{E} = 14.8$ V, $R_1 = 13.4\ \Omega$, $R_2 = 20.5\ \Omega$, and $R_3 = 9.80\ \Omega$.
    a. What is the current through each resistor?
    b. What is the voltage across each resistor?
    c. How much power is consumed by each resistor and by the emf device?

d. If R_3 is replaced by a new resistor that has twice the resistance, answer parts (a) through (c) for this new circuit. Check your new answers against your old answers.

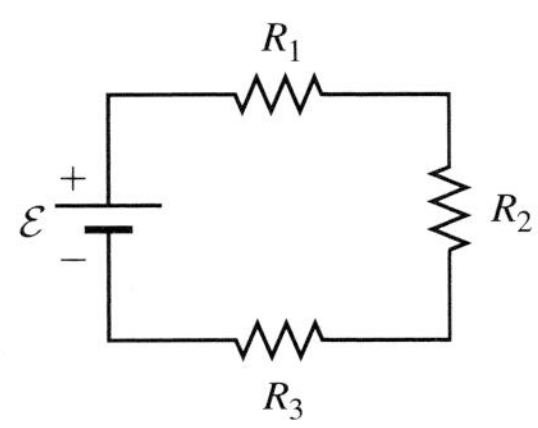

16. **N** In Figure P29.15, three resistors are connected to an ideal emf device. The resistances are $R_1 = 13.4\ \Omega$, $R_2 = 20.5\ \Omega$, and $R_3 = 9.8\ \Omega$. The current through the last resistor is 7.55 mA.
    a. What is the current through the other two resistors?
    b. What is the terminal potential of the emf device?

FIGURE P29.15 Problems 15 and 16.

17. **C** Some instructors use the following analogy to describe how the electric potential changes throughout a circuit: The flow of charge from higher to lower electric potential is like the flow of water down an incline from a greater height to a lower level. Consider an ideal emf device connected to a single resistor.
    a. Using the analogy, what happens to the water level as it crosses the emf device from the lower electric potential to the higher electric potential?
    b. What happens to the water level as it moves along the wire that connects the emf device and the resistor?
    c. Describe what happens to the water level as the resistor is crossed in the direction of the current flow.
    d. What is the "height" of the water level after it crosses the resistor compared to that at the end of the emf device with the lower electric potential?

18. **E** Students at one university have purchased large "handheld" calculators capable of doing many calculations. The professors at the same university use an older-style calculator with many fewer functions. The students' calculators need new batteries about twice per semester. The professors' calculators need new batteries about once every other year. Both types of calculators are powered by four AAA 1.5-V batteries connected in series. Assume students and professors use their calculators about the same amount of time each week. Model the calculators as simple circuits consisting of a resistor connected to the four batteries. Estimate these ratios:
    a. the resistance in a professor's calculator to the resistance in a student's calculator, and
    b. the current in a professor's calculator batteries to the current in a student's calculator batteries.
    Explain your assumptions.

Problems 19 and 20 are paired.

19. **N** An ideal emf device with $\mathcal{E} = 9.00$ V is connected to two resistors in series. One of the resistors has a resistance of 145 Ω, and the other has unknown resistance R. If the current through the emf device is 0.0155 A, what is the resistance R?

20. **A** An ideal emf device with emf $\mathcal{E}$ is connected to two resistors in series. One of the resistors has resistance R_1 and the other has unknown resistance R. If the current through the emf device is I, find an expression for the unknown resistance R in terms of the other quantities.

29-3 Resistors in Series

21. **CASE STUDY** Having fixed Carol's hot water heater problem, the students decide to work on a home space heater. They model a home space heater as a couple of resistors connected to a 120-V (DC) power supply. (Of course, the power supply is actually AC, but that does not affect the analysis in this problem.) The heater has two settings—one for 1300 W and the other for 1500 W.

a. **N** If the heater consumes 1500 W, what is its resistance?
b. **N** If the heater consumes 1300 W, what is its resistance?
c. **C** Are the resistors connected in series? Explain.

22. **C** Three series resistors (22.5 Ω, 45.0 Ω, and 90.0 Ω) are connected to an ideal emf source, so that a current I flows through the 22.5-Ω resistor first. The resistors are reordered so that the current flows through the 90.0-Ω resistor first, followed by the 22.5-Ω and the 45.0-Ω resistors.
    a. What is the effect of the new order on the current I?
    b. What is the effect on the voltage drop across each resistor? Explain your answers.

Problems 23 and 31 are paired.
23. **A** Six resistors with resistances $7R$, $6R$, $2R$, R, $R/2$, and $R/4$ are connected in series. What is the equivalent resistance of this combination?

Problems 24 and 25 are paired.
24. The emf devices and lightbulbs in Figure P29.24 are identical.
    a. **A** Find an expression for the current in each bulb.
    b. **C** List the bulbs in order from brightest to dimmest. Explain your answer.

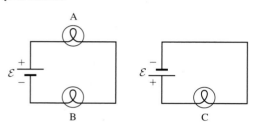

FIGURE P29.24 Problems 24 and 25.

25. **N** The emf devices and lightbulbs in Figure P29.24 are identical. The terminal potential of the emf device is 4.50 V. The current through bulb A is 9.00 mA.
    a. Find the current through bulb B.
    b. What is the current through bulb C?
26. **N** When three resistors with resistances in the ratio 1:2:3 are connected in series with a 12.0-V ideal battery, there is 6.00 A of current in the circuit. Determine
    a. the resistance of and
    b. the voltage drop across each of the three resistors.

29-4 Kirchhoff's Junction Rule

27. **N** Determine the currents through the resistors R_2, R_5, R_6, and R_7 in the set of junctions and branches shown in Figure P29.27. *Hint*: Use Kirchhoff's junction rule, be sure to consider the branches where a current is shown, and assume the branches that appear disconnected are connected to other parts of the circuit.

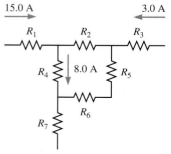

FIGURE P29.27

Problems 28, 29, 32, and 39 are grouped.
28. **C** The emf devices in the circuits shown in Figure P29.28 are identical.
    a. Redraw circuit 1 in Figure P29.28, including (and labeling) the current in each branch of the circuit. Apply the junction rule to write an equation in terms of the currents you have labeled.
    b. Repeat part (a) for circuit 2 in Figure P29.28.
    c. Simplify your equations if necessary, and compare your answers for each circuit. What can you say about the two circuits?

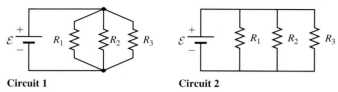

Circuit 1   Circuit 2

FIGURE P29.28 Problems 28, 29, 32, and 39.

29. **A** The emf devices in the circuits shown in Figure P29.28 are identical. The resistances are $R_1 = R$, $R_2 = 2R$, and $R_3 = 3R$.
    a. If the current through each emf device is I_0, find an expression for the current through each resistor in circuit 1.
    b. Repeat part (a) for circuit 2.
30. **C** Some people might describe Kirchhoff's junction rule as a statement that "the charge in the circuit is conserved." Describe what people are referring to when they make this statement. In what way is the charge conserved in the junction rule?

29-5 Resistors in Parallel

31. **A** Six resistors with resistances $7R$, $6R$, $2R$, R, $R/2$, and $R/4$ are connected in parallel. What is the equivalent resistance of this combination?
32. The emf in circuit 2 in Figure P29.28 is 22.5 V. The resistors have resistances $R_1 = 47.6$ Ω, $R_2 = 98.3$ Ω, and $R_3 = 50.0$ Ω.
    a. **N** What is the current in the emf device?
    b. **C** If a fourth resistor is added in parallel with the first three resistors, will the current in the emf device increase, decrease, or stay the same? Explain.
    c. **C** If, instead, a fourth resistor is added in series with the emf device, will the current in the emf device increase, decrease, or stay the same? Explain.
33. **C** The French natural philosopher Abbé Nollet (1700–1770) was interested in studying the strength of an electric shock as it traveled a great distance. He assembled a large group of monks into a circle reported to be a mile in circumference. Each monk was in contact with the next monk by a short piece of wire held in their hands. Between one pair of monks was a Leyden jar that they could touch and discharge, causing a (hopefully mild) shock. Imagine that, instead, Nollet connected the two "end monks" to a Volta pile (a battery) as shown in Figure P29.33A. Alternatively, Nollet could have made a mile-long chain of the same monks using an identical Volta pile as shown in Figure P29.33B. In which circuit is the current through the Volta pile greater? In which circuit is the current through a particular monk greater? Explain your answers.

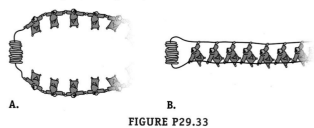

A.   B.

FIGURE P29.33

34. You are building a device that requires a 37.5-Ω resistor. You have only a box of a dozen 50.0-Ω resistors.
   a. C Draw the network of resistors that best meets your design requirement.
   b. N What is the resistance of your network?
35. A Figure P29.35 shows a combination of six resistors with identical resistance R. What is the equivalent resistance between points a and b?

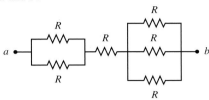

FIGURE P29.35

Problems 36 and 37 are paired.
36. A Each resistor shown in Figure P29.36 has resistance R. An ideal emf device ($\mathcal{E}$) is connected to points a and b via two leads (not shown in the figure). Find an expression for the current through the emf device.
37. N Each resistor shown in Figure P29.36 has a resistance of 100.0 Ω. An ideal emf device (120.0 V) is connected to points a and b via two leads (not shown in the figure). Find the current that flows through the emf device.

FIGURE P29.36 Problems 36 and 37.

38. C You have access to any number of 5.00-Ω and 75.0-Ω resistors. How can an effective resistance of 30.0 Ω be obtained by using these resistors?

29-6 Circuit Analysis

39. N The emf devices in the circuits shown in Figure P29.28 are identical and have a terminal potential of 7.50 V. The resistances are $R_1 = R$, $R_2 = 2R$, and $R_3 = 3R$, with $R = 15.0$ Ω.
   a. Find the current through the emf device and each resistor in circuit 1.
   b. Repeat part (a) for circuit 2.
40. N The emf in Figure P29.40 is 4.54 V. The resistances are $R_1 = 13.0$ Ω, $R_2 = 26.0$ Ω, and $R_3 = 39.0$ Ω. Find
   a. the current in each resistor,
   b. the power consumed by each resistor, and
   c. the power supplied by the emf device.

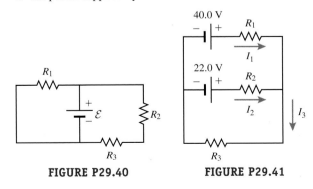

FIGURE P29.40 **FIGURE P29.41**

41. N Figure P29.41 shows three resistors ($R_1 = 14.0$ Ω, $R_2 = 8.00$ Ω, and $R_3 = 10.0$ Ω) and two batteries connected in a circuit.
   a. What is the current in each of the resistors?
   b. How much power is delivered to each of the resistors?

42. N Figure P29.42 shows five resistors and two batteries connected in a circuit. What are the currents I_1, I_2, and I_3?

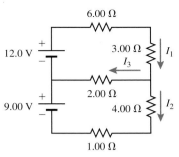

FIGURE P29.42

Problems 43 and 44 are paired.
43. N The emfs in Figure P29.43 are $\mathcal{E}_1 = 6.00$ V and $\mathcal{E}_2 = 12.0$ V. The resistances are $R_1 = 15.0$ Ω, $R_2 = 30.0$ Ω, $R_3 = 45.0$ Ω, and $R_4 = 60.0$ Ω. Find the current in each resistor when the switch is
   a. open and
   b. closed.

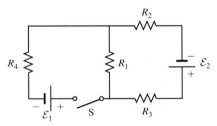

FIGURE P29.43 Problems 43 and 44.

44. N The emfs in Figure P29.43 are $\mathcal{E}_1 = 6.00$ V and $\mathcal{E}_2 = 12.0$ V. The resistances are $R_1 = 15.0$ Ω, $R_2 = 30.0$ Ω, $R_3 = 45.0$ Ω, and $R_4 = 60.0$ Ω. Find the power consumed by each resistor when the switch is
   a. open and
   b. closed.
45. N Figure P29.45 shows five resistors connected between terminals a and b.
   a. What is the equivalent resistance of this combination of resistors?
   b. What is the current through each resistor if a 24.0-V battery is connected across the terminals?

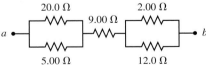

FIGURE P29.45

46. N Figure P29.46 shows a circuit with a 12.0-V battery connected to four resistors. How much power is delivered to each resistor?

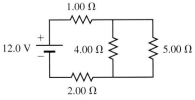

FIGURE P29.46

Problems 47 and 48 are paired.

47. **N** Two ideal emf devices are connected to a set of resistors as shown in Figure P29.47. If $\mathcal{E}_1 = 6.00$ V, $R_1 = 10.00$ Ω, $R_2 = 5.00$ Ω, $R_3 = 15.00$ Ω, $R_4 = 20.00$ Ω, and the current through R_4 is 0.250 A, what is the emf $\mathcal{E}_2$?

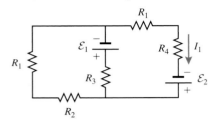

FIGURE P29.47 Problems 47 and 48.

48. **A** Two ideal emf devices are connected to a set of resistors as shown in Figure P29.47. Find an expression for the emf $\mathcal{E}_2$ in terms of $\mathcal{E}_1$, R_1, R_2, R_3, R_4, and the current through R_4, labeled I_1.

49. **N** Three resistors with resistances $R_1 = R/2$ and $R_2 = R_3 = R$ are connected as shown, and a potential difference of 225 V is applied across terminals a and b (Fig. P29.49).
    a. If the resistor R_1 dissipates 75.0 W of power, what is the value of R?
    b. What is the total power supplied to the circuit by the emf?
    c. What is the potential difference across each of the three resistors?

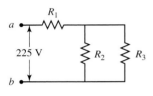

FIGURE P29.49

29-7 DC Multimeters

50. **E** An E-meter (Fig. P29.50) is an electric device patented by L. Ron Hubbard of the Church of Scientology (U.S. patent 3,290,589 on December 6, 1966). Originally the church used the device for medical purposes. This use was challenged in court, which ruled that the device is not medically or scientifically capable of improving the health or bodily functions of anyone. The church now states that the E-meter is strictly for the guidance of ministers of the church in confessionals and pastoral counseling. The E-meter, much like a lie detector, measures changes in the electrical resistance of human skin and is thus an ohmmeter. The resistivity of human skin is about 5×10^5 Ω·m. Assume the E-meter uses a 6.0-V DC emf device. Estimate the current passing through the person shown in the figure holding the metal cylinders. What do you imagine the person might feel during the test?

FIGURE P29.50 An E-meter in use.

51. **N** Consider a simple circuit consisting of a 15.0-V power supply connected to a 30.0-Ω resistor.
    a. What would an ideal voltmeter measure for the voltage across the resistor?
    b. An ideal voltmeter is imagined to have infinite internal resistance so that no current passes through it. A real voltmeter has a high internal resistance of 50,000 Ω. What is the current through this real voltmeter? How does it compare to the current through the 30.0-Ω resistor?

52. **N** Figure P29.52 shows a circuit in which the ideal ammeter has a reading of 2.00×10^{-3} A and the ideal voltmeter has a reading of 9.00 V. What are
    a. the unknown resistance R,
    b. the emf $\mathcal{E}$ of the battery, and
    c. the potential difference across the 1.65-kΩ resistor?

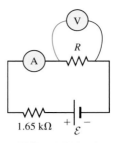

FIGURE P29.52

29-8 RC Circuits

53. **N** In Example 27.8 (page 851), we modeled the human nerve cell as a capacitor. Each nerve cell is a collection of nodes, and signals must be transmitted across each node. Each node may be modeled as a discharging RC circuit with resistance $R = 41 \times 10^6$ Ω and capacitance $C = 1.5 \times 10^{-12}$ F. Find the time constant for discharging a node. How does your answer compare to the measured firing times of about 50 μs?

54. **G** A charging RC circuit consists of a 12.0-V emf device, a 60.0-Ω resistor, and a 150-μF capacitor.
    a. Plot the charge stored by the capacitor as a function of time.
    b. On your graph, indicate the times when $t = \tau$, 2τ, 3τ, and 4τ. Record the charge stored by the capacitor at each of these times. Check your results algebraically.

55. **N** A 650.0-Ω resistor is connected across the terminals of a 12.0-nF capacitor that carries an initial charge of 7.40 μC.
    a. What is the magnitude of the maximum current in the resistor?
    b. What is the current in the resistor 5.00 μs after the circuit is completed and the capacitor begins to discharge through the resistor?
    c. How much charge remains in the capacitor 5.00 μs after the circuit is completed?

56. **N** At time $t = 0$, an RC circuit consists of a 12.0-V emf device, a 60.0-Ω resistor, and a 150.0-μF capacitor that is fully charged. The switch is thrown so that the capacitor begins to discharge.
    a. What is the time constant τ of this circuit?
    b. How much charge is stored by the capacitor at $t = 0.5\tau$, 2τ, and 4τ?

57. **N** A 210.0-Ω resistor and an initially uncharged 6.00-μF capacitor are connected in series to a 12.0-V emf source. A switch is closed to complete the circuit at $t = 0$.
    a. What is the time constant of this circuit?
    b. What is the maximum charge on the capacitor?
    c. What is the charge on the capacitor at $t = 3\tau$?

Problems 58 and 59 are paired.

58. **A** A real battery with internal resistance r and emf $\mathcal{E}$ is used to charge a capacitor with capacitance C. A resistor R is put in series with the battery and the capacitor when charging.
    a. Write an expression for the time constant of this circuit.
    b. Find an expression for the time when the capacitor has reached half its maximum charge.
    c. Find an expression for the current through the capacitor at this time.

59. **N** A real battery with internal resistance 0.500 Ω and emf 9.00 V is used to charge a 50.0-μF capacitor. A 10.0-Ω resistor is put in series with the battery and the capacitor when charging.
    a. What is the time constant for this circuit?
    b. What is the time when the capacitor has reached half its maximum charge?
    c. What is the current through the capacitor at this time?

60. **N** Figure P29.60 shows a simple RC circuit with a 2.50-μF capacitor, a 3.50-MΩ resistor, a 9.00-V emf, and a switch.

What are
a. the charge on the capacitor,
b. the current in the resistor,
c. the rate at which the capacitor is storing energy, and
d. the rate at which the battery is delivering energy exactly 7.50 s after the switch is closed?

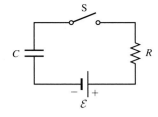

FIGURE P29.60

General Problems

61. **N** When connected in parallel, two resistors have an equivalent resistance of 125 Ω. The same two resistors have an equivalent resistance of 820 Ω when connected in series. What is the resistance of each of the resistors?

62. **A** Two lightbulbs have powers P_1 and P_2 when separately connected across an ideal battery with potential difference ΔV. The bulbs are then connected in series across the same battery. Determine the total power consumption of the bulbs in this configuration.

63. **N** A toy robot requires four D batteries rated at 1.50 V, and each delivers a charge of 1.20×10^3 milliampere-hour (mAh) or 4.32×10^3 C. If the internal resistance of the robot is 45.0 Ω, how long will one set of four batteries last?

64. **A** Ralph has three resistors, R_1, R_2, and R_3, connected in series. When connected to an ideal emf source $\mathcal{E}_1$, current I_1 flows through the resistors.
    a. If the resistors are instead connected to a second source with $\mathcal{E}_2 = 2\mathcal{E}_1$, what is the new current through the resistors in terms of the first current?
    b. Show that, if each resistance is doubled and the resistors are connected in series to the second emf source, the current through the resistors is equal to I_1.

65. **N** The reading on the ammeter in Figure P29.65 is 3.00 A, and the current runs from right to left through the ammeter. What is the value of current
    a. I_1 and
    b. I_2?
    c. What is the emf $\mathcal{E}$?

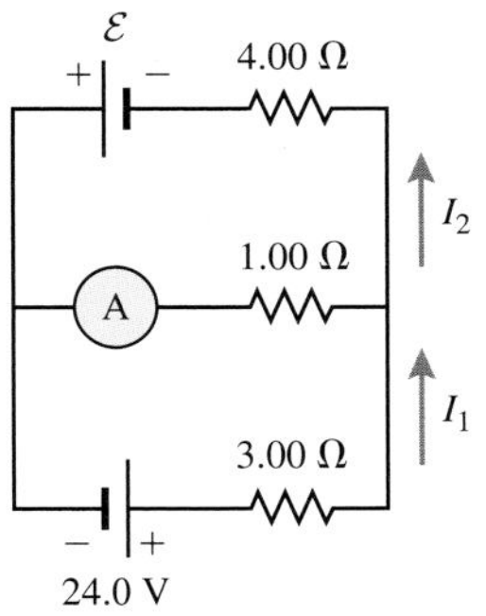

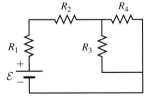

FIGURE P29.65

Problems 66, 67, and 68 are grouped.

66. **A** An ideal emf device is connected to a set of resistors as shown in Figure P29.66. Find an expression for the current through the resistor R_3 in terms of the emf and the resistances.

67. **N** An ideal emf device (24.0 V) is connected to a set of resistors as shown in Figure P29.66. If $R_1 = 22.5$ Ω, $R_2 = 52.5$ Ω, $R_3 = 125$ Ω, and $R_4 = 75.0$ Ω, find the current through resistor R_3.

FIGURE P29.66 Problems 66, 67, and 68.

68. **N** An ideal emf device (24.0 V) is connected to a set of resistors as shown in Figure P29.66. If $R_1 = 22.5$ Ω, $R_2 = 52.5$ Ω, $R_3 = 125$ Ω, and $R_4 = 75.0$ Ω, what is the voltage drop across each resistor?

69. **N** A real battery with emf 12.0 V and internal resistance 2.00 Ω is connected to a resistance of 4.00 Ω.
    a. What is the current through the battery?
    b. What is the terminal voltage of the battery?

70. **N** What is the equivalent resistance between points a and b of the six resistors shown in Figure P29.70?

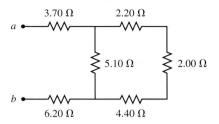

FIGURE P29.70

71. **N** Three batteries and four resistors are connected in a circuit as shown in Figure P29.71.
    a. What is the current in each of the resistors?
    b. What is the potential difference across the 175-Ω resistor?

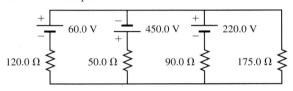

FIGURE P29.71

72. **A** A capacitor with initial charge Q_0 is connected across a resistor R at time $t = 0$. The separation between the plates of the capacitor changes as $d = d_0/(1 + t)$ for $0 \leq t < 1$ s. Find an expression for the voltage drop across the capacitor as a function of time.

73. **N** Batman is attempting to diffuse a bomb created by The Riddler and opens it to find a set of resistors connected as shown in Figure P29.73. He can tell from the coloring on several of the resistors that some are equivalent to each other, and he determines that $R_1 = 50.00$ Ω, $R_2 = 100.0$ Ω, $R_3 = 150.0$ Ω, and $R_4 = 300.0$ Ω. However, he needs to know the resistance of R_5 before he can begin diffusing the bomb. Using his Bat-Ohm-Meter, he is able to determine that the equivalent resistance between points a and b is 200.0 Ω. What is the resistance of the unknown resistor?

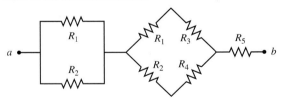

FIGURE P29.73

Problems 74 and 75 are paired.

74. **A** A capacitor with capacitance C is charged to a total charge Q. The capacitor is then connected in parallel to two resistors (R_1 and R_2) that are also connected in parallel.
    a. Write an expression for the time constant of this circuit.
    b. Find an expression for the time when the capacitor has lost half of its charge.
    c. Find an expression for the current through the capacitor at this time.

75. **N** A capacitor with capacitance 50.0 μF is charged to a total charge of 200.0 μC. The capacitor is then connected in parallel to two resistors (10.0 Ω and 30.0 Ω) that are also connected in parallel.
    a. What is the time constant for this circuit?
    b. What is the time when the capacitor has lost half of its charge?
    c. What is the current through the capacitor at that time?

76. N Figure P29.76 shows three identical resistors with $R = 125\ \Omega$ connected between terminals a and b. Each resistor can safely withstand a maximum power of 17.5 W.
  a. What is the upper limit on the safe voltage that can be applied across the terminals?
  b. How much power is delivered to each of the three resistors at the maximum voltage found in part (a)?
  c. How much total power is delivered to the combination of three resistors?

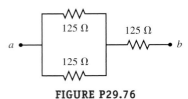

FIGURE P29.76

77. N Figure P29.77 shows a circuit with two batteries and three resistors.
  a. How much current flows through the 2.00-Ω resistor?
  b. What is the potential difference between points a and b in the circuit?

78. A In the RC circuit shown in Figure P29.78, an ideal battery with emf $\mathcal{E}$ and internal resistance r is connected to capacitor C. The switch S is initially open and the capacitor is uncharged. At $t = 0$, the switch is closed.
  a. Determine the charge q on the capacitor at time t.
  b. Find the current in the branch $b-e$ at time t. What is the current as t goes to infinity?

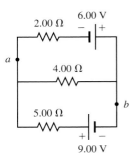

FIGURE P29.77

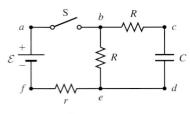

FIGURE P29.78

79. N A 12.0-V battery is used to charge a 4.00-μF capacitor in a simple RC circuit. After 5.00 s have elapsed, there is a potential difference of 6.60 V across the capacitor. What is the resistance of the resistor in this circuit?

80. A Calculate the equivalent resistance between points P and Q of the electrical network shown in Figure P29.80.

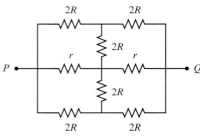

FIGURE P29.80

81. A In Figure P29.81, N real batteries, each with an emf $\mathcal{E}$ and internal resistance r, are connected in a closed ring. A resistor R can be connected across any two points of this ring, causing there to be n real batteries in one branch and $N - n$ resistors in the other branch. Find an expression for the current through the resistor R in this case.

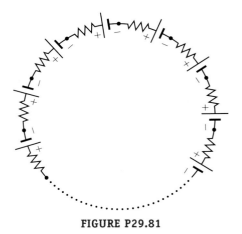

FIGURE P29.81

82. A By integrating, show that

$$\frac{1}{RC}\int_0^t dt = \int_0^q \frac{dq}{C\mathcal{E} - q}$$

leads to

$$-\frac{t}{RC} = \ln\left(\frac{C\mathcal{E} - q}{C\mathcal{E}}\right)$$

83. A Show that the time constant RC has the SI units of seconds.

Problems 84, 85, 86, and 87 are grouped.

84. A Figure P29.84 shows a circuit that consists of two identical emf devices. If $R_1 = R_2 = R$ and the switch is open, find an expression (in terms of R and $\mathcal{E}$) for the current I that is in the branch from point a to b.

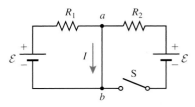

FIGURE P29.84 Problems 84–87.

85. A Figure P29.84 shows a circuit that consists of two identical emf devices. If $R_1 = R_2 = R$ and the switch is closed, find an expression (in terms of R and $\mathcal{E}$) for the current I that is in the branch from point a to b.

86. A Figure P29.84 shows a circuit that consists of two identical emf devices. If $R_1 = R$ and $R_2 = 2R$ and the switch is open, find an expression (in terms of R and $\mathcal{E}$) for the current I that is in the branch from point a to b.

87. A Figure P29.84 shows a circuit that consists of two identical emf devices. If $R_1 = R$ and $R_2 = 2R$ and the switch is closed, find an expression (in terms of R and $\mathcal{E}$) for the current I that is in the branch from point a to b.

Problems 88, 89, 90, 91, and 92 are grouped.

88. **A** Figure P29.88 shows a circuit that consists of two identical emf devices. If $R_1 = R_2 = R$ and the switch is open, find an expression (in terms of R and $\mathcal{E}$) for the current I_1 that is in resistor 1.

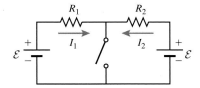

FIGURE P29.88 Problems 88–92.

89. **A** Figure P29.88 shows a circuit that consists of two identical emf devices. If $R_1 = R_2 = R$ and the switch is closed, find an expression (in terms of R and $\mathcal{E}$) for the current I_1 that is in resistor 1.

90. **A** Figure P29.88 shows a circuit that consists of two identical emf devices. If $R_1 = R$ and $R_2 = 2R$ and the switch is open, find an expression (in terms of R and $\mathcal{E}$) for the current I_1 that is in resistor 1.

91. **A** Figure P29.88 shows a circuit that consists of two identical emf devices. If $R_1 = R$ and $R_2 = 2R$ and the switch is closed, find an expression (in terms of R and $\mathcal{E}$) for the current I_1 that is in resistor 1.

92. **A** Figure P29.88 shows a circuit that consists of two identical emf devices. If $R_1 = R$ and $R_2 = 2R$ and the switch is open, find an expression (in terms of R and $\mathcal{E}$) for the current I_2 that is in resistor 2.

93. **A** Show that for a discharging RC circuit (Fig. 29.44) the charge $q(t)$ on the capacitor as a function of time is given by $q(t) = q_{max} e^{-t/\tau}$ (Eq. 29.16). *Hint*: What term in Equation 29.14 is zero for the discharging RC circuit?

94. **C** CASE STUDY Avi, Cameron, and Shannon turn their attention to the water heater problem. The symptom is that the hot water is merely warm and the supply runs out too quickly. From their research, the students know that the hot water heater has two heating coils that act as large resistors. Current in the coils increases their thermal energy, some of which flows into the water through conduction and convection. The power supply is AC, but the students have a digital multimeter that measures the rms voltage across any circuit element (rms = root mean square; Section 20-2). For the rest of the analysis, the circuit may be treated as a DC circuit with an emf equal to the rms voltage. The students use their multimeter to make two measurements: (1) The rms voltage across each coil is found to be 247 V. (2) They turn off power to the water heater and then measure the resistance across each coil. One coil's resistance is 13 Ω. When they try to measure the other resistance, the ohmmeter gives an error message. The students want to know if the coils are connected in parallel or in series and what the error message means.

Shannon: I think the coils are in parallel like the headlights in the car (Fig. 29.20).

Cameron: No. I think they are in series. If the coils are in series, the error message means the resistance of one has dropped to nearly zero, and that would explain the problem. The one working coil is just not enough to heat the whole tank of water.

Shannon: I agree that one coil must not be working, and it has to be the one that gives us the error message. I don't think the coils can be in series. It's like the headlights in a car. If one blows, both lights go out when they are in series.

Cameron: That's because when a lightbulb burns out, the filament breaks, so there is no current passing through it. I don't think the coil is broken like that. It's just lost its resistance.

Avi: There's an easy way to tell if they are in series or in parallel—from the voltage. If the coils have the same voltage, they're in parallel.

Cameron: If they're in parallel, and the resistance of one of them is essentially zero, then all the current would go through that one. The other coil would get no current. The water would never heat up because the good coil never gets hot.

Shannon: I think the error message might mean that the coil has *infinite* resistance. That makes more sense. If the resistance is very low, as it is for a wire, the meter would give you a zero or a really small number. I don't think it would give an error message.

Figure P29.94 shows the circuits described by Shannon and Cameron. Which circuit correctly fits the hot water heater's symptoms? Explain your answer.

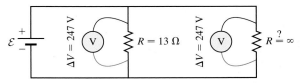

A. Shannon's circuit diagram

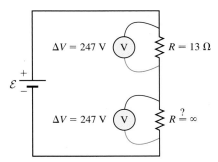

B. Cameron's circuit diagram

FIGURE P29.94

95. **A** Show that for a discharging RC circuit, the current $I(t)$ as a function of time is $I(t) = -I_0 e^{-t/\tau}$ (Eq. 29.17). Explain the leading negative sign.

PART FOUR
Magnetism

30 Magnetic Fields and Forces

Key Questions

What are the sources of magnetic fields?

What properties of matter make it subject to a magnetic force?

How do we calculate magnetic fields and magnetic forces?

30-1 Another fundamental force 935
30-2 Revealing magnetic fields 936
30-3 Ørsted's discovery 939
30-4 The Biot-Savart law 940
30-5 Using the Biot-Savart law 942
30-6 The magnetic dipole moment and modeling atoms 948
30-7 Ferromagnetic materials 951
30-8 Magnetic force on a charged particle 952
30-9 Motion of charged particles in a magnetic field 954
30-10 Case study: The Hall effect 957
30-11 Magnetic force on a current-carrying wire 960
30-12 Force between two long, straight, parallel wires 963
30-13 Current loop in a uniform magnetic field 964

❶ Underlying Principles
Magnetic fields and magnetic forces

✪ Major Concepts
1. Magnetic dipole and magnetic monopole
2. Biot-Savart law
3. Magnetic dipole moment
4. Ferromagnetic materials and magnetic domains
5. Magnetic force on a moving charged particle
6. Lorentz force
7. Magnetic force on a current-carrying wire

▶ Special Cases
1. Magnetic field due to
   a. Long, straight wire
   b. Circular wire (or current loop)
   c. Magnetic dipole
2. Magnetic force between two parallel current-carrying wires
3. Torque on a magnetic dipole
4. Magnetic potential energy stored in a dipole-magnetic field system

⊙ Tools
1. Magnetic field lines
2. Simple right-hand rule
3. Right-hand rule for direction of magnetic dipole moment

In the fall of 2007, Vikki Ortiz was on her way to a Milwaukee restaurant when her keys fell into a sewer grate. Vikki recounted her story in a blog read by Joselyn McKinley and her coworker, Dave Dulek. Joselyn and Dave's job is to rescue items from sewers. They used a large permanent magnet on a rope to retrieve the keys (Fig. 30.1).

The word *magnet* comes from the name Magnesia—a city in ancient Greece where naturally occurring magnets known as *lodestones* were found. Many ancient peoples believed lodestones were magical and even dangerous because they could attract bits of iron and attract or repel other lodestones. By the 12th century, the Chinese used lodestones as compasses. In the 1600s, magnets were studied by the scientific community. For 200 years, scientists tried to explain why magnets exert forces on bits of iron, push or pull on each other, and rotate to point toward the Earth's North Pole. We begin this part of the textbook by describing properties of magnets, magnetic fields, and magnetic forces.

30-1 Another Fundamental Force

If you play with permanent magnets, you will find that a magnet attracts certain metal objects, such as paper clips and Canadian coins. You will also learn that not all metal objects are attracted by a magnet. For example, a magnet will not pick up American coins. And, if you have two magnets, certain arrangements of the two magnets result in their mutual attraction while other arrangements result in their mutual repulsion. This repulsion can produce a stunning effect as shown in Figure 30.2; the upper magnets hover.

One of the amazing things we can see in Figure 30.2 is that, although none of the magnets are in contact with each other, they clearly exert a force on each other because the upper magnets are hovering. Magnet A experiences a downward force due to gravity, balanced by an upward **magnetic force** exerted by magnet B. Because the magnets are not in contact, we conclude that the magnetic force is a *field force*, like gravity. Physicists have identified four fundamental forces—the electromagnetic force, the gravitational force, the strong force, and the weak force (Fig. 23.2, page 684). The electromagnetic force is a combination of the electrostatic force, which we studied in Part III, and the magnetic force. All the fundamental forces are field forces.

The source of a gravitational field is an object that has mass, and the source of an electric field is an object that has an excess of either positively or negatively charged particles. The source of a **magnetic field** is an object that has a net motion of charged particles—or, in the case of permanent magnets, the primary source is the "motion" of the electrons bound to their atoms. Like the gravitational and electric fields, the magnetic field is a vector field, having both a magnitude and a direction at each point in space.

CASE STUDY The Hall Effect

In this case study, we tie up a major loose end that first appeared in our study of electricity (Chapters 23 and 24). Ben Franklin came up with his own model for electrostatics, in which all objects are full of an "electric fluid." When Franklin rubbed one object against another, such as silk against glass, he thought electric fluid was transferred from one object to the other. Franklin called the object with excess electric fluid *positive*, and the object with a deficit of fluid *negative*. However, Franklin could not perform an experiment to tell him which way the fluid flowed, so he made an arbitrary choice. He said some of the electric fluid is transferred from the silk to the glass after rubbing. The silk then has a deficit of electric fluid and is *negative*. The glass has an excess of electric fluid and is *positive*. Subsequent scientists followed Franklin's arbitrary choice, but today we know that when glass is rubbed against silk, electrons (not electric fluid) are transferred *from* the glass *to* the silk. Because of Franklin's arbitrary choice, we say that electrons are negative and protons are positive.

Choosing electrons to be negative leads to a complication, and a bit of an inconvenience, when we study current in a conductor, such as a wire. If we think of

FIGURE 30.1 Using a magnet to retrieve keys from a city sewer.

MAGNETIC FIELDS AND MAGNETIC FORCES ❶ **Underlying Principle**

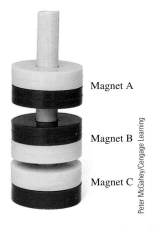

FIGURE 30.2 Each magnet consists of a blue side and a yellow side. The blue sides are mutually repulsive and the yellow sides are mutually repulsive. So magnet A is repelled by and hovers above magnet B, and magnet B is repelled by and hovers above magnet C. The blue side of one magnet attracts the yellow side of another magnet (not shown).

The words *force* and *field* are similar. Read carefully so you don't confuse them.

electrons as positive charge carriers, the current in a wire is in the same direction as the electron flow. However, because we think of electrons as negative charge carriers, we use the convention that current is the result of *imaginary* positively charged particles flowing in the opposite direction. By contrast, 18th- and 19th-century experimenters assumed that current is the result of the motion of *real* positive particles. Their assumption would not be challenged by any of the circuit analyses we performed earlier in this book.

So, how do we know that, in fact, the negative particles—the electrons—are the particles that flow in a conductor? The answer comes from observing a current-carrying conductor in a magnetic field. This experiment was performed by the American physicist Edwin Hall in 1879. His observation, now called the *Hall effect*, showed that the current in a conductor results from the motion of negative charge carriers, not positive ones. To understand the Hall effect, we must first learn about magnetic fields and then about the magnetic force exerted on moving charged particles.

A.

B.

FIGURE 30.3 A. Normally a Canadian coin is not a magnet. **B.** When the coin is near the toy magnet, the coin becomes magnetic and can pick up a paper clip.

30-2 Revealing Magnetic Fields

We know that Canadian coins and paper clips are not normally magnetic; Canadian coins cannot be used to pick up paper clips under normal circumstances (Fig. 30.3A). When a Canadian coin is near a magnet, however, it acts like a magnet and is able to pick up a paper clip (Fig. 30.3B). Just as when we hold an electric dipole near a conductor and the conductor becomes polarized like the dipole, we infer that when a Canadian coin is near a magnet, the magnet causes changes inside the coin that make it act like a magnet. We will learn more about how objects (like the Canadian coin) are magnetized in Section 30-7.

From our experience with magnets, we know that some objects (paper clips and Canadian coins) are attracted by magnets and others (American coins) are not. This difference is due to their composition. Because iron, for example, is easy to magnetize, it responds to the presence of a magnetic field. A *bar* magnet is an example of a permanent magnet made of a magnetic material, such as iron. The iron may be bent into other shapes, such as a horseshoe, the letter C, or a donut. The geometry of these magnets is more complicated, but the basic physics is the same—so we focus much of our attention on bar magnets.

Small, needle-shaped shavings of iron known as *iron filings* line up with a magnetic field to reveal the **magnetic field lines**. Figure 30.4 shows iron filings near a bar magnet that form long arcs running from one end of the magnet to the other. The iron filings also indicate relative magnetic field strengths. Regions with a high concentration of iron filings have a high concentration of magnetic field lines, and correspond to places where the magnetic field is particularly strong. For a bar magnet, the field is particularly strong near the ends of the bar, known as **poles**.

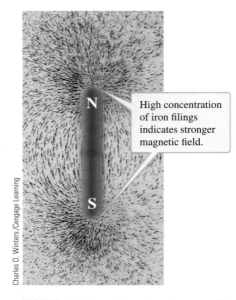

FIGURE 30.4 Iron filings trace the magnetic field near a bar magnet. (This magnet is painted red and green instead of blue and yellow. The colors are chosen for aesthetic reasons only.)

Iron filings are helpful in indicating the relative strength of the magnetic field, but to find its direction we need to use a compass. Imagine making a compass needle from a bar magnet. To make the magnet into a useful compass, it should be marked in terms of north and south. Therefore, you need to know which way is north. When the magnet is allowed to rotate, one end of the bar will always point north—that is, in the general direction of the Earth's North Pole. This end is labeled N for *north pole*, and the other end is labeled S for *south pole*. In making two compass needles, you would find the north pole of one compass repels the other's north pole. (Their south poles also repel each other.) You would also see that the north pole of one compass attracts the other magnet's south pole. In general, *like magnetic poles repel and opposite poles attract*.

Just as iron filings align with the magnetic field, compass needles align with magnetic field lines (Fig. 30.5A). By design, the north pole of a compass needle

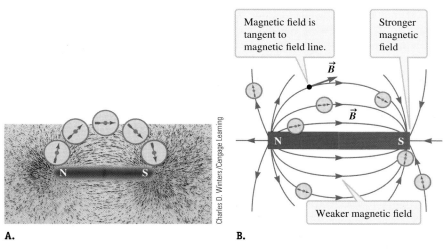

FIGURE 30.5 A. Compasses indicate the direction of the magnetic field. **B.** Magnetic field lines around a bar magnet.

points in the direction of the magnetic field. So, according to the compasses in Figure 30.5A, the magnetic field of a bar magnet points along the arc from its north pole to its south pole (Fig. 30.5B).

The symbol for the magnetic field is an uppercase $\vec{B}$.

Magnetic Monopoles and Dipoles

Compare Figure 30.5B to the *electric* field lines around an *electric* dipole in Figure 30.6A. The electric field around an electric dipole looks like the magnetic field around a bar magnet. The bar magnet is an example of a **magnetic dipole**. Although an electric dipole can be used as an analogy for a bar magnet, electric and magnetic dipoles are fundamentally different. The electric dipole's field is a result of stationary charged particles, whereas the bar magnet's field is a result of moving charged particles.

MAGNETIC DIPOLE ✪ **Major Concept**

This fundamental difference between an electric dipole and a bar magnet is revealed if you imagine breaking each one. First, imagine an electric dipole made up of two small spheres, one with charge $+q$ and the other with charge $-q$, held in place by a thin insulating rod (Fig. 30.6A). If you cut the rod in two and remove the negatively charged sphere, the electric field lines will point radially outward in all directions, as expected for a single positively charged sphere (Fig. 30.6B).

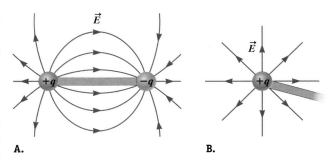

FIGURE 30.6 A. The electric field of an electric dipole is similar to the magnetic field of a magnetic dipole. **B.** The electric field of a broken dipole is simply the electric field of a single charged particle.

Now imagine cutting a bar magnet in two. You might expect that if you cut the bar into two equal pieces, the magnetic field lines of each piece would point radially outward (or inward). However, that is not what you find. Figure 30.7 shows iron filings around the two pieces of a broken bar magnet. Each piece has the same magnetic field pattern as the whole bar. In each piece, the iron filings form arcs that run from one end of the bar fragment to the other. A bar magnet is called a magnetic *di*pole because it has *two* poles. Even after breaking the bar magnet, we still observe two poles *in each fragment*, with magnetic field lines running from one pole to the other. Furthermore, both pieces are magnetic dipoles, no matter where you cut the bar magnet. Even if you slice off just a small portion of one end, the resulting pieces will both have magnetic field lines running from one end to the other. Both pieces are magnetic dipoles.

A magnetic field is fundamentally different from an electric field: The source of a magnetic field never produces field lines that simply point outward or inward. Instead, the magnetic field lines always loop around, as they do for a magnetic dipole. According to physics theories, a source with only one magnetic pole and field lines pointing either outward or inward is possible; such a source is known as a

FIGURE 30.7 The magnetic field of a broken bar magnet looks like the magnetic fields of two smaller bar magnets.

MAGNETIC MONOPOLE
✪ Major Concept

MAGNETIC FIELD LINES ⊙ Tool

Recall: Gravitational field lines originate from an object that has mass. Electric field lines originate from objects with excess positive charge and terminate on objects with excess negative charge.

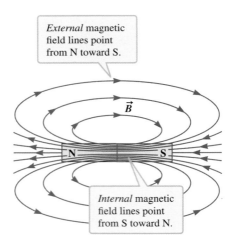

FIGURE 30.8 Magnetic field lines in and around a bar magnet. (The magnetic field lines exist in three dimensions, but only two dimensions are shown.)

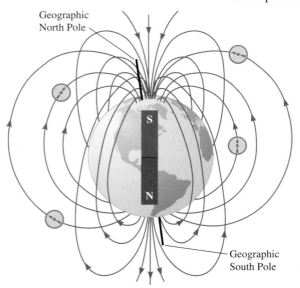

FIGURE 30.9 The Earth's magnetic field may be modeled as a bar magnet. The south magnetic pole is about 11° from the geographic North Pole. So a compass needle points toward the south magnetic pole, which is often good enough for navigation.

magnetic monopole. However, no one has ever found evidence of a magnetic monopole.[1,2] So, for the rest of our discussion, we will assume they don't exist, and electric dipoles will provide a good analogy for many magnetic sources.

Internal and External Magnetic Field Lines

Magnetic field lines indicate both the strength and direction of the magnetic field (Fig. 30.5B). The relative strength of a magnetic field is indicated by the concentration of field lines. The direction of the magnetic field is indicated by an arrow on the field line. The magnetic field at a particular point is tangent to the line and in the direction indicated by the arrow. Outside of a bar magnet, the magnetic field lines point *away from* the north pole and *toward* the south pole.

Because no magnetic monopole is known to exist, magnetic field lines, in contrast to gravitational and electric field lines, never originate from or terminate on a source. Instead, magnetic field lines always form closed loops.

In Figure 30.5A, it may look like the magnetic field lines (as revealed by the iron filings) start at the north pole and end at the south pole. However, these are just the *external* portions of the actual magnetic field lines. The magnetic field lines continue through the body of the bar magnet from the south pole to the north pole (Fig. 30.8). So, for a bar magnet, the *external* magnetic field lines point from the north pole toward the south pole, and the *internal* magnetic field lines point from the south pole toward the north pole. Notice that the magnetic field lines inside the bar magnet are densely packed, indicating that the internal magnetic field is very strong.

The Earth's Magnetic Field

The Earth's core is made up of molten iron and nickel. The swirling, convective motion of these molten metals gives rise to the Earth's magnetic field, which forms a magnetic dipole that may be modeled as a bar magnet (Fig. 30.9).

In Figure 30.5A, the north poles of the compass needles point toward the south pole of the bar magnet. Therefore, a compass's "north" points to the *south* magnetic pole of the Earth. In other words, the Earth's geographic North Pole is a *magnetic south* pole. If this seems confusing, remember how we imagined constructing a compass out of a bar magnet: We allowed the magnet to rotate so that it aligned with the Earth's magnetic field, and then we labeled as N the pole that pointed toward the Earth's geographic North Pole. Remember also that, in general, opposite magnetic poles attract.

There are a couple other complications. First, the Earth's geographic North Pole and magnetic south pole are not in exactly the same position. The geographic North Pole lies along the Earth's axis of rotation, as found by observing the North Star (Polaris). The North Star does not appear to move during the course of the night, so the Earth's rotation axis passes through the North Star. The magnetic south pole is found by using a compass and is about 11° from the geographic North Pole. Because most of us live in the midlatitudes, this slight discrepancy doesn't matter when we use a compass to navigate.

The other complication is that the Earth's magnetic poles are not fixed. The poles switch roughly every 300,000 years, so the north magnetic pole will one day be very close to the geographic North Pole, and the south magnetic pole will be very close to the geographic South Pole. The magnetic poles also wander several tens of kilometers daily.

---

[1] On Valentine's Day (February 14, 1982), Blas Cabrera reported finding a magnetic monopole (*Science* 216: 1082–1088, June 1982). Today this is known as the Valentine's Day Monopole. No one was able to confirm the discovery.

[2] In 2009, two independent research groups reported that they had created magnetic monopoles in an artificial substance called a spin ice. The monopoles they made were not like elementary particles; instead, they were an unbound pair of north and south poles in the spin ice.

CONCEPT EXERCISE 30.1

What would the magnetic field lines of an isolated north magnetic pole look like?

30-3 Ørsted's Discovery

Although magnets such as lodestones were studied and used for many centuries, the mechanism for generating a magnetic field was not known until 1820. While Hans Christian Ørsted (Oersted), a science professor at Copenhagen University, was giving an electricity and magnetism demonstration, he placed a compass near a current-carrying wire. Ørsted (and his audience) observed that the compass needle was deflected. Months later, Ørsted conducted the experiment more carefully and got similar results. His experiment connects the theories of electricity and magnetism by leading to the idea that *moving charged particles can be the source of a magnetic field.*

You may see a version of Ørsted's demonstration in your class (Fig. 30.10). Within a simple circuit, consisting of a wire connected to a switch and an emf device, a portion of the wire is arranged to be stiff and vertical. A small horizontal platform surrounds this portion of the wire, and several small compasses rest on the

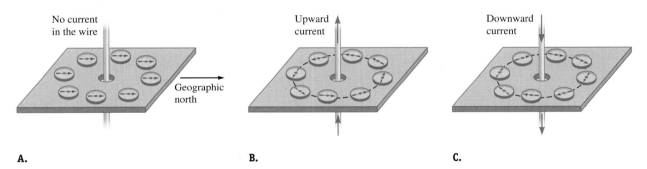

FIGURE 30.10 A. When there is no current in the wire, the compasses point in the general direction of the Earth's geographic North Pole (the Earth's magnetic south pole). **B.** When there is an upward current in the wire, the compass needles are deflected so that they point around a circle. **C.** When there is a downward current in the wire, the compass needles are deflected in the opposite direction from their deflection in part B.

platform. When the switch is open so that there is no current in the wire, the compass needles point in the general direction of the Earth's geographic North Pole, as usual (Fig. 30.10A). When the switch is closed, so that there is an upward current in the vertical portion of the wire, the compass needles are deflected, pointing counterclockwise around a circle, as seen from above the platform (Fig. 30.10B). If the emf device is reversed so that the current in the vertical wire is downward, the compass needles are deflected so that they point clockwise around the circle (Fig. 30.10C).

Consider again the upward current in Figure 30.10B. If we move the platform up or down, we find that the compass needles still point in a counterclockwise loop around the wire. If we move the compasses closer to or farther from the wire, we also find that they point in a counterclockwise loop. So, the magnetic field around a long, straight, current-carrying wire wraps around the wire in cylindrical sheets (Fig. 30.11).

There is a **simple right-hand rule (SRHR)** for finding the direction of the magnetic field produced by a current-carrying wire. Imagine grabbing the wire with your right hand and pointing your right thumb in the direction of the current. The current's direction is the direction in which real (or fictitious) positive particles move. Your fingers wrap around the wire in the direction of the magnetic field (Fig. 30.11). The SRHR works even if the wire is curved. Just imagine grabbing a small portion of wire, and your fingers will wrap in the direction of the magnetic field produced by that small portion.

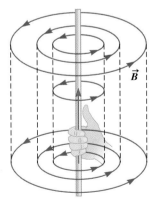

FIGURE 30.11 Using the SRHR: Place the thumb of your right hand in the direction of the current. Your fingers will wrap around the wire in the direction of the magnetic field $\vec{B}$. Here, $\vec{B}$ is counterclockwise as viewed from above.

SIMPLE RIGHT-HAND RULE (SRHR)
◉ Tool

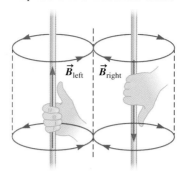

FIGURE 30.12 A wire is bent so that it forms a long, thin rectangle. The top and bottom of the rectangle are not shown in the figure. Between the left portion and the right portion of this long rectangular wire, the magnetic fields add and point into the page.

The magnetic field at any point near a curved current-carrying wire is the vector sum of the magnetic fields produced by each small segment. For example, imagine a wire bent into a long, thin rectangle. Figure 30.12 shows the left and right sides of the rectangle. The left side carries an upward current, and the right side carries a downward current. Use the SRHR twice—once for the left side and again for the right—to find the magnetic field midway between the two sides of the wire. According to the SRHR, the magnetic field due to each side of the wire points into the page, so the total magnetic field midway between the two portions points into the page (Fig. 30.12).

Compare the magnetic field produced by a bar magnet (Fig. 30.8) to the magnetic field produced by a long, straight, current-carrying wire (Fig. 30.11). Both sources produce closed magnetic field lines. The magnetic field lines associated with the bar magnet form squished loops that bunch together inside the bar, passing through the ends of the magnet at its poles. The magnetic field lines associated with the long, straight wire form concentric circles (along concentric cylinders). One major difference between a bar magnet and a long, straight wire is that the magnetic field lines of the bar magnet run into and out of the ends of the bar magnet. The places where the lines pass through the magnet are the poles. Because the loops around the wire do not pass through the wire, the wire does not have poles. Only when the field lines pass through a source, as in the case of the bar magnet, do we find it convenient to describe those places as poles.

The long, straight, current-carrying wire is an example of an **electromagnet**. The magnetic field produced by an electromagnet results from a gross motion of charged particles, such as the current in a wire. Electromagnets are often formed by coiling wire into a long cylinder, like a Slinky. By contrast, a permanent magnet is the source of a magnetic field even if the magnet carries no current. A permanent magnet's field results from the motion (and quantum-mechanical properties) of electrons within atoms, not from the gross motion of electrons throughout the magnet.

CONCEPT EXERCISE 30.2

In a wire made of copper or aluminum, the charged particles that move are really electrons. However, the conventional current is said to be due to fictitious positive particles that move in the opposite direction from the actual electrons. The simple right-hand rule is based on this convention. How does the SRHR have to be modified if we use the average motion of actual electrons instead of the usual convention for current?

30-4 The Biot-Savart Law

Biot-Savart rhymes with "Leo Guitar."

So far, our study of the magnetic field has been qualitative. Eventually, we want to calculate the magnetic *force* exerted on various *subjects*, but first we need to learn to calculate the magnetic *field* produced by various *sources*.

Soon after Ørsted's discovery, two French physics professors—Jean-Baptiste Biot and Felix Savart—experimented with current-carrying wires. By using a compass to measure the magnetic field around these wires, they discovered an empirical law known as the **Biot-Savart law** (Problem 94), which we can use to calculate the magnetic field.

The Magnetic Field Due to a Moving Charged Particle

Consider first the magnetic field that results from the motion of a single charged particle, where we are interested in finding the magnetic field at point P (Fig. 30.13). The vector $\vec{r}$ points from the charged particle to point P. According to the Biot-Savart law, the magnetic field $\vec{B}$ produced by a single particle with charge q moving at velocity $\vec{v}$ is given by

FIGURE 30.13 A charged particle moves at velocity $\vec{v}$, creating a magnetic field.

BIOT-SAVART LAW FOR A SINGLE CHARGED PARTICLE ⊙ **Major Concept**

$$\vec{B} = \left(\frac{\mu_0}{4\pi}\right) q \frac{\vec{v} \times \vec{r}}{r^3} \qquad (30.1)$$

The SI unit for the magnitude of the magnetic field (also called the field strength) is the **tesla** (T), where

$$1\,\text{T} \equiv 1\,\frac{\text{N}\cdot\text{s}}{\text{C}\cdot\text{m}} = 1\,\frac{\text{N}}{\text{A}\cdot\text{m}}$$

To help develop your intuition about magnetic field strengths, the field strengths for several sources are listed in Table 30.1. The constant μ_0 is called the **permeability of free space**, with the exact value

$$\mu_0 = 4\pi \times 10^{-7} \frac{\text{T}\cdot\text{m}}{\text{A}} \qquad (30.2)$$

Notice that when we substitute μ_0 into the Biot-Savart law (Eq. 30.1), the constant in parentheses is $10^{-7}\,\text{T}\cdot\text{m/A}$.

TABLE 30.1 Typical magnetic field strengths.

Source	B (T)
Interstellar clouds in the Milky Way galaxy	2×10^{-10}
Magnetic field produced by the human body	3×10^{-10}
Earth's magnetic field near its surface	5×10^{-5}
Sun's magnetic field near its surface	2×10^{-4}
Refrigerator magnet	5×10^{-3}
Magnet used in MRI[a]	2
World's strongest magnet[b]	45
Magnetic field near the surface of a neutron star	10^8 to 10^{10}

[a]MRI stands for *m*agnetic *r*esonance *i*maging, also known as NMR (*n*uclear *m*agnetic *r*esonance).
[b]As of 2009, this record is held by the hybrid magnet at the National High Magnetic Field Laboratory (Florida State University, Los Alamos National Laboratory, and University of Florida).

The direction of the magnetic field is determined by the sign of the charge and by the cross product in Equation 30.1. Point the fingers of your right hand in the direction of the first vector $\vec{v}$; then close your hand so that you "push" $\vec{v}$ into $\vec{r}$ (Section 12-6). For a positive charge, your thumb then points in the direction of $\vec{B}$. The magnetic field at point P is directed into the page (Fig. 30.13).

The magnitude of the cross product is given by Equation 12.22:

$$|\vec{v} \times \vec{r}| = vr\sin\varphi \qquad (30.3)$$

where φ is the angle between $\vec{v}$ and $\vec{r}$. We can find the magnitude of the magnetic field B at point P by substituting Equation 30.3 into Equation 30.1 and canceling a power of r from the numerator and denominator:

$$B = \left(\frac{\mu_0}{4\pi}\right) q\,\frac{v \sin\varphi}{r^2} \qquad (30.4)$$

Equation 30.4 shows that the strength of the magnetic field due to a single moving charged particle obeys an inverse-square law, just like the electric field due to a single stationary charged particle, $\vec{E}(r) = kQ_s\hat{r}/r^2$ (Eq. 24.3), or the gravitational field outside a massive spherical object, $\vec{g}(r) = -GM\hat{r}/r^2$ (Eq. 7.13).

The Magnetic Field Due to a Current

Now let's find the magnetic field due to a segment of current-carrying wire. In the arbitrarily shaped wire in Figure 30.14, the current is roughly from left to right. Imagine marking off segments of the wire small enough to be considered straight lines of length $d\ell$. Suppose we want to find the magnetic field at a particular

BIOT-SAVART LAW FOR A CURRENT

⭐ **Major Concept**

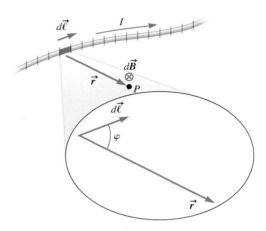

FIGURE 30.14 A current-carrying wire is broken into many small segments. The Biot-Savart law gives the magnetic field produced by each small segment at point P. According to the cross product $d\vec{\ell} \times \vec{r}$ the magnetic field $d\vec{B}$ produced by the blue segment at point P points into the page. Check this using the SRHR.

point, such as P. The Biot-Savart law for the magnetic field $d\vec{B}$ produced by one such small segment of wire is a slightly modified version of Equation 30.1:

$$d\vec{B} = \left(\frac{\mu_0}{4\pi}\right) I \frac{d\vec{\ell} \times \vec{r}}{r^3} \qquad (30.5)$$

Let's apply the Biot-Savart law to the segment shown in blue in Figure 30.14. The length of the segment is described by the vector $d\vec{\ell}$, whose magnitude $d\ell$ is the linear size of the segment, measured in meters. The direction of $d\vec{\ell}$ is given by the current, which is upward and to the right for this segment. The vector $\vec{r}$ extends from the segment to the point P, where we would like to know the magnetic field. The cross product in the Biot-Savart law (Eq. 30.5) has magnitude

$$|d\vec{\ell} \times \vec{r}| = (d\ell)(r) \sin \varphi \qquad (30.6)$$

where φ is the angle between $d\vec{\ell}$ and $\vec{r}$. The magnitude dB of the magnetic field produced by the blue segment at point P is found by substituting Equation 30.6 into Equation 30.5:

$$dB = \left(\frac{\mu_0}{4\pi}\right) I \frac{\sin \varphi}{r^2} d\ell \qquad (30.7)$$

Equation 30.7 shows again that the magnetic field strength due to a *small* segment of a current-carrying wire is inversely proportional to r^2.

The cross product in the Biot-Savart law gives the direction of $d\vec{B}$. According to the right-hand rule for cross products (Section 12-6), the direction of $(d\vec{\ell} \times \vec{r})$ in Figure 30.14 (and therefore of the resulting vector $d\vec{B}$) is into the page. This direction must be consistent with the SRHR from Section 30-3. To check, point your right thumb in the direction of the current (upward and to the right). Then your fingers wrap around the wire in the direction of the magnetic field. Your fingers go into the paper for points like P below the wire. So the SRHR is consistent with the right-hand rule for cross products. Because the cross product and the SRHR give the same results, you can use either one to find the magnetic field direction. Then use Equation 30.7 to find the magnitude dB. When using the Biot-Savart law to find the magnetic field for a current-carrying wire, we must add the contributions of all the wire segments. In the next section, we will learn some specific steps for doing this.

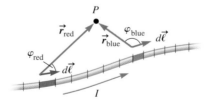

FIGURE 30.15 Which segment (red or blue) produces the stronger magnetic field at point P? *Hint:* $\varphi_{red} + \varphi_{blue} = 180°$

CONCEPT EXERCISE 30.3

In the next section, we'll learn how to calculate the total magnetic field at point P in Figure 30.15. For now, consider two segments—the blue one and the red one. Each segment has the same length $d\ell$. The red segment is farther from P, so $r_{red} > r_{blue}$; also $\varphi_{red} + \varphi_{blue} = 180°$. Use Equation 30.7 to state which segment produces the stronger magnetic field at point P.

30-5 Using the Biot-Savart Law

Using the Biot-Savart law to find the magnetic field involves a procedure similar to the one we used to find the electric field (Section 24-5). In both, we (1) imagine dividing the source into a number of small pieces, (2) find the field produced by a small piece, and (3) integrate over the entire source. In this section, we introduce a detailed problem-solving strategy and then use it to find the magnetic field produced by sources with different shapes.

PROBLEM-SOLVING STRATEGY

Using the Biot-Savart Law

:• INTERPRET and ANTICIPATE

We start with a diagram of the source—a current-carrying wire—and the point at which you want to find the magnetic field. There are four elements to this diagram:
1. A **coordinate system** and other **geometric details**.
2. **Labeled segments** of the source. Imagine slicing the current-carrying wire into a number of small segments. It usually helps to draw one or two of these segments, exploiting the symmetry of the problem when possible. Draw $d\vec{\ell}$, $\vec{r}$, and φ for each segment.
3. The **magnetic field vectors** produced by the segments at the point of interest. You may use either the SRHR or the cross product $(d\vec{\ell} \times \vec{r})$ from the Biot-Savart law to determine the direction of $d\vec{B}$ for each segment.
4. The **net magnetic field vector** at the point of interest. Find this graphically in order to determine whether any components cancel and to anticipate your result.

:• SOLVE

Step 1. Use the Biot-Savart law to find a **mathematical expression** for $d\vec{B}$, the small magnetic field produced by a single current segment at the point of interest. You may need only a component of $d\vec{B}$.

Step 2. Integrate $d\vec{B}$ to find $\vec{B}$ at the point of interest.

EXAMPLE 30.1 A Straight Wire

Figure 30.16 shows a straight wire of length L connected to a battery. The current in the wire is toward the right. Find an expression for the magnetic field at point P, directly above the midpoint of the wire.

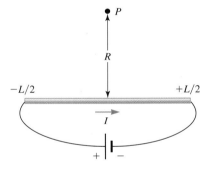

FIGURE 30.16

:• INTERPRET and ANTICIPATE

Start with a sketch. In this problem, we deal only with the straight wire, so we don't include the curved wire or the battery in our sketch (Fig. 30.17). We have included a **coordinate system** and some **geometric details**. Imagine slicing the wire into segments, exploit symmetry, and choose one segment from each side. We have **labeled segments** G and H. Draw $d\vec{\ell}$, $\vec{r}$, and φ for each segment. Because the wire lies along the x axis, we have labeled the length $d\vec{\ell}$ of each segment as $d\vec{x}$. Because $d\vec{x}$ is the same for both segments, we do not need a subscript. The **magnetic field vectors** produced by segments G and H both point out of the page in the positive z direction. So their **net magnetic field vector** $d\vec{B}$ also points out of the page, and we expect $\vec{B}$ to point in the positive z direction (Fig. 30.17).

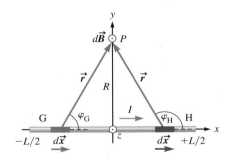

FIGURE 30.17

:• SOLVE

Step 1 Use the Biot-Savart law to find a **mathematical expression** for $d\vec{B}$. By symmetry, both segments G and H produce the same magnetic field $d\vec{B}_G = d\vec{B}_H$, so we can consider either one.

From Figure 30.17, $d\vec{B}$ is in the positive z direction, and the magnitude of the magnetic field is given by Equation 30.7. Substitute into Equation (1), replacing $d\ell$ with dx to match Figure 30.17.

$$d\vec{B} = (dB)\hat{k} \qquad (1)$$

$$d\vec{B} = \left(\frac{\mu_0}{4\pi} I \frac{\sin \varphi}{r^2} dx\right)\hat{k} \qquad (2)$$

Example continues on page 944 ▶

Step 2 Integrate Equation (2) over the entire length of the wire from $x = -L/2$ to $x = +L/2$.	$\int d\vec{B} = \int_{-L/2}^{L/2} \left(\frac{\mu_0}{4\pi} I \frac{\sin \varphi}{r^2} dx \right) \hat{k}$
In the many small segments that make up the wire, the vector $\vec{r}$ that runs from each segment to the point P varies in both magnitude and direction. So we cannot pull r or the angle φ outside the integral. We can pull out only the current and the constants.	$\vec{B} = \frac{\mu_0}{4\pi} I \left[\int_{-L/2}^{L/2} \left(\frac{\sin \varphi}{r^2} dx \right) \right] \hat{k}$ (3)

The variables x, r, and φ are interconnected. To find the connection, consider segment H and the triangle shown in Figure 30.18.

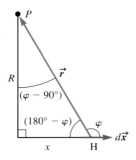

FIGURE 30.18

Use the Pythagorean theorem to express r in terms of x and the constant R. Next, use the trigonometric identity $\sin \alpha = \cos(\alpha - 90°)$. (See Appendix A.) Use this identity and Equation (4) to write $\sin \varphi$ in terms of x and R.	$r = (x^2 + R^2)^{1/2}$ (4) $\sin \varphi = \cos(\varphi - 90°) = \frac{R}{r} = \frac{R}{(x^2 + R^2)^{1/2}}$ (5)
Substitute Equations (4) and (5) into Equation (3).	$\vec{B} = \frac{\mu_0}{4\pi} I \left[\int_{-L/2}^{L/2} \frac{R}{(x^2 + R^2)^{1/2}} \frac{dx}{(x^2 + R^2)} \right] \hat{k} = \frac{\mu_0}{4\pi} I \left[\int_{-L/2}^{L/2} \frac{R \, dx}{(x^2 + R^2)^{3/2}} \right] \hat{k}$
The distance R between point P and the middle of the wire is constant, so we can pull it outside the integral.	$\vec{B} = \frac{\mu_0}{4\pi} IR \left[\int_{-L/2}^{L/2} \frac{dx}{(x^2 + R^2)^{3/2}} \right] \hat{k}$
The remaining integral can be found in Appendix A.	$\vec{B} = \frac{\mu_0}{4\pi} IR \left[\frac{x}{R^2 (x^2 + R^2)^{1/2}} \right]_{-L/2}^{L/2} \hat{k}$
Evaluate between the limits and simplify.	$\vec{B} = \frac{\mu_0}{4\pi} IR \left[\frac{L/2}{R^2 (L^2/4 + R^2)^{1/2}} - \frac{-L/2}{R^2 (L^2/4 + R^2)^{1/2}} \right] \hat{k}$ $\vec{B} = \frac{\mu_0 I}{2\pi R} \left[\frac{L}{(L^2 + 4R^2)^{1/2}} \right] \hat{k}$ (30.8)

∴ CHECK and THINK
As expected, the magnetic field at point P is in the positive z direction (out of the page).

A Very Long, Straight Wire

Equation 30.8 is an exact expression for the magnetic field at point P due to a straight current-carrying wire of length L (Fig. 30.16). We can model a long, straight wire as *infinitely* long when point P is very close to the wire. The limit of Equation 30.8 when L is very long compared to R is $\vec{B} \to (\mu_0 I / 2\pi R) \hat{k}$. The direction of the magnetic field at point P in Figure 30.16 is out of the page in the positive z direction. (See Figure 30.17 for the coordinates.) According to the SRHR, the magnetic field wraps in concentric cylinders around an infinitely long, straight wire (Fig. 30.11). In other words, because the magnetic field does not point in the z direction for all points around the wire, it is often best to use the SRHR to find the field direction, and to remember that the magnitude of the magnetic field around an infinitely long wire is

$$B = \frac{\mu_0 I}{2\pi r} \quad (30.9)$$

where, to be consistent with the equations for gravitational and electric fields, we use a lowercase "r" and r is the shortest distance from the wire to the point where the field is measured.

MAGNETIC FIELD DUE TO A LONG, STRAIGHT WIRE ▶ Special Case

Notice that the magnetic field due to a long, straight wire has an inverse r dependence, not an inverse r^2 dependence.

EXAMPLE 30.2 Special Case: A Current Loop

In addition to a long, straight, current-carrying wire, another important special case is the magnetic field produced by current in a circular loop of wire. Figure 30.19 shows a loop of wire connected to a battery. Model the loop as a circle, and show that the magnetic field at point P on the axis that runs through the center of the circle is given by

$$\vec{B} = \frac{\mu_0 I R^2}{2(R^2 + y^2)^{3/2}} \hat{j} \quad (30.10)$$

MAGNETIC FIELD DUE TO A CURRENT LOOP ▶ Special Case

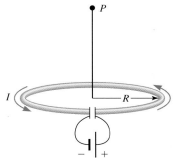

FIGURE 30.19

INTERPRET and ANTICIPATE

Sketch the current distribution (Fig. 30.20), choose a **coordinate system**, and add **geometric details**. In our coordinate system, the circular wire lies in the xz plane. Point P is on the y axis at a height y above the center of the circle. Imagine a cone with its vertex at point P and its base on the circle. The angle between the vertical and the wall of the cone is α. Now imagine dividing the loop into segments, with each segment so small that it is approximately straight. Two such **segments labeled** 1 and 2 are shown. The length element $d\vec{\ell}_1$ for segment 1 is in the positive z direction, and $d\vec{\ell}_2$ is in the negative z direction. Vectors $\vec{r}_1$ and $\vec{r}_2$ are in the xy plane, perpendicular to $d\vec{\ell}_1$ and $d\vec{\ell}_2$, respectively. Thus, $\varphi_1 = \varphi_2 = 90°$ (not shown on the figure). Sketch the **magnetic field vectors** produced by segments 1 and 2. Then add these to find the **net magnetic field vector**. It may be easiest to find the direction of $d\vec{B}$ using the cross product ($d\vec{\ell} \times \vec{r}$) in the Biot-Savart law. Consider segment 2, with $d\vec{\ell}_2$ in the negative z direction. Point your right fingers into the page; then close your hand so that you push $d\vec{\ell}_2$ into $\vec{r}_2$. Your right thumb should be parallel to the page, pointing upward and to the right (perpendicular to $\vec{r}_2$). This is the direction of $d\vec{B}_2$. The magnetic field $d\vec{B}_1$ due to segment 1 is also in the plane of the page, pointing upward and to the left (perpendicular to $\vec{r}_1$). Their magnitudes are equal: $dB_1 = dB_2$. They lie in the plane of the page, each having both an x and a y component. When these two vectors are added, their x components cancel and their resultant, $d\vec{B}_1 + d\vec{B}_2$, points in the positive y direction.

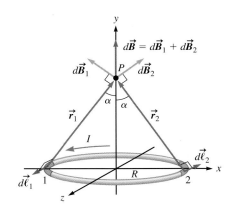

FIGURE 30.20

SOLVE
Step 1 Use the Biot-Savart law to find a **mathematical expression** for $d\vec{B}$. Any current segment will do, so we no longer need the subscripts. Remember that the horizontal components will cancel.

For all segments, $\varphi = 90°$ (the angle between $d\vec{\ell}$ and $\vec{r}$ in the Biot-Savart law; Eq. 30.5). The magnitude dB comes from Equation 30.7.	$dB = \left(\dfrac{\mu_0}{4\pi}\right) I \dfrac{\sin \varphi}{r^2} d\ell \quad (30.7)$ $dB = \left(\dfrac{\mu_0}{4\pi}\right) I \left(\dfrac{1}{r^2}\right) d\ell \quad (1)$

Example continues on page 946 ▶

Let's arbitrarily choose segment 2 (Fig. 30.21). We need to integrate only the y component of $d\vec{B}$. The horizontal components (x and z) cancel.	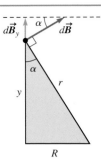 **FIGURE 30.21**
Apply trigonometry to the upper right triangle in Figure 30.21 to find the y component of $d\vec{B}$.	$dB_y = dB \sin \alpha$ (2)
Substitute Equation (1) into Equation (2).	$dB_y = \left(\dfrac{\mu_0}{4\pi}\right) I \dfrac{\sin \alpha}{r^2} d\ell$ (3)
Use the lower triangle highlighted in yellow in Figure 30.21 to write $\sin \alpha$ in terms of R and y.	$\sin \alpha = \dfrac{R}{r}$ $r = \sqrt{R^2 + y^2}$ (4) $\sin \alpha = \dfrac{R}{\sqrt{R^2 + y^2}}$ (5)
Substitute Equations (4) and (5) into Equation (3).	$dB_y = \left(\dfrac{\mu_0 I}{4\pi}\right)\left(\dfrac{1}{R^2 + y^2}\right)\left(\dfrac{R}{\sqrt{R^2 + y^2}}\right) d\ell$ $dB_y = \left(\dfrac{\mu_0 I}{4\pi}\right) \dfrac{R}{(R^2 + y^2)^{3/2}} d\ell$
Step 2 Integrate. To sum all the segments that make up the circular wire, we must integrate over the entire circumference. Thus, our limits of integration run from $\ell = 0$ to $\ell = 2\pi R$.	$\int dB_y = \int_0^{2\pi R} \left(\dfrac{\mu_0 I}{4\pi}\right) \dfrac{R}{(R^2 + y^2)^{3/2}} d\ell$
The parameters R and y are constant over the entire circle (Fig. 30.20), so we can pull them outside the integral along with the current and other constants.	$\int dB_y = \left(\dfrac{\mu_0 I}{4\pi}\right) \dfrac{R}{(R^2 + y^2)^{3/2}} \int_0^{2\pi R} d\ell$
Complete the integral, and evaluate between the limits.	$B_y = \left(\dfrac{\mu_0 I}{4\pi}\right) \dfrac{R}{(R^2 + y^2)^{3/2}} \ell \Big|_0^{2\pi R} = \left(\dfrac{\mu_0 I}{2}\right) \dfrac{R^2}{(R^2 + y^2)^{3/2}}$
Finally, write the magnetic field at point P in component form.	$\vec{B} = B_y \hat{j}$ $\vec{B} = \dfrac{\mu_0 I R^2}{2(R^2 + y^2)^{3/2}} \hat{j}$ ✓ (30.10)
:• CHECK and THINK The magnetic field at point P is in the positive y direction. A check of the SI units shows that they are teslas (T), as expected. In the next section, we will use Equation 30.10 to develop a model for bar magnets.	$[\vec{B}]_{\text{SI units}} = \dfrac{\text{T} \cdot \text{m}}{\text{A}} \dfrac{\text{A} \cdot \text{m}^2}{\text{m}^3}$ $[\vec{B}]_{\text{SI units}} = \dfrac{\text{T} \cdot \cancel{\text{m}} \cancel{\text{A}} \cdot \cancel{\text{m}^2}}{\cancel{\text{A}} \ \cancel{\text{m}^3}} = \text{T}$

EXAMPLE 30.3 A Bent Wire

Figure 30.22 shows a bent wire connected to an emf device. The orange segment of the wire consists of two straight portions and a portion that is a quarter of a circle of radius R. Point P is at the center of the complete circle. Find an expression for the magnetic field at point P. Assume the portions of the wire shown in black are far from point P and contribute a negligible magnetic field.

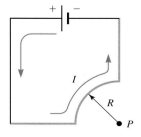

FIGURE 30.22

:• INTERPRET and ANTICIPATE

Sketch the current distribution (Fig. 30.23), choose a **coordinate system**, and add **geometric details**. The origin of the coordinate system is at point P (not labeled). The z axis points out of the page. Choose three **segments**: one for the horizontal portion, one for the curved portion, and one for the vertical portion—**labeled** 1, 2, and 3, respectively. The $\vec{r}$ vectors point from the length elements to the origin. For each portion of the wire, the angle φ between $d\vec{\ell}$ and $\vec{r}$ is constant. For the horizontal portion (segment 1), both $d\vec{\ell}_1$ and $\vec{r}_1$ are in the positive x direction, so $\varphi_1 = 0$. For the vertical portion (segment 3), $d\vec{\ell}_3$ points in the positive y direction and $\vec{r}_3$ points in the negative y direction, so $\varphi_3 = 180°$. The curved portion (segment 2) is part of a circle to which $d\vec{\ell}_2$ is tangent. Because $\vec{r}_2$ points toward the center of that circle, $\varphi_2 = 90°$. Use the cross product $(d\vec{\ell} \times \vec{r})$ in the Biot-Savart law to find the direction of $d\vec{B}$ for each segment. Because the cross product is zero for vectors that are parallel or antiparallel, segments 1 and 3 do not contribute to the magnetic field at point P. This leaves only segment 2. Place your right fingers in the direction of $d\vec{\ell}_2$ and close your hand in the direction of $\vec{r}_2$; your thumb points into the page. Thus, $d\vec{B}_2$ is in the negative z direction.

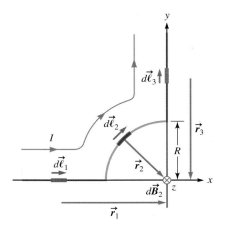

FIGURE 30.23

Only the curved portion of the wire contributes to the field at point P. To check our answer, we can use the results for a complete circular wire from Example 30.2.

:• SOLVE

Step 1 Use the Biot-Savart law (Eq. 30.7) to find a **mathematical expression** for dB. For every segment that makes up the curved portion, $\varphi = 90°$ and $r = R$.

$$dB = \left(\frac{\mu_0}{4\pi}\right) I \frac{\sin\varphi}{r^2} d\ell \qquad (30.7)$$

$$dB = \left(\frac{\mu_0}{4\pi}\right) I \frac{\sin 90°}{R^2} d\ell = \left(\frac{\mu_0}{4\pi}\right) \frac{I}{R^2} d\ell$$

Step 2 Integrate over the curved portion, whose length is a quarter of the circumference of a circle: $2\pi R/4$. The integration limits are from $\ell = 0$ to $\ell = \pi R/2$.

$$\int dB = \int_0^{\pi R/2} \left(\frac{\mu_0}{4\pi}\right) \frac{I}{R^2} d\ell$$

Both I and R are constant for the entire curved portion, so we can pull them, as well as the other constants, outside the integral. Complete the integration.

$$\int dB = \left(\frac{\mu_0}{4\pi}\right) \frac{I}{R^2} \int_0^{\pi R/2} d\ell$$

$$B = \left(\frac{\mu_0}{4\pi}\right) \frac{I}{R^2} \ell \Big|_0^{\pi R/2} = \frac{\mu_0 I}{8R}$$

Taking direction into account, write an expression for the magnetic field at point P.

$$\vec{B} = -\frac{\mu_0 I}{8R} \hat{k} \qquad (1)$$

:• CHECK and THINK

To check the magnitude of this answer, use the magnetic field of the complete circular wire found in Example 30.2. First, set $y = 0$ in Equation 30.10 to find an expression for the magnetic field at the center of the complete circle.

$$B(y) = \frac{\mu_0 I R^2}{2(R^2 + y^2)^{3/2}} \qquad (30.10)$$

$$B(0) = \frac{\mu_0 I R^2}{2(R^2)^{3/2}} = \frac{\mu_0 I R^2}{2R^3} = \frac{\mu_0 I}{2R} \qquad (2)$$

Example continues on page 948 ▶

Equation (2) is the magnetic field at the center of a complete circle of current, so we should divide by 4 to find the magnetic field due to a quarter of a circle. The result agrees with the magnitude of Equation (1).

$$B_{\text{quarter}} = \frac{1}{4}\left(\frac{\mu_0 I}{2R}\right) = \frac{\mu_0 I}{8R} \checkmark$$

30-6 The Magnetic Dipole Moment and Modeling Atoms

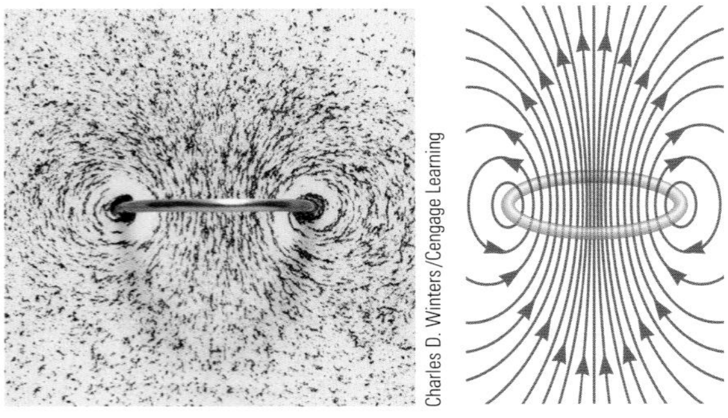

The magnetic field due to a current in a circular wire is such an important special case (Example 30.2) that it needs further exploration. We can find the magnetic field for any point on the axis perpendicular to the plane of the circle from

$$\vec{B} = \frac{\mu_0 I R^2}{2(R^2 + y^2)^{3/2}} \hat{j} \quad (30.10)$$

Finding the magnetic field at other points requires numerical integration. However, we can find the magnetic field lines experimentally by sprinkling iron filings around a current-carrying circle of wire (Fig. 30.24A). Compare these field lines in Figure 30.24B to the magnetic field lines for the bar magnet in Figure 30.8. The magnetic field lines look similar, and both the circular current loop and the bar magnet are magnetic dipoles.

In Section 24-4, we wrote an expression for the electric field at a distance x from an electric dipole compared to its size d:

$$\vec{E} \approx \frac{2k\vec{p}}{x^3} \quad (24.6)$$

where $\vec{p}$ is the electric dipole moment. By analogy, we can write an expression for the magnetic field at a distance y from a magnetic dipole compared to its size R. Start with Equation 30.10 and take the limit in the case that $y \gg R$:

$$\vec{B} \approx \frac{\mu_0 I R^2}{2y^3} \hat{j} \quad (30.11)$$

Also, by analogy, it is useful to define a **magnetic dipole moment** $\vec{\mu}$, often referred to as the **magnetic moment**. The magnetic dipole moment—like the electric dipole moment—is a vector. For a current loop, its magnitude is given by

$$\mu \equiv IA \quad (30.12)$$

where A is the area of the loop (of any shape). The SI units of the magnetic moment are $A \cdot m^2$. (The symbol for the permeability of free space μ_0 is similar to the symbol for the magnetic dipole moment μ. By convention, the subscript "0" denotes the permeability of free space.)

For a circular loop (Fig. 30.25), the area is $A = \pi R^2$, and the magnitude of the magnetic dipole moment is

$$\mu = I\pi R^2 \quad (30.13)$$

The direction of the magnetic moment is perpendicular to the plane of the loop, in the same direction as the magnetic field along the loop's axis. For the circular loop in Figure 30.25, the magnetic field and magnetic moment are in the positive y direction:

$$\vec{\mu} = (I\pi R^2)\hat{j} \quad (30.14)$$

There is a right-hand rule for finding the direction of the magnetic moment and therefore the direction of the magnetic field along the loop's axis. Wrap the fingers of your right hand in the direction of the current, and your thumb points in the direction of the magnetic moment (Fig. 30.25).

FIGURE 30.24 A. Iron filings around a circular wire show the magnetic field produced by current in the wire. **B.** The magnetic field lines associated with a circular current loop look similar to the magnetic field lines of a bar magnet.

MAGNETIC DIPOLE MOMENT
⊙ Major Concept

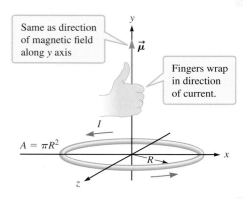

FIGURE 30.25 Find the direction of the magnetic dipole moment by using your right hand.

RIGHT-HAND RULE FOR DIRECTION OF $\vec{\mu}$ ⊙ Tool

We can write an expression for the magnetic field along the axis of a distant magnetic dipole by substituting $\vec{\mu}/\pi = IR^2 \hat{\jmath}$ (from Eq. 30.14) into Equation 30.11:

$$\vec{B} \approx \left(\frac{\mu_0}{2\pi}\right) \frac{\vec{\mu}}{y^3} \quad (30.15)$$

MAGNETIC FIELD DUE TO A MAGNETIC DIPOLE ▶ Special Case

Compare Equation 30.15 for the magnetic field of a magnetic dipole to $\vec{E} \approx 2k\vec{p}/x^3$ (Eq. 24.6), for the electric field of an electric dipole. Mathematically, these expressions are similar: Both show that the field far from a dipole decreases as the cube of the distance. Equation 30.15 holds for loops of wire of any shape, as long as y is large compared to the size of the loop.

Modeling Atoms

Ørsted's discovery that a current-carrying wire deflects a compass needle showed that moving charges are the source of a magnetic field. However, a bar magnet (Fig. 30.8) is also a source of a magnetic field, yet there is no current. How does such a permanent magnet produce a field?

In a permanent magnet, the magnetic field is caused by electrons that are bound to their particular atoms, so those electrons do not generate a current. A full explanation of how these electrons produce a magnetic field requires knowledge of both relativity and quantum mechanics at a level beyond the scope of this textbook. Here we will develop a model much like the microscopic models we used to describe friction (Section 6-2), solid matter (Section 14-4), fluids (Section 15-1), gases (Chapter 20), and—most recently—current in a conductor (Section 28-1). Although our model cannot be completely accurate, it will provide a reasonable explanation as well as a mental image.

In a solid, such as a permanent magnet, each atom is fixed in a lattice structure while its electrons move around the atom's nucleus. Imagine the motion of electrons around the nucleus as the motion of planets around the Sun. Each planet orbits the Sun and spins (rotates) on its own axis. By analogy, imagine each electron orbiting the nucleus of its atom while spinning on its own axis (Fig. 30.26).

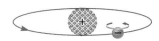

FIGURE 30.26 A mental image of an electron orbiting an atomic nucleus while spinning on its axis. (Not to scale.)

It is important to remember that this is just a mental image. In classical mechanics, an electron is considered a particle, and because particles have no spatial extent, they cannot spin (Section 2-1). In fact, the term *spin* as used in quantum mechanics does not really refer to the rotation of an electron. Instead, spin is a quantum property of an electron, just as color is a property of a shirt. With this warning in mind, we use the mental image of an electron spinning on its axis while orbiting its atomic nucleus to explain how permanent magnets work.

The Magnetic Moment of Electrons and Atoms

An electron in a circular orbit is much like the current in a circular wire (Fig. 30.25). So the orbiting electron is much like a dipole, and the magnetic field far from any particular atom is given by Equation 30.15:

$$\vec{B}_{\text{dipole}} \approx \left(\frac{\mu_0}{2\pi}\right) \frac{\vec{\mu}}{y^3}$$

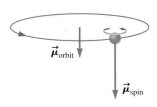

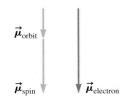

FIGURE 30.27 **A.** The magnetic moments resulting from the orbital motion and the spin of an electron. **B.** The net magnetic moment. In this special case, the orbital and spin magnetic moments are parallel.

The subscript reminds us that this is the field of a magnetic dipole. Because the electron carries a negative charge, when you use your right hand to find the direction of the orbiting electron's magnetic dipole moment, you must reverse the result. Wrap the fingers of your right hand in the direction of the electron's orbit in Figure 30.27A. Your thumb then points upward. But because the electron's charge is negative, you must flip your hand over, so now your thumb points downward—in the direction of the magnetic dipole moment. From Equation 30.15, we know that $\vec{B}_{\text{dipole}}$ (on the dipole's axis) is in the same direction as $\vec{\mu}$.

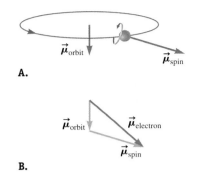

FIGURE 30.28 A. The electron's spin axis is often tilted with respect to its orbital axis. **B.** The net magnetic moment in this general case.

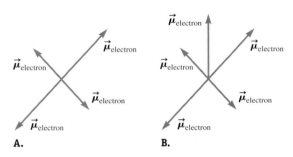

FIGURE 30.29 In these schematic representations of atoms, only the electrons' magnetic moments are shown. **A.** An atom with no net magnetic moment. **B.** An atom with a net upward magnetic moment.

Figure 30.27A shows two magnetic dipole moments for the electron. One of these ($\vec{\mu}_{\text{orbit}}$) is the result of the electron's orbital motion around the atomic nucleus. The other ($\vec{\mu}_{\text{spin}}$) is the result of the electron's spin around its axis. The net magnetic moment for an electron is the vector sum: $\vec{\mu}_{\text{electron}} = \vec{\mu}_{\text{orbit}} + \vec{\mu}_{\text{spin}}$. In this case, $\vec{\mu}_{\text{electron}}$ is in the same direction as $\vec{\mu}_{\text{orbit}}$ and $\vec{\mu}_{\text{spin}}$.

Figure 30.27 shows a special case in which the electron spins and orbits around parallel axes, so that $\vec{\mu}_{\text{spin}}$ is parallel to $\vec{\mu}_{\text{orbit}}$. Just as it is rare to find a planet whose orbital axis of rotation is parallel to its spin axis of rotation, this situation is also rare for electrons. Instead, the spin axis of the electron is often tilted with respect to the electron orbit, so that $\vec{\mu}_{\text{spin}}$ is tilted with respect to $\vec{\mu}_{\text{orbit}}$ (Fig. 30.28). The electron's net magnetic moment, the vector sum $\vec{\mu}_{\text{electron}} = \vec{\mu}_{\text{orbit}} + \vec{\mu}_{\text{spin}}$, points downward and to the right in this case. Its magnitude is slightly smaller than in the case where $\vec{\mu}_{\text{orbit}}$ and $\vec{\mu}_{\text{spin}}$ are parallel (Fig. 30.27).

Except for hydrogen, atoms are made up of many electrons, each orbiting and spinning in different directions. The net magnetic moment of the atom is the vector sum of all the electron magnetic moments: $\vec{\mu}_{\text{atom}} = \Sigma \vec{\mu}_{\text{electron}}$. In Figure 30.29A, an atom with four electrons has no net magnetic moment because the magnetic moments of the four electrons cancel: $\vec{\mu}_{\text{atom}} = \Sigma \vec{\mu}_{\text{electron}} = 0$. When an atom has an even number of electrons, their magnetic moments often cancel. In Figure 30.29B, an atom with an odd number of electrons has a net magnetic moment that is nonzero and upward.

An object is made up of many atoms; the net magnetic moment of the whole object is the vector sum of the atomic magnetic moments. In a nonmagnetic material, the vector sum of the atomic magnetic moments is zero. (The protons in an atom also have a dipole magnetic moment, but usually the net magnetic moment of the protons is small compared to the net magnetic moment of the electrons. So, the protons' magnetic moments can usually be ignored.)

EXAMPLE 30.4 The Orbital Magnetic Moment of an Electron

Assume an electron's orbit is a circle of radius $R \approx 10^{-10}$ m—about the size of an atom. Estimate the magnitude of the orbital magnetic dipole moment μ_{orbit} for an electron.

INTERPRET and ANTICIPATE
Figure 30.28A models our task. Before using $\mu_{\text{orbit}} = IA$ (Eq. 30.12), we need to find the area and current associated with the electron's orbit. We assume the electron moves in uniform circular motion, with the centripetal force provided by the electrostatic force (Coulomb's law).

SOLVE

The orbital period is the circumference divided by the electron's speed.	$T = \dfrac{2\pi R}{v}$	(1)
Find an expression for the current in terms of the electron's speed and orbital radius. Substitute Equation (1) into Equation (2).	$I = \dfrac{\Delta q}{\Delta t} = \dfrac{e}{T}$	(2)
	$I = \dfrac{ev}{2\pi R}$	(3)
Substitute Equation (3) into Equation 30.13 to get an expression for the magnitude of the orbital dipole moment.	$\mu_{\text{orbit}} = I\pi R^2$	(30.13)
	$\mu_{\text{orbit}} = \left(\dfrac{ev}{2\pi R}\right)\pi R^2 = \dfrac{evR}{2}$	(4)

To use Equation (4), we need to find the electron's speed v from its centripetal acceleration. The centripetal force is equal to the electrostatic force between the electron and the other charged particles in the atom.	$F_c = m_e a_c$ $F_E = \dfrac{kq_1 q_2}{R^2} = m_e a_c$

The centripetal force on the electron is the net force exerted by the other charged particles. In a neutral atom, the number of orbiting electrons equals the number of protons in the nucleus. The outermost electron feels the force from all the inner electrons as well as from the protons. (For instance, the outermost electron in iron feels the pull from 26 protons and the push from 25 electrons. Those electrons effectively shield the pull from 25 out of 26 protons. The net result is that the outermost electron feels the pull from approximately one proton.)

Estimate the magnitude of the centripetal force by using $q_1 = |-e|$ for the electron and $q_2 = e$ for the net charge of the protons and the other electrons.	$\dfrac{k(e)(e)}{R^2} = \dfrac{ke^2}{R^2} = m_e a_c$
Find the electron's speed from its centripetal acceleration, using $a_c = v^2/R$ (Eq. 4.36).	$a_c = \dfrac{ke^2}{m_e R^2} = \dfrac{v^2}{R}$ $v = \left(\dfrac{ke^2}{m_e R}\right)^{1/2}$ (30.16)
Substitute numbers and find the speed. Compare your answer to the speed of light (3×10^8 m/s). The electron is roughly 200 times slower than light (a reassuring answer).	$v = \left[\dfrac{(8.99 \times 10^9 \text{ N} \cdot \text{m}^2/\text{C}^2)(1.6 \times 10^{-19} \text{ C})^2}{(9.11 \times 10^{-31} \text{ kg})(10^{-10} \text{ m})}\right]^{1/2}$ $v = 1.6 \times 10^6$ m/s
Finally, substitute values into Equation (4). Remember that this is an order-of-magnitude estimate.	$\mu_{\text{orbit}} = \dfrac{(1.6 \times 10^{-19} \text{C})(1.6 \times 10^6 \text{ m/s})(10^{-10} \text{ m})}{2}$ $\mu_{\text{orbit}} = 1.3 \times 10^{-23} \text{ A} \cdot \text{m}^2 \sim 10^{-23} \text{ A} \cdot \text{m}^2$

CHECK and THINK

Our order-of-magnitude estimate has the correct SI units. How can such a small magnetic moment produce a magnetic field that we can detect on the macroscopic scale? Remember that a large number of electrons must have their orbital dipole magnetic moments aligned to produce a significant macroscopic magnetic field.

30-7 Ferromagnetic Materials

Permanent magnets—such as refrigerator magnets, bar magnets, and lodestones—are made from **ferromagnetic materials**, such as iron, iron alloys, cobalt, and nickel, and are known as **ferromagnets**. Each atom in a ferromagnetic material has a net nonzero magnetic moment (Fig. 30.29B). Of course, the entire object would have a net magnetic moment of zero if the atoms' magnetic moments canceled. But, in ferromagnetic materials, atoms interact strongly with their neighbors, such that the magnetic moments of nearby atoms line up. The interactions among atoms decrease with distance, so the magnetic moments are aligned over a small region known as a **magnetic domain**, or simply **domain**. A domain has between 10^{17} and 10^{21} atoms. In an unmagnetized ferromagnetic object, the magnetic moments $\vec{\mu}_{\text{domain}}$ of the domains are randomly oriented so that the net magnetic moment of the entire object is zero: $\vec{\mu}_{\text{net}} = \sum \vec{\mu}_{\text{domain}} = 0$, as shown in Figure 30.30.

Canadian coins are made from ferromagnetic materials. Normally a coin is unmagnetized, so Figure 30.30 is a good model for the coin's domains. What happens when the coin is in a magnetic field? Imagine each domain as a small bar magnet. In an unmagnetized coin, these atomic-sized bar magnets are randomly oriented (Fig. 30.31A). When an exterior magnet's north pole is held near the coin, the

FERROMAGNETIC MATERIALS AND MAGNETIC DOMAINS

○ Major Concepts

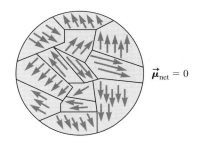

FIGURE 30.30 In an unmagnetized ferromagnetic object, the domains' magnetic moments are randomly oriented, so that the material has no net magnetic moment: $\vec{\mu}_{\text{net}} = 0$.

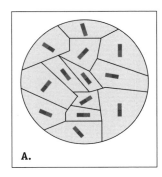

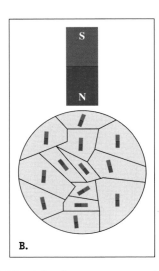

FIGURE 30.31 A. In an unmagnetized ferromagnetic material, each domain is modeled as a randomly oriented bar magnet. **B.** A Canadian coin is in the magnetic field of a bar magnet. In our model, the small bar magnets associated with the atoms inside the coin twist.

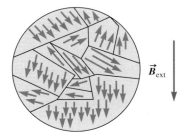

FIGURE 30.32 In a magnetized ferromagnetic material, the sum of the atomic magnetic moments is nonzero. In this case, the net dipole magnetic moment is roughly downward.

MAGNETIC FORCE ON A MOVING CHARGED PARTICLE
✪ **Major Concept**

atomic-sized bar magnets inside the coin twist so that their south poles point toward the exterior magnet's north pole (Fig. 30.31B). As a result, the coin is now essentially a magnet, with its south pole near the exterior magnet's north pole; thus, the two objects are mutually attracted.

Additionally, when a ferromagnetic material is in a magnetic field, domains that are aligned or nearly aligned with the external magnetic field grow, while the other domains shrink. Figure 30.32 shows the domains in a ferromagnet that is in a downward external magnetic field $\vec{B}_{ext}$. Compare Figure 30.32 to Figure 30.30, which has no external magnetic field. The domains that have a downward-pointing magnetic moment are larger in Figure 30.32 than in Figure 30.30, resulting in a net magnetic moment (roughly) downward in the same direction as the external magnetic field. The ferromagnet is now a "permanent" magnet and will respond as such.

A "permanent" magnet will not necessarily be magnetized forever. Permanent magnets maintain their magnetization for some period of time after an external magnetic field is removed. In the case of a Canadian coin, this effect does not last long because the domains are easily agitated and quickly become randomly oriented again, returning $\vec{\mu}_{net}$ to zero. There are three ways to demagnetize a ferromagnet:

1. Strike the object, for example by hitting the ferromagnet with a hammer or dropping it onto a hard surface.
2. Raise the object's temperature. Above a threshold temperature known as the Curie temperature, a ferromagnet becomes demagnetized.

Methods 1 and 2 both increase the ferromagnet's thermal energy, which also increases the random motion of atoms so that the domains become randomly oriented, thus demagnetizing the object.

3. Place the ferromagnet in a magnetic field $\vec{B}_{ext}$ (with a component) pointing in the direction opposite its $\vec{\mu}_{net}$. This method may create a "permanent" magnet with $\vec{\mu}_{net}$ reversed with respect to the original direction.

30-8 Magnetic Force on a Charged Particle

For two objects to interact through a particular force, they must have a particular property in common (mass for gravity, net charge for electric force). Because both objects have this property in common, in principle either one can be thought of as the source or as the subject. In practice, there is usually a good reason to choose one object over the other to be the source. For two objects to interact through the magnetic force, each must have a net motion of charged particles or a net magnetic moment. The magnetic force exerted between permanent magnets, each with a net magnetic moment, is beyond the scope of this textbook. Instead, we focus on the magnetic force exerted on subjects that have a net motion of charged particles. The source may be a permanent magnet or another object with a net motion of charged particles.

Start by considering a subject that is a single charged particle moving at velocity $\vec{v}$ (Fig. 30.33A). An unknown source creates a magnetic field vector $\vec{B}$ at the location of the subject. The magnetic force $\vec{F}_B$ on a particle with charge q is

$$\vec{F}_B = q(\vec{v} \times \vec{B}) \tag{30.17}$$

Because Equation 30.17 involves a cross product, we can find the magnitude of the magnetic force using Equation 12.22:

$$F_B = |q|vB \sin \varphi \tag{30.18}$$

where φ is the angle between $\vec{v}$ and $\vec{B}$ (Fig. 30.33A). From our experience with cross products (Section 12-6), we can rewrite the magnitude of the force F_B in terms of $B_\perp$ (the component of the magnetic field that is perpendicular to the subject's velocity; Fig. 30.33B). Substitute $B_\perp = B \sin \varphi$ into Equation 30.18:

$$F_B = |q|vB_\perp \tag{30.19}$$

Further, we can rewrite F_B in terms of $v_\perp$ (the component of the subject's velocity that is perpendicular to the magnetic field; Fig. 30.33C):

$$F_B = |q|v_\perp B \qquad (30.20)$$

From Equations 30.18 through 30.20, we see that for a particular particle moving through a particular magnetic field, the magnetic force F_B is at a maximum when the particle is moving perpendicular ($\varphi = 90°$) to the magnetic field.

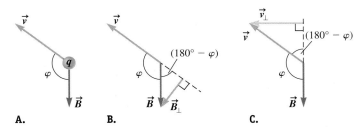

FIGURE 30.33 A. A charged particle is the subject of a magnetic field. We draw the subject's velocity vector $\vec{v}$ because it is important in determining the magnetic force. The cross product $(\vec{v} \times \vec{B})$ is out of the page. If q is positive, the magnetic force $q(\vec{v} \times \vec{B})$ is out of the page. If q is negative, the magnetic force $q(\vec{v} \times \vec{B})$ is into the page. **B.** The magnetic force is given by $F_B = |q|vB_\perp$. **C.** The magnetic force is also given by $F_B = |q|v_\perp B$.

There are four conditions that result in no magnetic force on the subject: (1) Perhaps it is obvious, but there is no magnetic force if there is no magnetic field; if $\vec{B} = 0$, then $\vec{F}_B = 0$. (2) There is no magnetic force exerted on a neutral particle; if $q = 0$, then $\vec{F}_B = 0$. (3) There is no magnetic force if the particle is at rest; if $\vec{v} = 0$, then $\vec{F}_B = 0$. (4) There is no magnetic force if the particle is moving parallel or antiparallel to the magnetic field; if $\varphi = 0$ or $\varphi = 180°$, then $\vec{F}_B = 0$.

The least intuitive aspect of the magnetic force is its direction. The magnetic *force* is *not* parallel or antiparallel to the magnetic *field*. The cross product in Equation 30.17, $\vec{F}_B = q(\vec{v} \times \vec{B})$, means the magnetic force $\vec{F}_B$ is perpendicular to both $\vec{B}$ and $\vec{v}$. The direction of the cross product $\vec{v} \times \vec{B}$ is given by the usual right-hand rule for cross products (Section 12-6), but the direction of $\vec{F}_B$ also depends on the sign of the particle's net charge. If the particle is positively charged, $\vec{F}_B$ is in the same direction as $\vec{v} \times \vec{B}$. In that case, when you use the right-hand rule to find the direction of the cross product, your thumb points in the direction of the magnetic force. If the particle is negatively charged, $\vec{F}_B$ is in the opposite direction from $\vec{v} \times \vec{B}$, and when you use the right-hand rule to find the direction of the cross product, you must flip your thumb over by 180° to find the direction of the magnetic force.

If you forgot how to find the direction of a cross product, return to Fig. 12.23 on page 346.

The Lorentz Force

We can summarize the main ideas of electricity (Part III) with a simple statement: *Charged particles interact through an electric force.* In addition, if the source and subject are both *moving* charged particles, they exert a magnetic force on each other. So we can make another simple statement: *Moving charged particles interact through both a magnetic force and an electric force.*

Of course, the motion of a particle depends on the observer's reference frame. If a charged source is moving in the observer's frame, there is both a magnetic field $\vec{B}$ and an electric field $\vec{E}$, and the subject (of charge q) may experience both an electric force $\vec{F}_E = q\vec{E}$ (Eq. 24.2) and a magnetic force $\vec{F}_B = q(\vec{v} \times \vec{B})$ (Eq. 30.17). We can combine these two equations to express the total electromagnetic force experienced by a particle with charge q moving with velocity $\vec{v}$:

$$\vec{F}_L = \vec{F}_E + \vec{F}_B = q(\vec{E} + \vec{v} \times \vec{B}) \qquad (30.21)$$

LORENTZ FORCE ✪ Major Concept

To use Equation 30.21, the field vectors $\vec{E}$ and $\vec{B}$, and the subject's velocity $\vec{v}$, must all be observed in the same reference frame. It is possible that in this frame, the velocity of the subject may be zero—in which case the Lorentz force is due solely to the electric force, even if a nonzero magnetic field is present.

Equation 30.21 is called the *Lorentz force* in honor of the Dutch physicist Hendrik Anton Lorentz (1853–1928).

CONCEPT EXERCISE 30.4

Cosmic rays are high-energy charged particles produced by astronomical objects. Many of the cosmic rays that make their way to the Earth are trapped by the Earth's magnetic field and never reach the surface. These trapped cosmic rays are found in the Van Allen belts—donut-shaped zones over the Earth's equator (Fig. 30.34). These cosmic rays are mostly protons with energies of about 30 MeV. The inset in the figure shows a cosmic ray proton as it is about to enter the Earth's magnetic field. The cosmic ray's velocity is initially perpendicular to the field. Three students discuss what happens to the incoming cosmic ray. Decide which student or students are correct.

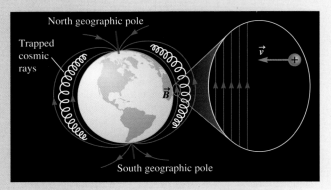

FIGURE 30.34 The Van Allen belts are donut-shaped zones of trapped cosmic rays above the Earth's surface. *Inset*: What happens to this cosmic ray as it enters the Earth's magnetic field?

Shannon: The velocity is perpendicular to the magnetic field, so the cosmic ray just passes through the field and hits the Earth's atmosphere.

Avi: What you are saying is that the magnetic field exerts no force on the cosmic ray. Actually, it exerts a huge force because the velocity is perpendicular to the magnetic field. The force will be into the page.

Cameron: Avi is right. The cosmic ray proton is going to feel a huge magnetic force. Because it is positively charged, it will be pushed upward along the magnetic field lines.

Shannon: I never said the force was zero. There is a force, but the force is perpendicular to the magnetic field lines. In this case, that's to the left—toward the Earth.

Avi: The force is perpendicular to the magnetic field, but it also has to be perpendicular to the velocity. Because $\vec{B}$ and $\vec{v}$ are both in the plane of the page, the force must be perpendicular to the page.

30-9 Motion of Charged Particles in a Magnetic Field

In this section, we explore the motion of charged particles in a magnetic field. When a particle is moving either parallel or antiparallel to the magnetic field (Fig. 30.35A), $v_\perp = 0$ and there is no magnetic force acting on the particle. In this case, the magnetic field does not accelerate the particle, and the particle continues to move at constant speed in a straight line (assuming no other forces are acting).

When a particle is moving perpendicular to a uniform magnetic field (Fig. 30.35B), the particle experiences a magnetic force that is always perpendicular to the particle's velocity (Eq. 30.17). In this case (assuming no other forces are acting), the particle moves with uniform speed in a circle of radius r because the particle's acceleration is perpendicular to its velocity (Section 4-6).

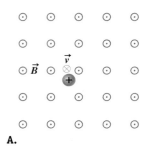

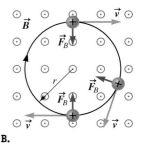

FIGURE 30.35 A. A particle moves at constant velocity antiparallel to a magnetic field. **B.** A particle moves in uniform circular motion in a magnetic field. The magnetic field does not change the particle's speed in either case.

The radius of the circle depends on the strength of the magnetic field. The required centripetal force F_c is supplied by the magnetic force F_B: $F_c = F_B$. The centripetal force is given by $F_c = mv^2/r$ (Eq. 6.7), and the magnetic force comes from $F_B = |q|vB \sin \varphi$ (Eq. 30.18) with $\varphi = 90°$:

$$m\frac{v^2}{r} = |q|vB \quad (30.22)$$

Solve Equation 30.22 for r:

$$r = \frac{mv}{|q|B} \quad (30.23)$$

What if a particle's initial velocity makes some arbitrary angle with respect to a uniform magnetic field? In that case, it is best to break up the velocity into two components. One component $\vec{v}_\perp$ is perpendicular to the magnetic field, and the other component $\vec{v}_\parallel$ is parallel to the magnetic field, so the total velocity is $\vec{v} = \vec{v}_\perp + \vec{v}_\parallel$. This expression is convenient because there is no magnetic force parallel to the magnetic field, so there is no acceleration parallel to the field, $\vec{a}_\parallel = 0$, and the parallel velocity $\vec{v}_\parallel$ is constant. However, the particle experiences a centripetal acceleration, which is constant in magnitude, and the particle's path is a helix (Fig. 30.36). The cross section of the helical path is a circle of radius r. The helix is a combination of linear motion and uniform circular motion. The radius of the helix is given by $r = mv_\perp/|q|B$ (Eq. 30.23), where v has been replaced by $v_\perp$.

The path of a particle in a nonuniform magnetic field can be more complicated. For example, the magnetic field shown in Figure 30.37 is weaker in the central region than it is near the top or bottom. The particle's path spirals outward (has a larger radius) as the particle moves toward the central region, and inward (with smaller radius) as it approaches the top or bottom. The particle spirals back and forth between the top and bottom, and it is effectively trapped. This magnetic field configuration is called a *magnetic bottle*. Magnetic bottles are used to contain very hot plasmas.

It is important to notice that whether the magnetic field is uniform or nonuniform, only the *direction* of the velocity can be changed by the magnetic field. The particle's *kinetic energy remains constant*. Put another way, the magnetic field can never do work on the particle.

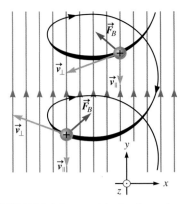

FIGURE 30.36 A positively charged particle moves in a helical path in a region with a uniform magnetic field. The particle's speed is constant.

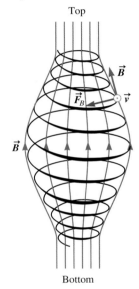

FIGURE 30.37 A positively charged particle spirals in a region with a nonuniform magnetic field. Use the right-hand rule to show that the magnetic force near the bottom points somewhat upward, and the magnetic force near the top points somewhat downward. The particle is trapped. When the particle reaches the top, it will spiral back down, and when it reaches the bottom, it will spiral back up.

CONCEPT EXERCISE 30.5

The Earth's Van Allen belts (Fig. 30.34) are a natural magnetic bottle, trapping cosmic rays that spiral back and forth between the North and South Poles. Use the shape of a cosmic ray's path to describe the variation in the magnetic field strength.

EXAMPLE 30.5 A Velocity Selector

Many laboratory experiments require a beam of charged particles, all moving with the same velocity. A *velocity selector* uses electric and magnetic fields to filter the particles by velocity. Figure 30.38 shows the basic parts of the device. The source on the left supplies a beam of charged particles with a range of velocities. The beam then passes through a region of space with a uniform electric field and a uniform magnetic field. The electric field for this velocity selector points downward, and the magnetic field points into the page. Show that only positively charged particles with speed

$$v = \frac{E}{B} \quad (1)$$

can pass straight through the velocity selector and emerge from the opening on the right.

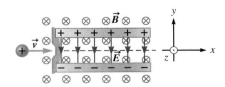

FIGURE 30.38 A velocity selector.

Example continues on page 956 ▶

:• **INTERPRET and ANTICIPATE**
Draw a free-body diagram (Fig. 30.39) for a particle that travels at constant velocity through the velocity selector. There is an electric field and a magnetic field, so the Lorentz force $\vec{F}_L = \vec{F}_E + \vec{F}_B = q(\vec{E} + \vec{v} \times \vec{B})$ (Eq. 30.21) applies. The electric force on a positively charged particle is downward in the negative y direction. The magnetic force is upward in the positive y direction. Because the particle travels at constant velocity, its acceleration is zero.

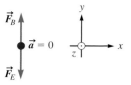

FIGURE 30.39

:• **SOLVE**

Because the particle's acceleration is zero, the (total) Lorentz force must be zero.	$\vec{F}_L = \vec{F}_E + \vec{F}_B = m\vec{a}$ $\vec{F}_E + \vec{F}_B = 0 = -F_E \hat{j} + F_B \hat{j}$ $F_B = F_E$ \hfill (2)
The velocity of the particle is perpendicular to the magnetic field, so $\varphi = 90°$ in $F_B = \|q\|vB \sin \varphi$ (Eq. 30.18). The magnitude of the electric force is given by Equation 24.2.	$F_B = \|q\|vB \sin \varphi = \|q\|vB$ $F_E = \|q\|E$ \hfill (24.2)
Substitute F_B and F_E into Equation (2), and solve for v.	$\|q\|vB = \|q\|E$ $v = \dfrac{E}{B}$ ✓

:• **CHECK and THINK**
This is exactly what we set out to derive. Our answer does not depend on the particle's charge. What if the particle had a negative charge? Then the magnetic force would be in the negative y direction and the electric force would be in the positive y direction. The two forces would cancel for a particle traveling at the speed given by Equation (1). In practice, the experimenter varies the ratio E/B in order to select a desired particle speed. In the next example, we show how a velocity selector may be used to determine a particle's mass.

EXAMPLE 30.6 A Mass Spectrometer

A *mass spectrometer* (Fig. 30.40) is a device used to measure the mass of charged particles, such as ions. Knowing the mass often allows the researcher to identify the type of ion. A mass spectrometer uses a velocity selector, shown on the left in the figure (region 1). The particles leave the velocity selector with speed $v = E/B_1$ and enter a region with another uniform magnetic field (region 2). In region 2, the magnetic field has magnitude B_2 directed into the page. There is no electric field in this region. The particles' velocities are perpendicular to the magnetic field, so the particles travel along a circular arc and strike a photographic plate (or other detector) at point P. The radius r of the semicircular trajectory is easily measured. The particle's mass is determined from the radius measurement.

Assume a beam of ions has been singly ionized so that the particles are missing just one electron. The velocity selector's electric field is $E = 3.54 \times 10^3$ N/C, and the magnetic field in region 1 is $B_1 = 0.035$ T. The magnetic field in region 2 is $B_2 = 0.075$ T, and the radius of the semicircular path is 39.1 cm. What is the particle mass (in atomic mass units)?

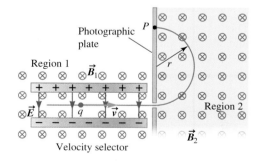

FIGURE 30.40 A mass spectrometer.

:• **INTERPRET and ANTICIPATE**
This problem combines the velocity selector from Example 30.5 with the circular motion of a charged particle in a uniform magnetic field. Once we know the particle's velocity when it enters region 2, we can find its mass. Because the required centripetal force is proportional to the particle's mass, we expect a more massive particle to move in a larger circular path.

SOLVE

First find the particle's velocity. There are two magnetic fields; here we need the magnetic field in the velocity selector (region 1).

$$v = \frac{E}{B_1} = \frac{3.54 \times 10^3 \text{ N/C}}{0.035 \text{ T}} = 1.01 \times 10^5 \text{ m/s}$$

Now solve Equation 30.23 for mass. Here we need the magnetic field in region 2.

$$r = \frac{mv}{|q|B_2} \quad (30.23)$$

$$m = \frac{|q|B_2 r}{v}$$

Substitute numerical values. The ion is missing an electron, so it has excess positive charge e.

$$m = \frac{|1.6 \times 10^{-19} \text{ C}|(0.075 \text{ T})(0.391 \text{ m})}{1.01 \times 10^5 \text{ m/s}} = 4.6 \times 10^{-26} \text{ kg}$$

Because this mass is so small, it is convenient to convert to atomic mass units (Section 19-6).

$$m = (4.6 \times 10^{-26} \text{ kg})\left(\frac{1 \text{ u}}{1.66 \times 10^{-27} \text{ kg}}\right) = 28 \text{ u}$$

CHECK and THINK

Let's see if we can identify this ion. It has all of its protons and neutrons, and it is missing only one electron, so it has most of its mass. According to the periodic table (Appendix B), silicon (Si) has an atomic mass of about 28 u. It seems likely that this ion is silicon, Si^+.

30-10 Case Study: The Hall Effect

Throughout Part III, we assumed a current in a metal conductor was due to the drifting of negatively charged particles (electrons) in the direction opposite to the current. If we put the conductor in a magnetic field, we can show that our assumption is correct. This experiment was performed by the American physicist Edwin Hall in 1879 and is called the **Hall effect**.

To show the Hall effect, a rectangular conductor is connected to a battery in a region with a uniform magnetic field (Fig. 30.41). The downward current through the conductor could be the result of either positively charged particles drifting downward or negatively charged particles drifting upward. An ammeter that measures the current cannot tell us the sign of the charges or the direction of the drifting particles. However, because the conductor is in a magnetic field, the magnetic force on the drifting particles will reveal which particles are really in motion. Try the next two concept exercises.

CONCEPT EXERCISE 30.6

CASE STUDY Hall Effect; Positive Charge Carriers

If the current in the rectangular conductor (Fig. 30.41) is the result of positively charged particles drifting downward in the negative y direction, what is the direction of the magnetic force on these particles?

CONCEPT EXERCISE 30.7

CASE STUDY Hall Effect; Negative Charge Carriers

If the current in the rectangular conductor (Fig. 30.41) is the result of negatively charged particles drifting upward in the positive y direction, what is the direction of the magnetic force on these particles?

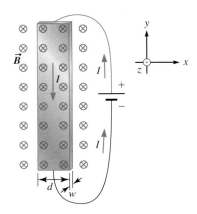

FIGURE 30.41 A rectangular conductor of width d and thickness w is attached to a battery. The current in the circuit is counterclockwise, so the current through the conductor is downward. The conductor is in a region with a uniform magnetic field pointing into the page.

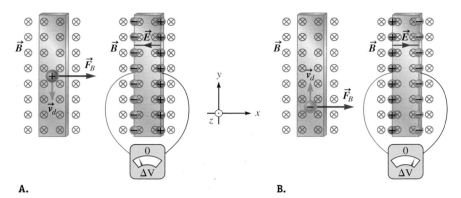

FIGURE 30.42 The Hall effect may be used to show that electrons flow in a conductor. **A.** If positively charged particles flow downward in the *negative y* direction, there is a buildup of positively charged particles on the right wall of the conductor. **B.** If negatively charged particles flow upward in the *positive y* direction, there is a buildup of negatively charged particles on the right wall of the conductor.

Figure 30.42 shows the two possible ways to have a downward current in the rectangular conductor. The current can be the result of positively charged particles drifting downward (Fig. 30.42A) or of negatively charged particles drifting upward (Fig. 30.42B). A positive particle drifting downward experiences a magnetic force to the right. After some time, excess positive charge builds up on the right wall of the conductor, leaving excess negative charge on the left wall. In contrast, a negative particle drifting upward also experiences a magnetic force to the right, and after some time there is a buildup of negative charge on the right wall and of positive charge on the left wall.

We can use a voltmeter to distinguish between these situations and thus to determine the sign of the flowing particles. The potential difference between the two walls of the conductor is called the **Hall voltage**, ΔV_{Hall}. If you attach the red lead to the left side of the conductor and the black lead to the right side, you find

$$\Delta V_{\text{Hall}} = V_{\text{red}} - V_{\text{black}} < 0$$

in the case of positive flowing charges (Fig. 30.42A). In the case of negative flowing charges (Fig. 30.42B), you find

$$\Delta V_{\text{Hall}} = V_{\text{red}} - V_{\text{black}} > 0$$

The Hall effect shows that negative charges (electrons) flow in a conductor because the Hall voltage is observed to be positive.

The Hall effect can also be used to find the number density of conduction electrons as well as their drift speed v_d. Once positive charge builds up on the left wall and negative charge on the right wall, an electric field in the conductor points from left to right in the positive x direction (Fig. 30.42B, right). An electron drifting upward is now subject to a magnetic force in the positive x direction and also to an electric force in the negative x direction. This situation is similar to that in the velocity selector (Example 30.5). When the electric and magnetic forces have the same magnitude, there is no net acceleration in the x direction and $F_B = F_E$. Using $F_B = |q|vB \sin \varphi$ (Eq. 30.18), with $\varphi = 90°$ because the drift velocity is perpendicular to the magnetic field, and $F_E = |q|E$ (Eq. 24.2), we find

$$|q|v_d B = |q|E \tag{30.24}$$

The charge cancels out:

$$v_d B = E \tag{30.25}$$

The magnitude of the electric field E comes from the Hall voltage (and Eq. 26.17):

$$\Delta V_{\text{Hall}} = Ed$$

$$E = \frac{\Delta V_{\text{Hall}}}{d} \tag{30.26}$$

where d is the width of the conductor (Fig. 30.41). Substitute Equation 30.26 into Equation 30.25:

$$v_d B = \frac{\Delta V_{\text{Hall}}}{d} \tag{30.27}$$

The drift speed may be written in terms of the current:

$$I = nev_d A \tag{28.6}$$

where n is the number density of the conduction electrons and A is the cross-sectional area of the conductor. For a rectangular conductor (Fig. 30.41), the cross-sectional area is $A = dw$, and Equation 28.6 becomes

$$I = nev_d dw \qquad (30.28)$$

Solve for v_d:

$$v_d = \frac{I}{nedw} \qquad (30.29)$$

Substitute Equation 30.29 into Equation 30.27:

$$\left(\frac{I}{nedw}\right)B = \frac{\Delta V_{\text{Hall}}}{d}$$

$$n = \frac{IB}{ew\Delta V_{\text{Hall}}} \qquad (30.30)$$

Equation 30.30 gives the number density n of conduction electrons. The drift speed comes from Equation 30.27:

$$v_d = \frac{\Delta V_{\text{Hall}}}{Bd} \qquad (30.31)$$

Because the drift speed is very low (between 10^{-5} m/s and 10^{-4} m/s), the Hall voltage is typically small and hard to measure accurately. Using Equation 30.31 directly can lead to inaccuracies in the drift speed measurement. To improve accuracy, experimenters slide the entire conductor in the direction of the current—opposite the direction of the electron's drift velocity. If the conductor in Figure 30.42B slides downward at exactly the drift speed, the conduction electrons have no net velocity with respect to the magnetic field. So the magnetic field exerts no force on the electrons, no charge builds up on the walls of the conductor, and the Hall voltage is zero. By sliding the conductor at increasing speeds until the Hall voltage reaches zero, the drift speed can be determined accurately.

EXAMPLE 30.7 CASE STUDY Hall Voltage

Suppose the rectangular conductor in Figure 30.41 is made of copper, with width $d = 2.54$ cm and thickness $w = 0.500$ cm. The number density of conduction electrons in copper is 8.42×10^{28} m^{-3} (Table 28.2, page 872). The magnetic field is that of the Earth, $B = 0.50 \times 10^{-4}$ T, in a direction perpendicular to the conductor. The current is measured to be 205 A.

A Find the expected Hall voltage. Answer in **CHECK and THINK**: If your voltmeter's error is $\pm 1\,\mu$V, can you measure the Hall voltage?

:• INTERPRET and ANTICIPATE
First find the drift speed in copper and then, using this information, find the Hall voltage. We expect a small voltage.

:• SOLVE
Find the drift speed from Equation 30.29. Substitute numerical values.

$$v_d = \frac{I}{nedw} \qquad (30.29)$$

$$v_d = \frac{205 \text{ A}}{(8.42 \times 10^{28} \text{ m}^{-3})(1.60 \times 10^{-19} \text{ C})(0.0254 \text{ m})(0.00500 \text{ m})}$$

$$v_d = 1.20 \times 10^{-4} \text{ m/s}$$

Example continues on page 960 ▶

Solve Equation 30.31 for the Hall voltage, and substitute numerical values.

$$v_d = \frac{\Delta V_{\text{Hall}}}{Bd} \quad (30.31)$$

$$\Delta V_{\text{Hall}} = Bdv_d = (0.50 \times 10^{-4}\,\text{T})(0.0254\,\text{m})(1.20 \times 10^{-4}\,\text{m/s})$$

$$\Delta V_{\text{Hall}} = 1.5 \times 10^{-10}\,\text{V} = 1.5 \times 10^{-4}\,\mu\text{V}$$

:• **CHECK and THINK**

As expected, the Hall voltage is very small. A voltmeter with an error of $\pm 1\,\mu\text{V}$ cannot distinguish this small potential difference from zero, and you cannot measure potential difference with it. In part B, we imagine using an electromagnet to increase the Hall effect.

B Instead of using the Earth's magnetic field in Figure 30.41, imagine we use a large electromagnet, producing a field with magnitude $B = 2.5$ T. What is the Hall voltage in this case?

:• **INTERPRET and ANTICIPATE**

The only change from part A is the stronger magnetic field, so we can start with $v_d = \Delta V_{\text{Hall}}/Bd$ (Eq. 30.31).

:• **SOLVE**

Solve for the Hall voltage, and substitute numerical values.

$$\Delta V_{\text{Hall}} = Bdv_d = (2.5\,\text{T})(0.0254\,\text{m})(1.20 \times 10^{-4}\,\text{m/s})$$

$$\Delta V_{\text{Hall}} = 7.6 \times 10^{-6}\,\text{V} = 7.6\,\mu\text{V}$$

:• **CHECK and THINK**

A voltmeter with an error of $\pm 1\,\mu\text{V}$ can distinguish this small potential difference from zero. It measures perhaps $7\,\mu\text{V}$ or $8\,\mu\text{V}$. The Hall voltage is still very small, requiring a very large current (205 A) for this measurement. The experiment could be further improved by using a larger sample of copper. The width of the sample in this problem is only 1 in. The Hall voltage ΔV_{Hall} increases linearly as the width d increases.

30-11 Magnetic Force on a Current-Carrying Wire

MAGNETIC FORCE ON A CURRENT-CARRYING CONDUCTOR

✪ **Major Concept**

Equation 30.17, $\vec{F}_B = q(\vec{v} \times \vec{B})$, is a convenient way to find the magnetic force on a single charged particle. In many practical circumstances, however, the subject is not a single moving particle, but instead a current in a conductor. So, in this section, we derive an expression for the magnetic force exerted on a current-carrying wire.

DERIVATION Magnetic Force on a Current-Carrying Conductor

Figure 30.43 shows a segment of a current-carrying wire in a region with a uniform magnetic field $\vec{B}$. We define a length vector $\vec{\ell}$ equal in magnitude to the length of the wire segment and pointing in the direction of the current. We show that the magnetic force on the straight current-carrying conductor is

$$\vec{F}_B = I(\vec{\ell} \times \vec{B}) \quad (30.32)$$

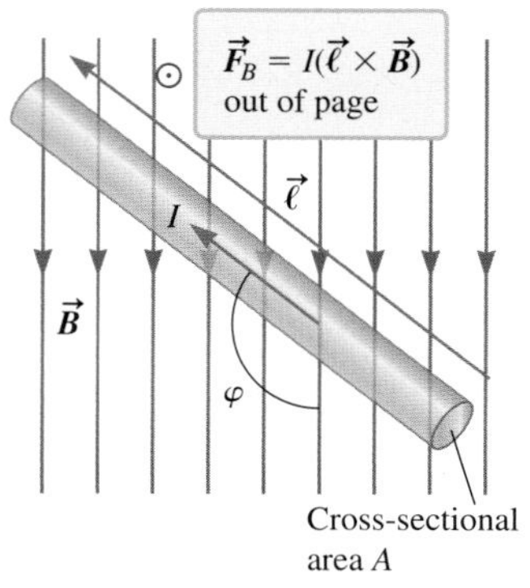

FIGURE 30.43 A straight segment of a current-carrying wire in a uniform magnetic field.

To derive an expression for the magnetic force exerted on the current-carrying wire, we first find the magnetic force on a single charged particle moving in the wire. We could consider the motion of an actual conduction electron in the direction opposite the current, but it is slightly easier to deal with the equivalent positive charge e moving at the drift velocity $\vec{v}_d$ in the direction of the current (Fig. 30.44).

FIGURE 30.44 One equivalent positive charge moving in the direction of the current.

The magnetic force on this single particle is given by Equation 30.17.	$\vec{f}_B = e(\vec{v}_d \times \vec{B})$	(30.17)

The magnetic force on the entire segment of wire is the sum of the magnetic forces on all the equivalent positively charged particles. So we need to know the number of conduction electrons. The number density of conduction electrons is n and the volume of the segment is $A\ell$, so the number of conduction electrons is $nA\ell$.

The total magnetic force on the segment is the number of conduction electrons times the force on a single particle.	$\vec{F}_B = (nA\ell)\vec{f}_B = nA\ell e(\vec{v}_d \times \vec{B})$ $\vec{F}_B = A\ell(ne\vec{v}_d \times \vec{B})$	(1)
We can write Equation (1) in terms of the current density: $\vec{J} = ne\vec{v}_d$ (Eq. 28.5).	$\vec{F}_B = A\ell(\vec{J} \times \vec{B})$	(2)
The current density has magnitude $J = I/A$, and in this case its direction is the same as that of $\vec{\ell}$. The term in parentheses is a unit vector pointing in the same direction as $\vec{\ell}$.	$\vec{J} = \dfrac{I}{A}\left(\dfrac{\vec{\ell}}{\ell}\right)$	(3)
Substitute Equation (3) into Equation (2). The cross-sectional area A and the magnitude ℓ of the length vector cancel.	$\vec{F}_B = A\ell\left(\dfrac{I\vec{\ell}}{A\ell} \times \vec{B}\right) = I(\vec{\ell} \times \vec{B})$ ✓	(30.32)

:• **COMMENTS**
1. The magnitude of the force is given by

$$F_B = I\ell B \sin\varphi \qquad (30.33)$$

where φ is the angle between $\vec{\ell}$ and $\vec{B}$. The direction of the force is given by the right-hand rule for cross products.

2. If the conductor is not straight, we find the magnetic force by imagining the conductor broken into many small segments that can each be considered a straight line with a length vector $d\vec{\ell}$ (Fig. 30.45). The magnetic force $d\vec{F}_B$ on any one of these segments is

$$d\vec{F}_B = I(d\vec{\ell} \times \vec{B}) \qquad (30.34)$$

The total force comes from integrating $d\vec{F}_B$ over the entire length of the conductor.

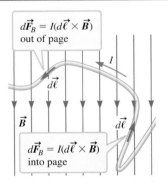

FIGURE 30.45 Imagine breaking up a curved current-carrying wire into small straight segments, like the two shown here. Then integrate the magnetic force on each segment.

EXAMPLE 30.8 Stereo Speakers and the Direction of Magnetic Force

Figure 30.46A shows the basic components of a stereo speaker. A permanent magnet shaped like a donut with a cylinder in the hole creates a radial magnetic field. In the space between the magnet's cylinder and its donut is a coil of wire called the *voice coil*. The signal from the stereo's amplifier controls the amount and direction of current in the voice coil. The permanent magnet exerts a force on the coil, causing it to move in and out. The motion of the voice coil results in the motion of the diaphragm, causing a sound wave in the surrounding air.

The magnitude and direction of the magnetic force on the voice coil depend on the current in it. Figure 30.46B shows a front view of these components. From this perspective, the voice coil looks like a single circular current loop. At the instant shown, the current is 1.5 A counterclockwise, and the magnitude of the magnetic field is 0.50 T (at the location of the coil). The radius of the voice coil is $R = 2.0$ cm. Find the magnetic force exerted on one loop of the voice coil for the instant shown in Figure 30.46B.

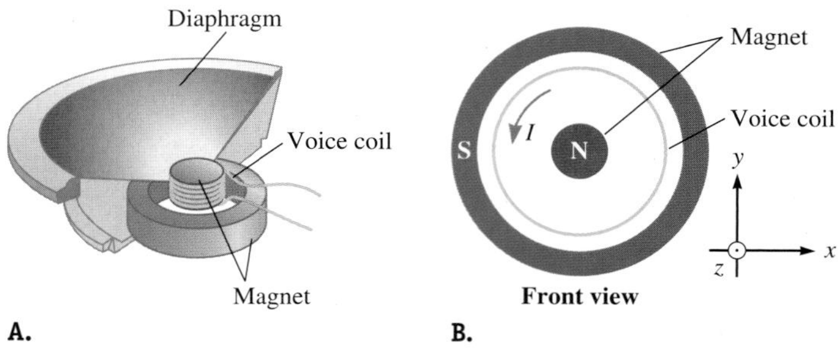

FIGURE 30.46 A. Parts of a stereo speaker. **B.** Magnet and voice coil of a stereo speaker. Find the direction of the magnetic force on the voice coil.

INTERPRET and ANTICIPATE

Because the wire loop of the coil is curved, break it up into small, straight segments. Find the magnitude of the force on one segment, and then integrate over the entire circular loop. Draw the magnetic field lines $\vec{B}$ due to the permanent magnet, pointing from the north pole to the south pole (Fig. 30.47). Next, find the magnetic force $d\vec{F}_B$ on a small, straight segment. We arbitrarily choose a segment of the coil at the bottom and draw $d\vec{\ell}$ for that segment, to the right in the direction of the current. Pointing the fingers of your right hand toward the right in the direction of $d\vec{\ell}$, imagine pushing $d\vec{\ell}$ into $\vec{B}$, which is downward near this segment. Your thumb should point into the page, so $d\vec{F}_B$ for this segment is into the page (in the negative z direction). For any other segment, $d\vec{F}_B$ is also in the negative z direction. So, the total magnetic force $\vec{F}_B$ on the coil is in the negative z direction.

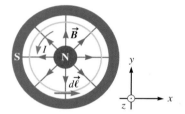

FIGURE 30.47

SOLVE

Start with Equation 30.34 for the segment shown in Figure 30.47. The magnetic field $\vec{B}$ is perpendicular to $d\vec{\ell}$, so the angle φ between $\vec{B}$ and $d\vec{\ell}$ is 90°.

$$d\vec{F}_B = I(d\vec{\ell} \times \vec{B}) \qquad (30.34)$$

$$dF_B = I(d\ell)(B)\sin\varphi = I(d\ell)(B)$$

Find the total magnetic force by integrating over the entire circumference of the circular loop, from 0 to $2\pi R$. The current and the magnetic field strength are constant over the integral.

$$\int dF_B = \int I(d\ell)(B)$$

$$F_B = IB\int_0^{2\pi R} d\ell = IB2\pi R$$

Substitute numerical values. Remember that the direction is into the page.

$$F_B = 2\pi(1.5\,\text{A})(0.50\,\text{T})(2.0 \times 10^{-2}\,\text{m})$$

$$\vec{F}_B = -9.4 \times 10^{-2}\hat{k}\,\text{N}$$

CHECK and THINK

We found the magnetic force at one instant. When the current reverses direction, the force will be outward, and so on. So the speaker vibrates in and out to create a sound wave.

30-12 Force Between Two Long, Straight, Parallel Wires

In Section 30-11, we studied the magnetic force exerted on a current-carrying wire, regardless of the source of the magnetic field. In this section, we consider the special case in which the source of the magnetic field is another long, straight, current-carrying wire. Picture two wires hanging vertically, with no current in either wire (Fig. 30.48A). When the wires are connected to a battery, so that they both have current, they exert an attractive magnetic force on each other (Fig. 30.48B).

Figure 30.49 shows the force between two straight, parallel wires carrying current I_1 in wire 1 and I_2 in wire 2. Each wire produces a magnetic field that exerts a force on the other wire. We can arbitrarily consider wire 1 to be the source of the magnetic field and wire 2 to be the subject. Using the simple right-hand rule, we find that the direction of the magnetic field produced by wire 1 is clockwise (Fig. 30.49A). The magnetic field $\vec{B}_1$ at the location of wire 2 is tangent to that circle in the negative y direction. From the side view (Fig. 30.49B), the magnetic field $\vec{B}_1$ at the location of wire 2 is directed into the page (in the negative y direction).

Now use Equation 30.32, $\vec{F}_B = I(\vec{\ell} \times \vec{B})$, and the right-hand rule for cross products to find the direction of the magnetic force exerted on wire 2 by the magnetic field $\vec{B}_1$. The magnetic force on wire 2 due to $\vec{B}_1$ is to the left (negative x direction), so wire 2 is attracted to wire 1. Next, let's find the magnitude of that attractive force using $F_B = I\ell B \sin \varphi$ (Eq. 30.33). The current we need is the subject's current (in this case I_2), and the magnetic field we need is the source's magnetic field B_1. The angle φ between the length vector $\vec{\ell}$ and the magnetic field $\vec{B}_1$ is 90°, so the magnitude of the force exerted by wire 1 on wire 2 is

$$F_{[1 \text{ on } 2]} = I_2 \ell B_1 \sin \varphi = I_2 \ell B_1 \sin 90°$$

$$F_{[1 \text{ on } 2]} = I_2 \ell B_1 \quad (30.35)$$

Wire 1 is long and straight, so substitute the expression $B_1 = (\mu_0 I_1)/(2\pi r)$ (Eq. 30.9) for a long, straight wire into Equation 30.35 in order to write the force exerted by wire 1 on wire 2 in terms of their currents:

$$F_{[1 \text{ on } 2]} = \left(\frac{\mu_0}{2\pi}\right)\frac{I_1 I_2 \ell}{r} \quad (30.36)$$

Equation 30.36 is the force exerted by wire 1 on wire 2. By Newton's third law, the force exerted by wire 2 on wire 1 must have the same magnitude but the opposite direction:

$$\vec{F}_{[2 \text{ on } 1]} = -\vec{F}_{[1 \text{ on } 2]}$$

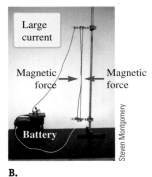

FIGURE 30.48 A. When there is no current in the wires, they hang vertically. **B.** When there is current in the wires, they exert an attractive magnetic force on each another. Both are bowed inward.

Using the notation from Section 5-9, we denote the magnetic force exerted by wire 1 on wire 2 by $\vec{F}_{[1 \text{ on } 2]}$.

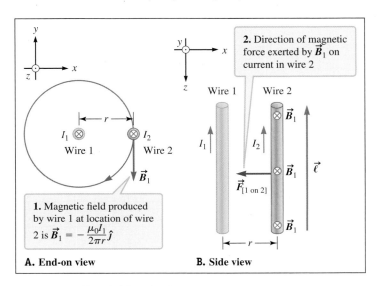

FIGURE 30.49 Two parallel, straight, current-carrying wires seen **A.** end-on and **B.** from the side. To find the magnetic field produced by wire 1, point your right thumb into the page in the direction of I_1 (in the negative z direction). Your fingers wrap clockwise as shown by the circular field line in part A. The magnetic force exerted by wire 1 on wire 2 is to the left (negative x direction) by $\vec{F}_{[1 \text{ on } 2]} = I(\vec{\ell} \times \vec{B})$.

964 CHAPTER 30 Magnetic Fields and Forces

Thus, Equation 30.36 gives the magnitude of the force exerted on either wire 1 or wire 2. Because the wires are very long, it is often convenient to calculate the magnitude of the force per unit length f on either wire:

$$f = \frac{F_{[1\text{ on }2]}}{\ell} = \frac{F_{[2\text{ on }1]}}{\ell}$$

$$f = \left(\frac{\mu_0}{2\pi}\right)\frac{I_1 I_2}{r} \tag{30.37}$$

MAGNETIC FORCE (PER UNIT LENGTH) BETWEEN TWO PARALLEL CURRENT-CARRYING WIRES   ▶ **Special Case**

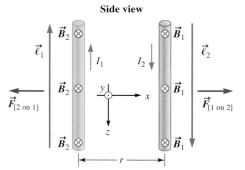

FIGURE 30.50 When they carry currents in opposite directions, long, straight, parallel wires repel each other.

We have shown that when two long, straight, parallel wires *carry currents in the same direction*, they are *attracted* to each other, and the magnitude of the force per unit length is given by Equation 30.37.

What if two parallel wires carry oppositely directed currents, as in Figure 30.50? The magnetic field produced by wire 1 at the location of wire 2 is into the page (negative y), just as it was in Figure 30.49B. In this case, however, the current in wire 2 is downward, so $\vec{\ell}_2$ is in the positive z direction. Applying the right-hand rule for the cross product in $\vec{F}_B = I(\vec{\ell} \times \vec{B})$ shows that the force $\vec{F}_{[1\text{ on }2]}$ exerted by $\vec{B}_1$ on wire 2 is in the positive x direction. By Newton's third law, $\vec{F}_{[2\text{ on }1]}$ is in the negative x direction. Thus, when parallel wires *carry currents in opposite directions*, the wires *repel* each other.

CONCEPT EXERCISE 30.8

When we think about two charged particles, it is helpful to remember the phrase *opposites attract and likes repel*. Come up with a similar phrase for the magnetic force between two parallel current-carrying wires.

30-13 Current Loop in a Uniform Magnetic Field

We have studied the magnetic force exerted on a single charged particle and on a straight current-carrying wire. In this section, we turn our attention to a current-carrying loop of wire. Current loops play an important role in practical devices such as galvanometers and electric motors, and they are also important in modeling permanent magnets (Section 30-7).

To keep things simple for now, consider a source that produces a uniform magnetic field, and a rectangular current loop carrying a current I as the subject (Fig. 30.51A). Our goal is to find the net force and the net torque on the loop. The top and bottom sides of the loop have length ℓ, and the left and right sides have length w. The loop is tilted with respect to the magnetic field, which is best seen from the side (Fig. 30.51B).

FIGURE 30.51 A current loop in a uniform magnetic field. **A.** Front view showing forces on all four sides. The net force on the loop is zero. **B.** Side view. The side labeled *right* is behind the left side and is not visible from this view. The forces exerted on the right and left sides are not shown in this view. There is a net torque on the loop.

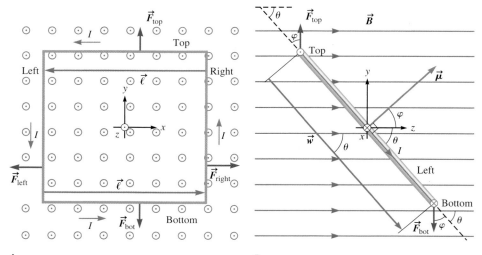

The direction of the magnetic force comes from Equation 30.32, $\vec{F}_B = I(\vec{\ell} \times \vec{B})$, and the right-hand rule for cross products. The magnetic field is perpendicular to the length vectors $\vec{\ell}$ for the top and bottom portions of the loop, so $\varphi = 90°$ and the magnitude of the magnetic force on the top and bottom is

$$F_{top} = F_{bot} = I\ell B \tag{30.38}$$

For the top, $\vec{\ell}$ is in the negative x direction, so the magnetic force is upward in the positive y direction:

$$\vec{F}_{top} = I\ell B \hat{\jmath} \tag{30.39}$$

For the bottom, $\vec{\ell}$ is in the positive x direction, so the magnetic force is downward in the negative y direction:

$$\vec{F}_{bot} = -I\ell B \hat{\jmath} \tag{30.40}$$

The magnetic field makes an angle θ with the length vectors $\vec{w}$ for the left and right portions of the loop (Fig. 30.51B). The magnitude of the magnetic force on the right and left is $F_{right} = F_{left} = IwB\sin\theta$. After finding the direction of these forces, we have

$$\vec{F}_{left} = -IwB\sin\theta\,\hat{\imath} \tag{30.41}$$

and

$$\vec{F}_{right} = IwB\sin\theta\,\hat{\imath} \tag{30.42}$$

The total magnetic force $\vec{F}_B$ exerted on the loop is the sum of the magnetic forces exerted on each of the four sides:

$$\vec{F}_B = \vec{F}_{top} + \vec{F}_{bot} + \vec{F}_{right} + \vec{F}_{left}$$

Substitute Equations 30.39 through 30.42:

$$\vec{F}_B = (I\ell B\hat{\jmath} - I\ell B\hat{\jmath}) + (IwB\sin\theta\,\hat{\imath} - IwB\sin\theta\,\hat{\imath})$$
$$\vec{F}_B = 0$$

We considered a rectangular loop, but our result holds for a loop of any shape: *A uniform magnetic field exerts no net magnetic force on a loop of current.*

The forces $\vec{F}_{top}$ and $\vec{F}_{bot}$ exert a net torque, which rotates the loop clockwise as seen from the side (Fig. 30.51B). Let's calculate the net torque around the loop's center of mass (the origin of our coordinate system). The magnitude of the torque exerted by a force F is given by

$$\tau = rF\sin\varphi \tag{12.16}$$

where $\vec{r}$ is the position vector from the rotation axis to the point at which $\vec{F}$ is applied and φ is the angle between $\vec{F}$ and $\vec{r}$. The distance from the axis of rotation to either the top or bottom of the loop is $r = w/2$. The magnitude of each force is given by Equation 30.38, so the magnitude of the torque exerted by $\vec{F}_{top}$ or $\vec{F}_{bot}$ is

$$\tau_{top} = \tau_{bot} = \left(\frac{w}{2}\right)(I\ell B)\sin\varphi \tag{30.43}$$

The direction of the torque is given by $\vec{\tau} = \vec{r} \times \vec{F}$ (Eq. 12.23) and the right-hand rule for cross products. For both $\vec{F}_{top}$ and $\vec{F}_{bot}$, the torque is into the page in the positive x direction (Fig. 30.51B). The total torque exerted on the loop is

$$\tau = 2\tau_{top} = 2\left(\frac{w}{2}\right)(I\ell B)\sin\varphi$$

$$\vec{\tau} = [I(\ell w)B\sin\varphi]\hat{\imath} \tag{30.44}$$

The term in parentheses is the area A of the loop:

$$\vec{\tau} = [(IA)B\sin\varphi]\hat{\imath} \tag{30.45}$$

The total torque on the loop is more conveniently expressed in terms of the loop's magnetic dipole moment of magnitude $\mu \equiv IA$ (Eq. 30.12). The torque becomes

$$\vec{\tau} = [\mu B \sin \varphi]\hat{\imath} \qquad (30.46)$$

The direction of the magnetic dipole moment $\vec{\mu}$ is found by wrapping the fingers of your right hand in the direction of the current; your thumb then points in the direction of the magnetic moment (Fig. 30.51B). The angle between $\vec{\mu}$ and $\vec{B}$ is φ, so the term in brackets in Equation 30.46 is the magnitude of the cross product of $\vec{\mu}$ and $\vec{B}$. Thus, the total torque is written as

TORQUE ON A MAGNETIC DIPOLE
○ Special Case

$$\vec{\tau} = \vec{\mu} \times \vec{B} \qquad (30.47)$$

Equation 30.47 is the torque exerted on a current loop (a magnetic dipole) by a uniform magnetic field. From Section 12-6, the direction of the torque is perpendicular to the plane in which the object rotates. The torque in Figure 30.51B is in the positive x direction, and the loop rotates clockwise in the yz plane. It is helpful to remember: *The torque attempts to align $\vec{\mu}$ with $\vec{B}$*.

Magnetic Potential Energy

The torque on a magnetic dipole causes it to rotate back and forth. Just as we described the back-and-forth motion of a pendulum in terms of changing kinetic and potential energies, we can associate kinetic and potential energies with the rotation of a current loop. To arrive at an expression for the magnetic potential energy, we can use our work in Section 24-9 on the electric dipole. A magnetic dipole rotating in a uniform magnetic field $\vec{B}$ is analogous to an electric dipole rotating in a uniform electric field $\vec{E}$. The torque on such an electric dipole is given by

$$\vec{\tau} = \vec{p} \times \vec{E} \qquad (24.22)$$

where $\vec{p}$ is the electric dipole moment. Comparing Equation 30.47 to Equation 24.22, we find

$$\vec{p} \to \vec{\mu} \quad \text{and} \quad \vec{E} \to \vec{B}$$

From Chapter 24, the electric potential energy stored in a dipole–electric field system is

$$U = -\vec{p} \cdot \vec{E} = -pE \cos \varphi \qquad (24.26, 24.25)$$

where the reference configuration is set so that when the electric dipole is perpendicular to the electric field ($\varphi = 90°$), the potential energy is zero. By analogy, the magnetic potential energy of a dipole–field system is

MAGNETIC POTENTIAL ENERGY
○ Special Case

$$U = -\vec{\mu} \cdot \vec{B} = -\mu B \cos \varphi \qquad (30.48)$$

where the reference configuration is set so that when the magnetic dipole moment is perpendicular to the magnetic field ($\varphi = 90°$), the potential energy is zero. The lowest potential energy is $U_{\min} = -\mu B$, when $\varphi = 0$ and the dipole moment is aligned with the magnetic field. The highest potential energy is $U_{\max} = +\mu B$, when $\varphi = 180°$ and the dipole moment is opposite to the magnetic field.

Figure 30.52 shows a current loop as it rotates clockwise in a uniform magnetic field. **1** Initially, the magnetic dipole is at rest, the kinetic energy is zero, and the potential energy is negative. Because the torque is into the page, the magnetic dipole rotates clockwise. **2** As the angle φ decreases, the potential energy decreases and the kinetic energy increases. **3** When the magnetic dipole moment is aligned with the magnetic field, the potential energy is at its minimum, the kinetic energy is at its maximum, and there is no torque. **4** Because the magnetic dipole has angular momentum, it continues to rotate in the same clockwise direction. As the angle φ increases, the potential energy increases (becomes less negative), the kinetic energy decreases, and the torque is now out of the page. **5** When the angle φ reaches its maximum, the magnetic dipole stops momentarily, the potential energy is at its maximum, the kinetic energy is zero, and the torque is still out of the page. So, the dipole begins to rotate in the other direction. In the absence of dissipative forces, the magnetic dipole would rotate back and forth as energy is converted between its potential and kinetic forms.

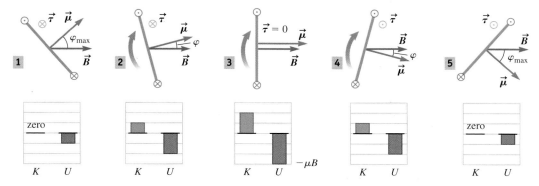

FIGURE 30.52 A current loop rotates clockwise, starting from rest.

We have considered a single current loop. If a wire is curled into a coil with N turns, the magnitude of the magnetic dipole moment is

$$\mu = NIA \quad (30.49)$$

The torque in a uniform magnetic field is still given by $\vec{\tau} = \vec{\mu} \times \vec{B}$ (Eq. 30.47), and the potential energy of the coil–field system is given by $U = -\vec{\mu} \cdot \vec{B}$ (Eq. 30.48).

Galvanometers

The essential element of an analog multimeter is a d'Arsonval **galvanometer** (Section 29-8), which consists of a rectangular coil of wire wrapped around an armature free to rotate between the poles of a permanent magnet (Fig. 30.53). The permanent magnet is bent as shown, and the armature is usually an iron cylinder. The result is that the magnetic field is nearly radial. The field lines point from the north to the south poles, crossing in the center. The field exerts a torque $\vec{\tau}_B$ on the coil when it carries a current. A spring then exerts a restoring torque $\vec{\tau}_S$ on the armature, so that the armature–coil assembly rotates until the two torques cancel. A needle attached to the armature rotates with it, and a scale printed on the galvanometer indicates the current in the coil.

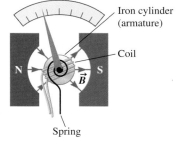

FIGURE 30.53 Basic parts of a d'Arsonval galvanometer.

Figure 30.54 shows just the coil and the radial magnetic field in a cross-sectional view like the one in Figure 30.51B. In the galvanometer, however, the magnetic field is radial, so the magnetic field that penetrates the metal of the coil (yellow in Fig. 30.54) is parallel to the plane of the coil. When there is no current in the coil, the magnetic dipole moment $\vec{\mu} = 0$ and the magnetic torque $\vec{\tau}_B = 0$ (Fig. 30.54A). This is also the relaxed position of the spring (not shown in this figure), so $\vec{\tau}_S = 0$. The current in the coil causes the armature to rotate until the magnetic torque is balanced by the spring's torque: $\tau_B = \tau_S$. The magnetic field passing through the metal of the coil is perpendicular to the coil's magnetic dipole moment $\vec{\mu}$.

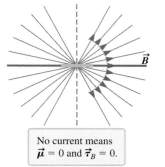 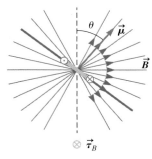 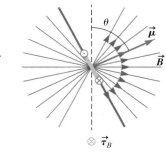

A.   B.   C.

FIGURE 30.54 A. No current in the coil. The magnetic dipole moment is zero and the magnetic torque is zero. **B.** Small current in the coil results in a small magnetic dipole moment and a magnetic torque directed into the page. **C.** Larger current means a large magnetic dipole moment and a greater magnetic torque. In all cases, $\tau_B = \tau_S$ and the coil is at rest.

EXAMPLE 30.9 Building a Galvanometer

Suppose you want to make a galvanometer using a square coil with sides of length 2.00 cm. The magnitude of the magnetic field is 0.250 T. The maximum current to be read is 1.00×10^{-4} A when the deflection of the needle is 45.0°. If the spring has torsion spring constant $\kappa = 3.00 \times 10^{-6}$ N·m/rad, how many turns does the coil need to have?

Example continues on page 968

: **INTERPRET and ANTICIPATE**
The armature stops rotating when the magnetic torque is balanced by the spring's torque. We need to find the torque exerted by both the magnetic and spring forces when the armature has rotated $\theta = 45.0°$. Figure 30.54B provides a rough sketch.

: **SOLVE**

The magnitude of the spring's torque comes from Equation 16.34. The spring is relaxed when the needle points straight up (indicating zero current), so θ is measured from the vertical position.	$\tau_S = \kappa\theta$	(16.34)
The magnitude of the magnetic torque comes from Equation 30.46.	$\vec{\tau} = [\mu B \sin\varphi]\hat{\imath}$ $\tau_B = \mu B \sin\varphi$	(30.46)
The magnetic field passing through the metal of the coil is perpendicular to the coil's dipole moment (Fig. 30.54), so $\varphi = 90°$.	$\tau_B = \mu B \sin 90° = \mu B$	(1)
Substitute the magnetic dipole moment for a coil (Eq. 30.49) into Equation (1). Write the area of the square in terms of the length ℓ of one of its sides.	$\mu = NIA$ $\tau_B = NIAB = NI\ell^2 B$	(30.49) (2)

The armature stops rotating when $\tau_B = \tau_S$, so set Equation (2) equal to Equation 16.34 and solve for N. Substitute numerical values, including $\theta = 45° = \pi/4$ rad.

$$\tau_B = \tau_S$$
$$NI\ell^2 B = \kappa\theta$$
$$N = \frac{\kappa\theta}{I\ell^2 B} = \frac{(3.00 \times 10^{-6}\,\text{N}\cdot\text{m/rad})(\pi/4\,\text{rad})}{(100 \times 10^{-6}\,\text{A})(2.00 \times 10^{-2}\,\text{m})^2(0.250\,\text{T})}$$
$$N = 236$$

: **CHECK and THINK**
Dimensional analysis shows that the result is a pure number, as it should be. The galvanometer is cleverly designed. If the magnetic field were not (nearly) radial, then as the armature rotated, the magnitude of the magnetic field's torque would vary depending on $\sin\varphi$. In other words, the amount of current in the coil would no longer be directly proportional to the angle θ of the coil's deflection. In Problem 71, you can see how a galvanometer would work if the magnetic field were uniform instead of radial.

DC Motors

The direct current (DC) motor is a practical application of the magnetic torque exerted on a current loop. A motor's job is to convert electrical energy into mechanical energy. A current loop—known as a *rotor*—is rotated by the torque exerted by a permanent magnet (Fig. 30.55). The rotor is connected to a DC power supply, such as battery, through a *commutator*. The commutator is a conductor—usually shaped like a ring or disk—that is split by an insulator, such as air. At the moment shown in Figure 30.55A, the commutator is in contact with the brushes (also conductors): The current in the yellow part of the rotor is in the negative z direction, and the current in the orange part is in the positive z direction. Further, the magnetic field is in the positive x direction, $\vec{B} = B\hat{\imath}$, and the magnetic dipole moment is in the negative y direction, $\vec{\mu} = -\mu\hat{\jmath}$. By $\vec{\tau} = \vec{\mu} \times \vec{B}$ (Eq. 30.47), the torque is

$$\vec{\tau} = -\mu B(\hat{\jmath} \times \hat{\imath}) = -\mu B(-\hat{k}) = \mu B \hat{k}$$

The z axis coincides with the rotor's rotation axis, and the torque is in the positive z direction. So, if you viewed the rotor from the commutator in the negative z direction, you would see the rotor turn counterclockwise.

When the rotor has turned 90°, so that the orange portion is directly above the yellow portion (Fig. 30.55B), the split in the commutator is centered on the two brushes. Each brush is in contact with both halves of the commutator. The commutator now provides a low-resistance path for the current, so the current passes through the commutator instead of through the rotor. In this position, the rotor has no current and no magnetic moment, so no torque is exerted on it. However, the rotor does have angular momentum and rotational inertia, so it continues to rotate despite the small amount of friction between the brushes and the commutator.

Soon, each brush is back in contact with only one half of the commutator, so the current must pass through the rotor again. In Figure 30.55C, the current in the yellow part of the rotor is in the positive z direction, and the current in the orange part is in the negative z direction, because the rotor–commutator assembly is oriented 180° from its position in Figure 30.55A. However, the magnetic moment is still in the negative y direction, and the torque is again in the positive z direction. In other words, in this position, the rotor continues to rotate in the same counterclockwise direction, as seen from the left in the figure. The split-ring commutator is designed to keep the current flowing clockwise as seen from above, even as the rotor flips over. The resulting current ensures that the magnetic moment always points in the same direction (negative y).

A real DC motor is more complicated. Its rotor actually contains a coil with many turns. This construction increases its magnetic dipole moment and the torque exerted on the rotor. To increase the torque further, a strong magnet is used—often an electromagnet instead of a permanent magnet. Finally, the magnitude of the torque exerted on the rotor depends on the angle between $\vec{\mu}$ and $\vec{B}$, which increases and decreases as the rotor rotates. To maintain a more constant torque, a real rotor consists of several coils oriented at different angles.

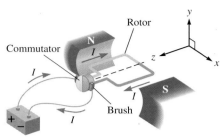
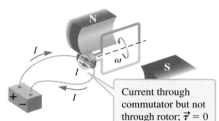
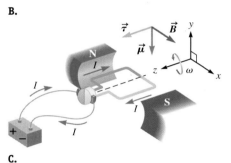

FIGURE 30.55 Components of a simple DC motor. **A.** The current is clockwise as seen from above. The magnetic moment is in the negative y direction, and the torque is in the positive z direction. **B.** When the rotor is vertical, there is no current in it. Instead, the current goes directly through the commutator. Because the rotor has angular momentum, it continues to rotate counterclockwise as seen from the left. **C.** The rotor has flipped over but the current is still clockwise as in A.

SUMMARY

❶ Underlying Principles

Magnetic fields are produced by moving charged particles and in magnetized materials by the orbital motion and "spin" (a quantum property) of subatomic particles, such as electrons. No matter what the source, magnetic fields may exert **magnetic forces** on subjects that are moving charged particles or magnetized materials.

✪ Major Concepts

1. A **magnetic dipole** has both a north and a south pole. A **magnetic monopole** has either a north or a south pole, but not both. So far, no magnetic monopole has been observed.

2. **Biot-Savart law:** The magnetic field $\vec{B}$ produced by a single particle with charge q moving at velocity $\vec{v}$:

$$\vec{B} = \left(\frac{\mu_0}{4\pi}\right) q \frac{\vec{v} \times \vec{r}}{r^3} \quad (30.1)$$

The Biot-Savart law for the magnetic field $d\vec{B}$ produced by a small segment of a wire:

$$d\vec{B} = \left(\frac{\mu_0}{4\pi}\right) I \frac{d\vec{\ell} \times \vec{r}}{r^3} \quad (30.5)$$

3. The **magnetic dipole moment** $\vec{\mu}$ (also called the **magnetic moment**) is a vector of magnitude

$$\mu \equiv IA \quad (30.12)$$

for a loop of current I and area A. The vector direction is given by a right-hand rule (tool 3, next page). The magnetic moment direction is the same as that of the magnetic field at the center of the current loop.

4. Each atom in a **ferromagnetic material** has a net nonzero magnetic moment and interacts strongly with its neighbors, causing the magnetic moments of nearby atoms to line up. A region in which the atoms' magnetic moments are aligned is called a **magnetic domain**.
5. Magnetic force on a moving charged particle:
$$\vec{F}_B = q(\vec{v} \times \vec{B}) \quad (30.17)$$

6. The **Lorentz force** is the total electromagnetic force experienced by a charged particle:
$$\vec{F}_L = \vec{F}_E + \vec{F}_B = q(\vec{E} + \vec{v} \times \vec{B}) \quad (30.21)$$

7. Magnetic force on a current-carrying wire:
$$\vec{F}_B = I(\vec{\ell} \times \vec{B}) \quad (30.32)$$

▶ Special Cases

1. Magnetic field due to
   a. an **infinitely long, straight, current-carrying wire**:
   $$B = \frac{\mu_0 I}{2\pi r} \quad (30.9)$$
   b. a **circular current loop** of radius R (on axis):
   $$\vec{B} = \frac{\mu_0 I R^2}{2(R^2 + y^2)^{3/2}} \hat{j} \quad (30.10)$$
   where $\hat{j}$ points along the loop axis.
   c. a **magnetic dipole** for distant points on the dipole axis:
   $$\vec{B} \approx \left(\frac{\mu_0}{2\pi}\right) \frac{\vec{\mu}}{y^3} \quad (30.15)$$

2. Magnetic force (per unit length) between two parallel current-carrying wires:
$$f = \left(\frac{\mu_0}{2\pi}\right) \frac{I_1 I_2}{r} \quad (30.37)$$
The force is attractive if the currents are in the same direction and repulsive if the currents are in the opposite directions.

3. **Torque on a magnetic dipole** in a uniform magnetic field:
$$\vec{\tau} = \vec{\mu} \times \vec{B} \quad (30.47)$$

4. Magnetic potential energy stored in a dipole–magnetic field system:
$$U = -\vec{\mu} \cdot \vec{B} = -\mu B \cos\varphi \quad (30.48)$$

⊙ Tools

1. **Magnetic field lines** indicate both the strength and direction of the magnetic field (Fig. 30.5B). Magnetic field lines always form closed loops. Outside a bar magnet, field lines point *away from* the north pole and *toward* the south pole. Inside a bar magnet, field lines point from the south pole toward the north pole (Fig. 30.8). Field lines are densely packed in regions where the magnetic field is relatively strong. The field direction at a particular point is tangent to the field line.
2. The **simple right-hand rule** (SRHR) gives the direction of the magnetic field produced by a current-carrying wire. Point your right thumb in the direction of the current; then your fingers wrap around the wire in the direction of the magnetic field (Fig. 30.11).
3. To find the direction of the magnetic moment $\vec{\mu}$ and of the magnetic field $\vec{B}$ along the axis of a current loop, use the **right-hand rule for the direction of the magnetic dipole moment**: Wrap the fingers of your right hand in the direction of the current; then your thumb points in the direction of $\vec{\mu}$ and $\vec{B}$ (Fig. 30.25).

PROBLEM-SOLVING STRATEGY Using the Biot-Savart Law $d\vec{B} = \left(\frac{\mu_0}{4\pi}\right) I \frac{d\vec{\ell} \times \vec{r}}{r^3}$

∶• INTERPRET and ANTICIPATE
Draw a diagram with these four elements:
1. A **coordinate system** and other **geometric details**.
2. **Labeled segments** of the source.
3. The **magnetic field vectors** produced by the segments at the point of interest.
4. The **net magnetic field vector** at the point of interest.

∶• SOLVE
1. Use the Biot-Savart law to find a **mathematical expression** for $d\vec{B}$, the small magnetic field produced by a single current segment at the point of interest.
2. Integrate $d\vec{B}$ to find $\vec{B}$ at the point of interest.

PROBLEMS AND QUESTIONS

A = algebraic   C = conceptual   E = estimation   G = graphical   N = numerical

30-1 Another Fundamental Force

1. **C** A yoga teacher tells her students to imagine their hands are magnets pulling on each other. What are the problems with this metaphor? What is a better metaphor?
2. **C** One end of a bar magnet is brought close to the end of a metal bar. This metal bar might be attracted, be repelled, or experience no magnetic force. In which of these three cases can you determine that the second metal bar is also a magnet? In which cases can you determine that the second bar is definitely not a magnet? Explain your reasoning.
3. **C** We have studied three field forces so far—gravity, the electric force, and the magnetic force. Which of these forces is associated with each of the following sources according to a stationary observer: **a.** a neutron at rest, **b.** a proton at rest, and **c.** a proton moving at velocity $\vec{v}$?

30-2 Revealing Magnetic Fields

4. **C** Why are there no gravitational dipoles?
5. **C** Because there are no magnetic monopoles, magnetic field lines must form closed loops. Consider the magnetic field lines for a bar magnet in Figure 30.8. **a.** Do all the lines form closed loops? **b.** Does the line that passes through the axis of the magnet form a closed loop? Explain.
6. **G** Copy Figure P30.6 and sketch the magnetic field lines that result from the bar magnets shown there.

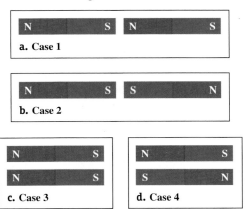

FIGURE P30.6

7. **G** A student attempts to draw the magnetic field lines near one end of a bar magnet that is held near a coin. The student produces Figure P30.7. Identify which aspects of the drawing are not consistent with the rules for correctly drawing magnetic field lines.

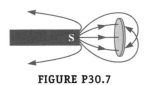

FIGURE P30.7

30-3 Ørsted's Discovery

8. **C** In your lab, you have a vertical wire that carries a current from the floor to the ceiling. A compass is moved in a plane perpendicular to the wire. When there is no current in the wire, the compass needle points parallel to the north-south direction. Assume that when there is a current in the wire, the magnetic field created by this current is much greater than the Earth's magnetic field. **a.** Describe where the compass should be placed relative to the wire so that the compass needle is undeflected when the current is turned on. **b.** Describe where the compass should be placed relative to the wire so that the compass needle is deflected by 90° when the current is turned on.
9. **C** Figure P30.9 shows very long current-carrying wires. Using the coordinate system indicated (with the z axis out of the page), state the direction of the magnetic field at point P in each case.

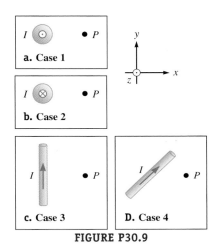

FIGURE P30.9

10. **C** Figure P30.10 shows a circular current-carrying wire. Using the coordinate system indicated (with the z axis out of the page), state the direction of the magnetic field at points A and B.

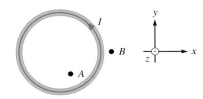

FIGURE P30.10

11. **C** Figure P30.11 shows three configurations of wires and the resultant magnetic fields due to current in the wires. What is the direction of the current that gives the resultant magnetic field shown in each case?

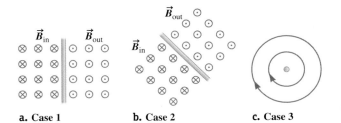

FIGURE P30.11

30-4 The Biot-Savart Law

12. **N Review** A proton is accelerated from rest through a 5.00-V potential difference. **a.** What is the proton's speed after it has been accelerated? **b.** What is the maximum magnetic field that this proton produces at a point that is 1.00 m from the proton?

13. **A** An electron moves in a circle of radius r at uniform speed around a single stationary proton. Find an expression for the magnitude of the magnetic field at the center of the circle in terms of μ_0, e, the speed v, and r.

Problems 14 and 15 are paired.

14. **N** One common type of cosmic ray is a proton traveling at close to the speed of light. If the proton is traveling downward, as shown in Figure P30.14, at a speed of 1.00×10^7 m/s, what are the magnitude and direction of the magnetic field at point B?

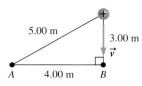

FIGURE P30.14 Problems 14 and 15.

15. **N** For the proton in Problem 14, what are the magnitude and direction of the magnetic field at point A?

30-5 Using the Biot-Savart Law

16. **C** Consider two very long, parallel wires that carry the same current but in opposite directions. Is the resulting magnetic field 0 everywhere? Does your answer depend on the distance between the wires?

17. **N** A straight wire carries a current of 5.00 A. What is the magnitude of the magnetic field it produces at a perpendicular distance of 50.0 cm from the wire?

Problems 18, 19, and 20 are grouped.

18. **A** Two long, straight, parallel wires are shown in Figure P30.18. The current in the wire on the left is double the current in the wire on the right. Find an expression for the magnetic field at points A and B. Use the indicated coordinate system to write your answer in component form.

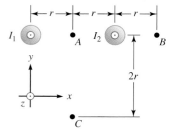

FIGURE P30.18
Problems 18–20.

19. **A** Two long, straight, parallel wires carry current as shown in Figure P30.18. If the currents are equal, find an expression for the magnetic field at points A and B. Use the indicated coordinate system to write your answer in component form.

20. **A** Two long, straight, parallel wires carry current as shown in Figure P30.18. If the currents are equal, find an expression for the magnetic field at point C. Use the indicated coordinate system to write your answer in component form.

21. **N** A counterclockwise current of 7.00 A flows in a square metal loop of side $d = 15.0$ cm. What is the magnitude of the magnetic field at the center of the loop?

22. **A** Two long, straight wires carry the same current as shown in Figure P30.22. One wire is parallel to the z axis and the other wire is parallel to the x axis as shown. Find an expression for the magnetic field at the origin.

23. **N** Figure P30.23 shows two infinitely long, straight, current-carrying wires located at $x = -20.0$ cm and at $x = +30.0$ cm. The left-hand wire carries current $I_1 = 2.00$ A in the negative z direction (into the page), and the current I_2 in the right-hand wire is unknown. If the magnitude of the total magnetic field due to the two wires at the origin is 4.00 μT, what are the two possible values for the magnitude and direction of the current in the right-hand wire?

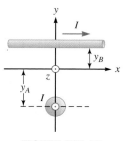

FIGURE P30.22

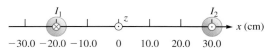

FIGURE P30.23

24. **A** A wire is bent in the form of a square loop with sides of length L (Fig. P30.24). If a steady current I flows in the loop, determine the magnitude of the magnetic field at point P in the center of the square.

25. **N** A conducting loop in the shape of an equilateral triangle with sides $L = 3.00$ cm carries a current of 8.00 A (Fig. P30.25). What is the magnitude of the magnetic field at the center O of the loop?

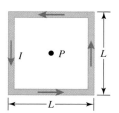

FIGURE P30.24

26. **A** Derive an expression for the magnetic field produced at point P due to the current-carrying wire shown in Figure P30.26. The curved parts of the wire are pieces of concentric circles. Point P is at their center.

27. **N** Two circular coils of radii 10.0 cm and 40.0 cm are connected in series to a battery. Find the ratio of the magnitudes of the magnetic fields at their centers.

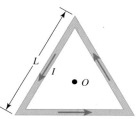

FIGURE P30.25

28. **N** Figure P30.28 shows three long, straight, parallel current-carrying wires with $I_1 = I_2 = I_3 = 5.00$ A into the page (the negative $\hat{k}$ direction) at the positions shown in the xy plane, with $d = 10.0$ cm. What are the magnitude and direction of the magnetic field at **a.** point A, **b.** point B, and **c.** point C?

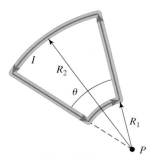

FIGURE P30.26

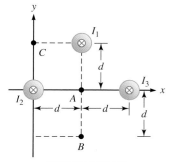

FIGURE P30.28

29. **N** Figure P30.29 shows two current-carrying loops with $I_1 = 4.00$ A clockwise and $I_2 = 9.00$ A counterclockwise, placed with their centers at the origin of the xy plane. **a.** If $r_1 = 10.0$ cm and $r_2 = 16.0$ cm, what are the magnitude and direction of the magnetic field due to the two loops at the origin? **b.** If r_1 is held constant at 10.0 cm, what would r_2 have to be for the magnetic field at the origin to be 0?

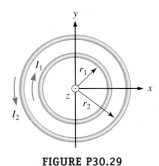

FIGURE P30.29

30-6 The Magnetic Dipole Moment and Modeling Atoms

30. **E** Model the Earth's magnetic field as a dipole produced by the motion of the molten iron and nickel core (1200 km $< R <$ 3500 km). Near the surface, the field strength is 0.5×10^{-4} T. Estimate the Earth's magnetic moment.

31. **N** A magnetic field of magnitude 0.450 T is directed parallel to the plane of a circular loop of radius 75.0 cm. A current of 9.00 mA is maintained in the loop. What is the magnetic moment of the loop?

32. **E** The world's strongest magnet according to Table 30.1 has a field strength of about 45 T. Imagine using a loop of wire and a battery to create a similar magnetic field at the center of the loop. Estimate the current and the size of the loop that you need to create this magnitude field. Can this be done in your lab room?

33. **N** A square loop of wire with side length 0.205 m carries a current of 1.50 A. What is the magnitude of the magnetic field along the axis of the square loop, 2.50 m from the center?

30-7 Ferromagnetic Materials

34. **C** The magnetic moments within several domains of a ferromagnetic material at different times are shown in Figure P30.34. What could have happened in the time between figures A and B to cause the change?

 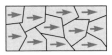

A.   **B.**

FIGURE P30.34

35. **C** Normally a refrigerator is not magnetized. If you tried to stick a bar magnet to a refrigerator, would you expect both poles to be attracted to the refrigerator, neither pole to be attracted, or just one pole to be attracted? Explain.

30-8 Magnetic Force on a Charged Particle

36. **C** For each case, determine the fundamental force or forces through which each pair of particles interacts: **a.** a stationary proton and a stationary neutron, **b.** a moving proton and a stationary neutron, **c.** a stationary proton and a stationary electron, and **d.** a moving proton and a moving electron. Explain your answers, including anything special about each situation.

37. **N** At the instant shown in both sketches in Figure P30.37, a positively charged particle ($q = 15.5$ mC) travels with speed $v = 356$ m/s in a magnetic field with $B = 0.650$ T. The angle θ is 32.0°. **a.** Find the magnetic force (magnitude and direction) exerted on the particle in both cases. **b.** If the positively charged particle is replaced by a negatively charged particle ($q = -15.5$ mC), how do your results change?

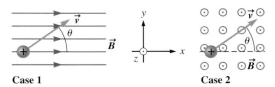

Case 1   Case 2

FIGURE P30.37

Problems 38 and 39 are paired.

38. **A** The magnetic field in a region is given by $\vec{B} = B_x\hat{\imath} + B_y\hat{\jmath}$. At some instant, a charged particle's velocity is given by $\vec{v} = v_x\hat{\imath} + v_y\hat{\jmath} + v_z\hat{k}$. Find an expression for the magnetic force exerted on the particle at that instant. In thinking about your result, find what happens to this force if **a.** $v_x = v_y = 0$ or **b.** $v_z = 0$.

39. **N** The magnetic field in a region is given by $\vec{B} = (0.750\,\hat{\imath} + 0.250\,\hat{\jmath})$ T. At some instant, a particle with charge $q = 10.0$ mC has velocity $\vec{v} = (35.6\hat{\imath} + 107.3\hat{\jmath} + 43.5\hat{k})$ m/s. What is the magnetic force exerted on the particle at that instant? (Give your answer in component form.)

40. **C** A positively charged particle starts at rest in a region of space that has both a uniform electric field and a uniform magnetic field. Describe the resulting motion of the particle if **a.** both the electric field and the magnetic field point to the right ($+\hat{\imath}$) in this region of space, and **b.** the electric field points to the right ($+\hat{\imath}$) and the magnetic field points out of the page ($+\hat{k}$).

41. **N** A charged particle enters a region of space with a uniform magnetic field $\vec{B} = 1.88\hat{\imath}$ T. At a particular instant in time, it has velocity $\vec{v} = (1.44 \times 10^6\hat{\imath} + 2.50 \times 10^6\hat{\jmath})$ m/s. Based on the observed acceleration, you determine that the force acting on the particle at this instant is $\vec{F} = 1.50\,\hat{k}$ N. What are **a.** the sign and **b.** the magnitude of the charged particle?

42. **N** The velocity vector of a singly charged helium ion ($m_{He} = 6.64 \times 10^{-27}$ kg) is given by $\vec{v} = 4.50 \times 10^5\,\hat{\imath}$ m/s. The acceleration of the ion in a region of space with a uniform magnetic field is 8.50×10^{12} m/s² in the positive y direction. The velocity is perpendicular to the field direction. What are the magnitude and direction of the magnetic field in this region?

43. **N** The magnetic field in a region of space is given by $\vec{B} = 0.540\hat{\jmath}$ T. The velocity vector of an electron moving at 7.55×10^6 m/s makes an angle of 45.0° with this field. What are the magnitudes of **a.** the magnetic force on the electron and **b.** the electron's acceleration?

30-9 Motion of Charged Particles in a Magnetic Field

44. **C** Can you use a mass spectrometer to measure the mass of a proton? Can you use a mass spectrometer to measure the mass of a neutron?

45. **N** In a laboratory experiment, a beam of electrons is accelerated from rest through a 154-V potential difference. The beam then

enters a uniform magnetic field and follows a circular path of radius $r = 19.3$ cm in the field region. **a.** What is the angle between the magnetic field and the electrons' velocity? **b.** What is the magnitude of the magnetic field?

46. A Two particles A and B with equal charges accelerated through potential differences V and $3V$, respectively, enter a region with a uniform magnetic field. The particles move in circular paths of radii R and $2R$, respectively. Determine the ratio of the masses of particles A and B.

47. N A proton is launched with a speed of 3.00×10^6 m/s perpendicular to a uniform magnetic field of 0.300 T in the positive z direction. **a.** What is the radius of the circular orbit of the proton? **b.** What is the frequency of the circular movement of the proton in this field?

48. C A proton moves in a helical path of constant radius in a magnetic field. **a.** What can you conclude about the magnitude and direction of the magnetic field? **b.** If, instead, the radius of the helical path decreases, what can you conclude about the magnetic field in that region?

49. A A proton and a helium nucleus (consisting of two protons and two neutrons) pass through a velocity selector and into a mass spectrometer. The radius of the proton's circular path is r_p. Find an expression for the radius r of the helium nucleus's path in terms of r_p. (You may assume the mass of a proton is roughly equal to the mass of a neutron, and the helium nucleus has the same speed as the proton.)

50. N Two ions are accelerated from rest in a mass spectrometer operating with potential difference ΔV. The first ion, with mass m_1, is singly ionized and is deflected into a semicircle of radius R_1 by the uniform magnetic field in the mass spectrometer. A second, doubly-ionized ion with mass m_2 is deflected into a semicircle with twice the radius of the first ion. What is the ratio m_2/m_1?

51. N A uniform electric field of magnitude 124 kV/m is directed upward in a region of space. A uniform magnetic field of magnitude 0.60 T perpendicular to the electric field also exists in this region. A beam of positively charged particles travels into the region. Determine the speed of the particles at which they will not be deflected by the crossed electric and magnetic fields.

30-10 Case Study: The Hall Effect

52. C What are the major differences between Ben Franklin's experiments and the Hall effect? Why can't we use the potential difference measured across a circuit element, such as a resistor, to find the sign of the charge carriers?

53. N A rectangular silver strip is 2.50 cm wide and 0.050 cm thick. It is in a magnetic field perpendicular to its surface (Fig. 30.41, page 957). The magnetic field is uniform, with a magnitude of 1.75 T. The strip carries a current of 6.45 A. According to Table 28.2, the number density of charge carriers in silver is 5.86×10^{28} m^{-3}. Find the Hall voltage for this strip.

Problems 54 and 55 are paired.

54. C An archeologist finds a painted rectangular metal artifact. She believes it was once part of a necklace and it may be made of lead, but she must be careful not to damage the paint in her examination. (She cannot dip the artifact in a fluid or take a sample of the metal.) Explain how she can use the Hall effect to find the number density of charge carriers and Table 28.2 to determine whether the artifact is made of lead.

55. N The artifact in Problem 54 is flat, with a thickness of 1.01 cm and a width of 1.139 cm. The archeologist places it in a strong magnetic field of 5.00 T that is perpendicular to the artifact (Fig. 30.41, page 957). The accuracy of her voltage measurement is ± 50.0 nV. To minimize the possibility of damaging

the artifact, she wishes to put the minimum current through it. **a.** If the artifact is made of lead, estimate the current required. **b.** Starting at the minimum, the researcher slowly turns up the current. When the current reaches six times the minimum, she measures a Hall voltage of 160.0 nV. What is the number density of charge carriers? Is the artifact made of lead? If not, use Table 28.2 to find its most likely (pure) composition.

30-11 Magnetic Force on a Current-Carrying Wire

56. N For both sketches in Figure P30.56, there is a 3.54-A current, a magnetic field strength $B = 0.650$ T, and the angle θ is 32.0°. Find the magnetic force per unit length (magnitude and direction) exerted on the current-carrying conductor in both cases.

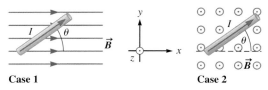

FIGURE P30.56

57. N A 1.40-m section of a straight wire oriented along the z axis carries 11.0 A of current in the positive z direction. The wire is placed in a region with a uniform magnetic field of 4.00 T in the positive y direction. What are the magnitude and direction of the magnetic force on the wire?

58. E Professor Edward Ney was the founder of infrared astronomy at the University of Minnesota. In his later years, he wore an artificial pacemaker. Always an experimentalist, Ney often held a strong laboratory magnet near his chest to see what effect it had on his pacemaker. Perhaps he was using the magnet to throw switches that control different modes of operation. An admiring student (without an artificial pacemaker) thought it would be fun to imitate this great man by holding a strong magnet to his own chest. The natural pacemaker of the heart (known as the sinoatrial node) carries a current of about 0.5 mA. Estimate the magnetic force exerted on a natural pacemaker by a strong magnet held to the chest. How do you think the student might have felt during the experiment? Explain your geometric assumptions. *Hints*: See Table 30.1 (page 941) to estimate the magnetic field, and assume the field is roughly uniform. Use Figure P30.58 to estimate the size of the sinoatrial node; your heart is about the size of your fist.

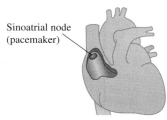

FIGURE P30.58

59. N A current of 5.64 A flows along a wire with $\vec{\ell} = (1.382\hat{i} - 2.095\hat{j})$ m. The wire resides in a uniform magnetic field $\vec{B} = (-0.300\hat{j} + 0.750\hat{k})$ T. What is the magnetic force acting on the wire?

60. N A wire with a current of $I = 8.00$ A directed along the positive y axis is embedded in a uniform magnetic field perpendicular to the current in the wire. If the wire experiences a magnetic force of 50.0 mN/m in the positive z direction, what are the magnitude and direction of the magnetic field in this region?

61. N A 3.60-m section of a straight wire carrying a current of 2.40 A is embedded in a 1.30-T uniform magnetic field. What is the magnitude of the magnetic force on the wire if the angle between the current in the wire and the direction of the magnetic field is **a.** 45.0°, **b.** 135.0°, and **c.** 90.0°?

62. **N** The triangular loop of wire shown in Figure P30.62 carries a current of 0.125 A, and a uniform magnetic field of 0.250 T points toward the right. Determine the force on each segment of the wire (indicate magnitude and direction) and the net force on the triangular loop.

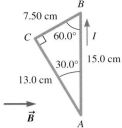

FIGURE P30.62

30-12 Force Between Two Long, Straight, Parallel Wires

Problems 63, 64, and 65 are grouped.

63. Three long, straight wires are seen end-on in Figure P30.63. The distance between the wires is $r = 0.256$ m. Each carries current $I_A = I_B = I_C = 1.63$ A. Assume the only forces exerted on wire A are due to wires B and C.
    a. **G** Draw a free-body diagram for wire A.
    b. **N** Find the magnetic force per unit length exerted on wire A.

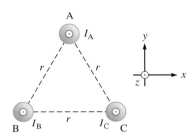

FIGURE P30.63 Problems 63–65.

64. **N** Consider the wires described in Problem 63. Find the magnetic force per unit length exerted on wire B.
65. **N** Consider the wires described in Problem 63, except now wire C carries current $I_C = 3.26$ A. Find the force per unit length exerted on **a.** wire A and **b.** wire B.
66. Two long, straight, parallel wires carry identical currents of 6.30 A in the positive x direction. The separation between the wires is 8.50 cm.
    a. **C** Is the force between the wires attractive or repulsive?
    b. **N** What is the magnitude of the force per unit length that each wire exerts on the other?

Problems 67, 68, and 84 are grouped.

67. **A** Three parallel current-carrying wires are shown in Figure P30.67. Find an expression for the net magnetic force per unit length on the middle wire due to the other wires.

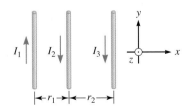

FIGURE P30.67 Problems 67, 68, and 84.

68. **N** Three parallel current-carrying wires are shown in Figure P30.67. The currents in each wire are $I_1 = 0.500$ A, $I_2 = 0.750$ A, and $I_3 = 0.250$ A. If the wires are separated by distances $r_1 = 0.185$ m and $r_2 = 0.290$ m, find the net magnetic force per unit length on the middle wire due to the other wires.

30-13 Current Loop in a Uniform Magnetic Field

69. **N** A current loop freely rotates in a uniform magnetic field ($B = 0.250$ T). The maximum torque on the loop is 0.540 N·m. What is the magnetic dipole moment of the loop?
70. **E** A student wishes to make a compass out of a wire and a 9-V battery. She wraps copper wire around a plastic water bottle (Fig. P30.70). The bottle hangs from a thread so that it is horizontal and free to rotate. Estimate the maximum torque exerted on her homemade compass due to the Earth's magnetic field.

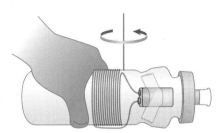

FIGURE P30.70

71. **C** In Figure 30.54 (page 967), we saw that the magnetic field inside a galvanometer is radial. Why is this an important part of its design? What would happen if the magnetic field were uniform instead? In particular, what could you say about the torque on the coil if the field were uniform?
72. **N** A straight conductor between points O and P has a mass of 0.50 kg and a length of 1.0 m and carries a current of 12 A (Fig. P30.72). It is hinged at O and is placed in a plane perpendicular to a magnetic field of 2.0 T. If the conductor begins from rest, determine the angular acceleration of the conductor due to the magnetic force. Ignore the effect of the gravitational force on the conductor.

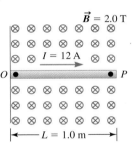

FIGURE P30.72

73. **N** A circular coil 15.0 cm in radius and composed of 145 tightly wound turns carries a current of 2.50 A in the counterclockwise direction, where the plane of the coil makes an angle of 15.0° with the y axis (Fig. P30.73). The coil is free to rotate about the z axis and is placed in a region with a uniform magnetic field given by $\vec{B} = 1.35\hat{j}$ T. **a.** What is the magnitude of the magnetic torque on the coil? **b.** In what direction will the coil rotate?

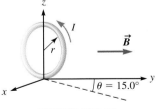

FIGURE P30.73

74. **N** A circular conductor with radius $r = 25.0$ cm, carrying a current of 3.40 A, is embedded in a 0.650-T uniform magnetic field. **a.** What is the magnitude of the maximum torque on the conductor? **b.** What are the minimum and maximum values of the potential energy of the conductor–magnetic field system?

General Problems

75. G, E The formula for the magnetic field along the axis of a magnetic dipole moment given by Equation 30.11,

$$\vec{B} \approx \frac{\mu_0 I R^2}{2y^3}\hat{j}$$

is valid far from the loop ($y \gg R$), but how far is far enough? Plot the magnetic field versus distance y for this expression, along with the exact expression

$$\vec{B} = \frac{\mu_0 I R^2}{2(R^2 + y^2)^{3/2}}\hat{j}$$

from Equation 30.10 for a loop that has a radius of 2.5 cm and a current of 0.5 A. Use your graph to estimate the distance at which the approximate value equals the exact value. Express your estimate in terms of loop's diameter.

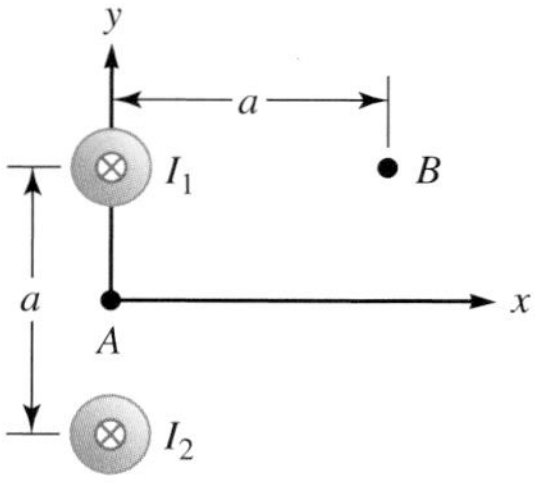

76. N Figure P30.76 shows two long, straight wires carrying currents $I_1 = 2.00$ A and $I_2 = 8.00$ A into the page. The wires are separated by a distance $a = 1.00$ m.
a. What are the magnitude and direction of the magnetic field at point A, halfway between the two wires? **b.** What are the magnitude and direction of the magnetic field at point B, at a distance $a = 1.00$ m to the right of the top wire?

77. N Figure P30.77 shows two long current-carrying wires in the xy plane with $I_1 = 2.00$ A to the left and $I_2 = 7.00$ A downward. The wires are not in contact with each other. What are the magnitude and direction of the magnetic field due to the two wires **a.** at point A, a distance $a = 60.0$ cm and $b = -50.0$ cm in the x and y directions from the intersection of the wires, and **b.** at $z = -50.0$ cm behind the intersection of the wires?

78. N Two long, straight, current-carrying wires run parallel to the y axis. The first wire is located at ($x = 0, z = 0$) with current $I_1 = 12.0$ A in the positive y direction, and the second wire, located at ($x = 95.0$ cm, $z = 0$), carries a current of 24.0 A in the negative y direction. At what location in the xy plane is the magnetic field due to the two wires equal to 0?

Problems 79, 80, and 81 are grouped.

79. A Consider a regular polygon with N sides carrying a steady current I. Determine the magnetic field at the center of such a polygon. Assume R is the perpendicular distance from the center of the polygon to the sides of the polygon. Assume the positive z axis points out of the page.

80. A In Problem 79, assume the regular polygon has a very large number of sides. Mathematically, this corresponds to taking the limit as $N \to \infty$. Obtain the analytical expression for the magnetic field at the center of such a polygon. Compare your result with the magnetic field at the center of a circular loop. Assume the positive z axis points out of the page.

81. N A hexagonal loop of side 6.0 cm carries a current of 2.0 A (Fig. P30.81). Determine the magnetic field at the center of

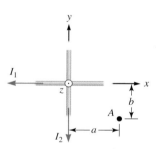

FIGURE P30.76

FIGURE P30.77

the loop. Assume the positive z axis points out of the page. *Hint*: Consider the results of Problems 79 and 80.

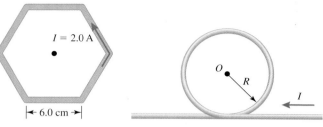

FIGURE P30.81   FIGURE P30.82

82. N A section of a long, straight wire is twisted into a coplanar loop of radius $R = 45.0$ cm (Fig. P30.82). If the current in the wire is $I = 5.00$ A, what are the magnitude and direction of the magnetic field at point O, the center of the loop? (The wire is sheathed in insulation.)

83. Two infinitely long current-carrying wires run parallel in the xy plane and are each a distance $d = 11.0$ cm from the y axis (Fig. P30.83). The current in both wires is $I = 5.00$ A in the negative y direction.
a. G Draw a sketch of the magnetic field pattern in the xz plane due to the two wires. What is the magnitude of the magnetic field due to the two wires
b. N at the origin and
c. A as a function of z along the z axis, at $x = y = 0$?

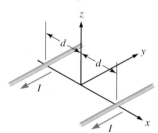

FIGURE P30.83

84. C Consider the current-carrying wires depicted in Figure P30.67. If the only sources of magnetic fields are the three wires, is it possible that the net magnetic force per unit length on the middle wire is 0? If yes, explain how this is possible. If no, explain what might be changed to make it possible for the net magnetic force per unit length to be 0.

85. C Each part of Figure P30.85 shows the velocity vector of a positively charged ion and the magnetic force it is experiencing at an instant in time. What is the direction of the magnetic field that causes the force on the ion in each of the cases shown in the figure?

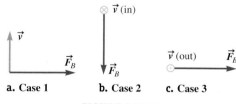

a. Case 1   b. Case 2   c. Case 3

FIGURE P30.85

Problems 86 and 87 are paired.

86. N An electron with velocity $\vec{v} = (3.14 \times 10^5 \hat{i} - 5.78 \times 10^5 \hat{j})$ m/s is moving into a region of space with a uniform magnetic field $\vec{B} = (-0.500\hat{i} + 0.250\hat{j})$ T. Find the resultant magnetic force on the electron at the instant it enters the field.

87. **A** A charged particle with charge q and velocity $\vec{v} = v_x\hat{i} + v_y\hat{j}$ is moving into a region of space with a uniform magnetic field $\vec{B} = B_x\hat{i} + B_y\hat{j}$. Find the resultant magnetic force on the particle the instant it enters the field.

88. **N** A magnetic field of magnitude 0.450 T is directed parallel to the plane of a circular loop of radius 75.0 cm. A current of 9.00 mA is maintained in the loop. What is the magnitude of the torque the magnetic field exerts on the loop?

89. **N** A potential difference of 5.10×10^3 V is used to accelerate a proton initially at rest in a region of space with a 0.500-T uniform magnetic field. What are the **a.** minimum and **b.** maximum magnetic forces the proton can experience in this region?

90. **N** A mass spectrometer (Fig. 30.40, page 956) operates with a uniform magnetic field of 20.0 mT and an electric field of 4.00×10^3 V/m in the velocity selector. What is the radius of the semicircular path of a doubly ionized alpha particle ($m_a = 6.64 \times 10^{-27}$ kg)?

91. **A** Three long, current-carrying wires are parallel to one another and separated by a distance d. The magnitudes and directions of the currents are shown in Figure P30.91. Wires 1 and 3 are fixed, but wire 2 is free to move. Wire 2 is displaced to the right by a small distance x. Determine the net force (per unit length) acting on wire 2 and the angular frequency of the resulting oscillation. Assume the mass per unit length of wire 2 is λ and $x \ll d$.

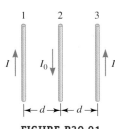

FIGURE P30.91

92. **N** A long, straight wire carrying a current of 2.0 A is placed in the plane of a conducting strip of width 8.0 cm (Fig. P30.92). The strip carries a current of 4.0 A. The distance from the wire to the near edge of the strip is 4.0 cm. Calculate the attractive force per unit length between the wire and the strip.

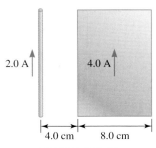

FIGURE P30.92

93. **A** A current-carrying conductor PQ of mass m and length L is placed on an inclined plane with angle of inclination θ (Fig. P30.93). A uniform magnetic field B is directed upward as shown. Assume friction is negligible. **a.** Determine the magnitude and direction of the current in the conductor so that it remains in equilibrium. **b.** If the direction of the current is reversed, will the conductor still be in equilibrium? If not, find the magnitude of the initial acceleration of the conductor.

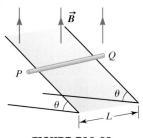

FIGURE P30.93

94. This problem is designed to show how the Biot-Savart empirical law may be found using simple laboratory equipment.[1] A wire is bent into a square U-shape and a current is set up in the wire (Fig. P30.94). The short segment of the wire has been aligned with the Earth's magnetic field $\vec{B}_\oplus$, so that when the current is turned off, the compass needle points in the x direction. The long segments of the wire produce magnetic fields that are nearly vertical at the location of the compass, so the needle is not deflected by these magnetic fields. When the current is on, the short segment produces a magnetic field $\vec{B}$ in the positive y direction (at the compass's location). The needle is deflected so that it points along the total magnetic field $\vec{B}_{tot}$, making an angle θ with the x axis. The compass rests on thin wooden sheets of known thickness. By varying the number of sheets, you can measure the angle θ as a function of distance r from the short segment of wire. (r comes from multiplying the thickness of a sheet by the number of sheets.)

a. **A** Assume the segment's magnetic field obeys the inverse-square Biot-Savart law, and derive an expression for $\tan\theta$ as a function of r.

b. **G** Plot $\tan\theta$ as a function of $1/r^2$ for the data in Table P30.94. Assume the error in $\tan\theta$ is ± 0.01. On this graph, sketch the expression you found in part (a). Assume the expression agrees with the data at $r = 8$ cm. (You may use centimeters on your graph.)

c. **C** Explain the discrepancy between the data and the expression predicted by the Biot-Savart law. How can you improve the experiment? *Hint*: How long is the segment described by the Biot-Savart law?

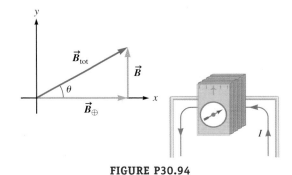

FIGURE P30.94

TABLE P30.94

θ (°)	r (cm)
22.9	2
15.1	2.67
11.1	3.33
7.88	4
5.71	4.67
3.96	5.33
2.86	6
2.86	6.67
1.76	7.33
1.76	8

95. **N** A proton enters a region with a uniform electric field $\vec{E} = 5.0\,\hat{k}$ V/m and a uniform magnetic field $\vec{B} = 5.0 \times 10^{-4}\,\hat{k}$ T. The proton has initial velocity $\vec{v}_0 = 2.5 \times 10^5\,\hat{i}$ m/s. How far along the z axis does the proton travel after it undergoes three complete revolutions?

---

[1] Based on *The Biot-Savart Law: From Infinitesimal to Infinite* by Jeffrey A. Phillips and Jeff Sanny (*The Physics Teacher* 46: 44–47, Jan. 2008).

31 Gauss's Law for Magnetism and Ampère's Law

Key Questions

How do we state formally that no magnetic monopoles have been observed?

How do we use Ampère's law and Maxwell–Ampère's law to find the magnetic fields produced by different sources?

31-1 Measuring the magnetic field 979

31-2 Gauss's law for magnetism 980

31-3 Ampère's law 984

31-4 Special case: Linear symmetry 987

31-5 Special case: Solenoids 992

31-6 Special case: Toroids 996

31-7 General form of Ampère's law 999

❶ Underlying Principles

1. Gauss's law for magnetism
2. Ampère's law (Ampère–Maxwell's law)

These laws are two of the four Maxwell equations.

✷ Major Concepts

Displacement current

◐ Special Cases

1. Sources with linear symmetry
2. Ideal solenoid
3. Toroid

FIGURE 31.1 Because Maxwell's four simple equations describe diverse phenomena concerning electricity, magnetism, and light, physicists believe they are perfect for T-shirts.

After doing hundreds of physics homework problems, you might be surprised to learn that physicists strive to make their theories as simple and eloquent as possible. An eloquent theory is an underlying principle usually expressed by a single equation or sentence that holds for a wide range of phenomena. Physicists believe there is beauty in a simple theory. Newton's laws of motion are considered beautiful because his three simple laws apply to so many situations.

Just as Newton's laws of motion are the principles that underlie classical mechanics, there are four laws that underlie electricity, magnetism, and light. These four laws are known as **Maxwell's equations**; they are considered so beautiful that they appear on T-shirts (Fig 31.1).

Newton's discovery relied on the careful observations and experiments of Galileo. Maxwell's work also depended on the careful observations and experiments of other scientists. In fact, Maxwell's equations were actually discovered by scientists who preceded him; each equation is known by the name of the scientist who originally discovered it. One such equation is Gauss's law for electricity (Chapter 25). In this chapter, we study two more of Maxwell's equations—Gauss's law for magnetism and Ampère's law.

Chapter 32 will introduce the last of Maxwell's equations—Faraday's law. Maxwell's amazing achievement was recognizing that these four laws combine to form the underlying principle for light (Chapter 34). Maxwell's equations form one of the most eloquent theories of physics.

31-1 Measuring the Magnetic Field

In 1820, Hans Christian Ørsted discovered that a compass needle is deflected by the current in a wire. Within a few months, French prodigy André Marie Ampère (1775–1836) conducted his own experiments and concluded that moving charged particles are the source of any magnetic field, including fields due to permanent magnets. Ampère's law is one of the four underlying principles of electricity and magnetism.

Today, we know that the magnetic field of a permanent magnet is largely due to the spin of electrons—an intrinsic property.

CASE STUDY Measuring the Magnitude of the Magnetic Field

In Chapter 30, we used the Biot-Savart law to find expressions for the magnetic field. Ampère's law as introduced in this chapter also allows us to find expressions for the magnetic field produced by various highly symmetrical sources. Much of this chapter focuses on finding mathematical expressions for the *magnitude* of the magnetic field for several different sources using Ampère's law. How do we know these expressions are correct? Of course, they must be verified by experiments. The goal of this case study is to confirm the expressions derived from Ampère's law using a compass and the simple techniques used to discover Biot-Savart's empirical law (similar to Problem 30.94, page 977).

To find the magnitude of the magnetic field, we measure the compass needle's deflection from a reference position. The Earth's magnetic field $\vec{B}_\oplus$ provides a natural reference position. When no other magnetic field is present, a compass needle points toward the Earth's south magnetic pole (the North geographic pole); let's call this the positive y direction (Fig. 31.2A). Now, imagine there is a second magnetic field $\vec{B}$ directed along the positive x axis (Fig. 31.2B). We can find the magnitude B of this magnetic field by measuring the deflection θ of the compass needle:

$$B = B_\oplus \tan \theta \quad (31.1)$$

where $B_\oplus$ is the magnitude of the Earth's magnetic field, about 0.5×10^{-4} T. Equation 31.1 holds only when the magnetic field $\vec{B}$ is perpendicular to the Earth's magnetic field $\vec{B}_\oplus$. (If $\vec{B}$ is not perpendicular to $\vec{B}_\oplus$ or some other known magnetic field, finding the magnitude B is more complicated, but we will not consider such cases.) The goals of this case study are to connect theory (Ampère's law) with observation, and to visualize the magnetic field through its effect on compass needles and graphs of its magnitude versus position.

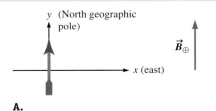

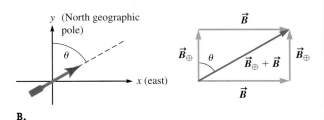

FIGURE 31.2 A. The Earth's magnetic field $\vec{B}_\oplus$ provides a reference position for a compass needle. The compass needle points toward the Earth's magnetic south pole (North geographic pole) when no other magnetic fields are present. **B.** When there is another magnetic field $\vec{B}$, the compass needle points in the direction of the total magnetic field $\vec{B}_\oplus + \vec{B}$.

CONCEPT EXERCISE 31.1

CASE STUDY Measuring the Magnetic Field Near a Bar Magnet

Suppose you have a bar magnet (Fig. 30.4, page 936) and you want to find the magnetic field strength along the magnet's axis as a function of distance from one of the poles. Explain how you could use a small compass to achieve your goal.

EXAMPLE 31.1 CASE STUDY: Playing with Magnets

From the information in Table 30.1 (page 941), calculate the deflection of a compass needle that results **A** from being near a refrigerator magnet and **B** from being near an MRI magnet. Assume the magnetic field in each case is perpendicular to the Earth's magnetic field. Report your results to three significant figures.

INTERPRET and ANTICIPATE
The magnetic field in both cases is perpendicular to the Earth's magnetic field (Fig. 31.2B).

SOLVE
Find the deflection θ from Equation 31.1.

$$B = B_\oplus \tan\theta \quad (31.1)$$

$$\theta = \tan^{-1}\left(\frac{B}{B_\oplus}\right)$$

A According to Table 30.1, the magnetic field near a typical refrigerator magnet is $B = 5 \times 10^{-3}$ T.

For a refrigerator magnet,

$$\theta = \tan^{-1}\left(\frac{5 \times 10^{-3}\ \text{T}}{5 \times 10^{-5}\ \text{T}}\right) = 89.4°$$

B Near a typical MRI magnet, the magnetic field is $B = 2$ T.

For an MRI magnet,

$$\theta = \tan^{-1}\left(\frac{2\ \text{T}}{5 \times 10^{-5}\ \text{T}}\right) = 90.0°$$

CHECK and THINK
This example illustrates a problem with using the Earth's magnetic field to set a reference position. The magnetic field of a refrigerator magnet is much weaker than that of an MRI magnet, yet both produce almost the same deflection of the compass needle. It would be difficult to measure the difference between the two deflections. One improvement is to use a stronger magnetic field than the Earth's as the reference. Ideally, it is best to choose a reference magnetic field that is comparable to the field you want to measure.

31-2 Gauss's Law for Magnetism

Gauss's law for magnetism and Ampère's law are both similar to Gauss's law for electricity, so let's review a few important ideas about Gauss's law from Chapter 25.

Quick Review of Gauss's Law for Electricity

We begin by imagining an electric field source hidden inside a box. We can get some idea of what is inside the box (a Gaussian surface) by suspending a small, positively charged sphere from a thin thread at various places around the box (Fig. 31.3). According to Gauss's law for electricity, the net charge contained in the surface is proportional to the electric flux Φ_E through the closed surface:

$$\Phi_E = \oint \vec{E} \cdot d\vec{A} = \frac{q_{in}}{\varepsilon_0} \quad (25.11)$$

We see that the box in Figure 31.3A contains a positively charged source, so the electric flux through the Gaussian surface is positive and the electric field lines emanate from the box. The box in Figure 31.3B contains a negatively charged source, so the electric flux through the Gaussian surface is negative and the electric field lines terminate in the box. Finally, if the positively charged sphere is not deflected no matter where it is placed outside the box, we conclude that the net charge inside is zero ($q_{in} = 0$), the electric field is zero ($\vec{E} = 0$) everywhere, and the electric flux through the Gaussian surface is zero ($\Phi_E = 0$).

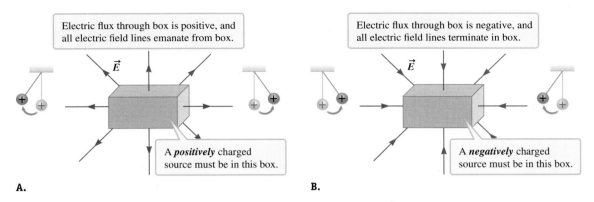

FIGURE 31.3 A small, positively charged sphere is used to gather information about the contents of a box. According to Gauss's law for electricity, the net charge contained in the box is proportional to the electric flux through the box. **A.** The sphere is repelled by the source, so the box contains a positively charged source and the net electric flux is positive. **B.** The sphere is attracted by the source, so the box contains a negatively charged source and the net electric flux is negative.

What can we conclude if the sphere is both attracted to and repelled by the contents of the box (Fig. 31.4)? First, because the sphere is deflected, we conclude that the electric field is nonzero, pointing leftward at the two positions shown. Second, the contents of the box cannot be either positive or negative because the sphere is both attracted to and repelled by the box, so the contents must be neutral. If the contents of the box are neutral, how can the positively charged sphere be deflected? One possible answer (Fig. 31.5A) is that the box is empty and there is a positively charged source located on the right, deflecting the positively charged ball toward the left. Another possible answer (Fig. 31.5B) is that the box contains an electric dipole. When the positively charged sphere is on the right side of the box, it is closer to the negative side of the dipole, so it is attracted to the box. When the positively charged sphere is on the left side of the box, it is closer to the positive side of the dipole, so it is repelled. In both cases, there are just as many electric field lines entering the box as leaving the box, and so the electric flux through the box is zero.

FIGURE 31.4 The positively charged sphere is deflected toward the box at some points and away from the box at other points.

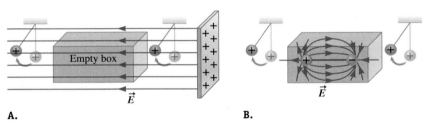

FIGURE 31.5 The net electric flux is zero in both cases.

Magnetic Monopoles Revisited

A magnetic monopole is an idealized magnetic source that is either a north pole or a south pole, analogous to an electric source that has either a net positive or a net negative charge. There is no conclusive evidence for the existence of magnetic monopoles (Section 30-2). Magnets always have both a north and a south pole. The simplest combination of a north pole and a south pole is a magnetic dipole, analogous to an electric dipole.

Gauss's law for magnetism is a formal mathematical statement that there are no known magnetic monopoles. Just as Gauss's law for electricity involves electric flux Φ_E, Gauss's law for magnetism involves magnetic flux Φ_B. The mathematical expression for magnetic flux is similar to the expression for electric flux (Eq. 25.6). The magnetic flux through a surface is given by

$$\Phi_B = \int \vec{B} \cdot d\vec{A} \qquad (31.2)$$

The SI unit for magnetic flux is the weber (Wb), named for Wilhelm Weber (1804–1891), a colleague of Gauss:

$$1 \text{ Wb} = 1 \text{ T} \cdot \text{m}^2 \qquad (31.3)$$

GAUSS'S LAW FOR MAGNETISM
ⓘ Underlying Principle

Gauss's law for magnetism involves the magnetic flux through a closed (Gaussian) surface. *Because there are no (known) magnetic monopoles, the magnetic flux through a Gaussian surface is zero.* Mathematically, **Gauss's law for magnetism** is

$$\Phi_B = \oint \vec{B} \cdot d\vec{A} = 0 \qquad (31.4)$$

The circle on the integral symbol in Equation 31.4 indicates that the integral is taken over the entire closed surface.

To understand Gauss's law for magnetism, imagine using a compass to probe the contents of a box. The compass's north pole is attracted to south poles and repelled by other north poles. Because there are no magnetic monopoles, we will never find a situation analogous to Figure 31.3 in which the positively charged sphere is always either repelled by or attracted to the source in the box.

The compass's north pole is attracted to the box when it is placed at some points and repelled by the box at other points, analogous to Figure 31.4. Figure 31.6A shows that when the compass is to the right of the box, its north pole is attracted to the box, and when the compass is to the left of the box, the compass's north pole is repelled. One possibility for what causes the deflection of the compass needle in this case is that there is nothing in the box; the deflection is caused by an external magnetic source such as a loop of current (Fig. 31.6B). Figure 31.6C shows a second possibility; there is a dipole in the box—in this case, a bar magnet. The north poles of the compass needles are attracted to the south pole of the bar magnet on the right and repelled by the north pole on the left. In both cases (current loop outside the box and bar magnet inside the box), the number of magnetic field lines entering the box is equal to the number of field lines leaving the box. If we calculate the magnetic flux through the entire closed Gaussian surface, we find there is no net magnetic flux through the box (Eq. 31.4).

Comparing Gauss's law for magnetism to Gauss's law for electricity shows that the electric flux through a Gaussian surface may be positive, negative, or zero depending on the net charge of the source enclosed by the surface, but the magnetic flux through a Gaussian surface is always zero. This difference arises because electric monopoles exist (sources with a net charge), whereas magnetic monopoles are not known to exist. If magnetic monopoles are discovered, Gauss's law for magnetism will have to be modified; the right side of Equation 31.4 would then be proportional to the net magnetic "charge" enclosed in the Gaussian surface.

It is possible to choose a Gaussian surface near an electric dipole so that the electric flux through that surface is not zero. For example, imagine drawing a small Gaussian surface that encloses only the positive charge of the dipole. All the electric field lines pass outward through this Gaussian surface, so the net electric flux through it is positive. There is no analogy in the case of a magnetic dipole. Suppose you draw a small Gaussian surface around the north pole of the bar magnet in Figure 31.6C. Just as many magnetic field lines enter the Gaussian surface from the right as leave the surface on the left. The net magnetic flux is zero no matter how small a portion of the bar magnet you place in your Gaussian surface. This is the same as saying that if you break a bar magnet, you have two new complete bar magnets, each with a north and a south pole (Fig. 30.7, page 937).

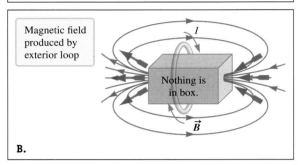
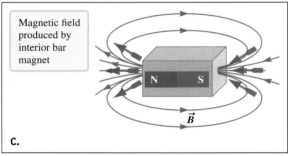

FIGURE 31.6 A. Analogous to Figure 31.4, we can use a compass needle to probe the magnetic contents of a box. **B.** One possible explanation for the deflection of these compass needles: an external source. This is analogous to Figure 31.5A. **C.** Another possibility: a bar magnet inside the box. The compass needles' north poles are attracted to the south pole of the bar magnet and repelled by the bar magnet's north pole. This is analogous to Figure 31.5B.

CONCEPT EXERCISE 31.2

Does Gauss's law for magnetism state that the magnetic flux $\Phi_B = \int \vec{B} \cdot d\vec{A}$ (Eq. 31.2) is always zero? Explain.

EXAMPLE 31.2 Magnetic Flux

Find the magnetic flux through the rectangular loop lying near the long, straight wire carrying current $I = 2.50$ A (Fig. 31.7). According to the simple right-hand rule (Fig. 30.11, page 939), this current produces a magnetic field in the negative z direction on the right side of the wire. The magnetic field is stronger near the wire and gets weaker farther away according to Equation 30.9, $B = (\mu_0 I)/(2\pi r)$.

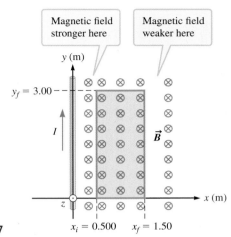

FIGURE 31.7

:• **INTERPRET and ANTICIPATE**
The current-carrying wire and the rectangular loop lie in the xy plane. The current is in the positive y direction.

:• **SOLVE**
Because the magnetic field is not uniform over the rectangular loop, we cannot pull $\vec{B}$ outside the integral in Equation 31.2. Instead, we must imagine cutting the rectangle into small vertical slices (Fig. 31.8). The direction of $d\vec{A}$ is perpendicular to the surface, so $d\vec{A}$ is in either the positive or negative z direction. Choose the negative z direction so that $d\vec{A}$ makes the smallest angle with the magnetic field.

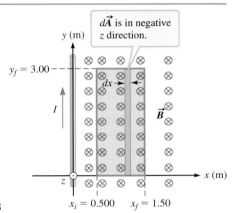

FIGURE 31.8

Because $d\vec{A}$ is parallel to $\vec{B}$, the dot product in the integral is simplified.	$\Phi_B = \int \vec{B} \cdot d\vec{A} = \int B \, dA$
The width of each slice is dx and the height is y_f, so the area of a slice is $dA = y_f \, dx$. Integrate from the left side of the rectangle ($x_i = 0.500$ m) to the right side ($x_f = 1.50$ m).	$\Phi_B = \int_{x_i}^{x_f} B y_f \, dx$
The magnetic field for a long, straight wire is given by $B = \mu_0 I / 2\pi r$ (Eq. 30.9; replace r with x).	$\Phi_B = \int_{x_i}^{x_f} \left(\frac{\mu_0 I}{2\pi x}\right) y_f \, dx = \frac{\mu_0 I y_f}{2\pi} \int_{x_i}^{x_f} \left(\frac{dx}{x}\right)$
Integrate and substitute values.	$\Phi_B = \frac{\mu_0 I y_f}{2\pi} \ln\left(\frac{x_f}{x_i}\right) = (2 \times 10^{-7} \text{ T} \cdot \text{m/A})(2.50 \text{ A})(3.00 \text{ m}) \ln\left(\frac{1.50 \text{ m}}{0.500 \text{ m}}\right)$ $\Phi_B = 1.65 \times 10^{-6}$ Wb

:• **CHECK and THINK**
The SI units work out correctly (1 T·m² = 1 Wb). The mathematical process for calculating magnetic flux is the same as the process for calculating electric flux (Section 25-2). Whereas the magnetic flux through any Gaussian surface is zero, the magnetic flux through an open surface is not necessarily zero. Magnetic flux through an open surface is very important in the next chapter, which will focus on the fourth of Maxwell's equations—Faraday's law.

31-3 Ampère's Law

Gauss's law for electricity gave us an alternative to calculating the electric field using the "brute force" method of applying Coulomb's law (Chapter 24). In practice, we used Gauss's law to find expressions for the electric field created by highly symmetrical charged sources such as an infinite line, sphere, or infinitely long cylinder (Chapter 25).

Using Coulomb's law to calculate the electric field is a lot like using the Biot-Savart law to calculate the magnetic field (Chapter 30). In both cases, we imagine slicing the source into a number of small pieces, calculating the electric or magnetic field due to each small piece, and then integrating over the entire source. What alternative do we have to the Biot-Savart law? Gauss's law for magnetism cannot be used to calculate the magnetic field because the right side of Gauss's law for magnetism equals zero. So, we get no information about the field.

However, Ampère came up with a law—like Gauss's law for electricity—that can be used to calculate the magnetic field for sources that have a high degree of symmetry. **Ampère's law** is expressed mathematically as

AMPÈRE'S LAW

❶ Underlying Principle

$$\oint \vec{B} \cdot d\vec{\ell} = \mu_0 I_{\text{thru}} \tag{31.5}$$

Path integrals were introduced in Section 9-4 on work and then used again to calculate potential energy and potential.

The symbol $d\vec{\ell}$ is used in Biot-Savart's law to represent a piece of a wire.

The integral in Equation 31.5 is a **path integral**, and $d\vec{\ell}$ is a small piece of that path. In Ampère's law, the path integral is around a closed path known as an **Ampèrian loop** as indicated by the circle on the integral symbol. Like a Gaussian surface, an Ampèrian loop is imaginary, so we are free to choose a loop of any shape or size. We are also free to choose the direction over which we integrate. The path integral of a vector over a closed path is sometimes called the **circulation integral** or, simply, the **circulation**.

EXAMPLE 31.3 A Rectangular Loop

Consider a uniform magnetic field $\vec{B}$ and complete the circulation integral $\oint \vec{B} \cdot d\vec{\ell}$. Integrate counterclockwise over the rectangular Ampèrian loop as indicated by the direction of the $d\vec{\ell}$ vectors in Figure 31.9.

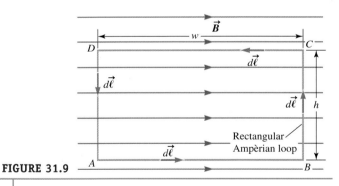

FIGURE 31.9

:• **INTERPRET and ANTICIPATE**

Because the rectangular loop consists of four straight paths, we can break up the circulation integral into four parts: $\vec{B} \cdot d\vec{\ell}$ is constant over each path, which makes the integration easy. Now we complete the four separate integrals along the four straight paths.

$$\oint \vec{B} \cdot d\vec{\ell} = \int_A^B \vec{B} \cdot d\vec{\ell} + \int_B^C \vec{B} \cdot d\vec{\ell} + \int_C^D \vec{B} \cdot d\vec{\ell} + \int_D^A \vec{B} \cdot d\vec{\ell}$$

:• **SOLVE**

From A to B: Both the magnetic field $\vec{B}$ and the direction of the path $d\vec{\ell}$ point toward the right, so the dot product is simply the product $Bd\ell$. The magnetic field is constant over this path, so we can pull B outside the integral and complete the integral. The length ℓ from A to B is the width w of the rectangle (Fig. 31.9).

$$\int_A^B \vec{B} \cdot d\vec{\ell} = \int_A^B B\,d\ell = B \int_A^B d\ell = B\ell \Big|_A^B$$

$$\int_A^B \vec{B} \cdot d\vec{\ell} = Bw \tag{1}$$

From C to D: This integral is similar to the first integral, so we solve it next. The magnetic field $\vec{B}$ points toward the right and the path $d\vec{\ell}$ points toward the left. Because $\vec{B}$ and $d\vec{\ell}$ are in opposite directions, the dot product is negative. As before, we can pull B outside the integral.	$\int_C^D \vec{B} \cdot d\vec{\ell} = -\int_C^D B d\ell = -B\int_C^D d\ell = -B\ell\Big|_C^D$ $\int_C^D \vec{B} \cdot d\vec{\ell} = -Bw$ (2)
From B to C and from D to A: For the path from B to C, $d\vec{\ell}$ is upward and $\vec{B}$ is toward the right, and for the last path, $d\vec{\ell}$ is downward and $\vec{B}$ is toward the right. Because $\vec{B}$ is perpendicular to $d\vec{\ell}$ for both these paths, the dot products and therefore the integrals are zero.	$\int_B^C \vec{B} \cdot d\vec{\ell} = 0$ (3) $\int_D^A \vec{B} \cdot d\vec{\ell} = 0$ (4)
The integral over the Ampèrian loop is found by adding the results for the four paths (Eqs. 1–4).	$\oint \vec{B} \cdot d\vec{\ell} = Bw + 0 + (-Bw) + 0$ $\oint \vec{B} \cdot d\vec{\ell} = 0$

:• **CHECK and THINK**

For this rectangular loop, the circulation is zero. This result is consistent with Ampère's law, $\oint \vec{B} \cdot d\vec{\ell} = \mu_0 I_{thru}$ (Eq. 31.5), which says the circulation is proportional to the current I_{thru} through the loop. There is no current through this Ampèrian loop, so the circulation is zero.

Finding the Current Through a Loop

To find the current I_{thru}, imagine stretching a soap film over the loop (see Fig. 36.12, page 1162). A wire that goes through the film contributes to the current I_{thru} through the loop. The current may be negative or positive depending on the direction of the current relative to the direction of the Ampèrian path. The shape of the loop and the path direction are chosen for convenience. Once the path direction in the Ampèrian loop has been chosen, use your right hand to find the direction of positive current through that loop. Wrap the fingers of your right hand in the direction of the path, and then your thumb points in the direction of positive current. For example, in Figure 31.10A, your thumb points in the positive x direction, so I_1 is positive and I_2 is negative. The current through this Ampèrian loop is $I_{thru} = I_1 - I_2$. The third wire carries current I_3. Because this wire does not go through the "soap film," it does not contribute to I_{thru}.

In Figure 31.10A, the soap film is in the same plane as the Ampèrian loop. However, the right-hand rule holds even if the imaginary film is distended. For example, the Ampèrian loop in Figure 31.10B is a circle lying in the yz plane.

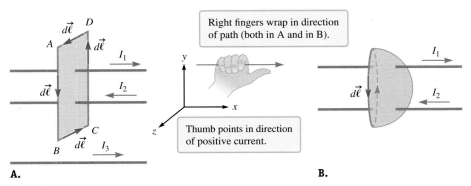

FIGURE 31.10 A. The net current through the rectangular Ampèrian loop is $I_{thru} = I_1 - I_2$. **B.** Imagine a soap film over this circular loop is stretched into a hemisphere. (Dashed portion of the path is *behind* the plane of the page.) The net current through the circular Ampèrian loop is still $I_{thru} = I_1 - I_2$.

Imagine the soap film bubbles outward as shown, with two wires penetrating it. The direction of the Ampèrian path is similar to the direction in Figure 31.10A as indicated by the $d\vec{\ell}$ vector, so your thumb points in the positive x direction as before. The current through this Ampèrian loop is $I_{\text{thru}} = I_1 - I_2$, the same as in Figure 31.10A.

CONCEPT EXERCISE 31.3

Two circular Ampèrian loops lie in the xz plane. Curved wires carrying current I penetrate the imaginary soap films as shown in Figure 31.11. Find the current through each Ampèrian loop.

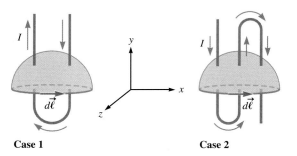

Case 1   Case 2

FIGURE 31.11

EXAMPLE 31.4 Finding Current

The circulation integral around the Ampèrian loop shown in case 2 of Figure 31.11 is $\oint \vec{B} \cdot d\vec{\ell} = -4.40 \times 10^{-6}$ T·m. What is the current I in the bent wire?

:• INTERPRET and ANTICIPATE

Ampère's law quickly leads to the answer, independent of the shape or size of the loop. All we need to know is that the circulation is proportional to the current through the film, $I_{\text{thru}} = I - 2I = -I$.

:• SOLVE

Start with Ampère's law (Eq. 31.5), and then divide the circulation by μ_0.

$$\oint \vec{B} \cdot d\vec{\ell} = \mu_0 I_{\text{thru}} \quad (31.5)$$

$$I_{\text{thru}} = \frac{1}{\mu_0} \oint \vec{B} \cdot d\vec{\ell} = \frac{-4.40 \times 10^{-6} \text{ T·m}}{(4\pi \times 10^{-7} \text{ T·m/A})}$$

$$I_{\text{thru}} = -3.50 \text{ A}$$

By the right-hand rule for finding the current through a loop (Fig. 31.10), we have $I_{\text{thru}} = -I$.

$$I_{\text{thru}} = -I = -3.50 \text{ A}$$

$$I = 3.50 \text{ A}$$

:• CHECK and THINK

It may seem odd that the circulation integral is negative, but the current we found is positive. Remember that we found the sign of the current through the loop consistent with the direction of the path using the right-hand rule.

PROBLEM-SOLVING STRATEGY

Finding the Magnetic Field Using Ampère's Law

We can use Ampère's law to find the magnetic field for a number of different sources, much like using Gauss's law for electricity to find the electric field. The steps here are similar to those we used in Section 25-3.

:• INTERPRET and ANTICIPATE
Step 1 Sketch the magnetic field lines. You may need to draw more than one perspective. Your sketch will help you choose the Ampèrian loop and anticipate your results.
Step 2 Choose an Ampèrian loop. You must choose a closed path that exploits the symmetry of the situation and also choose the direction of the path. If possible, choose the direction so that $\vec{B}$ and $d\vec{\ell}$ are parallel. In practice, pick an Ampèrian loop so that each piece of the path has an angle φ equal to 0, 90°, or 180° between $\vec{B}$ and $d\vec{\ell}$. The most commonly used Ampèrian loops are circles and rectangles.

:• SOLVE
Step 3 Do the **circulation integral** in $\oint \vec{B} \cdot d\vec{\ell} = \mu_0 I_{thru}$ (Eq. 31.5). Because the Ampèrian loop exploits the symmetry of the situation, the magnetic field is often a constant over each portion of the integral, and so the integral is usually simple. It may help to sketch a few length vectors $d\vec{\ell}$.
Step 4 Determine the amount of **current** I_{thru} through the Ampèrian loop. Imagine a soap film stretched across the loop. In most cases, you may assume the film is taut and in the same plane as the loop. Wrap the fingers of your right hand in the direction of the path you have chosen; then your thumb points in the direction of positive current. Add the currents through the film to find I_{thru}.

31-4 Special Case: Linear Symmetry

In the next three sections, we use Ampère's law to find expressions for the magnetic field strength as a function of position for sources with commonly found shapes. We begin with a long, straight, current-carrying wire.

EXAMPLE 31.5 $\vec{B}$ Due to a Long, Straight Wire Using Ampère's Law

In Section 30-5, we found the magnetic field due to a long, straight wire carrying current I by applying the Biot-Savart law:

$$B = \frac{\mu_0 I}{2\pi r} \qquad (30.9)$$

where r is the shortest distance from the wire to the point where the field is measured. The magnetic field wraps around the wire as shown in Figure 31.12. Use Ampère's law to re-derive Equation 30.9 for a long, straight wire.

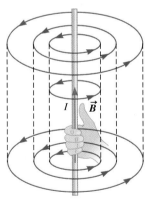

FIGURE 31.12

:• INTERPRET and ANTICIPATE
Step 1 Sketch the magnetic field lines. In this case, it is helpful to draw an end-on sketch of the wire coming out of the page and the magnetic field lines as concentric circles around the wire (Fig. 31.13).

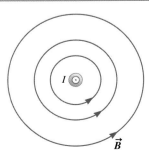

FIGURE 31.13

Example continues on page 988 ▶

Step 2 Choose an Ampèrian loop. A circular Ampèrian loop exploits the symmetry of the source (Fig. 31.14). We choose a path in the same direction as the magnetic field, so the angle φ between $\vec{B}$ and $d\vec{\ell}$ is zero all around the path. The radius of the circle is r.

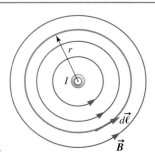

FIGURE 31.14

SOLVE

Step 3 Do the **circulation integral**. Because $\vec{B}$ and $d\vec{\ell}$ are parallel all around the circular loop, their dot product is $\vec{B} \cdot d\vec{\ell} = B\,d\ell$. The magnetic field is constant over the entire path.

$$\oint \vec{B} \cdot d\vec{\ell} = \oint B\,d\ell = B \oint d\ell$$

The integral $\oint d\ell$ covers the entire length of the path, which is the circumference $2\pi r$ of the circle.

$$\oint \vec{B} \cdot d\vec{\ell} = B(2\pi r)$$

Step 4 Find current I_{thru}. Imagine a soap film stretched taut across the circular Ampèrian loop (Fig. 31.14). The only current penetrating the film is the current I in the wire. Wrap the fingers of your right hand in the counterclockwise direction of the path; then your thumb points out of the page and the current through the Ampèrian loop is $I_{\text{thru}} = +I$. Substitute $I_{\text{thru}} = I$ into Equation (1).

$$\oint \vec{B} \cdot d\vec{\ell} = B(2\pi r) = \mu_0 I_{\text{thru}}$$

$$B = \frac{\mu_0 I_{\text{thru}}}{2\pi r} \tag{1}$$

$$B = \frac{\mu_0 I}{2\pi r} \checkmark \tag{30.9}$$

CHECK and THINK

This result is exactly what we found using the Biot-Savart law, but using Ampère's law is much simpler. Why would we ever use the Biot-Savart law? The Biot-Savart law can be applied to a source of any shape, but Ampère's law is best used to find the magnetic field produced by a source with a high degree of symmetry.

CONCEPT EXERCISE 31.4

CASE STUDY Magnetic Field Due to a Long, Straight Wire

In a laboratory, you measure the magnitude of the magnetic field generated by a long, straight wire, and you plot your results—B as a function of position r. Which of the graphs in Figure 31.15 best represents the magnetic field due to a long, straight wire?

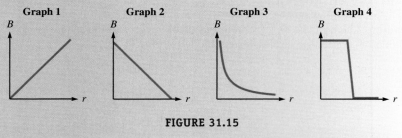

FIGURE 31.15

EXAMPLE 31.6 CASE STUDY Designing an Experiment for a Long, Straight Wire

When designing an experiment, you need to start with your expected results. Suppose you want to confirm $B = (\mu_0 I)/(2\pi r)$ (Eq. 30.9) for a long, straight wire. You decide your work will be easy if the current in the wire is exactly 1 A. Find the magnetic field strength at $r = 0.001$ m, 0.01 m, 0.1 m, 1.0 m, and 10 m from the wire. Imagine using a compass to measure the magnetic field strength at these positions. Assume the magnetic field at each position of the compass is perpendicular to the Earth's magnetic field, and find the deflection of the compass needle at each position. In **CHECK and THINK**: Sketch the deflection of the needle for the first three positions, and comment on your experimental design.

INTERPRET and ANTICIPATE
The magnetic field is stronger near the wire than it is farther away, so the compass needle will be deflected more when it is close to the wire. The question is: Will we be able to discern the deflection both near the wire and far from it?

SOLVE
To tabulate values of B, start by substituting $I = 1$ A into Equation 30.9.

$$B = \frac{\mu_0 I}{2\pi r} = \frac{(4\pi \times 10^{-7}\,\text{T}\cdot\text{m/A})(1\,\text{A})}{2\pi r}$$

$$B = \frac{2 \times 10^{-7}\,\text{T}\cdot\text{m}}{r} \tag{1}$$

We must also tabulate the angles θ (compass needle deflections), so solve Equation 31.1 for θ. We plan to use the Earth's magnetic field as the reference.

$$B = B_\oplus \tan\theta \tag{31.1}$$

$$\theta = \tan^{-1}\left(\frac{B}{B_\oplus}\right) = \tan^{-1}\left(\frac{B}{0.5 \times 10^{-4}\,\text{T}}\right) \tag{2}$$

Now substitute each value of r into Equation (1) to calculate B. Once B is calculated, substitute into Equation (2) to find θ for each distance r.

r (m)	B (T)	θ (°)
0.001	2.00×10^{-4}	76
0.01	2.00×10^{-5}	22
0.1	2.00×10^{-6}	2.3
1.0	2.00×10^{-7}	0.23
10	2.00×10^{-8}	0.023

CHECK and THINK
The sketch (Fig. 31.16) comes from the first three entries in the table. The Earth's magnetic field is in the positive y direction (straight up in this figure), and the current points out of the page. The top point (at 1 mm) is so close to the wire that the compass needle looks like it is touching the wire in the sketch. So, one problem is that we probably cannot get a compass needle that close to the wire, and we'll have to start at about 1 cm away from the wire. Another problem is that by 10 cm away, the compass needle deflection is nearly indistinguishable from zero. We will have difficulty measuring this slight deflection and will certainly not be able to measure the deflection at 1.0 m or 10 m. So we might hope to make about five reasonable measurements between about 1 cm and 10 cm. In Problem 24, you will be asked to make other improvements to this experiment.

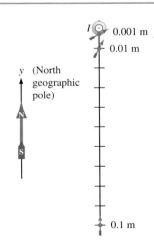

FIGURE 31.16

EXAMPLE 31.7 A Layered Cylindrical Wire

A very long, cylindrical wire has an insulating core surrounded by a cylindrical conductor carrying current I (Fig. 31.17). The current is uniformly distributed in the conducting cylinder. The radius of the insulating core is r_1, and the outer radius of the conducting cylinder is r_2. Find expressions for the magnetic field strength **A** outside the conducting cylinder ($r > r_2$), **B** inside the conducting cylinder but outside the insulator ($r_2 > r > r_1$), and **C** inside the insulator ($r < r_1$).

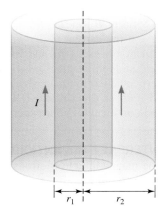

FIGURE 31.17 Current is only in the conducting cylindrical shell, not in the insulating core.

INTERPRET and ANTICIPATE

We can use Ampère's law and the problem-solving strategy to find expressions for B. The geometry resembles that of a long, straight wire, so start with an end-on view showing all three regions. We need three different Ampèrian loops—one for each region in which we must find B.

Step 1 Sketch magnetic field lines. Figure 31.18 is an end-on sketch showing the geometry of the situation. By the symmetry of the source, the magnetic field lines are circular in all regions. Because the conductor carries current I out of the page, the magnetic field lines are counterclockwise by the simple right-hand rule. (For now, don't worry if we find that there is no magnetic field in one or more regions.)

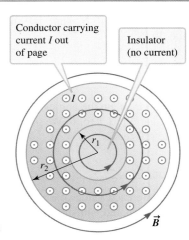

FIGURE 31.18

A Outside the hollow conducting cylinder.
Step 2 Choose an Ampèrian loop in the region outside the conductor ($r > r_2$). Because this region is much like the region outside a long, straight wire, a circular path in the direction of the magnetic field works well. The loop has radius $r > r_2$ (Fig. 31.19).

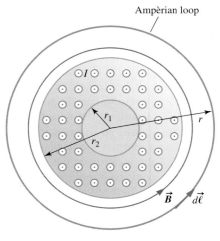

FIGURE 31.19

SOLVE
Step 3 Do the circulation integral. This calculation is exactly the same as for a long, straight wire.

$$\oint \vec{B} \cdot d\vec{\ell} = \oint B\, d\ell = B \oint d\ell = B(2\pi r)$$

Step 4 Find current I_{thru}. Just as for a long, straight wire, the current I penetrates an imaginary soap film stretched taut across the circular loop. Wrap the fingers of your right hand in the counterclockwise direction of the path; then your thumb points out of the page. Because I is out of the page, the current through the Ampèrian loop is $I_{thru} = +I$. Substitute $I_{thru} = I$ into Equation (1).

$$\oint \vec{B} \cdot d\vec{\ell} = B(2\pi r) = \mu_0 I_{thru}$$

$$B = \frac{\mu_0 I_{thru}}{2\pi r} \quad (1)$$

$$B = \frac{\mu_0 I}{2\pi r} \text{ for } r > r_2$$

:• **CHECK and THINK**
The magnetic field outside the conductor is exactly what we found for the magnetic field outside a long, straight, current-carrying wire.

B Inside the conducting cylinder but outside the insulator.

:• **INTERPRET and ANTICIPATE**
Step 2 Choose an Ampèrian loop in the region inside the conductor ($r_2 > r > r_1$). Again, a circular path in the direction of the field works well (Fig. 31.20). The loop has radius r, where $r_2 > r > r_1$.

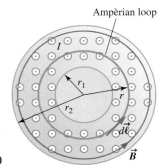

FIGURE 31.20

:• **SOLVE**
Step 3 Do the circulation integral. This calculation is the same as for part A.

$$\oint \vec{B} \cdot d\vec{\ell} = \oint B d\ell = B \oint d\ell$$

$$\oint \vec{B} \cdot d\vec{\ell} = B(2\pi r) \quad (2)$$

Step 4 Find current I_{thru}. Now we find a departure from the case of a long, straight wire. The current through the Ampèrian loop is *not* I; only a fraction of I penetrates an imaginary film covering the loop (Fig. 31.20). To find that fraction, begin with the current density (Eq. 28.3).

$$J = \frac{I}{A} \quad (28.3)$$

We need the cross-sectional area A of the conductor: a disk with a hole in it. The required area is that of the disk (radius r_2) minus that of the hole (radius r_1).

$$A = \pi r_2^2 - \pi r_1^2 = \pi(r_2^2 - r_1^2)$$

Substitute A into $J = I/A$, Equation 28.3.

$$J = \frac{I}{\pi(r_2^2 - r_1^2)} \quad (3)$$

There is no current through the insulator. The "film" area $A_{current}$ that is penetrated by the current is the area inside the Ampèrian loop (radius r) minus the area of the hole (radius r_1). (Keep in mind that $A_{current}$ does not equal the area A of the conductor.)

$$A_{current} = \pi r^2 - \pi r_1^2$$

$$A_{current} = \pi(r^2 - r_1^2) \quad (4)$$

The current I_{thru} penetrating that film is the current density J times $A_{current}$. Substitute Equation (3) for J and Equation (4) for $A_{current}$. The numerator and denominator do not cancel unless the Ampèrian loop is the size of the cylinder.

$$I_{thru} = J A_{current}$$

$$I_{thru} = \left[\frac{I}{\pi(r_2^2 - r_1^2)}\right][\pi(r^2 - r_1^2)]$$

$$I_{thru} = I \frac{(r^2 - r_1^2)}{(r_2^2 - r_1^2)} \quad (5)$$

Example continues on page 992 ▶

Substitute Equations (2) and (5) into Ampère's law, and solve for B.

$$\oint \vec{B} \cdot d\vec{\ell} = \mu_0 I_{thru}$$

$$B(2\pi r) = \mu_0 I \frac{(r^2 - r_1^2)}{(r_2^2 - r_1^2)}$$

$$B = \frac{\mu_0 I}{2\pi} \frac{(r^2 - r_1^2)}{(r_2^2 - r_1^2)} \frac{1}{r} \text{ for } r_2 > r > r_1 \quad (6)$$

:• **CHECK and THINK**
Equation (6) must give the same answer as Equation 30.9 at the outer surface of the conductor ($r = r_2$). Substitute $r = r_2$ into both equations; the results are the same.

From Equation 30.9, $B = \dfrac{\mu_0 I}{2\pi r_2}$

From Equation (6),
$$B = \frac{\mu_0 I}{2\pi} \frac{(r_2^2 - r_1^2)}{(r_2^2 - r_1^2)} \frac{1}{r_2} = \frac{\mu_0 I}{2\pi} \frac{1}{r_2}$$

C Inside the insulator

:• **INTERPRET and ANTICIPATE**
Step 2 Choose an Ampèrian loop. Now the Ampèrian loop is a circle inside the insulator ($r < r_1$; Fig. 31.21).

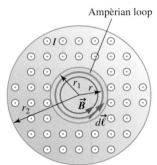

FIGURE 31.21

:• **SOLVE**
Step 3 Do the **circulation integral**. This calculation is the same as in parts A and B.

$$\oint \vec{B} \cdot d\vec{\ell} = B(2\pi r)$$

Step 4 Find **current** I_{thru}. Again imagine a film stretched over the Ampèrian loop; the conducting cylinder does not penetrate the film. Therefore, $I_{thru} = 0$, and by Ampère's law, the magnetic field inside the insulator is zero. (Now we know we didn't need the inner magnetic field line in Fig. 31.18 or Fig. 31.21.)

$$\oint \vec{B} \cdot d\vec{\ell} = \mu_0 I_{thru}$$
$$B(2\pi r) = \mu_0(0)$$
$$B = 0 \text{ for } r < r_1$$

:• **CHECK and THINK**
To check our result, we can calculate the magnetic field at the interface between the insulator and the conductor ($r = r_1$) using Equation (6). As expected, the magnetic field at $r = r_1$ is zero. In Problem 26, you will put all three parts of this example together to graph the magnetic field in all three regions.

$$B = \frac{\mu_0 I}{2\pi} \frac{(r_1^2 - r_1^2)}{(r_2^2 - r_1^2)} \frac{1}{r_1^2} = 0$$

31-5 Special Case: Solenoids

Today, we know that the magnetic field of a permanent magnet is largely due to the spin of electrons—an intrinsic property.

Just one week after Ampère theorized that moving charged particles are the source of all magnetic fields, the French physicist Dominique François Arago (1786–1853) presented a practical demonstration. Arago showed that a current-carrying coil of wire acts like a bar magnet by attracting bits of iron filings and that when the current is cut off, the coil's magnetic properties disappear. This demonstration led Ampère to try to magnetize an iron rod by placing it inside the coil. Ampère found that when the iron rod is removed from the coil, it is a permanent magnet.

A **solenoid** is a coil that consists of many windings or loops of wire. Solenoids are found in the circuits of many common electrical devices. Before we use Ampère's law to find the magnetic field strength due to a current-carrying solenoid, let's try to visualize the solenoid's magnetic field.

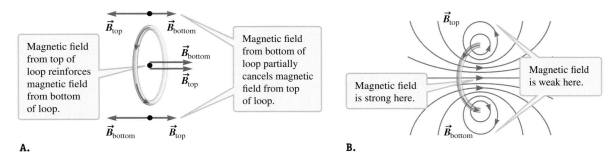

FIGURE 31.22 A. The current in the top portion of the loop points out of the page, and the current in the bottom portion points into the page. The direction of the magnetic field at the three points (above the loop, in the middle, and below the loop) comes from the simple right-hand rule. **B.** The lines are close together in the center of the loop because the magnetic field is strong there. Outside the loop, where the magnetic field is weaker, the lines are more widely spaced.

Each winding of a solenoid is similar to a single current loop. Figure 31.22A shows a perspective drawing of a loop that is perpendicular to the page. From this perspective, there is a current coming out of the page at the top of the loop and an equal current going into the page at the bottom. Consider a point directly above the loop; the top portion of the loop produces a magnetic field toward the left and the bottom portion produces a magnetic field toward the right. The point above the loop is closer to the top portion, so the magnetic field produced by the top portion is stronger than the bottom portion's magnetic field. The two vectors partially cancel out, resulting in a magnetic field pointing leftward for this point above the loop. A similar argument about the magnetic fields partially cancelling out can be made for the point directly below the loop, again resulting in a magnetic field that points toward the left. However, in the center of the loop, both the top portion and the bottom portion produce magnetic fields that point toward the right. The result is a strong magnetic field in the center of the loop and a weaker magnetic field outside (Fig. 31.22B).

A solenoid consists of many windings. First, consider the magnetic field produced by two loops placed near each other (Fig. 31.23). Between two top or two bottom portions, the curved magnetic field lines point in roughly opposite directions, indicating that the magnetic fields nearly cancel in these regions. The nearly straight field lines near the center of the loops are roughly in the same direction, so the magnetic fields add together and the result is that the field is strong in this region. The net result is that the magnetic field lines near the center of the two loops are tightly packed, nearly uniformly spaced, and nearly horizontal, pointing toward the right, whereas outside the loops, the magnetic field lines are widely spaced, indicating that the magnetic field is weak.

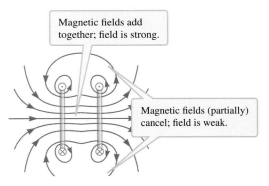

FIGURE 31.23 The magnetic fields of two loops must be added together. The combined magnetic field of two loops is strong near the center and weak outside.

Now let's look at a loosely wound solenoid (Fig. 31.24). The four turns of this solenoid produce a weak magnetic field on the outside, but inside the solenoid the magnetic field is strong and uniform. Compare the magnetic field of two loops (Fig. 31.23) to that of the loosely wound solenoid (Fig. 31.24). The solenoid has a more uniform interior magnetic field and a weaker exterior field.

In general, as the number of turns increases and as the turns are more closely packed, the interior magnetic field becomes stronger and more uniform while the exterior field becomes weaker. An **ideal solenoid** is infinitely long with tightly packed turns, a uniform interior magnetic field, and no exterior magnetic field (Fig. 31.25). In practice, a long solenoid with tightly packed turns can be approximated as an ideal solenoid in the region far from either end. Because the magnetic field inside the solenoid is uniform, the ideal solenoid is a practical device that we will encounter over and over again in electromagnetism. Throughout this textbook, you may assume all solenoids are ideal (even if they have a finite length). The direction of the magnetic field inside the solenoid is found by wrapping the fingers of your right hand in the direction of the current; then your thumb points in the field's direction.

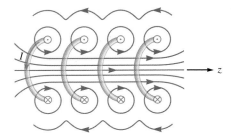

FIGURE 31.24 This solenoid has four loosely wound turns. The magnetic field in the inside is strong and nearly uniform in the z direction. The magnetic field outside is weak.

FIGURE 31.25 An ideal solenoid consists of tightly packed turns and is infinitely long. Only a few loops of wire are shown here. In many figures the loops are not drawn and the current is only indicated by the circled dots and circled crosses.

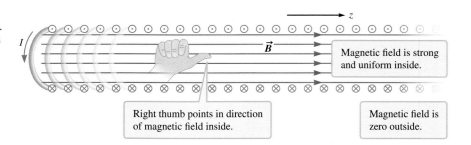

Right thumb points in direction of magnetic field inside.

Magnetic field is strong and uniform inside.

Magnetic field is zero outside.

EXAMPLE 31.8 $\vec{B}$ for a Solenoid

We have reasoned that the magnetic field outside an ideal solenoid is zero. Now use Ampère's law to show that the magnetic field strength inside an ideal solenoid with n turns per unit length is given by

$$B = \mu_0 n I \quad (31.6)$$

INTERIOR MAGNETIC FIELD OF IDEAL SOLENOID ▶ Special Case

INTERPRET and ANTICIPATE

Steps 1 and 2 Sketch the **magnetic field lines**, and choose an **Ampèrian loop**. Figure 31.26 shows a slice through the solenoid. Because the interior magnetic field forms horizontal parallel lines, a rectangular Ampèrian loop exploits the symmetry. The width of the rectangle is w. We have chosen a counterclockwise path so that the part of the loop inside the solenoid is lined up with the interior magnetic field.

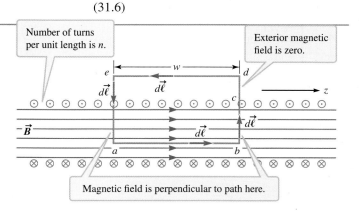

Number of turns per unit length is n.

Exterior magnetic field is zero.

Magnetic field is perpendicular to path here.

FIGURE 31.26

SOLVE
Step 3 Do the **circulation integral**. Because the loop is a rectangle, break up the integral into four pieces, one for each side of the rectangle.

$$\oint \vec{B} \cdot d\vec{\ell} = \int_a^b \vec{B} \cdot d\vec{\ell} + \int_b^d \vec{B} \cdot d\vec{\ell} + \int_d^e \vec{B} \cdot d\vec{\ell} + \int_e^a \vec{B} \cdot d\vec{\ell} \quad (1)$$

The path from a to b is parallel to the magnetic field, and the magnetic field is uniform along that part of the path. The dot product is $B d\ell$. The path length is the width of the rectangle.

$$\int_a^b \vec{B} \cdot d\vec{\ell} = \int_a^b B d\ell = B \int_a^b d\ell$$

$$\int_a^b \vec{B} \cdot d\vec{\ell} = Bw \quad (2)$$

The magnetic field is not constant over the path from b to d. Inside the solenoid from b to c, the magnetic field has some uniform value B, and outside the solenoid from c to d, the magnetic field is zero. We break up this straight path into two pieces. The integral over the path outside the solenoid (from c to d) is zero because the magnetic field is zero. The integral over the path inside the solenoid is also zero because the magnetic field is perpendicular to the path. So, the entire integral from b to d is zero.

$$\int_b^d \vec{B} \cdot d\vec{\ell} = \int_b^c \vec{B} \cdot d\vec{\ell} + \int_c^d \vec{B} \cdot d\vec{\ell}$$

$$\int_b^d \vec{B} \cdot d\vec{\ell} = \int_b^c B d\ell \cos 90° + \int_c^d 0$$

$$\int_b^d \vec{B} \cdot d\vec{\ell} = 0 \quad (3)$$

The integral from d to e is zero because the exterior magnetic field is zero.

$$\int_d^e \vec{B} \cdot d\vec{\ell} = 0 \quad (4)$$

Finally, the integral from e back to a is similar to the integral from b to d. Outside the solenoid, the magnetic field is zero, and inside the solenoid, the magnetic field is perpendicular to the path. So, the integral from e to a is zero.	$\int_e^a \vec{B} \cdot d\vec{\ell} = 0 \quad (5)$
Substitute Equations (2) through (5) into the circulation integral (Eq. 1).	$\oint \vec{B} \cdot d\vec{\ell} = Bw + 0 + 0 + 0 = Bw \quad (6)$
Step 4 Find **current** I_{thru}. Imagine a soap film stretched taut across the rectangular Ampèrian loop in Figure 31.26. If N turns penetrate the film and each turn carries current I, the total current that penetrates the film is NI. Because an ideal solenoid is infinitely long, it has an infinite number of turns. We should characterize an ideal solenoid in terms of the number of turns per unit length (n, measured in m^{-1}) instead of the total number of turns. The number N of turns that penetrate the film is then the width w of the rectangle times n. Finally, we must determine whether the current through the loop is positive or negative. Wrap the fingers of your right hand in the direction of the path from a around to e; then your thumb points out of the page. Because the top of each turn carries current out of the page, the current through the film is positive.	$I_{thru} = NI$ $I_{thru} = nwI \quad (7)$
Substitute Equations (6) and (7) into Ampère's law, and solve for B.	$\oint \vec{B} \cdot d\vec{\ell} = \mu_0 I_{thru}$ $Bw = \mu_0 nwI$ $B = \mu_0 nI \quad \checkmark \quad (31.6)$

:• **CHECK and THINK**

At the minimum, you need to confirm that our expression has the dimensions of magnetic field. Notice that Equation 31.6 does not have any spatial dependence. (It does not involve r, x, y, or z.) This result confirms that inside an ideal solenoid, the magnetic field is uniform. Throughout this textbook, you may use Equation 31.6 to find the interior magnetic field of any solenoid, whether or not it is described as an *ideal* solenoid, and assume the exterior magnetic field is zero.

CONCEPT EXERCISE 31.5

CASE STUDY Designing an Experiment for an Ideal Solenoid

Suppose you want to confirm $B = 0$ outside and $B = \mu_0 nI$ (Eq. 31.6) inside an ideal solenoid using a compass. Your ideal solenoid has exactly 5000 turns per meter. You align your solenoid so that its magnetic field is perpendicular to the Earth's magnetic field. The compass is easiest to read when its deflection is between 10° and 80°. What range of current will work best for your experiment?

EXAMPLE 31.9 Homemade Solenoid

Suppose you want to make your own solenoid using a 9-V battery and copper wire. You wrap the wire in one neat layer around a portion of a pencil. Estimate the magnetic field strength inside your solenoid. Assume the internal resistance of the battery is about 2 Ω.

:• **INTERPRET and ANTICIPATE**

Once we know the current in the wire and the number of turns per unit length, we can find the magnetic field inside the solenoid. The number of turns per unit length comes from estimating the parameters—the radius of the pencil, the length of the solenoid, and the thickness of the wire. The current depends on these geometric factors and also on the resistivity of copper. Start by estimating the geometric factors based on your experience with pencils and wires.

:• SOLVE

Using a ruler shows that a pencil is about 1 cm in diameter, so a pencil's radius is $r = 0.5$ cm. Now imagine a wire about as thick as a paper clip, which is slightly thinner than 1 mm. Finally, imagine wrapping the wire around the pencil. You will probably leave a little room at the ends for convenience, so the solenoid is about 10 cm long.	Pencil's radius $r = 0.005$ m Wire's thickness $t = 0.0008$ m Solenoid's length $L = 0.1$ m
The total number of turns N is found by dividing the length of the solenoid by the thickness of one turn (the thickness of the wire). If you had the solenoid in your hands, you could count the number of turns.	$N = \dfrac{L}{t} = \dfrac{0.1\,\text{m}}{0.0008\,\text{m}}$ $N = 125$ turns
Find the number of turns per unit length by dividing the number of turns N by the length of the solenoid.	$n = \dfrac{N}{L} = \dfrac{125}{0.1\,\text{m}} = 1250\,\text{m}^{-1}$ (1)
To find the current in the wire, we first need to find the wire's resistance, so we need the length ℓ of the wire. We can estimate that length as the circumference C of one turn multiplied by the total number of turns N. The circumference of one turn is roughly the circumference of the pencil.	$\ell = NC = N2\pi r$ $\ell = (125)2\pi(0.005\,\text{m}) = 4\,\text{m}$
We also need the cross-sectional area of the wire, which we find from its thickness.	$A = \pi(t/2)^2 = \pi(0.0008\,\text{m}/2)^2 = 5 \times 10^{-7}\,\text{m}^2$
The resistivity of copper is listed in Table 28.2 (page 872; $\rho = 1.68 \times 10^{-8}\,\Omega \cdot$m). Use Equation 28.24 to find the wire's resistance.	$R_{\text{wire}} = \rho\dfrac{\ell}{A}$ (28.24) $R_{\text{wire}} = (1.68 \times 10^{-8}\,\Omega \cdot \text{m})\left(\dfrac{4\,\text{m}}{5 \times 10^{-7}\,\text{m}^2}\right) = 0.13\,\Omega$
The circuit's total resistance is the sum of the battery's internal resistance and the wire's resistance. We can see that the wire's resistance does not contribute very much.	$R = R_{\text{internal}} + R_{\text{wire}} = (2 + 0.13)\,\Omega$ $R \approx 2\,\Omega$
Find the current in the wire from Equation 29.1; the sign does not matter in this problem.	$\Delta V_R = -IR$ (29.1) $I = \dfrac{\Delta V_R}{R} = \dfrac{9\,\text{V}}{2\,\Omega} = 4.5\,\text{A} \approx 4\,\text{A}$ (2)
Finally, to find the magnetic field inside the homemade solenoid, substitute values for n (Eq. 1) and I (Eq. 2) into Equation 31.6.	$B = \mu_0 nI$ (31.6) $B = (4\pi \times 10^{-7}\,\text{T}\cdot\text{m/A})(1250\,\text{m}^{-1})(4\,\text{A})$ $B = 6 \times 10^{-3}\,\text{T}$

:• CHECK and THINK

The magnetic field produced inside a crude homemade solenoid is about two orders of magnitude stronger than the Earth's magnetic field. If such a strong magnetic field were perpendicular to the Earth's magnetic field, it would deflect a compass needle by (essentially) 90°.

FIGURE 31.27 A toroid is a solenoid bent into a donut shape.

31-6 Special Case: Toroids

Another important source of a magnetic field is a toroid (Fig. 31.27). You can make a toroid by bending a solenoid into a donut. Toroids—like solenoids—are found in the circuits of many common devices. In this section, we use Ampère's law to find an expression for the magnetic field produced by an ideal toroid.

Figure 31.28 is a cross-sectional sketch of an ideal toroid; each turn is set neatly around the donut shape and there are no wires leading in or out. Take a look at the exterior regions—inside the donut hole and outside the toroid. Because there is no magnetic field outside a solenoid, we expect there is no magnetic field in either of these exterior regions. Let's use Ampère's law to confirm our expectations. First, consider an Ampèrian loop inside the donut hole. Imagine a film stretched across this loop. No wire penetrates the film, so there is no current through the Ampèrian loop: $I_{thru} = 0$. According to Ampère's law, $\oint \vec{B} \cdot d\vec{\ell} = \mu_0 I_{thru}$ (Eq. 31.5), the circulation must be zero no matter what loop shape we choose. The only way for the circulation integral to be zero independent of the path shape is for the magnetic field to be zero. We conclude that the magnetic field in the donut hole is zero.

Next, consider an Ampèrian loop outside the toroid. Again, imagine a film stretched across the loop. In this case, each turn of the toroid penetrates the film twice—once coming out of the page and once going into the page. The net result is that the total current through the loop is zero: $I_{thru} = 0$. As in the case of the Ampèrian loop in the donut hole, the circulation integral must be zero no matter what loop shape we choose. We conclude that the magnetic field outside the toroid must also be zero.

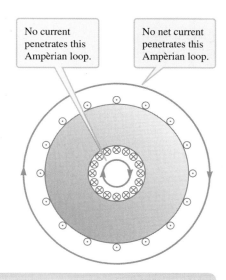

FIGURE 31.28 Circular Ampèrian loops inside the donut hole and outside the toroid.

EXAMPLE 31.10 $\vec{B}$ Inside an Ideal Toroid

Now we turn our attention to the magnetic field inside a toroid. Consider an ideal toroid that has N turns and a current I in the wire. Use Ampère's law to show that the magnetic field strength inside an ideal toroid is given by

$$B = \frac{\mu_0 NI}{2\pi r} \quad (31.7)$$

INTERIOR MAGNETIC FIELD OF IDEAL TOROID ● Special Case

Steps 1 and 2 Sketch the magnetic field lines, and **choose an Ampèrian loop**. The direction of the magnetic field is found using the same right-hand rule as for a solenoid. For any single turn, wrap the fingers of your right hand in the direction of the current, and then your thumb points in the direction of the magnetic field. The magnetic field lines bend into concentric circles on the inside, so a circular Ampèrian loop of radius r exploits the symmetry. We have chosen a clockwise path lined up with the magnetic field (Fig. 31.29).

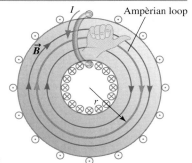

FIGURE 31.29

SOLVE
Step 3 Do the **circulation integral**. Because the magnetic field is parallel to and constant over the whole path, we can pull B outside the integral. The path integral is just the circumference of the Ampèrian loop.

$$\oint \vec{B} \cdot d\vec{\ell} = \oint B d\ell = B \oint d\ell$$

$$\oint \vec{B} \cdot d\vec{\ell} = B(2\pi r) \quad (1)$$

Step 4 Find **current** I_{thru}. Imagine a soap film stretched taut across the circular Ampèrian loop in Figure 31.29. All the turns penetrate the film. Each turn carries a current I, so the total current that penetrates the film is NI. Next, determine whether the current through the loop is positive or negative. Wrap the fingers of your right hand clockwise in the direction of the path, and your thumb points into the page. Because each turn carries current into the page, the current through the film is positive.

$$I_{thru} = NI \quad (2)$$

Substitute Equations (1) and (2) into Ampère's law, and solve for B.

$$\oint \vec{B} \cdot d\vec{\ell} = \mu_0 I_{thru}$$

$$B(2\pi r) = \mu_0 NI$$

$$B = \frac{\mu_0 NI}{2\pi r} \checkmark \quad (31.7)$$

Example continues on page 998 ▶

CHECK and THINK
Compare Equation 31.7, $B = (\mu_0 NI)/(2\pi r)$, for the magnetic field inside a toroid to Equation 31.6, $B = \mu_0 nI$, for the magnetic field inside a solenoid. The magnetic field inside a toroid depends on the position and is strongest near the donut hole but gets weaker at farther positions. By contrast, the magnetic field inside a solenoid is constant throughout the interior region.

EXAMPLE 31.11 CASE STUDY Plotting B for a Toroid

The current in a toroid is 1.0 A, and it consists of 66 turns. Suppose the radius of the donut hole is 1.2 cm and the outer radius of the toroid is 3.7 cm. Plot the toroid's magnetic field strength as a function of r from $r = 0$ to $r = 5.0$ cm.

INTERPRET and ANTICIPATE
This is an application of the equations we found for the magnetic field produced by a toroid. In the regions outside the toroid, the magnetic field strength is zero, and inside the toroid, the magnetic field depends on the position r.

SOLVE
Calculate and tabulate several values (Table 31.1). We need Equation 31.7 only for points inside the toroid—that is, for points between $r = 1.2$ cm and 3.7 cm. So, substitute all values except r and remember to work in SI units.

$$B = \frac{\mu_0 NI}{2\pi r} \quad (31.7)$$

$$B = \frac{(4\pi \times 10^{-7}\,\text{T}\cdot\text{m/A})(66)(1.0\,\text{A})}{2\pi r}$$

$$B = \frac{1.32 \times 10^{-5}}{r}\,\text{T}$$

TABLE 31.1 Field strength vs. r for toroid.

$r(\times 10^{-2}\,\text{m})$	$B(\times 10^{-4}\,\text{T})$	$r(\times 10^{-2}\,\text{m})$	$B(\times 10^{-4}\,\text{T})$	$r(\times 10^{-2}\,\text{m})$	$B(\times 10^{-4}\,\text{T})$
1.2	11.0	2.0	6.6	2.8	4.7
1.3	10.2	2.1	6.3	2.9	4.6
1.4	9.4	2.2	6.0	3.0	4.4
1.5	8.8	2.3	5.7	3.1	4.3
1.6	8.3	2.4	5.5	3.2	4.1
1.7	7.8	2.5	5.3	3.3	4.0
1.8	7.3	2.6	5.1	3.4	3.9
1.9	6.9	2.7	4.9	3.5	3.8

Plot the values from Table 31.1 as shown in Figure 31.30. Include $B = 0$ for $r < 1.2$ cm and for $r > 3.7$ cm on the graph.

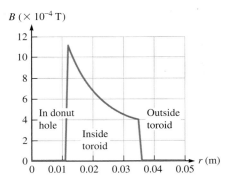

FIGURE 31.30

CHECK and THINK
Figure 31.30 clearly shows that the magnetic field inside the toroid is strongest near the donut hole. Another way to visualize this is with magnetic field lines (Fig. 31.31). The field lines inside the toroid are not evenly spaced; the concentric circles are more concentrated near the donut hole and get farther apart with increasing distance.

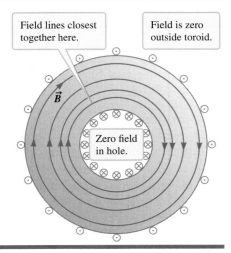

FIGURE 31.31

31-7 General Form of Ampère's Law

Maxwell's greatest insight was to combine four seemingly separate equations to describe electricity, magnetism, and light (Fig. 31.1). Each of these equations had been discovered by another researcher, and each carries the name of its discoverer. We have studied two and a half of Maxwell's equations so far: Gauss's law for electricity and Gauss's law for magnetism. The "half" is Ampère's law, which is incomplete in the form $\oint \vec{B} \cdot d\vec{\ell} = \mu_0 I_{\text{thru}}$ (Eq. 31.5). Maxwell discovered that it is missing a term, without which he would not have been able to discover that light is intimately connected to electricity and magnetism. In this section, we present Maxwell's addition to Ampère's law.

The need for Maxwell's addition to Ampère's law is best understood by considering a simple circuit that consists of an emf and a capacitor (Fig. 31.32). Imagine the switch S has been closed and the capacitor is charging. The top plate of the capacitor becomes positively charged and the bottom plate becomes negatively charged, so there is a downward-pointing electric field between the plates. As more charge builds up on the plates, the electric field grows stronger.

Suppose you wish to use Ampère's law to calculate the magnetic field at a point P; you need to use an Ampèrian loop that passes through that point (Fig. 31.33). Let's try to determine the current through the loop. If you imagine a film stretched taut across the loop (Fig. 31.33A), you find that the current through the loop is just the current in the wire: $I_{\text{thru}} = I$. However, if you imagine the film is bowed outward like a soap film so that the film encompasses the capacitor's top plate (Fig. 31.33B), there is no current through the loop.

This ambiguity over the current through the Ampèrian loop leads to a contradiction in the magnetic field at point P. If the film is taut (Fig. 31.33A), the current through the Ampèrian loop is nonzero and the magnetic field at point P is also nonzero. But if the film is bowed out (Fig. 31.33B), the current through the loop is zero and, according to Ampère's law $\oint \vec{B} \cdot d\vec{\ell} = \mu_0 I_{\text{thru}}$ (Eq. 31.5), the magnetic field must be zero. Because the magnetic field cannot be both nonzero and zero at the same time, something must be wrong with Ampère's law. Maxwell realized he could correct Ampère's law by adding a new term to the right side:

$$\oint \vec{B} \cdot d\vec{\ell} = \mu_0 I_{\text{thru}} + (\text{new term})$$

Even if $I_{\text{thru}} = 0$, as it does in the case of the bowed-out film, the new term would still give a nonzero value and then the magnetic field would be nonzero.

To see what the new term involves, take another look at Figure 31.33B. Although there is no current through the film, something else penetrates it: the electric field. The new term involves the *electric flux* (Eq. 25.3, p. 759) through the film. The general form of Ampère's law is known as Ampère–Maxwell's law:

$$\oint \vec{B} \cdot d\vec{\ell} = \mu_0 I_{\text{thru}} + \mu_0 \varepsilon_0 \frac{d\Phi_E}{dt} \quad (31.8)$$

For reasons that will become clear after its derivation, the new term (excluding the μ_0) is known as the **displacement current**. The displacement current involves the *rate of change* of the electric flux. If the electric flux is not changing, the displacement current is zero. Of course, when a capacitor is charging, the electric field is increasing, so the electric flux is changing.

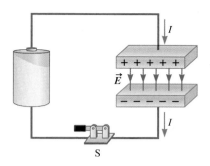

FIGURE 31.32 The switch is closed, there is current in the circuit, and the capacitor is charging. While a capacitor is charging, an electric field between the plates grows in strength.

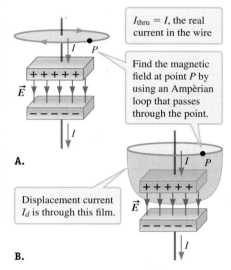

FIGURE 31.33 A. An Ampèrian loop with a film stretched taut across the loop. **B.** An Ampèrian loop with the film bowed outward like a soap bubble.

AMPÈRE–MAXWELL'S LAW
❶ Underlying Principle

DISPLACEMENT CURRENT
✪ Major Concept

DERIVATION   Displacement Current

We can gain a greater insight into the displacement current I_d by deriving it. We show that for the case of a parallel-plate capacitor with no dielectric between its plates,

$$I_d = \varepsilon_0 \frac{d\Phi_E}{dt} \quad (31.9)$$

Derivation continues on page 1000 ▶

At any particular instant, the charge stored by a capacitor is given by Equation 27.1.	$Q = C\Delta V$ (27.1)
The potential difference ΔV increases as charge continues to build up on the plates, and so does the magnitude of the electric field. (Remember that d is the distance between the plates.)	$\Delta V = Ed$ (26.17)
For a parallel-plate capacitor without a dielectric, the capacitance comes from Equation 27.10.	$C = \dfrac{\varepsilon_0 A}{d}$ (27.10)
Substitute Equations 27.10 and 26.17 into Equation 27.1.	$Q = \left(\dfrac{\varepsilon_0 A}{d}\right)(Ed) = \varepsilon_0 EA$
Substitute the electric flux $\Phi_E = EA$. This is the flux through the bowed-out film (Fig. 31.33B), and A is the area perpendicular to $\vec{E}$.	$Q = \varepsilon_0 \Phi_E$ (1)
The charge on the capacitor is increasing. The rate of charge increase is the current $I = dQ/dt$.	$I_d = \varepsilon_0 \dfrac{d\Phi_E}{dt}$ ✓ (31.9)

:• **COMMENTS**
The **displacement current** I_d is a fictitious current through the bowed-out film in Figure 31.33B. There is no real current between the plates of a capacitor. The displacement current is really a changing electric field between the capacitor's plates, which provides us with a helpful way to think about how the changing electric field between the plates can create a magnetic field at point P.

AMPÈRE–MAXWELL'S LAW

❶ **Underlying Principle**

Ampère–Maxwell's law can be written in terms of two currents—the real current I_{thru} and the displacement current I_d through the loop:

$$\oint \vec{B} \cdot d\vec{\ell} = \mu_0(I_{\text{thru}} + I_d) = \mu_0 I_{\text{tot}} \qquad (31.10)$$

where $I_{\text{tot}} = I_{\text{thru}} + I_d$ is the total (real plus displacement) current through the Ampèrian loop. The value of the displacement current is the same as the real current in the wire: $I_d = I$. So, whether we imagine a film stretched taut (Fig. 31.33A) or bowed out (Fig. 31.33B), we find the same value for the magnetic field.

For the taut film (Fig. 31.33A), the general form of Ampère's law is

$$\oint \vec{B} \cdot d\vec{\ell} = \mu_0 I + 0 \qquad (31.11)$$

where $I_{\text{thru}} = I$ and there is (almost) no electric flux through the taut film. For the bowed-out film (Fig. 31.33B), there is no real current through the film, so $I_{\text{thru}} = 0$. Maxwell–Ampère's law becomes

$$\oint \vec{B} \cdot d\vec{\ell} = 0 + \mu_0 \varepsilon_0 \dfrac{d\Phi_E}{dt}$$

The second term on the right includes the displacement current (Eq. 31.9):

$$\oint \vec{B} \cdot d\vec{\ell} = 0 + \mu_0 I_d$$

and the displacement current equals the real current in the wire:

$$\oint \vec{B} \cdot d\vec{\ell} = 0 + \mu_0 I \qquad (31.12)$$

This is exactly what we found (Eq. 31.11) for the taut film (Fig. 31.33A).

What about the direction of the magnetic field produced by the displacement current? The displacement current points in the same direction as the electric field if the electric flux is increasing. If the electric flux is decreasing, the displacement current is in the opposite direction as the electric field. To find the direction of the magnetic field, use the simple right-hand rule as you would for a real current. Point your right thumb in the direction of the displacement current, and your fingers naturally wrap in the direction of the magnetic field.

Finally, it is sometimes helpful to work with the **displacement current density**. Like the real current density, the displacement current density J_d is the displacement current divided by the cross-sectional area A:

$$J_d = \frac{I_d}{A} = \frac{1}{A}\left(\varepsilon_0 \frac{d\Phi_E}{dt}\right) = \frac{1}{A}\left[\varepsilon_0 \frac{d(EA)}{dt}\right]$$

$$J_d = \varepsilon_0 \frac{dE}{dt} \quad (31.13)$$

Newton and Maxwell both made major contributions in physics. Newton's laws are the basis of classical mechanics, and Maxwell's equations are the basis of electricity, magnetism, and light. Both scientists used their imagination to discover fundamental laws of nature. Newton imagined a world without friction, which enabled him to discover the law of inertia. Maxwell imagined a displacement current between the plates of a capacitor, allowing him to generalize Ampère's law and take a necessary step toward his greatest discovery that light is an electromagnetic phenomenon (Chapter 34).

EXAMPLE 31.12 CASE STUDY Magnetic Field Inside a Capacitor

A parallel-plate capacitor consists of two circular disks of radius R (Fig. 31.34). The capacitor is charging, and there is no dielectric between the plates.

A Find an expression for the magnetic field at point P between the plates of the capacitor. Point P is at a distance r from the axis, where $r < R$. The current in the wire is I.

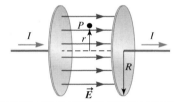

FIGURE 31.34

INTERPRET and ANTICIPATE

The same four steps we used to find the magnetic field with Ampère's law can be applied when using Ampère–Maxwell's law, provided we find the total (real and displacement) current through the Ampèrian loop.

Steps 1 and 2 Sketch the magnetic field lines, and choose an Ampèrian loop.
An end-on view is convenient for showing the magnetic fields. Imagine rotating Figure 31.34 so that the electric field is directed out of the page (Fig. 31.35). Because the electric flux increases while the capacitor is charging, the displacement current is in the same direction as the electric field. Use the simple right-hand rule to find the direction of the magnetic field. Point your thumb in the direction of the displacement current (out of the page); then your fingers wrap counterclockwise. To keep the figure clean, only one magnetic field line is shown in Figure 31.35. A circular Ampèrian loop exploits the symmetry of the problem. The path is in the same direction as the magnetic field.

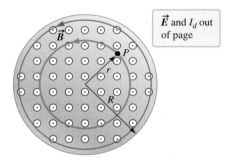

FIGURE 31.35

SOLVE
Step 3 Do the **circulation integral**. The magnetic field is parallel to the path and constant over the whole path. The path integral is just the circumference of the Ampèrian loop.

$$\oint \vec{B} \cdot d\vec{\ell} = \oint B\, d\ell = B \oint d\ell$$

$$\oint \vec{B} \cdot d\vec{\ell} = B(2\pi r) \quad (1)$$

Example continues on page 1002 ▶

Step 4 Find **current** I_{tot}. Imagine a soap film stretched taut across the circular Ampèrian loop in Figure 31.35. No wire penetrates the film, so there is no real current through the loop. However, electric field lines penetrate the loop. Because the electric field is increasing, there is a displacement current through the loop.	$I_{tot} = I_{thru} + I_d = 0 + I_d$ $I_{tot} = I_d$ \hfill (2)
The displacement current density J_d is the displacement current through the whole capacitor $(I_d)_{cap}$ divided by the area of the capacitor.	$J_d = \dfrac{(I_d)_{cap}}{\pi R^2}$
The displacement current through the capacitor equals the current in the wire in Figure 31.34. Write the displacement current density in terms of I.	$(I_d)_{cap} = I$ $J_d = \dfrac{I}{\pi R^2}$ \hfill (3)
The displacement current I_d through the Ampèrian loop is less than the displacement current through the whole capacitor $(I_d)_{cap}$. To find the displacement current through the Ampèrian loop, multiply the displacement current density J_d times the area A of the loop. Substitute Equations (3) and (2).	$I_d = J_d A = J_d \pi r^2 = \dfrac{I}{\pi R^2} \pi r^2$ $I_{tot} = I_d = I \dfrac{r^2}{R^2}$ \hfill (4)
Substitute Equations (1) and (4) into Ampère–Maxwell's law (Eq. 31.10), and solve for B.	$\oint \vec{B} \cdot d\vec{\ell} = \mu_0(I_{thru} + I_d) = \mu_0 I_{tot}$ $B(2\pi r) = \mu_0 I \dfrac{r^2}{R^2}$ $B = \dfrac{\mu_0 I}{2\pi} \dfrac{r}{R^2}$ \hfill (31.14)

:• **CHECK and THINK**

Equation 31.14 predicts that inside a charging capacitor, the magnetic field increases linearly with distance r from the axis. Measurements of such magnetic fields confirm Maxwell's modification to Ampère's law.

B Suppose you wish to confirm the displacement current term that Maxwell added to Ampère's law. You could use the deflection of compass needles to measure the magnetic field at several points between the plates of a large capacitor to see that it is given by Equation 31.14. The capacitor consists of two large circular plates of radius $R = 0.30$ m. You make your measurements using a very large current of 100 A. You have four compasses located at $r = 0, 0.10$ m, 0.20 m, and 0.30 m from the axis of the capacitor. Assume the magnetic field at these locations is perpendicular to the Earth's magnetic field. Find the deflection of the compass needles. Compare your results to the deflection of compass needles at the same distances due to a long, straight wire leading to the capacitor.

:• **INTERPRET and ANTICIPATE**

The goal of this part of the problem is to compare Equation 30.9, $B = (\mu_0 I)/(2\pi r)$, for the long, straight wire to Equation 31.14 for the region inside the capacitor. The latter must be derived with Maxwell's modification of Ampère's law. Ampère's law in its original form would conclude that the magnetic field inside the capacitor is zero.

:• **SOLVE**
The deflection angle comes from Equation 31.1. Substitute expressions for the magnetic fields produced by a long, straight wire (Equation 30.9) and by a capacitor, $B = (\mu_0 I r)/(2\pi R^2)$. Substitute numerical values to complete Table 31.2.

r (m)	B (T), capacitor	θ (°), capacitor	B (T), wire	θ (°), wire
0	0	0 ↑	→∞	90 →
0.10	2.2×10^{-5}	24 ↗	2.0×10^{-4}	76 ↗
0.20	4.4×10^{-5}	42 ↗	1.0×10^{-4}	63 ↗
0.30	6.7×10^{-5}	53 ↗	6.7×10^{-5}	53 ↗

TABLE 31.2

$$\theta = \tan^{-1}\left(\frac{B}{B_\oplus}\right)$$

For a straight wire,

$$\theta = \tan^{-1}\left(\frac{\mu_0 I}{2\pi r B_\oplus}\right) = \tan^{-1}\left(\frac{0.4}{r}\right)$$

For a capacitor,

$$\theta = \tan^{-1}\left(\frac{\mu_0 I}{2\pi R^2 B_\oplus} r\right)$$

$$\theta = \tan^{-1}(4.44 r)$$

:• **CHECK and THINK**
The last entry in the table shows that at the edge of the capacitor ($r = R$), the magnetic field produced by the displacement current equals the magnetic field produced by the real current. This is exactly what we would expect.

Compare the deflection angles. For the region between the plates of a charging (or discharging) capacitor, the deflection angle grows as we look at increasing distance because the magnetic field is strongest near the edge of the capacitor. The opposite is true for the long, straight wire because the magnetic field is strongest near the wire.

It might seem that Maxwell invented the displacement current just to fix a contradiction in Ampère's law and that his additional term does not correspond to any real phenomenon. However, experiments such as the one in Example 31.12 show that the displacement current term is a necessary part of Ampère's law. Maxwell's term predicts that there is nonuniform, nonzero magnetic field in the empty space between the capacitor's plates. The specific deflection angles predicted at each position in Example 31.12, part B, give us a way to test Maxwell's addition to Ampère's law. If Maxwell is correct about the displacement current, the magnetic field and the deflection angle should get larger as the compass is moved outward. Experiments have shown that Maxwell's addition of the displacement current to Ampère's law is correct. In fact, the displacement current is not only necessary but also fundamental because it led Maxwell to make a more important discovery about light (Chapter 34).

SUMMARY

❶ Underlying Principles

Two of Maxwell's equations:
1. According to **Gauss's law for magnetism**, because there are no (known) magnetic monopoles, the magnetic flux through a Gaussian surface is zero. Mathematically, Gauss's law for magnetism is

$$\Phi_B = \oint \vec{B} \cdot d\vec{A} = 0 \qquad (31.4)$$

The circle on the integral symbol in Equation 31.4 indicates that the integral is taken over the entire closed surface.

2. **Ampère's law** is expressed mathematically as

$$\oint \vec{B} \cdot d\vec{\ell} = \mu_0 I_{\text{thru}} \qquad (31.5)$$

The general form of Ampère's law, known as **Ampère–Maxwell's law**, is

$$\oint \vec{B} \cdot d\vec{\ell} = \mu_0 I_{\text{thru}} + \mu_0 \varepsilon_0 \frac{d\Phi_E}{dt} = \mu_0(I_{\text{thru}} + I_d)$$

(31.8) and (31.10)

Major Concepts

The **displacement current** is given by

$$I_d = \varepsilon_0 \frac{d\Phi_E}{dt} \quad (31.9)$$

The displacement current is a fictitious current; it represents not a real current but rather a changing electric field. The displacement current is a helpful way to think about how a changing electric field can create a magnetic field.

Special Cases

1. A **long, straight wire** exhibits linear symmetry. Ampère's law confirms that the magnetic field strength due to a long, straight, current-carrying wire is given by

$$B = \frac{\mu_0 I}{2\pi r} \quad (30.9)$$

2. The magnetic field inside an **ideal solenoid** is uniform; its strength is given by

$$B = \mu_0 n I \quad (31.6)$$

Outside the solenoid, the magnetic field is zero.

3. The magnetic field inside a **toroid** is given by

$$B = \frac{\mu_0 N I}{2\pi r} \quad (31.7)$$

Outside the toroid and inside the donut hole, the magnetic field is zero.

PROBLEM-SOLVING STRATEGY: Finding the Magnetic Field Using Ampère's Law

INTERPRET and ANTICIPATE
1. Sketch the magnetic field lines.
2. Choose an Ampèrian loop.

SOLVE
3. Do the circulation integral in

$$\oint \vec{B} \cdot d\vec{\ell} = \mu_0 I_{\text{thru}} \quad (\text{Eq. 31.5}).$$

4. Determine the amount of **current** I_{thru} through the Ampèrian loop.

PROBLEMS AND QUESTIONS

A = algebraic C = conceptual E = estimation G = graphical N = numerical

31-1 Measuring the Magnetic Field

1. **N CASE STUDY** Suppose you want to use a compass needle to measure the magnitude of a particular magnetic field. You plan to orient the field so it is perpendicular to the Earth's magnetic field. You have made a slight mistake, however, and instead the Earth's magnetic field makes an angle of 85° with respect to the field you want to measure. What is the percentage error in your observation?

Problems 2, 3, and 4 are grouped.

2. **C Review** Suppose you want to use a small, positively charged ball suspended by a light thread to map out the electric field in the space around a charged source. Describe the reaction of the ball as you place it at the two locations in front of the infinitely large, positively charged sheet shown in Figure P31.2A. Be sure to relate your description to what you know about the sheet's electric field.

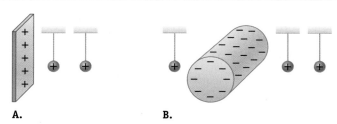

FIGURE P31.2 Problems 2 and 3.

3. **C Review** Suppose you want to use a small, positively charged ball suspended by a light thread to map out the electric field in the space around a charged source. Describe the reaction of the ball as you place it at the three locations in the space around the infinitely long, negatively charged rod shown in Figure P31.2B. Be sure to relate your description to what you know about the rod's electric field.

4. **C Review** Suppose you want to use a small, positively charged ball suspended by a light thread to map out the electric field in the space around a charged source. The deflection of the ball is used to find both the magnitude and direction of the electric field. What important role does the Earth's gravity play in your experiment?

Problems 5 and 6 are paired.

5. **N** In a laboratory exercise, you place a compass 2.00 cm from a wire as shown in Figure P31.5. Using a variable power supply, you slowly increase the current through the wire starting with $I = 0$. Assume the magnitude of the Earth's magnetic field is 5.0×10^{-5} T and that it points to the right in the figure. **a.** If you can detect a deflection of only 2°, what is the minimum current you can detect? **b.** Very large currents cause the compass needle to essentially align with the field due to the wire. Assuming you can't distinguish currents that lead to more than 88° of deflection, what is the largest current you can detect?

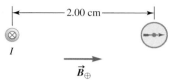

FIGURE P31.5 Problems 5 and 6.

6. **G** Plot the deflection angle of the compass needle in Problem 5 versus the current through the wire from 0 to 200 A. Is the compass needle a more sensitive measure for small currents or for large currents? Explain your answer.

31-2 Gauss's Law for Magnetism

7. **A** The magnetic field in some region is given by $\vec{B} = B_x\hat{i} + B_y\hat{j}$. The areas A_x, A_y, A_z, B_x, and B_y are constants. What is the magnetic flux through each of these areas: **a.** $\vec{A} = A_x\hat{i}$, **b.** $\vec{A} = A_y\hat{j}$, **c.** $\vec{A} = A_z\hat{k}$?

8. **C** Suppose the magnetic monopole is discovered. What change would be needed in Gauss's law for magnetism, if any? Write the new law, explaining your changes, if any.

9. **N** A loop of area 0.800 m² is in a uniform magnetic field $B = 0.652$ T. The magnetic flux through the loop is 0.240 Wb. What is the angle between the normal to the plane of the loop and the field?

10. **E** What is the Earth's magnetic flux through **a.** a basketball, **b.** a hula hoop standing up perpendicularly on its rim at the North Pole, and **c.** a hula hoop lying on the ground at the North Pole?

11. **N** A uniform magnetic field $\vec{B} = (0.0030\hat{i} + 0.0090\hat{j})$ T passes through a cube with side length $s = 10.0$ cm as shown in Figure P31.11. The edges of the cube align with the x, y, and z axes. Calculate the flux through each of the six surfaces of the cube, and show that Gauss's law for magnetism (Eq. 31.4) holds.

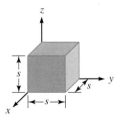

FIGURE P31.11

12. **N** A uniform magnetic field in a certain region of space has a magnitude of 0.25 T and points in a direction 30.0° from the positive x axis, 60.0° from the negative y axis, and 90.0° from the positive z axis. In each of the following cases, sketch the surface and determine the magnetic flux through the surface: **a.** $\vec{A} = 0.010\hat{j}$ m² and **b.** $\vec{A} = 1.0\hat{k}$ m².

31-3 Ampère's Law

Problems 13, 14, 15, and 16 are grouped.

13. **G** Figure P31.13 shows a uniform magnetic field. **a.** Can you find an Ampèrian loop that gives a circulation integral of zero? If so, draw the loop and the field. If not, explain why not. **b.** Can you find an Ampèrian loop that gives a nonzero circulation integral? If so, draw the loop and the field. If not, explain why not.

FIGURE P31.13 Problems 13 and 15.

14. **G** Figure P31.14 shows a magnetic field that is strongest near the center of the figure and weaker farther out. **a.** Can you find an Ampèrian loop that gives a circulation integral of zero? If so, draw the loop and the field. If not, explain why not. **b.** Can you find an Ampèrian loop that gives a nonzero circulation integral? If so, draw the loop and the field. If not, explain why not.

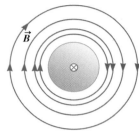

FIGURE P31.14 Problems 14 and 16.

15. **G** Figure P31.13 shows a uniform magnetic field. **a.** Can you find a (nonzero area) loop through which the magnetic flux is zero? If so, draw the loop and the field. If not, explain why not. **b.** Can you find a loop through which the magnetic flux is nonzero? If so, draw the loop and the field. If not, explain why not.

16. **G** Figure P31.14 shows a magnetic field that is strongest near the center of the figure and weaker farther out. **a.** Can you find a (nonzero area) loop through which the magnetic flux is zero? If so, draw the loop and the field. If not, explain why not. **b.** Can you find a loop through which the magnetic flux is nonzero? If so, draw the loop and the field. If not, explain why not.

17. **N** A 2.50 cm × 2.50 cm square Ampèrian loop exists in the xy plane in a region of space with a uniform magnetic field $\vec{B} = (1.25\hat{i} + 1.75\hat{j})$ T. Two sides of the loop are parallel to the x axis, and two sides are parallel to the y axis. The integration path is such that side 1 is traversed in the positive x direction, side 2 in the negative y direction, side 3 in the negative x direction, and side 4 in the positive y direction. Calculate the contribution to the circulation integral due to each segment of the loop, and determine the net current through the loop that must be present.

Problems 18 and 19 are paired.

18. **N** Suppose the current in each wire shown in Figure P31.18 is $I = 0.50$ A. What is the result of the circulation integral around the Ampèrian loop in parts A and B of the figure?

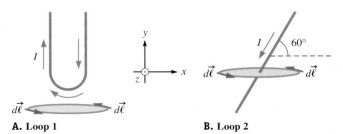

FIGURE P31.18 Problems 18 and 19.

19. **N** Suppose the circulation integral around Ampèrian loop 2 in Figure P31.18B is 3.142×10^{-7} T·m. What is the current through the loop?

20. **N** Figure P31.20 shows four current-carrying wires that are perpendicular to the page, where $I_1 = 1.0$ A, $I_2 = 2.0$ A, $I_3 = 3.0$ A, and $I_4 = 4.0$ A. Four Ampèrian loops are also shown. Find the current through each loop.

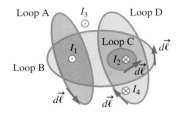

FIGURE P31.20

21. **A** A thin, infinite conducting sheet lies in the xy plane. It carries current per unit length J in the positive y direction. Find an expression for the magnitude of the magnetic field near the infinite sheet. (*Hint*: First, find the current through an Ampèrian loop perpendicular to the plane of the sheet using $I_{thru} = \int \vec{J} \cdot d\vec{\ell}$.)

Problems 22 and 23 are paired.

22. **A** A steady current I flows through a wire of radius a. The current density in the wire varies with r as $J = kr$, where k is a constant and r is the distance from the axis of the wire. Find expressions for the magnitudes of the magnetic field inside and outside the wire as a function of r. (*Hint*: Find the current through an Ampèrian loop of radius r using $I_{thru} = \int \vec{J} \cdot d\vec{A}$.)

23. **A** A steady current I flows through a wire of radius a. The current density in a wire varies with r as $J = kr^2$, where k is a constant and r is the distance from the axis of the wire. Find expressions for the magnitudes of the magnetic field inside and outside the wire as a function of r. (*Hint*: Find the current through an Ampèrian loop of radius r using $I_{thru} = \int \vec{J} \cdot d\vec{A}$.)

31-4 Special Case: Linear Symmetry

24. **C** **CASE STUDY** In Example 31.6, we discovered several problems in our experiment designed to measure the magnetic field due to a long, straight wire. How can this design be improved by more effectively using the same equipment?

25. **N** A magnetic field of 4.00 μT is measured at a distance of 25.0 cm from a long, straight wire with a current of 5.00 A. What is the distance from the wire at which a field of 0.500 μT will be measured?

26. **G** **CASE STUDY** A very long cylindrical wire has an insulating core as shown in Figure 31.17 (page 990). The core is surrounded by a cylindrical conducting shell with an inner and outer radius, carrying a current $I = 1.00$ A. The current is uniformly distributed in the conducting shell. Suppose the radius of the insulator and the inner radius of the shell is $r_1 = 0.00500$ m and the outer radius of the conductor is $r_2 = 0.0500$ m. Plot B versus r for this source. Compare your graph to that for a long, straight wire. *Hint*: Refer to Example 31.7 (page 990) and Concept Exercise 31.4 (page 988).

Problems 27, 28, and 29 are grouped.

27. A coaxial cable (sometimes called a "coax") is a long, layered cylindrical wire. You might find a coax attached to the back of your TV. Figure P31.27 is a cross-sectional view of a coax. The innermost layer is a long, straight conductor carrying current I into the page. The next layer is an insulator surrounded by another conductor carrying current I out of the page. The last layer is another insulator. The radius of each layer is labeled in the figure, and the current density is constant in each separate layer.
   a. **A** Find expressions for the magnitudes of the magnetic field in each of the four layers.

   b. **C** Compare your results to the magnetic field produced by a long, straight wire. Explain the advantage of using a coax.

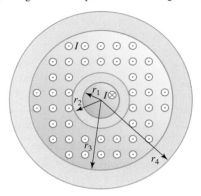

FIGURE P31.27 Problems 27, 28, and 29.

28. **G** Sketch a plot of the magnitude of the magnetic field as a function of position r for a coax (Fig. P31.27).

29. **N** A coaxial cable (Fig. P31.27) carries a current of 3.25 A. The radii are $r_1 = 0.500$ cm, $r_2 = 1.00$ cm, $r_3 = 1.50$ cm, and $r_4 = 2.00$ cm.
   a. Where is the magnetic field strongest?
   b. What is the magnitude of the magnetic field at this point?

30. The flow of ionic fluids can be viewed as a current, similar to electrons through a wire.
   a. **A** If a fluid with charge density ρ (in C/m^3) is pumped through a cylindrical tube with radius R (in m) at speed v (in m/s), write an expression for the magnetic field versus the distance r from the center of the tube, both inside and outside the tube.
   b. **G** Sketch the magnitude of the magnetic field versus distance from the center of the cylindrical tube (B vs. r), both inside and outside the tube.

Problems 31 and 32 are paired.

31. **N** Figure P31.31 shows a cylindrical conducting shell carrying current I to the right. The nonuniform current density in the conductor is given by $J = c(r - R_1)$ for the region $R_1 < r < R_2$, where c is a constant. What is the magnitude of the magnetic field at a radial distance r_A within the conductor? (*Hint*: Find the current through an Ampèrian loop of radius r using $I_{thru} = \int \vec{J} \cdot d\vec{A}$.)

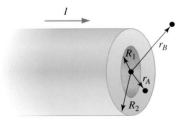

FIGURE P31.31 Problems 31 and 32.

32. **N** What is the magnitude of the magnetic field at a radial distance r_B from the center of a cylindrical conducting shell carrying current I (Fig. P31.31) in which the current density is given by $J = c(r - R_1)$ for the region $R_1 < r < R_2$, where c is a constant? (*Hint*: Find the current through an Ampèrian loop of radius r using $I_{thru} = \int \vec{J} \cdot d\vec{A}$.)

31-5 Special Case: Solenoids

33. **N** The magnitude of the magnetic field inside a very long solenoid is 1.0 T. If the current in the solenoid is 1.0 A, what is the number of turns in one meter?

34. **N** A solenoid of length 2.00 m and radius 1.00 cm carries a current of 0.100 A. Determine the magnitude of the magnetic field inside if the solenoid consists of 2000 turns of wire.

35. **N** The magnitude of the magnetic field at the center of a 25.0-cm-long solenoid with 650.0 turns is 2.00 mT. What is the current in the windings of the solenoid?

36. **N** **CASE STUDY** You want to measure magnetic fields in the range between 10 mT and 100 mT using a compass. You cannot use the Earth's magnetic field as a reference because it is too weak. Instead, you will use a solenoid to create a uniform magnetic field. If you want to keep the current smaller than 2.0 A, what is a reasonable value for the number of turns per unit length?

37. **N** A solenoid in your car is 33 cm long and has a diameter of 7.5 cm. It is made of copper wire with a radius of 0.51 mm. The turns are closely packed and form a single layer. Assume the resistivity of copper is $1.68 \times 10^{-8} \, \Omega \cdot m$. **a.** When the solenoid is connected to your battery with $\mathcal{E} = 11.6$ V, what is the magnetic field inside the solenoid? **b.** If, instead, the turns form two layers, how does your answer change?

Problems 38 and 39 are paired.

38. **N** The magnetic field inside a solenoid is 0.235 T. If the current is 4.53 A, how many turns are there per unit length?

39. **N** The magnetic field inside a solenoid 23.7 cm long is 0.235 T. If the current is 4.53 A and the diameter of the solenoid is 2.37 cm, what are the total number of turns and the length of the wire?

40. **N** A square conducting loop with side length $a = 1.25$ cm is placed at the center of a solenoid 40.0 cm long with 300 turns and aligned so that the plane of the loop is perpendicular to the long axis of the solenoid. According to an observer, the current in the single turn of the loop is 0.800 A in the counterclockwise direction, and the current in the windings of the solenoid is 8.00 A in the counterclockwise direction. **a.** What is the magnetic force on each side of the square loop? **b.** What is the net torque acting on the square loop?

41. **N** Aluminum wire 1.00 mm in radius is to be used to construct a solenoid 15.0 mm in radius so that the magnitude of the magnetic field at its center is 50.0 mT and the total resistance of the solenoid is 3.00 Ω when it carries a current of 1.00 A. How long does this solenoid need to be? The resistivity of aluminum is $2.655 \times 10^{-8} \, \Omega \cdot m$.

Problems 42 and 43 are paired.

42. **C** A straight current-carrying wire with current I_w resides inside a solenoid with n turns per unit length. A current I_s flows through the coils of the solenoid. **a.** Under what orientations will the wire experience no magnetic force? Explain your answer. **b.** Under what orientations will the wire experience a magnetic force? Explain your answer.

43. **A** A straight current-carrying wire with current I_w resides inside a solenoid with n turns per unit length. A current I_s flows through the coils of the solenoid. Find an expression for the magnitude of the maximum magnetic force per unit length the wire could experience inside the solenoid (the straight wire can be threaded through the coils of the solenoid, such that any angle is possible between the straight wire and the axis of the solenoid).

Problems 44 and 46 are paired.

44. **N** You have a spool of copper wire 5.00 mm in diameter and a power supply. You decide to wrap the wire tightly around a soda can that is 12.0 cm long and has a diameter of 6.50 cm, forming a coil, and then slide the wire coil off the can to form a solenoid. If the power supply can produce a maximum current of 125 A in the coil, what is the maximum magnetic field you would expect to produce in this solenoid? Assume the resistivity of copper is $1.68 \times 10^{-8} \, \Omega \cdot m$.

31-6 Special Case: Toroids

45. **N** The magnitude of the magnetic field inside a toroid with 1200 turns at a distance of 24 cm from the center is measured to be 0.020 T. Determine the current in the toroid.

46. **N** The solenoid in Problem 44 is wrapped into a donut to form a toroid. The solenoid (12.0 cm long) is wrapped as tightly as possible, so that the inner circumference of the toroid is the same as the length of the solenoid. What is the range of magnetic fields inside the toroid when the maximum current of 125 A is sent through the wire?

47. **N** A toroid with an inner radius of 50.0 cm and outer radius of 75.0 cm is wound with 760.0 turns of wire carrying a current of 955 A. What is the magnitude of the magnetic field along **a.** the inner radius of the toroid and **b.** the outer radius of the toroid?

48. **E** Today, nuclear power plants are based on fission reactions—reactions that split atomic nuclei. Another type of nuclear reaction is fusion, in which atomic nuclei are fused together. Fusion reactions power the Sun and may hold the key to Earth's power needs. JET (Joint European Torus) is a large device designed to study fusion reactions (Fig. P31.48). Inside JET's large donut-shaped apparatus, hydrogen plasma (a gas of free electrons and protons) is accelerated, reaching temperatures of about 100×10^6 K. Such a hot plasma is contained using a magnetic field. JET operates in short pulses, each several tens of seconds in duration. During a single pulse, JET uses 500 MW, about half of which goes to the toroid. The toroid gets very hot and must be water-cooled. The toroidal magnetic field keeps the plasma moving in its donut-shaped path. The maximum field strength is 4.5 T. If the magnetic field is produced by a 768-turn toroid, estimate the current in the toroid.

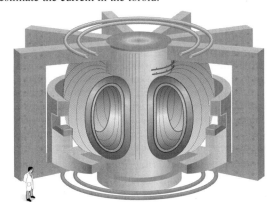

FIGURE P31.48 The Joint European Torus.

49. **N** A toroid with 150 turns of wire and inner and outer radii of 0.115 m and 0.295 m, respectively, carries a current of 1.46 A. At what location within the toroid is the magnitude of the magnetic field equal to half the value at the inner wall (the inner radius)?

Problems 50 and 51 are paired.

50. **N** The inner and outer radii of a toroid are 1.25 cm and 2.75 cm. The toroid has 64 turns and carries a current of 2.15 A. What are the strongest and weakest magnetic fields produced by this toroid?

51. **N** A toroid has an outer radius of 5.75 cm. The strongest magnetic field produced by the torus is 0.632 T and the weakest magnetic field is 0.316 T. What is the toroid's inner radius?

31-7 General Form of Ampère's Law

52. C In what way is the term *displacement current* misleading? In what way is the term useful?

53. N A parallel-plate capacitor consists of two rectangular plates, each of area $A = 36.7$ cm^2. There is a vacuum between the plates, which are separated by 2.14 mm. When a current of 0.0935 A is charging the capacitor, find the rate at which the magnitude of the electric field is changing.

54. N A parallel-plate capacitor is made using circular plates of radius 10.0 cm separated by a distance of 0.50 cm. It is connected to a battery, and the voltage between the capacitor plates increases at a rate of 125,000 V/s. What is the displacement current between the capacitor plates?

Problems 55, 56, and 57 are grouped.

55. C Review A capacitor with circular plates of radius a separated by distance d is being charged by a battery with emf $\mathcal{E}$. A vacuum exists between the plates of the capacitor. The constructed circuit has resistance R in series with the capacitor, and the capacitor is initially uncharged when a switch is closed, completing the circuit and causing a current to flow (an RC series circuit). What happens to the magnitude of the magnetic field at a location between the plates at a distance r from the axis through the center of the plates, as time goes on while the capacitor is charging? Explain.

56. A Review A capacitor with circular plates of radius a separated by distance d is being charged by a battery with emf $\mathcal{E}$. A vacuum exists between the plates of the capacitor. The constructed circuit has resistance R in series with the capacitor, and the capacitor is initially uncharged when a switch is closed, completing the circuit and causing a current to flow (an RC series circuit). Find an expression for the magnitude of the magnetic field between the plates at a distance r from the axis through the center of the plates. Assume $r < a$.

57. N Review A capacitor with circular plates of radius 0.0534 m separated by a distance of 5.00×10^{-4} m is being charged by a battery with an emf of 12.0 V. A vacuum exists between the plates of the capacitor. The constructed circuit has a resistance of 325 Ω in series with the capacitor, and the capacitor is initially uncharged when a switch is closed, completing the circuit and causing a current to flow (an RC series circuit). Find an expression for the magnitude of the magnetic field between the plates at a distance of 0.0250 m from the axis through the center of the plates, when one time constant has passed since the switch was closed.

General Problems

58. N A 2.50 cm × 2.50 cm square Ampèrian loop exists in the xy plane in a region of space with a magnetic field. The magnetic field has a y component that varies along the x direction and an x component that varies along the y direction as described by $\vec{B} = (0.125y\hat{i} - 0.125x\hat{j})$ T. Calculate the contribution to the circulation integral due to each segment of the loop, and determine the net current through the loop that must be present.

59. N A uniform magnetic field $\vec{B} = 5.44 \times 10^4 \hat{i}$ T passes through a closed surface with a slanted top as shown in Figure P31.59. **a.** Given the dimensions and orientation of the closed surface shown, what is the magnetic flux through the slanted top of the surface? **b.** What is the net magnetic flux through the entire closed surface?

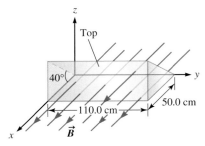

FIGURE P31.59

60. N Review Figure P31.60 shows four parallel wires arrayed at the indicated coordinates in the xy plane carrying identical currents of 8.00 A. The currents in wires A and B are in the positive z direction (out of the page), and the currents in wires C and D are in the negative z direction (into the page). What are the magnitude and direction of the magnetic field at the origin?

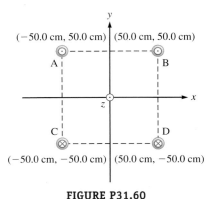

FIGURE P31.60

61. N A solenoid 1.25 m long with a current of 5.00 A in its windings produces a magnetic field at its center of 2.00 T. How many windings are in this solenoid?

62. C Review the new equations presented in this chapter. Use your own judgment to order the equations worth remembering according to their importance. Explain your reasoning.

63. N A current of 5.00 A flows through a 10.0-cm-long solenoid that has 55 turns. What is the magnitude of the field at the center of the solenoid?

64. N Figure P31.64 shows a cube with side length $d = 10.0$ cm with its center at the origin. The cube is subject to a uniform magnetic field $\vec{B} = (3\hat{i} - 5\hat{j} - 2\hat{k})$ T. **a.** What is the magnetic flux through the top face of the cube? **b.** What is the total magnetic flux through all six faces of the cube?

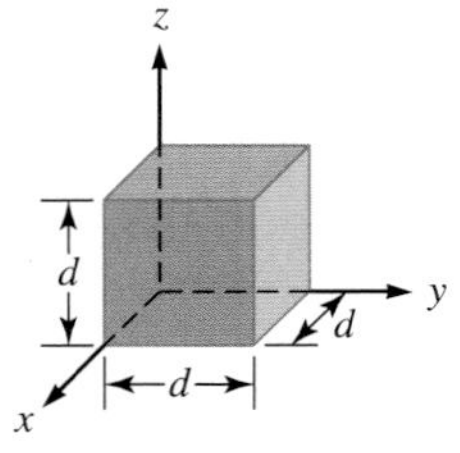

FIGURE P31.64

65. N A coaxial cable is constructed from a central cylindrical conductor of radius $r_A = 1.00$ cm carrying current $I_A = 8.00$ A in the positive x direction and a concentric conducting cylindrical shell with inner radius $r_B = 14.0$ cm and outer radius $r_C = 15.0$ cm with a current of 20.0 A in the negative x direction (Fig. P31.65). What are the magnitude and direction of the magnetic field **a.** at point O, a distance of 10.0 cm from the center of the coaxial cable along the y axis, and **b.** at point P, a distance of 20.0 cm from the center of the coaxial cable along the y axis?

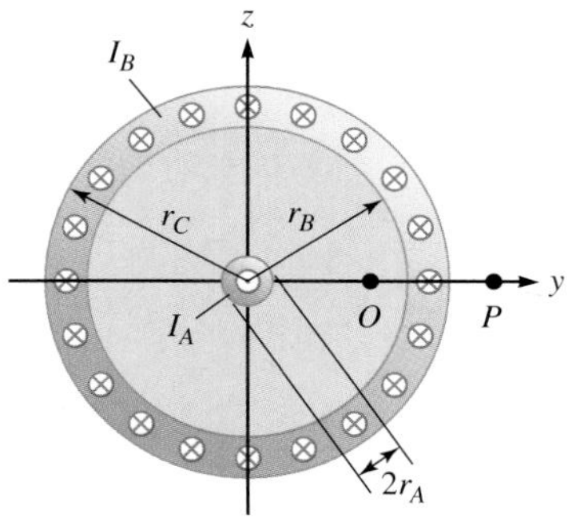

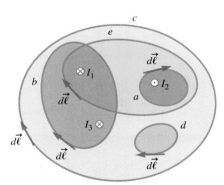

FIGURE P31.65

66. **N Review** In a laboratory vacuum chamber, a long, straight, horizontal wire with a current of 55.0 μA in the positive x direction is used to "levitate" a doubly ionized helium nucleus moving parallel to the wire with a speed of 5.10×10^5 m/s in the negative x direction. What is the distance above the wire where the magnetic force of the wire balances the gravitational force on the helium nucleus?

67. **N** Cross-sectional views of three long conductors carrying currents $I_1 = 2.0$ A, $I_2 = 1.0$ A, and $I_3 = 3.0$ A are shown in Figure P31.67. Determine the closed line integrals $\oint \vec{B} \cdot d\vec{\ell}$ around the five Ampèrian loops a through e as shown in the figure.

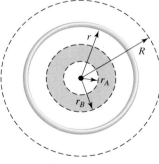

FIGURE P31.67

68. **N** A closed loop encloses several current-carrying wires with currents in different directions. The line integral $\oint \vec{B} \cdot d\vec{\ell}$ around the curve is 2.0×10^{-3} T·m. Determine the net current through the loop.

69. **N** Figure P31.69 shows a cross-sectional view of a 510-turn solenoid 50.0 cm in length and with radius $r = 2.00$ cm. The solenoid carries a current of 4.00 A. What is the magnetic flux through **a.** the circular area of radius $R = 3.50$ cm, and **b.** the gray-shaded annulus with $r_A = 5.00$ mm inner radius and $r_B = 15.0$ mm outer radius?

FIGURE P31.69

Problems 70 and 71 are paired.

70. **C** A square loop of wire with side length s carries current I_1 and resides inside a solenoid with n turns per unit length. The solenoid carries current I_2. Under what orientation does the square loop experience the maximum possible torque? Explain your answer.

71. **A** A square loop of wire with side length s carries current I_1 and resides inside a solenoid with n turns per unit length. The solenoid carries current I_2. Find an expression for the magnitude of the maximum possible torque experienced by the square loop of wire.

72. **A** A straight wire carrying current I_1 resides inside a toroid, oriented along the radial direction (running from the inner wall to the outer wall). The toroid has N turns and carries current I_2. The inner and outer radii of the toroid are r_i and r_o, respectively.
    a. A Find an expression for the magnitude of the magnetic force per unit length on the straight wire at r_i.
    b. A Find an expression for the magnitude of the magnetic force per unit length on the straight wire at r_o.
    c. C Given your answers to parts (a) and (b), what will happen to the wire if it is subject to only these forces initially?

73. **A** An infinite slab with thickness extending from $z = -L$ to $z = L$ carries a uniform current density $\vec{J} = J\hat{\imath}$ as shown in Figure P31.73. Find an expression for the magnetic field as a function of z inside, below, and above the slab. (*Hint*: First, find the current through an Ampèrian loop perpendicular to the plane of the sheet using $I_{\text{thru}} = \int \vec{J} \cdot d\vec{A}$.)

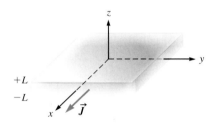

FIGURE P31.73

74. **A** Two long coaxial solenoids each carry current I but in opposite directions as shown in Figure P31.74. The inner solenoid of radius R_1 has n_1 turns per unit length, and the outer solenoid with radius R_2 has n_2 turns per unit length. Find expressions for the magnetic fields inside the inner solenoid, between the two solenoids, and outside both solenoids.

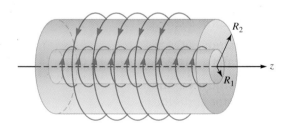

FIGURE P31.74

75. **A** A time-dependent current $I = I_0 \cos(\omega t)$ induces a time-dependent electric field

$$\vec{E}(s,t) = \frac{\mu_0 I_0 \omega}{2\pi} \sin(\omega t) \ln(R/s) \hat{k}$$

in a region of space. Obtain the total displacement current through the end of a cylinder with radius R that has its axis aligned with the z axis.

32 Faraday's Law of Induction

Key Questions

How can a magnetic field become a source of electric current?

How does Faraday's law lead to a practical method of generating electrical power?

- 32-1 Another kind of emf 1011
- 32-2 Faraday's law 1013
- 32-3 Lenz's law 1016
- 32-4 Lenz's law and conservation of energy 1019
- 32-5 Case study: Slide generator 1020
- 32-6 Case study: AC generators 1026
- 32-7 Case study: Faraday's generator and other DC generators 1030
- 32-8 Case study: Power transmission and transformers 1033

❶ Underlying Principles

1. Faraday's law, one of Maxwell's equations
2. Lenz's law

✪ Major Concepts

Eddy current

◐ Special Cases

1. Slide generator
2. AC generator
3. DC generator
4. Transformer

Michael Faraday was a great experimenter. After Hans Christian Ørsted discovered that electric currents are the source of magnetic fields, Faraday suspected that magnetic fields might be the source of electric currents because he believed the laws of nature are symmetrical. Many scientists look for symmetry in nature and are often rewarded with major new discoveries. Faraday's careful laboratory observations led him to discover a connection between magnetic fields and electric currents. His discovery is now known as **Faraday's law**.

Faraday saw great practical applications of his law and built the first electric generator. Today, the operation of all electric generators and electric motors is based on Faraday's law.

But that is not all. Faraday's law is our last of the four Maxwell's equations. Maxwell started with Faraday's law and Ampère's law and derived an equation showing that light is an electromagnetic wave. Faraday's law was a key component of Maxwell's major theoretical discovery that light is a combination of electricity and magnetism (Chapter 34). In this chapter, we focus on Faraday's law and its immediate practical application in our everyday lives.

32-1 Another Kind of Emf

There are two different kinds of electric field sources. We studied the first type, including sources with excess charge, such as the spheres and rods used by Ben Franklin and his contemporaries. We also studied the practical application of charged sources such as Leyden jars and batteries, which have excess positive charge on one terminal and excess negative charge on the other. It is convenient to characterize these sources in terms of the electric potential difference between their terminals, known as their *emf* (Section 29-1). As you know, a current results when the terminals of such an emf are connected by a conductor (Fig. 32.1).

About four decades after Franklin died, the British physicist Michael Faraday (1791–1867) discovered another kind of electric field source. This source is a *changing magnetic flux* Φ_B. Like a battery, a changing magnetic flux creates current in a conductor. We usually use the phrases "the battery *sets up* a current" and "the changing magnetic flux *induces* a current."

Because Faraday's law is based on his laboratory observations, let's begin by considering a few experiments. Figure 32.2 shows a wire connected to an ammeter, but—unlike the wire in Figure 32.1—there is no battery connected to the wire. A solenoid passes through the rectangular loop with its long axis perpendicular to the page. The solenoid is connected to an external circuit (not shown) so that it creates a uniform magnetic field $\vec{B}$ pointing out of the page. Recall that outside an ideal solenoid, the magnetic field is zero. The magnetic flux through the rectangular loop is given by Equation 31.2

$$\Phi_B = \int \vec{B} \cdot d\vec{A} \qquad (31.2)$$

$$\Phi_B = BA \qquad (32.1)$$

where A is the cross-sectional area of the solenoid.

In Figure 32.2A, the current in the solenoid is constant, so the solenoid's magnetic field is steady. The magnetic flux through the rectangular loop (Eq. 32.1) is therefore constant. We observe no current in the rectangular loop; the ammeter needle remains stationary at 0. As you may expect, the magnetic field of the solenoid has no effect on the rectangular wire.

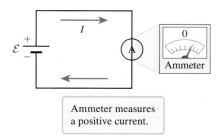

FIGURE 32.1 A battery represents one type of source of an electric field. The top terminal has excess positive charge, and the bottom terminal has excess negative charge. The circuit includes an ammeter, and the ammeter shows there is positive (clockwise) current.

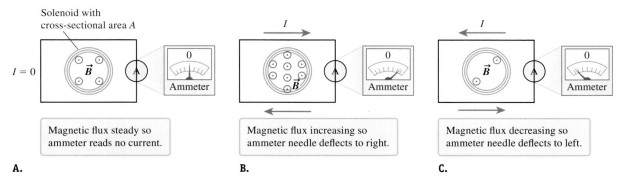

FIGURE 32.2 A. A solenoid generates a magnetic field pointing out of the page. The magnetic field is steady, and there is no current in the wire. **B.** The magnetic field increases, changing the magnetic flux through the rectangular loop. A current appears in the rectangular loop, and the ammeter's needle deflects to the right as it did with a battery in the circuit (Fig. 32.1). **C.** The magnetic field decreases, again changing the magnetic flux through the rectangular loop. A current appears in the rectangular loop, but this time the ammeter's needle deflects to the left, as if you had reversed the battery in Figure 32.1.

The situation is very different when the solenoid's magnetic field is changing. If the external current in the solenoid increases, so that its magnetic field $\vec{B}$ increases, then according to $\Phi_B = BA$ (Eq. 32.1), the magnetic flux Φ_B also increases. In this case, a clockwise current appears in the rectangular loop and the ammeter's needle is deflected to the right (Fig. 32.2B). This is exactly what we found when a battery was in the circuit (Fig. 32.1). To the ammeter, there is no difference between having a battery in the circuit and having a changing magnetic flux through the circuit. This is true of any device you might put in the circuit. For example, if you replaced the ammeter with a lightbulb, the bulb would light up whether you used a battery (Fig. 32.1) or there was a changing magnetic flux (Fig. 32.2B).

What happens if the current in the solenoid decreases rather than increases? In that case, the magnetic field $\vec{B}$ still points out of the page, but it decreases (grows weaker) and the magnetic flux Φ_B also decreases. The decreasing magnetic flux causes a counterclockwise current in the loop, and so this time, the ammeter needle deflects to the left (Fig. 32.2C). This is exactly what would happen if we reversed the battery in Figure 32.1.

From the experiments in Figure 32.2, we may guess that a changing magnetic flux is required to produce a current in the loop. So far, we have imagined that the changing magnetic flux is produced by changing the magnetic field of a solenoid. A permanent magnet can also be used to create a changing magnetic flux. If you hold a bar magnet stationary near a wire loop, there is no current in the loop because the magnetic flux Φ_B through the loop does not change. However, if you move the bar magnet toward or away from the loop, the magnetic flux through the loop changes and a current appears. The direction of the current depends on which pole of the magnet is closest to the loop and whether you move the bar magnet toward or away from the loop. For example, if you point the north pole at the loop and move the magnet toward it, the ammeter needle in Figure 32.3A deflects to the left. If you move the magnet away from the loop as in Figure 32.3B, the ammeter needle deflects to the right.

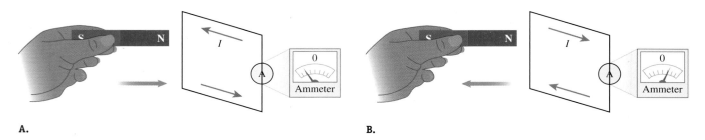

FIGURE 32.3 A. A bar magnet is pushed toward the loop as shown, and the ammeter needle deflects to the left. **B.** The bar magnet is pulled away from the loop, and the ammeter needle deflects to the right.

CASE STUDY AC/DC: The War of Currents

A changing magnetic flux (and Faraday's law) is the basis of the AC (alternating current) power supplies that you use every time you plug a device into a wall outlet. When the electrical infrastructure was being developed in the late 1800s, however, AC power was controversial. Some people supported developing a DC (direct current) infrastructure, while others recommended using AC. The debate was so controversial that it is described as a *war* over current. In this case study, we look at the roles played by three great inventors—Thomas Edison, Nikola Tesla, and George Westinghouse—in this war.

Edison—probably best known for inventing the incandescent bulb—believed DC was superior because it was safer. He built many steam-powered DC generators. Tesla and Westinghouse favored AC, but neither one was as famous in the field of electricity as Edison. Tesla was a young Serbian immigrant who worked as a junior engineer for Edison. Westinghouse had invented the air brake used on trains. Tesla believed DC generators and motors failed more often than their AC counterparts.

Another argument involved the location of generators. Edison's DC generators had to be located in metropolitan areas, so they had to be powered by steam engines. Both Tesla and Westinghouse recognized that AC generators could be located far from cities and so could be powered by waterfalls such as Niagara Falls (Fig. 32.4).

By the late 1880s, Edison's DC generators were used in major cities to light homes and businesses and to run motors. Westinghouse had built many AC generators, but at that time only DC motors were commercially viable, so his AC

generators were used only for lighting. In 1888, Edison wrote an 84-page book entitled *WARNING,* in which he cautioned the public not to invest in AC power. He said AC power was not as useful as DC power because there were no AC motors and AC power was deadly.

The public was moved by safety concerns. Harold Brown, an independent New York engineer, proposed a law limiting AC power to 300 V. If such a law had been passed, Westinghouse would have had to close all his operations. In this chapter, we explore the technological developments that led to the AC infrastructure we use today.

FIGURE 32.4 George Westinghouse used Niagara Falls to generate AC power. The power was used in Buffalo, New York, more than 20 miles away. This photo shows the current dam built in the 1960s.

CONCEPT EXERCISE 32.1

To calculate the magnetic flux through the rectangular loop in Figure 32.2, we used the cross-sectional area A of the solenoid in $\Phi_B = BA$ (Eq. 32.1). Why didn't we use the area of the rectangular loop?

32-2 Faraday's Law

The experiments described in Section 32-1 show that a changing magnetic flux may induce a current in a conductor, like the current set up by a battery. Both are measured by an ammeter, and both can be used to light a bulb. Just as it is convenient to characterize a battery in terms of the emf between its terminals, it is convenient to characterize a changing magnetic flux in terms of the emf it produces. One statement of **Faraday's law** is that *a changing magnetic flux through a conductor induces an emf in that conductor.* For a single conducting loop, Faraday's law is expressed mathematically as

$$|\mathcal{E}| = \left|\frac{d\Phi_B}{dt}\right| \qquad (32.2)$$

FARADAY'S LAW
❶ **Underlying Principle**

where Φ_B is the magnetic flux through the loop (Eq. 31.2), and $|\mathcal{E}|$ is the absolute value of the emf induced in the conductor. (In the next section, we'll learn how to find the sign of the emf, which allows us to remove the absolute-value signs from Equation 32.2.)

Let's consider a simple example of a uniform magnetic field and a single rectangular loop (Fig. 32.5). To calculate the magnetic flux, we need to find the direction of the loop's area vector. Recall that the area vector is perpendicular to the loop, so it can point either up and to the right or down and to the left (Section 25-2). Although either choice is acceptable, choosing the direction that is closest to the magnetic field is convenient and the one we make in this book—in this case, up and to the right (Fig. 32.5). The magnetic field $\vec{B}$ and the angle φ between $\vec{A}$ and $\vec{B}$ are constants. So Equation 31.2 for the magnetic flux becomes

$$\Phi_B = \int (B)(dA) \cos\varphi = B \cos\varphi \int dA$$

The integral is just the area A of the loop:

$$\Phi_B = BA \cos\varphi \qquad (32.3)$$

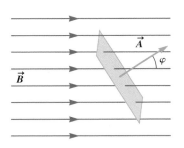

FIGURE 32.5 A rectangular conducting loop is perpendicular to the page and tilted with respect to the uniform magnetic field.

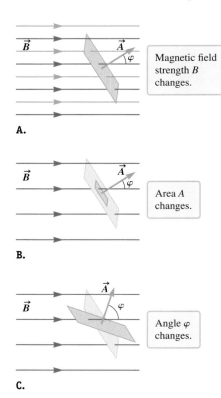

FIGURE 32.6 Three ways for the magnetic flux through the loop to change: **A.** The magnetic field gets weaker. **B.** The loop (made of a flexible material) shrinks. **C.** The loop rotates counterclockwise.

To find the absolute value of the emf induced in the loop, substitute Equation 32.3 for Φ_B into Faraday's law (Eq. 32.2):

$$|\mathcal{E}| = \left|\frac{d\Phi_B}{dt}\right| = \left|\frac{d(BA\cos\varphi)}{dt}\right|$$

Because of the time derivative in this expression, if B, A, and φ are all constants in time, no emf is induced in the loop. At least one of these parameters must change in order to induce an emf (Fig. 32.6). According to Faraday's law, during any one of these changes, an emf is induced in the loop. If the changes stop happening, the magnetic flux becomes constant and there is no longer an emf.

The changes in Figure 32.6 all show a reduction in the magnetic flux. Perhaps it is obvious from the linear dependence of B and A in $\Phi_B = BA\cos\varphi$ (Eq. 32.3) that a weaker magnetic field B and a smaller loop area A both reduce Φ_B. Because of the cosine dependence on φ, the magnetic flux Φ_B is greatest when the area vector is lined up with the magnetic field so that $\varphi = 0$. When the loop rotates counterclockwise, thus increasing φ, the magnetic flux is reduced (Fig. 32.6C). If the loop keeps rotating counterclockwise, the magnetic flux will become negative for the angle $90° < \varphi < 270°$, and the flux will be positive and increase again as the area vector moves between $\varphi = 270°$ and $360°$. Changing the angle between the area vector and the magnetic field is an important way of generating AC and DC power, so it is important to the case study.

In Figure 32.6, we imagined an emf induced in a single loop. If we replace the single loop with a coil of N turns, each turn has an induced emf. The turns are in series with one another, so the total induced emf is the sum of the individual emfs in each loop, or N times the emf in each loop. Faraday's law for a coil of N turns comes from multiplying Faraday's law for a single loop by N:

$$|\mathcal{E}| = N\left|\frac{d\Phi_B}{dt}\right| \qquad (32.4)$$

where $|\mathcal{E}|$ is the (absolute value of the) net emf induced in the coil.

It is sometimes convenient to use the term **flux linkage** for the combination $N\Phi_B$. Then we can state Faraday's law (Eq. 32.4) in terms of the flux linkage: *The emf induced in a coil is equal to the time rate of change of the flux linkage.*

CONCEPT EXERCISE 32.2

Figure 32.7 shows three cases involving a single, stationary circular loop of wire. For each case, decide whether an emf is induced in the circular loop. Explain your reasoning.

Case 1: A bar magnet twists back and forth near the loop.
Case 2: A bar magnet swings toward and away from the loop but never penetrates it.
Case 3: A bar magnet is at rest so that one end penetrates the loop.

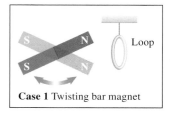
Case 1 Twisting bar magnet

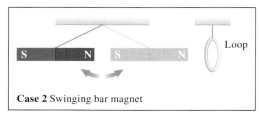
Case 2 Swinging bar magnet

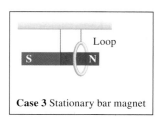
Case 3 Stationary bar magnet

FIGURE 32.7

EXAMPLE 32.1 Lighting a Flashlight the Hard Way

A small coil has area $A = 2.83 \times 10^{-3}$ m² and $N = 2550$ turns. A small bulb is attached to the coil, and the coil rests inside a long solenoid with $n = 3.65 \times 10^5$ turns per meter. The axis of the coil is aligned with the solenoid's axis. The current in the solenoid increases according to $I = \frac{1}{2}t$, where I is in amperes and t is in seconds. Find the absolute value of the emf induced in the coil, and describe the appearance of the bulb.

: INTERPRET and ANTICIPATE

Sketch the problem as in Figure 32.8. In this case, the area vector is aligned with the magnetic field, so $\varphi = 0$. To use Faraday's law, first find an expression for the magnetic flux Φ_B through the loop (Eq. 31.2). Often this can be done by inspection. After finding Φ_B, substitute it into Faraday's law (Eq. 32.4) to find the induced emf. We expect to find a numerical value in volts.

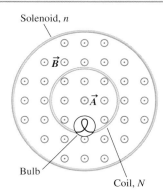

FIGURE 32.8 Cross-sectional view of the solenoid and coil.

: SOLVE

Use Equation 31.2 with $\varphi = 0$. The coil is in the uniform magnetic field of the solenoid. The result looks familiar (Eq. 32.1). We find this result whenever a loop is aligned with a uniform magnetic field.

$$\Phi_B = \int \vec{B} \cdot d\vec{A} = \int B(\cos 0) dA \quad (31.2)$$

$$\Phi_B = B \int dA = BA \quad (32.1)$$

Because the current in the solenoid is changing, write the magnetic flux in terms of that changing current. The solenoid's magnetic field is given by Equation 31.6, and the current is $I = \frac{1}{2}t$. Substitute both of these expressions in the flux equation (Eq. 32.1).

$$B = \mu_0 n I = \mu_0 n \left(\tfrac{1}{2}t\right) \quad (31.6)$$

$$\Phi_B = \left[\mu_0 n \left(\tfrac{1}{2}t\right)\right] A \quad (1)$$

Substitute the magnetic flux (Eq. 1) in Faraday's law (Eq. 32.4).

$$|\mathcal{E}| = N \left|\frac{d\Phi_B}{dt}\right| = N \left|\frac{d[\mu_0 n (\tfrac{1}{2}t)] A}{dt}\right| \quad (32.4)$$

Pull the constants outside the derivative and rearrange terms. In this case, the derivative is 1, so the emf is constant.

$$|\mathcal{E}| = \tfrac{1}{2}\mu_0 n N A \left|\frac{dt}{dt}\right| = \tfrac{1}{2}\mu_0 n N A$$

Substitute numerical values. This result means the bulb is lit with a constant brightness much as it is when connected to the batteries in your flashlight.

$$|\mathcal{E}| = (\tfrac{1}{2} \text{A/s})(4\pi \times 10^{-7} \text{ T·m/A})(2550)(3.65 \times 10^5 \text{ m}^{-1})(2.83 \times 10^{-3} \text{ m}^2)$$

$$|\mathcal{E}| = 1.66 \frac{\text{T·m}^2}{\text{s}}$$

: CHECK and THINK

One way to check our results is to make sure the answer has the expected SI units—volts. The tesla is defined in Section 30-4 as $1 \text{ T} \equiv 1 \text{ (N·s)}/(\text{C·m})$. A joule is a newton-meter, and a volt is a joule per coulomb (Section 26-4). So, our result is in volts as expected.

To help with future applications of Faraday's law, recall from Equation 31.3 that the SI unit of magnetic flux is the weber (Wb). So, a weber per second is a volt.

$$1 \frac{\text{T·m}^2}{\text{s}} = 1 \left(\frac{\text{N·s}}{\text{C·m}}\right)\left(\frac{\text{m}^2}{\text{s}}\right) = 1 \frac{\text{N·m}}{\text{C}}$$

$$1 \frac{\text{T·m}^2}{\text{s}} = 1 \frac{\text{J}}{\text{C}} = 1 \text{ V}$$

$$1 \frac{\text{Wb}}{\text{s}} = 1 \text{ V} \quad (32.5)$$

32-3 Lenz's Law

The experiment in Figure 32.2 shows that a changing magnetic flux induces an emf in the loop, which we observe by measuring the current. If the conductor has resistance R, the current is given by

$$I = \frac{\mathcal{E}}{R}$$

where the emf is found from Faraday's law. Although current and electric potential are both scalars, a direction is associated with them. Current is the flow of positively charged particles from high potential to low potential. We think of an emf as pointing in the same direction as the current. For example, the current in Figure 32.2B is clockwise, so we say the emf induced in the loop is also clockwise. Likewise, the current in Figure 32.2C is counterclockwise, so the emf in the loop is also counterclockwise. The Russian physicist Heinrich Friedrich Emil Lenz (1804–1865) discovered a rule—known as **Lenz's law**—for finding the direction of the induced emf.

Because the induced current and the induced emf are in the same direction, it is common to use the phrases "direction of the induced current" and "direction of the induced emf" interchangeably.

Observed Current Directions

Let's see what can be deduced from the four simple experiments in Figure 32.9. The first two experiments (parts A and B of Fig. 32.9) are the same as those in Figure 32.2, with the magnetic field out of the page. The second two experiments (parts C and D of Fig. 32.9) are similar, but with the solenoid's magnetic field into the page. We have listed three facts about each experiment—the direction of the magnetic field $\vec{B}$, the direction of the change in the magnetic field $d\vec{B}/dt$, and the direction of the observed current.

FIGURE 32.9 Lenz's law experiments. Your thumb always points in the direction opposite $d\vec{B}/dt$. An ammeter in each circuit indicates the direction of the current we observe. Notice that there is no correlation between your thumb's direction and that of the magnetic field $\vec{B}$.

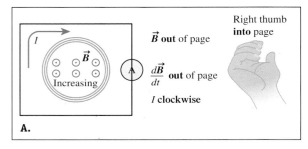

A.

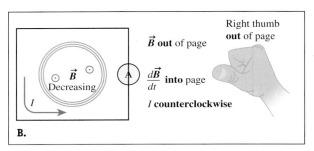

B.

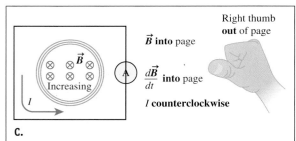

C.

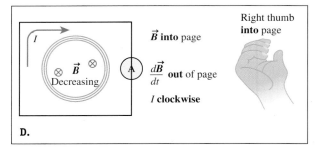

D.

Why is the direction of $d\vec{B}/dt$ important? From Faraday's law, the change in the magnetic field is what induces the current in these cases. The direction of $d\vec{B}/dt$ depends on the direction of the magnetic field $\vec{B}$ and whether that field is increasing or decreasing. When the magnetic field is increasing, $d\vec{B}/dt$ is in the same direction as $\vec{B}$. When the magnetic field is decreasing, $d\vec{B}/dt$ is in the direction opposite $\vec{B}$. You should verify this for the four experiments in Figure 32.9.

Another Right-Hand Rule

The observed current is in the plane of the page circulating either clockwise or counterclockwise, whereas $\vec{B}$ and $d\vec{B}/dt$ are perpendicular to the page. Again, a right-hand rule helps associate the direction of the current with a direction either into or out of the page. We can begin with the same right-hand rule we used to find the magnetic moment of a loop or the magnetic field produced by a loop near its center (Fig. 30.25, page 948): Wrap the fingers of your right hand in the direction of the current; then your thumb points in the direction of the loop's magnetic moment and the magnetic field that the loop produces along an axis passing through its center.

In Figure 32.9, the fingers of the right hand shown in each part wrap in the direction of the observed induced current. There is a correlation between the thumb directions and the change in the magnetic field's direction $d\vec{B}/dt$. The thumbs always point in the direction opposite $d\vec{B}/dt$. (Check the experiments in Figure 32.9 for yourself.) So, by studying the relationships in the figure, we can come up with a rule for finding the direction of the induced current. Point your right thumb in the direction *opposite* $d\vec{B}/dt$, and your fingers wrap in the direction of the induced current (and the induced emf). This is very close to what Lenz deduced, although his law is stated in more general terms.

According to **Lenz's law**, *the induced current produces a magnetic field that always acts to oppose the change in the magnetic flux that created the induced current.* Let's see how the law fits a couple of the experiments in Figure 32.9. In Figure 32.9A, the solenoid's magnetic field $\vec{B}_{sol}$ points out of the page through the loop and is increasing. Because the magnetic field is changing, a current is induced in the rectangular loop. The induced current is the source of a second magnetic field $\vec{B}_{loop}$. Near the center of the loop, the second magnetic field $\vec{B}_{loop}$ points either out of the page or into the page depending on the direction of the induced current. According to Lenz's law, $\vec{B}_{loop}$ *opposes the change in* $\vec{B}_{sol}$. Because $d\vec{B}_{sol}/dt$ points out of the page, $\vec{B}_{loop}$ must point into the page. When you point your thumb in the direction *opposite* $d\vec{B}/dt$, you are also pointing your thumb in the direction of $\vec{B}_{loop}$—the magnetic field created by the induced current. That's why your fingers correctly indicate the direction of the induced current.

As a second example, consider the experiment in Figure 32.9D. The solenoid's magnetic field $\vec{B}_{sol}$ points into the page and is decreasing. To oppose this change in $\vec{B}_{sol}$, the loop's magnetic field $\vec{B}_{loop}$ near the center of the loop must also point into the page. Point your right thumb in the direction of $\vec{B}_{loop}$ (into the page). Your fingers then wrap clockwise in the direction of the induced current. It may help to remember this: **If $\vec{B}_{sol}$ is decreasing, then $\vec{B}_{loop}$ is in the same direction as $\vec{B}_{sol}$, but if $\vec{B}_{sol}$ is increasing, then $\vec{B}_{loop}$ is in the direction opposite $\vec{B}_{sol}$.**

In our discussion of Lenz's law so far, we have focused on a changing magnetic field. There are two other ways for the magnetic flux Φ_B to change: The area A can change or the angle φ can change. How should we apply Lenz's law in these two cases?

First, consider a loop with a changing area in a uniform and constant magnetic field $\vec{B}_{sol}$. The key to finding the direction of the induced current is to first figure out the direction of the loop's magnetic field $\vec{B}_{loop}$. If the loop area **is shrinking**, the magnetic flux Φ_B is decreasing, so $\vec{B}_{loop}$ must be in the **same direction** as the original magnetic field $\vec{B}_{sol}$ (Fig. 32.10). If the area is **growing** instead, the magnetic flux Φ_B is increasing, so the induced current creates a magnetic field $\vec{B}_{loop}$ in the **opposite direction** as $\vec{B}_{sol}$. Once you know the direction of $\vec{B}_{loop}$, point your right thumb in that direction and your fingers wrap or point in the direction of the induced current (and induced emf).

For the case of a changing angle, consider a loop wrapped around a ball that rolls in a uniform and constant magnetic field $\vec{B}_{sol}$ (Fig. 32.11). Because the angle between the area vector of the loop and the magnetic field changes, a current is induced in the loop. We can apply Lenz's law at two separate instants to find the direction of the induced current. In the first instant (Fig. 32.11A), the loop's area vector $\vec{A}$ is initially aligned with the magnetic field vector $\vec{B}_{sol}$. So, initially the magnetic flux through the loop is at its maximum. After the ball rolls to the right, the area vector points downward and to the right, so the magnetic flux through the loop has decreased. To visualize this, imagine looking along the magnetic field from the left side of the page. You would see the solenoid's magnetic field $\vec{B}_{sol}$ pointed away from you. Initially the loop would look like a large circle, but a moment later it would look like a squashed oval, exactly as it looks in Figure 32.10B. Because the magnetic flux decreases, the loop's magnetic field must point in the same direction as the original magnetic field—away from you if you are looking from the left in the figure. To a reader looking straight at the page, the field points to the right. The only way to be consistent with the idea that the loop's magnetic field must have a component to

LENZ'S LAW
ⓘ Underlying Principle

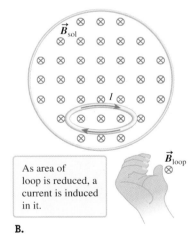

FIGURE 32.10 A. A thin, flexible loop is in the plane of the page, and a uniform and constant magnetic field $\vec{B}_{sol}$ points into the page. **B.** If you reduce the area of the loop, the magnetic flux Φ_B through the loop decreases. According to Lenz's law, the induced current creates a magnetic field $\vec{B}_{loop}$ that opposes the change. In this case, the change is a reduction in magnetic flux, so to oppose the change, $\vec{B}_{loop}$ must point into the page in the same direction as $\vec{B}_{sol}$. Point your right thumb in the direction of $\vec{B}_{loop}$ (into the page), and your fingers wrap clockwise. So the induced current and emf in the loop are clockwise.

FIGURE 32.11 A. The magnetic flux through the loop is reduced from a maximum, so the loop's magnetic field points in roughly the same direction as the original magnetic field $\vec{B}_{sol}$. Point your right thumb in the direction of $\vec{B}_{loop}$, and your fingers wrap in the direction of the induced current.
B. The magnetic flux increases from a minimum, so the loop's magnetic field points in roughly the opposite direction as the original magnetic field $\vec{B}_{sol}$. As before, if your right thumb points in the direction of $\vec{B}_{loop}$, your fingers wrap in the direction of the induced current.

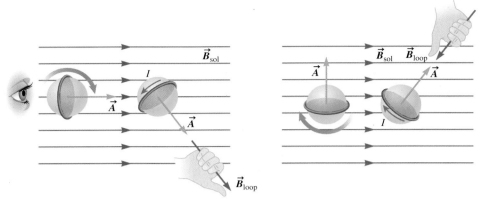

A. Φ_B is maximum initially

B. $\Phi_B = 0$ initially

the right is for $\vec{B}_{loop}$ to point downward and to the right as shown in Figure 32.11A. Therefore, $\vec{B}_{loop}$ has a component in the same direction as $\vec{B}_{sol}$ when rotation of the loop causes a decrease in the magnetic flux Φ_B.

In the second instant (Fig. 32.11B), the area vector $\vec{A}$ is initially perpendicular to the magnetic field vector $\vec{B}_{sol}$. So, initially the magnetic flux $\Phi_B = 0$ through the loop. The ball continues to roll to the right so that the loop's area vector points upward and to the right. The magnetic flux through the loop has increased. From your perspective at the left side of the page, initially the loop would look like a straight line, and a moment later it would look like a squashed oval, as if someone had stretched it open. Because the magnetic flux through the loop increases, the loop's magnetic field must point in the direction opposite the original magnetic field—toward you. For readers looking at the page, this is to the left. The only way to be consistent with the idea that the loop's magnetic field must have a component to the left is for $\vec{B}_{loop}$ to point downward and to the left as shown in Figure 32.11B. So, $\vec{B}_{loop}$ has a component in the direction opposite $\vec{B}_{sol}$ when rotation of the loop causes an increase in the magnetic flux Φ_B.

Combining Faraday and Lenz

We have been using a verbal statement of Lenz's law, but his law can also be expressed mathematically. The idea that an induced emf *opposes* a change in the magnetic flux means Lenz's law can be expressed as a negative sign in Faraday's law:

$$\mathcal{E} = -N\frac{d\Phi_B}{dt} \quad (32.6)$$

FARADAY-LENZ'S LAW
❶ Underlying Principle

Although Lenz's law is the negative sign in Equation 32.6, his name is often dropped and this equation is usually referred to as **Faraday's law**.

CONCEPT EXERCISE 32.3

Find and draw the direction of $\vec{B}_{loop}$ for each case in Figure 32.12. The original magnetic field $\vec{B}_{sol}$ in each case is produced by a solenoid (not shown) and is *decreasing*. All the loops are circular and seen in perspective.

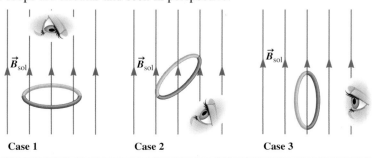

FIGURE 32.12   Case 1   Case 2   Case 3

CONCEPT EXERCISE 32.4

For each case in Figure 32.12, indicate the direction of the induced current as seen by the observer represented by the eye.

32-4 Lenz's Law and Conservation of Energy

Both Faraday's law and Lenz's law were discovered through careful observations made in a laboratory. Although neither law was derived from previously accepted laws, these laws are consistent with existing laws. In this section, we show that Lenz's law is really a statement of the conservation of energy.

First, imagine pushing a bar magnet in a frictionless environment. If the magnet is the system and you are outside the system, you do positive work on the bar magnet. If no other forces act on the system, the positive work you do must cause the bar magnet to speed up. Its kinetic energy increases:

$$W = \Delta K$$

Now, imagine pushing the bar magnet along a rough surface so that kinetic friction acts on the magnet. If the magnet and the surface are in the system, the positive work you do goes into increasing the magnet's kinetic energy as before, but it also goes into the thermal energy of the magnet and surface:

$$W = \Delta K + \Delta E_{th}$$

In order for energy to be conserved, kinetic friction must oppose the force you exert on the bar magnet. Suppose for a moment that weren't true. Then, when you give the bar magnet a small push, doing a small amount of positive work, kinetic friction would act in the same direction as your push. The magnet would continue to speed up and gain kinetic energy, and the system's thermal energy would also continue to increase. In this imaginary scenario, you would not need to put gasoline into your car. Instead, you could get someone to give your car a small push, and it would continue to speed up throughout your journey. In fact, stopping would be a big problem.

Finally, imagine pushing the bar magnet in a frictionless environment but toward a coil as shown in Figure 32.13A. Here, the magnet's north pole points toward the coil. The magnetic flux through the coil changes, inducing a current. According to Lenz's law, the coil's induced magnetic field opposes the change. A convenient way to visualize this is to remember that a coil's magnetic field looks similar to that of a bar magnet. So, you can imagine the coil as a bar magnet with its north pole directed toward the actual bar magnet's north pole (Fig. 32.13B). The coil's north pole repels the bar magnet's north pole, much like the situation in which you push the bar magnet along a rough surface; in that case, kinetic friction opposes your force. The work you do goes into the bar magnet's kinetic energy and the system's thermal energy.

If we consider both magnets to be part of the system, the work you do on the system goes into the bar magnet's kinetic energy and the electrical energy that induces the current in the coil and creates its magnetic field. Like friction, the coil's magnetic field must resist your force on the bar magnet, so that field must be oriented as shown in Figure 32.13B, with its north pole pointing toward the bar magnet. Suppose Lenz's law were reversed, so that the coil's south pole was pointing toward the bar magnet's north pole. Now, when you give the bar magnet a small push, doing a small amount of positive work, the coil's magnetic field would act in the same direction as your push. The bar magnet would continue to speed up, gaining kinetic energy. More energy would also go into the coil's current, creating an even stronger magnet. So the work you did on the system would be much smaller than the increase in the magnet's kinetic energy and the coil's electrical energy. In this imaginary scenario, you would not need to worry about powering electrical circuits or propelling objects. If you wished to light up a bulb, you could connect the bulb to a coil like the one in Figure 32.13. Then simply give a bar

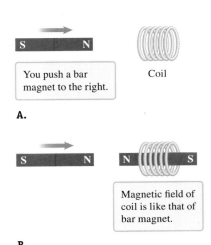

FIGURE 32.13 A. When you push a bar magnet toward a coil, **B.** the coil creates a second magnetic field that pushes back.

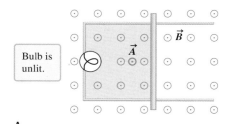

A.

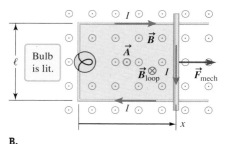

B.

FIGURE 32.14 A. A slide generator consists of a U-shaped conductor and a conducting bar in a uniform and constant magnetic field. **B.** When the bar slides to the right, there is a clockwise current in the conductors and the bulb lights up. (The mechanical force is balanced by a magnetic force; Example 32.2.)

SLIDE GENERATOR ▷ Special Case

CONCEPT EXERCISE 32.5

Suppose you have two bar magnets, with one fixed to a tabletop. You briefly push the other magnet toward the fixed magnet. There is negligible friction between the sliding magnet and the tabletop. You orient the sliding magnet so that its north end points toward the fixed magnet's south pole. Does the sliding magnet move at constant speed, speed up, or slow down? How is your answer consistent with the principle of conservation of energy?

32-5 Case Study: Slide Generator

Soon after Faraday discovered his law, he realized its enormous practical applications. In 1831, he invented an electric generator, a device that converts mechanical power to electrical power. We saw a simple example of a generator in Figure 32.13; the mechanical power you put in to move the bar magnet is converted to electrical power. Of course, having people move bar magnets around is not practical. Our case study focuses on the war over AC and DC power, and in this section we consider a simple device that highlights the basic features of any generator.

Figure 32.14 shows a **slide generator**, which consists of two conductors in a uniform magnetic field. One conductor looks like a sideways U. To visualize the current in the circuit, we have included a lightbulb in this conductor. The other conductor is a bar of length ℓ that is free to slide along the tracks of the U-shaped conductor without losing electrical contact. We'll assume the kinetic friction between the two conductors is negligible.

The bar and the U-shaped conductor form a rectangle of area $A = \ell x$. As the bar slides to the right, (1) the position x of the bar increases, (2) the area of the rectangle (shown in gray) grows, and (3) the magnetic flux through the rectangle increases. So, an emf is induced in the rectangular loop. According to Lenz's law, the loop's magnetic field must be directed into the page, and the current in the loop is clockwise (Fig. 32.14B). Our next steps are to derive the emf and current induced in, and the power required and produced by, a slide generator.

DERIVATION Motional Emf

The (absolute value of the) emf induced in the rectangular loop in Figure 32.14 is sometimes called a **motional emf** because the bar must be in motion ($v \neq 0$). Like the emf of a battery, the motional emf is constant as long as the speed v is constant, and the resulting current can operate useful devices such as the lightbulb.

The bar in Figure 32.14 is pulled to the right by a mechanical force $\vec{F}_{mech}$, and it moves to the right at constant velocity $\vec{v} = d\vec{x}/dt$. We will use Faraday's law to show the motional emf is

$$\mathcal{E} = B\ell v \qquad (32.7)$$

and the current in the conductor is

$$I = \frac{B\ell v}{R} \quad (32.8)$$

We are not concerned about direction here, so there is no need to include Lenz's law.

As in Example 32.1, the magnetic field is uniform over the entire area of the loop, and the area vector $\vec{A}$ is parallel to the magnetic field $\vec{B}$.	$\Phi_B = \int \vec{B} \cdot d\vec{A}$ $\Phi_B = BA \quad (32.1)$
Express the area A in terms of the length of the bar ℓ and its position x. As the bar slides to the right, x increases.	$\Phi_B = B(\ell x)$
Substitute the expression for the magnetic flux Φ_B into Faraday's law (Eq. 32.4).	$\mathcal{E} = N\dfrac{d\Phi_B}{dt} = N\dfrac{d(B\ell x)}{dt}$
There is only one loop, so $N = 1$. The magnetic field B and the length ℓ of the bar are constant.	$\mathcal{E} = \dfrac{d(B\ell x)}{dt} = B\ell \dfrac{dx}{dt}$
The derivative dx/dt is the constant speed v of the bar.	$\mathcal{E} = B\ell v \quad\checkmark \quad (32.7)$
The current in the conductor is found by dividing the motional emf by the conductor's total resistance R.	$I = \dfrac{B\ell v}{R} \quad\checkmark \quad (32.8)$

:• **COMMENTS**
Finding R in Equation 32.8 can be complicated. The bar, the legs of the U-shaped conductor, and the lightbulb are all connected in series (Fig. 32.14), so R is found by adding up the resistances of each of these elements. However, the resistance of the legs of the U-shaped conductor depends on the position x of the bar. When the bar is far along the tracks, the current must pass through a long portion of the U-shaped conductor's legs and the resistance is higher because $R = \rho L/A$ (Eq. 28.24), where L is the total length of the conductor through which current is flowing and A is its cross-sectional area.

EXAMPLE 32.2 Mechanical Power Supplied to a Slide Generator

Find an expression for the mechanical power put into the slide generator when the bar slides at constant speed in Figure 32.14.

:• **INTERPRET and ANTICIPATE**
A free-body diagram for the bar is helpful (Fig. 32.15). Two forces are exerted on the bar: the mechanical force $\vec{F}_{mech}$ to the right (as might be exerted by a person pulling the bar) and the magnetic force. The current and $\vec{\ell}$ are downward, and the magnetic field is out of the page. According to $\vec{F}_B = I(\vec{\ell} \times \vec{B})$ (Eq. 30.32) and the right-hand rule for cross products, the magnetic force on the bar is to the left.

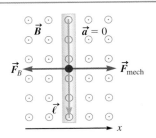

FIGURE 32.15

:• **SOLVE** Find the magnitude of the magnetic force. The angle between $\vec{\ell}$ and $\vec{B}$ is 90°.	$F_B = I\ell B \sin\varphi = I\ell B \sin 90°$ $F_B = I\ell B$

Example continues on page 1022 ▶

Because the bar is moving at constant velocity, its acceleration is zero and the mechanical force must be balanced by the magnetic force.	$F_{mech} = F_B$ $F_{mech} = I\ell B$ (1)
The power P_{mech} put in by the mechanical force is given by $P_{mech} = \vec{F}_{mech} \cdot \vec{v}$ (Eq. 9.38). Both the mechanical force $\vec{F}_{mech}$ and the bar's velocity $\vec{v}$ are to the right in the positive x direction, so the angle between them is zero.	$P_{mech} = F_{mech} v \cos 0 = F_{mech} v$
Substitute Equation (1) for the mechanical force.	$P_{mech} = I\ell B v$
Substitute Equation 32.8 for the current.	$P_{mech} = \left(\dfrac{B\ell v}{R}\right)\ell B v$ $P_{mech} = \dfrac{(B\ell v)^2}{R}$ (32.9)

∴ CHECK and THINK

As always, you should verify that our expression has the correct dimensions (energy per time). In the next example, we'll find the power supplied by the generator and think about both results together.

EXAMPLE 32.3 Electrical Power Supplied by a Slide Generator

Equation 32.9 is the power supplied *by the mechanical force*. Electrical power is, in turn, supplied *by the generator* and consumed by the conductors and the bulb. Find an expression for the power consumed by the conductors and the bulb. In **CHECK and THINK**, compare the power supplied by the mechanical force to the power P_{res} consumed.

∴ INTERPRET and ANTICIPATE

Because the conductors and the bulb have resistance, they must dissipate power. Our job is to find an expression for that power in terms of B, ℓ, v, and R so that we can compare P_{mech} to P_{res}.

∴ SOLVE

The conductors and the bulb have total resistance R, and the power they use is found from Equation 28.34.	$P_{res} = I^2 R$ (28.34)
Substitute $I = B\ell v/R$ (Eq. 32.8) for the current.	$P_{res} = \left(\dfrac{B\ell v}{R}\right)^2 R = \dfrac{(B\ell v)^2}{R}$ (32.10)

∴ CHECK and THINK

Equation 32.10 is the power used by the conductors and bulb. It is exactly what we found for the power supplied by the mechanical force: $P_{mech} = P_{res}$. Having the power supplied by the mechanical force equal to the power used by the conducting loop is consistent with the principle of conservation of energy. In the absence of dissipative forces such as kinetic friction, the mechanical work per unit time supplied to the generator equals the energy per unit time that is used by the electric circuit. So, if you want to light a 100-W bulb, you must supply 100 W of mechanical power.

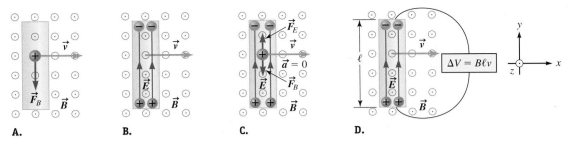

FIGURE 32.16 A. A conducting bar moves at constant velocity in a uniform magnetic field, leading to a downward magnetic force on the positive (imaginary) particles in the conductor. **B.** Positive charge builds up on the bottom of the bar and negative charge builds up on the top, resulting in an upward-pointing electric field. **C.** Now, the upward electric force on a positive particle is balanced by the downward magnetic force. **D.** When the particles are in equilibrium, the electric potential between the bottom and top is equal to the emf found using Faraday's law for a slide generator.

Another Look at the Motional Emf

The motional emf that is induced in the slide generator (Fig. 32.14) does *not* require the U-shaped conductor or Faraday's law. All that is necessary is the movement of the conducting bar at constant velocity through the uniform magnetic field (Fig. 32.16). Because the bar is moving, conducting electrons are moving with respect to the magnetic field, so a magnetic force is exerted on them. For convenience, we'll think about the magnetic force on the equivalent imaginary positive particles.

The bar moves at constant velocity in the positive x direction, so the imaginary positive particle's velocity $\vec{v}$ is in the positive x direction (Fig. 32.16A). The magnetic field $\vec{B}$ is out of the page in the positive z direction. According to $\vec{F}_B = q(\vec{v} \times \vec{B})$ (Eq. 30.17) and the right-hand rule for cross products, the magnetic force $\vec{F}_B$ on this imaginary positive particle is downward in the negative y direction. As a result of the magnetic force, positive charge builds up on the lower end of the conductor and negative charge on the upper end (Fig. 32.16B). This charge separation results in an electric field $\vec{E}$ in the bar pointing in the positive y direction—from the positive particles toward the negative ones. Once the electric field in the moving bar is established, an imaginary positive particle experiences two forces (Fig. 32.16C): a magnetic force $\vec{F}_B$ in the negative y direction and an electric force $\vec{F}_E$ in the positive y direction. The electric field reaches its maximum strength when these two forces are in balance. Once they are equal in magnitude, there is no net force on the imaginary charged particle, so it experiences no acceleration:

$$F_E = F_B$$

Equations 24.18 and 30.18 give the magnitudes of the electric and magnetic forces, respectively:

$$qE = qvB$$

The charge q cancels, so the maximum electric field at equilibrium is

$$E = vB \tag{32.11}$$

The electric field in the bar is uniform, so the electric potential difference between the bottom and the top of the bar (Fig. 32.16D) comes from modifying Equation 26.17:

$$\Delta V = V_{\text{bottom}} - V_{\text{top}} = E\ell$$

Substitute Equation 32.11 for the electric field:

$$\Delta V = Bv\ell = B\ell v \tag{32.12}$$

This is exactly what we found for the motional emf (Eq. 32.7) induced in the rectangular loop of the slide generator using Faraday's law. You could imagine using wires to connect the top and bottom of the bar (like the positive and negative terminals of a battery) to a lightbulb; then the bulb would be lit just as it is in Figure 32.14B.

EXAMPLE 32.4 Space Tether

Suppose you want to use a slide generator like the one in Figure 32.14 to light a small flashlight bulb. Normally, two 1.5-V batteries are used to power the bulb. The flashlight case is about 20 cm long, so let's assume the bar in Figure 32.14 is $\ell = 20$ cm long.

A If you use a uniform magnetic field of 0.15 T, at what speed must the bar be moved to produce the usual emf across the bulb?

INTERPRET and ANTICIPATE
This is a direct application of the slide generator. We expect to find a numerical result in m/s.

SOLVE

Solve Equation 32.7 for speed v, and then substitute values. The emf across the bulb when it is connected to two batteries is the sum of the emfs of both batteries (3.0 V).	$\mathcal{E} = B\ell v$ (32.7) $v = \dfrac{\mathcal{E}}{B\ell} = \dfrac{3.0\,\text{V}}{(0.15\,\text{T})(0.20\,\text{m})}$ $v = 100\,\dfrac{\text{V}}{\text{T}\cdot\text{m}}$ (1)
We need to obtain the expected SI units. Use Equation 32.5 to convert.	$1\,\dfrac{\text{T}\cdot\text{m}^2}{\text{s}} = 1\,\dfrac{\text{Wb}}{\text{s}} = 1\,\text{V}$ (32.5) so $1\,\dfrac{\text{V}}{\text{T}\cdot\text{m}} = 1\,\dfrac{\text{Wb/s}}{\text{T}\cdot\text{m}} = 1\,\dfrac{\text{T}\cdot\text{m}^2}{\text{T}\cdot\text{m}\cdot\text{s}} = 1\,\text{m/s}$
We have arrived at the correct units for speed. Substitute them into Equation (1).	$v = 100\,\text{m/s}$

CHECK and THINK
We just found that the bar must move at 100 m/s or approximately 224 mph, which is not practical. You might wonder whether slide generators have any use, but NASA engineers think such generators might be used in satellites. Part B is based on an experiment conducted by NASA.

B In 1996, NASA tested a large-scale slide generator. A space shuttle deployed a tether—a long conducting wire—with a spherical satellite at the end. The tether was moved through the Earth's magnetic field at a speed of 7.7 km/s, inducing an emf in the wire. The tether broke when it was 19.6 km long. Although the magnetic field is not uniform over the length of the wire, assume it is uniform with a magnitude of 2.5×10^{-5} T. Estimate the maximum motional emf between the ends of the tether.

INTERPRET and ANTICIPATE
This is another application of motional emf. In this case, we can think of the tether as the bar in Figure 32.16.

SOLVE

Equations 32.12 and 32.7 are equivalent. Use either to find the motional emf $\mathcal{E}$. Remember to convert the speed and length to SI units. Use Equation 32.5 to express the emf in volts.	$\mathcal{E} = B\ell v$ (32.12) $\mathcal{E} = (2.5 \times 10^{-5}\,\text{T})(19.6 \times 10^3\,\text{m})(7.7 \times 10^3\,\text{m/s})$ $\mathcal{E} = 3.8 \times 10^3\,\dfrac{\text{T}\cdot\text{m}^2}{\text{s}} = 3.8 \times 10^3\,\dfrac{\text{Wb}}{\text{s}}$ $\mathcal{E} = 3.8 \times 10^3\,\text{V}$

CHECK and THINK
This particular slide generator could be a very practical device. A space shuttle or satellite can produce an enormous motional emf that can then be used to produce a current in any sort of useful device. In fact, the tether actually broke because of the current. The tether was made of

many layers, with the innermost layer—a conducting wire—surrounded by an insulator full of small air pockets. Normally air is a good insulator, but due to the large motional emf, the air became a conductor and allowed for a complete circuit. The current in the circuit melted the tether.

Eddy Currents

Just as kinetic friction can slow down moving objects, so can currents induced as described by Faraday's law. Figure 32.17 shows a wire conducting loop and the poles of a C-shaped magnet. Between the poles of the magnet, the magnetic field is nearly uniform. The loop is made from a nonmagnetic material. When the loop is at rest, the magnet has no effect on the loop; there is no induced current and no magnetic force. Now imagine pulling the loop to the right (Fig. 32.17A). At first there is still no induced current and no magnetic force on the loop because the magnetic flux through it is constant. However, once the left edge of the loop begins to move through the magnetic field, the magnetic flux through the loop decreases, causing an induced current in the loop (Fig. 32.17B). According to Lenz's law, the loop's magnetic field $\vec{B}_{loop}$ must point downward near its center—in the same direction as the magnet's field. Point your right thumb downward, and your fingers wrap clockwise as seen from above the loop.

Now imagine pulling two loops to the right between the magnet's poles (Fig. 32.18). Both loops experience a change in the magnetic flux. For the loop on the right, the magnetic flux is decreasing. By Lenz's law, as argued above, $\vec{B}_{right}$ must point downward and the induced current in the right loop must be clockwise as seen from above. For the loop on the left, however, the magnetic flux is increasing. This time, according to Lenz's law, its magnetic field $\vec{B}_{left}$ must point in the opposite direction—upward—as the magnet's field. From the right-hand rule, the induced current in the left loop must be counterclockwise as seen from above (Fig. 32.18).

Finally, imagine pulling a conducting metal sheet to the right through the magnet's poles (Fig. 32.19). When the sheet is at rest, no magnetic force is exerted on it. However, when the sheet is moving with respect to the magnet, the conduction electrons experience a magnetic force like the force experienced by the charged particles in the bar in Figure 32.16. As a result, **eddy currents** develop in the metal sheet. The eddy currents on the right side swirl around clockwise, and the eddy currents on the left side swirl around counterclockwise, as we found for the two loops in Figure 32.18. Eddy currents—like the currents in the two loops—obey Lenz's law by opposing the change that created them. So, if you are pulling the sheet to the right, the magnetic force on the sheet pulls the sheet back toward the left.

A practical application of eddy currents is **magnetic braking**, which is used to damp the motion of delicate instruments and to stop some vehicles such as trains. You may have encountered a magnetic brake in your laboratory course. Typically,

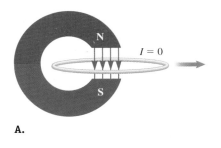

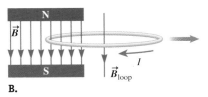

FIGURE 32.17 **A.** A loop is pulled to the right between the poles of a C-shaped magnet. There is no change in magnetic flux, so there is no current through the loop. **B.** Once the loop's left edge crosses the magnetic field lines, the magnetic flux begins to decrease and a current is induced in the loop.

EDDY CURRENT ✪ **Major Concept**

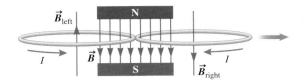

FIGURE 32.18 Two loops are pulled to the right between the poles of a C-shaped magnet. The magnetic flux through the right loop decreases, so it produces a magnetic field that points in the same direction as that of the magnet. The induced current in the right loop is clockwise as seen from above. The magnetic flux through the left loop increases, so it produces a magnetic field that points in the opposite direction as that of the magnet. The induced current in the left loop is counterclockwise as seen from above. (The loops are insulated from one another.)

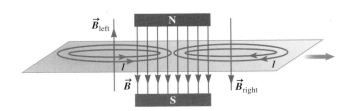

FIGURE 32.19 A conducting metal sheet is pulled to the right through the poles of a C-shaped magnet. Eddy currents are induced in the sheet. The eddy currents on the right are clockwise as seen from above, and the eddy currents on the left are counterclockwise as seen from above. Compare the eddy currents to the currents in the two loops in Figure 32.18.

the magnetic brake looks like a pendulum with a flat metal bob that swings between the poles of a magnet. Magnetic braking causes the pendulum's motion to damp, and the pendulum stops quickly. A train's magnetic brakes are electromagnets that are attached to the train and straddle the track. Normally there is no current or field in the electromagnets. When the train conductor wants to slow down or stop the train, current in the electromagnets produces a magnetic field, inducing eddy currents in the train track. Magnetic force produced by eddy currents in the track slows the train down. There is no kinetic friction between the brakes and the track, so the magnetic brakes do not wear out as quickly as normal brakes.

However, eddy currents heat up the train tracks. In terms of energy conservation, the kinetic energy of the train's motion must be transformed into some other form of energy when the train stops. If kinetic friction between the wheels and the track stops the train, both the wheels and the track gain thermal energy. If magnetic brakes are used, the track's thermal energy increases because the current in the conducting tracks encounters resistance. The power that heats the conductor is $P = I^2R$. Our case study focuses on electric generators. Even the most efficient generators lose energy due to eddy currents heating the conductors.

CONCEPT EXERCISE 32.6

A horizontal metal disk is balanced on a pivot. A strong bar magnet is moved in circles just above the surface of the disk. The disk is made of a nonmagnetic, conducting material, yet it begins to spin, following the magnet's circles. Explain how this is possible.

32-6 Case Study: AC Generators

The slide generator (Fig. 32.14) has two features that are common to all the generators we consider in this chapter: (1) The magnetic field is uniform and constant and may be due to a solenoid or a permanent magnet. Because a permanent magnet does not require a current to maintain its magnetic field, permanent magnets are used in most generators. (2) A change in the conductor induces the emf. Two features of the conductor can change—the area A and the angle φ. In the slide generator, the area A changes (grows or shrinks). We found, however, that the slide generator is usually not practical because it requires long tracks and a bar moving at very high speed. Another way to improve the slide generator is to increase the number of turns.

A more practical approach is to design a generator in which φ changes. In this section and the next, we look at two generators—AC and DC—based on a changing φ. The current in the AC (alternating current) generator increases, decreases, and changes direction. The current in the DC (direct current) generator is in a constant direction and is nearly constant in magnitude. Today, we use AC in all wall outlets. However, the decision to go with AC instead of DC was at the heart of a debate between Edison and his rivals, Westinghouse and Tesla.

Features and Operation of an AC Generator

Figure 32.20 shows the basic features of an AC generator. A coil rotates between the poles of one or more permanent magnets. Two wires protrude from the coil. The wire shown in blue is connected to the blue slip ring. The wire shown in red passes through the blue slip ring without making contact and is attached to the red slip ring. The slip rings rotate with the coil, and each ring is in contact with a stationary conducting brush. A wire is attached to the brushes to complete the circuit. Of course, the circuit can involve a number of useful devices. For simplicity, we imagine there is a lightbulb in the circuit.

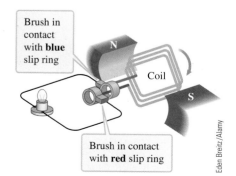

FIGURE 32.20 A basic AC generator.

Figure 32.21 shows how the coil rotation leads to a current. For simplicity, the coil is shown as a loop with only one turn. The loop is between the poles of a magnet (not shown); this permanent magnet's field lines are assumed to be horizontal and point from left to right. From our perspective (near the bulb), the loop rotates clockwise. In Figure 32.21A, the loop is not yet vertical. In Figure 32.21B, the loop is vertical so that the magnetic flux Φ_B through the loop is maximum. The magnetic flux increases as the loop rotates from time A to B and according to Lenz's law, the loop's magnetic field $\vec{B}_{\text{loop}}$ in part A must have a component in the direction opposite the permanent magnet's field $\vec{B}$. To find the direction of the current in the loop, point your right thumb in the direction of $\vec{B}_{\text{loop}}$ and trace the current through the bulb. (Follow it out through the red slip ring.) In part A, the current is from right to left through the bulb, the magnetic flux is increasing, and the emf is negative.

For a moment in Figure 32.21B, the magnetic flux does not change; its time derivative is zero, and so the induced emf is zero. Of course, the loop continues to rotate, and the magnetic flux decreases. By the time shown in Figure 32.21C, the loop's magnetic field $\vec{B}_{\text{loop}}$ must have a component in the same direction as the magnet's field $\vec{B}$. Point your right thumb in the direction of $\vec{B}_{\text{loop}}$, and follow the current out through the blue slip ring. In part C, the current through the bulb is from left to right, the magnetic flux is decreasing, and the emf is positive.

Now we are ready to look at one complete cycle. To keep the drawing simple, we show only the loop, the permanent magnet's field $\vec{B}$, and the loop's area vector $\vec{A}$. Figure 32.22 shows the loop end-on as it rotates clockwise from (1) vertical, going upside down and back to (9) vertical. At each moment, the magnetic flux is given by $\Phi_B = BA\cos\varphi$. As the loop rotates from (1) to (9), the magnetic flux through the loop decreases and then increases. The sign of the induced emf comes from Lenz's law and so is the opposite of $d\Phi_B/dt$. At times 1, 5, and 9, the flux is momentarily constant, and according to Faraday's law, the induced emf is zero at those times.

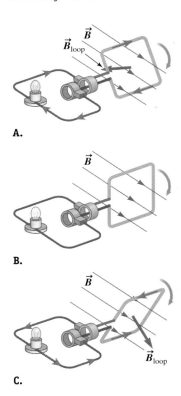

FIGURE 32.21 Imagine observing an AC generator in motion from a position near the bulb. **A.** The magnetic flux Φ_B is increasing, so $\vec{B}_{\text{loop}}$ points roughly opposite to $\vec{B}$ and the current through the bulb is to the *left*. **B.** Φ_B is at its maximum and momentarily constant, so $\vec{B}_{\text{loop}} = 0$, and there is no current. **C.** Φ_B is decreasing, so $\vec{B}_{\text{loop}}$ points roughly in the direction of $\vec{B}$ and the current through the bulb is to the *right*.

	Configuration	Magnetic flux Φ_B	Change in flux $\dfrac{d\Phi_B}{dt}$	Induced emf $\mathcal{E}$
1 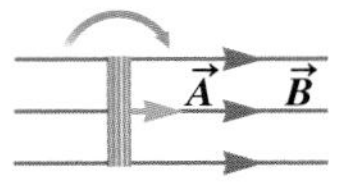		Positive, maximum	Momentarily zero (constant flux)	Zero
2 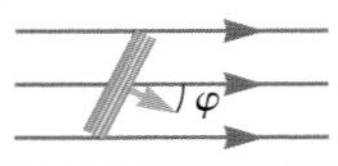		Positive	Decreasing (negative)	Positive
3 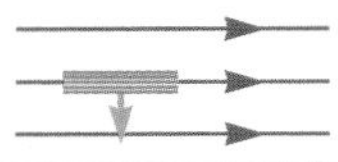		Zero	Decreasing (negative)	Positive
4 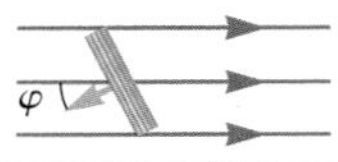		Negative	Decreasing (negative)	Positive
5 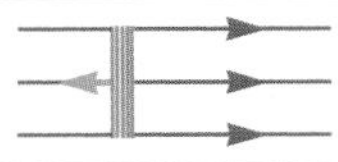		Negative, maximum	Momentarily zero (constant flux)	Zero
6 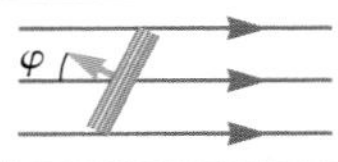		Negative	Increasing (positive)	Negative
7 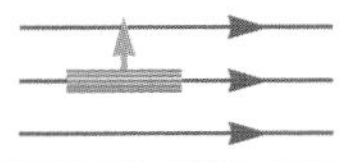		Zero	Increasing (positive)	Negative
8 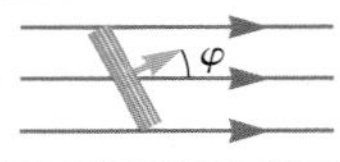		Positive	Increasing (positive)	Negative
9 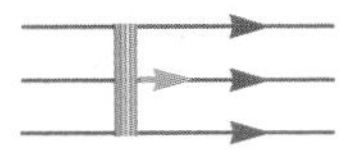		Return to positive, maximum	Momentarily zero (constant flux)	Zero

FIGURE 32.22 An AC generator loop shown schematically during one complete cycle.

DERIVATION | Time-Varying AC Emf and Current

We show here that the induced emf in an AC generator (Fig. 32.20) is

$$\mathcal{E} = NBA\omega \sin \omega t = \mathcal{E}_{max} \sin \omega t \quad (32.13)$$

AC GENERATOR ▶ Special Case

and the current is

$$I = \frac{NBA\omega}{R} \sin \omega t = I_{max} \sin \omega t \quad (32.14)$$

Then we use Figure 32.22 to connect these results to the motion in the generator.

First, we need an expression for the magnetic flux through the loop or coil. Because the magnetic field is uniform over the loop, the magnetic flux comes from Equation 32.3.	$\Phi_B = BA \cos \varphi$ $\Phi_B = (\Phi_B)_{max} \cos \varphi$ where $(\Phi_B)_{max} = BA$	(32.3) (1)
The coil in the generator rotates at constant angular speed ω, so we write φ in terms of ω and substitute $\varphi = \omega t$ into Equation (1).	$\Phi_B = (\Phi_B)_{max} \cos \omega t$	(32.15)
Now substitute Equation 32.15 for magnetic flux in Faraday–Lenz's law (Eq. 32.6).	$\mathcal{E} = -N\dfrac{d\Phi_B}{dt}$ $\mathcal{E} = -N\dfrac{d(BA \cos \omega t)}{dt}$	(32.6)
The magnetic field B and area A are constants, so we pull them outside the derivative.	$\mathcal{E} = -NBA\dfrac{d(\cos \omega t)}{dt}$ $\mathcal{E} = NBA\omega \sin \omega t = \mathcal{E}_{max} \sin \omega t$ ✓ where $\mathcal{E}_{max} = NBA\omega$	(32.13) (32.16)
The current is found by dividing the emf by the total resistance R in the circuit.	$I = \dfrac{NBA\omega}{R} \sin \omega t = I_{max} \sin \omega t$ ✓ where $I_{max} = \dfrac{\mathcal{E}_{max}}{R} = \dfrac{NBA\omega}{R}$	(32.14) (32.17)

∴ COMMENTS

The presence of a sine function in Equations 32.13 and 32.14 means the emf and current increase, decrease, and change sign periodically. Because the current *alternates*, such a generator is called an **alternating current** generator.

The coil's angular speed ω is numerically equivalent to the angular frequency ω of the emf and current oscillation that appear in Equations 32.13 and 32.14. Although *angular* frequency appears in these equations, *frequency f* is commonly used. Recall the relationship among angular frequency, frequency, and period from Equation 16.2: $\omega \equiv 2\pi f = 2\pi/T$. Standard AC power in the United States and Canada oscillates at $f = 60$ Hz. In many other countries, the standard frequency is $f = 50$ Hz.

EXAMPLE 32.5 | Plotting an AC Generator's Emf

Whenever a complicated function is involved, it is helpful to make a graph. Sketch the magnetic flux Φ_B (Eq. 32.15) and induced emf $\mathcal{E}$ (Eq. 32.13) as functions of time t for one cycle. Then connect these functions to the motion of the generator by including the numbers representing the time sequence of events in Figure 32.22, and check for consistency.

INTERPRET and ANTICIPATE

It is straightforward to sketch $\Phi_B = (\Phi_B)_{max} \cos \omega t$ and $\mathcal{E} = \mathcal{E}_{max} \sin \omega t$ because they are the familiar cosine and sine graphs. Our real work is checking these graphs for consistency with the motion of the AC generator.

SOLVE

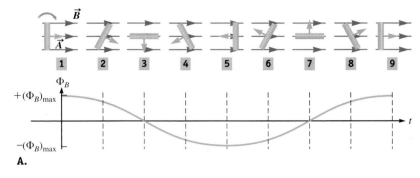

FIGURE 32.23 A. Magnetic flux though the loop and B. emf versus time for an AC generator. The configuration of the coil at selected instants in time is shown above the graphs.

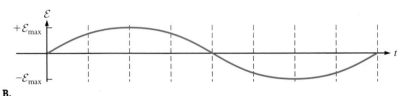

CHECK and THINK

First, compare the sketch of the generator at the nine times to the graph of magnetic flux (Fig. 32.23A). We see that the flux starts at its maximum at 1, decreases to its minimum at 5, and increases to its maximum again at 9. Now, let's check that the two graphs are consistent. At times 1, 5, and 9, the flux is at an extreme value, so the time derivative is zero. We should find that the induced emf $\mathcal{E}$ is zero at these three times, and this is exactly what is shown (Fig. 32.23B). Further, between times 1 and 5, the flux is decreasing, and because the slope of the graph is negative (Fig. 32.23A), according to Lenz's law, we expect the induced emf to be positive during this period. Again, this is exactly what we see in Figure 32.23B. Finally, consider the magnetic flux between times 5 and 9. The flux is increasing, and the slope of the graph is positive. According to Lenz's law, we expect the induced emf to be negative during this period. Once again, this result is consistent with the graph of emf versus time.

CONCEPT EXERCISE 32.7

Suppose the lightbulb in Figure 32.20 is fluorescent. Is it always illuminated while the generator is running? Explain.

EXAMPLE 32.6   CASE STUDY   AC Generator

The circular coil of an AC generator has 225 turns and an area of 2.00×10^{-2} m² (radius about 8 cm). The generator's magnetic field is 0.10 T. The maximum induced emf is $\mathcal{E}_{max} = 170$ V—typical of the AC power in American houses. What is the angular speed of the coil?

INTERPRET and ANTICIPATE

We expect to find a numerical answer in rad/s that is consistent with the frequency ($f = 60$ Hz) of the current's oscillation in a typical American house.

Example continues on page 1030 ▶

SOLVE
Find $\mathcal{E}_{max}$ from Equation 32.16. Solve for ω and substitute values.

$$\mathcal{E}_{max} = NBA\omega \quad (32.16)$$

$$\omega = \frac{\mathcal{E}_{max}}{NBA} = \frac{170\,\text{V}}{(225)(0.10\,\text{T})(2.00 \times 10^{-2}\,\text{m}^2)}$$

$$\omega = 378 \frac{\text{V}}{\text{T}\cdot\text{m}^2}$$

CHECK and THINK
We must make sure the units are correct and the value is consistent with what we expect for a typical American home. Equation 32.5 helps when reducing the units. Be careful; radians are dimensionless. *Angular* frequency and *angular* velocity are both in rad/s. Convert to frequency for the final answer. Our result is consistent with the frequency of current oscillation in a typical American house.

$$1\frac{\text{V}}{\text{T}\cdot\text{m}^2} = 1\frac{\text{V}}{\text{Wb}} = 1\frac{\text{Wb/s}}{\text{Wb}} = 1\,\text{s}^{-1}$$

$$\omega = 378\,\text{rad/s}$$

$$f = \frac{\omega}{2\pi} = \frac{378\,\text{rad/s}}{2\pi\,\text{rad}} = 60\,\text{Hz}$$

Compare these results to those for the slide generator in Example 32.4. We found an impractical result in the case of a slide generator, but the AC generator in this example has an achievable magnetic field, coil size, and number of turns. Now imagine turning the coils at 378 rad/s (60 rps). Is that also achievable? Of course, that is much too fast to do by hand, but another source of mechanical power such as a steam engine or a waterfall could do the trick. Edison's DC generators had to be located in cities, where it is unlikely to find a waterfall, so they were powered by steam engines.

The AC Motor

The AC motor is an AC generator running in reverse. In the motor, an AC current causes the coil to rotate. That mechanical rotation can be used to do any number of things, such as run an electric blender, vacuum cleaner, or power saw. Today many of our devices are based on AC motors running off of AC generators, but in the late 1880s DC motors (Fig. 30.55, page 969) were the standard. Edison worked to improve the DC generator and believed DC generators should become the standard. The DC generator is based on Faraday's generator (Section 32-7).

32-7 Case Study: Faraday's Generator and Other DC Generators

When presenting a new discovery, Faraday liked to quote Franklin: "Endeavor to make it useful." Within months of conducting the experiments that led to Faraday's law, he built a very useful device—the first electromagnetic generator as shown in Figure 32.24A.

Faraday's generator consists of a copper disk and a permanent magnet. A portion of the disk slips between the poles of the magnet as a person turns the hand crank. In Figure 32.24B, the disk is shown rotating clockwise and the permanent magnet's field points into the page. Conduction electrons move perpendicular to the magnetic field, so they experience a magnetic force. Consider an equivalent positive particle in Figure 32.24B. At the moment shown, the particle moves upward, so the magnetic force is radial toward the edge of the disk. The result is a radial current directed away from the shaft and toward the edge of the disk. One conducting brush is in contact with the shaft and another with the edge of the disk. Trace the current in the circuit through the bulb from left to right. As long as the crank is turned in the same direction, the direction of the current does not change. So, Faraday's generator is a DC (direct current) generator.

Faraday's generator was a technological breakthrough, but his actual design was impractical because it did not generate much power. In the 1870s, the Belgian engineer Zénobe-Théophile Gramme invented a practical DC generator and at the same time a DC motor (Fig. 30.55, page 969). A DC generator is a DC motor running in

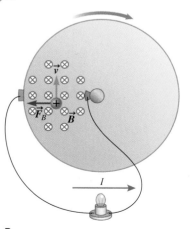

FIGURE 32.24 A. Faraday's electric generator (1831). **B.** As the copper disk is rotated, the magnetic field forces imaginary positive charges outward, creating a radial current from the shaft toward the edge of the disk.

reverse: Both consist of a coil or loop that rotates in a magnetic field. In a DC motor, a DC power supply such as a battery produces a current in the coil. A permanent magnet exerts a torque on the coil, so electrical power is converted to mechanical power.

In a DC generator, a coil is mechanically rotated (Fig. 32.25). Because the angle φ between the coil's area and the magnetic field changes, current is induced in the coil. So, mechanical power is converted to electrical power.

Compare the AC generator in Figure 32.21 to the DC generator in Figure 32.25. The main difference is that in the AC generator the coil is connected to two separate rings, and in the DC generator the coil is connected to a single **split ring commutator**. The split ring commutator ensures that the direction of the current is constant. Because the commutator is an external device, Tesla argued, all DC generators and motors are really AC. He said the commutator is a complicated device that often causes trouble. In fact, the commutator often sparked and wore out easily. Because the commutator is fragile and must be replaced often, Tesla wanted to eliminate it from the system, leaving an AC generator.

Nevertheless, it is important to understand how DC generators operate. To see how the commutator works, consider a few moments of the coil's rotation. The coil in Figure 32.25 is represented by a single loop rotating clockwise (as seen from the bulb). The loop is between the poles of a permanent magnet so that the magnetic field is horizontal and points roughly from left to right (from a perspective near the bulb). Two conducting brushes maintain contact with the split ring commutator as the loop rotates.

In Figure 32.25A, the loop is horizontal. There is no magnetic flux through the loop, but the flux increases as the coil rotates. The loop's magnetic field $\vec{B}_{\text{loop}}$ must have a component in the direction opposite the permanent magnet's field $\vec{B}$, and the current through the bulb is from right to left. In Figure 32.25B, the loop is vertical and the magnetic flux is at its maximum. For the moment, there is no change in flux and no induced current. Finally, in part C, the loop is once again horizontal. Compare part A to part C, paying close attention to the color of the loop and the commutator. In part A, the yellow part is near the magnet's north pole; in part C, the yellow part is near the magnet's south pole and the loop has flipped over. However, the situation in both parts is similar. There is no magnetic flux through the loop, but the flux increases as the coil rotates. So the loop's magnetic field $\vec{B}_{\text{loop}}$ in part C must have a component in the direction opposite the permanent magnet's field $\vec{B}$, and the current through the bulb is from right to left exactly as in part A.

In an AC generator as shown in Figure 32.22, at times 3 and 7, when the loop flips, the induced current reverses direction. Figure 32.25 shows that the split ring commutator in the DC generator keeps the current in the same direction when the loop flips.

Time-Varying DC Emf and Current

A DC generator is not the same as a battery. In a battery, the current or emf is constant in both direction and magnitude. In a DC generator, the direction of the current or emf is constant, but the magnitude is not constant. In Figure 32.25A and C, there is nonzero current, but in part B, the current is zero.

The procedure for finding the induced emf in the DC generator is the same as the procedure for finding the induced emf in the AC generator with one exception. The emf in the DC generator does not reverse direction; it is always positive, so we take the absolute value of the emf for the AC generator in $\mathcal{E} = NBA\omega \sin \omega t = \mathcal{E}_{\max} \sin \omega t$ (Eq. 32.13). Because $NBA\omega$ is positive, take the absolute value of the sine function:

$$\mathcal{E} = NBA\omega |\sin \omega t| = \mathcal{E}_{\max}|\sin \omega t| \quad (32.18)$$

The current is found by dividing the emf by the total resistance in the circuit:

$$I = \frac{NBA\omega}{R}|\sin \omega t| = I_{\max}|\sin \omega t| \quad (32.19)$$

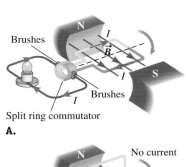

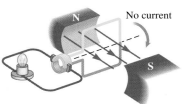

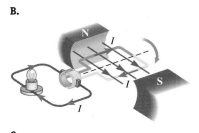

FIGURE 32.25 Compare this DC generator to the AC generator in Figure 32.21. **A.** The magnetic flux Φ_B is increasing, so $\vec{B}_{\text{loop}}$ points roughly opposite to $\vec{B}$ and the current through the bulb is to the *left*. **B.** Φ_B is at its maximum and momentarily constant, so $\vec{B}_{\text{loop}} = 0$, and there is no current. **C.** Φ_B is increasing, so $\vec{B}_{\text{loop}}$ points roughly opposite to $\vec{B}$ and the current through the bulb is to the *left* (same as in part A).

DC GENERATOR   ▶ **Special Case**

1032 CHAPTER 32 Faraday's Law of Induction

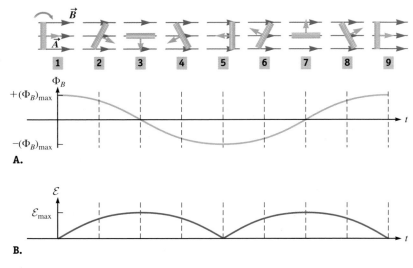

FIGURE 32.26 Graphs of **A.** the magnetic flux through the loop and **B.** the emf induced in a DC generator during one cycle. To make the comparison with the same graphs for an AC generator (Fig. 32.23) easy, the same time sequence 1 through 9 is shown.

Figure 32.26 shows the magnetic flux $\Phi_B = (\Phi_B)_{max} \cos \omega t$ (Eq. 32.15) and the induced emf $\mathcal{E}$ (Eq. 32.18) plotted as functions of time t for one cycle. Consider Φ_B at times 1, 5, and 9. At these times, the flux is at an extreme value, so the time derivative is zero (Fig. 32.26A). We find that the induced emf $\mathcal{E}$ is zero at these three times, just as in Figure 32.23 for the AC generator. Between times 1 and 5, the flux is decreasing. The slope of the graph is negative and the induced emf is positive during this period, exactly as for the AC generator. Finally, consider the magnetic flux between times 5 and 9. The flux is increasing and the slope of the graph is positive, but because of the split ring commutator, the emf remains positive. Compare this sequence, 5 through 9, to the same time interval in Figure 32.23; the result is the opposite of what we found for the AC generator.

A battery produces an emf that is constant in magnitude, but the simple DC generator in Figure 32.25 produces an emf that varies between zero and $\mathcal{E}_{max}$ (Fig. 32.26B). Many people—including Edison—made improvements to the DC generator. For example, including a loop perpendicular to the loop in Figure 32.25 results in an emf that varies less. By the late 1800s, the DC generator was efficient and widely used to light bulbs and run DC motors.

EXAMPLE 32.7 Gramme's DC Generator

Gramme's original DC generator was cranked by hand. It consisted of a coil with thirty turns and a diameter of roughly 51 cm. Assume the field of the permanent magnet was $B = 0.15$ T, and estimate the rotational speed needed to achieve an average emf of 110 V—the same as Edison required of his generator.

:• INTERPRET and ANTICIPATE
A key to solving this problem is finding an expression for the *average* emf from an equation that gives the *instantaneous* emf $\mathcal{E}$ as a function of time (Eq. 32.18).

:• SOLVE

Finding $\mathcal{E}_{avg}$ from $\mathcal{E} = \mathcal{E}_{max}\|\sin \omega t\|$ (Eq. 32.18) means finding the average of the function $\|\sin \omega t\|$, which is the same as the average of $\sin \omega t$ over half a period. (The sine function is positive from 0 to $T/2$.) We can integrate $\sin \omega t$ over half a period and then divide by that time interval, $T/2$.	$\|\sin \omega t\|_{avg} = \dfrac{1}{T/2}\displaystyle\int_0^{T/2} \sin \omega t\, dt = \dfrac{2}{T}\displaystyle\int_0^{T/2} \sin \omega t\, dt$
Solve the integral by substitution (Appendix A).	$\|\sin \omega t\|_{avg} = -\dfrac{2}{T}\left[\dfrac{1}{\omega}\cos \omega t\right]_0^{T/2}$
Eliminate ω in favor of T.	$\omega = \dfrac{2\pi}{T}$
	$\|\sin \omega t\|_{avg} = -\dfrac{2}{T}\left[\dfrac{T}{2\pi}\cos\dfrac{2\pi}{T}t\right]_0^{T/2} = -\dfrac{1}{\pi}\left[\cos\dfrac{2\pi}{T}t\right]_0^{T/2}$
Evaluate between limits.	$\|\sin \omega t\|_{avg} = -\dfrac{1}{\pi}\left[\cos\dfrac{2\pi}{T}\dfrac{T}{2} - \cos 0\right] = -\dfrac{1}{\pi}[\cos \pi - \cos 0] = \dfrac{2}{\pi}$ (1)

Start with Equation 32.18 and use Equation (1) to find an expression for the average emf.	$\mathcal{E}_{\text{avg}} = NBA\omega \lvert \sin\omega t \rvert_{\text{avg}} = NBA\omega \left(\dfrac{2}{\pi}\right)$
Solve for ω and substitute values. Assume the loop is circular, so from the given diameter, its area is $A = \pi r^2 = 0.20\,\text{m}^2$.	$\omega = \dfrac{\pi \mathcal{E}_{\text{avg}}}{2NBA} = \dfrac{\pi(110\,\text{V})}{2(30)(0.15\,\text{T})(0.20\,\text{m}^2)}$ $\omega \approx 190\,\text{rad/s} \approx 1800\,\text{rev/min}$

:• **CHECK and THINK**

Turning the crank 1800 times per minute (about 30 rps) is difficult for a person to do by hand for any substantial amount of time. DC generators are not usually hand-operated. Edison used a 500-hp steam engine to run the DC generator he used to light up hundreds of bulbs in the fall of 1882 in Lower Manhattan. This was one of the first steps in replacing gas-burning lamps with incandescent bulbs.

32-8 Case Study: Power Transmission and Transformers

Figure 32.27 illustrates how electrical power is transferred from a power plant to your home. When was the last time you saw a power plant? Probably on some long drive you took out of town. In the 1880s, however, Edison's DC generators were located in cities and towns because it wasn't practical to transmit electrical power more than about half a mile. Tesla pointed out to Edison that AC power plants could be located far outside the city, but Edison replied that he was not interested in high-voltage AC power because it was deadly. A person who touched any part of Edison's low-voltage DC system—from the generators to an incandescent bulb—would experience only a minor shock. Ultimately, Tesla won the argument; AC power lines run across the world. Yet Edison was right; high-voltage AC power lines are deadly. So, why is AC the standard used today? The answer has to do with the physics of power transmission as discussed in this section.

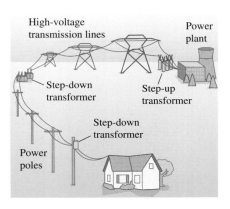

FIGURE 32.27 Electrical power is transmitted from the power plant to your home.

Root Mean Square Emf, Current, and Power

The instantaneous emf produced by an AC generator is given by $\mathcal{E} = NBA\omega \sin \omega t$ (Eq. 32.13), and the peak or maximum emf is given by $\mathcal{E}_{\text{max}} = NBA\omega$ (Eq. 32.16). The emf oscillates between $\mathcal{E}_{\text{max}}$ and $-\mathcal{E}_{\text{max}}$, so the average emf (over a whole number of periods) is zero: $\mathcal{E}_{\text{avg}} = 0$. An average emf of zero is not helpful in characterizing AC power sources. From Section 20-2, we know that the *root-mean-square* (rms) value is more useful in such circumstances. To find the root-mean-square emf $\mathcal{E}_{\text{rms}}$, we follow three steps:

1. *Square* the emf (Eq. 32.13):
$$\mathcal{E}^2 = (NBA\omega)^2 \sin^2 \omega t$$

2. Calculate the average over a whole number of periods, also known as the *mean*, of the emf squared:
$$(\mathcal{E}^2)_{\text{avg}} = [(NBA\omega)^2]_{\text{avg}} [\sin^2 \omega t]_{\text{avg}}$$

The average of the constant $[(NBA\omega)^2]_{\text{avg}}$ is $(NBA\omega)^2$. The average of the sine squared term is $\tfrac{1}{2}$ (Problem 55). So, the mean of the emf squared is
$$(\mathcal{E}^2)_{\text{avg}} = \dfrac{1}{2}(NBA\omega)^2$$

3. Take the *square root* of the average of the emf squared:
$$\mathcal{E}_{\text{rms}} = \sqrt{(\mathcal{E}^2)_{\text{avg}}} = \sqrt{\dfrac{1}{2}(NBA\omega)^2}$$

So the root-mean-square emf is

$$\mathcal{E}_{rms} = \frac{1}{\sqrt{2}} NBA\omega = \frac{\mathcal{E}_{max}}{\sqrt{2}} \quad (32.20)$$

In the United States and Canada, the standard rms voltage is $\mathcal{E}_{rms} = 120$ V. So the peak emf is

$$\mathcal{E}_{max} = \sqrt{2}\mathcal{E}_{rms} = \sqrt{2}(120\text{ V}) = 170\text{ V}$$

Now, let's consider the AC current given by $I = I_{max}\sin\omega t$ (Eq. 32.14), where the peak current is $I_{max} = \mathcal{E}_{max}/R = NBA\omega/R$ (Eq. 32.17). The average current (over a whole number of periods) is zero: $I_{avg} = 0$. The rms current is found by dividing the root-mean-square emf by the resistance:

$$I_{rms} = \frac{\mathcal{E}_{rms}}{R} = \frac{1}{\sqrt{2}}\frac{NBA\omega}{R}$$

$$I_{rms} = \frac{1}{\sqrt{2}}\frac{\mathcal{E}_{max}}{R}$$

Substitute Equation 32.17 to find a simple expression for the rms current:

$$I_{rms} = \frac{I_{max}}{\sqrt{2}} \quad (32.21)$$

Finally, we need expressions for the power supplied by an AC generator. The power supplied by any emf device is given by $P = I\mathcal{E}$ (Eq. 28.31), so the instantaneous power supplied by an AC generator comes from substituting Equations 32.14 and 32.13:

$$P = I_{max}\mathcal{E}_{max}\sin^2\omega t \quad (32.22)$$

Equation 32.22 says that the power's oscillation depends on a sine *squared*, so the average power is not zero. The average power is

$$P_{avg} = [I_{max}\mathcal{E}_{max}]_{avg}[\sin^2\omega t]_{avg}$$

$$P_{avg} = \frac{1}{2} I_{max}\mathcal{E}_{max} \quad (32.23)$$

By substituting Equations 32.20 and 32.21, we can write the average power in terms of the rms current and rms voltage:

$$P_{avg} = \frac{1}{2}(\sqrt{2}I_{rms})(\sqrt{2}\mathcal{E}_{rms})$$

$$P_{avg} = I_{rms}\mathcal{E}_{rms} \quad (32.24)$$

CONCEPT EXERCISE 32.8

The standard rms voltage in European countries is $\mathcal{E}_{rms} = 240$ V. What is the peak voltage in these countries?

Power Transmission in Wires

Edison's DC generators had a half-mile limit. Imagine the inconvenience of filling a city such as Manhattan with DC generators every mile or so. Why is the location of commercial DC generators so severely limited?

The problem with DC generators is that power is lost due to the resistance in the transmission wires. Normally, we approximate wires as having zero resistance, so we find that no power is lost. However, resistance is directly proportional to the length of a resistor or wire ($R = \rho\ell/A$; Eq. 28.24), so the resistance in long transmission lines cannot be ignored.

As a concrete example, suppose a DC generator supplies $P_{gen} = 150$ kW of power to a town just 2 or 3 miles away. The transmission lines have resistance $R = 0.25\ \Omega$.

Figure 32.28A is a simple model for the DC generator and the transmission lines. If the DC generator's emf $\mathcal{E} = 240$ V, the current in the transmission lines is given by $I = P_{gen}/V$ (Eq. 28.33):

$$I = \frac{150 \times 10^3 \text{ W}}{240 \text{ V}} = 625 \text{ A}$$

The power lost due to resistance in the transmission lines is found from the current in the wires and their resistance, $P_{loss} = I^2 R$ (Eq. 28.34):

$$P_{loss} = (625 \text{ A})^2 (0.25 \text{ }\Omega) \approx 98 \text{ kW}$$

A.

More than 65% of the power generated is lost in the transmission wires.

Edison's transmission lines were limited to about half a mile in order to reduce the transmission wire's resistance and reduce the power lost. He could, of course, also have reduced the resistance by making the transmission wires thicker, but this requires more copper (or whatever metal is used) and increases the cost.

In our example, the resistance is not very high (0.25 Ω), and reducing it much won't significantly reduce the power lost. In $P_{loss} = I^2 R$ (Eq. 28.34), the power lost depends linearly on the wire's resistance but on the *square* of the current. The best way to reduce the power lost is to reduce the current in the wire, but this means transmitting the power at a higher voltage, exactly as Tesla proposed to do with his AC generator.

B.

What happens if we replace the DC generator with an AC generator in this example? As before, the AC generator supplies average power $P_{gen} = 150$ kW to a town just 2 or 3 miles away. Figure 32.28B is a simple model for the AC generator and the transmission lines; the symbol for an AC generator is a wavy line inside a circle $-\bigcirc\!\!\!\!\sim-$. The transmission lines have the same resistance $R = 0.25$ Ω. If the AC generator's rms emf is $\mathcal{E}_{rms} = 24,000$ V (100 times greater than the DC generator's emf), the rms current in the transmission lines is given by Equation 32.24:

$$I_{rms} = \frac{P_{gen}}{\mathcal{E}_{rms}} = \frac{150 \times 10^3 \text{ W}}{24,000 \text{ V}} = 6.25 \text{ A}$$

The average power lost in a resistor carrying an AC current is

$$P_{loss} = I_{rms}^2 R \qquad (32.25)$$

So the power lost in the transmission wires is

$$P_{loss} = (6.25 \text{ A})^2 (0.25 \text{ }\Omega) \approx 9.8 \text{ W}$$

You may be wondering why an AC generator is necessary. As long as the voltage is high, the current in the transmission wires will be small and the power loss will be small. Why can't we just use a very high voltage DC generator? Part of the answer has to do with Edison's objection—high voltage is deadly. We don't really want to have 24,000 V between the terminals of every outlet in our house.

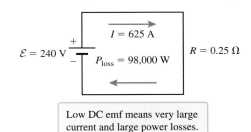

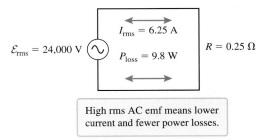

FIGURE 32.28 A. A DC emf device (such as a DC generator) powers a town and loses a lot of power. **B.** An AC generator powers the same town with less power lost in the wires. (These circuits only represent the resistance in the transmission wires and do not include the resistance of the devices powered in the town.)

Transformers

Westinghouse—Edison's rival—understood the real benefit of AC power after he learned about a European invention known today as a transformer. A **transformer** is a device that changes the voltage of alternating current; it does not work on direct current. Westinghouse's great insight was to realize that the transformer would allow AC generators to be located far from metropolitan areas so they could be powered by waterfalls. Additionally, one generator could serve the needs of a large number of people. In Figure 32.27, three transformers are shown. The **step-up transformer** near the power plant increases the voltage so the power loss in transmission is minimal. The two **step-down transformers** lower the voltage for transmission in town and into your home.

A transformer consists of two coils—the input or primary coil and the output or secondary coil. Westinghouse and his employees designed the modern transformer so that most of the magnetic flux produced by the primary coil passes through the

TRANSFORMER ⓞ **Special Case**

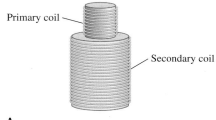

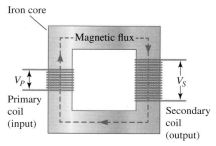

A.

B.

FIGURE 32.29 Two designs for transformers. **A.** The primary coil is inside the secondary coil. **B.** Both coils are wound around an iron core. The primary coil has N_P turns and carries a current I_P; the secondary coil has N_S turns and carries a current I_S.

Usually the voltages in Equation 32.27 are rms voltages.

secondary coil. Figure 32.29 shows two designs. In Figure 32.29A, the primary coil is inside the secondary coil, and in Figure 32.29B, the coils are both wrapped around an iron core.

The primary coil in Figure 32.29B is connected to an AC generator. Because the current in the coil oscillates, the magnetic field created by the coil oscillates. The coil's magnetic field creates an oscillating magnetic field in the iron core. Because the secondary coil is wrapped around that core, the magnetic flux through the secondary coil oscillates. According to Faraday's law (Eq. 32.6), the changing magnetic flux in the secondary coil induces an emf and current in the secondary coil.

The relative voltage V across the two coils depends on the relative number of turns:

$$\frac{V_S}{V_P} = \frac{N_S}{N_P} \quad (32.26)$$

where the subscript P stands for primary and the subscript S for secondary. In a step-up transformer, the voltage across the secondary coil is higher than the voltage across the primary coil, so the secondary coil has more turns than the primary coil: $N_S > N_P$. In a step-down transformer, the voltage across the secondary coil is lower than the voltage across the primary coil, so the secondary coil has fewer turns than the primary coil: $N_S < N_P$.

In an ideal transformer, there is no power loss from the primary to the secondary coil. In well-designed real transformers, only about 1% of the power is lost. So, if we assume the power is constant, we can find the relative current in the secondary coil:

$$P_{in} = P_{out}$$
$$I_P V_P = I_S V_S$$

Rearrange and use Equation 32.26:

$$\frac{I_S}{I_P} = \frac{V_P}{V_S} = \frac{N_P}{N_S} \quad (32.27)$$

By Equation 32.27, a step-up transformer increases the voltage $V_S > V_P$ and decreases the current $I_S < I_P$. Step-up transformers are used just before the current is sent through long power lines to reduce the power loss in those wires (Fig. 32.30A). Step-down transformers are used when a lower voltage is required. You can find step-down transformers in substations in towns, on utility poles near buildings, and in household devices (Fig. 32.30B and C).

FIGURE 32.30 A. A step-up transformer just outside a power plant. **B.** Step-down transformers used in a town. **C.** Transformers used inside the home.

A.

B.

C.

CONCEPT EXERCISE 32.9

Why can't you use a transformer to change the voltage of a direct current produced by **a.** a battery and **b.** a DC generator (Figs. 32.25 and 32.26)?

EXAMPLE 32.8 Big Metal Boxes in Town

You have probably seen substations (step-down transformers) like the one in Figure 32.27 in your town. The input voltage at a substation is rather high—assume 650 kV. The output voltage is typically lower than 10,000 V—assume 6.5 kV. The current in high-voltage transmission lines is typically 2 kA. For a simple ideal transformer, find the current that leaves the substation.

INTERPRET and ANTICIPATE
This is a straightforward application of Equation 32.27, giving us a chance to build our intuition about the magnitude of the voltage and current in a typical power distribution grid.

SOLVE
The primary coil is connected to the high-voltage lines, and the secondary coil is the output of the substation. Solve for the secondary current I_S.

$$\frac{I_S}{I_P} = \frac{V_P}{V_S} \quad (32.27)$$

$$I_S = \frac{V_P}{V_S} I_P = \left(\frac{650\,\text{kV}}{6.5\,\text{kV}}\right)(2 \times 10^3\,\text{A})$$

$$I_S = 2 \times 10^5\,\text{A} = 200\,\text{kA}$$

CHECK and THINK
This may seem like a huge current. In fact, it is. Besides stepping down the voltage, the substation splits the current so it can be directed along different distribution wires to serve many regions in the town.

Who Won the War?

A turning point in the war over current was the 1893 World's Fair in Chicago. Westinghouse and Edison both submitted bids to supply electricity for the fair. Westinghouse was awarded the contract, and the fair was run on AC power. By this time, Tesla's AC motor was working. Millions of visitors saw how electric motors could be used to do tasks that were being done by people or work animals. Westinghouse also demonstrated the flexibility of AC power. His original contract required him to light 92,000 bulbs. However, by the time the fair was running, about 180,000 bulbs were needed, and his AC generator was easily able to supply the extra power. Finally, in 1893, Westinghouse won the contract to build the first great hydroelectric power plant using Niagara Falls as the source of mechanical power. The AC power generated was then transmitted to Buffalo, demonstrating that Tesla's AC system had been perfected.

SUMMARY

❶ Underlying Principles

1. **Faraday's law**: A changing magnetic flux through a conductor induces an emf in that conductor.
2. **Lenz's law** follows from the principle of conservation of energy: The induced current produces a magnetic field that always acts to oppose the change that created the induced current. Lenz's law can be expressed mathematically as a minus sign in Faraday's law. The combination is known as **Faraday–Lenz's law**:

$$\mathcal{E} = -N\frac{d\Phi_B}{dt} \quad (32.6)$$

1038 CHAPTER 32 Faraday's Law of Induction

◆ Major Concepts

When a conductor such as a metal sheet is moving with respect to a magnetic field, the conduction electrons in the conductor experience a magnetic force. As a result, **eddy currents** develop in the conductor.

◆ Special Cases

1. The motional emf generated by a metal bar in a **slide generator** is
$$\mathcal{E} = B\ell v \quad (32.7)$$
2. The emf produced by an **AC generator** is
$$\mathcal{E} = NBA\omega \sin \omega t = \mathcal{E}_{max} \sin \omega t \quad (32.13)$$
3. The emf produced by a **DC generator** is
$$\mathcal{E} = NBA\omega |\sin \omega t| = \mathcal{E}_{max} |\sin \omega t| \quad (32.18)$$
4. A **transformer** is a device that changes the voltage of alternating current. It consists of two coils—the input or primary coil and the output or secondary coil. The relative current and voltage depend on the relative number of turns:
$$\frac{I_S}{I_P} = \frac{V_P}{V_S} = \frac{N_P}{N_S} \quad (32.27)$$

PROBLEMS AND QUESTIONS

A = algebraic   C = conceptual   E = estimation   G = graphical   N = numerical

32-1 Another Kind of Emf

1. A constant magnetic field of 0.275 T points through a circular loop of wire with radius 3.50 cm as shown in Figure P32.1.
   a. **N** What is the magnetic flux through the loop?
   b. **C** Is a current induced in the loop? Explain.
2. **C** In each of the following cases, a conducting loop is coaxial with a solenoid. Determine whether an emf is induced in the loop when the current in the solenoid is a. constant, b. increasing, and c. decreasing.
3. **A** A solenoid's long axis is perpendicular to the page and to the rectangular loop shown in Figure P32.3. The loop is completely inside the solenoid. The cross-sectional area of the solenoid is A_{sol}, and the area of the loop is A_{loop}. The solenoid's magnetic field $\vec{B}$ points out of the page. Find the magnetic flux Φ_B through the loop.
4. **C** Suppose magnetic monopoles exist. Would a current of monopoles be the source of an electric field, a magnetic field, or both? Explain.

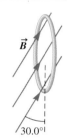

FIGURE P32.1

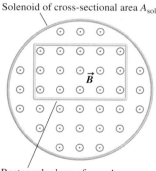

FIGURE P32.3

32-2 Faraday's Law

5. **N** The intensity of the Earth's magnetic field near the equator is 35.0 μT. A circular coil with 40 turns and a radius of 75.0 cm is placed so its axis points along the direction of the Earth's magnetic field. The coil is then rotated through an angle of 225° in 50.0 ms. What is the magnitude of the average emf generated in the circular coil?

6. **C** Figure P32.6 shows three situations involving a single circular loop of wire. For each case, decide whether an emf is induced in the circular loop. Explain your reasoning.

   Case 1: The loop lies in the plane of the page near a solenoid with its long axis perpendicular to the page. The solenoid's magnetic field is increasing.

   Case 2: A solenoid carries a constant current. The loop falls straight down inside the solenoid.

   Case 3: A solenoid carries a constant current. The loop is wrapped around a ball. The ball and loop roll along the inside of the solenoid.

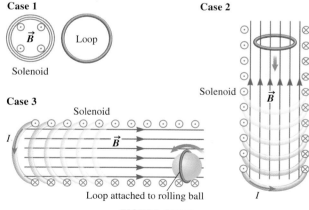

FIGURE P32.6

7. **A** A rectangular loop of length L and width W is placed in a uniform magnetic field $\vec{B}$ with its plane perpendicular to the field (Fig. P32.7). Determine the time-averaged induced emf if the loop rotates with constant angular velocity ω through an angle of 180° around an axis passing through the loop's center a. perpendicular to the loop and b. parallel to its width.

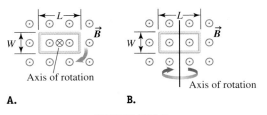

FIGURE P32.7

8. G The magnetic field through a square loop of wire with sides of length 3.00 cm changes with time as shown in Figure P32.8, where the sign indicates the direction of the field relative to the axis of the loop. Plot the emf induced in the loop versus time.

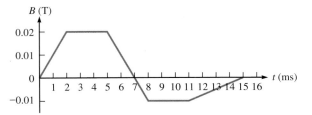

FIGURE P32.8

9. N A light conducting string is formed into a circular hoop with initial radius 20.0 cm. The hoop is placed in a 2.00-T uniform magnetic field so that its plane is perpendicular to the field, and the string is drawn to collapse the hoop to zero area in 125 ms. What is the average emf generated in the hoop during its collapse?

10. N A wire is formed into a square loop with sides $d = 18.0$ cm and positioned in a spatially uniform magnetic field with its plane perpendicular to the direction of the field. What is the magnitude of the average emf induced in the loop if the magnitude of the magnetic field is increased by 30.0 mT per second?

Problems 11 and 12 are paired.

11. Suppose a uniform magnetic field is perpendicular to the $8\frac{1}{2} \times$ 11-in. page of your homework and a rectangular metal loop lies on the page. The loop's sides line up with the edges of the page. The magnetic field is changing with time as described by $B = 3.75 \times 10^{-3} t$, where B is in teslas and t is in seconds.
  a. C Is the magnetic field increasing or decreasing?
  b. N Find the magnitude of the emf induced in the loop.

12. Suppose a uniform magnetic field is perpendicular to the 8½ × 11-in. page of your homework and a rectangular metal loop is perpendicular to the page such that one of its sides bisects the page into two long strips. The loop has the same dimensions as the page. The magnetic field is changing with time as described by $B = 3.75 \times 10^{-3} t^{-2}$, where B is in teslas and t is in seconds.
  a. C Is the magnetic field increasing or decreasing?
  b. N Find the emf induced in the loop.

13. N A square conducting loop with side length $a = 1.25$ cm is placed at the center of a solenoid 40.0 cm long with a current of 4.30 A flowing through its 420 turns, and it is aligned so that the plane of the loop is perpendicular to the long axis of the solenoid. The radius of the solenoid is 5.00 cm. **a.** What is the magnetic flux through the loop? **b.** What is the magnitude of the average emf induced in the loop if the current in the solenoid is increased from 4.30 A to 10.0 A in 1.75 s?

Problems 14 and 15 are paired.

14. A The magnetic field in a region of space is given by $\vec{B} = B_x \hat{\imath} + B_y \hat{\jmath}$. A coil of N turns is oriented so that its cross-sectional area is in the x direction: $\vec{A} = A_x \hat{\imath}$. A small bulb is connected across the ends of the coil. The total resistance of the coil and the bulb is R. Find an expression for the current through the bulb if $B_x(t) = B_0(t/t_0)^2$ and $B_y = B_0$, where B_0 and t_0 are constants.

15. A The magnetic field in a region of space is given by $\vec{B} = B_x \hat{\imath} + B_y \hat{\jmath}$. A coil of N turns is oriented so that its cross-sectional area is in the x direction: $\vec{A} = A_x \hat{\imath}$. A small bulb is connected across the ends of the coil. The total resistance of the coil and the bulb is R. Find an expression for the current through the bulb if $B_x(t) = B_0$ and $B_y = B_0(t/t_0)^2$, where B_0 and t_0 are constants.

16. N A coil of 255 turns and area 0.425 m² rotates in a uniform magnetic field $B = 0.325$ T such that the angle between the magnetic field and the area vector is given by $\varphi(t) = 6.35t$ rad. If the coil starts from rest at $t = 0$, what is the magnitude of the emf induced in the coil at $t = 25.8$ s?

32-3 Lenz's Law

17. C In Figure P32.17, a bar magnet oscillates back and forth near a stationary loop. Consider a half-cycle starting with the bar magnet at its farthest point from the loop. Describe the current induced in the loop from the perspective of an observer on the opposite side from the magnet (looking from the right in the figure).

FIGURE P32.17

18. G Imagine dropping a bar magnet through a horizontal conducting loop so that the north pole falls through the loop first. Assume a clockwise current as seen from above the loop is positive. Sketch the current as a function of time from this perspective. Remember that gravity accelerates the bar magnet. Explain the major features of your graph.

19. N A square loop with side length 5.00 cm is on a tabletop in a uniform magnetic field with the field pointing perpendicular to the loop, downward into the table. The field decreases from 0.500 T to 0.100 T in 0.15 s, and the loop has a resistance of 18.0 Ω. What are the induced current and its direction, as viewed from above?

20. A A thin copper rod of length L rotates with constant angular velocity ω about a point O, in a plane perpendicular to a uniform magnetic field $\vec{B}$ as shown in Figure P32.20. Determine the induced emf across its ends. Consider that the emf produced between the point O and a small segment of the rod, $d\vec{\ell}$, is given by $d\mathcal{E} = Bv\,d\ell$.

FIGURE P32.20

21. N Figure P32.21 shows a circular conducting loop with a 5.00-cm radius and a total resistance of 1.30 Ω placed within a uniform magnetic field pointing into the page. **a.** What is the rate at which the magnetic field is changing if a counterclockwise current $I = 4.60 \times 10^{-2}$ A is induced in the loop? **b.** Is the induced current caused by an increase or a decrease in the magnetic field with time?

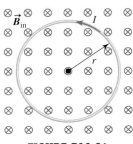

FIGURE P32.21

1040 CHAPTER 32 Faraday's Law of Induction

22. **G** Two solenoids are placed next to each other as shown in Figure P32.22. The solenoid on the left is connected to a battery and a switch. The switch is initially positioned such that the battery is not connected. It is then switched to include the battery, held in place for a minute, and then returned to the initial position. Sketch the current through the second solenoid versus time. Indicate the direction of current flow in the second solenoid as viewed from the first solenoid. Assume the solenoids have a nonzero resistance.

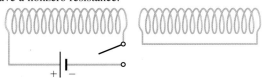

FIGURE P32.22

23. **A** A square loop with side length L, mass M, and resistance R lies in the xy plane. A magnetic field $\vec{B} = B_0(y/L)\hat{k}$ is present in the region of the space near the loop. Determine the magnitude and direction of the induced current in the loop as the loop starts moving at velocity $\vec{v} = v_y \hat{j}$.

24. **C** A small loop is arranged coaxially with a larger loop. The larger loop has a battery in series with a variable resistor such that a current is flowing clockwise in the larger loop as viewed from your vantage point. In what direction will the induced current flow in the small loop when the resistance is decreased?

25. **N** A coil with cross-sectional area 1.50×10^{-2} m^2 and 225 turns is positioned in a uniform magnetic field so that the plane of the coil is perpendicular to the direction of the field. The time-varying magnitude of the magnetic field is given by $B = 4.00 - 0.0200t - 0.00500t^2$, with B in teslas and t in seconds. What is the emf induced in the coil when $t = 2.00$ s?

26. **G** A rectangular loop lies in the plane of the page. In a small region of space, there is a uniform magnetic field that is perpendicular to the page as shown in Figure P32.26. **a.** The loop moves at a constant velocity from region L through region C to region R. Sketch a plot of the current in the loop versus time. Assume clockwise current is positive from your perspective. **b.** How does your answer change if the loop moves at a constant velocity from region R through C to L?

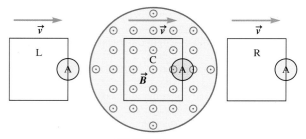

FIGURE P32.26

27. **N** A vibrating copper wire is magnetized so that it produces a variable magnetic field given by $B = 25.0 - 1.55\cos(67.0\pi t)$ perpendicular to the plane of a 14-turn coil with a cross-sectional area of 0.500 cm^2 placed nearby. How does the emf induced in the coil vary with time?

28. **A** A solenoid of area A_{sol} produces a uniform magnetic field (Fig. P32.28; shown in cross section). The solenoid's magnetic field points out of the page and is decreasing according to $B = B_0(t_0/t)^2$. A single conducting loop of area A_{loop} and resistance R is coaxial with the solenoid, with $A_{\text{loop}} < A_{\text{sol}}$. Find an expression for the current in the loop. What does the sign of your answer mean?

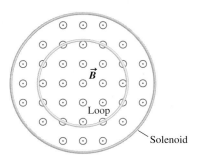

FIGURE P32.28

29. **C** Two circular conductors are perpendicular to each other as shown in Figure P32.29. Suppose conductor B carries a current. Will a current be induced in conductor A if there is a change in the current in conductor B? (The loops are insulated from one another.)

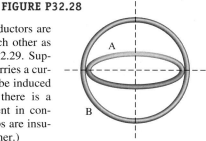

FIGURE P32.29

32-4 Lenz's Law and Conservation of Energy

Problems 30 and 31 are paired.

30. **C** Two circular conducting loops labeled A and B are close together and their axes are parallel. Loop A is connected to a power supply, and the current in A is increasing linearly with time. Loop B is not connected to a power supply, but the current in B runs through a bulb. **a.** Describe the current induced in loop B and the brightness of the bulb as a function of time. **b.** Explain how your answer is consistent with the principle of conservation of energy. **c.** If the current in loop A is decreasing instead of increasing, how do your answers change?

31. **C** Two circular conducting loops labeled A and B are close together and their axes are parallel. Loop A is connected to a power supply, and the current in A is increasing. Loop B is not connected to a power supply. Loop B is free to move toward or away from A. **a.** Does loop B move toward A, away from A, or not at all? Explain. **b.** If the current in loop A is decreasing instead of increasing, how does your answer change?

Problems 32 and 33 are paired.

32. **C** You are attending a magic show at a local theater, and the magician has the stage cleared except for a cylindrical copper pipe 2.0 m long and a small wooden block. The magician drops the wooden block through the top of the pipe, and it falls out the bottom about 0.7 s after it is released. You think this is what would have happened if the magician had simply dropped the wooden block from the same height. The magician then taps his wand three times against the pipe, while uttering an incantation, and drops a similar-looking block through the pipe. This time, however, the block takes significantly longer to fall out the bottom. You suspect the magician used a different block. What might have been different about this block? Explain how this trick could be done with a magnet and a copper pipe.

33. **C** Joanna first drops a magnet through a plastic pipe and then through a copper pipe. The pipes are the same length, but the magnet takes a much longer time to fall through the copper pipe. This means the final velocity of the magnet is much lower in one case than the other, yet the magnet began with the same total mechanical energy in each case. No nonconservative forces are exerted on the magnet–pipe system in either case. What happened to the supposed "missing energy" in the case where the magnet fell through the copper pipe?

32-5 Case Study: Slide Generator

34. **E** Estimate the magnitude of the emf induced across the ends of a commercial jet airplane's wings as a result of its motion through the Earth's magnetic field. Would it be practical to use this induced emf to power devices on the plane? Explain.

35. A slide generator has a movable bar with a length of 0.355 m on a U-shaped conductor and is in a magnetic field with a magnitude of 1.50 T, perpendicular to the plane of the generator.
    a. **N** How fast must the bar move to create an induced emf with a magnitude of 12.0 V?
    b. **C** Does it matter whether the bar is moving in one direction or the other along the U-shaped conductor? What is different when the bar moves in one direction versus the other?

Problems 36 and 37 are paired.

36. **A** Find an expression for the current in the slide generator in Figure 32.14 (page 1020) as a function of x. The U-shaped conductor and the bar have cross-sectional area A and resistivity ρ. The bar's length is ℓ, and the magnetic field is B. The bar is pulled at constant speed v. *Hint*: Your expression does not involve R.

37. **N** The slide generator in Figure 32.14 (page 1020) is in a uniform magnetic field of magnitude 0.0500 T. The bar of length 0.365 m is pulled at a constant speed of 0.500 m/s. The U-shaped conductor and the bar have a resistivity of $2.75 \times 10^{-8}\ \Omega \cdot m$ and a cross-sectional area of $8.75 \times 10^{-4}\ m^2$. Find the current in the generator when $x = 0.650$ m.

38. **C** **CASE STUDY** In 1887, Tesla was in need of funding to work on his AC system. He was introduced to an investor. The investor said he wanted Tesla to impress him, but the investor refused to watch a demonstration of the AC system. Tesla remembered that Christopher Columbus was able to make a great impression and gain an audience with Queen Isabella by balancing an egg on its end. Columbus did this by cracking the shell slightly. Isabella—of course—funded Columbus's expedition. So, Tesla said he would balance an egg on end without cracking its shell. Tesla installed a rotating magnetic field below a wooden tabletop. His hard-boiled egg was coated in copper. Not only was the egg standing on end, but it was rapidly rotating. The investor provided Tesla with the funds needed to start his own company—the Tesla Electric Company. Explain Tesla's demonstration.

39. A thin conducting bar (60.0 cm long) aligned in the positive y direction is moving with velocity $\vec{v} = (1.25\ m/s)\hat{\imath}$ in a region with a spatially uniform 0.400-T magnetic field directed at an angle of 36.0° above the xy plane.
    a. **N** What is the magnitude of the emf induced along the length of the moving bar?
    b. **C** Which end of the bar is positively charged?

40. **A** A stiff spring with a spring constant of 1200.0 N/m is connected to a bar on a slide generator as shown in Figure P32.40. Assume the bar has length $\ell = 60.0$ cm and mass $m = 0.75$ kg, and it slides without friction. The bar connects to a U-shaped wire to form a loop that has width $w = 40.0$ cm and total resistance 25 Ω and that sits in a uniform magnetic field $B = 0.35$ T. The bar is initially pulled 5.0 cm to the left and released so that it begins to oscillate. What is the induced current in the loop as a function of time, $I(t)$? (Ignore any effects due to the magnetic force on the oscillating bar.)

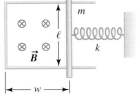

FIGURE P32.40

32-6 Case Study: AC Generators

41. **N** A generator spinning at a rate of 1.20×10^3 rev/min produces a maximum emf of 45.0 V. At what angular speed does this generator produce a maximum emf of 64.0 V?

42. **C** Suppose you have a simple homemade AC generator like the one in Figure 32.20 (page 1026) whose emf is about half of what you need to power a particular motor. How can you easily modify the generator to suit your needs? Explain your modifications specifically.

43. **N** A simple AC generator has a coil of exactly 250 turns (Fig. 32.20, page 1026). It rotates at 60.0 Hz in a magnetic field of 0.215 T. If the generator must produce a maximum emf of 170.0 V, what is the area of the coil? If the coil is square, what is the length of each side? Compare the size of the coil to some easy-to-visualize object.

44. Tesla's AC generator had three phases. To understand the basic idea behind his generators, imagine adding two loops to the AC generator in Figure 32.20 (page 1026). The three loops are set so that each is 120° from the other two.
    a. **G** For each loop, make a graph of $\mathcal{E}$ versus t similar to the one in Figure 32.23B (page 1029).
    b. **C** Imagine powering a motor with the three-phase generator. In what way is a three-phase generator superior to a one-phase generator?

45. **N** The magnetic flux through each turn of a 140-turn coil is given by $\Phi_B = 8.75 \times 10^{-3} \sin \omega t$, where ω is the angular speed of the coil and Φ_B is in webers. At one instant, the coil is observed to be rotating at a rate of 8.90×10^2 rev/min. a. What is the induced emf in the coil as a function of time for this angular speed? b. What is the maximum induced emf in the coil for this angular speed?

46. **C** Given a spool of wire, discuss the feasibility of using a rotating loop of wire in the Earth's magnetic field (about 10^{-4} T) to create a hand-powered AC generator to produce household voltage (170-V peak voltage with a frequency of 60 Hz).

47. **N** A square coil with a side length of 12.0 cm and 34 turns is positioned in a region with a horizontally directed, spatially uniform magnetic field of 82.0 mT and set to rotate about a vertical axis with an angular speed of 1.20×10^2 rev/min. a. What is the maximum emf induced in the spinning coil by this field? b. What is the angle between the plane of the coil and the direction of the field when the maximum induced emf occurs?

48. **N** A 44-turn rectangular coil with length $\ell = 15.0$ cm and width $w = 8.50$ cm is in a region with its axis initially aligned to a horizontally directed uniform magnetic field of 745 mT and set to rotate about a vertical axis with an angular speed of 64.0 rad/s. a. What is the maximum induced emf in the rotating coil? b. What is the induced emf in the rotating coil at $t = 1.00$ s? c. What is the maximum rate of change of the magnetic flux through the rotating coil?

32-7 Case Study: Faraday's Generator and Other DC Generators

49. **C** Explain Tesla's statement that all generators are really AC generators.

50. **C** Suppose you light a bulb with a battery, then Faraday's generator, then a DC generator, and finally an AC generator. Describe the brightness of the bulb as a function of time in each case. (Does the light flicker?) Do the lights in your room flicker? Explain.

51. **N, C** A typical flashlight uses two 1.5-V batteries. Could you crank Gramme's DC generator by hand to light such a bulb? Your answer should involve a short calculation similar to Example 32.7 (page 1032).

52. C An AC generator, a DC generator, and a DC battery with the same peak voltage are each connected to a lightbulb. Sketch the current through the lightbulb when it is connected to each generator separately, and rank them from the largest average current to the smallest average current.

32-8 Case Study: Power Transmission and Transformers

53. N The maximum value of the emf in the primary coil ($N_P = 1000$) of a transformer is 175 V. **a.** What is the maximum induced emf in the secondary coil ($N_S = 250$)? **b.** What is the ratio of the current in the primary coil to the current in the secondary coil?

Problems 54 and 55 are paired.

54. A Find the average of $f(x) = \sin x$ between $x = 0$ and 2π. What does your answer imply about the emf produced by an AC generator, averaged over one period?

55. A Find the average of $f(x) = \sin^2 x$ between $x = 0$ and 2π. What does your answer imply about the power supplied by an AC generator, averaged over one period?

56. N You have a 750-W hair dryer that is designed to work on the American power grid ($\mathcal{E}_{rms} = 120.0$ V, 60.0 Hz). **a.** What is the rms current in the dryer? **b.** If you connected the dryer to the European power grid ($\mathcal{E}_{rms} = 240.0$ V, 50.0 Hz), what would be the current in the dryer? What would happen to the dryer and why? **c.** If you used a transformer to connect the dryer to the European grid, what ratio of N_S/N_P would be required?

57. N An electric toothbrush charger is one example of a small transformer used in your home. The secondary coil in the toothbrush gets an induced emf from the primary coil in the charger when they are aligned. (The toothbrush coil is placed within the primary coil when it is plugged in.) **a.** If the maximum value of the emf in the primary coil ($N_P = 1000$) is 112 V, what is the maximum induced emf in the secondary coil ($N_S = 150$)? **b.** A maximum current of 0.050 A flows through the primary coil. What is the average power delivered to the toothbrush?

Problems 58 and 59 are paired.

58. N A step-down transformer has 65 turns in its primary coil and 10 turns in its secondary coil. The primary coil is connected to standard household voltage ($\mathcal{E}_{rms} = 120.0$ V, 60.0 Hz). **a.** What is the rms voltage in the secondary coil? **b.** If, instead, the transformer is connected to a 6-V DC battery, what is the rms voltage in the secondary coil after the first few milliseconds?

59. N Suppose the secondary coil in Problem 58(a) is connected to a 12.0-Ω resistor. **a.** What is the maximum current in the resistor? **b.** What is the average power dissipated by the resistor?

General Problems

60. C A conducting loop is placed over a lit incandescent bulb so that the filament is surrounded. Determine whether an emf is induced in the loop **a.** if the bulb is in a flashlight powered by a battery and **b.** if the bulb is in a lamp plugged into a wall outlet. Explain your answers.

Problems 61 and 62 are paired.

61. N A circular loop with a radius of 0.25 m is rotated by 90.0° over 0.200 s in a uniform magnetic field with $B = 1.50$ T. The plane of the loop is initially perpendicular to the field and is parallel to the field after the rotation. **a.** What is the average induced emf in the loop? **b.** If the rotation is then reversed, what is the average induced emf in the loop?

62. C A circular loop with a radius of 0.25 m is rotated by 90.0° over 0.200 s in a uniform magnetic field with $B = 1.50$ T. The plane of the loop is initially perpendicular to the field and is parallel to the field after the rotation. When the rotation is complete, the process is reversed and the loop is rotated back in the other direction, again in 0.200 s. Considering the induced emf and current, what is different about the two rotations? Are they indistinguishable?

63. N A 75-turn square coil constructed from aluminum wire is placed in a uniform 1.20-T magnetic field with the plane of the coil making an angle of 45.0° with the field direction. During the next 45.0 ms, the magnitude of the field is reduced to 0.300 T, resulting in an induced emf in the coil with magnitude 1.45 V. What is the total length of aluminum wire used to construct the coil?

64. C A bar magnet is dropped through a loop of wire as shown in Figure P32.64. **a.** What is the direction of the induced current as the magnet is approaching the loop, as viewed from above where the magnet begins? **b.** What is the direction of the induced current after the magnet falls through and is receding from the loop, as viewed from above where the magnet began?

FIGURE P32.64

65. N An airplane with wings 10.0 m long on either side flies south, parallel to the Earth's surface, at 250.0 m/s. The horizontal component of the Earth's magnetic field is 2.0×10^{-5} Wb/m², and the dip angle of the Earth's magnetic field is 60.0°. Calculate the induced emf between the wing tips. (The dip angle is the angle the field makes with the horizontal plane in which the airplane resides.)

66. N A helicopter has blades 4.0 m long (beginning at the rotor and extending outward) rotating at 300.0 rpm (revolutions per minute). Determine the maximum voltage that might develop between the two ends of the blade while they are rotating in the Earth's magnetic field (approximately 10^{-4} T). Assume the speed of the blade is determined by the linear speed of its center of mass and its mass is uniformly distributed.

67. N A circular coil with 75 turns and radius 12.0 cm is placed around an electromagnet that produces a uniform magnetic field through the coil and perpendicular to the plane of the coil. As the electromagnet powers up, the field it produces increases linearly from 0 to a maximum of 3.50 T in 0.110 s. If the total resistance of the coil is 5.00 Ω, what is the magnitude of the average current induced in the circular coil as the electromagnet powers up?

68. C Each of the three situations in Figure P32.68 shows a resistor in a circuit in which currents are induced. Using Lenz's law, determine whether the current in each situation is from a to b or from b to a. **a.** If the current I in the wire in Figure P32.68A is increased from zero to I, what is the direction of the current induced across the resistor R? **b.** The switch in Figure P32.68B is initially closed and is thrown open at $t = 0$. What is the direction of the current induced across the resistor R immediately afterward? **c.** A bar magnet is brought close to the circuit shown in Figure P32.68C. What is the direction of the current induced across the resistor R?

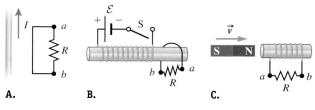

FIGURE P32.68

69. **N** A square loop with sides 1.0 m in length is placed in a magnetic field perpendicular to the plane of the loop. Half the area of the loop lies outside the magnetic field. The magnetic field varies with time as $B(t) = (0.010 - 2.00t)$ T, where t is in seconds. The loop also has a battery of emf 12.0 V as shown in Figure P32.69. Determine the resultant emf of the circuit.

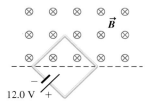

FIGURE P32.69

70. **C** You are working on an AC generator that consists of a coil rotating in a uniform magnetic field, and you discover that it is no longer creating the desired amount of power after being dropped on the floor. The coil has resistance R with N turns and cross-sectional area A, and the magnitude of the magnetic field is B. As you consider the cause of the power loss, your friend suggests that the coil may have been knocked slightly off-axis so that the axis of the coil is no longer pointing in the same direction as the static magnetic field, as desired. Could the coil being slightly off-axis explain the power loss? If no, what else might explain the loss of power? If yes, what can you do to compensate for the power loss due to the off-axis coil, assuming you are not able to adjust the coil or the magnetic field? Justify your answers.

71. **N** Two frictionless conducting rails separated by $\ell = 55.0$ cm are connected through a 2.00-Ω resistor, and the circuit is completed by a bar that is free to slide on the rails (Fig. P32.71). A uniform magnetic field of 5.00 T directed out of the page permeates the region. **a.** What is the magnitude of the force $\vec{F}_p$ that must be applied so that the bar moves with a constant speed of 1.25 m/s to the right? **b.** What is the rate at which energy is dissipated through the 2.00-Ω resistor in the circuit?

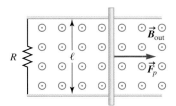

FIGURE P32.71

72. **C** CASE STUDY Imagine a glorious day after you've finished school; you are working as a scientist or engineer in a large research laboratory. Most likely you won't always agree with all the people who work on your team. What would you do if you disagreed with your boss, and your boss was the well-known American hero Thomas Edison? Think of Tesla. In 1882, Edison was known worldwide as the great inventor of the telegraph, phonograph, and incandescent bulb, while Tesla was a 26-year-old immigrant to the United States. Tesla originally worked for Edison, but they didn't get along and Tesla left in 1885. One of the major disagreements between the two was over the motor: Edison favored the DC motor and Tesla the AC version. In 1888, Tesla gave a presentation to the American Institute of Electrical Engineers, arguing for AC motors and generators. Present a case in favor of the AC motor and generator over their DC counterparts. Be fair in your presentation, listing the pros and cons of both AC and DC.

73. **N** The two free ends of a 320-turn circular coil with a radius of 9.00 cm are connected to a 2.40-Ω resistor. The coil is placed in a region with a uniform 3.00-T field that is initially in the upward direction so that its axis is parallel to the field. The coil then rotates about an axis perpendicular to the field by 180.0° during a time interval Δt. How much charge enters the resistor during the time interval Δt?

74. **A** Figure P32.74 shows an N-turn rectangular coil of length a and width b entering a region of uniform magnetic field of magnitude B_{out} directed out of the page. The velocity of the coil is constant and is upward in the figure. The total resistance of the coil is R. What are the magnitude and direction of the magnetic force on the coil **a.** when only a portion of the coil has entered the region with the field, **b.** when the coil is completely embedded in the field, and **c.** as the coil begins to exit the region with the field?

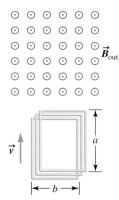

FIGURE P32.74

75. **N** A rectangular conducting loop with dimensions $w = 32.0$ cm and $h = 78.0$ cm is placed a distance $a = 5.00$ cm from a long, straight wire carrying current $I = 7.00$ A in the downward direction (Fig. P32.75). **a.** What is the magnitude of the magnetic flux through the loop? **b.** If the current in the wire is increased linearly from 7.00 A to 15.0 A in 0.230 s, what is the magnitude of the induced emf in the loop? **c.** What is the direction of the current that is induced in the loop during this time interval?

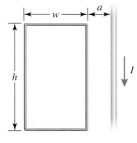

FIGURE P32.75

76. **C** Sadly, in an effort to retain the value of his patents involving the distribution of power with DC current, Edison electrocuted many animals using AC current in a series of public appearances attempting to discredit the use of AC power generation. The crowds at the time were unaware of the governing relationships among emf, current, and power, and these demonstrations caused many people to fear AC power. Given what you know about emf, current, and power, explain the faulty logic used to condemn AC power based on these demonstrations. Should people fear DC power as well? Or, should people fear neither?

77. **N** A conducting rod is pulled with constant speed v on a smooth conducting rail as shown in Figure P32.77. A constant magnetic field $\vec{B}$ is directed into the page. If the speed of the bar is doubled, by what factor does the rate of heat dissipation change?

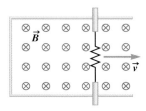

FIGURE P32.77

78. **A** A circular loop of radius a and resistance R is placed in a changing magnetic field so that the field is perpendicular to the plane of the loop. The magnetic field varies with time as $B(t) = B_0 e^{-t}$, where B_0 is a constant. Determine the electrical power in the circuit when $t = 0$.

79. **N** A conducting single-turn circular loop with a total resistance of 5.00 Ω is placed in a time-varying magnetic field that produces a magnetic flux through the loop given by $\Phi_B = a + bt^2 - ct^3$, where $a = 4.00$ Wb, $b = 11.0$ Wb/s^{-2}, and $c = 6.00$ Wb/s^{-3}. Φ_B is in webers, and t is in seconds. What is the maximum current induced in the loop during the time interval $t = 0$ to $t = 3.50$ s?

80. **A** A metal rod of mass M and length L is pivoted about a hinge at point O as shown in Figure P32.80. The axis of rotation passes through O into the page. A constant magnetic field $\vec{B}$ is applied into the page. Find the ratio of the maximum electric field inside the rod to the applied magnetic field when the rod is rotated with angular speed ω. Assume the speed of the rod is determined by the linear speed of its center of mass, and its mass is uniformly distributed.

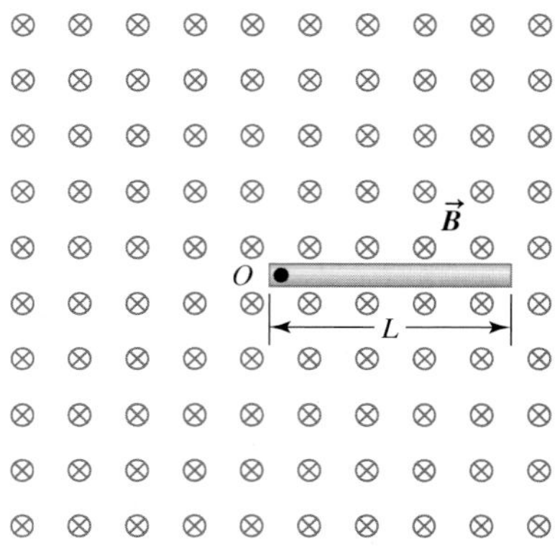

FIGURE P32.80

Inductors and AC Circuits

33

❗ Underlying Principles

There are no new principles in this chapter; instead, we apply:
1. Faraday's law, one of Maxwell's equations
2. Lenz's law

✚ Major Concepts

1. Inductor
2. Inductance
3. Energy density stored by a magnetic field
4. Capacitive reactance
5. Inductive reactance
6. Impedance
7. Resonance

▶ Special Cases

1. Inductance of an ideal solenoid
2. Inductor rule
3. RL circuit
4. LC circuit
5. AC circuit with resistance
6. AC circuit with capacitance
7. AC circuit with inductance
8. Filter circuits
9. RLC (AC) circuit

◉ Tools

Phasor diagrams

Key Questions

What practical devices exploit Faraday's law?

How is a pendulum like a circuit?

How do we analyze AC circuits?

- 33-1 **Inductors and inductance** 1046
- 33-2 **Back emf** 1049
- 33-3 **Special case: Resistor–inductor (*RL*) circuit** 1052
- 33-4 **Energy stored in a magnetic field** 1055
- 33-5 **Special case: Inductor–capacitor (*LC*) circuit** 1057
- 33-6 **Special case: AC circuit with resistance** 1062
- 33-7 **Special case: AC circuit with capacitance** 1064
- 33-8 **Special case: AC circuit with inductance** 1068
- 33-9 **Special case: AC circuit with resistance, inductance, and capacitance** 1072

Once an infrastructure has been built, creative people develop technologies that make use of that infrastructure. When cell phone towers were new, cell phones were large, clunky devices used only to make (expensive) calls. Now cell phones are compact, and we use them to send text messages and pictures to our friends. We use them to check our e-mail and get directions to a new restaurant.

The same sort of development took place after the AC power grid was established. At first, there were only a few AC-powered devices, but today we use AC circuits all the time. The AC electricity grid powers dishwashers, refrigerators, and vacuum cleaners. The development of AC circuits led to the development of communication devices such as radios, TVs, and cell phones. In this chapter, we study basic AC circuits.

INDUCTOR ⊗ **Major Concept**

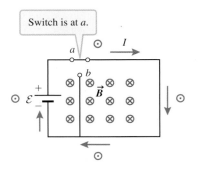

A.

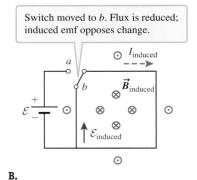

B.

FIGURE 33.1 A. When the switch is closed at *a*, there is a magnetic flux through the loop. **B.** When the switch is moved from *a* to *b*, the magnetic flux is reduced, so an emf is induced in the loop in the same direction as the battery's emf. The energy to induce this emf was stored in the magnetic field. We use an arrow (→) to indicate the direction of the *induced* emf. The arrow points in the direction the induced emf would tend to drive the current.

33-1 Inductors and Inductance

Every time you open or close a switch, you turn on or shut off a magnetic field. Consider the circuit in Figure 33.1A, in which the terminals of a battery are connected by wires through a closed switch at point *a*. As long as the switch is at point *a*, the current produces a nonzero magnetic flux through the circuit. When you close the switch at point *b*, you disconnect the battery from the circuit so that the circuit is simply a loop (Fig. 33.1B), and you turn off a magnetic field, so the magnetic flux through the loop decreases to zero. According to Lenz's law, an emf is induced that opposes the change. So, the induced emf must point in the same direction as the battery's emf (Fig. 33.1). This emf is not very strong and the induced current does not last very long, but this circuit illustrates an important point: Whenever you flip a switch, you induce an emf. Any emf—whether it is due to a battery or is induced—supplies energy in a circuit and moves charged particles. In a battery, the energy ultimately comes from chemical reactions inside the battery cell. So, where does the induced emf get its energy? The answer is *from the magnetic field*.

The circuit in Figure 33.1 is not well designed to take advantage of the energy stored in the magnetic field because it is the change in the magnetic flux that induces the emf, and a single loop cannot "hold" much magnetic flux. An **inductor** is a circuit element that is designed to hold more magnetic flux and store magnetic field energy. One type of inductor is a solenoid. Their simplicity makes solenoids easier to study than more complicated inductors. The solenoid is such an important type of inductor that the symbol for any inductor is a curly line that looks like a solenoid (—⁀⁀⁀—).

We encountered a similar circuit element—the capacitor—in our study of DC circuits. A capacitor "holds" charge on its plates and stores potential energy in an electric field between its plates. A parallel-plate capacitor is analogous to a solenoid inductor (Table 33.1).

An ideal solenoid's magnetic field is uniform in its interior (Section 31-5), analogous to the uniform electric field between the plates of a parallel-plate capacitor. The electric field E between the plates of a particular parallel-plate capacitor is proportional to the potential difference V_C between its plates: $E = V_C/d$ (Eq. 27.21). Likewise, the magnetic field B inside a particular solenoid is proportional to the current I in the solenoid's wire: $B = \mu_0 nI$ (Eq. 31.6). Compare Equations 27.21 and 31.6: The magnetic field inside the ideal solenoid is analogous to the electric field between the plates of a capacitor, $E \to B$, and the current in the solenoid is analogous to the voltage across the capacitor, $V_C \to I$.

To find an expression for the energy stored by an inductor, let's review the energy stored by a capacitor. Imagine two parallel-plate capacitors, each connected to identical batteries and fully charged so that each capacitor has the same potential difference V_C across its plates. The capacitor with the greater capacitance stores more

TABLE 33.1 Analogy between capacitors and inductors. Refer to this table often as you read the text.

Capacitor		Inductor	
⊣⊢		—⁀⁀⁀—	
$E = \dfrac{V_C}{d}$ (parallel-plate capacitor)	(27.21)	$B = \mu_0 nI$ (solenoid)	(31.6)
Electric field E between plates		Magnetic field B in coils	
Voltage V_C across plates		Current I in wire	
Charge Q stored by capacitor		Magnetic flux Φ_B through inductor	
Capacitance C		Inductance L	
$Q = CV_C$	(27.1)	$\Phi_B = LI$	(33.1)
$C = \dfrac{\varepsilon_0}{d} A$ (parallel-plate capacitor)	(27.10)	$L = \mu_0 n^2 \ell A = \dfrac{\mu_0 N^2}{\ell} A$ (solenoid)	(33.5)
$U_E = \dfrac{1}{2} C V_C^2$	(27.3)	$U_B = \dfrac{1}{2} L I^2$	(33.3)

charge on its plates, $Q = CV_C$ (Eq. 27.1), and more energy in its electric field, $U_E = \frac{1}{2}CV_C^2$ (Eq. 27.3).

Now, let's continue to use the capacitor as an analogy to come up with an expression for the energy stored in the magnetic field of an inductor. A capacitor holds charge on its plates, and an inductor holds magnetic flux through its coils. The charge stored Q on the plates of the capacitor is analogous to the total magnetic flux (or *flux linkage*) Φ_{tot} through the inductor, $Q \to \Phi_{tot}$. So, just as Q is proportional to V_C (Eq. 27.1), the magnetic flux through the inductor is proportional to the current:

$$\Phi_{tot} = LI \tag{33.1}$$

where the constant of proportionality L is the **inductance**. The inductance in Equation 33.1 is sometimes called the **self-inductance** because the flux Φ_{tot} through the inductor is the result of the current I in the same ("self") inductor. Comparing $\Phi_{tot} = LI$ (Eq. 33.1) to $Q = CV_C$ (Eq. 27.1) shows that inductance L is analogous to capacitance, $C \to L$. The SI unit for inductance is named the *henry* in honor of the American physicist Joseph Henry. The henry is abbreviated by an uppercase H, and

INDUCTANCE ⊗ Major Concept

$$1\,\text{H} = 1\frac{\text{Wb}}{\text{A}} = 1\frac{\text{T}\cdot\text{m}^2}{\text{A}} \tag{33.2}$$

Applying this analogy ($V_C \to I$ and $C \to L$) to $U_E = \frac{1}{2}CV_C^2$ (Eq. 27.3), we see that the energy stored in the magnetic field of an inductor is

$$U_B = \frac{1}{2}LI^2 \tag{33.3}$$

Now, let's review capacitance so we can use the analogy to understand the inductance L. The capacitance of a capacitor depends only on geometry. For an air-filled parallel-plate capacitor, $C = \varepsilon_0 A/d$ (Eq. 27.10), where A is the plate's area and d is the separation between the plates. Again, for two parallel-plate capacitors connected to identical batteries and with equal separation d between their plates, the capacitor with the larger plate area A has a greater capacitance. Because the capacitors are connected to identical batteries, the capacitor with the larger plates stores more charge.

Like capacitance, inductance depends only on the geometry of the inductor. For a solenoid, the inductance is proportional to its cross-sectional area A:

$$L \propto A \tag{33.4}$$

Imagine two solenoids, each with the same current I through their wires. The solenoid with the larger area A has a greater inductance L and, according to $\Phi_{tot} = LI$ (Eq. 33.1), has a greater magnetic flux through its loops. Also, because $U_B = \frac{1}{2}LI^2$ (Eq. 33.3), the inductor with the larger area stores more energy.

CONCEPT EXERCISE 33.1

Suppose the switch in Figure 33.1 is closed at point b for a long time so that there is no current in the circuit. Then the switch is closed at point a. Is there an induced emf? If not, explain why not. If so, what is the direction of the induced emf and why?

EXAMPLE 33.1 Inductance of an Ideal Solenoid

An ideal solenoid has n turns per unit length and cross-sectional area A. Show that its inductance is given by

INDUCTANCE OF AN IDEAL SOLENOID
▶ Special Case

$$L = \mu_0 n^2 \ell A = \frac{\mu_0 N^2}{\ell}A \tag{33.5}$$

where ℓ is the length of the solenoid and N is the number of turns.

Example continues on page 1048 ▶

INTERPRET and ANTICIPATE

To solve this problem, you must first find an expression for the total magnetic flux Φ_{tot} when current I is in the solenoid.

SOLVE

Figure 31.25 (page 994) shows an ideal solenoid. When the solenoid carries a current, the resulting magnetic field is uniform and perpendicular to its cross section. The magnetic flux through each loop is BA. The total magnetic flux or flux linkage is the flux through each loop times the number of loops N.	$\Phi_{tot} = NBA$
Write the number of loops in terms of the length ℓ of the solenoid and its number of turns per unit length: $N = n\ell$.	$\Phi_{tot} = (n\ell)BA$ (1)
When the current in the solenoid is I, the magnetic field inside the solenoid is given by Equation 31.6.	$B = \mu_0 n I$ (31.6)
Substitute Equation 31.6 into Equation (1). Equation (2) is the flux linkage.	$\Phi_{tot} = (n\ell)(\mu_0 n I)A = \mu_0 n^2 \ell A I$ (2)
Solve Equation 33.1 for L.	$\Phi_{tot} = LI$ (33.1) $L = \dfrac{\Phi_{tot}}{I}$
Substitute Equation (2) for Φ_{tot}. Also use $N = n\ell$ to express L in terms of N.	$L = \dfrac{\mu_0 n^2 \ell A I}{I}$ $L = \mu_0 n^2 \ell A = \dfrac{\mu_0 N^2}{\ell} A$ ✓ (33.5)

CHECK and THINK

Equation 33.5 is the inductance of a solenoid. As expected, and analogous to the capacitor, the inductance depends only on geometry—the number of turns per unit length n, the length ℓ, and the cross-sectional area A. We will often use Equation 33.5, but because an ideal solenoid is very long, it is sometimes more convenient to have an expression for the inductance per unit length $\mathscr{L}$ (Eq. 33.6).	$\mathscr{L} \equiv \dfrac{L}{\ell} = \mu_0 n^2 A$ (33.6)

FIGURE 33.2 Jimi Hendrix in concert.

CASE STUDY: Listening to Hendrix

In the late 1960s, Jimi Hendrix (1942–1970) was extremely popular (Fig. 33.2). He was known for his unique electric guitar sound. Imagine a college student in the 1960s listening to Hendrix while studying physics. This case study is about three of the devices that made this activity possible. The first device is Hendrix's electric guitar, the second is the student's radio (the type used then), and the third is the radio's speaker. All three devices depend on inductors.

The strings of an electric guitar are made of a magnetic material such as steel. Below each string are pickups that respond to the motion of the string with an induced current (Fig. 33.3A). The pickup closest to the guitar's neck is normally more sensitive to the string's low-frequency vibrations, and the pickup farthest from the neck is normally more sensitive to high frequencies. As shown in Figure 33.3B, a pickup is a small inductor consisting of a coil wrapped around a permanent magnet. The permanent magnet causes magnetic moments in the steel string to line up so that it becomes locally magnetized. When the string vibrates, the magnetic flux through the pickup coil oscillates. According to Faraday's law, the changing magnetic flux induces a current in the coil that oscillates at the same frequency as the string. The current can be amplified and put through speakers so we can hear the music.

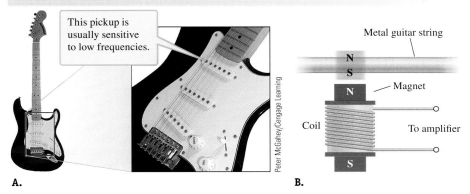

FIGURE 33.3 A. Some electric guitars have three sets of pickups. **B.** Each pickup is a coil of wire wrapped around a permanent magnet. The magnet causes the magnetic dipole moments in the steel string to line up so that a portion of the string is also magnetized.

CONCEPT EXERCISE 33.2

CASE STUDY Hendrix's Guitar

Jimi Hendrix—like other great guitar players—wanted to create his own unique sound. So, he rewrapped the pickup coils on his guitar, changing the number of turns. This changes the pickups' relative sensitivity to the string vibrations. Today, enthusiastic musicians make their own custom pickups. They might decrease the number of turns on the low-frequency pickup and increase the number on the high-frequency pickup. Model the pickup coil as a solenoid. Suppose you increase the number of turns by 10% on one particular pickup. Assume the cross-sectional area and length of the solenoid are essentially constant. By what amount does the inductance change?

33-2 Back Emf

To analyze practical circuits that involve inductors, we must understand how inductors behave when they carry a current. To do that, we must first introduce a DC power supply (Fig. 33.4). A DC power supply is much like a battery in that the potential difference between its terminals does not change sign. The advantage of a DC power supply over a battery is that you can adjust the terminal voltage by just turning a knob. The DC power supply's symbol looks like the symbol for a battery except an arrow is drawn across it ( ), indicating that the terminal voltage is variable.

Imagine connecting an inductor to a DC power supply (Fig. 33.5). The power supply sets up a clockwise current in the circuit. If that current is constant, the inductor acts like a piece of curly wire. A solenoid inductor may, for example, have a downward-pointing magnetic field in its interior (not shown in Fig. 33.5). Because the current is constant, the magnetic flux through the inductor is constant.

If the current is changing, however, the magnetic flux through the inductor changes and an emf $\mathcal{E}_L$ is induced. In Figure 33.5A, the current is increasing as you increase the voltage on the power supply. According to Lenz's law, the induced emf $\mathcal{E}_L$ must oppose the change—in this case, an increasing clockwise current—and the induced emf's direction is counterclockwise (as indicated by the upward arrow next to the inductor in Fig. 33.5). Put simply, the inductor "wants" to create a counterclockwise current to oppose the increasing clockwise current, so the

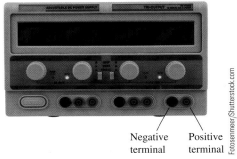

FIGURE 33.4 A variable DC power supply allows you to adjust the voltage across its terminals.

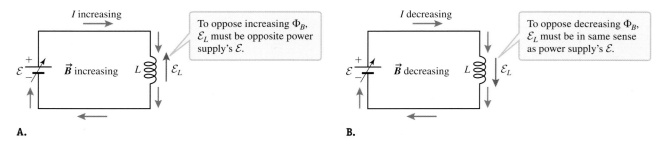

FIGURE 33.5 A variable DC power supply is connected to an inductor. **A.** As the power supply's emf increases, the current, the magnetic field, and the magnetic flux through the inductor all increase. To oppose that change, the inductor's emf $\mathcal{E}_L$ must be in the opposite sense as the power supply's emf. **B.** As the power supply's emf decreases, the magnetic flux through the inductor decreases. To oppose that change, the inductor's emf $\mathcal{E}_L$ must be in the same sense as the power supply's emf.

inductor acts like a second power supply working against the actual power supply. Because of the induced emf, the current does not increase as quickly as it would if the inductor were replaced by a simple straight wire.

If we decrease the power supply's voltage, the current is still clockwise but decreasing (Fig. 33.5B). Now the magnetic flux through the inductor decreases. In order to oppose the change, the induced emf must work with the power supply's emf. In other words, the inductor "wants" to maintain the clockwise current. So, the induced emf $\mathcal{E}_L$ is clockwise. Because of the induced emf, the current does not decrease as quickly as it would if the inductor were replaced by a simple straight wire.

We have found the direction of the induced emf; we can find the magnitude from Faraday's law. Equation 32.4 expresses Faraday's law in terms of the flux linkage (total magnetic flux):

$$\mathcal{E}_L = \frac{d(N\Phi_B)}{dt} = \frac{d\Phi_{tot}}{dt} \tag{32.4}$$

The flux linkage is replaced using $\Phi_{tot} = LI$ (Eq. 33.1):

$$\mathcal{E}_L = \frac{d(LI)}{dt}$$

The inductance L depends only on geometry, so it is constant:

$$\mathcal{E}_L = L\frac{dI}{dt} \tag{33.7}$$

Any real inductor has some resistance because the wire it is made from has resistance. Our focus is on **ideal inductors**, however, which have no resistance. A real inductor can be modeled as an ideal inductor in series with a resistor.

When analyzing DC circuits, we listed rules for finding the potential difference across various circuit elements (Section 29-1). These rules are helpful when we apply Kirchhoff's loop rule using the steps in Section 29-2. A similar rule for ideal inductors combines the directions we found with Lenz's law (Fig. 33.5) with the magnitude we found with Faraday's law (Eq. 33.7). Imagine connecting a voltmeter across an inductor so that we measure the potential difference in the direction of the current. The red lead is downstream, and the black lead is upstream (Fig. 33.6). If the current is decreasing (Fig. 33.5B), the induced emf $\mathcal{E}_L$ is downward from the black lead to the red lead. Compare this situation to that shown in Figure 29.2E (page 898), in which a voltmeter measures a *positive* potential across an emf device. In the case of the inductor, when we measure the voltage in the direction of the current and the current is decreasing ($dI/dt < 0$), the voltage is positive. Its magnitude is given by $\mathcal{E}_L = L(dI/dt)$, Equation 33.7. Combining the direction with the magnitude, we write an expression for the potential difference:

$$\Delta V_L = V_{red} - V_{black} = -L\frac{dI}{dt} \tag{33.8}$$

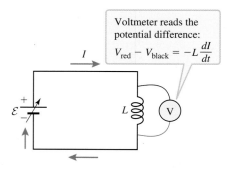

FIGURE 33.6 A voltmeter is connected across the inductor. If the current is decreasing, the voltmeter reports a positive potential difference. If the current is increasing, the voltmeter reports a negative potential difference.

where the red lead is downstream and the black lead is upstream so that the measurement is made in the direction of the current. Because $dI/dt < 0$ when the current is decreasing, the potential difference ($V_{\text{red}} - V_{\text{black}}$) is positive.

Equation 33.8 is the **inductor rule**; it holds even when the current is increasing. If the current is increasing (Fig. 33.5A), the induced emf points upward from the red lead to the black lead. In that case, our work with emf devices (Chapter 29) tells us the voltmeter measures a negative potential difference. This result is consistent with Equation 33.8 for an increasing current, $dI/dt > 0$. There are three things to keep in mind when using the inductor rule:

1. The potential difference across an inductor is zero if the current is constant ($dI/dt = 0$).
2. The negative sign in $\Delta V_L = -L(dI/dt)$ (Eq. 33.8) implies that we are measuring the potential difference in the direction of the current.
3. The negative sign also indicates that the voltage across the inductor always opposes the change in the current. The inductor voltage is referred to as the **back emf** because it is fighting *back* against the changing current. Thinking of the back emf is often the easiest way to reason out the sign of the potential difference across the inductor.

INDUCTOR RULE ◯ Special Case

(Strictly speaking, Kirchhoff's loop rule holds only in the case of conservative fields. The inductor rule is a way to preserve the use of Kirchhoff's loop rule; what follows in this chapter gives the right equations but misses on conceptual details. See Section 34-2 and Problems 13, 14, and 68 at the end of Chapter 34.)

Transformers Revisited

After electrical power is transported over great distances at high voltages and low current, transformers (Fig. 32.30, page 1036) step down the power to lower voltages for use in your home. Ultimately, the development of transformers led to the success of AC over DC power supplies (Case Study, Chapter 32). In Section 32-8, we stated the transformer equation

$$\frac{V_S}{V_P} = \frac{N_S}{N_P} \qquad (32.26)$$

without deriving it. Now we can derive this equation.

Start by analyzing the primary circuit, where the primary circuit consists of an AC generator and an inductor (Fig. 33.7A). Ideally, the resistance in the primary circuit is negligible. As in any circuit analysis, use Kirchhoff's loop rule to find an expression for the potential differences around the closed loop. This is an AC circuit, so the current's direction oscillates. To apply Kirchhoff's loop rule, imagine an instant when the current is clockwise. Starting in the bottom left corner and working our way around in the direction of the current, we first encounter the AC generator, whose emf is momentarily pointing upward (the negative terminal is below the positive one momentarily). From the emf rule, the potential difference is positive. Although the emf increases and decreases periodically, we can use the rms value $\mathcal{E}_{\text{rms}}$. Because there is no (or very little) resistance, the only other circuit element is the inductor. According to the inductor rule, the potential difference measured in the direction of the current is given by $\Delta V_L = -L(dI/dt)$ (Eq. 33.8). Kirchhoff's loop rule gives

$$\mathcal{E}_{\text{rms}} - L\frac{dI}{dt} = 0$$

To be consistent with the notation in Section 32-8, we write $\mathcal{E}_{\text{rms}} = V_P$, where P stands for primary, and solve for V_P:

$$V_P = L\frac{dI}{dt}$$

Eliminate the inductance L using Equation 33.7, $\mathcal{E}_L = L(dI/dt)$, and Equation 32.4, Faraday's law, $\mathcal{E} = N(d\Phi_B/dt)$:

$$V_P = L\frac{dI}{dt} = N_P\frac{d\Phi_B}{dt} \qquad (33.9)$$

where N_P is the primary solenoid's number of turns.

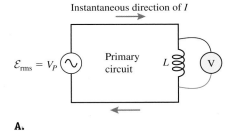

A.

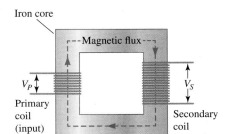

B.

FIGURE 33.7 A transformer has two circuits. **A.** The primary circuit consists of an inductor and an AC power supply. **B.** The iron core and both solenoids are shown.

Both solenoids (primary and secondary) are wound around an iron core (Fig. 33.7B). In an ideal transformer, the iron core ensures that the time-varying magnetic flux is the same through both the primary and secondary solenoids. The flux induces an emf in the secondary circuit given by Faraday's law:

$$V_S = N_S \frac{d\Phi_B}{dt} \tag{33.10}$$

where V_S is the rms induced emf in the secondary circuit and N_S is the number of turns in the secondary solenoid.

Because the time-varying magnetic flux $d\Phi_B/dt$ is the same through both solenoids, it can be eliminated from Equations 33.9 and 33.10 to arrive at the transformer equation:

$$\frac{d\Phi_B}{dt} = \frac{V_S}{N_S} = \frac{V_P}{N_P}$$

$$\frac{V_S}{V_P} = \frac{N_S}{N_P} \tag{32.26}$$

CONCEPT EXERCISE 33.3

Figure 33.8 shows an inductor that is part of a larger circuit.

a. If the voltmeter reads zero, what do you know about the current in the inductor?
b. If the voltmeter reads a positive value, what do you know about the current in the inductor?

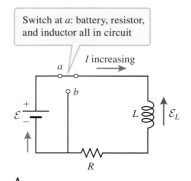

FIGURE 33.8

RL CIRCUIT ▶ Special Case

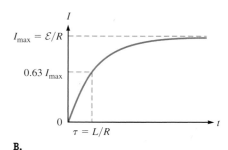

FIGURE 33.9 **A.** When the switch has just closed at position *a*, the current is increasing. **B.** Current versus time when the switch is at *a*.

33-3 Special Case: Resistor–Inductor (*RL*) Circuit

All real wires have resistance, so any real inductor has resistance, and circuits that involve resistance and inductance (*RL* circuits) are common. Figure 33.9A shows a circuit that consists of an emf device such as a battery, a switch, an inductor, and a resistor. The resistor may represent the inductor's own resistance, or it may be an actual resistor. In either case, the inductor alone is modeled as an ideal inductor with no other resistance.

Initially the switch is open, so there is no current in the circuit. The switch is then closed at position *a* (Fig. 33.9A). Because the current increases from zero, there is a back emf $\mathcal{E}_L$ across the inductor as shown. To find an expression for the increasing current, apply Kirchhoff's loop rule starting in the bottom left corner and working clockwise in the direction of the current. The voltage across the battery is positive going from the negative to the positive terminal; the voltage across the inductor is given by Equation 33.8, $\Delta V_L = -L(dI/dt)$, because we are measuring in the direction of the current, and the voltage across the resistor is negative because we are measuring in the direction of the current. According to Kirchhoff's loop rule,

$$\mathcal{E} - L\frac{dI}{dt} - IR = 0 \tag{33.11}$$

Equation 33.11 is a differential equation that is mathematically equivalent to the differential equation we found for the resistor–capacitor (*RC*) circuit (Section 29-8):

$$\mathcal{E} - R\frac{dq}{dt} - \frac{q}{C} = 0$$

We found the solution to this equation:

$$q(t) = C\mathcal{E}(1 - e^{-t/\tau}) \tag{29.10}$$

where the capacitive time constant is given by

$$\tau \equiv RC \tag{29.13}$$

We want to solve Equation 33.11 for I, and because we already solved the equivalent equation, we can just translate variables by comparing the two differential equations:

$$RC \text{ circuit} \rightarrow RL \text{ circuit}$$
$$\mathcal{E} \rightarrow \mathcal{E}$$
$$R \rightarrow L$$
$$q \rightarrow I$$
$$C \rightarrow \frac{1}{R}$$

Using this translation of variables, we can solve Equation 33.11:

$$I(t) = \frac{\mathcal{E}}{R}(1 - e^{-t/\tau}) \qquad (33.12)$$

where the inductive time constant is

$$\tau \equiv \frac{L}{R} \qquad (33.13)$$

If L is in henrys and R is in ohms, the inductive time constant τ is in seconds.

Figure 33.9B is a plot of Equation 33.12. Contrary to what you might think, when you close the switch at a, the current does not immediately reach its maximum value $I_{max} = \mathcal{E}/R$. Because the back emf opposes the increasing current, the current builds up to its maximum value over some time interval. At time $t = 0$, when the switch is first closed, the current is zero. The current gradually increases, and at $t = \tau$, the current has reached 63% of its maximum value. According to Equation 33.12, the current approaches its maximum value as $t \rightarrow \infty$. In practice, the current reaches its maximum value after a long time compared to τ, and then the back emf vanishes.

After the switch has been in position a for a while, it is switched to position b (Fig. 33.10A), so that only the resistor and inductor are in the circuit. The current does not immediately drop to zero because the back emf opposes the decrease in current. Kirchhoff's loop rule gives an expression for the decreasing current. Because the current is in the same direction as in Figure 33.9A, we start with $\mathcal{E} - L(dI/dt) - IR = 0$ (Eq. 33.11). This time, we must set $\mathcal{E} = 0$ because there is no battery in the circuit (Fig. 33.10A):

$$0 - L\frac{dI}{dt} - IR = 0$$

$$L\frac{dI}{dt} + IR = 0 \qquad (33.14)$$

Equation 33.14 is the same differential equation we found for a discharging capacitor in an RC circuit (Problem 29.93):

$$R\frac{dq}{dt} + \frac{q}{C} = 0$$

whose solution is

$$q(t) = q_{max} e^{-t/\tau}$$

As before, the capacitive time constant is given by $\tau \equiv RC$ (Eq. 29.13). We can use the same translation of variables to write the solution to Equation 33.14 in the case of an RL circuit:

$$I(t) = I_{max} e^{-t/\tau} \qquad (33.15)$$

where the inductive time constant is given by $\tau \equiv L/R$ (Eq. 33.13).

Figure 33.10B is a plot of Equation 33.15. When you first move the switch to position b, the current is at its maximum value. If the switch has been at position a for a long enough time, that maximum current is $I_{max} = \mathcal{E}/R$. If the switch has not

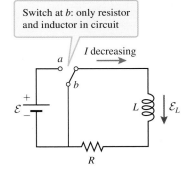

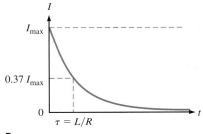

FIGURE 33.10 A. When the switch is moved to position b, the current decreases. **B.** Current versus time when the switch is at b.

been at position a long enough, the maximum current in Equation 33.15 is smaller: $I_{max} < \mathcal{E}/R$. Once again, after the battery is cut out of the circuit (switch at b), the current does not immediately drop to zero. Instead, the current gradually decreases because the back emf opposes the change. At time $t = \tau$, the current has dropped to 37% of its maximum value. Eventually the current reaches zero, at which time the back emf is also zero.

> **CONCEPT EXERCISE 33.4**
>
> In what ways is the RL circuit like an RC circuit (Section 29-8)?

EXAMPLE 33.2 **Half the Maximum**

It might seem odd that at $t = \tau$, the current in Figure 33.9B is at 63% of its maximum, but the current is at only 37% of its maximum in Figure 33.10B.

A At what time does the current reach half of its maximum when the switch is at a in Figure 33.9A? Give your answer in terms of τ.

: INTERPRET and ANTICIPATE
We need an expression for t when the current has risen to half its maximum value. From the graph in Figure 33.9B, our answer should be less than one time constant: $t < \tau$.

: SOLVE To find t when $I = \frac{1}{2}I_{max}$, substitute this value of the current into Equation 33.12.	$I(t) = \dfrac{\mathcal{E}}{R}(1 - e^{-t/\tau})$ (33.12) $\frac{1}{2}I_{max} = \dfrac{\mathcal{E}}{R}(1 - e^{-t/\tau})$
Because $I_{max} = \mathcal{E}/R$, we can eliminate $\mathcal{E}$ and R from the equation.	$\frac{1}{2}I_{max} = I_{max}(1 - e^{-t/\tau})$ $\frac{1}{2} = (1 - e^{-t/\tau})$
Solving for t involves taking a natural logarithm.	$e^{-t/\tau} = 1 - \frac{1}{2} = \frac{1}{2}$ $\ln e^{-t/\tau} = \ln \frac{1}{2}$ $-t/\tau = -0.693$ $t = 0.693\tau$

: CHECK and THINK
As expected, the time for the current to reach half its maximum value is less than one time constant: $t < \tau$.

B At what time does the current reach half its maximum when the switch is at b in Figure 33.10A? Give your answer in terms of τ.

: INTERPRET and ANTICIPATE
This is similar to part A, except now we need to look at the plot in Figure 33.10B to anticipate the result. Our answer here should also be less than one time constant: $t < \tau$.

: SOLVE Again, find t when $I = \frac{1}{2}I_{max}$. Substitute this value of the current into Equation 33.15, remembering that I_{max} does not necessarily equal $\mathcal{E}/R$.	$I(t) = I_{max} e^{-t/\tau}$ (33.15) $\frac{1}{2}I_{max} = I_{max} e^{-t/\tau}$

As before, solving for t involves taking a natural logarithm.	$e^{-t/\tau} = \frac{1}{2}$ $\ln e^{-t/\tau} = \ln \frac{1}{2}$ $-t/\tau = -0.693$ $t = 0.693\tau$

:• **CHECK and THINK**

This is exactly what we found in part A. The time for the current to drop to half its original value is sometimes called the **half-life**. The term *half-life* is used in any situation that involves exponential decay. For example, you might have heard of carbon dating for determining the age of fossils and archeological specimens. Carbon dating works because the amount of carbon-14 in a substance decreases when the organism dies and stops metabolizing carbon. The half-life of carbon-14 is about 5700 years.

33-4 Energy Stored in a Magnetic Field

Using the analogy between capacitors and inductors (Table 33.1), we found an expression for the energy stored by an inductor: $U_B = \frac{1}{2}LI^2$ (Eq. 33.3). In this section, we use our work on *RL* circuits to derive this expression, and then we use this expression to derive a more general expression for the energy stored by a magnetic field.

DERIVATION Energy Density Stored in a Magnetic Field

Consider the *RL* circuit in Figure 33.9A when the switch has just been thrown to position a. First, we derive $U_B = \frac{1}{2}LI^2$ (Eq. 33.3), and then we will show that in the case of a solenoid of length ℓ and cross-sectional area A,

$$U_B = \frac{1}{2}\frac{B^2}{\mu_0}(\ell A) \qquad (33.16)$$

Finally, we will show that the energy density (energy per unit volume) stored in a magnetic field is

$$u_B = \frac{U_B}{V} = \frac{1}{2}\frac{B^2}{\mu_0} \qquad (33.17)$$

ENERGY DENSITY STORED BY A MAGNETIC FIELD ✪ **Major Concept**

Start with Kirchhoff's loop rule (Eq 33.11) applied to this circuit (Fig. 33.9A).	$\mathcal{E} - L\frac{dI}{dt} - IR = 0$ (33.11) $\mathcal{E} = IR + L\frac{dI}{dt}$ (1)
Multiply each side of Equation (1) by I.	$\mathcal{E}I = I^2R + LI\frac{dI}{dt}$ (2)
The left side of Equation (2) is the power supplied by the battery. The first term on the right is the power dissipated by the resistor. The last term is the power stored in the inductor's magnetic field. Thus, part of the power supplied by the battery is dissipated by the resistor, and the rest is stored in the inductor's magnetic field.	$\begin{pmatrix}\text{power} \\ \text{supplied} \\ \text{by battery}\end{pmatrix} = \begin{pmatrix}\text{power} \\ \text{dissipated} \\ \text{by resistor}\end{pmatrix} + \begin{pmatrix}\text{power} \\ \text{stored} \\ \text{by inductor}\end{pmatrix}$ $P_\mathcal{E} \quad = \quad P_R \quad + \quad P_L$
Consider the power stored by the inductor (last term in Eq. 2). This is the rate at which the energy is stored in the inductor's magnetic field.	$P_L = \frac{dU_B}{dt} = LI\frac{dI}{dt}$ (3)

Derivation continues on page 1056 ▶

1056 CHAPTER 33 Inductors and AC Circuits

Cancel dt in Equation (3).	$dU_B = LI\,dI$
Integrate the left side from 0 to U_B (the stored energy when the current is I), and integrate the right side from 0 to the current I.	$\int_0^{U_B} dU_B = \int_0^I LI\,dI$
The inductance L depends only on geometry, not current, so pull L outside the integral. Complete the integral and substitute limits to find Equation 33.3 as before.	$U_B = L\int_0^I I\,dI = \tfrac{1}{2}LI^2$ (33.3)
Equation 33.3 is the energy stored in the magnetic field of an inductor of any shape. To find the energy stored in the special case of a solenoid, substitute the solenoid's inductance as given by $L = \mu_0 n^2 \ell A$ (Eq. 33.5).	$U_B = \tfrac{1}{2}(\mu_0 n^2 \ell A) I^2$
Use $I = B/(\mu_0 n)$ from Equation 31.6 to eliminate I, and simplify to find the energy stored in the magnetic field of a solenoid.	$U_B = \tfrac{1}{2}(\mu_0 n^2 \ell A)\left(\dfrac{B}{\mu_0 n}\right)^2$ $U_B = \dfrac{1}{2}\dfrac{B^2}{\mu_0}(\ell A)$ ✓ (33.16)
The energy density u_B stored in the solenoid's magnetic field is found by dividing by its volume: $V = \ell A$.	$u_B = \dfrac{U_B}{V} = \dfrac{1}{2}\dfrac{B^2}{\mu_0}$ ✓ (33.17)

:• **COMMENTS**
Equation 33.17 was derived for the special case of a solenoid but, because the expression is independent of geometry, it is true in general; it gives the magnetic energy density in any region that has a magnetic field.

CONCEPT EXERCISE 33.5

Skim through Chapter 27 to find an expression analogous to Equation 33.17 for the energy density stored in an electric field. Report your answer and explain.

EXAMPLE 33.3   The Earth's Magnetic Field

Estimate the energy density of the Earth's magnetic field near its surface.

:• **INTERPRET and ANTICIPATE**
We expect a (scalar) numerical answer with the SI units of joules per cubic meter.

:• **SOLVE**

This is a straightforward application of Equation 33.17. The magnetic field near the Earth's surface is about 0.5×10^{-4} T.	$u_B = \dfrac{1}{2}\dfrac{B^2}{\mu_0}$ (33.17) $u_B = \dfrac{1}{2}\dfrac{(0.5 \times 10^{-4}\,\text{T})^2}{(4\pi \times 10^{-7}\,\text{T}\cdot\text{m/A})}$ $u_B = 9.9 \times 10^{-4}\,\text{T}\cdot\text{A/m}$

:• **CHECK and THINK**

Make sure these units are right. It helps to think about the magnetic force exerted on a current-carrying conductor $\vec{F}_B = I(\vec{\ell} \times \vec{B})$ (Eq. 30.32) to write the SI units.	$1\,\text{N} = 1\,\text{T}\cdot\text{m}\cdot\text{A}$

Then use $1\,\text{N}\cdot\text{m} = 1\,\text{J}$ to find an expression for the SI units of energy density. The SI units we found are correct.

$1\,\text{N}\cdot\text{m} = 1\,\text{J} = 1\,\text{T}\cdot\text{m}^2\cdot\text{A}$

$1\,\text{J/m}^3 = 1\,\text{T}\cdot\text{m}^2\cdot\text{A/m}^3 = 1\,\text{T}\cdot\text{A/m}$

$u_B = 9.9 \times 10^{-4}\,\text{J/m}^3$

This problem helps build intuition about the energy stored in a magnetic field. Near the surface of the Earth, the magnetic energy density is about $10^{-3}\,\text{J/m}^3$. Compare that value to the energy density in gasoline (about $10^{10}\,\text{J/m}^3$). In Problem 18, you will calculate the energy density stored in the magnetic field near a neutron star.

33-5 Special Case: Inductor–Capacitor (LC) Circuit

LC CIRCUIT ▶ Special Case

So far we have taken a close look at two circuits—the RC circuit and the RL circuit—in which the current changes by either increasing or decreasing and finally reaching a steady value, but the current does not alternate in these circuits. What would happen if we built a circuit that has an inductor and a capacitor but no resistor? The answer is that energy would be conserved and transferred back and forth between the inductor and the capacitor, and the current would alternate.

Figure 33.11 shows an ideal LC circuit consisting of an inductor and a capacitor. Ideally, the circuit has no resistance, but of course in any real circuit there is resistance in the wires. We consider that complication in Section 33-9. The capacitor is fully charged, and the switch is open. The capacitor stores energy in its electric field. There is no magnetic field because there is no current. When the switch is closed, the capacitor discharges, resulting in a counterclockwise current from the positive plate toward the negative plate. Once there is a current through the inductor, it stores energy in its magnetic field. Eventually, the capacitor is completely discharged and all the energy is stored in the inductor's magnetic field. A current remains in the circuit, so the capacitor charges up again and the cycle repeats. Energy is transferred back and forth between the capacitor's electric field and the inductor's magnetic field, much like the energy that is transformed from gravitational potential energy into kinetic energy in the case of an oscillating pendulum. We say the circuit *oscillates*.

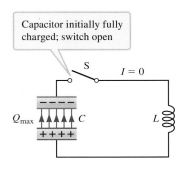

FIGURE 33.11 The capacitor has been charged by a battery that is not shown. The charged capacitor is then connected to an inductor through a switch.

Figure 33.12 shows nine key instants in one of the circuit's cycles. For each instant, an energy bar graph shows the relative amounts of energy stored in the capacitor's electric field and the inductor's magnetic field. When the capacitor is fully charged (times 1, 5, and 9), the total energy is stored in the capacitor's electric field and given by $E_{tot} = Q_{max}^2/2C$ (Eq. 27.2). When the capacitor is momentarily discharged, the current has reached it maximum value I_{max} and the magnetic field is at its maximum strength (times 3 and 7). Then the total energy is stored in the inductor's magnetic field and given by $E_{tot} = \frac{1}{2}LI_{max}^2$ (Eq. 33.3). When the capacitor is partially charged and there is current in the circuit (times 2, 4, 6, and 8), energy is stored in both the capacitor's electric field and the inductor's magnetic field. The total energy is the sum of the energy stored in the two devices: $E_{tot} = (Q^2/2C) + \frac{1}{2}LI^2$. As the circuit oscillates, energy is transferred back and forth but not lost. Thus, the total energy E_{tot} is a constant. Further, the total energy equals the maximum energy stored by the capacitor's electric field:

$$E_{tot} = \frac{1}{2}\frac{Q^2}{C} + \frac{1}{2}LI^2 = \frac{1}{2}\frac{Q_{max}^2}{C} = (U_E)_{max} \quad (33.18)$$

The total energy also equals the maximum energy stored by the inductor's magnetic field:

$$E_{tot} = \frac{1}{2}\frac{Q^2}{C} + \frac{1}{2}LI^2 = \frac{1}{2}LI_{max}^2 = (U_B)_{max} \quad (33.19)$$

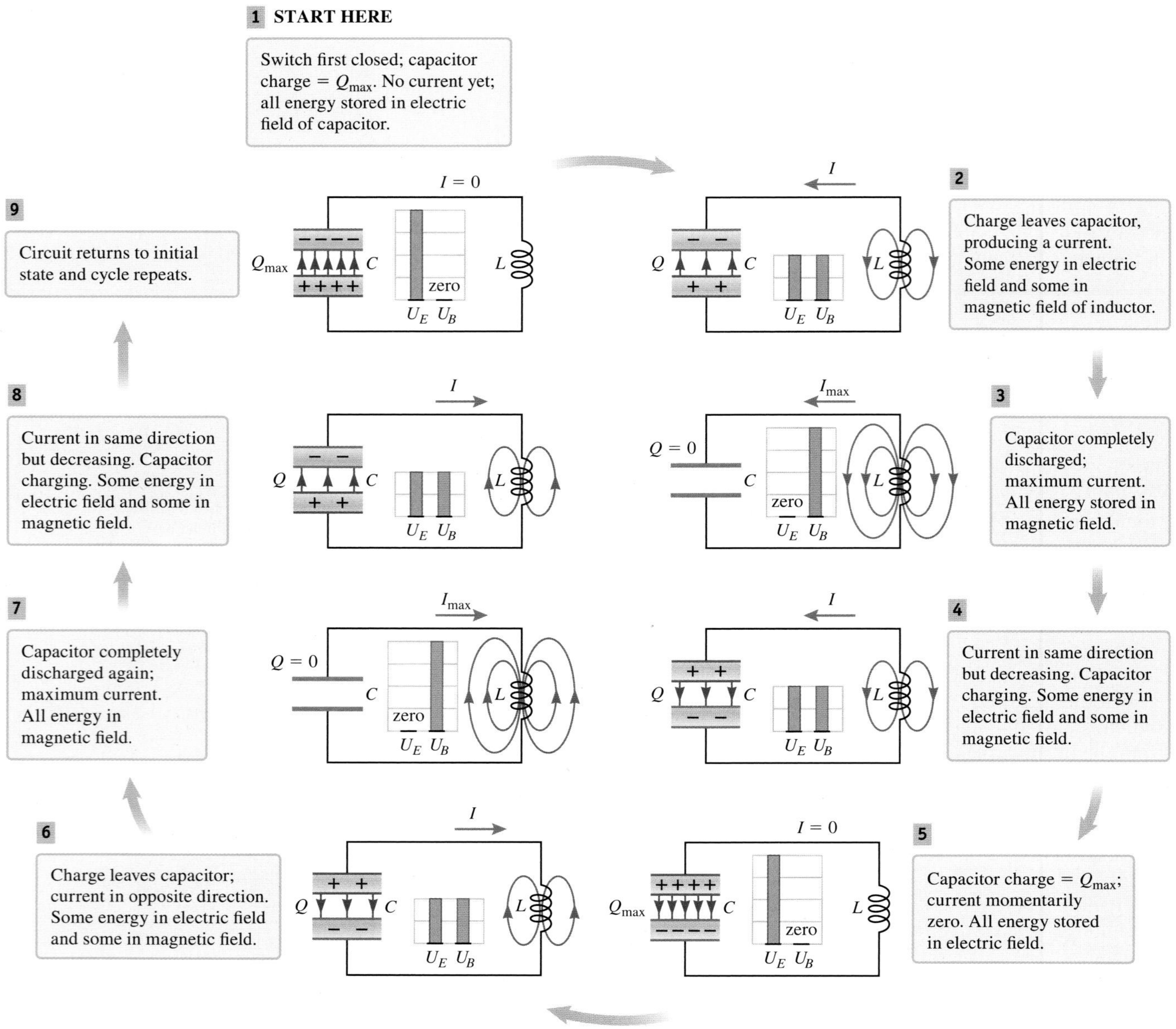

FIGURE 33.12 One complete cycle of an LC circuit. The cycle begins at time **1** with the capacitor fully charged, so its lower plate is positive. At time **5**, the circuit is halfway through its cycle; the capacitor is fully charged, but now the top plate is positive. By time **9**, the circuit has returned to its original state, after which the cycle repeats.

DERIVATION Q and I for the LC Circuit

We will show that the charge stored by a capacitor in an LC circuit oscillates according to

$$Q(t) = Q_{max} \cos(\omega t + \varphi) \tag{33.20}$$

$$\text{where } \omega = \sqrt{\frac{1}{LC}} \tag{33.21}$$

Also, we'll show that the current in the circuit is given by

$$I(t) = -I_{max} \sin(\omega t + \varphi) \tag{33.22}$$

$$\text{where } I_{max} = \omega Q_{max} \tag{33.23}$$

33-5 Special Case: Inductor–Capacitor (LC) Circuit

Start with the expression for constant total energy.	$E_{tot} = \frac{1}{2}\frac{Q^2}{C} + \frac{1}{2}LI^2$
The time derivative of a constant is zero. Taking the time derivative of both sides results in a differential equation set equal to zero.	$\frac{dE_{tot}}{dt} = \frac{d}{dt}\left(\frac{1}{2}\frac{Q^2}{C} + \frac{1}{2}LI^2\right) = 0$
Pull out the constants.	$\frac{1}{2}\left(\frac{1}{C}\frac{dQ^2}{dt} + L\frac{dI^2}{dt}\right) = 0$ (33.24)
Apply the chain rule (Appendix A) to each derivative.	$\frac{dQ^2}{dt} = \frac{dQ^2}{dQ}\frac{dQ}{dt} = 2Q\frac{dQ}{dt}$ and similarly for current I
Substitute these results into Equation 33.24.	$\frac{1}{2}\left[\frac{1}{C}\left(2Q\frac{dQ}{dt}\right) + L\left(2I\frac{dI}{dt}\right)\right] = 0$ $\frac{Q}{C}\left(\frac{dQ}{dt}\right) + LI\frac{dI}{dt} = 0$
Use $I = dq/dt$ (Eq. 28.1) to cancel the current I from both terms.	$\frac{Q}{C}(I) + L(I)\frac{dI}{dt} = 0$
Use Equation 28.1 again to write a differential equation for Q.	$\frac{Q}{C} + L\frac{d}{dt}\left(\frac{dQ}{dt}\right) = \frac{Q}{C} + L\frac{d^2Q}{dt^2} = 0$
Rearranging this differential equation shows that it is mathematically equivalent to Equation 16.25—the equation for a simple harmonic oscillator that consists of a particle of mass m attached to a spring of spring constant k. The particle's position is $y(t)$.	$-\frac{Q(t)}{LC} = \frac{d^2Q(t)}{dt^2}$ (33.25) $-\frac{k}{m}y(t) = \frac{d^2y(t)}{dt^2}$ (16.25)
We can use the solution to Equation 16.25 here. Compare Equation 33.25 to Equation 16.25 to come up with a translation of variables.	$y(t) \to Q(t)$ $\frac{k}{m} \to \frac{1}{LC}$
We found that the particle's position oscillates sinusoidally (Eq. 16.3) and the angular frequency ω depends on the mass and the spring constant. The initial phase (the phase constant) is the phase $(\omega t + \varphi)$ when $t = 0$. See Section 16-2.	$y(t) = y_{max}\cos(\omega t + \varphi)$ (16.3) $\omega = \sqrt{\frac{k}{m}}$ (16.26)
To solve the LC circuit's differential equation, substitute the translation of variables into Equations 16.3 and 16.26.	$Q(t) = Q_{max}\cos(\omega t + \varphi)$ ✓ (33.20) $\omega = \sqrt{\frac{1}{LC}}$ ✓ (33.21)
Take the time derivative to find the current in the circuit.	$I(t) = \frac{dQ(t)}{dt} = \frac{d}{dt}[Q_{max}\cos(\omega t + \varphi)]$ $I(t) = -\omega Q_{max}\sin(\omega t + \varphi)$ $I(t) = -I_{max}\sin(\omega t + \varphi)$ ✓ (33.22) where $I_{max} = \omega Q_{max}$ ✓ (33.23)

Derivation continues on page 1060 ▶

COMMENTS

While deriving expressions for the charge stored by the capacitor (Eq. 33.20) and the current (Eq. 33.22), we found that the *LC* circuit's differential equation is mathematically equivalent to the particle–spring oscillator's equation. The charge stored by the capacitor and the position of an oscillating particle are conceptually connected.

Figure 33.13 shows graphs of Q and I versus t. Compare this to a particle connected to a spring and oscillating back and forth from $+y_{max}$ to $-y_{max}$ as in Figure 16.5A (page 453). In a similar fashion, the charge on the bottom plate in Figure 33.12 is initially $+Q_{max}$. That charge is reduced to zero, and then the charge switches sign and increases in magnitude until it reaches $-Q_{max}$. The charge is reduced in magnitude again, reaches zero, and then increases until it reaches $+Q_{max}$ again. Like the particle attached to the spring, the charge on the capacitor oscillates back and forth from $+Q_{max}$ to $-Q_{max}$ as shown in Figure 33.13A, and current is analogous to the particle's velocity (Fig. 33.13B).

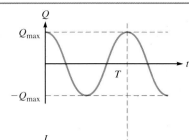

A.

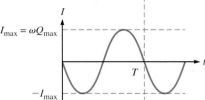

B.

FIGURE 33.13 Charge and current versus time for an *LC* circuit.

CONCEPT EXERCISE 33.6

Take another look at the simple harmonic oscillator (Section 16-2, page 453). How is $I(t)$ analogous to $v(t)$?

Graphs of U_E and U_B for the *LC* Circuit

In Figure 33.12, we see that the energy U_E stored by the capacitor's electric field and the energy U_B stored by the inductor's magnetic field both oscillate. We can also express this fact mathematically. We found that the charge stored by the capacitor oscillates according to $Q(t) = Q_{max} \cos(\omega t + \varphi)$ (Eq. 33.20), and we know that the energy stored in the electric field depends on the stored charge, $U_E = Q^2/2C$ (Eq. 27.2). So, as you will show in Problem 27, the energy stored in the electric field oscillates according to

$$U_E(t) = \frac{Q_{max}^2}{2C} \cos^2(\omega t + \varphi) \tag{33.26}$$

The same sort of mathematical reasoning applies to the energy stored in the magnetic field. As you will show in Problem 28, the energy stored in the magnetic field is given by

$$U_B(t) = \frac{Q_{max}^2}{2C} \sin^2(\omega t + \varphi) \tag{33.27}$$

At any moment, the total energy stored by the electric and magnetic fields is the sum of Equations 33.26 and 33.27:

$$E_{tot} = U_E + U_B = \frac{Q_{max}^2}{2C} \cos^2(\omega t + \varphi) + \frac{Q_{max}^2}{2C} \sin^2(\omega t + \varphi)$$

The total energy is a constant given by

$$E_{tot} = \frac{Q_{max}^2}{2C} = \frac{1}{2}LI_{max}^2 \tag{33.28}$$

as you will show in Problem 29.

The transformation of energy between the electric and magnetic fields is illustrated in the graphs of Equations 33.26 and 33.27 shown in Figure 33.14. Both energies are always positive, and at any instant the sum of the two is a constant given by Equation 33.28.

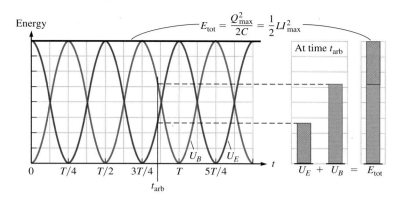

FIGURE 33.14 Energy stored in the electric field U_E and in the magnetic field U_B for the LC circuit. At any arbitrary time t_{arb}: $E_{tot} = U_E(t_{arb}) + U_B(t_{arb})$.

EXAMPLE 33.4 Energy Evenly Split Between the Fields

Consider the circuit in Figure 33.11 in which the capacitor is initially fully charged. The switch is closed at time $t = 0$. What is the first time after the switch is closed that half the energy is in the electric field of the capacitor and half is in the magnetic field of the inductor? Express your answer in terms of the period T of the oscillation.

:• **INTERPRET and ANTICIPATE**
The three graphs in Figures 33.13 and 33.14 show one and a half periods. Figure 33.14 shows that energy is equally split in less than a quarter of the period. We might guess that the split is halfway between 0 and $T/4$.

:• **SOLVE**

Set either U_E or U_B equal to half the total energy, and solve for t. We arbitrarily choose U_E (Eq. 33.26).	$U_E(t) = \dfrac{Q_{max}^2}{2C}\cos^2(\omega t + \varphi) = \dfrac{1}{2}E_{tot}$
Eliminate Q_{max} with $E_{tot} = Q_{max}^2/2C$ (Eq. 33.28).	$E_{tot}\cos^2(\omega t + \varphi) = \dfrac{1}{2}E_{tot}$
Cancel E_{tot}. The capacitor's charge is initially Q_{max}, so the phase constant $\varphi = 0$ (Fig. 33.13A).	$\cos^2(\omega t) = \dfrac{1}{2}$
Solve for ωt.	$\cos(\omega t) = \pm\dfrac{1}{\sqrt{2}}$ $\omega t = \cos^{-1}\left(\pm\dfrac{1}{\sqrt{2}}\right)$ $\omega t = \dfrac{\pi}{4}, \dfrac{3\pi}{4}, \dfrac{5\pi}{4}, \dfrac{7\pi}{4}, \dfrac{9\pi}{4}, \ldots$
We need only the first time after the switch is closed.	$\omega t = \pi/4$
Now use $\omega = 2\pi/T$ (Eq. 16.2) to find t in terms of the period.	$t = \dfrac{\pi}{4}\dfrac{1}{\omega} = \dfrac{\pi}{4}\dfrac{T}{2\pi} = T/8$

:• **CHECK and THINK**
As expected, the time for the energy to be evenly split between the capacitor and the inductor is less than a quarter of a period.

33-6 Special Case: AC Circuit with Resistance

AC CIRCUIT WITH RESISTANCE
▶ Special Case

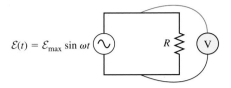

FIGURE 33.15 An AC generator and a resistor. A voltmeter measures the voltage across the AC generator and across the resistor. The voltage across the resistor equals the voltage across the generator.

In Chapter 29, we applied the principles of electricity to practical direct current (DC) circuits. In this chapter, we apply these principles to alternating current (AC) circuits. In the remainder of this chapter, we will consider circuits that involve an AC power source such as an AC generator (Fig. 32.20, page 1026). The symbol for an AC power source is ⌒. We start by considering circuits that contain an AC power source and just one other circuit element—a resistor, a capacitor, or an inductor. In the final section, we will consider an AC circuit that has all three of these elements. Our primary goal in each case is to find a connection between the voltage across the circuit element and the current in the circuit.

Figure 33.15 shows an AC power supply connected to a resistor. The AC power supply's emf oscillates according to $\mathcal{E}(t) = \mathcal{E}_{max} \sin \omega t$ (Eq. 32.13). The voltage across the generator equals the voltage across the resistor:

$$V_R(t) = \mathcal{E}(t) = \mathcal{E}_{max} \sin \omega t \qquad (33.29)$$

where V_R is the potential difference across the resistor.

To find the current in the circuit, apply Kirchhoff's loop rule at one particular instant in which the current is clockwise (not shown on the figure). Applying the loop rule and the resistance rule clockwise around the circuit in the direction of the current, we have

$$\mathcal{E}(t) - IR = 0 \qquad (33.30)$$

The current as a function of time is given by

$$I(t) = \frac{\mathcal{E}(t)}{R} = \frac{\mathcal{E}_{max} \sin \omega t}{R}$$

The maximum current is

$$I_{max} = \frac{\mathcal{E}_{max}}{R} \qquad (33.31)$$

and the current as a function of time may be written in terms of I_{max}:

$$I(t) = I_{max} \sin \omega t \qquad (33.32)$$

Figure 33.16A is a graph of the voltage across the resistor as a function of time, and Figure 33.16B is a graph of the current as a function of time. The current and the voltage across the resistor are **in phase**, which means the current and voltage rise and fall together at the same time.

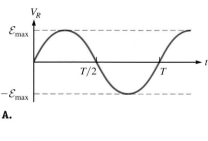

A.

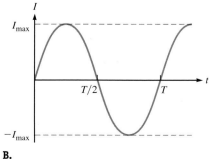

B.

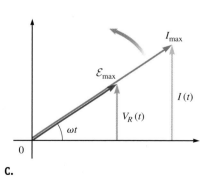

C.

FIGURE 33.16 A. Voltage across the resistor in Figure 33.15. **B.** Current in the circuit. **C.** A phasor diagram shows that I and V_R are in phase because their phasors lie on top of one another as they rotate at angular speed ω around the origin.

Phasor Diagrams

In our study of AC circuits, we compare the time dependence of the current to that of the voltage across various circuit elements. We can make this comparison visually by plotting current and voltage as functions of time as in parts A and B of Figure 33.16.

A **phasor diagram** is another visual tool for making such a comparison. In a phasor diagram, a quantity such as current or voltage is represented by a vector known as a phasor. The length of the phasor represents the quantity's maximum value. To represent the time dependence of the quantity, the phasor rotates around the origin at an angular speed ω. So, at any instant t, the phasor makes an angle of ωt with the horizontal axis. The instantaneous value of the quantity is projected onto either the horizontal axis or the vertical axis. If you choose to represent the time variation using the cosine function, the projection is onto the horizontal axis. In this textbook, we have chosen to use the sine function, so the phasor's projection onto the vertical axis gives the quantity's instantaneous value. Figure 33.16C is the phasor diagram representing the current and voltage across the resistor for the AC circuit with resistance (Fig. 33.15).

PHASOR DIAGRAMS ⦿ **Tool**

Although current and voltage may be represented by phasors, they are not vectors.

EXAMPLE 33.5 — Power in an AC Circuit with Resistance

The AC generator in Figure 33.15 supplies energy, and the resistor dissipates energy through heat. Find an expression for the (instantaneous) power P dissipated by the resistor as a function of time. Plot P versus t. Indicate the average power on the plot.

:• **INTERPRET and ANTICIPATE**

The current through the resistor and the voltage across the resistor oscillate with time. The power supplied to or consumed by any circuit element at any moment is given by $P = I\Delta V$ (Eq. 28.33), so multiply the expression for current by the expression for voltage. We expect to find that the power also oscillates.

:• **SOLVE**

Substitute $I(t) = I_{max} \sin \omega t$ and $V_R(t) = \mathcal{E}_{max} \sin \omega t$ (Eqs. 33.32 and 33.29) into Equation 28.33.

$$P(t) = I(t)V_R(t) \quad (28.33)$$
$$P(t) = (I_{max} \sin \omega t)(\mathcal{E}_{max} \sin \omega t)$$
$$P(t) = I_{max} \mathcal{E}_{max} \sin^2 \omega t$$

The factor $I_{max} \mathcal{E}_{max}$ is the maximum power, and it is convenient to write the expression in terms of P_{max}.

$$P(t) = P_{max} \sin^2 \omega t \quad (33.33)$$
where $P_{max} = I_{max} \mathcal{E}_{max}$

The graph of $P(t)$ versus t (Eq. 33.33) is always positive in Figure 33.17 because the sine squared is always positive.

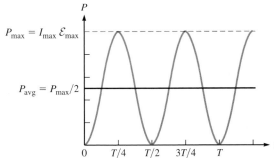

FIGURE 33.17 Power dissipated by the resistor in Figure 33.15.

The average power can be found from the rms current and rms voltage (Eq. 32.24). The rms voltage V_{rms} across the resistor equals the rms voltage $\mathcal{E}_{rms}$ across the power supply.

$$P_{avg} = I_{rms}V_{rms} \quad (32.24)$$
$$P_{avg} = I_{rms}\mathcal{E}_{rms} \quad (1)$$

Example continues on page 1064 ▶

Express the rms current and voltage in terms of their maximum values (Eqs. 32.21 and 32.20).	$I_{rms} = \dfrac{I_{max}}{\sqrt{2}}$ (32.21) $\mathcal{E}_{rms} = \dfrac{\mathcal{E}_{max}}{\sqrt{2}}$ (32.20)
Substitute these expressions for I_{rms} and V_{rms} into Equation (1).	$P_{avg} = \left(\dfrac{I_{max}}{\sqrt{2}}\right)\left(\dfrac{\mathcal{E}_{max}}{\sqrt{2}}\right) = \dfrac{I_{max}\mathcal{E}_{max}}{2}$
The numerator is the maximum power dissipated by the resistor, so the average power equals half the maximum power. A horizontal line on Figure 33.17 shows the average power dissipated by the resistor.	$P_{avg} = \dfrac{P_{max}}{2}$

CHECK and THINK

Figure 33.17 shows that the power oscillates as we expected. Suppose we were asked to find the power *supplied by the generator*. We would again use Equation 28.33 ($P = I\Delta V$). Because the voltage across the power supply equals the voltage across the resistor and the current through the power supply equals the current through the resistor, we would again find Equation 33.33 for the power supplied by the generator. Therefore, *all the power supplied by the generator is dissipated by the resistor.*

AC CIRCUIT WITH CAPACITANCE

▶ Special Case

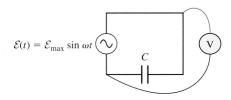

FIGURE 33.18 An AC generator and a capacitor. A voltmeter measures the voltage across the AC generator and across the capacitor. The voltage across the capacitor equals the voltage across the generator.

33-7 Special Case: AC Circuit with Capacitance

A purely capacitive AC circuit consists of an AC power source as before, but the resistor is replaced by a capacitor (Fig. 33.18). Again, imagine measuring the voltage across the power supply, which equals the voltage across the capacitor. As in the circuit described in Section 33-6,

$$V_C(t) = \mathcal{E}(t) = \mathcal{E}_{max} \sin \omega t \quad (33.34)$$

where V_C is the potential difference across the capacitor.

The voltage across a capacitor is proportional to the charge stored on its plates, $Q = CV_C$ (Eq. 27.1). So, the charge stored by the capacitor oscillates because V_C oscillates. To find an expression for $Q(t)$, substitute Equation 33.34 into Equation 27.1:

$$Q(t) = C\mathcal{E}_{max} \sin \omega t \quad (33.35)$$

We also need the current as a function of time, which we find by taking the time derivative of Equation 33.35:

$$I(t) = C\omega\mathcal{E}_{max} \cos \omega t \quad (33.36)$$

Equation 33.36 meets our goal of finding an expression for the current. In contrast to the purely resistive circuit in Section 33-6, the capacitive circuit has current and voltage that are *not* in phase: If the voltage is described by a sine function, the current is described by a cosine. In order to compare the phases, we should express both current and voltage in terms of either the sine function or the cosine function. We arbitrarily choose sine and, using the trigonometric identity $\cos \theta = \sin(\theta + 90°)$ from Appendix A, we rewrite Equation 33.36 as

$$I(t) = C\omega\mathcal{E}_{max} \sin(\omega t + 90°) \quad (33.37)$$

Angular frequency ω is in rad/s, so we often need to express the phase constant in radians: $90° = \pi/2$ rad.

The maximum current is the amplitude in front of the sine function:

$$I_{max} = C\omega\mathcal{E}_{max} \quad (33.38)$$

In Equation 33.37, the 90° is called the **phase constant**.

One more simplification is typically made. Compare Equation 33.38 to the current in the resistor circuit, $I_{max} = \mathcal{E}_{max}/R$ (Eq. 33.31). In both circuits, the maximum current is proportional to the generator's maximum emf $\mathcal{E}_{max}$. To make the equations look more similar and to understand the role of $C\omega$, a new quantity called the **capacitive reactance** X_C is defined as

$$X_C \equiv \frac{1}{\omega C} \qquad (33.39)$$

So, the current in the circuit shown in Figure 33.18 is given by

$$I(t) = \frac{\mathcal{E}_{max}}{X_C} \sin(\omega t + 90°) \qquad (33.40)$$

where the maximum current is given by

$$I_{max} = \frac{\mathcal{E}_{max}}{X_C} \qquad (33.41)$$

CAPACITIVE REACTANCE

⭐ **Major Concept**

Capacitive reactance has the SI units of ohms, and it is mathematically analogous to resistance. Conceptually, the reactance impedes the current. A large reactance means a small maximum current (Eq. 33.41).

Whereas resistance depends only on a particular resistor's geometry and composition, capacitive reactance depends on both the capacitor's capacitance *and the angular frequency of the generator*. When the angular frequency is low, the capacitive reactance is large (Fig. 33.19). So, at low angular frequencies, the current is strongly impeded and the maximum current is relatively low. This idea should make sense because if the AC generator were replaced by a DC generator, the angular frequency would be zero, and in this case, the capacitor prevents charge from flowing once it has been fully charged. When the angular frequency is high, the capacitive reactance is small. So, at high angular frequencies, the current is only weakly impeded and the maximum current is relatively high. As the angular frequency approaches infinity, the capacitive reactance approaches zero and the capacitor acts like an ideal wire.

Figure 33.20 shows the voltage across the capacitor and the current in the circuit as functions of time. The two sine functions are out of phase. The current reaches its peak a quarter of a cycle, or 90°, before the voltage reaches its peak. So we say *the current leads the voltage by 90°*, or *the voltage across a capacitor lags the current by 90°*.

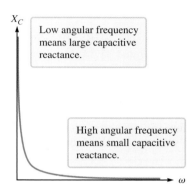

FIGURE 33.19 Capacitive reactance as a function of angular frequency.

CONCEPT EXERCISE 33.7

Read the dialogue and decide which student is correct. Give your reasons.

Shannon: You can only use the capacitive reactance to find the maximum current.

Cameron: No, you can use capacitive reactance exactly as you use resistance. So, to get the current you simply divide the voltage by X_C. It doesn't have to be the maximum current.

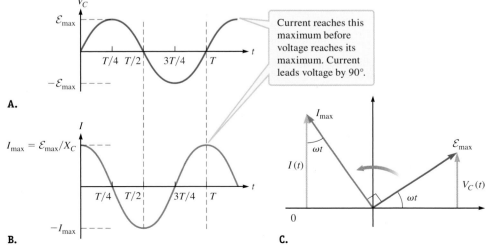

FIGURE 33.20 A. Voltage across the capacitor in Figure 33.18. **B.** Current in the circuit. **C.** Phasor diagram shows that I leads V_C by 90° as the two phasors rotate counterclockwise.

EXAMPLE 33.6 Power in an AC Circuit with Capacitance

The capacitor in Figure 33.18 stores energy in its electric field and releases that energy periodically. Find an expression for the (instantaneous) power P transferred to and from the capacitor as a function of time. Copy the graph of V_C as a function of time (Fig. 33.20A). Then plot P versus t, indicating the average power, and comment on the relationship between the two graphs.

:• INTERPRET and ANTICIPATE
The process for finding the power associated with this circuit is much like finding the power in the purely resistive circuit (Example 33.5). Again, we expect to find that the power transferred to and from the capacitor oscillates.

:• SOLVE

Substitute $I(t) = (\mathcal{E}_{max}/X_c)\sin(\omega t + 90°)$ and $V_C(t) = \mathcal{E}_{max}\sin\omega t$ (Eqs. 33.37 and 33.34) into Equation 28.33.	$P(t) = I(t)V_C(t)$ (28.33) $P(t) = \left[\dfrac{\mathcal{E}_{max}}{X_C}\sin(\omega t + 90°)\right](\mathcal{E}_{max}\sin\omega t)$ $P(t) = \dfrac{\mathcal{E}_{max}^2}{X_C}\sin(\omega t + 90°)\sin\omega t$
A few trigonometric identities help to simplify this expression. First, use $\sin(\omega t + 90°) = \cos\omega t$.	$P(t) = \dfrac{\mathcal{E}_{max}^2}{X_C}\cos\omega t \sin\omega t$
Second, use $\sin 2\theta = 2\sin\theta\cos\theta$ (Appendix A).	$P(t) = \dfrac{1}{2}\dfrac{\mathcal{E}_{max}^2}{X_C}\sin 2\omega t$
The factor in front of the sine function is the maximum power, and it is convenient to write the expression in terms of P_{max}.	$P(t) = P_{max}\sin 2\omega t$ (33.42) where $P_{max} = \dfrac{1}{2}\dfrac{\mathcal{E}_{max}^2}{X_C}$ (33.43)

The graph of $P(t)$ versus t (Fig. 33.21B) shows that the power can be either positive or negative. The power is positive when energy is being transferred *to* the capacitor and negative when energy is being transferred *from* the capacitor. The average power (over one cycle) is zero as shown in the figure.

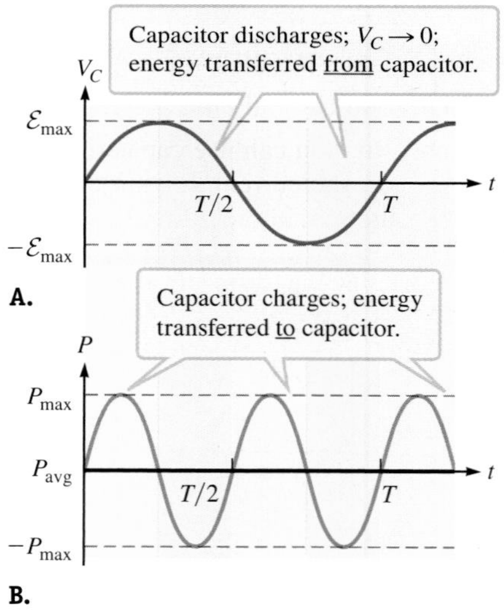

FIGURE 33.21 A. Voltage across the capacitor in Figure 33.18. **B.** Power transferred to and from the capacitor. In both parts, the yellow regions correspond to charging and the blue regions to discharging.

:• CHECK and THINK
Figure 33.21 shows that the power oscillates as we expected. When charge is building up on the capacitor, $|V_C|$ increases. The energy stored in the capacitor's electric field increases and power is transferred *to* the capacitor (positive). When charge is leaving the capacitor, the voltage between its plates approaches 0. The energy stored in the capacitor's electric field decreases and power is transferred *from* the capacitor (negative).

Special Case: RC Filters

The dependence of capacitive reactance on the generator's angular frequency has important practical applications, one of which is frequency filter circuits. A **filter circuit** produces negligible voltage for certain frequencies and nonnegligible voltage for other frequencies. To understand the basic idea of a filter, imagine the AC generator in Figure 33.18 is replaced by a power supply that produces an emf oscillating at many frequencies at once. For instance, the power might be supplied by a person singing into a microphone. The sound wave is made up of many frequencies (Chapter 18), so the resulting emf is made up of many frequencies. Many performers like to adjust their singing electronically by filtering out some frequencies that their voice produces. A filter circuit is inserted between the microphone and the speaker.

Let's use mathematics to see how a filter works. Because any real circuit has some resistance, consider a filter that consists of a resistor and a capacitor (Fig. 33.22A). We seek an expression for the maximum voltage across the capacitor as a function of angular frequency.

The current oscillates:

$$I(t) = I_{max} \sin \omega t \tag{33.44}$$

The voltage across the resistor is in phase with the current:

$$V_R(t) = RI_{max} \sin \omega t \tag{33.45}$$

The voltage across the capacitor lags the current by 90°:

$$V_C(t) = X_C I_{max} \sin\left(\omega t - \frac{\pi}{2}\right) \tag{33.46}$$

where we have written the phase constant in radians. The AC power supply's emf is not necessarily in phase with the current or with the voltage across the other circuit elements. We express its phase with an arbitrary constant φ:

$$\mathcal{E}(t) = \mathcal{E}_{max} \sin(\omega t + \varphi) \tag{33.47}$$

According to Kirchhoff's loop rule, at any instant t, the voltage across the AC power supply must equal the voltage across the resistor plus the voltage across the capacitor:

$$\mathcal{E}(t) = V_R(t) + V_C(t) \tag{33.48}$$

Substitute Equations 33.45, 33.46, and 33.47 into Equation 33.48:

$$\mathcal{E}_{max} \sin(\omega t + \varphi) = RI_{max} \sin \omega t + X_C I_{max} \sin\left(\omega t - \frac{\pi}{2}\right) \tag{33.49}$$

To solve for I_{max}, we must eliminate the phase constant φ. One way to do that is to evaluate Equation 33.49 at two special times. Let one such time be $t = 0$:

$$\mathcal{E}_{max} \sin(0 + \varphi) = RI_{max} \sin 0 + X_C I_{max} \sin\left(0 - \frac{\pi}{2}\right)$$

$$\mathcal{E}_{max} \sin \varphi = -X_C I_{max} \tag{33.50}$$

Let the other time be when $\omega t = \pi/2$:

$$\mathcal{E}_{max} \sin\left(\frac{\pi}{2} + \varphi\right) = RI_{max} \sin \frac{\pi}{2} + X_C I_{max} \sin(0)$$

$$\mathcal{E}_{max} \cos \varphi = RI_{max} \tag{33.51}$$

where we used the trigonometric identity $\sin(\pi/2 + \varphi) = \cos \varphi$. Now, square Equations 33.50 and 33.51 and add the results:

$$\mathcal{E}_{max}^2 \sin^2 \varphi + \mathcal{E}_{max}^2 \cos^2 \varphi = (-X_C I_{max})^2 + (RI_{max})^2$$

RC FILTERS ▶ Special Case

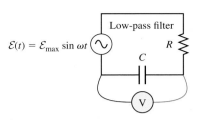

A.

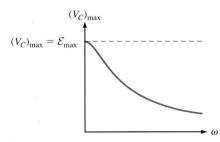

B.

FIGURE 33.22 A. A low-pass RC filter allows low frequencies to pass through the capacitor. **B.** Maximum voltage $(V_C)_{max}$ across the capacitor in an RC filter is frequency-dependent.

Use the trigonometric identity $\sin^2\varphi + \cos^2\varphi = 1$ to solve for I_{max}:

$$\mathcal{E}_{max}^2 = (X_C^2 + R^2)I_{max}^2$$

$$I_{max} = \frac{\mathcal{E}_{max}}{\sqrt{X_C^2 + R^2}} \quad (33.52)$$

Now that we have the maximum current, multiply by the capacitive reactance to find an expression for the maximum voltage across the capacitor:

$$(V_C)_{max} = X_C I_{max} = X_C \frac{\mathcal{E}_{max}}{\sqrt{X_C^2 + R^2}} \quad (33.53)$$

Finally, we substitute $X_C \equiv 1/\omega C$ (Eq. 33.39) for capacitive reactance so that our expression is in terms of ω:

$$(V_C)_{max} = \frac{1}{\omega C} \frac{\mathcal{E}_{max}}{\sqrt{(1/\omega^2 C^2) + R^2}} \quad (33.54)$$

Equation 33.54 meets our goal and is best understood from a graph of $(V_C)_{max}$ versus angular frequency ω (Fig. 33.22B). The voltage across the capacitor is very low at high frequencies. At low frequencies, the voltage across the capacitor nearly equals the voltage across the AC power supply. If the AC power supply is replaced by a microphone and a singer, the signal from the microphone has both high and low frequencies. Only the low-frequency signal produces a significant voltage across the capacitor. So, if the voltmeter in Figure 33.22A is replaced by a speaker, only the low-frequency signal passes to the speaker. If the singer makes some high squeaky notes, those are filtered out of the sound you hear. Because the low frequencies are passed through the filter, it is called a **low-pass filter**.

This same circuit can also be used as a **high-pass filter** (Fig. 33.23A), allowing the high-frequency signal to pass through. According to $\mathcal{E}(t) = V_R(t) + V_C(t)$ (Eq. 33.48), when the voltage across the capacitor is low (when the frequency is high), the voltage across the resistor must be high. The maximum voltage across the resistor is found by multiplying the maximum current by the resistance:

$$(V_R)_{max} = RI_{max} = R\frac{\mathcal{E}_{max}}{\sqrt{X_C^2 + R^2}} \quad (33.55)$$

Again we substitute $X_C \equiv 1/\omega C$ (Eq. 33.39) for capacitive reactance so that our expression is in terms of ω:

$$(V_R)_{max} = \frac{\mathcal{E}_{max} R}{\sqrt{(1/\omega^2 C^2) + R^2}} \quad (33.56)$$

Figure 33.23B shows a graph of $(V_R)_{max}$ versus angular frequency ω. The voltage across the resistor is very low at low frequencies, but at high frequencies, the voltage across the resistor nearly equals the voltage across the AC power supply. Again, imagine the AC power supply is replaced by a microphone and a person singing. Only the high-frequency signal produces a significant voltage across the resistor. So, if the voltmeter in Figure 33.23A is replaced by a speaker, only the high-frequency signal passes to the speaker. A high-pass filter forms the basic circuit for the *tweeter* in a stereo speaker.

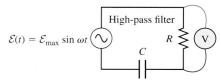

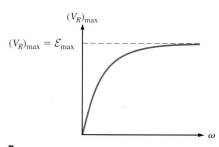

FIGURE 33.23 A. A high-pass RC filter allows high frequencies to pass through the capacitor. **B.** Maximum voltage $(V_R)_{max}$ across the resistor in an RC filter is frequency-dependent.

AC CIRCUIT WITH INDUCTANCE
▶ Special Case

33-8 Special Case: AC Circuit with Inductance

To form a purely inductive circuit, we replace the capacitor (Fig. 33.18) with an inductor (Fig. 33.24). The voltage across the power supply equals the voltage across the inductor:

$$V_L(t) = \mathcal{E}(t) = \mathcal{E}_{max} \sin \omega t \quad (33.57)$$

where V_L is the potential difference across the inductor. For all three circuits so far (purely resistive, capacitive, and inductive), the voltage across the circuit element is equal to the voltage across the power supply.

Consider an instant when the current in the circuit is clockwise so that the red lead of the voltmeter is on the upstream side of the inductor. We are measuring the voltage in the direction *opposite* the current, so according to the inductor rule (Eq. 33.8),

$$V_L(t) = V_{red} - V_{black} = L\frac{dI}{dt} \quad (33.58)$$

The minus sign is dropped because the voltmeter leads have been switched and $V_{red} - V_{black} = -(V_{black} - V_{red})$.

To find the current in the circuit, set Equation 33.57 equal to Equation 33.58:

$$L\frac{dI}{dt} = \mathcal{E}_{max} \sin \omega t$$

Isolate the current on one side of the equation:

$$dI = \frac{\mathcal{E}_{max}}{L}(\sin \omega t)dt$$

Then integrate:

$$\int dI = \int \frac{\mathcal{E}_{max}}{L}(\sin \omega t)dt$$

$$I = \frac{\mathcal{E}_{max}}{L}\int (\sin \omega t)dt$$

The integral of the sine function can be found in Appendix A:

$$I = -\frac{\mathcal{E}_{max}}{\omega L} \cos \omega t + \text{constant}$$

The constant of integration represents the current when the first term is zero. In other words, it is the DC current. Because there is no DC current in this circuit, we set the constant of integration to zero:

$$I = -\frac{\mathcal{E}_{max}}{\omega L} \cos \omega t$$

To compare the phase of the current to the phase of the voltage, we should express both current and voltage in terms of either the sine function or the cosine function. As in the case of the capacitor circuit, we have arbitrarily chosen the sine. Use a trigonometric identity (Appendix A) to rewrite the current as

$$I = \frac{\mathcal{E}_{max}}{\omega L} \sin(\omega t - 90°) \quad (33.59)$$

The maximum current is the amplitude in front of the sine function:

$$I_{max} = \frac{\mathcal{E}_{max}}{\omega L} \quad (33.60)$$

Analogous to the capacitive reactance X_C, the **inductive reactance** X_L is defined as

$$X_L \equiv \omega L \quad (33.61)$$

The current in the circuit shown in Figure 33.24 is thus

$$I(t) = \frac{\mathcal{E}_{max}}{X_L} \sin(\omega t - 90°) \quad (33.62)$$

where the maximum current is

$$I_{max} = \frac{\mathcal{E}_{max}}{X_L} \quad (33.63)$$

Inductive reactance has the SI units of ohms and is mathematically analogous to resistance and capacitive reactance. Like capacitive reactance, inductive reactance impedes the current; a large reactance means a small maximum current.

Also like capacitive reactance, inductive reactance depends on both the inductor's geometry *and the angular frequency of the generator*. Figure 33.25 shows a plot of the inductive reactance X_L versus the generator's angular frequency ω. Inductive reactance depends linearly on angular frequency, so when the angular frequency is

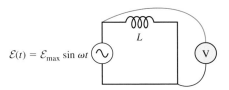

FIGURE 33.24 An AC generator and an inductor. A voltmeter measures the voltage across the AC generator and across the inductor. The voltage across the inductor equals the voltage across the generator.

INDUCTIVE REACTANCE

⊕ **Major Concept**

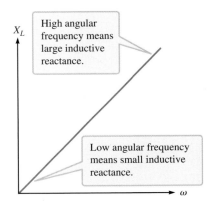

FIGURE 33.25 Inductive reactance as a function of angular frequency. Compare with Figure 33.19.

low, the inductive reactance is also low. Thus, at low angular frequencies, the current is only weakly impeded and the maximum current is large. This idea makes sense because if the AC generator were replaced by a DC generator, the angular frequency would be zero, and in that case, the inductor would not produce a back emf. Instead, the inductor would be modeled as an ideal wire, allowing current to flow freely. When the angular frequency is high, the inductive reactance is large; current is strongly impeded and the maximum current is low.

Figure 33.26 shows the voltage across the inductor and the current in the circuit as functions of time. The two sine functions are out of phase. The current reaches its peak a quarter of a cycle, or 90°, *after* the voltage reaches its peak. So, we say *the current lags the voltage by 90°*, or *the voltage across an inductor leads the current by 90°*.

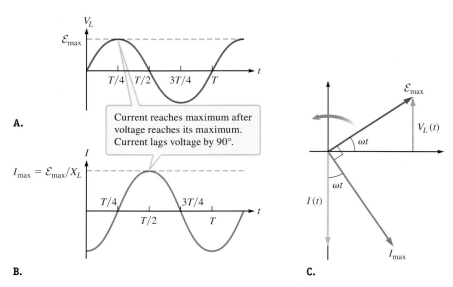

FIGURE 33.26 A. Voltage across the inductor in Figure 33.24. **B.** Current in the circuit. Compare with Figure 33.20. **C.** A phasor diagram shows that V_L leads I by 90°. Compare with Figure 33.20 for a capacitive circuit. One way to remember the difference between capacitive and inductive circuits is with the mnemonic CIVIL—in a **C**apacitive circuit, **I** (current) leads **V** (voltage); **V** leads **I** in an inductive (**L**) circuit.

Special Case: *RL* Filters

Because inductive reactance depends on the generator's angular frequency, an inductive circuit can be used as a filter. Figure 33.27A shows a basic *RL* filter consisting of a resistor, an inductor, and an AC power supply with an adjustable frequency. We can find an expression for the maximum voltage across the inductor as a function of the angular frequency by using a process much like the one we used in Section 33-7 for the *RC* filter. Start by finding an expression for I_{max} (Problem 46):

$$I_{max} = \frac{\mathcal{E}_{max}}{\sqrt{X_L^2 + R^2}} \qquad (33.64)$$

Multiply the maximum current by the inductive reactance to find an expression for the maximum voltage across the inductor:

$$(V_L)_{max} = X_L I_{max} = X_L \frac{\mathcal{E}_{max}}{\sqrt{X_L^2 + R^2}} \qquad (33.65)$$

Substitute $X_L \equiv \omega L$ (Eq. 33.61) for inductive reactance so that our expression is in terms of ω:

$$(V_L)_{max} = L\omega \frac{\mathcal{E}_{max}}{\sqrt{L^2\omega^2 + R^2}} \qquad (33.66)$$

RL FILTERS ⊙ Special Case

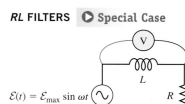

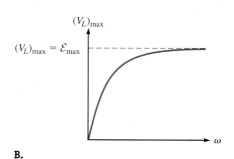

FIGURE 33.27 A. A high-pass *RL* filter allows high frequencies to pass through the inductor. **B.** Maximum voltage $(V_L)_{max}$ across the inductor of an *RL* filter is frequency-dependent. Compare with Figure 33.23 for a high-pass *RC* filter.

Figure 33.27B is a graph of the maximum voltage across the inductor (Eq. 33.66) as a function of ω. The voltage across the inductor is very low at low frequencies. But at high frequencies, the voltage across the inductor is high, nearly equal to the voltage across the AC power supply. Because the high frequency is passed through the inductor, the circuit in Figure 33.27A is a **high-pass filter**.

This same circuit can also be used as a **low-pass filter** (Fig. 33.28A), allowing the high-frequency signal to pass through. The maximum voltage across the resistor is found by multiplying the maximum current by the resistance:

$$(V_R)_{max} = R \frac{\mathcal{E}_{max}}{\sqrt{X_L^2 + R^2}} \qquad (33.67)$$

Substitute $X_L \equiv \omega L$ (Eq. 33.61) for inductive reactance:

$$(V_R)_{max} = \frac{\mathcal{E}_{max} R}{\sqrt{L^2\omega^2 + R^2}} \qquad (33.68)$$

Figure 33.28B is a graph of $(V_R)_{max}$ as a function of ω, showing that the voltage across the resistor is very low at high frequencies. At low frequencies, the voltage across the resistor nearly equals the voltage across the AC power supply.

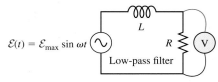

A.

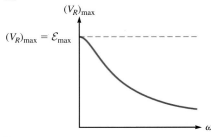

B.

FIGURE 33.28 A. A low-pass RL filter allows low frequencies to pass through the resistor. **B.** Maximum voltage $(V_R)_{max}$ across the resistor of an RL filter is frequency-dependent. Compare with Figure 33.22 for a low-pass RC filter.

EXAMPLE 33.7   CASE STUDY   A Stereo Speaker

In the 1960s, students listened to Jimi Hendrix play the guitar through stereo speakers at live concerts, broadcast on the radio, or from recordings. In any case, the input to the speakers corresponded to the notes Hendrix played on his guitar (made up of many frequencies). To help the listener hear the full range of frequencies, a good speaker is made up of two individual speakers. The smaller speaker—known as the **tweeter**—is good at producing high-frequency sound waves. The larger speaker—the **woofer**—is good at producing low-frequency sound waves. Ideally, a high-frequency voltage should be passed to the tweeter and a low-frequency voltage should be passed to the woofer. This is done with a high-pass RC filter and a low-pass RL filter (Fig. 33.29).

A Which circuit in Figures 33.22, 33.23, 33.27, and 33.28 corresponds to the tweeter circuit, and which corresponds to the woofer circuit? Which plot of maximum voltage versus angular frequency in these figures corresponds to the voltage across the tweeter? Which corresponds to the voltage across the woofer?

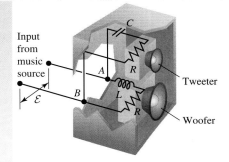

FIGURE 33.29 A stereo speaker, simplified.

INTERPRET and ANTICIPATE
Your job is to match the simple circuit diagrams in the figures listed to the drawing in Figure 33.29 and then identify the correct voltage-versus-angular frequency graphs.

SOLVE
Tweeter. The tweeter circuit in Figure 33.29 consists of a capacitor and a resistor as in Figures 33.22 and 33.23. The tweeter speaker is represented by a resistor. So, the tweeter circuit corresponds to Figure 33.23A, which has the voltmeter across the resistor (a high-pass filter).

Because the tweeter is a high-pass RC circuit and its voltage corresponds to the voltage across the resistor, the graph of $(V_R)_{max}$ versus ω in Figure 33.23B is the one we want. As this graph shows, the voltage across the tweeter is high at high frequencies.

Woofer. Like the circuits in Figures 33.27 and 33.28, the woofer circuit consists of an inductor and a resistor. The woofer speaker is represented by the resistor. So, the woofer circuit corresponds to Figure 33.28A, which has the voltmeter across the resistor.

The voltage across the resistor $(V_R)_{max}$ shown in Figure 33.28B corresponds to the voltage across the woofer. As this graph shows, the voltage across the woofer is high at low frequencies.

B Assume the tweeter's resistance equals the woofer's resistance. The **angular crossover frequency** is the angular frequency at which the maximum voltage across the tweeter equals the maximum voltage across the woofer. (This is sometimes defined in terms of the current through each element, but if the resistance is the same in each, the definitions are equivalent.) Roughly speaking, the tweeter produces sounds higher than the crossover frequency and the woofer produces sounds lower than the crossover frequency. Find an expression for the crossover frequency in terms of L and C. In the **CHECK and THINK** step answer this: How can you change the crossover frequency so that the tweeter picks up lower frequencies?

:• INTERPRET and ANTICIPATE
In part A, we identified the graphs of maximum voltage across the tweeter and the woofer. We can plot these curves on the same axes (Fig. 33.30). To find an expression for the crossover frequency, set the maximum voltage across the resistor in the high-pass RC filter equal to the voltage across the resistor in the low-pass RL filter.

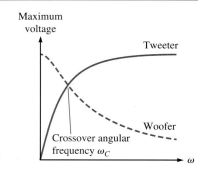

FIGURE 33.30

:• SOLVE
Set $(V_R)_{max}$ from Equation 33.56 equal to $(V_R)_{max}$ from Equation 33.68, and solve for the angular crossover frequency ω_c.

$$\frac{\mathcal{E}_{max} R}{\sqrt{(1/\omega_c^2 C^2) + R^2}} = \frac{\mathcal{E}_{max} R}{\sqrt{L^2 \omega_c^2 + R^2}}$$

$$(1/\omega_c^2 C^2) + R^2 = L^2 \omega_c^2 + R^2$$

$$1/\omega_c^2 C^2 = L^2 \omega_c^2 \qquad \omega_c^4 = \frac{1}{L^2 C^2}$$

$$\omega_c = \frac{1}{\sqrt{LC}}$$

:• CHECK and THINK
The crossover frequency depends on the inductance and the capacitance. Sometimes, when you listen to music through speakers, it sounds distorted. If you want the tweeter to pick up lower frequencies, you need the crossover frequency to decrease, so you should increase the capacitance. (Problem 54 asks you to think about what happens to the voltage across the woofer as a result.)

RLC (AC) CIRCUIT ⭕ Special Case

33-9 Special Case: AC Circuit with Resistance, Inductance, and Capacitance

Our last circuit (Fig. 33.31) consists of an AC power supply, an inductor, a resistor, and a capacitor and is usually referred to as an RLC circuit. RLC circuits play an important role in communications equipment such as radios, televisions, and telephones.

Current and Voltage in an RLC Circuit

We seek expressions for the current and the AC power supply's emf. The current varies sinusoidally:

$$I(t) = I_{max} \sin \omega t \qquad (33.69)$$

We want to find I_{max} in terms of $\mathcal{E}_{max}$, R, X_C, and X_L. Equation 33.69 for $I(t)$ is the same expression we had in the resistor circuit, but it is different from the expression we had for the capacitor and inductor circuits. Figure 33.32A is a graph of Equation 33.69.

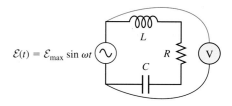

FIGURE 33.31 An AC generator, a resistor, an inductor, and a capacitor. A voltmeter measures the voltage across the AC generator and across the three other circuit elements. The voltage across the generator equals the sum of the voltages across the three other circuit elements.

To find I_{max}, we need the circuit voltages. As for the three circuits we have presented earlier, imagine measuring the voltage across the AC generator. In this case, the voltage across the AC generator is the sum of the voltages across the three other circuit elements:

$$\mathcal{E}(t) = V_L(t) + V_R(t) + V_C(t) \quad (33.70)$$

We need expressions for each of these terms. From our earlier work, we know how each voltage is related to the current. First, the voltage across the resistor is in phase with the current (Fig. 33.32B), so we can write an expression for V_R:

$$V_R(t) = RI_{max} \sin \omega t \quad (33.71)$$

where the maximum voltage across the resistor is RI_{max}. Second, the voltage across the capacitor lags the current by 90° (or $\pi/2$), so we can write an expression for V_C:

$$V_C(t) = X_C I_{max} \sin\left(\omega t - \frac{\pi}{2}\right) = -X_C I_{max} \sin\left(\omega t + \frac{\pi}{2}\right) \quad (33.72)$$

where we have used $\sin(\alpha + \pi) = -\sin \alpha$ and the maximum voltage across the capacitor is $X_C I_{max}$. (Eq. 33.72 is plotted in Fig. 33.32C.) Third, the voltage across the inductor leads the current by 90° (or $\pi/2$), so we can write an expression for V_L:

$$V_L(t) = X_L I_{max} \sin\left(\omega t + \frac{\pi}{2}\right) \quad (33.73)$$

where the maximum voltage across the capacitor is $X_L I_{max}$. (Eq. 33.73 is plotted in Fig. 33.32D.)

Our next step is to add the voltages (Eqs. 33.71, 33.72, and 33.73) across the three circuit elements to find the voltage across the AC generator:

$$\mathcal{E}(t) = RI_{max} \sin \omega t - X_C I_{max} \sin\left(\omega t + \frac{\pi}{2}\right) + X_L I_{max} \sin\left(\omega t + \frac{\pi}{2}\right)$$

Combine like terms:

$$\mathcal{E}(t) = RI_{max} \sin \omega t + (X_L - X_C) I_{max} \sin\left(\omega t + \frac{\pi}{2}\right) \quad (33.74)$$

We need an expression for $\mathcal{E}(t)$, the AC generator's emf, in terms of $\mathcal{E}_{max}$. Figure 33.33 helps organize the information graphically. Figure 33.33A is a graph of

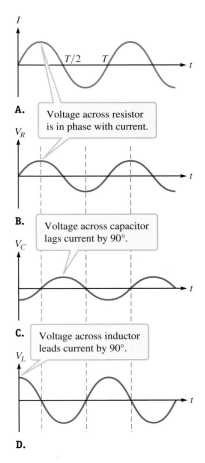

FIGURE 33.32 A. Current in an *RLC* circuit (Fig. 33.31). **B.** Voltage across the resistor. **C.** Voltage across the capacitor. **D.** Voltage across the inductor.

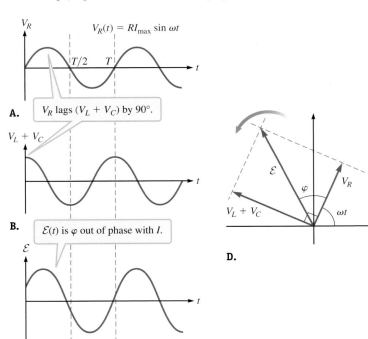

FIGURE 33.33 A. Voltage across the resistor from Figure 33.32B. **B.** Voltage across the inductor plus voltage across the capacitor: $V_L + V_C = (X_L - X_C)I_{max} \sin(\omega t + 90°)$. **C.** Voltage across the power supply. **D.** The corresponding phasor diagram. The current I (not shown) is in phase with V_R and so lies on its phasor. The voltage across the power supply is the sum of the three other voltages: $\mathcal{E}(t) = V_L(t) + V_R(t) + V_C(t)$.

$V_R(t)$, and part B is a graph of $V_L(t) + V_C(t)$. These graphs show that the voltage across the resistor lags the sum $V_L(t) + V_C(t)$ by 90°. When we add these values to find $\mathcal{E}(t)$, the result is out of phase with the current by some amount φ. To see this, compare Figure 33.33C for $\mathcal{E}(t)$ with the graph for the current in Figure 33.32A. We express the AC generator's emf as

$$\mathcal{E}(t) = \mathcal{E}_{max} \sin(\omega t + \varphi) \tag{33.75}$$

Substitute Equation 33.75 into Equation 33.74:

$$\mathcal{E}_{max} \sin(\omega t + \varphi) = RI_{max} \sin \omega t + (X_L - X_C)I_{max} \sin\left(\omega t + \frac{\pi}{2}\right) \tag{33.76}$$

To find I_{max}, evaluate Equation 33.76 at two different times in order to eliminate φ. First, at $t = 0$:

$$\mathcal{E}_{max} \sin \varphi = (X_L - X_C)I_{max} \tag{33.77}$$

Then, when $\omega t = \pi/2$:

$$\mathcal{E}_{max} \sin\left(\frac{\pi}{2} + \varphi\right) = RI_{max} \sin \frac{\pi}{2} + (X_L - X_C)I_{max} \sin\left(\frac{\pi}{2} + \frac{\pi}{2}\right)$$

$$\mathcal{E}_{max} \cos \varphi = RI_{max} + (X_L - X_C)I_{max} \sin \pi$$

$$\mathcal{E}_{max} \cos \varphi = RI_{max} \tag{33.78}$$

There are two unknowns (I_{max} and φ) and two equations (33.77 and 33.78). To solve for I_{max}, square these equations and add them:

$$(\mathcal{E}_{max} \sin \varphi)^2 + (\mathcal{E}_{max} \cos \varphi)^2 = [(X_L - X_C)I_{max}]^2 + (RI_{max})^2$$

$$\mathcal{E}_{max}^2 = [(X_L - X_C)^2 + R^2]I_{max}^2$$

$$I_{max} = \frac{\mathcal{E}_{max}}{\sqrt{(X_L - X_C)^2 + R^2}} \tag{33.79}$$

To find φ, divide Equation 33.77 by Equation 33.78:

$$\frac{\mathcal{E}_{max} \sin \varphi}{\mathcal{E}_{max} \cos \varphi} = \frac{(X_L - X_C)I_{max}}{RI_{max}}$$

$$\frac{\sin \varphi}{\cos \varphi} = \tan \varphi = \frac{(X_L - X_C)}{R}$$

$$\varphi = \tan^{-1}\left(\frac{X_L - X_C}{R}\right) \tag{33.80}$$

Our goal of finding an expression for the current is met by $I(t) = I_{max} \sin \omega t$ (Eq. 33.69) along with Equation 33.79. Likewise, $\mathcal{E}(t) = \mathcal{E}_{max} \sin(\omega t + \varphi)$ (Eq. 33.75) along with Equation 33.80 meets the second goal of finding the AC generator's emf.

Impedance

The denominator in Equation 33.79 is defined to be the **impedance Z**:

IMPEDANCE ✪ Major Concept

$$Z \equiv \sqrt{(X_L - X_C)^2 + R^2} \tag{33.81}$$

So, the maximum current is

$$I_{max} = \frac{\mathcal{E}_{max}}{Z} \tag{33.82}$$

The term *impedance* is descriptive. If a circuit has a large impedance, the maximum current is small. Like resistance, impedance has SI units of ohms. The impedance depends on the inductive reactance, the capacitive reactance, and the resistance, so *the impedance also depends on the AC generator's angular frequency*. Using the expressions for reactance $X_C \equiv 1/\omega C$ and $X_L \equiv \omega L$ (Eqs. 33.39 and 33.61), we can write impedance explicitly in terms of ω:

$$Z = \sqrt{\left(\omega L - \frac{1}{\omega C}\right)^2 + R^2} \qquad (33.83)$$

The inductance L, capacitance C, and resistance R depend on the geometry (size, shape, and composition) of each circuit element.

Resonance in an *RLC* Circuit

The dependence of impedance on angular frequency has important consequences. Consider a mechanical analog: a pendulum swinging back and forth. If there were no air resistance or friction, the pendulum would swing back and forth forever, repeatedly transforming gravitational potential energy into kinetic energy, and vice versa. This is analogous to the *LC* circuit (Fig. 33.12). Energy is repeatedly transferred from the electric field to the magnetic field, and vice versa, forever, in the absence of resistance. Of course, friction and air resistance cannot be completely eliminated, so any real pendulum loses mechanical energy and eventually stops. Friction and air resistance are analogous to resistance in a circuit. Because resistance can never be eliminated, the *LC* circuit loses energy and eventually the current stops. If you want to keep a pendulum from stopping, you can drive the pendulum by pushing it periodically. From Section 16-11, the pendulum responds best (that is, with maximum amplitude) if you drive it at its natural frequency, known as the **resonance frequency**. The same is true for a circuit. If you want to keep the current in an *RLC* circuit from fading to 0, you can use an AC generator to drive the current. The current will display the greatest response if the AC generator's driving frequency is set to the circuit's resonance frequency.

The maximum current I_{max} depends on the emf's driving frequency (according to Eqs. 33.82 and 33.83). The highest value of I_{max} occurs when the driving frequency equals the resonance angular frequency ω_0, the value of ω that minimizes the impedance Z. The minimum value of the impedance is $Z_{min} = R$. When $Z = R$, the first two terms in Equation 33.83 cancel:

$$\omega_0 L = \frac{1}{\omega_0 C}$$

Solve for ω_0:

$$\omega_0 = \frac{1}{\sqrt{LC}} \qquad (33.84)$$

Equation 33.84 is the resonance or natural angular frequency of the circuit. When the AC power supply's driving frequency equals the resonance frequency, we say the circuit is in **resonance** and the maximum value of the current is

$$I_{max}(\omega_0) = \frac{\mathcal{E}_{max}}{Z_{min}} = \frac{\mathcal{E}_{max}}{R} \qquad (33.85)$$

Figure 33.34 is a graph of I_{max} as a function of ω (Eqs. 33.82 and 33.83). The peak of the graph occurs at the resonance frequency ω_0. Imagine the AC generator in Figure 33.31 has a knob that allows you to adjust the angular frequency (the rate at which the coil in the generator rotates). As you slowly turn the knob up from $\omega = 0$, the maximum current in the circuit increases. When $\omega = \omega_0$, the maximum current has reached its highest value. As you continue to increase ω, the maximum current drops in value.

RESONANCE ⊕ **Major Concept**

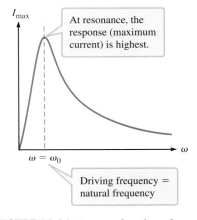

At resonance, the response (maximum current) is highest.

Driving frequency = natural frequency

FIGURE 33.34 I_{max} as a function of angular frequency ω for an *RLC* circuit.

EXAMPLE 33.8 Power Factor

An ideal *LC* circuit does not dissipate energy. Of course, any real circuit has some resistance, so energy is actually dissipated. Suppose you have an *RLC* circuit with an AC power supply that adds energy at the same rate that the resistor dissipates energy. Show that the average power dissipated by the resistor is

$$P_{avg} = I_{rms}\mathcal{E}_{rms}\cos\varphi \qquad (33.86)$$

(Problem 58 asks you to show that this is equal to the average power supplied.)

INTERPRET and ANTICIPATE
Start with a general expression for the average power dissipated by a resistor, and then make the appropriate substitutions for the specific case of the RLC circuit in Figure 33.31.

SOLVE

Start with Equation 32.25 for the average power dissipated by a resistor.	$P_{avg} = I_{rms}^2 R$	(32.25)
Solve Equation 33.78 for R.	$\mathcal{E}_{max} \cos \varphi = R I_{max}$ $R = \dfrac{\mathcal{E}_{max}}{I_{max}} \cos \varphi$	(33.78)
Use $\mathcal{E}_{rms} = \mathcal{E}_{max}/\sqrt{2}$ and $I_{rms} = I_{max}/\sqrt{2}$ (Eqs. 32.20 and 32.21) to write the maximum emf and maximum current in terms of their rms values.	$R = \dfrac{\sqrt{2}\mathcal{E}_{rms}}{\sqrt{2}I_{rms}} \cos \varphi$ $R = \dfrac{\mathcal{E}_{rms}}{I_{rms}} \cos \varphi$	(1)
Substitute Equation (1) into Equation 33.25.	$P_{avg} = I_{rms}^2 \left(\dfrac{\mathcal{E}_{rms}}{I_{rms}} \cos \varphi \right)$ $P_{avg} = I_{rms}\mathcal{E}_{rms} \cos \varphi$ ✓	(33.86)

CHECK and THINK
This is what we set out to show. The factor $\cos \varphi$ is called the **power factor**; it is needed because the current in an RLC circuit is *not* in phase with the AC power supply's emf. Equation 33.86 shows that the average power dissipated (and supplied) depends on the phase.

Let's look at three special cases: (1) If the inductance and the capacitance are zero (if the inductor and the capacitor are replaced by wires), the circuit in Figure 33.31 consists of only a power supply and a resistor and the phase is zero. When $\varphi = 0$, the average power dissipated is at its maximum: $P_{avg} = I_{rms}\mathcal{E}_{rms}$. (2) If the RLC circuit is in resonance, $X_L = X_C$ and again the phase is zero. The average power dissipated (and supplied) is $P_{avg} = I_{rms}\mathcal{E}_{rms}$. (3) If the resistance is zero (an ideal situation), $\varphi = 90°$ and no power is dissipated: $P_{avg} = I_{rms}\mathcal{E}_{rms} \cos 90° = 0$.

EXAMPLE 33.9 CASE STUDY Broadcast Radio

If you lived in the San Francisco area in the 1960s, you may have found a Hendrix song playing on 104.5 FM (KFOG). You tuned your radio to 104.5 MHz. A radio receiver is basically an RLC circuit except that the AC generator in Figure 33.31 is replaced by the signal picked up by the antenna. Of course, many radio stations are sending out signals at many different frequencies. Ideally, you want your radio to be sensitive to only one signal at a time. When you tune your radio to a certain station, you are adjusting the RLC circuit's resonance frequency so that it equals the broadcast frequency of that particular station. The resonance frequency depends on the inductance and the capacitance (Eq. 33.84), so altering either one allows you to tune your radio. In old radios and in radios built at home, typically the inductance is fixed and the capacitance is varied. Figure P27.36 (page 859) shows a variable capacitor consisting of several parallel plates that can be rotated by turning a knob. The position of the plates determines the geometry of the capacitor and its capacitance. Assume a simple radio can be modeled as an RLC circuit with an inductance of 1.64×10^{-7} H, a resistance of $0.0346 \, \Omega$, and a variable capacitor.

A What capacitance is needed so that your radio is tuned to 104.5 MHz (KFOG)?

33-9 Special Case: AC Circuit with Resistance, Inductance, and Capacitance

:• **INTERPRET and ANTICIPATE**
Your radio receiver's resonance frequency f_0 must be set equal to the broadcast frequency f of the radio station. All the equations we have worked with so far involve angular frequency ω, but the radio station's broadcast frequency f is given, so we must convert as needed.

:• **SOLVE**
Find an expression for the resonance frequency f_0 from $\omega_0 = 1/\sqrt{LC}$ (Eq. 33.84) and $\omega = 2\pi f$ (Eq. 16.2).

$$2\pi f_0 = \omega_0 = \frac{1}{\sqrt{LC}}$$

Solve for capacitance.

$$4\pi^2 f_0^2 = \frac{1}{LC}$$

$$C = \frac{1}{4\pi^2 f_0^2 L} = \frac{1}{4\pi^2 (104.5 \times 10^6 \,\text{Hz})^2 (1.64 \times 10^{-7}\,\text{H})}$$

$$C = 1.41 \times 10^{-11}\,\text{F} = 14.1\,\text{pF}$$

:• **CHECK and THINK**
This is the capacitance when you are listening to KFOG. When you want to listen to another station, you tune that station in by changing the capacitance.

B If the radio signal produces a maximum emf of 17.5 mV, what is the maximum current in the circuit?

:• **INTERPRET and ANTICIPATE**
The radio is tuned so that it is in resonance with the broadcaster's signal. You can model this as an RLC circuit in which the angular frequency of the AC power supply equals the resonance angular frequency: $\omega = \omega_0$. In that case, the capacitive reactance equals the inductive reactance, $X_C = X_L$, and the impedance equals the resistance, $Z = R$. Then the maximum current depends only on the resistance and the power supply's maximum emf.

:• **SOLVE**
When an RLC circuit is in resonance, its maximum current is given by Equation 33.85.

$$I_{\max}(\omega_0) = \frac{\mathcal{E}_{\max}}{R} \tag{33.85}$$

$$I_{\max}(\omega_0) = \frac{17.5 \times 10^{-3}\,\text{V}}{0.0346\,\Omega} = 5.06 \times 10^{-1}\,\text{A}$$

$$I_{\max}(\omega_0) = 506\,\text{mA}$$

:• **CHECK and THINK**
As shown in Figure 33.34, the maximum current is largest when the circuit is in resonance. If you tune your radio to another station, you change the resonance frequency of the circuit. KFOG still broadcasts its signal, but the current it produces in your radio receiver is much smaller. As long as that current is small compared to the current produced when you are tuned to the station you want to listen to, you won't get interference.

C There is another radio station in the San Francisco area at 104.9 FM (KCNL). Suppose this station produces a stronger signal so that the maximum emf in your radio is 35.0 mV (two times stronger than the station at KFOG). When your radio is tuned to KFOG (104.5 MHz), what is the maximum current produced by the KCNL signal? Do you expect KCNL to cause much interference?

:• **INTERPRET and ANTICIPATE**
Your radio is tuned so it is in resonance with the KFOG signal, not the KCNL signal. Because the radio is not in resonance with KCNL, the impedance depends on the resistance, the capacitive reactance, and the inductive reactance. The maximum current then depends on the maximum emf and the impedance.

Example continues on page 1078 ▶

SOLVE

Find KCNL's broadcast angular frequency.

$$\omega = 2\pi f = 2\pi(104.9 \times 10^6 \text{ Hz}) = 6.591 \times 10^8 \text{ Hz}$$

The capacitive reactance is given by Equation 33.39. Use the capacitance from part A (because the radio is still tuned to KFOG), but use the angular frequency of KCNL's broadcast. We retain an extra significant figure to avoid rounding error.	$X_C = \dfrac{1}{\omega C}$ (33.39) $X_C = \dfrac{1}{(6.591 \times 10^8 \text{ Hz})(1.41 \times 10^{-11} \text{ F})} = 107.6\, \Omega$
The inductive reactance is given by Equation 33.61. The inductance is constant for this circuit. Again, use the angular frequency of KCNL's broadcast.	$X_L = \omega L$ (33.61) $X_L = (6.591 \times 10^8 \text{ Hz})(1.64 \times 10^{-7} \text{ H}) = 108.1\, \Omega$
Because the circuit is not in resonance with KCNL, we must use Equation 33.81 to find the circuit's impedance for KCNL's frequency.	$Z = \sqrt{(X_L - X_C)^2 + R^2}$ (33.81) $Z = \sqrt{(108.1\,\Omega - 107.6\,\Omega)^2 + (0.0346\,\Omega)^2} = 0.5\,\Omega$
The maximum current for KCNL's broadcast is given by $I_{max} = \mathcal{E}_{max}/Z$ (Eq. 33.82).	$I_{max} = \dfrac{35.0 \times 10^{-3}\text{ V}}{0.5\,\Omega} = 70 \times 10^{-3}\text{ A} = 70\text{ mA}$

CHECK and THINK

KCNL's maximum emf is two times greater than KFOG's, but when your radio is tuned to KFOG (104.5 MHz), the current produced by KCNL (104.9 MHz) is about seven times smaller than the current produced by KFOG. You are not likely to hear any interference due to KCNL. You could have enjoyed listening to Hendrix on KFOG without worry.

SUMMARY

Underlying Principles

There are no new principles in this chapter.

Major Concepts

1. An **inductor** is a circuit element designed to store energy in a magnetic field.
2. **Inductance** L (sometimes called **self-inductance**) is the constant of proportionality in

   $$\Phi_{tot} = LI \quad (33.1)$$

   The total flux Φ_{tot} through the inductor is the result of the current I in the wire of the same (self) inductor. The SI unit for inductance is the **henry**:

   $$1\text{ H} = 1\dfrac{\text{Wb}}{\text{A}} = 1\dfrac{\text{T} \cdot \text{m}^2}{\text{A}} \quad (33.2)$$

3. **Energy stored by an inductor:** $U_B = \tfrac{1}{2}LI^2$ (33.3)
   Energy density stored by a magnetic field:

   $$u_B = \dfrac{U_B}{V} = \dfrac{1}{2}\dfrac{B^2}{\mu_0} \quad (33.17)$$

4. **Capacitive reactance:** $X_C = 1/\omega C$ (33.39)
5. **Inductive reactance:** $X_L = \omega L$ (33.61)
6. **Impedance:** $Z \equiv \sqrt{(X_L - X_C)^2 + R^2}$ (33.81)
7. **Resonance** occurs when the AC power supply's driving frequency equals the resonance or natural (angular) frequency. The resonance angular frequency is given by

   $$\omega_0 = \dfrac{1}{\sqrt{LC}} \quad (33.84)$$

   At resonance, the maximum current is given by

   $$I_{max}(\omega_0) = \dfrac{\mathcal{E}_{max}}{Z_{min}} = \dfrac{\mathcal{E}_{max}}{R} \quad (33.85)$$

Special Cases

1. **Inductance of an ideal solenoid:**

$$L = \mu_0 n^2 \ell A = \frac{\mu_0 N^2}{\ell} A \qquad (33.5)$$

2. **Inductor rule:** $\Delta V_L = V_{red} - V_{black} = -L\frac{dI}{dt}$ (33.8)

   where the red voltmeter lead is downstream and the black lead is upstream so that the measurement is made in the direction of the current.

3. An **RL circuit** consists of an inductor, a resistor, and a DC power supply that can be switched into or out of the circuit (Fig. 33.9). When the DC power supply is in the circuit, the current is

$$I(t) = \frac{\mathcal{E}}{R}(1 - e^{-t/\tau}) \qquad (33.12)$$

   When the DC power supply is switched out of the circuit, the current decays:

$$I(t) = I_{max} e^{-t/\tau} \qquad (33.15)$$

   where τ is the inductive time constant:

$$\tau \equiv \frac{L}{R} \qquad (33.13)$$

4. An **LC circuit** is an ideal circuit that consists of an inductor and a capacitor (no resistance). Energy is transferred back and forth between the magnetic field of the inductor and the electric field of the capacitor (Fig. 33.12). The electric field energy is

$$U_E(t) = \frac{Q_{max}^2}{2C} \cos^2(\omega t + \varphi) \qquad (33.26)$$

   The magnetic field energy is

$$U_B(t) = \frac{Q_{max}^2}{2C} \sin^2(\omega t + \varphi) \qquad (33.27)$$

   The current in the circuit and the charge on the capacitor both oscillate:

$$I(t) = -I_{max} \sin(\omega t + \varphi) \qquad (33.22)$$

$$Q(t) = Q_{max} \cos(\omega t + \varphi) \qquad (33.20)$$

   where the angular frequency is

$$\omega = \sqrt{\frac{1}{LC}} \qquad (33.21)$$

5. An **AC circuit with resistance** consists of an AC power supply and a resistor (Fig. 33.15). The voltage across the generator equals the voltage across the resistor:

$$V_R(t) = \mathcal{E}(t) = \mathcal{E}_{max} \sin \omega t \qquad (33.29)$$

   The current in the circuit and the voltage across the resistor are **in phase**:

$$I(t) = I_{max} \sin \omega t \qquad (33.32)$$

   where the maximum current is

$$I_{max} = \frac{\mathcal{E}_{max}}{R} \qquad (33.31)$$

6. An **AC circuit with capacitance** consists of an AC power supply and a capacitor (Fig. 33.18). The voltage across the power supply equals the voltage across the capacitor:

$$V_C(t) = \mathcal{E}(t) = \mathcal{E}_{max} \sin \omega t \qquad (33.34)$$

   The current in the circuit is

$$I(t) = \frac{\mathcal{E}_{max}}{X_C} \sin(\omega t + 90°) \qquad (33.40)$$

   where the maximum current is

$$I_{max} = \frac{\mathcal{E}_{max}}{X_C} \qquad (33.41)$$

   The current leads the voltage by 90°, or the voltage across a capacitor lags the current by 90°.

7. An **AC circuit with inductance** consists of an AC power supply and an inductor (Fig. 33.24). The voltage across the power supply equals the voltage across the inductor:

$$V_L(t) = \mathcal{E}(t) = \mathcal{E}_{max} \sin \omega t \qquad (33.57)$$

   The current in the circuit is

$$I(t) = \frac{\mathcal{E}_{max}}{X_L} \sin(\omega t - 90°) \qquad (33.62)$$

   where the maximum current is

$$I_{max} = \frac{\mathcal{E}_{max}}{X_L} \qquad (33.63)$$

   The current lags the voltage by 90°, or the voltage across an inductor leads the current by 90°.

8. a. An **RC filter circuit** consists of a resistor and a capacitor (Fig. 33.22A). Low frequencies are passed through the capacitor (low-pass filter):

$$(V_C)_{max} = X_C \frac{\mathcal{E}_{max}}{\sqrt{X_C^2 + R^2}} \qquad (33.53)$$

   High frequencies are passed through the resistor (high-pass filter):

$$(V_R)_{max} = R \frac{\mathcal{E}_{max}}{\sqrt{X_C^2 + R^2}} \qquad (33.55)$$

   b. An **RL filter circuit** consists of a resistor and an inductor (Fig. 33.27A). High frequencies are passed through the inductor (high-pass filter):

$$(V_L)_{max} = X_L \frac{\mathcal{E}_{max}}{\sqrt{X_L^2 + R^2}} \qquad (33.65)$$

   Low frequencies are passed through the resistor (low-pass filter):

$$(V_R)_{max} = R \frac{\mathcal{E}_{max}}{\sqrt{X_L^2 + R^2}} \qquad (33.67)$$

9. An **RLC circuit** consists of an inductor, a resistor, a capacitor, and an AC power supply (Fig. 33.31). The current is not in phase with the AC power supply's emf. If the current is given by

$$I(t) = I_{max} \sin \omega t \qquad (33.69)$$

the AC generator's emf is
$$\mathcal{E}(t) = \mathcal{E}_{max} \sin(\omega t + \varphi) \quad (33.75)$$
where φ is the phase constant:
$$\varphi = \tan^{-1}\left(\frac{X_L - X_C}{R}\right) \quad (33.80)$$
The maximum current is
$$I_{max} = \frac{\mathcal{E}_{max}}{Z} \quad (33.82)$$

The **resonance angular frequency** is
$$\omega_0 = \frac{1}{\sqrt{LC}} \quad (33.84)$$
When the AC power supply's driving frequency equals the resonance frequency, we say the circuit is in **resonance**, and the maximum value of the current is
$$I_{max}(\omega_0) = \frac{\mathcal{E}_{max}}{Z_{min}} = \frac{\mathcal{E}_{max}}{R} \quad (33.85)$$

⦿ Tool

In a **phasor diagram**, a quantity such as current or voltage is represented by a vector known as a phasor. (For examples, see Figures 33.16C, 33.20C, and 33.26C.) The length of the phasor represents the quantity's maximum value. To represent the time dependence of the quantity, the phasor rotates around the origin at an angular speed ω. At any instant t, the phasor makes an angle ωt with the horizontal axis. In this textbook, we have chosen to use the sine function for AC currents and voltages, so the phasor's projection onto the vertical axis gives the quantity's instantaneous value.

PROBLEMS AND QUESTIONS

A = algebraic   C = conceptual   E = estimation   G = graphical   N = numerical

33-1 Inductors and Inductance

1. **C** Two solenoids have the same number of turns per unit length, but one is short and wide and the other is long and narrow. **a.** Which inductor has the greater inductance per unit length L/ℓ? Explain. **b.** How is it possible for the two to have the same inductance L?
2. **E** Estimate the inductance of a typical (metal) Slinky.
3. **C** A solenoid is constructed by wrapping a wire of a fixed length d to form N loops of radius R such that the total length of the solenoid created is ℓ. A battery is connected to drive a current I through the solenoid. Which of the following would increase the inductance of the solenoid: **a.** squeezing the solenoid so that the loops have the same radius but are closer together, so that the total length of the solenoid is shorter; **b.** unwrapping the solenoid and using the same total length of wire d to form a new solenoid of the same length ℓ but with the coil wrapped less tightly so that the radius is larger and the number of loops is smaller; or **c.** increasing the voltage of the battery so that the current is three times larger? Justify your answer.

33-2 Back Emf

4. **C** The back emf in a circuit can sometimes be much greater than the emf supplied by the battery. **a.** Would you expect the inductance of such a circuit to be very high or very low? Explain. **b.** Suppose the inductance in some circuit is fixed. There is a steady current in the circuit, and you wish to cut that current. Explain how you could avoid inducing a large emf in the circuit.
5. A 15.0-mH inductor is connected to a DC power supply. The power supply's emf is fixed at 5.00 V.
   a. **N** If you turn down the power supply's current at a constant rate, $dI/dt = -50.0$ A/s, what is the magnitude of the back emf?
   b. **N** At what rate would you need to turn down the current so that the back emf is 5.00 V?
   c. **C** If you had a lightbulb in this circuit, how would it glow in the two cases? (Ignore the resistance of the bulb.)
6. **G** A voltage source is connected directly to an 80.0-mH inductor, and the current through the inductor is a triangular wave as shown in Figure P33.6. Plot the voltage across the inductor during the time interval from 0 to 16 s.

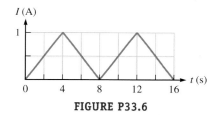

FIGURE P33.6

7. **N** What is the inductance of a coil in which the average emf induced is 23.0 mV when the current in the coil is increased from 4.00 A to 7.00 A in 0.330 s?
8. **N** A 45.0-cm-long solenoid is 8.00 cm in diameter and has 690 turns. **a.** What is the inductance of the solenoid? **b.** What is the rate of change of the current dI/dt required to produce an emf of 44.0 μV in the solenoid?

33-3 Special Case: Resistor–Inductor (RL) Circuit

9. **N** A series circuit contains a 4.00-H inductor, a 5.00-Ω resistor, and a 9.00-V battery. The current is initially zero when the circuit is connected at $t = 0$. At what time will the current reach **a.** 33.3% and **b.** 95.0% of its final value?
10. **C** When you close the switch on a flashlight, does the bulb instantaneously light up? When you open the switch on a flashlight, does the bulb instantaneously shut off? Explain your answers.

11. A battery with emf $\mathcal{E}$ is connected in series with an inductance L and a resistance R.
    a. **A** Assuming the current has reached steady state when it is at 99% of its maximum value, how long does it take to reach steady state, assuming the initial current is zero?
    b. **N** If an emergency power circuit needs to reach steady state within 1.0 ms of turning on and the circuit has a total resistance of 75 Ω, what values of the total inductance of the circuit are needed to satisfy the requirement?
12. **N** At one instant, a current of 6.0 A flows through part of a circuit as shown in Figure P33.12. Determine the instantaneous potential difference between points A and B if the current starts to decrease at a constant rate of 1.0×10^2 A/s.

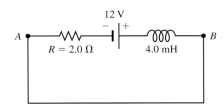

FIGURE P33.12

13. **N** The time constants for a series RC circuit with a capacitance of 5.00 μF and a series RL circuit with an inductance of 2.00 H are identical. a. What is the resistance R in the two circuits? b. What is the common time constant for the two circuits?
14. **N** After being closed for a long time, the switch S in the circuit shown in Figure P33.14 is thrown open at $t = 0$. In the circuit, $\mathcal{E} = 24.0$ V, $R_A = 4.00$ kΩ, $R_B = 7.00$ kΩ, and $L = 589$ mH. a. What is the emf across the inductor immediately after the switch is opened? b. When does the current in the resistor R_B have a magnitude of 1.00 mA?

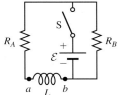

FIGURE P33.14

Problems 15, 16, and 17 are grouped.
15. In Figure 33.9A (page 1052), the switch is closed at a at $t = 0$.
    a. **A** Find an expression for the total charge that passes through the resistor in one time constant.
    b. **A** After the switch is left at a for many time constants, it is switched to b (Fig. 33.10A, page 1053). Find an expression for the total charge that passes through the resistor in one time constant.
    c. **C** Compare your results (ignoring any sign difference) and comment.

33-4 Energy Stored in a Magnetic Field

16. **G** In Figure 33.9A (page 1052), the switch is closed at a at $t = 0$. Find an expression for the power dissipated by the resistor as a function of time, and sketch your result. Is the power lost greater as soon as the switch is closed or a long time after it has been closed? Does your answer make sense?
17. In Figure 33.9A (page 1052), the switch is closed at a at $t = 0$.
    a. **A** Find an expression for the total energy dissipated by the resistor in one time constant.
    b. **A** After the switch is left at a for many time constants, it is switched to b. Find an expression for the total energy dissipated by the resistor in one time constant.
    c. **C** Compare your results and comment.
18. **E** Use Table 30.1 (page 941) to estimate the energy density stored in the magnetic field near a neutron star.
19. **N** Find the energy stored in a 4.0-mH inductor when the current is 4.0 A.
20. **N** If a high-voltage power line 25 m above the ground carries a current of 1.00×10^3 A, estimate the energy density of the magnetic field near the ground and compare it to the energy density of the Earth's magnetic field.

33-5 Special Case: Inductor–Capacitor (*LC*) Circuit

21. **N** What is the inductance of an LC circuit with $C = 4.50$ μF oscillating at 76.0 Hz?
22. **N** Figure 33.12 (page 1058) shows an LC circuit whose capacitor is initially ($t = 0$) fully charged. Find the phase constant for this common situation, and write an expression for $Q(t)$.
23. **A** In the LC circuit in Figure 33.11, the inductance is $L = 19.8$ mH and the capacitance is $C = 19.6$ mF. At some moment, $U_B = U_E = 17.5$ mJ. a. What is the maximum charge stored by the capacitor? b. What is the maximum current in the circuit? c. At $t = 0$, the capacitor is fully charged. Write an expression for the charge stored by the capacitor as a function of time. d. Write an expression for the current as a function of time.
24. **G** An LC circuit is an ideal circuit. Any real circuit has resistance, so energy is dissipated. This is analogous to the damped oscillator in Section 16-10. Use this analogy and Figure 16.22 (page 473) to sketch the charge stored by the capacitor and the current in a real LC circuit with resistance as functions of time.
25. **N** A 2.0-μF capacitor is charged to a potential difference of 12.0 V and then connected across a 0.40-mH inductor. What is the current in the circuit when the potential difference across the capacitor is 6.0 V?
26. **N** Figure P33.26 shows a circuit with $\mathcal{E} = 9.00$ V, $R = 6.00$ Ω, $L = 75.0$ mH, and $C = 2.55$ μF. After a long time interval at the position a shown in the figure, the switch S is thrown to position b at time $t = 0$. What is the maximum a. charge on the capacitor and b. current in the inductor for $t > 0$? c. What is the frequency of oscillation of the resulting LC circuit for $t > 0$?

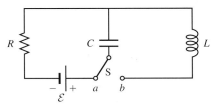

FIGURE P33.26

Problems 27, 28, and 29 are grouped.
27. **A** For an LC circuit, show that the energy stored in the electric field oscillates according to Equation 33.26, $U_E(t) = (Q_{max}^2/2C) \cos^2(\omega t + \varphi)$.
28. **A** For an LC circuit, show that the energy stored in the magnetic field oscillates according to Equation 33.27, $U_B(t) = (Q_{max}^2/2C) \sin^2(\omega t + \varphi)$.
29. **A** For an LC circuit, show that the total energy stored in the electric and magnetic fields is a constant given by Equation 33.28, $E_{tot} = Q_{max}^2/2C = \frac{1}{2} L I_{max}^2$.
30. **A** In Example 33.4 (page 1061), we found that the energy in an LC circuit is evenly split between the two fields at $t = T/8$. a. Find the charge and current at that time. b. Find the energy stored in each circuit element to confirm that the energy is evenly split at this time.
31. **N** A fully charged capacitor and a 0.20-H inductor are connected to form a complete circuit. If the circuit oscillates with a frequency of 1.2×10^3 Hz, determine the capacitance of the capacitor.

33-6 Special Case: AC Circuit with Resistance

32. **N** A 60-W lightbulb is used in an American desk lamp. **a.** What is the average power dissipated by the filament? **b.** What is the rms current in the filament? **c.** What is the maximum current in the filament? **d.** What is the maximum power dissipated by the filament? (Give your answers to two significant figures.)

33. **N** The rms current in a 29.0-Ω resistor connected to an AC source is 3.00 A. **a.** What is the rms voltage across this resistor? **b.** What is the peak voltage of the AC source? **c.** What is the average power dissipated by the resistor? **d.** What is the maximum current across this resistor?

34. **C** Suppose you connect a small lightbulb across a DC power supply that has an emf of 5.0 V. You then connect the bulb across a 60.0-Hz AC power supply that has an rms voltage of 5.0 V. Does the bulb dissipate more power on average when it is connected to the DC power supply or the AC power supply? Is the bulb brighter when it is connected to the DC power supply or the AC power supply? Explain.

Problems 35 and 36 are paired.

35. An AC generator delivers an alternating current $I(t) = (2.0\text{ A})\sin[(120\pi\text{ rad/s})t]$ to a single resistor in series with the generator.
    a. **N** What is the rms value of the current in the circuit?
    b. **N** If the resistor has a resistance of 100.0 Ω, what is the rms value of the source emf?
    c. **N** What is the maximum value of the source emf?
    d. **A** Write a function that describes the source emf as a function of time.

36. **G** An AC generator delivers an alternating current $I(t) = (2.0\text{ A})\sin[(120\pi\text{ rad/s})t]$ to a single resistor in series with the generator. Given a resistance of 100.0 Ω, draw a phasor diagram for this circuit, including the current, the potential difference across the resistor, and the source emf. Draw your diagram with the current phasor pointing toward the right along the horizontal axis.

37. **N** A simple AC circuit contains an AC source with output voltage $V_{max}\cos\omega t$ and a resistor with resistance $R = 150.0\ \Omega$. **a.** What is the angular frequency of the AC source if the first time $V_R = 0.400V_{max}$ is at $t = 1.00 \times 10^{-3}$ s? **b.** When is $V_R = 0.750V_{max}$ for the first time?

33-7 Special Case: AC Circuit with Capacitance

38. **C** Avi and Cameron are working on homework problems. They need to find the current in an AC circuit with capacitance at a particular time t. Decide who is right and why. If neither is right, make your own correct statement.

    Avi: Capacitive reactance is like resistance. To find the current at some particular time, you just divide the emf by the capacitive reactance X_C.
    Cameron: That works only if you want to find the maximum current. You need to divide the emf by the resistance R. Then multiply it by $\sin(\omega t + \pi/2)$.

39. **N** An 8.00-mF capacitor is connected across an AC source with an rms voltage of 68.0 V oscillating with a frequency of 50.0 Hz. What are the **a.** capacitive reactance of, **b.** rms current in, and **c.** maximum current in this circuit?

40. **C** Suppose you have an AC circuit with capacitance (Fig. 33.18, page 1064) and you wish to increase the maximum current. The AC power supply is adjustable. Without changing the capacitor, how can you increase the maximum current? (There is more than one answer.)

Problems 41, 42, and 43 are grouped.

41. An AC generator delivers an rms current of 3.50 A to a 6.25-μF capacitor connected in series. The frequency of the source emf is 60.0 Hz.
    a. **N** What is the capacitive reactance of the circuit?
    b. **N** What is the rms emf of the generator?
    c. **N** What are the maximum values of the current and the source emf?
    d. **A** Write a function for the potential difference across the capacitor as a function of time.

42. **G** An AC generator delivers an rms current of 3.50 A to a 6.25-μF capacitor connected in series. The frequency of the source emf is 60.0 Hz. Draw a phasor diagram for this circuit, including the current, the potential difference across the capacitor, and the source emf. Draw your diagram with the current phasor pointing toward the right along the horizontal axis.

43. **C** An AC generator delivers an rms current of 3.50 A to a 6.25-μF capacitor connected in series. The frequency of the source emf is 60.0 Hz. When the potential difference across the capacitor is at its maximum positive value, what is the value of the current? How does the potential difference compare to the source emf at this time? Justify and explain your answer.

44. **C** In an *ideal* AC circuit with capacitance, there is no resistance. Is any energy dissipated? How about in a *real* AC circuit with capacitance?

45. A radio telescope is designed to pick up natural signals at 1420 MHz. Unfortunately, it also picks up artificial signals at 1475 MHz. You want to use an *RC* filter to remove the signal above 1450 MHz.
    a. **C** Explain how an *RC* filter can be used to accomplish your task.
    b. **N** If the capacitance is 245 nF and the resistance is 34.7 Ω, what fraction of the artificial signal's maximum voltage $\mathcal{E}_{max}$ will the radio telescope pick up? Does your result tell you that the telescope won't suffer from artificial interference?

33-8 Special Case: AC Circuit with Inductance

46. **A** Follow the reasoning used for the *RC* filter to show that in the case of an *RL* filter, the maximum current is given by Equation 33.64:

$$I_{max} = \frac{\mathcal{E}_{max}}{\sqrt{X_L^2 + R^2}}$$

47. **N** A 68.0-mH inductor is connected across an AC source that has an rms voltage of 120.0 V oscillating at a frequency of 110.0 Hz. What are the **a.** inductive reactance of, **b.** rms current in, and **c.** maximum current in this circuit?

48. **C** Avi and Cameron are discussing another problem. Decide who is right and why.

    Avi: Inductive reactance is just like resistance. To find the current, you just need to divide by the inductive reactance. So the current at $t = 0$ is zero because the emf is zero.
    Cameron: That is only good for finding the maximum current.

Problems 49, 50, and 51 are grouped.

49. An AC generator with an rms emf of 15.0 V is connected in series with a 0.54-H inductor. The frequency of the source emf is 70.0 Hz.
    a. **N** What is the inductive reactance of the circuit?
    b. **N** What is the rms current in the circuit?
    c. **N** What are the maximum values of the current and the source emf?
    d. **A** Write an expression for the potential difference across the inductor as a function of time.

50. **G** An AC generator with an rms emf of 15.0 V is connected in series with a 0.54-H inductor. The frequency of the source emf is 70.0 Hz. Draw a phasor diagram for this circuit, including the current, the potential difference across the inductor, and the source emf. Draw your diagram with the current phasor pointing toward the right along the horizontal axis.

51. **C** An AC generator with an rms emf of 15.0 V is connected in series with a 0.54-H inductor. The frequency of the source emf is 70.0 Hz. When the current through the inductor is a maximum, what is the potential difference across the inductor? How does this compare to the source emf at this time? Justify and explain your answer.

52. **N** When connected to a 90.0-Hz AC source, an inductor has an inductive reactance of 33.0 Ω. What is the maximum current in this inductor if it is connected to an AC source that has an rms voltage of 120.0 V and a frequency of 110.0 Hz?

53. **N** An inductor ($L = 37.8$ mH) is connected to an AC power supply that has a maximum emf of 16.4 V and an angular frequency of 62.8 rad/s. At time $t = 0$, the emf is zero and increasing. **a.** What is the current at $t = 0$? **b.** What is the current across the inductor at $t = 25.0$ s? **c.** What is the maximum current in the circuit?

54. **C** **CASE STUDY** What happens to the woofer's voltage in Example 33.7 (page 1071) when the crossover frequency decreases?

55. **N** Design a high-pass filter using a 1.5-kΩ resistor and a 0.75-mF capacitor, and determine the approximate frequency above which voltages are able to pass through the circuit. Assume this is the frequency at which the output voltage is at least half the input voltage.

33-9 Special Case: AC Circuit with Resistance, Inductance, and Capacitance

56. **N** A radio tuner circuit is created with a 1.50-Ω resistor and a 2.50-μH inductor in series with a variable capacitor. **a.** What capacitance is needed for this circuit to have a resonance frequency of 88.7 MHz, the frequency of a local radio station? **b.** How much interference is there from the next lowest FM station at 88.1 MHz? That is, assuming the circuit is tuned to 88.7 MHz as in part (a), how much smaller is the current due to a signal at 88.1 MHz, assuming the signal strength is the same for both stations?

57. **N** A series *RLC* circuit with a resistance of 120.0 Ω has a resonant angular frequency of 4.0×10^5 rad/s. At resonance, the voltages across the resistor and inductor are 60.0 V and 40.0 V, respectively. **a.** Determine the values of L and C. **b.** At what frequency does the current in the circuit lag the voltage by 45°?

58. **A** Start with Equations 33.69 and 33.75, $I(t) = I_{max} \sin \omega t$ and $\mathcal{E}(t) = \mathcal{E}_{max} \sin(\omega t + \varphi)$, and show that the average power supplied is given by Equation 33.86, $P_{avg} = I_{rms}\mathcal{E}_{rms} \cos \varphi$.

59. **N** An AC source with $V_{rms} = 110.0$ V and $I_{rms} = 12.0$ A is connected to a series *RLC* circuit in which the current leads the voltage by 23.5°. What are the **a.** total resistance R and **b.** net reactance $(X_L - X_C)$ of this circuit?

60. **N** An AC source of angular frequency ω is connected to a resistor R and a capacitor C in series. The maximum current measured is I_{max}. While the same maximum emf is maintained, the angular frequency is changed to $\omega/3$. The measured current is now $I/2$. Determine the ratio of the capacitive reactance to the resistance at the initial frequency ω.

Problems 61 and 62 are paired.

61. An *RLC* series circuit is constructed with $R = 100.0$ Ω, $C = 6.25$ μF, and $L = 0.54$ H. The circuit is connected to an AC generator with a frequency of 60.0 Hz that delivers a maximum current of 2.00 A to the circuit.
a. **N** What is the impedance of this circuit?
b. **N** What are the maximum potential differences across each of the three circuit elements (R, L, and C)?
c. **N** What is the phase angle between the source emf and the current?
d. **A** Write expressions for the source emf and the current as functions of time.

62. **G** An *RLC* series circuit is constructed with $R = 100.0$ Ω, $C = 6.25$ μF, and $L = 0.54$ H. The circuit is connected to an AC generator with a frequency of 60.0 Hz that delivers a maximum current of 2.00 A to the circuit. Draw a phasor diagram for this circuit, including the current, the potential difference across each of the circuit elements, and the source emf. Draw your diagram with the current phasor pointing toward the right along the horizontal axis.

63. **N** A series *RLC* circuit driven by a source with an amplitude of 120.0 V and a frequency of 50.0 Hz has an inductance of 787 mH, a resistance of 267 Ω, and a capacitance of 45.7 μF. **a.** What are the maximum current and the phase angle between the current and the source emf in this circuit? **b.** What are the maximum potential difference across the inductor and the phase angle between this potential difference and the current in the circuit? **c.** What are the maximum potential difference across the resistor and the phase angle between this potential difference and the current in this circuit? **d.** What are the maximum potential difference across the capacitor and the phase angle between this potential difference and the current in this circuit?

Problems 64, 65, and 66 are grouped.

64. **N** In an *RLC* circuit (Fig. 33.31, page 1072), the resistance is 325 Ω, the inductance is 126 mH, and the capacitance is 13.7 μF (13.7×10^{-6} F). The angular frequency is 377 rad/s and $\mathcal{E}_{max} = 18.8$ V. **a.** What is the impedance? **b.** What is the maximum current? **c.** What is the phase constant φ of the power supply's emf with respect to the current?

65. **N** In an *RLC* circuit (Fig. 33.31, page 1072), the resistance is 325 Ω, the inductance is 126 mH, and the capacitance is 13.7 mF (13.7×10^{-3} F). The angular frequency is 377 rad/s and $\mathcal{E}_{max} = 18.8$ V. **a.** What is the impedance? **b.** What is the maximum current? **c.** What is the phase constant φ of the power supply's emf with respect to the current?

66. **G** In an *RLC* circuit (Fig. 33.31, page 1072), the resistance is 325 Ω, the inductance is 126 mH, the angular frequency is 377 rad/s, and $\mathcal{E}_{max} = 18.8$ V. Draw a phasor diagram if the capacitance is **a.** 13.7 μF and **b.** 13.7 mF.

General Problems

67. **N** How much energy is stored in the magnetic field of an 18.0-cm-long solenoid that has 154 turns and a radius of 0.900 cm and carries a current of 1.65 A?

68. **C** In circuits with nonzero inductance, interrupting the circuit by throwing a switch open can lead to a large spark at the switch. Why is a large spark possible, even though we generally assume a broken circuit always has zero current?

69. **N** An inductor ($L = 0.35$ H) and two resistors ($R_1 = R_2 = 2.0$ Ω) are connected to a battery with an emf of 12.0 V as shown in Figure P33.69.
a. If the switch S is closed at time $t = 0$, determine the potential drop across the inductor at time $t = 0.040$ s. **b.** After a steady state is reached,

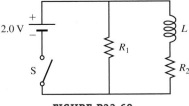

FIGURE P33.69

the switch is opened. What are the direction and magnitude of the current through R_1 at a time 0.040 s after the switch is opened?

70. **N** An ideal solenoid with 855 turns is 0.100 m long and has a cross-sectional area of 3.00×10^{-3} m². **a.** What is the inductance of this solenoid? **b.** If we connect this solenoid in series with a 12.0-V battery and a 185-Ω resistor, what is the maximum value of the current? **c.** Assuming the current is zero when $t = 0$ as the circuit elements are connected, at what time will the current reach 50% of its maximum value?

Problems 71 and 72 are paired.

71. **N** Figure P33.71 shows a series *RLC* circuit with a 25.0-Ω resistor, a 430.0-mH inductor, and a 24.0-μF capacitor connected to an AC source with $V_{max} = 60.0$ V operating at 60.0 Hz. What is the maximum voltage across the **a.** resistor, **b.** inductor, and **c.** capacitor in the circuit?

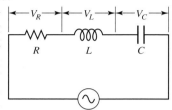

FIGURE P33.71 Problems 71 and 72.

72. **G** Figure P33.71 shows a series *RLC* circuit with a 25.0-Ω resistor, a 430.0-mH inductor, and a 24.0-μF capacitor connected to an AC source with $V_{max} = 60.0$ V operating at 60.0 Hz. Draw a phasor diagram for this circuit, including the current, the potential difference across each of the circuit elements, and the source emf. Draw your diagram with the current phasor pointing toward the right along the horizontal axis.

73. **N** The time variation of the current in a 37.0-mH inductor is given by $I = 3.00 + 4.00t - 2.00t^2$, with *I* in amperes and *t* in seconds. What is the magnitude of the emf induced in the inductor at **a.** $t = 2.00$ s and **b.** $t = 5.00$ s? **c.** For what value of *t* is the emf induced in the inductor 0?

74. **N** A 22.5-μF capacitor is charged by a 6.00-V battery and then connected to a 75.0-mH inductor. At what frequency does the current oscillate, and what is the maximum current?

Problems 75 and 76 are paired.

75. **N** In a series *RLC* circuit with a maximum current of 0.250 A, an AC source with $V_{max} = 115$ V operating at 60.0 Hz is connected to a 325-mH inductor, a 7.50-μF capacitor, and a resistor with unknown resistance *R*. What are the **a.** inductive reactance, **b.** capacitive reactance, and **c.** impedance of this circuit? **d.** What is the resistance *R* of the resistor in this circuit? **e.** What is the phase angle between the current and the source voltage?

76. **G** In a series *RLC* circuit with a maximum current of 0.250 A, an AC source with $V_{max} = 115$ V operating at 60.0 Hz is connected to a 325-mH inductor, a 7.50-μF capacitor, and a resistor with unknown resistance *R*. Draw a phasor diagram for this circuit, including the current, the potential difference across each of the circuit elements, and the source emf. Draw your diagram with the current phasor pointing upward along the vertical axis.

77. **N** A 45.0-V battery with an internal resistance of 13.0 Ω is connected to a 7.40-H inductor. The current is zero at $t = 0$. At what rate is the current in the inductor increasing at **a.** $t = 0$ and **b.** $t = 2.00$ s?

78. **A** Two coaxial cables of length ℓ with radii *a* and *b* are carrying currents in opposite directions as shown in Figure P33.78. Determine the inductance of the system. *Hint*: Use Ampère's law to write an expression for the magnetic field in the region between the cables, a distance *r* from the axis of the cables. Then calculate the magnetic flux through a narrow rectangular region between the cables such that the field is perpendicular to the area everywhere.

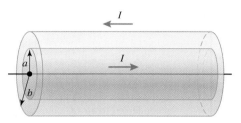

FIGURE P33.78

79. **N** An 11.0-μF capacitor is connected across an AC source with $V_{max} = 90.0$ V oscillating at 60.0 Hz. What is the maximum current in the capacitor?

Problems 80 and 81 are paired.

80. **G** A 20.0-Ω resistor, a 200.0-mH inductor, and a 200.0-μF capacitor are connected to a variable-frequency AC voltage source one at a time. The voltage source has a frequency *f* that ranges from 5.0 Hz to 100.0 Hz with a peak voltage of 10.0 V. Plot the peak current versus frequency for each circuit element.

81. **N** A 20.0-Ω resistor, a 200.0-mH inductor, and a 200.0-μF capacitor are each connected to a 120.0-V rms, 60.0-Hz household voltage one at a time. What is the peak current through each circuit element?

82. **A** A conducting bar of mass *M* and length ℓ is given an initial speed v_0 on a smooth horizontal conducting rail as shown in Figure P33.82. Assume the system has a constant inductance *L*, and note that the applied magnetic field $\vec{B}$ is into the page as shown in the figure. Find the maximum distance the bar travels before it stops. *Hint*: Use the relationship $B\ell(dx/dt) = L(dI/dt)$.

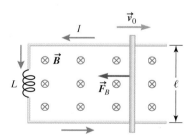

FIGURE P33.82

83. **N** An electronic device is composed of an *RLC* circuit with frequency $f = 60.0$ Hz. The capacitor in the circuit goes bad and must be replaced, but the engineer does not know the capacitance. If she wants the impedance to be 1000.0 Ω but knows only that $R = 500.0$ Ω and $L = 2.50$ H, find the two values of capacitance that will solve her dilemma. (There are two answers to this problem!)

84. **N** An inductor of inductance 2.0 mH is connected across a charged capacitor ($C = 4.0$ μF). The maximum value of the charge Q_{max} on the capacitor is 2.0×10^2 μC. **a.** When the charge on the capacitor is 1.0×10^2 μC, what is the value of $|dI/dt|$? **b.** When the charge on the capacitor is 2.0×10^2 μC, what is the value of the current?

Maxwell's Equations and Electromagnetic Waves

34

❶ Underlying Principles

1. General form of Faraday's law
2. Maxwell's equations
3. Lorentz force
4. Wave equation for an electromagnetic wave

✪ Major Concepts

1. Transverse electric and magnetic waves
2. Energy transferred by electromagnetic waves
3. Intensity
4. Momentum transferred by electromagnetic waves
5. Pressure exerted by electromagnetic waves
6. Polarization of electromagnetic waves and Malus's law

Key Questions

How can we express Faraday's law in a more general form?

What does Faraday's law tell us about an *induced* electric field?

How do Maxwell's equations describe electromagnetic waves?

- 34-1 **Light: One last classical topic** 1086
- 34-2 **Generalized form of Faraday's law** 1087
- 34-3 **Five equations of electromagnetism** 1092
- 34-4 **Electromagnetic waves** 1093
- 34-5 **The electromagnetic spectrum** 1101
- 34-6 **Energy and intensity** 1103
- 34-7 **Momentum and radiation pressure** 1107
- 34-8 **Polarization** 1112

How do scientists discover new laws of nature? There are many answers to this question, but one is based on the idea that nature is made up of symmetrical or parallel principles.

Maxwell reached his great insight by recognizing the symmetry in nature. Faraday formulated his law about 30 years before Maxwell modified Ampère's law. (We studied the two laws out of chronological order.) So Faraday's law was well known by the time Maxwell was just a young boy. Maxwell used Faraday's law and the idea that nature is based on parallel principles to expand Ampère's law. In doing so, he discovered the fundamental nature of light.

In this chapter, we gain a deeper understanding of electricity and magnetism by looking at all four of Maxwell's equations together. We also combine two of his equations to come up with a wave equation for electromagnetic waves.

34-1 Light: One Last Classical Topic

We are near the end of our study of classical physics. In fact, we have only one phenomenon left to discuss—electromagnetic waves, including light. Light is a very important topic, and the whole next part of this text is devoted to it.

All the classical physics in the preceding 33 chapters can be summed up in a few fundamental principles:

1. Newton's three laws of motion
   a. The law of inertia
   b. $\vec{F}_{\text{tot}} = d\vec{p}/dt$
   c. $\vec{F}_{[\text{B on A}]} = -\vec{F}_{[\text{A on B}]}$
2. Newton's law of universal gravity (one of the four fundamental forces)
3. The first law of thermodynamics (the principle of conservation of energy)
4. The second law of thermodynamics (the entropy of an isolated system never decreases; $\Delta S \geq 0$)
5. Maxwell's four equations and the Lorentz force (the second of the four fundamental forces)

The crowning achievement of classical physics is to describe in these few fundamental principles the enormous range of phenomena that we have studied so far and encounter so often in our daily lives. These phenomena include train crashes, the flight of airplanes, music, refrigerators, cosmic rays, electric generators, electric guitars, and nerve signals in the human body. The Universe would be a very different place if each of these phenomena was governed by different principles. Imagine designing a spacecraft if the physical principles that are true on the Earth were not valid in space.

Missing from this impressive list of phenomena, however, is light. By the early 1800s, scientists modeled light as a wave. That model led to more questions: What is waving? What kind of wave is it—longitudinal or transverse? In the mid-1800s, Maxwell showed that light is a transverse electromagnetic wave, which means electric and magnetic fields oscillate perpendicular to the direction of the wave's propagation (Section 34-4). Two other important properties of light were discovered in the 1900s: (1) Light waves require no medium; they can travel in a vacuum. By contrast, sound requires a medium such as air; there is no sound in a vacuum. (2) Light is also modeled as a particle known as a **photon**. Today, we say that light is both a wave and a particle: Sometimes we model light as a wave and sometimes as a particle, but photons are a modern concept, so we will not study them until Chapter 39.

CASE STUDY Part 1: The Power of Light

Sunlight enables life to exist on the Earth, but you probably never thought sunlight could save the Earth from destruction by a wandering comet or asteroid. The solar system contains a large number of these objects and sometimes one of them hits a planet, as in 1994 when Comet Shoemaker-Levy 9 collided with Jupiter. The comet had broken into many pieces, so there were multiple collisions; Figure 34.1 shows four impact sites along a diagonal line running upward and to the right. After the collision, Jupiter had scars that took about six months to heal. Many of the scars were several times the size of the Earth.

The Earth is not immune to such impacts. It is likely that the dinosaurs were wiped out by an impact 65 million years ago. Astronomers search for near-Earth objects such as comets and asteroids that might hit the Earth and do serious damage. If they find such an object, we could try to deflect it. If the object were predicted to strike in a few decades, we might

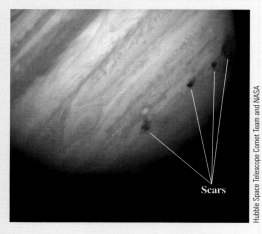

FIGURE 34.1 Jupiter was hit by Comet Shoemaker-Levy 9. Four impact sites are seen in this Hubble Space Telescope image.

consider detonating a nuclear weapon near the object. If we had even more warning, we might try to use the Sun's light to push the object off course. We could attach a solar sail to the object to harness the pressure of the Sun's radiation.

It might seem incredible that light could push a giant rock off course. This is possible, however, and it is just such an effect that limits the amount of light a star can give off without blowing itself apart. In this case study, we will look at (1) saving the Earth from a collision by using sunlight and (2) the maximum luminosity of stars.

CASE STUDY   Part 2: Meteor Shower Revisited

In the middle of August, three physics professors—Black, Noir, and Kuro—watched the Perseid meteor shower. On their drive home, Black asked whether either of the other two professors knew the size of a piece of the debris. Neither of them knew the answer, but among the three professors, they knew enough information to estimate an answer. They worked in the dark without a calculator and came up with a good estimate using two different approaches. We described the first approach based on conservation of energy in Chapter 9. Here we consider their second approach, based on the light from the meteor.

The Perseids peak on August 12. The shooting stars appear to come from the constellation Perseus.

CONCEPT EXERCISE 34.1

One way to think about the fundamental principles of physics is to imagine an extraterrestrial alien society. If the alien society could discover the principle, it must be a fundamental principle of the Universe. If the aliens could discover the principle only by coincidence (if they happened to live on a planet with properties identical to those of the Earth), the principle is not fundamental. Avi and Cameron discuss the fundamental nature of the constants g and G. With whom do you agree and why?

Avi: Little g is gravitational acceleration near the Earth. It isn't even good on the Moon, so it isn't fundamental. Big G is the constant in Newton's law of universal gravity. So it must be universal.

Cameron: Both constants depend on the SI system. They aren't the same even if we use U.S. customary units. Neither one is fundamental.

34-2 Generalized Form of Faraday's Law

In order to study light, we need to revisit Faraday's law. Faraday's law in the form presented in Chapters 32 and 33 is very useful in practical situations such as finding the current induced in a conductor. For example, in Figure 34.2A, a single conducting loop is in the plane of the page and a magnetic field (uniform over the loop's area) points into the page. (The magnetic field may be generated by a large solenoid.) If the magnetic field gets stronger (because the current in the solenoid increases), the magnetic flux through the loop increases. According to Faraday's law, the changing magnetic flux induces a current in the loop like the current that would be set up with a battery that has emf $\mathcal{E}$. So, a practical way to write Faraday's law for a single loop is in terms of the emf *induced* by the changing magnetic flux: $\mathcal{E} = -d\Phi_B/dt$ (Eq. 32.6 with $N = 1$).

To study light, however, we need a more general form of Faraday's law that involves the electric field instead of an induced emf. According to Faraday's experiments, a changing magnetic flux through a loop induces a current in the loop. You can think of that current as being set up by an emf, but you can also think of it as being set up by an electric field. To get positive charge to move in a counterclockwise circle (Fig. 34.2A), the electric field must form a counterclockwise circle (Fig. 34.2B). In fact, the conducting loop is *not* necessary, and the electric field is

A.

Practical way to look at Faraday's law: Changing magnetic flux induces current in conducting loop, so an emf is induced in conductor.

B.

General way to look at Faraday's law: Changing magnetic flux induces an electric field whether a conductor is present or not.

FIGURE 34.2 A. The changing magnetic flux induces an emf in the conducting loop. **B.** The changing magnetic flux induces an electric field. This kind of electric field is nonconservative.

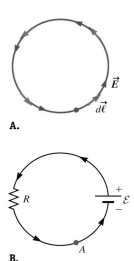

FIGURE 34.3 A. A particle moves in a circular path along a single electric field line. Each segment of the path is parallel to the electric field. **B.** A positively charged particle starts at point A and moves counterclockwise around the circuit. The particle gains energy when it goes through the battery and loses the same amount of energy when it passes through the resistor.

An *induced* electric field is nonconservative, so its path integral depends on the particular path.

found throughout the indicated region of space. The electric field lines are represented by a number of circles.

This should seem odd for electric field lines, which we said must originate from a positively charged particle and/or terminate on a negatively charged particle (Sections 24-3 and 24-4). The electric field lines in Figure 34.2B neither originate nor terminate. Instead, they wrap around on themselves. And, because the conductor is unnecessary, there are no charged particles in Figure 34.2B.

What is going on here? The rules in Chapter 24 were established for electrostatic fields—that is, for electric fields whose source is stationary, electrically-charged particles. The electric field in Figure 34.2B is generated by the changing magnetic field. (The magnetic field is the result of moving charges, so the ultimate source of this electric field is also those moving charged particles.) There are two types of electric fields that result from two types of electric sources: (1) Stationary charged particles produce an electrostatic field, and (2) a changing magnetic field produces an *induced* electric field.

The closed loops of the induced electric field lines in Figure 34.2 mean that the induced electric field is *nonconservative*. To see this, imagine a positively charged particle moves in a circular path following one of the field lines (Fig. 34.3A). The work done by the electric field on this positively charged particle is given by $W = \oint \vec{F}_E \cdot d\vec{\ell}$ (similar to Eq. 9.21), where the circle on the path integral means it is taken over a closed path of which $d\vec{\ell}$ is a segment. (The original form of Eq. 9.21 contains $d\vec{r}$. We have replaced it with $d\vec{\ell}$ so that we use the same symbols as in Ampère's law, Eq. 31.5.) The particle's charge is q and the electric field around the path is $\vec{E}$, so the force exerted by the electric field is given by $\vec{F}_E = q\vec{E}$ (Eq. 24.2). Substitute Equation 24.2 into Equation 9.21:

$$W = q \oint \vec{E} \cdot d\vec{\ell} \tag{34.1}$$

where the charge q is constant as the particle moves around the closed path. At any point along the path, the electric field is parallel to the path segment (Fig. 34.3A), so at every point the dot product $\vec{E} \cdot d\vec{\ell}$ is positive. Therefore, the entire integral and thus the work around the closed path are positive:

$$W = q \oint \vec{E} \cdot d\vec{\ell} > 0 \tag{34.2}$$

When the positive particle in Figure 34.3A completes one trip around its closed path, positive work has been done on it and its energy is greater than when it started. This result shows that the induced electric field fails the (second) test for a conservative force (Section 9-5). According to this test, the work done by a conservative force around a closed path is zero.

Compare the round trip of a particle in Figure 34.3A to that of a particle moving around the simple circuit in Figure 34.3B. Suppose a positive particle starts at point A in the circuit and moves in a counterclockwise direction (Fig. 34.3B). The particle passes through the battery from the negative to the positive terminal. Chemical reactions inside the battery do work on the particle, increasing its energy by $q\mathcal{E}$. Then the particle passes through the resistor, losing an amount of energy qV_R equal to the amount of work done on it by the battery: $q\mathcal{E} - qV_R = 0$. So, when the particle returns to point A, it has exactly the same energy it started with. This is really a statement of Kirchhoff's loop rule: The sum of the voltages around a closed loop is zero: $\mathcal{E} - V_R = 0$.

Kirchhoff's rule doesn't hold when an emf is induced in a conducting loop (Fig. 34.2A). From Faraday's law, the voltage around this loop is not zero; it is given by $\mathcal{E} = -d\Phi_B/dt$ (Eq. 32.6 with $N = 1$). When a positively charged particle moves once around the loop, positive net work W is done on it by the electric field:

$$W = q\mathcal{E} \tag{34.3}$$

Equation 34.3 must equal Equation 34.1, so

$$W = q\mathcal{E} = q \oint \vec{E} \cdot d\vec{\ell}$$

$$\mathcal{E} = \oint \vec{E} \cdot d\vec{\ell} \qquad (34.4)$$

Because the potential difference around the closed loop is *not* zero when a changing magnetic field is present, we cannot use Kirchhoff's loop rule. (See Problems 13, 14, and 68.) Faraday's law still holds, however.

Now, to find a general expression for Faraday's law, substitute Equation 34.4 into $\mathcal{E} = -d\Phi_B/dt$ (Eq. 32.6):

$$\oint \vec{E} \cdot d\vec{\ell} = -\frac{d\Phi_B}{dt} \qquad (34.5)$$

We can rewrite Equation 34.5 by replacing the magnetic flux with $\Phi_B = \int \vec{B} \cdot d\vec{A}$ (Eq. 31.2):

$$\oint \vec{E} \cdot d\vec{\ell} = -\frac{d}{dt}\int \vec{B} \cdot d\vec{A} \qquad (34.6)$$

GENERAL FORM OF FARADAY'S LAW
❶ **Underlying Principle**

Equation 34.6 says that a changing magnetic flux produces an electric field. It does not matter whether there is a conductor present, or free charges, or simply a vacuum.

The negative sign in Equations 34.5 and 34.6 is Lenz's law. Conceptually, Lenz's law says the induced electric field is in the direction that opposes the change in magnetic flux. In practice, Lenz's law tells us how to find the direction of the induced electric field. The left side $\left(\oint \vec{E} \cdot d\vec{\ell}\right)$ is a circulation integral (a path integral around a closed loop), for which you must choose the shape and direction of the path. Once you have picked the direction, wrap the fingers of your right hand in that direction. Your thumb then points in the direction of the positive area vector $\vec{A}$. If the magnetic field points in the same direction as $\vec{A}$, the magnetic flux $\left(\int \vec{B} \cdot d\vec{A}\right)$ is positive. If the magnetic flux *increases*, the right side of Equation 34.6 is negative (due to Lenz's law). In order for the left side to be negative also, the electric field must point *opposite* the path's direction you chose so that $\vec{E} \cdot d\vec{\ell} < 0$. In Concept Exercise 34.2, you are asked to work out the direction of the electric field for other situations.

As a simple rule of thumb, you can imagine the path you have chosen is made of a conducting loop. Use your usual Lenz's-law method to find the direction of the current induced in that (imaginary) conducting loop. The electric field points in the same direction as that (imaginary) induced current.

PROBLEM-SOLVING STRATEGY

Finding the Electric Field Using Faraday's Law

Because Ampère's law,

$$\oint \vec{B} \cdot d\vec{\ell} = \mu_0 \varepsilon_0 \frac{d}{dt}\int \vec{E} \cdot d\vec{A}$$

(see Eqs. 25.6 and 31.8 with $I_{\text{thru}} = 0$) is similar to Faraday's law,

$$\oint \vec{E} \cdot d\vec{\ell} = -\frac{d}{dt}\int \vec{B} \cdot d\vec{A} \qquad (34.6)$$

we can adapt the four steps from Chapter 31 (page 987) to come up with four steps for solving problems with Faraday's law. These steps are useful when we need to find the electric field induced by a changing magnetic flux.

∴ INTERPRET and ANTICIPATE
Step 1 Sketch electric field lines, and apply Lenz's law to find the direction of the electric field, if possible.
Step 2 Choose a closed path. You must choose a shape that exploits the symmetry of the situation and also choose the direction of the path. If possible, choose the direction so that $\vec{E}$ and $d\vec{\ell}$ are parallel.

∴ SOLVE
Step 3 Do the circulation integral $\oint \vec{E} \cdot d\vec{\ell}$. The electric field is often a constant over each portion of the integral, so the integral is usually very simple.
Step 4 Find the change in magnetic flux $\frac{d}{dt}\int \vec{B} \cdot d\vec{A}$, and **substitute into Faraday's law** (Eq. 34.6).

CONCEPT EXERCISE 34.2

a. Suppose you choose a direction for the path in Equation 34.6,

$$\oint \vec{E} \cdot d\vec{\ell} = -\frac{d}{dt} \int \vec{B} \cdot d\vec{A}$$

and you find that the magnetic field points in the same direction as the positive area vector. If the magnetic flux is *decreasing*, is the electric field in the same direction as the path or in the opposite direction?

b. Now you choose a new direction for the path, and the magnetic field points in the direction opposite the positive area vector. If the magnetic flux is *increasing*, is the electric field in the same direction as the path or in the opposite direction?

c. Again you choose a path such that the magnetic field points in the direction opposite the positive area vector. If the magnetic flux is *decreasing*, is the electric field in the same direction as the path or in the opposite direction?

EXAMPLE 34.1 A Circular Region

A magnetic field points out of the page and perpendicular to it. The field is uniform in space over a circular region of radius R and zero everywhere else. The field increases at a rate dB/dt.

A Find an expression for the electric field $E(r)$ at a distance r from the center of the circular region, where $r < R$.

• INTERPRET and ANTICIPATE
Use the four steps of the problem-solving strategy. We expect the electric field to depend on the rate at which the magnetic field changes. If the magnetic field changes rapidly, a strong electric field should be generated.

Step 1 Sketch electric field lines. The magnetic field points out of the page and is uniform in space over a circular region of radius R and zero everywhere else (Fig. 34.4). From the symmetry of the problem, the electric field must form concentric circles as shown. (For now, don't worry about the spacing of the electric field lines.) To find the direction of the induced electric field, imagine one of the electric field lines is a circular conductor. The current in the imaginary conductor must be clockwise to oppose the changing magnetic flux, so the induced electric field is clockwise.

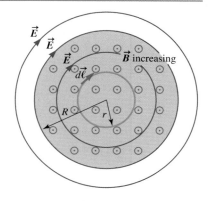

FIGURE 34.4

Step 2 Choose a closed path. To exploit the symmetry, we choose a circular path of radius r that runs along an electric field line. We have decided to integrate clockwise around the path in the direction of the electric field. Our path is shown in Figure 34.4.

• SOLVE
Step 3 Do the circulation integral. This is the integral on the left side of Faraday's law (Eq. 34.6). Because we choose to integrate clockwise in the direction of the electric field, the dot product is $E\,d\ell$. The electric field is constant over the path; that is why a circular path is a good choice. The path integral $\oint d\ell$ is the circumference of the circle.

$$\oint \vec{E} \cdot d\vec{\ell} = \oint E\,d\ell = E\oint d\ell$$

$$\oint \vec{E} \cdot d\vec{\ell} = E(2\pi r) \qquad (1)$$

Step 4 Find the change in magnetic flux. First use Equation 31.2 to find the magnetic flux through the circular path. To find the direction of the area vector, wrap the fingers of your right hand clockwise in the direction of the path. Your thumb then points into the page, so the area vector points into the page while the magnetic field points

$$\Phi_B = \int \vec{B} \cdot d\vec{A} \qquad (31.2)$$

out of the page. The dot product is therefore negative. The magnetic field is uniform in space through the circular path of radius $r < R$. The integral $\int dA$ is then the area of the circular path.	$\Phi_B = -\int B\,dA = -B\int dA = -BA$ $\Phi_B = -B\pi r^2$ (2)
Now **substitute** Equations (1) and (2) **into Faraday's law** (Eq. 34.6). The negative signs cancel on the right.	$\oint \vec{E}\cdot d\vec{\ell} = -\dfrac{d}{dt}\int \vec{B}\cdot d\vec{A}$ (34.6) $E(2\pi r) = -\dfrac{d}{dt}(-B\pi r^2) = \pi r^2 \dfrac{dB}{dt}$
Solve for E as a function of r.	$E(r) = \dfrac{r}{2}\dfrac{dB}{dt}$

:• **CHECK and THINK**
As expected, the strength of the electric field depends on the rate at which the magnetic field changes. If the magnetic field changes rapidly, the electric field is strong. The electric field is zero at the center of the region ($r = 0$) and gets stronger as r increases. At the outer border of the region ($r = R$), the electric field is $E(R) = (R/2)(dB/dt)$. We'll use this value to check our result in part B.

B Find an expression for the electric field $E(r)$ at a distance r from the center of the circular region, where $r > R$.

:• **INTERPRET and ANTICIPATE**
This part is similar to part A, so we can use our previous work to take a few shortcuts. At $r = R$, both expressions should give the same answer.

Step 1 Sketch electric field lines. The electric field lines are the same as in part A (Fig. 34.5). (Now we know the electric field increases linearly as a function of r, as we roughly represent in our spacing of the field lines.)

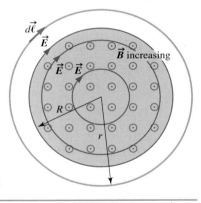

FIGURE 34.5

Step 2 Choose a closed path. The circular path must lie in the region we are interested in, so this time our circular path has radius $r > R$.

:• **SOLVE** **Step 3 Do the circulation integral.** The circulation integral is identical to the one in part A.	$\oint \vec{E}\cdot d\vec{\ell} = E(2\pi r)$ (1)
Step 4 Find the change in magnetic flux. This step is different from the step in part A because the magnetic field is not uniform in space throughout the circular path. It is uniform out to radius R but zero in the donut-shaped region between R and r. So we break up the flux integral (Eq. 31.2) into two pieces covering (1) the inner region over which the magnetic field is spatially uniform and (2) the donut-shaped region over which the magnetic field is zero.	$\Phi_B = \int \vec{B}\cdot d\vec{A}$ (31.2) $\Phi_B = \left(\int \vec{B}\cdot d\vec{A}\right)_{\text{inner}} + \left(\int \vec{B}\cdot d\vec{A}\right)_{\text{donut}}$
The magnetic flux over the donut-shaped region is zero because the magnetic field is zero.	$\Phi_B = \left(\int \vec{B}\cdot d\vec{A}\right)_{\text{inner}} + 0 = \left(\int \vec{B}\cdot d\vec{A}\right)_{\text{inner}}$

Example continues on page 1092 ▶

Finding the magnetic flux through the inner region is like finding the flux through the circular path in part A. The area vector points into the page and the magnetic field points out of the page, so the dot product is negative. The magnetic field through the inner region is uniform. The remaining integral is then just the area of the inner region of radius R.	$\Phi_B = \int \vec{B} \cdot d\vec{A} = -\int B\,dA = -B\int dA$ $\Phi_B = -BA = -B\pi R^2 \qquad (3)$
Now **substitute** Equations (1) and (3) **into Faraday's law** (Eq. 34.6). Solve for E as a function of r.	$\oint \vec{E} \cdot d\vec{\ell} = -\dfrac{d}{dt}\int \vec{B} \cdot d\vec{A} \qquad (34.6)$ $E(2\pi r) = -\dfrac{d}{dt}(-B\pi R^2) = \pi R^2 \dfrac{dB}{dt}$ $E(r) = \dfrac{R^2}{2r}\dfrac{dB}{dt}$
:• **CHECK and THINK** To check the result, find E at the border of the region, where $r = R$. The electric fields from parts A and B are the same at that distance, as required.	$E(R) = \dfrac{R^2}{2R}\dfrac{dB}{dt} = \dfrac{R}{2}\dfrac{dB}{dt} \quad \checkmark$

MAXWELL'S EQUATIONS
❶ **Underlying Principle**

34-3 Five Equations of Electromagnetism

The entire theory of electricity and magnetism boils down to just five equations, four of which are Maxwell's equations. James Clerk Maxwell (1831–1879) was born in Edinburgh, Scotland. Like Newton about 200 years earlier, Maxwell went to Cambridge University and worked on many of the problems of his time. He correctly described the nature of Saturn's rings, and he is one of the founders of the kinetic theory of gases (Chapter 20), but his greatest work was on electricity, magnetism, and light. Maxwell believed electricity and magnetism were part of a single unified principle. In 1864, he published *A Dynamical Theory of the Electromagnetic Field*, in which he wrote his four now-famous equations in mathematical form. To physicists, Maxwell's equations are beautiful (Fig. 31.1, page 978) because four relatively simple equations describe a wide range of phenomena. In this section, we review Maxwell's four equations and consider the achievements of Maxwell and the body of classical physics. We state Maxwell's four equations next (in the absence of a dielectric or magnetic material).

Gauss's Law for Electricity

The electric flux through a closed surface is proportional to the charge inside that surface:

$$\Phi_E = \oint \vec{E} \cdot d\vec{A} = \dfrac{q_{\text{in}}}{\varepsilon_0} \qquad (25.11)$$

Conceptually, Gauss's law for electricity says that particles with excess charge are the source of electrostatic fields, and electric field lines emerge from positive particles and terminate on negative ones.

Gauss's Law for Magnetism

The magnetic flux through a closed surface is zero:

$$\Phi_B = \oint \vec{B} \cdot d\vec{A} = 0 \qquad (31.4)$$

Gauss's law for magnetism says that so far, the existence of magnetic monopoles has not been confirmed. If magnetic monopoles are discovered, the 0 on the right side of the equation will have to be replaced so that the net magnetic flux through a closed surface is proportional to the net magnetic "charge" enclosed. Conceptually, Gauss's law for magnetism says that magnetic fields always form closed loops such as those created by magnetic dipoles (Fig. 30.8 on page 938).

Faraday's Law

The circulation integral of the electric field is proportional to the changing magnetic flux through the closed path:

$$\oint \vec{E} \cdot d\vec{\ell} = -\frac{d\Phi_B}{dt} \tag{34.5}$$

where the negative sign is Lenz's law. Conceptually, Faraday's law says a changing magnetic flux is another source of an electric field. We have referred to this as the *induced* electric field to distinguish it from the electrostatic field. Lenz's law states that the direction of the induced electric field is such that it opposes the change that created it. According to Faraday's law, we can simply say a changing magnetic field is the source of an electric field.

Ampère–Maxwell's Law

Ampère's law says that the circulation integral of the magnetic field is proportional to the current that passes through the closed path. Maxwell's addition to Ampère's law adds a second term so that the circulation integral also depends on the changing electric flux through the path:

$$\oint \vec{B} \cdot d\vec{\ell} = \mu_0 I_{\text{thru}} + \mu_0 \varepsilon_0 \frac{d\Phi_E}{dt} \tag{31.8}$$

Conceptually, Ampère–Maxwell's law says that a current and a changing electric *flux* are the sources of a magnetic field. More generally, we can say a changing electric *field* is the source of a magnetic field. Because magnetic monopoles have not been confirmed, this is the only source of a magnetic field.

The magnetic field of a permanent magnet is largely due to the spin of the electrons. (See Section 30-6.)

When Gauss, Faraday, and Ampère developed their laws, they did not express them in the mathematical form we use today. Not only was Maxwell the first to think of these four laws as being part of the same physical principle, but he was also the first to write them down mathematically in a form close to how we have given them here. This mathematical description is important in seeing how these laws come together to describe light (Section 34-4).

The Fifth Equation

An additional equation gives the Lorentz force:

$$\vec{F}_L = \vec{F}_E + \vec{F}_B = q(\vec{E} + \vec{v} \times \vec{B}) \tag{30.21}$$

LORENTZ FORCE
❶ **Underlying Principle**

The Lorentz force is the vector sum of the force exerted by an electric field plus the force exerted by a magnetic field on a charged particle. The electric force is either parallel (in the case of positive particles) or antiparallel (in the case of negative particles) to the electric field. The magnetic force is exerted only on particles that have velocity $\vec{v}$ that is not parallel (or antiparallel) to the magnetic field. The magnetic force is perpendicular to both $\vec{v}$ and $\vec{B}$.

These five equations are the fundamental equations of electricity and magnetism. All the new equations introduced in Chapters 23 through 34 are based on these five equations and other fundamental principles such as conservation of energy. Even light can be understood in terms of these few fundamental equations.

34-4 Electromagnetic Waves

In this section, we use Maxwell's equations to derive a wave equation for electromagnetic waves, including light. Maxwell presented his derivation in the late 1870s, but the experimental evidence confirming his prediction of electromagnetic waves came about 10 years later. Let's consider some experimental evidence first.

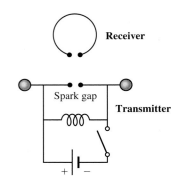

FIGURE 34.6 Hertz built the first radio transmitter and receiver. Both are essentially *LC* circuits.

Hertz's Experiments

In 1887, the German physicist Heinrich Rudolf Hertz (1857–1894) developed and performed an experiment that confirmed Maxwell's prediction. Figure 34.6 shows a very crude sketch of Hertz's basic apparatus, which consists of two parts. The *transmitter* is a circuit that produces an electromagnetic oscillation. The oscillation causes electromagnetic waves to travel outward, and the *receiver* detects those waves. The transmitter and receiver (both essentially *LC* circuits) are not in contact. The main parts of the transmitter are an induction coil and a capacitor that looks like a rod with a small gap. Hertz needed a very high potential difference between the two sides of the gap so that the air in between would become a conductor and a spark could jump across. So, Hertz used two large spheres to increase the capacitance of the circuit, allowing for a high voltage across the gap. A power supply (consisting of three batteries) was required to charge up the capacitor. The receiver is a simple loop of wire with a small gap. The loop has some inductance, and the gap is like the plates of a capacitor. The receiver does not have a separate power supply.

When the transmitter is turned on, a strong electric field builds up in its gap. The electric field ionizes the air, and acceleration of the resulting free electrons causes more ionization until the air conducts a spark. The plates of the capacitor charge and discharge periodically, while the sparks in the gap oscillate at the natural frequency of the *LC* circuit. (Recall that in an *LC* circuit, the electric field between the capacitor plates oscillates back and forth at the natural frequency; see Fig. 33.12, page 1058.) The electric field oscillation creates an electromagnetic wave. This is similar to the oscillation of a stereo speaker, which is analogous to the transmitter. The speaker's back-and-forth motion creates a sound wave in the surrounding air. Your eardrum is the receiver; the sound wave causes your eardrum to move back and forth at the same frequency as the speaker's frequency.

Hertz found that sparks in the transmitter's gap sometimes set off sparks in the receiver's gap. The receiver has its own natural frequency. Sparks in the receiver were present only when the natural frequency of the transmitter matched the natural frequency of the receiver.

Hertz also found that he could reflect the wave (the spark "signal" from the transmitter) off a metal plate just as a sound wave is reflected off a cliff. Recall that a standing wave can be formed by using the reflection of a traveling wave. Hertz was able to set up a standing electromagnetic wave similar to a sound wave in a pipe. By moving the plate from which the wave reflected, Hertz could find the nodes of the standing wave, which allowed him to determine n—the harmonic number and wavelength. Because the frequency of the wave was the natural frequency of the *LC* circuit, Hertz could find the speed of the wave using Equation 17.8 $(v = \lambda f)$. His result was a speed very close to that of light $(c = 3.00 \times 10^8 \text{ m/s})$.

Hertz's confirmation of Maxwell's theory came about eight years after Maxwell died. Not only was it an important confirmation of the existence of electromagnetic waves, but it also led to important new technologies. Hertz's setup is essentially a radio transmitter and receiver. Unfortunately, Hertz died in 1894 and did not see the important communication devices that resulted from his work. However, his experiment provided the insight needed by the Italian electrical engineer Guglielmo Marconi (1874–1937), who invented the first wireless communication device (a wireless telegraph).

Electromagnetic Wave Transmission

The panels in Figure 34.7 show a transmitter on the left and a receiver on the right, each represented by a simple capacitor. No other parts of the circuit are shown. We wish to follow the electric field from the transmitter to the receiver. One way to visualize an electric field is to imagine the electric force exerted on a positive test particle, so imagine a whole line of test particles along the *x* axis from the transmitter to the receiver. Suppose the transmitter is charged up so the bottom plate is positive and the top plate is negative. Then an upward-pointing electric field pulse moves from

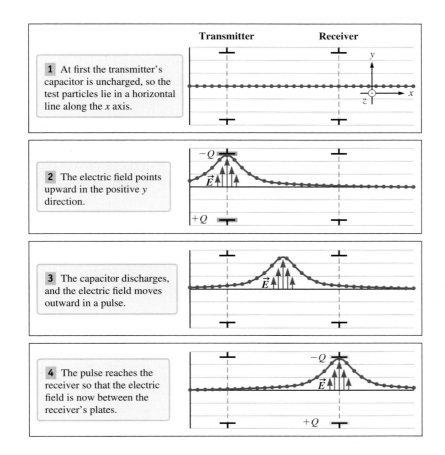

FIGURE 34.7 An electric pulse moves from the transmitter to the receiver.

The displacement current points in the same direction as the electric field if the electric flux is increasing.

the transmitter to the receiver. The magnetic field is not shown in Figure 34.7, but we can picture it. According to Ampère's law, the changing electric field causes a magnetic field. To find the direction of the magnetic field, point your right thumb in the direction of the displacement current (the y direction). Then your fingers wrap around in the direction of the magnetic field. So the magnetic field lies in the xz plane and is perpendicular to the electric field. Both fields move outward from the transmitter.

The Electromagnetic Wave Equation

To derive a wave equation for any possible electromagnetic wave is beyond the scope of this textbook. Instead, we will derive the wave equation for a plane wave traveling in the positive x direction (Fig. 34.8). The electric field oscillates parallel to the y axis, while the magnetic field oscillates parallel to the z axis, perpendicular to the electric field. The wave in Figure 34.8 is **linearly polarized**, which means the electric field is always parallel to a single axis (and therefore the magnetic field is always perpendicular to that chosen axis).

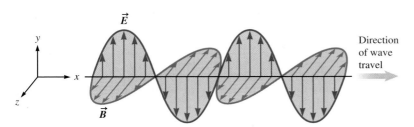

FIGURE 34.8 A linearly polarized electromagnetic wave (pulse) travels in the positive x direction. For all electromagnetic waves, *the electric and magnetic fields are perpendicular to each other and to the direction of the wave's propagation*. (However, $\vec{E}$ along y and $\vec{B}$ along z were chosen arbitrarily here.)

DERIVATION Wave Equation for Linearly Polarized Electromagnetic Waves

We will use the last two of Maxwell's equations—Faraday's law (Eq. 34.5) and Ampère–Maxwell's law (Eq. 31.8)—to derive the wave equation for the electric field of a linearly polarized wave in a vacuum:

$$\frac{\partial^2 E}{\partial x^2} = \mu_0 \varepsilon_0 \frac{\partial^2 E}{\partial t^2} \quad (34.7)$$

WAVE EQUATION FOR AN ELECTROMAGNETIC WAVE
❶ Underlying Principle

For homework (Problem 26), you will show that

$$\frac{\partial^2 B}{\partial x^2} = \mu_0 \varepsilon_0 \frac{\partial^2 B}{\partial t^2} \quad (34.8)$$

In a vacuum, you can assume there is no charge and therefore no current: $I_{\text{thru}} = 0$. Also assume the space between the transmitter and receiver is empty.

Apply Faraday's law (Eq. 34.5) to a linearly polarized wave (Fig. 34.8), using the four steps from this chapter's problem-solving strategy (page 1089) as a guide.

$$\oint \vec{E} \cdot d\vec{\ell} = -\frac{d\Phi_B}{dt} \quad (34.5)$$

Sketch electric field lines. The electric and magnetic field lines are shown in Figure 34.8. Consider a small portion of this wave. Figure 34.9 shows the electric field at one instant in a small, narrow rectangle that lies in the xy plane. The rectangle has height h and width dx. For this particular region at this particular instant, the electric field is weaker on the left, $\vec{E}(x)$, than it is on the right, $\vec{E}(x + dx)$.

Choose a closed path. A rectangular path exploits the symmetry of the problem. The electric field is zero along the horizontal portions of the path, and if we choose to integrate counterclockwise, the electric field is parallel to the path on the right side of the rectangle and antiparallel on the left side.

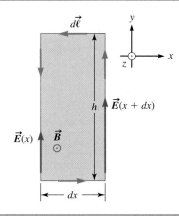

FIGURE 34.9 A small portion of the wave in Figure 34.8, showing the plane of the electric field.

Do the circulation integral. The horizontal portions of the path do not contribute to the circulation integral. The dot product over the right side of the path is positive and over the left side is negative. We take the electric field to be constant over each individual path segment. The integrals $\int d\ell$ are both equal to the height h of the rectangle.

$$\oint \vec{E} \cdot d\vec{\ell} = \left(\int \vec{E} \cdot d\vec{\ell}\right)_{\text{right}} + \left(\int \vec{E} \cdot d\vec{\ell}\right)_{\text{left}}$$

$$\oint \vec{E} \cdot d\vec{\ell} = E_{\text{right}} \int d\ell - E_{\text{left}} \int d\ell$$

$$\oint \vec{E} \cdot d\vec{\ell} = [E(x + dx)]h - [E(x)]h$$

The term in square brackets is the derivative of the electric field with respect to x times the differential dx. We use the partial derivative because the electric field is a function of time and position, and our derivative is taken only over position x while the time t is held constant.

$$\oint \vec{E} \cdot d\vec{\ell} = h[E(x + dx) - E(x)]_{t\,\text{constant}}$$

$$\oint \vec{E} \cdot d\vec{\ell} = h\left(\frac{\partial E}{\partial x}\right)dx \quad (34.9)$$

Find the change in magnetic flux. Figure 34.9 shows one magnetic field line perpendicular to the electric field and pointing out of the page. Wrap the fingers of your right hand counterclockwise in the direction of the path; then the magnetic field is in the same direction as the (positive) area vector. If we assume the magnetic field is uniform over this very small rectangle, the magnetic flux is easy to find. The area A is the area of the rectangle.

$$\Phi_B = \int \vec{B} \cdot d\vec{A} = BA$$

$$\Phi_B = B(h\,dx) \quad (34.10)$$

Substitute Equations 34.9 and 34.10 **into Faraday's law.** The height h and width dx of the rectangle are constants. The magnetic field is a function of position and time. Because we need to take the derivative with respect to time only, we again use a partial derivative, this time holding x constant. Equation 34.11 is the first of two partial differential equations we need.	$\oint \vec{E} \cdot d\vec{\ell} = -\dfrac{d\Phi_B}{dt}$ $h\left(\dfrac{\partial E}{\partial x}\right)dx = -\dfrac{d}{dt}B(h\,dx) = -h\left(\dfrac{\partial B}{\partial t}\right)dx$ $\left(\dfrac{\partial E}{\partial x}\right) = -\left(\dfrac{\partial B}{\partial t}\right)$ (34.11)
We are about halfway through our derivation. The next step is to apply Ampère–Maxwell's law (Eq. 31.8), setting $I_{\text{thru}} = 0$ because there is no current in a vacuum. Equation 34.12 is mathematically similar to Faraday's law, so the procedure is much like what we did above and follows the same four steps.	$\oint \vec{B} \cdot d\vec{\ell} = \mu_0 I_{\text{thru}} + \mu_0\varepsilon_0 \dfrac{d\Phi_E}{dt}$ (31.8) $\oint \vec{B} \cdot d\vec{\ell} = \mu_0\varepsilon_0 \dfrac{d\Phi_E}{dt}$ (34.12)

Figure 34.10 is similar to Figure 34.9 and combines the first two steps. **Sketch magnetic field lines, and choose a closed path.** The rectangle lies in the xz plane; its length is dx and its width is w. For this particular region at this particular instant, the magnetic field is weaker on the left, $\vec{B}(x)$, than it is on the right, $\vec{B}(x+dx)$. Our path is a rectangle as shown. Wrap the fingers of your right hand in the direction of the path; the electric field then points in the same direction as the (positive) area vector.

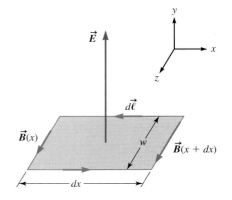

FIGURE 34.10 A small portion of the wave in Figure 34.8, showing the plane of the magnetic field.

The circulation integral is similar to the circulation integral over the electric field. Only the left and right portions of the path contribute to the integral. The dot product on the path's left is positive and on the right is negative. The magnetic field is constant over an individual path segment. The integrals $\int d\ell$ are both equal to the width w of the rectangle.	$\oint \vec{B} \cdot d\vec{\ell} = \left(\int \vec{B} \cdot d\vec{\ell}\right)_{\text{right}} + \left(\int \vec{B} \cdot d\vec{\ell}\right)_{\text{left}}$ $\oint \vec{B} \cdot d\vec{\ell} = -B_{\text{right}}\int d\ell + B_{\text{left}}\int d\ell$ $\oint \vec{B} \cdot d\vec{\ell} = [-B(x+dx)]w + [B(x)]w$
The term in square brackets is the derivative of the magnetic field with respect to x times the differential dx. As before, this is a partial derivative.	$\oint \vec{B} \cdot d\vec{\ell} = -w[B(x+dx) - B(x)]_{t\,\text{constant}}$ $\oint \vec{B} \cdot d\vec{\ell} = -w\left(\dfrac{\partial B}{\partial x}\right)dx$ (34.13)
Find the change in electric flux. The electric field is perpendicular to the rectangle and parallel to the area vector. If we assume the electric field is uniform over this very small rectangle, the electric flux is easy to find.	$\Phi_E = \int \vec{E} \cdot d\vec{A} = EA$ $\Phi_E = E(w\,dx)$ (34.14)
Substitute Equations 34.13 and 34.14 into Ampère–Maxwell's law as expressed in Equation 34.12. The length dx and width w of the rectangle are constants. The electric field is a function of position and time. Because we need to take the derivative with respect to time only, we again use a partial derivative, holding x constant. Equation 34.15 is the second of the two partial differential equations we need.	$\oint \vec{B} \cdot d\vec{\ell} = \mu_0\varepsilon_0 \dfrac{d\Phi_E}{dt}$ (34.12) $-w\left(\dfrac{\partial B}{\partial x}\right)dx = \mu_0\varepsilon_0 \dfrac{d}{dt}E(w\,dx)$ $-w\left(\dfrac{\partial B}{\partial x}\right)dx = \mu_0\varepsilon_0 w\left(\dfrac{\partial E}{\partial t}\right)dx$ $\left(\dfrac{\partial B}{\partial x}\right) = -\mu_0\varepsilon_0\left(\dfrac{\partial E}{\partial t}\right)$ (34.15)

Derivation continues on page 1098 ▶

We are not done because both equations (Eqs. 34.11 and 34.15) involve B and E in the same expression. We need one equation for E. There is a bit of a trick to finding it.

Take the partial derivative of Equation 34.11 with respect to x. On the left side, we find the second partial derivative of E with respect to x. On the right side, we switch the order of differentiation.	$\frac{\partial}{\partial x}\left(\frac{\partial E}{\partial x}\right) = -\frac{\partial}{\partial x}\left(\frac{\partial B}{\partial t}\right)$ $\frac{\partial^2 E}{\partial x^2} = -\frac{\partial}{\partial t}\left(\frac{\partial B}{\partial x}\right)$
Substitute Equation 34.15 for $\partial B/\partial x$. This eliminates B and gives us an equation that involves only E.	$\frac{\partial^2 E}{\partial x^2} = -\frac{\partial}{\partial t}\left[-\mu_0\varepsilon_0\left(\frac{\partial E}{\partial t}\right)\right]$ $\frac{\partial^2 E}{\partial x^2} = \mu_0\varepsilon_0\frac{\partial^2 E}{\partial t^2}$ ✓ (34.7)
We can find a similar wave equation for the magnetic field using a similar trick (Problem 26).	$\frac{\partial^2 B}{\partial x^2} = \mu_0\varepsilon_0\frac{\partial^2 B}{\partial t^2}$ (34.8)

⁖ COMMENTS

Equation 34.7 is the **wave equation** for the electric field, a differential equation in which the second spatial derivative is proportional to the second time derivative. Likewise, Equation 34.8 is the wave equation for the magnetic field.

A general wave equation is given by Equation 17.33, where v is the propagation speed of the wave.	$\frac{\partial^2 y}{\partial x^2} = \frac{1}{v^2}\frac{\partial^2 y}{\partial t^2}$ (17.33)

Properties of Electromagnetic Waves

We have just shown that if we combine Ampère's law and Faraday's law, we come up with an electromagnetic wave—an electric field and a magnetic field that oscillate periodically as depicted in Figure 34.8. By comparing Equation 17.33 to either Equation 34.7 or 34.8, we find that the speed of the wave comes from the coefficient on the right side of the wave equation:

$$\frac{1}{v^2} = \mu_0\varepsilon_0$$

$$v = \frac{1}{\sqrt{\mu_0\varepsilon_0}} \tag{34.16}$$

The entire wave moves along the x axis (in a direction perpendicular to both fields) at a speed v given by Equation 34.16. Let's calculate a numerical value for the speed by substituting for the constants. We leave it as a homework problem (Problem 18) to show that the units work out to m/s.

$$v = \frac{1}{\sqrt{\mu_0\varepsilon_0}} = \frac{1}{\sqrt{\left(4\pi \times 10^{-7}\frac{\text{T}\cdot\text{m}}{\text{A}}\right)\left(8.854 \times 10^{-12}\frac{\text{C}^2}{\text{N}\cdot\text{m}^2}\right)}} = 2.998 \times 10^8 \text{ m/s}$$

$$v = \frac{1}{\sqrt{\mu_0\varepsilon_0}} = c \tag{34.17}$$

The value is known as the **speed of light** c.

Because the electromagnetic wave equations (Eqs. 34.7 and 34.8) are mathematically identical to the corresponding equation for a wave on a string (Eq. 17.33), the solutions to electromagnetic wave equations are the same as the solution to the transverse mechanical wave equation $y(x, t) = y_{max} \sin(kx - \omega t)$ (Eq. 17.4) with a simple translation of variables:

$$E(x, t) = E_{max} \sin(kx - \omega t) \qquad (34.18)$$

$$B(x, t) = B_{max} \sin(kx - \omega t) \qquad (34.19)$$

As with the oscillation of beads on a string, both fields ($\vec{E}$ and $\vec{B}$) are perpendicular to the direction of propagation, so this is a **transverse wave**. But now, instead of beads on a string oscillating up and down, electric and magnetic fields oscillate in both magnitude and direction (Fig. 34.8). The electric field is parallel to the y axis. It grows and shrinks periodically just as a bead on a string moves up and down along the y axis. The magnetic field is parallel to the z axis and also grows and shrinks periodically. Compare Equations 34.18 and 34.19; the speed v, wave number k, and angular frequency ω are the same for the electric and magnetic field oscillations, so they have the same frequency f and period T. Like a mechanical wave, the speed, frequency, angular frequency, angular wave number, and wavelength are connected by

$$v = \frac{\omega}{k} = \lambda f = c \qquad (34.20)$$

TRANSVERSE ELECTRIC AND MAGNETIC WAVE ◯ **Major Concept**

Table 34.1 summarizes what we know about electromagnetic waves by comparing them to transverse waves on a string.

TABLE 34.1 Comparison between transverse wave on a string and electromagnetic wave.

	Transverse wave on a string	Linearly polarized electromagnetic wave in a vacuum
Snapshot—a picture of the wave taken at one instant	Figure 17.8A, page 491	Figure 34.8, page 1095
Wave equation(s)	$\dfrac{\partial^2 y}{\partial x^2} = \dfrac{1}{v^2} \dfrac{\partial^2 y}{\partial t^2}$ (17.33)	$\dfrac{\partial^2 E}{\partial x^2} = \mu_0 \varepsilon_0 \dfrac{\partial^2 E}{\partial t^2}$ (34.7) and $\dfrac{\partial^2 B}{\partial x^2} = \mu_0 \varepsilon_0 \dfrac{\partial^2 B}{\partial t^2}$ (34.8)
Solution to wave equation(s)	$y(x, t) = y_{max} \sin(kx - \omega t)$ (17.4) The string oscillates along the y direction, while the wave moves in the positive x direction.	$\vec{E}(x, t) = [E_{max} \sin(kx - \omega t)]\hat{\jmath}$ (34.18) $\vec{B}(x, t) = [B_{max} \sin(kx - \omega t)]\hat{k}$ (34.19) The electric field oscillates along the y direction and the magnetic field oscillates along the z direction, while the wave moves in the positive x direction.
Propagation speed—speed of the wave in the x direction	$v = \dfrac{\omega}{k} = \lambda f$ (17.7) and (17.8)	$v = \dfrac{\omega}{k} = \lambda f = c$ (34.20)

The Connection Between $\vec{E}$ and $\vec{B}$

Consider the wave on the string in Figure 17.8A (page 491). In that case, the string is connected to an oscillator. The amplitude y_{max} of the wave depends on the amplitude of the oscillator. Similarly, the amplitude E_{max} of the electric field depends on the maximum electric field produced by the transmitter. That same transmitter creates both an electric and a magnetic field, so there is a connection between the amplitude of the electric part of the wave (E_{max}) and the magnetic part (B_{max}).

Here's how we can find that connection. Start by taking the partial derivative of E (Eq. 34.18) with respect to x while holding t constant:

$$\frac{\partial E(x, t)}{\partial x} = \frac{\partial}{\partial x} E_{max} \sin(kx - \omega t) = k E_{max} \cos(kx - \omega t) \qquad (34.21)$$

Next, take the partial derivative of B (Eq 34.19) with respect to t while holding x constant:

$$\frac{\partial B(x, t)}{\partial t} = \frac{\partial}{\partial t} B_{max} \sin(kx - \omega t) = -\omega B_{max} \cos(kx - \omega t) \quad (34.22)$$

Now substitute Equations 34.21 and 34.22 into $(\partial E/\partial x) = -(\partial B/\partial t)$ (Eq. 34.11):

$$kE_{max} \cos(kx - \omega t) = -[-\omega B_{max} \cos(kx - \omega t)] = \omega B_{max} \cos(kx - \omega t)$$

Solve for E_{max}:

$$kE_{max} = \omega B_{max}$$

$$E_{max} = \frac{\omega}{k} B_{max}$$

Use $\omega/k = c$ (Eq. 34.20) to eliminate the angular frequency and angular wave number:

$$E_{max} = cB_{max} \quad (34.23)$$

Equation 34.23 is the connection between the amplitude of the electric field and that of the magnetic field. Inserting Equation 34.23 into Equation 34.18 gives

$$E(x, t) = c[B_{max} \sin(kx - \omega t)]$$

The term in square brackets is the magnetic field (Eq. 34.19), so

$$E(x, t) = cB(x, t) \quad (34.24)$$

At every moment, the magnitude of the electric field is c times the magnitude of the magnetic field. Although c is a very large number, the electric and magnetic fields are measured in different units, so you cannot conclude that the electric field is stronger than the magnetic field. Both fields contribute equally to the energy of the electromagnetic wave.

Let's look at the energy density stored in each field to show that they are equal. The energy density stored in an electric field in a vacuum ($\kappa = 1$) is $u_E = \frac{1}{2}\varepsilon_0 E^2$ (Eq. 27.22). Substitute $E = cB$ (Eq. 34.24):

$$u_E = \frac{1}{2}\varepsilon_0 c^2 B^2 \quad (34.25)$$

Now, we use the propagation speed $c = 1/\sqrt{\mu_0 \varepsilon_0}$ (Eq. 34.17):

$$u_E = \frac{1}{2}\varepsilon_0 \frac{1}{\mu_0 \varepsilon_0} B^2 = \frac{1}{2}\frac{B^2}{\mu_0}$$

This is the same as the energy density stored in the magnetic field, $u_B = B^2/2\mu_0$ (Eq. 33.17):

$$u_E = u_B \quad (34.26)$$

So, the electric and magnetic fields contribute an equal amount of energy to an electromagnetic wave.

CONCEPT EXERCISE 34.3

The electric part of an electromagnetic wave is given by
$E(x, t) = 0.75 \sin(0.30x - \omega t)$ V/m in SI units.

    a. What are the amplitudes E_{max} and B_{max}?
    b. What are the angular wave number and the wavelength?
    c. What is the propagation velocity?
    d. What are the angular frequency, frequency, and period?

34-5 The Electromagnetic Spectrum

In 1864, in front of the Royal Society of London, Maxwell declared that light and other forms of radiation are electromagnetic disturbances in the form of waves that propagate according to the laws of electricity and magnetism. In Section 34-4, we started with Maxwell's equations for electricity and magnetism and derived an equation for an electromagnetic wave. We showed that the wave moves at the speed of light.

The frequency and wavelength of any electromagnetic wave in a vacuum are given by $\lambda f = c$ (Eq. 34.20). The term **electromagnetic spectrum** refers to the continuum of electromagnetic waves arranged in order by frequency (and wavelength). In practice, the electromagnetic spectrum is divided into a few frequency bands. These bands blend together; there is no strict boundary between bands, and the terms given to the bands often vary from one branch of science and technology to the next. The electromagnetic spectrum and bands shown in Figure 34.11 and listed in Table 34.2 are commonly used in introductory physics. (You may find slight discrepancies when you look at other sources.)

All electromagnetic waves are produced by accelerating charged particles. Hertz's experiment is just one example in which electrons oscillating in an LC circuit create an electromagnetic wave. The bands in the electromagnetic spectrum are convenient because they correspond to how the charged particles are accelerated and to methods for detecting those waves. We will briefly describe the bands listed in Table 34.2.

The terms *electromagnetic wave* and *electromagnetic radiation* are interchangeable, but from now on we save the term *visible light* for electromagnetic waves that healthy human eyes can sense.

Unless we specify otherwise, assume the electromagnetic wave is in a vacuum with propagation speed c.

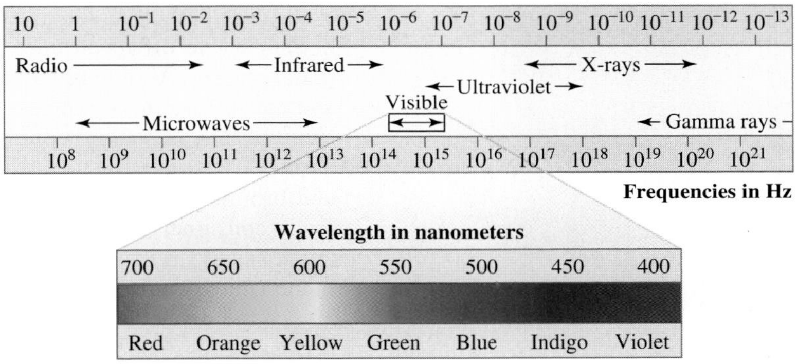

FIGURE 34.11 The electromagnetic spectrum is separated into bands for convenience. The bands overlap and do not have strict boundaries. It may help to memorize the visible colors in order from long to short wavelength by thinking of the name "Roy G. Biv," which stands for Red, Orange, Yellow, Green, Blue, Indigo, and Violet. (Values here may differ slightly from those in Table 34.2.)

TABLE 34.2 The electromagnetic spectrum broken into convenient bands. Values are approximate.

Name of band	Wavelength λ (m)	Frequency f (Hz)
Radio	$> 10^{-2}$	$< 10^{11}$
Microwave	10^{-4}–1	10^{9}–10^{13}
Infrared (IR)	10^{-6}–10^{-4}	10^{12}–10^{14}
Visible light	10^{-7}–10^{-6}	10^{14}–10^{15}
Ultraviolet (UV)	10^{-9}–10^{-7}	10^{15}–10^{18}
X-rays	10^{-12}–10^{-9}	10^{17}–10^{20}
Gamma rays	$< 10^{-10}$	$> 10^{19}$

Radio waves are generated by the acceleration of electrons in a circuit and are used for communication. In practice, the electrons oscillate in the antenna of a transmitter (Fig. 34.12). Radio waves are also produced by astronomical sources. For example, neutron stars produce radio waves when high-speed electrons are accelerated by very strong magnetic fields (Table 30.1). The resulting radiation covers many parts of the electromagnetic spectrum, but it is often easiest to detect in the radio band. A radio telescope receiver is much like the circuits that detect artificial radio waves.

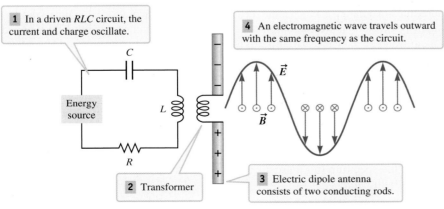

FIGURE 34.12 The basic design of a radio transmitter. **1** An LC circuit is driven to compensate for energy loss due to internal resistance and radiation. **2** The circuit is coupled to an antenna by a transformer, causing charge to oscillate in the antenna. **3** The current in the antenna oscillates. Effectively, the antenna has a dipole moment that oscillates in magnitude and direction. The dipole's electric field also oscillates. The changing current creates an oscillating magnetic field. **4** The changing electric and magnetic fields travel out into space as an electromagnetic wave.

FIGURE 34.13 Microwave radiation detected from the early Universe. The image is a map of the whole sky. The wavelengths of the microwaves are used to determine temperature. The red regions are slightly hotter than the blue regions.

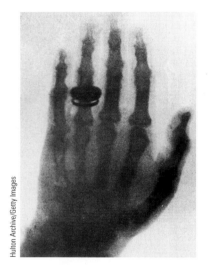

FIGURE 34.14 An X-ray image of Bertha Röntgen's hand.

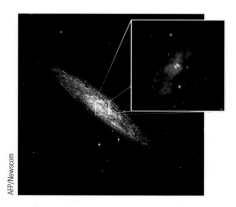

FIGURE 34.15 Chandra, a space-based telescope, took this X-ray image (*inset*) of the central region of a starburst galaxy (NGC 253). The X-ray image may be evidence of intermediate-mass black holes in the galaxy.

Microwaves are also generated by electronic devices and by natural sources. Their production and detection are similar to those of radio waves, and both wave bands are used for communication. Microwaves are better for certain applications such as **radar** (*ra*dio *d*etection *a*nd *r*anging), which was developed during World War II to detect aircraft and ships. Microwaves are transmitted, reflected by objects such as aircraft, and then detected by the radar receiver. In another common application, microwaves are used to cook food because they are absorbed by water and fat. One of the most important sources of microwaves is the Universe (Fig. 34.13). Early in the formation of the Universe, light was generated, and we observe that radiation today as microwaves that have a wavelength of about 1 mm coming from all directions. By studying this radiation, cosmologists can measure the age of the Universe and predict its fate.

Infrared radiation is produced by the motion within molecules when they vibrate and rotate. We often associate infrared radiation with a warm object such as a stovetop. The temperature of the object is a measure of its thermal energy. When you hold your hand near the stovetop, infrared radiation causes the molecules in your hand to move more quickly (Section 21-10), so your hand feels warmer.

Visible light is produced by the motion of electrons in atoms and molecules. Electrons in atoms orbit the nucleus at different distances. When an electron moves from a high orbit to a lower orbit, visible light is given off. Human eyes have two types of detectors—known as cones and rods—to detect visible light. Most of the Sun's electromagnetic radiation is in the form of visible light, and our eyes are sensitive to this radiation band because humans evolved on the Earth's surface. White light is made up of all the colors of the rainbow. Red light has the longest wavelength; violet has the shortest.

Ultraviolet (UV) radiation is produced in much the same way as visible light. Very hot stars give off much ultraviolet radiation (as does the Sun, but not as much as hotter stars). UV radiation causes human skin to tan or burn, promotes skin cancer, and causes cataracts, which cloud our vision. Clothing, sunscreen, and sunglasses should be used for protection. In addition, ozone (O_3) in the Earth's atmosphere absorbs UV radiation, but ozone is destroyed by chemical reactions with chlorine and bromine. Chlorofluorocarbons (CFCs), formerly present in refrigerants and in many consumer products, are a reservoir for chlorine. In the mid-1980s, the ozone layer was greatly depleted and a hole was discovered over Antarctica. Every country in the United Nations agreed to phase out the use of CFCs to protect the ozone layer. Today, the ozone layer has been greatly restored.

X-rays are produced when high-energy electrons bombard metal so that the metal stops the electrons. Because the electrons decelerate, electromagnetic radiation is created. X-rays can penetrate soft tissues in the body but not bone. The German physicist Wilhelm Conrad Röntgen won the first Nobel Prize in physics in 1901 for his discovery of X-rays. One of the first X-ray photos he took was of his wife Bertha's hand (Fig. 34.14). Her bones and ring are clearly visible. Today, X-rays are used both as a medical diagnostic tool and as a therapy for cancer.

Gamma rays are given off during certain nuclear reactions, such as fusion reactions in the Sun's core and fission reactions on the Earth. Gamma rays can cause significant biological damage, so—like X-rays—gamma rays can be used to destroy cancerous cells. However, gamma rays can also be very dangerous. Exposure to gamma rays can cause intestinal damage, cancer, and death. In the 1930s, women who painted radium watch and clock dials were exposed to gamma radiation, and many of them developed leukemia and breast cancer. Many survivors of the atomic bombs dropped over Hiroshima and Nagasaki also developed cancer. Both X-rays and gamma rays are produced by astronomical sources (Fig. 34.15). This radiation does not penetrate the Earth's atmosphere, so X-ray and gamma-ray detectors are put into orbit.

CONCEPT EXERCISE 34.4

Consider the spectrum in Figure 34.11 and propose an explanation of the terms in-fra*red* and ultra*violet*.

CONCEPT EXERCISE 34.5

The most abundant element in the Universe is hydrogen (the subject of the case study in Chapter 42). Neutral hydrogen gives off electromagnetic waves with a wavelength of 21 cm. What is the frequency of this radiation? In what part of the electromagnetic spectrum is it found?

34-6 Energy and Intensity

Like all waves, electromagnetic waves transport energy along the direction of propagation. In 1884, the British physicist John Henry Poynting (1852–1914) published a paper entitled *On the Transfer of Energy in the Electromagnetic Field*. In this paper, he showed that the flow of energy at a point can be expressed in terms of the electric and magnetic fields at that point. Mathematically, the energy transferred per unit time is given by the magnitude of the **Poynting vector** $\vec{S}$:

$$\vec{S} \equiv \frac{1}{\mu_0}(\vec{E} \times \vec{B}) \quad (34.27)$$

The magnitude S of the Poynting vector is the **power flux**, the energy transported per unit time per unit area (which is the same as the power per unit area) of a surface perpendicular to the direction of the Poynting vector. The SI units for power flux are W/m². The power flux depends on the position x and time t because the electric and magnetic fields depend on x and t.

To make this idea more concrete, let's use Equations 34.18 and 34.19 for a plane wave (Fig. 34.8) to find an expression for the Poynting vector. First, calculate the cross product:

$$(\vec{E} \times \vec{B}) = [E_{max} \sin(kx - \omega t)]\hat{\jmath} \times [B_{max} \sin(kx - \omega t)]\hat{k}$$

$$(\vec{E} \times \vec{B}) = [E_{max} B_{max} \sin^2(kx - \omega t)]\hat{\jmath} \times \hat{k}$$

Because $\hat{\jmath} \times \hat{k} = \hat{\imath}$, the Poynting vector is

$$\vec{S}(x, t) = \frac{1}{\mu_0}[E_{max} B_{max} \sin^2(kx - \omega t)]\hat{\imath} \quad (34.28)$$

The magnitude S of the Poynting vector at one instant is plotted in Figure 34.16. Compare the power flux S to the electric and magnetic fields in the wave at the same instant. At points on the x axis where the electric and magnetic fields are at their maxima, the power flux is at its maximum. At points where the electric and magnetic fields are zero, the power flux is also zero. The power flux is never negative, and when the electric and magnetic fields are pointing in the negative direction, the power flux is still positive, so the Poynting *vector* still points in the positive x direction.

ENERGY TRANSFERRED BY ELECTRO-MAGNETIC WAVES

✪ Major Concept

The Poynting vector "points" in the direction of energy transport.

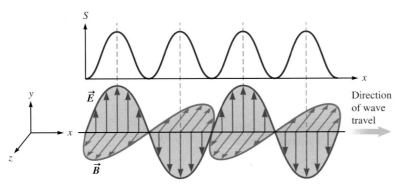

FIGURE 34.16 The power flux depends on time and position. This is the power flux at one particular time. The power flux is greatest when the fields are at their maxima. The direction of the fields does not affect the magnitude of the power flux.

The Poynting vector tells us about the power flux as a function of time and position. However, we are usually more interested in the intensity (power per unit area) of the radiation. From Section 17-7, the intensity I is the time-averaged power per unit area (flux):

$$I \equiv S_{av}$$

The average of $\sin^2(kx - \omega t)$ over a period is $\frac{1}{2}$ (Problem 76), so the intensity I is

$$I = \frac{1}{2\mu_0} E_{max} B_{max} \qquad (34.29)$$

It is often convenient to express the intensity in terms of either E_{max} or B_{max} by using $E = cB$ (Eq. 34.24):

INTENSITY ⭐ **Major Concept**

$$I = \frac{E_{max}^2}{2\mu_0 c} = \frac{c}{2\mu_0} B_{max}^2 \qquad (34.30)$$

Recall that the intensity from a point source such as a distant lightbulb, radio antenna, or galaxy decreases as the distance r from the source increases:

$$I = \frac{P_{av}}{4\pi r^2} \qquad (17.23)$$

where P_{av} is the average power emitted by the source.

Intensity is also related to the average energy density stored by the electric and magnetic fields, as we can show by starting with an expression for the total energy density:

$$u = u_E + u_B$$

The energy density stored by the electric field is equal to the energy density stored by the magnetic field, $u_E = u_B$ (Eq. 34.26), so the total energy density is

$$u = 2u_E = 2u_B$$

We can write the energy density in terms of the electric field or the magnetic field. Arbitrarily choosing the magnetic field, we substitute $u_B = B^2/2\mu_0$ (Eq. 33.17):

$$u = 2\left(\frac{B^2}{2\mu_0}\right) = \frac{B^2}{\mu_0}$$

Next, we substitute $B(x, t) = B_{max} \sin(kx - \omega t)$ (Eq. 34.19):

$$u = \frac{1}{\mu_0}\left[B_{max}^2 \sin^2(kx - \omega t)\right]$$

The energy density u depends on time. We need the energy density averaged over one or more periods and, as before, the average of $\sin^2(kx - \omega t)$ is $\frac{1}{2}$. So the average energy density is

$$u_{av} = \frac{1}{2\mu_0} B_{max}^2 \qquad (34.31)$$

If we had worked in terms of the electric field, we would have found

$$u_{av} = \frac{1}{2}\varepsilon_0 E_{max}^2 = \frac{1}{2\mu_0 c^2} E_{max}^2 \qquad (34.32)$$

Substitute Equation 34.31 or 34.32 into Equation 34.30, and the result is

$$I = cu_{av} \qquad (34.33)$$

So the intensity is proportional to the average energy density.

CONCEPT EXERCISE 34.6

What is the direction of the Poynting vector for the wave shown in Figure 34.8?

EXAMPLE 34.2 Buy More Sunscreen

A salesperson once suggested that the author of this book should wear sunscreen on her face every day whether she planned to be out in the Sun or not. The salesperson argued that indoor lighting can cause as much damage as the Sun to a person with fair skin. Let's analyze this claim.

The author does have fair skin, and on a sunny day she will burn in about 20 minutes without protection. The author's face has an area of roughly 2×10^{-2} m². The Sun's power output is 3.85×10^{26} W. Estimate the amount of energy from the Sun that the author's face is exposed to in 20 minutes. While writing this textbook, the author sits about 2 m from a 150-W lightbulb. Model the lightbulb as a point source, and estimate the time the author would need to sit near the bulb in order to absorb as much energy from the bulb as she does from the Sun in 20 minutes.

INTERPRET and ANTICIPATE
Our first step is to find the intensity of sunlight on Earth. Once we know that, we can find the power delivered to the author's face by multiplying the intensity by the area of the face. Finally, find the energy delivered in the 20-minute period by multiplying the power by the time of exposure. This will tell us how much energy her face must be exposed to in order to burn. Then find the intensity of the lightbulb's radiation at her face and the power her face is exposed to from that bulb. To find the time required in the problem, divide the energy her face is exposed to in 20 minutes of sunlight by the power delivered by the bulb. We expect the answer will be greater than 20 minutes.

SOLVE
Find the intensity of sunlight at the position of the Earth using Equation 17.23. The distance between the Earth and the Sun is 1 AU = 1.50×10^{11} m.

$$I = \frac{P_{av}}{4\pi r^2} \quad (17.23)$$

$$I = \frac{(3.85 \times 10^{26}\,\text{W})}{4\pi(1.50 \times 10^{11}\,\text{m})^2}$$

$$I = 1360\,\text{W/m}^2$$

The power P delivered to the author's face is the intensity times the area of her face. We retain an extra significant figure for now.

$$P = IA \quad (34.34)$$

$$P = (1360\,\text{W/m}^2)(2 \times 10^{-2}\,\text{m}^2)$$

$$P = 27\,\text{W}$$

The energy (heat) Q delivered to her face in 20 minutes (1200 s) is the power times the time of exposure. (Q stands for energy here, not charge.)

$$Q = (27\,\text{W})(1200\,\text{s})$$

$$Q = 3 \times 10^4\,\text{J} \quad (1)$$

Now we turn our attention to the 150-W lightbulb. Find its intensity at a distance of 2 m as given by Equation 17.23.

$$I = \frac{P_{av}}{4\pi r^2} \quad (17.23)$$

$$I = \frac{(150\,\text{W})}{4\pi(2\,\text{m})^2} = 3\,\text{W/m}^2$$

Again, the power P delivered to the author's face is the intensity times her face's area.

$$P = IA = (3\,\text{W/m}^2)(2 \times 10^{-2}\,\text{m}^2)$$

$$P = 0.06\,\text{W} \quad (2)$$

The time for her face to be exposed to as much energy from the lightbulb as it is in a 20-minute sunbath is the energy Q (Eq. 1) divided by the power P (Eq. 2).

$$\Delta t = \frac{Q}{P} = \frac{3 \times 10^4\,\text{J}}{0.06\,\text{J/s}}$$

$$\Delta t = 5 \times 10^5\,\text{s} = 140\,\text{h}$$

CHECK and THINK
It takes about $3\frac{1}{2}$ workweeks to be exposed to as much energy from the 150-W bulb as in just 20 minutes in sunlight. Of course, during those workweeks, a person spends about 8 hours per day in total darkness. Clearly, exposure to sunlight is much more hazardous than exposure to normal indoor lighting. Nevertheless, over the course of a lifetime, a person's face is exposed to a great deal of electromagnetic energy.

EXAMPLE 34.3 — CASE STUDY — Meteor Estimate Revisited

In Chapter 9's case study, three physics professors estimated the size of a meteor. They worked in the dark without a calculator, no steps were written down, and their estimate was made to one significant figure. Using the conservation of energy approach (Example 9.12, page 272), they found a radius $R_{mtr} = 0.3$ mm. Now consider another approach based on light from the meteor. All three professors agree that the meteor looks about as bright as an ordinary faint star. Professor Black knows that if the Sun were about 30 light-years from the Earth, it would look like an ordinary star. Professor Noir read that meteors are visible when they are about 100 km above us. They all know that the Sun's radius is about 110 times the Earth's radius, or about $R_\odot \approx 6.6 \times 10^5$ km. Finally, they know that a meteor gets hot when it travels through the Earth's atmosphere. They assume it gets about as hot as the Sun's surface. They model both the Sun and the meteor as black bodies. Use their values and observations to estimate the radius of a meteor.

:• INTERPRET and ANTICIPATE
This problem brings together concepts from many chapters, including some ideas from Chapter 21. We expect our estimate to be consistent with the estimate from Chapter 9 based on conservation of energy.

:• SOLVE

The meteor is the same temperature as the Sun's surface. From Equation 21.34, the power P of radiation given off by an object is proportional to the surface area of the object and its temperature to the fourth power. Because the meteor and the Sun are assumed to be the same temperature, the ratio of their emitted powers depends only on their radii. (Because they are modeled as good emitters, both objects have $\varepsilon \approx 1$; see Section 21-10.)

$$P = \frac{Q}{\Delta t} = 4\pi R^2 \varepsilon \sigma T^4 \qquad (21.34)$$

$$P_\odot = 4\pi R_\odot^2 \varepsilon \sigma T_\odot^4$$

$$P_{mtr} = 4\pi R_{mtr}^2 \varepsilon \sigma T_\odot^4$$

$$\frac{P_{mtr}}{P_\odot} = \frac{R_{mtr}^2}{R_\odot^2} \qquad (1)$$

The professors observed that the meteors look as bright as ordinary stars. Professor Black knows that the Sun would look like an ordinary star if it were 30 ly from us. So the intensity of the meteor at altitude $r_{mtr} = 100$ km equals the intensity the Sun would have if it were $r_\odot = 30$ ly away. Use $I = P_{av}/4\pi r^2$ (Eq. 17.23) with $r_\odot = 30$ ly and $r_{mtr} = 100$ km.

$$\frac{P_\odot}{4\pi r_\odot^2} = \frac{P_{mtr}}{4\pi r_{mtr}^2} \qquad (2)$$

Solve Equations (1) and (2) for the radius of the meteor, R_{mtr}.

$$\frac{P_{mtr}}{P_\odot} = \frac{4\pi r_{mtr}^2}{4\pi r_\odot^2} \qquad \frac{R_{mtr}^2}{R_\odot^2} = \frac{r_{mtr}^2}{r_\odot^2}$$

$$R_{mtr} = \frac{r_{mtr}}{r_\odot} R_\odot \qquad (3)$$

To convert 30 ly to kilometers, remember that $c = 3 \times 10^5$ km/s and there are roughly 3×10^7 s in a year. Substitute numerical values into Equation (3), including $r_\odot$ in kilometers.

$$r_\odot = 30 \text{ ly} \approx 30 \text{ yr} \times \frac{3 \times 10^5 \text{ km}}{\text{s}} \times \frac{3 \times 10^7 \text{ s}}{\text{yr}}$$

$$r_\odot \approx 3 \times 10^{14} \text{ km}$$

$$R_{mtr} = \left(\frac{100 \text{ km}}{3 \times 10^{14} \text{ km}}\right)(6.6 \times 10^5 \text{ km})$$

$$\boxed{R_{mtr} = 2.2 \times 10^{-7} \text{ km} \approx 0.2 \text{ mm}}$$

:• CHECK and THINK
This result is very close to the estimate from Chapter 9. This example illustrates how physicists model a complicated problem to make an estimate. It also shows the importance of knowing a couple of numerical facts. Whenever you must make a detailed calculation, start with an estimate as a way to check your answer. This is particularly important if you need to calculate a value that is completely unknown.

34-7 Momentum and Radiation Pressure

In Chapter 17, we mentioned that waves transport momentum and we developed equations for wave pressure. Electromagnetic waves also transport momentum and exert pressure. We are familiar with the pressure exerted by a sound wave on the eardrum, but the pressure exerted by light on our skin may not be so obvious. In this section, we estimate the (very small) pressure exerted by sunlight on skin. The pressure exerted on a large solar sail attached to a deadly asteroid may be high enough to save humanity.

First, let's derive an expression for the momentum an electromagnetic wave imparts to a single particle; our results can be extended to any object.

DERIVATION   **Momentum Delivered by an Electromagnetic Wave**

Consider a simple situation in which a single positively charged particle (charge $+q$) is at rest when an electromagnetic wave strikes it, delivering energy Q. The wave is linearly polarized as shown in Figure 34.17. We will show that the momentum p delivered to the particle is proportional to the energy Q:

$$p = Q/c \qquad (34.35)$$

MOMENTUM TRANSFERRED BY (ABSORBED) ELECTROMAGNETIC WAVE ✪ **Major Concept**

Don't confuse q for charge with Q for energy.

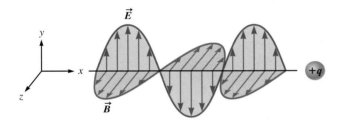

FIGURE 34.17 An electromagnetic wave traveling to the right encounters a single positively charged particle.

Consider one instant when the electric field $\vec{E}$ at the particle's location is in the positive y direction and the magnetic field $\vec{B}$ is in the positive z direction. The electric force on the particle is then in the positive y direction as given by Equation 24.2.	$\vec{F}_E = qE\hat{\jmath}$ (24.2)
This force causes the particle of mass m to accelerate (from rest) in the positive y direction.	$a_y = \dfrac{qE}{m}$
To find the particle's speed, assume it has been in contact with the wave for a short time t, and during that time, the electric force on it is constant.	$v_y = a_y t = \dfrac{qE}{m} t$ (34.36)
Because the particle is moving in the positive y direction, the magnetic field exerts a force on it in the positive x direction according to the right-hand rule for cross products (Eq. 30.17). The velocity is perpendicular to the magnetic field and its magnitude is given by Equation 34.36.	$\vec{F}_B = q(\vec{v} \times \vec{B})$ (30.17) $\vec{F}_B = qv_y B\hat{\imath} = \dfrac{q^2 E}{m} tB\hat{\imath}$
The magnetic force is in the direction of the wave's propagation, and we can think of the wave as delivering momentum in the x direction. To find the momentum p_x, combine Equations 11.3 and 11.4 (page 309). The particle is initially at rest, so its change in momentum is just its final momentum in the x direction, p_x.	$\Delta \vec{p} = \displaystyle\int_0^t \vec{F}_B \, dt$ $(p_f - p_i)_x = \dfrac{q^2 EB}{m} \displaystyle\int_0^t t\, dt = \dfrac{q^2 EB}{2m} t^2$
Use $E = cB$ (Eq. 34.24) to eliminate B.	$p_x = \dfrac{q^2 E^2}{2mc} t^2$

Derivation continues on page 1108 ▶

Substitute Equation 34.36, which is v_y. The particle started from rest, so the term $\frac{1}{2}mv_y^2$ is the increase in its kinetic energy.	$p_x = \dfrac{m}{2c}\left(\dfrac{q^2E^2}{m^2}t^2\right) = \dfrac{1}{c}\left(\dfrac{1}{2}mv_y^2\right)$
So, the momentum p delivered to the particle is proportional to the energy Q delivered to the particle by the electromagnetic radiation.	$p = \dfrac{Q}{c}$ ✓ (34.35)

:• **COMMENTS**
Although we considered a very simple object—a single charged particle—Equation 34.35 is still valid for more complicated, extended objects.

Reflected Versus Absorbed Waves

Equation 34.35 applies when the electromagnetic wave is *totally absorbed* by an object. It is analogous to a sticky ball thrown at a vertical wall; the ball's momentum is transferred to the wall. Now imagine the electromagnetic radiation is reflected from the surface. This case is similar to a ball that bounces off a wall so that its final momentum has the same magnitude as its initial momentum: $\vec{p}_f = -\vec{p}_i$. In that case, the change in the ball's momentum is

$$\Delta\vec{p} = \vec{p}_f - \vec{p}_i$$
$$\Delta\vec{p} = -\vec{p}_i - \vec{p}_i = -2\vec{p}_i$$

So the momentum delivered to the wall is $2\vec{p}_i$. If an electromagnetic wave is completely reflected from an object, the momentum delivered to the object is

MOMENTUM TRANSFERRED BY (REFLECTED) ELECTROMAGNETIC WAVE ✪ Major Concept

$$p = 2\dfrac{Q}{c} \quad (34.37)$$

If the electromagnetic wave is partially absorbed by the object, the momentum delivered is somewhere between $p = Q/c$ and $p = 2Q/c$.

Radiation Pressure

To find the pressure exerted by electromagnetic radiation on an object, start with Newton's second law written in terms of momentum $\sum \vec{F} = d\vec{p}/dt$ (Eq. 10.2). In this case, the only force we are interested in is the force in the x direction as exerted by the electromagnetic wave shown in Figure 34.17. We start with a wave that has been totally absorbed, so its momentum is given by $p = Q/c$ (Eq. 34.35) and the force is

$$F_x = \dfrac{d}{dt}\left(\dfrac{Q}{c}\right) = \dfrac{1}{c}\dfrac{dQ}{dt}$$

The rate at which the energy Q is delivered is the power, which is the intensity times the cross-sectional area A (Eq. 34.34):

$$F_x = \dfrac{1}{c}\dfrac{dQ}{dt} = \dfrac{\text{power}}{c} = \dfrac{IA}{c}$$

The force divided by the area is the pressure P:

In Equation 34.38, uppercase P stands for pressure, not power; lowercase p still stands for momentum.

$$\dfrac{F_x}{A} = P = \dfrac{I}{c} \quad (34.38)$$

Equation 34.38 is the pressure exerted by electromagnetic radiation when it is absorbed by an object. If the radiation is completely reflected, the pressure is twice as great:

PRESSURE EXERTED BY ELECTROMAGNETIC WAVES ✪ Major Concept

$$P = 2\dfrac{I}{c} \quad (34.39)$$

If the radiation is partially absorbed by the object, the pressure exerted on the object is somewhere between $P = I/c$ and $P = 2I/c$.

EXAMPLE 34.4 That Doesn't Hurt a Bit

On a day at the beach, you orient your face so it points straight up toward the Sun. Assume the intensity of sunlight is 1000 W/m² and the area of your face is $A = 2 \times 10^{-2}$ m². Estimate the force exerted by sunlight on your face.

INTERPRET and ANTICIPATE
Because you have pointed your face toward the Sun, the force exerted by sunlight is perpendicular to your face. Find that force by multiplying the radiation pressure by the area of your face. We expect a small number because we don't normally feel this force.

SOLVE
Sunlight is partially reflected from your skin, so the pressure exerted by sunlight on your face is somewhere between $P = I/c$ and $P = 2I/c$. Because we are only making an estimate, let's assume the sunlight is totally reflected. This assumption gives the greatest possible force.

$$P = 2\frac{I}{c} \qquad (34.39)$$

The force is the pressure times the area of your face. (We have retained an extra significant figure because the leading number is 1.)

$$F = PA = \left(2\frac{I}{c}\right)A = 2\left(\frac{1000 \text{ W/m}^2}{3.00 \times 10^8 \text{ m/s}}\right)(2 \times 10^{-2} \text{ m}^2)$$

$$F = 1.3 \times 10^{-7} \text{ N}$$

CHECK and THINK
As expected, this is a very small force—about equal to the weight of a couple of hairs on your head. So it is not surprising that we don't normally think about the force exerted by sunlight. In the next examples, we will show how this force can have an enormous effect.

EXAMPLE 34.5 CASE STUDY A Stellar Weight Loss Plan

Deep inside a star, nuclear fusion reactions generate huge amounts of power. The energy makes its way to the star's surface through convection and radiation, but in this problem, we consider only radiation. Radiation exerts outward pressure on the outer layers of a star, which are also pulled inward by the gravitational attraction of the inner layers. Normally, a star is in equilibrium, so the gravitational attraction is balanced by the outward pressure. However, there is a limit to the power a star can give off without blowing off its outer layers. More massive stars generate more power and run a greater risk of shedding material. Figure 34.18 shows the most massive star known in our Milky Way galaxy, Eta Carinae. Notice the bright spot near the middle of the image and the two enormous lobes of material that the star has blown off.

The maximum amount of power that an object can give off in the form of radiation without losing mass is called the **Eddington luminosity**. (*Luminosity* is another term for power.) The outer layer of stars is made up of mostly ionized hydrogen (protons). We'll model this hydrogen as particles that have an effective cross-sectional area given by $\sigma_T = 6.65 \times 10^{-29}$ m² and the mass of a hydrogen atom. (The light interacts strongly with the electrons, pushing them outward; the electrons attract the protons, dragging them outward as well.) Assume the radiation from lower layers is totally absorbed.

A Derive an expression for the Eddington luminosity L_{edd} for a star of mass M.

B Once you have that expression, calculate L_{edd} for the Sun and for a star with 100 times the mass of the Sun.

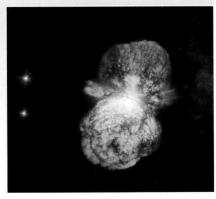

FIGURE 34.18 Eta Carinae may be the most massive star in our Milky Way galaxy. It gives off so much radiation that it is blowing off its outer layers.

Example continues on page 1110 ▶

:• INTERPRET and ANTICIPATE

To make this problem manageable, think about the forces exerted on a single particle in the outer layer of a star. Figure 34.19 shows a free-body diagram for that single particle. It is pulled toward the center of the star by gravity ($\vec{F}_G$) and pushed outward by radiation ($\vec{F}_{\text{rad}}$). At the Eddington limit, the particle does not accelerate and these two forces are balanced.

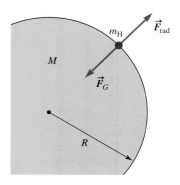

FIGURE 34.19

A SOLVE

The magnitude of the gravitational force on the particle equals the force due to radiation pressure.	$F_G = F_{\text{rad}}$
The gravitational force is given by Newton's law of universal gravity (Eq. 7.4). The force exerted by radiation is the pressure times the particle's cross-sectional area σ_T (Eq. 34.38). Here, the center-to-center distance in the law of universal gravity is the radius R of the star.	$F_G = G\dfrac{Mm_H}{R^2}$ (7.4) $G\dfrac{Mm_H}{R^2} = \left(\dfrac{I}{c}\right)\sigma_T$ (1)
The intensity is the power divided by the surface area of the star. In this case, that power is the Eddington luminosity L_{edd}.	$I = \dfrac{L_{\text{edd}}}{4\pi R^2}$
Substitute I into Equation (1) and solve for L_{edd}.	$G\dfrac{Mm_H}{R^2} = \dfrac{1}{c}\left(\dfrac{L_{\text{edd}}}{4\pi R^2}\right)\sigma_T$ $L_{\text{edd}} = \left(\dfrac{4\pi c G m_H}{\sigma_T}\right)M$ (2)

:• CHECK and THINK

The term in parentheses in Equation (2) is a constant; only M changes as we consider different stars. Finding a value for that constant will be convenient for the calculations in part B and allows us to check the units. We find the constant term has dimensions of power per mass, so the expression for the Eddington luminosity has dimensions of power as expected.

$$\left(\dfrac{4\pi c G m_H}{\sigma_T}\right) = \dfrac{4\pi (3.00 \times 10^8\,\text{m/s})(6.67 \times 10^{-11}\,\text{N}\cdot\text{m}^2/\text{kg}^2)(1.67 \times 10^{-27}\,\text{kg})}{6.65 \times 10^{-29}\,\text{m}^2}$$

$$\left(\dfrac{4\pi c G m_H}{\sigma_T}\right) = 6.31\,\dfrac{\text{N}\cdot\text{m/s}}{\text{kg}} = 6.31\,\dfrac{\text{W}}{\text{kg}} \quad (3)$$

B SOLVE

For our calculations, it is convenient to convert from kilograms to solar mass units, $M_\odot$. Do this by multiplying Equation (3) by the mass of the Sun ($M_\odot = 1.98 \times 10^{30}\,\text{kg}$).	$L_{\text{edd}} = 1.25 \times 10^{31}\,M$ (in watts when M is in solar mass units)
Now we are ready to calculate the Eddington luminosity for the Sun, $M = 1$.	$L_{\text{edd}} = 1.25 \times 10^{31}\,\text{W}$ (Sun)
For the 100-solar-mass star, $M = 100$.	$L_{\text{edd}} = 1.25 \times 10^{33}\,\text{W}$ (100-solar-mass star)

:• **CHECK and THINK**

The Eddington luminosity we found for the Sun is about 30,000 times greater than its actual luminosity ($L_\odot = 3.84 \times 10^{26}$ W), so the Sun's radiation is not blowing off its outer layers. Our Sun is stable (for now). However, the Eddington luminosity we found for the 100-solar-mass star is very close to the actual luminosity observed for such stars. The light given off by these stars is actually destroying them. They are blowing off their outer layers, as in the case of Eta Carinae (Fig. 34.18). Because the amount of radiation given off is greater for stars of greater mass, there is a limit to the ultimate mass of a star. Stellar masses are not greater than about 120 times the mass of the Sun because if a more massive star existed, its radiation would blow it apart. This limit was first theorized by Roberta Humphreys and Kris Davidson and is known as the *Humphreys-Davidson limit*.

EXAMPLE 34.6 **CASE STUDY** **Solar Sailing Saves the Earth?**

In 1950, an asteroid (1950DA) was discovered. It was observed again on December 31, 2000, and appears to be on a collision course with the Earth. The asteroid has about a 1-in-300 chance of colliding with us in about 800 years. Don't panic; we'll have a lot of time to figure out what we should do. One idea is to use a solar sail, harnessing the Sun's light to push the asteroid off course. Ideally, the sail would add little mass to the asteroid but would increase its area and the force due to the Sun's radiation. The mass of 1950DA is unknown, but its diameter is about 1 km. Suppose its density is 2500 kg/m³, the same as the density of Eros (another near-Earth asteroid). Assume the only forces acting on 1950DA are the gravitational attraction of the Sun and the force from radiation on its solar sail. What is the minimum area of the sail needed to accelerate the asteroid away from the Sun? Because of uncertainties in the mass, estimate to just one significant figure.

:• **INTERPRET and ANTICIPATE**

This problem is similar to Example 34.5. The free-body diagram is the same; there is a gravitational force toward the Sun and a force due to radiation pressure away from the Sun. We want the acceleration to be away from the Sun, so the force due to radiation pressure must be greater than the gravitational force. If the acceleration is zero, the two forces have the same magnitude. Use this condition to set a lower limit to the size of the sail. If the sail is larger, the force due to radiation will be greater in magnitude than that due to gravity.

:• **SOLVE**

First, estimate the mass of 1950DA. Its exact shape does not matter, and we can say that the volume is roughly the diameter cubed. (Retain a couple extra significant figures to avoid rounding errors.)

$$V \sim d^3 \sim (1000\,\text{m})^3 = 10^9\,\text{m}^3$$
$$m = \rho V \sim (2500\,\text{kg/m}^3)(10^9\,\text{m}^3)$$
$$m \sim 2.5 \times 10^{12}\,\text{kg}$$

Consider the critical case in which the gravitational force is balanced by the radiation force. As in Example 34.5, the gravitational force is given by Equation 7.4 and the force due to radiation is the pressure times the area. In this case, the pressure is given by $P = 2I/c$ (Eq. 34.39) for a perfectly reflecting sail.

$$F_G = F_{\text{rad}}$$
$$G\frac{M_\odot m}{r^2} = 2\frac{I}{c}A \qquad (1)$$

Find the intensity from $I = P_{\text{av}}/4\pi r^2$ (Eq. 17.23), where the average power is the Sun's luminosity.

$$I = \frac{L_\odot}{4\pi r^2} \qquad (2)$$

Substitute Equation (2) into Equation (1) and solve for A, the area of the sail. Notice that the distance r cancels out.

$$G\frac{M_\odot m}{r^2} = \frac{2}{c}\left(\frac{L_\odot}{4\pi r^2}\right)A \qquad A = \frac{2\pi c G M_\odot m}{L_\odot}$$

$$A = \frac{2\pi(3.0 \times 10^8\,\text{m/s})(6.7 \times 10^{-11}\,\text{N}\cdot\text{m}^2/\text{kg}^2)(2 \times 10^{30}\,\text{kg})(2.5 \times 10^{12}\,\text{kg})}{4.0 \times 10^{26}\,\text{W}}$$

$$A = 2 \times 10^{15}\,\text{m}^2$$

Example continues on page 1112 ▶

:• CHECK and THINK

The area we just found is enormous (about four times the Earth's surface area), but our goal need not be to eject the asteroid completely from the solar system. In other words, we don't have to create a force that exceeds the Sun's gravitational force. Like other objects in the solar system, the asteroid is in orbit around the Sun; we only need to change that orbit so it doesn't cross the Earth's orbit. A more realistic plan is to send an impactor propelled by a solar sail on a collision course with the asteroid, change its orbit, and avoid a collision with the Earth (Problem 75). Of course, any real solution must take into account the gravitational force exerted by other objects such as the Earth, the Moon, and other planets. Fortunately, we have 35 generations to come up with a plan.

34-8 Polarization

Due to the reflections off the surface of the water, you cannot see into the pot in Figure 34.20A. However, if you hold a filter between your eye and the pot, you can see there is a stone under the water (Fig. 34.20B). Why does this work? In this section, we study another property of electromagnetic radiation—polarization—and find out how filters can be constructed based on this property.

FIGURE 34.20 With the right filter, you can see into the water.

A. No filter.   **B.** Notice the hand holding a filter.

Polarized Versus Unpolarized Radiation

First consider the electromagnetic wave in Figure 34.8. The electric field oscillates in the xy plane, and the magnetic field oscillates in the xz plane. For the rest of our discussion, we will focus on only the electric field; the magnetic field is always perpendicular to the electric field. Because the electric field oscillates in only one plane, we say the wave is **plane polarized**. A transverse wave on a string, such as in Figure 17.8A (page 491), is another example of a plane-polarized wave. The string oscillates in a single plane. Imagine watching one of the beads in Figure 17.8A from a point along the direction of propagation (an end-on view). You would see the bead oscillating up and down along a single line as in Figure 17.8B. So, we also say the wave is **linearly polarized**. The electromagnetic wave in Figure 34.21 is also linearly polarized because if you view the wave end-on, you see the electric field oscillating along a single line—the y axis.

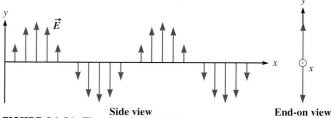

FIGURE 34.21 The electric field of a linearly polarized electromagnetic wave oscillates in one plane. In the end-on view, you see the electric field oscillating up and down along the y axis.

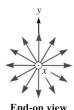

FIGURE 34.22 The electric field of unpolarized radiation oscillates in all planes perpendicular to the direction of propagation. In this end-on view, you see the electric field oscillating up and down along all radial directions.

The wave on the string in Figure 17.8A is linearly polarized because the oscillator that generated the wave oscillates up and down along a single line. In an ordinary radiation source such as a lightbulb, light is emitted by the motion of many electrons oscillating randomly. The net effect is that the electric field oscillates in randomly and rapidly changing planes. Such radiation is said to be **unpolarized** because it does not have a single plane of polarization. When we look at unpolarized radiation end-on, we see electric field vectors oscillating along all the lines perpendicular to the x axis (Fig. 34.22).

The oscillation of any one of these electric field vectors can be divided into two perpendicular components—the horizontal z component and the vertical y component. Consider one such vector $\vec{E}$ that oscillates along a line pointing from the lower left to the upper right. We divide this vector into its components at several times (Fig. 34.23). The components oscillate as a result of the electric field vector's oscillation. So, one way to think about unpolarized light is as the sum of light polarized in two perpendicular planes, each of which carries half of the whole wave's intensity.

Polarizers

A **polarizer** is a filter that allows only one plane of polarization to pass through. The polarized lenses in your good sunglasses are polarizers. Figure 34.24 shows unpolarized light that passes through a polarizer. Initially this light can be modeled as the sum of the light polarized in two perpendicular directions. It is helpful to imagine that one of these directions is parallel to the polarization that the filter allows through, and the other is perpendicular. The polarizer in Figure 34.24 allows the light's vertically polarized component to pass through but absorbs its horizontal component. Because half the intensity is carried in the horizontal component, only half the intensity gets through.

Let's see how a polarizer works. In the late 1920s, the American inventor Edwin Herbert Land (1909–1991) invented the sheet polarizer, which is the basis of such devices as polarized lenses in sunglasses. The polarizer consists of long chains of hydrocarbon molecules arranged in parallel rows, as represented by the horizontal parallel lines shown on the polarizer in Figure 34.24. The spacing between the rows must be smaller than the wavelength of the light to be filtered (a few hundred nanometers for visible light). The hydrocarbon molecules are coated in iodine. The conduction electrons in the coated molecules are able to move along the entire chain, so each long row of molecules acts like a thin

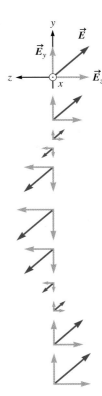

FIGURE 34.23 Any oscillating electric field vector can be broken into two vectors that oscillate perpendicular to each other. In this case, an electromagnetic wave travels in the x direction—out of the page. The electric field oscillates along a line that runs from the lower left to the upper right. The electric field is broken into two perpendicular components along y and z, and is shown at 11 different times. So, we can think of unpolarized light as equivalent to two waves with the same amplitude and perpendicular polarizations, varying independently.

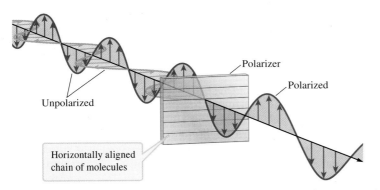

FIGURE 34.24 Only the electric field is represented here. A polarizing filter has all of its molecules aligned in the same direction. The electric field with the same orientation as the filter is absorbed by the molecules. The radiation that passes through the polarizing filter is polarized perpendicular to the chain of molecules and is reduced in intensity.

conducting wire or antenna. When unpolarized light meets the polarizer in Figure 34.24, the horizontal component of this light causes the conduction electrons to oscillate horizontally along the chain of molecules. Their oscillation results in a linearly polarized wave 180° out of phase with the horizontally polarized component of the incident light. Because they are 180° out of phase, the two waves cancel each other. In effect, the horizontal component is absorbed and only the vertical component passes through the polarizer.

Often polarizers are described in terms of their **transmission axis**, which is parallel to the component that passes through the filter. For the polarizer in Figure 34.24, the transmission axis is vertical. The transmission axis is always perpendicular to the rows (long chains) of molecules.

What happens when light passes through more than one polarizer? In Figure 34.25, unpolarized light first passes through a polarizer with a vertical transmission axis. After doing so, it is vertically polarized and has half the intensity of the original

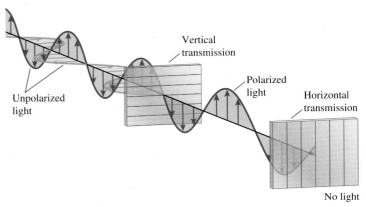

FIGURE 34.25 No light passes through two polarizers with perpendicular transmission axes.

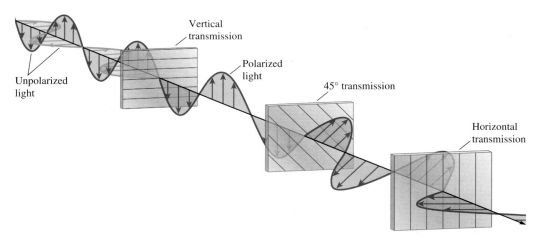

FIGURE 34.26 Some light passes through all three polarizers, each rotated 45° from the previous one. (Magnitudes are not to scale.)

unpolarized light. Now, if that vertically polarized light passes through a second polarizer with a horizontal transmission axis, no light gets through the second polarizer. In other words, the transmitted light has zero intensity.

What happens if you insert a third polarizer at some angle between vertical and horizontal—if, for example, the transmission axis is 45° from vertical? The answer depends on where you put the third polarizer. If you put it at the end, so the light encounters the polarizers in order from vertical transmission to horizontal transmission and then to 45° transmission, there is no change because no light reaches the last polarizer. What is really amazing is that if you put the 45° transmission polarizer between the vertical and horizontal transmission polarizers, some light passes through all three polarizers (Fig. 34.26). How can this happen? The first polarizer allows vertically polarized radiation through. But this vertically polarized radiation may also be broken into two components. One component is tilted by 45° and is parallel to the molecules in the second filter. The other component is perpendicular to these molecules, so it is parallel to the second polarizer's transmission axis. The second polarizer—tilted by 45°—allows radiation tilted by 135° (= 45° + 90°) relative to the original horizontal to pass through. This is the light that passes between the second and third polarizers. This radiation may in turn be broken into two components, one vertical and the other horizontal. The final polarizer allows the horizontally polarized radiation to pass. This is the radiation that passes through all three filters.

This experiment with three polarizers shows that the intensity of the radiation that passes through a filter depends on the angle between the radiation's plane of polarization and the polarizer transmission axis. This relationship was described by a French military engineer who worked for Napoleon, Etienne-Louis Malus (1775–1812). According to Malus's law, the intensity I_f of the linearly polarized radiation that passes through a polarizer is given by

MALUS'S LAW ✪ Major Concept

$$I_f = I_i \cos^2\varphi \qquad (34.40)$$

where I_i is the intensity of the polarized radiation incident on the polarizer and φ is the angle between the light's plane of polarization and the polarizer's transmission axis.

CONCEPT EXERCISE 34.7

Explain why the term *all polarized* might be better than *unpolarized*.

EXAMPLE 34.7 Three Polarizers

What percentage of the incident light's intensity in Figure 34.26 passes through all three polarizers?

: INTERPRET and ANTICIPATE
There are two keys to solving this problem: (1) Half the unpolarized light's intensity passes through the first polarizer, and (2) we can use Malus's law for the other two polarizers because the light that passes through them is linearly polarized.

: SOLVE

Let I_v be the intensity of unpolarized light that passes through the vertical polarizer. This light has half the intensity I_0 of the unpolarized light.	$I_v = \dfrac{1}{2} I_0$	(1)
The vertically polarized light encounters the second polarizer, whose transmission axis is tilted 45° with respect to the vertical (or 135° from the horizontal). Either angle may be used because in Malus's law the cosine function is squared. The intensity I_t of the light that passes through the tilted polarizer is given by Equation 34.40, with Equation (1) substituted for the light incident on this polarizer.	$I_t = I_v \cos^2 \varphi$ (34.40) $I_t = \dfrac{1}{2} I_0 \cos^2 45° = \dfrac{1}{2} I_0 \left(\dfrac{1}{2}\right)$ $I_t = \dfrac{1}{4} I_0$	(2)
Use Malus's law one more time to find the intensity I_f of the light that passes through the final polarizer. The light emerging from the second polarizer is tilted 45° with respect to the horizontal, but the last polarizer's transmission axis is horizontal. Substitute Equation (2) for the intensity incident on this polarizer.	$I_f = I_t \cos^2 \varphi$ (34.40) $I_f = \dfrac{1}{4} I_0 \cos^2 45° = \dfrac{1}{4} I_0 \left(\dfrac{1}{2}\right) = \dfrac{1}{8} I_0$	
Take the ratio I_f/I_0. We are asked for a percentage, so multiply the result by 100.	$\dfrac{I_f}{I_0} = 0.125 = \boxed{12.5\%}$	

: CHECK and THINK
To think about these results, use Malus's law for two perpendicular polarizers (Fig. 34.25). In this case, half the intensity gets through the first polarizer, but no light gets through the second polarizer because $\cos 90° = 0$. If we put a third polarizer between these two perpendicular polarizers, some light (12.5%) passes through all three polarizers. These results are amazing, but the order of the polarizers matters. If the tilted polarizer is placed third, after the horizontal polarizer, no light is incident on the tilted polarizer and so no light gets through.

SUMMARY

❶ Underlying Principles

1. **General form of Faraday's law:**

$$\oint \vec{E} \cdot d\vec{\ell} = -\dfrac{d\Phi_B}{dt} = -\dfrac{d}{dt}\int \vec{B} \cdot d\vec{A} \quad (34.5) \text{ and } (34.6)$$

The induced electric field is nonconservative, so $\mathcal{E} = \oint \vec{E} \cdot d\vec{\ell} \neq 0$.

2. **Maxwell's equations**, which are named for the scientists who first discovered them, describe all of electricity and magnetism. When combined, they describe electromagnetic radiation. The four equations are:

  a. Gauss's law for electricity:

$$\Phi_E = \oint \vec{E} \cdot d\vec{A} = \dfrac{q_{\text{in}}}{\varepsilon_0} \quad (25.11)$$

  b. Gauss's law for magnetism:

$$\Phi_B = \oint \vec{B} \cdot d\vec{A} = 0 \quad (31.4)$$

c. Faraday's law:
$$\oint \vec{E} \cdot d\vec{\ell} = -\frac{d\Phi_B}{dt} \quad (34.5)$$

d. Ampère–Maxwell's law:
$$\oint \vec{B} \cdot d\vec{\ell} = \mu_0 I_{\text{thru}} + \mu_0 \varepsilon_0 \frac{d\Phi_E}{dt} \quad (31.8)$$

3. **Lorentz force** exerted on a charged particle by electric and magnetic fields:
$$\vec{F}_L = \vec{F}_E + \vec{F}_B = q(\vec{E} + \vec{v} \times \vec{B}) \quad (30.21)$$

4. Electromagnetic waves satisfy a **general wave equation** of the form
$$\frac{\partial^2 y}{\partial x^2} = \frac{1}{v^2} \frac{\partial^2 y}{\partial t^2} \quad (17.33)$$

The wave equation for the electric field is
$$\frac{\partial^2 E}{\partial x^2} = \mu_0 \varepsilon_0 \frac{\partial^2 E}{\partial t^2} \quad (34.7)$$

and the wave equation for the magnetic field is
$$\frac{\partial^2 B}{\partial x^2} = \mu_0 \varepsilon_0 \frac{\partial^2 B}{\partial t^2} \quad (34.8)$$

⭐ Major Concepts

1. Electromagnetic waves are **transverse**. If a wave propagates along the positive x direction, its electric and magnetic fields are given by
$$\vec{E}(x, t) = [E_{\max} \sin(kx - \omega t)] \hat{j} \quad (34.18)$$
$$\vec{B}(x, t) = [B_{\max} \sin(kx - \omega t)] \hat{k} \quad (34.19)$$
where $E(x, t) = cB(x, t) \quad (34.24)$

2. Mathematically, the **energy transferred** by an electromagnetic wave per unit time is given by (the magnitude of) the Poynting vector $\vec{S}$:
$$\vec{S} \equiv \frac{1}{\mu_0} (\vec{E} \times \vec{B}) \quad (34.27)$$

3. The **intensity** of an electromagnetic wave is the time average of S: $I \equiv S_{\text{av}}$. In terms of the maximum electric and magnetic fields, the intensity I is
$$I = \frac{1}{2\mu_0} E_{\max} B_{\max} \quad (34.29)$$

The intensity may also be expressed in terms of either $E_{\max}$ or $B_{\max}$ alone:
$$I = \frac{E_{\max}^2}{2\mu_0 c} = \frac{c}{2\mu_0} B_{\max}^2 \quad (34.30)$$

For a distant point source of electromagnetic radiation, the intensity decreases as the distance r from the source increases:
$$I = \frac{P_{\text{av}}}{4\pi r^2} \quad (17.23)$$

where P_{av} is the average power emitted by the source. The intensity is also proportional to the average field energy density:
$$I = cu_{\text{av}} \quad (34.33)$$

4. The **momentum** p delivered by an electromagnetic wave is proportional to the energy Q delivered. When the electromagnetic wave is totally absorbed by an object, the momentum is given by
$$p = \frac{Q}{c} \quad (34.35)$$

When the electromagnetic wave is completely reflected from the object, the momentum delivered to the object is given by
$$p = 2\frac{Q}{c} \quad (34.37)$$

5. The **pressure** exerted by an electromagnetic wave that is totally absorbed by an object is
$$P = \frac{I}{c} \quad (34.38)$$

If the radiation is completely reflected, the pressure is
$$P = 2\frac{I}{c} \quad (34.39)$$

6. **Malus's law:** The intensity I_f of the linearly polarized radiation that passes through a polarizer is given by
$$I_f = I_i \cos^2 \varphi \quad (34.40)$$

where I_i is the intensity of the polarized radiation incident on the polarizer and φ is the angle between the light's plane of polarization and the polarizer's transmission axis.

PROBLEMS AND QUESTIONS

A = algebraic   C = conceptual   E = estimation   G = graphical   N = numerical

34-1 Light: One Last Classical Topic

1. **C** What do we mean when we say that light is sometimes modeled as a particle and sometimes modeled as a wave?

34-2 Generalized Form of Faraday's Law

2. **C** Suppose a positive particle moves once around the circuit in Figure 34.3B (page 1088). The amount of energy it gains going through the battery equals the amount of energy it loses going through the resistor, so when it returns to its initial position, its energy is unchanged. Now think about a positive particle that goes around the loop in Figure 34.3A. When this particle moves once around, it has gained energy. **a.** Where does that energy come from? **b.** Is this a source of *free* energy? Explain.

3. **N** A circular coil of radius 0.50 m is placed in a time-varying magnetic field $B(t) = (5.80 \times 10^{-4}) \sin[(12.6 \times 10^2 \, \text{rad/s})t]$ where B is in teslas. The magnetic field is perpendicular to the plane of the coil. Find the magnitude of the induced electric field in the coil at $t = 0.001$ s and $t = 0.01$ s.

4. **C** An end-on view of a coil is shown in Figure P34.4. A current in the coil is increasing toward a steady-state value. Indicate the direction (into or out of the page) of the magnetic field near the center of the coil. Indicate the direction (clockwise or counterclockwise) of the electric field inside the coil.

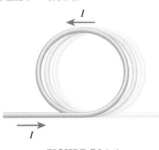

FIGURE P34.4

5. A solenoid with n turns per unit length has radius a. It is connected to a power supply that drives a current increasing linearly with time as $I(t) = Ct$, where C is a constant. Assume the magnetic field is uniform inside the solenoid.
   a. **A** Find an expression for the magnetic field inside the solenoid as a function of time, and determine the magnitude of the induced electric field just inside the solenoid.
   b. **N** What is the magnitude of the electric field induced just inside a solenoid that has exactly 10 turns/cm and a radius of 1.5 cm if the current increases at a rate of 0.50 A/s?

Problems 6 and 7 are paired.

6. **G** Suppose the magnetic field in Figure 34.4 (page 1090) increases at a constant rate dB/dt. Sketch $E(r)$ from $r = 0$ to $r = 2R$.

7. **A** If $B(t) = B_0 e^{-t/\tau}$ in Figure 34.4 (page 1090), find an expression for the electric field at $r = R$. Does the electric field depend on time?

34-3 Five Equations of Electromagnetism

8. **C** If you discover magnetic monopoles, what changes (if any) do you need to make to Maxwell's equations?

9. **N** A capacitor with square plates, each with an area of 36.0 cm² and plate separation $d = 2.54$ mm, is being charged by a 265-mA current. **a.** What is the change in the electric flux between the plates as a function of time? **b.** What is the magnitude of the displacement current between the capacitor's plates?

10. **C** Three students are discussing Maxwell's equations.
    Avi: If you don't have any electric charges at all, Maxwell's equations still tell you how electric and magnetic fields behave, right?
    Cameron: That doesn't make sense to me. Electric fields are made by charges, and magnetic fields are made by *moving* charges. So we have to have charges to describe electromagnetism. I don't see how Maxwell's equations can be used without electric charges.
    Shannon: Well, Maxwell's equations are still true, but the right sides of all the equations are zero anyway, so the electric and magnetic fields have to be zero.
    With which student do you agree? Justify your answer.

11. **N** An electric field with initial magnitude 140 V/m directed into the page and increasing at a rate of 12.0 V/(m·s) is confined to the area of radius $R = 17.0$ cm in Figure P34.11. If $r = 42.0$ cm, what are the magnitude and direction of the magnetic field at point A?

12. **C** Suppose a positive particle is initially at rest in Figure 34.4 (page 1090). Then the magnitude of the magnetic field begins to change. **a.** Will the particle be accelerated? **b.** If so, will the particle remain in the plane of the page? Explain your answers.

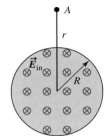

FIGURE P34.11

Problems 13 and 14 are paired.

13. **A** A circular conductor (Fig. P34.13) encloses a uniform magnetic field that is perpendicular to the page (not shown), pointing outward and increasing. The changing magnetic flux induces an emf $\mathcal{E}$ in the conductor and a clockwise current. The magnetic field exists only in the circular region enclosed by the conductor. If we measure the potential difference between points A and B using the two voltmeters as shown, we find $V_1 \neq V_2$. To see how that is possible, set up the path integral from A to B along path 1 to find V_1, and then set up the path integral from A to B along path 2 to find V_2. (These two paths pass through their respective voltmeters.) Finally, by combining these integrals, show that $V_2 - V_1 = \mathcal{E}$.

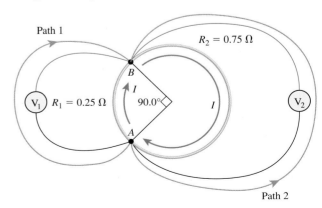

FIGURE P34.13 Problems 13 and 14.

14. **N** A circular conductor encloses a uniform magnetic field that is perpendicular to the page (not shown), pointing inward and increasing. The magnetic field outside the circular loop is zero. Suppose the changing magnetic flux induces a 1.0-V emf in the conductor and the conductor's resistance is 1.0 Ω. Imagine connecting two voltmeters to the conductor (Fig. P34.13). The clockwise distance from A to B is one-fourth the circumference. Because the resistance of a wire is proportional to its length, this portion of the loop has resistance $R_1 = 0.25$ Ω. Likewise, the clockwise distance from B to A is three-fourths the circumference, so this portion's resistance is $R_2 = 0.75$ Ω. It may seem incredible, but what a voltmeter measures depends on its placement; in Problem 13, we showed that $V_2 - V_1 = \mathcal{E}$. Find the voltage measured by both meters. *Hint*: Kirchhoff's loop rule (and other rules from Chapter 29) may be applied to a loop that does not enclose the changing magnetic flux.

15. **N** What is the acceleration of a proton moving with velocity $\vec{v} = 15.0\hat{\jmath}$ m/s through a region that has magnetic field $\vec{B} = 1.20\hat{\imath}$ T and electric field $\vec{E} = (3.00\hat{\jmath} + 2.00\hat{k})$ V/m?

16. **C** Write Maxwell's equations for the case in which there are no charges present, and explain in words the meaning of each equation.

34-4 Electromagnetic Waves

17. **N** Consider electromagnetic waves in free space. What is the wavelength of a wave that has a frequency of **a.** 2.00×10^{11} Hz and **b.** 8.00×10^{16} Hz?

18. **A** Show that the SI units of $1/\sqrt{\mu_0 \varepsilon_0}$ are m/s.

19. **N** The human eye can see light that has a maximum frequency of 7.69×10^{14} Hz. What is the corresponding wavelength of the light?

20. **N** An electromagnetic wave is given in SI units by $E(x, t) = 3.75 \sin(0.60x - \omega t)$ V/m. **a.** What is the angular frequency? **b.** What is the magnetic field at $x = 2.0$ m and $t = 3.0$ s?

21. **N** Ultraviolet (UV) radiation is a part of the electromagnetic spectrum that reaches the Earth from the Sun. It has wavelengths shorter than those of visible light, making it invisible to the naked eye. These wavelengths are classified as UVA, UVB, or UVC, with UVA the longest of the three at 320 nm to 400 nm. Both the U.S. Department of Health and Human Services and the World Health Organization have identified UV as a proven human carcinogen. Many experts believe that, especially for fair-skinned people, UV radiation frequently plays a key role in melanoma, the deadliest form of skin cancer, which kills more than 8000 Americans each year. UVB has a wavelength between 280 nm and 320 nm. Determine the frequency ranges of UVA and UVB.

22. **G** An electromagnetic wave is given in SI units by $E(x, t) = 3.75\sin(kx - 0.094t)$ V/m. Plot the magnetic energy density and the electrical energy density associated with this wave as a function of time t.

23. **N** What is the frequency of the blue-violet light of wavelength 405 nm emitted by the laser-reading heads of Blu-ray disc players?

24. **A** Write equations for both the electric and magnetic fields for an electromagnetic wave in the red part of the visible spectrum that has a wavelength of 710 nm and a peak electric field magnitude of 2.5 V/m.

25. **N** The amplitude of the electric field of an electromagnetic wave traveling in a vacuum is measured to be 4.3×10^2 V/m. What is the amplitude of the magnetic field in this wave?

26. **A** Start with Equation 34.11,

$$\left(\frac{\partial E}{\partial x}\right) = -\left(\frac{\partial B}{\partial t}\right)$$

and Equation 34.15,

$$\left(\frac{\partial B}{\partial x}\right) = -\mu_0 \varepsilon_0 \left(\frac{\partial E}{\partial t}\right)$$

and show that the wave equation for a linearly polarized magnetic wave in a vacuum is given by Equation 34.8,

$$\frac{\partial^2 B}{\partial x^2} = \mu_0 \varepsilon_0 \frac{\partial^2 B}{\partial t^2}$$

27. **N** WGVU-AM is a radio station that serves the Grand Rapids, Michigan, area. The main broadcast frequency is 1480 kHz. At a certain distance from the radio station transmitter, the magnitude of the magnetic field of the electromagnetic wave is 3.0×10^{-11} T. **a.** Calculate the wavelength. **b.** What is the angular frequency? **c.** Find the wave number of the wave. **d.** What is the amplitude of the electric field at this distance from the transmitter?

28. **N** Suppose the magnetic field of an electromagnetic wave is given by $B = (1.5 \times 10^{-10}) \sin(kx - \omega t)$ T. **a.** What is the maximum energy density of the magnetic field of this wave? **b.** What is maximum energy density of the electric field?

29. **N** The magnetic field of a plane electromagnetic wave is given by $B_z = (68.0) \sin(kx - 2.60 \times 10^6 t)$ nT. **a.** What is the amplitude of the electric field of this wave? **b.** What is the frequency f? **c.** What is the wavelength λ?

30. **A** Write equations for both the electric and magnetic fields for an electromagnetic wave (an X-ray) that has an angular frequency of 7.5×10^{18} Hz and a peak magnetic field magnitude of 10^{-10} T.

31. **N** A cell phone sends and receives electromagnetic waves. The quality of the phone's reception depends on the strength of the electric field. The stronger the electric field around a cellular telephone, the better the reception. In this scenario, however, there is a higher chance that the user's body will absorb the electric field signal, slowly leading to possible harmful effects like cancer. A guideline has been established limiting the electric field strength near a cell phone to protect the user. The "maximum permissible exposure" in this context is considered to be 100 V/m. If the amplitude of the electric field near a cell phone is 41 V/m, what is the amplitude of the magnetic field? How does it compare to the magnitude of the Earth's magnetic field near the surface, which is 5.0×10^{-5} T?

32. **N** An electromagnetic wave is traveling in the positive x direction with the electric field oscillating along the z direction. If the wavelength is 555 nm, $E_{\max} = 0.050$ V/m, and the wave is in a vacuum, write equations describing the electric and magnetic fields that make up the electromagnetic wave as functions of x and t.

Problems 33 and 34 are paired.

33. **A** By substitution, show that $E(x, t) = E_{\max} \sin(kx - \omega t)$ is a solution to Equation 34.7:

$$\frac{\partial^2 E}{\partial x^2} = \mu_0 \varepsilon_0 \frac{\partial^2 E}{\partial t^2}$$

where

$$c = \frac{1}{\sqrt{\mu_0 \varepsilon_0}} = \frac{\omega}{k}$$

34. **A** By substitution, show that $B(x, t) = B_{\max} \sin(kx - \omega t)$ is a solution to Equation 34.8:

$$\frac{\partial^2 B}{\partial x^2} = \mu_0 \varepsilon_0 \frac{\partial^2 B}{\partial t^2}$$

34-5 The Electromagnetic Spectrum

35. **C** Can you hear radio waves? Explain.

36. **C** When electromagnetic radiation shines through openings with a size that is comparable to the wavelength of the radiation,

diffraction becomes important, leading to variations in intensity due to interference of the waves. What type of electromagnetic radiation would lead to diffraction when shined on a lattice of atoms separated by 0.2 nm? What type of radiation would diffract through a row of skyscrapers separated by a couple hundred meters? Explain your answers.

37. **C** Which waves travel faster—radio waves or gamma rays?

Problems 38 and 39 are paired.

38. **C** Astronomers often speak of light from a distant source as being *red-shifted*, where the measured wavelength is made longer than that emitted by the source by some interfering effect (typically a Doppler effect). Consider the electromagnetic spectrum, and explain why astronomers might use this term to describe these wavelength measurements.

39. **C** Astronomers often speak of light from a distant source as being *red-shifted*, where the measured wavelength is made longer than that emitted by the source by some interfering effect (typically a Doppler effect). If someone described a wavelength measurement as being *blue-shifted* instead, what can you say about the original wavelength and frequency of the wave before the blue-shifting occurred?

34-6 Energy and Intensity

40. **E** Estimate the amount of energy from sunlight that your face absorbs over your entire lifetime. *Hints*: See Example 34.2. The intensity of sunlight at the Earth's surface is roughly 1000 W/m². (This intensity is lower than the 1360 W/m² used in Example 34.2 because of atmospheric absorption, scattering, and other factors.)

41. **N** Find the intensity of the electromagnetic wave described in each case: An electromagnetic wave with **a.** a wavelength of 710 nm and a peak electric field magnitude of 2.5 V/m and **b.** an angular frequency of 7.5×10^{18} rad/s and a peak magnetic field magnitude of 10^{-10} T.

42. An electric field $\vec{E} = (23.0\hat{i} - 55.0\hat{j} + 17.0\hat{k})$ V/m and a magnetic field $\vec{B} = (5.25\hat{i} + 4.05\hat{j} + 6.00\hat{k})$ nT are measured in a location in free space.
    a. **N** What is the Poynting vector at this location?
    b. **A** Show that the electric and magnetic fields are perpendicular to each other at this location.

43. **N** You wish to send a probe to the Moon. The probe has a radio transmitter that you test in the laboratory. When the probe is 10 m from your receiver, the intensity is I_0. When the probe is on the Moon, it sends radio waves to you. If you want to receive radio waves of the same intensity, how much stronger must the probe's electric field be when it is on the Moon?

44. **A** The intensity in Equation 34.29, $I = (1/2\mu_0)E_{max}B_{max}$, and in Equation 34.30, $I = E_{max}^2/(2\mu_0 c) = (cB_{max}^2)/(2\mu_0)$, is written in terms of the maximum electric and magnetic fields. Rewrite these equations in terms of the rms field values.

45. **N** The state-of-the-art digital air surveillance radar used in civil aviation emits its signal equally in all directions (isotropically) with an average power of 22.0 kW. What is the average intensity near an aircraft on final approach to an airport, 2.00 km from the radar tower?

46. **N** At an instant in time, the electric and magnetic fields of an electromagnetic wave are given by $\vec{E} = -4.00 \times 10^{-3}\hat{k}$ V/m and $\vec{B} = -1.33 \times 10^{-11}\hat{i}$ T. Find the Poynting vector for this wave.

47. **N** The electric field of an electromagnetic wave traveling in the vacuum of space is described by $E = (5.00 \times 10^{-3})(kx - \omega t)$ V/m.
    a. What is the maximum value of the associated magnetic field for this electromagnetic wave? **b.** What is the average energy density of the wave?

48. **E** You may have heard that our telecommunication signals are traveling out in space and that some alien society may be watching old TV episodes or listening to your phone calls. A powerful cell phone puts out about 3 W. The closest star to our solar system is about 4 light-years away. Think about a call you made four years ago. The electromagnetic wave from that call is just reaching our nearest neighbor star. How strong is the intensity at that distance? Compare your answer to the intensity of the signal picked up by the nearest cell phone tower on the Earth.

49. **C** Determine the direction of energy flow for the following four cases, where $\vec{E}$ and $\vec{B}$ are the electric and magnetic fields, respectively: **a.** $\vec{E} = E\hat{j}, \vec{B} = B\hat{k}$; **b.** $\vec{E} = -E\hat{i}, \vec{B} = B\hat{k}$; **c.** $\vec{E} = -E\hat{i}, \vec{B} = -B\hat{j}$; **d.** $\vec{E} = -E\hat{i}, \vec{B} = -B\hat{k}$.

50. **N** A circular mirror 78.0 cm in radius is used to reflect sunlight onto a small plate with a diameter of 3.40 cm that completely absorbs the light. The intensity of sunlight at the Earth's surface is 980 W/m². **a.** What is the intensity of sunlight incident on the absorbing plate? **b.** What is the amplitude of the electric field at the absorbing plate? **c.** What is the amplitude of the magnetic field at the absorbing plate?

Problems 51 and 52 are paired.

51. **A** A current of magnitude I flows through a coiled wire. The coil forms a cylinder of length L and radius R. The potential difference between the two ends of the wire is V. Determine the magnitude of the Poynting vector.

52. **A** A current of magnitude I flows through a coiled wire. The coil forms a cylinder of length L and radius R. The potential difference between the two ends of the wire is V. Integrate the Poynting vector over the cross-sectional area of the wire to obtain the energy per unit time (E/t) passing through the surface of the wire $(E/t = \int \vec{S} \cdot d\vec{A})$.

34-7 Momentum and Radiation Pressure

53. **N** Optical tweezers use light from a laser to move single atoms and molecules around. Suppose the intensity of light from the tweezers is 1.00×10^3 W/m², the same as the intensity of sunlight at the surface of the Earth. **a.** What is the pressure on an atom if light from the tweezers is totally absorbed? **b.** If this pressure were exerted on a hydrogen atom, what would be its acceleration? Assume the cross-sectional area is 6.65×10^{-29} m².

54. **N** The intensity of sunlight on the Earth is about 1.00×10^3 W/m². The average orbital radius of Mercury is about 40% of the Earth's orbital radius. Using this information, determine the approximate intensity of the sunlight and the radiation pressure on Mercury.

55. **N** What is the radiation pressure on a perfectly reflecting mirror due to a 0.600-W laser beam of radius 1.50 mm that is normally incident on the mirror?

56. Enrique claims he can push a toy cart across a frozen pond using a flashlight and a sail. He says that if he spreads the sail out to its maximum area of 64 cm² and attaches it to the 75-g toy cart, the light from the flashlight alone will push the cart. Angelique objects to Enrique's claim, noting that the surrounding sunlight is more intense than the light from the flashlight.
    a. **C** Evaluate Angelique's claim. Does the surrounding sunlight matter when the flashlight shines on the sail? In other words, does the sunlight cause a resistive force when Enrique tries to move the cart? Explain.
    b. **E, N** Estimate or research the intensity of light from a typical household flashlight when it is held about 0.01 m away from the sail. If we assume the light is perfectly reflected by the sail, what is the magnitude of the force exerted on the cart by the flashlight?
    c. **C** Given that there is a small amount of friction between the cart and the frozen pond, will Enrique observe the cart moving if the experiment is performed? Explain your answer.

57. N The accepted value of the intensity of sunlight at the Earth's surface is 1.36 kW/m². **a.** Determine the pressure exerted by sunlight if it strikes a perfect absorber. **b.** Determine the pressure exerted by sunlight if it strikes a perfect reflector.

58. C The expressions for the momentum transferred by radiation (Eqs. 34.35 and 34.37) and for the pressure exerted (Eqs. 34.38 and 34.39) may seem odd. To become more comfortable with them, check their dimensions.

59. N A perfectly reflecting circular mirror 15.0 cm in radius is placed in the path of a plane electromagnetic wave with an intensity of 3.65 W/m² traveling in the positive z direction. **a.** How much momentum does the wave impart to the mirror per second? **b.** What is the force exerted on the mirror by the wave?

60. N A budding magician holds a 5.00-mW laser pointer, wondering whether he could use it to keep an object floating in the air with the radiation pressure. This might be an idea for a new trick! Assuming the laser pointer has a circular beam 3.00 mm in diameter and the magician rigs up a totally reflecting sail on which to shine the laser, what is the maximum weight the magician could suspend with this technique?

34-8 Polarization

61. N Some unpolarized light has an intensity of 1365 W/m² before passing through three polarizing filters. The transmission axis of the first filter is vertical. The second filter's transmission axis is 30.0° from vertical. The third filter's transmission axis is 40.0° from vertical. What is the intensity of the light that emerges from the three filters?

Problems 62 and 63 are paired.

62. N Suppose you have a polarized light source in your lab that has an intensity of 1409 W/m², where the electric field oscillates in the z direction. The light travels toward a workbench where you can send it through either one of two polarizing filters. **a.** If the light passes through a filter with its transmission axis at an angle of 45.0° from the z direction, what is the intensity of the transmitted light? **b.** If, instead, the light passes through a filter with its transmission axis at an angle of 90.0° from the z direction, what is the intensity of the transmitted light?

63. N The light in Problem 62 travels toward a workbench and encounters two polarizing filters. **a.** If the light passes first through a filter with its transmission axis at an angle of 45.0° from the z direction, and then through a filter with its transmission axis at an angle of 90.0° from the z direction, what is the final intensity of the transmitted light? **b.** If, instead, the light passes first through a filter with its transmission axis at an angle of 90.0° from the z direction, and then through a filter with its transmission axis at an angle of 45.0° from the z direction, what is the final intensity of the transmitted light?

64. C You have a set of four polarizing filters, and you are free to orient their transmission axes however you prefer. If you begin with an unpolarized light wave, is it possible to halve the intensity of the wave each time it passes through a polarizer, so that the final intensity is 1/16 the original intensity? How could this be done?

65. N Unpolarized light passes through three polarizing filters. The first filter has its transmission axis parallel to the z direction, the second has its transmission axis at an angle of 30.0° from the z direction, and the third has its transmission axis at an angle of 60.0° from the z direction. If the light that emerges from the third filter has an intensity of 250.0 W/m², what is the original intensity of the light?

General Problems

66. N The average Earth–Sun distance is 1.00 astronomical unit (AU). At how many AUs from the Sun is the intensity of sunlight 1/25 the intensity at the Earth?

67. The magnetic field of an electromagnetic wave is given by $B(x, t) = (4.0 \times 10^{-8}) \sin[(1.4 \times 10^4 \,\text{rad/m})x + \omega t]$ T.
a. C In which direction is the wave traveling?
b. N Determine the wave number, wavelength, and frequency of the wave.
c. A Write an equation for the electric field as a function of x and t.

68. A As mentioned in Section 34-2, Kirchhoff's loop rule holds only for conservative fields, and the inductor rule is a way to preserve using the loop rule. To see that we get the same mathematical relationship $\mathcal{E} - L(dI/dt) - IR = 0$ as we did in Section 33-3 when we (mis-)applied Kirchhoff's loop rule to a circuit consisting of a battery, a resistor, and an inductor, correctly apply Faraday's law (Equation 34.6)

$$\oint \vec{E} \cdot d\vec{\ell} = -\frac{d}{dt}\int \vec{B} \cdot d\vec{A}$$

to the circuit shown in Figure P34.68. This circuit is essentially a resistor, a battery, and a one-loop inductor. There is negligible resistance in the wires, and the total inductance is L. When the switch closes at $t = 0$, the current is counterclockwise, while the magnetic field points out of the page and increases. *Hints*: Choose an outward-pointing $d\vec{A}$, and choose $d\vec{\ell}$ in the same direction as the current. In the **CHECK and THINK** step, describe the conceptual difference between (mis-)applying Kirchhoff's loop rule and applying Faraday's law.

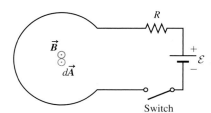

FIGURE P34.68

69. N What is the amplitude of the magnetic field a distance of 3.50 km away from a radio station that is broadcasting isotropically (in all directions) with a power of 43.0 kW?

70. N The magnetic field of an electromagnetic wave is given by $B = 1.5 \times 10^{-10} \sin(kx - \omega t)$ T. **a.** If the wavelength is 752 nm, what are the frequency, angular frequency, and wave number of this wave? **b.** What is the maximum total energy density of this wave?

71. N Household solar panels typically have an efficiency of 15.0%. If the intensity of sunlight is assumed to be a constant 9.80×10^2 W/m², what should be the total area of rooftop solar panels on a house in order to supply an average of 33.0 kW each day? Efficiency denotes the percentage of incident solar energy that is converted to electricity by the solar panel.

72. N A plane electromagnetic sinusoidal wave with a wavelength of 34.0 mm and a magnetic field oscillating in the yz plane with an amplitude of 96.0 nT is traveling in the positive z direction (Fig. P34.72). **a.** What is the frequency of this wave? **b.** What are the magnitude and direction of the electric field when $\vec{B} = B_{max}\hat{j}$? **c.** Expressing the magnitude of the electric field as $E(z, t) = E_{max} \sin(kz - \omega t)$, obtain an expression for E that includes numerical values for E_{max}, k, and ω.

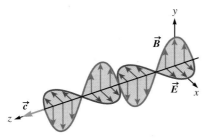

FIGURE P34.72

73. The electric field of an electromagnetic wave traveling in a vacuum is described by the equation $E = (5.00 \times 10^{-3})\sin(kz - \omega t)$ V/m, where the electric field oscillates along the y direction.
    a. **C** In what direction is this wave traveling?
    b. **N** If the wavelength is 454 nm, what are the frequency, angular frequency, and wave number for this wave?
    c. **C** Along which direction does the associated magnetic field oscillate?
    d. **N** Write an equation describing the associated magnetic field as a function of z and t.

74. **E CASE STUDY** The region around the Earth is filling up with space junk such as old satellites. One idea for cleaning up space involves using sails that create drag (Fig. P34.74). Perhaps one day satellites will be equipped with sails that are deployed at the end of their missions. The NASA mission NanoSail-D was launched in 2010 to test this idea. This problem compares the drag on a solar sail due to the Earth's upper atmosphere with the force exerted on the sail by the Sun's radiation and with the gravitational force exerted on the satellite by the Earth. Nano-Sail-D orbited at an altitude of 6.5×10^5 m, and its solar sail had an area of 10 m². Assume the satellite's total mass was about 4 kg and it was in a circular orbit. Also assume the Earth's atmosphere at that altitude has a density of about 5×10^{-14} kg/m³ and a drag coefficient $C \approx 1$, and the Earth's gravitational force provided the centripetal acceleration. Finally, assume the sail perfectly reflected the sunlight. Estimate the magnitude of the three forces exerted on the satellite.

FIGURE P34.74 An artist's conception of a solar sail.

75. **E CASE STUDY** In Example 34.6 (page 1111), we imagined equipping 1950DA, an asteroid on a collision course with the Earth, with a solar sail in hopes of ejecting it from the solar system. We found that the enormous size required for the solar sail makes the plan impossible at this time. Of course, there is no need to eject such an object from the solar system; we only need to change the orbit. A much more pressing problem is Apophis, a 300-m asteroid that may be on a collision course with the Earth and is due to come by on April 13, 2029. It is unlikely to hit the Earth on that pass, but it will return again in 2036. If Apophis passes through a 600-m keyhole on its 2029 pass, it is expected to hit the Earth in 2036, causing great damage. There are plans to deflect Apophis when it comes by in 2029. For example, we could hit it with a 10- to 150-kg impactor accelerated by a solar sail. The impactor is launched from the Earth to start orbiting the Sun in the same direction as the Earth and Apophis. The idea is to use a solar sail to accelerate the impactor so that it reverses direction and collides head-on with Apophis at 80–90 km/s and thereby keeps Apophis out of the keyhole. Consider the momentum in the impactor's orbit (Fig. P34.75) when the solar sail makes an angle of $\theta = 60°$ with the tangent to its orbit. Current solar sails may be about 40 m on a side, but the hope is to construct some that are about 160 m on a side. Estimate the impactor's tangential acceleration when it is about 1 AU from the Sun. Keep in mind that the sail is neither a perfect absorber nor a perfect reflector, and a heavier impactor would presumably be equipped with a larger sail. Don't be surprised by what may seem like a very small acceleration.

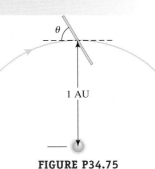

FIGURE P34.75

76. **A** Show that over one full cycle, the average value of $\sin^2\theta$ is $\frac{1}{2}$.

77. **N** A plane sinusoidal electromagnetic wave traveling in the positive z direction has a wavelength of 625 nm and a magnetic field amplitude of 48.0 μT. Obtain expressions for the electric and magnetic fields of this wave, assuming the electric field oscillates along the y direction.

78. **C** We know from Chapter 27 that insulating materials have a different electric permittivity (ε) than does a vacuum (ε_0). This difference can be accounted for by defining a *dielectric constant* κ such that $\varepsilon = \kappa\varepsilon_0$. How do you suspect the speed of an electromagnetic wave is affected when it enters an insulating material? Does the wave speed up, slow down, or remain at the same speed it had while traveling in a vacuum? Justify your answer.

79. **N** In 2010, the Japanese IKAROS satellite became the first to demonstrate the viability of using solar sails as a means of propulsion. The 315-kg spacecraft used a perfectly reflecting polyimide sail of area 2.00×10^2 m² and successfully reached the planet Venus. The intensity of the solar radiation near Venus is 2.62×10^3 W/m². Ignore the gravitational effects from the Sun and other bodies. **a.** What is the magnitude of the force exerted on the sail by the solar radiation near Venus? **b.** What is the acceleration of the spacecraft near Venus?

Problems 80, 81, and 82 are grouped.

80. **A** Consider two circular regions of the same cross-sectional area A with uniform magnetic fields, one with the field pointing into the page and the other pointing out of the page (Fig. P34.80). Initially, both fields have magnitude B_0, and the magnitude of each is increasing at the same rate dB/dt. Calculate $\oint \vec{E} \cdot d\vec{\ell}$ using **a.** circular path 1 and **b.** circular path 2.

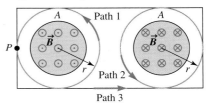

FIGURE P34.80 Problems 80, 81, and 82.

81. **A** Consider two circular regions of the same cross-sectional area A with uniform magnetic fields, one with the field pointing into the page and the other pointing out of the page (Fig. P34.80). Initially, both fields have magnitude B_0, and the magnitude of each is decreasing at the same rate dB/dt. Calculate $\oint \vec{E} \cdot d\vec{\ell}$ using rectangular path 3.

82. Consider two circular regions of the same cross-sectional area with uniform magnetic fields, one with the field pointing into the page and the other pointing out of the page (Fig. P34.80). Initially, both fields have magnitude B_0, and the magnitude of each is decreasing at the same rate dB/dt, so that the magnitude of either at time t is B.
    a. **A** What is the magnetic flux through circular path 1?
    b. **A** What is the magnetic flux through rectangular path 3?
    c. **C** Explain why it is difficult to use either path 1 or path 3 to calculate the electric field at point P. If you could overcome that difficulty (perhaps by using a computer to integrate), would your result depend on which path you used? Explain.

83. **N** A linearly polarized sinusoidal electromagnetic wave moving in the positive x direction has a wavelength of 5.80 mm. The magnetic field of the wave oscillates in the xz plane and has an amplitude of 8.70 μT. The magnitude of the electric field vector for this wave can be written as $E = E_{max} \sin(kx - \omega t)$. What are **a.** the value of E_{max}, **b.** the wave number k, **c.** the angular frequency ω, **d.** the plane in which the electric field oscillates, **e.** the average Poynting vector, **f.** the radiation pressure exerted by this wave on a perfectly reflecting lightweight solar sail, and **g.** the acceleration of the solar sail if its dimensions are 5.00 m $\times$ 8.00 m and its mass is 34.5 g?

84. **C** In Section 34-1, we summarized classical mechanics. You may have noticed that conservation of momentum (and angular momentum) is not on that list. How is this principle already included in the list?

PART FIVE
Light

Diffraction and Interference

35

❶ Underlying Principles

Huygens's principle

✪ Major Concepts

1. Diffraction
2. Interference
3. Conditions for constructive and destructive interference
4. Coherence

◐ Special Cases

1. Young's double-slit experiment (position of fringes; intensity)
2. Single-slit diffraction (position of dark fringes; intensity)
3. Double-slit diffraction (intensity)

⊙ Tools

Visual representations of light

Key Questions

What evidence supports the wave model for light?

How do we find the location and intensity of the fringes in interference and diffraction patterns?

35-1 **Light is a wave**   1124

35-2 **Sound wave interference revisited**   1126

35-3 **Young's experiment: Position of the fringes**   1128

35-4 **Single-slit diffraction**   1132

35-5 **Young's experiment: Intensity**   1137

35-6 **Single-slit diffraction intensity**   1140

35-7 **Double-slit diffraction**   1144

Many people think vision is the most important human sense. It is certainly a highly developed sense. About 25% of the human brain is used for vision, so people have a preference for *visible* electromagnetic radiation. This entire part of the book is dedicated to the physics of visible light.

A doctoral graduate student in physics must pass a qualifying exam before being allowed to pursue her dissertation. As part of her qualifying exam, this book's author had to pass an oral exam during which she was asked, "How do you know that light is a wave?" How would you answer? You might be tempted to derive the wave equation from Maxwell's equations (Section 34-4). But such a derivation is based on a *theory*, whereas the question asks for *evidence*.

Before James Clerk Maxwell was born, the British physicist Thomas Young (1773–1829) considered this question. By the time Young was 19, he was highly proficient in Latin, which enabled him to master important scientific

FIGURE 35.1 Sunlight through windows produces crisp shadows that look like stars and hexagons.

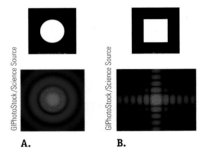

FIGURE 35.2 The top row shows two apertures (not to scale). The bottom row shows the resulting patterns made by the light on a screen. When laser light passes through a small aperture, the pattern of light on a screen does not simply have the same shape as the aperture. **A.** The aperture is circular, but the light pattern shows concentric circles around a central disk. **B.** The aperture is a square, but the light pattern is a cross pattern of rectangles.

DIFFRACTION ⭐ **Major Concept**

FIGURE 35.3 A needle illuminated by a bright red laser. Notice the diffraction pattern both inside the eye and around the outside of the needle in its shadow.

works, including Newton's *Principia* and *Opticks*. Young is best known for his 1803 experiment that provided strong evidence that light is a wave. Whereas a theory can be disproved by a single experiment, one experiment cannot prove a theory. A single experiment can only confirm or be consistent with a theory. If a number of experiments confirm a theory, the theory is accepted as a good model of nature. The wave model of light did not become widely accepted until another experiment, performed by Francois Arago in 1819 (10 years before Young died), confirmed Young's results. This chapter presents evidence that supports the wave model for light.

35-1 Light Is a Wave

Contemporary scientists model light both as a photon (particle) and as a wave, choosing the model that works best in any particular situation (Part VI). Several hundred years ago, scientists believed that light must be either a collection of particles or a wave, but not both. On one side of the debate were the French philosopher and mathematician René Descartes (1596–1650), the English physicist Robert Hooke (1635–1703), and the Dutch mathematician and astronomer Christiaan Huygens (1629–1695), who all believed that light is a wave. On the other side was Isaac Newton (1642–1727), who believed that light is made up of particles.

Figure 35.1 shows sunlight passing through an open window. The sharp shadows cast by the edges of the (glassless) window suggest that Newton was right: Light is made up of particles. Imagine that, instead of sunlight passing through the window, there was a steady stream of particles, such as paint droplets from a spray can, all moving in the same direction. The paint droplets that made it though the holes of the windows would paint the wall in the shape of the window. So, we expect particles passing through an aperture to trace the shape of the aperture. You might think observations of light passing through an aperture, such as those in Figure 35.1, would settle the debate in favor of the particle model.

If we look at laser light passing through a small aperture, however, we find evidence that does not support the particle model. In the top row of Figure 35.2 are two apertures with different shapes. In the bottom row are the resulting patterns of laser light projected onto a screen. These complicated patterns, known as **diffraction patterns**, do *not* simply trace the shapes of the apertures and cannot be explained by the particle model. Instead, diffraction patterns are a result of the interference of light *waves*.

Diffraction occurs when light passes through a narrow aperture or near the edge of an obstacle; the diffracted light bends and forms a diffraction pattern of bright and dark fringes. You may not notice diffraction because it is prominent only when the size of the aperture (or obstacle) is small enough compared to the wavelength of light. Because visible light has a wavelength of a few hundred nanometers, the aperture must be fairly small for us to observe diffraction. (For many of our applications in this chapter, we consider apertures and obstacles that are roughly a millimeter in size.)

Diffraction is not unique to light waves; we encountered the phenomenon when we studied mechanical waves. For example, the diffraction of sound waves allows you to hear around obstacles. If you sit in the back of a room so crowded that you cannot see the speaker standing at the front, you can still hear his voice. Because sound waves have wavelengths of roughly a meter, you can hear around obstacles, such as people, that are roughly a meter across. Because the wavelength of light is short compared to the width of a person, light essentially does not diffract around people. As a result, you cannot see around the people sitting in front of you.

Similarly, a diffraction pattern of bright and dark bands (or fringes) may be seen around the edges of an obstacle such as a sewing needle (Fig. 35.3). The pattern around the needle looks like thin fringes of light and darkness outlining the needle, and some of the bright fringes appear in the region that we normally think of as the needle's shadow.

Creating diffraction patterns does not require special equipment. You can make your own diffraction pattern by holding your hand in front of a bright light source, such as an open window or lightbulb. Extend two fingers so that they are parallel to each other with a gap in between. As you bring your fingers close together, look at the space between them, and in it you will see one or more dark lines running parallel to your fingers.

Both diffraction and **interference** are wave phenomena that result from the superposition of waves. Because both phenomena produce a characteristic pattern of bright and dark fringes, the distinction between interference and diffraction is arbitrary and not consistently made. However, a common distinction between the two phenomena is that diffraction involves only a single aperture (or source), whereas interference involves two or more such apertures.

Throughout this chapter, we'll consider the details of interference and diffraction experiments that produce such patterns of bright and dark fringes and thus confirm the wave model. Before we get into details, however, we need to consider some convenient ways to visually represent light as a wave.

INTERFERENCE ✪ Major Concept

CONCEPT EXERCISE 35.1

Perhaps Newton never observed a diffraction pattern. If he had, what might he have concluded about light? Why do you suppose he was unable to see a diffraction pattern?

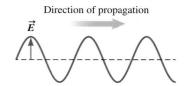

FIGURE 35.4 This representation of light looks like a wave on a string. You can imagine it is the oscillating electric field, so that the amplitude of the "string wave" represents the maximum electric field magnitude. The magnetic field oscillates in the plane that is perpendicular to both the electric field and the direction of propagation.

Representing Light as a Wave

From Chapter 34, we know that light is a transverse electromagnetic wave with an electric field oscillating perpendicular to a magnetic field and both fields oscillating perpendicular to the propagation direction (Fig. 34.8, page 1095). When we use this wave model, it is inconvenient and unnecessary to sketch both fields and the direction of propagation. Instead, we sometimes draw a single sinusoidal curve (Fig. 35.4), which closely resembles a transverse traveling wave on a string (Fig. 17.9, page 492). You can think of such a sketch as representing the oscillating electric field. Then you can imagine that the magnetic field oscillates perpendicular to the electric field—but there is no need to draw it.

However, a sketch of a single curve fails to represent a wave that may be two- or three-dimensional. To represent two-dimensional waves, such as waves on the surface of water, we imagine looking at the wave from above and draw solid lines to represent the wave crests (also called the **wave fronts**) and dashed lines to represent the wave troughs or valleys. Figure 35.5A shows a **plane parallel wave**. Each wave crest appears as a straight line parallel to all other crests. If you imagine viewing Figure 35.5A from the side, it would look like Figure 35.4, so a plane parallel wave is really a one-dimensional wave. Figure 35.5B shows a **circular wave** in which the wave crests form concentric circles. Circular waves travel in two dimensions. Because there is no good way to represent a three-dimensional wave on a two-dimensional surface, we must draw a two-dimensional slice through the wave. Figure 35.5B is exactly what we would draw to represent a three-dimensional **spherical wave** in which the wave crests form concentric spheres. Spherical waves are given off by sources such as the Sun and lightbulbs.

VISUAL REPRESENTATIONS OF LIGHT ◉ Tool

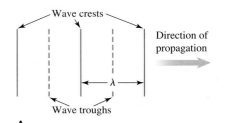

A.

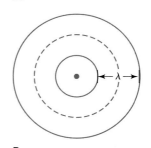

B.

FIGURE 35.5 A. Representation of a plane parallel wave. **B.** Representation of a two-dimensional circular wave or a three-dimensional spherical wave.

Propagation of Light

Long before Maxwell discovered that light is an electromagnetic wave, researchers had studied the propagation of light beams. Huygens came up with a way to model the propagation of light. To understand Huygens's model, we use an analogy with waves on the surface of water. First, imagine a small ball bobbing up and down, creating circular waves on the water as in Figure 17.19A (page 502). Then imagine a rod bobbing up and down, creating parallel waves on the water as in Figure 17.21B. Huygens's idea is that if you replace the rod with a large number of small bobbing balls, the circular waves produced by the balls will add together to produce the plane parallel wave. Huygens extended this idea beyond such bobbing, or oscillation, to

HUYGENS'S PRINCIPLE
❶ Underlying Principle

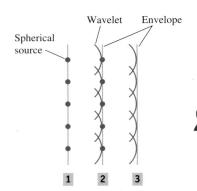

FIGURE 35.6 Huygens's principle applied to a plane parallel wave. The wave front starts at **1** and is made up of point sources creating spherical wavelets. The wave front at time **2** is the envelope of all these wavelets. This process is repeated to find the wave front at time **3** and at later times not shown.

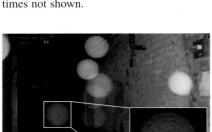

FIGURE 35.7 Some people believe orbs are ghosts. They are really diffractions of light due to dust or other small particles near the camera. *Inset:* A close-up of an orb reveals rings.

describe the forward motion of any wave front. Today we call this **Huygens's principle**: *Each point on a primary wave front serves as the source of a spherical wave, called a wavelet, that advances with a speed and frequency equal to those of the primary wave. The primary wave front at some later time is the envelope of these wavelets* (Fig. 35.6).

CASE STUDY Part 1: Ghosts

A group of college students discusses a photograph.

Shannon: Check this out. I took a picture of ghosts (Fig. 35.7).

Avi: I've seen this kind of picture before. I think people call these things orbs. They're caused by dust, or something like that.

Shannon: No, I thought of that. I cleaned the lens and took another picture, and there were more orbs.

Avi: The dust I mean isn't in the camera. It's in the air.

Shannon: Are you kidding? Dust in the air would be too small, and it wouldn't have rings. Take a close look at the orb. It has a light spot in the middle and dark rings like this (Fig. 35.7, inset).

Cameron: I've seen that kind of pattern before. It kind of looks like the fringes around this sewing needle (Fig. 35.3).

CASE STUDY Part 2: Science Competition

In 1819, the French Academy of Sciences held an essay competition to explain properties of light. At the time, most members of the Academy strongly supported the particle model for light, and they expected the competition to disprove the wave model. A physicist and military engineer named Augustin-Jean Fresnel submitted a theory based on the wave model. The French mathematician Siméon-Denis Poisson (1781–1840) was one of the judges and a supporter of the particle model. He read Fresnel's essay and thought he had found what he needed to disprove the wave model. Poisson showed that Fresnel's work predicted the existence of a bright spot in the center of the shadow cast by a round object, such as a disk or sphere. Poisson argued that, because it is absurd to expect to see a bright spot in the middle of a shadow, light must be made up of particles, not waves. In this chapter's two-part case study, we'll learn why the orb in Figure 35.7 has rings and find out whether Fresnel or Poisson was right.

CONCEPT EXERCISE 35.2
CASE STUDY Hiding?

You are kneeling behind a large sandbag in a game of paintball. On the other side of this obstacle, a friend fires paint in your direction. Do you get hit? Now imagine floating behind a rock in the ocean, with water waves moving toward the rock. Do you oscillate up and down when the wave arrives at your location? Explain your answers. Later in the case study, we'll use these situations as analogies for light that encounters an obstacle.

35-2 Sound Wave Interference Revisited

Finding the connection between two seemingly unrelated phenomena helps us learn something new and gain a deeper insight into both phenomena. In this section, we put aside our study of light for the moment in order to review sound wave interference.

Then we'll be more prepared to study the interference of light waves. Sound and light are similar, but they differ in two major ways: First, sound waves are longitudinal waves, whereas light is a transverse wave. Second, sound needs a medium, such as air, whereas light needs no medium.

We are interested in the interference pattern produced by two identical harmonic waves. Harmonic waves are mathematically simple, represented by a single sine or cosine function. In particular, we'll review sound waves emitted by speakers at a single tone (or frequency). Such a sound is not like music or speech; rather, it is something like an emergency alarm.

Section 18-3 presented three important properties of waves: (1) When two or more waves are present, the resulting wave is the **superposition** (addition) of those individual waves. (2) Constructive interference results when two identical harmonic waves are in phase; the resulting wave has twice the amplitude of either of the original waves (Fig. 18.15A, page 527). (3) Destructive interference results when two identical harmonic waves are 180° out of phase (Fig. 18.15B).

Consider the ideal situation in which two speakers produce identical harmonic sound waves in a large room as in Example 18.2. The waves that emerge from each speaker are three-dimensional hemispheres (Fig. 35.8). There are points of constructive and destructive interference throughout the room. In this ideal situation, places of constructive interference are loud and places of destructive interference are silent.

Whether there is constructive or destructive interference at a particular point depends on how far the point is from each speaker. If the difference in the distance traveled by two waves is an integer multiple of the wavelength, the result is constructive interference. The condition for constructive interference is

$$\Delta d = n\lambda \qquad (n = 0, 1, 2, 3, \ldots) \qquad (18.2)$$

where Δd is the difference in the distances traveled by the two waves. This condition is met by point C in Figure 35.8. To see this, use the wavelengths indicated to find the distance. C's distance from speaker 1 is $d_1 = 2\lambda$, and C's distance from speaker 2 is $d_2 = \lambda$. The difference $\Delta d = \lambda$, so $n = 1$ and there is constructive interference at point C.

If one wave travels an extra $\lambda/2$, the two waves are 180° out of phase and interfere destructively. The distance between point D and speaker 1 is $3\lambda/2$, and D's distance from speaker 2 is 2λ (Fig. 35.8). So, point D is $\lambda/2$ farther from speaker 2 than it is from speaker 1. As a result, waves arrive at point D 180° out of phase, and point D is a place of destructive interference. More generally, if the difference in the distances traveled by two waves is a half-integer number of wavelengths, the result is destructive interference:

$$\Delta d = \left(n + \frac{1}{2}\right)\lambda \qquad (n = 0, 1, 2, 3, \ldots) \qquad (18.3)$$

CONCEPT EXERCISE 35.3

In Figure 35.8, the speakers are 2λ apart. What would happen at points C and D if speaker 2 was moved so that the speakers' separation increased to $5\lambda/2$?

Young's Double-Slit Experiment

Young's experiment, shown in Figure 35.9, looks much like Figure 35.8 with the two speakers replaced by two slits. Each slit is a source of identical light waves. As in Figure 35.8, these light waves interfere constructively or destructively depending on location. We experience constructive interference of sound waves as a loud sound and destructive interference as silence. In Young's experiment with light waves, we see places of constructive interference as bright fringes and places of destructive interference as dark fringes on a screen. Particles do not undergo interference, but waves do. So, Young's experiment supports the wave model. Young's work was not well received in England because Newton was a figure of national pride and opposition to his theories was difficult to accept.

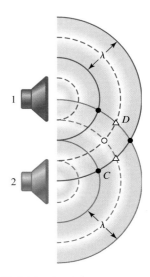

FIGURE 35.8 The filled circles (dots) are places where two peaks meet. The open circle is a place where two valleys meet. Both are places of constructive interference. The triangles are places where a valley meets a peak. These are places of destructive interference.

CONDITIONS FOR CONSTRUCTIVE AND DESTRUCTIVE INTERFERENCE
✪ Major Concepts

Notice that we modified Equation 18.3, but the two expressions are equivalent.

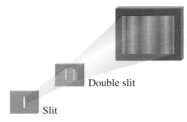

FIGURE 35.9 Young's experiment shows that light makes an interference pattern. Light passes through two small apertures. Usually the apertures are narrow slits, so Young's experiment is sometimes called the *double-slit experiment*. Young needed to use the single slit in the foreground to make sunlight that passes through the double-slit coherent. In contemporary experiments using laser light, this single slit is not necessary. (This diagram is a schematic showing only a crude representation of Young's experiment.)

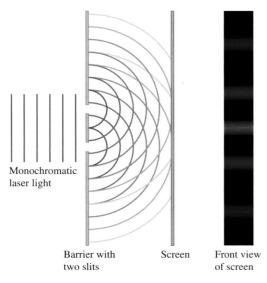

FIGURE 35.10 Schematic diagram of Young's experiment, as seen from above. The pattern on the screen cannot be seen from this perspective, but it is shown to the right. A front view of the screen is placed at the location of the screen in subsequent figures.

COHERENCE ⊙ Major Concept

Figure 35.10 is a schematic drawing of Young's experiment, shown from above the apparatus. Bright light encounters a barrier that has two slits. The resulting interference pattern is projected onto the screen on the right. From this perspective, you can't see the pattern, so we have included a front view of the pattern on the screen. The brightest fringe is in the center of the pattern; the farther fringes are somewhat dimmer. The bright fringes gradually fade into the dark fringes with no sharp cutoff.

The diagram also shows the wave fronts. Before the light encounters the slits, it is a plane wave. According to Huygens's principle, we can think of the wave front as comprised of spherical wavelets. When the wave front encounters the barrier, only a couple of wavelets emerge. These hemispherical waves are like the sound waves that emerge from a speaker. As a result, the wave pattern from the two slits (Fig. 35.10) is like the pattern from the speakers (Fig. 35.8). So, the conditions for constructive and destructive interference in Young's experiment are also given by Equations 18.2 and 18.3.

In our daily lives we are often exposed to two light sources, so why don't we see interference patterns all the time? The answer is that four conditions must be met for us to see a fringe pattern like the one in Figure 35.10. First, the waves must be in the same medium, so they have the same propagation speed. Second, they must have the same frequency or, equivalently, the same wavelength. Third, they may differ by a (nonzero) phase constant φ, but that difference must remain constant in both space and time. The combination of the second and third conditions is called **coherence**. Two waves are said to be **coherent** if they are monochromatic sources of the same frequency and there is a constant phase difference between them. Finally, if two waves that are 180° out of phase are to cancel, the two waves must also have the same amplitude.

We don't normally see interference patterns because two ordinary light sources can maintain a constant phase difference for only about 10^{-8} s or less. Any interference pattern they produce is shifted around rapidly and randomly, so, on average, you cannot detect any interference. Also, different parts of the same light source, such as two pieces of the same lightbulb filament, do not produce coherent light waves. The atoms in any one source normally emit light of many different frequencies and amplitudes. Even if the source produces monochromatic (single frequency) light, the light waves from different parts of the source do not maintain a constant phase difference. The incoherence of light normally destroys the interference pattern. For the same reason, we normally don't notice interference when listening to music in an auditorium: The speakers in such a case do not emit coherent sound waves.

In your laboratory or classroom, you will likely see a demonstration of Young's experiment using a laser. A laser produces monochromatic, coherent light. Young conducted his experiment in 1803, long before the laser was invented. He needed a source of bright light, though, so he used the Sun, but sunlight is neither monochromatic nor coherent. To create a nearly coherent source, Young placed another small aperture between the Sun and the double slits (Fig. 35.9). The light that makes it through this first slit is essentially a single source of light, which Young then split into two coherent sources. In other words, the light that emerges from the double slits is coherent. Because Young used sunlight, the interference pattern he saw showed all the colors arranged in bright and dark fringes. If he had used a monochromatic light source instead of the Sun, he would have seen the familiar monochromatic bright and dark pattern (Fig. 35.10). For the rest of our discussion of Young's experiment, we will assume that the light source is monochromatic and coherent; you can imagine it is a laser.

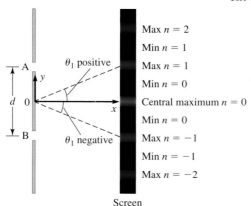

FIGURE 35.11 Places of constructive interference are bright and are known as maxima. Dark regions are places of destructive interference and are known as minima. Angles measured counterclockwise from the x axis are positive, and angles measured clockwise are negative.

35-3 Young's Experiment: Position of the Fringes

In this section, we find expressions for the positions of the bright and dark fringes in Young's double-slit experiment. Figure 35.11 shows the coordinate system and the conventions we use to label these fringes. The origin of the coordinate system is the point on the barrier directly between the two slits. A positive x axis points from the origin to the screen. At the point where the x axis touches the screen ($y = 0$), there is a bright fringe. The bright fringes are also called **maxima**, and the bright fringe at $y = 0$ is called the **central maximum**. Each maximum is assigned an integer n starting with the central maximum, for which $n = 0$. Maxima on the positive y axis are assigned positive integers, and maxima on the negative y axis are assigned negative integers.

The dark fringes are called **minima** and are also assigned integers n, so you must determine from the context whether n refers to a maximum or a minimum. The labeling of dark fringes corresponds to that of bright fringes. Each dark fringe is surrounded by two bright fringes. The label n of the dark fringe is the same as the lower absolute magnitude of the labels of the two adjacent bright fringes. Consider the dark fringe just above the central maximum; it is surrounded by two bright fringes ($n = 0$ and $n = 1$). It is labeled $n = 0$, the same as the neighboring bright fringe with the lower absolute magnitude.

The position of each fringe is indicated by the angle θ measured from the x axis, normally with a subscript corresponding to the integer n. So, the position of the $n = 1$ maximum is given by the positive angle θ_1, and the position of the $n = -1$ minimum is given by the negative angle θ_1.

The central maximum is at $\theta_0 = 0$. The distance d_A from slit A to the central maximum is equal to the distance d_B from slit B to the central maximum: $d_A = d_B$ (Fig. 35.12). So, the waves that emerge from these slits arrive in phase at the position of the central maximum. Other fringes are not equidistant from both slits. Whether there is a bright fringe or a dark fringe at some location depends on the difference between the distances traveled by the light waves that emerge from each slit.

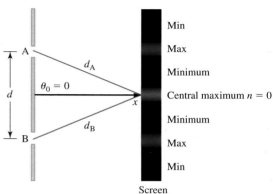

FIGURE 35.12 The distance from slit A to the central maximum equals the distance from slit B to the central maximum: $d_A = d_B$.

DERIVATION Bright and Dark Fringe Positions

We will show that the positions of the maxima in Young's double-slit experiment are given by

$$d \sin \theta_n = n\lambda \quad (n = 0, \pm 1, \pm 2, \pm 3, \dots) \quad (35.1)$$

and the positions of the minima are given by

$$d \sin \theta_n = \left(n + \frac{1}{2}\right)\lambda \quad (n = 0, \pm 1, \pm 2, \pm 3, \dots) \quad (35.2)$$

where d is the separation between the slits and λ is the wavelength of the light.

In Figure 35.13, the distance from the barrier to the screen is x. We assume that the distance x is much greater than the separation d between the slits. Let's start by arbitrarily choosing a maximum at a positive position above the central maximum. The distance from slit A to this maximum is shorter than the distance from slit B to this maximum: $d_B > d_A$. This extra distance, the path-length difference $\Delta d = d_B - d_A$, determines whether the waves arrive in phase or out of phase.

We must relate Δd to the angle θ using an approximation. As long as the distance to the screen is greater than the distance between the slits, we can approximate the two paths (one from each slit) to any point on the screen as nearly parallel (Fig. 35.14).

YOUNG'S EXPERIMENT—POSITION OF FRINGES
○ **Special Case**

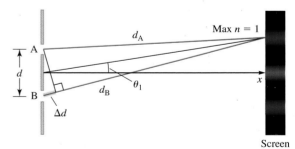

FIGURE 35.13

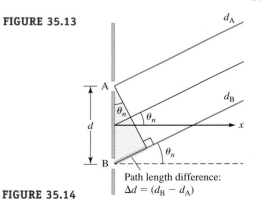

FIGURE 35.14

Derivation continues on page 1130 ▶

1130 CHAPTER 35 Diffraction and Interference

Angle θ_n to the nth bright fringe is measured from the x axis and also equals the angle at vertex A of the highlighted right triangle in Figure 35.14. Use this triangle to relate the angle θ to the path-length difference Δd.	$\sin \theta_n = \dfrac{\Delta d}{d}$ $\Delta d = d \sin \theta_n$ (35.3)
Setting the condition for constructive interference (Eq. 18.2) equal to Equation 35.3, we find an expression for the position of the positive maxima. To include the fringes below the central maximum at negative positions, we add $\pm$ to indicate that n may be positive or negative. Choose the positive sign for fringes at positive y positions (above the central maximum) and the negative sign for fringes at negative y positions (below the central maximum).	$\Delta d = n\lambda \quad (n = 0, 1, 2, 3, \ldots)$ (18.2) $\Delta d = d \sin \theta_n = n\lambda \quad (n = 0, 1, 2, 3, \ldots)$ $d \sin \theta_n = n\lambda \quad (n = 0, \pm1, \pm2, \pm3, \ldots)$ ✓ (35.1)
From here, it is easy to find an expression for the positions of the minima. The path-length difference is still given by Equation 35.3, but there is destructive interference at the minima, so we set Equation 35.3 equal to Equation 18.3 for destructive interference. Again, we include $\pm$ to indicate that n may be positive or negative.	$\Delta d = \left(n + \dfrac{1}{2}\right)\lambda \quad (n = 0, 1, 2, 3, \ldots)$ (18.3) $\Delta d = d \sin \theta_n = \left(n + \dfrac{1}{2}\right)\lambda$ $d \sin \theta_n = \left(n + \dfrac{1}{2}\right)\lambda \quad (n = 0, \pm1, \pm2, \pm3, \ldots)$ ✓ (35.2)

:• COMMENTS
In Equations 35.1 and 35.2, d (the distance between slits) and λ should be in the same length units. However, it is not necessary or convenient to use meters; we typically use nanometers instead.

CONCEPT EXERCISE 35.4
When Young performed his double-slit experiment, he used sunlight. Explain why the interference pattern he saw showed a rainbow of colors.

EXAMPLE 35.1 A Birthday Present

You are given a red-light laser pointer for your birthday, but you don't know what wavelength of light it emits. You take the laser into your physics laboratory to perform a double-slit experiment. The slits are 0.350 mm apart. The distance from the barrier to the screen is 2.50 m. You measure the distance from the central maximum to the $n = 2$ maximum and find that it is 9.04 mm. What is the wavelength of your laser pointer's light?

:• INTERPRET and ANTICIPATE
The location of the bright fringes depends on the wavelength of the light, so if we know the position of one of the bright fringes (with respect to the central maximum), we can find the wavelength. Because the laser light is red, we expect to find an answer between 620 nm and 780 nm.

Sketch the situation (Fig. 35.15). The linear position y_2 of the $n = 2$ maximum is known from the experiment and is indicated on the figure. This corresponds to the angular position θ_2.

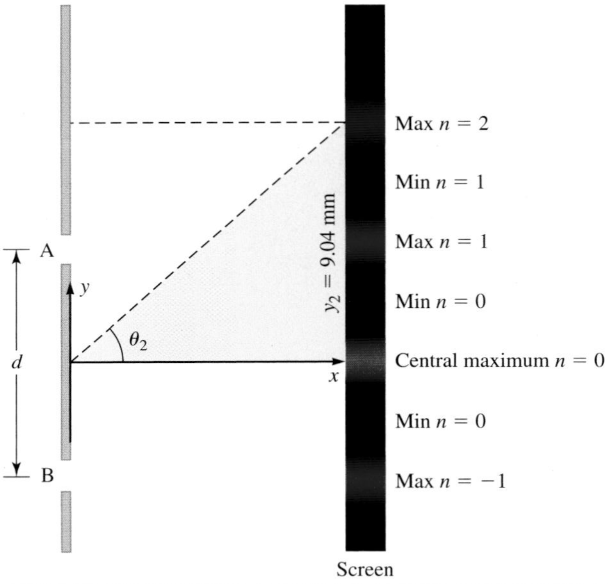

FIGURE 35.15

35-3 Young's Experiment: Position of the Fringes

:• SOLVE

Find the angular position from the linear position using the highlighted right triangle (Fig. 35.15). Both y_2 and x must be in the same length units; we use millimeters here. (We keep an extra significant figure in this intermediate step.)	$\tan\theta_2 = \dfrac{y_2}{x}$ $\theta_2 = \tan^{-1}\left(\dfrac{y_2}{x}\right) = \tan^{-1}\left(\dfrac{9.04\text{ mm}}{2500\text{ mm}}\right) = 0.2072°$
Solve Equation 35.1 for wavelength, setting $n = 2$.	$d\sin\theta_n = n\lambda = 2\lambda$ (35.1) $\lambda = \dfrac{d\sin\theta_n}{2} = \dfrac{(0.350\text{ mm})\sin 0.2072°}{2} = 6.33\times 10^{-4}\text{ mm}$ $\lambda = 633\text{ nm}$

:• CHECK and THINK

As expected, the wavelength is in the red range. Now let's think about improving this experiment. The distance was measured between the central maximum and the $n = 2$ maximum. This distance is fairly small and would have been even smaller if the $n = 1$ maximum were used instead. To make the distance easier to measure, the screen could be moved farther away. Also, it may be easier to measure the distance between extreme maxima, such as between $n = -3$ and $n = 3$.

EXAMPLE 35.2   Small Angles

Often the angular position of a fringe is a very small angle. Find an approximate expression for the y position in this case. Then repeat Example 35.1 using this approximate position equation and compare your results in the **CHECK and THINK** step.

:• INTERPRET and ANTICIPATE

Figure 35.15 shows the geometry of the situation, but because we are considering small angles, θ_2 is greatly exaggerated. The angle in Example 35.1 was very small ($\theta_2 = 0.207° = 3.16 \times 10^3$ rad), so a small-angle approximation should produce a result very close to the exact wavelength we found before.

:• SOLVE

From the highlighted triangle in Figure 35.15, we found an exact expression involving θ_n. When this angle is small, the tangent of the angle approximately equals the sine of the angle.	$\tan\theta_n = \dfrac{y_n}{x}$ $\sin\theta_n \approx \dfrac{y_n}{x}$ (1)
Substitute Equation (1) into Equation 35.1 for a bright fringe.	$d\sin\theta_n = n\lambda$ ($n = 0, \pm 1, \pm 2, \pm 3, \ldots$) (35.1) $d\left(\dfrac{y_n}{x}\right) \approx n\lambda$
Solve for y_n.	$y_n \approx \dfrac{nx\lambda}{d}$ ($n = 0, \pm 1, \pm 2, \pm 3, \ldots$) (35.4)

:• CHECK and THINK

To see whether Equation 35.4 leads to a different wavelength than in Example 35.1, solve Equation 35.4 for λ and substitute values, including $n = 2$. It makes sense that our approximate answer matches our exact answer because in this case angle θ_2 is very small.	$\lambda \approx \dfrac{y_n d}{nx} \approx \dfrac{(9.04\text{ mm})(0.350\text{ mm})}{2(2500\text{ mm})} \approx 6.33\times 10^{-4}\text{ mm}$ $\lambda \approx 633\text{ mm}$

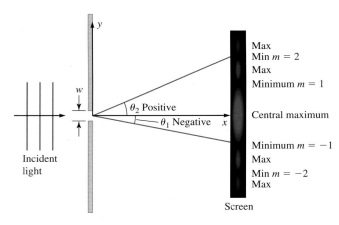

FIGURE 35.16 Coherent light incident from the left encounters a single slit of width w. A diffraction pattern is seen on a screen on the right. The origin of the coordinate system is at the center of the slit. The dark fringes are labeled by positive and negative integers.

35-4 Single-Slit Diffraction

As stated earlier, the distinction between interference and diffraction is arbitrary and not consistently made. However, a common distinction between the two phenomena is that diffraction involves only a single wave front, whereas interference involves a small number of coherent sources. So, diffraction is observed when coherent light passes through a single slit, whereas interference is observed when coherent light passes through two slits. The pattern observed in Young's double-slit experiment is called an interference pattern. However, we'll see that it is a combination of interference and diffraction. First, we need to take a close look at diffraction. To keep things relatively simple, we focus on the diffraction pattern produced by a single slit. In this section, we find a mathematical expression for the positions of the dark fringes produced by the slit.

Figure 35.16 is a schematic drawing of a single-slit diffraction experiment with a slit of width w. The coordinate system and labeling are similar to those for Young's double-slit experiment (Fig. 35.11). At the point ($y = 0$) where the x axis touches the screen, there is a bright fringe, the **central maximum**. The other bright fringes are sometimes referred to as **secondary maxima** or **side lobes**.

We can derive the position of the dark fringes (the minima) by following a procedure similar to that for the double-slit experiment. Here we label each minimum with an integer m. Minima at positive y positions have positive-integer labels, and those at negative y positions have negative-integer labels. This labeling convention is different from that in Figure 35.11 for two slits; in that case, there was an $n = 0$, but here there is no $m = 0$.

The position of each dark fringe is indicated by the angle θ measured from the x axis (positive if measured counterclockwise; negative if measured clockwise). Normally we use a subscript on θ that corresponds to the integer m.

Before we derive the position of the dark fringes, let's see why there is a central maximum. The key is to use Huygens's principle: Each point on a primary wave front serves as the source of a spherical wavelet that advances with a speed and frequency equal to those of the primary wave. The primary wave front at some later time is the envelope of these spherical waves (Fig. 35.6). We can model the wave front that enters the slit as the source of four spherical wavelets (Fig. 35.17). The four waves arrive in phase at point C on the screen, and the superposition of these four wavelets produces constructive interference. The choice to break up the wave front into four spherical waves is arbitrary. We could have chosen any number and found the same result: Constructive interference at C produces a central maximum. (The mathematics leading to the position of the other maxima is left for a more advanced course.)

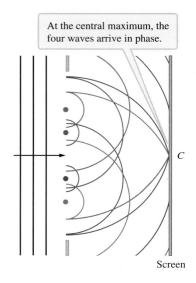

FIGURE 35.17 From Huygens's principle, the wave front is broken arbitrarily into four spherical waves represented by four colors. These waves arrive in phase at point C on the screen.

DERIVATION Dark Fringe Positions in a Single-Slit Diffraction Pattern

We will show that the angular positions θ_m of the dark fringes in a single-slit diffraction pattern are given by

$$w \sin \theta_m = m\lambda \quad (m = \pm 1, \pm 2, \pm 3, \ldots) \quad (35.5)$$

SINGLE-SLIT DIFFRACTION—POSITION OF DARK FRINGES ▶ Special Case

We begin by finding expressions for the first and second dark fringes, and then extrapolate from these to find an expression for the position of any dark fringe.

Finding the positions of the dark fringes produced by a single slit is much like finding the positions of the fringes produced by a double slit. Break the slit into two zones A and B, each of width $w/2$. Use Huygens's principle to break up the wave front into as many spherical wavelets as we have zones. The source of one wavelet is at the top of zone A, and the source of the other is at the top of zone B. Figure 35.18 does not show the sources or the wavelets, but instead shows the distance from each source to the position of the first dark fringe on the screen. The distance d_B from source B is greater than the distance d_A from source A.	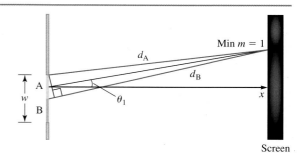 **FIGURE 35.18**
As in the case of Young's double-slit experiment, the path-length difference $d_B - d_A$ determines the position of the dark fringes. We again make the approximation that the distance x from the slit to the screen is much greater than the width w of the slit: $x \gg w$. In this approximation, the paths from each zone to any point on the screen are nearly parallel (Fig. 35.19).	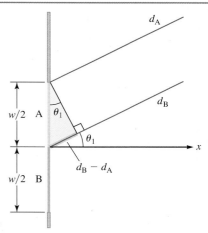 **FIGURE 35.19**
Use the highlighted right triangle in Figure 35.19 to find an expression for the path-length difference Δd.	$\Delta d \equiv d_B - d_A = \dfrac{w}{2}\sin\theta_1$
In order for there to be a dark fringe at the corresponding screen position, the two waves must arrive 180° out of phase. This occurs when the path-length difference equals half a wavelength. Equation (1) may be used to find the angular position θ_1 of the first dark fringe above or below the central maximum. (The angle below the central maximum is negative.)	$\Delta d = \dfrac{\lambda}{2}$ $\dfrac{w}{2}\sin\theta_1 = \dfrac{\lambda}{2}$ $w\sin\theta_1 = \lambda$ (1)
Find the position of the second dark fringe by splitting the slit into four zones, each with width $w/4$ and labeled A through D. We imagine four spherical wave sources, one at the top of each zone. The path length from each of these sources to the second dark fringe is labeled d with a subscript corresponding to the zone. As before, if $x \gg w$, the paths are approximately parallel. The path-length difference between the paths from adjacent zones is Δd.	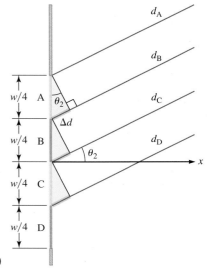 **FIGURE 35.20**

Derivation continues on page 1134 ▶

Find this path-length difference Δd by applying trigonometry to any one of the right triangles highlighted in Figure 35.20, such as the top triangle.	$\Delta d \equiv d_B - d_A = \dfrac{w}{4}\sin\theta_2$
Because θ_2 is the position of a dark fringe, the path-length difference must be half a wavelength. Equation (2) may be used to find the angular position θ_2 of the second dark fringe above or below the central maximum.	$\Delta d = \dfrac{\lambda}{2} = \dfrac{w}{4}\sin\theta_2$ $w\sin\theta_2 = 2\lambda$ (2)
We can generalize this procedure to find the angular position θ_m of any dark fringe by dividing the slit into an even number of zones n_{zones}. The path-length difference between the paths from adjacent zones is found by applying trigonometry to a right triangle.	$\Delta d = \dfrac{w}{n_{zones}}\sin\theta_m$
From our procedure for finding the positions of the first and second dark fringes, we see that the number of zones n_{zones} is two times the label m for the fringe.	$n_{zones} = 2m$ $\Delta d = \dfrac{w}{2m}\sin\theta_m$
At the position of any dark fringe, the path-length difference must be half a wavelength.	$\Delta d = \dfrac{\lambda}{2} = \dfrac{w}{2m}\sin\theta_m$ (3)
Rearrange Equation (3) and note that m is an integer other than 0 because there must be at least two zones.	$w\sin\theta_m = m\lambda \quad (m = \pm 1, \pm 2, \pm 3, \ldots)$ ✓ (35.5)

:• **COMMENTS**
Equation 35.5 closely resembles

$$d\sin\theta_n = n\lambda \quad (n = 0, \pm 1, \pm 2, \pm 3, \ldots) \tag{35.1}$$

which gives the angular position of the *bright* fringes produced in the double-slit interference pattern. So, you need to find a way to tell them apart. Here are two differences: (1) Equation 35.5 is used to find the angular position of the *dark* fringes, not the *bright* ones; and (2) in Equation 35.5, the first dark fringe is labeled $m = \pm 1$, whereas the first bright fringe in Equation 35.1 is labeled $n = 0$.

Effect of Slit Width on the Diffraction Pattern

The diffraction pattern and, in particular, the width of the central maximum depend on the slit width. Let's say that the width Δy of the central maximum is the distance between the two first dark fringes (Fig. 35.16); that is,

$$\Delta y = 2y_1 = 2x\tan\theta_1 \tag{35.6}$$

where θ_1 is positive. For a slit illuminated by monochromatic light of wavelength λ, let's see what happens if the slit width equals the wavelength: $w = \lambda$. In that case, Equation 35.5 (with $m = \pm 1$) becomes

$$\sin\theta_1 = \pm\dfrac{\lambda}{w} = \pm\dfrac{w}{w} = \pm 1 \tag{35.7}$$

and so

$$\theta_1 = \pm 90°$$

When we substitute 90° into Equation 35.6, the central maximum's width is infinite: $\Delta y = 2x\tan 90° \to \infty$. In other words, if the slit width equals the light's wavelength, the first dark fringes are infinitely far from the central maximum, which completely covers the viewing screen. You will not see a diffraction *pattern*; instead, you will see that the center of the screen is very bright and the brightness fades away from the center.

By combining $\Delta y = 2x \tan \theta_1$ (Eq. 35.6) and $\sin \theta_1 = \pm \lambda/w$, we can plot Δy as a function of λ/w as in Figure 35.21. For monochromatic light, the horizontal axis is a measure of the slit width. The widest slit is on the far left, and the narrowest slit ($w = \lambda$) for which a pattern may be observed on a flat screen is on the far right. This graph shows that for a wide slit, the central maximum is narrow, so the first dark fringes are very close together. By contrast, for a narrow slit, the central maximum is wide and the first dark fringes are far apart. And, as we already know, for the narrowest slit ($w = \lambda$), the central maximum is infinite.

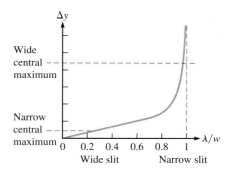

FIGURE 35.21 The position of the first dark fringe as a function of λ/w. A wide slit is on the left of the graph, and the slit width gets narrower toward the right. If λ is constant, the horizontal axis depends only on the slit width.

CONCEPT EXERCISE 35.5

When we studied Young's double-slit experiment, we mostly ignored the dark fringe pattern produced by diffraction. Use Figure 35.21 to describe situations in which that omission makes sense. Think especially about the single slit used in front of the double slit in Young's experiment (Fig. 35.9).

EXAMPLE 35.3 A Wide Slit

We just found that with a narrow slit, diffraction spreads out light into a wide central maximum, so we don't usually notice the dark fringe pattern produced. We also don't notice the diffraction pattern when light passes through a slit that is wide compared to the wavelength. Now consider a narrow beam of coherent, monochromatic light of wavelength $\lambda = 550$ nm passing through a single slit of width $w = 1.00$ cm. (This slit is nearly 20,000 times wider than the wavelength.) The light illuminates a screen at $x = 1.00$ m from the slit. On the screen, we see a bright central spot (Fig. 35.22, top). There are actually 20 dark fringes on either side of the central maximum, but you cannot see these fringes because they are too close together. The inset in Figure 35.22 is a magnified sketch of the diffraction pattern, showing the 20 fringes on either side of the central bright spot. (Even after magnification, the pattern is blurred.) Estimate the magnification required to see these fringes at the scale indicated in the figure. Proceed by finding the y positions of the $m = \pm1, \pm2, \pm3, \pm10,$ and ±20 dark fringes, and then find the magnification needed to display the images on this scale.

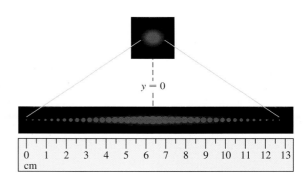

FIGURE 35.22 A bright central spot from a green laser and a magnified view of its diffraction pattern. The distance from the slit to the screen is x (out of the page, not labeled in this figure).

:• **INTERPRET and ANTICIPATE**

Because we normally don't see the dark fringes when light passes through a wide slit, we expect that all 20 fringes are close to the central maximum. The angular position should be very small, so we can use the small-angle approximation: $\sin \theta \approx \tan \theta \approx \theta$ for θ in radians.

:• **SOLVE**

Use trigonometry to write an expression for the y position of the mth dark fringe.	$y_m = \pm x \tan \theta_m$	(1)
Equation 35.5 gives the angular position of the mth dark fringe.	$w \sin \theta_m = m\lambda \quad (m = \pm1, \pm2, \pm3, \ldots)$	(35.5)
Apply the small-angle approximation to Equations (1) and 35.5.	$y_m \approx \pm x \theta_m$ $w \theta_m \approx m\lambda$	(2) (3)

Example continues on page 1136 ▶

Combine Equations (2) and (3) by eliminating θ_m.	$y_m \approx \pm mx\dfrac{\lambda}{w}$
The wavelength λ, slit width w, and distance to the screen x have the same values for all dark fringes. Substitute these values. It is convenient to work in millimeters.	$y_m \approx \pm m(1000\text{ mm})\left(\dfrac{5.50 \times 10^{-4}\text{ mm}}{10.0\text{ mm}}\right)$ $y_m \approx \pm m(5.50 \times 10^{-2}\text{ mm})$
Substitute the 10 values of m. We show five examples here.	$\begin{array}{cc} m & y_m \text{ (mm)} \\ \pm 1 & \pm 5.50 \times 10^{-2} \\ \pm 2 & \pm 0.110 \\ \pm 3 & \pm 0.170 \\ \pm 10 & \pm 0.550 \\ \pm 20 & \pm 1.10 \end{array}$
According to the values we just found, there are 40 dark fringes in a region only 2.20 mm across. The scale in Figure 35.22 shows fringes spread over about 12.7 cm when magnified. Find the magnification M by dividing the distance seen in the figure by the actual distance.	$M = \dfrac{127\text{ mm}}{2.20\text{ mm}} = 57.7$ $M \approx 60\times$

:• CHECK and THINK

As expected, the 40 dark fringes are all close to (less than 1.10 mm from) the central maximum. To see the fringes spread out as they are in the inset of Figure 35.22, you must magnify the diffraction pattern by about 60 times, which requires a microscope. This explains why we normally don't notice the diffraction pattern when light passes through a very wide slit (compared to its wavelength): The dark fringes are so close together that they blur into a seemingly dark region around the central maximum.

EXAMPLE 35.4 CASE STUDY What Is an Orb?

Shannon: Of course, I know I didn't take a picture of a ghost, but I still don't think I took a picture of dust. Dust should just look like a small spot.

Cameron: Yeah, the orb is big and has a very regular ring pattern.

Avi: I see that too. But what is causing the pattern? How exactly did you take the picture?

Shannon: I used my digital camera. I wiped the lens clean, so I'm sure there was no dust on the lens. I had the flash on because it was dark in the basement.

Assume the orb is due to dust in the air. Come up with a simple, conceptual reason why the orb has a ring pattern.

:• SOLVE

Orbs are best seen when a flash photo is taken by a camera whose flash is close to the lens, as it often is on contemporary digital cameras. The illuminated dust particle reflects light. The small dust is effectively the source of (nearly) coherent light—something like the sunlight that passed through the small slit in Young's experiment (Fig. 35.9). That light must pass through the camera's circular aperture. Notice that the diffraction pattern for a circular aperture (Fig. 35.2A) looks much like the close-up of the orb in Shannon's photo (Fig. 35.7, inset). So, it is likely that the ring pattern is caused by diffraction.

:• CHECK and THINK

There is usually dust in the air, yet we rarely see orbs in photographs. However, orbs are becoming more common as people use digital photography. Orbs are best seen in digital photography because orbs are most visible in the UV spectrum. Film is not very sensitive to UV, but the detectors used in digital cameras are. Also, orbs are more likely to be noticed if the dust is relatively close to the camera's lens because then the lens acts as a magnifier and the orb may appear very large on the photo.

35-5 Young's Experiment: Intensity

In Section 35-4, we derived expressions for the positions of the bright and dark fringes. In this section, we derive an expression for the intensity of the pattern. Examine the pattern in Figure 35.10; the bright fringes are *not* uniformly bright and the dark fringes are not uniformly dark. Instead, the intensity gradually changes from bright to dark and back again. So, we expect the mathematical description of the intensity to be a function that gradually varies from high intensity to zero intensity and back again.

DERIVATION Intensity of the Double-Slit Interference Pattern

Again, consider Young's double-slit experiment in which the slits are separated by distance d, the light has wavelength λ, and θ is the angular position shown in Figure 35.11. We show that the intensity of the interference pattern is given by

$$I = I_{max} \cos^2 \frac{\varphi}{2} = \frac{2E_0^2}{\mu_0 c} \cos^2 \frac{\varphi}{2} \qquad (35.8)$$

YOUNG'S EXPERIMENT— INTENSITY ⊙ Special Case

where

$$\varphi = \frac{2\pi}{\lambda}\Delta d = \frac{2\pi}{\lambda} d \sin\theta \qquad (35.9)$$

For this derivation, we need only consider the electric field part of the electromagnetic waves. Let the field from slit A have subscript A and the field from slit B have subscript B. The light that passes through the two slits is coherent, so those two sets of waves have the same amplitude E_0 and angular frequency ω. The only difference between the two waves that hit any particular point on the screen is a constant phase difference φ due to the path-length difference.	$E_A = E_0 \sin \omega t$ $E_B = E_0 \sin(\omega t + \varphi)$
According to the principle of superposition, the electric field at the location of the screen is the sum of the individual waves.	$E_{tot} = E_A + E_B = E_0[\sin \omega t + \sin(\omega t + \varphi)]$
Apply the trigonometric identity $\sin\alpha \pm \sin\beta = 2\sin\frac{1}{2}(\alpha \pm \beta)\cos\frac{1}{2}(\alpha \mp \beta)$ (Appendix A) to simplify the expression for E_{tot}.	$E_{tot} = E_0[2\sin\frac{1}{2}(2\omega t + \varphi)\cos\frac{1}{2}(\varphi)]$ $E_{tot} = 2E_0 \cos\frac{\varphi}{2} \sin\left(\omega t + \frac{\varphi}{2}\right) \qquad (1)$
Equation (1) describes an electric wave whose maximum value E_{max} depends on the relative phase φ between the original two waves.	$E_{max} \equiv 2E_0 \cos\frac{\varphi}{2} \qquad (2)$

Derivation continues on page 1138 ▶

Using our definition for E_{max} (Eq. 2), we rewrite Equation (1) in a more familiar form.	$E_{tot} = E_{max} \sin\left(\omega t + \dfrac{\varphi}{2}\right)$
The intensity of an electromagnetic wave is given by Equation 34.30. Substitute for E_{max} from Equation (2).	$I = \dfrac{E_{max}^2}{2\mu_0 c}$ (34.30) $I = \dfrac{1}{2\mu_0 c}\left(2E_0 \cos\dfrac{\varphi}{2}\right)^2 = \dfrac{1}{2\mu_0 c}\left(4E_0^2 \cos^2\dfrac{\varphi}{2}\right)$ $I = \dfrac{2E_0^2}{\mu_0 c}\cos^2\dfrac{\varphi}{2}$ (3)
We define the maximum intensity I_{max} and write Equation (3) in terms of I_{max}.	$I_{max} \equiv \dfrac{2E_0^2}{\mu_0 c}$ $I = I_{max}\cos^2\dfrac{\varphi}{2} = \dfrac{2E_0^2}{\mu_0 c}\cos^2\dfrac{\varphi}{2}$ ✓ (35.8)
We still need to show that $\varphi = (2\pi/\lambda)\Delta d$ (Eq. 35.9). Let's look at a few specific cases to see if we can find a general trend. When there is no path-length difference, the two waves are in phase. When the path-length difference is half a wavelength, the two waves are 180° (π rad) out of phase. When the path-length difference is a whole wavelength, the two waves are back in phase, with a phase difference of 360° (2π).	$\begin{array}{cc} \Delta d & \varphi \text{ (radians)} \\ \hline 0 & 0 \\ \lambda/2 & \pi \\ \lambda & 2\pi \end{array}$
Reviewing this short table, we find that, in general, the phase difference (in radians) is $2\pi/\lambda$ times the path-length difference, which is $\Delta d = d\sin\theta$ (Eq. 35.3).	$\varphi = \dfrac{2\pi}{\lambda}\Delta d = \dfrac{2\pi}{\lambda}d\sin\theta$ ✓ (35.9)

:• COMMENTS
The expression for intensity in Equation 35.8 involves a cosine function, so as expected, the intensity varies from zero to a maximum value I_{max}. Note that the intensity cannot be negative. Because the cosine function is squared, the intensity varies from zero to I_{max} as a function of φ (or, of θ).

Finding the Maxima and Minima from the Intensity

The intensity expression $I = I_{max}\cos^2(\varphi/2)$ is consistent with the expressions we found for the locations of the bright and dark fringes. In the center of a bright fringe, the intensity must be at its maximum. This occurs when $\cos(\varphi/2) = \pm 1$. So there is constructive interference when

$$\dfrac{\varphi}{2} = 0, \pm\pi, \pm 2\pi, \pm 3\pi, \ldots$$

$$\varphi = 0, \pm 2\pi, \pm 4\pi, \pm 6\pi, \ldots$$

Now let's use $\varphi = (2\pi/\lambda)\Delta d = (2\pi/\lambda)d\sin\theta$ (Eq. 35.9) to write this condition for constructive interference as

$$d\sin\theta = \Delta d = \dfrac{\lambda}{2\pi}\varphi$$

$$d\sin\theta = \Delta d = 0, \pm\lambda, \pm 2\lambda, \pm 3\lambda, \ldots$$

This is exactly the constructive interference condition we found earlier: $d \sin \theta_n = n\lambda$ (Eq. 35.1).

Likewise, in the center of a dark fringe, the intensity must be zero, which occurs when $\cos(\varphi/2) = 0$. So there is destructive interference when

$$\frac{\varphi}{2} = \pm\frac{\pi}{2}, \pm\frac{3\pi}{2}, \pm\frac{5\pi}{2}, \pm\frac{7\pi}{2}, \ldots$$

$$\varphi = \pm\pi, \pm 3\pi, \pm 5\pi, \pm 7\pi, \ldots$$

Again, when we use $\varphi = (2\pi/\lambda)\Delta d = (2\pi/\lambda) d \sin \theta$ (Eq. 35.9), this condition for destructive interference is

$$d \sin \theta = \Delta d = \frac{\lambda}{2\pi}\varphi$$

$$d \sin \theta = \Delta d = \pm\frac{\lambda}{2}, \pm\frac{3\lambda}{2}, \pm\frac{5\lambda}{2}, \pm\frac{7\lambda}{2}, \ldots$$

This is the same destructive interference condition we found earlier: $d \sin \theta_n = (n + \frac{1}{2})\lambda$ (Eq. 35.2).

Plotting Intensity

It is helpful to plot the intensity as a function of either the phase difference φ or the path-length difference $\Delta d = d \sin \theta_n$. Figure 35.23 is a graph of $I = I_{max} \cos^2(\varphi/2)$ (Eq. 35.8). The horizontal axis has been labeled in terms of both the phase difference and the path-length difference using $\varphi = (2\pi/\lambda)\Delta d$ (Eq. 35.9).

Also included on this graph is the intensity of a single source $I_A = E_0^2/(2\mu_0 c)$ (Eq. 34.30). The single source's intensity is a horizontal line on this graph because it does not depend on φ. The maximum intensity of the double-slit pattern is four times greater than the intensity of the single source. You might have expected the maximum intensity through two slits to be *twice* as great as the intensity through a single slit. The electric field's amplitude is twice as great, but the intensity depends on the electric field squared, so it is four times greater.

You may have noticed something odd about the expression for intensity derived here. According to $I = I_{max} \cos^2(\varphi/2)$ or Figure 35.23, the intensity of every fringe is exactly the same. So all the bright fringes should be equally bright no matter how far they are from the central maximum. But this is not what we observe. Instead, the central maximum is the brightest fringe, and the fringes fade as a function of their distance from the central maximum (Fig. 35.10). Our derivation did not agree with this observation because we ignored the diffraction of light through the individual slits. We will consider double-slit diffraction in Section 35-7. We can ignore diffraction and use $I = I_{max} \cos^2(\varphi/2)$ (Eq. 35.8) as a good approximation as long as we consider only the fringes near the central maximum.

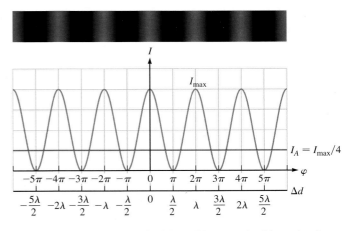

FIGURE 35.23 A graph of intensity for the double-slit interference pattern (blue) and for a single source (red).

EXAMPLE 35.5 Half the Maximum

For Young's double-slit experiment, find the first three values of phase difference φ and path-length difference Δd for which the intensity is half its maximum. Include positive and negative values of each quantity.

Example continues on page 1140 ▶

INTERPRET and ANTICIPATE

The interference pattern's intensity varies from zero at a dark fringe to I_{max} at a bright fringe. We have already found the phase differences and path-length differences that produce dark and bright fringes, as shown on the graph in Figure 35.23. We can find the phase differences and path-length differences where the intensity is half its maximum by using either algebra or the graph of intensity. We use the graph in Figure 35.23 to anticipate our algebraic solution.

Imagine a horizontal line at $I = I_{max}/2$ in Figure 35.23, and read off the first three values of φ and Δd that intersect this line. Next, we use algebra to check our results against the values we found from the graph.	$\varphi = \pm\dfrac{\pi}{2},\ \pm\dfrac{3\pi}{2},\ \text{and}\ \pm\dfrac{5\pi}{2}$ $\Delta d = \pm\dfrac{\lambda}{4},\ \pm\dfrac{3\lambda}{4},\ \text{and}\ \pm\dfrac{5\lambda}{4}$

SOLVE

Set the intensity in Equation 35.8 equal to the intensity ($I = I_{max}/2$) we're interested in.	$I = I_{max} \cos^2 \dfrac{\varphi}{2}$ (35.8) $\dfrac{I_{max}}{2} = I_{max} \cos^2 \dfrac{\varphi}{2}$
The maximum intensity cancels. The cosine function is squared, so taking its square root gives two possible solutions.	$\dfrac{1}{2} = \cos^2 \dfrac{\varphi}{2}$ $\cos \dfrac{\varphi}{2} = \pm \dfrac{1}{\sqrt{2}}$
Write down the first three solutions in radians.	$\dfrac{\varphi}{2} = \pm\dfrac{\pi}{4},\ \pm\dfrac{3\pi}{4},\ \text{and}\ \pm\dfrac{5\pi}{4}$ $\varphi = \pm\dfrac{\pi}{2},\ \pm\dfrac{3\pi}{2},\ \text{and}\ \pm\dfrac{5\pi}{2}$
Use these solutions for φ and Equation 35.9 to find the first three solutions for Δd.	$\varphi = \dfrac{2\pi}{\lambda} \Delta d$ (35.9) $\Delta d = \dfrac{\lambda}{2\pi} \varphi = \pm\dfrac{\lambda}{4},\ \pm\dfrac{3\lambda}{4},\ \text{and}\ \pm\dfrac{5\lambda}{4}$

CHECK and THINK

The solution we found algebraically is exactly the same as the values we read off the graph in Figure 35.23. Remember that the intensity of the double-slit interference pattern varies smoothly from zero to I_{max} and back again. Sometimes it is easy to forget this, and instead to think that the intensity is either I_{max} at a bright fringe or zero at a dark fringe, with no transitions in between.

35-6 Single-Slit Diffraction Intensity

In Section 35-5, we found the intensity of the interference pattern produced by Young's double-slit experiment. Now let's find the intensity of the single-slit diffraction pattern.

Figure 35.24 shows a single-slit pattern with four points, c, p, d, and q. Within the diffraction pattern, light intensity depends on position. The maximum intensity I_{max} is in the center of the pattern at $\theta = 0$. The intensity drops off for points that are farther from the center, such as point p. The intensity is zero at the first dark fringes on either side of the central maximum. Then the intensity increases, but never again reaches I_{max}. So, the intensity at point q is lower than I_{max}. Mathematically, the **intensity of the single-slit diffraction pattern** as a function of angular position θ is given by

SINGLE-SLIT DIFFRACTION—INTENSITY
◉ Special Case

$$I = I_{max} \left(\dfrac{\sin \alpha}{\alpha}\right)^2 \quad (35.10)$$

where α is defined as

$$\alpha \equiv \dfrac{1}{2}\varphi = \dfrac{\pi w}{\lambda} \sin \theta \quad (35.11)$$

where w is the slit width and φ is described below.

35-6 Single-Slit Diffraction Intensity

In this section, we do not formally derive the intensity expression (Eqs. 35.10 and 35.11), but we explore the concepts behind such a derivation. We also explain how the intensity expression is related to the expression for the position of the dark fringes. Finally, we plot the intensity for different slit widths.

Where the Intensity Expression Comes From

First, we show how the intensities at the four labeled points in Figure 35.24 are determined. As when we found the position of the dark fringes, we use Huygens's principle and divide the slit into a number of zones. Consider the four zones shown in Figure 35.20. The electric (and magnetic) field at any point on the screen is the result of the superposition of the four spherical wavelets. The relative phase of those four wavelets depends on the distance each wave must travel to get to that particular point on the screen. So the total electric field E_{tot} at each point is found by adding up the electric fields of the four waves. To get the intensity, we must find E_{tot}^2 because the intensity is proportional to the square of the electric field's amplitude $I \propto E_{max}^2$ (Eq. 34.30).

Table 35.1 illustrates how the relative phase of the four wavelets determines the intensity at each of the four points. For each point, we graph the electric fields of the four wavelets, their total electric field, and the square of the total electric field. Table 35.1 also includes three parameters for each point. The first is the sine of the angular position θ because it enters into Equation 35.11. The second parameter is the phase difference φ between the wavelet from zone A and the one from zone D (Fig. 35.20). The third parameter is α as given by Equation 35.11.

At the central point c in Figure 35.24, the angular position is $\theta = 0$. The four spherical waves arrive at c in phase and interfere constructively, so the total electric field is very strong. The intensity at the central point is the maximum intensity in the pattern, I_{max}.

Now let's skip to the first dark fringe—point d, at the positive angular position given by $\sin \theta = \lambda/w$. According to Equation 35.11, α is given by

$$\alpha = \frac{\pi w}{\lambda}\left(\frac{\lambda}{w}\right) = \pi$$

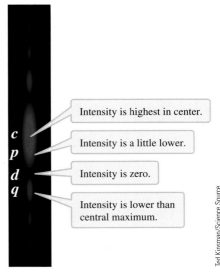

FIGURE 35.24 A diffraction pattern produced on a screen by laser light that passes through a single narrow slit. The label c stands for *center* and d for *dark*.

- Intensity is highest in center.
- Intensity is a little lower.
- Intensity is zero.
- Intensity is lower than central maximum.

TABLE 35.1 Finding the intensity of points c, p, d, and q in Figure 35.24. (For point c, the four waves overlap and cannot be seen separately.)

Point c	Point p	Point d	Point q
$\sin\theta = 0$	$\sin\theta = 3\lambda/8w$	$\sin\theta = \lambda/w$	$\sin\theta = 3\lambda/2w$
$\varphi = 0$	$\varphi = 3\pi/4$	$\varphi = 2\pi$	$\varphi = 3\pi$
$\alpha = 0$	$\alpha = 3\pi/8$	$\alpha = \pi$	$\alpha = 3\pi/2$
$I = I_{max}$	$I \approx 0.60 I_{max}$	$I = 0$	$I \approx 0.05 I_{max}$

Also by Equation 35.11, the total phase difference φ between the wavelet from zone A and the one from zone D is $\varphi = 2\alpha = 2\pi$.

Let's think about what this total phase difference of $\varphi = 2\pi$ means for the four wavelets that arrive at d. Suppose the time dependence of the wave from A is given by

$$E_A(t) = E_{max} \sin(\omega t)$$

If the total phase difference across the slit is $\varphi = 2\pi$, the phase difference between each of the four adjacent zones must be $2\pi/4 = \pi/2$. So the time dependence of the waves from the other three zones must be

$$E_B(t) = E_{max} \sin\left(\omega t + \frac{\pi}{2}\right)$$

$$E_C(t) = E_{max} \sin(\omega t + \pi)$$

$$E_D(t) = E_{max} \sin\left(\omega t + \frac{3\pi}{2}\right)$$

These four electric fields plotted in Table 35.1 for point d interfere destructively and cancel, so that their superposition E_{tot} is zero. Thus, E_{tot}^2 is also zero and because the intensity is proportional to E_{max}^2, the intensity is zero, exactly as you would expect for a dark fringe.

Now let's consider point p in Figure 35.24, between the center c and the first dark fringe d. Point p falls within the central maximum, but it is not at the center of this maximum. We have chosen point p at an angular position given by $\sin\theta = 3\lambda/8w$. According to Equation 35.11, the total phase difference is $\varphi = 3\pi/4$, and $\alpha = 3\pi/8$. The wavelets do not interfere completely constructively or completely destructively. Compare the column for point p in Table 35.1 to the graphs of E_{tot}^2 for points c, d, and q, and you'll see that the intensity for p is between zero and I_{max} (about 60% of I_{max}).

Finally, consider point q in the center of the second maximum. It is halfway between the first ($\sin\theta_1 = \lambda/w$) and the second ($\sin\theta_2 = 2\lambda/w$) dark fringes, so its angular position is given by $\sin\theta = 3\lambda/2w$. According to Equation 35.11, the total phase difference is $\varphi = 3\pi$, and $\alpha = 3\pi/2$. The four wavelets that arrive at point q are not in phase. Furthermore, the intensity at point q is lower than it is at either c or p (about 5% of I_{max}).

We have explored the intensity at four points (c, d, p, and q) in the diffraction pattern by dividing the slit into four zones (A through D) of width $w/4$. In a thorough derivation of Equations 35.10 and 35.11, the slit is divided into infinitely many infinitesimal zones. The total electric field at an arbitrary point is then found from the superposition of the electric fields from these infinitesimal zones. Even without doing such a formal derivation, however, we can see how the relative intensity in the diffraction pattern is determined.

Dark Fringes

Our next goal is to connect the expression for intensity with the position of the dark fringes. To do this, we first find the values of α that correspond to zero intensity by setting $I = 0$ in Equation 35.10:

$$I = I_{max}\left(\frac{\sin\alpha}{\alpha}\right)^2 = 0$$

This expression is zero when $\sin\alpha = 0$, so

$$\alpha_m = m\pi \quad (m = \pm 1, \pm 2, \pm 3, \ldots) \tag{35.12}$$

Substitute Equation 35.12 into Equation 35.11 to find an expression for the angular positions θ where the intensity is zero:

$$\alpha_m = \frac{\pi w}{\lambda}\sin\theta_m = m\pi \quad (m = \pm 1, \pm 2, \pm 3, \ldots)$$

$$w\sin\theta_m = m\lambda \quad (m = \pm 1, \pm 2, \pm 3, \ldots) \tag{35.5}$$

This is exactly what we found for the position of the dark fringes (Eq. 35.5).

Why didn't we start with $m = 0$ in Equation 35.12? Because if $\alpha = 0$, both the numerator and the denominator in Equation 35.10 are zero. To find the value of the fraction in this case, take the limit using L'Hôspital's rule:

$$\lim_{\alpha \to 0}\left(\frac{\sin \alpha}{\alpha}\right) = \lim_{\alpha \to 0}\left(\frac{\cos \alpha}{1}\right) = 1$$

So, at $\alpha = 0$ (corresponding to $\theta = 0$), the intensity equals its maximum $I(0) = I_{max}$, which obviously is not the location of a *dark* fringe. Instead, the most intense spot is in the center, at $\theta = 0$ (Fig. 35.24).

Visualizing Intensity

To visualize the intensity of the diffraction pattern produced by a variety of slit widths, imagine shining laser light of wavelength λ through three different slits. Figure 35.25 shows graphs of I versus θ for these three situations. The vertical axis is the relative intensity I/I_{max}, so its maximum value is 1. Each part of the figure extends only 30° to either side of the central point. The vertical and horizontal scales are thus the same in each case.

The diffraction pattern in Figure 35.25A comes from a slit whose width equals the wavelength of the light ($w = \lambda$). We found in this case that the first dark fringe is at $\theta = 90°$, so the screen is completely covered by the central maximum. At the edges of this small portion of the screen that extends to $\theta = \pm 30°$, the intensity drops to 40% of its maximum value.

The pattern in Figure 35.25B comes from the same laser, using a slit that is three times wider ($w = 3\lambda$). The central maximum in this case extends only to about $\theta = \pm 20°$, and at these points we see the first dark fringes. We can also see about half of a side lobe on either side of the central maximum. Notice that the side lobe is much dimmer than the central maximum.

Finally, Figure 35.25C shows the pattern produced by a slit that is six times wider than the first slit ($w = 6\lambda$). The central maximum is now much thinner, extending to only roughly $\theta = \pm 10°$. We now see two side lobes on either side of the central maximum, narrower than those in Figure 35.25B.

Comparing the three graphs in Figure 35.25 leads to three general statements. First, a wide slit produces a narrow central maximum. Second, a wide slit produces more side lobes (and dark fringes) on the screen. Third, a wide slit produces narrow side lobes. In fact, at the limit of a very wide slit, the pattern we see is like the pattern due to diffraction around an edge, as around the outside of the sewing needle shown in Figure 35.3.

CONCEPT EXERCISE 35.6

In Figure 35.25, we imagined using a single light source with a fixed wavelength and three different slits of widths λ, 3λ, and 6λ. We could have equally imagined using a single slit of width w and varying the wavelength. In this case, order the graphs from shortest wavelength to longest. If light of all three wavelengths passed simultaneously through a single slit, which wavelength would produce the broadest central maximum?

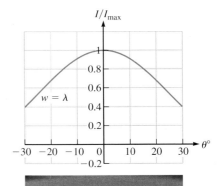

A.

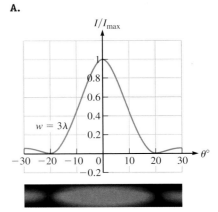

B.

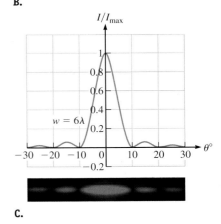

C.

FIGURE 35.25 In all three cases, a graph of intensity versus angular position is shown above the pattern that is actually observed on the screen. These patterns were made using a slit whose width **A.** equals the wavelength of the light; **B.** is three times the wavelength of the light; and **C.** is six times the wavelength of the light.

EXAMPLE 35.6 Half as Intense

Red light of wavelength $\lambda = 625$ nm passes through a single slit of width $w = 1875$ nm. Estimate to three significant figures the (positive) angular position θ where light in the diffraction pattern equals half of its maximum.

Example continues on page 1144 ▶

:• **INTERPRET and ANTICIPATE**
If we could set I in Equation 35.10 equal to $I = I_{max}/2$ and then solve for α, we could use Equation 35.11 to find the angular position θ. However, we cannot isolate α in Equation (1). Instead of solving for α, we must estimate its value by making a good guess at what the answer might be. We can use Figure 35.25 as a basis for our initial guess. This also sets our expectation for the final estimate.

$$I = I_{max}\left(\frac{\sin \alpha}{\alpha}\right)^2 \quad (35.10)$$

$$\frac{I}{I_{max}} = \frac{1}{2} = \left(\frac{\sin \alpha}{\alpha}\right)^2 \quad (1)$$

:• **SOLVE**
In Figure 35.25, the horizontal axis is angular position θ. Once we read off our initial guess for θ, we can use Equation 35.11 to make an initial guess for α. In this problem, the slit width is three times the wavelength. From Figure 35.25B, read off θ for $I/I_{max} = 0.5$.

$$\theta \approx 8° = 0.14 \text{ rad}$$

$$\alpha = \frac{\pi w}{\lambda} \sin \theta \quad (35.11)$$

$$\alpha = \frac{\pi(1875 \text{ nm})}{(625 \text{ nm})}(\sin 0.14) = 1.3 \text{ rad}$$

Our initial guess corresponds to $\alpha = 1.3$ rad. To find a better estimate, calculate $[(\sin \alpha)/\alpha]^2$ for values near the initial guess. The goal is to find the value of α that comes closest to $[(\sin \alpha)/\alpha]^2 = 1/2$ as required in Equation (1). Make a table to help organize the work. Our initial guess gives a result that is a little greater than $1/2$. When we try a smaller α, we find that the result is greater than $1/2$. So we try a larger α, and we find that the result is also greater than $1/2$ but getting closer. An even larger value for α shows that we have gone too far because the result is smaller than $1/2$.

α (rad)	$\left(\frac{\sin \alpha}{\alpha}\right)^2$
1.3 initial guess	0.549 too high
1.25 low guess	0.576 worse
1.35 high guess	0.522 better
1.40 high guess	0.495 too low

Now we know that α is between 1.35 and 1.40. Make a new table of guesses in this range. We were asked to find an estimate to three significant figures. We check our answer to four significant figures to confirm that we don't need to round up to 1.40. So our best estimate is $\alpha = 1.39$ rad.

α (rad)	$\left(\frac{\sin \alpha}{\alpha}\right)^2$
1.36	0.517 better
1.38	0.506 better
1.39	0.500 great!
1.392	0.4997 too low

Use Equation 35.11 to find θ from our estimate of α.

$$\alpha = \frac{\pi w}{\lambda} \sin \theta \quad (35.11)$$

$$\theta = \sin^{-1}\left(\frac{\alpha \lambda}{\pi w}\right) = \sin^{-1}\left[\frac{(1.39 \text{ rad})(625 \text{ nm})}{\pi(1875 \text{ nm})}\right]$$

$$\theta = 0.148 \text{ rad} = \boxed{8.48°}$$

:• **CHECK and THINK**
As expected, the intensity drops to about half its maximum at $\theta \approx 8°$. The entire central maximum is about 40° wide (Fig. 35.25B). Because the intensity drops by 50% at roughly $\theta \approx 8°$ from the center, the inner 16° of the maximum is fairly bright compared to the outer 24°. (As a final note, some problem-solvers may use a computer-based approach instead of the method used here.)

35-7 Double-Slit Diffraction

In Section 35-3, we imagined that the slits in Young's double-slit experiment were very narrow compared to the wavelength of light ($w \ll \lambda$). In that case, the diffraction pattern from each slit was wide enough to cover the entire screen, so the interference pattern of light from the two slits dominated what we saw. In particular, the intensity of the bright fringes was nearly constant. In effect, we ignored the diffraction of light through the individual slits.

In this section, we examine the pattern produced by two slits that are wide enough that we cannot ignore diffraction. The resulting pattern is the combination of diffraction through the individual slits and interference between the two slits. This sort of pattern is often referred to as **double-slit diffraction**, a shorter way of describing the pattern produced by the interference of light from two slits and the diffraction due to each slit. We limit our study to the intensity produced by two slits of equal width.

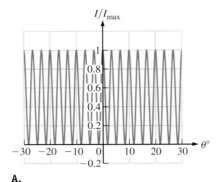
A.

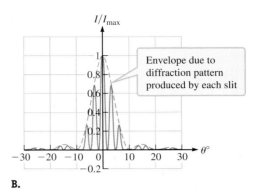
B.

FIGURE 35.26 A. A double-slit interference pattern produced by very narrow slits. **B.** A double-slit interference pattern produced by slits of width $w = 6\lambda$.

Interference from light passing through two very narrow slits ($w \ll \lambda$) produces bright fringes with equal intensities (Fig. 35.26A). Now imagine repeating this experiment with wider slits, six times the wavelength. In this case, diffraction through each slit is important. The intensity due to diffraction through either slit *alone* would look like Figure 35.25C. But the intensity produced by double-slit diffraction looks like a *combination* of double-slit interference (Fig. 35.26A) and single-slit diffraction (Fig. 35.25C). Figure 35.26B shows that single-slit diffraction defines an envelope for the intensity we see through both slits; the fringes are in the same location as the fringes in Figure 35.26A, but their intensities are limited by the diffraction pattern in Figure 35.25C.

For easy comparison, Figure 35.27 shows both a graph of intensity and the pattern you would see on the screen. The thin fringes are due to interference of light from the two slits. Even if the slits were very narrow, the thin fringes would still be there. The fringes are also grouped in a series of broad bands due to the diffraction pattern produced by each slit. If you covered up one of the slits, the thin fringes would disappear, but the broad bands would still be there.

Mathematically, the intensity of double-slit diffraction is given by

$$I = I_{max}\left(\frac{\sin \alpha}{\alpha}\right)^2 \cos^2 \beta \quad (35.13)$$

where $\alpha = (\pi w/\lambda) \sin \theta$ (Eq. 35.11) and β is given by

$$\beta = \frac{\varphi}{2} = \frac{\pi d}{\lambda} \sin \theta \quad (35.14)$$

where d is the center-to-center separation between the slits. We expect from Figure 35.26 that the intensity of double-slit diffraction is a combination of the intensities of single-slit diffraction and double-slit interference. To check, compare Equation 35.13 to Equation 35.10 for single-slit diffraction. Then compare Equation 35.13 to $I = I_{max} \cos^2 \beta$ (Eq. 35.8 with $\beta = \varphi/2$) for double-slit interference. You can see that the $[(\sin \alpha)/\alpha]^2$ term comes from single-slit diffraction, while the $\cos^2 \beta$ term comes from double-slit interference.

Equation 35.13 actually describes the intensity of either the double-slit interference pattern or the single-slit diffraction pattern, so you don't have to remember three separate equations. To see this, first consider double-slit interference. We must imagine that the slits are very narrow ($w \ll \lambda$), so that $\alpha \to 0$ (Eq. 35.11). From L'Hôspital's rule, when $\alpha \to 0$,

$$\frac{\sin \alpha}{\alpha} \to \cos \alpha \to 1$$

So, when the slits are very narrow, Equation 35.13 reduces to

$$I = I_{max} \cos^2 \beta$$

which is exactly the intensity of the double-slit interference pattern (Eq. 35.8). To find the single-slit diffraction pattern, we imagine that the two slits overlap,

DOUBLE-SLIT DIFFRACTION—INTENSITY
◉ Special Case

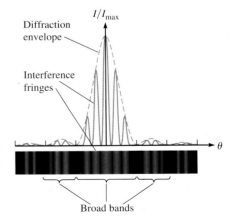

FIGURE 35.27 The relationship between the graph of intensity and the pattern you see on the screen.

which means $d \to 0$ and $\beta \to 0$ (Eq. 35.14). Substitute $\beta = 0$ into Equation 35.13 to find

$$I = I_{max}\left(\frac{\sin\alpha}{\alpha}\right)^2 \cos^2 0 = I_{max}\left(\frac{\sin\alpha}{\alpha}\right)^2$$

This is the same as Equation 35.10 for the single-slit diffraction pattern.

EXAMPLE 35.7 Double-Slit Interference Versus Diffraction

You are working in a laboratory with a single laser of wavelength λ. The laser is used to illuminate one or two slits, and interference or diffraction patterns are observed as a result. Give your results to three significant figures.

A If there are two very thin slits ($w \ll \lambda$) separated by $d = 18\lambda$, at what (positive) angular position is the $m = 1$ bright fringe? Give your answer in radians and degrees. What is the relative intensity I/I_{max} of this fringe?

:• INTERPRET and ANTICIPATE
Because the slits are so narrow, we can ignore diffraction. This situation is exactly like double-slit interference (Section 35-3).

:• SOLVE
Use Equation 35.1 to find the positive position of the $m = 1$ bright fringe.

$$d \sin\theta_m = m\lambda \quad (35.1)$$
$$d \sin\theta_1 = \lambda$$
$$\theta_1 = \sin^{-1}\left(\frac{\lambda}{d}\right) \quad (1)$$
$$\theta_1 = \sin^{-1}\left(\frac{1}{18}\right)$$
$$\theta_1 = 5.56 \times 10^{-2} \text{ rad} = 3.18°$$

To find the relative intensity, substitute Equation (1) into Equation 35.14 to find β when $m = 1$.

$$\beta = \frac{\pi d}{\lambda}\sin\theta_1 \quad (35.14)$$
$$\beta = \pi$$

Now substitute $\beta = \varphi/2 = \pi$ into Equation 35.8.

$$I = I_{max}\cos^2\beta$$
$$\frac{I}{I_{max}} = \cos^2\beta = \cos^2\pi$$
$$\frac{I}{I_{max}} = 1$$

:• CHECK and THINK
It makes sense that $\beta = \pi$ because $\beta = \varphi/2$ and the phase difference must be $\varphi = 2\pi$ in order for there to be constructive interference at the first bright fringe. It also makes sense that the relative intensity of the bright fringe is 1 because the intensity of all fringes in a double-slit interference pattern is the same (Fig. 35.26A).

B Now the laser is used to illuminate a single slit of width $w = 6\lambda$. What is the relative intensity at the angular position you found in part A?

:• **INTERPRET and ANTICIPATE**
The key to solving this problem is to find α from θ_1, found in the preceding part. The relative intensity is only 1 at the central maximum ($\theta = 0$), so we expect that the relative intensity at θ_1 should be less than 1.

:• **SOLVE**
Use Equation 35.11 to find α.

$$\alpha = \frac{\pi w}{\lambda} \sin \theta \qquad (35.11)$$

$$\alpha = 6\pi \sin 3.18° = 1.046 \text{ rad}$$

Substitute α into Equation 35.10 to find the relative intensity for a single slit.

$$I = I_{max}\left(\frac{\sin \alpha}{\alpha}\right)^2 \qquad (35.10)$$

$$\frac{I}{I_{max}} = \left(\frac{\sin 1.046 \text{ rad}}{1.046 \text{ rad}}\right)^2 = 0.684$$

:• **CHECK and THINK**
The intensity at this particular location is about 68% of the maximum, consistent with our expectations.

C The laser is used to illuminate two slits of width $w = 6\lambda$ and separation $d = 18\lambda$. What is the relative intensity at the angular position you found in part A?

:• **INTERPRET and ANTICIPATE**
Both interference and diffraction play important roles in this case. Because we are considering the same angular position, we can use the values we found for α and β in parts A and B.

:• **SOLVE**
Substitute α and β in Equation 35.13 to find the relative intensity for double-slit diffraction.

$$I = I_{max}\left(\frac{\sin \alpha}{\alpha}\right)^2 \cos^2 \beta \qquad (35.13)$$

$$\frac{I}{I_{max}} = \left(\frac{\sin 1.046}{1.046}\right)^2 \cos^2 \pi = 0.684$$

:• **CHECK and THINK**
Our results are pretty amazing. With the use of wider slits, the intensity at this particular spot ($\theta \approx 3°$) is actually decreased to about 68% of what it was when we used two narrow slits (part A).

EXAMPLE 35.8 CASE STUDY Poisson's Spot

Recall that Poisson was an advocate for the particle model of light. When he analyzed Fresnel's competition essay, he thought he found a fatal flaw in Fresnel's use of the wave model. Fresnel's work predicted a bright spot at the center of the shadow cast by an object with a circular cross section. François Arago—another judge—tested Poisson's prediction experimentally. This experiment has been repeated many times; you may even do it in your laboratory (Fig. 35.28). At the center of the shadow is a bright spot as predicted. Although this phenomenon is often called Poisson's spot, Poisson probably was not happy to have seen it because it supported the wave model of light. The spot is sometimes called Fresnel's spot because it is a direct consequence of his work, and sometimes Arago's spot because Arago devised the experiment that confirmed its existence.

Come up with a simple, conceptual reason for the light in the center of the shadow in Figure 35.28. *Hint*: See Concept Exercise 35.2.

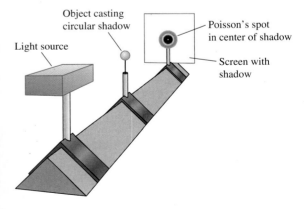

FIGURE 35.28 A bright spot is clearly seen in the center of the shadow of a circular object.

INTERPRET and ANTICIPATE
The key to explaining the bright spot lies in modeling light as a wave. If light obeyed only the particle model, like paint pellets hitting a sandbag, the light would be stopped and there would be no light in the shadow. But when you float at a point behind a rock in the ocean and waves arrive, you oscillate up and down because the water waves diffract around the rock and reach you.

SOLVE
Figure 35.29 shows the diffraction of a wave (such as light or water) around an object with a circular cross section. The waves that emerge around the edges of the object interfere constructively in the center. If these were light waves, you would see a bright spot in the center of the object's shadow.

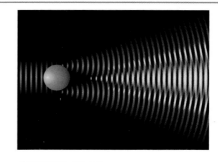

FIGURE 35.29 Diffraction around an object causes a bright spot in its shadow.

CHECK and THINK
The spot in the center of the shadow in Arago's experiment and the interference pattern observed in Young's experiment are convincing evidence that light should be modeled as a wave.

SUMMARY

❶ Underlying Principles

Huygens's principle: Each point on a primary wave front serves as the source of a spherical wave, called a wavelet, that advances with a speed and frequency equal to those of the primary wave. The primary wave front at some later time is the envelope of these wavelets (Fig. 35.6).

✪ Major Concepts

1. **Diffraction** occurs when waves pass through a narrow opening or near the edge of an obstacle. A diffraction pattern consists of bright and dark regions.
2. Like diffraction, **interference** is a wave phenomenon that results from the superposition of waves, producing a characteristic pattern of bright and dark fringes. A common distinction between the two phenomena is that diffraction involves only a single aperture (or source), whereas interference involves two or more such apertures.
3. **Conditions for constructive interference:**
   For two waves to interfere constructively, they must be in phase, so the difference between their path lengths must be given by

$$\Delta d = n\lambda \quad (n = 0, 1, 2, 3, \ldots) \quad (18.2)$$

Conditions for destructive interference:
For two waves to interfere destructively and cancel, they must be 180° out of phase, so the difference between their path lengths must be given by

$$\Delta d = \left(n + \frac{1}{2}\right)\lambda \quad (n = 0, 1, 2, 3, \ldots) \quad (18.3)$$

4. Two light waves are said to be **coherent** if they are monochromatic sources of the same frequency with a constant phase difference φ between them; φ must be a constant in both space and time. In order to see an interference pattern, the light involved must be coherent.

▶ Special Cases

1. In **Young's double-slit experiment**, the **position θ_n of the bright fringes (maxima)** is given by

$$d \sin \theta_n = n\lambda \quad (n = 0, \pm 1, \pm 2, \pm 3, \ldots) \quad (35.1)$$

where d is the distance between the slits and λ is the wavelength of the monochromatic light source. The **position θ_n of the dark fringes (minima)** is given by

Special Cases cont'd

$$d \sin \theta_n = \left(n + \frac{1}{2}\right)\lambda \quad (n = 0, \pm 1, \pm 2, \pm 3, \ldots) \quad (35.2)$$

The **intensity of the fringes in Young's double-slit experiment** is given by

$$I = I_{max} \cos^2 \frac{\varphi}{2} = \frac{2E_0^2}{\mu_0 c} \cos^2 \frac{\varphi}{2} \quad (35.8)$$

where

$$\varphi = \frac{2\pi}{\lambda} \Delta d = \frac{2\pi}{\lambda} d \sin \theta \quad (35.9)$$

2. In **single-slit diffraction**, the angular **position** θ_m **of the dark fringes** is given by

$$w \sin \theta_m = m\lambda \quad (m = \pm 1, \pm 2, \pm 3, \ldots) \quad (35.5)$$

and the **intensity** is given by

$$I = I_{max} \left(\frac{\sin \alpha}{\alpha}\right)^2 \quad (35.10)$$

where $\alpha \equiv \varphi/2 = (\pi w/\lambda) \sin \theta$ (Eq. 35.11).

3. The **intensity of double-slit diffraction** is given by

$$I = I_{max} \left(\frac{\sin \alpha}{\alpha}\right)^2 \cos^2 \beta \quad (35.13)$$

where $\alpha = (\pi w/\lambda) \sin \theta$ (Eq. 35.11) and $\beta = \varphi/2 = (\pi d/\lambda) \sin \theta$ (Eq. 35.14).

Tools

Visual representations of light
1. Draw a single sinusoidal curve to represent the oscillating electric field (Fig. 35.4).
2. Draw solid lines to represent the wave crests and dashed lines to represent the wave troughs or valleys (Fig. 35.5).

PROBLEMS AND QUESTIONS

A = algebraic   **C** = conceptual   **E** = estimation   **G** = graphical   **N** = numerical

35-1 Light Is a Wave

1. **C** As shown in Figure P35.1, spray paint can be used with a stencil to produce an image that has sharp edges. Should you model the paint as waves or as particles? Explain.

35-2 Sound Wave Interference Revisited

2. **G** Draw two harmonic waves that are 90° out of phase. Then draw the superposition of these two waves.

3. **C** Two speakers produce identical harmonic waves as in Figure P35.3. If the two waves are 90° out of phase when they arrive at your ear, what is the minimum distance between the speakers in terms of the wavelength?

FIGURE P35.1

FIGURE P35.3

4. You are seated on a couch equidistant between two speakers. The speakers are 2.0 m apart and you are seated 4.0 m away from the point between the speakers as shown in Figure P35.4. Both speakers play an A note (a constant frequency of 440 Hz) in phase and at the same volume. As you move left and right along the couch, you notice that the volume alternates between minimum and maximum due to interference effects. Assume the speed of sound is 343 m/s.
   a. **C** When you are seated at the center of the couch, is the sound a maximum or minimum in volume?
   b. **N** If you slide along the couch to the left or right, approximately how far do you need to move until you reach the next location that has the same volume? (*Hint*: You may wish to obtain an exact expression and then use a spreadsheet or trial and error to find an approximate solution.)

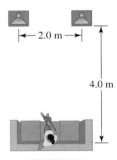

FIGURE P35.4

35-3 Young's Experiment: Position of the Fringes

5. **N** The first bright fringe of an interference pattern occurs at an angle of 11.5° from the central fringe when a double slit is illuminated by a 415-nm blue laser. What is the spacing of the slits?

6. **C** In Young's double-slit experiment, the positions of the bright fringes are given by $d \sin \theta_n = n\lambda$, where $n = 0, \pm 1, \pm 2, \pm 3, \ldots$ (Eq. 35.1). Mathematically, there are an infinite number of bright fringes. However, there is a physical

limit to the number of bright fringes that appear on the screen. Why is there a physical limit?

7. **N** A student shines a red laser pointer with a wavelength of 675 nm through a double-slit apparatus in which the two slits are separated by 75.0 μm. He observes the diffraction pattern on the wall 1.50 m away. What is the distance between the central bright fringe and either of the neighboring bright fringes on the wall?

8. **N** Monochromatic light is incident on a pair of slits that are separated by 0.200 mm. The screen is 2.50 m away from the slits.
   a. If the distance between the central bright fringe and either of the adjacent bright fringes is 1.67 cm, find the wavelength of the incident light.
   b. At what angle does the next set of bright fringes appear?

9. **N** In a double-slit experiment, the wavelength of monochromatic light used is 489.7 nm and the distance between the slits is 7.500 μm. How many bright fringes are created by the light passing through the slits?

10. **N** In a Young's double-slit experiment with microwaves of wavelength 3.00 cm, the distance between the slits is 5.00 cm and the distance between the slits and the screen is 100.0 cm. Determine the number of bright fringes on the screen and their distances from the central bright fringe on the screen. Assume the screen is long enough to at least show the first order bright fringes.

11. **N** A beam from a helium-neon laser with wavelength 635 nm strikes two slits separated by 0.240 mm. What is the distance between the first and third dark fringes on a screen located 4.20 m from the slits?

Problems 12, 13, and 14 are grouped.

12. **N** The wavelength of light emitted by a particular laser is 633 nm. This laser light illuminates two slits that are 50.0 μm apart.
    a. What is the angular separation between the $n = 2$ and $n = -2$ maxima?
    b. If a screen is placed 1.50 m from the slits, what is the distance between these two maxima?
    c. If you want these two maxima to be 1.00 cm apart, how far from the slits do you need to place the screen?

13. The wavelength of light emitted by a particular laser is 633 nm. This laser light illuminates two slits that are 50.0 μm apart, and a screen is 1.50 m from the slits.
    a. **N** What is the (linear) distance on the screen between the central maximum and the $n = 2$ maximum?
    b. **N** What is the (linear) distance on the screen between the central maximum and the $n = 20$ maximum?
    c. **C** Are the maxima evenly spaced? Explain.

14. The wavelength of light emitted by a particular laser is 633 nm. This laser light illuminates two slits that are 50.0 μm apart, and a screen is 1.50 m from the slits.
    a. **N** What is the (linear) distance on the screen between the central maximum and the $n = 2$ minimum?
    b. **N** What is the (linear) distance on the screen between the central maximum and the $n = 20$ minimum?
    c. **C** Are the minima evenly spaced? Explain.

15. **N** Light from a sodium vapor lamp ($\lambda = 589$ nm) forms an interference pattern on a screen 0.80 m from a pair of slits in a double-slit experiment. The bright fringes near the center of the pattern are 0.35 cm apart. Determine the separation between the slits. Assume the small-angle approximation is valid here.

16. **A** In a Young's double-slit experiment with two sources of different wavelengths, the eighth maximum of wavelength λ_1 is at a distance y_1 from the central maximum and the sixth maximum of wavelength λ_2 is at a distance y_2 from the central maximum. Find an expression for the ratio y_1/y_2.

17. **N** When a Young's double-slit experiment is carried out with 620.0-nm light, first-order bright fringes near the center of the pattern appear on a screen placed 2.50 m away each with a spacing of 1.75 cm from the central bright fringe. What is the distance between the slits?

18. **C** After shining a red laser pointer through a double-slit apparatus and observing the interference pattern on the wall, three students discuss what would happen if they used light of a shorter wavelength, such as that produced by a green laser pointer.

    Avi: It's still coherent, monochromatic light, so I expect the pattern to look the same.
    Cameron: But the wavelength is shorter, so the pattern will be smaller. I mean the peaks will not be spread out as far from the center.
    Shannon: I think these interference effects work the other way, though—an inverse relationship. So I think the interference pattern will actually be more spread out on the wall.

    Which student do you agree with, and why?

19. **N** Susan designs a double-slit experiment that uses coherent light with a wavelength of 408.0 nm and a slit separation of 0.125 mm. Peter sets up a second double-slit experiment with a slit separation of 0.250 mm. If Peter wants to produce an interference pattern that matches Susan's, what wavelength of coherent light must he use?

20. **N** In a Young's double-slit experiment, 586-nm-wavelength light is sent through the slits. A screen is held at a distance of 1.50 m from the slits. The second-order maxima appear at an angle of 2.50° from the central bright fringe. How far apart do the first-order ($m = 1$) and second-order ($m = 2$) maxima appear on the screen?

21. **N** Two slits separated by 0.130 mm are illuminated by visible light, forming an interference pattern on a screen 3.50 m away. What is the wavelength of the light if the first bright fringe of the interference pattern is 0.235° from the central fringe?

22. **N** Red light with a wavelength of 715.5 nm and violet light with a wavelength of 412.5 nm are used simultaneously in a double-slit experiment. The first maximum of the violet light is 1.975 mm from the central maximum. What is the distance between the central maximum and the first maximum of the red light?

35-4 Single-Slit Diffraction

23. **N** What is the maximum width of a single slit for which 454.6-nm light from an argon laser does not produce any diffraction minima?

24. **C** Figure P35.24 shows the diffraction patterns produced by a slit of varying width. What is the relative width of the slit in each case, from narrowest to widest?

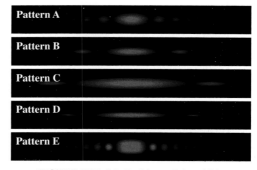

FIGURE P35.24 Problems 24 and 32.

25. **N** Monochromatic light of wavelength 529 nm is incident on a single slit. The second-order diffraction minimum is at an angle of 7.50×10^{-3} rad. What is the width of the slit?

26. **C** One day you are running late to class and realize as you hurry down the hall that you can hear your instructor even though you are not yet in front of the doorway. However, you also notice that your instructor's voice sounds lower and more muffled than

normal. Why do you hear the lower frequencies (longer wavelengths) from your instructor's voice, but not the higher frequencies (shorter wavelengths)? How does the width of the doorway affect what you hear?

27. **N** A thread must have a uniform thickness of 0.525 mm. To check the thickness of the thread, you can illuminate it with a laser of wavelength 625.8 nm. A diffraction pattern like the one produced by a single slit forms on a screen.
    a. If the screen is 3.00 m from the thread, how far apart are the fifth-order minima from one another?
    b. If the thread's thickness increases by 20%, how far apart will the fifth-order minima be?

28. **N** Microwaves with a wavelength of 3.56 mm are passed through a single slit. If only five dark fringes appear on either side of the central maximum, what is the minimum width of the slit?

29. **N** What is the width of the central maximum in the diffraction pattern formed on a viewing screen placed 1.60 m away from a single slit of width 0.185 mm illuminated by a 410.0-nm blue laser?

30. **N** A radio wave of wavelength 21.5 cm passes through a window of width 96.3 cm.
    a. What is the angular separation of the first-order minima?
    b. How does your answer change for a radio wave of twice the wavelength?

31. **N** For single-slit diffraction, what ratio λ/w produces a first dark fringe at exactly $\theta_1 = 30°$?

32. **C** Suppose the diffraction patterns produced in Figure P35.24 are made by a single slit of fixed width. The slit is illuminated in turn by monochromatic light of varying wavelengths. What are the relative wavelengths of the light, from shortest to longest?

33. **N** A single slit is illuminated by light consisting of two wavelengths. One wavelength is 540.0 nm. If the first minimum of one color is located at the second minimum of the other color, what are the possible wavelengths of the other color? Is the light visible?

34. **N** A long, narrow slit 3.0 μm wide is illuminated by light of wavelength 1.20×10^3 nm. Determine the angle corresponding to the first minimum of intensity in the diffraction pattern.

35. **N** A single slit 0.450 mm wide is illuminated with monochromatic light, forming a diffraction pattern on a screen placed 2.15 m away. What is the wavelength λ of the incident light if the width of the central maximum is 5.55 mm?

36. **N** A diffraction pattern is produced by passing He-Ne laser light of wavelength 632.8 nm through a single slit. The pattern shown is viewed on a screen 2.0 m behind the slit, where the second-order minima are separated by 15.2 cm. Find the width of the slit.

35-5 Young's Experiment: Intensity

37. **N** Two identical coherent waves with different intensities interfere with each other. The intensity of the second wave is double the intensity of the first wave. Determine the ratio of the maximum intensity to the minimum intensity when the two waves interfere.

38. **C** If one slit in the double-slit experiment is blocked so that no light passes through it, what happens to the maximum intensity observed?

Problems 39 and 40 are paired.

39. **N** A double-slit experiment is conducted, and at some point on the screen the intensity is found to be 75.0% of its maximum. What is the minimum phase difference (in radians and degrees) that produces this result?

40. **N** A double-slit experiment is conducted with light of wavelength 550.0 nm, and at some point on the screen the intensity is found to be 75.0% of its maximum. The slits are 75.0×10^3 nm apart, and the screen is 1.75 m from the slits. What is the minimum separation between this point and the central maximum?

41. **N** A 520.0-nm light source illuminates two slits with a separation of 6.00×10^{-4} m, forming an interference pattern on a screen placed 4.20 m away from the slits. At a point a distance of 3.75 mm from the central maximum, **a.** what is the phase difference between the two waves originating at the slits, and **b.** what is the intensity compared to that of the central maximum?

Problems 42 and 43 are paired.

42. **A** There isn't a sharp edge between bright and dark fringes, so how do we define the width of a fringe? Assume the angular width of a bright fringe corresponds to the angular distance between points that are two-thirds of the maximum intensity. Find an algebraic expression in terms of d and λ for the angular width $\Delta\theta$ of a bright fringe. Use the small-angle approximation.

43. **N** Assume the angular width of a bright fringe corresponds to the angular distance between points that are two-thirds of the maximum intensity. A certain double-slit experiment uses green light at 555.5 nm with slit separation d_{green}. You want to design another double-slit experiment using red light at 725.0 nm. If the red fringes in your experiment are to have the same angular width as the fringes in the experiment that used green light, what slit separation must you use? Express your answer in terms of d_{green}.

44. **N** Light of wavelength 455 nm is incident on two slits separated by $d = 3.75$ mm, forming an interference pattern on a screen placed 1.85 m away. At what distance y from the central maximum is the intensity exactly half that of the maximum?

45. **N** Light of wavelength 589 nm is used to illuminate two slits with a separation of 0.270 mm. What is the percentage of the maximum intensity a distance of 0.950 cm away from the central maximum if the screen on which the interference pattern forms is 1.25 m away from the slits?

Problems 46 and 47 are paired.

46. **A** In a Young's double-slit experiment, light of wavelength λ is sent through the slits. The intensity I at angle θ_0 from the central bright fringe is lower than the maximum intensity I_{max} on the screen. Find an expression for the spacing between the slits in terms of λ, θ_0, I, and I_{max}.

47. **N** In a Young's double-slit experiment, 586-nm-wavelength light is sent through the slits. The intensity at an angle of 2.50° from the central bright fringe is 80.0% of the maximum intensity on the screen. What is the spacing between the slits?

48. **G** Red light with a wavelength of 715.5 nm and violet light with a wavelength of 412.5 nm are used simultaneously in a double-slit experiment. The slits are 651.5 nm apart. On the same graph, sketch the ratio I/I_{max} as a function of θ for both colors over the range $-50° < \theta < 50°$. Comment on the relative position and width of the maxima.

35-6 Single-Slit Diffraction Intensity

Problems 49 and 50 are paired.

49. **C** Figure P35.49 shows the intensity of the diffraction patterns produced by a slit of varying width. Rank the relative widths of the slit in each case, from narrowest to widest.

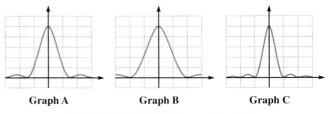

FIGURE P35.49 Problems 49 and 50.

50. **C** Figure P35.49 shows the intensity of the diffraction patterns produced by a slit of fixed width. The slit is illuminated in turn by monochromatic light of varying wavelengths. Rank the relative wavelengths of the light in each case, from shortest to longest.

51. **N** Assume the maxima produced by a single-slit diffraction pattern have angular locations α that are halfway between the adjacent minima of angular location a. Find the intensity of the first three maxima as a fraction of the maximum intensity.

52. **N** Light with a wavelength of 426 nm is incident on a single slit with a width of 4.5 μm. If the intensity of the central bright fringe is 655 W/m², find the intensity at these angular positions from the fringe: **a.** 20.0°, **b.** 40.0°, and **c.** 60.0°.

53. **N** Light of wavelength 750.0 nm passes through a single slit of width 2.00 μm, and a diffraction pattern is observed on a screen 25.0 cm away. Determine the relative intensity of light I/I_{max} at 15.0 cm away from the central maximum.

54. **N** Monochromatic light of wavelength 414 nm is incident on a single slit of width 32.0 μm. The distance from the slit to the screen is 2.60 m. Consider a point at $y = 16.5$ mm from the center of the central maximum. What is the ratio of the intensity at that point to the maximum intensity?

55. **N** For the following questions, assume the noncentral maxima in a single-slit diffraction pattern are exactly halfway between the minima immediately adjacent to them. What percentage of the central maximum's intensity is **a.** the first-order maximum's intensity, and **b.** the second-order maximum's intensity?

56. **N** Light with a wavelength of 585 nm is incident on a single slit with a width of 3.21 μm.
    a. What is the angular position of the second-order minimum?
    b. What is the ratio of the intensity at the angular position of the second-order minimum and the maximum intensity (I/I_{max})?

35-7 Double-Slit Diffraction

Problems 57 and 58 are paired.

57. **N** Light of wavelength 515 nm is incident on two slits of width 1.545 μm. The slits are 6.18 μm apart. What is the relative intensity at $\theta = 15.0°$?

58. **N** Light of wavelength 515 nm is incident on two slits of width 1.545 μm. The slits are 6.18 μm apart. How many bright fringes are within the central maximum of the diffraction pattern?

59. **A** Two slits are separated by distance d and each has width w. If $d = 2w$, how many bright fringes are within the central maximum of the diffraction pattern?

60. **N, C** Suppose you want exactly 11 bright fringes inside the central maximum of a double-slit diffraction pattern. What must be true of the slit width and separation?

General Problems

61. **C** Two very small identical lightbulbs are connected to the same light fixture. Explain why no interference pattern is seen.

62. **C** If you spray paint through two slits, what pattern results? How does it compare to the pattern seen in Young's experiment?

63. **N** Light of wavelength 570.0 nm incident on a single slit forms a diffraction pattern on a screen placed 1.10 m from the slit. What is the width of the slit if the first and fifth dark fringes are separated by 5.30 mm?

64. **C** Much of this chapter is devoted to Young's double-slit experiment, which we will revisit in later chapters. Why is this experiment so important?

65. **N** A screen is located 2.0 m from a single slit of width 4.0 μm. If light from a helium-neon laser of wavelength 632.8 nm shines on the slit, determine the position of the first minimum on the screen.

Problems 66 and 67 are paired.

66. **N** The two slits of a double-slit apparatus each have a width of 0.120 mm and their centers are separated by 0.720 mm. What orders are missing in the diffraction pattern?

67. **N** The two slits of a double-slit apparatus each have a width of 0.120 mm and their centers are separated by 0.120 mm. What orders are missing in the diffraction pattern?

68. **N** A single slit with width 0.314 mm is illuminated with light of wavelength 490.0 nm, and a diffraction pattern with a central maximum of width 6.60 mm forms on the viewing screen.
    a. What is the distance from the slit to the viewing screen?
    b. What is the width of the first side maximum in this diffraction pattern?

69. **N** An exterior sliding door of a building 1.20 m in width opens to let in coherent microwaves with a wavelength of 28.0 cm, creating a diffraction pattern on the opposite wall of the lobby located 11.2 m from the door. If we assume the building is opaque to this wavelength, what is the distance between the second-order minimum and the central maximum on the lobby wall?

70. A single slit of width w is illuminated by light of wavelength λ so that the first-order minimum occurs at an angle θ_1.
    a. A Find an expression for the angular position of the first-order minimum in terms of θ_1 if the width of the slit is halved.
    b. A, C How does your answer change if the small-angle approximation is valid in this case?

71. **N** Monochromatic light incident on a slit of width 0.635 mm produces a diffraction pattern on a screen 1.55 m away from the slit. If the third-order dark fringe of the pattern is 3.04 mm from the center of the pattern, what is the wavelength of the light incident on the slit?

72. **C** When you arrive at a concert, you are dismayed to find that your seats are directly behind a pillar 1 m wide. (That's why the tickets were so cheap.) Your friend says not to worry and assures you that your enjoyment of the concert will not be affected. She says, "The sound will bend around the pillar, and you will still be able to hear everything." Another concertgoer suggests that you will hear the concert but everything will sound "bassier," or that you will hear primarily the lower frequencies (longer wavelengths). Explain how the pillar will affect what you hear and why.

73. **N** A narrow slit is illuminated with light of wavelength 750.0 nm, 1.0 m in front of a screen. The first minima on either side of the central maximum of the diffraction pattern observed are separated by 3.0 mm. Determine the width of the slit.

74. **N** Sound with a wavelength of 2.29 m is incident on an opening in a wall that acts as a single slit with a width of 4.59 m. The maximum intensity of the sound after passing through the opening is 1.00×10^{-6} W/m². The locations of absolute silence would be found in a way similar to finding the angular positions of dark fringes for light passing through a single slit. What is the angular position for the first-order ($m = 1$) location of silence in this example?

75. **N** When a double slit is illuminated with 622-nm light, the eighth interference fringe is observed a distance of 5.92 mm from the central maximum on a screen 3.60 m away. What is the separation of the two slits?

76. **N** Light with a wavelength of 550.0 nm is incident on a pair of slits with a separation of 0.350 mm.
    a. Find the angles corresponding to the locations of the first three orders of fringes away from the central bright fringe.
    b. If the screen is 2.00 m away from the slits, what is the distance between the first-order ($n = 1$) and second-order ($n = 2$) bright fringes?

77. **N** Consider two monochromatic sources A and B with wavelength λ such that A is initially ahead of B in phase by 66°. The waves interfere at a certain point after having traveled different paths, where B travels $\lambda/4$ farther than A. Determine the phase difference between the waves when they interfere at this point.

78. **C** Another way to construct a double-slit experiment is to use a Lloyd's mirror (Fig. P35.78). Light from the single slit strikes the screen and interferes with the light that has reflected from the mirror. Explain why at the center of the fringes there is a dark fringe instead of a bright fringe.

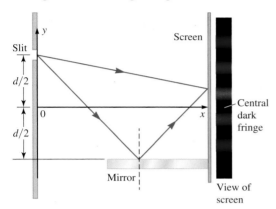

FIGURE P35.78

79. **N** Sound waves with frequency 1850 Hz are incident on a pair of slits placed 38.0 cm apart. Use 343 m/s for the speed of sound.
    a. What is the angular separation between the central maximum and an adjacent maximum?
    b. What would the slit separation have to be for the interference pattern of microwaves with wavelength 2.65 cm to have the same angular separation between the central maximum and an adjacent maximum?

Problems 80 and 81 are paired.

80. **G** Table P35.80 presents data gathered by students performing a double-slit experiment. The distance between the slits is 0.0700 mm, and the distance to the screen is 2.50 m. The distance y from the central maximum to other maxima is given. Plot the data and find the wavelength of the laser's light.

81. **N** Table P35.80 presents data gathered by students performing a double-slit experiment. The distance between the slits is 0.0700 mm, and the distance to the screen is 2.50 m. The intensity of the central maximum is 6.50×10^{-6} W/m². What is the intensity at $y = 0.500$ cm?

82. **N** A pair of slits separated by 0.340 mm and placed 2.10 m away from a viewing screen is illuminated by a source that produces both 580-nm and 620-nm light, resulting in overlapping interference patterns. What is the shortest distance from the central maximum for which bright fringes for the two wavelengths coincide?

TABLE P35.80

n	y (cm)
-4	-7.52
-3	-5.11
-2	-3.52
-1	-1.70
0	0.00
1	1.63
2	3.40
3	4.80
4	6.98

83. **N** Monochromatic light waves of wavelengths 400.0 nm and 560.0 nm are incident simultaneously and normally on a Young's double-slit apparatus. The separation between the slits is 0.100 mm, and the distance between the slits and the screen is 1.00 m. Determine the distance between the first two adjacent dark fringes on the screen.

84. **N** In a Young's double-slit experiment, the intensity at a point on the screen is measured to be 55.0% of the maximum.
    a. What is the minimum phase difference between the two sources, measured in radians, that would give this intensity?
    b. If the path difference between the light from the two slits is 106 nm, what is the wavelength of light being used in the experiment?

85. **N** Two transmitters for rival radio stations both transmitting at 20.0 MHz are located 45.0 m apart on a hilltop. Due to a tuning error, the second transmitter transmits 180° out of phase with the first. If a technician walks from the first transmitter to the second along the line joining them, what is the shortest distance he must walk to reach a point where the transmissions from the two towers are in phase?

86. **G** Three waves, which can each be represented as $y(t) = A \sin(\omega t + \varphi)$, reach the same point in space. For times $0 < t < 10.0$ s, plot the superposition of the three waves for the following conditions and comment on the qualitative differences.
    a. The waves are coherent: $A = 2.0$ mm and $f = 3.0$ Hz for each wave, but they have phases of $\varphi = 0$, 0.30 rad, and 0.80 rad.
    b. The waves are incoherent: The parameters of the three waves are (1) $A = 2.0$ mm, $f = 3.0$ Hz, and $\varphi = 0$; (2) $A = 1.0$ mm, $f = 4.0$ Hz, and $\varphi = 0.30$ rad; and (3) $A = 1.5$ mm, $f = 5.0$ Hz, and $\varphi = 0.80$ rad.

87. **N** A source of 485-nm light is used to perform Young's double-slit experiment. The separation of the slits is 0.185 mm, and the screen is placed 2.30 m away from the slits.
    a. What is the distance between the central maximum and an adjacent maximum in the interference pattern on the screen?
    b. What is the distance between the first and third dark bands in the interference pattern on the screen?

Problems 88 and 89 are paired.

88. **A** Show that the resultant intensity from light waves with amplitudes E_1 and E_2 coming through two slits of a Young's double-slit apparatus is given by
$$I = \left(\frac{1}{2\mu_0 c}\right)[E_1^2 + E_2^2 + 2E_1 E_2 \cos \varphi]$$
where φ is the phase difference between the two interfering waves.

89. **A** One of the slits in a Young's double-slit apparatus is wider than the other, so that the amplitude of the light that reaches the central point of the screen from one slit alone is twice that from the other slit alone. Determine the resultant intensity as a function of the direction θ on the screen, the wavelength λ of the incident light, the incident intensity I_0, and the slit separation d.

90. A laser shines through a double-slit apparatus, creating bright and dark fringes.
    a. **N** If we define the bright fringes as the regions between locations where the intensity falls below 20%, what is the ratio of the width of the bright fringes to the width of the dark fringes?
    b. **G** Plot the intensity of the interference pattern for a few peaks around $\varphi = 0°$ and indicate the positions for which the intensity is higher than 20% of the maximum value to confirm your answer to part (a).

36 Applications of the Wave Model

Key Questions

How do thin films generate interference patterns?

How did Michelson and Morley use interference to show that light does not require a medium?

What instruments exploit the wave nature of light?

❶ **Underlying Principles**

No new principles are introduced in this chapter.

✪ **Major Concepts**

1. Rayleigh's criterion
2. Dispersion
3. Resolving power

▷ **Special Cases**

1. Circular aperture resolution
2. Thin-film interference
3. Diffraction grating (position and half-width of lines)

36-1 Implications of the wave model 1155

36-2 Circular aperture diffraction 1155

36-3 Thin-film interference 1158

36-4 Diffraction gratings 1166

36-5 Dispersion and resolving power of gratings 1170

36-6 Case study: Michelson's interferometer 1174

Rainbows are commonly observed in the skies of Hawaii because brief rain showers are often followed by sunny skies. The author of this book spent much of her early childhood in the Midwest, however, where rain showers are followed by dreary skies. She rarely saw rainbows in the midwestern skies. But, after a rainstorm, she often saw rainbows in parking lots because many vehicles at that time left films of gasoline or oil floating in shallow puddles. Of course, this is harmful to the environment and today's vehicles are much cleaner. Nevertheless, the rainbows observed in a thin film of oil are beautiful (Fig. 36.1). The wonderful colors we see in such polluted water are best explained by the wave model of light. Chapter 35 presented evidence that supports the wave model, and this chapter is about applying that model to various phenomena and using it to design instruments to study the Universe.

FIGURE 36.1 Oil on water creates beautiful colors.

36-1 Implications of the Wave Model

Because the nature of light was hotly debated, performing experiments that supported the wave model was important in the process of scientific discovery. Such experiments often lead to more discoveries and inventions. For example, after Ben Franklin discovered that lightning is a giant electrical spark, he went on to invent the lightning rod (case study in Chapter 24). Applying the wave model allows us to explain phenomena and build new instruments. The case study in this chapter involves both the invention of a new scientific instrument and the important discovery made with that instrument.

CASE STUDY The Michelson-Morley Experiment

This case study involves the most important *failed* experiment in history. In the 19th century, most scientists thought light is a wave that propagates in a medium they called the **luminiferous ether** (usually referred to simply as the **ether**). Today we know that no such medium exists and that light travels in a vacuum. But 19th-century scientists reasoned that because light travels through the open space between the Sun and the Earth, as well as through substances such as water and glass, the ether permeates everything. These scientists wanted to characterize and measure its properties.

Luminiferous means "light-bearing" in Latin, and the Greek term *ether* designates an unknown medium.

The speed of a wave through a medium depends on the properties of that medium. For example, the speed of sound depends on the medium's bulk modulus B (or compressibility): $v_s = \sqrt{B/\rho}$ (Eq. 17.13). The speed of sound is greater in a less compressible fluid (one with a greater bulk modulus). Because the speed of light is very high, scientists reasoned that the ether was nearly incompressible (had a very great bulk modulus).

If the ether was also everywhere, then everything—you, racehorses, and the Earth—moved through the ether. If the ether created a drag or frictional force on moving objects, we would notice the resulting deceleration. For example, the Earth has been orbiting the Sun for 4.5 billion years. If the ether exerted a dissipative force on the Earth, the Earth would have spiraled in toward the Sun. Because no such dissipative force was observed, 19th-century scientists concluded that the ether was frictionless.

The ether was also supposedly at rest, and thus it should have been possible to measure the speed of objects moving with respect to the ether. In other words, the ether defined an absolute reference frame against which to measure all motion. So, in the late 1800s, Albert Michelson and Edward Morley set out to measure the speed of the Earth through the ether. Like Thomas Young's experiment, Michelson and Morley's experiment depended on observing the fringes produced by interference.

Much to their surprise, Michelson and Morley were unable to measure the speed of the Earth with respect to the ether. Instead, they found *no* evidence that the ether existed. Michelson was bothered by this failed experiment for the rest of his life, but today we understand that this is the most important experimental failure in history. The simplest explanation for the failure is that the ether does not exist and light waves do not require a medium to propagate through.

36-2 Circular Aperture Diffraction

In many practical situations, light must pass through a circular or nearly circular aperture, such as the shutter on your camera or the aperture in a telescope. Diffraction occurs whenever light passes through an aperture or near the edge of an object. In the case study in Chapter 35, we considered diffraction of light through a circular aperture and around an object with a circular cross section. In this section, we take a more mathematical approach.

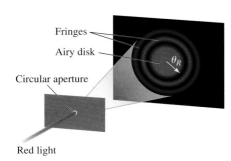

FIGURE 36.2 When light passes through a circular aperture, the diffraction pattern has a central maximum, known as an Airy disk, and fringes that form concentric circles.

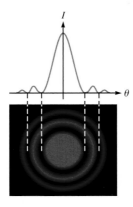

FIGURE 36.3 Light intensity as a function of angular position for diffraction through a circular aperture.

RAYLEIGH'S CRITERION
✪ **Major Concept**

CIRCULAR APERTURE RESOLUTION
▶ **Special Case**

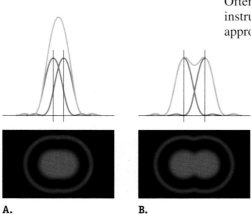

FIGURE 36.4 A. These two sources appear to be a single extended source. **B.** These two sources fit Rayleigh's criterion; they are barely resolved.

In Figure 36.2, light from a single source passes through a circular aperture. The diffraction pattern that results has a circular central maximum called an **Airy disk**. The angular radius θ_R of the Airy disk is given by

$$\sin \theta_R = 1.22 \frac{\lambda}{d} \tag{36.1}$$

where λ is the wavelength of the light and d is the diameter of the aperture. Both the wavelength and the diameter must be measured in the same length units. Often the angular radius is small, so it is approximately given by

$$\theta_R \approx 1.22 \frac{\lambda}{d} \tag{36.2}$$

where θ_R is in radians. The central maximum is surrounded by bright and dark fringes that form concentric circles, often unnoticed because they are much fainter than the central maximum. Figure 36.3 shows a graph of the light intensity across the diffraction pattern. The central maximum is many times more intense than the bright fringes.

If the fringes are so faint compared to the central maximum, why do we care about diffraction through a circular aperture? Diffraction is important when the patterns of two or more sources overlap. Figure 36.4 shows graphs of intensity and the diffraction patterns that result when two point sources emit light through a circular aperture. In Figure 36.4A, the point sources are close together and their diffraction patterns overlap. The total intensity has a single central hump, so it looks similar to the intensity from a single source (Fig. 36.3). If these two sources were two very distant stars, the resulting image through a telescope would appear to be a single star (perhaps slightly elongated). We would say that the images of the two stars are **not spatially resolved**. **Resolution** is the ability of an aperture to separate the diffraction patterns produced by two sources. Resolution depends on the wavelength of the light passing through the aperture and on the aperture's size.

Figure 36.4B shows the diffraction patterns of two sources that are just barely resolved. On the intensity graph, the total intensity has a slight dip in the center, so it does not look like the intensity from a single source; instead, it hints at two sources. The image has two lobes, again hinting that there are two sources.

According to **Rayleigh's criterion**, two sources are barely resolved when the central maximum of one diffraction pattern is centered on the first minimum of the other diffraction pattern (Fig. 36.4B). That is, the two images are barely resolved when they are separated by the angular radius of the Airy disk. So, we can use Equation 36.1 to write an expression for the minimum angular separation θ_{min} between two images that are barely resolved:

$$\sin \theta_{min} = 1.22 \frac{\lambda}{d} \tag{36.3}$$

Often θ_{min} is called the **diffraction limited resolution**, or simply the **resolution**, of an instrument with a circular aperture. Usually θ_{min} is small, and we use the small-angle approximation to write

$$\theta_{min} \approx 1.22 \frac{\lambda}{d} \tag{36.4}$$

where θ_{min} is in radians. Often the angle is *very* small and so is more conveniently expressed in arc seconds, in which case Equation 36.4 becomes

$$\theta_{min} \approx 251643 \frac{\lambda}{d} \text{ arcsec} \tag{36.5}$$

Why Is There a Factor of 1.22?

We cannot derive Equation 36.3 or 36.4 without using special mathematical functions (known as Bessel functions) that are beyond the scope of this textbook. However, we can give a plausible argument using a couple of approximations and come close to the exact answer.

Let's start by finding an expression for the resolution θ_{min} of a (rectangular) slit of width w, assuming its height is much greater than its width. According to Rayleigh's criterion, two sources whose light passes through the single slit will be resolved if the central maximum of one falls no closer to the central maximum of the other than the first dark fringe of the other's diffraction pattern. So, their minimum separation is equal to the position of the first dark fringe (Eq. 35.5 with $m = 1$):

$$\sin \theta_{min} = \frac{\lambda}{w}$$

For the rest of this discussion, we will assume that the resolution θ_{min} is small and use the small-angle approximation:

$$\theta_{min} \approx \frac{\lambda}{w} \text{ (radians)} \quad (36.6)$$

For a long, narrow slit, the diffraction pattern is perpendicular to the long axis of the slit (Fig. 35.16). Circles are not well modeled by a single narrow slit, and we improve our approximation by using a square aperture. Figure 36.5 shows the diffraction pattern that is formed by a square aperture: The pattern is stretched out along both the x and y directions. The resolution in each of these directions is given by Equation 36.6:

$$(\theta_{min})_x = (\theta_{min})_y \approx \frac{\lambda}{w}$$

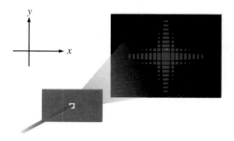

FIGURE 36.5 The diffraction pattern produced by a square aperture.

Let's model a circular aperture as a square whose width equals the diameter of the circle: $w = d$ (Fig. 36.6A). The resolution (in any direction) is given by Equation 36.6, with $w = d$:

$$\theta_{min} \approx \frac{\lambda}{d}$$

From this simple model, we see why θ_{min} in Equation 36.4 depends directly on λ and inversely on d. However, Figure 36.6A is not a good model for a circular aperture because the area of the square is greater than the area of the circle. Let's try a second model, using a square whose area equals the area of the circle (Fig. 36.6B). To find the resolution, we first need to find an expression for the width of this second, smaller square. Because its area is equal to the area of the circle,

$$w^2 = \pi \left(\frac{d}{2}\right)^2$$

so the width of the square in Figure 36.6B is

$$w = \frac{\sqrt{\pi}}{2} d \quad (36.7)$$

To find the resolution, substitute Equation 36.7 into $\theta_{min} \approx \lambda/w$ (Eq. 36.6):

$$\theta_{min} \approx \frac{2}{\sqrt{\pi}} \frac{\lambda}{d} \approx 1.13 \frac{\lambda}{d}$$

Now we have found a slightly closer result; instead of the factor of 1.22, our factor is 1.13. Although we did not get the exact result (Eq. 36.4), we have a way to understand where the equation comes from. The circle is modeled as a square whose width is smaller than the diameter of the circle in order to ensure that their areas are equal. We don't get the exact factor of 1.22 because the square is not quite the right shape. Its corners extend beyond the circle, and its sides fall within the circle. A better model is beyond the mathematics of this textbook.

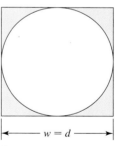

A.

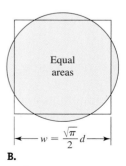

B.

FIGURE 36.6 Model a circular aperture as a square **A.** whose width equals the diameter of the circle, and **B.** whose area equals the area of the circle.

CONCEPT EXERCISE 36.1

When it comes to telescopes, bigger is better. By *bigger*, we mean a larger-diameter primary lens or mirror. How does the diameter of a telescope affect its resolution?

EXAMPLE 36.1   Your Eye Versus Your Telescope

For your birthday, your very nice aunt gives you a 4-in. refracting telescope. The diameter of a fully open pupil is about 8 mm. How does the resolution of your telescope compare to the resolution of your fully open pupil? Assume the minimum angle is small.

INTERPRET and ANTICIPATE
Resolution depends on the diameter of the aperture and the wavelength of the light. For a telescope, the diameter of the objective lens is all that matters. The phrase "4-in. refracting telescope" means that the objective lens has a 4-in. diameter. Because the diameter of the telescope is so much larger than the diameter of your pupil (while the wavelength of the light remains the same), we expect the resolution of the telescope to be much better than that of your eye. It is convenient to work in millimeters (4 in. = 102 mm).

SOLVE
Because the angle is small, we can use Equation 36.4 for the resolution. We need only the ratio of resolutions, so wavelength cancels out.

$$\theta_{min} \approx 1.22 \frac{\lambda}{d} \quad (36.4)$$

$$\frac{(\theta_{min})_{tele}}{(\theta_{min})_{eye}} = \left(1.22\frac{\lambda}{d}\right)_{tele} \bigg/ \left(1.22\frac{\lambda}{d}\right)_{eye}$$

$$\frac{(\theta_{min})_{tele}}{(\theta_{min})_{eye}} = \frac{d_{eye}}{d_{tele}} = \frac{8 \text{ mm}}{102 \text{ mm}} = 0.08$$

CHECK and THINK
The resolution of the telescope is 0.08 times the resolution of your eye. That means the telescope can produce separate images of sources that are 0.08 times closer together than the sources resolved by your eye. Let's look at a few specific values. Your eye can just barely resolve two sources that are separated by about 16 arcsec—the width of a human hair held at arm's length. The telescope can resolve two sources that are about 1.3 arcsec apart, the width of a human hair held at 13 arm's lengths.

36-3 Thin-Film Interference

THIN-FILM INTERFERENCE

Special Case

You have probably seen the rainbow of colors in the thin film of a soap bubble or an oil slick (Fig. 36.1). These colors are a result of wave interference. When light is incident on a thin film, such as oil floating on water, it undergoes two (or more) reflections before it enters your eye, and these reflected waves interfere with one another. To understand how a thin film can cause interference, we first need to take a closer look at the speed of light and what happens when light encounters the boundary between two media such as air and oil.

Speed of Light

All electromagnetic waves propagate at the speed of light c in a vacuum. If an electromagnetic wave is traveling through a transparent medium such as glass or water, its speed is lower than c because the electric and magnetic fields interact with the atoms and molecules in the medium. To see this mathematically, first remember that when we derived the wave equation for an electromagnetic wave in a vacuum, we found that the speed is given by $v_{vac} = 1/\sqrt{\mu_0 \varepsilon_0}$ (Eq. 34.16). Next, remember that we take the effect of a dielectric into account by replacing the permittivity constant ε_0 with $\kappa \varepsilon_0$ in an equation where κ is the dielectric constant (Section 27-8). Similarly, we replace the permeability constant μ_0 with $\kappa_m \mu_0$. Let's model a transparent medium as a dielectric. Then the speed of light in a medium is found by replacing ε_0 with $\kappa \varepsilon_0$ and μ_0 with $\kappa_m \mu_0$ in Equation 34.16:

$$v = \frac{1}{\sqrt{\kappa_m \mu_0 \kappa \varepsilon_0}} = \frac{v_{vac}}{\sqrt{\kappa_m \kappa}} = \frac{c}{\sqrt{\kappa_m \kappa}}$$

For most dielectrics, $\kappa_m \approx 1$ and $\kappa > 1$, so the speed of light in a medium is lower than the speed of light in a vacuum.

The **index of refraction** n is the (dimensionless) ratio of the speed of light in a vacuum to the speed of light in the medium:

$$n = \frac{c}{v} \quad (36.8)$$

The indices of refraction for various media are listed in Table 36.1. We'll take a closer look at the index of refraction in Section 38-1. For now, the index of refraction for a vacuum is exactly 1 and for air is nearly 1. So the speed of light in air is nearly c. All other media have a higher index of refraction, meaning that light propagates more slowly in a medium than it does in air. (Remember that higher n means lower v.) When light is transmitted from a medium with a low index of refraction to one with a higher index of refraction, the light slows down.

Because the speed of light depends on its wavelength and frequency according to $v = \lambda f$ (Eq. 17.8), if there is a change in speed, there must be a change in either frequency or wavelength, or both. When light is transmitted from one medium to another, the frequency does not change, however, so the wavelength in the medium must change. If the speed of light in a vacuum is $c = \lambda_0 f$, then by substituting $v = c/n$ into $v = \lambda f$, we find

$$\frac{c}{n} = \frac{\lambda_0 f}{n} = \lambda f$$

where λ_0 is the wavelength in a vacuum and λ is the wavelength in the medium. So the wavelength in the medium is

$$\lambda = \frac{\lambda_0}{n} \quad (36.9)$$

Equation 36.9 shows that the wavelength in the medium is shorter than in a vacuum because the index of refraction for all media is greater than 1.

Reflection and Phase Changes

The reflection of light has some similarities to the reflection of a mechanical wave on a rope. Figure 36.7 shows a pulse on a rope that is fixed to the pole on the right. With respect to the incoming pulse in Figure 36.7A, the reflected pulse in Figure 36.7B is

TABLE 36.1 Index of refraction.

Medium	n
Air	1.0002926
Cubic zirconia	2.14
Diamond	2.417
Fused quartz	1.458
Heavy flint glass	1.890
Ice	1.3049
Quartz	1.54
Water	1.333
Zinc crown glass	1.517

Notes:
1. The index of refraction depends on wavelength. The values in the table are good for yellow light at $\lambda = 589$ nm. You may use these values in this chapter unless you are told otherwise.
2. You may notice that $n \approx \sqrt{\kappa}$, but don't try to find the index of refraction by using the values for the dielectric constants in Table 27.2. Those values were measured using constant electric fields, and κ is usually much lower when the fields oscillate rapidly.

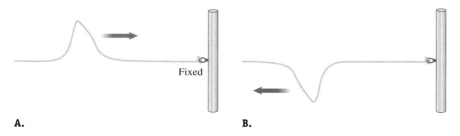

FIGURE 36.7 When a pulse reflects from a fixed end, the reflected pulse is 180° out of phase with the incoming pulse.

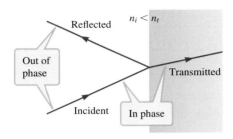

FIGURE 36.8 When light is reflected from a boundary and $n_i < n_t$, the reflected wave is 180° out of phase with the incident wave. We have sketched light as a ray here (see Chapter 37). We are still using the wave model because we are interested in the relative phase of the light, but it is more convenient in this figure to draw rays.

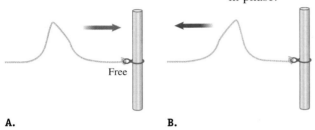

FIGURE 36.9 When a pulse reflects from a free end, the reflected pulse is upright.

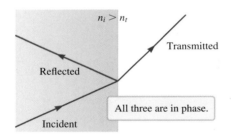

FIGURE 36.10 When light is reflected from a boundary and $n_i > n_t$, the reflected wave is in phase with the incident wave.

both inverted and flipped from left to right. This difference between incoming and reflected pulses can be described in terms of their relative phase. Imagine two such pulses approaching each other on the same rope. Where they overlap completely, the pulses cancel and the rope remains flat (Fig. 18.6, page 524). Because the reflected wave and the incident wave interfere destructively, they are 180° out of phase.

This situation is analogous to light incident in a medium with index of refraction n_i and reflected from a medium with index of refraction n_t, where $n_i < n_t$, such as when light in air is reflected from glass. *An electromagnetic wave reflected from the boundary of a medium with a higher index of refraction than that of the medium in which it is incident is 180° out of phase with the incident wave* (Fig. 36.8).

Now consider a pulse on a rope whose end is free (Fig. 36.9). As before, the incident pulse moves to the right, but now the reflected pulse is flipped from left to right while remaining upright. To find the relative phase of the two pulses, imagine that they approach each other on the same rope. Where they overlap completely, the pulses add together and the amplitude doubles (Fig. 18.8, page 525). Because the reflected wave and the incident wave interfere constructively, they are in phase.

This situation is analogous to light incident in a medium with index of refraction n_i and reflected from a medium with index of refraction n_t, where $n_i > n_t$, such as when light in glass is reflected at an air-glass boundary. *An electromagnetic wave reflected from the boundary of a medium with a lower index of refraction than that of the medium in which it is incident is in phase with the incident wave* (Fig. 36.10).

To keep the two possibilities straight, remember the phrase *low to high is out*, which corresponds to Figure 36.8. By contrast, in both situations ($n_i < n_t$ and $n_i > n_t$), the transmitted wave is in phase with the incident wave.

Thin Films: Conditions for Constructive and Destructive Interference

Now we are ready to analyze the interference produced by a thin film and come up with mathematical expressions for the conditions of constructive and destructive interference. The film has index of refraction n_f and width w (Fig. 36.11), and it is surrounded by a medium with index of refraction n_i. The film's index of refraction is higher than that of the surrounding medium: $n_f > n_i$. This film could be soap and the surrounding medium air, which is a special case of thin-film interference. There are many other possibilities: For example, the film could be surrounded by a material that has a higher index of refraction, or there could be one medium on one side of the film and another medium on the other side. Some of these other special cases are explored in the examples and homework problems.

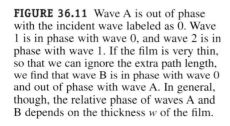

FIGURE 36.11 Wave A is out of phase with the incident wave labeled as 0. Wave 1 is in phase with wave 0, and wave 2 is in phase with wave 1. If the film is very thin, so that we can ignore the extra path length, we find that wave B is in phase with wave 0 and out of phase with wave A. In general, though, the relative phase of waves A and B depends on the thickness w of the film.

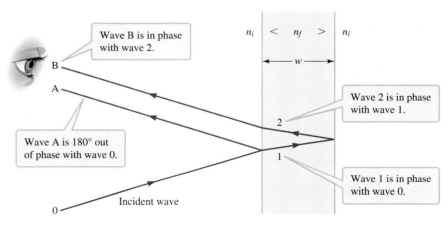

Consider one wave incident in air (Fig. 36.11). We'll trace its progress as it encounters first the front side of the film and then the back side. When the wave encounters the front side of the film, it is both reflected and refracted (transmitted). The reflected wave is labeled A. Because the wave is incident in the medium with the lower index of refraction ($n_i < n_f$), the reflected wave is 180° out of phase with the incident wave (Fig. 36.8).

The transmitted wave is labeled 1 and is in phase with the incident wave. When wave 1 encounters the back side of the thin film, it is both transmitted into the surrounding medium and reflected back into the film. We'll ignore the light transmitted into the surrounding medium and instead follow the reflected wave, labeled 2. Wave 2 is in phase with wave 1 because wave 2 is reflected from the back side of the film. For this boundary, the incident medium is the film and the transmitted medium is the air; because the incident medium has a greater index of refraction, the reflected wave does not experience a phase shift. Next, wave 2 encounters the front side of the film and is both reflected and transmitted. This time we ignore the reflection and follow the transmitted wave, labeled B. Like all transmitted waves, wave B is in phase with its incident wave (wave 2 in this case).

When your eye intercepts waves A and B, you see a bright fringe if the two waves are in phase and a dark fringe if they are 180° out of phase. The relative phase of A and B depends on two factors: (1) Waves A and B result from two different reflections. Reflection from the front of the film introduces a 180° phase shift, but reflection from the back side does not. (2) Wave B in effect travels farther than wave A because B includes the paths of waves 1 and 2 through the film. Let's separate these two factors for the moment by imagining that the film is very thin, so that the extra path length is negligible. In this case, waves A and B must be 180° out of phase because A is 180° out of phase with the incident wave (labeled 0) while B is in phase with the incident wave (labeled 0).

Now let's take into account the extra path length traveled by B. If the waves are nearly perpendicular to the film, the distance traveled either by wave 1 or by wave 2 is equal to the width of the film. So the extra path length is twice the width w of the film:

$$\Delta d = 2w \qquad (36.10)$$

We just stated that if the film is very thin so that there is no path-length difference, then A and B are 180° out of phase. But if the extra path length is a half-wavelength, this path-length difference puts the two waves back in phase. In fact, as long as the extra path-length difference is an odd number of half-wavelengths, A and B are in phase. We can express this condition for constructive interference mathematically as

$$\Delta d = \left(m + \frac{1}{2}\right)\lambda_f \quad (m = 0, 1, 2, 3, \ldots) \qquad (36.11)$$

In some equations, we represent a whole number with the letter m instead of n to avoid confusion with the index of refraction.

Recall that the wavelength λ_f in the film is given by $\lambda_f = \lambda_0/n_f$ (Eq. 36.9), where λ_0 is the wavelength in a vacuum (or in air). It is most convenient to write the condition for constructive interference in terms of the width of the film and the wavelength of light in a vacuum. Substituting Equations 36.9 and 36.10 into Equation 36.11 gives

$$2w = \left(m + \frac{1}{2}\right)\frac{\lambda_0}{n_f} \quad (m = 0, 1, 2, 3, \ldots) \qquad (36.12)$$

Now we find the condition for destructive interference. In this case, the path-length difference must be a whole number of wavelengths. Mathematically, we have

$$\Delta d = m\lambda_f \quad (m = 0, 1, 2, 3, \ldots) \qquad (36.13)$$

As before, we write the condition for destructive interference in terms of the width of the film and the wavelength of light in a vacuum:

$$2w = m\frac{\lambda_0}{n_f} \quad (m = 0, 1, 2, 3, \ldots) \qquad (36.14)$$

Equations 36.12 and 36.14 may be applied to situations other than the one depicted in Figure 36.11, in which a thin film is surrounded by a medium with a lower index of refraction. These equations apply whenever the reflection at one boundary produces a 180° phase shift and the reflection at the other boundary produces no phase shift.

1162 CHAPTER 36 Applications of the Wave Model

In other situations, the equations may need to be switched so that Equation 36.14 is for constructive and Equation 36.12 is for destructive interference.

We derived Equations 36.12 and 36.14 for the reflected waves A and B, ignoring the waves that are transmitted all the way through the soap film. The transmitted waves also produce an interference pattern, however, so you can see these patterns from either side of the film. The derivation of the conditions for constructive and destructive interference for the transmitted waves is similar to the derivation done here (Problem 12).

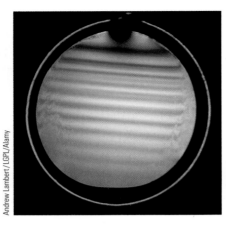

FIGURE 36.12 A thin-film interference pattern.

CONCEPT EXERCISE 36.2

We derived the conditions for constructive and destructive interference for a thin film such as the soap bubble in Figure 36.12. Use these conditions to explain why a colorful pattern is seen.

EXAMPLE 36.2 Newton's Rings

Thin-film interference may be used to test the shape of a lens. Consider a plano-convex glass lens, which has one plane surface and one convex surface as shown in Figure 36.13A. To test its shape, the lens is placed on top of a flat slab of glass so that there is a thin film of air between the lens and the slab. Monochromatic light is incident from above the lens, producing an interference pattern as shown from above in Figure 36.13B. If the lens is perfect, the pattern is a series of concentric rings called Newton's rings. (If a different pattern is seen, the lens has a defect.) Find an approximate expression for the radius r_m of the mth dark ring in terms of the lens's radius of curvature R and the wavelength of the incident light. Assume the index of refraction for air is 1 and the radius of a fringe is much smaller than the radius of curvature of the lens ($r_m \ll R$). In the **CHECK and THINK** step, explain why there is a dark spot at the center of the pattern.

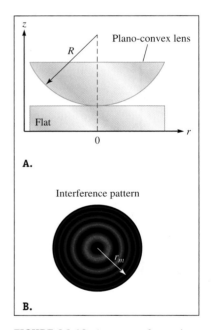

FIGURE 36.13 Apparatus for testing the shape of a lens.

:• INTERPRET and ANTICIPATE

When solving problems that involve thin-film interference, start by deciding whether the conditions for constructive (Eq. 36.12) or destructive (Eq. 36.14) interference apply to the situation. These equations apply whenever the reflection at one boundary produces a 180° phase shift and the reflection at the other boundary produces no phase shift. In our current problem, the film is air surrounded by glass, so we can use Figure 36.11, but now $n_i > n_f$. When the incident wave is reflected from the first boundary, there is no phase shift because the light is incident in a medium (glass) with a higher index of refraction. Wave 1 travels through the film of air. When wave 1 is reflected from the next boundary, there is a 180° phase shift because the light is incident in a medium with a lower index of refraction. So wave 2 is 180° out of phase with wave 1. There is no phase shift as wave B is transmitted in the lens. So, as in the case of the soap film in air (Fig. 36.11), for the air film in glass, one reflection has no phase shift and the other has a 180° phase shift. Thus, Equations 36.12 and 36.14 are the conditions for constructive and destructive interference, respectively. Because we are interested in one of the dark rings, which are the result of destructive interference, we should apply the condition for destructive interference to find its radius.

:• SOLVE

Draw a sketch showing the relevant geometry (Fig. 36.14). The lens's radius of curvature is R, and the radius of an arbitrary dark ring is r_m. The width of the air film is w at the location of that ring.

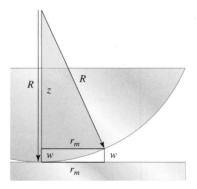

FIGURE 36.14

In Figure 36.14, we have defined z such that $z + w$ is the radius of curvature.	$z + w = R$ (1)
Square Equation (1).	$(z + w)^2 = R^2$ $z^2 + 2zw + w^2 = R^2$ (2)
Apply the Pythagorean theorem to the highlighted right triangle with sides z, r_m, and R.	$z^2 + r_m^2 = R^2$ (3)
Eliminate R^2 from Equations (2) and (3); then solve for r_m^2.	$r_m^2 = R^2 - z^2 = z^2 + 2zw + w^2 - z^2$ $r_m^2 = 2zw + w^2$ (4)
We are trying to derive an approximate expression for the situation where $r_m \ll R$. In Figure 36.14, when $r_m \ll R$, $z \approx R$ and $w \ll z$. We can approximate Equation (4) by replacing z with R and ignoring the last term w^2 because this term is much smaller than $2zw$.	$r_m^2 \approx 2Rw$ (5)
Now apply the condition for destructive interference (Eq. 36.14). In this case, the film is air with index of refraction $n_f \approx 1$. We drop the zero subscript because the wavelength of light is roughly the same in air as it is in a vacuum.	$2w = m\dfrac{\lambda_0}{n_f}$ $m = 0, 1, 2, 3, \ldots$ (36.14) $2w = m\lambda$ $w = \dfrac{m\lambda}{2}$ (6)
Substitute Equation (6) into Equation (5), and solve for r_m.	$r_m^2 \approx 2Rw \approx 2R\dfrac{m\lambda}{2}$ $r_m \approx \sqrt{m\lambda R}$

:• **CHECK and THINK**

This is the equation we were asked to derive. It gives the radius of each dark ring as a function of the wavelength and the lens's radius of curvature. By measuring r_m, you can make sure the lens has the expected radius of curvature R.

There is a dark spot in the center of the interference pattern because the lens is in contact with the flat glass slab. Light incident at the center undergoes two reflections: one involving a 180° phase shift and the other involving no phase shift. But, because no extra distance is traveled, the two reflected waves are out of phase and undergo destructive interference.

EXAMPLE 36.3   Like Oil and Water

A thin film with of an index of refraction of 1.20 lies on top of water with an index of refraction of 1.33. (See Fig. 36.1 for an example.)

A Explain why the conditions for constructive and destructive interference (Eqs. 36.12 and 36.14) are not applicable here. How can these equations be modified to fit this situation?

:• **INTERPRET and ANTICIPATE**

The key to determining whether Equations 36.12 and 36.14 apply to a situation is to see whether one reflection causes a 180° phase shift while the other causes no phase shift. (Consult Figure 36.11.) In this case, the incident light is initially in air. At the boundary between the air and the oily film, the reflected wave is 180° out of phase with the incident wave because the oil has a higher index of refraction than the air. At the next boundary between the oily film and the water, there is also a 180° phase shift because the water has a higher index of refraction than the oil. Because there are *two* 180° phase shifts, Equations 36.12 and 36.14 do **not** represent constructive and destructive interference, respectively.

Example continues on page 1164 ▶

SOLVE

Now our task is to modify these equations to fit this situation. First imagine the film is so thin that there is no extra path-length difference. In this case, the two reflected waves are in phase because both are 180° out of phase with the incident wave. So, if the path-length difference is zero or a whole number of wavelengths, the two reflected waves will remain in phase and interfere constructively. If the extra path-length difference is a half-integer number of wavelengths, the two reflected waves will be out of phase and interfere destructively. All the conditions that held when we derived Equations 36.12 and 36.14 have been switched, so all we need to do is switch the two equations.

Equation 36.12 is now the condition for destructive interference in the case of an oily film that has air on one side and water on the other.	Destructive condition in this case: $$2w = \left(m + \frac{1}{2}\right)\frac{\lambda_0}{n_f} \quad (m = 0, 1, 2, 3, \dots) \quad (36.12)$$
Equation 36.14 is the condition for constructive interference in this case.	Constructive condition in this case: $$2w = m\frac{\lambda_0}{n_f} \quad (m = 0, 1, 2, 3, \dots) \quad (36.14)$$

CHECK and THINK

One of the most important lessons of this problem is that you must not use Equations 36.12 and 36.14 blindly. Always consider first whether or not each reflection produces a phase shift. Note that the conditions involve the index of refraction of the film. You don't need to know the index of refraction for the surrounding media; you just need to know whether the surrounding media have higher or lower indices of refraction than the film.

B If the film is 465 nm thick and you are viewing it from directly above, for what wavelengths of visible light is the reflection brightest because of constructive interference? (Assume the Sun is directly above the film.)

INTERPRET and ANTICIPATE

In this part of the problem, we apply the condition for constructive interference that we found in part A.

SOLVE Solve the condition for constructive interference for wavelength. Then substitute numerical values for all variables except m. Keep an extra significant figure to avoid rounding errors.	$$2w = m\frac{\lambda_0}{n_f} \quad (m = 0, 1, 2, 3, \dots) \quad (36.14)$$ $$\lambda_0 = (2n_f w)\frac{1}{m} = 2(1.20)(465 \text{ nm})\frac{1}{m}$$ $$\lambda_0 = (1116 \text{ nm})\left(\frac{1}{m}\right)$$
Now, to find the wavelengths of visible light, start with $m = 1$ to find all the wavelengths between 400 nm and 780 nm. (Do not include $m = 0$ because that gives an infinite wavelength.)	$\lambda_0(1) = (1116 \text{ nm})\left(\frac{1}{1}\right) = 1116$ nm $\lambda_0(2) = (1116 \text{ nm})\left(\frac{1}{2}\right) = 558$ nm $\lambda_0(3) = (1116 \text{ nm})\left(\frac{1}{3}\right) = 372$ nm
The only wavelength that is in the visible range is green light.	$\lambda_0(2) = 558$ nm

CHECK and THINK
When you look from directly above, you see the film as a bright green patch. If you look at other patches (that are not directly below you), you see other colors because the path length for light through the film is longer.

C Now imagine viewing the film from directly below, as a scuba diver might do. For what wavelengths of visible light is the transmitted light brightest because of constructive interference?

INTERPRET and ANTICIPATE
Once again, we need to decide whether Equations 36.12 and 36.14 apply to this situation. This is our first problem involving transmitted light, so we start with the new sketch in Figure 36.15, keeping the vertical film orientation for easy comparison to Figure 36.11. Let the index of refraction of air be n_i (incident), the index of refraction of the film n_f (film), and the index of refraction of water n_t (transmitted). When the incident light in air encounters the film, it is reflected and transmitted, but we are not interested in the reflection that occurs at this boundary. The light transmitted into the film has no phase shift. Next, that light encounters the water and is again transmitted and reflected. The transmitted wave is labeled A, and the reflected wave is labeled 1. Wave 1 undergoes a 180° phase shift because the water has a higher index of refraction than the film. Next, wave 1 encounters the boundary between the film and air. Wave 1 is reflected and transmitted, but again we are not interested in the transmitted wave. Instead, follow the reflected wave, labeled 2. Wave 2 does not undergo a phase shift because the air has a lower index of refraction than the film. Finally, wave 2 encounters the film-water boundary, where once again it is reflected and transmitted. Now we are interested in the transmitted wave, labeled B. To sum up: Wave A has undergone no phase shifts due to reflection because it has undergone no reflection. Wave B has undergone one 180° phase shift due to reflection, although it has undergone two reflections. So Equations 36.12 and 36.14 may be used in this case, exactly as they appear.

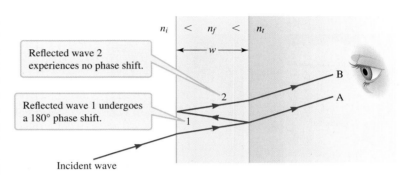

FIGURE 36.15

: SOLVE	
Now our work is much like that in part B, except we use Equation 36.12 for constructive interference. Solve for wavelength.	$2w = \left(m + \dfrac{1}{2}\right)\dfrac{\lambda_0}{n_f}$ $(m = 0, 1, 2, 3, \ldots)$ (36.12) $\lambda_0 = (2n_f w)\left(m + \dfrac{1}{2}\right)^{-1}$ $\lambda_0 = (1116 \text{ nm})\left(m + \dfrac{1}{2}\right)^{-1}$
To find the wavelengths of visible light, start with $m = 0$ and calculate all the wavelengths between 400 nm and 780 nm.	$\lambda_0(0) = (1116 \text{ nm})(\tfrac{1}{2})^{-1} = 2232$ nm $\lambda_0(1) = (1116 \text{ nm})(\tfrac{3}{2})^{-1} = 744$ nm $\lambda_0(2) = (1116 \text{ nm})(\tfrac{5}{2})^{-1} = 446$ nm $\lambda_0(3) = (1116 \text{ nm})(\tfrac{7}{2})^{-1} = 319$ nm
In this case, two colors—red and indigo—are visible.	$\lambda_0(1) = 744$ nm $\lambda_0(2) = 446$ nm

CHECK and THINK
An observer above the film sees green, while the diver below the surface sees burgundy (a blend of red and indigo).

36-4 Diffraction Gratings

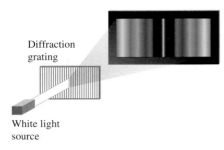

FIGURE 36.16 A diffraction grating produces a spectrum.

A **diffraction grating**, also known simply as a **grating**, is a device that has a large number of slits called **rulings**. Often diffraction gratings have thousands of rulings per millimeter. As in Young's double-slit experiment, light from each ruling arrives at points on a screen and interferes with light from the other rulings. Well-separated bright fringes, also known as **lines**, appear on the screen at places of constructive interference. Because the location of the lines depends on wavelength, the grating produces a "rainbow" (Fig. 36.16). This rainbow is called a **visible light spectrum**, a **color spectrum**, or just a **spectrum**.

The Hydrogen Spectrum

One of the most important uses of a diffraction grating is to spread polychromatic light out into a color spectrum. A complete full-color spectrum is generated when light from a **black body**, an *ideal* object that absorbs all incident radiation, passes through a diffraction grating. When a black body is in thermal equilibrium, it is an ideal emitter (with emissivity $\varepsilon = 1$; Section 21-10) that produces an *ideal* spectrum called a **black-body spectrum**. No real object can produce a perfect black-body spectrum, but when white light—for example, from a glowing tungsten filament—passes through a diffraction grating, a nearly full-color spectrum from red to violet appears on the screen (Fig. 36.16). Other light sources are not well modeled as black bodies. For example, a compact fluorescent bulb's spectrum is missing many colors (Fig. 36.17).

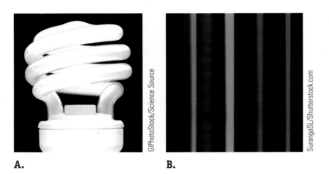

FIGURE 36.17 A. A compact fluorescent bulb. **B.** A compact fluorescent bulb's spectrum. The fluorescent bulb's spectrum is missing many colors.

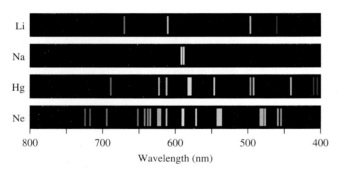

FIGURE 36.18 Each atom's spectrum is a unique set of colors.

The spectrum of each type of atom or molecule is unique, showing lines of only certain colors. Figure 36.18 shows the spectra produced by several different elements. For example, the spectrum of lithium (Li) has about four well-spaced lines, whereas Na has two lines near the center of the range of visible wavelengths. One practical application of a diffraction grating is to examine the spectrum of a light source in order to determine the source's composition. This is particularly important in astronomy because the great distances involved mean that nearly all of the information we have comes from the light emitted by the objects. Observations of spectra have determined that roughly 75% of the universe is hydrogen and most of the rest is helium. All the other elements are present in only trace amounts. The hydrogen spectrum is discussed in the case study in Chapter 42.

Figure 36.19A shows that hydrogen ordinarily produces four strong lines in the visible part of the electromagnetic spectrum: Hα (red), Hβ (aqua), Hγ (indigo), and Hδ (violet). You may also see another faint shorter wavelength line; we'll ignore this line in our discussion. Figure 36.19B shows the spectrum of hydrogen observed in a laboratory using a diffraction grating. There are several features to notice in this spectrum. First, we see two sets of bright fringes labeled with the appropriate value of m. Second, the four expected lines are blended together so we only see two separate lines on either side of the center. Third, the central maximum ($m = 0$) is pale pink because the different colored lines overlap in the center. Fourth, the short-wavelength lines are closer to the center than are the long-wavelength lines, so compared to the $m = 1$ lines, the $m = -1$ lines are in the reverse order. The general process of using a diffraction grating to observe the spectrum of a light source may be applied to any element or compound. Our next step is to derive an expression for the position of observed lines.

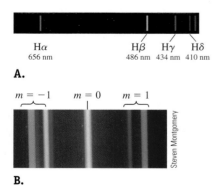

FIGURE 36.19 A. Hydrogen gives off four colored lines in the visible part of the electromagnetic spectrum. The wavelength of each line in nanometers is provided on the figure. **B.** The spectrum of hydrogen observed through a poor quality diffraction grating.

36-4 Diffraction Gratings

DERIVATION Position of the Maxima (Lines) in a Diffraction Grating

We show that the angular position θ of the lines produced by a diffraction grating is given by

$$d \sin \theta = m\lambda \quad (m = 0, \pm 1, \pm 2, \pm 3, \ldots) \quad (36.15)$$

where d is the separation of adjacent rulings and λ is the wavelength of the light. The procedure for finding an expression for the position of the lines produced by a diffraction grating is very similar to that for Young's double-slit experiment (page 1129).

DIFFRACTION GRATING—POSITION OF LINES ▶ Special Case

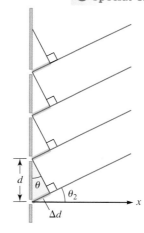

Consider a viewing screen far from the grating, so that the paths from the rulings to any point on the screen are approximately parallel (Fig. 36.20). A point on the screen has the angular position θ measured from the horizontal axis.

FIGURE 36.20

Consider the bottom two rulings and the paths from them to the screen in Figure 36.20. The path from the bottom ruling to the point on the screen is longer. Find the path-length difference Δd by applying trigonometry to the highlighted right triangle.	$\Delta d = d \sin \theta$ ✓ (1)
In order for a maximum to appear on the screen at the angular position θ, the waves must arrive in phase, so the path-length difference must be an integer number of wavelengths. Using this condition and Equation (1), we find an expression for the position of the lines.	$\Delta d = m\lambda \quad (m = 0, \pm 1, \pm 2, \pm 3, \ldots)$ $d \sin \theta = m\lambda \quad (m = 0, \pm 1, \pm 2, \pm 3, \ldots)$ ✓ (36.15)

:• COMMENTS

Each integer m corresponds to a different line, as labeled in Figure 36.21. The integers m are called the **order numbers**. We refer to the zeroth-order line, the first-order line, and so on.

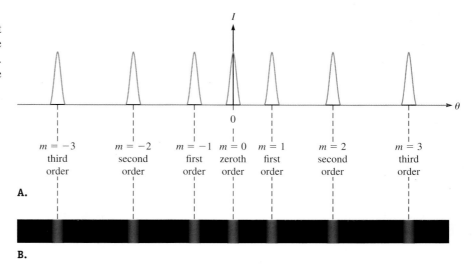

FIGURE 36.21 A. A graph of intensity versus angular position θ for monochromatic light passing through a diffraction grating. **B.** The corresponding pattern of lines produced by this diffraction grating. The lines are referred to by the order number m.

Width of the Maxima in a Diffraction Grating

Now suppose you don't know which element produced the observed spectrum in Figure 36.19B, so you compare this observed spectrum to the theoretical spectra shown in Figures 36.18 and 36.19A. It is hard to identify the element from its observed spectrum because the lines overlap and blend together. As a result, the diffraction grating used to produce the observed spectrum in Figure 36.19B is not very effective because the lines are not well separated. To characterize the line separation, we need an expression for the width of the lines (described in this section). In the next section, we show how to characterize the efficiency of diffraction gratings.

We want to quantify how effective a grating is at producing a spectrum. A good diffraction grating produces thin lines, which is important so that the colors are well separated. It is common to characterize the width of a line in terms of its **half-width** $\Delta\theta_{hw}$, which is measured from the line's center to the darkness found on either side (Fig. 36.22). The half-width of any line is given by

$$\Delta\theta_{hw} = \frac{\lambda}{Nd\cos\theta} \qquad (36.16)$$

where N is the total number of rulings in the grating (Problem 65).

Diffraction grating manufacturers often specify the total number N of rulings. We can find the ruling separation d in terms of N and the length ℓ of the grating:

$$d = \frac{\ell}{N} \qquad (36.17)$$

If the manufacturer specifies instead the linear density of the rulings n (number of rulings per unit length), we use $d = 1/n$.

DIFFRACTION GRATING—HALF-WIDTH OF LINES ⊙ Special Case

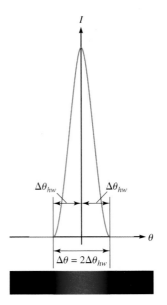

FIGURE 36.22 The width of a line is quantified in terms of the half-width from the center to the darkness on either side.

CONCEPT EXERCISE 36.3

Explain why it would be better to call a *diffraction* grating an *interference* grating.

EXAMPLE 36.4 A Tale of Two Gratings

Compare the hydrogen spectra produced by two different gratings by completing the specified tasks. Each grating has the same separation $d = 2000$ nm between rulings, but the first grating has 5 rulings and the second grating has 50 rulings. For each grating, find the $m = 1$ and $m = 2$ angular positions of the four hydrogen lines (Hα, Hβ, Hγ, and Hδ). Then find the half-widths of these (eight) lines. Sketch the lines, taking into account their positions and thicknesses.

:• **INTERPRET and ANTICIPATE**
This problem requires repeating the same sort of calculation several times, so you may wish to use a spreadsheet program. Because the second grating has more rulings, we expect it to produce spectral lines that are narrower than the lines produced by the first grating.

:• **SOLVE**
Equation 36.15 gives the angular position of the lines. Because we must calculate the angular positions for four different wavelengths, we substitute all the other values. Start with the first-order lines ($m = 1$).

$$d\sin\theta = m\lambda \qquad (36.15)$$

$$\theta = \sin^{-1}\left(\frac{m\lambda}{d}\right)$$

$$\theta_1 = \sin^{-1}\left(\frac{\lambda}{2.00\times 10^{-6}\,\text{m}}\right)$$

We also need the second-order lines ($m = 2$).

$$\theta_2 = \sin^{-1}\left(\frac{2\lambda}{2.00 \times 10^{-6}\,\text{m}}\right)$$

Notice that the angular positions do not depend on the number of rulings, so the values of θ_1 and θ_2 can be used for both gratings. Organize these results in a table.

TABLE 36.2 Angular positions of four hydrogen lines.

Line	λ ($\times 10^{-7}$ m)	θ_1 (rad)	θ_2 (rad)
Hα	6.56	0.334	0.716
Hβ	4.86	0.246	0.508
Hγ	4.34	0.219	0.449
Hδ	4.10	0.207	0.423

The next step is to find the half-widths of these eight lines from Equation 36.16. Because the half-width depends on the number of rulings, we treat each grating separately.

$$\Delta\theta_{hw} = \frac{\lambda}{Nd\cos\theta} \quad (36.16)$$

Substitute $N = 5$ and the value of d to find an expression for the half-widths of the lines produced by the first grating.

First grating, $N = 5$:

$$\Delta\theta_{hw} = \frac{\lambda}{(5)(2.00 \times 10^{-6}\,\text{m})\cos\theta}$$

$$\Delta\theta_{hw} = \frac{\lambda}{(1.00 \times 10^{-5}\,\text{m})\cos\theta}$$

Then use the wavelengths and the angular positions from Table 36.2 to find the half-width of each of the eight lines produced by the first grating. Again, display the results in a table. Because the half-width depends on θ, we must calculate $\Delta\theta_{hw}$ separately for θ_1 and θ_2.

TABLE 36.3 Half-widths for first grating, $N = 5$.

Line	$(\Delta\theta_{hw})_1$ (rad)	$(\Delta\theta_{hw})_2$ (rad)
Hα	0.0695	0.0870
Hβ	0.0501	0.0556
Hγ	0.0444	0.0482
Hδ	0.0419	0.0450

Sketch the lines produced by the first grating (Fig. 36.23). The positions of the eight lines are given in Table 36.2. The thicknesses of the lines come from Table 36.3. Notice that the lines from the two orders overlap.

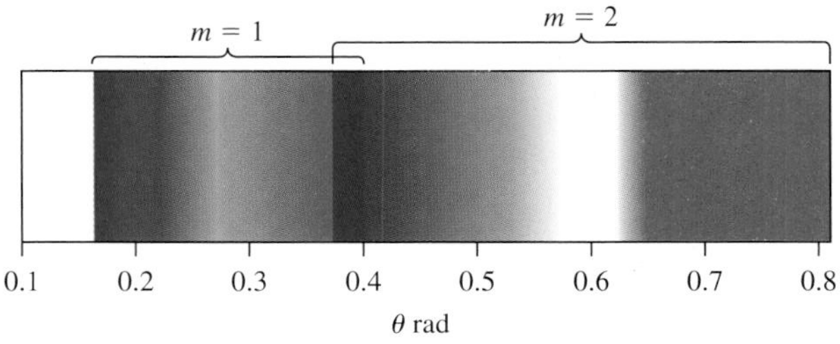

FIGURE 36.23 In a diffraction grating with only five rulings, the hydrogen lines overlap.

Because the second grating has 50 rulings, we must repeat the process of finding the half-widths of the lines it produces.

$$\Delta\theta_{hw} = \frac{\lambda}{(50)(2.00 \times 10^{-6}\,\text{m})\cos\theta}$$

$$\Delta\theta_{hw} = \frac{\lambda}{(1.00 \times 10^{-4}\,\text{m})\cos\theta}$$

Table 36.4 gives the half-widths of the eight lines produced by the second grating. Notice that these lines are thinner than the lines produced by the first grating.

TABLE 36.4 Half-widths for second grating, $N = 50$.

Line	$(\Delta\theta_{hw})_1$ (rad)	$(\Delta\theta_{hw})_2$ (rad)
Hα	0.00695	0.00870
Hβ	0.00501	0.00556
Hγ	0.00444	0.00482
Hδ	0.00419	0.00450

Example continues on page 1170 ▶

Sketch the lines produced by the second grating (Fig. 36.24). As for the first grating, the positions of the eight lines are given in Table 36.2. The thicknesses of the lines are listed in Table 36.4.

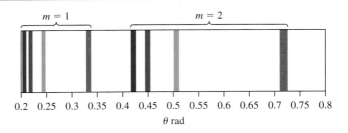

FIGURE 36.24 In a diffraction grating with 50 rulings, the hydrogen lines are well separated. Both the $m = 1$ and the $m = 2$ lines are shown.

CHECK and THINK
Compare Figures 36.23 and 36.24. The second grating has 10 times more rulings than the first grating and produces a spectrum of eight lines that are well separated. If you want to identify the element that produced the spectrum, you may compare Figure 36.24 to the theoretical spectrum in Figure 36.19A. It would be difficult to do the same with Figure 36.23 because the lines are blended together.

36-5 Dispersion and Resolving Power of Gratings

Figures 36.23 and 36.24 show how different diffraction gratings can produce spectra of different qualities. The effectiveness of a diffraction grating is quantified in terms of two parameters—*dispersion* and *resolving power*. **Dispersion** is a measure of the angular separation between the lines of a spectrum. **Resolving power** is a measure of the thickness of these lines. The gratings that produced the spectra in Figures 36.23 and 36.24 have the same dispersion, but the resolving power of the grating is greater in Figure 36.24. (The term *resolution* describes the spatial separation between the diffraction patterns produced by two different sources through a single aperture; *resolving power* describes the narrowness of the lines produced by a single source through a diffraction grating.)

Dispersion

The mathematical definition of dispersion D is

DISPERSION ⊕ Major Concept

$$D \equiv \frac{\Delta\theta}{\Delta\lambda} \quad (36.18)$$

where $\Delta\theta$ is the angular separation between two lines whose wavelength difference is $\Delta\lambda$. The SI units for dispersion are radians per meter, although sometimes it is more convenient to work in degrees per meter, arc seconds per meter, or arc minutes per meter.

If the dispersion of a grating is high, its lines have a high degree of separation. When designing a grating, you can increase the dispersion by using closer rulings. In Problem 66, you will show that the dispersion of a grating is given by

Don't confuse $\Delta\theta$ with $\Delta\theta_{hw}$: $\Delta\theta$ is the separation between two different lines, and $\Delta\theta_{hw}$ is the half-width (thickness) of a single line.

$$D \equiv \frac{\Delta\theta}{\Delta\lambda} = \frac{m}{d\cos\theta} \quad (36.19)$$

where m is the order number, θ is the angular position of the corresponding line, and d is the separation between the rulings. Equation 36.19 shows that the dispersion of a grating is greater (better) if the rulings are closely spaced (d is small) and if the order number m is large. The factor $\cos\theta$ in the denominator means that the dispersion also depends on the angular position of the lines. Lines that have a large (absolute) value for θ have a greater dispersion D.

CONCEPT EXERCISE 36.4

Find the dispersion for the ($m = 1$) Hα and Hβ lines produced by the first grating in Example 36.4. How does it compare to the dispersion for the second grating?

Resolving Power

The lines in Figures 36.23 and 36.24 have the same dispersion, but they are better separated in Figure 36.24 because the lines are narrower. The resolving power R is a measure of the lines' narrowness. The narrower the lines, the higher the resolving power R, which is defined as

$$R \equiv \frac{\lambda_{av}}{\Delta\lambda} \quad (36.20)$$

RESOLVING POWER ✪ **Major Concept**

where λ_{av} is the average wavelength of two lines that are just barely separated and $\Delta\lambda$ is the difference in their wavelengths.

DERIVATION Resolving Power of a Diffraction Grating

Next we show that the resolving power of a diffraction grating is given by

$$R \equiv \frac{\lambda_{av}}{\Delta\lambda} = Nm \quad (36.21)$$

where N is the number of rulings and m is the order number.

Solve the dispersion equation (Eq. 36.19) for $\Delta\theta$.	$D \equiv \dfrac{\Delta\theta}{\Delta\lambda} = \dfrac{m}{d\cos\theta} \quad (36.19)$ $\Delta\theta = \left(\dfrac{m}{d\cos\theta}\right)\Delta\lambda$
If the angular separation $\Delta\theta$ between two lines is just barely visible, the minimum of one line falls on the maximum of the other line, and their separation is their (average) half-width $\Delta\theta_{hw}$ (Eq. 36.16).	$\Delta\theta_{hw} = \Delta\theta$ $\dfrac{\lambda_{av}}{Nd\cos\theta} = \left(\dfrac{m}{d\cos\theta}\right)\Delta\lambda$
Solve for $\lambda_{av}/\Delta\lambda$.	$\dfrac{\lambda_{av}}{\Delta\lambda} = Nm$
Substitute the definition of the resolving power (Eq. 36.20).	$R \equiv \dfrac{\lambda_{av}}{\Delta\lambda} = Nm$ ✓ $\quad (36.21)$

:• **COMMENTS**
According to Equation 36.21, the resolving power is greater (better) if the diffraction has more rulings (if N is large) and if the order number m is also large.

CONCEPT EXERCISE 36.5

Why isn't the spectrum produced by Young's double-slit experiment effective at producing well-separated spectral lines? (For simplicity, think about the first-order separation of a red line and a violet line.)

EXAMPLE 36.5 A Tale of Two Gratings, Revisited

Let's apply the idea of resolving power to the two diffraction gratings in Example 36.4. What resolving power is required to just barely resolve the $m = 1$ Hγ (indigo) and Hδ (violet) lines? Find the resolving power of each grating. In the **CHECK and THINK** step, compare the resolving power of each grating to the resolving power required to just barely separate the lines.

Example continues on page 1172 ▶

INTERPRET and ANTICIPATE
From Figures 36.23 and 36.24, the first grating cannot resolve the lines but the second one can. Now we show this numerically.

SOLVE
Find the average wavelength of the $H\gamma$ and $H\delta$ lines and the difference between their wavelengths.

$$\lambda_{av} = \frac{(410 + 434)\,\text{nm}}{2} = 422\,\text{nm}$$

$$\Delta\lambda = (434 - 410)\,\text{nm} = 24\,\text{nm}$$

The resolving power required to just separate the lines is $R = \lambda_{avg}/\Delta\lambda$ (Eq. 36.20).

$$\frac{\lambda_{av}}{\Delta\lambda} = \frac{422\,\text{nm}}{24\,\text{nm}} = 17.5$$

Now, to find the first-order ($m = 1$) resolving power of the two gratings in Example 36.4, use Equation 36.21. The first grating has 5 rulings and the second grating has 50 rulings.

$$R = Nm \tag{36.21}$$

First grating:
$R_1 = (5)(1) = 5$

Second grating:
$R_2 = (50)(1) = 50$

CHECK and THINK
The resolving power R_1 of the first grating is lower than the required value of 17.5, so the two lines are blended together as shown in Figure 36.23. However, the resolving power R_2 of the second grating is several times higher than the required resolving power, so the lines are well separated as seen in Figure 36.24.

EXAMPLE 36.6 A Third Grating

Let's compare the two diffraction gratings from Example 36.4 to a third grating that has 50 rulings and a distance between rulings $d = 3000$ nm. For the first order ($m = 1$), find the angular positions and half-widths of the four hydrogen lines. Sketch this first-order spectrum and compare it to the spectrum found for the second grating in Example 36.4 (Fig. 36.24).

INTERPRET and ANTICIPATE
Use the same procedure as in Example 36.4. We expect this third grating to have the same resolution as the second grating because both have the same number of rulings ($N = 50$). However, we expect this grating to have a lower dispersion because the separation between the rulings is greater.

SOLVE
The angular positions of the lines come from Equation 36.15. Because we must calculate the angular positions for four different wavelengths, it is helpful to substitute all the other values, including $m = 1$.

$$d \sin\theta = m\lambda \tag{36.15}$$

$$\theta = \sin^{-1}\left(\frac{m\lambda}{d}\right) = \sin^{-1}\left(\frac{\lambda}{3.00 \times 10^{-6}\,\text{m}}\right)$$

Organize the results in a table.

TABLE 36.5 Angular positions of four hydrogen lines.

Line	λ ($\times 10^{-7}$ m)	θ_1 (rad)
$H\alpha$	6.56	0.221
$H\beta$	4.86	0.163
$H\gamma$	4.34	0.145
$H\delta$	4.10	0.137

The next step is to find the widths of these four lines. The half-width comes from Equation 36.16. It depends on the separation d between rulings and the angular position of the lines, so we don't expect to find the same half-widths that we found in Example 36.4.

$$\Delta\theta_{hw} = \frac{\lambda}{Nd\cos\theta} \quad (36.16)$$

$$\Delta\theta_{hw} = \frac{\lambda}{(50)(3.00\times 10^{-6}\,\text{m})\cos\theta} = \frac{\lambda}{(1.50\times 10^{-4}\,\text{m})\cos\theta}$$

Table 36.6 lists the half-widths of the four lines produced by the third grating.

TABLE 36.6 Half-widths for new grating.

Line	$(\Delta\theta_{hw})_1$ (rad)
Hα	0.00448
Hβ	0.00328
Hγ	0.00293
Hδ	0.00277

Sketch the lines produced by the third grating (Fig. 36.25). The positions of the lines are given in Table 36.5, and the thicknesses of the lines come from Table 36.6. We have included the $m=1$ portion of the second grating's spectrum for easy comparison. (To do this, we had to rescale the horizontal axis to match the scale of the third grating's spectrum.)

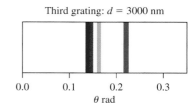

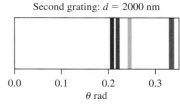

:• **CHECK and THINK**
A lower dispersion for the third grating means that the lines are closer to the central maximum at $\theta = 0$ and closer to one another. So the second grating is the most effective of the three. It has the greatest dispersion and the highest resolution.

FIGURE 36.25 The $m = 1$ hydrogen spectra for the third grating ($d = 3000$ nm) and the second grating ($d = 2000$ nm). Both gratings have 50 rulings.

X-ray Diffraction

So far, we have imagined using an artificially constructed diffraction grating to study light from a natural source such as the Sun. In such a case, we know the separation of adjacent rulings in the grating and we may wish to measure the wavelengths of light produced by the Sun. In this subsection, we see how artificially generated X-rays may be used to find the spacing of atoms in a solid, which acts as a natural three-dimensional diffraction grating.

X-rays are electromagnetic waves with wavelengths between about 1 pm and 1 nm (Section 34-5).

To get an idea of how this works, imagine finding a diffraction grating with no indication of the separation d between rulings. You want to use the diffraction grating, but first you must find d. So, you use a laser of known wavelength to create a diffraction pattern. By measuring the angular positions of the different order lines, you can deduce the ruling separation using $d\sin\theta = m\lambda$ (Eq. 36.15).

Finding the separation between atoms in a solid is somewhat more complicated. First, because the separation between atoms is on the order of 10^{-10} m, visible light has a wavelength about three orders of magnitude too long to create a diffraction pattern. So X-rays are used instead. Second, the solid is three-dimensional and the diffraction pattern is a result of the X-rays reflected from parallel planes (Fig. 36.26).

Consider an X-ray beam reflected from two such parallel planes (Fig. 36.27). The beam makes an angle θ with respect to the planes. The X-ray reflected from the lower plane travels farther than the X-ray reflected from the upper plane. We can find the path-length difference by examining the small highlighted right triangle in

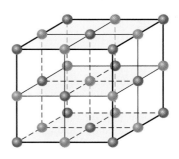

FIGURE 36.26 Atoms in a solid are arranged in a three-dimensional pattern.

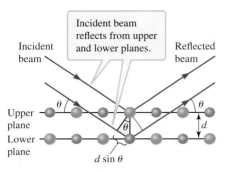

FIGURE 36.27 X-rays are reflected from two parallel planes separated by a distance d.

Figure 36.27. The hypotenuse of this triangle is d, and the extra path length traveled by the incident beam is $d \sin \theta$. Of course, this is also the extra path length traveled by the beam reflected from the lower plane, so the total path-length difference is $2d \sin \theta$. If this extra path length is an integer multiple of the X-ray wavelength, the two reflected beams interfere constructively. We've shown only two adjacent planes, but as long as the planes are parallel and evenly spaced, the same condition produces constructive interference for all planes. The condition for constructive interference is known as **Bragg's law** and is given by

$$2d \sin \theta = m\lambda \quad (m = 1, 2, 3, \dots) \tag{36.22}$$

The spacing d between atomic planes is found by using X-rays of known wavelength and measuring the angular positions θ for constructive interference spots observed at order m. To get the three-dimensional structure, the solid must be rotated to find the spacing between other parallel planes.

Bragg's law is named for William Lawrence Bragg (1890–1971), who shared the Nobel Prize in physics with his father William Henry Bragg (1862–1942) in 1915 for formulating the condition for constructive interference (Eq. 36.22). The two men continued to work together to find the structure of other substances, including sodium chloride—table salt.

36-6 Case Study: Michelson's Interferometer

The first American to win a Nobel Prize in science was Albert Abraham Michelson (1852–1931), who was born in a part of Prussia that is now in Poland and moved to the United States at age 2. Michelson graduated in 1873 from the United States Naval Academy and subsequently became a teacher there, where he began the experiments that led to his prize-winning work.

By the time Michelson was at the academy, Young had already demonstrated that light is a wave and Maxwell had derived the electromagnetic wave equation. So, like you, Michelson was taught that light is a wave. In Michelson's time, however, scientists thought that all waves must propagate in a medium. They called the medium in which light propagates the **ether**, reasoned to be weightless, (nearly) incompressible, frictionless, transparent, and omnipresent—permeating all space and matter. Physicists in the 1800s also reasoned that there was no chemical way to detect the ether, meaning it couldn't be captured in a bottle and normal chemical experiments could not determine its properties. One of the major differences between science and religion is that in science, something exists only if it is detectable. So, for ether to exist, there must be some way to detect it and determine its physical characteristics. Michelson designed an experiment to measure the speed of the Earth with respect to the ether. What he found instead is that the ether does not exist.

Swimming Race Analogy

Michelson designed an instrument that we now call a **Michelson interferometer**, intended to allow him and his collaborator Edward Morley to measure the speed of the Earth through the ether. Michelson thought the kinematics of relative motion (Sections 4-7 and 4-8) could be applied to the situation of the Earth moving relative to the ether. He explained his experiment using an analogy with swimmers racing in a river, so we'll use this analogy too.

Figure 36.28 shows a river flowing at constant uniform speed v to the right with respect to the river's banks. In this river, Simon and Greta are about to have a swimming race. Both start in the river at point S near one bank. Simon swims to point 1 on the opposite side of the river, while Greta swims to point 2 downstream. Then both swimmers reverse course and return to their starting position at S. The distance L from the starting point S to point 1 equals the distance from S to point 2.

Now, here is the unusual thing about this race: Both swimmers swim at the same speed c with respect to the river. (For the moment, c is not the speed of light, just the speed of the swimmers.) So, if the river were not flowing (if $v = 0$), the race would be a tie; both swimmers would arrive back at S at the same moment, and each swimmer's time would be $\Delta t = 2L/c$.

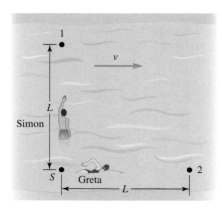

FIGURE 36.28 The river flows to the right at a uniform and constant speed v. The race begins and ends at point S. Simon swims to point 1 and returns to S. Greta swims to point 2 and returns to S. The distance L from S to 1 equals the distance from S to 2. Each swimmer's speed with respect to the water is c. Simon always wins as long as $0 < v < c$.

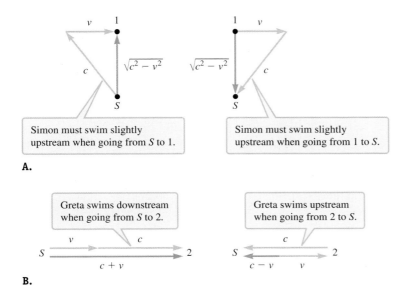

FIGURE 36.29 Each swimmer's speed with respect to the water is c. (The magnitude of each vector is written next to each vector.) **A.** Simon's velocity with respect to the river's banks must point directly toward 1 on his way out and directly toward S on his way back. So, Simon must swim slightly upstream, whether he is swimming to or from the starting position S. As a result, his speed relative to the banks is same whether he is swimming toward or away from S. **B.** Greta swims with the current when she goes from S to 2, so her speed relative to the banks equals her speed relative to the water plus the speed of the water. When she swims from 2 back to S, she swims against the current, so her speed relative to the banks equals her speed relative to the water minus the speed of the water.

However, if the river is moving at speed v, Simon always wins the race. To see why this is true, let's derive each swimmer's travel time. Simon must swim slightly upstream in order to compensate for the river's motion to the right, whether he is swimming toward or away from the starting position at S (Fig. 36.29A). We can use the Pythagorean theorem to determine Simon's speed relative to the banks:

$$v_{\text{Simon}} = \sqrt{c^2 - v^2} \tag{36.23}$$

where v is the speed of the water with respect to the banks.

By contrast, Greta's speed relative to the banks depends on whether she is swimming away from or toward the starting position. Her speed relative to the water is c in either case (Fig. 36.29B). But when she swims with the current away from S, her speed relative to the banks is higher because she is helped by the current. When she swims back toward the starting position, her speed relative to the banks is lower. So, when Greta swims from S to 2 (downstream), her speed with respect to the banks is given by

$$(v_{\text{Greta}})_{\text{down}} = c + v \tag{36.24}$$

When she swims from 2 toward S (upstream), her speed with respect to the banks is given by

$$(v_{\text{Greta}})_{\text{up}} = c - v \tag{36.25}$$

The swimmer who wins the race has the shortest round-trip time, which we need to find. Simon's speed relative to the banks is the same whether he is swimming toward or away from the starting point. The round-trip distance he travels (relative to the banks) is $2L$. So, his round-trip time is his round-trip distance divided by his speed v_{Simon} relative to the banks:

$$\Delta t_{\text{Simon}} = \frac{2L}{v_{\text{Simon}}} \tag{36.26}$$

Substitute Equation 36.23 into Equation 36.26:

$$\Delta t_{\text{Simon}} = \frac{2L}{\sqrt{c^2 - v^2}} \tag{36.27}$$

Greta's speed relative to the banks is faster when she is moving downstream, so her time to go from S to 2 is shorter than her time on the return trip. We must take this into account when we calculate her round-trip time. Her downstream time is the distance from S to 2 divided by her downstream speed:

$$\Delta t_{\text{down}} = \frac{L}{(v_{\text{Greta}})_{\text{down}}} = \frac{L}{c + v} \tag{36.28}$$

Her upstream time is the distance from 2 to S divided by her upstream speed:

$$\Delta t_{\text{up}} = \frac{L}{(v_{\text{Greta}})_{\text{up}}} = \frac{L}{c - v} \tag{36.29}$$

So, her total round-trip time is the sum of Equations 36.28 and 36.29:

$$\Delta t_{\text{Greta}} = \Delta t_{\text{down}} + \Delta t_{\text{up}} = \frac{L}{c + v} + \frac{L}{c - v}$$

$$\Delta t_{\text{Greta}} = \frac{2Lc}{c^2 - v^2} \tag{36.30}$$

One way to show that Simon always wins is to find an expression for the ratio of their round-trip times (Eq. 36.30 divided by Eq. 36.27):

$$\frac{\Delta t_{\text{Greta}}}{\Delta t_{\text{Simon}}} = \left(\frac{2Lc}{c^2 - v^2}\right)\left(\frac{\sqrt{c^2 - v^2}}{2L}\right)$$

$$\frac{\Delta t_{\text{Greta}}}{\Delta t_{\text{Simon}}} = \frac{1}{\sqrt{1 - (v/c)^2}} \tag{36.31}$$

where v is the speed of the river with respect to the banks and c is the speed of each swimmer with respect to the water. Now examine Equation 36.31 to see who wins. First, if $v = c$ (the swimmers' speed equals the river's speed), the ratio (Eq. 36.31) goes to infinity. Second, if $v > c$ (the swimmers are slower than the river), the ratio is an imaginary number. So we conclude that the river must be slower than the swimmers: $v < c$. Third, if the river does not move with respect to the banks ($v = 0$), the ratio is 1 and there is a tie as we already predicted. Finally, as long as $0 < v < c$, then $\Delta t_{\text{Greta}} > \Delta t_{\text{Simon}}$, and Greta's round-trip time is greater than Simon's so that Simon always wins.

This swimming analogy is meant to help us understand the light measured by Michelson's interferometer. There is one more useful idea we need from the analogy: Both swimmers have the same speed c with respect to the water, so if they were to swim the same distance in still water, their race would always be a tie. If Simon were to win in still water, you would conclude that he had a shorter distance to swim. So, another way to think about the race in moving water (Fig. 36.28) is to say that the moving water is *equivalent* to Simon (the perpendicular swimmer) having a shorter distance to swim than Greta (the parallel swimmer).

The Michelson–Morley Experiment

With the swimming analogy in mind, we return to light, the ether, and Michelson's interferometer. Michelson—like other scientists at the time—believed that the ether permeated all space, so that objects such as the Earth moved through the ether. His goal was to measure the speed of the Earth relative to the ether, which was thought to be motionless. If the Earth moved through the ether, we and everything else on the Earth would experience the ether as a river moving in the direction opposite to the Earth's motion. In our analogy, the ether is like the river flowing at constant speed v. To design his interferometer, Michelson needed a rough idea of the Earth's speed through the ether. He estimated that the speed should be similar to the Earth's orbital velocity around the Sun, or $v \approx 30$ km/s.

Figure 36.30 shows the basic design of Michelson's interferometer. We arbitrarily assume that the ether moves to the right. Light enters the interferometer from the source on the left. The light first encounters a glass plate with a half-silvered back, labeled S. This plate is called a **beam splitter** because it allows about half the light to pass through but reflects the other half. The splitter is tilted at 45° with respect to the incoming beam. The reflected light travels to the mirror at point 1, while the transmitted light travels to the mirror at point 2. In the swimming

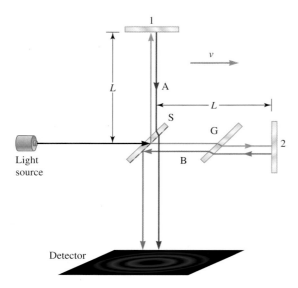

FIGURE 36.30 The ether flows to the right at a uniform and constant speed v. A half-silvered mirror S splits the light, so that half of the light travels along path A and the other half travels along path B. The light on path A travels to the mirror at 1, back through S, and to the detector. The light on path B travels to the mirror at 2. On its way both to and from 2, it passes through the glass slab G. When it returns to S, it is reflected toward the detector. The distance L from S to 1 equals the distance from S to 2. The speed of light with respect to the ether is c, and the light that follows A always gets to the detector before the light that travels along B. The result is an interference pattern produced at the detector. (The colors used are only for clarity; the light does not change color. Rays are slightly offset to avoid overlap.)

analogy, the splitter is at the starting position. The light is like the two swimmers racing from point S. The mirrors at points 1 and 2 are like the places where the swimmers turn around and head back toward S. The two paths labeled A and B are perpendicular to each other.

Let's follow the light along path A. After it is reflected from the mirror at point 1, it returns to S. Half of the light is reflected by the splitter, and half passes straight through. We are not interested in the reflected light in this case, so it is not shown. Instead, we follow the light that passes straight through S and goes to the detector at the bottom of the figure.

Next, follow the light along path B. After it is reflected from the mirror at 2, it returns to S. Half of that light passes straight through the splitter and back toward the light source. We are not interested in this light, so it is not shown. The other half of the light is reflected from the splitter and goes to the detector.

A glass slab labeled G is located along path B to make sure that the light passes through the same total thickness of glass along both paths. Before the incoming light is split, it must travel through the glass slab of the splitter in order to encounter the half-silvered surface on the back side. The light that travels along path A must pass back through that glass twice on its way to 1, and it must also pass back through the glass once on its way to the detector. However, the light that travels along path B passes through the splitter's glass only once. To make up for path A's two extra passes through the splitter's glass, a second glass slab G is inserted in path B with the same thickness and composition as the glass in the splitter. So, like the light that travels along path A, the light along path B travels through glass three times. (Slab G ensures that the race is fair!)

As in the case of the two swimmers, Michelson predicted that the light traveling on the perpendicular path (path A) always would arrive at the detector first. We can apply Equation 36.31 to Michelson's experiment by replacing Δt_{Simon} with Δt_A and Δt_{Greta} with Δt_B:

$$\frac{\Delta t_B}{\Delta t_A} = \frac{1}{\sqrt{1 - (v/c)^2}} \tag{36.32}$$

Now c is the speed of light, which Michelson had carefully measured in earlier experiments (at the Naval Academy). If he could also measure the ratio of the times $\Delta t_B / \Delta t_A$ that light took to travel over the two paths, he could solve for v, the speed of the Earth relative to the ether. Of course, the speed of light is very high compared to the anticipated speed of the Earth through the ether (30 km/s), so Michelson expected the ratio to be very close to 1:

$$\frac{\Delta t_B}{\Delta t_A} = \frac{1}{\sqrt{1 - \left(\frac{3 \times 10^4 \text{ m/s}}{3 \times 10^8 \text{ m/s}}\right)^2}} = 1.000000005 \tag{36.33}$$

The brilliance of Michelson's experimental design is that he did not need to measure the time delay directly. Another way to think about the time delay is in terms of a path-length difference. A shorter travel time for light along path A is equivalent to a shorter *effective* path length. As in Young's experiment, this path-length difference should result in a phase difference between the two beams of light that arrive at the detector, resulting in an interference pattern. If the interference pattern resembles a bull's eye and the path-length difference is $\lambda/2$, there should be destructive interference at the center of the pattern as shown by the dark spot in Figure 36.30.

We arbitrarily decided that the ether moves to the right in Figure 36.30, parallel to path B. That arbitrary decision led to the conclusion that path A is effectively shorter than path B. In the Michelson–Morley experiment, the entire interferometer was placed on a large piece of sandstone that floated in mercury (Fig. 36.31). This setup reduced errors introduced by the vibrations of moving vehicles and other nearby machinery and also allowed the experimenters to rotate the interferometer, thus changing the effective path lengths. For example, imagine that the two paths are turned so that they are both at 45° with respect to the ether's velocity. Then the two paths would have the same effective length and both beams would arrive at the detector at the same instant. As a result, there would be a bright spot at the center of the interference pattern due to the constructive interference of the two beams, instead of the dark spot shown in Figure 36.30.

By measuring the shift in the fringes of the interference pattern as the interferometer was rotated, Michelson hoped to measure the speed of the Earth through the ether. Let's see what he expected to find. The maximum path-length difference Δd occurs when one path is parallel to the ether's velocity as in Figure 36.30. Then

$$\Delta d = d_B - d_A \qquad (36.34)$$

It is helpful to rewrite Equation 36.34 as

$$\Delta d = d_A\left(\frac{d_B}{d_A} - 1\right) \qquad (36.35)$$

The ratio of the effective path lengths is the same as the ratio of the times:

$$\frac{d_B}{d_A} = \frac{c\Delta t_B}{c\Delta t_A} = \frac{\Delta t_B}{\Delta t_A} \qquad (36.36)$$

Substitute Equation 36.36 into Equation 36.35:

$$\Delta d = d_A\left(\frac{\Delta t_B}{\Delta t_A} - 1\right)$$

In the Michelson–Morley experiment, $d_A \approx 22$ m and $\Delta t_B/\Delta t_A$ is given by Equation 36.33, so the effective path-length difference is

$$\Delta d = (22 \text{ m})(1.000000005 - 1)$$
$$\Delta d = 1.1 \times 10^{-7} \text{ m} = 110 \text{ nm} \qquad (36.37)$$

As mentioned previously, Michelson and Morley rotated their interferometer. Suppose path A is originally perpendicular to the ether's velocity and then the interferometer is rotated by 90° so that path B is perpendicular. In effect, this is equivalent to lengthening path A and shortening path B by the same amount. By rotating the interferometer, the experimenters could double the difference between the two path lengths so that the pattern would shift by

$$m = \frac{2\Delta d}{\lambda} \qquad (36.38)$$

fringes. Because the Michelson–Morley experiment used visible light with a wavelength of 550 nm, they expected the pattern to shift by

$$m = \frac{2(110 \text{ nm})}{(550 \text{ nm})} = 0.4 \text{ fringe}$$

How can an interference pattern be observed to shift by a fraction of a fringe? Michelson's detector included a telescope (Fig. 36.31) that could detect shifts of as little as 0.01 fringe, so the experimenters should have easily been able to detect the shift they expected. Michelson and Morley ran their experiment at several different times of day and night, and several different times of year. The experiment was even tried at different locations on the Earth. But they never detected any shift in the pattern.

Michelson and Morley failed to detect the Earth's motion with respect to the ether. This is considered the most important failed experiment in history because it led to a major change in our understanding. Because the experiment found no shift in the interference pattern, it demonstrated that the Earth does not move with respect to the ether. In Chapter 39, we will argue that this experiment also demonstrated that the ether does not exist. Instead, light propagates in a vacuum. Although Albert Einstein claimed that he had not heard of Michelson and Morley's experiment before he developed his theory of special relativity, their experiment provides an important physical basis for Einstein's work.

Einstein and Michelson eventually became friends, but Michelson never felt comfortable with his failure. He never warmed to Einstein's theory, and when he died in 1931, he still believed that the ether existed.

FIGURE 36.31 In the Michelson–Morley experiment, the interferometer was placed on a stone slab floated on mercury so that it could be rotated. The apparatus of the actual interferometer includes a telescope for viewing the interference pattern.

SUMMARY

❶ Underlying Principles

No new principles are introduced in this chapter.

✪ Major Concepts

1. According to **Rayleigh's criterion**, two sources are barely resolved when the central maximum of one diffraction pattern is centered on the first minimum of the other diffraction pattern (Fig. 36.4).
2. For a grating, **dispersion** is a measure of the angular separation between the lines (maxima):

$$D \equiv \frac{\Delta\theta}{\Delta\lambda} = \frac{m}{d\cos\theta} \quad (36.18 \text{ and } 36.19)$$

where m is the order number, θ is the angular position of the corresponding line, and d is the separation between the rulings.

3. For a grating, the **resolving power** is a measure of the thickness of the lines:

$$R \equiv \frac{\lambda_{av}}{\Delta\lambda} = Nm \quad (36.21)$$

where λ_{av} is the average wavelength of two lines that are just barely separated, $\Delta\lambda$ is the difference in their wavelengths, N is the number of rulings, and m is the order number.

◐ Special Cases

1. The **resolution** of an instrument with a **circular aperture** is given by

$$\sin\theta_{min} = 1.22\frac{\lambda}{d} \quad (36.3)$$

2. When a **thin film** of thickness w is surrounded by a medium with a lower index of refraction ($n_f > n_i$), the condition for **constructive interference** is

$$2w = \left(m + \frac{1}{2}\right)\frac{\lambda_0}{n_f} \quad (m = 0, 1, 2, 3, \ldots) \quad (36.12)$$

and the condition for **destructive interference** is

$$2w = m\frac{\lambda_0}{n_f} \quad (m = 0, 1, 2, 3, \ldots) \quad (36.14)$$

3. For a **diffraction grating**, the angular **position** θ of the lines is given by

$$d\sin\theta = m\lambda \quad (m = 0, \pm 1, \pm 2, \pm 3, \ldots) \quad (36.15)$$

and each line's **half-width** is given by

$$\Delta\theta_{hw} = \frac{\lambda}{Nd\cos\theta} \quad (36.16)$$

PROBLEMS AND QUESTIONS

A = algebraic   C = conceptual   E = estimation   G = graphical

36-2 Circular Aperture Diffraction

1. **C** Many circular apertures are adjustable, such as the pupil of your eye or the shutter of a camera. Describe the change in the diffraction pattern as such an aperture decreases in size.
2. **C** Many of the images we regularly look at are digitized; that is, they are made up of many small individual dots. For example, a color laser printer produces images by printing many dots of various colors. If the printer is of high quality, we do not see the individual dots. Why not?
3. **N** The "hydrogen line" at 1420.4 MHz corresponds to the natural frequency of neutral hydrogen atoms and plays an important role in radio astronomy. What size dish is required so that a radio telescope receives this frequency with an angular resolution of 0.0500°?
4. **E** Post-Impressionist Georges Seurat is famous for a painting technique known as pointillism. His paintings are made up of different-colored dots, and each dot is roughly 2 mm in diameter. When viewed from far enough away, the dots blend together and only large-scale images are seen. Assuming that the human pupil is about 2.5 mm in diameter, determine an approximate minimum distance you need to stand from one of Seurat's paintings to see the dots blended together.

5. **E** Estimate the diffraction-limited resolution of the radio telescope in Arecibo, Puerto Rico. The radio telescope's diameter is 305 m, and it operates at a frequency of 300.0 MHz.

Problems 6 and 9 are paired.

6. **N** The resolution of a telescope on the Earth is not limited by diffraction. Instead, the Earth's atmosphere causes blurriness such that the best resolution on Earth is about 0.50 arcsec. The Hubble Space Telescope was placed into orbit to avoid such blurriness. Its resolution is determined by diffraction. The Hubble's mirror is 2.4 m in diameter. Estimate its diffraction-limited resolution. *Hint:* Use an estimate of 6.0×10^{-7} m (or 600 nm) for the wavelength of light in this calculation.

7. **N** A microscope with an objective lens of diameter 7.50 mm is used to view samples in light of wavelength 500.0 nm.
   a. What is the limiting resolution of the microscope with this objective lens?
   b. If the lights in the laboratory could be tuned to the range of visible wavelengths, which wavelength would produce the smallest angle of resolution?
   c. What would the angular resolution of the microscope be at the wavelength found in part (b)?
   d. What would the angular resolution of the microscope be at the wavelength found in part (b) if the microscope was immersed in water?

8. **C** In a particular microscope, you are just able to resolve two blood cells using a certain aperture. How is your ability to resolve the two cells altered if the diameter of the aperture is increased? What if it is decreased?

9. **N** Use the results from Problem 6 to find the distance between two objects on the Moon that can just barely be resolved by the Hubble Space Telescope.

10. **N** Later in this textbook, you will explore the intriguing idea that matter can exhibit wavelike behavior when moving at velocities near the speed of light. One example where this is used is in the imaging of materials in an electron microscope. A common wavelength for the electrons in a transmission electron microscope is 2.5×10^{-12} m. The diameter of the apertures is 5.0 μm. What is the diffraction-limited resolution of this microscope?

11. **N** Suppose that in a particular microscope, a diffraction-limited resolution of 0.25° is necessary to resolve two plant cells that are part of a specimen. If the microscope uses light with a wavelength of 525 nm, what is the minimum aperture diameter so that the cells can be resolved?

36-3 Thin-Film Interference

12. **A** For a thin film such as a soap bubble surrounded by a medium with a smaller index of refraction such as air, find the conditions for constructive and destructive interference produced by the transmitted light.

13. **N** The nonreflective coating on a camera lens with an index of refraction of 1.27 is designed to minimize the reflection of 589-nm light. If the lens glass has an index of refraction of 1.54, what is the minimum thickness of the coating that will accomplish this task?

14. **N** When you spread oil ($n_{oil} = 1.50$) on water ($n_{water} = 1.33$) and have white light incident from above, you might observe different reflected colors that depend on the thickness of the oil on the water at each point. Consider the wavelength of red light (750.0 nm) and the wavelength of violet light (380.0 nm), covering the visible spectrum. What is the minimum possible thickness of oil that would give rise to seeing **a.** the red light reflected and **b.** the violet light reflected?

15. Sometimes a nonreflective coating is applied to a lens, such as a camera lens. The coating has an index of refraction between the index of air and the index of the lens. The coating cancels the reflections of one particular wavelength of the incident light. Usually, it cancels green-yellow light ($\lambda = 550.0$ nm) in the middle of the visible spectrum.
   a. **A** Assuming the light is incident perpendicular to the lens surface, what is the minimum thickness of the coating in terms of the wavelength of light in that coating?
   b. **N** If the coating's index of refraction is 1.38, what should be the minimum thickness of the coating?

16. **C** Just outside of Esther's bedroom window is an illuminated sign. In red (716 nm) and blue (408 nm) letters it proclaims "Danger Ahead." Tired of light from this sign, Esther coats her glass ($n = 1.50$) window on the inside with a thin film ($n_f = 1.39$). Now, when she looks straight through her window, the sign reads "anger head." What color was emitted by the missing letters? Assume the film is the minimum thickness required to prevent the colored light transmission.

17. **N** An oil slick on water displays a variety of iridescent colors due to interference effects. Assuming the film has the minimum thickness to produce the colors observed, what thickness of the oil film will appear green ($\lambda = 550.0$ nm) or red ($\lambda = 700.0$ nm)? The index of refraction of the oil is 1.47.

Problems 18 and 19 are paired.

18. **N** A thin film ($n_f = 1.29$) coats a glass ($n = 1.56$) window on the inside to prevent the transmission of red light at 735 nm straight through the window. What is the minimum thickness of the film?

19. **N** A thin film ($n_f = 1.29$) coats a glass ($n = 1.56$) camera lens on the outside to prevent the reflection of red light at 735 nm from rays that are perpendicular to the surface. What is the minimum thickness of the film?

Problems 20 and 21 are paired.

20. **C** When you spread oil ($n_{oil} = 1.50$) on water ($n_{water} = 1.33$) and have white light incident from above, you might observe different reflected colors that depend on the thickness of the oil on the water at each point. Suppose we find a location where we see green light reflected from the thin film of oil. If we slowly increase the thickness of oil at that location, what happens to the observed reflected color?

21. **C** In Problem 20, what happens to the observed reflected color if we slowly decrease (rather than increase) the thickness of the oil at that location?

22. **E** Police measure the speed of moving vehicles with radar guns. A radar gun sends out radio waves with a typical frequency of 10 GHz. The radio waves reflect from vehicles. By measuring the Doppler shift of the reflected radio waves (Section 17-9), the radar detector measures the speed of the vehicle. A physics student wants to paint his car with a nonreflective coating that will prevent radar detectors from measuring his speed. Estimate the thickness of the coating, and describe your assumptions.

23. **N** An oil slick of thickness 325 nm and index of refraction $n_{oil} = 1.50$ forms on the surface of a still pond ($n_{water} = 1.33$). When it is viewed straight from above, what is the wavelength of visible light that is **a.** most strongly reflected and **b.** most strongly transmitted?

24. **N** An oil tanker bound for the Suez Canal runs aground in the Red Sea, spilling sweet light crude oil that forms a uniform thin film on the surface of the sea. After ten hours of weathering, the index of refraction of the crude oil has increased to 1.28, and the index of refraction of seawater at the tanker's location is 1.341. What is the minimum thickness of the oil film that will result in the strong reflection of normally incident 450-nm blue light?

36-4 Diffraction Gratings

25. **N** Light of wavelength 566 nm is incident on a grating. Its third-order maximum is at 15.7°. What is the density of rulings on this grating?

26. **N** A diffraction grating with a ruling separation of 3.00 μm is illuminated with light of wavelength 525 nm. What is the highest-order maximum visible?

27. **N** A diffraction grating is illuminated with a krypton laser operating at 416 nm. What is the spacing between rulings on the grating if the first-order maximum is observed at 18.8°?

28. **N** A grating has 3330 rulings per centimeter. When monochromatic light is incident on this grating, a fourth-order line is seen at 30.0°. What is the wavelength of the light?

29. **N** A diffraction grating is illuminated with green light of wavelength 530.0 nm, and a second-order maximum is observed at 27.5°. **a.** How many slits per centimeter does the diffraction grating have? **b.** What is the number of maxima that can be seen in the diffraction pattern?

30. **N** A diffraction grating is 5.00 cm wide and has 515 rulings per millimeter. If light of **a.** 450.0 nm and **b.** 650.0 nm is incident on the grating, what is the half-width of the central maximum in each case?

31. **N** A light source emits a mixture of wavelengths from 450.0 nm to 600.0 nm. When the light passes through a diffraction grating, two adjacent spectra barely overlap at an angle of 30.0°. How many rulings per meter are on the grating?

32. **A** A diffraction grating of length ℓ has N rulings. When the grating is illuminated with light of wavelength λ, the $m = 3$ maximum is diffracted at angle $\theta = 90.0°$. Find an expression for the wavelength of the light in terms of ℓ and N.

33. **N** A diffraction grating with 1.60×10^3 rulings per centimeter spreads white light into its spectral components. What is the angle at which the second-order maximum appears for green light with a wavelength of 530.0 nm?

34. **N** Light is incident on a grating. The third-order line is observed at 65°. At what angular position is the first-order line?

35. **N** A diffraction grating with 2.65×10^3 rulings per centimeter is illuminated by white light. **a.** How many orders of the entire visible light spectrum, from wavelength 400.0 nm to 780.0 nm, are visible? **b.** How many orders of violet light of wavelength 400.0 nm are visible?

36. **N** A diffraction grating with 3.65×10^3 rulings per centimeter is illuminated by light of wavelength 632.8 nm from a helium-neon laser. **a.** At what angle does the third-order maximum occur? **b.** At what angle would the third-order maximum occur if the experiment was carried out underwater? **c.** What is the relationship between the diffracted rays in parts (a) and (b)?

36-5 Dispersion and Resolving Power of Gratings

Problems 37 and 38 are paired.

37. **N** A grating has 3.200×10^4 rulings and is 5.00 cm wide. Light of wavelength 555 nm is incident on the grating. What is its dispersion for the first- and second-order lines?

38. **N** A grating has 3.200×10^4 rulings and is 5.00 cm wide. What is its resolving power for the first- and second-order lines?

Problems 39 and 40 are paired.

39. **N** A diffraction grating is exposed to light from a source that consists of two different wavelengths: 575 nm and 585 nm. What is the minimum number of rulings necessary to be able to resolve the **a.** $m = 1$ lines, **b.** $m = 2$ lines, and **c.** $m = 3$ lines?

40. **C** A diffraction grating is exposed to light from a source that consists of two different wavelengths: 575 nm and 585 nm. Explain why the resolving power of the grating must be higher to resolve the $m = 1$ lines versus the $m = 3$ lines.

41. **N** A diffraction grating with 3.52×10^3 rulings/cm and length 2.50 cm is illuminated by light with a wavelength of 425 nm. What is the dispersion of the first-order maximum?

Problems 42 and 43 are paired.

42. **A** A diffraction grating with n rulings per unit length and total length ℓ is illuminated by light with several similar wavelengths. Find an expression for the resolving power of this diffraction grating for the first-order maxima of two similar wavelengths of light in terms of n and ℓ.

43. **N** A diffraction grating with 2.00×10^3 rulings/cm and total length 3.00 cm is illuminated by light with several similar wavelengths. Find the resolving power of this diffraction grating for the first-order maxima of two similar wavelengths of light.

44. **N** A diffraction grating with 1.54×10^3 rulings/cm and length 3.10 cm is illuminated by light with a wavelength of 310.0 nm. What is the dispersion of the first three maxima that fall on the screen?

36-6 Case Study: Michelson's Interferometer

45. **N** **CASE STUDY** Michelson's interferometer played an important role in improving our understanding of light, and it has many practical uses today. For example, it may be used to measure distances precisely. Suppose the mirror labeled 1 in Figure 36.30 (page 1176) is movable. If the laser light has a wavelength of 632.5 nm, how many fringes will pass across the detector if mirror 1 is moved just 1.000 mm? If you can easily detect the passage of just one fringe, how accurately can you measure the displacement of the mirror?

46. **N** **CASE STUDY** Michelson's interferometer played an important role in improving our understanding of light, and it has many practical uses today. For example, it may be used to measure precisely the indices of refraction of various media. Suppose an initially evacuated tube of length 6.50 cm is placed in one of the paths in Michelson's interferometer. The tube is then filled slowly with a gas of an unknown index of refraction, and as a result 42 fringes shift. The laser light has a wavelength of 608.5 nm. You will find that the index of refraction is very close to 1. Give $n - 1$ to three significant figures.

47. **N** **CASE STUDY** The interferometer uses interference effects to produce an observable change with a small change in the optical path length of one path relative to the other. Michelson and Morley also used a telescope to zoom in to observe phase shifts of only 1% of the width of a fringe. **a.** In Figure 36.30, how far would mirror 2 need to move to shift the interference pattern by 1.0% of the width of a fringe, when used with light of wavelength 500.0 nm? **b.** If Michelson and Morley tried to measure the difference in travel times for light traveling along paths A and B, what time resolution would they need? That is, how much extra time is needed for the light to cover the distance found in part (a)? Assume the speed of light is 3.0×10^8 m/s.

48. **N** **CASE STUDY** The Michelson–Morley interferometer can also be used to determine the wavelength of light. Given light with an unknown wavelength, the position of one of the mirrors is shifted by 1.000 mm and the interference pattern shifts by exactly 4500 fringes. What is the wavelength of the light?

General Problems

Problems 49 and 50 are paired.

49. **C** Optical flats are flat pieces of glass used to determine the flatness of other optical components. They are placed at an angle above the component as shown in Figure P36.49A, and

monochromatic light is incident and observed from above, leading to interference fringes. Parts B and C of Figure P36.49 show the results of tests on two optical components. Which of the two is more flat? Explain.

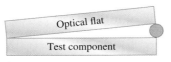

A.

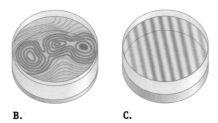

B. C.

FIGURE P36.49 Problems 49 and 50.

50. **E** Optical flats are flat pieces of glass used to determine the flatness of other optical components. They are placed at an angle above the component as shown in Figure P36.49A, and monochromatic light is incident and observed from above, leading to interference fringes. Figure P36.49C shows the results of one of these tests. What is the approximate difference in the gap thickness between the left and right sides of the optical flat and the component? Is it possible to determine from this figure alone which side has the greater gap thickness (left or right)?

Problems 51 and 52 are paired.

51. **N** A thin film of material is coated on a substrate to protect the substrate while it is immersed in oil (Fig. P36.51). Light incident in the oil with a wavelength of 485 nm is reflected from the thin film. What is the minimum thickness of the thin film that results in the reflection, given the incident wavelength? Assume $n_{oil} = 1.70$, $n_{film} = 1.40$, and $n_{substrate} = 1.25$.

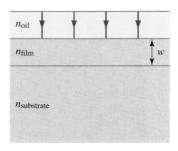

FIGURE P36.51 Problems 51 and 52.

52. **N** A thin film of material is coated on a substrate to protect the substrate while it is immersed in oil (Fig. P36.51). Light incident in the oil with a wavelength of 485 nm is transmitted through the thin film, not reflected. What is the minimum thickness of the thin film that results in the transmission, given the incident wavelength? Assume $n_{oil} = 1.70$, $n_{film} = 1.40$, and $n_{substrate} = 1.25$.

53. **N** Figure P36.53 shows two thin glass plates separated by a wire with a square cross section of side length w, forming an air wedge between the plates. What is the edge length w of the wire if 42 dark fringes are observed from above when 589-nm light strikes the wedge at normal incidence?

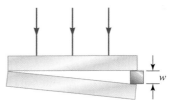

FIGURE P36.53

54. **N** Viewed from above, a thin film of motor oil with index of refraction $n = 1.28$ on a wet asphalt surface (water, $n = 1.33$) strongly reflects light of 595 nm wavelength and does not reflect any light at 476 nm. What is the minimum thickness of the film of motor oil?

55. **N** Newton's rings, discovered by Isaac Newton, are an interference pattern of dark and bright rings formed because of the air gap of increasing thickness w between a spherical surface and an adjoining flat surface (Example 36.2). Figure P36.55 shows a Newton's rings apparatus with a plano-convex lens with radius of curvature R atop a flat glass slab, both with index of refraction n_{glass}. In the configuration shown, the seventh bright fringe has a radius of 1.10 cm. After the air gap in the apparatus is filled with an unknown liquid, the radius of the seventh fringe decreases to 0.968 cm. What is the index of refraction of the unknown liquid?

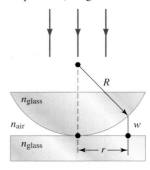

FIGURE P36.55

56. **C** A diffraction grating of length ℓ has N rulings. When the grating is illuminated with light of wavelength λ, only a certain number of maxima appear on the screen. How will the number of visible maxima change if the number of rulings is decreased or increased, while the length of the grating remains the same? Explain.

57. **N** What is the radius of the beam of an argon laser with wavelength 454.6 nm when viewed 50.0 km away from the laser if the laser's aperture has a radius of 3.00 mm?

58. **N** A diffraction grating with a length of 5.65 cm consists of 185 rulings and is illuminated with light that consists of several similar wavelengths. One of the lines corresponds to light with a wavelength of 489 nm. If this line is a first-order maximum, what are the wavelengths of the nearest first-order maxima that could be resolved, assuming they are present in the light?

59. **N** A diffraction grating with 428 rulings per millimeter is illuminated by light from a xenon ion laser. What is the wavelength of the laser light being used if the separation between the central and second-order maxima on a screen placed 2.10 m away from the grating is 67.5 cm?

60. **N** How many rulings must a diffraction grating have if it is just to resolve the sodium doublet (589.592 nm and 588.995 nm) in the second-order spectrum?

61. **N** Freeway exit signs are typically constructed with letters 25.0 cm in height and placed a similar distance apart. Assuming the diameter of the pupil is 3.00 mm in daylight and the sign is observed with light of wavelength 589 nm, at what distance can a driver distinguish the individual letters on a freeway sign?

62. **N** White light is incident on a diffraction grating that has 2.85×10^3 rulings per centimeter, producing first-order maxima at 6.55°, 8.75°, and 11.2°.
    a. What are the wavelengths responsible for these three first-order maxima?
    b. At which angles do these wavelengths produce second-order maxima?

63. **N** X-rays incident on a crystal with planes of atoms located 0.378 nm apart produce a diffraction pattern in which a first-order maximum is observed at an angle of 14.2°. **a.** What is the wavelength of the X-rays incident on the crystal? **b.** How many orders are visible in the diffraction pattern?

64. **N** Later in this book, you will explore the intriguing idea that matter can exhibit wavelike behavior when moving at velocities near the speed of light. One example where this is used is in the imaging of materials in an electron microscope. A common wavelength for the electrons in a transmission electron microscope is 2.5×10^{-12} m. Assume the diameter of the aperture is 500.0 μm. What is the required distance between the aperture and the specimen if we want to resolve two atoms separated by approximately 1.0×10^{-9} m, assuming the resolution is not dependent on other factors?

65. **A** Confirm Equation 36.16,

$$\Delta \theta_{hw} = \frac{\lambda}{Nd \cos \theta}$$

by showing that the half-width of the zeroth-order line produced by a diffraction grating is given by $\Delta \theta_{hw} = \lambda/Nd$. Do not begin with Equation 36.16, but instead use our work finding the first dark minimum produced by single-slit diffraction as a guide (Section 35-4). The angular position of the first dark minimum produced by a single slit is like the angular position of the dark fringes on either side of the zeroth-order line from a grating.

66. **A** Show that the dispersion of a diffraction grating is given by Equation 36.19,

$$D \equiv \frac{\Delta \theta}{\Delta \lambda} = \frac{m}{d \cos \theta}$$

where m is the order number, θ is the angular position of the corresponding line, and d is the separation between the rulings.

Problems 67 and 68 are paired.

67. **N** The fringe width β is defined as the distance between two consecutive maxima (or consecutive minima). In a Young's double-slit experiment, the fringe width obtained from a source of wavelength 500.0 nm is 3.90 mm. If the apparatus is immersed in a liquid that has index of refraction $n = 1.30$, what is the new fringe width? Assume the apparatus was originally in air and that the small-angle approximation applies to this situation.

68. **N** The fringe width β is defined as the distance between two consecutive maxima (or consecutive minima). When a Young's double-slit apparatus is completely immersed in a liquid, the fringe width decreases by 20%. Determine the liquid's index of refraction. Assume the apparatus was originally in air.

69. **N** A pair of closely spaced slits is illuminated with 600.0-nm light in a Young's double-slit experiment. During the experiment, one of the two slits is covered by an ultrathin Lucite plate with index of refraction $n = 1.485$. What is the minimum thickness of the Lucite plate that produces a dark fringe at the center of the viewing screen?

37 Reflection and Images Formed by Reflection

Key Questions

How do we find the location and magnification of an image formed by a mirror?

- 37-1 Geometric optics 1185
- 37-2 Law of reflection 1187
- 37-3 Images formed by plane mirrors 1190
- 37-4 Spherical mirrors 1195
- 37-5 Images formed by convex mirrors 1199
- 37-6 Images formed by concave mirrors 1204
- 37-7 Spherical aberration 1210

❶ Underlying Principles

Geometric optics

✪ Major Concepts

1. Law of reflection
2. Magnification
3. Virtual and real images
4. Mirror equation

▷ Special Cases

1. Camera obscura
2. Plane mirrors
3. Convex spherical mirrors
4. Concave spherical mirrors

⊙ Tools

1. Ray diagrams
2. Primary rays

Imagine this. You are having dinner with the President. It feels like some food is stuck to your front teeth, but you can't excuse yourself from the table. What can you do? You can try to catch a glimpse of your image in your knife or spoon. Your knife is a lot like your bathroom mirror. You are familiar with the image of yourself that you see in a mirror. You know that if you see food on your image's left front tooth, it is really on your right tooth. If you try using the bowl of your spoon as a mirror, your image is large but upside down. If you look at the back of your spoon, your image is upright but small. In this chapter, you will see how the shape of a mirror affects the image you see.

37-1 Geometric Optics

In earlier chapters, we modeled light as a wave and applied Huygens's principle. In this chapter and the next, we will model light as a ray and explore the images formed when light reflects or refracts. This chapter is devoted to reflection and the images formed by mirrors. You probably look into your bathroom mirror every morning. Mirrors have many other everyday uses, such as in the car, in a store as part of its security system, and in the dentist's office. Mirrors are also used in major telescopes, one of which inspires the case study for this chapter (Section 37-2).

Ray Model

In many circumstances, the wave nature of light is not relevant to the problem at hand; only its direction of propagation is relevant. In those cases, we crudely model light as a **ray**, drawn as a line perpendicular to the wave fronts with an arrowhead indicating the direction of the wave's motion. Rays are not vectors; to distinguish them from vectors, we draw the arrow for a ray on the line, not at its end. The electric and magnetic fields oscillate perpendicular to the ray, but they are not drawn. It is easy to represent a plane parallel wave with a single ray or several parallel rays (Fig. 37.1A). A spherical wave is represented by many rays (Fig. 37.1B).

The ray model ignores the wave qualities of light. It is a good model when light encounters only obstacles and apertures that are large compared to its wavelength. For practical purposes, the ray model works for obstacles and apertures that are at least 1 mm in diameter. In this chapter, we assume all mirrors and apertures are larger than 1 mm in diameter or width. Because rays reflected from such mirrors travel in straight lines, much of our analysis involves geometry. Situations that can be studied using the ray model fall into the realm of **geometric optics** (Chapters 37 and 38). Situations involving the wave qualities of light require *physical optics*. We used physical optics when we studied such phenomena as polarization, interference, and diffraction (Chapters 34, 35, and 36).

A narrow beam of light, such as from a laser, may be represented by a single ray or a small number of parallel rays (Fig. 37.1A). However, light reflected from a complicated object such as a flower must be represented by rays pointing in many different directions. When you view such an object, you are aware only of the rays that enter your eyes. To see an entire flower, you move your eyes and head to see more of the rays. Your brain uses the ray model to interpret the object's distance and size by assuming any ray that enters the eye must have traveled along a single straight path. The straight-ray assumption made by the human brain is sometimes exploited to create optical illusions (Fig. 37.2). In this chapter and the next, we will use the straight-ray assumption to find the images generated by mirrors and lenses.

Camera Obscura

To see how geometric optics works, let's consider a specific example—the camera obscura or pinhole camera. The term *camera obscura* is Latin for "dark chamber." This device dates back to antiquity. A camera obscura is a dark room with a small hole (the **aperture**) that allows light to enter from the **object** (Fig. 37.3). In optics, the object is anything from which light rays are emitted. It can be a source of light such as the Sun, but more often it is something that reflects light. People in the camera obscura look at an **image** (the apparent reproduction of the object) projected on a screen or wall on the opposite side of the room from the hole. In Figure 37.3, the image is formed by rays that pass through the aperture. A pinhole camera is a miniature camera obscura in which the room is replaced by a small box with a pinhole aperture. A larger hole cut into the box allows people to look at an image projected on its back wall.

When using geometric optics to analyze a situation, we draw a simple sketch called a **ray diagram** in which the object is usually represented by one or two arrows for simplicity. In Figure 37.3, we have overlaid the object with two perpendicular arrows. This practice enables us to see if the image is upright or upside down, and if it is flipped from left to right. Many rays are emitted by the object, but we need to draw only a few. Figure 37.3 shows three rays—one from the point of each arrow and the third from the bottom of the vertical arrow. Extending these rays all the way to the back wall shows that the image is both upside down and flipped (left to right).

GEOMETRIC OPTICS
❶ Underlying Principle

VISUAL REPRESENTATIONS OF LIGHT ◉ Tool

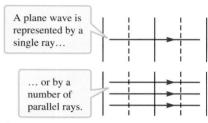

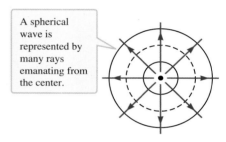

FIGURE 37.1 A. Representation of a plane parallel wave. **B.** Representation of a two-dimensional circular wave or a three-dimensional spherical wave. Compare to Figure 35.5.

FIGURE 37.2 Is there a piano in the room? Optical illusions are created because the human brain assumes all the rays that enter the eye have traveled along a straight line.

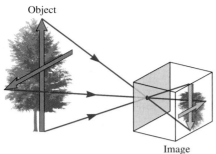

FIGURE 37.3 The camera obscura was a great form of entertainment. The image produced is upside down and reversed.

MAGNIFICATION ⊗ **Major Concept**

MAGNIFICATION OF A CAMERA OBSCURA ▷ **Special Case**

For a pinhole camera, both the object distance and the image distance are positive.

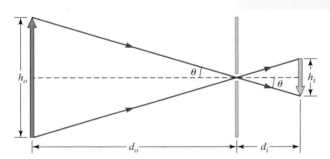

FIGURE 37.4 A ray diagram allows us to use geometry to find the magnification of a pinhole camera.

The image in a pinhole camera is usually blurry because every point on the object emits many rays that make it through the hole and to the back wall. These rays all come through at slightly different angles, so the image of any part of the object is slightly spread out over the whole image. You can improve a pinhole camera by making the hole smaller, but that also reduces the amount of light, which makes the image dimmer.

We are often interested in the **linear magnification** of an image—that is, its size relative to the object's size. Often the term *linear* is dropped. If the *o*bject's height is h_o and the *i*mage's height is h_i, the magnification M is

$$M = \frac{h_i}{h_o} \tag{37.1}$$

Magnification depends on geometric factors, such as distance. To find an expression for the magnification created by a camera obscura in terms of the distance d_o between the hole and the object and the distance d_i between the hole and the image, we start with the ray diagram in Figure 37.4. This diagram shows the object, the hole, the image, and two rays—one from the top of the arrow and the other from its base. We can use geometry to find the magnification. If the dashed line bisects the object and the image, we have

$$\tan\theta = \frac{h_o/2}{d_o} = \frac{h_i/2}{d_i}$$

Solving for the magnification h_i/h_o gives

$$M = \frac{h_i}{h_o} = -\frac{d_i}{d_o} \tag{37.2}$$

where we have inserted a negative sign by convention, the first of three we'll encounter in this chapter. The convention we use in this textbook is that if the image is inverted (upside down), its height is negative. Because the image's height is negative, the magnification is also negative. The negative sign in Equation 37.2 means the image in a pinhole camera or camera obscura is inverted.

For this pinhole camera, the image distance is less than the object distance, $d_i < d_o$, so the absolute value of the magnification is less than 1, $|M| < 1$. When $|M| < 1$, the image is smaller than the object. In everyday language, *magnification* means "making larger." But, in optics, magnification describes the relative size of the image, which may actually be smaller than the object.

EXAMPLE 37.1 Transit of Venus

On June 8, 2004, and then again on June 5, 2012, Venus crossed in front of the Sun. This event is called the transit of Venus. Venus had gone 122 years without such an event, and it will now be more than a century before the next one. Figure 37.5A shows a slightly distorted image of the Sun and Venus taken in 2012, and Figure 37.5B shows a sketch of Venus's path across the Sun. Estimate the magnification of the image in Figure 37.5A. If this image was made with a camera obscura, what was the distance from the hole to the image? The Sun's radius is 6.955×10^8 m.

FIGURE 37.5 A. Image of Venus in transit across the Sun. The image is distorted because the photographer had to stand to the side of the image in order not to block the sunlight. **B.** Sketch of Venus's path.

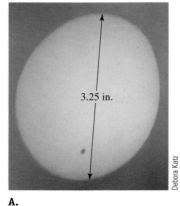
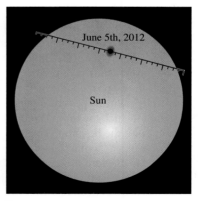

A.   B.

:• **INTERPRET and ANTICIPATE**
The photo shows that the image of the Sun has a diameter of 3.25 in., so we can find the image height h_i. The object is the Sun itself, whose radius is given inside the cover of this text. From these values, we can estimate the magnification. We expect the absolute value of the magnification to be much less than 1 because the Sun is very large and the image is very small.

:• **SOLVE**

Convert the image diameter to meters. This is the image height.	$h_i = 3.25$ in. $= 8.255 \times 10^{-2}$ m
The diameter of the Sun is the object height.	$h_o = 2(6.955 \times 10^8 \text{ m}) = 1.391 \times 10^9$ m
Compare the sketch to the image. The sketch shows Venus crossing the top half of the Sun (Fig. 37.5B). In the image, Venus is on the bottom (Fig. 37.5A). So the image is inverted, and the magnification is negative.	$M = \dfrac{h_i}{h_o} = \dfrac{-8.255 \times 10^{-2} \text{ m}}{1.391 \times 10^9 \text{ m}}$ $M = -5.93 \times 10^{-11}$
Now use Equation 37.2 to find the distance between the image and the hole. The object distance is the distance between the Earth and the Sun (1 AU). See Appendix B.	$M = \dfrac{h_i}{h_o} = -\dfrac{d_i}{d_o}$ (37.2) $d_i = -Md_o = -(-5.93 \times 10^{-11})(1.496 \times 10^{11} \text{ m})$ $d_i = 8.87$ m

:• **CHECK and THINK**
The Sun is much larger than its image in Figure 37.5, and as expected the absolute value of the magnification is much less than 1. If a camera obscura was used, the hole was about 9 m (or roughly 30 ft) from the screen. This is somewhat larger than the giant camera obscura in San Francisco, but much smaller than the world's largest camera obscura ($d_i \approx 55$ ft); that camera obscura was made from an F-18 fighter plane's hangar. Most likely this image was not made with a camera obscura, but with another sort of instrument such as a telescope.

37-2 Law of Reflection

Consider a typical room full of objects illuminated by any number of lightbulbs. You can see those objects because the light from the bulbs reflects from the surface of the objects. When light reflects from a nonflat object—such as a flower, a piano, or the many walls of a room—the many reflected rays travel in many different directions. However, let's consider a simple situation in which light reflects from a flat surface, such as laser light reflected from a polished flat mirror.

Figure 37.6A shows a light beam of width w making an angle γ_i with the surface of a medium. Because of the angle γ_i, the beam first encounters the medium point at A. To see what happens, we apply Huygens's principle (Section 35-1) and consider a wavelet centered on point A. Just as the wavelet at A is beginning to emerge, the other edge of incident beam is still a distance r from medium at point B. Of course, both the wavelet centered on A and the beam move at the speed of light. So, by the time beam has reached point B, the radius of the wavelet must equal the distance r. We've only considered two points A and B on the surface, and if we applied Huygens's principle to a great number of wavelets originating from points between A and B, we would find a plane parallel wave traveling away from the glass at an angle γ_r.

With a little trigonometry, we can find the relationship between the angle γ_i that the *i*ncident light makes with the surface and the angle γ_r that the *r*eflected beam makes. Figure 37.6B shows a simpler sketch, showing the geometry we need. The distance between points A and B is h. There are two right triangles with the same hypotenuse in common. Because each one has a leg of length r, the remaining leg

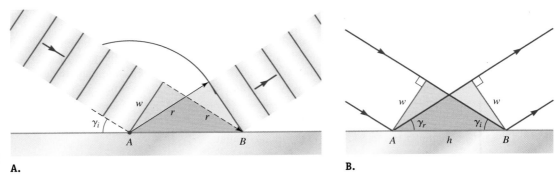

FIGURE 37.6
A. A beam of light reflects from a flat surface. **B.** Geometry of this reflection.

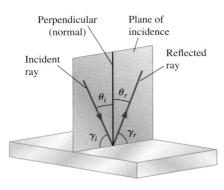

FIGURE 37.7 The law of reflection. Compare this figure to Figure 18.10.

LAW OF REFLECTION

⊛ **Major Concept**

on each triangle must also be equal. So the width of the reflected beam must be w. Use these triangles to find the sine of the two angles γ_i and γ_r:

$$\sin \gamma_i = \sin \gamma_r = \frac{w}{h}$$

Because $\sin \gamma_i = \sin \gamma_r$, the angles must be equal: $\gamma_i = \gamma_r$.

The angles γ_i and γ_r are measured between the beams and the surface. By custom, however, the angle used to express the *law of reflection* is measured between the beam and a line perpendicular to the surface. Figure 37.7 shows the angles θ_i and θ_r measured with respect to the perpendicular (called the *normal*). These angles θ_i and θ_r are referred to as the **angle of incidence** and the **angle of reflection**, respectively.

The **law of reflection** has two parts. The first part is that the angle of incidence equals the angle of reflection:

$$\theta_i = \theta_r \tag{37.3}$$

The second part is that the incident ray, the reflected ray, and the perpendicular (normal) line all lie in a single plane called the **plane of incidence** (Fig. 37.7). The law of reflection applies to curved surfaces as long as you consider a portion that is small enough to be considered flat.

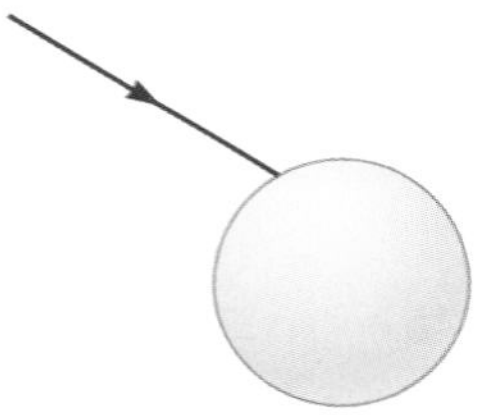

CONCEPT EXERCISE 37.1

A beam in air strikes a glass ball as shown in Figure 37.8. Draw the reflected ray.

FIGURE 37.8

EXAMPLE 37.2 The Ultimate in Security

Figure 37.9 shows the plan for a small apartment. A light source is in the left part of the apartment. A hole is in the wall on the right. Two mirrors are fixed on the long wall at the bottom of the plan. You would like to stand outside the apartment, look through the hole, and see light from the source. Where along the short wall should you place the third mirror? As part of your thinking, ask yourself: If you stood at the position of the light source, could you see out through the hole (assuming there was light outside the apartment)?

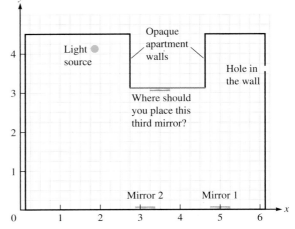

FIGURE 37.9 Apartment floor plan with three mirrors.

:• **INTERPRET and ANTICIPATE**
The physics behind this problem is the law of reflection, $\theta_i = \theta_r$. We can solve this problem by neatly drawing a few rays and then just reading the position of the third mirror off the diagram.

:• **SOLVE**
Start by drawing the line normal to each mirror, as shown in the middle of each mirror in Figure 37.10. When we find the position of the third mirror, we are really finding the position of the normal line we have drawn.

Also, draw the ray (labeled 1) from the last mirror at (5.00, 0). This is the ray you can see from that point on the mirror.

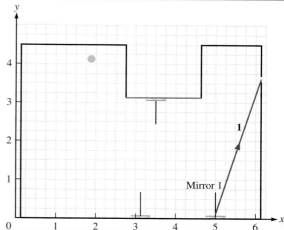

FIGURE 37.10

Apply the law of reflection ($\theta_i = \theta_r$) to this mirror and ray in order to draw the incident ray, labeled 2. The incident ray must come from the third mirror (on the green wall). By sliding the third mirror so that it can produce ray 2, we find the position of that mirror (Fig. 37.11).

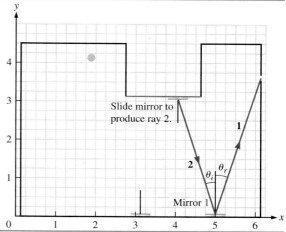

FIGURE 37.11

We are not done because it is possible that the light from the source cannot get to this third mirror. Repeating the process of using the law of reflection to find rays 3 and 4, we find that the light can make it from the source to all three mirrors and the hole (Fig. 37.12).

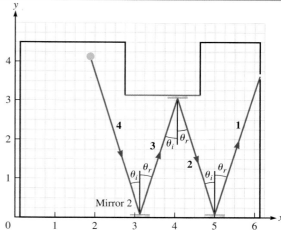

FIGURE 37.12

Examine Figure 37.12 to find the position of the mirror. Read off the position of the normal line with respect to the x axis.

The third mirror is at $x = 4.1$ m.

:• **CHECK and THINK**
If you tried to set up these mirrors on the scale shown, your placement would not need to be very precise because each mirror is about 0.5 m wide. So, if the center of the third mirror is not exactly at $x = 4.1$ m, this will still probably work. Also, there is nothing special that distinguishes reflected rays from incident rays. Suppose you stand in the room at the position of the light source. If the light source is off but there is light coming from the hole, you will be able to see that light. The "incident rays" shown in Figure 37.12 would be reflected rays, and the "reflected rays" would be incident rays. The W-shaped path would look the same, except the arrowheads would be reversed.

CASE STUDY The Hubble Space Telescope's Mirror

Smart people learn from their mistakes. This case study is about a mistake and how it was fixed. In April 1990, the Hubble Space Telescope (HST) was launched. About a month later, scientists discovered that the HST's images were blurry (Fig. 37.13A).

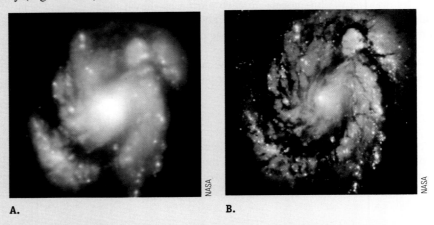

FIGURE 37.13 A. A picture taken by the Hubble Space Telescope before lenses were used to correct the shape of its mirror. **B.** A picture of the same astronomical object taken after the correction.

The largest telescopes are reflecting telescopes, which means they use mirrors, not lenses, to create images of astronomical objects. The HST—like large ground-based telescopes—uses an enormous dish-shaped mirror (the primary mirror) to focus light from distant objects. The HST's initial images were blurry because its primary mirror was incorrectly shaped. Three years later, this problem was corrected, and the images taken by the HST became much sharper (Fig. 37.13B). In this case study, we explore reflecting telescopes and how a misshaped primary mirror causes a blurry image.

37-3 Images Formed by Plane Mirrors

A mirror is a device that reflects more than 90% of the light incident on it, and the reflection is specular rather than diffuse. (We ignore the glass that usually covers the shiny reflective surface of a real mirror.) When you look at *diffuse* reflection from a surface such as the wall of a room, you can see reflected light no matter where you look. Even if you shine a laser at the wall, you can see the reflection (the spot) from many positions (Fig. 37.14A). When you look at a *specular* reflection, however, your eyes must be in the right place (determined by the law of reflection) in order to see the reflected light (Fig. 37.14B). The result is that walls do not form

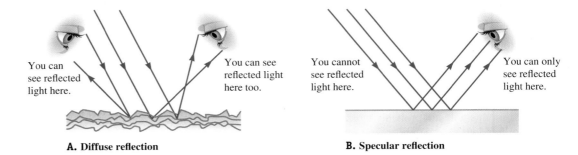

A. Diffuse reflection

B. Specular reflection

FIGURE 37.14 A. In diffuse reflection, you can see reflected light from many vantage points. **B.** In specular reflection, your eyes must be in the right place to see a reflection.

images, but mirrors do. For the rest of this chapter, we will study the images formed by mirrors of various shapes. In each case, our goal is to come up with expressions for the position and magnification of the image formed by the mirror. We begin with the familiar example of a **plane mirror**; such a mirror is flat and lies in a single plane.

PLANE MIRRORS ▶ Special Case

Ray Diagrams

We used a crude ray diagram when we studied the camera obscura. Now we describe the elements needed to sketch a ray diagram that can be used to find the image formed by a plane mirror (Fig. 37.15):

RAY DIAGRAMS ◉ Tool

1. A thick line representing the **mirror** seen edge-on. Light totally reflects from the thin layer on the front of the mirror. Assume an observer and an object are in front of the mirror. (Usually, the front side is shown on the left, but that is not necessary and is sometimes inconvenient.) It is not normally necessary to represent the observer.
2. An arrow representing the **object**.
3. A small number of **emitted rays** (usually four or fewer) coming from key points on the object, such as the tip and base of the arrow.
4. **Reflected rays.** For each emitted ray, use the law of reflection to draw the reflected ray.
5. An arrow representing the **image**. Consider two rays emitted from a single point, such as from the tip or the base of the arrow, and find the place where their reflected rays cross or would appear to cross. Because these rays were emitted from a single point, the image of that point is formed where the reflected rays cross (or seem to cross). Use this crossing point to sketch the image at that location. (In Figures 37.3 and 37.4, we considered rays emitted from two separate points of the object, so they do not cross to form a point on the image.)

Virtual and Real Images

In Figure 37.15, the two rays emitted from the tip of the object arrow are reflected by the mirror. These reflected rays enter the observer's eye without crossing, and the human brain assumes the rays are not bent. So, in effect, the brain traces these rays back along two straight lines that meet at an imaginary point *behind* the mirror, but there is no light behind the mirror. (Your bathroom mirror is probably hanging on a wall.) Nevertheless, your brain interprets the point where these rays appear to meet as the location of the image. We define a **real image** as one in which light rays from every point of an object **converge** or come together to create every point of the image. This convergence of light rays makes it possible to project a real image on a detector like film, a screen, or your retina. A movie camera lens and a camera obscura both produce real images. By contrast, we call the image produced by your bathroom mirror a **virtual image**. Light rays appear to **diverge** or spread from every point of a virtual image, just as they actually diverge from every point of an object. This divergence of light rays means it is not possible to project a virtual image on a screen, just as an object by itself does not project an image. A screen behind your bathroom mirror does not capture your image because there is no light there. A screen facing a bathroom mirror does not capture your image either. Nevertheless, because your eye has a lens, you see a virtual image of yourself in a plane mirror. Some curved mirrors produce real images (Section 37-4), but plane mirrors always produce virtual images.

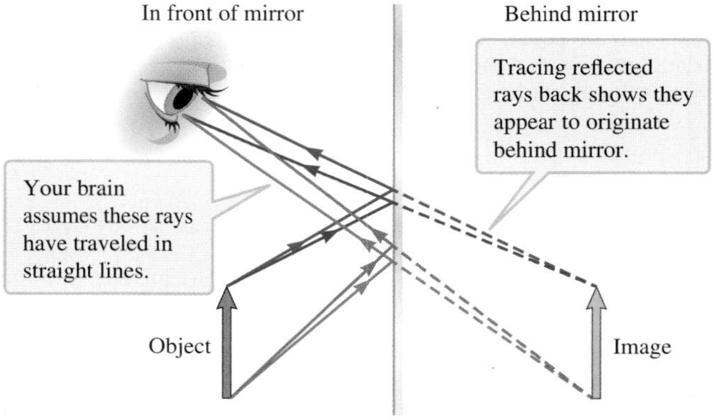

VIRTUAL AND REAL IMAGES
✦ Major Concept

FIGURE 37.15 We have drawn two rays emitted from the tip and two from the base of the object's arrow. For all four emitted rays here, we have used the law of reflection to draw the reflected rays. The image you see in a plane mirror is behind the mirror, but there is no light back there. Such an image is called a *virtual image*.

Finding the Image Formed by a Plane Mirror

Your experience looking at your image in a flat mirror provides you with a set of expectations. First, imagine sharing the mirror with a roommate who is standing behind you. Your roommate's image is behind your image, so we expect that the image distance d_i is proportional to the object distance d_o. Second, the image is

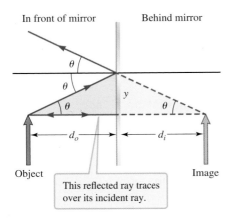

FIGURE 37.16 A simple ray diagram for a plane mirror. All the angles labeled θ are equal, so the highlighted triangles show that the image distance equals the object distance: $d_i = d_o$.

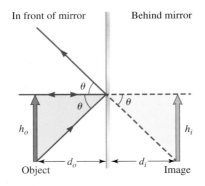

FIGURE 37.17 Another simple ray diagram for a plane mirror. All the angles labeled θ are equal, so the highlighted triangles show that the image height equals the object height: $h_i = h_o$.

FIGURE 37.18 The reflection of the left hand looks like the palm of the right hand.

upright, so the image height h_i is positive, and we expect the magnification M also to be positive. Third, when you look in a bathroom mirror, you don't look smaller or larger than you really are, so we expect the magnification to be $M = 1$.

The simplified ray diagram in Figure 37.16 shows two rays emitted from the tip of an object (represented by the large upright arrow in front of the mirror). One ray is perpendicular to the mirror, so according to the law of reflection, its reflected ray is also perpendicular. The other ray is also emitted from the tip, but drawn in some arbitrary direction toward the mirror. The law of reflection determines its reflected ray. The two reflected rays do not cross in front of the mirror. To find the image, trace these two rays back to the point behind the mirror where they seem to meet.

The two highlighted right triangles in Figure 37.16 share a common side of length y. The base of the triangle on the left is the object distance d_o, and the base of the triangle on the right is the image distance d_i. The tangent of θ for both of these triangles is

$$\tan\theta = \frac{y}{d_o} = \frac{y}{d_i}$$

So the image distance equals the object distance:

$$d_i = -d_o \quad (37.4)$$

using a second sign convention that when the image is located behind the mirror, its distance d_i is negative. Equation 37.4 meets our expectation that the image distance d_i is proportional to the object distance d_o.

Another simple ray diagram (Fig. 37.17) is helpful in finding the magnification. As before, we draw one ray perpendicularly from the tip of the object arrow to the mirror, so that ray's reflection traces back over the incident ray. The other ray is drawn from the base of the arrow, and it strikes the mirror at the same place as the first incident ray. It obeys the law of reflection, and the two reflected rays are traced back to points behind the mirror.

Take the tangent of θ for both highlighted triangles:

$$\tan\theta = \frac{h_o}{d_o} = \frac{h_i}{d_i}$$

According to Equation 37.4, the magnitudes of the object and image distances are equal, so the bases of the triangles are equal:

$$\frac{h_o}{d_o} = \frac{h_i}{d_o}$$

The image height thus equals the object height:

$$h_o = h_i$$

The magnification (Eq. 37.2) of a plane mirror is given by

$$M = \frac{h_i}{h_o} = -\frac{d_i}{d_o} = 1 \quad (37.5)$$

Because d_i is negative, Equation 37.5 meets our expectations that the magnification is positive and equal to 1.

Left Hand or Right Hand?

We just showed that your image in a bathroom mirror is upright, has a magnification of 1, is as far behind the mirror as you are in front of the mirror, and is virtual. However, the image you see every morning also appears to be reversed from left to right. The camera obscura causes a left-right reversal, and it also causes the image to be inverted (Fig. 37.3).

To see how a plane mirror can cause the left-right reversal without inverting the image, let's consider the reflection of a person's left hand. In Figure 37.18, the left hand has a ring on it, so you can keep track of which hand is which. The reflected image of the left hand in the mirror looks like the right hand, with its fingers pointing up, its palm facing out, and its thumb pointing toward the right. The only substantial difference is that the image of the left hand has a ring. So, you might say the mirror has made the left hand look like the right hand. This is the left-right reversal you see every time you look in the bathroom mirror.

However, describing what you see as a simple left-right reversal is not quite correct. Examine the actual left hand. The left fingers point up and the thumb points toward the right, as in the image of the left hand. If the person flipped her left hand so that the palm faced forward, the fingers would be up but the thumb would point toward the left. So the mirror doesn't really create a left-right reversal. The real difference is that the image of the left hand in the mirror shows the *palm* of the left hand, and the view of the actual left hand is of the back. In a sense, the mirror creates a *depth reversal*, not a left-right reversal. To make the left hand look like its image, each particle of the left hand would need to move straight through to the other side of the hand, so that the back of the hand would be behind the palm.

A depth reversal may sound complicated, but when we consider a moving object and its mirror image, the concept becomes clearer. Consider Charlie Chaplin's encounter with his twin in Figure 37.19A. When Chaplin moves his left hand toward his twin in the positive x direction, the twin creates a mirror image by moving his right hand toward Chaplin in the *negative x* direction. Now imagine the twin is replaced by a plane mirror (Fig. 37.19B). When Chaplin moves his left hand in the positive x direction, his mirror image moves its right hand in the negative x direction. To Chaplin, the twin in Figure 37.19A appears to be a mirror image because the twin imitates the depth reversal of the mirror image in Figure 37.19B.

CONCEPT EXERCISE 37.2

The terms *virtual image* and *real image* are not particularly descriptive or intuitive. Come up with your own terms, and explain your reasons for suggesting them.

A.

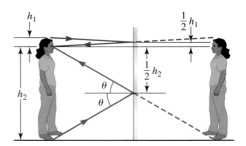

B.

FIGURE 37.19 Charlie Chaplin (1889–1977) was a British comedian. **A.** Charlie Chaplin and his twin create a mirror image. **B.** Charlie Chaplin and his mirror image move like Chaplin and his twin in part A.

EXAMPLE 37.3 Full-Length Mirror

People often purchase a full-length mirror that is at least as tall as they are, but such a large mirror is not necessary. Suppose a woman of height h wishes to see her entire image in a mirror. What is the minimum height of the plane mirror she needs?

INTERPRET and ANTICIPATE
When solving problems in geometric optics, start with a ray diagram. Because the object—the woman—is also the observer, we have sketched her instead of a simple arrow (Fig. 37.20). The key to solving this problem is to draw one ray emitted from the top of her head and another ray from her foot. These rays come from the extreme parts of her body. If she can see their reflections, she will be able to see the rays reflected from points between these extremes.

FIGURE 37.20 A ray diagram of a woman seeing her entire image in a mirror.

SOLVE
Label a few key distances. Figure 37.20 shows a horizontal line through the woman's eye and the image of her eye. The top of her head is a distance h_1 above her eye, and her foot is a distance h_2 below it, so her total height is the sum of these two distances.

$$h = h_1 + h_2 \tag{1}$$

Example continues on page 1194 ▶

After the emitted rays are chosen, find and draw the reflected rays. The ray emitted from her foot is incident on the mirror at angle θ, so its angle of reflection is also θ. In order for her to see the reflected ray, the incident ray must strike the mirror at a point halfway between her eye and her foot.	The bottom of the mirror is $h_2/2$ below the horizontal line that passes through her eye.
The same sort of geometry applies to the ray emitted from the top of her head. That ray must strike the mirror halfway between her eye and the top of her head.	The top of the mirror is $h_1/2$ above the horizontal line that passes through her eye.
Let h_{mirror} be the total height of the mirror that just enables the woman to see her entire reflection. This total height is $h_1/2$ (top of mirror to woman's eye level) plus $h_2/2$ (bottom of mirror to woman's eye level). Using Equation (1), we find the mirror must be half the height of the woman.	$h_{mirror} = \dfrac{h_1}{2} + \dfrac{h_2}{2} = \dfrac{h_1 + h_2}{2}$ $h_{mirror} = \dfrac{h}{2}$

:• **CHECK and THINK**
Notice that we did not need to consider how far the woman is from the mirror, so our answer does not depend on where she stands. She sees her entire body whether she is far from the mirror or close to it. However, we did consider where the mirror hangs on the wall. The top of the mirror must be almost level with the top of the woman's head. Why do people buy such large full-length mirrors (about their own height)? There are a number of reasons. If two people of different heights want to use a mirror of the size we calculated, it needs to be half the height of the taller person, and it needs to hang almost level with the top of that person's head. A second, much shorter person wouldn't be able to see his whole body. In the extreme case that one person is a tall adult and the other is a short child, the child's head may be below the bottom of the mirror, so he may not be able to see himself in the mirror at all.

EXAMPLE 37.4 True Mirror

Figure 37.21A shows the image of a message and a clock created by a plane mirror. The image in Figure 37.21B is not reversed because it is the image created by a second plane mirror perpendicular to the first. Two perpendicular mirrors are called a *true mirror* because the final image is not reversed. Use a ray diagram to show that a true mirror produces an image that is not reversed.

FIGURE 37.21 A. This is the image in the first mirror. **B.** A second mirror creates an image that is not reversed.

:• **INTERPRET and ANTICIPATE**
Start by sketching a side view of the two perpendicular mirrors and the object, which is represented by an arrow (Fig. 37.22). We can break down the next part of the job into steps. First draw a few emitted rays and their reflection from one of the two mirrors. In the second step, trace these reflected rays to the second mirror. Find the rays that reflect from the second mirror, and then trace these reflected rays back to the image formed by the second mirror.

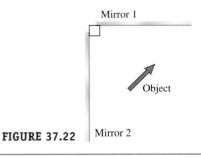

FIGURE 37.22

SOLVE
We arbitrarily choose the mirror shown at the top of Figures 37.22 and 37.23 as the first mirror. (In Problem 20, you show that the same result occurs if you choose the other mirror as the first mirror.) Consider three emitted rays; two are emitted from the tip of the arrow and the third from the base. Each ray is traced to mirror 1, and its reflection from mirror 1 is drawn. Then trace these three rays to points behind mirror 1 to show the image formed by mirror 1. We find that image 1 is reversed.

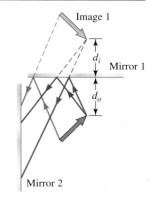

FIGURE 37.23 Image 1 is reversed.

In the second step, trace the rays reflected from mirror 1 to mirror 2 (Fig. 37.24). In doing this, we are treating image 1 as the object for mirror 2. For each of the three rays incident on mirror 2, we find the ray that reflects from that mirror. Tracing these reflected rays back to points behind mirror 2 shows the image formed by mirror 2; image 2 is the image of image 1.

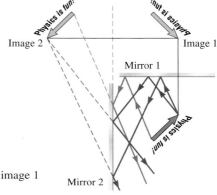

FIGURE 37.24 The image of image 1 is not reversed.

CHECK and THINK
Compare the final ray diagram (Fig. 37.24) to the photos in Figure 37.21. The words in the first image in mirror 1 are backward (reversed), which matches what we see in Figure 37.21A. The words in the second mirror (image 2) are not reversed, as in Figure 37.21B. So our ray diagram seems to match the images in Figure 37.21.

37-4 Spherical Mirrors

Mirrors can be bent into complicated shapes. When the mirror's geometry is complicated, however, it is difficult to calculate the position and magnification of the image formed by the mirror. To keep these tasks manageable, we will focus on *spherical mirrors*. A **spherical mirror's** surface forms either a sphere or part of a sphere. The reflective surface may be on the outside or the inside of the sphere. In Figure 37.25A, an artist sees his reflection in a mirror that forms a complete sphere, with the reflective surface on the outside. In Figure 37.25B, a photographer reaches for her reflection in a spherical mirror. The surface of her mirror does not form a complete sphere, and the reflective surface is on the inside of the sphere. The perimeter of a spherical mirror does not need to be a circle. Figure 37.26 shows a spherical mirror with a roughly rectangular perimeter. The surface of the mirror is part of a sphere, and you can imagine peeling it off a large sphere whose inside surface is reflective.

A plane mirror forms an upright virtual image that is the same size as the object and located as far behind the mirror as the object is in front of the mirror (Section 37-3). By contrast, the image formed by a spherical mirror may be upright or inverted; it may be larger or smaller than the object; and it may be virtual and behind the mirror, or real and in front of the mirror.

In this section, we use ray diagrams to help visualize the images produced by spherical mirrors and to derive algebraic expressions for the positions and magnifications of

FIGURE 37.25 A. The image of the artist in this convex spherical mirror is upright. **B.** The image of the photographer in this concave spherical mirror is inverted.

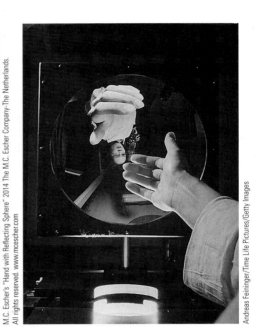

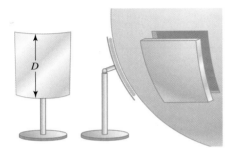

FIGURE 37.26 A spherical mirror does not have to be circular. This rectangular mirror is a spherical mirror. Its surface is a patch of a spherical surface. The inside surface of the sphere is reflective.

CONVEX AND CONCAVE SPHERICAL MIRRORS ▶ Special Case

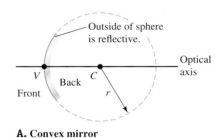

A. Convex mirror

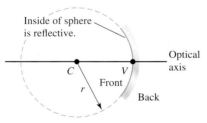

B. Concave mirror

FIGURE 37.27 A. For a *convex* spherical mirror, the outside surface is reflective. **B.** For a *concave* spherical mirror, the inside surface is reflective. To keep the two terms straight, remember that from the front a concave mirror looks like a *cave*. The dashed circles represent the complete spheres to which these mirrors correspond and are not normally included in a ray diagram.

the images. When representing a spherical mirror on a ray diagram, we draw an edge-on view that is either a circle or a circular arc (Fig. 37.27). The radius of curvature r is the radius of the circle. The center of the circle—labeled C—is called the mirror's **center of curvature**.

In Figure 37.27A, the reflective surface of the mirror is the outside the sphere, as with the mirror in Figure 37.25A. Such a mirror is called a **convex** mirror. When you stand in front of a convex mirror, the surface bulges out toward you. Convex mirrors are used in public places such as parking garages and stores because these mirrors have a wide field of view, allowing you to look for traffic around the corner of a garage or for shoplifters in a store. In Figure 37.27B, the reflective surface of the mirror is the inside of the sphere, as with the mirror in Figure 37.25B. Such a mirror is called a **concave** mirror. When you stand in front of a concave mirror, the surface looks like a bowl or *cave*. The image formed by a concave mirror may be larger than the object, so concave mirrors are used as shaving or makeup mirrors. Concave mirrors are also used in telescopes (Section 37-7).

It is important to distinguish between the radius of curvature and the size of the mirror. The radius of curvature is the radius r of the entire sphere (Fig. 37.27). It is a measure of how curved the mirror is. A small radius of curvature means a tightly curved mirror. A flat mirror has an infinite radius of curvature. The size of the mirror is its largest linear dimension; for example, D is the height of the rectangular mirror in Figure 37.26. If the mirror happens to have a circular perimeter, D is the diameter of that circle.

It is also important to distinguish between the center of curvature and the *center of the mirror*. The **center of the mirror** lies on its surface and is also called the **vertex**. The center of curvature does not lie on the mirror. In Figure 37.27, the center of curvature is labeled C and the vertex is labeled V. The **optical axis** is the line that runs through the center of curvature C and the vertex V; it is perpendicular to the mirror at V. Sometimes the optical axis is referred to as the **principal axis** or the **central axis**.

Finding the Focal Point

Consider a simple situation in which the object is very far from the mirror. For example, you can imagine the Sun is the object and the mirror is on the Earth. Because the Sun is very far away, any rays that strike the mirror must be traveling in the same direction and are therefore parallel. To keep things simple for now, consider only mirrors that are small compared to their radii of curvature: $D \ll r$. (We'll consider larger mirrors in Section 37-7.)

Now imagine orienting a convex mirror so that its optical axis is parallel to the Sun's rays (Fig. 37.28A). The law of reflection determines each ray's path after it strikes the mirror. We find that the reflected rays **diverge** (do not cross). No real image of the Sun forms in front of the mirror. However, if we trace each of these reflected rays back along a straight path, we see that they meet at a single point behind the mirror. If these reflected rays enter your eye, you see a virtual image of the Sun. It looks like the rays emerged from the point labeled *F*. The point where parallel rays form a virtual image is known as the **virtual focal point**. Only the reflections of parallel rays, or rays from a very distant object, appear to emerge from the virtual focal point.

Now consider what happens when parallel rays encounter a concave mirror. As before, these rays must be parallel to the optical axis. By the law of reflection, this time we find that the reflected rays meet or **converge** at a point in front of the mirror. This point—labeled *F*—is called the **real focal point**. A real image of the Sun forms at the real focal point; it is like the real image formed by a camera obscura. If you put a small piece of paper (or film) at the focal point so that it doesn't block much of the Sun's light from reaching the mirror, you see an image projected on the paper. (You may even be able to light the paper on fire.)

Because the law of reflection is symmetrical (the angle of incidence equals the angle of reflection), we can reverse the arrowheads on all the rays in Figure 37.28. So, if an incoming ray is aimed at the virtual focal point of a convex mirror, the reflected ray is parallel to the optical axis (Fig. 37.28A). Further, if an incident ray passes through the real focal point of a concave mirror, its reflection is parallel to the optical axis (by ray reversal of Fig. 37.28B). Flashlight and headlight manufacturers make use of this fact. If you examine a flashlight or headlight closely, you see that the bulb is placed at or near the focal point of a concave mirror.

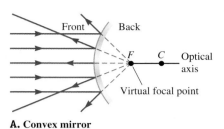

A. Convex mirror

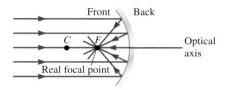

B. Concave mirror

FIGURE 37.28 These mirrors are small compared to their radii of curvature. **A.** Parallel rays reflected by the convex mirror appear to come from points behind the mirror. As in the case of a flat mirror, we see a virtual image located behind the mirror. **B.** Parallel rays incident on a concave mirror are focused in front of the mirror, forming a real image.

Primary Rays

There are four rays that are not only easy to draw but also particularly helpful in ray diagrams involving spherical mirrors. These four rays are known as the **primary rays** and are described in Table 37.1.

A ray diagram does not necessarily need to show all four primary rays. Often two rays are sufficient. Then you may wish to draw a third ray to check your work. Choosing which primary rays to draw sometimes depends on geometry. Some choices may make your work easier than others, but very often all rays are equally good choices. As with drawing free-body diagrams, practice will help you draw good ray diagrams.

PRIMARY RAYS ⊙ Tool

DERIVATION   Focal Length *f*

The **focal length** *f* of a mirror is the distance between the focal point and the vertex of the mirror (Fig. 37.29). We show that for a small convex mirror ($D \ll r$), the focal length is half the radius of curvature:

$$f = \frac{r}{2} \quad (37.6)$$

In Problem 27, you are asked to repeat this derivation for a concave mirror.

We can derive the focal length by drawing just one primary ray and using geometry. Rays 3 and 4 work equally well, but we have arbitrarily decided to use ray 4. Ray 4 is aimed at the virtual focus (Fig 37.30) and reflects at point *A*. Its reflected ray is parallel to the optical axis.

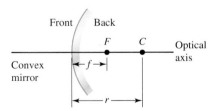

FIGURE 37.29 The focal length *f* is the distance between the focal point and the vertex.

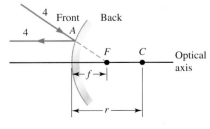

FIGURE 37.30

Derivation continues on page 1198 ▶

The line that runs through points A and C is perpendicular to the mirror (Fig. 37.31). The angles of incidence and reflection are measured with respect to that line. According to the law of reflection, these two angles θ are equal. In addition, these angles are equal to the two angles θ on the back side of the mirror. By the rules of geometry, the angle ACF must also be equal to θ. The triangle ACF has two equal angles, so it is an isosceles triangle. The two short sides must be equal. We have labeled their lengths ℓ.

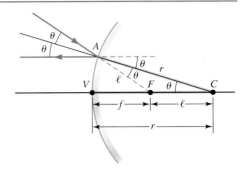

FIGURE 37.31 The triangle ACF is isosceles, so its two short sides have the same length ℓ.

The radius of curvature r is the distance ℓ from C to F plus the distance f from F to V.	$r = \ell + f$	(1)
Although Figure 37.31 is exaggerated, the mirror is small compared to its center of curvature ($r \gg D$), so the distance from the focal point F to any point on the mirror is roughly the same. Because the distance from F to V is the focal length f, the distance from F to any point on the mirror is roughly f. So the distance ℓ from F to A is roughly f.	$\ell \approx f$	(2)
Substitute Equation (2) into Equation (1), and solve for f.	$r \approx f + f = 2f$ $$f = \frac{r}{2} \checkmark$$	(37.6)

∴ COMMENTS

We dropped the approximately equal sign in the result, but keep in mind that Equation 37.6 is true only when the spherical mirror is small compared to its radius of curvature. This derivation shows the value of considering primary rays. In the next two sections, we will use primary rays to find the images formed by both convex and concave mirrors for objects that are at some finite distance from the mirror.

TABLE 37.1 Primary rays for spherical mirrors. Colors shown are arbitrary and do not relate to actual color of light.

Ray	Convex Mirror	Concave Mirror
Ray 1 strikes the mirror at its vertex. Because the optical axis is perpendicular to the mirror at its vertex, the incident and reflected rays are symmetrical with respect to the optical axis.		
For a convex mirror, **ray 2** is aimed at the center of curvature C. For a concave mirror, ray 2 passes through the center of curvature. For either type of mirror, ray 2 is perpendicular to the mirror surface and the reflected ray is back along ray 2.		

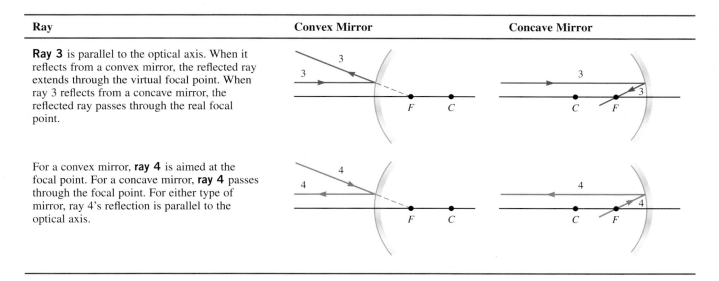

Ray	Convex Mirror	Concave Mirror
Ray 3 is parallel to the optical axis. When it reflects from a convex mirror, the reflected ray extends through the virtual focal point. When ray 3 reflects from a concave mirror, the reflected ray passes through the real focal point.		
For a convex mirror, **ray 4** is aimed at the focal point. For a concave mirror, **ray 4** passes through the focal point. For either type of mirror, ray 4's reflection is parallel to the optical axis.		

Sign Conventions

The third and final sign convention is for the focal length f and radius of curvature r. For a convex mirror, the center of curvature C and the focal point F are both behind the mirror (Figs. 37.28A and 37.29). In this case, the focal point is virtual, and by convention the focal length and radius of curvature are both negative. For a concave mirror, the center of curvature C and the focal point F are both in front of the mirror (Fig. 37.28B). In this case, the focal point is real, and by convention the focal length and radius of curvature are both positive. The three sign conventions for mirrors are summarized in Table 37.2.

TABLE 37.2 Sign conventions for mirrors.

Quantity	Positive	Negative
1. Image height h_i and magnification M	If image is **upright**	If image is **inverted**
2. Object distance d_o and image distance d_i	If object or image is **real**	If object or image is **virtual**
3. Radius of curvature r and focal length f	If mirror is **concave**	If mirror is **convex**

CONCEPT EXERCISE 37.3

Estimate the focal length of the mirror shown in Figure 37.25A.

37-5 Images Formed by Convex Mirrors

Figure 37.32 shows a convex mirror and an object whose distance from the mirror is d_o. Only an object that is very far from the mirror, so that its rays are parallel to the optical axis, forms a virtual image at the virtual focus. Thus, the object in Figure 37.32 does not form a virtual image at F. To get a rough idea of the location and size of the image, we make a ray diagram using primary rays 1 and 2. Both rays are drawn from a single point on the object (the tip of the object arrow). The two reflected rays do not intersect, but when we trace these rays back behind the mirror, their extensions intersect. Because these rays come from the tip of the object arrow,

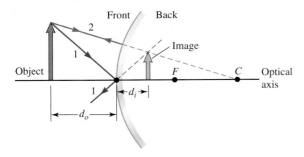

FIGURE 37.32 To find the location and size of the image created by this convex mirror, we draw two of the four primary rays. We arbitrarily choose rays 1 and 2. Ray 1 strikes the vertex of the mirror; ray 1 and its reflection are symmetrical with respect to the optical axis. Ray 2 is aimed at the center of curvature C, and its reflection is back along itself.

the image of the tip must be at this intersection. As with a plane mirror, there is no light behind the convex mirror, so this is a virtual image located a distance d_i behind the mirror.

Figure 37.32 also shows that the image points upward, just like the object. Because the image is *upright* (rather than inverted), the magnification is positive (Table 37.2). Additionally, the image is smaller than the object, so the absolute value of the magnification is less than 1 ($0 < M < 1$). Compare the image formed by the convex mirror (Fig. 37.32) to the image formed by a plane mirror (Fig. 37.17). Both images are formed behind the mirror, so both are virtual. The images formed in both mirrors are upright, so $M > 0$. However, the image and the object for a plane mirror have the same size, so $M = 1$, whereas the image is smaller than the object in a convex mirror, so $0 < M < 1$.

DERIVATION Magnification and the Mirror Equation for Spherical Mirrors

Ray diagrams provide a rough idea of the location and magnification of an image. We can get more precise results if we use geometry to derive algebraic expressions. Here we show that the magnification for a spherical mirror is given by

MAGNIFICATION OF SPHERICAL MIRROR ✪ Major Concept

MIRROR EQUATION ✪ Major Concept

$$M = \frac{h_i}{h_o} = -\frac{d_i}{d_o} \quad (37.7)$$

and we also derive the **mirror equation**, which relates the image distance d_i to the object distance d_o and to the focal length f of the mirror:

$$\frac{1}{d_o} + \frac{1}{d_i} = \frac{1}{f} \quad (37.8)$$

We will derive these equations for a convex mirror, but they also apply to a concave mirror (Problems 36 and 40).

Start with a simple ray diagram (Fig. 37.33) showing only rays 1 and 3. We'll apply the sign convention later; for now, we will use absolute values of all lengths.

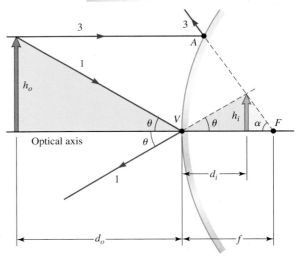

FIGURE 37.33 Geometry for deriving the mirror equation.

Use the two right triangles highlighted in yellow in Figure 37.33 to write expressions for $\tan \theta$.	$\tan \theta = \dfrac{\|h_o\|}{\|d_o\|} = \dfrac{\|h_i\|}{\|d_i\|}$ (1)
Rearrange Equation (1) to come up with an expression for the absolute value of the magnification: $\|M\| = \|h_i\|/\|h_o\|$.	$\|M\| = \dfrac{\|h_i\|}{\|h_o\|} = \dfrac{\|d_i\|}{\|d_o\|}$ (2)
The **magnification of a spherical mirror** is found by applying the first two sign conventions (Table 37.2) to Equation (1). The image height and object height are both positive because both object and image are upright, so the magnification must be positive. The image is virtual, so the image distance is negative. We must insert a negative sign in Equation (1) to be consistent with this convention.	$M = \dfrac{h_i}{h_o} = -\dfrac{d_i}{d_o}$ ✓ (37.7)

Now consider the right triangle highlighted in blue. The horizontal leg of this triangle has length $\lvert f\rvert - \lvert d_i\rvert$. Use this to find an expression for $\tan\alpha$.	$$\tan\alpha = \frac{\lvert h_i\rvert}{\lvert f\rvert - \lvert d_i\rvert} \qquad (3)$$

For a small spherical mirror (compared to the radius of curvature), point A where ray 3 strikes the mirror is almost directly above the vertex V (Fig. 37.34). Thus, AVF is approximately a right triangle.

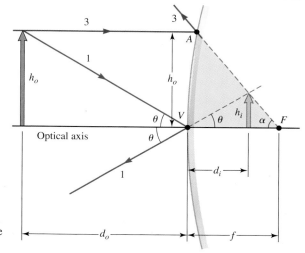

FIGURE 37.34 For a small spherical mirror, the distance from A to V is approximately the height of the object h_o.

Right triangle AVF highlighted in yellow yields another expression for $\tan\alpha$.	$$\tan\alpha = \frac{\lvert h_o\rvert}{\lvert f\rvert} \qquad (4)$$
Set Equation (3) equal to Equation (4) to find another expression for the absolute value of the magnification, $\lvert M\rvert$.	$$\tan\alpha = \frac{\lvert h_i\rvert}{\lvert f\rvert - \lvert d_i\rvert} = \frac{\lvert h_o\rvert}{\lvert f\rvert}$$ $$\lvert M\rvert = \frac{\lvert h_i\rvert}{\lvert h_o\rvert} = \frac{\lvert f\rvert - \lvert d_i\rvert}{\lvert f\rvert}$$
Use Equation (2) to eliminate the object and image heights.	$$\frac{\lvert d_i\rvert}{\lvert d_o\rvert} = \frac{\lvert h_i\rvert}{\lvert h_o\rvert} = \frac{\lvert f\rvert - \lvert d_i\rvert}{\lvert f\rvert} = 1 - \frac{\lvert d_i\rvert}{\lvert f\rvert}$$
Divide both sides by the image distance.	$$\frac{1}{\lvert d_o\rvert} = \frac{1}{\lvert d_i\rvert} - \frac{1}{\lvert f\rvert}$$ $$\frac{1}{\lvert d_o\rvert} - \frac{1}{\lvert d_i\rvert} = -\frac{1}{\lvert f\rvert} \qquad (5)$$
Now apply the second and third sign conventions (Table 37.2) to Equation (5). The object distance is positive because the object is real (in front of the mirror), but the image distance is negative because the image is virtual (behind the mirror). The focal length is also negative because the mirror is convex.	$$\frac{1}{d_o} + \frac{1}{d_i} = \frac{1}{f} \qquad (37.8)$$

:• COMMENTS

Equations 37.7 and 37.8, along with the sign conventions in Table 37.2, hold for all mirrors—even plane ones as shown in the next example. In applying geometric optics equations such as these, we can use any convenient length units because these equations do not involve fundamental constants.

CONCEPT EXERCISE 37.4

Show that all four primary rays produce the same results as those in Figures 37.32 and 37.33.

EXAMPLE 37.5 A Flat Mirror Is a Spherical Mirror

One way to think of a flat mirror is as a special case of a spherical mirror in which the radius of curvature is infinite. Start with the mirror equation and show that, in the limit where $r \to \infty$, the mirror equation is equal to $d_i = -d_o$ (Eq. 37.4) and the magnification is given by Equation 37.5:

$$M = \frac{h_i}{h_o} = -\frac{d_i}{d_o} = 1$$

INTERPRET and ANTICIPATE
Take a moment to see whether the problem makes sense. Imagine a mirror as a patch on a very large sphere (Fig. 37.26), and imagine that sphere growing in size while the size of the patch remains the same. As the sphere grows larger, the patch gets flatter. So, as $r \to \infty$ for a spherical mirror, it becomes a plane mirror.

SOLVE
Because a flat mirror does not have an inside and an outside, as a spherical mirror does, we can start with either a convex or a concave mirror. We'll choose the concave mirror arbitrarily. By Equation 37.6, as the radius of curvature approaches infinity, so does the focal length.

$$f = \frac{r}{2} \quad (37.6)$$

$$\lim_{r \to \infty}\left(f = \frac{r}{2}\right) \to \infty$$

$$f \to \infty$$

Take the limit of the mirror equation (37.8) as $f \to \infty$. The term $1/f$ goes to 0, and the image distance is the negative of the object distance (Eq. 37.4), as for a flat mirror.

$$\lim_{r \to \infty}\left(\frac{1}{d_o} + \frac{1}{d_i} = \frac{1}{f}\right) \to 0$$

$$\frac{1}{d_o} + \frac{1}{d_i} = 0, \text{ so } \frac{1}{d_i} = -\frac{1}{d_o}$$

$$d_i = -d_o \checkmark \quad (37.4)$$

The magnification follows from substituting $d_i = -d_o$ into Equation 37.2. As expected, the magnification is 1.

$$M = \frac{h_i}{h_o} = -\frac{d_i}{d_o} \quad (37.2)$$

$$M = -\frac{-d_o}{d_o} = 1 \checkmark$$

CHECK and THINK
We have shown that a flat mirror is a special case of a spherical mirror, with infinite radius of curvature and focal length. Here's a follow-up question: If an object is infinitely far from a flat mirror, where does its image form? Because the object is very far from the flat mirror, its rays are parallel. The reflected rays cross at the focal point, which is located at infinity. So, the image of an infinitely far object is located at infinity.

EXAMPLE 37.6 Find the Car's Image and Magnification

A car is 3.00 m from a convex mirror in a parking garage. The mirror's radius of curvature is 8.00 m.

A Find the image distance and the magnification. Is the image upright or inverted? Is the image real or virtual?

INTERPRET and ANTICIPATE
To anticipate our results, consider either Figure 37.33 or Figure 37.34. According to these figures, we expect the image to be behind the mirror, upright, and virtual. We'll use the mirror equation and the magnification equation to get exact results, and then confirm our expectations by using the sign conventions.

SOLVE

Find the focal length of this mirror from the radius of curvature (Eq. 37.6). Because the mirror is convex, the radius of curvature and the focal length are negative (Table 37.2).

$$f = \frac{r}{2} \quad (37.6)$$

$$f = \frac{-8.00 \text{ m}}{2} = -4.00 \text{ m}$$

Use the mirror equation (Eq. 37.8) to find the image distance. Be careful; the answer we want is often in the denominator. Solve for $1/d_i$ and then take the inverse (reciprocal). From the sign convention in Table 37.2, the object distance d_o is positive because the object is real. We get a negative image distance (image is virtual).

$$\frac{1}{d_o} + \frac{1}{d_i} = \frac{1}{f} \quad (37.8)$$

$$\frac{1}{d_i} = \frac{1}{f} - \frac{1}{d_o} = -\frac{1}{4.00 \text{ m}} - \frac{1}{3.00 \text{ m}} = -\frac{7.00}{12.0 \text{ m}}$$

$$d_i = -\frac{12.0 \text{ m}}{7.00} = -1.71 \text{ m}$$

To find the magnification, use Equation 37.7. The positive result tells us the image is upright.

$$M = -\frac{d_i}{d_o} \quad (37.7)$$

$$M = -\frac{-1.71 \text{ m}}{3.00 \text{ m}} = 0.571$$

CHECK and THINK

Our calculations are consistent with Figures 37.33 and 37.34. The image is smaller than the object (about 60% of the object's size). This result matches the common experience of looking at images in convex mirrors (Fig. 37.35). The image of the car is smaller than the arrow sign below the mirror.

FIGURE 37.35 A convex mirror allows you to see traffic around the corner.

B How do your answers change if the car is 5.00 m from the mirror?

INTERPRET and ANTICIPATE

The only difference from part A is that the car is farther from the mirror. So, our expectations are the same: negative image distance and positive magnification.

SOLVE

This is the same mirror with the same focal length as in part A; simply substitute the new object distance.

$$\frac{1}{d_i} = \frac{1}{f} - \frac{1}{d_o} = -\frac{1}{4.00 \text{ m}} - \frac{1}{5.00 \text{ m}} = -\frac{9.00}{20.0 \text{ m}}$$

$$d_i = -\frac{20.0 \text{ m}}{9.00} = -2.22 \text{ m}$$

Find the magnification from Equation 37.7 using the new object and image distances.

$$M = -\frac{d_i}{d_o} \quad (37.7)$$

$$M = -\frac{-2.22 \text{ m}}{5.00 \text{ m}} = 0.444$$

Example continues on page 1204 ▶

:• **CHECK and THINK**
As expected, the image is still behind the mirror ($d_i < 0$) and upright ($M > 0$). In part A, the car's distance to the mirror (3.00 m) is less than the focal length (4.00 m). In this part, the car's distance to the mirror (5.00 m) is greater than the focal length. When the car is closer, its image is closer and its magnification is greater.

C How do your answers change if the car is 4.00 m from the mirror?

:• **INTERPRET and ANTICIPATE**
In this part, the object distance is equal to the focal length of the mirror. This problem allows us to see whether this special case produces special results.

:• **SOLVE** Substitute the new value of the object distance.	$\dfrac{1}{d_i} = \dfrac{1}{f} - \dfrac{1}{d_o} = -\dfrac{1}{4.00\text{ m}} - \dfrac{1}{4.00\text{ m}} = -\dfrac{2.00}{4.00\text{ m}}$ $d_i = -\dfrac{4.00\text{ m}}{2.00} = -2.00\text{ m}$
Use Equation 37.7 again to find the magnification.	$M = -\dfrac{d_i}{d_o}$ (37.7) $M = -\dfrac{-2.00\text{ m}}{4.00\text{ m}} = 0.500$

:• **CHECK and THINK**
These results are interesting. When the object distance equals the focal length ($d_o = f$), the image distance is half the object distance ($d_i = 0.5d_o$) and the magnification $M = 0.5$. Combining these results with those of parts A and B, we can say that when the object distance is less than the focal length ($d_o < f$), the image distance is more than half the object distance ($d_i > 0.5d_o$) and $M > 0.5$. When the object distance is greater than the focal length ($d_o > f$), the image distance is less than half the object distance ($d_i < 0.5d_o$) and $M < 0.5$. You can add these observations to your list of expectations when solving problems involving a convex mirror.

37-6 Images Formed by Concave Mirrors

We found that images formed by plane and convex mirrors are behind the mirror, virtual, and upright. Concave mirrors are more complicated because the image may be behind the mirror, virtual, and upright, or it may be in front of the mirror, real, and inverted. The type of image depends on the position of the object. We must consider three possible positions for the object relative to the mirror: (1) closer than the focal point, (2) farther than the focal point, and (3) at the focal point.

The four primary rays for convex mirrors also apply to concave mirrors. Also, three equations

$$f = \frac{r}{2} \tag{37.6}$$

$$M = \frac{h_i}{h_o} = -\frac{d_i}{d_o} \tag{37.7}$$

and the mirror equation

$$\frac{1}{d_o} + \frac{1}{d_i} = \frac{1}{f} \tag{37.8}$$

hold for concave mirrors. The center of curvature C and the focal point F are both in front of the mirror (Fig. 37.28B). So, according to the third sign convention (Table 37.2), the radius of curvature and the focal length are both positive: $r > 0$ and $f > 0$.

Concave Mirror: Object Closer than the Focal Point

In Figure 37.36, the object distance is less than the mirror's focal length: $d_o < f$. We can use any two of the four primary rays to find the image formed by the mirror. We arbitrarily choose rays 1 and 3. Their reflected rays do not intersect in front of the mirror, but if we extend them backward, they intersect behind the mirror. The place where they intersect locates the tip of the image. The image is behind the mirror, virtual, and upright. However, unlike what happens with plane or convex mirrors, the image is larger than the object, so $M > 1$. Also, the image distance is greater than the object distance, $d_i > d_o$.

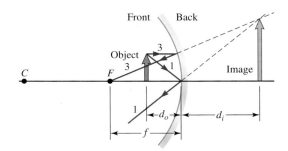

FIGURE 37.36 Ray 1 and ray 3 are used to find the image produced by a concave mirror when the object is closer than the focal point F. Ray 1 strikes the vertex; ray 1 and its reflection are symmetrical around the optical axis. Ray 3 is parallel to the optical axis. Its reflection passes through the focal point F. The reflected rays are traced backward, locating a virtual and upright image behind the mirror.

EXAMPLE 37.7 — Comparing Concave and Convex Mirrors, Part 1

Find the image distance and the magnification for a concave mirror (Fig. 37.36) with focal length $f = 4.00$ m. The object distance $d_o = 3.00$ m. These are the same values from Example 37.6, part A, so compare your results here to the results from that example.

: INTERPRET and ANTICIPATE

As in Example 37.6, apply the mirror equation and the magnification equation in conjunction with the sign conventions.

: SOLVE

Use the mirror equation (Eq. 37.8) to find the image distance. The focal length is positive because the mirror is concave, and the object distance is positive because the object is real (Table 37.2).

$$\frac{1}{d_o} + \frac{1}{d_i} = \frac{1}{f} \qquad (37.8)$$

$$\frac{1}{d_i} = \frac{1}{f} - \frac{1}{d_o} = \frac{1}{4.00 \text{ m}} - \frac{1}{3.00 \text{ m}} = -\frac{1.00}{12.0 \text{ m}}$$

$$d_i = -12.0 \text{ m}$$

Because the image distance is negative, the image is virtual and behind the mirror. The magnitude of the image distance is greater than that of the object distance. Compare this result to those for plane and convex mirrors. For a plane mirror, the image distance and the object distance have the same magnitude. For a convex mirror, the magnitude of the image distance is smaller than that of the object distance. In Example 37.6A, we found $d_i = -1.71$ m, so the image was close to that convex mirror. By contrast, the image is far from the concave mirror in this example.

Find the magnification using $M = -d_i/d_o$ (Eq. 37.7).

$$M = -\frac{d_i}{d_o} = -\frac{-12.0 \text{ m}}{3.00 \text{ m}} = 4.00$$

: CHECK and THINK

The magnification is positive, so the image is upright. The magnification is greater than 1, so the image is larger than the object. Contrast this with Example 37.6, part A, in which $M = 0.571$. The convex mirror produced a small image, but this concave mirror produces an image four times larger than the object. This magnification is why a concave mirror is useful as a shaving mirror, so the shaver can see a larger image of himself.

Concave Mirror: Object Farther than the Focal Point

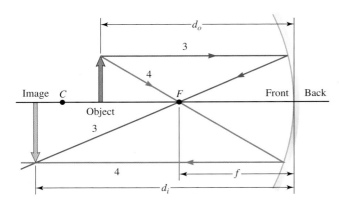

FIGURE 37.37 Ray 3 and ray 4 are used to find the image produced by a concave mirror when the object distance is greater than the focal length. Ray 3 is parallel to the optical axis. Its reflection passes through the focal point F. Ray 4 passes through the focal point on its way to the mirror. Its reflection is parallel to the optical axis. These reflected rays cross in front of the mirror, creating an image that is in front of the mirror, real, and inverted.

In Figure 37.37, the object distance is greater than the mirror's focal length: $d_o > f$. Again, we can use any two of the four primary rays to find the image formed by the mirror. We have arbitrarily chosen rays 3 and 4. Unlike the rays reflected by a flat mirror, these reflected rays *do* meet at a point in front of the mirror. Because the rays converge in front of the mirror, it is possible to project the image on a screen, so the image is *real*. The place where the rays intersect locates the tip of the image. This point is below the optical axis, so the image is *inverted*. These results contrast with those for plane and convex mirrors.

Also, the image in Figure 37.37 is larger than the object, $|M| > 1$, but this is not always the case for real images formed by concave mirrors. Whether $|M|$ is greater or less than 1 depends on the location of the object relative to the center of curvature. When the object is between the center of curvature C and the focal point F (Fig. 37.37), its image is larger than the object ($|M| > 1$). When the object is farther than the center of curvature, however, the image is smaller than the object ($|M| < 1$). We can see this from Figure 37.37 by imagining that the arrow labeled *Image* is actually the object. Then just reverse the arrowheads on the rays to find the location of the image. The image will be smaller than the object ($|M| < 1$).

EXAMPLE 37.8 Comparing Concave and Convex Mirrors, Part 2

Find the image distance and the magnification for a concave mirror (Fig. 37.37) with focal length $f = 4.00$ m and radius of curvature $r = 8.00$ m. The object distance is $d_o = 5.00$ m, so the object is between the focal point and the center of curvature. These are the same values used in Example 37.6, part B, so compare your results here to those results.

• INTERPRET and ANTICIPATE
As in Example 37.7, apply the mirror equation and the magnification equation in conjunction with the sign conventions, and compare your results to those in Example 37.6.

• SOLVE
Use the mirror equation (Eq. 37.8) to find the image distance. The focal length is positive because the mirror is concave, and the object distance is positive because the object is real (Table 37.2).

$$\frac{1}{d_o} + \frac{1}{d_i} = \frac{1}{f} \qquad (37.8)$$

$$\frac{1}{d_i} = \frac{1}{f} - \frac{1}{d_o} = \frac{1}{4.00 \text{ m}} - \frac{1}{5.00 \text{ m}} = \frac{1.00}{20.0 \text{ m}}$$

$$d_i = 20.0 \text{ m}$$

• CHECK and THINK
Because the image distance is positive, the image is real and in front of the mirror. Also, the image distance is greater than the object distance. In Example 37.6, part B, we found $d_i = -2.22$ m, so the image was close to and behind that convex mirror. In this example, where the object is between the focal point and the center of curvature of a concave mirror, the image is far from and in front of that mirror.

• SOLVE
Now, let's find the magnification using $M = -d_i/d_o$ (Eq. 37.7).

$$M = -\frac{d_i}{d_o} = -\frac{20.0 \text{ m}}{5.00 \text{ m}} = -4.00$$

:• **CHECK and THINK**
The magnification is negative, so the image is inverted. The absolute value of the magnification is greater than 1, so the image is larger than the object. Contrast this with Example 37.7, in which we found $M = 4.00$ also for a concave mirror. In that case, however, the object was closer to the mirror than the focal point. In this case, the object is farther from the mirror than the focal point, and although the image is four times larger than the object, this mirror would not be particularly good for shaving because the image is inverted.

EXAMPLE 37.9 Comparing Concave and Convex Mirrors, Part 3

Find the image distance and the magnification for a concave mirror with focal length $f = 4.00$ m and radius of curvature $r = 8.00$ m. The object distance is $d_o = 9.00$ m, so the object is to the left of the center of curvature in Figure 37.37.

:• **INTERPRET and ANTICIPATE**
As in Example 37.8, apply the mirror equation and the magnification equation in conjunction with the sign conventions. But now imagine the rays in Figure 37.37 are reversed, so the image should be real, in front of the mirror, and inverted. The image will be smaller than the object.

:• **SOLVE**
Use the mirror equation (Eq. 37.8) to find the image distance. The focal length is positive because the mirror is concave, and the object distance is positive because the object is real (Table 37.2).

$$\frac{1}{d_o} + \frac{1}{d_i} = \frac{1}{f} \quad (37.8)$$

$$\frac{1}{d_i} = \frac{1}{f} - \frac{1}{d_o} = \frac{1}{4.00 \text{ m}} - \frac{1}{9.00 \text{ m}} = \frac{5.00}{36.0 \text{ m}}$$

$$d_i = 7.20 \text{ m}$$

:• **CHECK and THINK**
As expected, the image distance is positive, so the image is real and in front of the mirror. In this case, the image distance is less than the object distance.

:• **SOLVE**
Find the magnification using $M = -d_i/d_o$ (Eq. 37.7).

$$M = -\frac{d_i}{d_o} = -\frac{7.20 \text{ m}}{9.00 \text{ m}} = -0.800$$

:• **CHECK and THINK**
The magnification is negative, so as expected, the image is inverted. The absolute value of the magnification is less than 1. So, when the object is farther away than the center of curvature of a concave mirror, the image is smaller than the object.

Concave Mirror: Object at the Focal Point

Finally, what can we say about the image when the object is exactly at the focal point of a concave mirror? To answer this question, start with the mirror equation, with $d_o = f$:

$$\frac{1}{d_i} = \frac{1}{f} - \frac{1}{d_o} = \frac{1}{f} - \frac{1}{f} = 0$$

When the object is at the focal point of the concave mirror, the image distance is infinite:

$$d_i \to \infty$$

To understand what this result means for the image, let's review how images are formed. Each point of an object emits rays in all directions. A real image is formed when those rays from each point converge on a detector such as a screen, so that each point of the object is represented by a point in the image. Likewise, a virtual image is composed of points from which rays appear to diverge, and which correspond to points in the object. For an image to form, the reflected rays from a single point must either meet or appear to meet, and their intersection is located at d_i. So, when the object is located at the focal point of a concave mirror, its image is located at infinity, $d_i \to \infty$, which means the reflected rays from each point meet at infinity. But rays that "meet at infinity" are really parallel rays; they don't meet at a finite distance, and so no image forms.

CONCEPT EXERCISE 37.5

Fill in the table by inserting a greater than (>), less than (<), or equals (=) sign in each blank.

Plane mirror	$\|d_i\|$ ___ $\|d_o\|$	M ___ 0	$\|M\|$ ___ 1
Convex mirror with $d_o < \|f\|$	$\|d_i\|$ ___ $\|d_o\|$	M ___ 0	$\|M\|$ ___ 1
Convex mirror with $d_o > \|f\|$	$\|d_i\|$ ___ $\|d_o\|$	M ___ 0	$\|M\|$ ___ 1
Concave mirror with $d_o < \|f\|$	$\|d_i\|$ ___ $\|d_o\|$	M ___ 0	$\|M\|$ ___ 1
Concave mirror with $\|r\| > d_o > \|f\|$	$\|d_i\|$ ___ $\|d_o\|$	M ___ 0	$\|M\|$ ___ 1
Concave mirror with $d_o > \|r\|$	$\|d_i\|$ ___ $\|d_o\|$	M ___ 0	$\|M\|$ ___ 1

CONCEPT EXERCISE 37.6

CASE STUDY Why Use a Secondary Mirror?

As shown in Figure 37.38, reflecting telescopes usually have two mirrors—the primary and the secondary. Light from an astronomical object strikes the primary mirror. The reflected rays then strike the secondary mirror. In the case of a telescope like the HST, rays from the secondary mirror pass through a hole in the primary mirror to an instrument such as a digital camera. The focal length of the HST's primary mirror is 57.6 m. If the secondary mirror were removed, where would the images of distant stars form? The length of the HST's tube is 13.2 m. Explain the statement "A secondary mirror allows the telescope to be compact."

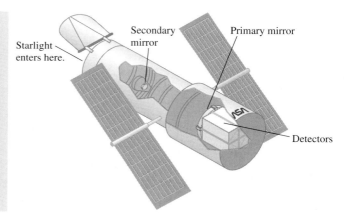

FIGURE 37.38 Diagram of the Hubble Space Telescope.

CONCEPT EXERCISE 37.7

CASE STUDY Where Are Those People?

Figure 37.39 shows the HST's primary mirror in a clean room before the telescope was launched. You can see an image of four people in the mirror, but you cannot see them in the room. Roughly how far are they from the mirror? Is their image virtual or real? In this exercise, you may model the HST's mirror as concave and spherical.

FIGURE 37.39 The HST's primary mirror has a hole in the center. It has been covered in this photo.

EXAMPLE 37.10 CASE STUDY Hubble's Image of the Moon

In April 1999, the HST took a picture of the Moon (Fig. 37.40). The large impact crater in the photo is called Copernicus. Use the information in Concept Exercise 37.6 to find the image distance and magnification, given also the Earth–Moon distance 3.82×10^8 m. The diameter of the Copernicus crater is about 93 km. What is the diameter of its image? In this problem, model the HST as a concave spherical mirror. To check your results, answer these questions: Is the image upright or inverted? Is it real or virtual?

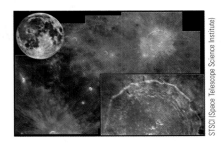

FIGURE 37.40 An HST image of the Moon. The large crater is called Copernicus. The photo of the whole Moon was taken by a ground-based telescope. The lower right image is a close-up of Copernicus.

• INTERPRET and ANTICIPATE
Figure 37.37 is a good ray diagram for this situation because the Moon is very far from the focal point. This diagram shows that the image is inverted and real. So we expect that the magnification is negative and the image distance is positive.

• SOLVE
The object distance is roughly the distance between the Earth and the Moon. The object distance is positive because the object is real. The focal length is 57.6 m (Concept Exercise 37.6). The focal length is positive because the mirror is concave.

$$d_o = 3.82 \times 10^8 \text{ m}$$
$$f = 57.6 \text{ m}$$

When using the mirror equation (Eq. 37.8) for a distant object, we find the image distance is roughly equal to the focal length. This makes sense because the rays from the Moon are roughly parallel, so they focus at the focal point.

$$\frac{1}{d_o} + \frac{1}{d_i} = \frac{1}{f} \quad (37.8)$$

$$\frac{1}{d_i} = \frac{1}{f} - \frac{1}{d_o} = \frac{1}{57.6 \text{ m}} - \frac{1}{3.82 \times 10^8 \text{ m}} \approx \frac{1}{57.6 \text{ m}}$$

$$d_i = 57.6 \text{ m}$$

• CHECK and THINK
As expected, the image distance is positive, so the image is real and in front of the mirror. A telescope requires a real image in order to record that image as a photograph. The image must be projected onto a film or digital detector.

• SOLVE
Find the magnification from Equation 37.7.

$$M = -\frac{d_i}{d_o} \quad (37.7)$$

$$M = -\frac{57.6 \text{ m}}{3.82 \times 10^8 \text{ m}} = \boxed{-1.51 \times 10^{-7}}$$

• CHECK and THINK
As expected, the magnification is negative, so the image is inverted.

• SOLVE
To find the size of the crater's image, use Equation 37.7 again, but as the ratio of image height to object height.

$$M = \frac{h_i}{h_o} \quad (37.7)$$

$$h_i = h_o M = (93 \text{ km})(-1.51 \times 10^{-7})$$

$$h_i = -1.4 \times 10^{-5} \text{ km} = -1.4 \text{ cm}$$

• CHECK and THINK
The negative image height means the image is inverted. And, of course, it makes sense that the crater's image is much smaller than the crater.

EXAMPLE 37.11 Cosmetic Mirror

A mirror manufacturer reports that its cosmetic mirror has a magnification of 5×. Estimate the radius of curvature and the size of the image of a person's eye.

INTERPRET and ANTICIPATE
The mirror must be concave because neither a plane mirror nor a convex mirror produces an image that is larger than the object. For the image to be useful, it must be upright. So the person's face must be closer to the mirror than the focal point (Fig. 37.36). The key to this problem is estimating the distance between the person's face and the mirror. How close is your face to a shaving or cosmetic mirror? Perhaps 5–10 in. Let's say about 7 in. (18 cm).

SOLVE
We have estimated the object distance $d_o = 18$ cm. Use the magnification equation (Eq. 37.7) to find the image distance.

$$M = -\frac{d_i}{d_o} \quad (37.7)$$

$$d_i = -Md_o = -(5)(18 \text{ cm}) = \boxed{-90 \text{ cm}}$$

CHECK and THINK
The image distance is negative, as expected for a virtual image. The image is farther from the mirror than the object, consistent with Figure 37.36.

SOLVE
Because we know the object distance and the image distance, we can find the focal length of the mirror from the mirror equation (Eq. 37.8).

$$\frac{1}{d_o} + \frac{1}{d_i} = \frac{1}{f} \quad (37.8)$$

$$\frac{1}{f} = \frac{1}{18 \text{ cm}} + \frac{1}{-90 \text{ cm}}$$

$$f = 22.5 \text{ cm}$$

The radius of curvature is twice the focal length (Eq. 37.6). As with all estimates, we report our result to one significant figure.

$$f = \frac{r}{2} \quad (37.6)$$

$$r = 2f = 2(22.5 \text{ cm}) = \boxed{50 \text{ cm}}$$

To estimate the size of the image of an eye, start by estimating the size of your eye. Perhaps your eye is about an inch in diameter, or about 2 cm. Use Equation 37.7 (image height to object height) to find the size of the eye's image.

$$M = \frac{h_i}{h_o} \quad (37.7)$$

$$h_i = Mh_o = (5)(2 \text{ cm}) = \boxed{10 \text{ cm}}$$

CHECK and THINK
Typically a cosmetic mirror is about 20 cm tall, so if it has a radius of curvature of 50 cm, the mirror has the shape of a very shallow bowl. Such a mirror would make an image of your eye that is 10 cm in diameter, or about half the size of your hand. An image that size is useful for applying makeup. The image of your eye in such a mirror is larger than the 1.4-cm image of the Copernicus crater produced by the HST. Unlike a cosmetic mirror, a telescope is designed to gather light, not to produce an enlarged image.

37-7 Spherical Aberration

In the past three sections, we considered only spherical mirrors that are small compared to their radii of curvature: $D \ll r$. We even modeled the HST's primary mirror as a small spherical mirror. In this section, we show that a large spherical mirror produces a blurry image and that the HST's mirror cannot always be modeled as a small spherical mirror.

Figure 37.41 shows two spherical mirrors—one concave and the other convex. Each is larger than its radius of curvature ($D > r$). Consider an object such as a star that is very far away so that its rays are parallel. The rays close to the optical axis come together in front of the concave mirror at the focal point (Fig. 37.41A). If the mirror were small ($D \ll r$), all the parallel rays that struck the mirror would be reflected through the focal point. However, this mirror is large, so reflections of the outer rays do not cross at the focal point, but instead cross at a point slightly closer to the mirror. Rays don't end at the point where they cross. If you place a small screen at the focal point to capture the image of a distant star, you see the light from the inner rays that cross at a single point, and you also see light from the outer rays above and below this point. Such a screen is in the **focal plane**, perpendicular to the optical axis and passing through the focal point. A distant star is a point source of light, and ideally its image on the focal plane should be a dot. But the image produced by a large concave mirror is blurry because it includes light from the outer rays.

Large convex mirrors also produce blurry images. Again, the rays from a distant object are parallel, and the inner rays seem to diverge from the focal point behind the mirror. But the outer rays seem to diverge from a point slightly closer to the mirror (Fig. 37.41B). If you look at the virtual image in this mirror, it also looks blurry.

A defect in an optical system that results in a poor image is called **aberration**. **Spherical aberration** results from rays far from the optical axis, or rays that make a large angle with the optical axis of a spherical mirror. Such rays strike points on the mirror that are far from its vertex. **Paraxial rays** are those that are nearly parallel and close to the optical axis, striking near the vertex of the mirror. We considered only small mirrors in earlier sections because any rays that strike a small mirror ($D \ll r$) must be paraxial.

For many practical applications, spherical aberration is avoided by making spherical mirrors small compared to their radii of curvature, such as a cosmetic mirror (Example 37.11). However, small spherical mirrors still suffer from some spherical aberration. In the case of a cosmetic or parking garage mirror, the aberration does not diminish the mirror's usefulness. But even a little spherical aberration can be a major problem in some applications, such as in astronomy. Astronomy requires the best images possible, so telescopes are designed to avoid all aberration, including spherical aberration.

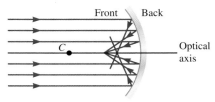

A. Concave mirror

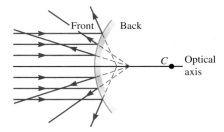

B. Convex mirror

FIGURE 37.41 A. Parallel rays incident on a concave mirror are focused in front of the mirror. The inner rays focus at a single point, but the outer rays focus at another point closer to the mirror. **B.** The rays reflected by a convex mirror appear to come from points behind the mirror. The inner rays appear to originate from one point, but the outer rays appear to originate from another point, closer to the mirror.

> **CASE STUDY** **Hubble's Primary Mirror**
>
> A reflecting telescope consists of two mirrors—the primary and the secondary (Fig. 37.38). The front of the primary mirror faces the sky and plays the most important role in gathering light from distant objects. For the moment, consider a special type of telescope design called a *prime focus*, which uses only a single (primary) mirror. Light from a distant object forms an image at the focal point of the mirror. The astronomer captures that image either on photographic film or on a sensor known as a charge-coupled device (CCD), like the sensor in a digital camera. Because the image must be projected on a film or sensor, the image must be real, so the primary mirror must be concave. If the primary mirror is spherical, the parallel rays do not form a single image in the focal plane.
>
> However, if the primary mirror has a parabolic or hyperbolic shape, even the rays far from the optical axis come together at the focal point (Fig. 37.42). (A slice through a parabolic mirror is a parabola.) Astronomical telescopes often have parabolic or hyperbolic primary and secondary mirrors to avoid spherical aberration. But there are drawbacks to these mirrors. If the rays from a distant object are not parallel to the optical axis, they do not form a clear image at the focal point. When a star is at the center of a telescope's field, it forms a clear image on the focal plane. However, stars that are elsewhere than at the center of the field form distorted images. These images look like comets' tails and are known as *comas*, so this sort of aberration is called **coma**. Parabolic and hyperbolic mirrors are also difficult to make, which led to the problem with the HST's primary mirror.
>
> The HST was designed to have two hyperbolic mirrors. The primary mirror was incorrectly shaped due to a testing procedure used during the manufacturing process. Its incorrect shape led to spherical aberration: Parallel rays did not come together at a single focal point, and the HST's images were blurry.

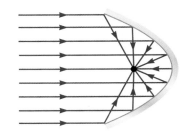

FIGURE 37.42 All rays focus at a single point for a parabolic mirror.

The problem was corrected by the addition of other optical systems. Once the actual shape of the primary mirror was measured, engineers designed a system of mirrors called COSTAR (*c*orrective *o*ptics *s*pace *t*elescope *a*xial *r*eplacement) to correct the problem. This system is like the corrective lenses given to a nearsighted person. COSTAR was installed in December 1993. Figure 37.13 shows that COSTAR was successful. There is an important lesson in this story: People make mistakes, but they can also correct those mistakes. Since 1993, the HST has contributed to major discoveries in our understanding of the Universe.

SUMMARY

① Underlying Principles

Geometric optics: When light encounters only obstacles and apertures that are at least 1 mm in diameter, the ray model is valid. According to this model, light travels in straight lines represented by rays. The ray model ignores the wave properties of light.

✪ Major Concepts

1. According to the **law of reflection**, the angle of incidence equals the angle of reflection:
$$\theta_i = \theta_r \tag{37.3}$$
The incident ray, the reflected ray, and the perpendicular (normal) line all lie in a single plane, called the plane of incidence (Fig. 37.7).

2. **Magnification** (also known as linear magnification) is given by
$$M = \frac{h_i}{h_o} = -\frac{d_i}{d_o} \tag{37.7}$$
where the sign conventions in Table 37.2 (page 1199) must be used.

3. Light rays from each point of an object converge to create each point of a **real image**, whereas light rays appear to diverge from each point of a **virtual image**. A real image can appear on a screen, but a virtual one cannot.

4. The **mirror equation** relates the image distance d_i to the object distance d_o and the focal length f of the mirror:
$$\frac{1}{d_o} + \frac{1}{d_i} = \frac{1}{f} \tag{37.8}$$
where the sign conventions in Table 37.2 must be used.

◐ Special Cases

1. A **plane mirror** is a flat mirror that has an infinite radius of curvature and produces an upright and virtual image with $M = 1$ (Fig. 37.17). The image distance is $d_i = -d_o$ (Eq. 37.4).

2. A **convex spherical mirror** has a negative (finite) radius of curvature and produces an upright and virtual image (Fig. 37.32).

3. A **concave spherical mirror** has a positive (finite) radius of curvature. It produces either an upright, virtual image (Fig. 37.36) or an inverted, real image (Fig. 37.37).

⊙ Tools

The elements of a **ray diagram** (Fig. 37.15) are:
1. A thick line representing the **mirror** seen edge-on
2. An arrow representing the **object**
3. A small number of **emitted rays** coming from key points on the object, such as the tip or base of the arrow
4. **Reflected rays** for each emitted ray
5. An arrow representing the **image**

When drawing a ray diagram, you may find two or more of the four **primary rays** described in Table 37.1 to be particularly helpful.

PROBLEMS AND QUESTIONS

A = algebraic   C = conceptual   E = estimation   G = graphical   N = numerical

37-1 Geometric Optics

1. **N** A camera obscura is used to form an image of a distant object. If the object is 10.0 m away from the aperture and the magnification of the image is −0.15, what is the image distance?

2. **C** Because you should never stare directly into the Sun, pinhole cameras are very handy for observing solar eclipses. Come up with a simple, inexpensive design for a pinhole camera that you could use to observe an eclipse. Describe your design and how you would use the camera. A sketch may be helpful.

3. An image is formed from an object using a camera obscura. The image is one-fifth the height of the object and is oriented upside-down compared to the orientation of the object.
   a. **N** What is the magnification of the image? Answer with two significant figures.
   b. **A** Write an expression for the image distance in terms of the object distance. Answer with two significant figures.

4. **E** The image of a person seen through a camera obscura seems to be 1.5 ft tall. If the image distance is roughly 7.5 ft, estimate the distance to the person.

5. **N** We have all enjoyed sitting in the shade of a leafy tree. If the leaves of the shade tree are not too densely packed, light from the Sun is able to filter through the tiny openings created by overlapping leaves. These openings essentially form tiny pinholes that project potentially hundreds of circular images of the Sun onto the forest floor. If one of the projected images of the Sun you observe on the forest floor is 4.1 cm in diameter, how high above the ground in the tree canopy is the pinhole that projected this image? *Hint*: You may need to look up the diameter of the Sun and the distance from the Sun to the Earth.

6. **C** Figure P37.6 shows the image of a person seen through a camera obscura. Sketch the person as if viewed from the camera's location.

FIGURE P37.6

37-2 Law of Reflection

7. **C** In this chapter, we study the reflection of visible light. Have we ignored other parts of the electromagnetic spectrum because radiation at other frequencies does not reflect? If so, explain why other wavelength radiation cannot reflect. If not, give an example of other wavelength radiation reflecting from some surface.

Problems 8 and 9 are paired.

8. **A** Two mirrors are perpendicular as shown in Figure P37.8. A narrow beam of light is incident on one mirror. The angle of incidence is θ_i. Find an expression for the angle of reflection θ_r from the second mirror.

9. **N** Two mirrors are perpendicular as shown in Figure P37.8. A narrow beam of light is incident on one mirror. The angle of incidence is 22.3°. Find the angle of reflection from the second mirror.

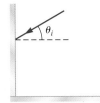

FIGURE P37.8 Problems 8 and 9.

10. **N** A light ray enters a region between two parallel mirrors at the angle shown in Figure P37.10. The mirrors are separated by 1.00 m and are 2.00 m long. How many times will the light ray be reflected before it exits the region between the mirrors?

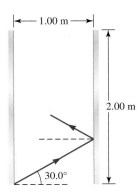

FIGURE P37.10

Problems 11 and 12 are paired.

11. **A** Two mirrors make a 45.0° angle as shown in Figure P37.11. A narrow beam of light is incident on one mirror. The angle of incidence is θ_i. Find an expression for the angle of reflection θ_r from the second mirror.

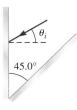

FIGURE P37.11 Problems 11 and 12.

12. **N** Two mirrors make a 45.0° angle as shown in Figure P37.11. A narrow beam of light is incident on one mirror. The angle of incidence is 22.3°. Find the angle of reflection from the second mirror.

13. **N** Figure P37.13 shows a beam of light incident on one of two parallel mirrors at an angle of 12.0°. How many total reflections will the beam undergo in the mirrors?

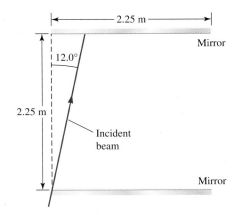

FIGURE P37.13

14. **N** One way to train a dog to stay when you leave the room is to use a mirror so that you can see the dog from outside the room. Many doors have shiny metal plates at the bottom that are convenient for this purpose (Fig. P37.14). At what angle should the door be placed with respect to the threshold in order for the trainer to see the dog? Use the dimensions given in the figure.

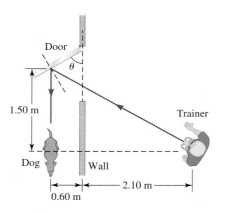

FIGURE P37.14

Problems 15 and 16 are paired.

15. **N** Light rays strike a plane mirror at an angle of 45.0° as shown in Figure P37.15. At what angle should a second mirror be placed so that the reflected rays are parallel to the first mirror?

16. **N** Determine the angle between two plane mirrors such that a ray of light incident on the first mirror and parallel to the second mirror is reflected from the second mirror and emerges parallel to the first mirror as shown in Figure P37.16.

17. **N** A monochromatic beam of light enters a square enclosure with mirrored interior surfaces at an angle of incidence θ_i (Fig. P37.17). If each of the mirrored walls (other than the one with the opening) reflects the beam only once, what is the value of θ_i for which the beam will exit the enclosure through the same hole?

37-3 Images Formed by Plane Mirrors

18. **C** In Figure P37.18, you can see the painting reflected in the mirror. Why can't we see an image of the painting on the floor?

19. **N** You and your roommate share a plane mirror. She stands behind you so that her face is 18 in. behind your face. Your face is 3.5 ft from the mirror. How far does the image of your roommate's face appear to be from you? Give your answer in feet.

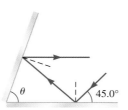

FIGURE P37.15

FIGURE P37.16

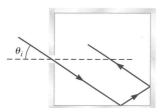

FIGURE P37.17

FIGURE P37.18

20. **G** Two perpendicular mirrors are used to produce an image that is not reversed (Fig. 37.21B). In Example 37.4 (page 1194), we used a ray diagram to show how this is possible. Repeat the example, but this time choose mirror 2 as the first mirror and mirror 1 as the second mirror.

21. **N** At a department store, you adjust the mirrors in the dressing room so that they are parallel and 5.0 ft apart. You stand 2.0 ft from one mirror and face it. You see an infinite number of reflections of your front and back.
    a. How far from you is the first "front" image?
    b. How far from you is the first "back" image?

22. **A** An object moves toward a plane mirror with speed v_p at an angle θ with respect to the normal to the interface as shown in Figure P37.22. Determine the relative velocity between the object and the image.

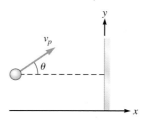

FIGURE P37.22

23. **N** A rectangular room used for ballet practice is 12.0 m long and 6.20 m wide. The ballet teacher has mounted a 1.45-m-wide mirror on one of the 12.0-m-long walls of the room so that she can at all times observe the students lined at the handrail along the opposite wall. At one instant, the teacher is facing away from the students, a distance of 1.10 m from the mirror. What is the extent of the line of students that the teacher can see while looking forward?

24. **A** Two perpendicular mirrors are used to produce a true image (Fig. P37.24). The distance from the tip of the arrow to mirror 1 is d_1, and the distance from the tip of the arrow to mirror 2 is d_2. Find an expression for the distance r between the tip of the arrow and its true image.

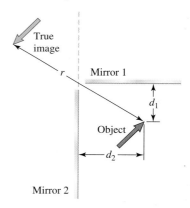

FIGURE P37.24

25. **N** Diane wants to buy a flat wall mirror in which she can see her full image. If she is 165 cm tall, what is the minimum height of the mirror she should purchase?

26. **N** A statue on a pedestal is placed in a room with flat mirrors on the parallel northern and southern walls, producing an infinite number of images in each mirror. If the walls are 7.00 m apart and the statue is 1.00 m from the northern wall, what are the distances from the statue to the first four images formed in the northern mirror?

37-4 Spherical Mirrors

27. A Derive $f = r/2$ (Eq. 37.6) for a concave spherical mirror.

37-5 Images Formed by Convex Mirrors

28. N An object is placed 30.0 cm in front of a convex mirror. If the focal length of the mirror is -20.0 cm, find the distance between the object and the final image.

29. N A convex mirror with a radius of curvature of -25.0 cm is used to form an image of an arrow that is 10.0 cm away from the mirror. If the arrow is 2.00 cm tall and inverted (pointing below the optical axis), what is the height of the arrow's image?

30. N The magnitude of the radius of curvature of a convex spherical mirror is 22.0 cm.
   a. What are the location and magnification of the image of an object placed 16.0 cm in front of the mirror? Is this image upright or inverted?
   b. What are the location and magnification of the image of an object placed 44.0 cm in front of the mirror? Is this image upright or inverted?

31. N When an object is placed 60.0 cm from a convex mirror, the image formed is half the height of the object. Where should the object be placed so that the height of the image becomes one-third the height of the object?

32. The image formed by a convex spherical mirror with a focal length of magnitude 12.0 cm is located one-fourth of the object–mirror distance from the mirror.
   a. N What is the distance of the object from the mirror?
   b. C Is the image formed by the mirror upright or inverted?
   c. N What is the magnification of this image?

33. N An object is placed 25.0 cm from the surface of a convex mirror and an image is formed. The same object is placed 20.0 cm in front of a plane mirror and, again, an image is formed. Suppose in each case the object is located at $x = 0$ and the mirrors lie along the positive x axis. If the images formed in the two mirrors lie at the same x coordinate, determine the radius of curvature of the convex mirror.

34. At an ice cream shop, a cylindrical mirror is used to give customers the impression that they are taller and thinner than they really are.
   a. C How should the cylinder be oriented? If the cylinder were tipped by 90°, how would the customers look?
   b. C Is the image upright or inverted? Real or virtual?
   c. E If the image of the customer's width is about 75% of the actual width, estimate the mirror's radius of curvature.

35. N A convex mirror magnifies an upright arrow so that the image of the arrow is one-third the height of the arrow. If the arrow is 35.0 cm away from the mirror, what is the focal length of the mirror?

37-6 Images Formed by Concave Mirrors

36. A Derive the mirror equation $1/d_o + 1/d_i = 1/f$ (Eq. 37.8) for a concave mirror.

37. N A dental hygienist uses a small concave mirror to look at the back of a patient's tooth. If the mirror is 1.33 cm from the tooth and the magnification is 2.00, what is the mirror's focal length?

38. C Come up with a procedure to find the focal length of a concave mirror, such as a spoon.

39. N The magnitude of the radius of curvature of a concave spherical mirror is 34.0 cm.
   a. What are the location and magnification of the image of an object placed 60.0 cm in front of the mirror? Is this image real or virtual? Is it upright or inverted?
   b. What are the location and magnification of the image of an object placed 34.0 cm in front of the mirror? Is this image real or virtual? Is it upright or inverted?

40. A Derive $M = h_i/h_o = -d_i/d_o$ (Eq. 37.7) for a concave mirror.

41. C A researcher wishes to capture the image of an object from a mirror using a CCD camera. Why is a concave mirror better suited to this task than a convex mirror?

Problems 42 and 43 are paired.

42. N A concave mirror with a radius of curvature of 25.0 cm is used to form an image of an arrow that is 10.0 cm away from the mirror. If the arrow is 2.00 cm long and inverted (pointing below the optical axis), what is the height of the arrow's image?

43. N A concave mirror with a radius of curvature of 25.0 cm is used to form an image of an arrow that is 30.0 cm away from the mirror. If the arrow is 2.00 cm tall and inverted (pointing below the optical axis), what is the height of the arrow's image?

44. C In your physics class, you may have seen a *mirage* demonstration (Fig. P37.44). One such demonstration uses a device that consists of two concave mirrors with the same radius of curvature. One mirror has a small hole near its vertex. An object, such as a strawberry, is placed at the vertex of the other mirror. The mirror with the hole is placed on top of the first mirror, so that their shiny surfaces point toward each other. The image of the strawberry hovers at the hole. Use a ray diagram to show how this works. *Hint*: The focal length of each mirror is about equal to the distance between their vertices.

FIGURE P37.44

45. N A concave spherical mirror with a focal length of magnitude 2.10 m is used to form an image of the Sun. The angular size of the Sun's disk on the Earth is 0.533°.
   a. What is the position of the solar image formed by the mirror?
   b. What is the radius of the solar image formed by the mirror?

46. The upright image formed by a concave spherical mirror with a focal length of 14.0 cm is 2.50 times larger than the object.
   a. N What is the distance of the object from the mirror?
   b. C Is the image formed by the mirror real or virtual?
   c. G Draw a ray diagram showing the locations of the object and the image by tracing at least three rays.

47. N A concave mirror magnifies an upright arrow so that the image of the arrow is three times taller and inverted. If the arrow is 35.0 cm from the mirror, what is the focal length of the mirror?

48. An object is located 12.0 cm in front of a concave spherical mirror that has a radius of curvature of 32.0 cm.
   a. N What are the location and magnification of the image?
   b. G Draw a ray diagram showing the locations of the object and the image by tracing at least three rays.

49. N The focal length of a concave mirror is 30.0 cm. Find the two positions of the object in front of the mirror so that the image height is three times greater in magnitude than the object height.

Problems 50 and 51 are paired.

50. **N** An object in front of a concave mirror has a real image that is 11.0 cm from the mirror. The mirror's radius of curvature is 20.0 cm.
    a. What is the object distance?
    b. What is the magnification?

51. **N** An object in front of a concave mirror has a virtual image that is 11.0 cm from the mirror. The mirror's radius of curvature is 20.0 cm.
    a. What is the object distance?
    b. What is the magnification?

52. **C** **CASE STUDY** The United States Naval Academy owns the primary mirror to a defunct telescope. The academy staff would like to put the mirror into a new telescope, and to do so they must construct a telescope tube with the correct dimensions. The mirror is 20 in. in diameter. The length of the tube depends in part on the focal length of the primary mirror. Of course, the secondary mirror will help to fold the optics, but no further planning can be done without knowing the focal length of the primary mirror. Come up with an experiment to find the focal length. The mirror's surface must remain clean and free of scratches.

53. **N** What must be the radius of curvature of a concave mirror to form an image of the Sun 2.0 cm in diameter?

37-7 Spherical Aberration

54. **C** How does the radius of curvature compare to the size of the convex mirror in Figure 37.25A (page 1196)? Explain why the image of the artist's face seems more realistic than the image of the ceiling.

General Problems

55. An object 3.50 cm high is located 18.0 cm in front of a convex spherical mirror that has a radius of curvature of magnitude 8.00 cm.
    a. **N** What is the location of the image?
    b. **N** What is the magnification of the image?
    c. **C** Is the image upright or inverted?
    d. **N** What is the height of the image?

56. A screen is located 6.75 m from a spherical mirror. An image with $h_i = -4.00 h_o$ is to be formed on the screen.
    a. **C** Should a concave or convex mirror be selected for this task?
    b. **N** What is the focal length of the mirror that will accomplish this task?
    c. **N** What should the object distance be in this case?

Problems 57, 58, 59, and 60 are grouped.

57. **G** You see the image of a sign through a camera obscura. The sign reads FOX AND SOX. Sketch the image.
58. **G** You see the image of a sign in a plane mirror. The sign reads FOX AND SOX. Sketch the image.
59. **G** You see the image of a sign produced by a convex mirror. The sign reads FOX AND SOX. Sketch the image.
60. **G** You see the image of a sign produced by a concave mirror. The sign reads FOX AND SOX. Sketch the image. There are two possible images; sketch them both and explain under what circumstances each one is possible.
61. **A** An object is placed midway between two concave spherical mirrors as shown in Figure P37.61. The distance between the mirrors is D, and they have the same focal length. Determine the value(s) of D in terms of the focal length f for which only one image is formed in each mirror.

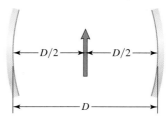

FIGURE P37.61

62. The upright image from a spherical mirror is 30.0% of the object's size and located 33.0 cm away from the object.
    a. **C** Is this a convex or concave mirror?
    b. **N** What is the distance between the object and the mirror?
    c. **N** What is the focal length of the mirror?

63. **N** A critical characteristic of light rays that are reflected diffusely from an object such as your friend's nose is the extent to which adjacent rays diverge from each other. Imagine holding a gumball 2.0 cm in diameter in front of your friend's nose as shown in Figure P37.63. By what angle do the two tangential rays shown in the diagram diverge from each other when the center of the gumball is **a.** $D = 5.00$ cm, **b.** $D = 20.0$ cm, **c.** $D = 100.0$ cm, and **d.** $D = 100.0$ km from the tip of your friend's nose?

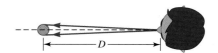

FIGURE P37.63

64. **C** An arrow points above the optical axis, and a spherical mirror forms an image of that arrow. Is it possible to determine initially whether or not the mirror forming the image is convex or concave? Why or why not? Does your answer change if you are told that the image is also upright, pointing above the optical axis?

Problems 65 and 66 are paired.

65. **N** The height of an inverted image formed by a concave spherical mirror is 5.00 times the height of the object. The object and the image are separated by 48.0 cm. What is the focal length of the mirror?

66. **N** The height of an image formed by a convex spherical mirror is 40.0% of the object's height. The object and the image are separated by 48.0 cm. What is the focal length of the mirror?

Problems 67 and 68 are paired.

67. **E** Observe your reflection in the back of a spoon. From that observation, estimate the radius of curvature of the spoon. *Hint*: Model the spoon as a spherical mirror.

68. **E** In Problem 67, you used the back of a spoon to see your face. Now imagine you flip the spoon over to see your face in its bowl. Use your estimate from Problem 67 to find the diameter of your face's image in the bowl of the spoon.

69. **N** A small convex mirror and a large concave mirror are separated by 1.00 m, and an object is placed 1.40 m to the left of the concave mirror (Fig. P37.69). The concave mirror forms an image of this object at distance $d_i = 25.0$ cm. This image is then reflected in the convex mirror, which forms an image a distance of 8.00 cm behind the convex mirror. What is the focal length of the small convex mirror?

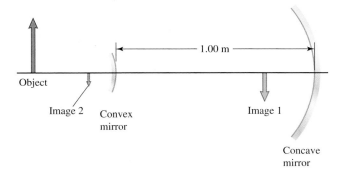

FIGURE P37.69

Problems 70 and 71 are paired.

70. Two plane mirrors are facing each other, placed on opposite walls of a room. The mirrors are separated by a distance D. A vase is placed at the midpoint between the mirrors. An infinite number of images of the vase are created in each mirror. (Try this at home!) The first image created in each mirror is of the vase itself, and those images are separated by a total distance $2D$.
  a. **G** Sketch the scenario and the first image created in each mirror to verify the distance between the images.
  b. **A** The first image in each mirror creates a second image in each opposite mirror. How far apart is this second set of images?

71. **A** For the scenario in Problem 70, derive an expression in terms of D for the distance between any corresponding pair of images in both mirrors. You should use a variable like n to indicate the pair of corresponding images to which you are referring. For example, the first set described here would be $n = 1$. The second set ($n = 2$) would be the set referred to in part (b) of Problem 70.

72. **C** A light ray is traveling along the positive y axis toward the origin. A rotating plane mirror is located at the origin and can be oriented to reflect the incoming ray. If you want the reflected ray to travel along the negative x axis, how should the mirror be oriented? How does the answer change if you want the reflected ray to travel along the positive x axis?

73. **A Fermat's principle of least time for reflection.** A ray of light traveling in a medium with speed v leaves point A and strikes a reflecting surface at point O, a horizontal distance x from point A as shown in Figure P37.73. The reflected ray reaches point B, where the horizontal distance between A and B is L.
  a. Derive an expression for the time t required for the light to travel from A to B in terms of the parameters labeled in the figure.
  b. Now take the derivative of t with respect to x. What is the condition for which the ray of light will take the shortest time to travel from A to B?

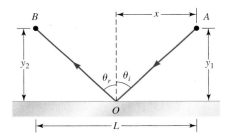

FIGURE P37.73

38 Refraction and Images Formed by Refraction

Key Questions

How do we use the law of refraction to find the location and magnification of images formed by refraction?

In what ways are thin lenses like spherical mirrors?

- 38-1 **Law of refraction** 1219
- 38-2 **Total internal reflection** 1221
- 38-3 **Dispersion** 1223
- 38-4 **Refraction at spherical surfaces** 1226
- 38-5 **Thin lenses** 1232
- 38-6 **Images formed by diverging lenses** 1238
- 38-7 **Images formed by converging lenses** 1240
- 38-8 **The human eye** 1243
- 38-9 **One-lens systems** 1248
- 38-10 **Multiple-lens systems** 1252

❶ Underlying Principles

No new principles are introduced in this chapter.

✪ Major Concepts

1. Law of refraction
2. Total internal reflection
3. Dispersion
4. Thin-lens equation
5. Magnification equation for thin lenses
6. Lens maker's equation
7. Angular magnification

▶ Special Cases

1. Diverging lens
2. Converging lens

⊙ Tools

Ray diagrams and primary rays for thin lenses

Your brain is working a lot harder than you think. When you look at your dog, the light reflected from your dog enters your eye. The lens in your eye bends or refracts this light to form an image on your retina. But the image is upside down! Your dog's feet are at the top of the image, and his head is at the bottom. This upside-down image is then carried to your brain for interpretation. The world appears right-side-up to you because your brain flips the image. Newborn babies see upside-down images, and their brains learn to flip the images in a few days. If you wore lenses that inverted the image projected onto your retina, your brain would take a few days to learn how *not* to flip the image.

This chapter is about refraction and images formed by refraction. Many real images formed by lenses are inverted.

38-1 Law of Refraction

LAW OF REFRACTION
✪ Major Concept

Chapter 37 was about mirrors and practical applications of reflection; this chapter is about lenses and practical applications of refraction. Any image you see is the result of refraction in your eye. If your vision is impaired, you may need corrective lenses. Our study of images formed by mirrors is helpful here because we use much of the same language and mathematics to study the images formed by refraction. Before we study the images formed by lenses, however, we need to consider what happens to a light ray when it passes from one transparent medium into another.

Recall that the speed of light in a medium depends on the medium's index of refraction n such that $v = c/n$ (Eq. 36.8). The index of refraction depends on the wavelength of the light and the type of medium (Table 38.1). So, the speed at which light propagates changes as it passes from one medium into another. Recall that when a wave's propagation speed changes, the wave's path bends or *refracts* (Section 17-8). So, light bends as it is transmitted from one medium into another because the speed of light depends on the medium.

TABLE 38.1 Index of refraction. (Unless otherwise specified, use n for yellow light.)

Medium	$n, \lambda = 589$ nm (yellow)	$n, \lambda = 486$ nm (blue)	$n, \lambda = 656$ nm (red)
Air	1.0002926		
Cubic zirconia	2.14		
Diamond	2.417		
Fused quartz	1.458		
Heavy flint glass	1.890	1.919	1.879
Ice	1.3049		
Quartz	1.54		
Water	1.333	1.337	1.331
Zinc crown glass	1.517	1.523	1.514

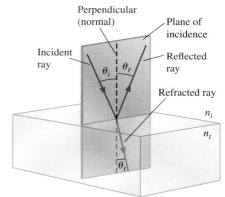

The term *refraction* is used to describe the bending of light as it propagates from an incident medium into a transmitted medium. The **law of refraction** has two parts. The first part states that the refracted ray, the incident ray, and the normal all lie in a single plane—the plane of incidence. Combining this law with the law of reflection, we can say that all three rays (incident, reflected, and refracted) lie in the plane of incidence (Fig. 38.1). The second part of the law of refraction is a mathematical relationship known as **Snell's law**, in honor of the Dutch mathematician and physicist Willebrord van Roijen Snell. You can derive Snell's law for the refraction of light by starting with $\sin \theta_2 = (v_2/v_1) \sin \theta_1$ (Eq. 17.26) and then substituting $v = c/n$ (Eq. 36.8) for the speed of light in each medium. In optics, it is convenient to use the subscripts t and i, and we write Snell's law as

$$n_t \sin \theta_t = n_i \sin \theta_i \quad (38.1)$$

where θ_t and θ_i are the *t*ransmitted and *i*ncident angles (Fig. 38.1), and n_t and n_i are the indices of refraction for the transmitted and incident media, respectively. According to Snell's law, when $n_t > n_i$, the light is bent toward the normal: $\theta_t < \theta_i$.

FIGURE 38.1 The angle of incidence θ_i and the angle of refraction θ_t are measured with respect to the line normal to the surface. The incident ray, the reflected ray, and the refracted ray all lie in the plane of incidence.

CONCEPT EXERCISE 38.1

Light travels from air into glass. Which sketch in Figure 38.2 correctly shows the incident, reflected, and refracted beams? *Hint*: Consider the law of reflection (Section 37-2).

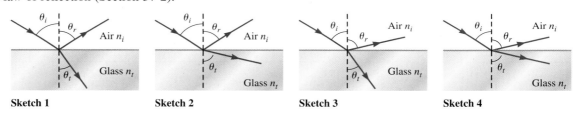

FIGURE 38.2   Sketch 1   Sketch 2   Sketch 3   Sketch 4

EXAMPLE 38.1 — Glass or Diamond?

Suppose you don't know whether an object is made of glass, cubic zirconia, or diamond. You test the material by shining a laser at its surface so that the incident beam makes a 15.0° angle with the normal. You find that the refracted beam makes a 6.15° angle with the normal. What is the object made of? What is the speed of light in that medium?

INTERPRET and ANTICIPATE
Use Snell's law to find the index of refraction of the object. Then use Table 38.1 to identify the material. Once you know the index of refraction, find the speed by dividing c by that index of refraction.

SOLVE
The light travels from air with an index of refraction n_i. We know the angle of incidence and the angle of refraction, so we use Snell's law (Eq. 38.1) to solve for n_t.

$$n_t \sin \theta_t = n_i \sin \theta_i \quad (38.1)$$

$$n_t = \frac{n_i \sin \theta_i}{\sin \theta_t}$$

The incident light is in air, so we find the index of refraction for air in Table 38.1. The angles are given in the problem statement.

$$n_t = \frac{(1.00029) \sin 15.0°}{\sin 6.15°} = 2.42$$

According to Table 38.1, this is the index of refraction for **diamond**.

To find the speed of light in diamond, use Equation 36.8.

$$n = \frac{c}{v} \quad (36.8)$$

$$v = \frac{c}{n_t} = \frac{3.00 \times 10^8 \text{ m/s}}{2.42} = \boxed{1.24 \times 10^8 \text{ m/s}}$$

CHECK and THINK
It makes sense that when light passes from air into diamond, the beam is bent toward the normal because the index of refraction of diamond is higher than the index of refraction of air. When light passes from diamond into air, the light is bent away from the normal. It also makes sense that the speed of light in diamond is much less than it is in a vacuum. In a medium with a high index of refraction, light travels slowly.

EXAMPLE 38.2 — From Water into Glass

Light travels from water into flint glass as shown in Figure 38.3. The angle of incidence is 25.7°. Find the angle of reflection and the angle of refraction. Draw a sketch showing the incident, reflected, and refracted beams.

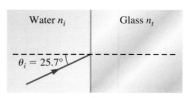

FIGURE 38.3

INTERPRET and ANTICIPATE
We need to use the law of reflection and the law of refraction to find the angles. We expect that the refracted beam is bent toward the normal because the light travels from a medium with a low index of refraction into a medium with a higher index of refraction. Once we've made our calculations, we represent the three beams with three rays.

SOLVE
Use Equation 37.3 to find the angle of reflection.

$$\theta_i = \theta_r \quad (37.3)$$

$$\theta_r = 25.7°$$

Use Snell's law (Eq. 38.1) to find the angle of refraction. The indices of refraction are listed in Table 38.1.

$$n_t \sin \theta_t = n_i \sin \theta_i \qquad (38.1)$$

$$\sin \theta_t = \frac{n_i}{n_t} \sin \theta_i$$

$$\theta_t = \sin^{-1}\left(\frac{n_i}{n_t} \sin \theta_i\right) = \sin^{-1}\left(\frac{1.333}{1.890} \sin 25.7°\right)$$

$$\theta_t = 17.8°$$

Use the angles to sketch the three beams (Fig. 38.4).

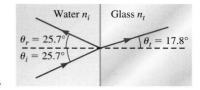

FIGURE 38.4

:• **CHECK and THINK**
As expected, the refracted beam is closer to the normal than the incident beam.

CASE STUDY **Devices You Use to See**

In this chapter's case study, we explore instruments with lenses that you can use to correct or improve your vision. If you are nearsighted or farsighted, you may wear glasses or contacts. When an object is so small you cannot see its details with your unaided eyes, you may use a lens called a *magnifying glass* to produce an enlarged image. If a magnifying glass cannot produce a large enough image, you may use a microscope. If you wish to see the faint light from a distant star, you may use a refracting telescope. All these instruments rely on refraction.

38-2 Total Internal Reflection

This chapter is about *refraction*, so it might seem odd that this section is about *reflection*. But you will soon see how refraction can lead to a particular kind of reflection.

Figure 38.5A shows light traveling from a medium with a high index of refraction into a medium with a lower index of refraction: $n_i > n_t$. According to Snell's law, $\sin \theta_t = (n_i/n_t) \sin \theta_i$ (Eq. 38.1), the refracted ray is bent away from the normal, so $\theta_t > \theta_i$ as shown in Figure 38.5A.

Also according to Snell's law, as the angle of incidence gets larger, the angle of refraction also gets larger. At the **critical incident angle**, the angle of refraction is 90° and the refracted ray is parallel to the surface (Fig. 38.5B). At any incident angle

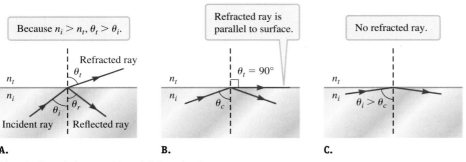

FIGURE 38.5 A. When the external transmitting index of refraction n_t is less than the internal incident index of refraction n_i, the refracted ray is bent away from the normal. **B.** When the angle of incidence equals the critical angle, the refracted ray is parallel to the surface. **C.** If the incident angle is greater than the critical angle, there is no refracted ray, only a reflected one.

TOTAL INTERNAL REFLECTION
⭐ Major Concept

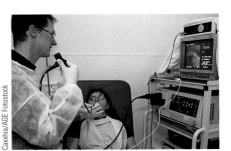

FIGURE 38.6 A doctor looks inside a patient with fiber optics.

larger than the critical angle, there is no refracted ray and no light is transmitted into the second medium (Fig. 38.5C).

Reflected rays are present in all three cases (Fig. 38.5), obeying the law of reflection $\theta_i = \theta_r$ (Eq. 37.3). If the angle of incidence is larger than the critical angle, no light is transmitted; instead, all light is reflected. This phenomenon is called **total internal reflection** and is possible only if light is incident in a medium with a higher index of refraction than that of the medium beyond the boundary. From Figure 38.5B, at the critical angle, the angle of refraction is 90°. According to Snell's law (Eq. 38.1), the critical angle is

$$n_t \sin 90° = n_i \sin \theta_c$$

$$\sin \theta_c = \frac{n_t}{n_i} \qquad (38.2)$$

Optical fibers are a practical application of total internal reflection. An optical fiber is made of a material, such as plastic or glass, that has a relatively high index of refraction. Many fibers are often bundled together in a device. Figure 38.6 shows a medical application of an optical fiber bundle that enables a doctor to look inside a patient. Other uses are in communication systems, such as telephones, televisions, and the Internet.

CONCEPT EXERCISE 38.2

Imagine looking at your own reflection in a pond. If your reflection is clear, is that a result of total internal reflection? Explain.

EXAMPLE 38.3 A Simple Optical Fiber

A laser is directed at an optical fiber (Fig. 38.7). The fiber has an index of refraction of 1.27 and is surrounded by air. For what range of the angle θ_i is the light totally internally reflected inside the fiber?

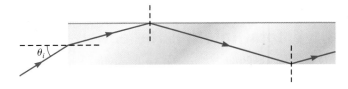

FIGURE 38.7

:• INTERPRET and ANTICIPATE
Begin with a sketch in which the light already inside the fiber is incident on the air–fiber boundary at the critical angle (Fig. 38.8). The light is refracted as it goes from the air into the fiber, but when the light gets to the second boundary, the refracted ray is parallel to the surface. We can use geometry and Snell's law to find θ_i for this critical case. Then any incident angle less than θ_i results in total internal reflection inside the fiber.

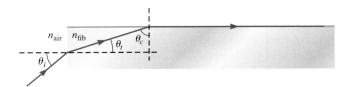

FIGURE 38.8

:• SOLVE
First, find the critical angle (Eq. 38.2) for the light that starts in the fiber. In that case, $n_i = n_{\text{fib}} = 1.27$; the medium beyond the boundary is air, so $n_t = n_{\text{air}} = 1.00029$ (Table 38.1).

$$\sin \theta_c = \frac{n_t}{n_i} \qquad (38.2)$$

$$\theta_c = \sin^{-1}\left(\frac{n_t}{n_i}\right) = \sin^{-1}\left(\frac{1.00029}{1.27}\right)$$

$$\theta_c = 52.0°$$

Because the triangle in Figure 38.8 is a right triangle, we know the two acute angles add up to 90°.	$\theta_c + \theta_t = 90°$ $\theta_t = 90° - \theta_c$ $\theta_t = 90° - 52.0° = 38.0°$
The incident angle θ_i is related to the refracted angle θ_t by Snell's law (Eq. 38.1). At the left boundary in Figure 38.8, light is incident in air, so $n_i = n_{air} = 1.00029$ and $n_t = n_{fib} = 1.27$.	$n_t \sin \theta_t = n_i \sin \theta_i$ (38.1) $\theta_i = \sin^{-1}\left(\dfrac{n_t \sin \theta_t}{n_i}\right)$ $\theta_i = \sin^{-1}\left(\dfrac{1.27 \sin 38.0°}{1.00029}\right) = 51.4°$
If the laser is aimed at any angle less than 51.4°, its light will not leak out the sides of the fiber. In the limiting case where $\theta_i = 0$, so that the incident beam is perpendicular to the end of the fiber, the laser light is not refracted. Instead, it travels in a straight line down the fiber's long axis.	$\theta_i \leq 51.4°$

:• CHECK and THINK

Our answer also shows that any ray whose incident angle is greater than about 52° is transmitted out the sides of the fiber. So optical fibers do not perfectly transmit light along their axes; they leak. To help reduce leakage, the fibers should curve gently. A fiber with a sharp right angle would leak a great deal of light.

38-3 Dispersion

In the mid-1600s, people didn't know that white light is a mixture of all colors. Some people believed color was a mixture of light and darkness. But Isaac Newton performed an experiment in 1666 showing that white light can be spread out to reveal all the colors and that all those colors can be recombined to form white light. In this section, we'll see how such an experiment is possible. (In Problem 27, you'll consider the details.)

The speed of an electromagnetic wave in a medium depends on the medium's index of refraction n (Eq. 36.8). The index of refraction depends in turn on the frequency, wavelength, or color of the light (Table 38.1). For many media, the index of refraction is highest for violet light and lowest for red light. So, when white light propagates from a medium with a low index of refraction, like air, into a medium with a high index of refraction, like glass, the violet light slows down more than the red light. In addition, light bends or refracts when it is transferred from one medium into another because the speed of light is different in the two media. In general, violet light is refracted more than red light (Fig. 38.9A). The result is that white light is spread out into a broader beam that is separated by color. This broad beam looks something like a rainbow and is called a **visible light spectrum**, a **color spectrum**, or simply a **spectrum** (Fig. 38.9B). The spreading out of light by color due to differences in the index of refraction is called **dispersion**.

DISPERSION ✪ Major Concept

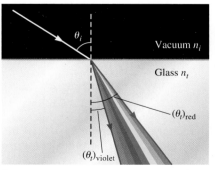

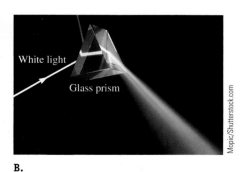

FIGURE 38.9 A. When white light (composed of red, orange, yellow, green, blue, indigo, and violet) goes from a vacuum into glass, the violet light is refracted more than the red light. The other colors are refracted in order by wavelength from red to violet. The spreading out of light by color is known as dispersion. **B.** A prism is used to spread white light into its color spectrum.

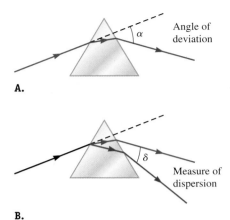

FIGURE 38.10 A. The angle between the original path of the light and the path of the light that emerges from the prism is the angle of deviation α. **B.** The angle between the red ray and the violet ray that emerge from the prism is the measure of dispersion δ.

A device known as a *prism* makes use of dispersion to create a color spectrum (Fig. 38.9B). Two angles—the **angle of deviation** α and the **measure of dispersion** δ—characterize the dispersion of light emerging from a prism (Fig. 38.10). The measure of dispersion (difference between the deviations for violet and red) depends on the difference between n for violet light and n for red light: The greater this difference, the greater the measure of dispersion.

Rainbows

Artificial devices such as prisms are not the only causes of dispersion. Dispersion also occurs naturally, creating beautiful phenomena such as rainbows (Fig. 38.11A). For you to see a rainbow, there must be water droplets in the air and the Sun must be behind you. Sunlight is refracted and reflected by the water droplets and into your eyes. These droplets are spread out in the air, and the position of each droplet determines which color enters your eyes (Fig. 38.11B). The highest droplets allow red light to enter your eyes, and the lowest ones allow violet light to do so. Drops in between allow other colors to enter your eyes.

Sometimes it is possible to see a second rainbow above the first. The second rainbow's colors are reversed so that violet is on top and red is on the bottom. The reversal occurs because the sunlight that reaches the secondary rainbow's droplets undergoes two reflections from the back side of each droplet (Fig. 38.11C). The net result is that the violet ray emerges below the red ray.

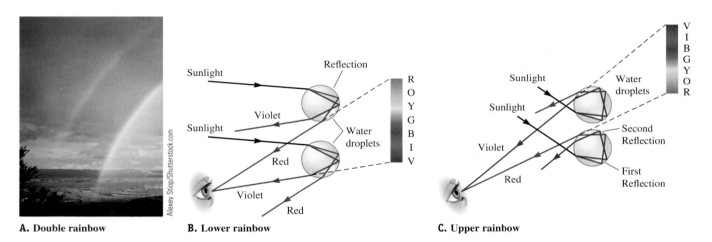

A. Double rainbow   **B. Lower rainbow**   **C. Upper rainbow**

FIGURE 38.11 A. You will see a rainbow only when there are water droplets in the air and the Sun is behind you. Sometimes, you may see a second rainbow above the first with the colors in reverse order. **B.** Because the index of refraction depends on the color of the light, red light is not bent as much as violet light. The red ray strikes the back of each droplet at a point slightly above where the violet ray strikes it. When the rays reach the near side of the droplet, they are refracted again as they leave the water and enter the air. **C.** The second rainbow's colors are reversed because two reflections take place inside each droplet. So, violet light is seen from the highest droplets and red light from the lowest droplets.

CONCEPT EXERCISE 38.3

In Figure 38.9A, white light has angle of incidence θ_i. Imagine $\theta_i = 0$ so that the light is perpendicular to the surface of the glass. Describe the spectrum in this case.

EXAMPLE 38.4 Dispersion in a Prism

Figure 38.12 shows a beam of white light that was originally in air and is incident on a (high-dispersion) prism. The prism's index of refraction for violet light is $n_{\text{violet}} = 1.511$ and for red light is $n_{\text{red}} = 1.487$. Assume the index of refraction is the same for all colors of light in air: $n_{\text{air}} = 1.000293$. The prism's cross section is an equilateral triangle. The beam is parallel to the base of the triangle. Find the measure of dispersion δ.

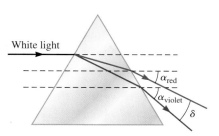

FIGURE 38.12

INTERPRET and ANTICIPATE

The measure of dispersion is the deviation of violet light minus the deviation of red light: $\delta = \alpha_{\text{violet}} - \alpha_{\text{red}}$ (Fig. 38.10B). We must find both angles of deviation before we can find the measure of dispersion. Because this is a *high-dispersion* prism, we expect the measure of dispersion to be fairly large so that the spectrum is easily seen.

Start by making a sketch of a single beam that enters the prism at L and exits at R (Fig. 38.13). The angles of incidence and refraction are labeled θ_1 through θ_4, and α is the angle of deviation between the original horizontal path of the light and the path of the ray after it has passed through the prism. From basic geometry, the internal angles of an equilateral triangle are each 60°. The only physics we need is Snell's law, but we also need to apply some geometry.

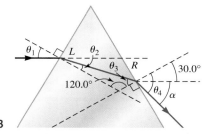

FIGURE 38.13

SOLVE

Apply Snell's law (Eq. 38.1) at the air-to-glass boundary at point L.

$$n_{\text{air}} \sin\theta_1 = n_{\text{glass}} \sin\theta_2 \quad (1)$$

Now apply Snell's law at the glass-to-air boundary at point R.

$$n_{\text{glass}} \sin\theta_3 = n_{\text{air}} \sin\theta_4 \quad (2)$$

Use geometry to find expressions involving $\theta_1, \theta_2, \theta_3, \theta_4$, and α (Problem 33). The angles here are derived from geometry, so they are exact.

$$\theta_2 + \theta_3 = 60° \quad (3)$$
$$\alpha = \theta_4 - 30° \quad (4)$$
$$\theta_1 = 30° \quad (5)$$

Normally, we do all the algebra before substituting values. However, in this problem, it helps to substitute values now and check our preliminary results for both violet light and red light. Rearrange Equation (1), and then substitute the indices of refraction and Equation (5) into Equation (1) to find θ_2.

$$\theta_2 = \sin^{-1}\left(\frac{n_{\text{air}} \sin\theta_1}{n_{\text{glass}}}\right)$$

For violet light,

$$\theta_2 = \sin^{-1}\left(\frac{1.000293 \sin 30°}{1.511}\right) = 19.330°$$

For red light,

$$\theta_2 = \sin^{-1}\left(\frac{1.000293 \sin 30°}{1.487}\right) = 19.654°$$

CHECK and THINK

It makes sense that the angle of refraction is greater for red light because this means red light is refracted farther from the normal (closer to the original path of the light). So, as expected, red light is bent less than violet light.

Example continues on page 1226 ▶

Use $\theta_2 + \theta_3 = 60°$ (Eq. 3) to find θ_3.	For violet light, $\theta_3 = 60° - 19.330° = 40.670°$
	For red light, $\theta_3 = 60° - 19.654° = 40.346°$
Next, rearrange Equation (2) to find θ_4.	$\theta_4 = \sin^{-1}\left(\dfrac{n_{\text{glass}} \sin \theta_3}{n_{\text{air}}}\right)$
	For violet light,
	$\theta_4 = \sin^{-1}\left(\dfrac{1.511 \sin 40.670°}{1.000293}\right) = 79.877°$
	For red light,
	$\theta_4 = \sin^{-1}\left(\dfrac{1.487 \sin 40.346°}{1.000293}\right) = 74.239°$
Use $\alpha = \theta_4 - 30°$ (Eq. 4) to find α. Round to the correct number of significant figures.	For violet light, $\alpha_{\text{violet}} = 79.877° - 30° = 49.88°$
	For red light, $\alpha_{\text{red}} = 74.239° - 30° = 44.24°$
The difference between the angles of deviation for violet light and red light is the measure of dispersion.	$\delta = \alpha_{\text{violet}} - \alpha_{\text{red}} = 49.88° - 44.24°$
	$\delta = 5.64°$

CHECK and THINK
The measure of dispersion is about 6°. As expected, this angle would be easy to see.

38-4 Refraction at Spherical Surfaces

In Section 38-3, we saw that refraction may be used to create a color spectrum. In the rest of this chapter, we'll study how refraction forms images. We start with a simple example: observing a fish's image in a still pond. Rays from a point on the fish—located a distance d_o below the water—are refracted at the boundary between the water and the air (Fig. 38.14). These refracted rays do not cross in the air. However, the diverging rays enter your eye, and as in the case of a plane mirror, your visual system in effect traces these rays back to a point a distance d_i below the water. Because the refracted rays are diverging, the resulting image cannot be projected onto a screen. Thus, the image you see is virtual; the refracted rays appear to diverge from the points that make up the image.

FIGURE 38.15 An image of a person is formed by this spherical globe.

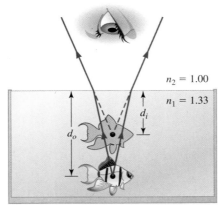

FIGURE 38.14 The virtual image of the fish is formed by refraction. This image appears above the actual fish. The eye is not drawn to scale.

Paraxial Assumption

The surface of the water in Figure 38.14 is flat. From Example 37.5, a plane mirror is a special case of a spherical mirror for which the radius of curvature is infinite. Likewise, the flat surface of the water is a special case of a spherical surface with an infinite radius of curvature. Let's look at the more general case of a spherical surface with a *finite* radius of curvature, which might form one surface of a lens.

In the image formed by the glass globe in Figure 38.15, the person's face near the center is very clear. However, around the globe's edge, you can see a very distorted image of the person's throat, hair, and background wall, an aberration we wish to avoid. For mirrors, we can ignore spherical aberration if we consider only small spherical mirrors so that the rays are paraxial. Similarly, in this chapter, you can assume the rays are paraxial unless otherwise indicated. In practice, this means our derivations hold for images formed near the optical axis.

DERIVATION Position of Image Formed by Refraction at a Spherical Surface

Light from an object travels in a medium with index of refraction n_i; it is then refracted in a medium with a higher index n_t (Fig. 38.16). The boundary between the media is spherical, and the object is small compared to the radius of curvature r, so the paraxial ray assumption holds. Call the region with the object the *front side*, so that the center of curvature C is on the back side. We will show that the object distance d_o and image distance d_i are related by

$$\frac{n_i}{d_o} + \frac{n_t}{d_i} = \frac{(n_t - n_i)}{r} \tag{38.3}$$

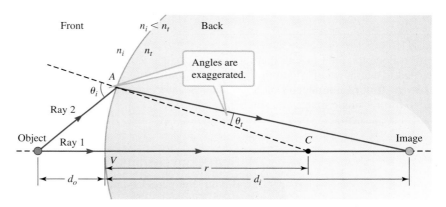

FIGURE 38.16 An image forms due to refraction at a spherical surface.

Consider two rays emitted by a single point on the object (Fig. 38.16). Ray 1 is emitted along the optical axis. It strikes the surface at the vertex V, and because it is perpendicular to the surface, it is not bent. Ray 2 strikes the surface at point A and is refracted toward the normal. The only physics we need for this derivation is Snell's law; the rest comes from geometry.

Apply Snell's law (Eq. 38.1) to ray 2 at point A.	$n_t \sin \theta_t = n_i \sin \theta_i$	(38.1)
For paraxial rays, the angles θ_t and θ_i are small. Use the small-angle approximation for the sine function.	$\sin \theta_t \approx \theta_t$ $\sin \theta_i \approx \theta_i$	
Write Snell's law in terms of this approximation.	$n_t \theta_t \approx n_i \theta_i$ radians	(1)

Our next steps involve geometry as shown by the two highlighted triangles and angles α, β, and γ in Figure 38.17.

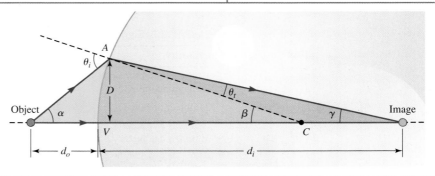

FIGURE 38.17

Derivation continues on page 1228 ▶

In Problem 36, you'll find two expressions relating α, β, and γ to the angles of incidence and reflection.	$\theta_i = \alpha + \beta$ (2) $\theta_t = \beta - \gamma$ (3)
Substitute Equations (2) and (3) into Equation (1).	$n_t(\beta - \gamma) = n_i(\alpha + \beta)$ $n_i\alpha + n_t\gamma = (n_t - n_i)\beta$ (4)
For paraxial rays, point A is almost directly above the vertex V. Call this vertical distance D (Fig. 38.17). Three right triangles share side D and can be used to find the tangents of α, β, and γ. For paraxial rays, these three angles must be small, so we use the small-angle approximation for the tangent function.	$\tan \alpha = \dfrac{D}{d_o} \approx \alpha$ $\tan \beta = \dfrac{D}{r} \approx \beta$ $\tan \gamma = \dfrac{D}{d_i} \approx \gamma$
Use these three approximate expressions to eliminate the angles α, β, and γ from Equation (4).	$n_i \dfrac{D}{d_o} + n_t \dfrac{D}{d_i} = (n_t - n_i)\dfrac{D}{r}$
The vertical distance D cancels.	$\dfrac{n_i}{d_o} + \dfrac{n_t}{d_i} = \dfrac{(n_t - n_i)}{r}$ ✓ (38.3)

:• COMMENTS

We can use Equation 38.3 to find that the image distance d_i depends only on the shape of the surface (that is, its radius of curvature), the index of refraction of both media, and the position of the object. It does not depend on any angle. So, we conclude that all paraxial rays come together to form a single image at a single position d_i.

Sign Conventions for Spherical Refractors

If you compare Equation 38.3 to the mirror equation $(1/d_o) + (1/d_i) = (2/r)$ (Eqs. 37.6 and 37.8), you find three similarities. First, in both cases, the object distance and the image distance are in the denominator of two separate terms. Second, in both cases, the radius of curvature plays the same role in determining the position of the image. Finally, both equations require a sign convention. The sign convention for a spherical refractor is similar to that for a spherical mirror.

The front of the refractor is the side from which the light originates. If the center of curvature is in the back, as in Figure 38.16, the surface is convex and the radius of curvature $r > 0$. If the center of curvature is in the front, the surface is concave and $r < 0$. If the object is in the front, $d_o > 0$ and the object is real. This is often the case, but it is possible for multiple refractions that the object is in the back—in which case $d_o < 0$ and the object is virtual. If the image is in the back, $d_i > 0$ and the image is real. As for spherical mirrors, we use the convention that an inverted image has a negative height and its magnification is negative. These sign conventions are summarized in Table 38.2.

TABLE 38.2 Sign conventions for spherical refracting surfaces.

Quantity	Positive	Negative
1. Image height h_i and magnification M	If image is **upright**	If image is **inverted**
2. Object distance d_o	If object is **real** (in front of surface)	If object is **virtual** (behind surface)
3. Image distance d_i	If image is **real** (behind surface)	If image is **virtual** (in front of surface)
4. Radius of curvature r	If surface is **convex**	If surface is **concave**

DERIVATION: Magnification by Refraction at a Spherical Surface

Next, we show that the magnification of the image created by refraction at a spherical surface is given by

$$M = \frac{h_i}{h_o} = -\frac{n_i \, d_i}{n_t \, d_o} \qquad (38.4)$$

We use the same two media and surfaces as in the preceding derivation, but now we must consider an extended object represented by an arrow (Fig. 38.18). As before, we consider only paraxial rays.

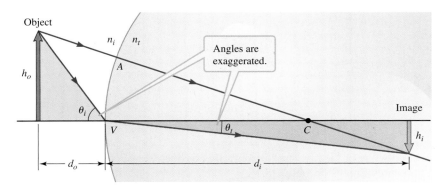

FIGURE 38.18

We draw two rays from the tip of the image. The first ray passes through the center of curvature, is perpendicular to the surface at point A, and is therefore not refracted. The other ray strikes the surface at the vertex V and is refracted toward the normal. The two rays cross to form a real inverted image in the second medium. We will apply the sign convention at the end of the derivation.

Use Snell's law (Eq. 38.1) for the ray at V.	$n_t \sin \theta_t = n_i \sin \theta_i$ (1)
Use the two right triangles highlighted in Figure 38.18 to write expressions for the tangents of θ_i and θ_t. These angles are small, so we can use the approximation that the tangent of a small angle equals the sine of the angle.	$\tan \theta_i = \dfrac{h_o}{d_o} \approx \sin \theta_i$ (2) $\tan \theta_t = \dfrac{h_i}{d_i} \approx \sin \theta_t$ (3)
Substitute Equations (2) and (3) into Equation (1).	$n_t \tan \theta_t \approx n_i \tan \theta_i$ $n_t \dfrac{h_i}{d_i} = n_i \dfrac{h_o}{d_o}$
The magnification is the image height divided by the object height (Eq. 37.2). By the first sign convention in Table 38.2, we insert a negative sign for an inverted image.	$M = \dfrac{h_i}{h_o} = -\dfrac{n_i \, d_i}{n_t \, d_o}$ ✓ (38.4)

:• COMMENTS

Compare the magnification in this case to the magnification of a spherical mirror, $M = -d_i/d_o$ (Eq. 37.7), and you notice a similarity. Both depend on $-d_i/d_o$, but in this case of refraction by a spherical surface, the magnification also depends on the indices of refraction. Throughout this chapter, many equations are similar to those we learned in Chapter 37 on reflection, except here the equations involve the index of refraction. Making comparisons like this will help you to keep the equations straight in your mind.

EXAMPLE 38.5 Looking at a Fish

In this problem, we find an expression for the image distance of the fish in Figure 38.14 in two different ways.

A Use the rays shown in Figure 38.14 to find an expression for d_i. Include the appropriate sign convention from Table 38.2 in your final expression.

INTERPRET and ANTICIPATE
Figure 38.19 is a simple ray diagram for one point on the fish. The angles must be small because the person is directly above the fish. Only slightly bent rays will make it into the person's eye.

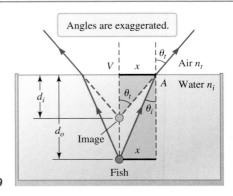

FIGURE 38.19

SOLVE
We need to consider the refraction of only one of the rays. Let's arbitrarily consider the ray that is refracted at point A and apply Snell's law (Eq. 38.1).

$$n_t \sin \theta_t = n_i \sin \theta_i \quad (38.1)$$

Use the highlighted right triangles to find expressions for $\tan \theta_i$ and $\tan \theta_t$. The angles are small, so the tangent of each angle is roughly equal to the sine of that angle.

$$\tan \theta_t = \frac{x}{d_i} \approx \sin \theta_t$$

$$\tan \theta_i = \frac{x}{d_o} \approx \sin \theta_i$$

Substitute these approximate expressions into Snell's law.

$$n_t \frac{x}{d_i} = n_i \frac{x}{d_o}$$

Solve for d_i.

$$d_i = \frac{n_t}{n_i} d_o$$

The fish is the object, so the front side of the boundary is below the surface of the water. Because the image is virtual (in front of the surface), the image distance is negative according to the sign convention (Table 38.2).

$$d_i = -\frac{n_t}{n_i} d_o$$

B Use Equation 38.3 and consider the flat water surface as a special case of a spherical surface in which the radius of curvature is infinite to derive an expression for d_i. Check your results by comparing parts A and B.

INTERPRET and ANTICIPATE
Figure 38.19 still applies to this problem, but now we think of the flat surface as a special case of a spherical surface.

SOLVE
Start with Equation 38.3 and take the limit as the radius of curvature goes to infinity. The right side goes to 0, so the left side must also approach 0.

$$\frac{n_i}{d_o} + \frac{n_t}{d_i} = \frac{(n_t - n_i)}{r} \quad (38.3)$$

$$\lim_{r \to \infty} \left[\frac{(n_t - n_i)}{r} \right] \to 0$$

$$\frac{n_i}{d_o} + \frac{n_t}{d_i} \to 0$$

Solve for d_i.	$\dfrac{n_i}{d_o} = -\dfrac{n_t}{d_i}$
	$d_i = -\dfrac{n_t}{n_i} d_o$

:• CHECK and THINK

As expected, this is the same expression we found in part A. This time we didn't have to insert a negative sign because the sign convention is built into Equation 38.3.

EXAMPLE 38.6 A Thick Lens

Suppose you have a nearly perfect spherical vase full of water and have discovered you can use the vase as a lens to magnify the writing on a sheet of paper. When you place a message on one side of the vase and look through the opposite side, you see an upright, slightly magnified image of the message (Fig. 38.20). Model the vase as a spherical refractor (a thick lens) with a radius of 6.0 cm and an index of refraction of 1.3. Assume the air has an index of refraction of 1.0. If you hold the message 3.0 cm from the vase, find the position of the final image and its magnification. (As shown in Figure 38.20, you can also use the thick lens to project a real image onto a screen; in Problem 39, you will find the distance and magnification for this image.)

FIGURE 38.20 An upright image of the message "Physics is Fun!" as seen through the vase. The image is greatly distorted, so the whole message cannot be read. An inverted image of a lamp is projected through the vase and onto a screen.

:• INTERPRET and ANTICIPATE

Both the vase of water and the globe in Figure 38.15 are thick lenses. Unlike in Example 38.5, where the light is refracted only after it passes from water into air, in this example the light is refracted twice through the thick lens. The refraction from one surface produces an image that becomes the object for the refraction from the second surface. We will therefore work this problem in two parts. We'll first find the image produced by the first surface. Then, using that image as an object, we'll find the final image produced by the second surface. Throughout this problem, we consider only paraxial rays and we model the vase of water as a single spherical refractor, ignoring the thin layer of glass.

:• SOLVE

In Figure 38.21, the sphere is divided into two surfaces labeled 1 and 2. The object (the message) is to the left, in front of both surfaces; your eye (not shown) is to the right, behind them. So, for the rays emitted by the object, surface 1 is convex and surface 2 is concave.

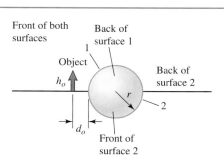

FIGURE 38.21

Solve Equation 38.3 for $1/d_i$.	$\dfrac{n_i}{d_o} + \dfrac{n_t}{d_i} = \dfrac{(n_t - n_i)}{r}$	(38.3)
	$\dfrac{1}{d_i} = \dfrac{1}{n_t}\left[\dfrac{(n_t - n_i)}{r} - \dfrac{n_i}{d_o}\right]$	(1)

Example continues on page 1232 ▶

Substitute values, taking into account the sign convention, to find the image distance produced by surface 1. Surface 1 is convex, so r is positive. The object is real and in front of surface 1, so d_o is positive.	$\dfrac{1}{d_i} = \dfrac{1}{1.3}\left[\dfrac{(1.3-1.0)}{6.0\text{ cm}} - \dfrac{1.0}{3.0\text{ cm}}\right]$ $d_i = -4.6$ cm
We find that the image produced by the first surface has a negative image distance, so this image is in front of surface 1 (Fig. 38.22). We can think of it as image 1, but we must also think of it as object 2 (the object for surface 2).	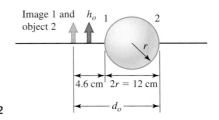

FIGURE 38.22

Before finding the position of the final image, let's find the magnification of image 1 using Equation 38.4. The result is positive, so the image is upright.	$M_1 = -\dfrac{n_i\, d_i}{n_t\, d_o}$ (38.4) $M_1 = -\dfrac{1.0}{1.3}\dfrac{(-4.6\text{ cm})}{(3.0\text{ cm})} = 1.2$
Now consider image 1 as the object for surface 2. Find the image distance with Equation (1). The object (image 1) is in front of surface 2 (Fig. 38.22), so the object distance is positive. This distance is 4.6 cm plus the diameter (12 cm) of the spherical lens. The radius of curvature is negative because surface 2 is concave. Be careful; now the incident ray is in the lens ($n_i = 1.3$) and the transmitted ray is in the air ($n_t = 1.0$).	$\dfrac{1}{d_i} = \dfrac{1}{n_t}\left[\dfrac{(n_t-n_i)}{r} - \dfrac{n_i}{d_o}\right]$ $\dfrac{1}{d_i} = \dfrac{1}{1.0}\left[\dfrac{(1.0-1.3)}{-6.0\text{ cm}} - \dfrac{1.3}{16.6\text{ cm}}\right]$ $d_i = -35$ cm

:• **CHECK and THINK**
The final image you see is about 35 cm in front of the second surface, or about 23 cm in front of the first surface. So the image is in front of the object (the message). But, from the perspective of the eye at the back of these surfaces, the image is behind the vase and virtual. You cannot project this image onto a screen. Instead, you see the image by looking through the vase (Fig. 38.20).

The magnification of image 2 is also given by Equation 38.4. It is the ratio of image 2's height to object 2's height.	$M_2 = -\dfrac{n_i\, d_i}{n_t\, d_o} = -\dfrac{1.3}{1.0}\dfrac{(-35\text{ cm})}{(16.6\text{ cm})} = 2.7$
To find the total magnification—the ratio of image 2's height to object 1's (the message's) height—multiply the magnifications caused by each surface.	$M = M_1 M_2 = (1.2)(2.7)$ $M = 3.2$

:• **CHECK and THINK**
The vase acts as a magnifying glass, producing an enlarged image. However, because the lens is thick, there is a great deal of aberration and the image is blurred around the edges. The image also suffers from *chromatic aberration* because the index of refraction depends on color, and different colors produce images at different places. Don't worry if you find it difficult to see the halo of colors around the image (Fig. 38.20); the photo is very small.

38-5 Thin Lenses

Unless otherwise specified, assume all lenses are thin.

A lens is an optical system with two refracting surfaces. For example, the vase of water in Figure 38.20 is a thick lens. Optical instruments such as cameras, microscopes, and the human eye are best modeled as thin lenses. The thickness of a **thin lens** is small compared to its radius of curvature. In the rest of this chapter, we study thin lenses. Our goal in this section is to find expressions for the image distance and the magnification produced by thin lenses. Our study of spherical mirrors provides a major shortcut in reaching these goals because the mirror equation (Eq. 37.8),

$$\frac{1}{d_o} + \frac{1}{d_i} = \frac{1}{f} \quad (38.5)$$

THIN-LENS EQUATION
✪ Major Concept

is the same as the **thin-lens equation**. Also (as you show in Problem 44), the magnification equation for a thin lens is the same as that for a spherical mirror (Eq. 37.7):

$$M = \frac{h_i}{h_o} = -\frac{d_i}{d_o} \quad (38.6)$$

LINEAR MAGNIFICATION OF LENSES
✪ Major Concept

However, thin lenses require the slightly different sign conventions given in Table 38.3. Also, the focal length of a lens depends on both its shape and the lens's index of refraction. The focal length of a thin lens is given by the *lens maker's equation*, which we derive next.

TABLE 38.3 Sign conventions for thin spherical lenses.

Quantity	Positive	Negative
1. Image height h_i and magnification M	If image is **upright**	If image is **inverted**
2. Object distance d_o	If object is **real** (in front)	If object is **virtual** (behind)
3. Image distance d_i	If image is **real** (behind)	If image is **virtual** (in front)
4. Radius of curvature r	If surface is **convex**	If surface is **concave**
5. Focal length f	If lens is **converging**	If lens is **diverging**

DERIVATION The Lens Maker's Equation

We derive the lens maker's equation, an expression for the focal length f of a thin lens:

$$\frac{1}{f} = (n-1)\left[\frac{1}{r_1} - \frac{1}{r_2}\right] \quad (38.7)$$

LENS MAKER'S EQUATION
✪ Major Concept

where n is the lens's index of refraction. Although the lens maker's equation holds for thin lenses, we begin by first considering a thick lens and then we move to a thin lens mathematically. The lens's surfaces are both spherical, but they do not necessarily have the same radius of curvature, so the focal length depends on r_1 and r_2 for each surface (Fig. 38.23). Assume the lens is in a vacuum with an index of refraction equal to 1. (Because the index of refraction of air is 1.0003, this assumption holds well for a lens in air.)

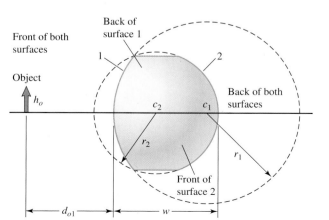

FIGURE 38.23

Derivation continues on page 1234 ▶

Our approach here is similar to that in Example 38.6 with the vase of water. An object is located in front of surface 1 (Fig. 38.23). Surface 1 produces an image that becomes the object for surface 2. The image produced by surface 2 is the final image produced by the lens.

Start by applying Equation 38.3 to surface 1. The subscript 1 is a reminder that we are working with surface 1. Light from the object passes first through the vacuum and then into the lens, so $n_i = 1$ and $n_t = n$ for the lens.	$\dfrac{n_i}{d_o} + \dfrac{n_t}{d_i} = \dfrac{(n_t - n_i)}{r}$	(38.3)
	$\dfrac{1}{d_{o1}} + \dfrac{n}{d_{i1}} = \dfrac{(n-1)}{r_1}$	(1)

Surface 1 produces one of three possible images. If the image is virtual and in front of both surfaces (Fig. 38.24), d_{i1} is negative. Because this image is also the object of surface 2, the object distance d_{o2} is positive and the object is real.

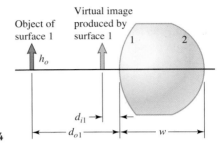

FIGURE 38.24

If surface 1 produces a real image inside the lens itself (Fig. 38.25), d_{i1} is positive. Because the image is in front of surface 2, as in the preceding case, the object for surface 2 is real.

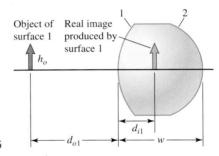

FIGURE 38.25

Finally, the image produced by surface 1 may be real and behind surface 2 (Fig. 38.26). Because this image is the object for surface 2 and is behind surface 2, the object is virtual and its distance is negative.

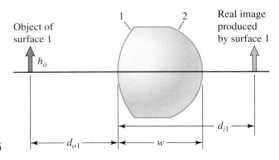

FIGURE 38.26

In all three cases, the object distance for surface 2 is given by Equation (2). (Test this for yourself in Problem 112.)	$d_{o2} = w - d_{i1}$	(2)
Now we're ready to apply Equation 38.3 to surface 2. This time the light is incident in the lens and then refracted into the vacuum. So, $n_i = n$ for the lens, and $n_t = 1$. The object distance d_{o2} is given by Equation (2).	$\dfrac{n_i}{d_o} + \dfrac{n_t}{d_i} = \dfrac{(n_t - n_i)}{r}$	(38.3)
	$\dfrac{n}{d_{o2}} + \dfrac{1}{d_{i2}} = \dfrac{(1-n)}{r_2}$	
	$\dfrac{n}{w - d_{i1}} + \dfrac{1}{d_{i2}} = \dfrac{(1-n)}{r_2}$	(3)
To find the lens maker's equation, add Equations (1) and (3). For a thin lens, $w \to 0$.	$\dfrac{1}{d_{o1}} + \dfrac{n}{d_{i1}} + \dfrac{n}{w - d_{i1}} + \dfrac{1}{d_{i2}} = \dfrac{(n-1)}{r_1} + \dfrac{(1-n)}{r_2}$	

The object distance d_{o1} is the distance of the original real object, while d_{i2} is the image distance of the final image. So the double subscripts are no longer necessary.	$\dfrac{1}{d_{o1}} + \dfrac{n}{d_{i1}} - \dfrac{n}{d_{i1}} + \dfrac{1}{d_{i2}} = \dfrac{(n-1)}{r_1} + \dfrac{(1-n)}{r_2}$ $\dfrac{1}{d_{o1}} + \dfrac{1}{d_{i2}} = (n-1)\left[\dfrac{1}{r_1} - \dfrac{1}{r_2}\right]$ $\dfrac{1}{d_o} + \dfrac{1}{d_i} = (n-1)\left[\dfrac{1}{r_1} - \dfrac{1}{r_2}\right]$ (4)
Compare Equation (4) to the thin-lens equation (Eq. 38.5). The left sides are equal; therefore, the right sides are equal. We have reached the lens maker's equation.	$\dfrac{1}{d_o} + \dfrac{1}{d_i} = \dfrac{1}{f}$ (38.5) $\dfrac{1}{f} = (n-1)\left[\dfrac{1}{r_1} - \dfrac{1}{r_2}\right]$ ✓ (38.7)

∴ COMMENTS
Compare the focal length of a thin lens (Eq. 38.8) to the focal length for a spherical mirror $f = r/2$ (Eq. 37.6). Both depend on the radius of curvature, but the focal length for a thin lens is more complicated. First, a thin lens is made up of two curved surfaces and so depends on two radii of curvature. Also, because the light is refracted by the thin lens, its focal length depends on its index of refraction. Why does the focal length of a thin lens have the "1" in the $(n-1)$ factor?

Sign Conventions for Thin Lenses

To use the thin-lens equation (38.5), the lens maker's equation (38.7), and the magnification equation (38.6), we need the sign conventions from Table 38.3. Because all these lens equations follow from refraction at spherical surfaces, the sign conventions for thin lenses are similar to those for spherical surfaces. First determine the front and back of the lens using the same convention as for spherical surfaces. The front of the lens is the side from which light originates; the other side is the back. Because the lens is thin, we don't concern ourselves with the inside.

Comparing Table 38.3 (sign conventions for thin lenses) to Table 38.2 for spherical refracting surfaces shows they have much in common. However, a refracting surface does not have a focal length.

The sign of the lens's focal length comes from the signs of the radii of curvature. Figure 38.27 shows two thin lenses. In both cases, the object is on the left, the left side is the front, and surface 1 is the surface that the object's light encounters first. In Figure 38.27A, surface 1 bulges out toward the object. Its center of curvature is behind the lens, so surface 1 is convex. According to the fourth sign convention in Table 38.3, this surface's radius of curvature is positive: $r_1 = |r_1|$. Surface 2 bulges away from the object. Its center of curvature is in front of the lens, so surface 2 is concave. According to the sign convention, its radius of curvature is negative: $r_2 = -|r_2|$.

Now let's find the sign of the focal length using the lens maker's equation (Eq. 38.7). The term $(n-1)$ is positive because the index of refraction n for the lens must be greater than 1. For the lens in Figure 38.27A, the term in square brackets is

$$\left[\dfrac{1}{|r_1|} - \dfrac{1}{-|r_2|}\right] = \left[\dfrac{1}{|r_1|} + \dfrac{1}{|r_2|}\right] > 0$$

so the focal length is positive. A lens with a positive focal length is called a **converging lens**, whereas a lens with a negative focal length is called a **diverging lens**.

Next, let's show that the lens in Figure 38.27B has a negative focal length and is a diverging lens. Because c_1 is in front of the lens, surface 1 is concave and its

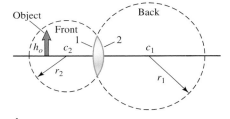

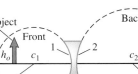

FIGURE 38.27 A thin spherical lens has two surfaces. The center of curvature for surface 1 is c_1. The center of curvature for surface 2 is c_2. **A.** Surface 1 is convex and surface 2 is concave. This is a converging lens, and so it has a positive focal length. **B.** Surface 1 is concave and surface 2 is convex. This is a diverging lens, so it has a negative focal length.

CONVERGING AND DIVERGING LENSES
◯ Special Case

radius of curvature is negative: $r_1 = -|r_1|$. Because c_2 is behind the lens, surface 2 is convex and its radius of curvature is positive: $r_2 = |r_2|$. So we have

$$\left[\frac{1}{-|r_1|} - \frac{1}{|r_2|}\right] = -\left[\frac{1}{|r_1|} + \frac{1}{|r_2|}\right] < 0$$

which means the focal length of the lens in Figure 38.27B is negative.

Converging Versus Diverging Lenses

Figure 38.28 illustrates the difference between converging and diverging lenses. In each case, parallel rays from a very distant object are incident on the lens. The converging lens (Fig. 38.28A) bends the rays so that they come together or *converge* at a point behind the lens. The point where they come together is the *focal point F*, and a real image of the distant object is formed there. The distance from the lens to the focal point is the focal length *f*. If you place a screen at *F*, you see the real image projected onto the screen.

When parallel rays encounter a diverging lens (Fig. 38.28B), the refracted rays separate or *diverge*. Diverging rays never cross, so no real image forms. However, the diverging rays appear to originate from a point in front of the lens. This point is the focal point *F* of the lens. If you look through the back of the lens, you see a virtual image at the focal point. In the next two sections, we'll explore the images produced by both types of lenses when the objects are *not* infinitely far away.

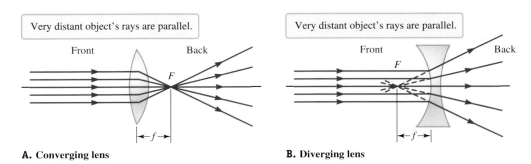

A. Converging lens   **B. Diverging lens**

FIGURE 38.28 A. When parallel rays are incident on a converging lens, they are bent toward one another and a real image forms at the focal point, sometimes called the *real focal point*. **B.** When parallel rays are incident on a diverging lens, they are bent away from one another and a virtual image forms at the focal point, sometimes called the *virtual focal point*.

EXAMPLE 38.7 Convex Meniscus Lens

A Figure 38.29 shows a convex meniscus lens and an object. If $|r_1| = 6.50$ cm and $|r_2| = 8.50$ cm, find the focal length and determine whether the lens is converging or diverging. The lens is made of glass with index of refraction $n = 1.55$.

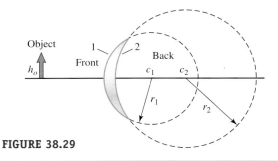

FIGURE 38.29

:• **INTERPRET and ANTICIPATE**
The key to solving this problem is to apply the sign conventions correctly (Table 38.3). When we find the focal length using the lens maker's equation, we can use the sign conventions again to determine whether the lens is converging or diverging.

SOLVE

The object is on the left, so the left side of the lens is the front. For both surfaces, the center of curvature is in the back, so both surfaces are convex and both radii of curvature are positive.

$r_1 = 6.50 \text{ cm}$
$r_2 = 8.50 \text{ cm}$

Substitute values into the lens maker's equation (Eq. 38.7). Remember to invert the equation to find f. The focal length is positive, so according to the sign convention, the lens is converging.

$$\frac{1}{f} = (n-1)\left[\frac{1}{r_1} - \frac{1}{r_2}\right] \quad (38.7)$$

$$\frac{1}{f} = (1.55 - 1)\left[\frac{1}{6.50 \text{ cm}} - \frac{1}{8.50 \text{ cm}}\right] \quad (1)$$

$\boxed{f = 50.2 \text{ cm, converging}}$

CHECK and THINK

Both surfaces are convex, so both have a positive radius of curvature. If the first surface had the larger radius of curvature, the lens would have had a negative focal length (diverging).

B Suppose you flip the lens around, back to front (Fig. 38.30). Find the focal length and determine whether the lens is still converging.

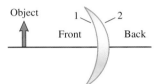

FIGURE 38.30

INTERPRET and ANTICIPATE

This is the same lens, so all the parameters are the same. However, surface 1 and surface 2 are switched, so $|r_1| = 8.50 \text{ cm}$ and $|r_2| = 6.50 \text{ cm}$.

SOLVE

The object is on the left, so the left side is the front. For both surfaces, the center of curvature is in the front, so both surfaces are concave. According to the fourth sign convention in Table 38.3, both radii of curvature must be negative.

$r_1 = -8.50 \text{ cm}$
$r_2 = -6.50 \text{ cm}$

Substitute values into the lens maker's equation (Eq. 38.7). Notice that Equation (2) is the same as Equation (1) in part A. The focal length is the same as before the lens was flipped.

$$\frac{1}{f} = (n-1)\left[\frac{1}{r_1} - \frac{1}{r_2}\right] \quad (38.7)$$

$$\frac{1}{f} = (1.55 - 1)\left[\frac{1}{-8.50 \text{ cm}} - \frac{1}{-6.50 \text{ cm}}\right]$$

$$\frac{1}{f} = (1.55 - 1)\left[\frac{1}{6.50 \text{ cm}} - \frac{1}{8.50 \text{ cm}}\right] \quad (2)$$

$\boxed{f = 50.2 \text{ cm, still converging}}$

CHECK and THINK

We found that the orientation of the lens does not determine whether the lens is converging or diverging.

Ray Diagrams for Thin Lenses

In the next two sections, we study the images that result when an object is placed at a finite distance from a thin lens. The process of finding the image distance and magnification is much like the process we used with spherical mirrors. Again, we use

RAY DIAGRAMS ⊙ Tool

ray diagrams to illustrate the problem. A **ray diagram** for a thin lens usually includes these five elements (Figs. 38.31 and 38.32; pages 1239 and 1240):

1. The **lens**, including a vertical line that runs through its middle. Sketch the rays as though they refract at this midline.
2. The **optical axis** is perpendicular to the line through the lens and passes through the lens's center.
3. The **focal points** F drawn on both sides of the lens.
4. An **object** represented as a simple shape, such as an arrow. The placement of the object determines the front of the lens. Because we read from left to right, we usually draw the object on the left side of the lens.
5. Two or three **primary rays** (described below) emerging from one point of the object, usually the tip of the arrow. Only two rays are needed to find the position, orientation, and magnification of the image. Sometimes a third ray is helpful to check your drawing.

Primary Rays for Thin Lenses

PRIMARY RAYS FOR THIN LENSES ⊙ Tool

In Section 37-4, we listed four primary rays that are helpful when working with spherical mirrors. Likewise, there are three primary rays that are helpful when you must find the position, magnification, and orientation of an image produced by a thin lens. These primary rays are described in Table 38.4.

TABLE 38.4 Primary rays for thin lenses.

Ray	Converging lens	Diverging lens
Ray 1 passes through the center of the lens. It is not deflected because any ray that passes through the center of the lens goes through two nearly parallel surfaces.		
Ray 2 is parallel to the optical axis. For a converging lens, ray 2's refraction passes through the focal point on the back of the lens. For a diverging lens, ray 2's refraction is bent away from the optical axis. When you extend the refracted ray backward, it passes through the focal point on the front of the lens. Notice that you draw the bend in the ray only where it strikes the vertical line through the middle of the lens.		
For a converging lens, **ray 3** passes through the focal point on the front of the lens. For a diverging lens, ray 3 is aimed at the focal point on the back of the lens. For both lenses, the refracted ray is parallel to the optical axis.		

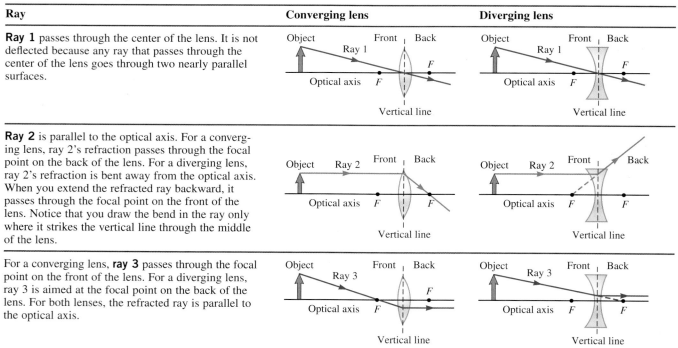

38-6 Images Formed by Diverging Lenses

We can use ray diagrams to find the position, orientation, and magnification of a diverging lens when an object is at some finite distance from it. The location of an object relative to the focal point of a diverging lens does not matter. Figure 38.31 shows an object at some arbitrary distance in front of a diverging lens along with the three primary rays. The three refracted rays that result do not cross behind the lens, so no real image is formed. However, if you look through the lens from the back, the three rays entering your eye can be traced to a point in front of the lens. The resulting image is upright, in front of the lens, virtual, and smaller than the object.

Let's connect our ray diagram (Fig. 38.31) to the thin-lens equation (Eq. 38.5). The focal length is negative because the lens is diverging, and the object distance is positive because the object is real and in front of the lens. The thin-lens equation (Eq. 38.5),

$$\frac{1}{d_i} = \frac{1}{-|f|} - \frac{1}{|d_o|}$$

shows that the image distance must be negative, which means the image is virtual and in front of the lens (Table 38.3) as shown in the figure.

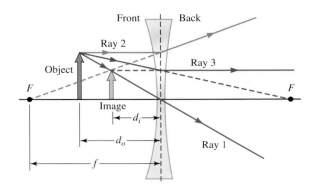

FIGURE 38.31 The three primary rays show that for a diverging lens, the image is virtual, upright, and small.

CONCEPT EXERCISE 38.4

The image produced by a diverging lens is upright and virtual no matter where the object is in front of the lens. We'll show in Section 38-7, however, that for a converging lens, the placement of the object relative to the focal length determines whether the image is virtual or real, upright or inverted, and larger or smaller than the object. Examine the convex and concave mirrors in Sections 37-5 and 37-6. Is a diverging lens more like a convex mirror or a concave mirror? Explain.

EXAMPLE 38.8 All Possible Regions for a Diverging Lens

A diverging lens has focal length $|f| = 10.0$ cm. Find the image distance and magnification if the object is

A 15.0 cm,
B 10.0 cm, and
C 5.00 cm in front of the lens.

• INTERPRET and ANTICIPATE
For a diverging lens, the image is upright, in front of the lens, and smaller than the object, no matter where the real object is placed. So we expect to find $0 < M < 1$ and $d_i < 0$ in all three cases.

• SOLVE

Isolate the image distance on one side of the thin-lens equation (Eq. 38.5). Substitute the focal length, which is negative because this is a diverging lens.	$\dfrac{1}{d_i} = \dfrac{1}{f} - \dfrac{1}{d_o} = -\dfrac{1}{10.0\,\text{cm}} - \dfrac{1}{d_o}$
A Find the image distance when $d_o = 15.0$ cm.	$\dfrac{1}{d_i} = -\dfrac{1}{10.0\,\text{cm}} - \dfrac{1}{15.0\,\text{cm}} = -\dfrac{5}{30.0\,\text{cm}}$ $d_i = -6.00\,\text{cm}$
Use $M = -d_i/d_o$ (Eq. 38.6) to find the magnification.	$M = -\dfrac{d_i}{d_o} = -\dfrac{-6.00\,\text{cm}}{15.0\,\text{cm}}$ $M = +0.400$

Example continues on page 1240 ▶

B Find the image distance when $d_o = 10.0$ cm.	$\dfrac{1}{d_i} = -\dfrac{1}{10.0 \text{ cm}} - \dfrac{1}{10.0 \text{ cm}} = -\dfrac{2}{10.0 \text{ cm}}$ $d_i = -5.00$ cm
Again, use $M = -d_i/d_o$ (Eq. 38.6) to find the magnification.	$M = -\dfrac{d_i}{d_o} = -\dfrac{-5.00 \text{ cm}}{10.0 \text{ cm}}$ $M = +0.500$
C Find the image distance when $d_o = 5.00$ cm.	$\dfrac{1}{d_i} = -\dfrac{1}{10.0 \text{ cm}} - \dfrac{1}{5.00 \text{ cm}} = -\dfrac{3}{10.0 \text{ cm}}$ $d_i = -3.33$ cm
Find the magnification as before.	$M = -\dfrac{d_i}{d_o} = -\dfrac{-3.33 \text{ cm}}{5.00 \text{ cm}} = +0.666$

:• **CHECK and THINK**

As expected, the image distance is negative in all three cases, so the image is in front of the lens and virtual. Also, the magnification is positive and less than 1, so the image is upright and smaller than the object in all three cases. When the object distance equals the focal length (part B), the image is half the size of the object. When the object is farther away than the focal length (part A), the image is less than half the size of the object. When the object is closer than the focal length (part C), the image is more than half the size of the object.

38-7 Images Formed by Converging Lenses

In this section, we explore converging lenses using ray diagrams. The image produced by a converging lens may be real or virtual, upright or inverted, depending on the placement of the object. As in the case of diverging lenses, our goal here is to find the position, orientation, and magnification of a converging lens when an object is at some finite distance from it. There are three regions in which to place the object: (1) farther away from the lens than the focal point, (2) closer to the lens than the focal point, and (3) at the focal point. For each of these regions, we'll draw a ray diagram, including the primary rays, and then we'll connect our ray diagram to the lens equation.

Converging Lens: Object Farther than Focal Point

Consider an object that is farther away from the front of a converging lens than the focal point (Fig. 38.32). The three primary rays refract and cross behind the lens, forming a real image. If you put a screen where the rays cross, you would see an inverted image. So, the image is inverted, behind the lens, and real.

Let's connect our ray diagram (Fig. 38.32) to the thin-lens equation (Eq. 38.5). The focal length is positive because the lens is converging, and the object distance is positive because the object is real and in front of the lens. So the thin-lens equation becomes

$$\dfrac{1}{d_i} = \dfrac{1}{+|f|} - \dfrac{1}{+|d_o|}$$

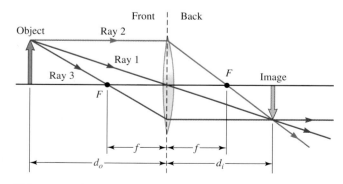

FIGURE 38.32 An object is farther from the converging lens than the focal point. The image is inverted, real, and behind the lens. In this figure, the image is smaller than the object.

Because the object is farther away from the lens than the focal point, $d_o > f$ and the first term on the right must be greater than the second term:

$$\frac{1}{|f|} > \frac{1}{|d_o|}$$

which means the image distance must be positive:

$$\frac{1}{d_i} > 0$$

A positive image distance (Table 38.3) means the image is real and behind the lens, in agreement with Figure 38.32.

The image may be smaller or larger than the object, depending on the object's position. To see why the object's position matters, imagine switching the situation shown in Figure 38.32 so that the arrow labeled "Image" is an object in front of the lens. Now the front of the lens is on the right, and the back of the lens is on the left. When you reverse the arrowheads on the primary rays, the rays come together at the arrow labeled "Object," which we now think of as the image of the arrow on the right. In this case, the image is larger than the object.

We can also do this switching mathematically. For a converging lens with an object farther from the lens than the focal point, the object distance, image distance, and focal length are all positive. So, if the object and image distances are switched in the thin-lens equation,

$$\frac{1}{d_o} + \frac{1}{d_i} = \frac{1}{d_i} + \frac{1}{d_o} = \frac{1}{f}$$

the equation still holds for the same focal length. We say the thin-lens equation is *symmetrical* with respect to such a switch of variables. However, switching these distances changes the magnification because $M = -d_i/d_o$ (Eq. 38.6) is not symmetrical. If $d_o > d_i$, as in Figure 38.32, $|M| < 1$ and the image is smaller than the object. If $d_o < d_i$, as if the arrowheads in Figure 38.32 were reversed, $|M| > 1$ and the image is larger than the object.

Converging Lens: Object Closer than Focal Point

Figure 38.33 shows an object that is closer to a converging lens than the focal point. Drawing the primary rays is a little trickier because to draw ray 3 you must start by extending the ray from the focal point on the front side. Of course, no ray comes from that focal point, so the portion of the ray from the focal point to the tip of the object is shown as a dashed line.

The three refracted rays do not cross behind the lens, so no real image forms. If you trace the three refracted rays backward, they intersect at a point in front of the lens. A virtual image forms at this point. The image is upright and larger than the object.

Now let's connect our ray diagram (Fig. 38.33) to the thin-lens equation (Eq. 38.5). The object distance is positive because the object is real and in front of the lens. The focal length is positive because the lens is converging. The object distance is less than the focal length: $d_o < f$. So,

$$\frac{1}{|f|} < \frac{1}{|d_o|}$$

The thin-lens equation becomes

$$\frac{1}{d_i} = \frac{1}{f} - \frac{1}{d_o} < 0$$

so the image distance is negative. According to the sign conventions in Table 38.3, the image is virtual and in front of the lens, in agreement with the ray diagram.

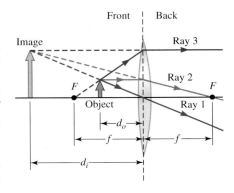

FIGURE 38.33 An object is closer to the converging lens than the focal point. The image is upright, virtual, and in front of the lens.

Converging Lens: Object at Focal Point

Finally, consider an object at the focal point of a converging lens, along with two primary rays (Fig. 38.34). We cannot include ray 3 because we cannot extend a ray through the focal point and the tip of the arrow. (The light would have to pass through the object itself.) However, we have included two other rays that also emerge from the tip of the object. All the refracted rays are parallel. Because they don't intersect, no real or virtual image forms.

Now let's apply the thin-lens equation (Eq. 38.5) to this situation by setting the object distance equal to the focal length ($d_o = f$):

$$\frac{1}{d_o} + \frac{1}{d_i} = \frac{1}{f} + \frac{1}{d_i} = \frac{1}{f}$$

The focal length cancels out:

$$\frac{1}{d_i} = 0$$

So the image distance must be infinite: $d_i \rightarrow \infty$. Our interpretation of this result is that the parallel refracted rays cross at infinity, which means the image forms at infinity. Compare this result to the situation when an object is placed at the focal point of a concave mirror (Section 37-6).

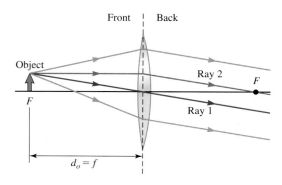

FIGURE 38.34 An object at the focal point does not form an image; the rays that emerge from the lens are parallel. Put another way, the image forms at infinity.

CONCEPT EXERCISE 38.5

Review the discussions of convex and concave mirrors in Sections 37-5 and 37-6. Is a converging lens more like a convex mirror or a concave mirror? Explain.

EXAMPLE 38.9 All Possible Regions for a Converging Lens

A converging lens has focal length $|f| = 10.0$ cm. Find the image distance and magnification if the object is

A 15.0 cm,
B 10.0 cm, and
C 5.00 cm in front of the lens.

INTERPRET and ANTICIPATE

For a converging lens, whether the image is upright or inverted, in front of or behind the lens, and larger or smaller than the object depends on where the real object is placed. If the object is farther away than the focal point, we expect an inverted real image to form behind the lens. That image may be larger or smaller than the object. If the object is at the focal length, we expect an image to form at infinity (no image). Finally, if the object is closer than the focal point, we expect an upright virtual image to form in front of the lens, larger than the object. The distances here are the same as in Example 38.8 for a diverging lens. To think about our results, we'll compare our answers in this example to what we found there.

SOLVE

Isolate the image distance on one side of the thin-lens equation (Eq. 38.5). Substitute for the focal length, which is positive because this is a converging lens.	$\dfrac{1}{d_i} = \dfrac{1}{f} - \dfrac{1}{d_o} = \dfrac{1}{10.0\,\text{cm}} - \dfrac{1}{d_o}$
A Find the image distance when $d_o = 15.0$ cm. This situation corresponds to Figure 38.32.	$\dfrac{1}{d_i} = \dfrac{1}{10.0\,\text{cm}} - \dfrac{1}{15.0\,\text{cm}} = \dfrac{1}{30.0\,\text{cm}}$ $d_i = 30.0$ cm

| Use $M = -d_i/d_o$ (Eq. 38.6) to find the magnification. | $M = -\dfrac{d_i}{d_o} = -\dfrac{-30.0 \text{ cm}}{15.0 \text{ cm}} = \boxed{-2.00}$ |

:• **CHECK and THINK**
The positive image distance means the image is behind the lens and real, as in Figure 38.32. The negative magnification means the image is inverted. Because $|M| > 1$, the image is larger than the object.

| **B** Find the image distance when $d_o = 10.0$ cm. | $\dfrac{1}{d_i} = \dfrac{1}{10.0 \text{ cm}} - \dfrac{1}{10.0 \text{ cm}} = 0$

 $d_i \to \infty$ |

:• **CHECK and THINK**
When the object is at the focal point, the image forms at infinity. In this case, it does not make sense to calculate a magnification because no actual image forms.

| **C** Find the image distance when $d_o = 5.00$ cm. | $\dfrac{1}{d_i} = \dfrac{1}{10.0 \text{ cm}} - \dfrac{1}{5.00 \text{ cm}} = -\dfrac{1}{10.0 \text{ cm}}$

 $d_i = -10.0 \text{ cm}$ |

| Again, use Equation 38.6 to find the magnification. | $M = -\dfrac{d_i}{d_o} = -\dfrac{-10.0 \text{ cm}}{5.00 \text{ cm}} = \boxed{+2.00}$ |

:• **CHECK and THINK**
In part C, we found that the image distance is negative, so the image is virtual and in front of the lens. The magnification is positive and greater than 1, so the image is upright and larger than the object, as in Figure 38.33.

Now let's compare our results for this converging lens to what we found for the diverging lens in Example 38.8. The diverging lens always produces an upright virtual image in front of the lens, and the image is always smaller than the object. When the object is at the focal point, there is nothing special about the image other than that it is exactly half the size of the object. By contrast, a converging lens does not produce an image if the object is at the focal point. If the object is farther than the focal point, a converging lens produces a real inverted image that may be larger or smaller than the object. Finally, if the object is closer than the focal point, the converging lens produces an upright virtual image that is always larger than the object. So, a converging lens can enlarge an image, whereas a diverging lens cannot. These facts will be important as we consider the applications of lenses in the case study.

38-8 The Human Eye

This chapter's case study is about devices with lenses that you use to see. Your vision begins with your eyes, so first we look at the how the human eye works. We then consider devices that use one lens, and finally devices that use two lenses.

Anatomy of the Human Eye

Figure 38.35 shows the anatomy of the human eye. The **sclera** is the white part of the eye; the **iris** is the colored part. The **pupil** is the dark spot in the center of the iris. It is actually a hole or aperture that allows light to pass into the eye's interior. The pupil's diameter changes in response to the amount of light. In bright light, the pupil's diameter may be as small as 2 mm, whereas in the dark, its diameter may

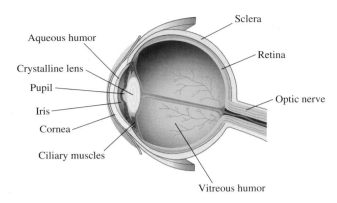

FIGURE 38.35 Anatomy of the human eye.

be 8 mm. The **cornea** is a clear protective covering over the iris and pupil that allows light to pass through. Behind the cornea is a compartment full of transparent fluid similar to water, called the **aqueous humor**. Behind this compartment is the **crystalline lens**, or simply the **lens**. Behind the lens is another compartment full of watery fluid called the **vitreous humor**. At the back of the eye is the light-sensitive **retina**. In response to light, the retina produces a signal that travels along the **optic nerve** to the brain.

Modeling the Eye

Now we develop a model for the optical system of the eye.

Light incident in air first encounters the cornea. The index of refraction of the cornea is about 1.376. Next, the light enters the aqueous humor, a fluid with an index of refraction of 1.336. Then the light passes through the lens, which has an index of refraction of 1.410. And, finally, the light travels through the vitreous humor to the retina. The vitreous humor is very similar to the aqueous humor, so its index of refraction is also 1.336.

The light thus passes through four boundaries—(1) from air to cornea, (2) from cornea to aqueous humor, (3) from aqueous humor to lens, and (4) from lens to vitreous humor—on its way to the retina. The biggest difference in the indices of refraction is between the air ($n_{air} = 1.00026$) and the cornea ($n_{cornea} = 1.376$), so the greatest refraction occurs at the air–cornea boundary. (This explains why it is difficult for you to see underwater. The index of refraction of the water is close to the index of refraction of the cornea, so light is not refracted as much when you look through water as when you look through air.) Although the light passes through four media before reaching the retina, we can still model the eye as a single lens. (A more detailed model would include two lenses—the cornea and the crystalline lens—and would take into account the watery fluid surrounding the crystalline lens.)

Does the eye produce a real or virtual image? Refracted rays of light must make contact with the retina in order to stimulate its photo sensors and create an electrical signal that passes on to the brain. So a real image must be projected onto the retina. Because a diverging lens does not form a real image, we should model the eye as a converging lens and the retina as a screen where a real image forms. Now, is the image inverted or upright? From Figure 38.32, the real image formed by a converging lens is inverted. Objects do not appear upside down to you because the electrical signal that transmits the image is processed by your brain, which has learned to invert that image. As an analogy, consider a digital camera. The image that forms on the light sensor in the camera may be inverted, but the software in the camera displays an upright image to you.

How the Human Eye Focuses

The image distance in your eye is fixed because the distance between your lens and retina does not change. In a normal human eye, the image distance is about an inch (2.2 cm to 2.6 cm). Of course, people like to look at objects at various distances d_o. Because d_i is fixed, the focal length of the lens must be adjusted to form a real image on the retina.

How can the focal length of the lens be adjusted? The answer comes from the lens maker's equation:

$$\frac{1}{f} = (n - 1)\left[\frac{1}{r_1} - \frac{1}{r_2}\right] \tag{38.7}$$

We cannot change the index of refraction of our lens. However, the crystalline lens is flexible and is held in place by ligaments attached to the muscles of the **ciliary**, a ring-like structure surrounding it (Fig. 38.35). The contraction (or relaxation) of the ciliary muscles changes the shape (r_1 and r_2) and therefore the focal length of the lens.

When its ciliary muscles are relaxed, a normal eye is able to focus on objects at infinity. In practical terms, these are objects that are about 5 m or more from the eye.

The rays from such distant objects are parallel. So, when the eye focuses on a distant object, its (converging) lens forms an image on the focal plane and $d_i = f$. In a normally working eye, the focal plane is the retina, typically located 2.5 cm behind the lens (Fig. 38.36A). The *far-point distance* d_{far} is the maximum distance at which an eye can form an image of an object on its retina. The far point of a normal eye is at infinity, and its lens has focal length $f = 2.5$ cm when it focuses on distant objects.

There are two common disorders that alter the far-point distance. Figure 38.36B shows a myopic or nearsighted eye. In a myopic eye, the distance between the lens and the retina is farther than in a normal eye. Parallel rays from a distant object converge at a point in front of the retina. As a result, the person's far point is closer than infinity. In a hyperopic or farsighted eye (Fig. 38.36C), by contrast, the distance between the lens and the retina is shorter than in a normal eye. As a result, the image of a distant object forms behind the retina.

To see a close object, on the other hand, the eye's ciliary muscles must contract (Fig. 38.37). For example, a comfortable reading distance is $d_o = 25$ cm. The image distance is still fixed at $d_i = 2.5$ cm as before. By the thin-lens equation, the focal length must now be

$$\frac{1}{25 \text{ cm}} + \frac{1}{2.5 \text{ cm}} = \frac{1}{f}$$
$$f = 2.3 \text{ cm}$$

which is shorter than the relaxed focal length ($f = 2.5$ cm). Contracting the ciliary muscles in order to look at close objects for a long time, such as when you read, can cause eyestrain. Some doctors recommend that you briefly focus on distant objects every time you reach the end of a page.

The closest distance at which an eye can focus is called the *near-point distance* d_{near}. As a person grows older, the eye sheds cells that build up on the crystalline lens, causing the lens to grow thicker and become harder to flex. Older people thus find it more difficult to focus on nearby objects. The near point moves farther away with increasing age, an effect called *presbyopia*. Table 38.5 lists typical near-point distances as a function of age. When you are in your twenties, your near-point distance is about 10 cm, so you can hold a menu 10 cm from your face and read it comfortably. By the time you are in your sixties, you must hold the menu 1 m to 2 m from your face in order to read it. Because your arms are simply not long enough, you must wear corrective lenses.

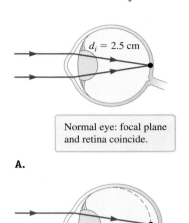
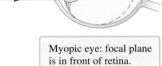
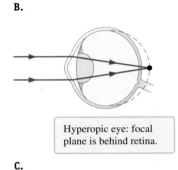

FIGURE 38.36 When an eye focuses on a distant object, the muscles are relaxed and the focal length of the lens is at its maximum. **A.** For a normal eye, the image is formed on the retina. **B.** For a myopic eye, the image is formed in front of the retina. **C.** For a hyperopic eye, the image is formed behind the retina.

TABLE 38.5 Typical near point as a function of age.

Age in years	Near point d_{near} in cm
10	8
20	10
30	13
40	20
50	42
60	100

Corrective Lenses

We have discussed three problems with the human eye. Myopia occurs when the far-point distance d_{far} is too small, and the image of a distant object forms in front of the retina. Hyperopia and presbyopia occur when the near-point distance d_{near} is too great, and the image of a near object forms behind the retina (Fig. 38.36C)— because of either the shape of the eye (hyperopia) or the inflexibility of its lens

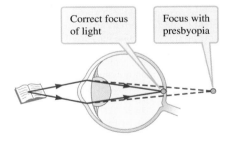

FIGURE 38.37 Reading text at a close distance requires the ciliary muscles to contract so that the focal length of the lens is shortened. As people age, it becomes more difficult to shorten that focal length. When an older person tries to read text up close, the image forms at a point behind her retina.

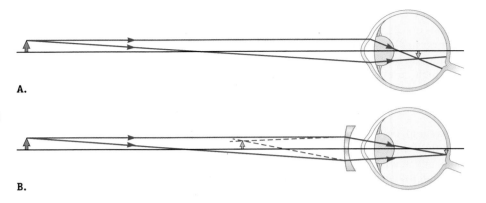

FIGURE 38.38 A. In a myopic eye, the image forms in front of the retina. The eye's lens over-refracts the rays. The result is that the far point is too close to the person's eye. B. A diverging corrective lens compensates for this over-refraction by spreading out the rays and producing an image closer than the object. The person looks at this closer image.

(presbyopia). All three problems can be fixed with corrective lenses such as eyeglasses or contact lenses.

In Figure 38.38A, an uncorrected myopic eye forms an image of a very distant object in front of the retina. In effect, the lens has over-refracted the rays. If the object were closer, the lens would refract the rays by the right amount and the image would form on the retina. A corrective lens is needed that will create an image of a very distant object closer to the eye, at the person's natural far point. Because the person will look at this image through the corrective lens, the image must be upright and virtual (Fig. 38.38B). A diverging lens always produces an upright virtual image that is closer to the lens than the object is (Fig. 38.31) and thus corrects for myopia. The diverging lens spreads out the rays to correct for the over-refracting converging lens of the myopic eye.

In both a hyperopic and a presbyopic eye, the near-point distance is too great and the image forms behind the retina (Fig. 38.39A). In effect, the eye's lens does not refract the rays sufficiently. If the object were farther, the lens would refract the rays by the right amount and the image would form on the retina. A corrective lens must create an image that is farther from the eye. Because the person looks at this image through the lens, this image must be upright and virtual (Fig. 38.39B). A converging lens produces an upright virtual image that is farther than the object as long as the object is closer than the focal point (Fig. 38.33). The normal reading distance is 25 cm, so as long as the focal length of the corrective converging lens is greater than 25 cm, the lens produces an upright virtual image. By bending the rays together, the converging lens corrects the hyperopic or presbyopic eye's under-refraction of the rays, allowing an image to form on the retina.

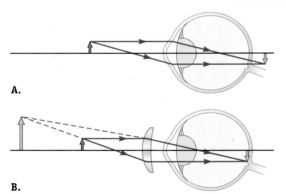

FIGURE 38.39 A. In a hyperopic or presbyopic eye, the image forms behind the retina because the eye's lens cannot refract the rays sufficiently. As a result, the near point is too far from the person's eye. B. A converging corrective lens compensates for this under-refraction by bringing the rays closer together, producing an image farther away than the object. The person looks at this far image.

In physics, *power* can also mean energy per unit time. Don't confuse the two usages.

Instead of specifying the focal length of a corrective lens, it is common to specify its *power*. The **power** D of a lens is the reciprocal of the focal length:

$$D = \frac{1}{f} \tag{38.8}$$

The SI unit of focal length is the meter, so the SI unit of power is m^{-1}, but the power of a lens is usually expressed in diopters, where

$$1 \text{ diopter} = 1 \text{ m}^{-1} \tag{38.9}$$

For example, a prescription for a corrective lens may be $D = +5$ diopters, corresponding to a focal length of $f = 0.2$ m. Because the power is positive, the focal length is positive, and this prescription is for a converging lens. If the power is negative, the lens is diverging.

EXAMPLE 38.10 CASE STUDY How to Read Contact Lens Prescriptions

Sophia and Ezra both wear contact lenses. Sophia's corrective lenses have a power of -4.50 diopters; Ezra's lenses have a power of $+2.75$ diopters.

A Determine whether each person has myopia or hyperopia. Then find the focal length of each person's corrective lenses.

B Assuming a nearsighted person can comfortably see infinitely distant objects when wearing corrective lenses, what is the nearsighted person's uncorrected far-point distance d_{far}? Give your answer in centimeters.

C Assume the farsighted person's corrective lenses allow him or her to just barely read at 25 cm. What is that person's near-point distance d_{near} without corrective lenses?

INTERPRET and ANTICIPATE

In the case of myopia, the person's far-point distance is too short; a nearsighted person can see things that are up close but not far away. So the corrective lenses must produce an image that is closer than the object. In the case of hyperopia, the person's near-point distance is too great; a farsighted person can see things that are far away but not up close. So the corrective lenses must produce an image that is farther than the object. With these ideas in mind, we can reason out whether the corrective lenses must be converging or diverging in each case and, from that, the sign of the focal length and the power. We expect the nearsighted person has a far point closer than 5 m and the farsighted person has a near point farther than 25 cm.

SOLVE

A A person with myopia needs diverging lenses to create an image that is closer than the object. A diverging lens has a negative focal length and power.	Sophia's prescription is for lenses that have a negative power; Sophia is myopic.
A person with hyperopia needs converging lenses to create an image that is farther away than the object. A converging lens has a positive focal length and power.	Ezra's prescription is for lenses that have a positive power; Ezra is hyperopic.
Now let's find the focal length of Sophia's contacts. The focal length is the reciprocal of the power (Eq. 38.8).	$D = \dfrac{1}{f}$ (38.8) $f = \dfrac{1}{D} = \dfrac{1}{-4.50 \text{ diopters}} = -0.222 \text{ m}$ $f = -22.2$ cm for Sophia
Find the focal length of Ezra's contacts in the same way.	$D = \dfrac{1}{f}$ $f = \dfrac{1}{D} = \dfrac{1}{2.75 \text{ diopters}} = 0.364 \text{ m}$ $f = 36.4$ cm for Ezra
B Next, we find Sophia's uncorrected far-point distance. Sophia's contacts produce an image at her far point. Assume the object is very far away, $d_o \to \infty$. In this case, the image distance is negative because the image is virtual and in front of the lens (Fig. 38.38B). Sophia's uncorrected far point is the absolute value of this image distance.	$\dfrac{1}{\infty} + \dfrac{1}{d_i} = \dfrac{1}{f}$ $d_i = f = -22.2$ cm $d_{far} = 22.2$ cm for Sophia

Example continues on page 1248 ▶

C Ezra's contact lenses allow him to read words that are 25 cm from his eyes, so this is the object distance d_o. These lenses produce an image located at his natural near point. The image distance is negative because the image is in front of the contact lens (Fig. 38.39B). This image is the object of his eye's lens. Ezra's uncorrected near-point distance is the absolute value of the image distance.

$$\frac{1}{d_o} + \frac{1}{d_i} = \frac{1}{f}$$

$$\frac{1}{d_i} = \frac{1}{f} - \frac{1}{d_o} = \frac{1}{36.4 \text{ cm}} - \frac{1}{25 \text{ cm}}$$

$$d_i = -79.8 \text{ cm}$$

$$d_{\text{near}} = 80 \text{ cm for Ezra}$$

:• CHECK and THINK

As expected, the nearsighted Sophia has a far-point distance less than 5 m. Her uncorrected eyes cannot focus on objects that are farther than about 22 cm. Also, as expected, the farsighted Ezra has a near-point distance greater than 25 cm. His uncorrected eyes cannot focus on objects that are closer than about 80 cm. If Ezra suffers from presbyopia rather than hyperopia, he is probably about 55 years old (Table 38.5).

38-9 One-Lens Systems

In this section, we look at two practical devices that require only a single lens. The first device is a basic camera. Cameras were originally used by artists, but other individuals quickly saw their value. Today, the camera plays an important role in astronomy, biology, police work, medicine, and communications. The second device is a **simple magnifier**, or just **magnifier**. In ordinary language, it is more commonly called a magnifying glass. You have probably used a magnifier to look at coins, stamps, fingerprints, or jewelry.

FIGURE 38.40 Daguerre's camera.

FIGURE 38.41 A contemporary camera has an adjustable aperture, much like the pupil of your eye.

The Camera

The digital camera you carry in your pocket can trace its evolution back to a modified pinhole camera (Section 37-1). You can make a pinhole camera (a small camera obscura) by poking a small hole in a box. Light from outside the camera forms an inverted real image on a wall or screen inside the device. Artists, such as the scene painter Louis Daguerre (1787–1851), used the camera obscura to trace the image of a subject on paper or on another medium. Daguerre was also a physicist who wanted to find another way to preserve nature in an image. His primary contributions to photography were a photosensitive medium (iodized silver on a glass plate), a development process (exposure to mercury vapors), and—most important—a process for fixing the image on the medium in order to make it permanent (putting the plate in a salt solution). Today's film cameras contain film coated in a photosensitive medium, and part of the development process fixes the image so that further exposure to light does not cause the image to disappear.

Daguerre used a modified pinhole camera (Fig. 38.40). Its chamber had a somewhat large hole or aperture at one end, with a lens behind the hole to keep the image sharp. The image was projected onto a glass plate coated with iodized silver. The aperture in a camera is adjustable, so it can be widened or narrowed depending on the lighting conditions (Fig. 38.41). The photosensitive iodized silver is analogous to the photosensitive cells of the retina. A real image must form on the glass plate (or film or digital detector) in a camera, as on the retina. So a camera's lens must be converging, and the image formed on the plate or film is inverted. After the image is printed, we simply invert the photograph again in order to view it.

In the human eye, the image distance is fixed because the distance between the lens and retina does not change. You are able to focus on objects at various distances because your crystalline lens is flexible, able to change shape and therefore focal length. By contrast, the camera's lens is not flexible; its focal length is fixed by its shape (and the material it is made of). In order to form an image on the plate, then, the image distance must be flexible. As shown in Figure 38.40, Daguerre's camera consisted of two open-ended boxes, one holding the lens and the other the plate. Daguerre could slide the two boxes relative to each other to change the image distance, creating a sharp image. In a contemporary camera, the lens sits in a cylinder that moves in and out in order to adjust the image distance.

A contemporary camera usually has a lens made of several optical elements that collectively act like a single converging lens. These optical elements are designed to work together to reduce aberrations (defects that result in a poor image; Section 37-7). Aberrations are not a result of poor construction but are a natural consequence of how light is refracted through lenses. Like spherical mirrors, spherical lenses suffer from spherical aberration when the rays that reach them are not paraxial.

Lenses also suffer from **chromatic aberration**, which occurs because the index of refraction of light depends on its color. Violet light is refracted more than red light (Fig. 38.9), so when white light is refracted by a lens, the different colors that make up white light focus at different positions (Fig. 38.42A). When colors do not focus at exactly the same position, the resulting photograph shows a halo of colors. Figure 38.42B shows how chromatic aberration may be partially compensated for with a second lens. In this case, red and yellow light form an image at a single point, but blue light still does not focus at this point.

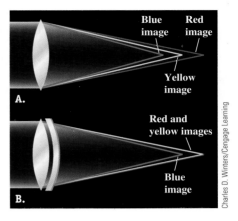

FIGURE 38.42 A. A single converging lens suffers from chromatic aberration. Blue light forms an image in front of yellow light, which forms an image in front of red light. **B.** A second lens is added, and the chromatic aberration is improved. Now the red and yellow images are formed at the same position.

EXAMPLE 38.11 Making Your Own Daguerre Camera

You can build a simple camera out of a pair of open boxes provided one box can easily slide inside the other. Wrap the seam between the boxes in dark fabric to keep the light out. Place light-sensitive paper at the back wall of one box, and poke a small hole in the front wall of the other box. Attach a lens in front of the hole.

Suppose you use a converging lens with a 30.0-cm focal length. Find the closest and farthest useful distances between the front and back walls of the two boxes. Be practical in your design. Assuming you will typically take pictures of objects that are 5 m from the camera, what is the typical useful distance between the front and back walls?

:• INTERPRET and ANTICIPATE
The thin-lens equation applies in this case. The lens is converging and the object is in front of it, so both the focal length and the object distance are positive. The image distance, which is also positive, is the distance between the front and back walls. So, for the first question, think about finding the full useful range of image distances. For the second question, solve for the image distance when the object distance is 5 m.

:• SOLVE
The image distance is shortest when the object is at infinity. In that case, the image distance equals the focal length. So, the closest useful distance between the front and back walls equals the focal length.

$$\frac{1}{d_o} + \frac{1}{d_i} = \frac{1}{f} \quad (38.5)$$

$$\frac{1}{\infty} + \frac{1}{d_i} = \frac{1}{f}$$

$$d_i = f = 30.0 \text{ cm}$$

(closest useful distance between walls)

Example continues on page 1250 ▶

A converging lens does not produce a real image if the object is at the focal point or closer (Figs. 38.33 and 38.34). Assume that, for practical purposes, you don't expect to photograph objects that are closer than 60.0 cm, or twice the focal length of the lens. (60.0 cm is about 2 ft—shorter than your arm. You probably won't take a picture of a friend from less than an arm's length away.)	$\frac{1}{d_i} = \frac{1}{f} - \frac{1}{d_o}$ $\frac{1}{d_i} = \frac{1}{30.0 \text{ cm}} - \frac{1}{60.0 \text{ cm}} = \frac{1}{60.0 \text{ cm}}$ $d_i = 60.0 \text{ cm}$ (farthest useful distance between walls)
But it seems reasonable to take pictures of objects that are about 5 m from the camera. We find that this image distance is very close to the focal length of the lens.	$\frac{1}{d_i} = \frac{1}{f} - \frac{1}{d_o} = \frac{1}{30.0 \text{ cm}} - \frac{1}{500 \text{ cm}}$ $d_i = 32 \text{ cm}$ (typical useful distance)

:• CHECK and THINK

Now we see why early cameras were so large. For them to focus on close objects, the distance between the lens and the glass plate had to be large. Notice that the image distance nearly equals the focal length when the object is at 5 m. This gives us a practical measure for what we mean when we say an object is infinitely far. For a lens with a 30.0-cm focal length, an object at 5 m is, practically speaking, infinitely far away.

A.

B.

FIGURE 38.43 **A.** A magnifier produces an upright virtual image. **B.** The angular size θ_o is the angle subtended by the object at your eye. When the flower is close to your eye, it subtends a larger angle, so it appears larger than it does when it is far from your eye.

CASE STUDY The Magnifying Glass

Now let's turn our attention to the second example of a single-lens device—a magnifier (Fig. 38.43A). A magnifying lens differs from a camera lens in that when you look through a magnifying lens, you always see an enlarged image. This image must be upright and on the same side of the lens as the object. This image is virtual, but it cannot be made by a diverging lens because such a lens always produces an image that is smaller than the object. A converging lens produces a virtual upright image that is larger than the object when the object is closer to the lens than the focal point (Fig. 38.33).

An object looks bigger to you if its image fills up more space on your retina. Mathematically, we describe the apparent size of an object in terms of its angular size θ_o as defined in Figure 38.43B. An object that is close to your eye has a large angular size and appears large. However, you cannot focus on objects closer to your eye than your near point d_{near}. So there is a limit to the angular size of a particular object (Fig. 38.44A). If the object's height is h_o, the largest angle the object can subtend is

$$(\theta_o)_{\text{max}} = \frac{h_o}{d_{\text{near}}} \qquad (38.10)$$

where $(\theta_o)_{\text{max}}$ is small so that $\tan(\theta_o)_{\text{max}} \approx \sin(\theta_o)_{\text{max}} \approx (\theta_o)_{\text{max}}$.

If you wish to see a larger image of the object, you can use a magnifier to make an enlarged image that is far from your eye. Ideally, the image forms at infinity so that your ciliary muscles can relax. To achieve this, the object must be placed at one focal point of the lens with your eye at the other focal point. In Figure 38.44B, we have drawn two rays that emerge from an object—one from the top and the other from the bottom. Both rays are parallel to the optical axis and bend toward the focal point of the lens. This figure is not a ray diagram; in a ray diagram, the rays emerge from a single point of an object. Compare Figure 38.44B to the ray diagram for an object at the focal point of a converging lens (Fig. 38.34).

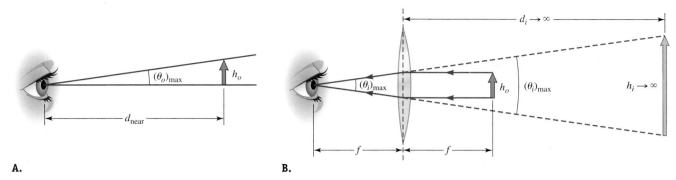

FIGURE 38.44 A. For an unaided eye, the maximum angular size of an object is seen when the object is at the person's near point. **B.** When you use a magnifier, the maximum angular size of the image is seen when the object is at one focal point of the converging lens and your eye is at the other focal point. Then the image is at infinity, so its height is infinite. For practical purposes, an image with infinite height at infinity is indistinguishable by a normal eye from a large image at a distance greater than 5 m whose rays are paraxial when they reach the eye.

As shown in the ray diagram, rays from a single point on the object, such as the tip, do not come together at the focus. Instead, the rays emerge from the lens parallel to one another; the image forms at infinity, and its height is infinite. Such an image cannot be drawn, so in Figure 38.44B we have drawn a large image at a great distance.

Because the rays in Figure 38.44B are from the top and bottom of the object, the angle between the refracted rays is the angular size of the image θ_i. In this ideal geometry, where the image is understood to have an infinite height and to be located at infinity, the angular size of the image must be at its maximum $(\theta_i)_{max}$. Assuming $(\theta_i)_{max}$ is small enough so that $\tan(\theta_i)_{max} \approx \sin(\theta_i)_{max} \approx (\theta_i)_{max}$, we can express the image's angular size as

$$(\theta_i)_{max} = \frac{h_o}{f} \tag{38.11}$$

The **angular magnification** m of a magnifier is the ratio of the image's angular size with the magnifier to the object's angular size without it:

$$m \equiv \frac{\theta_i}{\theta_o} \tag{38.12}$$

Using the maximum values (Eqs. 38.11 and 38.12) for these angles, we find

$$m = \frac{(\theta_i)_{max}}{(\theta_o)_{max}} = \left(\frac{h_o}{f}\right)\left(\frac{d_{near}}{h_o}\right)$$

$$m = \frac{d_{near}}{f} \tag{38.13}$$

ANGULAR MAGNIFICATION
✪ Major Concept

Equation 38.13 gives the angular magnification for a simple magnifier. Because it is common to assume the near point of a person's eye is 25 cm, the angular magnification of a magnifier is often expressed as

$$m = \frac{25 \text{ cm}}{f} \tag{38.14}$$

Throughout this textbook, we do not drop the term *angular*, but we usually drop the term *linear*.

It is important to distinguish between the angular magnification m and the (linear) magnification M, the ratio of image height to object height (Eq. 37.1). In the case of a magnifier, the image height is infinite, so the linear magnification is infinite. However, the image does not look infinitely large. Often the terms *angular* and *linear* are dropped when magnification is specified. When you buy a magnifier, the manufacturer usually reports the angular magnification as a number followed by ×, which stands for *times*. For example, binoculars that have a magnification of 50× produce an image whose angular size is 50 times larger than the angular size of the object.

According to $m = d_{near}/f$ (Eq. 38.13), the shorter the focal length, the higher the angular magnification. However, there is a practical limit to angular magnification due to spherical aberration. With no correction for spherical aberration, the highest angular magnification for a single-lens magnifier is about $4\times$. Such magnification is practical for many purposes, but not for all. Coin collectors and jewelers often require a higher magnification. Correcting for spherical aberration can produce a magnifier with about $20\times$ angular magnification. If you need an even higher magnification, you must use a microscope (Section 38-10).

CONCEPT EXERCISE 38.6

CASE STUDY An Emergency Magnifier

You are in the woods taking photos with your camera when you come across an interesting and very rare bird's feather. Your camera has a removable lens. Can you use that lens as a magnifier to take a closer look at the feather? If so, explain what you need to do. If not, explain why not.

CONCEPT EXERCISE 38.7

CASE STUDY Focal Length and Angular Magnification

A magnifier has an angular magnification of $3\times$. What is the focal length of the lens?

38-10 Multiple-Lens Systems

In this section, we study two devices that require two lenses. The first is a **compound microscope**, or simply a **microscope**. The other is a **refracting telescope**. The purpose of a microscope is to produce a greatly enlarged image. Many people think a telescope serves the same purpose, but that is not the case. The purpose of a telescope is to gather a great deal of light from very distant objects. However, both devices are made of two converging lenses.

CASE STUDY The Compound Microscope

Suppose you are using a magnifier to see some fine detail on a small object but the image is simply not large enough. Fortunately, you have a second magnifier, so you can use both of them to see the detail. This is the basic idea behind a microscope (Fig. 38.45A). The object is placed on a slide and illuminated from below by a light source. Just above the object is the **objective lens**, or simply the **objective**. (As a memory aid, the *objective* lens is close to the *object*.) At the top of the microscope is a second lens called the **eyepiece**. You place your eye near the eyepiece.

Figure 38.45B shows a ray diagram for the two lenses that make up the microscope. The focal length of the objective lens is f_o, and the object distance is greater than the focal length: $d_o > f_o$. Because the objective is a converging lens and the object is farther away than its focal point, the lens forms a real inverted image at the image distance d_i (as in Fig. 38.32). To make sure this real image is larger than the object, the object must be close to the focal point of the lens. Then this enlarged real image becomes the object for the second lens—the eyepiece.

The job of the eyepiece is simply to act as a magnifier of the real image produced by the objective. Like any simple magnifier, the eyepiece should produce a virtual enlarged image. From Section 38-9, we know that a converging lens works as a

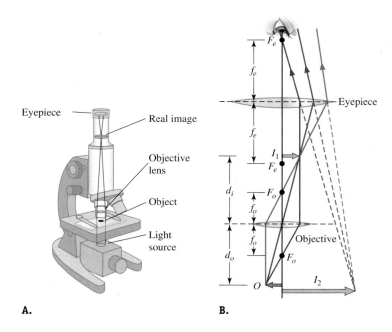

FIGURE 38.45 A. Parts of a compound microscope. **B.** A ray diagram for two lenses that make up the microscope. I_1 is the image produced by the objective; it is the object of the eyepiece. I_2 is the image that you observe through the eyepiece.

magnifier when the object is just inside its focal point (Fig. 38.44). So the real image produced by the objective lens must form just inside the focal point of the eyepiece.

The total magnification of a microscope depends on the linear magnification of the objective lens and the angular magnification of the eyepiece. The linear magnification of the objective is given by

$$M = \frac{h_i}{h_o} = -\frac{d_i}{d_o} \qquad (38.6)$$

Because the object and image distances are both positive for the objective, the magnification is negative and the real image I_1 produced by the objective is inverted (Fig. 38.45B). The magnification is highest when the object distance is short and the image distance is long. Because we need the objective to produce a real image, however, the object cannot be closer to the lens than the focal point. So, in a well-designed microscope, the object distance is only slightly longer than the focal length of the objective. Let's make the approximation that the object distance equals the focal length of the objective lens. Then $M = -d_i/d_o$ (Eq. 38.6) becomes

$$M_o = -\frac{d_i}{f_o} \qquad (38.15)$$

By contrast, the angular magnification of the eyepiece is given by $m_e = d_{\text{near}}/f_e$ (Eq. 38.13). The total angular magnification of the microscope is found by multiplying the linear magnification of the objective by the angular magnification of the eyepiece:

$$m_{\text{micro}} = M_o m_e = -\frac{d_i}{f_o} \frac{d_{\text{near}}}{f_e} \qquad (38.16)$$

The negative sign means that the image you see through a microscope is inverted. But, if you imagine using a microscope to look at blood cells, for example, an inverted image is not a problem.

Often microscope manufacturers drop the negative sign in Equations 38.15 and 38.16. Also, they usually assume the near point is at 25 cm, so the angular magnification of a microscope is reported as

$$m_{\text{micro}} = (25 \text{ cm}) \frac{d_i}{f_o f_e} \qquad (38.17)$$

Equation 38.17 shows that the angular magnification is increased by using an eyepiece and an objective with short focal lengths.

When you use a microscope, you must be able to adjust its angular magnification. Suppose you want to look at a particular blood cell. You start by studying an image with low angular magnification, so that you can see all the cells in the sample. Then you make sure the cell you are interested in is in the center of the image, so that when you look with a higher angular magnification, that cell remains in the image.

According to Equation 38.17, the magnification of a microscope can be controlled by using eyepieces and objectives of varying focal lengths. If the manufacturer supplies only one eyepiece, the magnification is controlled by switching the objective lenses. As shown in Figure 38.45A, several objective lenses are often arranged on a wheel. You control the magnification by choosing a particular objective; the objective with the shortest focal length produces the highest magnification.

We modeled the microscope as a system of two thin converging lenses. However, like modern camera lenses, each lens of a modern microscope usually consists of several optical elements.

CASE STUDY The Refracting Telescope

A microscope provides a greatly enlarged image of a small close object. Contrast this to an astronomical telescope, which provides an image of a large distant object. The main job of the telescope is to gather more light than you could with your unaided eyes. Both microscopes and refracting telescopes are made of two converging lenses, but the arrangement of these lenses determines which instrument results.

Figure 38.46A shows the major components of a refracting telescope. As in a microscope, the lens closer to the object is the objective and the lens closer to your eye is the eyepiece. An object examined through a telescope is far away, so rays from the object are parallel and come together at the focal point of the objective lens (Fig. 38.46B). The image formed at the focal point is very small. The second lens—the eyepiece—magnifies that image. The image produced by the objective becomes the object for the eyepiece, which produces an enlarged virtual image.

Ideally, the image you see through the eyepiece is at infinity so that your ciliary muscles can relax, which means the image produced by the objective should be at the focal point of the eyepiece. So the distance between the objective lens and the eyepiece should be the sum of their focal lengths: $f_o + f_e$ in Figure 38.46B.

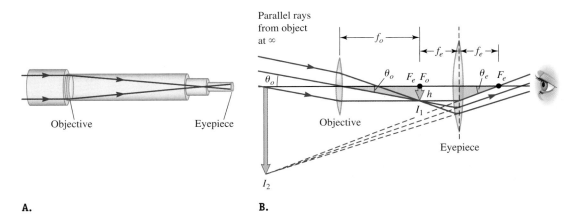

FIGURE 38.46 A. A refracting telescope has two lenses. **B.** A ray diagram for an astronomical telescope. I_1 is the image produced by the objective; it is the object of the eyepiece. I_2 is the image that you observe through the eyepiece.

Although the primary purpose of a telescope is to gather light and not to magnify, it does produce an enlarged image. This enlargement is useful when you want to see something far away, but not *astronomically* far. The total angular magnification of a telescope depends on the angular magnifications of the objective and the eyepiece. The angular size seen with the unaided eye is the same as the angle θ_o subtended by the object at the objective lens, and the angular size seen through the telescope is about equal to the angle θ_e subtended at the eyepiece (Fig. 38.46B). The angular magnification of the telescope is the angular size of the image seen with the telescope θ_e divided by the angular size of the object seen with the unaided eye θ_o:

$$m_{\text{tele}} = \frac{\theta_e}{\theta_o} \tag{38.18}$$

Using the two right triangles highlighted in Fig. 38.46B and some trigonometry, we find

$$\tan\theta_o = \frac{h}{f_o} \text{ and } \tan\theta_e = -\frac{h}{f_e}$$

The minus sign is introduced because the image seen through the eyepiece is inverted. (The image seen with the unaided eye is not inverted.) The angles are small, so we use the small-angle approximation for the tangent:

$$\theta_o \approx \frac{h}{f_o} \tag{38.19}$$

and

$$\theta_e \approx -\frac{h}{f_e} \tag{38.20}$$

Substitute Equations 38.19 and 38.20 into Equation 38.18 to find that h cancels out:

$$m_{\text{tele}} = -\frac{f_o}{f_e} \tag{38.21}$$

Now compare the angular magnification of a microscope, $m_{\text{micro}} = (25 \text{ cm})(d_i/f_o f_e)$ (Eq. 38.17), to the angular magnification of a telescope (Eq. 38.21). For both instruments, the angular magnification is inversely proportional to the focal length of the eyepiece. However, for the telescope the angular magnification is directly proportional to the focal length of the objective, whereas for the microscope it is inversely proportional. Often a microscope has only one eyepiece, and the magnification is controlled by changing the objective. The opposite is true for a telescope. Usually the telescope manufacturer provides only one objective, and that lens is not easily removed. However, the eyepieces are relatively inexpensive and easy to change. An astronomer generally starts by using a low-magnification (long focal length) eyepiece to see a large field of view. When she knows that the object she wishes to see is in the center of the image, she may switch to a higher-magnification (short focal length) eyepiece to see details. The larger image is fainter, however, because the same amount of light is spread out over a larger area of her retina. As in the case of microscopes, the negative sign is often dropped when a telescope's angular magnification is reported.

The primary purpose of an astronomical telescope is to collect light from objects too distant to see with unaided eyes. We describe this property of telescopes as their *light-gathering power*, or LGP. The LGP of a telescope depends on its diameter. As an analogy, suppose that instead of collecting photons from a distant object, you wish to collect rain. You will collect more rain in the same amount of time if you use a bowl with a large diameter rather than a tall glass with a small diameter. Similarly, the light-gathering power of a telescope is proportional to the area A of its objective lens:

$$\text{LGP} \propto A$$

Because the cross section of the lens is a circle, the LGP is proportional to the square of the radius or the diameter of the lens:

$$\text{LGP} \propto \pi R^2 \propto \pi \left(\frac{d}{2}\right)^2$$

$$\text{LGP} \propto d^2 \quad (38.22)$$

The LGP of a telescope is so important that telescopes are usually described in terms of their objective lens diameter. So, an astronomer may say he is going to use the "8-inch," which means he will use a telescope whose objective lens is 8 inches in diameter.

Refracting telescopes are great fun and were very important until the early 1900s. Refracting telescopes are not generally used by today's professional astronomers, however, because very large telescopes are needed to see very faint objects. Large lenses are heavy and also suffer from chromatic aberration. It is expensive to build very large refracting telescopes that are corrected for chromatic aberration. Instead, astronomers use large reflecting telescopes.

CONCEPT EXERCISE 38.8

CASE STUDY How Good Is Your Telescope?

When you talk to an astronomer about his telescope, why doesn't it make sense to ask about its magnification? Why does it make more sense to ask about the size of the objective lens?

EXAMPLE 38.12 CASE STUDY Your Eye Versus Your Telescope

For your birthday, your very nice aunt gives you a 4-inch refracting telescope. The objective lens has a focal length of 880 mm. The telescope has two eyepieces with focal lengths of 10 mm and 50 mm.

A How does the light-gathering power of your telescope compare to the light-gathering power of your fully open pupil?

:• INTERPRET and ANTICIPATE

The LGP depends on the diameter. For a telescope, the diameter of the objective is all that matters. The phrase "4-inch refracting telescope" means the objective has a 4-inch diameter. By contrast, the diameter of a fully open pupil is about 8 mm (Section 38-8). Because the diameter of the telescope is so much larger than the diameter of your pupil, we expect the LGP of the telescope to be much greater than that of your eye. It is helpful to work in millimeters: 4 in. = 102 mm.

:• SOLVE

The ratio of the LGP of the telescope to the LGP of your eye is proportional to the ratio of their diameters squared (Eq. 38.22).

$$\frac{(\text{LGP})_{\text{tele}}}{(\text{LGP})_{\text{eye}}} = \left(\frac{d_{\text{tele}}}{d_{\text{eye}}}\right)^2 = \left(\frac{102 \text{ mm}}{8 \text{ mm}}\right)^2$$

$$\frac{(\text{LGP})_{\text{tele}}}{(\text{LGP})_{\text{eye}}} = 13$$

:• CHECK and THINK

As expected, the LGP of the telescope is much higher (13 times higher) than the LGP of your eye. So, when you use a telescope, you can see much fainter objects than you can without it.

B What is the telescope's highest angular magnification?

:• INTERPRET and ANTICIPATE
The magnification depends on the focal length of the objective and the focal length of the eyepiece you choose. There is only one objective, but you can use either the 10-mm or the 50-mm eyepiece. Notice that we refer to an eyepiece by its focal length, not its diameter. The eyepiece with the shorter focal length produces the higher angular magnification. So, to find the telescope's highest angular magnification, you would choose the 10-mm eyepiece.

:• SOLVE
Use Equation 38.21 to find the angular magnification. We have dropped the minus sign, as is often done.

$$m_{\text{tele}} = \frac{f_o}{f_e} \quad (38.21)$$

$$m_{\text{tele}} = \frac{880 \text{ mm}}{10 \text{ mm}} = 88\times$$

:• CHECK and THINK
The highest angular magnification you can achieve with this instrument is $88\times$. You might be able to get a higher magnification by using an eyepiece with an even shorter focal length, but that would make the image fainter. A telescope manufacturer often specifies the highest practical angular magnification. If the magnification is too high, the image is too faint to be useful.

SUMMARY

❶ Underlying Principles

No new principles are introduced. This chapter uses geometric optics.

✪ Major Concepts

1. The first part of the **law of refraction** states that the refracted ray, the incident ray, and the normal all lie in a single plane—the plane of incidence. The second part is **Snell's law**:

$$n_t \sin \theta_t = n_i \sin \theta_i \quad (38.1)$$

where θ_t and θ_i are the *t*ransmitted and *i*ncident angles (Fig. 38.1), and n_t and n_i are the indices of refraction for the transmitted and incident media, respectively.

2. **Total internal reflection** is possible if light is incident in a medium that has a higher index of refraction than that of the medium beyond the boundary. Then, no light is transmitted if the angle of incidence is larger than the critical angle as given by

$$\sin \theta_c = \frac{n_t}{n_i} \quad (38.2)$$

3. The spreading out of light by color due to differences in the index of refraction is called **dispersion**.

4. The **thin-lens equation** is identical to the mirror equation:

$$\frac{1}{d_o} + \frac{1}{d_i} = \frac{1}{f} \quad (38.5)$$

5. The **magnification** by a thin lens is the same as for a spherical mirror:

$$M = \frac{h_i}{h_o} = -\frac{d_i}{d_o} \quad (38.6)$$

6. The **lens maker's equation** is

$$\frac{1}{f} = (n-1)\left[\frac{1}{r_1} - \frac{1}{r_2}\right] \quad (38.7)$$

Equations 38.5, 38.6, and 38.7 use the sign conventions given in Table 38.2.

7. The **angular magnification** is given by

$$m \equiv \frac{\theta_i}{\theta_o} \quad (38.12)$$

Special Cases

1. When a real object is in front of a **diverging lens**, the image formed is in front of the lens and virtual. The image is upright and smaller than the object.
2. When a real object is in front of a **converging lens**, the image formed depends on the position of the object relative to the focal point.
   a. A real inverted image is produced if $d_o > f$. If $d_o > 2f$, the image is smaller than the object. If $f < d_o < 2f$, the image is larger than the object.
   b. An upright virtual image is produced if $d_o < f$. The image is always larger than the object.

Tools

The elements of a **ray diagram** (Fig. 38.31) are:
1. The **lens**, including a vertical midline
2. The **optical axis**
3. The **focal points** F drawn on both sides of the lens
4. An arrow representing the **object**
5. Two or three **primary rays** (described in Table 38.4) emerging from one point of the object, usually the tip of the arrow

PROBLEMS AND QUESTIONS

A = algebraic   C = conceptual   E = estimation   G = graphical   N = numerical

38-1 Law of Refraction

1. **N** The Sun appears at an angle of 53.0° above the horizontal as viewed by a dolphin swimming underwater. What angle does the sunlight striking the water actually make with the horizon?

2. **C** In this chapter, we studied the refraction of visible light. Have we ignored other parts of the electromagnetic spectrum because radiation at other frequencies does not refract? If so, explain why radiation of other wavelengths cannot refract. If not, give an example of radiation that refracts at other wavelengths.

Problems 3 and 4 are paired.

3. **N** A light ray is incident on an interface between water ($n = 1.333$) and air ($n = 1.0002926$) from within the water. If the angle of incidence in the water is 30.0°, what is the angle of the refracted ray in the air?

4. **N** A light ray is incident on an interface between water ($n = 1.333$) and air ($n = 1.0002926$) from within the air. If the angle of incidence in the air is 30.0°, what is the angle of the refracted ray in the water?

Problems 5 and 6 are paired.

5. **A** A mirror is above a swimming pool and perpendicular to the surface of the water. A narrow beam is incident on the mirror (Fig. P38.5). Find an expression for the angle of refraction in the water.

6. **N** A mirror is above a swimming pool and perpendicular to the surface of the water. A narrow beam is incident on the mirror (Fig. P38.5). If $\theta_i = 22.3°$, find the angle of refraction in the water.

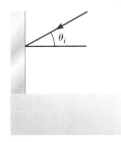

FIGURE P38.5 Problems 5 and 6.

7. **N** Figure P38.7 shows a monochromatic beam of light striking the right-hand face of a right-angle prism at normal incidence. If $\theta_t = 20.0°$, what is the index of refraction of the prism?

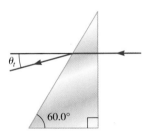

FIGURE P38.7

8. **N** A ray of light enters a liquid from air. If the angle between the incident and refracted rays is 150° and the angle between the reflected and refracted rays is 60°, find the refractive index of the liquid. Assume the refractive index of air is 1.00.

9. **N** A mirror is above a swimming pool and tilted toward the surface of the water. A narrow beam is parallel to the surface of the water and incident on the mirror (Fig. P38.9). The angle of incidence on the mirror's surface is 22.3°. Find the angle of refraction in the water.

FIGURE P38.9

10. **N** Figure P38.10 on the next page shows a monochromatic beam of light of wavelength 575 nm incident on a slab of crown glass surrounded by air. Use a protractor to measure the angles of incidence and refraction. **a.** What is the speed of the beam of

light within the glass slab? **b.** What is the frequency of the beam of light within the glass slab? **c.** What is the wavelength of the beam of light within the glass slab?

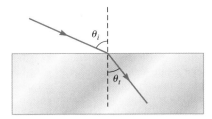

FIGURE P38.10

11. **N** Cone receptors in the human eye contain visual pigments that are sensitive to the color of light. In the 1800s, two physiologists, Thomas Young and Herman von Helmholtz, proposed that for each of the three fundamental color sensations in human vision (red, green, and blue), there is a different kind of color receptor. The excitation of a particular receptor leads to the sensation of the corresponding color. This theory, known as the Young–Helmholtz or trichromatic theory, is accepted today. The pigments are found in the aqueous humor inside the eyeball. The receptor responsible for red pigment shows peak sensitivity to light at a wavelength of about 580.0 nm in the aqueous humor. Using red light in air with a wavelength of 700.0 nm, determine the index of refraction of the aqueous humor and the speed and frequency of the light in this substance.

12. **C** You are camping in the woods and find yourself running low on food. You notice several fish in a small pool of clear water. After making a spear from a tree limb, you try to spear a meal by thrusting the spear into the water where you see a fish. Why is this a poor strategy for using the spear? Where should you target a particular fish if you hope to spear it?

Problems 13 and 107 are paired.

13. **N** A block is constructed from layers of cubic zirconia ($n = 2.14$), flint glass ($n = 1.80$), and quartz ($n = 1.54$) as shown in Figure P38.13. The block is surrounded by air. A ray of monochromatic light is incident on the cubic zirconia–flint glass interface with $\theta_i = 23.0°$. What is the refraction angle θ_t when the ray exits the slab at the bottom?

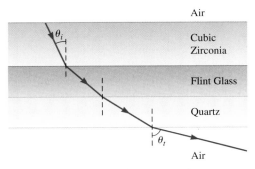

FIGURE P38.13 Problems 13 and 107.

38-2 Total Internal Reflection

14. **C** If you are underwater, can you look straight up and use total internal reflection to see your own face? Explain.

15. **N** What is the minimum value of the index of refraction of a right-angled, isosceles prism for which a light ray entering normally on one face will be totally internally reflected as shown in Figure P38.15? Assume the index of refraction of air is 1.00029.

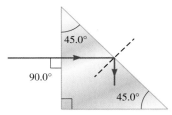

FIGURE P38.15

Problems 16 and 17 are paired.

16. **C** A fish is 3.25 m below the surface of still water (Fig. P38.16). You do not want the fish to see your fishing boat. Is it possible to place your boat so that total internal reflection keeps it hidden from the fish? If so, explain how this is done. If not, explain why not.

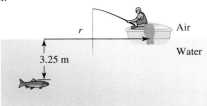

FIGURE P38.16 Problems 16 and 17.

17. **N** A fish is 3.25 m below the surface of still water. Because of total internal reflection, it is hidden from the view of a fisher in a boat on the water as long as the boat is outside a circle of radius r. The center of the circle is directly above the fish (Fig. P38.16). Find the minimum value of r.

18. **N** A beam of monochromatic light within a fiber optic cable is incident on one of the sides of the cable ($n = 1.485$) at an angle of incidence θ_i. Assume the fiber is surrounded by air ($n = 1.00029$). **a.** What is the critical angle for total internal reflection so that the beam stays within the fiber? **b.** What would the critical angle be if the fiber were completely immersed in water ($n = 1.33$)?

19. **N** A beam of light travels through a fiber with an index of refraction of 1.55. The fiber is surrounded by a medium with an index of refraction of 1.36. At what minimum angle of incidence is all the light reflected back into the fiber?

Problems 20 and 21 are paired.

20. Consider a light ray that enters a pane of glass with air on either side as shown in Figure P38.20. The light ray experiences refraction at the first interface when it enters the glass and again at the second interface when it exits the glass. Assume the index of refraction of the glass is 1.54.
    a. **N** If $\theta_1 = 53.0°$, find θ_2, θ_3, and θ_4.
    b. **C** Explain why it is impossible for a light ray striking the glass pane to experience total internal reflection at either interface.

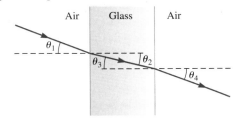

FIGURE P38.20

21. **N** Consider a light ray that enters a pane of glass with air on one side and water on the other side as shown in Figure P38.21. The light ray experiences refraction at the first interface when it enters the glass from the water and again at the second interface when it exits the glass into the air. Assume the index of refraction of the glass is 1.54. For a ray of light, find the angle of incidence θ_1 in the water such that the ray experiences total internal reflection when it strikes the glass–air interface on the other side.

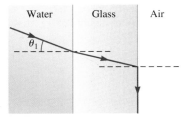

FIGURE P38.21

22. **N** Consider a beam of 486-nm blue light within a slab of material. The beam is incident on the slab's interface with air. What is the critical angle for total internal reflection if the slab is made of **a.** ice ($n = 1.309$), **b.** fluorite ($n = 1.434$), and **c.** diamond ($n = 2.417$)?

23. **N** A source of light S is placed at the bottom of a container holding a liquid that has an index of refraction of 1.67. A person is viewing the source from above the liquid while there is an opaque disk of radius 1.00 cm floating on the surface as shown in Figure P38.23. The center of the disk lies vertically above the source S. The liquid from the container is slowly drained out through a tap, while the center of the disk remains directly above the light source. What is the height of the liquid when the source can suddenly no longer be seen from above? Assume the index of refraction of air is 1.00029.

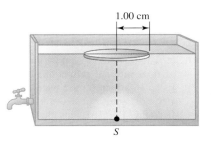

FIGURE P38.23

24. A light ray is traveling inside a block of glass. Assume the index of refraction of the glass is 1.54.
    a. N If the medium outside the block of glass is air, find the critical angle for the glass–air interface.
    b. C Any light ray that strikes the glass–air interface at an incident angle greater than the critical angle will experience total internal reflection. Will any ray that strikes the glass–air interface at an incident angle less than the critical angle necessarily experience transmission out of the block of glass on the other side? Explain.

25. **N** Figure P38.25 shows a beam of 656-nm red light striking the top surface of a slab of quartz ($n = 1.54$) surrounded by air at an angle of incidence θ_i. After refraction at the air–quartz interface, the beam strikes the right-hand edge of the slab at point P. What is the greatest angle θ_i for which total internal reflection will occur at point P?

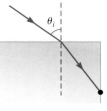

FIGURE P38.25

38-3 Dispersion

26. **N** Dispersion occurs when the index of refraction of a medium depends on the wavelength of incident light, resulting in different angles of refraction for each wavelength. A beam of white light is incident on a block of zinc crown glass with $\theta_i = 34.0°$. The index of refraction of this glass for 589-nm yellow light is 1.528, while the index of refraction for 486-nm blue light is 1.517. **a.** What is the refraction angle θ_t for yellow light in the glass? **b.** What is the refraction angle θ_t for blue light in the glass?

27. **C** Isaac Newton, working in a dark room, let a beam of sunlight pass through a hole in the window shade. His goal was to show that white light is composed of all the colors. He placed a prism in the path of the beam and saw that the white light spread out to form a color spectrum. To prove that the colors were not produced by the prism, he used a screen with a small slit to select one color of light. He then placed a second prism in the path of this monochromatic (one-color) beam. **a.** What colors emerged from the second prism? **b.** Next, Newton removed the filter (the slit) and used the second prism to form a beam of white light (Fig. P38.27). Why was this step necessary?

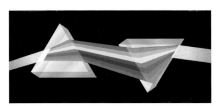

FIGURE P38.27

28. **C** When UV and IR radiation pass from a vacuum into some medium, which one is refracted more by the change in media? Explain.

29. The wavelength of light changes when it passes from one medium into another. Suppose green light at 550.0 nm passes from air into glass ($n = 1.5$).
    a. N What is its wavelength in glass?
    b. N What is its frequency?
    c. C Do you see the same color in both media?

30. **N** A new material is being considered for encapsulating and protecting a sensitive electronic device. If light of wavelength 500.0 nm in air ($n = 1.00029$) has a wavelength of 240.0 nm when it enters the new material, what is the index of refraction of the material?

31. Light is incident on a prism as shown in Figure P38.31. The prism, an equilateral triangle, is made of plastic with an index of refraction of 1.46 for red light and 1.49 for blue light. Assume the apex angle of the prism is 60.00°.
    a. C Sketch the approximate paths of the rays for red and blue light as they travel through and then exit the prism.
    b. N Determine the measure of dispersion, the angle between the red and blue rays that exit the prism.

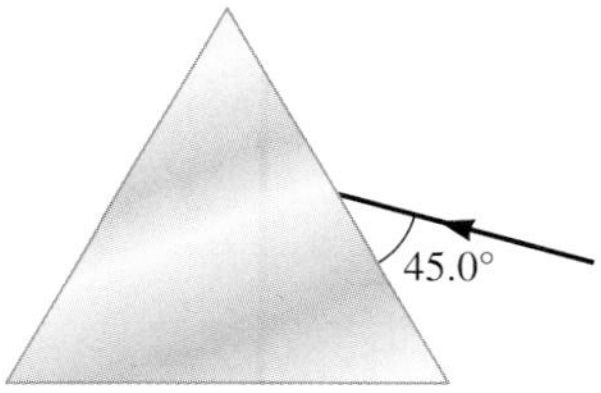

FIGURE P38.31

32. **C** Zak and Sallie are playing outside in a water sprinkler. Sallie can see a rainbow when she looks at the water, but Zak cannot. Where are the two children standing? Explain.

33. **A** Use geometry and Figure 38.13 (page 1225) to show these relationships: **a.** $\theta_2 + \theta_3 = 60°$, **b.** $\alpha = \theta_4 - 30°$, and **c.** $\theta_1 = 30°$.

34. **C** Light travels through air and enters a prism. How, if at all, do the speed, wavelength, frequency, and energy of the light change as it enters the prism?

38-4 Refraction at Spherical Surfaces

35. **N** A paperweight is made of a transparent material with an index of refraction of 1.75. The paperweight is a hemisphere of radius R. At its base is a flattened ladybug. Where is the image of the ladybug, and what is its magnification? Assume the index of refraction for air is 1.00.

36. **A** Use geometry and Figure 38.17 (page 1227) to show these relationships: **a.** $\theta_i = \alpha + \beta$ and **b.** $\theta_t = \beta - \gamma$.

37. **N** In a still pond, you see a fish whose distance from the surface of the water is 3.25 m. How far below the surface does the fish appear to you? If you were trying to catch the fish in a net, would you need to take the difference between the image distance and the object distance into account? For this problem, take the index of refraction for air to be 1.00 and for water to be 1.33.

38. **N** A Lucite slab ($n = 1.485$) 5.00 cm in thickness forms the bottom of an ornamental fish pond that is 40.0 cm deep. If the pond is completely filled with water, what is the apparent thickness of the Lucite plate when viewed from directly above the pond?

39. **N** Figure 38.20 (page 1231) showed an image produced by a vase of water projected onto a screen. Suppose the object is 1.0 m from surface 1. How far is the screen from surface 2, and what is the magnification of the final image?

40. **N** In Example 38.6, we found that a spherical vase of radius 6.0 cm full of water acts like a magnifying lens for an object 3.0 cm in front of it. Imagine taking the empty vase, capping the top to contain the air, and submerging it upside down in a pool to try to form a spherical "air lens" in a water environment. Determine the magnification of an object 3.0 cm in front of the spherical "air lens" surrounded by water to see whether the vase in these circumstances also acts like a magnifying lens.

41. **N** The end of a solid glass rod of refractive index 1.50 is polished to have the shape of a hemispherical surface of radius 1.0 cm. A small object is placed in air (refractive index 1.00) on the axis 5.0 cm to the left of the vertex. Determine the position of the image.

42. **A** Figure P38.42 shows a hemispherical material with radius of curvature R and refractive index 1.5. The flat portion is silvered, so it is a reflective surface. Find the necessary distance x of the silvered plane surface from the point P so as to form an image of a very distant object at that point.

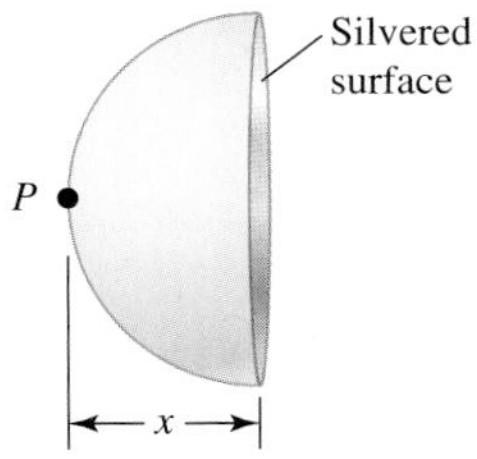

FIGURE P38.42

38-5 Thin Lenses

43. **N** Figure P38.43 shows a concave meniscus lens. If $|r_1| = 8.50$ cm and $|r_2| = 6.50$ cm, find the focal length and determine whether the lens is converging or diverging. The lens is made of glass with index of refraction $n = 1.55$. **CHECK and THINK**: How do your answers change if the object is placed on the right side of the lens?

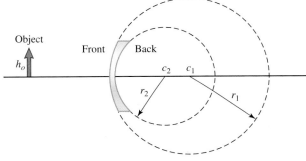

FIGURE P38.43

44. **A** Show that the magnification of a thin lens is given by $M = -d_i/d_o$ (Eq. 38.6). *Hint*: Follow the derivation of the lens maker's equation (page 1233) and start with a thick lens.

45. **N** A converging lens has a focal length of 30.0 cm and is made of glass ($n = 1.50$). If $|r_1| = |r_2| \equiv |r|$, what is $|r|$?

46. **C** Compare a converging lens to a diverging lens by filling in each blank with either **converging** or **diverging**: The focal length of a _____ lens is positive, and the focal point is behind the lens. Parallel rays form a real image at the focal point of a _____ lens. The focal length of a _____ lens is negative, and the focal point is in front of the lens. Parallel rays form a virtual image at the focal point of a _____ lens.

47. **N** A thin lens made of acrylic ($n = 1.49$) has $|r_1| = 17.5$ cm and $|r_2| = 7.5$ cm, where both surfaces of the lens have their radii of curvature on the same side of the lens. **a.** What is the focal length of this lens? **b.** Is the lens diverging or converging?

Problems 48 and 49 are paired.

48. **N** The radius of curvature of the left-hand face of a flint glass biconvex lens ($n = 1.60$) has a magnitude of 8.00 cm, and the radius of curvature of the right-hand face has a magnitude of 11.0 cm. The incident surface of a biconvex lens is convex regardless of which side is the incident side. What is the focal length of the lens if light is incident on the lens from the left?

49. **N** The radius of curvature of the left-hand face of a flint glass biconvex lens ($n = 1.60$) has a magnitude of 8.00 cm, and the radius of curvature of the right-hand face has a magnitude of 11.0 cm. The incident surface of a biconvex lens is convex regardless of which side is the incident side. What is the focal length of the lens if light is now incident on the lens from the right?

38-6 Images Formed by Diverging Lenses

50. **C** Devise an experiment that uses a small object (such as a coin) and a ruler to find the focal length of a diverging lens.

51. **N** A diverging lens with focal length $|f| = 15.5$ cm produces an image with a magnification of $+0.750$. What are the object and image distances?

52. **N** An object with a height of -0.050 m points below the principal axis (it is inverted) and is 0.150 m in front of a diverging lens. The focal length of the lens is -0.30 m. **a.** What is the image distance? **b.** What is the magnification? **c.** What is the image height?

53. **N** An object is placed 10.0 cm in front of a diverging lens of focal length -6.00 cm. Determine the image magnification and the image distance.

54. An object is placed 14.0 cm in front of a diverging lens with a focal length of -40.0 cm.
    a. N What are the location and magnification of the image?
    b. G Draw a ray diagram showing the locations of the object and the image by tracing at least two rays.

55. **N** An object is in front of a diverging lens with a focal length of -15.0 cm. The image seen has a magnification of 0.400.
    a. How far is the object from the lens? **b.** If the object has a height of -10.0 cm because it points below the principal axis (it is inverted), what is the image height h_i?

56. A student with a diverging lens is struggling to form an image on a screen. A small lightbulb is placed 25.0 cm in front of the lens, which has a focal length of 10.0 cm.
    a. G Draw a ray-tracing diagram to find the position of the image formed by the lens, and determine why the student is unable to project an image onto the screen.
    b. N Calculate the location and magnification of the image to confirm the accuracy of your drawing.

38-7 Images Formed by Converging Lenses

57. **N** An object is placed a distance of $4.00f$ from a converging lens, where f is the lens's focal length. **a.** What is the location of the image formed by the lens? **b.** Is the image real or virtual? **c.** What is the magnification of the image? **d.** Is the image upright or inverted?

58. A magnifying glass is a converging lens, which could be used to make a real image of an object. A small lightbulb is placed 25.0 cm in front of a convex lens with a focal length of 10.0 cm.
    a. G Draw a ray-tracing diagram to find the position of the image formed by the lens.
    b. N Calculate the location and magnification of the image to confirm the accuracy of your drawing.

59. **N** An object has a height of 0.050 m and is held 0.250 m in front of a converging lens with a focal length of 0.150 m. **a.** What is the magnification? **b.** What is the image height?

60. **E** Estimate the magnification and focal length of the pitcher of water shown in Figure P38.60.

FIGURE P38.60

61. **N** An object is in front of a converging lens with a focal length of 15.0 cm. The image seen has a magnification of -2.50. **a.** How far is the object from the lens? **b.** If the object has a height of -10.0 cm because it points below the principal axis (it is inverted), what is the image height h_i?

62. **C** Devise an experiment that uses a ruler to find the focal length of a converging lens.

63. **C** Explain the relative positions of the black and white backgrounds seen through the champagne glass shown in Figure P38.63.

FIGURE P38.63

38-8 The Human Eye

64. **E** Use a ruler and a small object (such as a coin) to find the near point of each of your eyes. Explain your procedure. How does your near point compare to the typical values in Table 38.5? Does each eye have the same near point? Estimate the focal length of each lens when you focus on an object at your near point. You may instead measure the near point of another person's eyes.

65. **C** Explain how wearing goggles while swimming enables you to see clearly given that your underwater vision is much less clear without goggles.

66. **C** **CASE STUDY** Two people are in the woods, and they want to start a fire by using their eyeglasses to focus sunlight onto some dry brush. One person is myopic, and the other is hyperopic. Whose glasses should they use, and why?

67. **C** **CASE STUDY** Take a look at the eyes of the cartoon character in Figure P38.67. Is she farsighted or nearsighted? Explain.

FIGURE P38.67

68. **C** **CASE STUDY** Ben Franklin invented bifocals (Fig. P38.68). The lower lenses are for reading, while the upper lenses are for looking at distant objects. What kind of lenses are in the bottom? What kind are in the top? Explain your answers.

FIGURE P38.68

69. **N** **CASE STUDY** Susan wears corrective lenses. The prescription for her right eye is -7.75 diopters, and the prescription for her left eye is -8.00 diopters. Is she nearsighted or farsighted? If she is nearsighted, what is the far point for each of her eyes? If she is farsighted, what is the near point for each of her eyes? **CHECK and THINK:** In which eye does she have better vision?

70. **A** Fill in the missing entries in Table P38.70.

TABLE P38.70

	Convex mirror or diverging lens, $f < 0$	Concave mirror or converging lens, $f < 0$		
Object distance	$d_o > 0$	$d_o > 2f > 0$	$2f > d_o > f > 0$	$d_o < f > 0$
Image distance	$0 > d_i > f$		$d_i > 2f > 0$	
Magnification				$|M| > 1$
Real or virtual image	Virtual		Real	
Upright or inverted image			Inverted	

38-9 One-Lens Systems

71. **N** A converging lens is made such that both sides have a radius of curvature of 7.50 cm. The glass used has an index of refraction of 1.510 for red light and 1.530 for blue light, leading to chromatic aberration. **a.** What is the focal length of the lens for both red and blue light? **b.** If this lens is used to form an image of an object that is 10.0 cm in front of the lens, how far apart are the images formed for red and blue light?

72. **C** Is it possible to look through a magnifier and see a real image?

73. **N** In Example 38.11 (page 1249), we imagined making a camera from a single converging lens with a 30.0-cm focal length. If you wanted to use this lens as a magnifier, what would be the achieved magnification if you use the lens to create a virtual image by placing it 25.0 cm from an object, where you might have placed your eye to see the object as clearly as possible?

74. **C** Suppose you want to use a single lens to project a slide (a transparent film) onto a screen. Do you need a converging or diverging lens? Explain.

38-10 Multiple-Lens Systems

75. **N** An object 2.50 cm tall is 15.0 cm in front of a thin lens with a focal length of 5.00 cm. A thin lens with a focal length of -12.0 cm is placed 2.50 cm beyond this converging lens as shown in Figure P38.75. Determine the final image height and the position of the final image relative to the second lens.

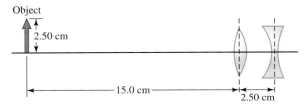

FIGURE P38.75

76. **N** Figure P38.76 shows an object placed a distance d_{o1} from one of two converging lenses separated by $s = 1.00$ m. The first lens has focal length $f_1 = 22.0$ cm, and the second lens has focal length $f_2 = 45.0$ cm. An image is formed by light passing through both lenses at a distance $d_{i2} = 15.0$ cm to the left of the second lens. **a.** What is the value of d_{o1} that will result in this image position? **b.** Is the final image formed by the two lenses real or virtual? **c.** What is the magnification of the final image? **d.** Is the final image upright or inverted?

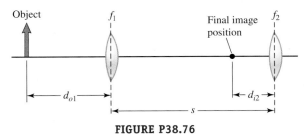

FIGURE P38.76

Problems 77 and 78 are paired.

77. **N** A system of lenses has light passing through a converging lens and then through a diverging lens. An object is 2.00 cm to the left of the converging lens, the lenses are 4.00 cm apart, and the focal length of the converging lens is 1.00 cm. **a.** Find the image distance for the image formed by the first lens. **b.** Find the focal length of the diverging lens if the final image position (after the light has gone through both lenses) is 0.50 cm to the left of the diverging lens.

78. **G** A system of lenses has light passing through a converging lens and then through a diverging lens. An object is 2.00 cm to the left of the converging lens, the lenses are 4.00 cm apart, and the focal length of the converging lens is 1.00 cm. Draw the ray diagram for the image formed by this system of lenses.

79. **N** An object is placed 22.0 cm to the left of a pair of lenses, with a converging lens of focal length $+15.0$ cm on the left, and a diverging lens of focal length -12.0 cm a distance of 40.0 cm to its right. **a.** What are the location and magnification of the final image formed by the two lenses? **b.** Is the final image formed by the two lenses upright or inverted?

Problems 80 and 81 are paired.

80. **CASE STUDY** A group of students is given two converging lenses. Lens A has a focal length of 12.5 cm, and lens B has a focal length of 50.0 cm. The diameter of each lens is 6.50 cm. The students are asked to construct a telescope from these lenses if possible, and they have this discussion:

Avi: To make a telescope, we pick lens B to be the objective and lens A to be the eyepiece. Lens B has the greater focal length, so it has to be the objective.

Cameron: Both lenses have the same diameter—6.50 cm. It doesn't matter which is the objective.

Shannon: It does matter because the magnification depends on their relative focal lengths. We still want to get the best magnification.

a. **C** What do you think?
b. **N** If a telescope can be constructed from these two lenses, describe its design. What are its LGP and angular magnification? Compare the LGP to the value for your fully open pupil.

81. **CASE STUDY** A group of students is given two converging lenses. Lens A has a focal length of 12.5 cm, and lens B has a focal length of 50.0 cm. The diameter of each lens is 6.50 cm. The students are asked to construct a microscope from these lenses that has the same magnification as the telescope in Problem 80 if possible, and they have this discussion:

Avi: These are the same lenses we used to make a telescope. So they won't work as a microscope. Microscopes are for looking at close objects; telescopes are for looking at far objects.

Cameron: All you need for a microscope are two converging lenses. I think the difference from a telescope is just that the order of the lenses is switched. A microscope is just a backward telescope.

Shannon: I think the order of the lenses doesn't matter because the magnification is inversely proportional to both focal lengths. I think we have to adjust the distance between the lenses.

a. **C** What do you think?
b. **N** If a microscope can be constructed with these two lenses, describe its design. What is the minimum separation of the lenses? Where must you place the object?

82. **N** Microscope objectives and eyepieces are often labeled simply with a magnification. A $4\times$ eyepiece and a $20\times$ objective result in a total magnification of $80\times$ for the system. Assuming these are simple converging lenses and the objective is brought to a distance of 8.5 mm from the specimen, what focal lengths are needed for the eyepiece and the objective to give the $80\times$ system magnification?

83. **N** Two lenses are placed along the x axis, with a diverging lens of focal length -8.00 cm on the left and a converging lens of focal length 16.0 cm on the right. When an object is placed 10.0 cm to the left of the diverging lens, what should the separation s of the two lenses be if the final image is to be focused at $x = \infty$?

84. **C CASE STUDY** What is the purpose of the tube used in the construction of a microscope or a telescope?

General Problems

85. **N** A monochromatic beam of light is incident on a slab of unknown material at angle $\theta_i = 51.0°$ (Fig. P38.85). If the angle at which the beam exits the slab is $\theta_t = 69.0°$, what is the index of refraction of the slab?

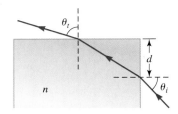

FIGURE P38.85

Problems 86 and 87 are paired.

86. **N** At a still mountain lake where the air temperature is 0.0°C, a plane sound wave with wavelength 602 mm is incident on the lake's surface with angle of incidence $\theta_i = 9.00°$. The lake's water is at 15.0°C. *Note*: Refer to Table 17.1 (page 498) for the speed of sound in air and in water at the given temperatures. **a.** What is the angle of refraction θ_t of the sound wave in the water? **b.** What is the wavelength of the sound wave in the water?

87. Consider the mountain lake in Problem 86. The angle of incidence of a beam of 656-nm red light on the lake's surface is $\theta_i = 9.00°$.
    a. N What is the angle of refraction θ_t of the ray of light in the water?
    b. N What is the wavelength of the ray of light in the water?
    c. C Compare the answers to parts (a) and (b) to the corresponding parts of Problem 86. How do the behaviors of sound waves and light waves differ during refraction?

88. **N** A thick glass container ($n = 1.523$) with inner and outer vertical walls has its interior filled with water ($n = 1.333$) and has a quarter lying on the bottom. A ray of light with wavelength 486 nm in the water has reflected from the quarter and strikes the interface between the water and the inner wall of the glass at an angle of incidence of 15.0°. **a.** What is the frequency of the light while in the water? **b.** What is the light ray's transmitted angle when it finally leaves the glass, passing into the air?

89. **N** A Pyrex ($n = 1.47$) pan is filled with water, and a monochromatic beam of light is directed at the water–Pyrex interface. The beam is observed to reflect off the bottom of the Pyrex such that it hits the water–Pyrex interface at an angle of incidence of 24.3°. What is the angle of refraction of the beam of light as it emerges into the water?

Problems 90 and 91 are paired.

90. **A** Prove that when a wave moves from medium 1 into medium 2, $n_1\lambda_1 = n_2\lambda_2$.

91. **N** When a light wave moves from one medium into another, the frequency of the wave stays the same. If a light wave with a wavelength of 489 nm moving in air ($n = 1.0002926$) enters water ($n = 1.333$), find **a.** the frequency of the light in the air, **b.** the wavelength of the light in the water, and **c.** the speed of the light in the water.

92. **N** A beam of monochromatic light is incident at $\theta_i = 33.0°$ on a block of Lucite ($n = 1.495$) surrounded by air. **a.** What is the angle of refraction θ_t? **b.** What would the angle of refraction be if the block were made of flint glass with a 71% lead content ($n = 1.805$)?

Problems 93, 94, and 95 are grouped.

93. **N** Light in air is incident on diamond, with $\theta_i = 34.5°$. **a.** What is the angle of reflection? **b.** What is the angle of refraction in the diamond?

94. **N** Light in water is incident on diamond, with $\theta_i = 34.5°$. **a.** What is the angle of reflection? **b.** What is the angle of refraction in the diamond?

95. **N** Light in air is incident on cubic zirconia, with $\theta_i = 34.5°$. **a.** What is the angle of reflection? **b.** What is the angle of refraction in the cubic zirconia?

96. **N** A monochromatic beam of light is incident on a large rectangular sapphire crystal ($n = 1.760$) with $\theta_i = 27.0°$. **a.** What is the angle of refraction θ_t in the sapphire crystal? **b.** At what angle is the beam incident on the back surface of the sapphire crystal? **c.** At what angle is the beam refracted at the crystal–air interface?

97. **N** A beam of monochromatic light strikes the interface between air and safflower oil ($n = 1.466$) at angle θ_i and is refracted (Fig. P38.97). The beam next strikes the safflower oil–water interface at angle $\gamma = 17.0°$ and is refracted at angle θ_t. **a.** What is angle θ_i? **b.** What is angle θ_t?

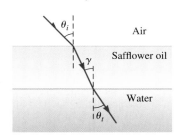

FIGURE P38.97

98. **A** Fermat's principle of least time for refraction. A ray of light traveling in a medium with speed v_1 leaves point A and strikes the boundary between the incident and transmitted media a horizontal distance x from point A as shown in Figure P38.98. The refracted ray travels with speed v_2 in the second medium, eventually reaching point B. The horizontal distance between points A and B is L. **a.** Calculate the time t required for the light to travel from A to B in terms of the parameters labeled in the figure. **b.** Now take the derivative of t with respect to x. What is the condition for which the ray of light will take the shortest time to travel from A to B?

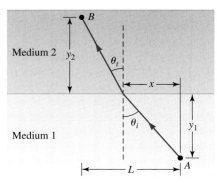

FIGURE P38.98

99. **N** A monochromatic beam of light is incident on a slab of Lucite ($n = 1.495$) surrounded by air. What is the incident angle θ_i that results in refraction angle $\theta_t = \theta_i/2$? *Hint*: Use the trigonometric identity $\sin 2\theta = 2 \sin \theta \cos \theta$.

100. **N** When the index of refraction of a medium depends on the wavelength of the incident light, each wavelength refracts at a different angle, causing dispersion that can be quantified as the difference in the refraction angles of two wavelengths (called the measure of dispersion). A beam of light containing

red and yellow wavelengths strikes a block of heavy flint glass at $\theta_i = 66.0°$. If the index of refraction of the glass for 656-nm red light is 1.879 and the index of refraction for 589-nm yellow light is 1.919, what is the dispersion angle of the block of flint glass for these wavelengths?

101. **N** An outdoor hot tub is cylindrical with depth d and radius 1.25 m and is completely filled with water. In the morning, the bottom of the hot tub begins to receive sunlight when the Sun is 24.0° above the horizon. What is the depth d of the hot tub?

Problems 102 and 103 are paired.

102. **N** A quarter lies on the bottom of a glass ($n = 1.54$) that contains water ($n = 1.33$). Light rays bounce off the quarter in all directions. **a.** What is the critical angle for light traveling from the glass into the water? **b.** What is the critical angle for light traveling from the glass into the air ($n = 1.00029$)?

103. **N** A quarter lies on the bottom of a glass ($n = 1.54$) that contains water ($n = 1.33$). Light rays bounce off the quarter in all directions. What is the minimum angle of incidence for a ray of light from the quarter in the water that will result in the ray being at the critical angle when it arrives at the glass–air interface?

104. **N** Figure P38.104 shows a monochromatic beam of light incident on one end of a cylinder with index of refraction $n = 1.38$ at an angle of incidence θ_i. What is the maximum angle for which the beam of light will be totally internally reflected by the cylindrical wall and exit at the left-hand end?

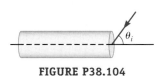

FIGURE P38.104

Problems 105 and 106 are paired.

105. **N** Curved glass–air interfaces like those observed in an empty shot glass make it possible for total internal reflection to occur at the shot glass's internal surface. Consider a glass cylinder ($n = 1.54$) with an outer radius of 2.50 cm and an inner radius of 2.00 cm as shown in Figure P38.105. Find the minimum angle θ_i such that there is total internal reflection at the inner surface of the shot glass.

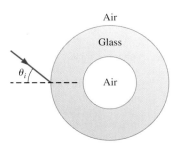

FIGURE P38.105 Problems 105 and 106.

106. **N** In Problem 105, we discovered how a light ray that enters a cylindrical glass container can experience total internal reflection at the internal glass–air interface when the outer angle θ_i is larger than a particular value. This phenomenon is very difficult to observe for thin containers because the outer angle is very close to 90°, so not much light experiences total internal reflection. However, we can observe this interesting example of total internal reflection in a thin-walled glass cylinder ($n = 1.54$) if we submerge it in water. Consider a cylinder with an outer radius of 2.50 cm and inner radius of 2.00 cm. Find the minimum angle θ_i at which total internal reflection is observed. The interior will still be air but the glass will be surrounded by water on the outside.

107. **N** A block is constructed from layers of cubic zirconia ($n = 2.14$), flint glass ($n = 1.80$), and quartz ($n = 1.54$) as shown in Figure P38.13. The block is surrounded by air. For what values of the incident angle θ_i does total internal reflection occur at the quartz–air interface?

108. **N** A hiker stands at the edge of a clear alpine lake that is 5.00 m deep. **a.** What is the apparent depth of the lake? **b.** Returning in the summer, the hiker finds the lake surface 2.00 m lower than before. What is the apparent depth of the lake now?

Problems 109 and 110 are paired.

109. **N** A light source forms parallel beams that strike the flat face of a transparent hemisphere of flint glass ($n = 1.65$) at normal incidence (Fig. P38.109). At what distance behind the hemisphere do the paraxial rays of the light source focus if the magnitude of the hemisphere's radius of curvature is 8.50 cm?

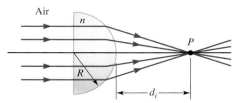

FIGURE P38.109

110. **N** A point source of light is placed at point P a distance d_o in front of the curved face of a transparent hemisphere of flint glass ($n = 1.65$) as in Figure P38.110. What should the object distance d_o be if the rays emerging from the flat surface of the hemisphere are to form an image at infinity? The radius of curvature is 8.50 cm.

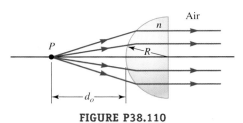

FIGURE P38.110

111. **N** A group of children watches a playful dolphin swim toward the underwater viewing window at an aquarium. If the dolphin's speed in the water is 1.00 m/s, what is its apparent speed as seen by the children?

112. **C** Return to the derivation of the lens maker's equation (page 1234), and explain why $d_{o2} = w - d_{i1}$ (Eq. 2) holds in all three cases shown in Figures 38.24, 38.25, and 38.26.

Problems 113 and 114 are paired.

113. An object is 20.0 cm to the left of a converging lens with a focal length of 30.0 cm. A second lens is placed 15.0 cm to the right of the converging lens. The second lens is a diverging lens with focal length −20.0 cm.
 a. N Find the final image position.
 b. C Describe where the final image is located.

114. **N** An object is 20.0 cm to the left of a converging lens with a focal length of 30.0 cm. A second lens is placed 15.0 cm to the right of the converging lens. The second lens is a diverging lens with a focal length of −20.0 cm. **a.** If the object has a height of 17.0 cm and is upright, find the magnification due to each lens and the magnification of the final image. **b.** What is the final image height?

115. **N** The magnification of an upright image that is 34.0 cm away from its object is 0.400. What type of lens is used to form the image, and what is its focal length?

116. **C** What is the physical difference between a real and a virtual image? Describe a method for using a converging lens to demonstrate the difference to a friend.

117. N The plano-convex lens made of glass ($n = 1.52$) shown in Figure P38.117 has a focal length of 20.0 cm. Calculate the radius of curvature of the curved surface of the lens.

FIGURE P38.117

118. C You are handed two lenses, one of which is a diverging lens and the other a converging lens. Explain how you would tell the difference between these lenses if you were able to **a.** feel them with your hands, **b.** try to project an image of a lightbulb onto a piece of paper, or **c.** look through each of them at an object on your desk.

Problems 119 and 120 are paired.

119. N Consider a converging lens with a 16.0-cm focal length. How far is the object from the lens if a real image is formed **a.** 22.0 cm behind the lens and **b.** 42.0 cm behind the lens?

120. N Consider a converging lens with a 16.0-cm focal length. How far is the object from the lens if a virtual image is formed **a.** 22.0 cm in front of the lens and **b.** 42.0 cm in front of the lens?

121. N Consider a diverging lens with focal length $|f| = 14.0$ cm. **a.** What is the location of the image if $d_o = 28.0$ cm? What is the magnification of the image? Is the image real or virtual? Is it upright or inverted? **b.** What is the location of the image if $d_o = 6.00$ cm? What is the magnification of the image? Is the image real or virtual? Is it upright or inverted?

122. A In Example 38.7, we found that when a convex meniscus lens is flipped front to back, it remains a converging lens with the same focal length. Is this true of all lenses, or is it a particular feature of the convex meniscus lens? Starting with the lens maker's equation, determine which variables change when a lens is flipped and whether the results of Example 38.7 are true for all lenses.

123. N An object is placed 32.0 m in front of a converging lens with focal length $f = +18.0$ cm, forming a real image. The object is moved away from the lens with an initial speed of 3.00 m/s. **a.** With what initial speed does the image formed by the lens move? **b.** What is the direction of motion of the image—toward the lens or away from the lens?

Problems 124, 125, 126, and 127 are grouped.

124. Consider a prism with apex angle β. A light ray is incident at angle θ_1 as shown in Figure P38.124. The refractive index of the prism relative to air is n, and assume the index of refraction of the surrounding air is 1.
 a. A Show that the total deviation α of a ray from its original path is given by

$$\alpha = \theta_1 - \beta + \sin^{-1}\left[\sin \beta \sqrt{n^2 - \sin^2 \theta_1} - \cos \beta \sin \theta_1\right]$$

 b. G Plot the angle of deviation α versus the angle of incidence θ_1 for $n = 1.5$ and the apex angle $\beta = 60.0°$.
 c. N For what angle θ_1 is α minimized? What is the significance of this minimum?

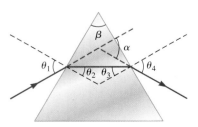

FIGURE P38.124

125. A In Problem 124, we found that the deviation α is minimized when the ray of light passes through the prism symmetrically.
 a. Show that the minimum deviation α_{min} satisfies the equation

$$n = \sin\left(\frac{\beta + \alpha_{min}}{2}\right) \Big/ \sin\left(\frac{\beta}{2}\right)$$

 b. Assume the apex angle of the prism is small, so that $\sin \beta \approx \beta$. Obtain the minimum deviation angle.

126. A A combination of two prisms in which the deviation produced by the first prism is equal and opposite to that produced by the second prism is called a **direct vision prism**. This combination produces dispersion without deviation. If n_1 and n_2 are the refractive indices of the two prisms and β_1 is the apex angle of the first prism, determine the apex angle β_2 of the second prism so that the net deviation is zero. Assume the angles of the prisms are small, so that $\sin \beta_1 \approx \beta_1$ and $\sin \beta_2 \approx \beta_2$. *Hint*: You may use the result from part (b) in Problem 125.

127. N Light enters a prism of crown glass and refracts at an angle of 5.00° with respect to the normal at the interface. The crown glass has a mean index of refraction of 1.51. It is combined with one flint glass prism ($n = 1.65$) to produce no net deviation. **a.** Find the apex angle of the flint glass. **b.** Assume the index of refraction for violet light ($\lambda_v = 430$ nm) is $n_v = 1.528$ and the index of refraction for red light ($\lambda_r = 768$ nm) is $n_r = 1.511$ for crown glass. For flint glass using the same wavelengths, $n_v = 1.665$ and $n_r = 1.645$. Find the net dispersion.

128. A A converging lens with a focal length of 18.0 cm is placed at $y = 40.0$ cm on the y axis. An object originally at the origin is moved slowly along the y axis toward the lens until it is within 10.0 cm of the lens. What is the position of the image formed by the lens as a function of the position y of the object?

129. N An object is placed a distance of 10.0 cm to the left of a thin converging lens of focal length $f = 8.00$ cm, and a concave spherical mirror with radius of curvature $+18.0$ cm is placed a distance of 45.0 cm to the right of the lens (Fig. P38.129). **a.** What is the location of the final image formed by the lens–mirror combination as seen by an observer positioned to the left of the object? **b.** What is the magnification of the final image as seen by an observer positioned to the left of the object? **c.** Is the final image formed by the lens–mirror combination upright or inverted?

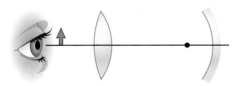

FIGURE P38.129

PART SIX
20th Century Physics

39

Relativity

❶ Underlying Principles

1. Einstein's postulates of special relativity
2. Einstein's postulates of general relativity

✪ Major Concepts

1. Lorentz transformations
2. Length contraction
3. Time dilation
4. Relativistic Doppler effect
5. Perpendicular velocity transformation
6. Parallel velocity transformation
7. Mass transformation
8. Relativistic momentum
9. Mass–energy equivalence principle
10. Curvature of space
11. Gravitational Doppler shift and gravitational time dilation

◓ Special Cases

Galilean relativity

Key Questions

How is Einstein's theory of special relativity different from Galilean relativity? What are the strange consequences of special relativity?

How is general relativity different from special relativity?

- 39-1 It's in the eye of the observer 1268
- 39-2 Special case: Galilean relativity 1271
- 39-3 Postulates of special relativity 1275
- 39-4 Lorentz transformations 1276
- 39-5 Length contraction 1279
- 39-6 Time dilation 1281
- 39-7 The relativistic Doppler effect 1285
- 39-8 Velocity transformation 1288
- 39-9 Mass and momentum transformation 1290
- 39-10 Newton's second law and energy 1293
- 39-11 General relativity 1297
- 39-12 Gravitational lenses and black holes 1300

We expect that some measurements should not depend on the observer. For example, your physics class is 50 minutes long, you are 5 feet 8 inches tall, and the speed of light is 3.00×10^8 m/s. The truth, however, is that observers need to agree about only one of these three observations—the speed of light. The duration of your physics class and your height depend on the velocity of the observer. It may seem crazy, but an observer moving at a very high speed will say that your physics class lasts longer than 50 minutes and you are shorter than 5 feet 8 inches!

The idea that you and a moving observer won't agree on something as fundamental as your height or the duration of your physics class may seem unbelievable. But, early in the 20th century, Albert Einstein (1879–1955) found that measurements of space and time depend on the motion of the observer. This led to some amazing discoveries, such as (1) the observed mass of an object depends on its speed, (2) the ultimate speed limit is the speed of light, and (3) mass and energy are equivalent. In this chapter, we explore Einstein's fundamental ideas and their consequences.

39-1 It's in the Eye of the Observer

In this textbook, we have studied physics in nearly chronological order—from Galileo's work on kinematics in the late 1500s to Maxwell's study of light in the middle 1800s. In this last part, we turn our attention to discoveries made in the early 1900s—a departure from classical physics referred to as *modern physics* or *20th-century physics*. Much of Einstein's work in developing his theories of relativity is actually based on a careful study of kinematics, so we must return to the beginning and study motion from the perspectives of different observers.

Experimenting in a Noninertial Reference Frame

Recall that we defined an inertial reference frame as one in which Newton's first law holds (Section 5-5). An inertial reference frame may be at rest or may move at constant velocity relative to the observer, but it cannot accelerate. Imagine for a moment that all scientists, including Newton, lived their entire lives on cruise ships that sometimes accelerated dramatically (Fig. 5.6, page 125). It would be difficult for these scientists to discover the law of inertia because sometimes they would observe the dramatic acceleration of objects that had no net force exerted on them.

The idea of living your entire life on a cruise ship may seem ridiculous, but it is actually closer to your experience than you might think. All human beings live in a *noninertial* reference frame. The Earth spins on its axis and orbits the Sun. The entire solar system orbits the center of the Milky Way galaxy, and the entire galaxy is falling toward the Andromeda galaxy. All this motion means that all of our laboratories are always accelerating. For most of this book, we could ignore the Earth's acceleration, but in sensitive experiments we cannot. One of Newton's great insights was to imagine a truly inertial reference frame. He had to infer the existence of such a frame and of the law of inertia by extrapolating from the results of experiments made on the Earth. His thought experiment led scientists to wonder whether such an inertial frame exists physically.

Looking for the Inertial Reference Frame

In the 19th century, scientists believed that light propagated in a medium known as the ether, which was also supposed to be at rest. So, perhaps the ether was the object that defined the one true inertial reference frame. If we could measure the speed of objects moving with respect to the ether, we would find that these objects obey Newton's laws. However, the Michelson–Morley experiment (case study in Chapter 36) failed to detect the motion of the Earth with respect to the ether, showing that either the ether is attached to the Earth or it does not exist.

The Aberration of Starlight

The possibility that the ether is attached to the Earth was ruled out by data collected even before Michelson and Morley conducted their experiment. An important piece of evidence to disprove that possibility comes from the observation of a star. The British astronomer James Bradley (1693–1762) observed the star Gamma Draconis (γ Dra) near the zenith for more than a year. He found that γ Dra appeared to move with respect to the zenith and that its greatest angular displacement (20.5 arcsec) occurred in September and March. This apparent motion of a star is known as **stellar aberration** and is a direct consequence of the Earth's orbital motion.

Let's start with the assumptions that the ether exists and that it is attached to the Earth. In this case, the velocity of the Earth with respect to the ether is zero (consistent with Michelson and Morley's results). If you want to observe a star that is directly overhead, you point your telescope straight up at the zenith (Fig. 39.1A). If the Earth, the ether, and the star are all at rest in the same reference frame, light from the star travels straight down the tube of your telescope to your detector at the bottom. You would say the star is at the zenith because that is the direction in which you pointed your telescope to collect the starlight.

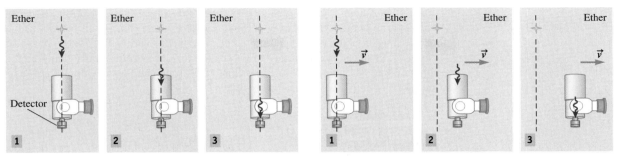

A. The Earth, ether, and star are in the same reference frame.

B. Only the Earth and ether are in the same reference frame.

FIGURE 39.1 **A.** The Earth, the ether, and the star are all at rest in the same frame. **B.** The ether is attached to the Earth, and both move to the right relative to the star. In both cases, **1** light (shown as a photon) is emitted by the star, **2** the photon enters the telescope, and **3** the photon is detected. In both cases, the telescope is aimed straight up and the star is detected at the zenith. *Neither of these scenarios is true!*

Now suppose the ether and the Earth are at rest with respect to each other in the same reference frame, which is moving to the right with respect to the star (Fig. 39.1B). You point your telescope straight up at the star. Light from the star enters the aperture of your telescope. As the photon drifts down toward your detector, the telescope (attached to the Earth) and the ether move to the right, so the light must also move to the right as it drifts down. The result is that you detect the light. Again, you would claim the star is at the zenith because that is the direction in which you must point your telescope. But, if this were the case, if the ether were attached to the Earth, Bradley would not have observed the aberration of γ Dra. No matter what the time of year, he could point his telescope toward the zenith to see γ Dra. Instead, Bradley had to tilt his telescope by as much as 20.5 arcsec to see γ Dra at certain times of the year.

To see why, imagine the ether exists but is not attached to the Earth. Instead, the Earth moves through the ether at roughly the same speed ($v \approx 30$ km/s) at which it orbits the Sun. Suppose you wish to observe γ Dra in March, when the Earth's velocity is to the right with respect to the ether. The star is still at the zenith, but you must point your telescope at the angle θ in order to see it (Fig. 39.2A), so you believe this is the angular position of the star. In time t, the photon travels a vertical distance ct, while the telescope travels a horizontal distance vt (Fig. 39.2B). So, the (small) angle θ is approximately

$$\theta \approx \frac{vt}{ct} = \frac{v}{c} = \frac{30 \times 10^3 \text{ m/s}}{3.00 \times 10^8 \text{ m/s}} \approx 20.6 \text{ arcsec}$$

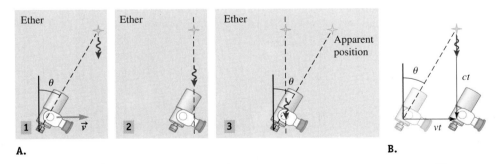

A.

B.

FIGURE 39.2 Suppose the Earth moves through the ether to the right at speed v. **A.** **1** A star emits a photon, **2** the photon enters the telescope, and **3** the photon is detected. If the Earth moves with respect to the ether, a star at the zenith appears to be at an angle θ with respect to the zenith. **B.** In time t, the photon travels distance ct and the detector travels distance vt. This scenario explains Bradley's observation, but the Michelson–Morley experiment requires that the ether does not exist.

This is in agreement with Bradley's observation of stellar aberration. Because Michelson and Morley failed to detect the velocity of the Earth with respect to the ether, we must conclude that the ether is attached to the Earth (Fig. 39.1B) or that it does not exist. Because of Bradley's observation, the ether cannot be attached to the Earth, and therefore the ether does not exist. This conclusion is an important underpinning of Einstein's theories of relativity.

CASE STUDY Truth Is Stranger than Fiction

Educated people acquire knowledge from many different sources. Sometimes, scientific discoveries inspire works of fiction, and we can learn about scientific breakthroughs through such works. Fiction is a great avenue for a first encounter with something new. It inspires our imagination, which makes the learning exciting. But there is a drawback to learning about science through fiction: Fiction writers are not obligated to be truthful. They can exaggerate or ignore a scientific truth in order to create their art. So, it can be hard for us to sort out what is true and what is made up.

In this chapter's case study, we use fiction to explore the strange consequences of Einstein's theory of relativity. If you first encountered these strange ideas in a work of fiction, you would probably think they were made up. But each one is a prediction that results from relativity and has been confirmed by observation. Einstein's theory of relativity demonstrates the old saying: *Truth is stranger than fiction.*

CONCEPT EXERCISE 39.1

Which of the following are (approximately) inertial reference frames? Explain your answer.

   a. Your classroom
   b. A railroad car moving at constant speed along a straight track
   c. A car moving at constant speed around a turn
   d. An airplane during takeoff

EXAMPLE 39.1 A Rotating Room

Aaron lives in a room that rotates with constant speed. The room has a coordinate system painted on its glass ceiling, and it is located entirely inside a stationary room in which Hannah lives (Fig. 39.3). Through the glass ceiling of his rotating room, Aaron can see a ball glued (off axis) to the ceiling of the stationary room. Draw the path of the ball as seen by Aaron through his rotating glass ceiling. What does he conclude about the law of inertia? A desk lamp in Aaron's room projects his coordinate system onto Hannah's ceiling.

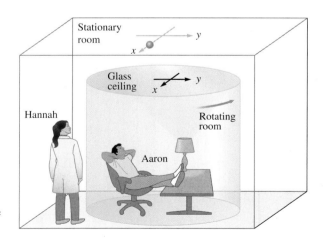

FIGURE 39.3 What path does the ball take in Aaron's frame?

INTERPRET and ANTICIPATE
Because it is accelerating, a rotating room is not an inertial reference frame (Section 5-5). So we expect to find that the ball, as observed from Aaron's frame, violates the law of inertia.

SOLVE

It is difficult to imagine immediately what Aaron sees from the rotating room. It is somewhat easier to imagine first what Hannah sees if Aaron's coordinate system is projected onto her stationary ceiling. To Hannah, the ball is at rest and the coordinate system rotates counterclockwise as shown at three time instants in Figure 39.4.

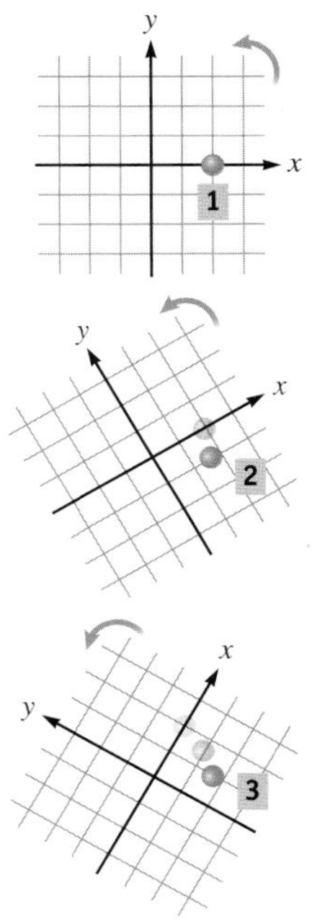

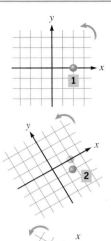

FIGURE 39.4 Hannah sees the ball remain fixed (bright red) and Aaron's coordinates moving (shown with images of the ball's previous position in Aaron's coordinates).

In the rotating room, Aaron sees a fixed coordinate system on his ceiling. He sees the ball moving clockwise in a circle around his coordinate system (Fig. 39.5). He cannot identify a source of centripetal force, however, or any net force on the ball. So, he concludes that the ball's motion violates the law of inertia.

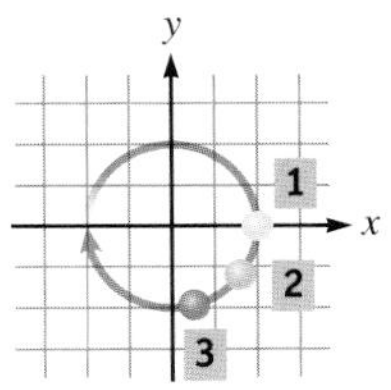

FIGURE 39.5 Aaron sees the ball moving in a circle.

CHECK and THINK

This is exactly what we would expect from an observer in an accelerating reference frame. Aaron sees an apparent violation of the law of inertia because, although the ball is stationary in an inertial frame, he is in a noninertial frame. If the ball had been glued to Aaron's glass ceiling, he would have seen the ball at rest and he would not have observed any violation of the law of inertia.

39-2 Special Case: Galilean Relativity

Einstein developed much of his theory of relativity by thinking about the kinematics observed from different reference frames. His ideas about space and time are a departure from the classical ideas of space and time used until now in this textbook. To help distinguish the classical model of relative motion from Einstein's model, we'll refer to the classical model as **Galilean relativity**.

GALILEAN RELATIVITY

▶ Special Case

We'll start with a review of Galilean relativity as in Section 4-7, using mostly scalar components instead of vector ones and with less cumbersome notation. Consider two reference frames, one of which is at rest and is referred to as the *laboratory* frame. Unless otherwise specified, the other frame is moving at constant velocity $\vec{v}_{rel}$ relative to the laboratory frame. To distinguish between the two coordinate systems, a prime (') is used on all the symbols for the moving frame, and no prime is used on the symbols for the laboratory frame (Fig. 39.6).

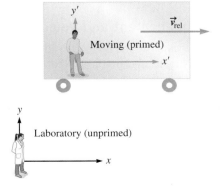

FIGURE 39.6 Two inertial frames with the laboratory frame at rest while the moving frame has velocity $\vec{v}_{rel}$ as measured by the observer in the laboratory frame. Each observer in each frame chooses a coordinate system such that he or she is at the origin. The two coordinate systems are parallel. We show them offset so that they don't overlap, but the primed frame is not actually located above the unprimed frame. In later figures, we just draw the coordinate systems; however, we will refer to the *laboratory observer* and the *primed* (or moving) *observer*.

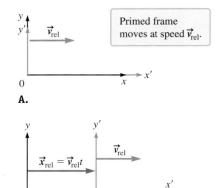

Position and Distance

Initially (at $t = 0$), the origins of two inertial frames overlap (Fig. 39.7A). The primed frame moves to the right at speed v_{rel}, and by time t, the observer in the laboratory frame finds that the primed frame's origin is at x_{rel} as shown in Figure 39.7B:

$$x_{rel} = v_{rel} t \tag{39.1}$$

Now suppose that at time t, both observers measure the position of a helicopter hovering in place relative to the laboratory (Fig 39.8A). They measure the same vertical position:

$$y = y' \tag{39.2}$$

However, they don't get the same horizontal position: $x = x' + x_{rel}$ (Eq. 4.45) and, using Equation 39.1, we find

$$x = x' + v_{rel} t \tag{39.3}$$

FIGURE 39.7 A. At $t = 0$, the two origins coincide. **B.** At a later time t, the observer in the laboratory frame finds that the moving frame's origin is at x_{rel}.

Equations 39.2 and 39.3 are called **transformation equations** because they transform the measurements made in one frame into the corresponding measurements made in the other frame. When a quantity is always the same in both frames, such as the vertical position of the helicopter, we say that quantity is *invariant*. Usually, we consider the simple situation in which one frame moves in the x direction relative to the other frame so that the perpendicular positions (y and z) are invariant.

Our next step is to find a transformation equation for a horizontal separation, such as between the helicopter and a tower (Fig. 39.8B). According to the laboratory observer, the separation is

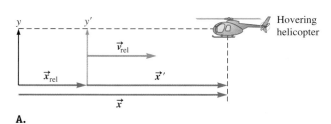

$$L = x_t - x_h \tag{39.4}$$

where the subscript t stands for *t*ower and h for *h*elicopter. To the primed observer, the separation is $L' = x'_t - x'_h$. The transformation equation for the separation distance is found by substituting Equation 39.3 into Equation 39.4 twice: once for the tower's position and once for the helicopter's position:

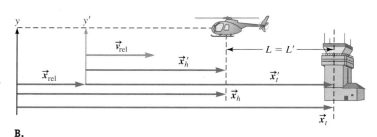

$$L = (x'_t + v_{rel} t) - (x'_h + v_{rel} t) = (x'_t - x'_h) + (v_{rel} t - v_{rel} t)$$

FIGURE 39.8 A. At t, both observers measure the position of a helicopter hovering relative to the laboratory. **B.** At t, both observers measure the distance $L = L'$ between a helicopter and a tower.

The relative velocity is constant, and the measurements of the helicopter and tower's positions were made simultaneously (same t), so the second term is zero. The first term is the distance measured in the primed frame:

$$L = L' \tag{39.5}$$

Equation 39.5 shows that the distance or length is invariant under Galilean relativity. However, measurements of the endpoint positions (those of the helicopter and tower in this case) must be made simultaneously.

Displacement and Velocity

Now suppose the helicopter flies to the right from position x_i at time t_i to x_f at time t_f, as measured by the laboratory observer (Fig. 39.9). We can find the initial and final positions measured by the primed observer by using Equation 39.3 twice. For the initial position,

$$x_i = x'_i + v_{rel} t_i \tag{39.6}$$

and for the final position,

$$x_f = x'_f + v_{rel} t_f \tag{39.7}$$

The displacement measured by the primed observer is

$$\Delta x' = x'_f - x'_i \quad (39.8)$$

The displacement of the helicopter measured by the laboratory observer is

$$\Delta x = x_f - x_i \quad (39.9)$$

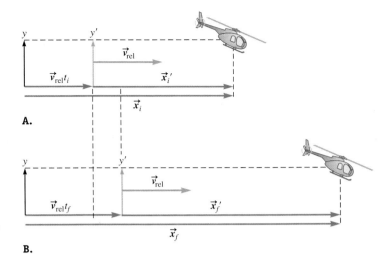

FIGURE 39.9 Both observers measure the position of a helicopter **A.** at t_i and **B.** at t_f.

We don't expect the displacement to be invariant, however, because in the time between the two measurements, the primed frame has moved. We expect the displacement $\Delta x'$ measured by the primed observer to depend on the elapsed time. Relate the two displacements by substituting Equations 39.6 and 39.7 into Equation 39.9:

$$\Delta x = (x'_f + v_{rel}t_f) - (x'_i + v_{rel}t_i) = (x'_f - x'_i) + v_{rel}(t_f - t_i)$$

The first term is the displacement as measured by the primed observer, so

$$\Delta x = \Delta x' + v_{rel}\Delta t \quad (39.10)$$

where $\Delta t = t_f - t_i$ is the elapsed time. Unlike a length, the displacement of the moving object is not invariant because it depends on the elapsed time as expected.

To find the velocity transformation equation, take the time derivative of Equation 39.3:

$$\frac{dx}{dt} = \frac{d}{dt}(x' + v_{rel}t) = \frac{dx'}{dt} + \frac{d(v_{rel}t)}{dt}$$

The relative velocity is a constant, so

$$v_x = v'_x + v_{rel} \quad (39.11)$$

where $v_x = dx/dt$ is the helicopter's velocity along the x direction measured in the laboratory frame and $v'_x = dx'/dt$ is its velocity along the same direction measured in the primed frame. Equation 39.11 shows that horizontal velocity is not invariant under Galilean relativity. However, the velocity in the perpendicular directions (y and z) is invariant: $v_y = v'_y$ and $v_z = v'_z$ (Problem 5).

Acceleration

Now suppose the helicopter is accelerating in some arbitrary direction. To find the transformation of acceleration in the x direction, take the time derivative of Equation 39.11:

$$\frac{dv_x}{dt} = \frac{dv'_x}{dt} + \frac{dv_{rel}}{dt}$$

The primed frame is an inertial frame, which means it is not accelerating. So v_{rel} is constant, $dv_{rel}/dt = 0$, and we find $dv_x/dt = dv'_x/dt$. Thus, acceleration in the parallel direction is invariant: $a_x = a'_x$. In Problem 5, you are asked to show that acceleration in the perpendicular directions is also invariant.

CONCEPT EXERCISE 39.2

Suppose the primed and laboratory observers want to measure the length of a rod that rests on the ground horizontally in the space between the helicopter and the tower (Fig. 39.8B). To derive the length transformation $L = L'$ (Eq. 39.5), we had to assume that the positions of the two ends were determined simultaneously. What happens to the length transformation equation if both observers measure the end below the helicopter at one time t_1 and the other end at a later time t_2?

EXAMPLE 39.2 How Fast Can You Bowl?

Aaron, a good bowler, is in the primed frame. He rolls a ball at 8.50 m/s (about 19 mph) in the x' direction. The bowling lane in the primed frame is 18.3 m (60 ft) long. The primed frame's speed with respect to Hannah in the laboratory frame is 25.0 m/s in the x direction. Assume the ball's speed is constant as it rolls down the lane.

A According to Hannah, what is the displacement of the ball?

INTERPRET and ANTICIPATE
During the time the ball rolls along the lane to the right (Fig. 39.10), Hannah sees Aaron's frame move to the right. So we expect her to see a greater displacement of the ball than he does.

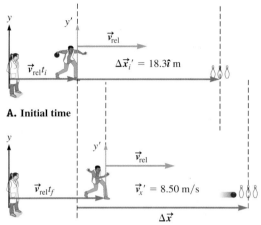

FIGURE 39.10

A. Initial time

B. Final time

SOLVE
First, find the time the ball takes to travel down the lane. Because the ball's velocity is constant, we find the time interval by dividing the displacement by the ball's speed as observed by Aaron.

$$\Delta t = \frac{\Delta x'}{v'_x} = \frac{18.3 \text{ m}}{8.50 \text{ m/s}} = 2.15 \text{ s}$$

Use Equation 39.10 to transform Aaron's observation of displacement into Hannah's observation of displacement.

$$\Delta x = \Delta x' + v_{\text{rel}} \Delta t \qquad (39.10)$$
$$\Delta x = 18.3 \text{ m} + (25.0 \text{ m/s})(2.15 \text{ s})$$
$$\Delta x = 72.1 \text{ m}$$

CHECK and THINK
As expected, the displacement measured by Hannah is greater than the displacement measured by Aaron.

B According to Hannah, what is the speed of the ball?

INTERPRET and ANTICIPATE
There are two ways to find the ball's constant speed in the laboratory frame. We can use the velocity transformation equation, or we can divide the ball's displacement by the time interval. We'll use the transformation equation and check our results by dividing the displacement by the time interval. Because Aaron's frame and the ball are both moving in the same direction, we expect that Hannah observes a higher speed than Aaron does.

SOLVE
The velocity transformation is given by Equation 39.11.

$$v_x = v'_x + v_{\text{rel}} \qquad (39.11)$$
$$v_x = 8.50 \text{ m/s} + 25.0 \text{ m/s} = 33.5 \text{ m/s}$$

CHECK and THINK
To check the answer, divide the displacement measured by Hannah by the time interval of the ball's travel. Both methods give the same speed for the ball in the laboratory frame, and, as expected, Hannah finds that the ball is moving faster than the speed Aaron observes in his moving frame.

$$v_x = \frac{\Delta x}{\Delta t} = \frac{72.1 \text{ m}}{2.15 \text{ s}}$$

$$v_x = 33.5 \text{ m/s} \checkmark$$

The second method assumes the time interval is the same in both the primed and laboratory frames ($\Delta t = \Delta t'$). Although this assumption holds in Galilean relativity, it does not hold in Einstein's theory of relativity.

39-3 Postulates of Special Relativity

A **postulate** is a presupposition or condition that underlies a line of reasoning. The postulate of Galilean relativity is that the laws of mechanics (Newton's laws) hold in all inertial reference frames. You might expect that the laws of electricity and magnetism (Maxwell's equations) also hold in all inertial reference frames. However, near the end of the 19th century, it seemed that Maxwell's equations did not hold in all inertial frames; these laws seemed to be true in only one reference frame—the ether's frame. But, as described in Section 39-1, the observation of stellar aberration along with Michelson and Morley's experiment revealed that the ether does not exist. In the early 1900s (with no knowledge of Michelson and Morley's experiment), Einstein, a theoretical physicist, postulated that all the laws of physics (mechanics, electricity, and magnetism) are true in all inertial reference frames. With this presupposition in mind, Einstein published a paper in 1905 describing a new theory of relativity. This theory is known as *special relativity* because it is a restricted (or special) case of relativity, in which the reference frames are inertial (nonaccelerating). In the next decade, Einstein developed *general relativity*—an unrestricted theory that allows for the acceleration of reference frames.

Einstein's theory of special relativity has two postulates:

1. All the laws of physics are true in all inertial reference frames. This means there is no special frame, such as the ether's frame.
2. The speed of light in a vacuum has the same value c as measured by all observers, regardless of the observer's velocity. Put another way, the speed of light is invariant.

In 1905, Einstein was working in the Swiss Patent Office in Bern.

EINSTEIN'S POSTULATES OF SPECIAL RELATIVITY ❶ Underlying Principle

Although the second postulate may seem to come out of nowhere, it is the fundamental condition that ensures that the laws of electricity and magnetism hold in all inertial frames. To see why, recall that Maxwell derived the wave equation for light by starting with Faraday's law and Ampère–Maxwell's law. Faraday's law says that a changing magnetic field is the source of an electric field, and Ampère–Maxwell's law says that a changing electric field is the source of a magnetic field. Because the wave equation for light results from the combination of these two laws, a light wave needs both a changing magnetic field to create an electric field and a changing electric field to create a magnetic field.

Suppose an observer in the laboratory frame sees a light wave traveling at c. Einstein imagined observing the same light wave from a primed frame moving at a relative speed $v_{rel} = c$ in the same direction as the wave. According to Galilean relativity (Eq. 39.11), the light wave would have no speed in the primed frame: $v'_x = v_x - v_{rel} = c - c$. Such a primed (moving) observer would claim that the light wave does not exist and would not see a changing magnetic or electric field. Of course, this observation by the moving observer would violate Einstein's first postulate because, in the laboratory frame, a light wave is observed as predicted by Maxwell's equations, and so the wave must exist in all frames. In order for Maxwell's equations to hold in both frames, both observers must see a light wave moving at the speed of light.

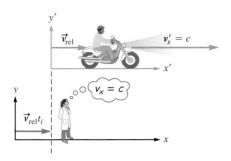

FIGURE 39.11 Aaron in the primed frame shines a light to the right. Observers in both the primed frame and the laboratory frame see the light moving at c.

Einstein's second postulate is consistent with Michelson and Morley's results (Section 36-6). Recall that in their experiment, light traveled along two perpendicular paths of equal length. Imagine two reference frames: each one moving with the light along each of the two paths. Their entire apparatus was on the moving Earth. But when they changed the orientation of this apparatus, they effectively changed the relative speed of the two frames. No matter what the orientation, no time delay was found between the two paths. Michelson and Morley found that the speed of light (in a vacuum) is constant, independent of the observer's speed.

Because our intuition is based on observing the motion of objects that have speeds much less than the speed of light, the second postulate leads to some counterintuitive results. For example, suppose Aaron in the primed frame turns on his headlight (Fig. 39.11). If Aaron could actually watch photons as they emerge from the headlight, he would see the light moving at speed $v'_x = c$. You might expect that Hannah in the laboratory frame would see the photons moving at $v_x = v_{rel} + c$. According to Einstein's second postulate, though, Hannah must still observe the light moving at $v_x = c$. The relative speed of the two frames does not matter. (Alternatively, you might think Hannah observes $v_x \approx c$ because the relative speed is low compared to c, so that when we add a low speed to the speed of light, we get approximately the speed of light. However, Einstein's second postulate is *not an approximation*; it says the speed of light is exactly c for all observers.)

39-4 Lorentz Transformations

One of the most fundamental ideas in Galilean relativity is that a time interval is invariant for all observers. So, one of the most counterintuitive implications of Einstein's second postulate is that time intervals are *not* invariant.

To see this, let's compare the motion of an object such as a helicopter (Fig. 39.9) to the motion of a photon. The displacement of the moving helicopter in the primed frame is $\Delta x'$ (Eq. 39.8), and its displacement as measured in the laboratory frame is greater: $\Delta x = \Delta x' + v_{rel}\Delta t$ (Eq. 39.10). Divide Equation 39.10 by Δt:

$$\frac{\Delta x}{\Delta t} = \frac{\Delta x'}{\Delta t} + v_{rel} \quad (39.12)$$

Now imagine replacing the helicopter moving to the right in Figure 39.9 with a photon. As in the case of the helicopter, the displacement of the photon in the primed frame is $\Delta x'$, and the displacement as measured in the laboratory frame is greater: $\Delta x > \Delta x'$. However, according to Einstein's second postulate, the speed of the photon as measured by either observer must be equal to c:

$$\frac{\Delta x}{\Delta t} = \frac{\Delta x'}{\Delta t'} = c$$

It is impossible for the time interval in the primed frame $\Delta t'$ to equal the time interval in the laboratory frame Δt because the displacements are not equal: $\Delta x' \neq \Delta x$. Furthermore, because the displacement is greater in the laboratory's frame, the time interval measured in that frame must be longer: $\Delta t > \Delta t'$. Compare this moving-photon analysis to the situation of the moving helicopter. According to Galilean relativity, the time interval of the helicopter's trip is the same in both frames, so Equation 39.12 shows that $\Delta x/\Delta t > \Delta x'/\Delta t'$; the helicopter's speed in the laboratory frame is higher than it is in the primed frame. By contrast, the speed of light is the same in both frames, and as a result the time interval for the photon's trip is longer in the laboratory's frame.

Because a time interval is *not* invariant in special relativity, we must come up with a transformation equation not only for position but also for time. These equations for transforming position and time are called the **Lorentz transformations**, named for the Dutch theoretical physicist Hendrik Lorentz. Lorentz did not start with Einstein's postulates; instead, he formulated these equations in an effort to understand Michelson and Morley's results. However, our arguments and derivations are based on Einstein's postulate that the speed of light is the same for all observers.

LORENTZ TRANSFORMATIONS
✪ Major Concept

We can use the Lorentz transformations to derive other important equations. For convenience, these transformation equations along with the Galilean transformations are listed in Table 39.1.

TABLE 39.1 Galilean and Lorentz transformation equations.

Galilean (valid for low relative velocity)				Lorentz			
From primed to lab		From lab to primed		From primed to lab		From lab to primed	
$x = x' + v_{rel}t$	(39.3)	$x' = x - v_{rel}t$		$x = \gamma(x' + v_{rel}t')$	(39.15)	$x' = \gamma(x - v_{rel}t)$	(39.16)
$y = y'$	(39.2)	$y' = y$		$y = y'$	(39.13)	$y' = y$	(39.13)
$z = z'$		$z' = z$		$z = z'$	(39.14)	$z' = z$	(39.14)
$t = t'$		$t' = t$		$t = \gamma\left(t' + \dfrac{v_{rel}x'}{c^2}\right)$	(39.19)	$t' = \gamma\left(t - \dfrac{v_{rel}x}{c^2}\right)$	(39.20)
				where $\gamma = \dfrac{1}{\sqrt{1 - (v_{rel}/c)^2}} = \dfrac{1}{\sqrt{1 - \beta^2}}$			(39.17)

The Correspondence Principle

The **correspondence principle** is based on the observation that Galilean relativity seems to hold in circumstances when the relative speed between frames is low compared to the speed to light. According to the correspondence principle, the Lorentz transformations must be identical to the Galilean transformations when the relative speed is low. Sometimes a high relative speed is referred to as a **relativistic speed**, shorthand for saying the relative speed is high enough that we must take Einstein's theory of special relativity into account because Galilean relativity is not a good approximation. We will use the correspondence principle to check the Lorentz transformations.

Lorentz Transformations for x, y, and z

Let's take a look at the Lorentz transformations for position (Table 39.1). Figure 39.7 establishes our coordinate systems. Because the primed frame is moving at constant velocity in one dimension (along x), there is no difference between the Galilean and Lorentz transformations for the perpendicular components of positions:

$$y = y' \tag{39.13}$$

and

$$z = z' \tag{39.14}$$

These perpendicular components are invariant, but the x component of position is not. Instead, the Lorentz transformation for x is given by

$$x = \gamma(x' + v_{rel}t') \tag{39.15}$$

and the inverse transformation (from the laboratory frame to the primed one) is given by

$$x' = \gamma(x - v_{rel}t) \tag{39.16}$$

This dimensionless constant γ is sometimes called the **gamma factor**, or the **Lorentz factor**, and is given by

$$\gamma = \frac{1}{\sqrt{1 - (v_{rel}/c)^2}} = \frac{1}{\sqrt{1 - \beta^2}} \tag{39.17}$$

where

$$\beta \equiv \frac{v_{rel}}{c} \tag{39.18}$$

There are three important features of the Lorentz factor γ. First, it depends only on the relative speed of the primed observer. It is constant in time (no t dependence) and uniform in space (no x, y, or z dependence). Second, the Lorentz factor is greater than or equal to 1: $\gamma \geq 1$. Third, if the relative velocity is low compared to the speed of light, $\lim_{\beta \to 0} \gamma \to 1$.

Because $\gamma \to 1$ when the relative velocity is low, the Galilean transformation for x is a very good approximation of the Lorentz transformation at low relative speed, as expected by the correspondence principle. (In Problem 72, you will use Einstein's second postulate to confirm Equation 39.17 for the gamma factor.)

Lorentz Transformation for t

Though it is counterintuitive to think so, time also depends on the motion of the observer. The Lorentz transformation for time is given by

$$t = \gamma\left(t' + \frac{v_{\text{rel}} x'}{c^2}\right) \quad (39.19)$$

The inverse transformation from the laboratory frame to the primed frame is given by

$$t' = \gamma\left(t - \frac{v_{\text{rel}} x}{c^2}\right) \quad (39.20)$$

The Lorentz time transformation must be the same as the Galilean transformation when the relative velocity is low. When $v_{\text{rel}} \ll c$, the gamma factor is approximately 1. Also, the second term in Equations 39.19 and 39.20 must be approximately zero because $\lim_{v_{\text{rel}} \to 0} v_{\text{rel}}/c^2 \to 0$. So, for low relative speed, the Lorentz transformation for time is approximately the same as in Galilean relativity. (In Problem 14, you are asked to confirm Equations 39.19 and 39.20.)

Simultaneity Is Relative

A related corollary to Einstein's second postulate is that simultaneity is not invariant: Two events that occur simultaneously in one reference frame are not necessarily simultaneous in another frame moving with respect to the first frame. So when we say two events occur simultaneously, we must state in which frame these events are observed. The idea that simultaneity is not invariant seems counterintuitive, because in our everyday experience, in which relative speeds are low, simultaneous events occur at roughly the same time in both frames.

EXAMPLE 39.3 **Displacement in the Lab Frame**

The helicopter in Figure 39.9 is at x_i' at t_i' and it moves to x_f' at t_f', as seen by the primed observer. Find an expression for the displacement observed in the laboratory frame in terms of the displacement $\Delta x'$ and the time interval $\Delta t'$ measured in the primed frame. Use the Lorentz transformation.

: INTERPRET and ANTICIPATE

The procedure for finding the displacement here is similar to the procedure in Example 39.2A, when we found a displacement under Galilean relativity. We expect the two results to be consistent when the relative velocity is low.

: SOLVE

The displacement in the laboratory frame is the difference between the final position x_f and the initial position x_i.	$\Delta x = x_f - x_i$

To find the transformation between the primed frame and the laboratory frame, substitute the transformation for each position using $x = \gamma(x' + v_{rel}t')$ (Eq. 39.15).	$\Delta x = \gamma(x'_f + v_{rel}t'_f) - \gamma(x'_i + v_{rel}t'_i)$
Regroup the terms.	$\Delta x = \gamma[(x'_f - x'_i) + v_{rel}(t'_f - t'_i)]$
Write the result in terms of the displacement and the time interval in the primed frame.	$\Delta x = \gamma(\Delta x' + v_{rel}\Delta t')$ (39.21)

CHECK and THINK
Equation 39.21 is the transformation of displacement to the laboratory frame, and the result depends on the displacement and the time interval in the primed frame. Compare that to the displacement transformation under Galilean relativity, $\Delta x = \Delta x' + v_{rel}\Delta t$ (Eq. 39.10), which also depends on the displacement in the primed frame. In Galilean relativity, however, the time interval is the same in both frames, $\Delta t' = \Delta t$, so we are free to write the Galilean transformation in terms of Δt.

Check to see that the correspondence principle holds. At a low relative speed, the Lorentz factor is approximately 1 and time is invariant, so the time interval is the same in both frames.	$\lim_{v_{rel} \to 0} \Delta x = \lim_{v_{rel} \to 0} \gamma(\Delta x' + v_{rel}\Delta t')$ $\Delta x \to \Delta x' + v_{rel}\Delta t$ ✓

39-5 Length Contraction

As discussed in the case study, the consequences of Einstein's theory of relativity may sound like science fiction or fantasy. For example, a consequence of the Lorentz transformation for position is that the length of an object depends on its motion relative to the observer. For a fictional take on this idea, think of the ancient Greek myth about a villain named Procrustes, who forced his victims to lie on a bed. If the victim was longer than the bed, Procrustes cut his victim down to make him fit. If the victim was shorter than the bed, Procrustes stretched his victim until he fit snugly in the bed.

Let's imagine a slight twist on this ancient myth. Suppose Procrustes's bed and our hero Theseus are in the primed frame moving at relative velocity v_{rel} with respect to the laboratory frame (Fig. 39.12). Theseus measures his height h' and the length of the bed L' in the primed frame, and he finds to his relief that they are equal: $L' = h'$. So Theseus believes he will fit in the bed. But Theseus's friend Ariadne is in the laboratory frame. Our challenge (in Example 39.4) is to see whether Ariadne agrees that Theseus will fit in the bed.

When you find the length of an object, you must measure the positions of its endpoints simultaneously. So, Theseus measures the positions of the bed's foot x'_f and head x'_h simultaneously. Then he subtracts to find the bed's length:

$$L' = x'_h - x'_f \quad (39.22)$$

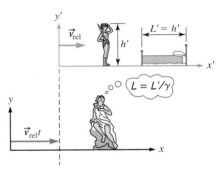

FIGURE 39.12 In the primed frame, the length of the bed equals Theseus's height. According to Ariadne in the laboratory frame, the bed is shorter than Theseus's height.

Theseus makes his measurement in the primed frame, where the bed is at rest. A measurement of length made in the same frame in which the object is at rest is called the **proper length**. So L' is the bed's proper length.

Ariadne also measures the positions of the bed's foot x_f and head x_h simultaneously and then subtracts:

$$L = x_h - x_f \quad (39.23)$$

To see how her measurement compares to the proper length, we substitute Equation 39.16, $x' = \gamma(x - v_{rel}t)$, for the positions in Equation 39.22:

$$L' = \gamma(x_h - v_{rel}t) - \gamma(x_f - v_{rel}t)$$

Her measurements are made simultaneously, so the terms involving $v_{rel}t$ cancel:

$$L' = \gamma(x_h - x_f)$$

Substitute Equation 39.23:

$$L' = \gamma L \tag{39.24}$$

or

$$L = \frac{L'}{\gamma} \tag{39.25}$$

LENGTH CONTRACTION

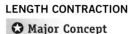

So Ariadne's measurement made in the laboratory frame is the proper length divided by the Lorentz factor. Because $\gamma \geq 1$, Ariadne's measurement is, alarmingly, shorter than the proper length! It is always true that the proper length is the longest measurement of an object's length. Because an observer moving relative to an object always finds it has a shorter length than does an observer in the object's frame, we say that moving relative to an object causes an observed **length contraction**.

Equations 39.24 and 39.25 indicate that, in Einstein's theory of special relativity, length depends on the relative motion of the observer. However, length is invariant in Galilean relativity (Eq. 39.5), and according to the correspondence principle, this is what we expect to find at low relative speeds. When $v_{rel} \ll c$, the Lorentz factor approaches 1 ($\gamma \to 1$) and

$$L' = \gamma L \to L$$

as expected.

EXAMPLE 39.4   CASE STUDY   Does Theseus Fit in the Bed?

Suppose Theseus's frame is moving at $0.6c$ relative to Ariadne's frame in Figure 39.12. Theseus measures his own height $h' = 2$ m (exactly), and he measures the length of the bed $L' = 2$ m. He is happy to find that his height equals the length of the bed. Theseus believes he is safe: Provided Procrustes is in his frame, Procrustes will not have to "adjust" Theseus's height.

A   What is Theseus's height as measured by Ariadne?

: INTERPRET and ANTICIPATE

Measuring Theseus's height means measuring the positions of his endpoints (feet and head) simultaneously. Because both are y coordinates and the y axes are perpendicular to the primed frame's velocity, we expect his height to be invariant.

: SOLVE	
Theseus measures the positions of his feet y'_f and head y'_h simultaneously and then subtracts to get his height h'.	$h' = y'_h - y'_f = 2$ m
Likewise, Ariadne measures the positions of Theseus's feet y_f and head y_h simultaneously and subtracts to get his height h.	$h = y_h - y_f$
Use the Lorentz transformation of y (Eq. 39.13) to transform Ariadne's measurement of Theseus's height.	$y = y'$ (39.13) $h = y'_h - y'_f$ $h = h' = 2$ m

: CHECK and THINK

As expected, Ariadne and Theseus measure his height to be the same. In general, length measurements that are perpendicular to the frame's motion are invariant.

B What is the length of the bed as measured by Ariadne?

:• **INTERPRET and ANTICIPATE**
Because the bed is horizontal (parallel to the primed frame's motion), we expect Ariadne's measurement to differ from Theseus's measurement. Theseus is in the same frame as the bed, so he measures the bed's proper length. We expect Ariadne's measurement to be shorter than 2 m because she is not in the bed's frame.

:• **SOLVE**
For problems involving special relativity, it often helps to find a value for the Lorentz factor first (Eq. 39.17).

$$\gamma = \frac{1}{\sqrt{1-(v_{\text{rel}}/c)^2}} = \frac{1}{\sqrt{1-\beta^2}} \quad (39.17)$$

$$\gamma = \frac{1}{\sqrt{1-0.6^2}} = 1.25$$

Find Ariadne's measurement from Equation 39.25. Because we already found the Lorentz factor, simply divide Theseus's measurement by that value.

$$L = \frac{L'}{\gamma} \quad (39.25)$$

$$L = \frac{2\text{ m}}{1.25} = 1.6\text{ m}$$

:• **CHECK and THINK**
As expected, Ariadne claims the bed is shorter than 2 m. To her alarm, it seems that Theseus won't fit in the bed (so Procrustes will cut him down to size). But Theseus thinks he will fit exactly. Who is right? Both are right. As long as Theseus remains vertical, they will both agree that his height is 2 m. However, Theseus finds the bed is 2 m long, while Ariadne finds it is only 1.6 m long. Still, Theseus believes he can safely lie in the bed—and he is right. As soon as he lies down, he will be horizontal, and Ariadne will find that his height is now 1.6 m—the same as the length of the bed. So neither observer will see the need for Procrustes to make any "adjustments" to Theseus. This is reassuring because it would violate the laws of physics to have Theseus cut down to size in one frame and remain intact in another frame.

39-6 Time Dilation

The Forever War by Joe Haldeman is a science fiction novel in which people must fight a war in a distant part of the galaxy. They must travel at nearly the speed of light to get to their battles. To these soldiers, each expedition lasts only a few months. However, when the solders return to the Earth, they find that many centuries have passed, and it is difficult for them to adjust to society in the future. This novel makes use of a key result of special relativity—namely, that the length of a time interval is relative, depending on the motion of the observer. This phenomenon is known as **time dilation**. In this section, we derive an expression for time dilation, and in Example 39.5, we apply it to Haldeman's characters.

DERIVATION The Time Dilation Equation

We will show that a time interval measured in the laboratory frame is given by

$$\Delta t = \gamma \Delta t' \quad (39.26)$$

TIME DILATION
✪ Major Concept

where $\Delta t'$ is known as the **proper time**. The proper time is defined as the time interval between two events that occur at the same position. Here, the proper time is measured in the primed frame.

Derivation continues on page 1282 ▶

Suppose in the primed frame Aaron fires a flashbulb at t_i'. A mirror is located directly above the bulb at h' (Fig. 39.13).

FIGURE 39.13 The observer in the primed frame sees a photon travel straight up and down.

Imagine Aaron can watch as a single photon travels up to the mirror, reflects from the mirror, and returns to the flashbulb at t_f'. The two events happen at the instant the photon leaves the bulb and the instant it returns to the bulb. Aaron measures the proper time between the two events because to him they both occur at the same position—the position of the bulb.	Proper time: $\Delta t' = t_f' - t_i'$
According to Aaron, the total path length is $2h'$. The proper time is the path length divided by the photon's speed c.	$\Delta t' = \dfrac{2h'}{c}$ $h' = \dfrac{c\Delta t'}{2}$ (1)
However, in the laboratory frame, Hannah does not measure the proper time because the bulb is not in the same position when the photon returns to it as when the photon leaves it. Instead, she sees the photon move along two legs of a triangle, with the displacement Δx of the bulb forming the triangle's base (Fig. 39.14). **FIGURE 39.14** The observer in the laboratory frame sees the photon trace out two legs of a triangle.	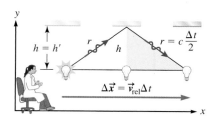
The displacement of the bulb depends on the relative speed of the primed frame. To Hannah, the bulb moves in the positive x direction at speed v_{rel}. In the time Δt it takes the photon to move from the bulb, reflect from the mirror, and return to the bulb, the bulb has been displaced by Δx.	$\Delta x = v_{rel}\Delta t$
The photon travels along the legs of the triangle of length r at speed c. The time it takes the photon to go from the bulb to the mirror is equal to the time it takes to go from the mirror to the bulb. Each is half the total travel time.	$r = c\dfrac{\Delta t}{2}$
The height of the triangle is h. Apply the Pythagorean theorem to the right triangle shaded blue in Figure 39.14. The length of the horizontal leg of this right triangle is half the displacement of the bulb.	$r^2 = h^2 + \left(\dfrac{\Delta x}{2}\right)^2$ $\left(c\dfrac{\Delta t}{2}\right)^2 = h^2 + \left(\dfrac{v_{rel}\Delta t}{2}\right)^2$ $h^2 = \left(c\dfrac{\Delta t}{2}\right)^2 - \left(\dfrac{v_{rel}\Delta t}{2}\right)^2$ (2)
The height of the triangle equals the vertical distance between the mirror and the bulb. There is no length contraction in the perpendicular direction.	$h = h'$ (3)
Square Equation (3) and substitute Equations (1) and (2).	$h^2 = h'^2$ $\left(c\dfrac{\Delta t}{2}\right)^2 - \left(\dfrac{v_{rel}\Delta t}{2}\right)^2 = \left(\dfrac{c\Delta t'}{2}\right)^2$
Solve for Δt^2.	$\Delta t^2(c^2 - v_{rel}^2) = c^2 \Delta t'^2$ $\Delta t^2 = \dfrac{c^2}{(c^2 - v_{rel}^2)}\Delta t'^2$

Write the factor $c^2/(c^2 - v_{rel}^2)$ in terms of the gamma factor (Eq. 39.17).	$\dfrac{c^2}{(c^2 - v_{rel}^2)} = \dfrac{1}{1 - (v_{rel}/c)^2} = \gamma^2$
Solve for Δt.	$\Delta t^2 = \gamma^2 \Delta t'^2$ $\Delta t = \gamma \Delta t'$ ✓ (39.26)

:• **COMMENTS**

Because $\gamma \geq 1$, the time interval Δt measured in the laboratory frame is longer than the proper time $\Delta t'$. To help remember this, we sometimes use the phrase *moving clocks run slowly*. (The proper clock is in the primed frame, and the laboratory frame moves relative to that frame.) Now use the correspondence principle to check Equation 39.26. There is no time dilation in Galilean relativity. If the relative speed is low, then $\gamma \to 1$ and $\Delta t \to \Delta t'$ as expected.

EXAMPLE 39.5   CASE STUDY   Back to the Future

In *The Forever War*, the soldiers are away from the Earth for about 2 months. When they return, they find that centuries have passed. Let's consider a slightly simpler problem based on those soldiers' experience. Suppose an observer in the primed frame claims that the time between two events at the same location is exactly 60 days (5.184×10^6 s). To keep this problem simple, you can imagine that the two events are the emission and return of the photon in Figure 39.13. An observer in the laboratory frame measures the time interval of the same two events as exactly 100 years (3.15576×10^9 s). What is the constant relative speed of the primed observer? Give your answer to eight significant figures in terms of the speed of light.

:• **INTERPRET and ANTICIPATE**

There is no time dilation if the relative speed is low. The time dilation in this problem is enormous, so we expect the relative speed to be very high.

:• **SOLVE** Solve Equation 39.26 for the Lorentz factor.	$\Delta t = \gamma \Delta t'$ (39.26) $\gamma = \dfrac{\Delta t}{\Delta t'}$
Substitute values. The proper time is the 60 days measured in the primed frame.	$\gamma = \dfrac{3.15576 \times 10^9 \text{ s}}{5.184 \times 10^6 \text{ s}} = 608.75$
Now solve $\gamma = \dfrac{1}{\sqrt{1-\beta^2}}$ (Eq. 39.17) for β.	$1 - \beta^2 = \dfrac{1}{\gamma^2}$ $\beta = \sqrt{1 - \dfrac{1}{\gamma^2}}$
Substitute for γ.	$\beta = \sqrt{1 - \dfrac{1}{608.75^2}} = 0.99999865 = v_{rel}/c$ $v_{rel} = 0.99999865 c$

Example continues on page 1284 ▶

CHECK and THINK
As expected, the primed frame's relative speed must be very relativistic—more than 99% of the speed of light. This is a real consequence of Einstein's theory of relativity. The only thing that makes this result fictional is that we have yet to come up with a spacecraft that can move at nearly the speed of light.

EXAMPLE 39.6 Muon Decay

Time dilation isn't found only in fiction; it has been confirmed by experiments involving subatomic particles known as muons. Muons can be created by high-energy particle accelerators, and they are unstable. When muons are at rest with respect to the laboratory, their half-life is roughly 2 μs. Muons are also made in the Earth's atmosphere by cosmic rays and subsequently move downward at roughly $0.98c$.

A How far can these muons move in the Earth's atmosphere during the time interval of their half-life?

INTERPRET and ANTICIPATE
We can consider the Earth and its atmosphere to be in the laboratory frame and the muons to be in the primed frame. The question is really asking us to find the distance the muons travel in the laboratory frame. We'll use a downward-pointing y axis because that is the direction in which the primed frame moves relative to the laboratory frame.

SOLVE
This is a straightforward application of kinematics. In the laboratory frame, the muons move downward at $0.98c$. We must find how far they travel in 2 μs.

$$\Delta y = v_{\text{rel}} \Delta t$$
$$\Delta y = (0.98)(3.00 \times 10^8 \text{ m/s})(2 \times 10^{-6} \text{ s})$$
$$\Delta y = 588 \text{ m} \approx 6 \times 10^2 \text{ m}$$

CHECK and THINK
If we put muon detectors in the laboratory frame separated by $\Delta y = 600$ m, we would expect to find that only half the muons would survive to reach the lower detector. However, if you really did this, you would find that *more* than half the muons would survive to reach the second detector due to time dilation. The muons' half-life in their own frame is 2 μs. The detectors are in the laboratory frame, though, and we should *not* expect the time interval to be the same in both frames.

B What is the half-life of these muons observed in the laboratory frame?

INTERPRET and ANTICIPATE
The half-life measured when the muons are *at rest* in a laboratory is the proper time for their half-life. No matter how fast the muons are moving, in their own frame their half-life is 2 μs. But, because of the relative speed of the laboratory frame, the half-life must be longer in the laboratory frame.

SOLVE
Find the value of the Lorentz factor (Eq. 39.17).

$$\gamma = \frac{1}{\sqrt{1 - \beta^2}} \quad (39.17)$$

$$\gamma = \frac{1}{\sqrt{1 - 0.98^2}} = 5.0$$

Use the time dilation equation (Eq. 39.26) to find the half-life of a muon as observed in the laboratory frame.	$\Delta t = \gamma \Delta t'$ (39.26) $\Delta t = (5.0)(2\,\mu s) \approx \boxed{10\,\mu s}$

:• **CHECK and THINK**
More than half the muons would survive to reach the lower detector because when muons move at $0.98c$ relative to detectors in the laboratory frame, their half-life is $10\,\mu s$, not $2\,\mu s$. So, if only half the muons are to reach the second detector, the detectors must be farther than 600 m apart to allow more time to pass.

C   How far apart should the detectors be placed for half the muons produced at the higher detector to reach the lower detector?

:• **INTERPRET and ANTICIPATE**
This is similar to part A, except we must use the half-life we expect to observe in the laboratory frame.

:• **SOLVE** In the laboratory frame, the muons move downward at $0.98c$. Find how far they travel in $10\,\mu s$.	$\Delta y = v_{rel}\Delta t$ $\Delta y = (0.98)(3.00 \times 10^8\,\text{m/s})(1 \times 10^{-5}\,\text{s})$ $\Delta y = 2940\,\text{m} \approx \boxed{3000\,\text{m}}$

:• **CHECK and THINK**
When the detectors are 3000 m apart, the lower detector finds half as many muons as the upper detector. You may be wondering what would happen if an observer moved with the muons. Would he also see half as many muons arrive at the second detector? The answer is yes. But, to this observer in the primed frame, the distance between the two detectors is contracted to 600 m, and the half-life of the muons is $2\,\mu s$. So the time interval between when the first detector encounters the muons and when the second detector encounters the muons is equal to their proper half-life in the primed frame.

39-7 The Relativistic Doppler Effect

You have probably noticed that as a police car using its siren passes you, you hear a change in the siren's frequency. This change in the frequency of sound waves is known as the **Doppler shift** (Section 17-9). The frequency and wavelength of light also depend on the relative motion of the source and the observer, so the color of an object (a light source) depends on the relative motion between the source and an observer. When a light source moves away from an observer, the wavelength is longer and the light looks redder. When a light source moves toward the observer, the wavelength is shorter and the light looks bluer.

DERIVATION   The Relativistic Doppler Formula

We will show that the relativistic Doppler formula for change in wavelength is

$$\lambda = \sqrt{\frac{1 \mp \beta}{1 \pm \beta}}\,\lambda' \quad \text{(radial motion)} \quad (39.27)$$

where λ' is the wavelength measured in the primed frame and λ is the wavelength measured in the laboratory frame. An observer in the primed frame sees that the frequency of light is f', the period (time interval between arriving wave fronts) is $T' = 1/f'$, and the wavelength is $\lambda' = c/f' = cT'$.

RELATIVISTIC DOPPLER EFFECT
★ Major Concept

Choose the top signs if the source is moving toward the observer; choose the bottom signs if the source is moving away from the observer.

Derivation continues on page 1286 ▶

A monochromatic light source is stationary in the primed frame. In the laboratory frame, Hannah is standing in front of the light source and sees the light source approaching her (Fig. 39.15). We use the phrase *radial motion* to describe motion that is directly toward or away from the observer; we have chosen to consider motion toward the observer, and we'll generalize our results at the end of the derivation.

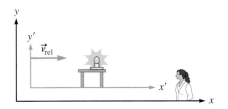

FIGURE 39.15 An observer in the laboratory frame sees a light source approaching her.

In the laboratory frame, the time between the emissions of each wave crest is T. But this is not the same as the period Hannah observes; because the source is moving toward her, the crests will reach her with a shorter period.

Figure 39.16 shows the motion of the source and the emission of its wave crests in the laboratory frame, similar to Figure 17.28 (page 510). At time 1, a wave crest is emitted. The source moves to the right at v_{rel}. By the time the second crest is emitted at time 2, the source has moved to $v_{rel}T$. By time 3, when a third crest is emitted, the second crest's radius has grown to cT and the first crest's radius has grown to $2cT$. The distance between the centers of the first and second crests is $v_{rel}T$. Along the direction of motion, the distance between the first and second crests (wave fronts) is the wavelength λ observed in the laboratory.

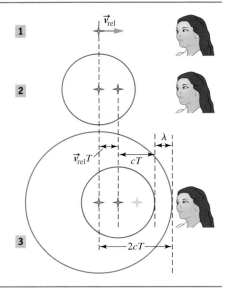

FIGURE 39.16 The colors of the source and wave crests have been chosen to help you connect the wave crest with the location of the source when the wave was emitted. For example, the large red wave crest shown at time **3** was emitted when the source was at the far left position at time **1**.

Use Figure 39.16 to find an expression for the wavelength λ.	$2cT = v_{rel}T + cT + \lambda$ $\lambda = cT - v_{rel}T = (c - v_{rel})T$ (1)
The period T' measured in the primed frame is the proper time between the emissions of the crests. To find T, take time dilation (Eq. 39.26) into account.	$T = \gamma T'$ (39.26) $T' = \dfrac{T}{\gamma}$ (2)
Write the period in the primed frame in terms of the wavelength observed in the primed frame.	$\lambda' = \dfrac{c}{f'} = cT'$ (3)
Substitute Equation (2) into Equation (3), and solve for T.	$T = \dfrac{\gamma}{c}\lambda'$ (4)
Substitute Equation (4) into Equation (1), and simplify by using $\gamma = \dfrac{1}{\sqrt{1-\beta^2}}$ and $\beta = \dfrac{v_{rel}}{c}$ (Eqs. 39.17 and 39.18).	$\lambda = (c - v_{rel})\left(\dfrac{\gamma}{c}\lambda'\right) = \gamma(1 - v_{rel}/c)\lambda'$ $\lambda = \dfrac{(1-\beta)}{\sqrt{1-\beta^2}}\lambda' = \dfrac{(1-\beta)}{\sqrt{(1-\beta)(1+\beta)}}\lambda'$ $\lambda = \sqrt{\dfrac{1-\beta}{1+\beta}}\lambda'$
We assumed Hannah is standing in front of the source, so that the source is moving toward her. If the source were moving away, the signs in front of β would be reversed. We write our final relativistic Doppler formula by writing these signs explicitly.	$\lambda = \sqrt{\dfrac{1 \mp \beta}{1 \pm \beta}}\lambda'$ ✓ (39.27)

The relativistic Doppler formula can also be written in terms of frequency (Problem 36).	$f = \sqrt{\dfrac{1 \pm \beta}{1 \mp \beta}} f'$ (39.28) Choose the top signs if the source is moving toward the observer; choose the bottom signs if the source is moving away from the observer.

:• **COMMENTS**
Equations 39.27 and 39.28 are used when the source is moving radially toward or away from the observer. There is also a transverse Doppler shift associated with relative motion that is perpendicular to the observer's line of sight. This transverse Doppler shift is due entirely to time dilation.

CONCEPT EXERCISE 39.3

Show that Equation 39.27 predicts that when a light source is moving toward an observer, the source appears bluer, and when the source is moving away, it appears redder.

EXAMPLE 39.7 Quasars

In 1960, Thomas Matthews and Allan Sandage discovered a starlike object whose spectrum could not be identified with any known substance. Sandage said, "The thing is exceedingly weird." When other similar objects were discovered, they were called quasi-stellar radio sources, or quasars, because they are sources of radio emission that appeared like stars in our own galaxy. Today, we know that quasars are much more powerful than stars, and they are among the most distant objects we can observe in the Universe. One particular quasar emits light with a wavelength of 121.6 nm in its own frame. On the Earth, this light is observed to have a wavelength of 885.2 nm. The quasar is moving either toward or away from the Earth. Decide which, and then find the quasar's speed relative to the Earth.

:• **INTERPRET and ANTICIPATE**
First, decide what is in the primed frame and what is in the laboratory frame. We need to find the quasar's speed relative to the Earth. This suggests the quasar is in the primed frame and the Earth is in the laboratory frame.

:• **SOLVE** The observer in the laboratory sees a longer wavelength than the one emitted in the primed frame. So the quasar must be moving away from the Earth, and we choose the bottom signs in Equation 39.27.	$\lambda = \sqrt{\dfrac{1+\beta}{1-\beta}}\,\lambda'$ (39.27)
Solving for β requires a few steps of algebra.	$\left(\dfrac{\lambda}{\lambda'}\right)^2 = \dfrac{1+\beta}{1-\beta}$ $-\beta[(\lambda/\lambda')^2 + 1] = 1 - (\lambda/\lambda')^2$ $\beta = \dfrac{(\lambda/\lambda')^2 - 1}{(\lambda/\lambda')^2 + 1}$

Example continues on page 1288 ▶

Substitute values. Use $\beta = v_{rel}/c$ (Eq. 39.18) to find the relative speed.

$$\beta = \frac{\left(\dfrac{885.2 \text{ nm}}{121.6 \text{ nm}}\right)^2 - 1}{\left(\dfrac{885.2 \text{ nm}}{121.6 \text{ nm}}\right)^2 + 1} = 0.9630$$

$$v_{rel} = 0.9630c$$

:• **CHECK and THINK**
When an object's speed is very relativistic, it is common to report that speed in terms of the speed of light, rather than in conventional units such as m/s. This quasar has a high recessional speed from the Earth. This fits with other observations made in astronomy: In general, the farther an object is from us, the faster it moves away. Quasars are very distant objects, so their relative speed is very high.

39-8 Velocity Transformation

Imagine that the primed observer and the laboratory observer measure the velocity of a moving object, while (as always) the primed observer is moving at $\vec{v}_{rel}$ along the x axis with respect to the laboratory. According to Galilean relativity, the two observers will agree on the component of the velocity perpendicular to $\vec{v}_{rel}$, while the parallel component's transformation is given by $v_x = v'_x + v_{rel}$ (Eq. 39.11).

The transformation of velocity is more complicated in special relativity, however. Because the observed velocity of an object depends on both its displacement and the time interval of measurement, a velocity transformation equation involves both length contraction and time dilation. There is no length contraction in the perpendicular directions, but there is time dilation, so the two observers will not measure the same perpendicular velocity. And, because there is length contraction in the parallel direction, the parallel velocity transformation involves both time dilation and length contraction. Our goal is to derive expressions for parallel and perpendicular velocity transformations. We can check our work in two ways. First, when the relative speed is low, our results should match the results of Galilean transformation. Second, if the moving object is a photon, we should find that both observers agree that its speed is c.

Problem 40 asks you to show that the transformation of a velocity component perpendicular to $\vec{v}_{rel}$ is given by

$$v_y = \frac{v'_y}{\gamma\left(1 + \dfrac{v_{rel}}{c^2} v'_x\right)} \quad (39.29)$$

PERPENDICULAR VELOCITY TRANSFORMATION ✪ Major Concept

and the inverse transformation is given by

$$v'_y = \frac{v_y}{\gamma\left(1 - \dfrac{v_{rel}}{c^2} v_x\right)} \quad (39.30)$$

We can write similar equations for the z component of velocity. The perpendicular velocity transformations depend on the parallel velocity component v_x.

DERIVATION Parallel Velocity Transformation

Here we show that the transformation of a velocity component parallel to $\vec{v}_{rel}$ is given by

$$v_x = \frac{(v'_x + v_{rel})}{\left(1 + \dfrac{v_{rel}}{c^2} v'_x\right)} \quad (39.31)$$

PARALLEL VELOCITY TRANSFORMATION ✪ Major Concept

39-8 Velocity Transformation

Consider an object that moves parallel to the primed frame's motion, such as the helicopter in Figure 39.9. The x component of its velocity in both frames is the time derivative of its parallel position. Length contraction and time dilation mean that both the numerator and the denominator of this derivative differ in the two frames.	$v_x = \dfrac{dx}{dt}$ (1) $v_x' = \dfrac{dx'}{dt'}$ (2)
First, we derive the transformation from the primed frame to the laboratory frame. Start by taking the differentials of Equations 39.15 and 39.19.	$x = \gamma(x' + v_{rel} t')$ (39.15) $dx = \gamma(dx' + v_{rel} dt')$ (3) $t = \gamma\left(t' + \dfrac{v_{rel} x'}{c^2}\right)$ (39.19) $dt = \gamma\left(dt' + \dfrac{v_{rel} dx'}{c^2}\right)$ (39.32)
Substitute Equations (3) and 39.32 into Equation (1).	$v_x = \dfrac{\gamma(dx' + v_{rel} dt')}{\gamma\left(dt' + \dfrac{v_{rel} dx'}{c^2}\right)}$
The Lorentz factor cancels. Divide the numerator and denominator by dt'.	$v_x = \dfrac{\left(\dfrac{dx'}{dt'} + v_{rel}\right)}{\left(1 + \dfrac{v_{rel}}{c^2}\dfrac{dx'}{dt'}\right)}$
Use Equation (2) to write the expression in terms of the velocity measured in the primed frame.	$v_x = \dfrac{(v_x' + v_{rel})}{\left(1 + \dfrac{v_{rel}}{c^2} v_x'\right)}$ ✓ (39.31)
Equation 39.31 is the parallel velocity transformation from the primed frame to the laboratory frame. Problem 38 asks you to show that Equation 39.33 is the corresponding transformation from the laboratory frame to the primed frame.	$v_x' = \dfrac{(v_x - v_{rel})}{\left(1 - \dfrac{v_{rel}}{c^2} v_x\right)}$ (39.33)
:• **COMMENTS** Check to see whether Equation 39.31 matches the Galilean transformation in the case of low relative speed: $v_{rel}/c^2 \to 0$. Our result approaches Equation 39.11 (the Galilean transformation).	$\lim\limits_{v_{rel} \to 0} v_x = \dfrac{(v_x' + v_{rel})}{\left(1 + \dfrac{v_{rel}}{c^2} v_x'\right)} \to v_x' + v_{rel}$ ✓
Next, imagine the helicopter is replaced by a photon moving to the right at $v_x' = c$. As expected, both observers see the photon moving at c.	$v_x = \dfrac{(c + v_{rel})}{\left(1 + \dfrac{v_{rel}}{c^2} c\right)}$ $v_x = \dfrac{(c + v_{rel})}{\left(1 + \dfrac{v_{rel}}{c}\right)} = \dfrac{c\left(1 + \dfrac{v_{rel}}{c}\right)}{\left(1 + \dfrac{v_{rel}}{c}\right)}$ $v_x = c$ ✓

EXAMPLE 39.8 A Photon Flies Straight Up

Suppose a photon flies straight up (along the y direction) at speed c in the primed frame, so $v_x' = 0$ and $v_y' = c$. As usual, the primed frame moves in the positive x direction at speed v_{rel} as observed from the laboratory frame. What is the speed of the photon observed in the laboratory frame?

INTERPRET and ANTICIPATE
Because of Einstein's second postulate, we expect the laboratory observer will observe the photon moving at c. This problem is tricky because the perpendicular velocity transformation depends on the horizontal motion of the photon. The situation is somewhat similar to the one in Figure 39.13, but there is no mirror, so the photon does not reverse direction. As in that case, because of the relative horizontal (parallel) motion of the primed frame, the laboratory observer sees the photon moving in both the y and x directions (as in Fig. 39.14). So we must find both v_x and v_y for the photon.

SOLVE
First, find the horizontal speed v_x in the laboratory frame (Eq. 39.31). The photon moves straight up in the primed frame, so $v_x' = 0$. The photon's horizontal speed in the laboratory frame equals the relative speed.

$$v_x = \frac{(v_x' + v_{\text{rel}})}{\left(1 + \frac{v_{\text{rel}}}{c^2} v_x'\right)} = \frac{(0 + v_{\text{rel}})}{\left[1 + \frac{v_{\text{rel}}}{c^2}(0)\right]}$$

$$v_x = v_{\text{rel}} \quad (1)$$

Next, find the photon's vertical speed in the laboratory frame (Eq. 39.29). Because the photon moves straight up at c in the primed frame, set $v_x' = 0$ and $v_y' = c$.

$$v_y = \frac{v_y'}{\gamma\left(1 + \frac{v_{\text{rel}}}{c^2} v_x'\right)} = \frac{c}{\gamma\left[1 + \frac{v_{\text{rel}}}{c^2}(0)\right]}$$

$$v_y = \frac{c}{\gamma}$$

Substitute for the gamma factor.

$$v_y = \frac{c}{\gamma} = c\sqrt{1 - \left(\frac{v_{\text{rel}}}{c}\right)^2} \quad (2)$$

The photon speed in the laboratory frame is found in the usual way from its x and y velocity components. Substitute Equations (1) and (2); then simplify.

$$v^2 = v_x^2 + v_y^2$$

$$v^2 = v_{\text{rel}}^2 + c^2\left[1 - \left(\frac{v_{\text{rel}}}{c}\right)^2\right]$$

$$v^2 = v_{\text{rel}}^2 + c^2 - v_{\text{rel}}^2 = c^2$$

$$v = c$$

CHECK and THINK
As expected, the speed of the photon as observed in the laboratory frame is c. We have confirmed that the perpendicular velocity transformation (Eq. 39.29) satisfies Einstein's second postulate.

39-9 Mass and Momentum Transformation

One strange consequence of special relativity is that the mass of an object also depends on the relative motion of the object and the observer. To see why this is true, we consider a pair of elastic collisions. Recall that in an elastic collision, both momentum and kinetic energy are conserved.

Figure 39.17 shows a collision between two perfectly elastic hockey pucks that are viewed in the same reference frame: a red puck of mass m_{rest} and a blue puck of mass m. Each puck is given the same initial speed v_{puck}. They collide elastically and each reverses direction, traveling afterward at the same speed v_{puck}. The change in momentum of the red puck is

$$\Delta \vec{p}_{red} = \vec{p}_f - \vec{p}_i = m_{rest} v_{puck} \hat{k} - (-m_{rest} v_{puck} \hat{k}) = 2 m_{rest} v_{puck} \hat{k} \quad (39.34)$$

The change in the blue puck's momentum is

$$\Delta \vec{p}_{blue} = -m v_{puck} \hat{k} - m v_{puck} \hat{k} = -2 m v_{puck} \hat{k} \quad (39.35)$$

Momentum is conserved, so we apply $\Delta \vec{p}_{red} = -\Delta \vec{p}_{blue}$ (Eq. 11.9) to Equations 39.34 and 39.35: $2 m_{rest} v_{puck} \hat{k} = 2 m v_{puck} \hat{k}$. We find that the pucks have the same mass:

$$m = m_{rest} \quad (39.36)$$

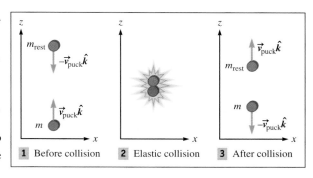

1 Before collision 2 Elastic collision 3 After collision

FIGURE 39.17 As observed in the laboratory frame, two pucks have the same initial speed. They collide, and afterward their speeds are unchanged.

This is not a surprising result; it is what we found for collisions in Section 11-5.

Now imagine that the blue puck and Paul are at rest in the primed frame, which moves to the right with speed v_{rel} relative to the laboratory frame. The red puck and Lil are in the laboratory frame. Paul and Lil agree to launch their pucks with the same speed v_{puck} parallel to the z axis, so that the two pucks collide. Of course, Paul must launch his puck *before* he is lined up with Lil's puck. As in the case of Figure 39.14, according to Lil in the laboratory frame, Paul's the blue puck traces out two legs of a triangle (Fig. 39.18). However, she sees her own red puck travel back and forth along the z axis at speed v_{puck}. She finds that the red puck's change in momentum is exactly what it was when both pucks were in the same frame:

$$\Delta \vec{p}_{red} = 2 m_{rest} v_{puck} \hat{k} \quad (39.34)$$

In the primed frame, Paul launched the blue puck at v_{puck} in the z' direction. So, in his frame, the puck's initial velocity components are $v'_x = 0$ and $v'_z = v_{puck}$, while the final velocity components are $v'_x = 0$ and $v'_z = -v_{puck}$. With $v'_x = 0$ in Equation 39.31, $v_x = (v'_x + v_{rel})/(1 + v_{rel} v'_x/c^2)$ (parallel velocity transformation), Lil finds that the x component of the blue puck's velocity is $v_{rel} \hat{i}$, which is unchanged by the collision. So, to Lil, the x component of the blue puck's momentum is unchanged: $(\Delta p_{blue})_x = m v_{rel} - m v_{rel} = 0$. However, she sees the blue puck reverse its motion in the z direction, so she finds a nonzero change in its momentum along z. Before we can calculate this change in momentum, we need to know the speed Lil observes for the blue puck. Use Equation 39.29, $v_z = v'_z/[\gamma(1 + v_{rel} v'_x/c^2)]$ (perpendicular velocity transformation), with $v'_x = 0$ and $v'_z = v_{puck}$, to find the initial velocity along z seen in Lil's laboratory frame:

$$v_z = \frac{v_{puck}}{\gamma} \quad (39.37)$$

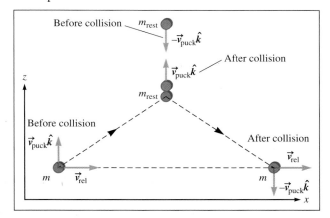

FIGURE 39.18 As seen from the laboratory frame, two pucks collide. In this frame, the red puck's velocity has no x component. In the primed frame (not shown), the blue puck's velocity has no x component.

Both observers agree that after the collision, the blue puck reverses its motion in the z direction with no change in speed. So Lil finds that the change in the blue puck's momentum is given by

$$\Delta \vec{p}_{blue} = \left(-m \frac{v_{puck}}{\gamma} - m \frac{v_{puck}}{\gamma}\right) \hat{k} = -2 m \frac{v_{puck}}{\gamma} \hat{k} \quad (39.38)$$

According to the conservation of momentum, $\Delta \vec{p}_{red} = -\Delta \vec{p}_{blue}$ (Eq. 11.9), and using Equations 39.34 and 39.38, we find

$$2 m_{rest} v_{puck} \hat{k} = -\left(-2 m \frac{v_{puck}}{\gamma} \hat{k}\right)$$

which reduces to

$$m = \gamma m_{rest} \quad (39.39)$$

MASS TRANSFORMATION
✪ Major Concept

Some physicists do not like to define *rest* mass; instead, they use the term *mass* (an invariant quantity). Then only other quantities such as momentum and energy are affected by relative motion.

RELATIVISTIC MOMENTUM
○ Major Concept

When the two pucks were in the same reference frame, we found that their masses were equal (Eq. 39.36). However, Equation 39.39 says that the mass of an object depends on its relative motion. The **rest mass** m_{rest} of any object must be measured in the object's frame. Because $\gamma \geq 1$, if the mass is measured in a frame in which the object is moving, its mass will be greater than its rest mass.

An implication of the experiment in Figure 39.18 is that we must modify how we express the momentum of an object. Suppose that in the laboratory frame, we observe a particle moving at velocity $\vec{v}$. If we think of the particle as being at rest in the primed frame, its momentum as observed in the laboratory frame is

$$\vec{p} = m\vec{v} = \gamma m_{rest} \vec{v} \qquad (39.40)$$

The Lorentz factor in this case depends on the speed v of the particle:

$$\gamma = \frac{1}{\sqrt{1-(v/c)^2}} = \frac{1}{\sqrt{1-\beta^2}} \qquad (39.41)$$

The conservation of momentum holds in all inertial reference frames as long as we use Equation 39.40 for momentum.

The Ultimate Speed Limit

Often in works of science fiction, there is some mention of traveling at or faster than the speed of light. According to the special theory of relativity, though, the speed of light is the ultimate speed limit. Nothing can travel faster than light. Specifically, any object that has mass must travel at less than the speed of light. To see why, substitute Equation 39.41 into Equation 39.39:

$$m = \frac{m_{rest}}{\sqrt{1-(v/c)^2}} = \frac{m_{rest}}{\sqrt{1-\beta^2}} \qquad (39.42)$$

As $\beta \to 1$ (or, equivalently, as $v \to c$), the object's mass approaches infinity: $m \to \infty$. Because it makes no sense to have an object with infinite mass (as measured in any reference frame), we conclude that any object that has mass cannot exceed the speed of light. Put another way, the object would have infinite momentum, and it would take an infinite amount of energy to accelerate the object to the speed of light. So, the speed of light is the ultimate speed limit and is reached only by massless particles such as photons.

EXAMPLE 39.9   **CASE STUDY**   **Try Working Out**

In Example 39.5, we estimated that the characters in *The Forever War* move at a very high speed ($\beta = 0.99999865$). If a typical soldier's mass is 85.0 kg in his own frame, what is his mass according to an observer in the laboratory frame?

: INTERPRET and ANTICIPATE
The smallest possible mass of the soldier is his rest mass. We expect his mass observed in the laboratory frame to be greater than his rest mass.

: SOLVE From Example 39.5, we know the Lorentz factor.	$\gamma = 608.75$
Use Equation 39.39 to find the mass observed in the laboratory frame.	$m = \gamma m_{rest}$ (39.39) $m = (608.75)(85.0\,\text{kg})$ $m = 5.17 \times 10^4\,\text{kg}$

: CHECK and THINK
The soldier's weight as observed in the laboratory frame is nearly 60 tons! Of course, in his own frame, his mass is still 85.0 kg.

39-10 Newton's Second Law and Energy

You have rearranged Newton's second law countless times to find an object's acceleration: $a = F/m$. However, according to special relativity, the mass of an object depends on its motion relative to an observer. Consider the seemingly simple scenario of observing an object that is initially at rest. You and the object are in the laboratory frame, and you find the object's mass equals its rest mass m_{rest}. Then a constant force is applied to the object, so the object accelerates while you remain in the laboratory frame. As the speed of the object increases, you see its mass increase: $m > m_{rest}$. By Newton's second law, then, its acceleration must decrease. However, a constant force should result in a constant acceleration, and there is a simple way around this contradictory observation. We just need to express Newton's second law in terms of momentum, $\vec{F}_{tot} = d\vec{p}/dt$ (Eq. 10.2), rather than in terms of mass and acceleration. This expression of Newton's second law holds even when objects are moving at relativistic speeds. The familiar expression $\vec{F}_{tot} = m\vec{a}$ is an approximation that holds when the object's speed is low enough that its mass equals its rest mass.

Just as the familiar form of Newton's second law is an approximation, the familiar form of kinetic energy, $K = \frac{1}{2}mv^2$, is also an approximation. In this section, we derive a more general equation for kinetic energy that applies in the case of relativistic motion. We also examine Einstein's most famous equation, $E = mc^2$.

DERIVATION Relativistic Kinetic Energy

We will show that the relativistic kinetic energy is given by

$$K = mc^2 - m_{rest}c^2 \qquad (39.43)$$

for a particle of rest mass m_{rest} whose mass is m when it is observed in some frame other than its own.

The classical kinetic energy of a particle depends on both its speed and its mass. According to special relativity, the mass of a particle depends on its speed. So, to find the change in a particle's kinetic energy, we must take into account both its change in speed and its change in mass. Throughout this derivation, we observe a particle from the laboratory frame.

Suppose a constant force in the x direction does work on a particle, with no potential energy or thermal energy involved, so that the work done by the force goes only into changing the particle's kinetic energy (Fig. 39.19).

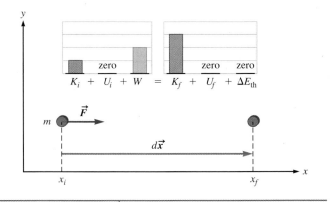

FIGURE 39.19 A constant force does work on a particle. The particle moves from x_i to x_f, and its kinetic energy increases.

When the particle moves a short distance dx, the work done is the force times this distance (Eq. 9.1). According to the work–kinetic energy theorem (Eq. 9.5), this work equals the increase in its kinetic energy dK.	$W = \vec{F} \cdot d\vec{x} = dK \qquad (1)$
The force is parallel to the displacement, so the dot product becomes $F\,dx$. This is the only force exerted on the particle, so substitute Newton's second law, $F = dp/dt$ (Eq. 10.2), into Equation (1). Simplify using $v = dx/dt$.	$dK = \dfrac{dp}{dt}dx = dp\dfrac{dx}{dt}$ $dK = v\,dp$

Derivation continues on page 1294 ▶

In the laboratory frame, we can express the momentum as $p = mv$ as long as we keep in mind that $m = \gamma m_{\text{rest}}$ (Eq. 39.39).	$dK = v\,d(mv)$
When we take the differential of mv, we must use the product rule because both m and v change in this case.	$dK = v(m\,dv + v\,dm)$ $dK = mv\,dv + v^2\,dm$ \hfill (2)
To take into account the change in the particle's mass, start with Equation 39.42 and isolate m^2c^2.	$m = \dfrac{m_{\text{rest}}}{\sqrt{1-(v/c)^2}}$ \hfill (39.42) $m^2 = \dfrac{m_{\text{rest}}^2}{1-(v/c)^2} = \dfrac{c^2}{c^2-v^2}m_{\text{rest}}^2$ $m^2c^2 = m^2v^2 + m_{\text{rest}}^2 c^2$
Differentiate using the product rule. The last term is zero because both c and m_{rest} are constants.	$d(m^2c^2) = d(m^2v^2) + d(m_{\text{rest}}^2 c^2)$ $2mc^2\,dm = 2m^2 v\,dv + 2mv^2\,dm + 0$
Cancel $2m$ from each term.	$c^2\,dm = mv\,dv + v^2\,dm$ \hfill (3)
The right sides of Equations (2) and (3) are identical, so we can set these equations equal to each other.	$dK = c^2\,dm$ \hfill (4)
Equation (4) shows that a change in mass dm results in a change in kinetic energy dK. We can integrate to find an expression for the kinetic energy. The lower limit on the integrals comes from the observation that the kinetic energy is zero when the particle is at rest in the laboratory frame, in which case the particle's mass equals its rest mass.	$\displaystyle\int_0^K dK = \int_{m_{\text{rest}}}^m c^2\,dm = c^2 m \Big|_{m_{\text{rest}}}^m$ $K = mc^2 - m_{\text{rest}}c^2$ ✓ \hfill (39.43)

:• COMMENTS

Equation 39.43 shows that the relativistic kinetic energy depends on the change in mass, $m - m_{\text{rest}}$. In Problem 52, you will show that this reduces to the familiar $K = \tfrac{1}{2}m_{\text{rest}}v^2$ in the case of a slow-moving particle.

The Famous Equation $E = mc^2$

Equation 39.43, $K = mc^2 - m_{\text{rest}}c^2$, has important implications for the natures of mass and energy. Because we derived this equation for the special case of a particle, we didn't consider potential energy or any internal energy. However, if the system had been more complicated, we would have found that Equation 39.43 holds for a change in energy of any form. Conceptually, a change in energy of any form is proportional to a change in mass. If a system's mass increases, its energy increases. Just as James Joule concluded that heat is a form of energy because it can increase a system's thermal energy (Section 21-1), we can conclude that mass is a form of energy.

This is a major change in our understanding of mass. So far, we have thought of mass as inertia (Section 5-4). The more mass an object has, the more it resists a change in velocity. Now we can think of mass as a form of energy. This idea is known as the **mass–energy equivalence principle**. Mathematically, we write that the total energy E of a system is

$$E = mc^2 \tag{39.44}$$

MASS–ENERGY EQUIVALENCE PRINCIPLE ⊕ Major Concept

where m is the observed mass of the system, $m = \gamma m_{\text{rest}}$ (Eq. 39.39). This principle allows us to write Equation 39.43 in a more general form by substituting Equation 39.44 and rearranging to find

$$E = mc^2 = m_{\text{rest}}c^2 + K \tag{39.45}$$

Equation 39.45 says that the total energy of a system is equal to its **rest mass energy** ($m_{rest}c^2$) plus its kinetic energy. Even if the system is at rest ($K = 0$), it still has energy associated with its rest mass.

There is another major consequence of Equation 39.44. Our previous most general statement of the conservation of energy principle was that the total energy of the universe must be conserved. Now, the mass–energy equivalence principle leads us to broaden the principle of energy conservation to include mass. In its broadest terms, the conservation of mass and energy says that **the total mass plus energy of the universe is constant**. In practical terms, the total mass and energy of an isolated system must be conserved. So, if an isolated system loses mass, it must gain energy in some other form, such as thermal energy.

CONCEPT EXERCISE 39.4

You have found a cold bowling ball at rest with effectively no thermal energy. If the system consists of only the ball, estimate its total energy

   a. according to classical mechanics and
   b. according to special relativity.

EXAMPLE 39.10 Fusion in the Sun

Sunlight is the ultimate source of energy on the Earth, and the Sun generates the energy for this light through nuclear fusion reactions in its core. For now, we need to know that the net (overall) fusion reaction in the Sun is

$$4H \rightarrow He$$

In this nuclear reaction, four hydrogen nuclei are fused together to make a single helium nucleus. A hydrogen nucleus is only one proton, and a helium nucleus is made up of two protons and two neutrons.

A Find the mass of the four hydrogen nuclei. Assuming the mass of the helium nucleus equals the mass of its four particles, find its mass. Report your answers to five significant figures.

INTERPRET and ANTICIPATE
We must calculate the total mass of four protons and of a helium nucleus (two protons plus two neutrons). Because the neutron is slightly more massive than the proton, we expect the helium nucleus to be slightly more massive than the four protons.

SOLVE

Look up the masses of the particles. Retain six significant figures to avoid rounding errors.	$m_p = 1.67262 \times 10^{-27}\,\text{kg}$ $m_n = 1.67493 \times 10^{-27}\,\text{kg}$
Multiply the proton's mass by 4 to find the mass of four protons.	$4m_p = 4(1.67262 \times 10^{-27}\,\text{kg})$ $4m_p = 6.6905 \times 10^{-27}\,\text{kg}$
Add the mass of two protons to the mass of two neutrons to find the mass of a helium nucleus.	$m_{He} = 2m_p + 2m_n$ $m_{He} = 2(1.67262 \times 10^{-27}\,\text{kg}) + 2(1.67493 \times 10^{-27}\,\text{kg})$ $m_{He} = 6.6951 \times 10^{-27}\,\text{kg}$

Example continues on page 1296 ▶

CHECK and THINK
Our answer for the mass of the helium nucleus is greater than our answer for the mass of four protons. However, experiments show that the mass of a helium nucleus is *less* than the mass of four protons. Helium has less mass than the sum of its parts because some mass is converted to energy in binding the four particles together.

B A helium nucleus actually has a smaller mass than you found in part A; its mass is 6.643×10^{-27} kg. When four hydrogen nuclei are fused into a single helium nucleus, mass is lost. This loss is known as a **mass deficit**. According to the principle of conservation of mass and energy, this mass deficit requires an increase of another form of energy. Find the energy associated with this mass deficit in joules.

INTERPRET and ANTICIPATE
The mass deficit is the difference between the mass of the helium nucleus and the mass of the four protons. In order for mass and energy to be conserved, this mass must be converted to another form of energy. We can find the amount of energy using the famous mass–energy equivalence principle.

SOLVE
To find the mass deficit, subtract the (actual) mass of the helium nucleus from the mass of the four protons.

$$\Delta m = 4m_p - m_{He}$$
$$\Delta m = (6.6905 \times 10^{-27}\,\text{kg}) - (6.643 \times 10^{-27}\,\text{kg})$$
$$\Delta m = 4.75 \times 10^{-29}\,\text{kg}$$

To find the amount of energy equivalent to this mass deficit, multiply by c^2 (Eq. 39.44).

$$\Delta E = \Delta m c^2 \quad (39.44)$$
$$\Delta E = (4.75 \times 10^{-29}\,\text{kg})(3.00 \times 10^8\,\text{m/s})^2$$
$$\Delta E = 4.28 \times 10^{-12}\,\text{J}$$

CHECK and THINK
The mass deficit due to the fusion of four protons into a helium nucleus is equivalent to a small amount of energy released by the Sun in the form of light.

C The Sun emits $P = 3.84 \times 10^{26}$ W. Assume all the energy associated with the mass deficit in part B goes into the Sun's power. How many fusion reactions take place per second in the Sun's core?

INTERPRET and ANTICIPATE
Because each fusion reaction releases very little energy and because the Sun's power is so great, we expect to find a great number of reactions per second.

SOLVE
Divide the Sun's power by the energy generated in each reaction to get the number n of reactions per second.

$$n = \frac{P}{\Delta E} = \frac{3.84 \times 10^{26}\,\text{J/s}}{4.28 \times 10^{-12}\,\text{J/reaction}}$$
$$n = 8.97 \times 10^{37}\,\text{reactions per second}$$

CHECK and THINK
As expected, the fusion reaction rate in the Sun is very high. The tiny mass deficit in each of these reactions supplies energy to generate sunlight, which is then transferred throughout the solar system. This sunlight can be seen even in distant parts of the galaxy, and it contributes to the glow of the Milky Way galaxy seen in distant parts of the Universe.

39-11 General Relativity

Einstein was not satisfied with restricting his theory of relativity to the special case of reference frames moving at constant velocity. In the decade that followed his groundbreaking work on special relativity, he developed a general theory of relativity that incorporated accelerating reference frames. In these last two sections, we take a brief look at general relativity, which includes a new theory of gravity.

Like special relativity, Einstein's general relativity is based on two postulates. The **first postulate of general relativity** is similar to the first postulate of special relativity: The laws of nature are the same in all reference frames, including accelerating ones. Before stating the second postulate, we must review a few things about mass, inertia, and gravity.

EINSTEIN'S FIRST POSTULATE OF GENERAL RELATIVITY
❶ Underlying Principle

The Principle of Equivalence

For most of this book, we have taken for granted that the inertial mass of an object is equivalent to its gravitational mass. Although this may seem obvious, it was a problem for Newton and other scientists, including Einstein.

The mass of either of the two objects in the law of universal gravity, $F_G = GM_{\text{grav}} m_{\text{grav}}/R^2$ (Eq. 7.4), is referred to as the object's **gravitational mass**. This is the property of the objects that creates a gravitational force between them. From Section 5-4, the mass in Newton's second law ($\vec{F}_{\text{tot}} = m_{\text{inert}} \vec{a}$) is the **inertial mass** of an object. Experimental evidence supports the idea that the gravitational mass of any object equals its inertial mass (Section 7-3). But, until Einstein, no one knew why they should be equal, so no one could explain why the gravitational force between two particles should depend on their inertial masses rather than on some other fundamental property.

Einstein came up with an explanation for why the inertial and the gravitational masses are equivalent. He said that experimenting in a gravitational field (such as on the surface of the Earth) is equivalent to experimenting in an accelerating elevator that is located out in space, far from any gravitational field.

Figure 39.20 shows one such pair of experiments. Aaron and Hannah perform the same experiment in separate small rooms, neither of which has a view of the outside universe. Aaron is on the Earth, and Hannah is in an elevator that is far from any gravitational field but is accelerating upward at g. Aaron drops a ball, and, as you have probably done in your own laboratory, he measures the downward acceleration of the ball and finds it is g.

Hannah in the elevator cannot tell that she is out in space. Her feet feel the normal force of the elevator, and it feels exactly like the normal force of the floor when she is standing on the Earth. Like Aaron, she drops a ball. She sees the ball fall to the floor of the elevator with an acceleration g. Because we can view this situation from an inertial frame outside the elevator, we see that in this frame, the ball hovers in place. There is no net force on the ball, so according to Newton's first law, the ball remains at rest. And, in our inertial frame, the floor of the elevator accelerates upward toward the ball. While we can see all of this, Hannah in the elevator cannot, so she cannot tell whether she is on the Earth or in an accelerating elevator. Likewise, Aaron cannot tell whether he is on the Earth or whether his small room is really an accelerating elevator.

In fact, Einstein said no experiment can be done to tell the difference between being in an accelerating frame and being in an equally strong gravitational field. This leads to the **second postulate of general relativity**: A gravitational field is equivalent to an accelerating reference frame without gravity. This second postulate is called the **principle of equivalence**, and it leads to some mind-blowing consequences.

FIGURE 39.20 A. Aaron—an experimenter on the Earth—releases a ball that falls downward with an acceleration g. **B.** Hannah—an experimenter in an elevator accelerating upward at g, far from any gravitational field—releases a ball. She observes the ball moving downward with an acceleration g. Einstein concluded that gravity is equivalent to an accelerating reference frame.

PRINCIPLE OF EQUIVALENCE (EINSTEIN'S SECOND POSTULATE OF GENERAL RELATIVITY)
❶ Underlying Principle

Gravity Is an Illusion

The first consequence has to do with gravity. According to Einstein, gravity is an illusion. An object with mass does not create a gravitational field; rather, it distorts or *curves* space. Humans see space as having three dimensions: (1) left and right, (2) forward and back, and (3) up and down. When Einstein says that an object with mass curves space, he means that the three-dimensional space we are used to thinking about is bent in a fourth spatial dimension.

CASE STUDY Flatlanders

No one—not even Einstein—can visualize four spatial dimensions. But, because visualizing a situation is an important part of understanding it, we'd like to have some picture of curved space in our heads. To create such a picture, we use an analogy with fewer dimensions. Instead of imagining the world through the eyes of three-dimensional beings, we imagine what two-dimensional beings would see. These two-dimensional beings are sometimes referred to as Flatlanders because of a novel by Edwin Abbott called *Flatland*. Flatlanders live on a two-dimensional surface. They describe the kinematics of everything in their universe in terms of just two dimensions: (1) left and right and (2) forward and back. Flatlanders may be able to imagine a third dimension (up and down) and work out the mathematical description of three-dimensional space, but they cannot picture it any more than we can picture four-dimensional space.

Suppose Flatlanders live on the two-dimensional surface of a smooth sphere. You can easily picture them living on such a sphere, but they would have no way to picture the curvature of their two-dimensional universe in three-dimensional space.

Now imagine two Flatlanders start at the equator and walk north on two different lines of longitude (Fig. 39.21). They see the equator as a straight line, and they see their two paths as perpendicular to this line, so they say their paths are parallel to each other. They know parallel lines never cross, so they expect to remain the same distance apart. However, when they reach the North Pole, they find they have arrived at the same position. They conclude that they were drawn away from their parallel straight-line paths by some attractive force that they call *gravity*. As three-dimensional beings, we could explain to the Flatlanders that (1) no force was involved, (2) their paths crossed because they were really walking along great circles (lines of longitude) on the surface of a three-dimensional sphere, and (3) the force they experienced and called "gravity" is an illusion because of their inability to picture three-dimensional space.

Einstein argued that we are like the Flatlanders. Gravity is an illusion; we invented it because, as three-dimensional beings, we cannot picture four spatial dimensions. Gravity is just our explanation for the curved motion of objects, such as the orbital motion of the Moon around the Earth. Although we cannot picture a fourth spatial dimension, it exists, and our three-dimensional universe is curved in this four-dimensional space due to the presence of massive objects such as the Earth. According to general relativity, the Earth does *not* exert a gravitational force on the Moon; instead, the Earth curves space and the shape of the Moon's path is a natural consequence of that curvature.

The idea that gravity is an illusion also explains why the gravitational mass is the same as the inertial mass. Or, said more precisely, if gravity is an illusion, the question is moot. There is no such thing as two different kinds of mass. Objects have only one kind of mass, and that mass curves space.

You might feel uncomfortable with the idea that gravity is an illusion. However, we can still use Newton's law of universal gravity in most practical situations, such as when calculating the orbits of satellites. Gravity may be an illusion, but Newton's description of this illusion works very well. Also, even when we talk about the theory of general relativity, we do not abandon the terms *gravity* and *gravitational*. They are shorthand for the illusion or effect we perceive due to the presence of an object with mass.

FIGURE 39.21 Flatlanders believe their universe is two-dimensional. When they walk on parallel paths, they find that they are drawn together. They conclude that a force must have pulled them together. We can see that their paths simply cross in three-dimensional space.

CURVATURE OF SPACE
⭐ Major Concept

Time Dilation and the Gravitational Doppler Shift

Consider an elevator in free space (Fig. 39.22) accelerating upward at g. On the floor of the elevator is a monochromatic light source emitting light at frequency f_{emt}. A detector on the ceiling of the elevator measures the frequency of the light that enters it. At some instant, the source emits a photon. In the time it takes the photon to cross from the floor to the ceiling, the elevator has moved upward. As we have seen before in both Galilean and special relativity, when the source and detector move toward or away from each other, there is a Doppler shift (Sections 17-9 and 39-7). Further, we saw that relative motion also produces time dilation (Section 39-6). According to the principle of equivalence, the acceleration due to gravity near a massive object is equivalent to an accelerating reference frame in empty space. So, the source on an elevator can be replaced by a source in the gravitational field of a massive object, and then the detector, located far from the object, must still measure a *gravitational* Doppler shift and *gravitational* time dilation. The **gravitational Doppler shift** and **gravitational time dilation** are given by

$$\frac{f_{obs}}{f_{emt}} = \frac{\Delta t_{emt}}{\Delta t_{obs}} = \left(1 - \frac{2GM}{c^2 r}\right)^{1/2} \quad (39.46)$$

GRAVITATIONAL DOPPLER SHIFT AND GRAVITATIONAL TIME DILATION ✪ **Major Concepts**

where the subscript "obs" indicates the *observed* measurement made by the detector and the subscript "emt" indicates the frequency emitted (or the time interval measured) in the source's frame. The mass of the object is M, and the source's distance from the object is r. The detector is assumed to be infinitely far from the object.

The negative sign in Equation 39.46 indicates that the observed frequency is lower than the emitted frequency. Expressed in terms of wavelength, the observed wavelength is longer than the emitted wavelength. So the observed light is redder than the emitted light. We say the light has been *gravitationally red-shifted*. Further, Equation 39.46 indicates that time passes more slowly near the massive object.

It is often convenient to approximate Equation 39.46. If $2GM/c^2 r \ll 1$, then

$$\frac{f_{obs}}{f_{emt}} = \frac{\Delta t_{emt}}{\Delta t_{obs}} \approx 1 - \frac{GM}{c^2 r} \quad (39.47)$$

Further, if the source and the detector are relatively close together and the gravitational field g is nearly uniform, it is convenient to write the Doppler shift as

$$\frac{f_{obs} - f_{emt}}{f_{emt}} = \frac{\Delta f}{f_{emt}} \approx -\frac{gh}{c^2} \quad (39.48)$$

where h is the distance between the source and the detector. (You can derive this equation in Problem 82.)

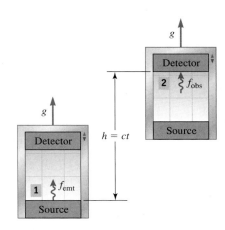

FIGURE 39.22 An elevator in free space accelerates upward at g. **1** A photon is released at $t = 0$. **2** The photon is detected at a later time t.

EXAMPLE 39.11 Harvard's Test of the Gravitational Doppler Shift

In 1960, a test of the gravitational Doppler shift was conducted at Harvard University. A γ–ray photon was emitted from the bottom of a tower of height 22.6 m. The photon was then detected at the top of the tower. The experimental result was $\Delta f/f_{emt} = -(2.57 \pm 0.26) \times 10^{-15}$. Does this agree with the prediction of general relativity?

:• INTERPRET and ANTICIPATE
We must calculate $\Delta f/f_{emt}$ based on the tower's height and the gravitational field of the Earth. Because the tower is on the Earth's surface, g is nearly uniform over the height of the tower: $g = 9.81 \text{ m/s}^2$.

:• SOLVE
Substitute into Equation 39.48.

$$\frac{\Delta f}{f_{emt}} \approx -\frac{gh}{c^2} \quad (39.48)$$

$$\left(\frac{\Delta f}{f_{emt}}\right)_{pred} \approx -\frac{(9.81 \text{ m/s}^2)(22.6 \text{ m})}{(3.00 \times 10^8 \text{ m/s}^2)^2}$$

$$\left(\frac{\Delta f}{f_{emt}}\right)_{pred} \approx -2.46 \times 10^{-15}$$

Example continues on page 1300 ▶

To make the comparison between the predicted value and the measured value, find the minimum and maximum of the measurement using the given experimental error.

$$-2.83 \times 10^{-15} < \left(\frac{\Delta f}{f_{emt}}\right)_{meas} < -2.31 \times 10^{-15}$$

The predicted value falls within the experimental range.

CHECK and THINK

The prediction and the measurement are consistent. This result is one confirmation of the gravitational Doppler shift. Such confirmations are comforting because there are so many counterintuitive consequences of general relativity. In the next section, we look at another test.

39-12 Gravitational Lenses and Black Holes

One of the most mind-blowing consequences of the principle of equivalence is that light is bent or refracted by a massive object, much as a lens refracts light. To see why this happens, consider Aaron in free space in an elevator accelerating upward at g. The elevator has large windows on opposite walls. A light source and a target are stationary outside the elevator, on opposite sides (Fig. 39.23A). The source emits light that hits the target. Hannah is in the frame of the source and target; to her, the light's path is horizontal. To Aaron, however, the light must move downward to hit the target. When the light was emitted, the source was lined up with his shoulders. In the time the light takes to cross the elevator and arrive at the target, the elevator moves upward, and the target ends up aligned with Aaron's feet. So Aaron must see the light fall from the height of his shoulders to the height of his feet as it moves across the elevator. The light's motion is analogous to that of a tennis ball launched horizontally at shoulder height on the Earth: The ball falls to the ground in a parabolic arc (Section 4-5). So the path of the light seen by Aaron is also a parabola (Fig. 39.23B).

According to the principle of equivalence, the accelerating elevator is the same as a gravitational field produced by an object with mass. So the curved path the light takes near a massive object is a natural consequence of the shape of space. No force is exerted on the light.

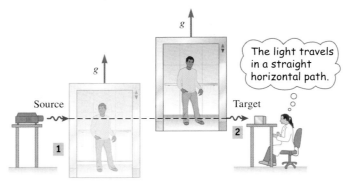

A. Path is straight in Hannah's frame.

B. Path is a parabola in Aaron's frame.

FIGURE 39.23 An elevator in free space accelerates upward at g. The elevator has windows on both sides. **A.** Light travels horizontally from the source to the target, passing through the elevator as seen in the laboratory frame. **B.** But in the moving frame, the light's path is a parabola.

Using an Eclipse to Test General Relativity

The bending of light near a massive object is not predicted by Newton's law of universal gravity. According to Newton's law, only objects that have mass can be affected by gravity. Because light has no mass, it should move in a straight line through a gravitational field. An observation that light is refracted by a massive object would be a confirmation of Einstein's theory of general relativity and a refutation of Newton's theory of gravity.

In 1919, a total solar eclipse was observed in Brazil and Principe in the Gulf of Guinea. During the eclipse, astronomers were able to take pictures of stars whose light passed near the Sun. They found that the positions of the stars appeared shifted by 1.64 arcsec compared to their normal positions. This shift was due to the refraction of starlight by the Sun's gravitational field, a confirmation of Einstein's theory of general relativity.

Gravitational Lenses

Since 1919, refraction of light by a massive object has been observed in many circumstances. The massive object that causes the light to refract is called a **gravitational lens** and is always located between the source and the observer. Rays from the source are refracted by the lens and come together at the observer's position. By tracing the rays back along straight paths, we find the image formed by the lens. If the source is directly behind the lens and the lens is symmetrical, the image is a ring (Fig. 39.24). Einstein described such a ring in 1936. The angular radius θ_r (in radians) of the ring depends on the mass M of the gravitational lens:

$$\theta_r = \sqrt{\frac{4GM}{c^2}\left(\frac{d_{src} - d_{lens}}{d_{src}d_{lens}}\right)} \quad (39.49)$$

where d_{src} and d_{lens} are the distances to the source and to the lens, respectively. By observing the radius of such a ring, astronomers are able to determine the mass of the gravitational lens. Often the gravitational lens is a galaxy, so this allows astronomers to measure the masses of entire galaxies. Such observations indicate the presence of dark matter (case study in Chapter 7).

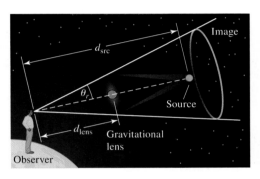

FIGURE 39.24 Any massive object between the source and the observer may refract the source's light and act as a gravitational lens. If the source is directly behind the lens, it is possible for the image to be a ring.

Black Holes

In a star like the Sun, the inward pull of its own gravity is balanced by the thermal pressure that results from nuclear fusion taking place in its core. But all stars eventually run out of the reactants needed for fusion. When that happens, gravity causes the star to collapse. The Sun will form a white dwarf—an object about the size of the Earth. The inward pull of gravity will eventually be balanced by the pressure exerted by a sea of electrons in the white dwarf. For a star that is several times as massive as the Sun, the electron pressure is not enough to keep it from collapsing under its own weight. A very massive star is eventually crunched into a single point known as a *singularity*. Anything that gets too close to a singularity cannot escape. By "too close" we mean inside a spherical region called the **event horizon** (Fig. 39.25). The singularity plus the event horizon are known as a **black hole**.

These dense objects are called *black holes* because anything, even light, that enters the event horizon cannot escape. The radius of the event horizon is called the **Schwarzschild radius** R_{sch} (Fig. 39.25). The Schwarzschild radius depends on the mass of the black hole. The correct way to find an expression for the Schwarzschild radius is to use general relativity to take into account the curvature of space and time dilation at the event horizon.

However, we can also find the correct expression by using classical mechanics. The escape speed from an object of mass M is $v_{esc} = \sqrt{2GM/R}$ (Eq. 8.17). At the event horizon, the escape speed equals the speed of light. We can find the Schwarzschild radius by setting v_{esc} equal to c:

$$c = \sqrt{\frac{2GM}{R_{sch}}}$$

FIGURE 39.25 Light that passes inside the event horizon never escapes. The radius of the event horizon is called the Schwarzschild radius.

So the Schwarzschild radius is given by

$$R_{sch} = \frac{2GM}{c^2} \quad (39.50)$$

Often, black holes are several times the mass of the Sun ($M_\odot$) and have Schwarzschild radii of just a few kilometers, so it is helpful to rewrite Equation 39.50 in the form

$$R_{sch} \approx 3\frac{M}{M_\odot}\text{ km} \quad (39.51)$$

where M is measured in solar mass units and the Schwarzschild radius is in kilometers.

CONCEPT EXERCISE 39.5

The Sun will not form a black hole. But if a black hole were discovered to have as great a mass as the Sun, what would its Schwarzschild radius be in kilometers?

SUMMARY

Underlying Principles

1. **Einstein's postulates of special relativity** All the laws of physics are true in all inertial reference frames. The speed of light c in a vacuum is invariant.
2. **Einstein's postulates of general relativity** The laws of nature are the same in all reference frames, including accelerating ones.

A gravitational field is equivalent to an accelerating reference frame without gravity. This second postulate is called the **principle of equivalence**.

Major Concepts

1. **Lorentz transformations** for position and time measured in two different reference frames hold in special relativity (Table 39.1, page 1277). These transformations are written in terms of the Lorentz (gamma) factor:

$$\gamma = \frac{1}{\sqrt{1 - (v_{rel}/c)^2}} = \frac{1}{\sqrt{1-\beta^2}} \quad (39.17)$$

where

$$\beta \equiv \frac{v_{rel}}{c} \quad (39.18)$$

2. **Length contraction** is described by

$$L = \frac{L'}{\gamma} \quad (39.25)$$

The object's proper length is measured in its own frame and is its longest observable length.

3. **Time dilation** is given by

$$\Delta t = \gamma \Delta t' \quad (39.26)$$

Moving clocks run slowly. The proper time is the time interval between two events that occur at the same position.

4. The **relativistic Doppler effect** is given by

$$\lambda = \sqrt{\frac{1 \mp \beta}{1 \pm \beta}}\lambda' \quad (39.27)$$

and

$$f = \sqrt{\frac{1 \pm \beta}{1 \mp \beta}}f' \quad (39.28)$$

Choose the top signs if the source is moving toward the observer; choose the bottom signs if the source is moving away from the observer.

5. The **perpendicular velocity transformation** is given by

$$v_y = \frac{v'_y}{\gamma\left(1 + \frac{v_{\text{rel}}}{c^2}v'_x\right)} \quad (39.29)$$

with a similar equation for z; this transforms a component of an object's velocity that is perpendicular to the relative motion between reference frames.

6. The **parallel velocity transformation** is given by

$$v_x = \frac{(v'_x + v_{\text{rel}})}{\left(1 + \frac{v_{\text{rel}}}{c^2}v'_x\right)} \quad (39.31)$$

This relationship holds for the component of an object's velocity that is parallel to the relative motion between reference frames.

7. **Mass transformation** is given by

$$m = \gamma m_{\text{rest}} \quad (39.39)$$

where the rest mass m_{rest} must be measured in the object's frame.

8. **Relativistic momentum** for a particle observed with velocity $\vec{v}$ in the laboratory frame is given by

$$\vec{p} = m\vec{v} = \gamma m_{\text{rest}}\vec{v} \quad (39.40)$$

where

$$\gamma = \frac{1}{\sqrt{1 - (v/c)^2}} \quad (39.41)$$

9. According to the **mass–energy equivalence principle**, mass is a form of energy given by

$$E = mc^2 \quad (39.44)$$

10. Gravity is an illusion. An object that has mass does not create a gravitational field. Instead, the object **curves space**.

11. The **gravitational Doppler shift** and **gravitational time dilation** are given by

$$\frac{\Delta t_{\text{emt}}}{\Delta t_{\text{obs}}} = \frac{f_{\text{obs}}}{f_{\text{emt}}} = \left(1 - \frac{2GM}{c^2 r}\right)^{1/2} \quad (39.46)$$

▷ Special Cases

Galilean relativity: The Galilean transformations hold when the relative speed is low (Table 39.1, page 1277). The relativistic transformations must be identical to the Galilean transformations in this special case.

PROBLEMS AND QUESTIONS

A = algebraic   C = conceptual   E = estimation   G = graphical   N = numerical

39-1 It's in the Eye of the Observer

1. **C** A simple pendulum surrounded by pegs is known as a Foucault pendulum (Fig. P39.1). Imagine watching such a pendulum placed at the Earth's North Pole. As the pendulum swings back and forth, it gradually hits each of the pegs. Explain why it hits all the pegs in 24 h. *Hint*: What would happen if the pendulum were in an inertial reference frame?

FIGURE P39.1

39-2 Special Case: Galilean Relativity

2. **N** Desmond and Lilani are traveling on different high-speed trains on tracks that run parallel to each other. Lilani's train is traveling west at 305 km/h, and Desmond's train is traveling east at 275 km/h. **a.** How far does Desmond travel in 30.0 s according to a stationary farmer near the tracks? **b.** How far does Lilani travel in 30.0 s according to a stationary farmer near the tracks? **c.** How far does Desmond travel in Lilani's frame of reference in 30.0 s? **d.** How far does Lilani travel in Desmond's frame of reference in 30.0 s?

Problems 3 and 4 are paired.

3. **N** Carrying a suitcase in an airport, you walk at about 2.0 mph. A fast-moving sidewalk (conveyor belt) moves at about 9.00 km/h and is 0.50 km long. **a.** If you stand still on the moving sidewalk, how long will it take you to cover the 0.50 km? **b.** If you walk next to the conveyer belt, how long will it take you to walk 0.50 km? **c.** If, instead, you walk on the sidewalk at your usual pace, how long will it take you to cover this distance? (Report your answers to the nearest second.)

4. **N** In an airport terminal, there are two fast-moving sidewalks (9.0 km/h); one carries its passengers south, and the other carries its passengers north. Each sidewalk is 0.50 km long. At the instant a woman steps onto the north end of the southbound sidewalk, a man steps onto the south end of the northbound sidewalk. He stands still with respect to the sidewalk, while she walks south at 5.0 km/h. **a.** How long after stepping onto the

sidewalks do they pass each other? (Report your answer to the nearest second.) **b.** How far does each person travel in that time? (Report your answer in kilometers.)

5. **A** A primed frame is moving in the positive x direction relative to the laboratory frame. Use $y = y'$ (Eq. 39.2) to show that both v_y and a_y are invariant under Galilean relativity.

Problems 6 and 7 are paired.

6. **N** Jason is driving north on a highway at 60.0 mph when he sees a car ahead. Kevin is driving that car north at 53.0 mph. **a.** What are the magnitude and direction of Kevin's velocity according to Jason? **b.** How far does Kevin travel in Jason's frame of reference in the time it takes Jason to travel 3.0 mi?

7. **N** Jason is driving north on a highway at 60.0 mph when he sees a car ahead. Kevin is driving that car south at 53.0 mph. **a.** What are the magnitude and direction of Kevin's velocity according to Jason? **b.** How far does Kevin travel in Jason's frame of reference in the time it takes Jason to travel 3.0 mi?

39-3 Postulates of Special Relativity

8. **C** Suppose there are two observers. Observer A is traveling toward a light source at $0.5c$, and observer B is traveling away from the same light source at $0.5c$. The source emits a brief pulse. **a.** What is the speed of the light pulse measured by each observer? **b.** At the instant the pulse is emitted, the two observers are the same distance from the source. Which observer will see the pulse first, as determined by a person in the frame of the source? Explain.

9. **C** One reference frame is moving relative to another, and it is said that both reference frames are inertial. The first postulate of special relativity is that the laws of physics hold in all inertial reference frames. What does it mean to say that a particular reference frame is "inertial"? In other words, what is required for a reference frame to be called "inertial"?

39-4 Lorentz Transformations

10. **N** A particle is at a point $x' = 12.0$ m, $y' = 4.00$ m, $z' = 6.00$ m at time $t' = 4.00 \times 10^{-4}$ s as measured in the primed frame. The particle is at rest in the primed frame. Assume the origins of the primed and unprimed frames were coincident at time $t = 0$. What are the coordinates (x, y, z, t) as measured in the laboratory frame if the particle is moving along the x and x' axes with speed **a.** $v_{rel} = 400.0$ m/s, **b.** $v_{rel} = -400.0$ m/s, and **c.** $v_{rel} = 2.0 \times 10^8$ m/s?

11. **N** In a stationary reference frame S, a green laser located at $x_1 = 1.00$ m is switched on at $t_1 = 5.00$ ns, and 11.0 ns later, a blue laser is switched on at $x_2 = 4.00$ m. A second reference frame S' is moving in the positive x direction so that the origins of the two reference frames coincide at time $t = 0$ as measured in both reference frames. To an observer in S', both lasers are switched on at the same location. **a.** What is the relative speed between the S and S' reference frames? **b.** What is the x coordinate of the two lasers in the S' reference frame when they are switched on? **c.** What is the time interval between the two lasers being switched on in the S' reference frame?

Problems 12 and 13 are paired.

12. **C** The primed frame has speed $0.75c$ in the x direction according to an observer in the laboratory frame. An observer in the primed frame sees a balloon hovering at $(x', y') = (6.0$ m, 12.0 m). What can you say about the position of the balloon as seen in the laboratory frame?

13. **N** The primed frame has speed $0.75c$ in the x direction according to an observer in the laboratory frame. At $t = t' = 0$, the origins of the two frames coincide. At $t' = 16.0$ s, an observer in the primed frame sees a balloon hovering at $(x', y') = (6.0$ m, 12.0 m). **a.** What can you say about the position of the balloon as seen in the laboratory frame? **b.** How much time has elapsed in the laboratory frame when the observer in the primed frame makes this observation?

14. **A** Starting with the Lorentz transformations relating x and x', confirm **a.** Equation 39.19, $t = \gamma(t' + v_{rel}x'/c^2)$, and **b.** Equation 39.20, $t' = \gamma(t - v_{rel}x/c^2)$.

Problems 15 and 16 are paired.

15. An electron is moving with a speed of 6.75×10^7 m/s as measured by an observer in a laboratory.
    a. N What is the Lorentz factor for the electron moving in the laboratory frame of reference?
    b. C Which frame of reference (the electron's or the laboratory's) would we label the x' frame and which the x frame?

16. **N** An electron is moving with a speed of 6.75×10^7 m/s as measured by an observer in a laboratory. The electron is at rest in the primed frame of reference and at its origin. If we assume the origins of the reference frames of the electron and the laboratory are coincident at $t = t' = 0$ and the electron is moving along the $+x$ axis, what is the position of the electron in the primed frame of reference when the position in the unprimed frame is **a.** 1.0 mm, **b.** 1.0 cm, **c.** 1.0 m, and **d.** 1.0 km?

39-5 Length Contraction

17. **N** A spaceship is moving at a relativistic speed. The length of the spaceship is measured to be exactly half its proper length. Determine the speed of the spaceship (in units of the speed of light, c) relative to the laboratory frame.

18. **C** In a factory, a rod lies on a conveyor belt so that its long axis is parallel to the direction of motion. A factory worker is watching the rod on the belt when the belt stops. Does the worker see the rod's length grow or shrink? Assume the worker can observe the change in length due to the velocity of the rod.

19. **N** CASE STUDY A space traveler aboard a spacecraft measures its length and finds it is 15.0 m long. An observer in another inertial reference frame sees that the spacecraft is 7.50 m long. What is the relative speed of the spacecraft in this frame?

20. **N** What is the speed with which a ruler 12.0 in. (30.48 cm) long, as measured in its own reference frame, must be moving if its apparent length is 20.0 cm?

21. **N** A meterstick in the primed frame makes a 45.0° angle with the x' axis. The meterstick is at rest in the primed frame. As seen by an observer in the laboratory frame, the primed frame's relative speed in the x direction is $0.975c$. What is the length of the meterstick as observed in the laboratory frame?

22. **N** An astronaut on the Moon observes a UFO flying by with a velocity of $0.75c$ as shown in Figure P39.22. **a.** If the long side of the UFO, when at rest, has a length of 265 m, what is the length of this side as perceived by the astronaut if it flies by as shown in case 1? **b.** What is the perceived length of the long side if it flies by the astronaut as shown in case 2? Explain.

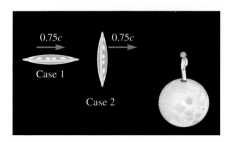

FIGURE P39.22

23. **N** As Brent's spacecraft overtakes Jasmin's identical spacecraft at high speed, Brent notifies Jasmin that his spacecraft is 35.0 m long; Jasmin's spacecraft appears to be 31.0 m long according to his observations. **a.** What is the length of Jasmin's spacecraft in her frame of reference? **b.** What is the length of Brent's spacecraft as observed by Jasmin? **c.** What is the relative speed between their spacecrafts?

Problems 24 and 25 are paired.

24. **N** A starship is 1025 ly from the Earth when measured in the rest frame of the Earth. The ship travels at a speed of $0.80c$ on its way back to the Earth. What is the distance traveled as measured by the crew of the starship?

39-6 Time Dilation

25. **N** A starship is 1025 ly from the Earth when measured in the rest frame of the Earth. The ship travels at a speed of $0.80c$ on its way back to the Earth. **a.** How much time will it take for the ship to return to the Earth as measured by the Earth observer? **b.** How much time will it take for the ship to return to the Earth as measured by the crew of the ship?

26. **N** Two atomic clocks are carefully synchronized on the Earth (considered an inertial frame). One remains in a laboratory at the airport. The other is carried on an airplane traveling at twice the speed of sound. When the plane returns, according to the laboratory clock, 3.50 h have passed. According to the clock that traveled on the plane, how much time has elapsed? What is the difference between the two clocks? Ignore the effects of acceleration on the clocks. *Hint*: Work to 10 significant figures throughout solving the problem, until expressing the time difference, and use the binominal theorem (Appendix A) to find the Lorentz factor.

27. **N** What is the speed with which a clock must be moving with respect to an observer at rest if its second hand sweeps through 45.0 s during each 60.0 s measured by the observer's clock?

28. **N** A neutron outside of a nucleus is an unstable particle. It decays in roughly 10.0 min. How fast would a neutron need to travel in order to be stable for 11.0 min as seen by an observer in the laboratory frame? How far would the neutron travel in that time according to the laboratory observer?

29. **A** The primed observer measures a time interval of $\Delta t'$, and the laboratory observer measures $\Delta t = 2\Delta t'$. What is the primed frame's relative speed?

30. **N** An astronaut takes her grandfather clock with her on her spaceship. Later, her ship is traveling by the Earth at a speed of $0.65c$. According to the astronaut, the pendulum on the clock has a frequency of 1.00 Hz. What is the frequency of the pendulum according to an observer on the Earth?

31. **N** Atomic clocks operate by measuring the resonances of certain atoms, most commonly cesium-133 (which is used to define the second). An atomic clock is held stationary on the ground, while a second identical atomic clock moves with a constant speed of 350.0 m/s for 2.00 h as measured by the stationary clock. What is the time difference between the two clocks after this 2.00-h interval?

32. **N** It takes 1.0 h to bake a casserole in a frame moving at a speed of 2.0×10^8 m/s. How long is the baking time as measured by a cook in the laboratory frame?

33. **N** Scientists are conducting timed experiments on two spacecraft and on the Earth. The experimenters on the Earth observe spacecraft A moving with speed $0.400c$ away from the Earth and toward spacecraft B, which is moving with speed $0.710c$ toward the Earth. The experiment requires a 2.00-h incubation period as measured by the crew on spacecraft A. **a.** How much time has elapsed on spacecraft B when the spacecraft A clock shows an elapsed time of 2.00 h? Assume that the relative speed between the two spacecraft is $0.864c$ (you will learn why in section 39-8). **b.** How much time has elapsed for the experimenters on the Earth when the spacecraft A clock shows an elapsed time of 2.00 h?

39-7 The Relativistic Doppler Effect

34. **C** A spiral galaxy is seen edge-on. From this perspective, half the disk moves toward the observer and half moves away. Describe the relative color of the two sides as seen by the observer.

35. **N** A supernova explosion sends out a spherical shell of hydrogen gas. The shell expands at the rate of 1000.00 km/s. The gas is transparent, so we can see the front of the shell moving toward us and the back moving away. In the frame of the gas, the hydrogen gives off radio waves with a frequency of 1420.04 MHz. What are the lowest and highest frequencies observed on the Earth?

36. **A** Beginning with Equation 39.27, derive Equation 39.28:

$$f = \sqrt{\frac{1 \pm \beta}{1 \mp \beta}} f' = \sqrt{\frac{c \pm v_{rel}}{c \mp v_{rel}}} f'$$

37. **N** A source of hydrogen gas gives off Hα (Fig. 36.19A, page 1166). The line is emitted with a wavelength of 656 nm and observed with a wavelength of 652 nm. What is the constant speed of the source relative to the observer? Is the source moving toward or away from the observer?

39-8 Velocity Transformation

38. **A** Show that Equation 39.33,

$$v'_x = \frac{(v_x - v_{rel})}{\left(1 - \frac{v_{rel}}{c^2} v_x\right)}$$

is the x velocity transformation from the laboratory frame to the primed frame.

39. **N** As measured in a laboratory reference frame, a linear accelerator ejects a proton with a speed of $0.780c$. Moments later, a muon is ejected at a speed of $0.920c$ as measured in the laboratory reference frame. What is the speed of the proton in a reference frame where the velocity of the muon is zero?

40. **A** Show that Equation 39.29,

$$v_y = \frac{v'_y}{\gamma \left(1 + \frac{v_{rel}}{c^2} v'_x\right)}$$

is the transformation of a velocity component perpendicular to $\vec{v}_{rel}$ and Equation 39.30,

$$v'_y = \frac{v_y}{\gamma \left(1 - \frac{v_{rel}}{c^2} v_x\right)}$$

is the inverse transformation from the laboratory frame to the primed frame.

41. **N** The components of the velocity of an electron measured in the primed frame are $v'_x = 5.00 \times 10^7$ m/s, $v'_y = 4.00 \times 10^7$ m/s, and $v'_z = 3.00 \times 10^7$ m/s. Determine the magnitude of the velocity in the laboratory frame. Assume the relative velocity between the primed and the laboratory frames is $2.00 \times 10^7 \hat{\imath}$ m/s.

Problems 42, 43, and 44 are grouped.

42. **N** In a laboratory, two particles are fired sequentially. Particle 1 has speed $0.80c$ and is traveling to the east. A moment later, particle 2 is launched with speed $0.70c$ and is also traveling to the east. What is the speed of particle 2 as measured in the reference frame of particle 1?

43. **N** In a laboratory, two particles are fired sequentially. Particle 1 has speed $0.80c$ and is traveling to the east. A moment later, particle 2 is launched with speed $0.70c$ but is traveling to the west. What is the speed of particle 2 as measured in the reference frame of particle 1?

44. **N** In a laboratory, two particles are fired sequentially. Particle 1 has speed $0.80c$ and is traveling to the east. A moment later, particle 2 is launched with speed $0.70c$. **a.** What is the speed of particle 1 as measured in the reference frame of particle 2, if particle 2 is also traveling to the east? **b.** What is the speed of particle 1 as measured in the reference frame of particle 2, if particle 2 is traveling to the west?

45. **N** **CASE STUDY** In a fictional spaceport, a fast-moving sidewalk moves at about $0.90c$. The sidewalk is 0.50 ly long. A man gets on the sidewalk, and a woman standing near the sidewalk watches him. In each case, find his speed in her frame in terms of c. **a.** He is at rest with respect to the sidewalk. **b.** He walks at 2.2 m/s with respect to the sidewalk. **c.** He uses fictional roller blades so that his speed relative to the sidewalk, and in the same direction, is $0.50c$.

39-9 Mass and Momentum Transformation

46. **N** What is the momentum of a proton ($m_p = 1.67 \times 10^{-27}$ kg) that has a speed of **a.** $0.0500c$, **b.** $0.550c$, and **c.** $0.950c$?

Problems 47 and 48 are paired.

47. **N** The rest mass of a baseball is 0.145 kg. Find the mass observed if the ball is moving at **a.** 90.0 mph (40.2 m/s)—a fast pitch, **b.** 250.0 km/s—the speed of the Sun around the center of the Milky Way galaxy, and **c.** $0.850c$ with respect to the laboratory.

48. **N** The rest mass of a baseball is 0.145 kg. Find the momentum observed if the ball is moving at **a.** 90.0 mph (40.2 m/s)—a fast pitch, **b.** 250.0 km/s—the speed of the Sun around the center of the Milky Way galaxy, and **c.** $0.850c$ with respect to the laboratory.

49. **N** Consider a proton moving with speed $0.00100c$. How fast does an electron have to move to have the same momentum as this proton?

Problems 50 and 51 are paired.

50. **N** In transmission electron microscopy (TEM), electrons are accelerated to have kinetic energies of hundreds of thousands of electron volts (1 eV = 1.602×10^{-19} J). Suppose an electron has a kinetic energy of 300.0 keV. **a.** What is the mass of the electron according to an observer in the lab? **b.** What is the momentum of the electron according to an observer in the lab? Answer using three significant figures.

39-10 Newton's Second Law and Energy

51. **N** In transmission electron microscopy (TEM), electrons are accelerated to have kinetic energies of hundreds of thousands of electron volts (1 eV = 1.602×10^{-19} J). Suppose an electron has a kinetic energy of 300.0 keV. **a.** What is the rest mass energy of the electron? **b.** What is the total energy of the electron? **c.** What is the speed of the electron? Answer using three significant figures.

52. **A** Show that $K = mc^2 - m_{rest}c^2$ (Eq. 39.43) reduces to the familiar $K = \frac{1}{2}m_{rest}v^2$ in the case of a slow-moving particle.

53. **N** The *top quark*, an elementary particle and a fundamental constituent of matter, was discovered in 1995 at Fermilab in Chicago. It has a mass of 172.9 GeV/c^2. What is the mass of the top quark in kilograms?

54. **N** Consider a proton moving with speed $0.2000c$. How fast would an electron have to move to have the same kinetic energy as this proton?

55. **N** A 15.00-kg satellite orbits the Earth at 8314 m/s. Calculate the satellite's kinetic energy using special relativity. How does your answer compare to what you find if you use classical physics?

56. **E** Assume the Sun's luminosity (the same as its power) is constant over its entire 10-billion-year life. Estimate the amount of mass it must consume in nuclear fusion to put out the required amount of energy. Give your answer in kilograms and in solar masses.

57. **N** Consider an electron moving with speed $0.980c$. **a.** What is the rest mass energy of this electron? **b.** What is the total energy of this electron? **c.** What is the kinetic energy of this electron?

58. **N** VY Canis Majoris, the largest known star in the Milky Way galaxy, is 4.30×10^5 times more luminous than the Sun. If this star has an average power output of 1.65×10^{32} W, how many kilograms of its mass are converted to energy per second?

39-11 General Relativity

Problems 59 and 60 are paired.

59. **N** Hα ($f_{emt} = 4.57 \times 10^{14}$ Hz) is emitted from the surface of the Sun. What is the frequency of the light when it is observed far from the Sun?

60. **N** Hα ($f_{emt} = 4.57 \times 10^{14}$ Hz) is emitted from the surface of a neutron star. Its radius is 6.96×10^4 m and its mass is 2 solar masses. What is the frequency of the light observed far from the neutron star?

61. **N** According to an observer on the Earth, a solar flare lasts for 3.25 min on the surface of the Sun. What is the lifetime of the flare as observed from the Sun?

62. **N** Imagine a star with a mass that is 1000 times that of the Sun and a radius that is 1/100 the radius of the Sun. **a.** If the star emits red light with a frequency of 3.2×10^{14} Hz, what is the frequency observed far away from the star? **b.** A dark spot forms on the surface of the star and lasts for 13.5 days as observed from the star. How long does the spot last according to observers far from the star?

39-12 Gravitational Lenses and Black Holes

63. **N** The Milky Way galaxy has a supermassive black hole at its center. Its mass is 3.6 million solar masses. What is its Schwarzschild radius? Give your answer in kilometers and in Earth radii.

64. **N** When a 10-solar-mass star runs out of reactants in its core, it sheds its outer layers in a supernova explosion, and the core collapses to form a black hole of about 4.0 solar masses. What is its Schwarzschild radius?

65. **N** A particular Einstein ring has an angular diameter of 2.1 arcsec. The distance to the source (a quasar) is 2.3×10^{26} m, and the distance to the gravitational lens (a galaxy) is 3.3×10^{25} m. What is the mass of the lens? Give your answer in kilograms and in solar masses. Compare your answer to the Andromeda galaxy, whose mass is about 710 billion solar masses.

General Problems

66. **N** At a distance of 6.00×10^2 ly, Kepler-22b is the nearest habitable planet found orbiting another star as of 2012. What would the speed of a spacecraft traveling to this planet have to

be so that the distance to the star is 1.50×10^2 ly in the reference frame of the people on the spacecraft?

67. **N** An event has a duration of 1.00 ms in the laboratory frame. How much time passes during this event for an observer moving past the lab at a speed of $0.733c$?

68. **N** The energy released in the first peacetime nuclear weapon test at Bikini Atoll in 1946 was approximately 6.3×10^{16} J. What mass of plutonium-239 would have the same rest mass energy as the amount of energy released in this explosion?

69. **N** While her spacecraft is traveling from the Earth to the Alpha Centauri star system at a speed of $0.910c$, Lea exercises by doing midair cartwheels, each taking 4.20 s to complete. According to an observer on the Earth, how long does each of Lea's cartwheels take to complete?

70. **C** Consider Example 39.1 (page 1270). Imagine one more scenario in which the ball hangs from the glass ceiling by a string. Aaron would see the string make an angle with respect to the vertical. What would Aaron conclude about Newton's first law?

71. **N** Joe and Moe are twins. In the laboratory frame at location S_1 (2.00 km, 0.200 km, 0.150 km), Joe shoots a picture for a duration of $t = 12.0$ μs. For the same duration as measured in the laboratory frame, at location S_2 (1.00 km, -0.200 km, 0.300 km), Moe also shoots a picture. Both Joe and Moe begin taking their pictures at $t = 0$ in the laboratory frame. Determine the duration of each event as measured by an observer in a frame moving at a speed of 2.00×10^8 m/s along the x axis in the positive x direction. Assume that at $t = t' = 0$, the origins of the two frames coincide.

72. **A** Start with the Lorentz transformation for x and x', and show that the Lorentz factor is given by Equation 39.17:

$$\gamma = \frac{1}{\sqrt{1 - (v_{rel}/c)^2}} = \frac{1}{\sqrt{1 - \beta^2}}$$

Hint: Imagine that at $t = t' = 0$, the origins of both coordinate systems overlap (Fig. 39.7A) and at that instant the observer in the primed frame fires a flashbulb, sending a photon to the right parallel to the x and x' axes.

73. **N** On its way back to the Earth, an interplanetary spaceship with a proper length of 180.0 m is observed by flight controllers to pass a stationary beacon on the Moon in 0.550 μs. What is the speed of the spacecraft in the Earth's reference frame?

74. **N** What is the speed of a proton that has a momentum 4.00 times greater than its classical momentum?

75. **N** A light in a spaceship moving at a relativistic speed appears red (wavelength 675 nm) to the passengers in the ship. But it appears yellow (wavelength 575 nm) to an observer on the Earth. What is the speed (in terms of the speed of light c) of the spaceship? Is it coming toward the Earth or moving away from the Earth?

76. **N** The Sun radiates energy with a power of about 3.8×10^{26} W. **a.** What is the rest mass energy of the Sun? **b.** How much mass is "lost" per second by the Sun if the power radiated is due to the conversion of mass to energy? **c.** Given your answer to part (b), what percentage of the Sun's mass is radiated each second?

77. **N** Far into the future, trains on the Earth may travel at speeds approaching the speed of light. Imagine such a train traveling at $0.875c$ and having a proper length of 145 m. The first of several tunnels along the train's path is 80.0 m long. Is the train ever completely in the tunnel according to an observer standing on a trackside platform near the tunnel? If so, by how much? If not, how much longer than the tunnel is the train?

78. **N** In December 2012, researchers announced the discovery of ultramassive black holes, with masses up to 40 billion times the mass of the Sun (seen as the bright spot at the center of the galaxy near the center of Fig. P39.78). **a.** What is the Schwarzschild radius of a black hole that has a mass 40 billion times that of the Sun? **b.** Suppose this black hole is 1.3 billion ly from the Earth. What is the angular radius of a galaxy that is 1.7 billion ly behind it, as viewed from the Earth?

FIGURE P39.78

Problems 79 and 80 are paired.

79. **A** A rectangular tank filled with a liquid of density ρ' is in a moving frame with its edges parallel to the coordinate axes. What is the density of the liquid as measured in the laboratory frame? Assume the relative speed along the x axis between the two frames is v_{rel}.

80. **A** A rectangular tank filled with a liquid of density ρ' is in a moving frame with its edges parallel to the coordinate axes. If the liquid in the tank is evaporating, show that in both the laboratory and the primed frames, the rate at which mass is being lost from the tank by evaporation is the same.

81. **N** How much work is required to increase the speed of a helium nucleus ($m_{He} = 6.64 \times 10^{-27}$ kg) from $0.750c$ to $0.990c$?

82. **A** Show that the gravitational Doppler shift formula is approximately given by Equation 39.48:

$$\frac{f_{obs} - f_{emt}}{f_{emt}} = \frac{\Delta f}{f_{emt}} \approx -\frac{gh}{c^2}$$

Hint: Start by considering the scenario illustrated in Figure 39.22.

83. **N** A heavy nucleus breaks up into two pieces while in motion. The first piece, of rest mass $m_1 = 1.45$ MeV/c^2, is observed to have momentum $\vec{p}_1 = -10.0 \,\hat{i}$ MeV/c, and the second piece, of rest mass $m_2 = 0.850$ MeV/c^2, is observed to have momentum $\vec{p}_2 = 5.75 \,\hat{j}$ MeV/c by an observer in the laboratory frame. **a.** What is the speed of the first piece, as viewed by an observer in the lab? **b.** What is the speed of the second piece, as viewed by an observer in the lab?

84. **N** In radioactive decay, a parent nucleus decays into one or more daughter nuclei. During one such reaction, the parent nucleus with rest mass 8.99×10^{-27} kg decays into two daughter nuclei with rest masses m_1 and m_2 moving with velocities $v_1 = -0.915c\,\hat{i}$ and $v_2 = 0.785c\,\hat{i}$. What are the rest masses of the two daughter nuclei?

85. **N** A train moving with constant velocity $0.775c$ approaches a station where a large mirror is hung next to the tracks. A playful boy on the train aims a laser pointer at the mirror when the train is 10.0 km from the station, as measured by observers at the station, and the beam is reflected by the mirror back toward the train. **a.** According to the boy, how much time elapses from when he switches on the laser pointer until the beam gets back to him? **b.** According to observers at the station, how much time elapses from when the boy switches on the laser pointer until the beam gets back to him?

40 The Origin of Quantum Physics

Key Questions

How is quantum physics different from classical physics?

What is the wave-particle duality?

❶ Underlying Principles

1. Planck's **quantum theory** and model of black-body radiation
2. **Wave-particle duality** of light and matter

✪ Major Concepts

1. Black body
2. Photon
3. Compton shift
4. De Broglie wavelength

40.1 Another Modern Idea 1309

40.2 Black-Body Radiation and the Ultraviolet Catastrophe 1309

40.3 The Photoelectric Effect 1316

40.4 The Compton Effect 1320

40-5 Wave-Particle Duality 1326

40-6 The Wave Properties of Matter 1328

By the late 1800s, many physicists thought that all the major questions of nature had been answered by Newton's laws of mechanics and Maxwell's equations for electricity and magnetism. Even controversial topics, such as the nature of light, had been worked out. They were satisfied that light is a wave, as predicted by Maxwell's equations and shown experimentally by Young and Arago. They believed that only a few minor questions remained to be answered and that experimentalists would have nothing else to do but make more accurate measurements, such as measuring the speed of light to six significant figures.

But as we saw in Chapter 39, they couldn't have been more wrong. Michelson and Morley didn't believe that their experiment would lead to groundbreaking results. They thought that they were just measuring the speed of the Earth with respect to the ether. They certainly didn't expect that their failure to find the ether would be explained by Einstein's theory that observations of space and time depend on the motion of the observer.

Those who thought that the major questions of nature had been answered were in for another big surprise in the early 1900s. In trying to understand something as seemingly mundane as the glow given off by a warm object, such as a wood-burning stove, another radical departure from classical physics was discovered. Today, we call this new departure the theory of *quantum physics* (or *quantum mechanics*).

40-1 Another Modern Idea

The basic idea behind quantum physics is that energy is not necessarily continuous; instead, energy comes in small packets called *quanta*. We have already seen a similar idea in our study of electromagnetism. Electric charge is quantized in that the elementary charge e is carried by either a proton ($+e$) or an electron ($-e$). An object that is positively charged carries an excess number n of protons, and its charge is $q = ne$, where n is an integer. Because it is impossible to find an object with a fractional number of excess electrons or protons, an object cannot have a charge of $\pm 1.75e$, for example. Since the elementary charge is so small ($e = 1.6 \times 10^{-19}$ C), the charge of a macroscopic object (at even a few nanocoulombs) appears continuous: The addition or subtraction of a single elementary charge has very little effect compared to the total charge.

The same is true for quantized energy. When we consider the energy of a baseball in flight, for example, to the batter the energy seems not quantized but continuous. Quantum physics is most applicable on the submicroscopic scale of atoms. So perhaps it is not surprising that quantum physics plays an important role in many of today's electronic devices, two of which are discussed in this chapter's case study.

It has been discovered that protons are not fundamental particles; instead they are made up of quarks. Quarks have a fractional charge (Chapter 43).

CASE STUDY Digital Cameras and Electron Microscopes

In this chapter, we'll return to the controversy over whether light should be modeled as a wave or a particle, and we'll find that it should be modeled as both. In addition, matter (such as electrons, protons, and even people) can be modeled both as waves and as particles. It might seem that how we choose to model a physical phenomenon is only important to theorists, who need words to describe the phenomenon. However, models lead to practical devices as well. For example, your digital camera is based on modeling light as a particle. By contrast, the high-powered electron microscope used in fields such as medicine, biology, and crystallography is based on modeling the electron as a wave.

CONCEPT EXERCISE 40.1

a. An atom is missing one electron. What is its net charge?
b. A pith ball's charge is 9 nC. How many electrons is it missing?
c. Use these two examples to explain why charge seems continuous on the macroscopic level of a pith ball but quantized on the level of a single atom.

40-2 Black-Body Radiation and the Ultraviolet Catastrophe

BLACK BODY ✪ Major Concept

The first steps leading to quantum mechanics were taken by the German physicist Max Karl Ernst Ludwig Planck (1858–1947). Like scientists before him, Planck was interested in the thermal radiation given off by an idealized object known as a **black body**. A black body is both a perfect emitter and a perfect absorber of radiation. So, for example, when a black body is cooler than its surroundings, by comparison it appears dark or black because it absorbs radiation of every color (or frequency).

Although no real object is a black body, many objects are well modeled as such. For example, consider a small hole in a cavity of an opaque vessel, such as an old-fashioned keyhole in a closet door. (The keyhole, not the closet, is the black body.) Any light that passes into the hole is reflected off the interior surfaces many times, but very little, if any, light may escape the hole (Fig. 40.1). So the hole has effectively absorbed all the radiation impinging on it and is well modeled as a black body.

A black body is also a perfect emitter, giving off electromagnetic waves at all frequencies. The power P emitted as thermal radiation (another name for black-body

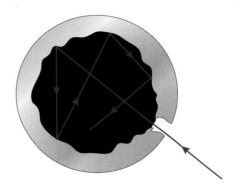

FIGURE 40.1 The hole in a cavity is well modeled as a black body.

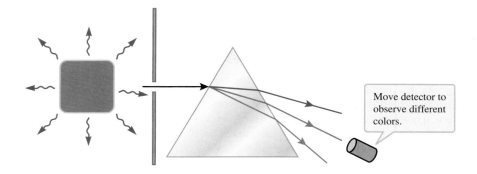

FIGURE 40.2 A prism is used to disperse the light from a hot lump of coal. By moving the detector, the intensity at each color may be observed.

Recall σ is Stefan–Boltzmann's constant and is found on the inside back cover.

radiation) depends on the temperature T and surface area A of the object: $P = \sigma \varepsilon A T^4$ (the Stefan–Boltzmann equation with $P = Q/\Delta t$, Eq. 21.34). The emissivity ε in this equation is a measure of a substance's ability to emit or absorb radiation. So a black body is an ideal object whose emissivity is one: $\varepsilon = 1$. A real object's emissivity is always between 0 and 1, but some objects and materials, such as coal ($\varepsilon = 0.95$), come very close to the ideal. For example, charcoal is well modeled as a black body, and Figure 21.30 (page 642) shows that when charcoal is cooler than its surroundings, it appears black. When it is hotter than its surroundings, however, it appears to glow brightly. For the rest of this section, we'll model such real objects as black bodies. (Often, real objects that are nearly black bodies are referred to as "gray bodies." However, in this book we'll use the term *black body* with the understanding that no real object is an ideal black body.) When an object is in thermal equilibrium with its surroundings, the thermal power it absorbs must equal the thermal power it emits.

When a black body gives off radiation, the emitted spectrum contains a complete rainbow of colors. (The white light source in Fig. 36.16 is well modeled as a black body, but the compact fluorescent bulb in Fig. 36.17 is not.) The intensity (power per unit area) emitted at each color depends on the black body's temperature. Suppose you wished to measure the intensity of radiation emitted by a black body, such as a lump of hot coal, as a function of wavelength. The hot coal emits visible light. So you might place the lump of coal behind a small slit, so that you get a nice beam of light. Then you could use a device such as a prism or a diffraction grating to disperse the light into its color components (Fig. 40.2). A detector could measure the intensity of the radiation it receives through the prism. By moving the detector to different positions, you could measure the intensity of light as a function of color or wavelength. To be more precise, the detector is placed at one position at a time and has a finite width, so it measures the intensity of light at λ (due to the position of the detector) with range of wavelengths $d\lambda$ (due to the detector's width). So the detector measures the intensity per unit wavelength band. There are several names for intensity per wavelength; in this book we use the term *spectral intensity* and the symbol I_λ.

A **black-body curve** is a graph of emitted spectral intensity I_λ versus wavelength. Figure 40.3 shows several black-body curves for temperatures ranging from 6000 K to 12,000 K. Notice that the warmest black body (which gives off the most power) has the shortest peak wavelength. The peak wavelength $\lambda_{\max}$ of the black-body curve is given by Wien's law:

$$\lambda_{\max} T = 2.898 \times 10^{-3}\,\text{m} \cdot \text{K} = 2.898 \times 10^{6}\,\text{nm} \cdot \text{K} \tag{40.1}$$

where T is the black body's temperature in kelvins. Figure 40.3 also shows a warm black-body curve with peak intensity in the green part of the visible spectrum. Because the curve is fairly flat, this black body looks nearly white—something like the Sun. Compare this curve for a warm black body to that shown for a hot black body, for which the most intense color is indigo. If you were to look at such a black body, it would look bluish, like the hot star Rigel in the constellation Orion (Fig. 40.4).

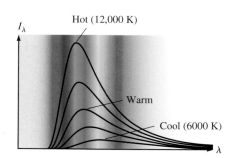

FIGURE 40.3 Graph of spectral intensity versus wavelength for five black bodies of different temperatures. All the curves have the same shape, differing only in scale.

The total intensity over all wavelengths (the area under the curve) depends on the temperature of the black body, so that a warmer black body emits a greater total intensity. (Often "total" is dropped from the term *total intensity* when the meaning can be understood from the context.) The total intensity is found by integrating the spectral intensity over all wavelengths:

$$I = \int_0^\infty I_\lambda d\lambda \tag{40.2}$$

Notice that the symbol for total intensity I has no subscript. The total intensity depends on the absolute temperature of the object. We find the total intensity of a thermal radiator by dividing the power it emits (Eq. 21.34) by its surface area:

$$I(T) = \varepsilon \sigma T^4 \tag{40.3}$$

As usual, for a black body the emissivity $\varepsilon = 1$.

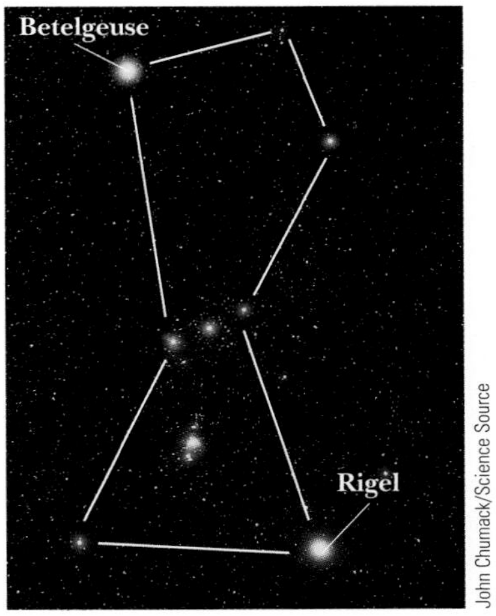

Modeling Black-Body Curves

Before Planck, scientists noticed that all black-body curves (Fig. 40.3) had the same shape (with different scalings). But no one came up with a model for matter and radiation that would reproduce the observed curves, although two important attempts were made. One attempt was made by the German physicist Wilhelm Wien (1864–1928), who noticed that black-body curves look much like the Maxwell–Boltzmann distribution for the speeds of particles in gas (Fig. 20.8, page 589). Wien reasoned that the temperature of a black body is related to the motion of the particles (such as the atoms or molecules) inside the black body. Think of it this way: An object that is in thermal equilibrium with its surroundings absorbs as much thermal energy as it emits. The thermal energy the object absorbs causes an increase in its particles' energy (atoms and molecules). Of course, the particles in a solid object oscillate around their equilibrium positions, and when their energy increases, their oscillations become more vigorous. (We measure this as an increase in the object's temperature.) The particles are made up of electrons and protons. These charged particles (typically electrons) are accelerated by the oscillations, and accelerating charged particles give off radiation. This emitted radiation may keep the black body in thermal equilibrium if the power emitted equals the power absorbed. The important point here is that the motion of the particles in a black body is responsible for the observed black-body radiation, so it made some sense for Wien to think that the Maxwell–Boltzmann distribution is related to the black-body curves. Using the Maxwell–Boltzmann distribution

$$f(v) = 4\pi \left(\frac{m}{2\pi k_B T}\right)^{3/2} v^2 e^{-(mv^2/2k_B T)} \tag{20.19}$$

as a guide, Wien derived a black-body intensity formula that was a good fit for the data at short wavelengths (Fig. 40.5). (At this point, you may wish to review Section 20-4 on the Maxwell–Boltzmann distribution function.)

FIGURE 40.4 The constellation Orion has two very bright stars—Rigel and Betelgeuse. Rigel's peak wavelength is actually in the ultraviolet; it looks blue to us because our eyes are not sensitive to UV.

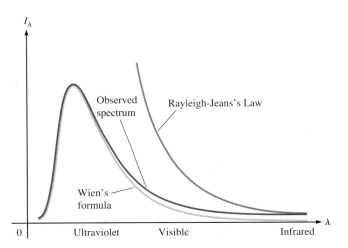

FIGURE 40.5 Neither Wien's formula nor Rayleigh–Jeans's law fits the black-body curve observed for an object at a given temperature. Wien's formula is best at short wavelengths, and Rayleigh–Jeans's law at long wavelengths.

Black-body curves may be observed by heating up an oven with a very small hole in one of its sides. The hole is similar to the one shown in Figure 40.1, but now the cavity (an oven) is hotter than its surroundings. The oven may be heated to different temperatures, and light emitted from the hole is detected using a prism (Fig. 40.2) to construct a black-body curve. Perhaps thinking about such an oven inspired the British scientists Lord Rayleigh (1842–1919) and Sir James Jeans (1877–1946) to come up with some ideas for black-body radiation. Instead of thinking about the atoms that make up the black body, as Wien did, Rayleigh focused on the electromagnetic waves in the cavity. He reasoned that the radiation inside the oven reflects from the walls and sets up standing waves. Like a wave on a string, each standing wave has a fundamental mode and an infinite number of higher harmonic modes. The spectral intensity is measured per waveband $d\lambda$. So Rayleigh found an expression for the number of modes per volume with wavelengths between λ and $\lambda + d\lambda$. According to classical mechanics, each mode contributes an average energy of $k_B T$. From the number density of modes per waveband and the average energy per mode, the energy density per waveband in the cavity can be found. And from this average energy density, Rayleigh–Jeans's law for the spectral intensity as a function of wavelength

$$I_\lambda = \frac{2\pi c}{\lambda^4} k_B T \quad (40.4)$$

is found. Rayleigh–Jeans's law is shown in Figure 40.5.

Take a close look at Figure 40.5, and you will see that Rayleigh–Jeans's formula misses one of the most important features of the actual black-body curves—the turnover. A black-body curve has a peak, but Rayleigh–Jeans's curve continues to climb at short wavelengths. To see what this means, imagine that you want to cook over a campfire. A campfire's radiation peaks somewhere in the infrared or visible red part of the spectrum. According to Rayleigh–Jeans's formula, though, even the most modest campfire would give off high-intensity radiation in the ultraviolet, X-ray, and gamma-ray range. You would suffer much more than a bad sunburn from such radiation. Because everyone knew that black-body radiation did not follow this intensity distribution, Rayleigh–Jeans's formula was said to result in an **ultraviolet catastrophe**.

Planck's Solution

In 1900, the German physicist Max Planck solved the ultraviolet catastrophe empirically by adjusting Wien's formula to fit the black-body curves at both long and short wavelengths. **Planck's formula** for the spectral intensity I_λ of a black body is:

$$I_\lambda(\lambda, T) = \frac{2\pi h c^2}{\lambda^5 (e^{hc/\lambda k_B T} - 1)} \quad (40.5)$$

Planck's formula depends on the temperature T of the black body. For any single curve in Figure 40.3, the temperature is constant. Each curve in Figure 40.3 is a graph of I_λ versus wavelength λ and is described by Equation 40.5 for a particular T. There are three physical constants in Planck's formula: the speed of light c, Boltzmann's constant k_B, and Planck's constant h with the value

$$h = 6.626 \times 10^{-34} \, \text{J} \cdot \text{s} = 4.136 \times 10^{-15} \, \text{eV} \cdot \text{s}$$

At first, Planck's formula was merely an empirical fit to the data. He knew that the formula described the observed black-body curves, but he didn't have a physical model or theory to explain why the formula worked. He wanted to start from some fundamental principle and derive Equation 40.5.

So Planck—like Wien—started with the kinetic theory of gases. He found that he could derive his formula if he (1) modeled the particles in a black body as oscillators and (2) postulated that the energy of each oscillator is **quantized**, existing in discrete bundles instead of being continuous. In this postulate, the smallest possible amount of energy $E_{\min}$ is called a **quantum of energy** and is proportional to the frequency of the oscillator:

$$E_{\min} = hf \quad (40.6)$$

If a molecule in the black body oscillates at frequency f, then its energy is an integer multiple of E_{min}:

$$E = nE_{min} = nhf \qquad n = 1, 2, 3 \ldots \qquad (40.7)$$

The idea that this energy is quantized is called **Planck's quantum theory,** and it marks a radical departure from the classical idea of energy. According to the classical model, an oscillator can have any energy. According to the quantum model, by contrast, an oscillator's energy comes in discrete bundles. However, no one realized that Planck had made such a radical statement until Einstein used the idea to explain the photoelectric effect (see Section 40-3).

To illustrate Planck's quantum theory, consider an analogy with two types of radios. The old radio in Figure 40.6A is tuned by turning a knob. In theory, the knob can be set to any position, and so the radio can be tuned to any frequency in its range. This is analogous to the classical model of an oscillator. According to classical mechanics, an oscillator can have any energy, just as the old radio can be tuned to any frequency. The radio in Figure 40.6B is a more contemporary digital radio. It is tuned by pressing a button. Each time the button is pressed the frequency is adjusted by discrete amount. So the frequency of a digital radio can only be set to certain discrete frequencies. This is analogous to Planck's quantum theory for an oscillator. According to this theory, an oscillator's energy can only take on certain discrete values.

PLANCK'S QUANTUM THEORY

❶ **Underlying Principle**

A.

B.

FIGURE 40.6 A. An analog radio can be tuned to any frequency. **B.** A digital radio can only be tuned to discrete frequencies.

CONCEPT EXERCISE 40.2

Planck's idea that energy is quantized is a radical departure from classical physics. You may find the radio analogy (Fig. 40.6) useful, but like all analogies this one is not perfect. Describe the ways in which this analogy fails and the ways in which it succeeds.

EXAMPLE 40.1 How hot is that flame?

We normally associate the color blue with cold objects and the color red with hot ones, but this association probably arises from our thinking of water as blue and cool and fire as red and hot. In reality, a blue black body is hotter than a red one. Take a close look at the flame in Figure 40.7, and you will see the flame is blue near the source and red on the far end. Model the flame as a black body, and estimate the temperature at each end. In the check and think step, consider that a candle's flame has a temperature of about 1000°C.

FIGURE 40.7

:• **INTERPRET and ANTICIPATE**
We expect to get a higher temperature for the blue part of the flame.

:• **SOLVE**
From the photos and Figure 34.11 (page 1101), we estimate the wavelengths based on the color of each flame. (It is okay to keep an extra significant figure for this step.)

| Blue: $\lambda_{blue} = 475$ nm |
| Red: $\lambda_{red} = 675$ nm |

Assume that the observed color is near the peak of the black-body spectrum for each flame; then the temperature can be estimated from Wien's law. Our estimates are only good to two significant figures.

$$\lambda_{max} T = 2.898 \times 10^6 \, \text{nm} \cdot \text{K} \tag{40.1}$$

$$T = \frac{2.898 \times 10^6 \, \text{nm} \cdot \text{K}}{\lambda_{max}}$$

Blue: $T_{blue} = \dfrac{2.898 \times 10^6 \, \text{nm} \cdot \text{K}}{475 \, \text{nm}} \approx 6.1 \times 10^3 \, \text{K}$

Red: $T_{red} = \dfrac{2.898 \times 10^6 \, \text{nm} \cdot \text{K}}{675 \, \text{nm}} \approx 4.3 \times 10^3 \, \text{K}$

:• **CHECK and THINK**
We found that the blue part is hotter than the red part. This result is satisfying because the blue part is closer to the source, and we expect the flame to be hotter near the source. However, there must be something wrong because the temperatures we found (about 5000°C) are considerably greater than the 1000°C we were expecting for a candle's flame. The problem is that we assumed the flame could be modeled as a black body, but a candle's flame is more complicated than a simple black body. The color (and temperature) variation depends on the chemical reactions that occur in various parts of the flame. In Section 42-3, we see how the structure of atoms affects the radiation we observe; not all radiation is well-modeled as black body radiation.

EXAMPLE 40.2 Cosmic Background Radiation

As we learned in Section 34-5, radiation was generated early in the history of the Universe. This radiation may be observed today. Our observations show that this radiation is well modeled as a black body with a temperature of 2.725 K. Sketch a graph of spectral intensity as a function of wavelength for this cosmic background radiation.

:• **INTERPRET and ANTICIPATE**
This problem gives us a chance to use Planck's formula $I_\lambda(\lambda, T) = \dfrac{2\pi hc^2}{\lambda^5 (e^{hc/\lambda k_B T} - 1)}$ (Eq. 40.5).

It is always helpful to plot complicated functions, as we are doing here.

:• **SOLVE**
Substitute all the constants into Planck's formula, as well as the given temperature. A spreadsheet program is helpful. A few values have been provided so that you can check your own results.

λ (m)	I_λ (W/m³)
1.00×10^{-4}	3.47×10^{-19}
2.00×10^{-4}	3.57×10^{-9}
3.00×10^{-4}	3.24×10^{-6}
4.00×10^{-4}	6.38×10^{-5}
5.00×10^{-4}	2.97×10^{-4}

Plotting spectral intensity on the vertical axis in milliWatts per meter cubed and wavelength on the horizontal axis in millimeters (Fig. 40.8) produces a curve that looks very much like the black-body curves in Figure 40.3.

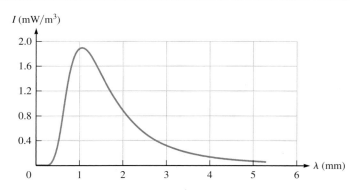

FIGURE 40.8

CHECK and THINK
To check our work, let's use Wien's law (Eq. 40.1) to find the peak wavelength for a black body with a temperature of 2.725 K.

$$\lambda_{max} T = 2.898 \times 10^{-3} \, \text{m} \cdot \text{K} \quad (40.1)$$

$$\lambda_{max} = \frac{2.898 \times 10^{-3} \, \text{m} \cdot \text{K}}{T} = \frac{2.898 \times 10^{-3} \, \text{m} \cdot \text{K}}{2.725 \, \text{K}}$$

$$\lambda_{max} = 1.06 \times 10^{-3} \, \text{m} = 1.06 \, \text{mm}$$

The peak wavelength we just found is consistent with our graph, which you can see peaks at just a little more than 1 mm. Further, we learned in Section 34-5 that the cosmic background radiation is observed in the microwave part of the electromagnetic spectrum. Compare Figure 40.8 to Table 34.2, and you will see that our entire graph falls in the microwave part of the spectrum. Finally, we note that cosmologists measure the background radiation, and then fit the spectrum to Planck's function to arrive at the corresponding black-body temperature of 2.725 K. (This is the reverse of what we did in this example.) Often, this temperature is referred as *the temperature of the Universe*, which may be a little misleading because not everything in the Universe is at the same temperature.

EXAMPLE 40.3 The Ultraviolet Catastrophe

Find the total intensity of a black body as predicted by Rayleigh–Jeans's law (Eq. 40.4).

INTERPRET and ANTICIPATE
We know that Rayleigh–Jeans's law predicts at short wavelengths a higher spectral intensity than is observed (Fig. 40.5). So we expect that the total intensity will be greater than what is observed.

SOLVE
The total intensity comes from integrating the spectral intensity over all wavelengths (Eq. 40.2).

$$I = \int_0^\infty I_\lambda \, d\lambda \quad (40.2)$$

Substitute Rayleigh–Jeans's formula (Eq. 40.4).

$$I = \int_0^\infty \left(\frac{2\pi c}{\lambda^4} k_B T \right) d\lambda$$

Pull out the constants, integrate, and substitute limits.

$$I = 2\pi c k_B T \int_0^\infty \frac{d\lambda}{\lambda^4} = -\frac{2\pi c k_B T}{3} \frac{1}{\lambda^3} \Big|_0^\infty$$

$$I = \frac{2\pi c k_B T}{3} \left(\frac{-1}{\infty} + \frac{1}{0} \right) \to \infty$$

CHECK and THINK
We just found that the total intensity predicted by Rayleigh–Jeans's formula is infinite, which, as expected, is greater than the total intensity observed in a black body. Had we integrated Planck's law instead, we would have found that the total intensity is given by $I(T) = \sigma T^4$ where $\sigma = \frac{2\pi^5 k_B^4}{15 h^3 c^2}$, a finite value consistent with Equation 40.3, when $\varepsilon = 1$.

EXAMPLE 40.4 — A Star's Radius

Have you ever wondered how astronomers measure the size of distant stars? They are too far away to resolve into a disk, so their radius is inferred from their luminosity and temperature. The star's temperature is easy to estimate from its spectrum, which looks like a black-body spectrum, but with some particular dark lines due to photon absorption in the star's atmosphere. These lines are used to estimate the temperature. The star's luminosity is another name for its total emitted power. Luminosity is harder to estimate than temperature because it comes from measuring the star's brightness and estimating its distance. Consider a star that has the same temperature as the Sun but is 1.0×10^4 times brighter than the Sun. Model the star and the Sun as black bodies, and find the star's radius in terms of the Sun's radius.

INTERPRET and ANTICIPATE
The luminosity or power emitted by a star is easily found from $P = \sigma \varepsilon A T^4$ (Eq. 21.34). By comparing the luminosity of the star to that of the Sun, we can find the ratio of their radii.

SOLVE

We are modeling the star and the Sun as black bodies, so we set the emissivity ε to 1.	$P = \sigma A T^4$
Both the star and the Sun are spheres. Substitute the surface area of a sphere into the luminosity expression.	$P = 4\pi R^2 \sigma T^4$
Write an expression for the ratio of the star's power to that of the Sun. We have used the usual symbol $\odot$ to stand for the Sun, and the symbol $*$ to stand for the star.	$\dfrac{P_*}{P_\odot} = \dfrac{4\pi R_*^2 \sigma T_*^4}{4\pi R_\odot^2 \sigma T_\odot^4}$ $\dfrac{P_*}{P_\odot} = \left(\dfrac{R_*}{R_\odot}\right)^2 \left(\dfrac{T_*}{T_\odot}\right)^4$
Use the fact that the star has the same temperature as the Sun to write an expression for the ratio of their radii.	$\dfrac{P_*}{P_\odot} = \left(\dfrac{R_*}{R_\odot}\right)^2$ $\dfrac{R_*}{R_\odot} = \sqrt{\dfrac{P_*}{P_\odot}}$
Substitute values.	$\dfrac{R_*}{R_\odot} = \sqrt{1.0 \times 10^4} = 1.0 \times 10^2$ $R_* = 1.0 \times 10^2 R_\odot$

CHECK and THINK
This very bright star is 100 times larger than the Sun. Such a large star is called a supergiant. Supergiants are "dying stars," which means they have run out of the fuel in their core and are quickly fusing other material before their final demise.

40-3 The Photoelectric Effect

Einstein's other 1905 paper was about Brownian motion.

For Einstein, 1905 was an *annus mirabilis*—a wonderful year—because in that year he published four major papers. Two of these were on special relativity (Sections 39-3 through 39-10), and another was on the quantization of light energy. This paper revived the age-old debate over whether light is best modeled as a wave or as a particle. In the early 1900s, it was generally agreed that the debate had been settled. Maxwell's equations predicted that light is a wave. Interference experiments (Young's double slit) and diffraction experiments (Arago's bright spot in an object's shadow) confirmed the wave model. However, Einstein found that light can also be modeled as a particle known as a *photon*. In this section, we'll explore how Einstein reasoned that light can also be modeled as a particle, and we'll describe the experimental confirmation of that model.

In his paper *Über einen die Erzeugung und Verwandlung des Lichtes betreffenden heuristischen Gesichtspunkt* [*On a Heuristic Viewpoint Concerning the Production and Transformation of Light*], Einstein reasoned that

1. according to Planck's quantum theory, a vibrating molecule's energy is quantized;
2. when such a molecule gives off light, it loses energy equal to the light's energy;
3. the energy the molecule loses must be quantized: $\Delta E = nhf$; so
4. the light's energy must be quantized.

As an example, suppose that the vibrating molecule initially had 10 quanta of energy $E_i = 10hf$. Then the molecule gives off light and loses energy so that it has seven quanta of energy $E_f = 7hf$. The light's energy must be equal to the energy lost by the molecule, so the light has three quanta of energy $3hf$.

This argument has broader implications for light. Because all light comes from the motion of charged particles, light's energy must depend on the energy lost by those particles. If the particle's energy is quantized, then the light's energy is quantized (discrete), and it is convenient to think of light as little bundles of energy called **photons**. A photon's energy depends on the frequency of light:

$$E = hf \qquad (40.8)$$

When a photon is absorbed by an atom (or a molecule), the atom's energy increases by $E = hf$ and the photon ceases to exist.

A test to see if light can be modeled as photons is made possible by a process called the **photoelectric effect**. When light shines on a metallic surface, electrons are sometimes released. You can think of it this way: An orbiting electron is bound to the nucleus of an atom like objects on the Earth are bound to its surface by gravity. If the light delivers enough energy to the electron, the electron may escape the pull of the nucleus. By conducting an experiment that involves the photoelectric effect, we can determine whether the light delivers energy in continuous amounts, the way a wave would, or whether it delivers energy in discrete amounts, as photons.

Figure 40.9 shows the basic photoelectric effect experiment. The circuit is made up of a capacitor, a variable DC power supply, and a couple of meters. The capacitor is charged so that the upper plate is negative and the lower plate is positive. The capacitor is fully charged when the potential across its plates equals the power supply's terminal voltage, and there is no current in the circuit. The capacitor's plates are enclosed in a vacuum tube with opaque walls.

A small transparent hole in the tube allows light from a monochromatic light source to strike only the lower (positive) plate, freeing some of its electrons. The free electrons are repelled by the upper (negative) plate and attracted to the lower (positive) plate, so most of the electrons quickly fall back to the lower plate. However, a few free electrons with high kinetic energy K make it to the negative plate. When an electron crosses from the lower plate to the upper plate, it loses kinetic energy, and the system (electric field plus electron) gains potential energy. The kinetic energy lost by the electron and gained as the system's potential energy is $|\Delta K| = |\Delta U| = |q\Delta V|$ (Eq. 26.7).

Electrons that make it across the gap in the capacitor will then travel through the circuit's wires and through the ammeter on their way to the lower plate. Their passage through the ammeter means that a current will be detected. The number of electrons that can make it through the gap to the negative plate depends on the potential difference across the capacitor. If that potential difference is low, then many electrons can get through, but if the potential difference is high, then very few can make it. Because the potential difference across the capacitor equals the power supply's voltage, the current through the ammeter will depend on that voltage.

The key to testing the photon model is to adjust the power supply's voltage so that the current just barely stops. Starting with a low voltage, we slowly turn the voltage

Similar experiments were conducted before Einstein's theoretical work.

PHOTON ⭐ **Major Concept**

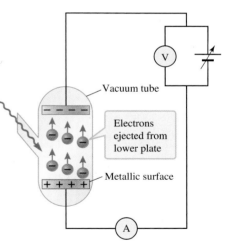

FIGURE 40.9 An experimental setup for testing the photoelectric effect.

knob up until the ammeter reads zero current. This voltage is called the *stopping potential* V_0. When the capacitor's potential difference equals the stopping potential, the free electrons with the most kinetic energy K_{max} stop just short of crossing from the lower plate to the upper plate. These electrons are ejected from the lower plate with a kinetic energy equal to the potential energy gained by the system:

$$K_{max} = eV_0 \tag{40.9}$$

Now let's think about a single electron bound to the nucleus of an atom on the surface of the lower plate. Such a bound electron must gain a certain minimum amount of energy called the **work function** W_0 in order to be freed. If the light delivers an amount of energy exactly equal to the work function, then the free electron will have no kinetic energy and will quickly fall back to the lower plate. If the light delivers an amount of energy E_{light} equal to the work function W_0 plus the kinetic energy K_{max} required to get from the lower plate almost to the upper plate, then this electron is one of the few that almost makes the trip across the gap:

$$E_{light} = K_{max} + W_0 \tag{40.10}$$

The work function W_0 is the minimum energy required to free an electron. But most electrons are more tightly bound and require more energy. After they are freed, these tightly bound electrons don't have enough kinetic energy to make it across the gap. So Equation 40.10 holds only for the mostly loosely bound electrons in the lower plate's surface.

Now we come to the measurement that tests Einstein's photon theory against the wave theory. The kinetic energy of the most energetic electrons is calculated from the stopping potential using $K_{max} = eV_0$ (Eq. 40.9), which is related to the energy delivered by the monochromatic light as:

$$eV_0 = K_{max} = E_{light} - W_0 \tag{40.11}$$

The light source can be tuned to adjust the frequency and the intensity independently. Although adjusting f or I may change the delivered energy E_{light}, the work function W_0 depends only on the type of substance that the plate is made of. The wave model and the photon model of light differ in predicting what happens as the light intensity and frequency are varied. In the wave model, the electrons crossing the gap (as measured by the current) depend on the intensity, but not on the frequency of the light. The photon model makes the opposite prediction: The electrons crossing the gap depend on the frequency, not the intensity, of the light. Table 40.1 compares the predictions made by the two models.

The American physicist Robert Andrews Millikan (1868–1953) was convinced that Einstein's photon theory was wrong. He believed that the particle-or-wave

TABLE 40.1 Results of photoelectric effect experiments as predicted by the wave and photon models of light.

Wave Model	Photon Model
K_{max} is proportional to I but not affected by f.	**K_{max} is proportional to f but not affected by I.**
The energy delivered by an electromagnetic wave is proportional to the wave's intensity, but **not** its frequency (Section 34-6). As the light intensity I is increased, E_{light} and therefore K_{max} will increase, but increasing f will have no effect.	Light's energy depends on its frequency, **not** on its intensity: $E = hf$ (Eq. 40.8). Assuming the frequency is above the threshold ($f > W_0/h$), increasing the light's frequency increases the kinetic energy of the electrons. However, increasing the light's intensity increases the number of electrons released but not the kinetic energy of each electron.
At low light intensity, energy is delivered slowly, so it takes some time before a single electron gains sufficient energy to cross the gap.	**Light intensity does not affect the time for electrons to arrive at the upper plate.**
Hence, there will be a time delay before electrons arrive at the upper plate.	There will in effect be no time delay. As long as a single photon has enough energy to liberate a single electron with enough kinetic energy to cross the capacitor gap, then it will do so, as it does when the intensity is high.
At very low frequency, current will be observed through the ammeter in Figure 40.9.	**At very low frequency, even if there is no potential difference across the capacitor, there will be no current through the ammeter.**
The frequency of the light does not affect the ejection of electrons in the wave model.	If the light energy is lower than the work function $E_{light} < W_0$, then no electrons will be ejected. This condition corresponds to $f < W_0/h$.

debate had been settled and that light clearly should be modeled as a wave. He spent nearly a decade building and conducting the photoelectric experiment. His results (Fig. 40.10) were consistent with the predictions made in the column on the right in Table 40.1, supporting Einstein's photon model. Despite these results, initially Millikan wasn't convinced. As late as 1927, he put out a textbook that only mentioned Einstein's theories as part of his biographical notes (and that included the ether). We might feel some sympathy for Millikan because it is difficult to accept a new theory. So it may help us to know that other experiments also confirmed the photon model. The next section describes another such historic experiment.

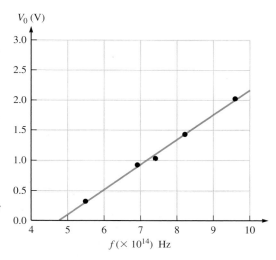

FIGURE 40.10 Stopping potential V_0 (see Fig. 41.5) as a function of frequency based on Millikan's 1916 results (*A Direct Photoelectric Determination of Planck's "h," Physical Review* 7, pp. 355–388). The vertical axis is proportional to K_{max} ($K_{max} = eV_0$). So this linear fit shows that K_{max} is proportional to f as predicted by the photon model (Table 40.1).

CONCEPT EXERCISE 40.3

Millikan used the data in Figure 40.10 to measure Planck's constant. Use the data to estimate h to two significant figures and compare your answer to the accepted value $h = 6.6 \times 10^{-34}$ J·s.

EXAMPLE 40.5 The Energy Range of Visible Photons

We are able to see light in the wavelength range from about 400 nm to about 700 nm. What is the corresponding range in the energy of photons that we can detect with our eyes? Report your answer to one significant figure in joules and in electron-volts.

:• **INTERPRET and ANTICIPATE**
An electromagnetic wave's frequency can be found from its wavelength. The photon's energy is directly proportional to the wave's frequency.

:• **SOLVE**
Find the frequencies using Equation 34.20. (It is okay to keep an extra significant figure in this step.)

$$\lambda f = c \quad (34.20)$$

$$f = c/\lambda$$

Red: $f_{red} = \dfrac{3.00 \times 10^8 \text{ m/s}}{700 \times 10^{-9} \text{ m}} = 4.3 \times 10^{14}$ Hz

Violet: $f_{violet} = \dfrac{3.00 \times 10^8 \text{ m/s}}{400 \times 10^{-9} \text{ m}} = 7.5 \times 10^{14}$ Hz

Find the corresponding energies from the frequencies.

$$E = hf \quad (40.8)$$

Red: $E_{red} = (6.626 \times 10^{-34} \text{ J·s})(4.3 \times 10^{14} \text{ Hz}) = \boxed{3 \times 10^{-19} \text{ J}}$

$E_{red} \approx \boxed{2 \text{ eV}}$

Violet: $E_{violet} = (6.626 \times 10^{-34} \text{ J·s})(7.5 \times 10^{14} \text{ Hz}) = \boxed{5 \times 10^{-19} \text{ J}}$

$E_{violet} \approx \boxed{3 \text{ eV}}$

:• **CHECK and THINK**
So we are able to detect photons with energies between approximately 2 and 3 eV.

FIGURE 40.11 A charge-coupled device (CCD) based on modeling light as photons. A CCD creates a digital image by collecting electrons that are excited into a higher energy state by the photons that strike the detector.

CASE STUDY Digital Cameras

As difficult as it may be to accept that light is modeled as particles, the design of today's digital cameras is based on the photon model of light. Some digital cameras use a semiconductor detector known as a charge-coupled device (CCD), such as the one shown in Figure 40.11. A CCD is made of a two-dimensional array of picture elements known as *pixels*. (Your camera's CCD may have around a million pixels.) When a photon strikes a pixel, an electron is excited into the semiconductor's conduction band. (Electrons in the conduction band are "free" to move throughout the conductor, much as conduction electrons are free to move throughout a conductor. See Chapter 28.) These electrons are collected and processed to form an image. The number of electrons collected from each pixel is proportional to the number of photons striking it.

Digital cameras are convenient and a lot of fun because you can see the image just moments after taking a picture. These cameras also play a very important role in contemporary astronomy. Early astronomers had to sketch what they saw, so the human eye was the original astronomical detector. But the eye can only detect about 1 out of 100 photons; we say that the human eye has a quantum efficiency of 1%. In the mid-1800s, astronomers began to take photographs using photographic plates, which are about equally as good at detecting photons as the human eye. But since the late 1900s, CCD cameras have been commonly used in astronomy. Not only do CCDs have almost 100% quantum efficiency, they also offer other advantages:

1. They are sensitive over a wide range of photon energies, from the infrared to X-rays.
2. They have a wide dynamic range, meaning that they can differentiate between very faint and very bright objects.
3. They have a linear response, meaning that if the number of photons doubles, the signal doubles.

CCDs are so important in astronomy today that it is probably fair to say that all contemporary astronomical images you have seen were taken with a CCD. So not only was the photon model of light revolutionary, but CCDs designed on the basis of that model have revolutionized our view of the universe.

40-4 The Compton Effect

One of the most important experiments confirming Einstein's photon model was carried out in 1923 by the American physicist Arthur Holly Compton (1892–1962). Four years later, Compton shared the Nobel Prize with the British physicist Charles Wilson (1869–1959) for this experiment, which discovered what is now known as the **Compton effect**. Compton was investigating the scattering of X-rays from elements such as carbon. He found two surprising results: (1) The scattered X-rays have a longer wavelength (lower frequency) than the original X-rays, and (2) the wavelength (or frequency) of the scattered X-rays depends on their scattering angle. These results are surprising because according to classical physics, scattering does not change the frequency of electromagnetic radiation.

Compton's work confirmed the photon model because he could explain his two experimental results if he treated the scattering event as an elastic collision between an incoming X-ray photon carrying a discrete amount of momentum and energy, and a free electron initially at rest. The electrons may be scattered at high speeds, so Compton needed the results from special relativity to analyze their motion after the collision. Provided the X-rays were treated as particles (photons), Compton could apply conservation of energy and momentum, as for any other elastic collision, to derive an expression for the change in the X-ray's wavelength.

DERIVATION The Relationship between Relativistic Energy and Momentum

Compton needed a useful relationship between total relativistic energy and momentum:
$$E^2 = p^2c^2 + m_{rest}^2 c^4 \qquad (40.12)$$
We will show that this relationship is true.

We can express any particle's momentum in terms of its energy. Start by writing the famous Equation 39.44 in terms of the rest mass and the Lorentz factor γ. (Use Equation 39.39, $m = \gamma m_{rest}$.)	$E = mc^2 \qquad (39.44)$ $E = \gamma m_{rest} c^2$
We need to use two algebraic tricks. First, square both sides.	$E^2 = \gamma^2 m_{rest}^2 c^4$
Second, add $\gamma^2 m_{rest}^2 c^2 v^2$ and subtract $\gamma^2 m_{rest}^2 c^2 v^2$ on the right side.	$E^2 = \gamma^2 m_{rest}^2 c^4 + \gamma^2 m_{rest}^2 c^2 v^2 - \gamma^2 m_{rest}^2 c^2 v^2$ $E^2 = \gamma^2 m_{rest}^2 c^2 (c^2 + v^2 - v^2)$ $E^2 = \gamma^2 m_{rest}^2 c^2 v^2 + \gamma^2 m_{rest}^2 c^2 (c^2 - v^2)$
Use $\gamma = \frac{1}{\sqrt{1-(v/c)^2}}$ (Eq. 39.41) to rewrite the second term on the right so that the Lorentz factor cancels out.	$E^2 = \gamma^2 m_{rest}^2 c^2 v^2 + \gamma^2 m_{rest}^2 c^4 [1 - (v/c)^2]$ $E^2 = \gamma^2 m_{rest}^2 c^2 v^2 + \gamma^2 m_{rest}^2 c^4 \left(\frac{1}{\gamma^2}\right)$ $E^2 = \gamma^2 m_{rest}^2 v^2 c^2 + m_{rest}^2 c^4 \qquad (1)$
Use the magnitude of relativistic momentum, $p = \gamma m_{rest} v$ (Eq. 39.40) to rewrite the first term on the right in Equation (1).	$\gamma^2 m_{rest}^2 v^2 c^2 = (\gamma m_{rest} v)^2 c^2 = p^2 c^2$ $E^2 = p^2 c^2 + m_{rest}^2 c^4 \checkmark \qquad (40.12)$

COMMENTS
Equation 40.12 is another expression for momentum and energy that holds under special relativity for any particle. Compton used it to derive an expression for the momentum of a photon.

Starting with Equation 40.12, we can find a simple expression for the momentum of a photon by noting that its rest mass is zero.
$$E_{photon}^2 = p_{photon}^2 c^2 + 0 = p^2 c^2$$
$$E_{photon} = p_{photon} c \qquad (40.13)$$

Solving for momentum and substituting $E_{photon} = hf$ (Eq. 40.8) for the energy of a photon results in
$$p_{photon} = \frac{E_{photon}}{c} = \frac{hf}{c}$$

Since $f = c/\lambda$ (Eq. 34.20),
$$p_{photon} = \frac{h}{\lambda} \qquad (40.14)$$

According to Equation 40.14, a photon's momentum depends only on its wavelength (or frequency). With this equation in hand, we are ready to examine the Compton effect.

1322   CHAPTER 40   The Origin of Quantum Physics

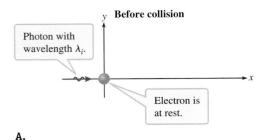

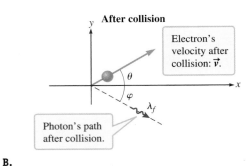

Figure 40.12 illustrates the event that produces the Compton effect. A photon of wavelength λ_i collides with an electron that is initially at rest (Fig. 40.12A). After the collision, the electron moves off at angle θ with respect to the x axis, while the scattered photon moves off at angle φ with respect to the x axis, with wavelength $\lambda_f \geq \lambda_i$ (Fig. 40.12B).

We seek to understand both of Compton's results. (1) Why does the photon's wavelength increase (and frequency decrease)? (2) Why does the scattered photon's wavelength (and frequency) depend on the scattering angle? As stated earlier, we find the answers by modeling light as a photon (carrying a discrete amount of momentum and energy) and then applying the principles of conservation of energy and momentum. The result will be an expression for the photon's change in wavelength $\Delta \lambda = \lambda_f - \lambda_i$, known as the *Compton shift*.

Our procedure resembles the analysis of two-dimensional elastic collisions in Section 11-6, but using special relativity to calculate the electron's energy and momentum. Because the photon has no mass, we must use $p = h/\lambda$ (Eq. 40.14) for its momentum.

FIGURE 40.12 The Compton effect. **A.** Before the collision, the electron is at rest and the photon approaches with wavelength λ_i along the x axis. **B.** After the collision, both the photon and the electron are in motion and the photon's wavelength has changed to λ_f.

DERIVATION   The Compton Shift

We will show that a photon scattering from an electron (Fig. 40.12) has a change in wavelength known as the *Compton shift*, given by:

$$\Delta\lambda = (\lambda_f - \lambda_i) = \lambda_C (1 - \cos\varphi) \qquad (40.15)$$

where λ_C is the **Compton wavelength** of a free electron:

$$\lambda_C \equiv \frac{h}{m_e c} = 2.426 \times 10^{-12}\,\text{m} = 2.426\,\text{pm} \qquad (40.16)$$

and m_e is the rest mass of an electron.

COMPTON SHIFT
⭐ Major Concept

∴ CONSERVATION of ENERGY
Begin with a bar chart (Fig. 40.13) for the two particles (photon and electron). The photon has no rest mass energy, so we need only one bar to represent its energy. The two bars for the electron represent its rest mass energy and kinetic energy.

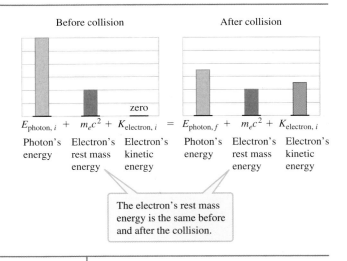

FIGURE 40.13

The electron's rest mass energy is the same before and after the collision, so the conservation of energy condition involves the photon's energy E_{photon} and the electron's kinetic energy K_{electron}. We assume the electron is initially at rest ($K_{\text{electron},i} = 0$).	$E_{\text{photon},i} + K_{\text{electron},i} = E_{\text{photon},f} + K_{\text{electron},f}$ $E_{\text{photon},i} = E_{\text{photon},f} + K_{\text{electron},f}$ (1)

40-4 The Compton Effect

Use $E = hf$ (Eq. 40.8) for the photon's total energy, then substitute $f = c/\lambda$ (Eq. 34.20).	$E_{photon} = hf = \dfrac{hc}{\lambda}$ (40.17)
Substitute Equation 40.17 for the photon's energy before and after the collision, and $K_{electron} = mc^2 - m_ec^2$ (Eq. 39.43) for the electron's kinetic energy after the collision into Equation (1). Then, substitute $m = \gamma m_{rest}$ (Eq. 39.39) for the scattered electron's observed mass.	$\dfrac{hc}{\lambda_i} = \dfrac{hc}{\lambda_f} + (mc^2 - m_ec^2) = \dfrac{hc}{\lambda_f} + (\gamma m_ec^2 - m_ec^2)$ $\dfrac{hc}{\lambda_i} = \dfrac{hc}{\lambda_f} + (\gamma - 1)m_ec^2$ (2)

:• CONSERVATION of MOMENTUM
As in Section 11-6, apply conservation of momentum separately in the x and y directions for the two-dimensional collision in Figure 40.12.

Before the collision, the photon's momentum is in the positive x direction. After the collision, both the photon and the electron have momentum in the positive x direction.	Along x: $p_{photon,i} = p_{photon,f}\cos\varphi + p_{electron,f}\cos\theta$ (3)
Before the collision, neither particle has momentum in the y direction. After the collision, the photon has momentum in the *negative* y direction, whereas the electron has momentum in the *positive* y direction.	Along y: $0 = -p_{photon,f}\sin\varphi + p_{electron,f}\sin\theta$ (4)
Substitute $p_{photon} = h/\lambda$ (Eq. 40.14) and $p_{electron} = \gamma m_e v$ (Eq. 39.40) into Equations (3) and (4).	$\dfrac{h}{\lambda_i} = \dfrac{h}{\lambda_f}\cos\varphi + \gamma m_e v\cos\theta$ (5) $0 = -\dfrac{h}{\lambda_f}\sin\varphi + \gamma m_e v\sin\theta$ (6)

The next steps involve many lines of algebra to eliminate v, γ, and θ from Equations (2)–(6). We'll do some of the work here and leave most of the steps for homework. The goal is to arrive at a single expression for $\Delta\lambda = \lambda_f - \lambda_i$ as a function of the scattering angle φ.

In Problem 83, you will eliminate θ from Equations (5) and (6) and arrive at Equation (7).	$\gamma^2 m_e^2 v^2 = \dfrac{h^2}{\lambda_i^2} - \dfrac{2h^2}{\lambda_i\lambda_f}\cos\varphi + \dfrac{h^2}{\lambda_f^2}$ (7)
In Problem 28 you will show that $\gamma^2 v^2 = c^2(\gamma^2 - 1)$. Use this equation to eliminate v from Equation (7).	$c^2(\gamma^2 - 1)m_e^2 = \dfrac{h^2}{\lambda_i^2} - \dfrac{2h^2}{\lambda_i\lambda_f}\cos\varphi + \dfrac{h^2}{\lambda_f^2}$ (8)
Now we must eliminate γ, which can be done by starting with Equation (2). The result is an expression for $\gamma^2 - 1$ (Problem 32).	$(\gamma^2 - 1) = \dfrac{h^2}{m_e^2c^2}\left(\dfrac{1}{\lambda_i^2} - \dfrac{2}{\lambda_i\lambda_f} + \dfrac{1}{\lambda_f^2}\right) + \dfrac{2h}{m_ec}\left(\dfrac{1}{\lambda_i} - \dfrac{1}{\lambda_f}\right)$ (9)
In Problem 84 you will substitute Equation (9) into Equation (8) and arrive near our goal of finding an expression for the Compton shift.	$(\lambda_f - \lambda_i) = \dfrac{h}{m_ec}(1 - \cos\varphi)$
The constant h/m_ec is the **Compton wavelength** λ_C of a free electron.	$\Delta\lambda = (\lambda_f - \lambda_i) = \lambda_C(1 - \cos\varphi)$ ✓ (40.15)

:• COMMENTS
The wave model for light predicts that when light scatters, $\Delta\lambda = 0$, so that the light does not change in wavelength (or frequency). But by using the photon model, we found a very different prediction for $\Delta\lambda$. The Compton shift (Eq. 40.15) predicts that scattered light should have a longer wavelength (that is, be redder) than incoming light. Additionally, the shift depends on the scattering angle φ.

The equation $\Delta\lambda = \lambda_C(1 - \cos\varphi)$ mathematically describes what Compton observed. Because $-1 \leq \cos\varphi \leq 1$, the Compton shift is always positive or zero: $\Delta\lambda \geq 0$. The wavelength of the scattered photon is the same or longer than that of the incoming photon. Put in terms of frequency and energy, the scattered photon has a frequency that is the same or lower, and an energy that is the same or lower, than the incoming photon.

To help visualize how the shift in wavelength depends on φ, Figure 40.14 diagrams three different scattering angles. The incoming photon and the electron are not shown. The smallest scattering angle (Fig. 40.14A) corresponds to a photon with the smallest Compton shift, the shortest wavelength, the highest frequency, and the most energy after the event. The greatest scattering angle (Fig. 40.14C) corresponds to the greatest Compton shift, the longest possible wavelength, the lowest frequency, and the least amount of energy after scattering.

Remember that the wave model predicts that all scattered radiation will have the same frequency (and wavelength) as the incoming radiation. By contrast, the Compton shift equation $\Delta\lambda = \lambda_C(1 - \cos\varphi)$ (Eq. 40.15) comes from modeling electromagnetic radiation as a photon with energy and momentum. Because Compton's experiment is well described by Equation 40.15, which was derived using the photon model, Compton's experimental results support the photon model for light.

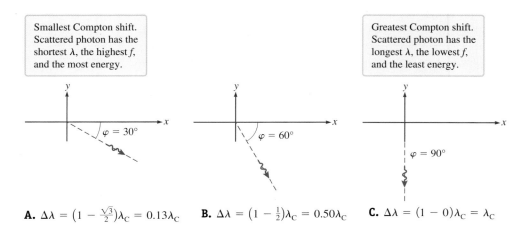

FIGURE 40.14 Three different Compton scattering angles φ are shown. Compare the wavelengths, frequencies, and energies of the scattered photons.

CONCEPT EXERCISE 40.4

In Figure 40.14, the three electrons are not shown. However, you can infer their change in energy from the photon's Compton shift. Which of the three electrons gains the most energy?

EXAMPLE 40.6 Plotting the Compton Shift

Graphs usually clarify complicated mathematical expressions. Make a graph of the relative Compton shift $\Delta\lambda/\lambda_C$ versus the scattering angle φ for $0 \leq \varphi \leq 180°$. In the **CHECK and THINK** step, comment on what is happening physically when $\varphi > 90°$.

:• **INTERPRET and ANTICIPATE**
From Figure 40.14, we expect the Compton shift to be smallest for small scattering angles.

:• **SOLVE**
Start with Equation 40.15 and isolate $\Delta\lambda/\lambda_C$ on one side.

$$\frac{\Delta\lambda}{\lambda_C} = (1 - \cos\varphi)$$

Substitute values from 0 to 180° for the scattering angle and plot the results (Fig. 40.15). A spreadsheet program may be helpful.

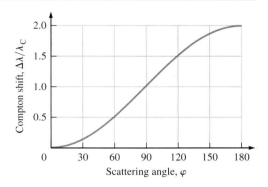

FIGURE 40.15

:• **CHECK and THINK**
As expected, the Compton shift increases as the scattering angle increases. When the scattering angle is greater than 90°, the scattered photon's motion begins to have a backward component. Consider the case when $\varphi = 180°$, so that the photon travels on a path back to its source. These photons have the greatest Compton shift, so they have the longest possible wavelength, lowest possible frequency, and least possible amount of energy after the scattering.

EXAMPLE 40.7 What Happens to the Electron?

The Compton shift equation tells us about the photon that is scattered by the electron, but what about the electron? Suppose an X-ray photon with a wavelength of 0.2400 nm is scattered off an electron that is initially at rest, as in Figure 40.12. The scattered X-ray beam is observed at an angle of 60.0°: down and to the right, as shown. Find the speed and momentum of the electron after the collision.

:• **INTERPRET and ANTICIPATE**
The X-ray photon loses energy and momentum as a result of its collision with the electron. The electron gains this energy and momentum. We need to find the electron's speed and momentum. Because speed is a scalar, we don't need to find the direction. So the easiest way to find the electron's speed is from its energy. Getting the electron's momentum is a little trickier because momentum is a vector quantity.

:• **SOLVE**
Start with the Compton shift to find the wavelength of the scattered photon.

$$\frac{\Delta\lambda}{\lambda_C} = \frac{\lambda_f - \lambda_i}{\lambda_C} = (1 - \cos\varphi) \quad (40.15)$$

$$\lambda_f = \lambda_i + \lambda_C(1 - \cos\varphi)$$

$$\lambda_f = 0.2400 \text{ nm} + 0.002426 \text{ nm}(1 - \cos 60.0°)$$

$$\lambda_f = 0.2412 \text{ nm}$$

Next use $E = hc/\lambda$ (Eq. 40.17) to find the change in the photon's energy.

$$\Delta E = E_f - E_i = h(f_f - f_i)$$

$$\Delta E = hc\left(\frac{1}{\lambda_f} - \frac{1}{\lambda_i}\right)$$

Example continues on page 1326 ▶

The energy lost by the photon goes into the kinetic energy of the electron. (Remember, the electron is initially at rest.)	$K_e = -\Delta E = hc\left(\dfrac{1}{\lambda_i} - \dfrac{1}{\lambda_f}\right)$
Let's start by assuming the speed is nonrelativistic. If we find a speed that is close to the speed of light, we will start over, taking special relativity into account.	$\dfrac{1}{2}m_e v_e^2 = hc\left(\dfrac{1}{\lambda_i} - \dfrac{1}{\lambda_f}\right)$ $v_e = \sqrt{\dfrac{2hc}{m_e}\left(\dfrac{1}{\lambda_i} - \dfrac{1}{\lambda_f}\right)}$ $v_e = \sqrt{\dfrac{2(6.626 \times 10^{-34}\,\text{J}\cdot\text{s})(2.998 \times 10^8\,\text{m/s})}{9.101 \times 10^{-31}\,\text{kg}}\left(\dfrac{1}{0.2400\,\text{nm}} - \dfrac{1}{0.2412\,\text{nm}}\right) \times 10^9\,\text{nm/m}}$ $v_e = 3.008 \times 10^6\,\text{m/s}$

:• **CHECK and THINK**
The speed we found is two orders of magnitude lower than the speed of light. We don't have to take special relativity into account.

:• **SOLVE**

The photon's momentum is initially in the x direction, and its magnitude is given by $p_\text{photon} = \dfrac{h}{\lambda}$ (Eq. 40.14).	$\vec{p}_i = \dfrac{h}{\lambda_i}\hat{\imath}$
We find the photon's change in momentum, remembering that momentum is a vector. Find the components separately. As shown in Figure 40.12, the photon's y momentum change is negative.	$\Delta p_x = p_f \cos\varphi - p_i = h\left(\dfrac{1}{\lambda_f}\cos\varphi - \dfrac{1}{\lambda_i}\right)$ $\Delta p_x = (6.626 \times 10^{-34}\,\text{J}\cdot\text{s})\left(\dfrac{1}{0.2412}\cos 60.0° - \dfrac{1}{0.2400}\right) \times 10^9\,\text{nm/m}$ $\Delta p_x = -1.3875 \times 10^{-24}\,\text{kg}\cdot\text{m/s}$ $\Delta p_y = p_f \sin\varphi = -\dfrac{h}{\lambda_f}\sin\varphi$ $\Delta p_y = -\dfrac{6.626 \times 10^{-34}\,\text{J}\cdot\text{s}}{0.2412 \times 10^{-9}\,\text{m}}\sin 60.0°$ $\Delta p_y = -2.3789 \times 10^{-24}\,\text{kg}\cdot\text{m/s}$
Because momentum is conserved, the loss in the photon's momentum must equal the gain in the electron's momentum. The electron is at rest before the collision, so the gain in momentum is the electron's momentum after the collision.	$\vec{p}_\text{electron} = (1.3875\hat{\imath} + 2.379\hat{\jmath}) \times 10^{-24}\,\text{kg}\cdot\text{m/s}$

:• **CHECK and THINK**
According to our expression for momentum, the electron moves up and to the right, consistent with Figure 40.12. Other than how we calculated the photon's energy and momentum, this problem is much like the two-dimensional collision problems we solved in Chapter 11 for particles. Put in simple terms, the electron cannot tell if it collided with another particle or with light. This example further shows why the Compton effect supports the photon model. We normally think that particles can collide but waves cannot. (Waves interfere; particles collide.)

40-5 Wave-Particle Duality

Let's review the debate over whether light should be modeled as a particle or a wave. In the 1600s, Newton was the strongest advocate of the particle model, whereas Huygens was the strongest advocate of the wave model. Because Newton's other theories (his three laws of dynamics and his law of universal gravity) were so successful, it was difficult for scientists to think that Newton could be wrong about light. Not until the late 1800s, after Young's interference experiment and Arago's diffraction experiment, was the wave model accepted.

Since Newton's and Huygens's time, scientists had believed that light must be modeled either as a wave or as a particle, but not as both. So in the 1800s, when the wave model was widely accepted, the particle model was widely rejected. Scientists believed that Young's and Arago's experiments disproved the particle model. However, the photoelectric effect and the Compton effect, observed in the early 1900s, could not be explained by the wave model.

Today we know of many experiments showing that light behaves as a wave and many other experiments showing that light behaves as a particle. So to have a full understanding of light, we must accept both models. This two-part description of light is known as **wave-particle duality**. Sometimes we must model light as a wave and at other times as a stream of photons.

WAVE-PARTICLE DUALITY OF LIGHT
❶ **Underlying Principle**

How do we know when to model light as a wave and when to model it as a photon? To explain the interference pattern produced by Young's double-slit experiment (Fig. 35.10, page 1128), we considered the path taken by two light waves that emerged from the two slits (Figs. 35.13 and 35.14, page 1129). Then we used the principle of superposition to find places of constructive and destructive interference. In general, we model light as a wave in any experiment where the *propagation* of light plays the key role. In these interference experiments, we consider the paths taken by the light waves and use the principle of superposition.

By contrast, in the Compton effect, light is modeled as a particle (a photon) that collides with another particle (an electron). The photon carries momentum and energy, which are conserved in its collision with the electron. In general, we model light as photons in any experiment where its *interaction* with matter plays a key role. These scattering experiments require that light be modeled as a particle having energy given by $E = hf = hc/\lambda$ (Eq. 40.17) and momentum given by $p = hf/c = h/\lambda$ (Eq. 40.14). In fact, these equations link the wave model of light (its frequency and wavelength) to the particle model (its energy and momentum).

We return to wave-particle duality in Chapter 41.

On the macroscopic level, we see some objects, such as basketballs and airplanes, that are best modeled as particles. We also experience phenomena, such as sound, that are best modeled as waves. The particle model and the wave model seem completely distinct. So when scientists were arguing about how to model light, it seemed that they were really arguing about its fundamental nature. Is light made up of a bunch of objects such as tiny balls, or is it a disturbance like a sound wave? When we conclude that our experiments show that we must model light *both* as a stream of particles and as a wave, how can we know what light is? The duality of light seems unsatisfactory.

Light's existence as both a wave and a stream of particles is dissatisfying because it doesn't fit our everyday experience. We can hold the basketball that we model as a particle in our hands, and we can watch it translate from place to place. By contrast, modeling sound as a wave is a little harder because we ordinarily can't see sound waves. However, we can make an analogy with water waves. By watching the water surface moving up and down in a water wave, we can imagine air molecules moving back and forth in a sound wave in a similar fashion. But it is even harder to digest wave-particle duality because we have no single macroscopic analogy. Instead, we must sometimes visualize light as a stream of tiny particles, and at other times as a waving disturbance. This is our best current description of light's nature.

Wave-particle duality is also counterintuitive because waves and particles are so different. For example, picture a water wave in a large pool; the wave spreads to take up the whole pool. If a second wave is introduced, the result is the superposition of both waves, which overlap and occupy the entire pool. By contrast, a basketball sitting on the floor of the court occupies a small place. If a second ball is placed on the floor of the court, you can be certain that its position is different from the position of the first ball. Two particles cannot occupy the same position; they would collide. So when we say that light is both a wave and a particle, we mean that sometimes we'll use the principle of superposition when two light waves overlap (as in Young's experiment), and at other times we'll treat light as composed of particles that can collide with other particles (as in Compton's experiment).

CONCEPT EXERCISE 40.5

What do we mean when we say that $E = hf = hc/\lambda$ (Eq. 40.17) "links" the wave model of light to the particle model? (How does the energy carried by a photon differ from the energy carried by a wave in the classical model?)

40-6 The Wave Properties of Matter

Our story of quantum physics continues with a French prince, Louis Victor Pierre Raymond de Broglie (1892–1987), who was originally a historian. He became interested in science during World War I. After the war, he earned his doctorate in physics in 1924. His doctoral thesis was revolutionary—in it, de Broglie proposed that the wave-particle duality extended beyond light. In fact, his thesis was so revolutionary that his professor wasn't sure what to make of it. So he sent a copy to Einstein, who said that de Broglie might be on to something.

As we have seen, exploiting the symmetry of nature sometimes leads to the discovery of new laws. De Broglie thought that if light behaves both as a wave and as a particle, then other things that we traditionally model as particles, such as electrons, protons, and even basketballs, might also be modeled as waves.

De Broglie said that the wavelength associated with such a particle comes from solving $p = h/\lambda$ (Eq. 40.14) for λ:

$$\lambda = \frac{h}{p} \qquad (40.18)$$

WAVE-PARTICLE DUALITY FOR MATTER
❶ Underlying Principle

DE BROGLIE WAVELENGTH
✪ Major Concept

The wavelength of a particle is called the **de Broglie wavelength**. The numerator is Planck's constant, which is a very small number. The denominator is large for macroscopic objects such as basketballs, so the wavelength of macroscopic objects is too small to be detected. Therefore, we don't notice the wave properties of such objects, and it is best just to model them as particles—as we have been doing.

If the particle is small, however, such as an electron, then it doesn't have a lot of mass, and its momentum is low too. In the case of a microscopic particle, the de Broglie wavelength is not negligible, and the particle's wave nature cannot be ignored.

Einstein pointed out that if de Broglie is correct, then electrons will show interference and diffraction patterns, similar to the interference and diffraction patterns observed by Young and Arago for light. The de Broglie wavelength of an electron is on the order of 10^{-10} m, or about 5000 times shorter than the wavelength of visible light. In order to observe an electron diffraction pattern like the one produced when light passes through a grating, you need a grating with rulings on the order of 10^{-10} m apart. Fortunately, the array of atoms in a solid can have spacings on this order of magnitude.

In 1927, Clinton Davisson (1881–1958) and Lester Germer (1896–1971) were working in their New York City laboratory. They had been scattering electrons from the surface of a nickel crystal, when by accident they discovered that the scattered electrons form a pattern of peaks and valleys. Figure 40.16 shows the electron diffraction pattern obtained in a contemporary experiment involving graphite instead of nickel. The underlying concept of this experiment is the same as that of Davisson and Germer's: The scattered electrons form a pattern like the diffraction pattern seen when light passes through a grating or a circular aperture (Fig. 36.2, page 1156). If you model the electrons as waves and the nickel crystal as a diffraction grating, then the pattern you see can be interpreted as a diffraction pattern. The spacing of fringes in the pattern can be used to find the electron's wavelength. When Davisson and Germer made this calculation, they found that the electron's wavelength is in agreement with the de Broglie wavelength.

Since this experiment in the 1920s, many others have been carried out showing that protons, neutrons, and other subatomic particles are well modeled as waves with wavelengths given by the de Broglie relation (Eq. 40.18). One of these experiments

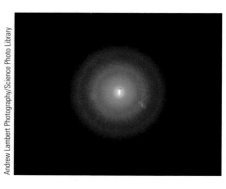

FIGURE 40.16 An electron diffraction pattern obtained from a thin film of polycrystalline graphite consists of concentric rings. In general the diffraction pattern observed depends on the structure of the sample; many observed patterns are more complicated that the simple concentric rings shown here.

40-6 The Wave Properties of Matter 1329

was carried out by Sir George Paget Thomson (1892–1975)—the son of J. J. Thomson. Since J. J. Thomson is famous for showing that the electron is a particle, it is ironic that his son successfully showed that the electron is a wave.

Wave-particle duality applies to objects that we traditionally model as particles, as well as to light. On the macroscopic scale, we continue to model basketballs as particles, but on the microscopic scale, we often need to model objects such as electrons, protons, and neutrons as waves instead of as particles. Sometimes we use the term **matter wave** to describe the wave associated with objects traditionally modeled as particles.

EXAMPLE 40.8 Your de Broglie Wavelength

A jogger with mass 65.0 kg is jogging at 4.00 m/s. What is the jogger's de Broglie wavelength?

: INTERPRET and ANTICIPATE
We model the jogger as a matter wave whose wavelength depends on his momentum. Since a person's jogging speed is much less than the speed of light, we do not need to use special relativity.

: SOLVE Find the magnitude of the momentum using Equation 10.1.	$p = mv$ $p = (65.0\,\text{kg})(4.00\,\text{m/s})$ $p = 260\,\text{kg}\cdot\text{m/s}$	(10.1)
Substitute this momentum into Equation 40.18 to find the de Broglie wavelength of the jogger.	$\lambda = \dfrac{h}{p}$ $\lambda = \dfrac{6.626 \times 10^{-34}\,\text{J}\cdot\text{s}}{260\,\text{kg}\cdot\text{m/s}}$ $\lambda = 2.55 \times 10^{-36}\,\text{m}$	(40.18)

: CHECK and THINK
The wavelength of a jogger is about 26 orders of magnitude smaller than the diameter of a single atom ($\sim 10^{-10}$ m). This is much too small to be detected, so the wave properties of a jogger can be safely ignored. This is why we are able to model objects as particles on the macroscopic scale.

EXAMPLE 40.9 An Electron's de Broglie Wavelength

In a laboratory experiment, an electron has a speed of 3.00×10^6 m/s. What is its de Broglie wavelength?

: INTERPRET and ANTICIPATE
As in the previous example, find the electron's de Broglie wavelength from its momentum. The electron's speed is $0.01c$—much less than the speed of light. Once again, we don't need special relativity.

: SOLVE Find the magnitude of the momentum using Equation 10.1.	$p = mv$ $p = (9.11 \times 10^{-31}\,\text{kg})(3.00 \times 10^6\,\text{m/s})$ $p = 2.73 \times 10^{-24}\,\text{kg}\cdot\text{m/s}$	(10.1)

Example continues on page 1330 ▶

Substitute this momentum into Equation 40.18 to find the electron's de Broglie wavelength.

$$\lambda = \frac{h}{p} \quad (40.18)$$

$$\lambda = \frac{6.626 \times 10^{-34}\,\text{J}\cdot\text{s}}{2.73 \times 10^{-24}\,\text{kg}\cdot\text{m/s}}$$

$$\lambda = 2.42 \times 10^{-10}\,\text{m} = 0.242\,\text{nm}$$

:• CHECK and THINK

The electron's de Broglie wavelength is about the size of an atom or of the space between atoms in a crystal. So when a beam of electrons is aimed at a crystal, the electrons form a diffraction pattern such as the one shown in Figure 40.16. In this case, the electrons behave like a wave, and we must use the wave model. In order to understand the structure of atoms (Chapter 42), we must model their electrons as matter waves.

CASE STUDY The Electron Microscope

De Broglie's idea that electrons may be modeled as waves may seem theoretical, with no practical applications. However, less than a decade after he put forth this idea, the first primitive electron microscope was built. An electron microscope uses electron waves in much the same way that an optical microscope uses light waves (Section 36-9). However, the optical lenses are replaced by magnetic (or electrostatic) fields that exert forces on the electrons, bringing them into focus.

There are two basic designs for an electron microscope (Fig. 40.17). In both designs, a beam of electrons is generated by an electron gun composed of a hot, V-shaped negative plate and a disk-shaped positive plate. Electrons are accelerated by the potential difference between the two plates. In a transmission electron microscope, the electron beam passes through the sample, and its image is projected onto a screen at the other end of the device (Fig. 40.17A). In a scanning electron microscope, the electron beam liberates electrons from the sample, and these electrons are then collected and used to form a three-dimensional image of the sample (Fig. 40.17B).

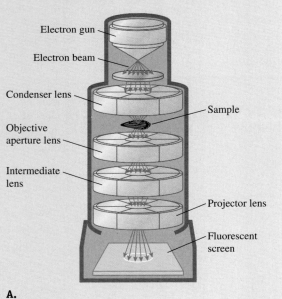
A.

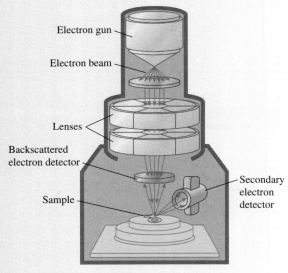
B.

FIGURE 40.17 Two electron microscope designs. In both designs, an electron gun generates a beam. **A.** In a transmission electron microscope, a series of lenses cause the electron beam to pass through the sample and form an image on a screen. **B.** In a scanning electron microscope, the electron beam liberates electrons from the sample. Some are scattered backward toward the electron gun, and others are scattered (roughly) perpendicular to the original beam. The two scattered beams of electrons are used to generate a three-dimensional image.

You have probably seen a photograph that has been over-magnified; the image looks like many individual squares, and it is hard to discern its subject. The useful magnification of an image is limited by how many individual pixels or resolution elements went into creating the image. For a microscope, the useful magnification depends on its resolution.

Recall that the diffraction-limited resolution of an instrument such as a microscope is proportional to the wavelength (Eq. 34.4). To construct a high-resolution, greatly magnified image, you should illuminate the sample with short wavelengths. As illustrated in the next example, it is more practical to do this with electron waves than with light.

EXAMPLE 40.10 — CASE STUDY: Comparing a Conventional Microscope to an Electron Microscope

In this example, we compare a conventional microscope that illuminates a sample with an electromagnetic wave to an electron microscope (of either design) that uses an electron wave. Suppose you wish to image a strand of DNA, which has a width of roughly 2 nm. You'd like a high-resolution, greatly magnified image, and you estimate that you require a wavelength of about 0.02 nm, so that there will be roughly 100 pixels or resolution elements across the DNA's width.

 Find the energy (in eV) of the electromagnetic wave used to illuminate the DNA in a conventional microscope. In the **CHECK and THINK** step, discuss the practical limitations of such a device.

:• INTERPRET and ANTICIPATE
Use the estimated wavelength to find the energy of the photons in the electromagnetic wave your microscope would employ.

:• SOLVE
The energy of the photons depends on their frequency, according to Equation 40.8. Their frequency is related to their wavelength by $\lambda f = c$ (Eq. 34.20). Convert the energy to eV.

$$E = hf \quad (40.8)$$

$$E = h\frac{c}{\lambda} = (6.63 \times 10^{-34}\,\text{J}\cdot\text{s})\left(\frac{3.00 \times 10^8\,\text{m/s}}{2 \times 10^{-11}\,\text{m}}\right)$$

$$E = 9.9 \times 10^{-15}\,\text{J}$$

$$1\,\text{eV} = 1.60 \times 10^{-19}\,\text{J}$$

$$E = 6 \times 10^4\,\text{eV}$$

:• CHECK and THINK
As shown in Table 34.2 (page 1101), electromagnetic radiation with a wavelength of 2×10^{-11} m is in the X-ray or gamma-ray band. Such high-energy radiation would pose a threat to the researchers, their microscope, and their sample. Using X-rays is not practical in a high-resolution microscope.

 Find the kinetic energy (in eV) of the electron wave used in an electron microscope. Assume that the electrons are not relativistic. In **CHECK and THINK**, compare your results to those in part A.

:• INTERPRET and ANTICIPATE
Find the momentum of the electrons from their wavelength, and find their kinetic energy from their momentum.

:• SOLVE
Use Equation 40.18 to find the electron momentum.

$$\lambda = \frac{h}{p} \quad (40.18)$$

$$p = \frac{h}{\lambda}$$

Example continues on page 1332 ▶

Write the expression for kinetic energy in terms of the momentum. Substitute values, then convert the result to eV.	$K = \dfrac{1}{2}m_e v^2 = \dfrac{(m_e v)^2}{2m_e}$ $K = \dfrac{p^2}{2m_e} = \dfrac{1}{2m_e}\left(\dfrac{h}{\lambda}\right)^2$ $K = \dfrac{1}{2(9.11 \times 10^{-31} \text{ kg})}\left(\dfrac{6.63 \times 10^{-34} \text{ J}\cdot\text{s}}{2 \times 10^{-11} \text{ m}}\right)^2$ $K = 6.0 \times 10^{-16} \text{ J} = 4 \times 10^3 \text{ eV}$

:• CHECK and THINK

The energy required by the electron microscope is about 15 times lower than that required by the conventional microscope. It will be easier to power an electron microscope and to control the high-energy electrons with magnetic (or electrostatic) fields.

Although the resolution of an electron microscope is not limited significantly by diffraction, like a conventional microscope, it is subject to spherical aberration, distortion, chromatic aberration, and other imaging defects. So even an electron microscope cannot achieve the resolution of 0.02 nm that we would like here. The best resolution achieved by an electron microscope is typically between 0.1 and 0.5 nm, and its highest useful magnification is typically between 10^4 and 10^5. Figure 40.18 is an image of a bundle of strands DNA taken with a transmission electron microscope. The bar in the bottom left corner is the scale; it represents 20 nm. This amazing image shows six DNA stands wrapped around a seventh strand. The inset shows a close-up of the helical structure, with red arrows pointing to the individual turns of the helix. The width of the DNA is 2 nm, so assuming that the resolution is 0.2 nm, we find that there are about 10 resolution elements across the width of a single strand of DNA. Without such detailed resolution, the helical structure would not be seen, the image would look like a smooth line without any structure.

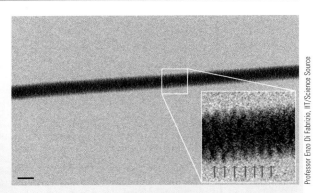

FIGURE 40.18

SUMMARY

❶ Underlying Principles:

1. According to **Planck's quantum theory**, the smallest possible amount of energy $E_{\min}$ (a **quantum of energy**) for an oscillator of frequency f is:

$$E_{\min} = hf \quad (40.6)$$

    Planck modeled the particles in a black body as oscillators and concluded that the energy of each oscillator is **quantized** and given by:

$$E = nE_{\min} = nhf \quad n = 1, 2, 3\ldots \quad (40.7)$$

    Planck's formula for the intensity I of a black body's **radiation** is:

$$I(\lambda, T) = \dfrac{2\pi hc^2}{\lambda^5(e^{hc/\lambda k_B T} - 1)} \quad (40.5)$$

    where **Planck's constant** is $h = 6.626 \times 10^{-34}$ J·s.

2. **Wave-particle duality**: A wave model and a particle model must be applied to both light and matter.

Major Concepts

1. In the photon model, light is modeled as bundles of energy or particles known as **photons**. A photon's energy depends on the frequency of the light:

$$E_{photon} = hf = \frac{hc}{\lambda} \quad (40.17)$$

A photon's momentum is given by:

$$p = \frac{hf}{c} = \frac{h}{\lambda} \quad (40.14)$$

2. When a photon is scattered (Fig. 40.12), it has a change in wavelength known as the **Compton shift**, given by:

$$\Delta\lambda = (\lambda_f - \lambda_i) = \lambda_C(1 - \cos\varphi) \quad (40.15)$$

where λ_C is the **Compton wavelength** of a free electron,

$$\lambda_C \equiv \frac{h}{m_e c} = 2.426 \times 10^{-12} \, \text{m} = 2.426 \, \text{pm} \quad (40.16)$$

3. The **de Broglie wavelength** associated with a particle is

$$\lambda = \frac{h}{p} \quad (40.18)$$

PROBLEMS AND QUESTIONS

A = algebraic   C = conceptual   E = estimation   G = graphical   N = numerical

40-1 Another Modern Idea

1. **C** Many of the images that we see today are composed of many small dots of color, or pixels. Normally, we don't notice that the images are made out of discrete pixels. Why do the images seem continuous? What would happen if the size of the pixels increased?

40-2 Black-Body Radiation and the Ultraviolet Catastrophe

2. **N** The surface temperature of the Sun is about 5.8×10^3 K. Model the Sun as a black body; find the peak wavelength of its radiation.

3. **N** A stove top burner operates at a temperature of 350.0°F (Fig. P40.3). Modeling the burner as a black body, what is the peak wavelength emitted by the burner?

FIGURE P40.3

Problems 4 and 5 are paired.

4. **C** Samantha tells George that stars that appear yellow are really hot and that the color is indicative of temperature. George says he's heard that red stars are often very big and should thus be hotter than yellow ones, which are often smaller. Samantha tells George that he is wrong about red stars being hotter and observes that some stars can even be blue. She says these would be even hotter than the yellow ones. George insists that she has it backwards because blue things should be cold. Who is correct and why?

5. **N** Calculate the peak wavelength and identify the color of the following stars based on their different temperatures. Treat them as black bodies: **a.** 4.00×10^3 K, **b.** 5.30×10^3 K, **c.** 6.40×10^3 K.

6. **E** Model yourself as a black body and estimate your peak wavelength.

7. **N** What is the temperature of a black body with a wavelength distribution peaking at 628 nm?

8. **A** Show that Planck's constant h has the same dimensions as angular momentum.

9. **N** An incandescent lightbulb filament reaches a temperature of around 4.14×10^3 K. Assuming that it radiates as a black body, at what wavelength is the peak intensity of the black-body radiation? In what region of the electromagnetic spectrum is this radiation?

10. **N** What is the frequency of the light associated with the peak wavelength emitted by a black body with a temperature of 4.83×10^3 K?

11. **N, C** A molecule vibrates with a minimum frequency of 9.50×10^{13} Hz. What is the minimum quantum of energy that this molecule can have? Can this molecule vibrate at 12.5×10^{13} Hz? If not, why not? If so, what is its energy when it vibrates at this frequency?

12. A black body has a temperature of 3575 K.
    a. N What is the peak wavelength of this black body?
    b. N What is the intensity of the light from the black body at this wavelength?
    c. C How would the answer to part (b) compare to the intensity of other wavelengths from the black body?

13. **N** The Sun, which can be modeled as a black body, radiates 3.85×10^{26} J of energy from its surface each second. The Sun's radius is 6.9551×10^5 km.
    a. What is the black-body temperature of the Sun's surface?
    b. What is the peak wavelength of the radiation emitted by the Sun?

14. **C** When you look at a flame or a fire, you may notice that different regions appear to glow with different colors (Fig. P40.14). In terms of black-body radiation, explain why different regions within the flame or fire appear to be different colors.

FIGURE P40.14

40-3 The Photoelectric Effect

15. **N** The work function of silicon is $W_0 = 4.85$ eV.
    a. What is the maximum wavelength of incident light for which photoelectrons will be released from silicon?
    b. What is the minimum frequency of incident light, called the cutoff frequency, for which photoelectrons will be released from silicon?
    c. What is the maximum kinetic energy of photoelectrons emitted by silicon if 7.00-eV photons strike its surface?

16. A material has a work function of 2.3 eV.
    a. **N** What frequency of light is necessary to eject electrons from its surface?
    b. **C** Assuming light of the frequency you found in part (a), does the number of ejected electrons increase if the intensity of the light is increased? How might you explain this in terms of the particle representation of light?
    c. **N** What is the stopping potential for this material if it is illuminated by light with a frequency of 6.50×10^{14} Hz?
    d. **N, C** What is the maximum kinetic energy of ejected electrons if light with a frequency of 3.50×10^{14} Hz is incident on this material? Explain.

Problems 17 and 18 are paired.

17. **N** The work function of gold is 5.1 eV. What is the minimum frequency of light that will produce the photoelectric effect in gold?
18. **G** The work function of gold is 5.1 eV. Plot the maximum speed of the ejected electrons from gold versus the frequency of incident photons for photon energies between 0 and 10 eV.
19. **N** Far ultraviolet light with a wavelength of 175 nm is incident on an unknown solid surface, which releases photoelectrons with a maximum speed of 7.70×10^5 m/s.
    a. What is the work function of the unknown solid?
    b. What is the minimum frequency of incident light for which photoelectrons will be released from this surface?

20. **N** For each entry in Table P40.20, find the photon's energy in eV.

TABLE P40.20

Color	Wavelength λ (nm)	Frequency f ($\times 10^{12}$ Hz)	Energy E (eV)
Red	620–780	380–480	
Orange	600–620	480–500	
Yellow	575–600	500–522	
Green	500–575	522–600	
Blue	450–500	600-670	
Indigo	420–450	670–710	
Violet	400–420	710–750	

Problems 21 and 22 are paired.

21. **N** A strange metallic rock is found and is being tested. Suppose that light with a frequency of 7.50×10^{14} Hz is incident upon the rock and a stopping potential of 1.00 V is needed to reduce the electron current to zero in a photoelectric experiment.
    a. What is the maximum kinetic energy of an electron ejected by this light from this material?
    b. What is the work function of this material?
22. **N** For the situation in Problem 21, what is the minimum frequency of light for which electrons are still ejected from the surface of this material?
23. **N** Iron has a work function of 4.50 eV. What is the minimum frequency of light needed to eject electrons from iron?
24. **C** What has more energy, an X-ray photon or a radio photon? Explain.
25. **N** **CASE STUDY** Larger diameter telescopes are more desirable, in part, because they are better at gathering photons. The light-gathering power of a telescope is proportional to the area of its primary mirror. Large telescopes are expensive, so they are generally owned and operated by large institutions. However, the quantum efficiency of CCDs means that small institutions with small telescopes are also equipped to detect faint astronomical objects. Consider a small telescope equipped with a CCD. To make a concrete comparison, consider the unlikely scenario of a large 5.00-m diameter telescope equipped with a photographic plate. Suppose that in the same time exposure, the two telescopes detect equally faint objects. If the photographic plate's quantum efficiency is 1.5% of the CCD's, what is the diameter of the small telescope?
26. **C** The work functions for sodium and platinum are 2.28 eV and 6.35 eV respectively. The same light source with a sufficient frequency to free electrons from each metal is shined on each surface.
    a. How does the number of electrons freed from each metal differ?
    b. How does the maximum kinetic energy of the electrons freed from each metal differ?

40-4 The Compton Effect

27. **N** Suppose that the ratio of the wavelengths of two photons is $\lambda_1/\lambda_2 = 4.00$. What is the ratio of their momenta, p_1/p_2?
28. **A** The derivation of the Compton shift in Section 40-4 takes many lines of algebra. You are asked to provide many of the missing steps. Show that $\gamma^2 v^2 = c^2(\gamma^2 - 1)$.

29. **N** X-rays of wavelength 1.250 Angstroms (10^{-10} m) incident on an aluminum target undergo Compton scattering.
  a. For a photon scattered at 90° relative to the incident beam, what is the wavelength of the scattered photon?
  b. What are the magnitude of the momentum and the energy of this scattered photon at this angle?

Problems 30 and 31 are paired.

30. **A** A photon undergoes a shift in wavelength due to a collision with an electron so that its wavelength changes by 10.0%. Find an expression for the scattering angle in terms of the initial wavelength of the photon.

31. **N** For the photon in Problem 30, what is the scattering angle if the initial wavelength was 15.0 pm?

32. **A** The derivation of the Compton shift in Section 40-4 takes many lines of algebra. You are asked to provide many of the missing steps. Show that

$$(\gamma^2 - 1) = \frac{h^2}{m_e^2 c^2}\left(\frac{1}{\lambda_i^2} - \frac{2}{\lambda_i \lambda_f} + \frac{1}{\lambda_f^2}\right) + \frac{2h}{m_e c}\left(\frac{1}{\lambda_i} - \frac{1}{\lambda_f}\right).$$

33. **N** What is the shift in wavelength of X-rays that are scattered from a target at an angle of 62.0° to the incident beam?

34. **A** Show that $\lambda_C \equiv h/(m_e c)$ has the dimensions of length.

Problems 35, 36, and 37 are grouped.

35. **N** Suppose that you perform the Compton experiment with X-rays of wavelength 1.984×10^{-11} m. What is the wavelength (in picometers) of the scattered X-rays observed at **a.** 0°, **b.** 45° and, **c.** 180°?

36. **N** Suppose that you perform the Compton experiment with X-rays of wavelength 1.984×10^{-11} m. What is the frequency (in exaHertz) of the scattered X-rays observed at **a.** 0°, **b.** 45° and, **c.** 180°?

37. **N** Suppose that you perform the Compton experiment with X-rays of wavelength 1.984×10^{-11} m. What is the energy (in keV) of the scattered X-rays observed at **a.** 0°, **b.** 45.00° and, **c.** 180.0°?

38. **N** An X-ray with a frequency of 8.7700×10^{16} Hz undergoes a Compton scattering process and is detected at an angle of 45.000° relative to its original velocity.
  a. What is the wavelength of the photon after it has scattered?
  b. What is the kinetic energy of the electron after its collision with the X-ray?

39. **N** A beam of 75.0-keV X-rays is Compton scattered by a target in such a way that the scattered X-rays are at an angle of 44.0° to incident beam.
  a. What is the wavelength for the scattered X-rays?
  b. What is the energy of the scattered X-rays?
  c. What is the energy of the electrons scattered by the incident X-rays?

40. **C** According to the theory behind Compton scattering, is there a difference expected for scattering from targets made from two different metals? Why or why not?

41. **N** An X-ray with a frequency of 4.500×10^{17} Hz undergoes a Compton scattering process. What is the wavelength of the X-ray that is scattered at **a.** 30.00°, **b.** 60.00°, and **c.** 90.00°?

40-5 Wave-Particle Duality

42. **C** Wave-particle duality is hard to accept. Let's work through an analogy that might make the duality more comprehensible. You know you cannot be both a human being and a chimpanzee. However, you know that you *can* be both a physics student and a server in a restaurant. Explain why you cannot be both a human being and a chimpanzee, but you can be both a student and a server. Use this as an analogy to explain why light can be both a particle and a wave.

43. **C** For each of the following experiments, or cases, identify whether the particle or wave nature of light would be observed: **a.** light is incident on a metal giving rise to free electrons, **b.** light is incident on a single slit giving rise to an interference pattern on a screen, **c.** light is incident on a pair of closely spaced slits giving rise to an interference pattern on a screen, and **d.** x-rays are scattered elastically by free electrons.

44. **C** We know that the intensity of light from a point source decreases as a function of distance r with a dependence of $1/r^2$. Explain that inverse square dependence using the photon model.

40-6 The Wave Properties of Matter

45. **N** Consider a photon and a proton carrying equal momenta of 2.13×10^{-25} kg·m/s.
  a. What is the wavelength of a photon with this momentum?
  b. What is the speed of a proton with this momentum?

Problems 46 and 47 are paired.

46. **E** Determine an estimate for the wavelength of a fastball thrown by a major league baseball pitcher.

47. **C** The space between slats on a fence is about 1.5 cm. Would it be possible to observe the diffraction of a baseball thrown at the fence? Explain why or why not.

48. **C** A number of times throughout this textbook, we have said that by exploiting this symmetry nature new physical laws are proposed. For examples see Chapter 34 (page 1085) and Section 40-6. What do we mean when we say that nature is symmetric? How does the term compare and contrast to our common usage?

49. **N** If an electron has a kinetic energy 3.40 eV, what is the de Broglie wavelength of the electron?

50. **C** Compare and contrast an electron with a photon.

51. **C** In a transmission electron microscope, it is possible to control the energy of the electrons when imaging. High-resolution imaging usually requires 300 keV electrons. A researcher might describe 100 keV electrons as being "too fat" for high-resolution imaging. What does the researcher mean when she says this?

52. **E** Similar to X-ray and electron diffraction, neutron diffraction can occur if the de Broglie wavelength of the neutrons is comparable to the interatomic spacing of the target. Approximately what speed would be required to produce neutron diffraction effects?

53. **N** A proton is observed with a speed of 7.98×10^5 m/s. What is the de Broglie wavelength of this proton?

54. **N** An alpha particle (a helium nucleus) moves with a speed of 2.46×10^5 m/s. What is the wavelength of this particle?

55. **N** A stream of neutrons moves with a speed of 1.23×10^3 m/s and is incident on two rows of atoms within a sample material with a spacing of 0.35 nm between the rows. The neutrons undergo diffraction as they pass between the rows of atoms. What is the width of the central bright region on a screen that is 2.4 cm away from the sample material?

56. **C** In Example 40.8 (page 1329), we found that a jogger's wavelength is around 10^{-36} m, and in Example 40.9 we found the wavelength of an electron is about 0.2 nm. Why isn't it possible to detect the wave properties of a jogger, whereas it is possible to detect the wave properties of an electron?

57. **N** Suppose that an electron and a proton are traveling at the same speed. What is the ratio λ_p/λ_e of their wavelengths?
58. **N** Find the wavelength and kinetic energy of each of the following objects, assuming they move with a speed of 1.23×10^6 m/s: **a.** an electron, **b.** a dust particle ($m = 1.54 \times 10^{-10}$ kg), and **c.** a baseball ($m = 0.145$ kg).

Problems 59 and 60 are paired.
59. **N** Find **a.** the energy of a photon with a wavelength of 3.56 nm, and **b.** the kinetic energy of an electron with a de Broglie wavelength of 3.56 nm. **c.** Compare your results by finding the ratio of your answers (part (a)/part(b)).
60. **N** Find **a.** the energy of a photon with a wavelength of 3.56 fm, and **b.** the kinetic energy of an electron with a de Broglie wavelength of 3.56 fm. **c.** Compare your results by finding the ratio of your answers (part (a)/part(b)).

General Problems

61. **N** What is the stopping potential for a samarium (Sm) surface with work function $W_0 = 2.70$ eV illuminated with 405-nm blue light?
62. **N** An electron moves with a speed of $0.60c$.
    a. Find the wavelength of the electron semiclassically by using $p = mv$ for the momentum of the electron.
    b. Find the wavelength of the electron using special relativity to express the momentum of the electron.
    c. Calculate the percent difference between your answers for parts (a) and (b).

Problems 63 and 64 are paired.
63. **N** A researcher fires a stream of X-rays with a frequency of 2.50×10^{18} Hz at a target where they undergo Compton scattering. A circular detector is set up to detect X-rays at all possible scattering angles. What are **a.** the maximum and **b.** the minimum wavelengths that are detected?
64. **N** Consider the stream of X-rays and the circular detector in Problem 63.
    a. An X-ray scatters, resulting in the maximum possible wavelength being detected. What is the magnitude of the momentum of the electron scattered during this process?
    b. An X-ray scatters, resulting in the minimum possible wavelength being detected. What is the magnitude of the momentum of the electron scattered during this process?
65. **N** How many photons are emitted in 3.0 ms by a HeNe laser ($\lambda = 633$ nm) if the laser has a power of 15 mW?
66. **C** **CASE STUDY** You are helping a friend buy a new digital camera. The salesperson insists that the more expensive camera is better because it has 2 megapixels. Your friend has limited funds and only plans to print relatively small (8×10 in) images. How do you advise your friend? Explain.
67. **N** A proton has a de Broglie wavelength of 74.50 fm. What is its speed?
68. **N** A cup containing 5.67 kg of water is placed in a microwave oven. The oven creates microwaves with a frequency of 2.45 GHz. The initial temperature of the water is 20.0°C, and the specific heat capacity of water is 4186 J/kg · K.
    a. How much energy is necessary to raise the water to its boiling point?
    b. How many microwave photons are necessary to bring the water to its boiling point, assuming all of the energy from each photon is absorbed by the water?
69. **N** An electric tea kettle is plugged in without water in the kettle, and the heating element of the kettle reaches a temperature of 175.0°C. What is the peak wavelength of the radiation from the heating element?
70. **N** What are the wavelength and the energy in electron volts of a photon with a frequency of **a.** 892 MHz, **b.** 14.8 GHz, and **c.** 7.99 THz (terahertz = 10^{12} Hz)?
71. **N** **CASE STUDY** In Example 40.10 (page 1331), we found that the kinetic energy of the electrons in an electron microscope is 4 keV. What potential difference must be provided by the electron gun (Fig. 40.17)?

Problems 72 and 73 are paired.
72. **N** Light of ever-increasing frequency is incident on a metal until electrons are observed to be freed from the surface. If free electrons are first observed when the light has a frequency of 5.45×10^{14} Hz, what is the work function of the metal?
73. **N** For the light incident on metal in Problem 72, what stopping potential is necessary to completely stop free electrons created by the light when the light's frequency is **a.** 7.45×10^{14} Hz and **b.** 9.45×10^{14} Hz?
74. **N** Electron microscopes have a much greater resolving power than optical microscopes because electrons have wavelengths that are much shorter than visible light wavelengths. What energy must electrons have to image atoms with a transmission electron microscope operating at a wavelength of 0.0300 nm?

Problems 75, 76, and 77 are grouped.
75. **N** If 5% of the power radiated by a 100-W bulb is emitted as visible light at 550 nm, determine the number of photons that are emitted per second at this wavelength. Answer with two significant figures.
76. **E** Estimate the number of photons that are emitted by the Sun per second.
77. **N** **CASE STUDY** Let's imagine replacing the electrons in an electron microscope (Fig. 40.17) with protons. If you desired the same resolution as in Example 40.10, what would be the necessary kinetic energy of the protons?
78. **N** A free electron initially at rest is scattered by a photon with an energy of 1.24 MeV in such a way that the scattering angle of the photon and the electron are equal (Fig. P40.78).
    a. What is the scattering angle φ?
    b. What are the energy and momentum of the photon after it is scattered?
    c. What are the kinetic energy and momentum of the electron after it is scattered?

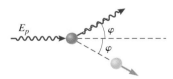

FIGURE P40.78

79. **C** Why can't we use $p = \gamma m_{rest} c^2$ to find the magnitude of a photon's momentum? *Hint*: What does this expression predict for photon's momentum?

Problems 80 and 81 are paired.
80. **C** A solar sail is a proposed propulsion device for space travel. The sail intercepts photons from the Sun similar to how a sail on a ship intercepts the wind on the ocean (Fig. P40.80). Should the side of the sail that faces the Sun be reflective or absorptive, assuming the goal is to get the most momentum possible from each photon collision with the sail? Explain.

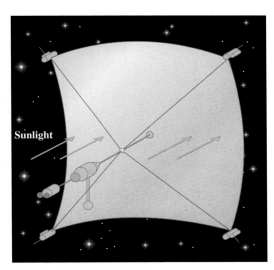

FIGURE P40.80

81. **N** Suppose you have a solar sail with an area of 1.00×10^4 m. In space near the Earth, the intensity of the sunlight is approximately 1.40 kW/m². Suppose the light consists entirely of green photons ($\lambda = 535$ nm).
   a. If the sail absorbs the photons as they strike the sail, what is the change in momentum of the solar sail each second?
   b. If the sail reflects the photons as they strike the sail, what is the change in momentum of the solar sail each second?

82. **N** A free electron initially at rest recoils with a speed of 1.750×10^6 m/s during the Compton scattering of an X-ray photon. The scattering angle of the photon is 21.50°.
   a. What is the wavelength of the incident X-ray photon?
   b. What is the scattering angle of the electron?

83. **A** The derivation of the Compton shift in Section 40-4 takes many lines of algebra. Start with Equations (5) and (6) from the derivation (page 1322), and eliminate θ from these equations to derive:

$$\gamma^2 m_e^2 v^2 = \frac{h^2}{\lambda_i^2} - \frac{2h^2}{\lambda_i \lambda_f} \cos \varphi + \frac{h^2}{\lambda_f^2}$$

This is Equation (7) in the derivation.

84. **A** The derivation of the Compton shift in Section 40-4 takes many lines of algebra. Substitute Equation (9) into Equation (8) from the derivation (page 1322), and simplify your expression until you derive the Compton shift expressed as:

$$\lambda_i \lambda_f \left(\frac{1}{\lambda_i} - \frac{1}{\lambda_f} \right) = (\lambda_f - \lambda_i) = \frac{h}{m_e c} (1 - \cos \varphi)$$

41 Schrödinger's Equation

Key Questions

If a particle is a wave, then what is doing the "waving"?

In what fundamental way are the predictions made by quantum mechanics different from those made by classical mechanics?

What does quantum mechanics predict for the energy and position of a trapped particle?

41-1 **The New Quantum Theory** 1339

41-2 **A Trapped Particle** 1340

41-3 **The Double-Slit Experiment Revisited: Probability Waves** 1343

41-4 **Schrödinger's Equation** 1346

41-5 **Special Case: A Particle in an Infinite Square Well** 1349

41-6 **Special Case: A Particle in a Finite Square Well** 1352

41-7 **Barrier Tunneling** 1355

41-8 **Special Case: Quantum Simple Harmonic Oscillator** 1359

41-9 **Heisenberg's Uncertainty Principle** 1362

Underlying Principles

1. Schrödinger's equation
2. Bohr's correspondence principle
3. Heisenberg's uncertainty principle

Major Concepts

1. Probability waves and probability density
2. Normalization condition
3. Boundary conditions
4. Barrier tunneling

Special Cases:

1. Particle in an infinite square well
2. Particle in a finite square well
3. Quantum simple harmonic oscillator

Tools

Energy-level diagram

You and your friend are up late studying for physics finals. Suddenly, he jumps up and says that he can walk through walls! You think he must be really worried about his exams tomorrow; but then he explains himself. He says that atoms are mostly empty space. The nucleus is a tightly concentrated ball in the center of the atom, and the electrons hover around it in a kind of cloud. So both he and the wall are mostly empty space. He should be able to pass his atoms through the wall's atoms like a comb passes through hair.

You laugh and say that he is wrong. When you press yourself against the wall, it exerts a normal force on you, owing to the way the atoms in your body interact with those in the wall. But he replies that your view is a *classical* view of the world, where everything is knowable and predictable. In modern physics, he says, nothing is knowable. You can only say that things are probable or improbable. So there really is a small probability that he can walk through walls.

Believe it or not, your friend is basically right. According to quantum physics, everything is subject to probability. However, probability matters a lot more on the microscopic scale. It plays a very important role for a single small particle, such as an electron; but on the macroscopic scale, probabilities calculated by quantum physics are essentially indistinguishable from the values determined by classical physics. So can your friend walk through walls? The probability of doing such a thing is so very low that you can safely say *no*—and go back to studying.

41-1 The New Quantum Theory

At the end of the previous chapter, we learned that de Broglie came up with a wave model for matter, such as an electron. This model marks the end of the first phase of the development of the *old* quantum physics theory. The old quantum theory was a great departure from classical physics, but it left many questions unanswered. Why should energy be quantized instead of continuous? And if an electron is a wave, what is doing the "waving"? (We could ask this about any particle.)

The *new* quantum theory was first developed by three independent theorists—Werner Heisenberg (1901–1976), Paul Dirac (1902–1984), and Erwin Schrödinger (1887–1961). Whole textbooks and courses could be devoted to the work of any one of these individuals; in this book, we primarily focus on Schrödinger's contribution.

The Austrian Erwin Schrödinger took a copy of de Broglie's thesis on a vacation during the winter of 1925. Schrödinger was bothered by the ad hoc nature of Planck's quantum theory (Section 40-2). He liked de Broglie's *matter wave* model, but he believed that de Broglie's theory was too vague. While he was on his vacation, he came up with a wave equation for these matter waves. This equation is known as **Schrödinger's (nonrelativistic) time-dependent wave equation**; in one dimension, Schrödinger's equation is:

SCHRÖDINGER'S TIME-DEPENDENT EQUATION ● Underlying Principle

$$-\frac{h^2}{8\pi^2 m}\frac{\partial^2 \Psi(x,t)}{\partial x^2} + U(x,t)\Psi(x,t) = \frac{ih}{2\pi}\frac{\partial \Psi(x,t)}{\partial t} \quad (41.1)$$

where U is potential energy. Schrödinger's equation is a wave equation for $\Psi(x, t)$, called the **wave function**. (The symbol Ψ is the uppercase Greek letter psi.) Compare Schrödinger's equation to the classical wave equation:

In Equation 41.1, h is Planck's constant, $i \equiv \sqrt{-1}$, and $\Psi(x, t)$ may be a complex quantity.

$$\frac{\partial^2 y(x,t)}{\partial x^2} = \frac{1}{v_x^2}\frac{\partial^2 y(x,t)}{\partial t^2} \quad (17.33)$$

The two equations are similar in that both involve partial derivatives with respect to x and t. Schrödinger's equation is more complicated than the classical wave equation because a classical wave only carries energy and momentum. Schrödinger's equation describes the propagation of a matter wave, which includes all the things we normally associate with a particle: energy, momentum, mass, and sometimes charge.

Schrödinger's equation is an underlying principle of quantum physics, much as Newton's laws are principles of classical physics. This equation will be our focus in this chapter. Although we cannot derive it, we'll show why it is plausible, and then we'll apply it to a number of situations. Schrödinger's equation makes predictions that are counterintuitive, and even seem impossible—such as the one in this chapter's case study.

CASE STUDY The Sun

The Sun generates energy through the nuclear fusion of hydrogen into helium (Example 39.10) in a three-part chain reaction. The first reaction is the most difficult because it requires fusing two hydrogen nuclei—that is, two protons—together. By Coulomb's law (Section 23-5), two protons repel one another, and the closer they get together, the greater their mutual repulsion. Nothing we have studied so far can explain how two protons can be smashed together. In fact, according to classical physics the reaction shouldn't take place—but we know it does. If it didn't, there would be no sunlight. We'll see how what seems to be impossible becomes possible in quantum physics.

41-2 A Trapped Particle

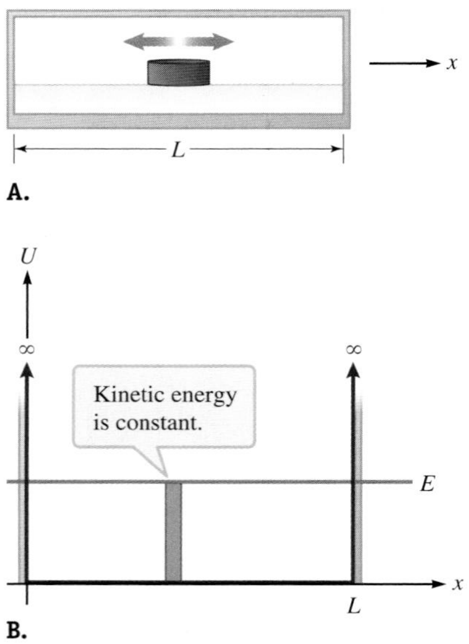

Let's start by studying a simple situation using a *quasi-quantum* approach. Consider a hockey puck trapped in a long narrow box, so that it can only move back and forth along a single line (Fig. 41.1A). The surface underneath the puck is ice. We model this as an ideal system consisting of the puck and the walls of the box, in which there are no dissipative forces. No work is done and no energy is lost by heat—so mechanical energy is conserved. Our goal is to use a quasi-quantum approach to find the system's energy, allowing us to see how problems in quantum mechanics are characterized. For this simple example, we'll get the same results we'll find later using Schrödinger's equation (Section 41-5). We call this situation "a particle in a one-dimensional box," because the puck can only move along one dimension. In this chapter we only consider the one-dimensional Schrödinger equation, so this problem is as general as we need. For this one-dimensional problem, we can ignore the two long walls. Because we only need to concern ourselves with the short walls at each end, we'll refer to these simply as "the walls." (The puck is never in contact with the ceiling, and there is no net vertical force exerted on the puck.)

When the puck is moving between the walls of the box, its kinetic energy is constant. Also, no net force is exerted on the system, so the system's potential energy is set to zero. In contrast, when the puck encounters a wall, it experiences a normal force that causes its momentum to be reversed instantaneously, without any change in magnitude. So at the walls, the system's potential energy is infinite. In quantum mechanics we work with potential energy instead of forces. Figure 41.1B shows an energy graph for this system. Between the walls, the system's total energy is finite—that is, too low for the puck to be found outside the box—so the walls are at turning points. If this system were being modeled according to classical mechanics, the puck could be at rest or moving at any speed within the box, and the system's energy E could have any value. But our quasi-quantum approach will show that the system's energy is restricted to certain values.

De Broglie proposed that a particle should be modeled as a wave with a wavelength given by $\lambda = h/p$ (Eq. 40.18). Because the puck is trapped in a box of length L, let's model the puck as a standing wave on a string of length L fixed at both ends, like a guitar string (Section 18-5, page 534). Recall that a standing wave on a string has a wave function of the form given by Equation 18.6, $y(x, t) = [2y_{\max} \sin(kx)] \cos(\omega t)$. We can separate this wave function's time part (the cosine term) from its space part (the sine term). We are only interested in the space part, which we write as

$$y(x) = A \sin kx \qquad (41.2)$$

where A is the amplitude of the space part. Because the ends of the guitar string are fixed, they cannot oscillate, so the standing wave that forms on the string

FIGURE 41.1 A. A hockey puck on ice is trapped inside a long narrow box. The puck can only move in one dimension, along the x axis. Only the short walls perpendicular to the x axis can limit the puck's one-dimensional motion. **B.** The potential energy curve is zero in the box and infinite at the walls.

Throughout our discussion of quantum mechanics, we'll use a puck as an example of a particle. This is simply a visual aid; you should imagine that the puck is so small that quantum-mechanical effects are important.

must have a node on each end. Therefore, only standing waves with wavelengths given by

$$\lambda_n = 2\frac{L}{n} \quad (n = 1, 2, 3 \ldots) \quad (18.9)$$

will fit on the string. This restriction on the wavelength can also be expressed as a restriction on the angular wave number $k = 2\pi/\lambda$ (Eq. 17.5) as

$$k_n = \frac{n\pi}{L} \quad (n = 1, 2, 3 \ldots) \quad (41.3)$$

We say that the wavelength and angular wave number are *quantized*, meaning that only certain separate values are allowed. So the family of possible standing waves is given by

$$y_n(x) = A_n \sin k_n x = A_n \sin \frac{n\pi x}{L} \quad (41.4)$$

where n is referred to as the "harmonic number" and each possible standing wave y_n is called the "nth harmonic." The amplitude of each harmonic is A_n.

Unless otherwise specified, n is a counting number: $n = 1, 2, 3 \ldots$

Now let's go back to modeling the trapped hockey puck. In Equation 41.4, y is the vertical position of the string at the horizontal position x. When we say that the puck is modeled as a wave on a string, we don't mean that its vertical position is waving. We really don't know what is waving (although we'll get a better idea of that in the next section); we just mathematically represent whatever is waving as ψ, and write (by analogy with Eq. 41.4):

$$\psi_n(x) = A_n \sin k_n x = A_n \sin \frac{n\pi x}{L} \quad (41.5)$$

Conceptually, Equation 41.5 says that the trapped puck is modeled as a family of standing waves. Instead of referring to n as the harmonic number, it is called the **quantum number**, and instead of referring to each possible standing wave ψ_n as the "nth harmonic," we refer to each ψ_n as a **quantum stationary state**, or simply a **state**. The lowest state is called the **ground state**, corresponding to $n = 1$.

The angular wave number k_n and the wavelength λ_n of the waves that represent the puck are quantized, just as they are for a standing wave on a string. Further, because de Broglie proposed that the momentum of a matter wave depends on its wavelength, the puck's momentum is quantized:

$$p_n = \frac{h}{\lambda_n} = \frac{nh}{2L} \quad (41.6)$$

Because the system's potential energy is zero inside the box (the only place the puck can be found), its mechanical energy is equal to the puck's kinetic energy. If the puck's mass is m, then the system's energy is $E = p^2/2m$. Since the puck's momentum is quantized, the system's energy is quantized:

$$E_n = \frac{(nh/2L)^2}{2m} = n^2\left(\frac{h^2}{8mL^2}\right) \quad (41.7)$$

Contrast this quasi-quantum analysis of a puck in a box with a classical analysis. According to classical mechanics, the puck can have any momentum and the system can have any energy. However, when we model the puck as a standing wave, we find it can only have certain momenta (Eq. 41.6), and the system can only have certain energies (Eq. 41.7). This sounds a lot like Planck's quantum hypothesis (Section 40-2): Planck proposed that the particles inside a black body have quantized energy levels. He didn't have an explanation for why this is true, but he found that it fit the data. De Broglie's model is a little better because it is based on the idea that nature is symmetric; because light is modeled as both a wave and a particle, matter should be modeled as both as well. The quantization of energy levels is a natural consequence of modeling a particle as a wave. However, de Broglie's model is not completely satisfying because we don't have a reason for modeling the trapped puck as a string fixed on two ends. Part of our goal in this chapter is to

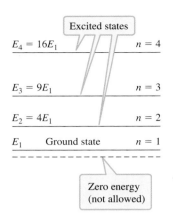

FIGURE 41.2 An energy-level diagram shows the relative energies possible for a system. In this case, the system is a particle in a box with infinitely high walls.

ENERGY-LEVEL DIAGRAM Tool

achieve a more complete understanding of quantum mechanics—then we will return to this trapped particle.

Energy Levels

There are a few more things we can learn from our quasi-quantum approach. First, because the lowest allowed value for the quantum number n is 1, the puck's lowest allowed momentum is $p_1 = h/2L$, which means the puck cannot be at rest. Second, the minimum value of the system's energy is $E_1 = h^2/8mL^2$. Both of these facts are true for all quantum-mechanical systems in which a particle is trapped, and the system's minimum energy E_1 is called its **zero-point energy** or **ground-state energy**. Third, if such a system is to gain or lose energy, it must do so in whole quantized steps, so that the system is in one of the states with an energy given by Equation 41.7.

The energy of a quantum system is often best represented visually by an **energy-level diagram**. The energy of each state is represented by a horizontal line labeled with its quantum number n, and the relative spacing of the lines represents the energy between each state. Figure 41.2 shows the energy levels for a particle trapped in a box (Eq. 41.7). For this system, the energy levels get farther apart for higher quantum numbers, which we can best see by writing the energy levels in terms of the zero-point energy: $E_n = n^2 E_1$.

EXAMPLE 41.1 A Hockey Puck in a Box

Suppose that a hockey puck's mass is 0.160 kg and it is in a box of length 6.35 m (Fig. 41.1). Find the system's zero-point energy and the energies of the first three excited states. In **CHECK and THINK**, comment on the difference between the third excited state and the ground state.

∴ INTERPRET and ANTICIPATE
A hockey puck in a box is a classical (macroscopic) system. So we expect to find results that seem familiar.

∴ SOLVE Find the zero-point energy by setting $n = 1$ in Equation 41.7.	$E_n = n^2 \left(\dfrac{h^2}{8mL^2} \right)$ (41.7) $E_1 = \dfrac{h^2}{8mL^2} = \dfrac{(6.63 \times 10^{-34} \text{ J} \cdot \text{s})^2}{8(0.160 \text{ kg})(6.35 \text{ m})^2}$ $E_1 = 8.52 \times 10^{-69}$ J
The first three excited states have quantum numbers 2, 3, and 4. Write the energy of the excited states in terms of E_1.	$E_n = n^2 E_1$ $E_2 = 2^2 E_1 = 3.41 \times 10^{-68}$ J $E_3 = 3^2 E_1 = 7.66 \times 10^{-68}$ J $E_4 = 4^2 E_1 = 1.36 \times 10^{-67}$ J

∴ CHECK and THINK
The lowest energy level is not zero, but it is so small that we would not distinguish it from zero. Also, keep in mind that we derived the energy levels for the ideal situation in which there are no dissipative forces, and that any real hockey puck would experience dissipative forces. Furthermore, the energy difference between the third excited state and the ground state is about 10^{-67} J. Again, this is too small to be distinguished from zero.

41-3 The Double-Slit Experiment Revisited: Probability Waves

Schrödinger was able to apply his equation to a number of situations, but he did not come up with a physical interpretation of the wave function Ψ. In other words, he was unable to answer the question "*What is doing the waving?*" To answer that question, we must think more about wave-particle duality, and return to the double-slit experiment.

When coherent light passes through two narrow slits, an interference pattern of bright and dark fringes is seen on a screen (Fig. 35.10, page 1128). The pattern is well described by the wave model for light (Section 35-1). We've also seen that objects that we have classically modeled as particles, such as electrons, can produce diffraction patterns (Fig. 40.16). The fringe pattern in part E of Figure 41.3 looks much like the interference pattern created by light passing through two slits, but in this case it was formed by electrons. What is really remarkable about Figure 41.3 is that the fringe pattern is built up one electron at a time. (Photons will also create an interference pattern that is built up one photon at a time.) How can an *interference* pattern be created when each particle (a photon or electron) encounters the slit by itself, with no other particles to interfere with it?

This question was answered by the German physicist Max Born (1882–1970) who, in doing so, came up with an interpretation for Ψ. He said that for matter waves there is no *physical* interpretation of the wave function Ψ itself, but we can think of Ψ as a **probability wave**. (It makes sense that there is no physical interpretation of Ψ because Ψ involves imaginary numbers.) Furthermore, he said that we don't actually detect Ψ; instead we detect $|\Psi|^2$, which is the **probability density** of finding the particle at a particular location and time. As we already stated, we will mostly consider one-dimensional situations, and so $|\Psi|^2$ is probability per unit length.

To understand the idea of a probability wave, let's return to the familiar interference pattern created by photons that are incident on a double slit. We know that one way to model light is as an electromagnetic wave, and that the bright and dark fringes are *not* a direct measure of the light's electric (or magnetic) field. Instead, the brightness of the fringes is a measure of the light's intensity, and intensity is proportional to the amplitude of the electric (or magnetic) field: $I \propto E^2 \propto B^2$ (Eq. 34.30). So the fringes are measuring E^2 (or B^2), but not E (or B) directly. When we see a bright fringe, we infer that amplitude of the electric field (and magnetic field) is great.

Now let's apply the photon model to the double-slit experiment. Imagine that we use narrow detectors, labeled A and B, to count the arrival of photons at two locations on the screen (Fig. 41.4). Let's also imagine that each detector makes a unique sound when it detects a particle: A chirps and B clicks. Because the individual photons arrive at random times, we hear chirps and clicks from the detectors at random time intervals. We cannot predict *when* a particle will arrive at either detector. However, detector A is located on a bright fringe, whereas detector B is located somewhere between a bright fringe and a dark fringe. The bright fringes are bright because they receive many more photons per area than do the other regions of the screen. That is, the probability of a particle arriving at detector A is much greater than the probability of it arriving at B—which means that in the same time interval we expect to hear more chirps from A than clicks from B. We have just reasoned that it is more likely that a particle will be detected at a bright place than at a dark place. Mathematically, we can say that the probability density (here, probability per unit area) of detecting a photon at a particular location on the screen is proportional to intensity, I. Because $I \propto E^2 \propto B^2$, we can argue that the probability density is proportional to E^2 (or B^2). Put simply in informal terms, the "things" that do the waving in an electromagnetic wave are E and B, and the probability density is proportional to their *squares*.

In Born's interpretation, however, Ψ is the "thing" that waves. Ψ is analogous to the electric (or magnetic field) of a light wave, but it has no physical interpretation.

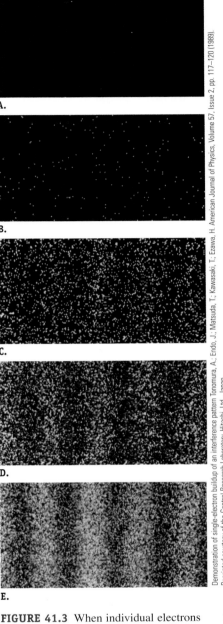

FIGURE 41.3 When individual electrons are successively incident on two slits, an interference pattern of bright and dark fringes builds up. Individual electrons alone do not reveal the pattern; see progression from part **A.** to part **E.**

PROBABILITY WAVE AND PROBABILITY DENSITY ⊙ **Major Concept**

Some authors refer to a probability wave as a wave function.

As described in Section 41-4, $|\Psi|^2 = \Psi \cdot \Psi^*$.

$|\Psi|^2$ is analogous to E^2 (or B^2).

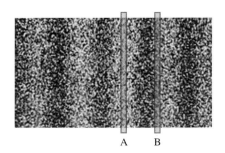

FIGURE 41.4 Detector A is located on a bright fringe; detector B is located between a bright and a dark fringe. (Demonstration of single-electron buildup of an interference pattern Tonomura, A.; Endo, J.; Matsuda, T.; Kawasaki, T.; Ezawa, H. American Journal of Physics, Volume 57, Issue 2, pp. 117–120 (1989). Reprinted courtesy of the Central Research Laboratory, Hitachi, Ltd., Japan.)

So by analogy with the intensity, the probability density is proportional to $|\Psi|^2$. Using this interpretation, we can understand how the interference pattern can be built up one electron (or one photon) at a time. A source generates a particle (an electron or a photon). Associated with *each* particle is a probability wave function Ψ, which we cannot measure directly. This wave travels from the source to the detector (the screen). When the wave arrives at the detector, it is detected as a particle (an electron or a photon). So the pattern on the screen is a measure of the probability density $|\Psi|^2$ of the particle arriving at a particular location. Of course, each particle arrives at some particular location and does not create the entire pattern by itself. The entire pattern is created only when a large number of particles have arrived (Fig. 41.3).

The idea that we can only measure the probability density $|\Psi|^2$ is a major departure from classical physics. According to classical physics, if we know the energy and momentum of a particle at some particular time, we can determine whether or not it will arrive at some detector. (This is just like calculating whether a basketball will drop through the basket.) According to classical mechanics, your answer is completely determined: Its probability is 100%.

Probability only enters into classical physics when we try to study a system with too many particles to count practically. If we are interested in whether a gas molecule will pass through a hole in the wall of its container, we answer with a probability only because there are too many particles to calculate the trajectory of each particle based on each particle's energy and momentum.

The probability that enters into quantum physics is different in nature. Quantum mechanics says that all of our calculations result in probabilities and that none of our answers are 100% determined. So instead of asking, "Will the basketball drop through the basket?" or "Will the gas molecule pass through the hole?" we should ask, "What is the probability of the basketball dropping through the basket or the molecule through the hole?" Quantum mechanics says that this is not a matter of the basketball player's skill or of having too many particles to keep track of. Probability is a fundamental principle of nature.

The idea that nature works in probabilities was very controversial when Born first proposed it. It took about 10 years to become generally accepted, and some scientists—such as Schrödinger, Bohr, and Einstein—never really accepted this interpretation of nature. Although the idea may be hard to accept, the probabilities calculated by using Schrödinger's equation work well to describe many experiments and physical phenomena that cannot be explained by the deterministic interpretation of classical mechanics.

CONCEPT EXERCISE 41.1

Three students were asked why the temperature of a gas is proportional to the *average* kinetic energy of the gas particles as opposed to the *exact* kinetic energy of the particles. (You may wish to review Chapter 20.) Read the discussion and decide who you think is correct. Explain your answer.

Avi: According to Born's interpretation, nature works in probabilities. We can only calculate averages and not exact values.

Cameron: We cannot observe the exact microscopic kinetic energy of every particle in a gas because there are too many particles. We are forced to work with averages when we have so many particles.

Shannon: We use an average because the gas molecules are very tiny, and quantum mechanics applies to tiny particles. That's why we are forced to use Born's interpretation.

EXAMPLE 41.2 Throwing Dice

A die is a cube whose six sides are marked by dots; one side has one dot, another two dots and so forth. If you roll a single die, the probability of rolling a 4 is 1/6 because there are six sides and only one has exactly four dots. If you roll two dice simultaneously, what is the probably of rolling a 4?

INTERPRET and ANTICIPATE
This is a classical mechanical problem, and if we knew exactly the initial conditions of the dice, we could use dynamics and kinematics to determine how the dice land. The only reason we can apply probability is that we don't know exactly how the two dice were tossed. This problem gives us a chance to review probability. The probability is found by determining how many possible ways the two dice may land such that the total is 4 out of all the possible ways the dice may land.

SOLVE
Make a sketch (Fig. 41.5), to find all ways the two dice may land. Imagine one die is blue and the other is red. The sketch shows that if blue die is a 1, the red die may be any value 1 through 6 for six possible combinations. The same reasoning can be used if the blue die is 2 for another six combinations, and so forth.

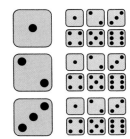

FIGURE 41.5   . . .   . . .

This sketch (Fig. 41.5) only shows the first three possible values for blue die, but there are six. The sketch shows 18 possible combinations, but that is only half of the total possible combinations.	$N_{tot} = 36$
Next make another sketch (Fig. 41.6) to consider all the ways the two dice may be tossed so they total 4.	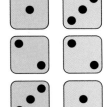 **FIGURE 41.6**
From Figure 41.6, we see there are three ways to roll a 4.	$N = 3$
The probability of rolling a 4 is the number of ways to roll a 4 divided by the number of possible ways the dice may land.	$P = \dfrac{N}{N_{tot}} = \dfrac{3}{36} = \dfrac{1}{12}$ $P \approx 8\%$

CHECK and THINK
So the probability of rolling a 4 when you roll just one die is twice as likely as when you roll two dice.

41-4 Schrödinger's Equation

With Born's probability interpretation in hand, we now turn our attention back to Schrödinger's equation. When we apply Schrödinger's equation to a situation, our goal is to come up with an expression for Ψ, which we can then use to find the probability density $|\Psi|^2$—a quantity that we can measure.

We must begin with a simplification. In Section 41-1, we presented the time-dependent version of Schrödinger's equation:

$$-\frac{h^2}{8\pi^2 m}\frac{\partial^2 \Psi(x,t)}{\partial x^2} + U(x,t)\Psi(x,t) = \frac{ih}{2\pi}\frac{\partial \Psi(x,t)}{\partial t} \quad (41.1)$$

In our study of classical waves, we learned that standing-wave solutions are somewhat easier to study because we can separate the space and time dependencies into two factors, such as $y(x,t) = [2y_{max}\sin(kx)]\cos(\omega t)$ in Equation 18.6. The same is true for the solutions to Schrödinger's equation; standing-wave solutions can be expressed as $\Psi(x,t) = \psi(x)e^{-i\omega t}$, where the lowercase Greek letter psi ψ represents the space part of the solution and $\omega = 2\pi E/h$. Finding general solutions for the time-dependent partial differential equation adds a layer of complication we do not explore in this book. Here we only consider standing-wave solutions for the time-independent, one-dimensional Schrödinger equation:

TIME-INDEPENDENT SCHRÖDINGER'S EQUATION ❶ **Underlying Principle**

$$-\frac{h^2}{8\pi^2 m}\frac{d^2\psi(x)}{dx^2} + U(x)\psi(x) = E\psi(x) \quad (41.8)$$

We don't need partial derivatives in the time-independent equation.

where E is the mechanical energy of the system.

Equation 41.8 holds for any system in which the potential energy $U(x)$ only depends on space (and not time) and the mechanical energy E is a constant. You will derive this time-independent equation in Problem 41.11. Conceptually, you can think of Schrödinger's equation as describing a particle modeled as a matter wave; the particle is part of a system whose mechanical energy is E and whose potential energy $U(x)$ does not change with time, but varies from place to place. From now on, when we refer to the Schrödinger equation, you can assume we mean the time-independent version (Eq. 41.8) unless otherwise specified.

A Plausible Equation

Schrödinger's equation is a departure from classical physics, so it cannot be derived from the principles of classical physics. However, we can show that it seems plausible. The argument that follows is a test that is just a little more thorough than checking the equation's dimensions. Recall that Schrödinger was looking for a theory to support de Broglie's matter wave model. Assume that $\psi(x)$ is a simple sine wave with a wavelength given by the de Broglie wavelength $\lambda = h/p$ (Eq. 40.18):

$$\psi(x) = \psi_{max}\sin kx = \psi_{max}\sin\frac{2\pi x}{\lambda}$$

Take the second derivative with respect to x:

$$\frac{d^2\psi}{dx^2} = -\psi_{max}\frac{4\pi^2}{\lambda^2}\sin\frac{2\pi x}{\lambda} = -\frac{4\pi^2}{\lambda^2}\psi \quad (41.9)$$

Use $K = p^2/2m$ to write the de Broglie wavelength in terms of the particle's (nonrelativistic) kinetic energy:

$$\lambda = \frac{h}{p} = \frac{h}{\sqrt{2mK}}$$

Substitute into Equation 41.9:

$$\frac{d^2\psi}{dx^2} = -4\pi^2\frac{2mK}{h^2}\psi = -\frac{8\pi^2 mK}{h^2}\psi \quad (41.10)$$

Finally, substitute $K = E - U$ and rearrange:

$$\frac{d^2\psi}{dx^2} = -\frac{8\pi^2 m(E-U)}{h^2}\psi$$

$$-\frac{h^2}{8\pi^2 m}\frac{d^2\psi(x)}{dx^2} + U(x)\psi(x) = E\psi(x) \checkmark$$

We have *not derived* Schrödinger's equation; instead we have simply shown that it is plausible that the equation describes a particle modeled as a matter wave with a de Broglie wavelength in a system with potential energy U.

Finding Solutions

Now we're ready to explore solutions to Schrödinger's equation. Each solution ψ depends on a particular form of the potential energy $U(x)$. Once we arrive at a solution, $|\psi|^2$ is the probability density of finding the particle at some position x. Because we are working with one-dimensional systems, the probability density is the probability per unit length, and the probability of finding the particle in a region dx is $|\psi|^2 dx$. Of course, the particle must be found somewhere along the x axis; mathematically, we express this commonsense idea as

$$\int_{-\infty}^{\infty}|\psi(x)|^2 dx = 1 \qquad 41.11$$

NORMALIZATION CONDITION
✪ **Major Concept**

Equation 41.11 is known as the **normalization condition**, and it is required in order to interpret $|\psi|^2$ as a probability density. In practice, we must measure the positions of a large number of identical systems with the same wave function ψ in order to come up with an average value for the position x. The average position of x is called the **expectation value** of x, and is given by:

$$\langle x \rangle \equiv \int_{-\infty}^{\infty} x|\psi(x)|^2 dx \qquad (41.12)$$

The symbol $\langle q \rangle$ means the expectation value of the quantity q. Don't be alarmed by this terminology: it is the same as the familiar phrase *average value* of q.

The process of solving Schrödinger's equation is mathematical. From our experience with physics, we know that often there are solutions that work mathematically but have no physical meaning. In order for a solution ψ to be physically meaningful so that we can interpret $|\psi|^2$ as a probability density, we require $\psi(x)$ to meet the following conditions.

1. $\psi(x)$ and $d\psi(x)/dx$ must be a continuous, single-valued function for all points along x. (Otherwise we couldn't take the second derivative $d^2\psi(x)/dx^2$ in Schrödinger's equation.)
2. $\psi(x)$ must equal zero if x is in a region where the particle cannot be found.
3. $\psi(x)$ must meet the normalization condition (Eq. 41.11). This is same as requiring that $\psi(x) \to 0$ as $x \to \pm\infty$ in order for the integral to converge.

In general, the solution $\psi(x)$ to Schrödinger's equation may be a complex number. Then the probability density is $|\psi|^2 = \psi \cdot \psi^*$, where ψ^* is called the **complex conjugate**. The complex conjugate is found from ψ by replacing i with $-i$. For example, if $\psi = \psi_{max} e^{ikx}$, then its complex conjugate is $\psi^* = \psi_{max} e^{-ikx}$, and $|\psi|^2 = \psi_{max}^2 e^{ikx} e^{-ikx} = \psi_{max}^2$. Much of the rest of this chapter is devoted to $\psi(x)$ and then to finding the probability density ψ^2 for a number of special cases. Our major goal is to find the energy levels of each system and present these on an energy-level diagram.

One way that we will check our results is with **Bohr's correspondence principle**. According to this principle, in the limit of very large quantum numbers, the classical calculation and the quantum calculation must yield the same results. This principle is much like what we do in special relativity, when we check to see that at low relative speed our results are consistent with Galilean relativity.

BOHR'S CORRESPONDENCE PRINCIPLE
❗ **Underlying Principle**

> **CONCEPT EXERCISE 41.2**
>
> Come up with at least one example from the previous 40 chapters where a mathematical solution does not make sense physically. *Hint*: You are likely to find an example from as early as Chapter 2.

EXAMPLE 41.3 — Average Position for a Particle in an Infinite One-Dimensional Well, Classical Approach

Consider the hockey puck trapped between the walls of the one-dimensional box in Figure 41.1. As described in Section 41-2, when the puck encounters a wall, it reverses direction instantaneously and there are no dissipative forces. So the puck's kinetic energy is constant and nonzero. Use classical mechanics to find the (average position) expectation value of x for a puck in an infinite well of width L. (In Example 41.4, we will take a quantum-mechanical approach to answer the same question.)

INTERPRET and ANTICIPATE
The expectation value is another way of saying the average value. Because the puck slides back and forth without losing speed, you might expect that its average position is in the center of the box.

SOLVE
The puck doesn't change speed, and so it is equally likely to be anywhere inside the box. So the probability density $|\psi|^2$ is a constant between the walls of the box and zero outside the box as shown in Figure 41.7.

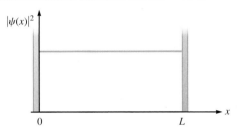

FIGURE 41.7

Of course, the puck must be located somewhere along the x axis inside the box. To express this commonsense idea, we apply the normalization condition (Eq. 41.11). In this case, we know $\|\psi\|^2$ is a constant between $x = 0$ and L and zero elsewhere.	$\int_{-\infty}^{\infty} \|\psi(x)\|^2 dx = 1$ (41.11) $\int_{-\infty}^{0} \|\psi(x)\|^2 dx + \int_{0}^{L} \|\psi(x)\|^2 dx + \int_{L}^{\infty} \|\psi(x)\|^2 dx = 1$ $0 + \|\psi(x)\|^2 \int_{0}^{L} dx + 0 = \|\psi(x)\|^2 L = 1$ $\|\psi(x)\|^2 = \dfrac{1}{L}$ (for $0 < x < L$)
Now we are ready to find the expectation value of x using Equation 41.12. (Keep in mind that the integrals from $x = -\infty$ to 0 and from $x = L$ to ∞ are zero because $\|\psi\|^2$ is zero in these regions. These two integrals are not shown here.)	$\langle x \rangle = \int_{-\infty}^{\infty} x \|\psi(x)\|^2 dx$ (41.12) $\langle x \rangle = \int_{0}^{L} x \|\psi(x)\|^2 dx = \int_{0}^{L} x \dfrac{1}{L} dx = \dfrac{1}{L}\left(\dfrac{x^2}{2}\right)\Big\|_{0}^{L}$ $\langle x \rangle = \dfrac{L}{2}$

CHECK and THINK
We found that the average position is in the center of the box as expected. In Example 41.4, we'll find the same thing using quantum mechanics.

41-5 Special Case: A Particle in an Infinite Square Well

Now we are ready to return to a particle trapped in a one-dimensional box, like the puck in Figure 41.1. This time, however, we'll solve Schrödinger's equation to find the system's energy levels. We'll gain a deeper insight into quantum mechanics because we'll also find expressions for where the particle is most likely to be found. Because we know that the quantum nature of a hockey puck in a box is difficult to observe, it is better to imagine a much smaller particle, such as a proton or an electron, that is "trapped in a box" owing to some electric potential energy well.

We use the term *potential energy well* to describe the general shape of a potential energy curve that rises on two sides, known as its **boundaries**. Figure 41.1B is a good illustration for a particle in an infinite *square* well. The system consists of the (unknown) source of potential energy and a particle that is trapped between $x = 0$ and $x = L$. The potential energy is $U = 0$ for $0 < x < L$ and $U = \infty$ for $x \leq 0$ and for $x \geq L$.

The particle cannot be outside the box, so $\psi(x) = 0$ for $x \leq 0$ and for $x \geq L$. The value that $\psi(x)$ must have at a boundary is known as the **boundary condition**. We'll use boundary conditions to help make a general solution to Schrödinger's equation fit a particular situation. The strict boundary conditions in this special case correspond to our physical intuitions about an infinite square well. But when we solve Schrödinger's equation in the more general case of a *finite* square well using more relaxed boundary conditions, we'll find that our intuition that the particle cannot be outside the infinite square well is correct.

BOUNDARY CONDITION
⊙ Major Concept

Inside the well $0 < x < L$, $U(x) = 0$, and Schrödinger's equation is

$$-\frac{h^2}{8\pi^2 m}\frac{d^2\psi(x)}{dx^2} = E\psi(x)$$

It is convenient to combine the constants as $k^2 \equiv 8\pi^2 mE/h^2$; then we write Schrödinger's equation as:

$$\frac{d^2\psi(x)}{dx^2} + k^2\psi(x) = 0 \quad (41.13)$$

The general solution to Equation 41.13 may be written as:

$$\psi(x) = A\sin kx + B\cos kx \quad (41.14)$$

We apply the boundary condition $\psi(0) = 0$:

$$\psi(0) = A\sin 0 + B\cos 0 = 0$$

and we find that $B = 0$. So now our solution is a little more specific; there is no cosine term. We apply the other boundary condition $\psi(L) = 0$:

$$\psi(L) = A\sin kL = 0$$

and we find that this boundary condition holds if k is quantized:

$$k_n = \frac{n\pi}{L} \quad (n = 1, 2, 3\ldots) \quad (41.15)$$

where we start with $n = 1$. Although $n = 0$ also fits the boundary condition, it cannot be normalized, Put another way, allowing $n = 0$ is the same as saying that the wave does not exist. So we reject this physically unacceptable solution.

Without finding our complete solution to Schrödinger's equation, we have already found that its energy levels are quantized. To see this, combine $k^2 \equiv 8\pi^2 mE/h^2$ with Equation 41.15:

$$E_n = \left(\frac{h^2}{8\pi^2 m}\right)\left(\frac{n^2\pi^2}{L^2}\right) = n^2\left(\frac{h^2}{8mL^2}\right) \quad (41.16)$$

PARTICLE IN AN INFINITE SQUARE WELL
⊙ Special Case

This is exactly what we found using our quasi-quantum approach (Eq. 41.7). Now, however, we have a better understanding of how we got to these quantized levels. Schrödinger's equation describes the mechanical energy of the system. The potential energy is zero in the box, and the particle cannot be found outside the box. When we express the boundary conditions mathematically, we find that the only solution to Schrödinger's equation requires the energy levels to be quantized. This is a general truth about trapped particles: The energy of of a trapped particle comes in discrete values.

Normalizing the Wave Function

Let's complete our solution by normalizing $\psi(x)$. So far, $\psi(x) = A \sin k_n x$ inside the well, where k_n is given by Equation 41.15. When we normalize the wave function, we'll find A. Conceptually, the normalization condition says that the particle must exist somewhere, and mathematically we have $\int_{-\infty}^{\infty} |\psi(x)|^2 dx = 1$ (Eq. 41.11). Substitute $\psi(x)$ into the normalization condition:

$$\int_{-\infty}^{\infty} |A \sin k_n x|^2 dx = 1$$

We break this integral into three parts, corresponding to potential energy values to the left of the well, inside the well, and to the right of the well:

$$\int_{-\infty}^{0} |A \sin k_n x|^2 dx + \int_{0}^{L} |A \sin k_n x|^2 dx + \int_{L}^{\infty} |A \sin k_n x|^2 dx = 1$$

We have already established that the particle cannot be outside the box, so the first and last integrals are zero:

$$\int_{0}^{L} |A \sin k_n x|^2 dx = 1$$

Using

$$\int \sin^2 ax \, dx = \frac{x}{2} - \frac{\sin 2ax}{4a} + C$$

from Appendix A to solve this integral, we find that $A^2(L/2) = 1$ (Problem 41.79). So $A = \sqrt{2/L}$ and

$$\psi(x) = \sqrt{\frac{2}{L}} \sin k_n x \tag{41.17}$$

Interpreting the Wave Function

Equation 41.17 is the solution to Schrödinger's equation for a particle in an infinite square well. (Of course, outside the well $\psi = 0$.) Although we have no physical interpretation for ψ, $|\psi|^2$ is the probability density, given by:

$$|\psi|^2 = \frac{2}{L} \sin^2 k_n x \tag{41.18}$$

Figure 41.8 shows the probability density function plotted for four different states.

For a moment, let's think classically about the hockey puck oscillating back and forth between the walls of its box (Fig. 41.1). You might ask, "At any instant, where is the puck most likely to be in the box?" The answer has to do with the puck's speed as it moves back and forth. If the puck maintains its speed, then at any instant it is equally likely to be anywhere inside the box (Example 41.3 and Problem 41.14). Of course, this is true for the puck in the box whether the system's energy is great or small.

By analogy, Equation 41.18 and Figure 41.8 are attempts to answer the question, "At any instant, where is the particle most likely to be in the well?" Perhaps the results for the first quantum states are the most surprising because these show that the particle is most likely to be in one, two, or three particular regions in the box. Only for a high quantum number state are the maxima so closely spaced that it becomes equally likely to find the particle anywhere in the box.

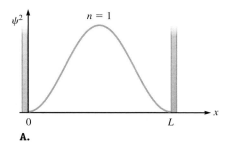
A.

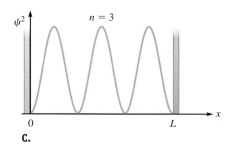
C.

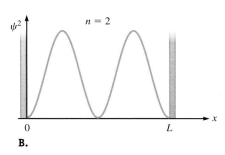
B.

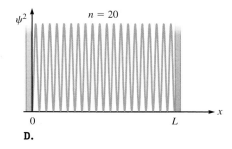
D.

FIGURE 41.8 The probability densities for a particle in a box for quantum states **A.** $n = 1$, **B.** $n = 2$, **C.** $n = 3$, and **D.** $n = 20$.

However, we need to be careful with our analogy. Our model of a puck sliding back and forth between the walls of a box is insufficient for a particle in quantum mechanics. Instead, we model a particle as a wave whose probability density is shown in Figure 41.8. Instead of visualizing a solid particle in motion, it is better to think of (the probability distribution of) the particle as a cloud. In its ground state, the cloud is most concentrated in the center of the box and thinly spread over the rest. In its $n = 2$ state, the cloud is concentrated in two regions and so on. For a very high quantum number, the cloud is well distributed throughout the box.

CONCEPT EXERCISE 41.3

Does Figure 41.8 show the motion of a trapped particle? If so, describe this motion in words. If not, describe what the figure is really showing about the particle.

EXAMPLE 41.4 Average Position for a Particle in an Infinite One-Dimensional Well, Quantum Approach

Find the expectation value of x for a particle in an infinite well of width L. This time apply the solution we found to Schrödinger's equation for this situation.

:• INTERPRET and ANTICIPATE
This problem is the same as Example 41.3, but here we take a quantum-mechanical approach. In the end we'll compare our results.

:• SOLVE
Start with Equation 41.12 for the expectation value of x. Change the limits of integration, since the particle cannot be found outside the box.

$$\langle x \rangle = \int_{-\infty}^{\infty} x|\psi(x)|^2 \, dx \qquad (41.12)$$

$$\langle x \rangle = \int_{0}^{L} x|\psi(x)|^2 \, dx$$

Substitute $\|\psi\|^2 = \frac{2}{L}\sin^2 k_n x$ (Eq. 41.18), which came from our solution to Schrödinger's equation.	$\langle x \rangle = \int_0^L x\left(\frac{2}{L}\sin^2 k_n x\right)dx$
This integral can be solved by parts.	$\langle x \rangle = \frac{2}{L}\left(\frac{x^2}{4} - \frac{x\sin 2k_n x}{4k_n} - \frac{\cos 2k_n x}{8k_n^2}\right)\Big|_0^L$
Substitute the limits and use $k_n = n\pi/L$ (Eq. 41.15). Notice that $\sin 2n\pi = 0$ and $\cos 2n\pi = 1$. (In quantum mechanics the energy is quantized.)	$\langle x \rangle = \frac{2}{L}\left(\frac{L^2}{4} - \frac{\cos 2k_n L}{8k_n^2} + \frac{1}{8k_n^2}\right)\Big|_0^L$ $\langle x \rangle = \frac{2}{L}\left(\frac{L^2}{4} - \frac{\cos 2n\pi}{8k_n^2} + \frac{1}{8k_n^2}\right)$ $\langle x \rangle = \frac{L}{2}$

:• **CHECK and THINK**
Many times quantum mechanics seems counterintuitive, but not this time. Here the average position is in the center of the box, just as you'd expect (and found in Example 41.3).

EXAMPLE 41.5 A Ground-State Energy Free Particle

Take the limit of $E_1 = h^2/8mL^2$ (Eq. 41.16 with $n = 1$) to find the zero-point energy of a free particle.

:• **INTERPRET and ANTICIPATE**
The key to this problem is knowing what limit we must take in order to go from the energy expression for a trapped particle to that for a free particle. There are only two parameters we can change: m and L. Changing the mass does not set a particle free, but increasing the width of the well does. An infinitely wide well means that the particle can go anywhere.

:• **SOLVE** Take the limit as $L \to \infty$.	$\lim_{L \to \infty} E_1 = \lim_{L \to \infty} \frac{h^2}{8mL^2} \to 0$

:• **CHECK and THINK**
When a particle is free (not part of any system), it can have any energy value. So its minimum energy is zero and it may be at rest. Only a confined particle is required to have nonzero energy and to be in motion.

41-6 Special Case: A Particle in a Finite Square Well

In the previous section, we applied quantum mechanics to a particle trapped in a square well with infinite sides. Now we consider a particle in a finite square well. Astonishingly, we'll find that particle may be found slightly outside of the box! This is something like watching a hockey puck ooze through the walls of its box. This bizarre behavior is predicted by the solution to Schrödinger's equation.

The finite square well looks much like the infinite square well; as in that case, the potential energy is $U = 0$ for $0 < x < L$ inside the well (Fig. 41.9). The difference is that the potential energy is finite and uniform outside the well: $U = U_0$ for $x \leq 0$ and for $x \geq L$. We are interested in a trapped particle, so we will only consider a system for which $E < U_0$.

Schrödinger's equation for the region inside the well is the same as it was in the case of an infinite well:

$$\frac{d^2\psi(x)}{dx^2} + k^2\psi(x) = 0 \quad (41.13)$$

where $k^2 \equiv 8\pi^2 mE/h^2$. The general solution must be the same as in the infinite well:

$$\psi(x) = A \sin kx + B \cos kx \quad (41.14)$$

But now the boundary conditions are different. We no longer require that $\psi(x) = 0$ for $x \leq 0$ and for $x \geq L$. Instead, we only need $\psi(x)$ and $d\psi/dx$ to be continuous functions at the boundaries. (Later in this chapter we will show that these boundary conditions are sufficient to produce the same results as in the case of the infinite square well when $U_0 \to \infty$.) With these more relaxed boundary conditions, $B \neq 0$. Both terms in Equation 41.14 hold inside the well, and together they describe an oscillating function.

Outside the well, Schrödinger's equation is:

$$-\frac{h^2}{8\pi^2 m}\frac{d^2\psi(x)}{dx^2} + U_0\psi(x) = E\psi(x)$$

We group terms to write:

$$\frac{d^2\psi(x)}{dx^2} - \frac{8\pi^2 m}{h^2}(U_0 - E)\psi(x) = 0$$

and define $\kappa^2 \equiv (8\pi^2 m/h^2)(U_0 - E)$, so that we have:

$$\frac{d^2\psi(x)}{dx^2} - \kappa^2\psi(x) = 0 \quad (41.19)$$

for Schrödinger's equation outside the well. But we are only considering the case of a particle trapped in the well, such that $E < U_0$ or $\kappa^2 > 0$. The general solution to Equation 41.19 in this case is:

$$\psi(x) = \alpha e^{-\kappa x} + \beta e^{\kappa x} \quad (41.20)$$

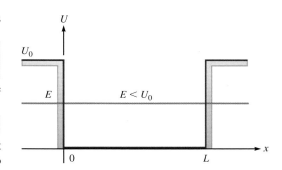

FIGURE 41.9 The potential energy is zero inside the well and finite outside.

In the region to the left of the well, where $x < 0$, we require $\psi(x) \to 0$ as $x \to -\infty$, and so we must set $\alpha = 0$ in this region and the solution reduces to $\psi_{\text{left}}(x) = \beta e^{\kappa x}$. In the region to the right of the well, where $x > L$, we require $\psi(x) \to 0$ as $x \to \infty$, and so we must set $\beta = 0$ in this region and the solution reduces to $\psi_{\text{right}}(x) = \alpha e^{-\kappa x}$.

In principle, we solve for A, B, α, and β by ensuring that $\psi(x)$ and $d\psi/dx$ are continuous at both boundaries. In practice, this is a tedious process, and unnecessary for the points we wish to make here. Instead we'll present the solution graphically for the first three quantum states in Figure 41.10A. Like the infinite square well (Fig. 41.10B), the solution inside the finite square well is sinusoidal. The major difference is that outside the well, the solution is a decaying exponential function. However, you can see a more subtle difference if you carefully compare the solutions in parts A and B of the figure: The wavelengths inside the finite square well are slightly

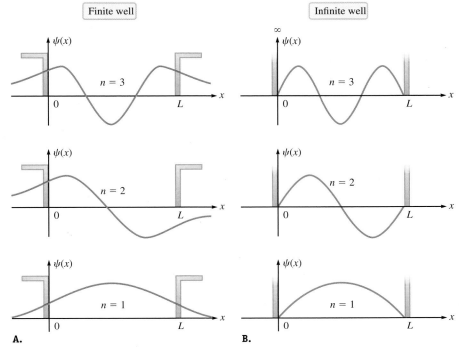

FIGURE 41.10 A. The wave functions for a finite square well extend beyond the boundaries of the well. B. The wave functions for an infinite square well are zero beyond the boundaries.

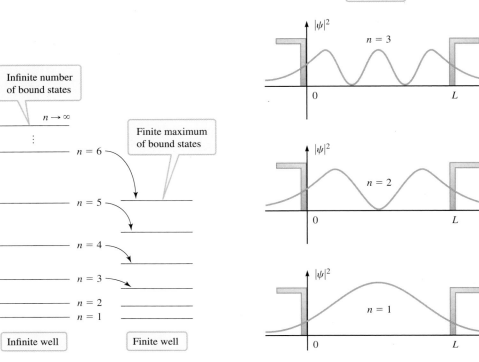

FIGURE 41.11 Compare the energy levels for two square wells of the same width. The well with a finite height has lower energy levels.

FIGURE 41.12 The probability densities for the first three quantum states of a particle in a finite square well.

PARTICLE IN A FINITE SQUARE WELL

⊙ Special Case

longer than the wavelengths inside the infinite square well. Because $E_n = p_n^2/2m$ and $p_n = h/\lambda_n$ (Eq. 41.6), the longer wavelengths for the finite square well mean that the energy levels there are a little lower (Fig. 41.11). Also, if the system's energy is greater than U_0, the particle is free. So there are a finite number of bound energy levels in the case of a finite square well.

Of course, the most amazing difference between the infinite and the finite square wells is that in the case of the finite well, the particle has a nonzero probability of being found outside the well (Fig. 41.12). According to classical physics, it would be impossible for the particle to be found outside the well. Because the system's energy E is less than the potential energy U_0, a particle outside the well would have negative kinetic energy. So the region beyond the boundaries of the well is called the *classically forbidden region*. The probability of finding the particle beyond the boundaries is small and drops off as the distance beyond the boundaries increases. The **penetration distance** Λ depends on κ:

$$\Lambda_n = \frac{1}{\kappa_n} = \frac{h}{2\pi\sqrt{2m(U_0 - E_n)}} \quad (41.21)$$

The penetration distance is a measure of how far the particle may be reasonably found in the classically forbidden region. (When $U_0 \to \infty$, that is, when the finite square well approaches an *infinite* square well, the penetration distance $\Lambda_n \to 0$ in agreement with our intuition that the particle cannot be found outside an infinite well.) This quantum-mechanical possibility, that a particle can be outside the boundaries of an finite square well, plays an important role in our case study.

CONCEPT EXERCISE 41.4

Use cloud imagery to describe Figure 41.12.

EXAMPLE 41.6 — The Penetration Distance of Particle in a Finite Square Well

When a particle in a finite square well is in its ground state, its penetration distance is Λ_1. The well's height is $U_0 = 5.55$ eV. When the same system is in an excited state with an energy of 5.35 eV, its penetration distance is $4\Lambda_1$. What is the system's ground-state energy (in eV)?

:• **INTERPRET and ANTICIPATE**
The penetration distance depends on the system's energy (Eq. 41.21). We expect the ground-state energy to be lower than the excited state's energy.

:• **SOLVE** We are given that when the system is in the excited state, the penetration distance is four times longer than when it is in the ground state.	$\Lambda_x = 4\Lambda_1$ (1)
Substitute Equation 41.21 into Equation (1) and simplify.	$\dfrac{h}{2\pi\sqrt{2m(U_0 - E_x)}} = \dfrac{4h}{2\pi\sqrt{2m(U_0 - E_1)}}$ $\dfrac{1}{\sqrt{(U_0 - E_x)}} = \dfrac{4}{\sqrt{(U_0 - E_1)}}$
Solve for E_1.	$E_1 = 16E_x - 15U_0$ $E_1 = 16(5.35 \text{ eV}) - 15(5.55 \text{ eV})$ $E_1 = 2.35 \text{ eV}$

:• **CHECK and THINK**
As expected, the ground-state energy is less than the excited state's energy.

41-7 Barrier Tunneling

We have just found that a trapped particle has a nonzero probability of being found outside its confines (Fig. 41.12). This amazing result contradicts classical mechanics, which is based on our everyday experience. However, our lives depend daily on reactions in the Sun that require particles to "slip out of their containers." As an analogy, let's consider a more ordinary experience from a classical perspective.

Isaac is playing an unusual game of catch with his mother. Isaac is at the bottom of a slide and his mother is behind the slide; Isaac rolls a ball up the slope (Fig. 41.13). Ideally, the ball makes it over the top of the slide and falls into his mother's hands. This is a problem that we can analyze with classical physics. Consider a system that consists of the ball of mass m and the Earth, and use conservation of energy. We'll ignore the rolling kinetic energy of the ball and any energy lost due to dissipative forces. If we assign the zero point of gravitational potential energy to the point at the bottom of the slide where Isaac releases the ball, then initially the only energy in the system is the ball's kinetic energy K_i. The potential energy at the top of the slide is mgy, where y is the height of the slide. So if the ball's kinetic energy at the bottom is greater than the potential energy at the top ($K_i > mgy$), then the ball will make it over the top and to Isaac's mother. If the ball's initial kinetic energy is less, so that the potential energy at its peak $K_i < mgy$, then the ball will get partway up the slide before rolling back down to Isaac. (Of course, if $K_i = mgy$, the ball will reach the top of the slide and stop there.) Classical physics is completely deterministic. If the ball has enough energy at the bottom

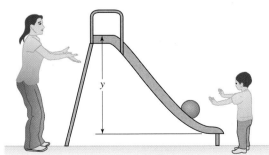

FIGURE 41.13 Isaac's ball will make it to his mother if the ball's kinetic energy at the bottom is greater than the system's potential energy when the ball is at the top of the slide.

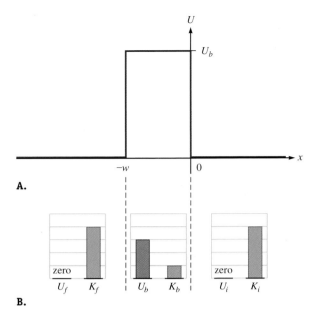

FIGURE 41.14 A. The potential energy curve and **B.** bar charts for a square barrier.

of the slide, we can say that there is a 100% probability of it getting to Isaac's mother. The slide is an example of a **barrier**. According to classical physics, if the system has enough energy, the ball will get to the other side of the barrier; if it does not, the ball will roll back to its starting point.

According to quantum physics, however, the question to ask is "What is the *probability* of the ball getting past the barrier to Isaac's mother?" We answer that question by solving Schrödinger's equation for this situation and then taking the absolute value of the wave function squared $|\Psi|^2$ to get the probability. Since this is a macroscopic problem, we would find that our answer would be (nearly) 100%. If $K_i < mgy$, for example, then we would find that the probability that the ball does not get to Isaac's mother is (nearly) 100%. But notice that we have said that the probability is *nearly* 100%. This means there is some very small nonzero probability that the ball will get to Isaac's mother even when $K_i < mgy$; in this improbable case, it would be as though the ball *passed through the slide* rather than going over it. This is why some people, like the student in this chapter's opening paragraph, like to say there is some probability that they can walk through walls.

In the case of macroscopic objects such as balls or people passing through walls, even quantum physics says that the probability of such an event is very small—so small that it is negligible. However, for microscopic particles such as electrons and protons, the probability that they can pass through a barrier becomes nonnegligible and may be very important.

Let's look at an example involving a proton moving at constant speed in the negative x direction. Initially, the proton is moving through some region of space ($x > 0$) where the potential is a constant that we set to zero. It is heading to a region ($-w < x < 0$) where an electric field sets up a constant positive potential. Beyond that region ($x < -w$), the potential returns to zero. The electric potential energy of this proton-electric field system is $U = eV$ (by Eq. 26.6). The system's potential energy graph looks like a rectangular barrier (Fig. 41.14A), so it is called a **potential energy barrier**. The height of the barrier is U_b, which is the maximum potential energy. When the proton is in the region to the right of the barrier ($x > 0$), the system's energy equals the proton's kinetic energy, as shown by the single bar in the bar chart (Fig. 41.14B). When the electron is in the region of the barrier ($-w < x < 0$), the system's energy is the sum of the barrier's potential energy and the proton's kinetic energy. In this region, there are two bars in the bar chart. When the proton is to the left of the barrier ($x < -w$), once again the system's energy equals the kinetic energy of the proton.

According to classical physics, the proton can only enter the region of the barrier if its initial kinetic energy is greater than the potential energy in the barrier, $K_i > U_b$. Put another way, the proton cannot make it into the region where $x < 0$ unless it has enough energy to cross over the top of the barrier. This is analogous to the example of the ball and the slide. The ball can only get over the slide if it has enough kinetic energy when it is released at the bottom. We use the word *barrier* because, according to classical physics, if $K_i < U_b$, the proton will be barred from the region to the left of the barrier's edge, $x < 0$.

According to quantum physics, however, the question when $K_i < U_b$ is, "What is the *probability* that the proton may be found in the region where $x < 0$?" Again, this question is answered by solving Schrödinger's equation for this situation and taking the absolute value of the wave function squared $|\psi|^2$ to get the probability. A graph of $|\psi|^2$ versus x is shown in Figure 41.15, placed over the potential energy graph. In the region in front of the barrier (where

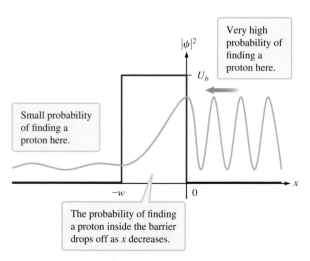

FIGURE 41.15 According to quantum mechanics, it is possible to find the proton on the back side of the barrier, even if $K_i < U_b$.

$x > 0$), there is a very high probability of finding the proton. The probability of finding the proton decreases inside the barrier and is very low to the left of the barrier. However, unlike in classical physics, when $K_i < U_b$, the probability of finding the proton to the left of barrier is not zero. According to classical physics, a proton (with $K_i < U_b$) that makes it into the region to the left of the barrier is like a person who has walked through a wall; this cannot happen. But according to quantum mechanics, it is possible.

This effect is known as **barrier tunneling** or **quantum tunneling**. The probability of tunneling through a rectangular barrier is given by:

$$P_\text{tunnel} \approx e^{-\gamma} \qquad (41.22)$$

where

$$\gamma^2 \equiv \frac{32\pi^2 mw^2(U_b - K_i)}{h^2} \qquad (41.23)$$

and w is the width of the barrier.

Scanning Tunneling Microscope

Barrier tunneling may sound like a great *theory* with no practical application. However, in 1986 the Swiss physicists Gerd Binnig and Heinrich Rohrer won the Nobel Prize for inventing a microscope based on quantum tunneling. This type of microscope is now called a **scanning tunneling microscope** (STM), and it allows for much greater magnification than either a conventional microscope (Fig. 38.45, page 1253) or either type of electron microscope (Fig. 40.17, page 1330). Recall that the electron microscope can resolve a strand of DNA (Example 40.10, page 1331). Contrast this microscope to the scanning tunneling microscope, which is able to resolve individual atoms! For example, using a STM we can see the individual iron atoms on the surface of copper (Fig. 41.16).

Figure 41.17 shows the basic operation of a STM. A pointy metal tip is placed near the sample. A potential difference is established between the tip and the surface of the sample. If the tip is at a lower potential than the sample, the electrons in the tip are attracted to the surface. However, there is a vacuum between the tip and the surface. The vacuum is an insulator. According to classical physics, electrons do not flow through an insulator. However, according to quantum mechanics, the vacuum is a barrier with the metal tip and the surface at the walls of the barrier similar to Figure 41.14. Further, quantum mechanics tells us that it is possible for an electron to tunnel through the barrier. The probability of tunneling depends on the width of the barrier (Eqs. 41.22 and 41.23). In this case the width of the barrier is the length of the path the electron traverses from the tip to the sample. A longer path means a lower tunneling probability (Fig. 41.17B). In fact, because of the exponential dependence in the probability (Eq. 41.22), the probability rapidly decreases for longer paths. The tip is scanned over the sample, changing the length of the path. By measuring the current that results from electron tunneling as the tip is scanned, an image of the surface is inferred. A computer is then used to create an image such as those in Figure. 41.16.

The strong electric field between the tip and the sample can also be used to move individual atoms along the surface as shown in Figure 41.16. So STMs can be used both to image individual atoms and to create nanoscale structures.

BARRIER TUNNELING
⭐ **Major Concept**

γ in Equation 41.23 is not the same as the Lorentz factor.

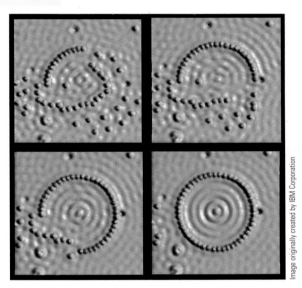

FIGURE 41.16 Iron atoms on the surface of copper are revealed by a scanning tunneling microscope. The microscope is used to push the atoms around, forming a circle.

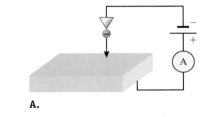

A.

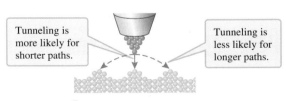

B.

FIGURE 41.17 A. In a scanning tunneling microscope, a metal point is separated from the sample by a vacuum. There is an electric potential difference between the point and the sample. **B.** Electrons tunneling through the vacuum (barrier).

CASE STUDY: Hydrogen Fusion in the Sun

Barrier tunneling happens all the time inside the Sun. The Sun generates the energy for its light (the ultimate source of all energy on the Earth) through nuclear fusion reactions in its core. From Example 39.10 (page 1295), the net fusion reaction in the Sun is

$$4H \rightarrow He$$

This is a nuclear reaction in which four hydrogen nuclei are fused together to make a single helium nucleus. A hydrogen nucleus is just a proton, whereas a helium nucleus is made up of two protons and two neutrons. Fusion in the Sun is a three-step chain reaction. In the first step, two protons must come together.

In classical physics, there are two problems with this first step. The first of them applies to any nucleus. According to Coulomb's law, the protons are repelled by one another with a force that grows stronger the closer together they are. So how can protons stay together in a nucleus? Why are many nuclei stable? The answer is that another force is involved called the *strong nuclear force* (or simply the *strong force*), a short-range attractive force that is stronger than Coulomb's repulsive force inside the nucleus. When the protons are very close together in the nucleus, we can ignore the Coulomb force.

The strong force explains why nuclei are stable. But there is a second problem: Outside the nucleus, Coulomb's repulsive force dominates. How can two protons ever get close enough together to form a nucleus? The answer has to do with barrier tunneling.

To keep things simple, imagine that one proton is at rest at the origin, $x = 0$. Another proton approaches it from the positive x direction. Figure 41.18 shows the potential energy of the two-proton system. When the protons are farther apart than x_0, the Coulomb force dominates and the electric potential energy falls off as shown in the right part of the graph. When the protons are closer together than x_0, the strong force dominates and the potential energy is a flat-bottomed well. In order for the protons to stick together, then, they must get closer together than x_0.

The electric potential curve forms a barrier shaped something like a playground slide. According to classical physics, the protons will get together if the moving proton's initial kinetic energy (when it is far away) is greater than the peak of the barrier's potential energy, $K_i > U_b$. In this case the barrier's potential energy is $U_b = 700$ keV. We can find the proton's initial kinetic energy from the temperature in the core of the Sun: about 1.5×10^7 K, so that the average kinetic energy of the protons is about 1 keV (Problem 41.1). Since K_i is about 700 times smaller than U_b, according to classical physics there is no chance the protons will get together.

You might argue that we have used the *average* kinetic energy for the protons. Of course, some protons have more kinetic energy, and perhaps some of them have more than 700 keV. The Maxwell–Boltzmann speed distribution equation (Eq. 20.19, page 589) predicts that the probability of a proton having enough kinetic energy to cross over the top of the barrier is less than 10^{-200}.

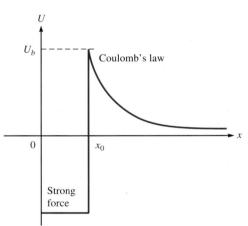

FIGURE 41.18 According to quantum mechanics, it is possible for two protons to come together even if the kinetic energy is less than peak of the potential energy curve $K_i < U_b$.

EXAMPLE 41.7 CASE STUDY: The Probability of Proton Tunneling in the Sun

Equation 41.22 is for a rectangular barrier. Although the barrier in Figure 41.18 is not rectangular, use Equation 41.22 to make an order-of-magnitude estimate of the probability of tunneling required for two protons to fuse together in the Sun. (Don't be surprised if your estimate is off.) Note that in Figure 41.18, $x_0 \sim 10^{-15}$ m (approximate diameter of a proton).

INTERPRET and ANTICIPATE
We expect to find a small but non-zero result.

:• SOLVE

We need the barrier potential energy and the initial kinetic energy in SI units.	$U_b = 700 \text{ keV} \sim 10^{-13} \text{ J}$ $K_i = 1 \text{ keV} \sim 10^{-16} \text{ J}$
It is also helpful to find their difference.	$U_b - K_i \sim (10^{-13} \text{ J} - 10^{-16} \text{ J})$ $U_b - K_i \sim 10^{-13} \text{ J}$
The strong force is important at x_0, so take this as the left edge of the barrier.	$x_0 \sim 10^{-15} \text{ m}$
At the right edge of the barrier, the proton's initial kinetic energy equals the system's potential energy given by Coulomb's law.	$\dfrac{ke^2}{x} = K_i$ $x = \dfrac{ke^2}{K_i} = \dfrac{(9 \times 10^9 \text{ N} \cdot \text{m}^2/\text{C}^2)(1.6 \times 10^9 \text{ C})^2}{10^{-16} \text{ N} \cdot \text{m}}$ $x \sim 10^{-12} \text{ m}$
If we simply subtract the left edge x_0 from the right edge x, we get an overestimate for the width of the barrier because the barrier is not a rectangle.	$x - x_0 = (10^{-12} \text{ m} - 10^{-15} \text{ m}) = 10^{-12} \text{ m}$
A slightly better estimate for the average width of the barrier is about an order of magnitude smaller.	$w \sim 10^{-13} \text{ m}$
Now we can use Equation 41.23 to estimate γ. (The subscript p on the m stands for *proton*.)	$\gamma^2 \equiv \dfrac{32\pi^2 m_p w^2 (U_b - K_i)}{h^2}$ $\gamma^2 \sim \dfrac{(300)(10^{-27} \text{ kg})(10^{-13} \text{ m})^2 (10^{-13} \text{ J})}{(7 \times 10^{-34} \text{ J} \cdot \text{s})^2} \sim 600$ $\gamma \sim 25$
The probability of tunneling comes from Equation 41.22	$P_{\text{tunnel}} \approx e^{-\gamma}$ $P_{\text{tunnel}} \sim e^{-25} \sim 10^{-11}$

:• CHECK and THINK

We modeled the barrier as rectangular. A better model finds a higher probability of about $P_{\text{tunnel}} \sim 10^{-10}$. It means that if 10^{10} pairs of protons attempt to tunnel, only one pair is successful. This might sound like a very low probability. However, it is a high enough probability to account for the 10^{38} fusion reactions (see Example 39.10, page 1295) that take place every second in the core of the Sun.

41-8 Special Case: Quantum Simple Harmonic Oscillator

QUANTUM SIMPLE HARMONIC OSCILLATOR ▶ Special Case

In classical mechanics, we found the simple harmonic oscillator to be a good model for a number of complex systems. Its role is equally important in quantum mechanics, and there is a particularly important reason for studying it. Planck's quantum hypothesis was that particles in a black body can be modeled as oscillators with the energy of each oscillator given by:

$$E = nE_{\min} = nhf \quad (n = 1, 2, 3 \ldots) \quad (40.7)$$

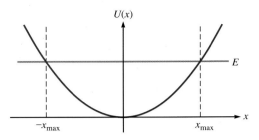

FIGURE 41.19 A simple harmonic oscillator's energy diagram.

Here k is the spring constant in Hooke's law.

where f is the frequency and $E_{min} = hf$. In Planck's model, the energy levels are evenly spaced. By contrast, when a particle is confined to either an infinite or a finite well, we have seen that the system's energy is quantized as in Planck's hypothesis, but the energy levels are not evenly spaced. Because the energy levels in a simple harmonic oscillator are evenly spaced, it makes a much better model for the molecules in a black body than for the particles in a well.

Our system consists of a particle connected to a spring that is fixed on its other end, so that the particle oscillates back and forth along an x axis. The spring obeys Hooke's law, so the magnitude of force on the particle is $F_H = kx$. The system's potential energy is $U(x) = \frac{1}{2}kx^2$. The natural frequency of the system is $\omega = \sqrt{k/m}$, and we use this equation to write the potential energy as $U(x) = \frac{1}{2}m\omega^2 x^2$. The system's energy is E, and classically the particle oscillates between $\pm x_{max}$ (Fig. 41.19).

Schrödinger's equation for the simple harmonic oscillator is

$$-\frac{h^2}{8\pi^2 m}\frac{d^2\psi(x)}{dx^2} + \frac{1}{2}m\omega^2 x^2 \psi(x) = E\psi(x) \tag{41.24}$$

The x^2 in the second term makes this a somewhat difficult equation to solve, so we will just give the solution for the ground state:

$$\psi_1(x) = A_1 e^{-(x/x_0)^2} \tag{41.25}$$

where $x_0^2 = h/\pi m\omega$ and A_1 is a constant that is found by normalization. In Problem 41.36, you will confirm that Equation 41.25 is a solution to Schrödinger's equation, and you will find that the ground-state energy is $E_1 = h\omega/4\pi$.

We can write the ground-state energy in terms of the frequency f as $E_1 = \frac{1}{2}hf$. The solutions for the excited states show that the energy is quantized and given by:

$$E_n = \left(n - \frac{1}{2}\right)hf \tag{41.26}$$

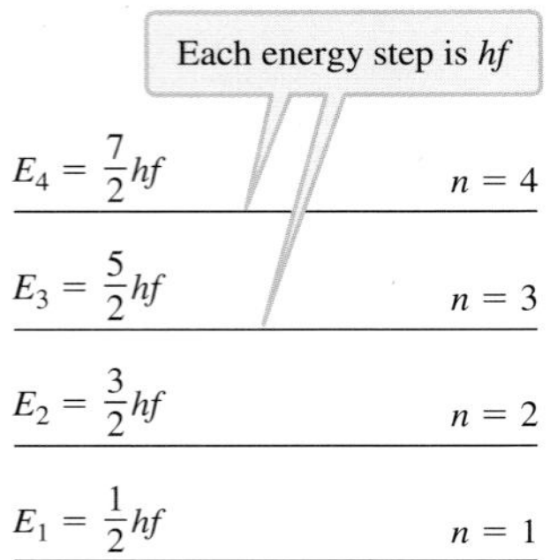

According to Equation 41.26, the energy levels of a simple harmonic oscillator are evenly spaced (Fig. 41.20). Compare these energy levels to those proposed by Planck (Eq. 40.6); the difference between adjacent levels is $\Delta E = hf$ in both cases. The only difference is that Planck's minimum energy $E_{min} = hf$ is twice the ground state energy $E_1 = \frac{1}{2}hf$ of the simple harmonic oscillator. Like any trapped particle, according to quantum mechanics our particle attached to a spring cannot be at rest; the system's minimum energy cannot be zero.

FIGURE 41.20 An energy-level diagram for a simple harmonic oscillator. The energy levels are evenly spaced and much like those in Planck's model for a black body.

EXAMPLE 41.8 — In Which State Is the Oscillator?

A simple harmonic oscillator in its ground state has a frequency of 3.450 MHz. Now its energy is found to be 2.400×10^{-26} J. What state is it in?

INTERPRET and ANTICIPATE
The frequency of the ground state is the frequency in the excited state. So we can use the given frequency and the energy of the excited state to find which state the oscillator is in.

SOLVE Solve Equation 41.26 for n.	$E_n = \left(n - \frac{1}{2}\right)hf \quad (41.26)$ $n = \dfrac{E_n}{hf} + \dfrac{1}{2}$
Substitute values. The answer must be an integer, so in this case we round up to the nearest whole number.	$n = \dfrac{(2.400 \times 10^{-26}\text{ J})}{(6.626 \times 10^{-34}\text{ J}\cdot\text{s})(3.450 \times 10^6\text{ Hz})} + \dfrac{1}{2}$ $n = 10.9988 = \boxed{11}$

CHECK and THINK

To check, let's find the difference between adjacent levels: $\Delta E = hf = (6.626 \times 10^{-34} \text{ J} \cdot \text{s})(3.450 \times 10^6 \text{ Hz}) = 2.286 \times 10^{-27}$ J. So the system's energy is about 2.400×10^{-26} J/2.286×10^{-27} J = 10.5 times greater than the difference between adjacent energy levels. The ground-state energy ($n = 1$) is 0.5 hf, and the system is 10 levels above the ground state, so $n = 1 + 10 = 11$, as we found.

EXAMPLE 41.9 Modeling Iodine

A number of times throughout this book, we have modeled molecular bonds as tiny springs connecting atoms. (For examples, see Figures 5.15, 14.21, and 21.25.) In particular, a vibrating diatomic molecule may be modeled as a simple harmonic oscillator (Fig. 41.21). In this problem, we connect the quantum-mechanical simple harmonic model to the classical model we have used for a molecular bond. An iodine molecule consisting of two iodine atoms is observed to have equally spaced energy levels of 4.24×10^{-21} J. Model the molecule as a particle-spring system, with one end of the spring fixed and the other attached to the oscillating particle. The particle's mass is 1.053×10^{-25} kg. (This the *reduced* or effective mass of the iodine molecule.) Find the spring constant.

Vibrate in and out.

FIGURE 41.21

INTERPRET and ANTICIPATE

Modeling the iodine molecule as a quantum simple harmonic oscillator is reasonable because the observed energy spacing is uniform as predicted by Equation 41.26. The frequency of a classical object-spring oscillator (Section 16-5) depends on the mass and spring constant. So, making the connection between the classical and quantum models means finding the frequency of the quantum oscillator and from that frequency finding the spring constant.

SOLVE

The difference in adjacent energy levels depends on frequency.

$$\Delta E = hf \quad (1)$$

The frequency of an object-spring oscillator is given by Equation 16.26 (modified by using $\omega = 2\pi f$.)

$$f = \frac{1}{2\pi}\sqrt{\frac{k}{m}} \quad 16.26$$

Combine Equations (1) and 16.26 to arrive at the spring constant.

$$\frac{\Delta E}{h} = \frac{1}{2\pi}\sqrt{\frac{k}{m}}$$

$$k = 4\pi^2 m \left(\frac{\Delta E}{h}\right)^2$$

$$k = 4\pi^2 (1.053 \times 10^{-25} \text{ kg}) \left(\frac{4.24 \times 10^{-21} \text{ J}}{6.626 \times 10^{-34} \text{ J} \cdot \text{s}}\right)^2$$

$$\boxed{k = 1.70 \times 10^2 \text{ N/m}}$$

CHECK and THINK

In this problem we have come full circle on our models of molecular bonds. We see why it was reasonable to model such bonds by tiny springs throughout this book. However, we must remember there is a major difference between a classical and quantum simple harmonic oscillator: In a classical oscillator the mechanical energy can have any value, but in a quantum oscillator the energy can only have certain discrete, evenly spaced values.

41-9 Heisenberg's Uncertainty Principle

Quantum mechanics is a great departure from classical mechanics largely because classical mechanics is deterministic and quantum mechanics is probabilistic. According to classical mechanics, both the location and momentum of a particle may be precisely determined. By contrast, quantum mechanics tells us that no matter how carefully we conduct an experiment, we cannot precisely determine a particle's position and momentum simultaneously. To see why quantum mechanics gives us such a different picture of nature, let's consider a free particle.

A Free Particle

So far we have only considered particles that have been trapped in potential wells. Now we'll consider a free particle traveling in the positive x direction that is not part of any system, so that $U(x) = 0$. Our plan is to solve for both the time-independent and the time-dependent wave functions. We then find the probability density. For a free particle, Schrödinger's equation is given by

$$-\frac{h^2}{8\pi^2 m}\frac{d^2\psi(x)}{dx^2} = E\psi(x) \quad (41.27)$$

The total energy of a free particle is its kinetic energy, which we express in terms of its momentum $E = p^2/2m$ and substitute into Equation 41.27:

$$-\frac{h^2}{8\pi^2 m}\frac{d^2\psi(x)}{dx^2} = \frac{p^2}{2m}\psi(x)$$

Rearrange this equation to isolate the second derivative on the left side:

$$\frac{d^2\psi(x)}{dx^2} = -\frac{4\pi^2 p^2}{h^2}\psi(x) \quad (41.28)$$

Use the de Broglie wavelength $\lambda = h/p$, and $\lambda = 2\pi/k$ to write the angular wave number in terms of momentum:

$$k = \frac{2\pi p}{h} \quad (41.29)$$

Substitute Equation 41.29 into Equation 41.28:

$$\frac{d^2\psi(x)}{dx^2} = -k^2\psi(x) \quad (41.30)$$

Mathematically, Equation 41.30 is identical to the differential equation for the position y of a simple harmonic oscillator, Equation 16.47: $d^2y(t)/dt^2 = -(k/m)y(t)$. Unfortunately, the solution given by Equation 16.3, $y(t) = y_{\max}\cos(\omega t + \varphi)$, is not general enough in this case. As before, we won't solve this differential equation; instead, we assert that the most general solution is

$$\psi(x) = \psi_0 e^{ikx} + \psi_1 e^{-ikx} \quad (41.31)$$

where ψ_0 and ψ_1 are arbitrary constants (Problem 41.40).

Although we are not normally interested in the time-dependent Schrödinger equation in this book, in this one case, the solution is interesting. We can easily find the time-dependent solution from $\Psi(x, t) = \psi(x)e^{-i\omega t}$; we just multiply the solution for $\psi(x)$ in Equation 41.31 by $e^{-i\omega t}$:

Remember $\omega = \dfrac{2\pi E}{h}$

$$\Psi(x, t) = \psi_0 e^{i(kx - \omega t)} + \psi_1 e^{-i(kx + \omega t)} = \Psi_0(x, t) + \Psi_1(x, t) \quad (41.32)$$

In Chapter 17, we wrote the wave function of a wave traveling in the positive x direction as Equation 17.4, $y(x, t) = y_{\max}\sin(kx - \omega t)$, and in the negative x direction as Equation 17.9, $y(x, t) = y_{\max}\sin(kx + \omega t)$. But using the sine function is an arbitrary choice in representing a wave. The key is that a function of the form $F(kx \pm \omega t)$ represents a wave traveling in either the positive or the negative x direction. $\Psi_0(x, t) = \psi_0 e^{i(kx - \omega t)}$ represents a wave traveling in the positive x direction, and $\Psi_1(x, t) = \psi_1 e^{-i(kx + \omega t)}$ represents a wave traveling in the negative x direction. We are interested in a free particle traveling in the positive x direction, so we focus our attention on $\Psi_0(x, t) = \psi_0 e^{i(kx - \omega t)}$.

Let's find this particle's probability density:

$$|\Psi_0|^2 = (\psi_0 e^{i(kx-\omega t)})(\psi_0 e^{-i(kx-\omega t)}) = \psi_0^2 e^{[i(kx-\omega t) - i(kx-\omega t)]} = \psi_0^2 e^0$$

$$|\Psi_0|^2 = \psi_0^2$$

We just found that the probability density is a constant that is independent of x. This means that the free particle is equally likely to be found anywhere along the x axis. The uncertainty Δx in the particle's position is infinite; the particle's position is *completely* uncertain. However, the particle's momentum is given by $p = hk/2\pi$ (Eq. 41.29), and because the angular wave number k is precisely known, the free particle's momentum is precisely known.

Uncertainty Is Inherent

Our inability precisely to know both the position and the momentum of a free particle is an example of Heisenberg's uncertainty principle. This uncertainty has nothing to do with the human ability to design an experiment; it is fundamental to nature. Our intuition that we can simultaneously measure position and momentum to an unlimited precision is based on our classical model of a particle, but the quantum model of a "particle" is of a probability *wave*. No forces are exerted on a free particle; its momentum is a constant that we can theoretically measure to any level of precision so that there is no inherent uncertainty in momentum $\Delta p = 0$. But because it is possible to measure its momentum precisely, when we do so, the idea of the particle having a position loses its meaning; the particle can be anywhere.

A free particle is an extreme example of precisely knowing a particle's momentum but having no way of knowing its position. What about a particle that is trapped? In such a case, we may know something about both its momentum and its position. But the more precisely we can know its position, the less precisely we can know its momentum.

To come up with a mathematical relationship for these uncertainties, let's think about how we could make such measurement. Suppose you wish to measure the position of a puck: You would probably reflect light off the puck to see it. Because of diffraction, the measurement of the puck's position is limited by the wavelength of the light used. So the uncertainty in the position measured is $\Delta x \approx \lambda$. You could, of course, improve the measurement by using very short wavelength radiation (or even an electron beam); but no matter how you design your experiment, there will be an inherent uncertainty in your position measurement. Next, you would like to measure the puck's momentum. You do this by measuring its position at two different times, calculating its velocity, and then multiplying by its mass. If the photons you use to illuminate the puck at these times have wavelength λ, then their momentum is h/λ. So the photons' interaction with the puck introduces an uncertainty in the puck's momentum: $\Delta p \approx h/\lambda$. If we multiply the uncertainty in the position by the uncertainty in the momentum, we find:

$$\Delta x \Delta p \approx (\lambda)\left(\frac{h}{\lambda}\right) \approx h \quad (41.33)$$

The German physicist Werner Heisenberg (1901–1976) worked this relationship out more carefully. According to **Heisenberg's uncertainty principle**, the standard deviation in position Δx and the standard deviation in momentum are given by:

$$\Delta x \Delta p \geq \frac{h}{4\pi} \quad (41.34)$$

HEISENBERG'S UNCERTAINTY PRINCIPLE ● Underlying Principle

Equivalently, Heisenberg's uncertainty principle may be expressed in terms of the standard deviation in energy and the standard deviation in time:

$$\Delta E \Delta t \geq \frac{h}{4\pi} \quad (41.35)$$

The equalities in Equations 41.34 and 41.35 are rarely realized. To make estimates, we usually use Equation 41.33 for position and momentum, and

$$\Delta E \Delta t \approx h \quad (41.36)$$

for energy and time.

Conceptually, $\Delta E \Delta t \approx h$ tells us that as the time taken to make an energy measurement increases, the uncertainty inherent in the result decreases. This equation has an even bolder implication. It says that the conservation of energy principle can be violated as long as the time period of this violation is short enough to be consistent with Equation 41.35.

Heisenberg's uncertainty principle gives us another way to understand the requirement that trapped particles can never be at rest; that is, their zero-point energy is not zero. The position of a particle trapped in a potential well has a relatively small uncertainty: within a box of width L, so we can say that $\Delta x \approx L$ (Fig. 41.1). Such a particle cannot be at rest because then its momentum would be precisely known, $\Delta p = 0$. Then $\Delta x \Delta p \approx (L)(0) = 0$, which violates Heisenberg's uncertainty principle.

Heisenberg's uncertainty principle also sheds light on the reactions in the core of the Sun (**CASE STUDY**). Fusion in the Sun requires protons to bind together, and for this to happen, quantum tunneling is required. Here is another way to look at tunneling. A proton may be characterized by its wavelength λ, and the uncertainty in its position must be at least on the order of its wavelength. As long as the barrier is no wider than a few times this wavelength, then tunneling is possible because the proton's position uncertainty includes both sides of the barrier. In other words, the proton is already on the other side of the barrier. It is like finding Isaac's ball is already on both sides of the slide (Fig. 41.13).

EXAMPLE 41.10 Science Fiction

As we saw in Chapter 39's **CASE STUDY** (page 1270), science fiction writers often use real physics in their stories. Let's test an idea a writer had. Feeding people on deep space missions is challenging. In a sense, the mission would need to somehow pack all the nutrition the crew needs for the entire voyage. The writer's solution to this problem is based on Heisenberg's uncertainty principle. According to the principle, the conservation of energy can be violated for a brief period of time given by $\Delta E \Delta t \approx h$. So a meal with energy ΔE could be created "out of nothing" and last a short time Δt. The real trick is that the person would need to eat the meal in that short time. Estimate Δt.

• INTERPRET and ANTICIPATE
The key to this problem is estimating the energy created out of nothing. This estimate comes from Einstein's famous equation $E = mc^2$ (Eq. 39.44). We find the mass of a typical meal, then the energy that must be created.

• SOLVE
Let's say the meal is a quarter-pound hamburger. The mass including the bun, French fries, and drink may be about 0.2 kg. We find the energy that must be created.

$$\Delta E = mc^2 \quad (39.44)$$
$$\Delta E = (0.2 \text{ kg})(3 \times 10^8 \text{ m/s})^2$$
$$\Delta E = 1.8 \times 10^{16} \text{ J}$$

According to Heisenberg's uncertainty principle, the mass can be created out of nothing as long as it disappears again in a short time given by Equation 41.36.

$$\Delta E \Delta t \approx h \quad (41.36)$$
$$\Delta t \approx \frac{h}{\Delta E} = \frac{6.63 \times 10^{-34} \text{ J} \cdot \text{s}}{1.8 \times 10^{16} \text{ J}} = 3.68 \times 10^{-50} \text{ s}$$
$$\Delta t \approx 4 \times 10^{-50} \text{ s}$$

• CHECK and THINK
This is a very short time. As a comparison, it takes about 300 milliseconds to blink your eye. So in a blink of an eye about 10^{45} such meals could come into existence and vanish. The writer should try to think of another solution for feeding the fictional crew. In case you are worried that we haven't done a good job of making our estimate because the person really only needs the chemical energy stored in the food (the calories), and not the mass, try Problem 41.80.

EXAMPLE 41.11 Hawking Radiation Is NOT Science Fiction

In the previous example, we imagined a whole meal being created from nothing. The result was that such a meal could only exist for a negligible amount of time.

But don't think that such events are impossible. The British physicist Stephen Hawking has a theory that predicts radiation from black holes. Recall from Section 39-12 that nothing can escape from the event horizon of a black hole. However, Hawking theorizes that black holes can radiate; we call this radiation *Hawking radiation*. His idea is that particle and antiparticle pair are created out of nothing near the event horizon of a black hole. Normally, when a particle and antiparticle are created, they soon come together and annihilate one another. But if in their short lifetime one of the particles falls inside the event horizon, the other particle will escape.

In principle we could see this escaped particle as Hawking radiation from the black hole. This radiation carries energy away from the black hole. Thus the black hole effectively loses energy and its mass is reduced, and we say the black hole *evaporates*. This evaporation process is very slow for solar mass black holes; such a black hole would take over 10^{67} years to evaporate. The universe is only 13.7 billion years old, and so no solar mass black holes have evaporated. However, much lower mass black holes known as *primordial black holes* may have had enough time to evaporate. Such small black holes would produce a final burst of high-energy Hawking radiation. Although no such radiation has been observed yet, it is theoretically possible.

Consider the creation of an electron-position pair near a black hole. Similar to the writer's idea about the meal in Example 41.10, one of the particles must be consumed by the black hole in their brief lifetime. Estimate the lifetime of such an electron-positron pair.

:• INTERPRET and ANTICIPATE
Like the previous example, we use Einstein's famous equation to estimate the energy that must be created.

:• SOLVE	
The mass of electron is the same as positron's mass.	$\Delta E = mc^2$ (39.44) $\Delta E = 2m_e c^2 = 2(9.11 \times 10^{-31}\text{ kg})(3.00 \times 10^8 \text{ m/s})^2$ $\Delta E = 1.64 \times 10^{-13}$ J
According to Heisenberg's uncertainty principle, the mass can be created out of nothing as long as it disappears again in a short time given by Equation 41.36.	$\Delta E \Delta t \approx h$ (41.36) $\Delta t \approx \dfrac{h}{\Delta E} = \dfrac{6.63 \times 10^{-34} \text{ J} \cdot \text{s}}{1.64 \times 10^{16} \text{ J}}$ $\Delta t \approx 4.04 \times 10^{-21}$ s

:• CHECK and THINK
This is still a very short time. However, it is 10^{29} longer than the length of time the meal would survive. According to quantum mechanics, such particle–antiparticle pairs are constantly created in empty space. And according to Hawking, there is a nonzero probability that in some cases one of the particles falls into the black hole and the other is set free.

SUMMARY

❶ Underlying Principles

1. **Schrödinger's equation** describes the propagation of a matter wave, and a matter wave carries all the things we normally associate with a particle—energy, momentum, mass, and sometimes charge.

   Schrödinger's (nonrelativistic) time-dependent wave equation is

   $$-\dfrac{h^2}{8\pi^2 m}\dfrac{\partial^2 \Psi(x,t)}{\partial x^2} + U(x,t)\Psi(x,t) = \dfrac{ih}{2\pi}\dfrac{\partial \Psi(x,t)}{\partial t} \quad (41.1)$$

❗ Underlying Principles cont'd

The time-independent, one-dimensional Schrödinger equation is

$$-\frac{h^2}{8\pi^2 m}\frac{d^2\psi(x)}{dx^2} + U(x)\psi(x) = E\psi(x) \quad (41.8)$$

2. According to **Bohr's correspondence principle** in the limit of very large quantum numbers, the classical calculation and the quantum calculation must yield the same results.

3. According to **Heisenberg's uncertainty principle**, the standard deviation in position Δx and the standard deviation in momentum Δp are given by:

$$\Delta x \Delta p \geq \frac{h}{4\pi} \quad (41.34)$$

and the standard deviation in energy ΔE and the standard deviation in time Δt by:

$$\Delta E \Delta t \geq \frac{h}{4\pi} \quad (41.35)$$

✪ Major Concepts

1. According to Born, there is no *physical* interpretation of the wave function Ψ itself, but we can think of Ψ as a **probability wave**. We don't actually detect Ψ. Instead we detect $|\Psi|^2$, which is the **probability density** of finding the particle at a particular location and time.
2. The **normalization condition** mathematically expresses this commonsense idea that the particle must be located somewhere along the x axis as

$$\int_{-\infty}^{\infty} |\psi(x)|^2 dx = 1 \quad (41.11)$$

the **normalization condition**, and it is required in order to interpret $|\psi|^2$ as a probability density.
3. The value that $\psi(x)$ must have at a boundary is known as the **boundary condition**. The boundary conditions are used to make a general solution to Schrödinger's equation fit a particular situation.
4. **Barrier tunneling** describes the prediction made by quantum mechanics that a particle may be found on the other side of a potential energy barrier that would be insurmountable according to classical mechanics. The probability that a particle with energy K_i will tunnel through a rectangular barrier of height U_b is given by:

$$P_{\text{tunnel}} \approx e^{-\gamma} \quad (41.22)$$

where

$$\gamma^2 \equiv \frac{32\pi^2 m w^2 (U_b - K_i)}{h^2} \quad (41.23)$$

and w is the width of the barrier.

▷ Special Cases

1. Schrödinger's equation for a **particle in an infinite square well** is

$$\frac{d^2\psi(x)}{dx^2} + k^2\psi(x) = 0 \quad (41.13)$$

and the specific solution is $\psi(x) = \sqrt{\frac{2}{L}}\sin k_n x$ (Eq. 41.17). The system's energy levels are

$$E_n = n^2\left(\frac{h^2}{8mL^2}\right) = n^2 E_1 \quad (41.16)$$

2. Schrödinger's equation for a **particle in a finite square well** is the same as that for an infinite well (Eq. 41.13). However, outside the well, Schrödinger's equation is

$$\frac{d^2\psi(x)}{dx^2} - \kappa^2\psi(x) = 0 \quad (41.19)$$

where $\kappa^2 = (8\pi^2 m/h^2)(U_0 - E)$. The solutions for the first three quantum states are presented graphically in Figure 41.10A.

The energy levels are a little lower in the finite well than in the infinite well (Fig. 41.11), and if the system's energy is greater than U_0, the particle is free.

A particle in a finite well may be found outside the well. The **penetration distance** Λ depends on κ:

$$\Lambda_n = \frac{1}{\kappa_n} = \frac{h}{2\pi\sqrt{2m(U_0 - E_n)}} \quad (41.21)$$

3. Schrödinger's equation for the **quantum simple harmonic oscillator** is:

$$-\frac{h^2}{8\pi^2 m}\frac{d^2\psi(x)}{dx^2} + \frac{1}{2}m\omega^2 x^2 \psi(x) = E\psi(x) \quad (41.24)$$

where the ground-state solution is:

$$\psi_1(x) = A_1 e^{-(x/x_0)^2} \quad (41.25)$$

The system's ground-state energy is $E_1 = h\omega/4\pi$ or $E_1 = \frac{1}{2}hf$, and the energies of the excited states are:

$$E_n = \left(n - \frac{1}{2}\right)hf \quad (41.26)$$

PROBLEMS AND QUESTIONS

A = algebraic   C = conceptual   E = estimation   G = graphical   N = numerical

41-1 The New Quantum Theory

1. **E** **CASE STUDY** The Sun's core temperature is about 1.5×10^7 K. Show that this means the average kinetic energy of the protons in the core is about 1 keV (order of magnitude only).
2. **C** In what way is de Broglie's matter wave theory ad hoc (makeshift)? In what way is Schrödinger's equation ad hoc, and in what way is it not ad hoc?

41-2 A Trapped Particle

Problems 3 and 4 are paired.

3. **N** Suppose that a hockey puck's mass is 0.160 kg, and it is in a box of length 6.35 m (Example 41.1). What excited state would give an energy of 8.52 J?
4. **C** Why do the energy levels of a hockey puck trapped in a box seem to be continuous?
5. **N** A particle of mass m is trapped in a one-dimensional box of length L. What combination of mL^2 is required for the zero-point energy to be 25.0 eV?
6. **N** Suppose we wanted to try to observe the quantization of energy for a 1.00-g peanut trapped in a box. If our detector has a resolution such that it is able to measure the energy difference between the ground state and first excited state, as long as the difference is greater than 0.00100 eV, what is the maximum length L of the box? Note that, of course, we cannot have a box this small. Thus, we do not observe the quantization of peanuts.
7. **N** A proton is trapped in a one-dimensional box of length L. If the zero-point energy is to be 25.0 eV, what is L?

41-3 The Double-Slit Experiment Revisited: Probability Waves

8. **E** Suppose that 6.5×10^6 particles created the interference pattern in Figure 43.4. There is a 13% probability of a particle being detected by device A and a 2.5% probability of a particle being detected by B. **a.** Estimate the number of particles detected by each detector. **b.** How many more particles were detected by A than by B?

Problems 9 and 10 are paired

9. Consider again the throwing of the dice in Example 41.2. In quantum mechanics, we will often speak of the current state of the system given a set of possible states. The sum of the faces of the two dice that are face-up after being thrown (or rolled) can be thought of as representing the state of the dice. This means the possible total, or state of the dice, could be 2, 3, 4, and so on, with 12 being the maximum value.
   a. **N** How many possible ways are there for the state of the dice to equal 10?
   b. **N** How many possible ways are there for the state of the dice to equal 6?
   c. **C** Which of these two states is more likely to occur? Explain your answer.

10. Consider the throwing of the dice in the previous problem. The number of possible ways a state can be achieved is called the *multiplicity* of the state and can be used to calculate the probability of the state occurring.
    a. **N** What is the multiplicity of the state where the total of the dice is equal to 7?
    b. **N** What is the total number of possible states of the dice when they are thrown? Be sure to consider that each possible state may have a multiplicity greater than one.
    c. **N** What is the probability of finding the dice in the state where their total is 7 after being thrown?
    d. **C** Is 7 the most likely result? Explain your answer.

41-4 Schrödinger's Equation

11. **A** Substitute $\Psi(x, t) = \psi(x)e^{-i\omega t}$ (where $\omega = 2(\pi)E/h$) into Equation 41.1, the time-dependent Schrödinger equation:

$$-\frac{h^2}{8\pi^2 m}\frac{\partial^2 \Psi(x,t)}{\partial x^2} + U(x,t)\Psi(x,t) = \frac{ih}{2\pi}\frac{\partial \Psi(x,t)}{\partial t}$$

and derive the time-independent version, Equation 41.8:

$$-\frac{h^2}{8\pi^2 m}\frac{\partial^2 \psi(x)}{\partial x^2} + U(x)\psi(x) = E\psi(x)$$

12. **C** A chapter of the Society of Physics Students at the United States Naval Academy designed a T-shirt with a pun based on quantum mechanics (Fig. P41.12). Explain the meaning of the T-shirt.

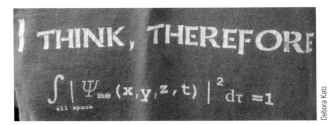

FIGURE P41.12

41-5 Special Case: A Particle in an Infinite Square Well

13. **N** A neutron is bound in an infinite square well with a width of 3.25 nm. Find the energies of the ground state and the first three excited states.
14. **C** We imagine that a puck is moving in a box with very stiff walls, so that the puck reverses direction without any time delay. Of course, in a real box the puck would take a moment to reverse direction. Where would you most likely find the puck in such a real situation? How does your answer change depending on the length of the box?
15. **N** A proton is bound in an infinite square well. A photon with a frequency of 2.34×10^{15} Hz is emitted when the proton makes a transition from the $n = 6$ excited state to the ground state. What is the width of the square well?
16. **G** The wave function at a particular moment in time for an electron in an infinite one-dimensional box of length L, as in Section 41-5, is

$$\psi(x) = \sqrt{\frac{2}{L}} \sin\left(\frac{2\pi x}{L}\right) \text{ for } 0 < x < L$$

Sketch the probability density for the electron in the box.

Problems 17 and 18 are paired

17. **N** An infinite square well of width L is used to confine a particle. The well extends from $x = 0$ to $x = L$. Assume the particle is in the ground state.
    a. What is the probability of detecting the particle between $x = 0$ to $x = L$?
    b. What is the probability of detecting the particle between $x = 0$ to $x = 0.50L$?
18. **N** An infinite square well of width L is used to confine a particle. The well extends from $x = 0$ to $x = L$. Assume the particle is in the first excited state ($n = 2$).
    a. What is the probability of detecting the particle between $x = 0$ to $x = L$?
    b. What is the probability of detecting the particle between $x = 0$ to $x = 0.50L$?

Problems 19, 20, and 21 are grouped.

19. **N** An electron is trapped in an infinite well of width 2.50 nm. What is its zero-point energy? Give your answer in joules and electron-volts.
20. **C** An electron is trapped in an infinite well of width 2.50 nm. Initially, the system is in the $n = 3$ quantum state; it then jumps to the $n = 5$ quantum state. Did the system lose or gain energy? Explain.
21. **N** An electron is trapped in an infinite well of width 2.50 nm. Initially, the system is in the $n = 3$ quantum state; it then jumps to the $n = 5$ quantum state. Determine the amount of energy lost or gained by the system. Give your answer in electron-volts.
22. **A** Write the wave functions for the stationary ground state and each of the first three stationary excited states for a particle in an infinite square well of width L_0.
23. **N** The ground state ($n = 1$) of a proton confined to an infinite square well of width L is 3.75 keV.
    a. What is the width L of the infinite square well?
    b. What is the energy difference between the $n = 3$ excited state and the ground state for this proton?

Problems 24 and 25 are paired.

24. **E** A proton in the nucleus of an atom can be modeled crudely as a proton trapped in an infinite well with a width on the order of 10^{-14} m. What is the order of magnitude of its zero-point energy? Give your answer in joules and electron-volts.
25. **E** An electron bound in an atom can be modeled crudely as an electron trapped in an infinite well with a width on the order of 10^{-10} m. What is the order of magnitude of its zero-point energy? Give your answer in joules and electron-volts. If you worked Problem 24, compare your results.

Problems 26 and 27 are paired.

26. **N** An electron is in an infinite potential well with a width of 1.50 nm.
    a. What is the wavelength of the photon emitted when the electron transitions from the fourth excited state, $n = 5$, to the ground state, $n = 1$?
    b. What is the maximum wavelength of a photon capable of transitioning the electron from the ground state to the third excited state, $n = 4$?
27. **N** A proton is in an infinite potential well with a width of 1.50 nm.
    a. What is the wavelength of the photon emitted when the proton transitions from the fourth excited state, $n = 5$, to the ground state, $n = 1$?
    b. What is the maximum wavelength of a photon capable of transitioning the proton from the ground state to the third excited state, $n = 4$?

41-6 Special Case: A Particle in a Finite Square Well

28. **C** A particle is in the ground state of a finite square well with energy $E_1 = 5.42$ eV. The walls of the well have a height 10.62 eV. If the particle absorbs a photon with an energy of 6.00 eV, what can we say about the state of the particle? How is this different than if the particle were in an infinite square well with the same ground-state energy and absorbed the same photon?
29. **N** An electron is bound in the ground state of a finite square well with $U_0 = 75$ eV.
    a. How much energy is required to free the electron from the well if the ground-state energy is 3.0 eV?
    b. If this transition is accomplished through the absorption of one photon of light, what is the maximum wavelength of that photon?

Problems 30 and 31 are paired.

30. **C** A particle is bound in the ground state of a finite square well. An external potential is applied such that the walls of the well increase in height, or the well becomes deeper. Describe what happens to the ground state energy of the particle as the wall height increases. What happens to the penetration distance?
31. **N** A proton is bound in a finite square well such that its energy E is 10.0% of the height of the well, U_0. Find the penetration distance of the particle when the height of the well is **a.** 5.0 eV, **b.** 50.0 eV, and **c.** 500.0 eV.
32. **A** Find the dimensions of the quantity in Equation 41.21,

$$\Lambda_n = \frac{h}{2\pi \sqrt{2m(U_0 - E_n)}}$$

Problems 33 and 34 are paired.

33. **N** A proton is in a finite square well with $U_0 = 35.5$ MeV. It is found that $\Lambda_2 = 2.35 \times 10^{-15}$ m. What is E_2? Give your answer in joules and in electron-volts.
34. **N** For the proton in Problem 33, given your answer, place a limit on the system's zero-point energy.
35. **N** The width of a finite square well can be varied such that the ground-state energy remains unchanged as we vary the depth of the square well. Suppose this is done, where the ground state energy of a proton is 3.50 eV and the various depths of the well

are **a.** 10.0 eV, **b.** 100.0 eV, and **c.** 1000.0 eV. Find the penetration distance for the proton in the ground state for each well depth.

41-7 Barrier Tunneling

36. **E** CASE STUDY The probability of two protons tunneling in the Sun's core is $P_{tunnel} \sim 10^{-10}$. This means that out of each 10^{10} pairs of protons, one pair tunnels successfully. Make an estimate showing that this probability can account for the 10^{38} fusion reactions that take place each second in the Sun's core. *Hint:* Assume that the Sun's core contains about 40% of its mass.

37. **N** CASE STUDY Use the Maxwell–Boltzmann speed distribution equation (Eq. 20.19, page 589) to find the probability that a proton has enough kinetic energy to cross over the top of the barrier in Figure 41.18. *Hint:* The Sun's core temperature is about 1.5×10^7 K, so the average kinetic energy of the protons in the core is about 1 keV (order of magnitude only). In this case the barrier's potential energy is $U_b = 700$ keV.

Problems 38, 39, and 40 are grouped.

38. **N** An electron with a kinetic energy of 45.34 eV is incident on a square barrier with $U_b = 54.43$ eV and $w = 2.400$ nm. What is the probability that the electron tunnels through the barrier?

39. **N** An electron with a kinetic energy of 45.34 eV is incident on a square barrier with $U_b = 54.43$ eV and $w = 2.400$ pm. What is the probability that the electron tunnels through the barrier?

40. **N** A proton with a kinetic energy of 45.34 eV is incident on a square barrier with $U_b = 54.43$ eV and $w = 2.400$ pm. What is the probability that the proton tunnels through the barrier?

41. **N** A proton is approaching a potential barrier with a height of 10.24 eV and a width of 4.560 pm. If there is a 25.00% chance of the proton tunneling through the barrier, what is the de Broglie wavelength of the proton?

42. **A** Derive an expression that relates the rate of change of the probability of tunneling for a particle incident on a potential barrier with height U_b and width w, to the rate of change of the kinetic energy of the particle.

43. **N** In a quantum wire, an electric current can be controlled by varying the height of a potential barrier. Suppose that the electrons each have a kinetic energy of 8.24 eV and that 1.33×10^4 electrons pass through the wire every second when there is no potential barrier. If the width of the barrier is 50.00 pm, what barrier energy height would cause the current to become 1.00% of its initial value?

41-8 Special Case: Quantum Simple Harmonic Oscillator

44. **A** Confirm that Equation 41.25, $\psi_1(x) = A_1 e^{-(x/x_0)^2}$, where $x_0^2 = h/(\pi m \omega)$ and A_1 is a constant that is found by normalization, is a solution to Schrodinger's equation (Eq. 41.24):

$$-\frac{h^2}{8\pi^2 m}\frac{d^2\psi(x)}{dx^2} + \frac{1}{2}m\omega^2 x^2 \psi(x) = E\psi(x)$$

In doing so, also show that the ground state energy is $E_1 = h\omega/4\pi$.

45. **N** The energy difference between two adjacent energy levels in a quantum simple harmonic oscillator is 3.58 eV. What is the energy of the ground state?

46. **A** Assume an electron is in the ground state of a quantum simple harmonic oscillator potential well with a ground-state energy of 4.65 eV. Write the wave equation for the stationary ground state. You do not need to determine the normalization constant, A_1.

47. **N** A particle is in the $n = 5$ excited state of a quantum simple harmonic oscillator well. A photon with a frequency of 1.25×10^{15} Hz is emitted as the particle moves to the $n = 3$ excited state. What is the minimum photon frequency required for this particle to make a quantum jump from the ground state of this well to the $n = 4$ excited state?

48. **N** A quantum simple harmonic oscillator has an angular frequency $\omega = 6.43 \times 10^{15}$ rad/s. Find the energy of the ground state and the first three excited states of the oscillator.

41-9 Heisenberg's Uncertainty Principle

49. **A** Substitute Equation 41.31, $\psi(x) = \psi_0 e^{ikx} + \psi_1 e^{-ikx}$, into Equation 41.30, $d^2\psi(x)/dx^2 = -k^2\psi(x)$, to show that it is a solution.

50. **A** Show that a free particle that is traveling in the negative x direction is equally likely to be found anywhere along the x axis.

51. **N** A hard-luck baseball player who can't seem to hit the fastball of an all-star pitcher claims that his difficulties rest in the physics of the ball being pitched. "Given the speed of the ball being 103 mph, the minimum uncertainty in the position of the ball is just too large to be able to know where and when to swing," he says. Assuming the mass of the baseball is 0.145 kg, find the minimum uncertainty in the position of the baseball. Is the player's claim legitimate?

52. **E** CASE STUDY In Chapter 42 we study atoms, and part of our case study is about hydrogen. Let's use Heisenberg's uncertainty principle to estimate the ground-state energy of hydrogen. In our model, the electron is confined in a one-dimensional well with a length about the size of hydrogen, so that $\Delta x = 0.0529$ nm. Estimate Δp, and then assume that the ground-state energy is roughly $\Delta p^2/2m_e$. Give your answer in joules and electron-volts.

Problems 53 and 54 are paired.

53. **N** An alpha particle is a helium nucleus consisting of two protons and two neutrons. It is moving with a speed of 1.10×10^3 m/s.
    a. What is the momentum of the alpha particle?
    b. If there is a 10.0% uncertainty in the momentum of this alpha particle, what is the minimum uncertainty in the position of the alpha particle?

54. **N Review** An alpha particle is a helium nucleus consisting of two protons and two neutrons. It is moving with a speed of 1.10×10^8 m/s.
    a. What is the momentum of the alpha particle?
    b. If there is a 10.0% uncertainty in the momentum of this alpha particle, what is the minimum uncertainty in the position of the alpha particle?

Problems 55 and 56 are paired.

55. **N Review** The lifetime of a muon in its own frame of reference is measured to be 3.10×10^{-6} s. What is the minimum uncertainty in the energy of the muon?

56. **Review** The lifetime of a muon moving relative to an observer is measured to be 15.6×10^{-6} s.
    a. **N** What is the minimum uncertainty in the energy of the muon according to the observer?
    b. **C** Consider the muon from Problem 55 and the uncertainty in its energy. Explain how the energy can be more certain for this moving muon than for a muon that is at rest for the same observer.

General Problems

57. **A** An infinite square well of width L confines an electron that transitions from the $n = 1$ ground state to the $n = 4$ state after absorbing a photon with wavelength λ_A. Assume the photon has just enough energy to accomplish this transition.
  a. What is the width L of the well confining the electron in terms of the mass of the electron, m_e, h, λ_A, and c?
  b. The electron next transitions from the $n = 4$ to the $n = 2$ state. What is the wavelength λ_B of the photon emitted during this transition in terms of λ_A?

58. **C** Although there is no physical interpretation for the wave function $\psi(x)$, it does have units. Consider the form of wave function for the case of the infinite square well, $\psi(x) = (\sqrt{2/L}) \sin(k_n x)$. What are the SI units of the wave function?

59. **N** An infinite square well of width $L = 155$ pm is used to confine an electron. The well extends from $x = 0$ to $x = L$. Assume the electron is in the ground state.
  a. What is the probability of detecting the electron between $x = 0$ to $x = 0.50L$?
  b. What is the probability of detecting the electron between $x = 0$ to $x = 0.25L$?

60. **N** An infinite square well of width $L = 155$ pm is used to confine an electron. The well extends from $x = 0$ to $x = L$. Assume the electron is in the first excited state ($n = 2$).
  a. What is the probability of detecting the electron between $x = 0$ to $x = 0.50L$?
  b. What is the probability of detecting the electron between $x = 0$ to $x = 0.25L$?

Problems 61, 62, and 63 are grouped.

61. **N** A proton is trapped in an infinite well of length 0.350 nm. Find the system's ground-state energy. Give your answer in joules and electron-volts.

62. **N** A proton is trapped in an infinite well of length 0.350 nm. Find the energy of the first four excited states ($n = 2$ to $n = 5$). Give your answers in joules and electron-volts.

63. **N** A proton is trapped in an infinite well of length 0.350 nm. Initially the system is in its ground state. It then absorbs a photon with an energy of 1.34×10^{-2} eV. What state is the system in as a result?

64. **C** A free particle with kinetic energy 185 eV travels in a region of space where $U = 0$. It passes over a square potential well of depth 152 eV.
  a. What happens to the total energy of the particle as it passes over the well?
  b. What happens to the kinetic energy of the particle as it passes over the well?

65. **C** Consider the quantum simple harmonic oscillator potential well. Would you expect there to be a penetration distance into the walls of the harmonic potential well? Explain your answer.

Problems 66 and 67 are paired

66. **N** A hockey puck ($m = 0.160$ kg) is placed in a two-dimensional box where the distance between one set of walls is 0.500 m and the distance between the other set of opposing walls is 0.250 m. Assume the walls in each dimension represent an infinite square well in each case. In this two-dimensional case, the state of the puck would be represented by the combination of standing waves in each of the two dimensions.
  a. What is the zero-point energy of the puck in each dimension?
  b. The actual zero-point energy is the sum of the zero point energy in each dimension. Find this zero-point energy for the puck in the two-dimensional box.

67. **N** A hockey puck ($m = 0.160$ kg) is placed in a three-dimensional box where the distance between one set of walls is 0.500 m, the distance between another set of opposing walls is 0.250 m, and the distance between the remaining set of walls is 0.350 m. Assume the walls in each dimension represent an infinite square well in each case. In this three-dimensional case, the state of the puck would be represented by the combination of standing waves in each of the three dimensions.
  a. What is the zero-point energy of the puck in each dimension?
  b. The actual zero-point energy is the sum of the zero-point energy in each dimension. Find this zero-point energy for the puck in the three-dimensional box.

Problems 68 and 69 are paired.

68. **C** A particle is in an infinite square well that has its width halved. Is the percentage change in the third excited state energy (E_4) the same as the percentage change in the ground-state energy E_1? Explain your answer.

69. **C** Consider two adjacent energy levels for a particle in an infinite square well, n and m, where $n > m$. When the width of the well is halved, does the energy difference $E_n - E_m$ stay the same, increase, or decrease? Explain your answer.

70. **A, C** The width of a potential barrier decreases at a constant rate (dw/dt) until the width becomes zero. Derive an approximate expression for the rate of change of the probability (dP_{tunnel}/dt) that a particle with mass m and kinetic energy K_i tunnels through the barrier with height U_i. Interpret the meaning of your answer.

71. **N** When transitioning from the $n = 4$ excited state to the ground state ($n = 1$), an isolated atom emits a 567-nm photon. When transitioning from the $n = 3$ excited state to the ground state, the same atom emits a 626-nm photon. What is the wavelength of the photon emitted by this atom when it transitions from the $n = 4$ excited state to the $n = 3$ excited state?

72. **A** In this chapter, you have seen the solution for the time-independent, one-dimensional Schrödinger equation for an infinite square well with a width from $x = 0$ to $x = L$. Though much more difficult to derive, Schrödinger's equation can be solved for an infinite square well that is symmetric about $x = 0$, running from $x = -L/2$ to $x = L/2$. In this case, the normalized results for the wave function are $\psi(x) = (\sqrt{2/L}) \sin(k_n x)$ when n is an odd number, and $\psi(x) = (\sqrt{2/L}) \cos(k_n x)$ when n is an even number. Show that $\psi(x) = (\sqrt{2/L}) \cos(k_n x)$ is a solution for the time-independent, one-dimensional Schrödinger equation.

73. **N** A particle incident on a potential barrier has a probability $P_{\text{tunnel}} = 50\%$ of tunneling through the barrier.
  a. If the width of the barrier is halved, what is the new probability of the particle tunneling through the barrier?
  b. If, instead, the width of the barrier is doubled, what is the new probability of the particle tunneling through the barrier?

74. **C** You may often encounter derivations, models, or solutions that claim to use a "semiclassical" approach. Describe what is meant by "semiclassical" in these instances.

75. **C** Shannon and Cameron have been studying all morning for tomorrow's physics final. Avi has slept in.

Cameron: Avi, so good of you to join us. Do you already know everything in all 43 chapters?

Avi: No, but I don't have to. All I need to know is what Heisenberg said, *nothing is certain, and anything is possible*. I can ace the exam without knowing anything else.

Shannon: Let me tell you just one thing about Heisenberg's uncertainty principle that might even help you tomorrow...

Complete Shannon's thought: What does Heisenberg's uncertainty principle really say?

Problems 76, 77 and 78 are grouped

76. **A** Although each stationary state represents a single-state solution to Schrödinger's time-independent equation for a particle in an infinite square well, a more general solution might include a mix of states, where there might be a different probability of finding the particle in each state. Suppose a linear combination of the ground state and the first excited state describes the current state of the particle, $\psi(x) = A_1 \sin(k_1 x) + A_2 \sin(k_2 x)$. The coefficients of each stationary state, A_1 and A_2, are related to the probability of each state occurring ($|A_1|^2$ and $|A_2|^2$ respectively). When a particle is measured, this kind of mixed state will "collapse" into one of the possible states – in this case, either the ground state or the first excited state. Substitute each individual state into the left side of the time-independent Schrödinger's equation (Eq. 41.8), with $U(x) = 0$ and show that in each case, $k_n = \sqrt{8\pi^2 m E_n / h^2}$.

77. **N** Consider the mix of stationary states described in Problem 76. If there is a 1 in 4, or 25%, chance of finding the particle in the first excited state ($n = 2$), what is the probability of finding the particle in the ground state?

78. **N** Consider the mix of stationary states described in Problem 76. Suppose there is a 1 in 4, or 25%, chance of finding the particle in the first excited state ($n = 2$).
    a. What is the value of A_2?
    b. What is the value of A_1?

79. **A** Start with the normalization condition $\int_0^L |A \sin k_n x|^2 dx = 1$ and show $A = \sqrt{2/L}$. Therefore, the wave function for a particle trapped in an infinite square well is given by $\psi(x) = \sqrt{\frac{2}{L}} \sin k_n x$ (Eq. 41.17).

80. **E** Suppose the science fiction writer in Example 41.10 only wishes to produce 2000 calories of pure energy (with no mass) for a crew member to consume. (We won't worry how the body can absorb the energy without the usual metabolic process.) Use Heisenberg's uncertainty principle to estimate how long such energy could be created from nothing before vanishing.

42 Atoms

Key Questions

How do we best model an atom?

What observations support our model?

42-1 Early Atomic Models 1373

42-2 Rutherford's Model of the Atom 1375

42-3 Bohr's Model and Atomic Spectra 1377

42-4 De Broglie's Theory and Atoms 1386

42-5 Schrödinger's Equation Applied to Hydrogen 1388

42-6 Magnetic Dipole Moments and Spin 1392

42-7 Other Atoms 1395

42-8 Organizing Atoms 1397

42-9 The Zeeman Effect 1401

42-10 Practical Devices 1405

❶ Underlying Principles

Pauli's exclusion principle

✪ Major Concepts

1. Thomson's plum pudding model
2. Rutherford's solar system model
3. Bohr's atomic model
4. De Broglie's model applied to an atom
5. Schrödinger's equation applied to an atom
6. Periodic table of the elements
7. Zeeman effect

In Chapter 41, we considered particles under ideal conditions, such as when they are trapped in an infinite well. Exploring those ideal conditions enabled us to tackle the mathematics more readily. So you might think that quantum mechanics is a theory having little to do with anything practical. In this chapter, we take the opposite approach and explore something very practical—atoms. Almost everything is made of atoms. (Some exceptions include plasmas, white dwarfs, and neutron stars.) In this chapter we'll use quantum mechanics to explain the structure of atoms. Quantum mechanics explains the organization of the periodic table of the elements and shows why some types of atoms are highly reactive and others are not reactive at all. Although the mathematics involved in modeling atoms with quantum mechanics is beyond the scope of this book, we'll imitate our practice in the previous chapter by deducing the steps that are required without actually performing all the calculations. However, you can expect to understand and use elsewhere the results derived here from quantum mechanics.

42-1 Early Atomic Models

You have probably encountered a solar system model of the atom from your early school days (Fig. 42.1). At the center of the atom, there is a nucleus consisting of protons and neutrons. Because it consists only of positive and neutral particles, this nucleus is positively charged. Negatively charged electrons orbit the nucleus, much as the planets orbit the Sun. The atom as a whole is neutral because the negative charge of all the electrons has the same magnitude as the positive charge of the nucleus. But because this solar system model does not take into account the wave properties of subatomic particles, it is an incorrect description of the atom. In this chapter, we take a closer look at the structure of the atom, one that goes beyond the solar system model. We begin with a brief history of atomic models.

The word *atom* comes from the Greek word *atomos*, which means *undivided*; so, an atom was thought to be an *indivisible* particle. The nature of matter was debated by ancient (fifth century BCE) Greek philosophers. The dominant idea at the time was that matter is continuous, but a few philosophers believed that matter was made up of small, discrete, indivisible particles called atoms. These atoms were thought to come in different sizes and shapes, which would determine the properties of matter on the macroscopic scale. Nevertheless, these atoms were also thought to be fundamental particles with no internal structure. This ancient atomic view of matter was largely ignored or rejected until the 1600s, when the modern scientific method was developed.

Evidence in favor of the modern atomic model was based on the study of gases (the kinetic theory of gases; Chapter 20) and chemical reactions. By the latter half of the 1800s, over 63 elements were identified and arranged by mass to form the beginning of the **periodic table** (Appendix B). By then scientists also knew that the chemical properties of an element correlate with its place in the periodic table. So the molecules that could form out of various elements were determined by the place of these elements in the periodic table. At first scientists didn't know what forces bind the atoms together to form molecules. Later, scientists such as Sir Humphry Davy (1778–1829) and his assistant Michael Faraday (1791–1867)—whose law is one of Maxwell's equations (Chapter 34 and page 1093)—discovered, through experimentation, that molecules are held together by electrical forces.

The idea that atoms have no internal structure persisted until the late 1800s, when the work of the British physicist Sir Joseph John Thomson (1856–1940) first showed that atoms are made up of smaller particles. Physicists at that time were trying to understand an experiment involving a capacitor in a vacuum tube (Fig. 42.2). The capacitor is charged, and then the tube is evacuated. As the air is removed from the tube, a blue glow appears between the capacitor's plates. As more air is removed, the glow turns pink. Physicists at the time called this glow (actually glowing particles) *cathode rays*. The nature of cathode rays was controversial. It was known that cathode rays could be deflected by a magnetic field, but in early experiments they were not deflected by an electric field. If cathode rays were made up of charged particles, then an electric field should deflect them.

At first, Thomson believed that cathode rays were charged particles and were not deflected by an electric field because the tube still contained too much air. By 1897, Thomson developed a much better vacuum system, and showed that cathode rays are indeed deflected by an electric field, as would be expected for a stream of negatively charged particles. Thomson concluded that the negative particles—electrons—emerge from the atoms that make up the metal of the capacitor plates. He then went on to change the type of metal used for the plates, with similar results. So he concluded that electrons are present in all atoms.

Based on his experiments, Thomson developed a model for the atom. Thomson thought that the atom consists of two parts: a positive homogeneous sphere and electrons embedded in this positive sphere. This was known as the plum pudding

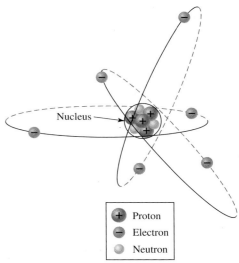

FIGURE 42.1 Solar-system model for an atom.

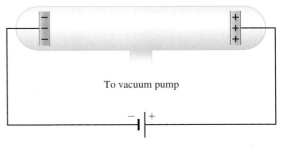

FIGURE 42.2 A capacitor in a tube can be used to observe several phenomena, such as cathode rays or the emissions from a gas such as hydrogen.

THOMSON'S PLUM PUDDING MODEL
✪ Major Concept

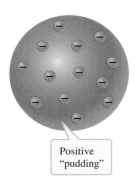

FIGURE 42.3 In the plum pudding model of an atom, negatively charged electrons are embedded in a positive "pudding." There is no nucleus, and the electrons do not orbit.

model (Fig. 42.3). Since you have probably never seen a plum pudding, it might be better to think of it as the chocolate chip ice cream model. The scoop of vanilla ice cream represents the positive sphere, and the chocolate chips represent the embedded electrons.

Although Thomson's model is consistent with his experiment, it cannot explain the spectral lines we observe from atoms (Section 36-4). Any acceptable atomic model must be able to explain the spectral lines. One part of this chapter's case study focuses on the spectrum produced by the simplest atom—hydrogen.

CONCEPT EXERCISE 42.1

What is misleading about the term *atom*? Come up with a term to replace *atom*. Explain your choice.

CASE STUDY Part 1: Lasers

As discussed in the opening paragraph, our focus in this chapter is atoms. Using quantum mechanics to understand atoms deepens our understanding of the universe and leads to practical devices. So the case study in this chapter is divided into two parts. One part is about a practical device—the laser. The other part is about hydrogen—the simplest and most abundant element in the universe. Lasers have become ubiquitous. You or a friend may have a laser on your key ring; your professor is likely to use a laser pointer during a lecture. You probably used a laser in your laboratory to produce interference or a diffraction pattern such as those shown in Figures 35.10, 35.16, 36.2, and 36.5. Lasers are used in medicine to correct vision, to remove tattoos, and to treat some cancers. In the future, lasers may be used to detect gravity waves from distant astronomical events such as the merging of black holes (Fig. 42.4). By the end of this chapter, you will understand the basic operation of a laser.

FIGURE 42.4 General relativity predicts that a gravity wave is triggered by the merging of two massive objects such as black holes or neutron stars. These gravity waves, like waves on the surface of water, travel out into space and cause distant objects to oscillate. Detectors on Earth use lasers to search for the slight oscillations, which would confirm the existence of gravity waves. As of press time, no gravity waves have been detected, but you may wish to check with such organizations as LIGO (Laser Interferometer Gravitational-Wave Observatory).

CASE STUDY Part 2: The Importance of the Hydrogen Spectrum

Because hydrogen is the most abundant element in the universe, astronomers spend a lot of time observing hydrogen, especially the hydrogen spectrum (Fig. 42.5). This spectrum has much to teach us: for example, it is used to find the temperature and size of stars, the mass of the Milky Way galaxy, and the velocity of distant quasars. Each kind of atom has a unique internal structure that determines its unique spectrum. In this case study, we'll focus on the structure of hydrogen atoms to see how that structure is responsible for hydrogen's spectrum. We'll end this part of the case study by seeing if observations of the Sun's spectrum can be used to measure the Sun's magnetic field.

Hα 656 nm   Hβ 486 nm Hγ 434 nm Hδ 410 nm

FIGURE 42.5 Hydrogen produces several colored lines in the visible part of the electromagnetic spectrum. The wavelength (in nanometers) of four dominant lines is provided on the figure. Hydrogen is the most abundant element in the Universe and so plays a major role in astronomy.

EXAMPLE 42.1 CASE STUDY The Hydrogen Spectrum

As shown in Figure 42.5, hydrogen produces four lines in the visible part of the spectrum: Hα (656 nm), Hβ (486 nm), Hγ (434 nm) and Hδ (410 nm). Find the energy associated with the photons that produce each of these lines.

INTERPRET and ANTICIPATE
A photon's energy is proportional to its frequency. The red line (Hα) has a lower frequency than the violet line (Hδ), so we expect the red line to be produced by photons that are less energetic than the ones that produce the violet line.

SOLVE
Use $f = c/\lambda$ (Eq. 34.20) to find the frequency of each line.

Line	$f (\times 10^{14}$ Hz)
Hα	4.57
Hβ	6.17
Hγ	6.91
Hδ	7.32

Use $E = hf$ (Eq. 40.8) to find the energy of each photon.

Line	$E (\times 10^{-19}$ J)
Hα	3.02
Hβ	4.09
Hγ	4.58
Hδ	4.85

The energies are very small. When the energy is so low, it is customary to work in electron volts (eV) instead of in SI units. The conversion factor is 1 eV = 1.6×10^{-19} J.

Line	E (eV)
Hα	1.89
Hβ	2.56
Hγ	2.86
Hδ	3.03

CHECK and THINK
As expected, the red line (Hα) is produced by photons having less energy than those producing the violet line (Hδ). In this chapter's case study, we'll show why these four unique lines are produced by hydrogen.

42-2 Rutherford's Model of the Atom

Thomson's plum pudding model only survived until about 1911, when Ernest Rutherford (1871–1937) designed an experiment that refuted it. Rutherford was born and began his education in New Zealand. In 1895 he won a scholarship, and decided to continue his education at the Cavendish laboratory in Cambridge, where Thomson was working. Thomson was known as a great teacher and mentor, and Rutherford flourished in Cambridge.

By studying the natural radioactive decay of uranium, Rutherford discovered that it emits two kinds of radiation. One kind is better than the other at penetrating a substance such as aluminum foil. Rutherford named the less-penetrating radiation **alpha rays**, and the more-penetrating radiation **beta rays**. Today we know that alpha rays are helium nuclei. Since helium nuclei consist of two protons and two neutrons, they have a charge of $+2e$ and are about 7000 times more massive than an electron.

Today we use the term *ray* to mean a stream of particles. For example, an alpha ray is a stream of alpha particles (helium nuclei).

FIGURE 42.6 Rutherford's experiment aimed a beam of alpha rays at a sheet of gold foil. If the beam hits the screen at A, an alpha particle experiences little or no deflection because it passes through the gold foil far from any atomic nucleus. If it hits the screen at B, the alpha particle is somewhat deflected by passing close to an atomic nucleus. If it hits the screen at C, the alpha particle must hit the atomic nucleus nearly head-on. Rutherford was surprised to detect the scattered beam at all angles around the foil. Rutherford's experiment was carried out by Hans Geiger and by an undergraduate named Ernest Marsden.

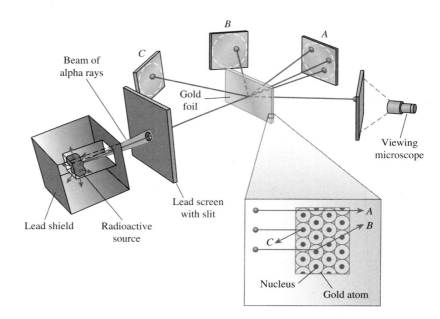

Rutherford aimed alpha rays at a piece of gold foil (Fig. 42.6). A screen coated with zinc sulfide was used to detect the scattered alpha rays. When the zinc sulfide is struck by an alpha particle, it emits light, which the experimenter observes with a small microscope. Rutherford moved his screen to a number of positions around the foil in order to see if the foil would deflect the alpha rays. If the plum pudding model was correct, then the alpha rays would not be deflected very much by the foil. Electrons in the foil (consisting of plum pudding gold atoms) could not cause much of a deflection because they are 7000 times less massive than alpha rays. These electrons would be accelerated by the alpha rays owing to the electromagnetic force, but the alpha rays themselves would only experience a negligible acceleration. By contrast, the positive "pudding" in the gold atoms would be too spread out to cause a significant net force on the alpha rays. The slight deflection caused by the many gold atoms in the foil would nearly cancel out, so that most of the alpha rays would be detected near point A in Figure 42.6.

Rutherford did find that many of the alpha particles were not deflected by the foil and arrived near point A. However, a significant number were greatly deflected and were detected at positions B and C. A few even backscattered off the foil and were detected near their source! Rutherford was surprised by his results; he said

> It was quite the most incredible event that has ever happened to me in my life. It was almost as incredible as if you fired a 15-inch shell at a piece of tissue paper and it came back and hit you.

Because the alpha particles were deflected at all angles, including 180°, Rutherford concluded that the positive part of the atom is concentrated at its center. His model of the atom consisted of a massive positively charged nucleus at the center surrounded by orbiting electrons, so that most of the atom is empty space. Rutherford's model is basically the solar system model that we discussed earlier (Fig. 42.1). This model accounts for the backscattering results as follows. When an alpha particle passes through atoms without coming close to a nucleus, it experiences almost no deflection and is detected at position A. But if the alpha particle comes close to a positive nucleus, then according to Coulomb's law the alpha particle will experience a repulsive force. The closer the alpha particle comes to a nucleus, the greater the force. This repulsive force causes the alpha particle to be deflected and in some cases even to go backward.

In Rutherford's model, the electrons must orbit the nucleus. If they did not orbit, then the attractive force between the nucleus and the electron would cause the electron to fall into the nucleus. If that were the case, then matter would be unstable. (As we'll see in the next sections, quantum mechanics does not require electrons to orbit in order for atoms to be stable.)

RUTHERFORD'S SOLAR SYSTEM MODEL
✪ **Major Concept**

EXAMPLE 42.2 Orbiting Electrons in Hydrogen

Let's examine Rutherford's model more closely. Consider hydrogen, the simplest atom, with a nucleus consisting of a single proton that is orbited by a single electron. Model the nucleus as a sphere with a radius of roughly 10^{-15} m, and assume the electron orbits the center of the nucleus in a circle with a radius of roughly 10^{-10} m. Find the magnitude of the force exerted on the electron by the nucleus and the electron's speed.

INTERPRET and ANTICIPATE
We could have done this problem in Chapter 23. Coulomb's law gives the force exerted on the electron, and because no other forces are present, this is the centripetal force. We can find the electron's speed from its centripetal acceleration.

SOLVE
Use Coulomb's law (Eq. 23.3) to find the force on the electron. As usual, quantity r is the distance from the center of the sphere to the particle; in this case r is the radius of the electron's orbit. For hydrogen, the nucleus and the electron have the same magnitude of charge, e.

$$F_E = k\frac{|q_1 q_2|}{r^2} \quad (23.3)$$

$$F_E = k\frac{e^2}{r^2}$$

$$F_E = (8.99 \times 10^9 \text{ N}\cdot\text{m}^2/\text{C}^2)\frac{(1.60 \times 10^{-19}\text{ C})^2}{(10^{-10}\text{ m})^2}$$

$$F_E = 2.3 \times 10^{-8} \text{ N}$$

This is the centripetal force, so the electron's speed is found using $F_c = mv^2/r$ (Eq. 6.7).

$$F_c = F_E = m_e \frac{v^2}{r}$$

$$v = \sqrt{\frac{rF_E}{m_e}} = \sqrt{\frac{(10^{-10}\text{ m})(2.3 \times 10^{-8}\text{ N})}{9.11 \times 10^{-31}\text{ kg}}}$$

$$v = 1.6 \times 10^6 \text{ m/s}$$

CHECK and THINK
The electron's orbital speed is about 0.5% the speed of light, so there is no need to use special relativity. Our estimate of this speed seems reasonable (not too fast, not too slow), which partly explains why the solar system model of the atom along with classical physics explains so many phenomena. However, Rutherford's solar system model cannot account for the hydrogen spectrum (Fig. 42.5), the subject of part 2 of our **CASE STUDY**.

42-3 Bohr's Model and Atomic Spectra

Rutherford's solar system model encountered problems immediately. First, since the electrons orbit around the nucleus, they are accelerating. Accelerating charged particles give off radiation, so the electrons should be giving off radiation continually. Second, since radiation carries energy away from its source, the electrons should lose energy. You would expect the electrons eventually to spiral into the nucleus, and matter would be unstable. However, no such radiation is observed, and matter is stable. Finally, each element is known to give off a unique set of discrete spectral lines. How can Rutherford's model of the atom explain these observations? A complete model of the atom must be able to account for the observed spectral lines.

The Hydrogen Spectrum

Let's begin by considering in some detail what researchers in the late 19th century were able to observe. One way to view the spectrum emitted by a gas is to use a tube like the one in Figure 42.2. Instead of maintaining a vacuum in the tube, replace the air with the rarefied gas that you wish to study, such as hydrogen. The capacitor is then connected to a high-voltage power supply. This excites the gas, the gas radiates, and the spectrum can be seen by using a diffraction grating (Section 36-4). From Figure 42.5, we see that the resulting visible spectrum of hydrogen consists of four lines. The wavelengths of these lines (656 nm, 486 nm, 434 nm, and 410 nm) may seem arbitrary or random. However, in 1885 a Swiss high school mathematics teacher named Johann Jakob Balmer (1825–1898) discovered their pattern. He realized that the wavelengths of the four lines could be found from the formula

$$\frac{1}{\lambda} = R\left(\frac{1}{2^2} - \frac{1}{n^2}\right) \quad (n = 3, 4, 5, 6) \tag{42.1}$$

where R is called the Rydberg constant and has the value:

$$R = (1.0973731568539 \pm 0.0000000000055) \times 10^7 \, \text{m}^{-1} \tag{42.2}$$

Today we refer to the series of hydrogen lines that fits Equation 42.1 as the **Balmer series**.

Four other hydrogen line series were discovered and named for the researchers who discovered them. The wavelengths of the lines that make up these other series fit an equation (known as the **Rydberg formula**) similar to Equation 42.1:

$$\frac{1}{\lambda} = R\left(\frac{1}{m^2} - \frac{1}{n^2}\right) \tag{42.3}$$

(In Eq. 42.3, $m < n$; you can remember that the variables are in alphabetical order, from low values to high.) The value of m is a constant (integer) for each series, and the lowest value of n is $(m + 1)$ for that series. For example, in the Balmer series $m = 2$, and the lowest value of n is 3. The other series and their values of m are listed in Table 42.1. Equation 42.3 is an empirical fit; it was first stated without a physical model to explain it. That physical model was later developed by Bohr.

TABLE 42.1 The hydrogen spectral line series.

Name of series	m
Lyman	1
Balmer	2
Pashen	3
Brackett	4
Pfund	5

CONCEPT EXERCISE 42.2

CASE STUDY The Balmer Series

Show that the Balmer formula correctly gives the four observed wavelengths Hα (656 nm), Hβ (486 nm), Hγ (434 nm) and Hδ (410 nm) in Figure 42.5.

Bohr's Atomic Model

BOHR'S ATOMIC MODEL
✪ **Major Concept**

Niels Hendrik David Bohr (1885–1962) was part of a very scientific family. Not only did he win the Nobel Prize in physics in 1922, his son Niels Aage Bohr also won the Nobel Prize in physics in 1975. Bohr worked with Rutherford for four years. For a moment, try to imagine being in Bohr's position in 1911. Rutherford's experiment had showed that the atom's positive charge is concentrated in a nucleus and that most of the atom is empty space. If you wanted to explore this model, you would probably focus your attention on the simplest atom—hydrogen. Bohr knew that the visible hydrogen spectrum showed a small number of lines whose wavelengths were given by Balmer's formula (Eq. 42.1). Bohr also knew that Planck had solved the "ultraviolet catastrophe" by hypothesizing that the energy of each oscillator is *quantized*. In fact, Bohr and Planck were in similar positions. Planck had discovered an empirical fit to the black-body curve, and he wanted to develop a physical model for black bodies (Section 40-2). Bohr knew that the hydrogen spectrum was described by Balmer's empirical fit, and he wanted a physical model for hydrogen that would allow him to derive that empirical fit from basic physical principles.

Bohr recognized the problems with Rutherford's atomic model, but he set out to fix these problems without completely overturning it. In both Rutherford's model and Bohr's model, the positive nucleus is at the center of the atom, and the electrons orbit this nucleus like the planets orbit the Sun.

However, Bohr's model differs from Rutherford's in two major ways, known as **Bohr's postulates**. His **first postulate** is that *the orbit of the electron is restricted to one of a number of allowed circular paths*. In making this postulate, Bohr was guided by Planck's success in modeling black bodies, which led Bohr to hypothesize that the angular momentum of the electron orbit is quantized. Specifically, Bohr proposed that the only allowable orbits have an angular momentum L given by:

$$L = n\left(\frac{h}{2\pi}\right) = n\hbar \qquad (n = 1, 2, 3 \ldots) \qquad (42.4)$$

where $\hbar = h/2\pi = 6.58211814 \times 10^{-16}$ eV·s. The integer n is called the **principal quantum number**.

Contrast this to the orbits of planets and other objects in the solar system. Any object orbiting the Sun can have any orbital path with any angular momentum. Comet and asteroid orbits have a wide range of semi-major axes, and thus a wide range of angular momenta. Bohr's first postulate is equivalent to saying that most of these paths are impossible and that all objects orbiting the Sun must follow certain allowed paths.

Bohr's second postulate is that *electrons in orbit around the nucleus do not radiate*. An electron only radiates when it moves from one allowed orbit to another allowed orbit that has a lower energy. Because the electron orbits are quantized, an electron cannot be found moving in between allowed orbits. We must imagine that the electron *jumps* discontinuously from one allowed orbit to another.

Although Bohr's first postulate is based on Planck's work on black body radiation, his second postulate involves a unique departure from classical physics. According to classical physics, a charged particle moving in a circular orbit must radiate. Bohr could not use classical physics to explain why the electrons orbiting the nucleus don't radiate, but this must be true if Rutherford's solar system model is basically correct.

Bohr's two postulates "fixed" the problems with Rutherford's model. First, an electron in a particular orbit does not radiate continually, which explains why we don't observe this sort of radiation coming from matter. Second, because an orbiting electron is not continually losing energy through radiation, it does not spiral into the nucleus. This fixes the problem with the stability of matter. Third, the atom only emits radiation when the electron jumps from one orbit to a lower energy orbit. When that happens, the atom loses a specific amount of energy in the form of a photon that has a specific wavelength. So we only observe photons of certain wavelengths (which in the case of hydrogen are given by Balmer's fit); as a result, the observed spectrum is made up of a few specific lines.

These two postulates were the only two departures Bohr made from classical physics. He was able to derive Balmer's empirical fit using classical physics and these postulates. We'll work through Bohr's derivation here in order to answer part of the question posed by our **CASE STUDY**—why does hydrogen produce just a handful of lines in the visible spectrum (Fig. 42.5)?

DERIVATION Spectral Lines of Hydrogen

We will use Bohr's model to derive the Rydberg formula,

$$\frac{1}{\lambda} = R\left(\frac{1}{m^2} - \frac{1}{n^2}\right) \quad (42.3)$$

which was originally found empirically for the hydrogen spectrum.

In Bohr's model, the nucleus is a proton with charge $+e$. The electron has charge $-e$, and its orbit is a circle of radius r. The electron's mass is given by its rest mass m_e.

As in Example 42.2, find the speed squared of the orbiting electron by equating the electric force (Eq. 23.3) exerted by the nucleus on the electron with the centripetal force (Eq. 6.7).	$F_E = F_c$ $k\dfrac{e^2}{r^2} = m_e \dfrac{v^2}{r}$ $v^2 = \dfrac{ke^2}{m_e r}$ (1)
Next we apply Bohr's first postulate: The angular momentum is quantized and given by Equation 42.4, where n is an integer.	$L = n\left(\dfrac{h}{2\pi}\right)$ (42.4) where $n = 1, 2, 3\ldots$
The magnitude of the angular momentum is given by $L = rp \sin \varphi$ (Eq. 13.25) and $p = mv$ (Eq. 10.1). Since the electron's velocity is perpendicular to its orbital radius, $\sin \varphi = 1$.	$L = rm_e v \sin \varphi$ $L = rm_e v$ (2)
Set Equation 42.4 equal to Equation (2) and solve for v. Remember n is an integer, so we find that the electron's speed is also quantized, with allowed speeds given by Equation (3).	$L = n\left(\dfrac{h}{2\pi}\right) = rm_e v$ (42.5) $v = \dfrac{nh}{2\pi m_e r}$ (3)
Substitute Equation (3) into Equation (1).	$\dfrac{n^2 h^2}{4\pi^2 m_e^2 r^2} = \dfrac{ke^2}{m_e r}$
Solve for r. We find that r is also quantized.	$r = \dfrac{n^2 h^2}{4\pi^2 k m_e e^2}$ (4)
The smallest possible orbital radius is known as the **Bohr radius** r_B, and we find it by substituting $n = 1$ as well as all of those constants into Equation (4) (Problem 42.9). The allowed radii are given by Equation 42.7.	$r_B = 5.29 \times 10^{-11}$ m (42.6) $r = n^2 r_B$ (42.7)
Now consider the energy of the atom, a system consisting of the nucleus and an electron. The nucleus is at rest, so the system's kinetic energy is due only to the electron's orbital speed. From Example 42.2, speed is not relativistic. Substitute Equation (1) for v^2.	$K = \dfrac{1}{2} m_e v^2$ $K = \dfrac{1}{2} m_e \left(\dfrac{ke^2}{m_e r}\right)$ $K = \dfrac{1}{2} \dfrac{ke^2}{r}$ (5)
The electric potential energy of the nucleus–electron system is given by $U_E = kQq/r$ (Eq. 26.3). By the usual convention, the potential energy is zero when the electron is infinitely far from the nucleus. For all other positions, the potential energy is negative.	$U_E = -\dfrac{ke^2}{r}$ (42.8)

Add Equations (5) and 42.8 to find the atom's total mechanical energy.	$E = K + U_E = \dfrac{1}{2}\dfrac{ke^2}{r} - \dfrac{ke^2}{r}$ $E = -\dfrac{1}{2}\dfrac{ke^2}{r}$
Substitute Equation (4) for r. The resulting Equation 42.9 says that the atom's energy is quantized, depending on n. Because of the negative sign, the higher the value of n, the greater (less negative) the atom's energy.	$E = -\dfrac{1}{2}ke^2\left(\dfrac{4\pi^2 km_e e^2}{n^2 h^2}\right)$ $E = -\dfrac{2\pi^2 k^2 m_e e^4}{h^2}\dfrac{1}{n^2}$ (42.9)
Now we apply Bohr's second postulate: The atom only emits radiation when the electron jumps from its orbit to a lower energy orbit. The higher energy orbit has a greater quantum number n, and the lower energy orbit has a lower quantum number m.	$\Delta E = E_n - E_m$ $\Delta E = -\dfrac{2\pi^2 k^2 m_e e^4}{h^2}\left(\dfrac{1}{n^2} - \dfrac{1}{m^2}\right)$ $\Delta E = \dfrac{2\pi^2 k^2 m_e e^4}{h^2}\left(\dfrac{1}{m^2} - \dfrac{1}{n^2}\right)$ (6)
When the atom loses energy ΔE, it emits radiation in the form of a photon whose energy, according to Einstein, is $E = hc/\lambda$ (Eq. 40.17). Set hc/λ equal to Equation (6) and find the wavelength of the photon emitted by solving for $1/\lambda$.	$hf = \Delta E$ $\dfrac{hc}{\lambda} = \dfrac{2\pi^2 k^2 m_e e^4}{h^2}\left(\dfrac{1}{m^2} - \dfrac{1}{n^2}\right)$ $\dfrac{1}{\lambda} = \left(\dfrac{2\pi^2 k^2 m_e e^4}{h^3 c}\right)\left(\dfrac{1}{m^2} - \dfrac{1}{n^2}\right)$
In Problem 42.8 you will show that the constant $2\pi^2 k^2 m_e e^4/h^3 c$ is equal to the Rydberg constant R (Eq. 42.2).	$\dfrac{1}{\lambda} = R\left(\dfrac{1}{m^2} - \dfrac{1}{n^2}\right)$ ✓ (42.3)

:• **COMMENTS**

We have shown that Equation 42.3 follows from Bohr's two postulates and classical physics. We can explain the visible hydrogen spectrum in terms of the Bohr atomic model, in which the hydrogen atom consists of a proton, which is the nucleus and an electron in orbit around the nucleus. The electron orbit is quantized: Only certain orbital radii (Eq. 42.7) or energies (Eq. 42.9) are allowed. The atom only radiates when the electron jumps from one orbit to another orbit with lower energy. The energy of the photon given off by the atom equals the energy difference between these two orbits. Because only certain orbital energies are allowed, only photons of certain specific energies can be emitted. The energy of the emitted photons corresponds to their wavelength or the color we see; thus only lines of those colors are found in the hydrogen spectrum.

Energy Levels in Hydrogen

In following Bohr's derivation, we found that the hydrogen atom's energy is quantized, as given by Equation 42.9. The absolute value of this energy is small, so it is more convenient to work in electron volts (eV) instead of in SI units (joules). Substituting values of the constants in Equation 42.9, we find that the energy levels of the hydrogen atom are given by:

$$E_n = -\dfrac{13.6\,\text{eV}}{n^2} \quad (n = 1, 2, 3\ldots) \quad (42.10)$$

All atoms and molecules have quantized energy levels, although Equation 42.10 only holds for the hydrogen atom. The energy levels of any atom or molecule are best visualized with an energy-level diagram (Section 41-2). Figure 42.7 is an energy-level diagram for hydrogen. The quantum number of the lowest energy level is $n = 1$; this level is known as the **ground state**. The atom cannot have less energy

It is also convenient to express Planck's constant in eV instead of in joules: $h = 4.1356 \times 10^{-15}$ eV·s

FIGURE 42.7 Energy-level diagram for hydrogen (not to scale). According to Bohr's model, the energy of a hydrogen atom increases as the size of the electronic orbit increases. The lines are labeled by Greek letters in Greek alphabetical order, starting with the lowest energy photon. Bohr's model is based on Rutherford's solar system model. The size of the electron's orbit corresponds to the atomic energy, as shown here for three levels.

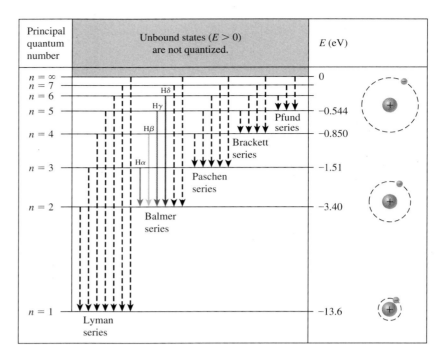

than its ground-state energy. For hydrogen, the ground-state energy is $E_1 = -13.6$ eV. Equation 42.10 can be written in terms of hydrogen's ground state energy:

$$E_n = \frac{E_1}{n^2} \tag{42.11}$$

Because the other energy levels are found by dividing by n^2, these energy levels are closer together for high quantum numbers than they are for low quantum numbers. The energy is negative, as is our usual convention for any *bound* system, such as the electrons bound to a nucleus to form an atom.

Energy-level diagrams are particularly useful for trying to visualize the processes that change an atom's energy state. For example, when a hydrogen atom loses energy, its electron moves into a lower orbit—one that is smaller and has less energy. Such a transition is represented by a vertical, downward-pointing arrow on an energy-level diagram. The arrow must begin on one energy level and end on a lower energy level. When an atom loses energy, it emits a photon whose energy equals the difference between the two energy levels. For example, the vertical arrow labeled Hα in Figure 42.7 begins on the $n = 3$ level and ends on the $n = 2$ level, representing the electron's transition from the larger $n = 3$ orbit to the smaller $n = 2$ orbit. The amount of energy lost by the hydrogen atom is the difference between the two levels:

$$\Delta E = E_2 - E_3 = -3.40 \text{ eV} - (-1.51 \text{ eV})$$

$$\Delta E = -1.89 \text{ eV}$$

The energy lost by the atom takes the form of a photon of energy 1.89 eV (Example 42.1), and wavelength:

$$\lambda = \frac{hc}{E} = \frac{(4.1356 \times 10^{-15} \text{ eV} \cdot \text{s})(2.9979 \times 10^8 \text{ m/s})}{1.89 \text{ eV}}$$

$$\lambda = 6.56 \times 10^{-7} \text{m} = 656 \text{ nm}$$

This is the wavelength observed for Hα. Figure 42.7 shows the transitions corresponding to all the hydrogen series listed in Table 42.1. The Balmer series (the visible light series shown in Fig. 42.5) corresponds to transitions *to* the $n = 2$ level.

Excited Atoms and Ions

You can imagine that when a hydrogen atom is in its ground state, its electron is in the closest orbit around the nucleus. The atom is stable and no photons are emitted. In order for an atom to emit a photon, it must first be in some energy level above the ground state. When an atom has more than its ground-state energy, we say that it is **excited**. Excitation requires that energy be transferred from the environment to the atom, usually by one of two mechanisms: (1) a collision with another object, such as another atom, a molecule, or even a free electron; or (2) absorption of a photon.

Because the atom's energy is always quantized, energy transferred to the atom must also be quantized (unless the atom becomes ionized). If a hydrogen atom is in its ground state, the environment could transfer $13.6 \text{ eV} - 3.40 \text{ eV} = 10.2 \text{ eV}$ to the atom, placing the atom in its $n = 2$ state. Ground-state hydrogen can absorb other amounts of energy as long as that energy is the difference between some energy level and the ground state. But it cannot absorb 1 eV, for example, because there is no available energy level at -12.6 eV.

However, the atom can absorb more than 13.6 eV. In that case, the electron is freed from the atom, and the atom becomes an ion. So the energies listed in Figure 42.7 can also be seen as the minimum amounts of energy the atom needs in each state in order to become an ion. They are called the **binding energy** or **ionization energy** for that state. If a hydrogen atom is in its $n = 2$ state, for example, it needs 3.40 eV in order to become an ion. If an atom gains an amount of energy that is greater than its binding energy, then the electron carries off the excess as kinetic energy. For example, if a ground-state hydrogen atom gains 15.6 eV, it becomes an ion, and the freed electron's kinetic energy is $15.6 \text{ eV} - 13.6 \text{ eV} = 2.0 \text{ eV}$.

Bohr's model works very well for the hydrogen atom and is even a good model for ions that resemble hydrogen (that is, those that have only one remaining electron). Bohr's model is not good for other atoms, but his two postulates were important steps toward discovering better models for atoms and molecules.

CASE STUDY Part 2: The Sun's Hydrogen Spectrum

You have probably seen the spectrum produced by sunlight as it passes through a prism; it appears to be a complete rainbow of colors. However, only an ideal black-body spectrum shows all the colors (Fig. 42.8A). A black-body spectrum is known as a *continuum spectrum*.

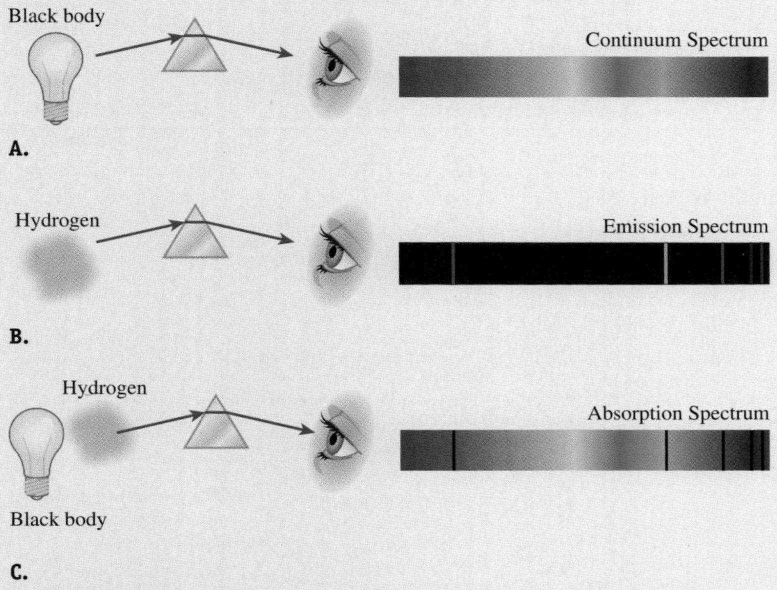

The Sun is not an ideal black body, so what colors does its spectrum consist of? The Sun is mostly made of hydrogen, so you might expect its spectrum to show the four bright Balmer lines in Figure 42.8B. Such a spectrum is known as an **emission spectrum** because these photons are emitted by atoms. However, what is actually observed looks like a continuum spectrum minus the Balmer lines (Fig. 42.8C). Such a spectrum is called an **absorption spectrum**.

To see why we observe an absorption spectrum from the Sun (and most other stars), imagine that some hydrogen gas (in a low energy state) is illuminated by white light from a black body (Fig. 42.8C); the hydrogen absorbs the photons whose energy corresponds to the difference between two hydrogen energy levels. The rest of the photons pass through the gas. So if you look at the spectrum from the perspective shown in Figure 42.8C, you will see the black-body spectrum, minus the photons that have been absorbed by hydrogen.

When we observe the Sun, we are looking at its outermost layer. The layers below radiate like a black body, and the outermost layer of hydrogen absorbs the photons that have energies corresponding to energy-level differences in the hydrogen atom. Normally, when we pass sunlight through a prism, we don't notice the few missing absorption lines. However, these absorption lines are present and with more sensitive equipment, astronomers regularly observe these lines to determine physical properties (such as temperature and magnetic field strength) of the Sun and other stars.

FIGURE 42.8 A. A black body is represented by the incandescent bulb. Its spectrum contains all the colors. **B.** A hydrogen emission spectrum is observed. **C.** A hydrogen absorption spectrum is observed when the light from a black body passes through hydrogen gas.

EXAMPLE 42.3 — CASE STUDY: Other Hydrogen Lines

Use the information in Figure 42.7 to find the wavelength of the first (lowest energy) line of the Lyman, Paschen, and Brackett series. In what part of the electromagnetic spectrum are these lines found?

INTERPRET and ANTICIPATE
Find the wavelength of the emitted photons from the energy differences shown in Figure 42.7. The greater the energy difference, the greater the energy of the photon. High-energy photons have short wavelengths. Of the three series, the Lyman transition has the greatest energy difference, so we expect its photon will have the shortest wavelength. Check the results by using the empirical fit (Eq. 42.3).

SOLVE

Find the energy difference for each of the lines. The subscripts L, P, and B refer to the first line of the *L*yman, *P*aschen and *B*rackett series, respectively.

$$\Delta E_L = E_2 - E_1 = -3.40 \text{ eV} - (-13.6 \text{ eV}) = 10.2 \text{ eV}$$

$$\Delta E_P = E_4 - E_3 = -0.850 \text{ eV} - (-1.51 \text{ eV}) = 0.66 \text{ eV}$$

$$\Delta E_B = E_5 - E_4 = -0.544 \text{ eV} - (-0.850 \text{ eV}) = 0.306 \text{ eV}$$

In each case, the energy lost by the atom equals the photon energy. Use $E = hc/\lambda$ (Eq. 40.17) to find the photon's wavelength.

$$\lambda = \frac{hc}{E} = \frac{(4.1356 \times 10^{-15} \text{ eV} \cdot \text{s})(2.998 \times 10^8 \text{ m/s})}{E} = \frac{1.240 \times 10^{-6} \text{ eV} \cdot \text{m}}{E}$$

$$\lambda_L = \frac{1.240 \times 10^3 \text{ eV} \cdot \text{nm}}{10.2 \text{ eV}} = 122 \text{ nm}$$

$$\lambda_P = \frac{1.241 \text{ eV} \cdot \mu\text{m}}{0.66 \text{ eV}} = 1.88 \, \mu\text{m}$$

$$\lambda_B = \frac{1.241 \text{ eV} \cdot \mu\text{m}}{0.306 \text{ eV}} = 4.06 \, \mu\text{m}$$

Use Figure 34.11 and Table 34.2 (page 1101) to identify the part of the electromagnetic spectrum in which each line is found.

$$\lambda_L = 1.22 \times 10^{-7} \text{ m in the UV}$$

$$\lambda_P = 1.88 \times 10^{-6} \text{ m in the IR}$$

$$\lambda_B = 4.05 \times 10^{-6} \text{ m in the IR}$$

CHECK and THINK
As expected, the Lyman line has the shortest wavelength. We can check our results using Equation 42.3.

$$\frac{1}{\lambda} = R\left(\frac{1}{m^2} - \frac{1}{n^2}\right) \tag{42.3}$$

$$\frac{1}{\lambda_L} = R\left(\frac{1}{1^2} - \frac{1}{2^2}\right) = \frac{3R}{4}$$

$$\lambda_L = \frac{4}{3R} = \frac{4}{3(1.097 \times 10^7 \text{ m}^{-1})} = 122 \text{ nm} \checkmark$$

$$\frac{1}{\lambda_P} = R\left(\frac{1}{3^2} - \frac{1}{4^2}\right) = \frac{7R}{144}$$

$$\lambda_P = \frac{144}{7R} = 1880 \text{ nm} \checkmark$$

$$\frac{1}{\lambda_B} = R\left(\frac{1}{4^2} - \frac{1}{5^2}\right) = \frac{9R}{400}$$

$$\lambda_B = \frac{400}{9R} = 4050 \text{ nm} \checkmark$$

EXAMPLE 42.4 Tungsten

Although Bohr's model doesn't work well for complicated atoms, it does fit ions that have been stripped of all but one electron. Consider such an ion of tungsten. If the ion goes from its first excited state ($n = 2$) down to the ground state, what is the energy of the photon emitted?

INTERPRET and ANTICIPATE
We start by modifying the expression (Eq. 42.9) for hydrogen's energy levels to find an expression for ionized tungsten's energy levels. Then we can find the energy between the first excited state and the ground state.

SOLVE

Revisit the derivation of the hydrogen spectral lines (Eq. 42.3). We started by applying Coulomb's law to the single proton and electron in a hydrogen atom. When we apply Coulomb's law to ionized tungsten, we must take into account that it has Z protons, where Z is its atomic number. So the speed of the electron (Eq. 1 in the derivation) increases by a factor of Z.	For hydrogen $$F_E = k\frac{q_1 q_2}{r^2} = k\frac{e^2}{r^2}$$ For other ions (with one electron) and Z protons $$F_E = k\frac{q_1 q_2}{r^2} = k\frac{Ze^2}{r^2}$$ $$v^2 = \frac{kZe^2}{m_e r} \quad (1)$$
This higher speed means that the kinetic energy is higher by a factor of Z.	For hydrogen $$K = \frac{1}{2}m_e v^2 = \frac{1}{2}\frac{ke^2}{r}$$ For other ions (with one electron) and Z protons $$K = \frac{1}{2}\frac{kZe^2}{r} \quad (2)$$
Continue to revisit the derivation, and you find that the radius of the electron's orbit (Eq. 4 in the derivation) is reduced by this factor of Z.	For hydrogen $$r = \frac{n^2 h^2}{4\pi^2 k m_e e^2}$$ For other ions (with one electron) and Z protons $$r = \frac{n^2 h^2}{4\pi^2 k m_e Ze^2} \quad (3)$$
There is another modification we must make to the derivation. The potential energy of the nucleus-electron system is modified by a factor of Z.	For hydrogen $$U_E = k\frac{Qq}{r} = -\frac{ke^2}{r}$$ For other ions (with one electron) and Z protons $$U_E = -\frac{kZe^2}{r} \quad (4)$$
Combine equations (2) through (4) to find an expression for the energy levels in an ion with only one electron remaining.	$$E = K + U = -\frac{1}{2}\frac{kZe^2}{r}$$ $$E = -\frac{1}{2}kZe^2\left(\frac{4\pi^2 k m_e Ze^2}{n^2 h^2}\right) = -\frac{2\pi^2 k^2 m_e Z^2 e^4}{h^2}\frac{1}{n^2}$$
Compare the expression here to Equation 42.9 for hydrogen's energy levels, and you see that they differ by a factor of Z^2.	For hydrogen $$E = -\frac{2\pi^2 k^2 m_e e^4}{h^2}\frac{1}{n^2} \quad (42.9)$$

We find a convenient expression for the energy levels in the ion by multiplying Equation 42.10 by Z^2.	$E_n = -\dfrac{13.6 \text{ eV}}{n^2}$ (42.10) For other ions (with one electron) and Z protons $E_{\text{ion}} = Z^2 E_{\text{Hydrogen}} = -\dfrac{Z^2(13.6 \text{ eV})}{n^2}$ (5)
From the periodic table (Appendix B), tungsten's atomic number is 74, which we substitute in Equation (5).	$E_{\text{Tungsten}} = -\dfrac{74^2(13.6 \text{ eV})}{n^2} = -\dfrac{74.5 \text{ keV}}{n^2}$ (6)
We use Equation (6) to write an expression for the difference in the tungsten ion's energy levels, which corresponds to Equation 4 in the derivation for the hydrogen spectral lines. Then we substitute for the energy levels. The photon carries away energy equal to the difference in these levels.	$\Delta E_{\text{Tungsten}} = 74.5 \text{ keV}\left(\dfrac{1}{m^2} - \dfrac{1}{n^2}\right)$ $\Delta E_{\text{Tungsten}} = 74.5 \text{ keV}\left(\dfrac{1}{1^2} - \dfrac{1}{2^2}\right) = 55.9 \text{ keV}$ $E_{\text{photon}} = 55.9 \text{ keV}$

∴ CHECK and THINK

When tungsten is stripped of all but one electron, it has 74 protons and 1 electron. We expect that when the ion is in its ground state, it should be much more tightly bound than when a hydrogen atom is in its ground state. This is exactly what we find when we compare Equation (6) in this example for the tungsten ion to Equation 42.10 for hydrogen. The numerator in these equations indicates the amount of energy it would take to free the electron from its ground state. In the case of hydrogen, this is 13.6 eV. For tungsten the energy requirement is roughly 74500 eV—a factor of $74^2 = 5476$ times greater.

42-4 De Broglie's Theory and Atoms

DE BROGLIE'S MODEL APPLIED TO AN ATOM ⊕ **Major Concept**

Although Bohr's postulates work well at creating a model for hydrogen, Bohr did not provide a theoretical basis for these postulates that would explain *why* the energy levels in an atom are quantized. This question is at least partially answered by de Broglie's idea that particles, such as electrons, are also waves (Section 40-6). Bohr did not have the benefit of de Broglie's thinking because de Broglie wrote his thesis about a decade after Bohr came up with his model of the atom.

An electron confined inside a hydrogen atom by the electric potential of its nucleus is like a particle confined in a potential well (Chapter 41). In both situations, the system's energy is quantized and the lowest energy level is nonzero. We modeled a particle confined in an infinite square well as a standing wave on a string with both ends fixed. We can use a similar model for an electron confined in a hydrogen atom.

First, write the de Broglie wavelength (Eq. 40.18) explicitly in terms of the speed v: $\lambda = h/p = h/(m_e v)$. Solve for $m_e v$:

$$m_e v = \dfrac{h}{\lambda} \quad (42.12)$$

Then substitute Equation 42.12 into Equation 42.5 for the angular momentum, $L = n(h/2\pi) = rm_e v$:

$$L = n\left(\dfrac{h}{2\pi}\right) = r\left(\dfrac{h}{\lambda_n}\right)$$

Keep in mind that L is used to represent both angular momentum and length. You may use dimensional analysis if you get confused.

Solve for the wavelength:

$$\lambda_n = \dfrac{2\pi r}{n} \quad (42.13)$$

Equation 42.13 shows that the wavelengths are quantized, and it also gives us a new model for the atom. Compare Equation 42.13 to Equation 18.9, $\lambda_n = 2L/n$, and you'll see that an electron confined in an atom behaves like a standing wave on a string of length $L = \pi r$ (or like a particle confined in a box of this length). Bohr's model predicts the hydrogen spectral lines, so the Rydberg formula derivation still holds—but with a new physical model behind it. Instead of interpreting r as the radius of the electron's orbit, we relate r to the length of the "box" in which the electron is confined. So we can replace the solar system model of the atom with the model of a standing wave on a string.

Because the numerator in Equation 42.13 is $2\pi r$—the circumference of a circle—we can think that the electron is confined in a circular "box" of radius r. However, a circular "box" is not the best visual analogy. Instead we consider the analogy of a standing wave on a string, and imagine a thin circular wire of radius r (Fig. 42.9). Bohr's model correctly gives the energy levels of the hydrogen atom (Fig. 42.7), but these levels are no longer said to correspond to the size of the electron's orbit. Instead, in de Broglie's model, the energy levels correspond to (circular) standing waves, with wavelengths given by Equation 42.13.

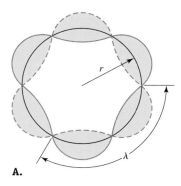

A.

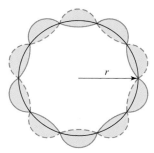

B.

FIGURE 42.9 An electron confined in an atom is like a standing wave on a circular wire. Here we show two wavelengths: **A.** $\lambda_3 = 2\pi r/3$ and **B.** $\lambda_5 = 2\pi r/5$.

CONCEPT EXERCISE 42.3

In what fundamental way is de Broglie's model of hydrogen different from Bohr's model? In what fundamental way is de Broglie's model of hydrogen the same as Bohr's model?

EXAMPLE 42.5 De Broglie's Model of Hydrogen

If a hydrogen atom's energy is -1.51 eV, what is the de Broglie wavelength of the atomic electron?

:• INTERPRET and ANTICIPATE
Use the energy-level diagram to find the principal quantum number. The quantum number determines the radius of the circular wire. (In Bohr's model, this is the radius of the electron's orbit.) The wavelength is determined by the quantum number and the radius of the circular wire.

:• SOLVE	
The energy-level diagram (Fig. 42.7) shows which quantum number corresponds to an energy of -1.51 eV.	$n = 3$
Find the orbital radius from Equation 42.7, where r_B is the Bohr radius and $n = 3$.	$r = n^2 r_B$ (42.7) $r = (3^2)(5.29 \times 10^{-11} \text{ m}) = 4.76 \times 10^{-10} \text{ m}$
The de Broglie wavelength is given by Equation 42.13, $\lambda_n = 2\pi r/n$.	$\lambda_3 = \dfrac{2\pi(4.76 \times 10^{-10} \text{ m})}{3}$ $\lambda_3 = 9.97 \times 10^{-10}$ m $\lambda_3 = 0.997$ nm

:• CHECK and THINK
This problem required a reinterpretation of Bohr's model. Instead of thinking of r as the radius of the electron's orbit, we think of it as the radius of the wire to which the "waving" electron is confined.

42-5 Schrödinger's Equation Applied to Hydrogen

Neither de Broglie's nor Bohr's model provides a theoretical basis for why the energy levels of hydrogen are quantized. However, such a basis comes from Schrödinger's equation. In Chapter 41, we learned that the procedure for applying Schrödinger's equation is to (1) substitute the potential energy function, (2) solve for a general expression for ψ, (3) apply the boundary conditions, and (4) interpret $|\psi|^2$ as the probability density. When we applied Schrödinger's equation to a trapped particle, we found that the energy levels are quantized. Because the electron is trapped in the potential well created by the proton, we expect that solving Schrödinger's equation for the hydrogen will yield quantized energy levels. And in particular, when we apply the boundary conditions, we will find an expression for its energy levels.

The potential energy for a hydrogen atom is given by $U_E = -ke^2/r$ (Eq. 42.8) and is plotted in Figure 42.10. The atom and its potential energy function are three dimensional, requiring a three-dimensional version of Schrödinger's equation, which is beyond the scope of this book. However, finding the solution to a three-dimensional version of Schrödinger's equation follows the same steps as solving the one-dimensional version, and we can understand the solution and use it without doing the complicated derivation.

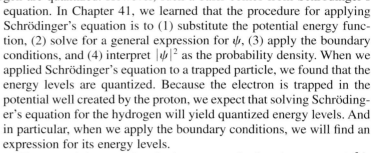

FIGURE 42.10 The energy graph for a hydrogen atom in its ground state. To show the spherical symmetry of the function, we have plotted (positive) r on both sides of the vertical (energy) axis.

Hydrogen's Quantum Numbers

When we solved the one-dimensional Schrödinger's equation, we applied one set of boundary conditions and found that one quantity—energy—is quantized and described by one quantum number. When Schrödinger's equation is solved and boundary conditions are applied in three dimensions, three quantities are quantized and are described by three quantum numbers.

The first quantum number is the **principal quantum number** n found in Bohr's model. This quantum number describes the quantization of energy; for hydrogen we have $E_n = E_1/n^2$ (Eq. 42.11), where hydrogen's ground-state energy is $E_1 = -13.6$ eV. The principal quantum number can be any positive integer: $n = 1, 2, 3 \ldots$

The second quantum number is the **orbital quantum number** ℓ. It describes quantization of the magnitude of the electron's orbital angular momentum L. The orbital quantum number is an integer whose range depends on the principal quantum number such that $\ell = 0, 1, 2, \ldots (n-1)$. The magnitude of the electron's orbital angular momentum is:

$$L = \sqrt{\ell(\ell+1)}\hbar \quad (42.14)$$

The third quantum number is the **orbital magnetic quantum number**, or simply the **magnetic quantum number** m. It describes quantization of the direction of the orbital angular momentum. The magnetic quantum number is an integer whose range depends on the orbital quantum number such that $m = -\ell, -(\ell-1), \ldots, 0, \ldots +(\ell-1), +\ell$. According to the uncertainty principle, the direction of $\vec{L}$ is only partially measurable; we can measure only one component, and then the others become uncertain. Traditionally, we imagine that the atom is in a magnetic field pointing in the positive z direction; then L_z is the component of $\vec{L}$ that we measure (Fig. 42.11). This z component is quantized and given by

$$L_z = m\hbar \quad (42.15)$$

The vector $\vec{L}$ can be anywhere on the surface of a cone that makes an angle θ with the z axis. This angle is quantized, and from Figure 42.11, we find

$$\cos\theta = \frac{L_z}{L} \quad (42.16)$$

Traditionally, the states with the same principal quantum number n are considered part of the same **shell**. You may have run across this terminology in chemistry. These shells are designated by the uppercase letters K, L, M, and so forth alphabetically,

SCHRÖDINGER'S EQUATION APPLIED TO AN ATOM ⊙ **Major Concept**

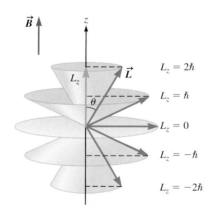

FIGURE 42.11 The magnetic field is in the z direction. When $\ell = 2$, $m = -2, -1, 0, 1$ and 2, corresponding to the five L_z listed here. The components of the angular momentum that are perpendicular to z are not known, so the angular momentum vector for any one value of m lies on the surface of a cone (or disk in the case of $m = 0$.)

Don't confuse m for magnetic quantum number with m for mass. It helps to keep in mind that the quantum numbers are unitless.

corresponding to $n = 1, 2, 3\ldots$. Further, these shells are divided into **subshells** based on orbital quantum number ℓ. These subshells are labeled with lowercase letters corresponding to particular values of ℓ (Table 42.2). This practice allows for a compact notation. For example, if hydrogen has $n = 3$ and $\ell = 2$, then we write its state as $3d^1$, where the superscript 1 tells us that one electron occupies this subshell.

TABLE 42.2 Subshell labels

ℓ	Lowercase letter label
0	s
1	p
2	d
3	f
4	g
5	h

After f, the labels proceed alphabetically.

CONCEPT EXERCISE 42.4

True or false? For an electron in an atom such as hydrogen, the angular momentum vector $\vec{L}$ can never be in the $\pm z$ direction if the magnitude of z component is given by $L_z = m\hbar$ (Eq. 42.15). Explain your answer.

Hydrogen's ground state

If hydrogen is in its ground state, then $n = 1$, and the maximum and only value of its orbital quantum number is $\ell = 1 - 1 = 0$ in this case. Substituting $\ell = 0$ in Equation 42.14, the electron's orbital angular momentum in the ground state is $L = \sqrt{0(0 + 1)}\hbar = 0$. Contrast this with Bohr's model, in which the ground-state electron has an angular momentum $L = \hbar$ (Eq. 42.4 with $n = 1$). To make this contrast more vivid, consider the classical "orbit" of any particle, such as a planet, that has no angular momentum. Such a particle travels on a straight line toward (or away) from the central object, eventually colliding with this object (or escaping from it). The classical system just described is unstable. However, hydrogen does exist in its ground state with no orbital angular momentum, and according to quantum mechanics such an "orbit" is stable.

You may notice something strange in this description. We went to great lengths in the previous section to discredit the solar system model and replace it with a standing-wave model. However, we now refer to the electron's angular momentum, implying that we are treating the electron as a particle in orbit around the nucleus. This is a common practice in the description of atoms. The solar system model is so familiar that we don't entirely abandon it; instead, we treat it as an analogy. Our terminology including the word "orbit" is based on that analogy, but remember that it is only an analogy. We find a more appropriate visual image for our description of the electron, after we consider the solutions to Schrödinger's equation.

We don't need to solve the three-dimensional version of Schrödinger's (time-independent) equation ourselves to appreciate and work with the solutions. In general, these three-dimensional solutions depend on three Cartesian coordinates x, y, and z. However, because of the spherical symmetry, the solutions are easier to find and understand using the spherical coordinates r, θ, and φ. We are only interested in the part of the solution that depends on r, the radial distance from the origin. This is called the *radial solution*, but from brevity we'll just refer to it as the *solution*.

The ground state ($n = 1$) solution for hydrogen is

$$\psi_{1s}(r) = \frac{1}{\sqrt{\pi r_B^3}} e^{-r/r_B} \quad (42.17)$$

where r_B is the Bohr radius.

As with any wave function, there is no physical interpretation for ψ, but we can interpret $|\psi|^2$ as the probability density. (Since an atom is three dimensional, this is the probability per unit volume.)

Figure 42.12 illustrates the probability density for an electron in its ground state in hydrogen. The cloud gives us a way to visualize the probability density. The probability density is greatest where the cloud is thickest. This figure stands in sharp contrast to the solar system model. The electron is not a particle in orbit around the nucleus. Instead, the electron is represented by a cloud that is thick near the nucleus and more diffuse at greater distances. Remember that in the ground state, the electron has no angular momentum. In the solar system model an electron with no angular momentum, trapped in the potential well of the nucleus,

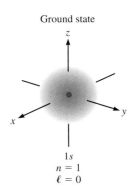

FIGURE 42.12 The probability density for a hydrogen atom in the ground state is spherically symmetric. The electron cloud is densest in the center, meaning that the electron is most likely to be found in the center.

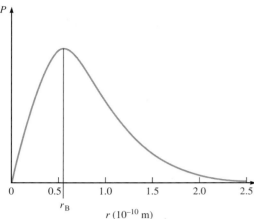

FIGURE 42.13 The radial probability per unit length for the electron in the ground state of the hydrogen atom. The peak occurs at the Bohr radius.

would end up in the nucleus. Contrast this to the quantum-mechanical model. In the ground state, the electron has no angular momentum, but the electron's position is uncertain. So we cannot say that the electron is in the nucleus. Instead we say that the probability density is greatest in the center of the atom.

To find the most likely location for the electron, we need the *radial* probability density $P(r)$, not the *volume* probability density $|\psi|^2$. The two probability densities are related by

$$P(r)dr = |\psi|^2 dV \tag{42.18}$$

The volume element dV is the volume of a spherical shell of thickness dr, so $dV = 4\pi r^2\, dr$. So the radial probability density is

$$P(r) = 4\pi r^2 |\psi|^2 \tag{42.19}$$

The radial probability density allows us to determine the probability $P(r)$ of finding the electron as function distance r from the nucleus. Problem 42.64 asks you to show that

$$P(r) = \frac{4}{r_B^3} r^2 e^{-2r/r_B} \tag{42.20}$$

which is plotted in Figure 42.13. The peak of the graph is located at the Bohr radius. So when hydrogen is in its ground state, the electron is most likely to be found at the Bohr radius—and we can see why Bohr's model predicts the behavior of hydrogen so well. However, quantum mechanics gives a probabilistic interpretation of Bohr's model for hydrogen; allows us to model more complicated atoms; explains the arrangement of atoms in the periodic table; and explains observed lines in hydrogen's (and other atom's) spectrum that cannot be accounted for by the Bohr model.

Hydrogen in an Excited State

So far we have considered hydrogen in its ground state. The wave functions and possibilities are more complicated for the excited states. For example, when hydrogen is in an excited state with $n = 2$, there are two possible values for ℓ and three possible values for m; these are listed in Table 42.3. So there are four possible states in which $n = 2$ (not taking the next quantum number m_s into account). These four states have the same principal quantum number n and make up a single shell with the same energy given by $E_2 = -13.6/(2^2)$ eV (Eq. 42.10 with $n = 2$). The energy of other atoms beyond hydrogen depends on both n and ℓ.

In principle, these four $n = 2$ states are described by four different wave functions and four separate probability densities. However, the probability densities are identical for two of these states: $n = 2$, $\ell = 1$ and $m = +1$ or -1. For hydrogen in the 2s state ($n = 2$, $\ell = 0$) the solution is

$$\psi_{2s}(r) = \frac{1}{\sqrt{2\pi r_B^3}}\left(\frac{1}{2} - \frac{r}{4r_B}\right)e^{-r/2r_B} \tag{42.21}$$

The other two wave functions depend on the azimuthal angle measured from the z axis in the xy plane and the polar angle measured from the z axis. So, to keep the mathematics of our discussion manageable, we omit the other two wave functions.

TABLE 42.3 Quantum numbers for hydrogen in the $n = 2$ state.

n	$\ell = 0, 1, 2 \ldots (n-1)$	$m = 0, \pm 1, \pm 2 \ldots \pm \ell$
2	0	0
2	1	+1
2	1	0
2	1	−1

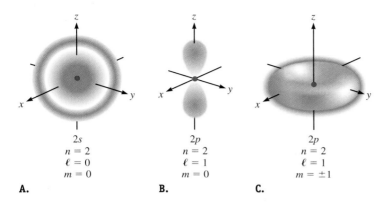

FIGURE 42.14 A. The probability density for a hydrogen atom in the $n = 2$, $\ell = 0$ state is spherically symmetric. Notice the white spherical shell region where $|\psi|^2 = 0$. **B.** The probability density in the $n = 2$, $\ell = 1$, $m = 0$-state is not spherically symmetric. **C.** The probability density in the $n = 2$, $\ell = 1$, $m = \pm 1$ state.

However, Figure 42.14 shows $|\psi|^2$ for all four states. The probability density for $n = 2$, $\ell = 0$ is made up of two components (Fig. 42.14A). One component is a sphere that is densest in the center. The second component is a spherical shell, and there is a spherical gap between the two components. Overall, the probability density for this state is spherically symmetric, much like the probability density for $n = 1$, $\ell = 0$ (Fig. 42.12). This is a general property of wave functions: All quantum states with $\ell = 0$ have spherically symmetric wave functions.

By contrast, the probability density for $n = 2$, $\ell = 1$, $m = 0$ (Fig. 42.14B) has two separate lobes and is symmetric, but not spherically symmetric. The probability density for $n = 2$, $\ell = 1$, $m = \pm 1$ (Fig. 42.14C) looks like a donut, again symmetric but not spherically symmetry. The donut display of this state is often shown in physics textbooks. In chemistry books, this same state is separated into two components, as shown in Figure 42.15. The display traditionally used in chemistry textbooks is helpful to chemists who are interested in studying how atoms bond, and the stretched out displays help us to understand in which directions these bonds are likely to form.

For more on symmetry see p. 756.

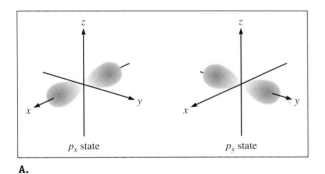

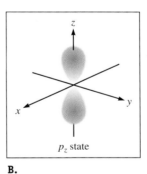

FIGURE 42.15 A. In chemistry books the probability density for the $n = 2$, $\ell = 1$, $m = \pm 1$ state is divided into p_x and a p_y component. **B.** Then the probability density of the $n = 2$, $\ell = 1$, $m = 0$ is referred to as the p_z. This is identical to Figure 42.14B.

EXAMPLE 42.6 Radial Probability density for Hydrogen in 2p states

Figure 42.16 shows the radial probability density for hydrogen in both the $n = 2$, $\ell = 0$ and $n = 2$, $\ell = 1$ subshells. The radial probability density for the $\ell = 1$ subshell peaks at $r_{max} = 4r_B$. The radial probability density for the $\ell = 0$ peaks at a greater distance; find this r_{max} in terms of the Bohr radius r_B.

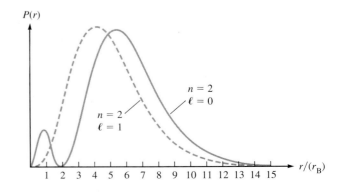

FIGURE 42.16

Example continues on page 1392 ▶

:• **INTERPRET and ANTICIPATE**
We know the wave function for this state. To find the radial probability density $P(r)$, we'll need to square this wave function and multiply the result by $4\pi r^2$. To find the maximum, we use the fact that the derivative of a function is zero at its extrema.

:• **SOLVE** We substitute the wave function (Eq. 42.21) into Equation 42.19 for the radial probability density.	$P(r) = 4\pi r^2 \|\psi\|^2$ (42.19) $P(r) = 4\pi r^2 \left[\dfrac{1}{\sqrt{2\pi r_B^3}} \left(\dfrac{1}{2} - \dfrac{r}{4r_B} \right) e^{-r/2r_B} \right]^2$
We need a few steps of algebra. Because we will be taking the derivative and setting it to zero, we can make the algebra a little easier if we replace the equals with a proportionality and drop the leading constants.	$P(r) \propto \left[\left(\dfrac{1}{2} - \dfrac{r}{4r_B} \right) r \right]^2 e^{-r/r_B}$ $P(r) \propto \left[\dfrac{1}{4}\left(2 - \dfrac{r}{r_B} \right) r \right]^2 e^{-r/r_B} \propto \left[\left(2 - \dfrac{r}{r_B} \right) r \right]^2 e^{-r/r_B}$ $P(r) \propto \left(2r - \dfrac{r^2}{r_B} \right)^2 e^{-r/r_B}$
Now take the derivative with respect to r and set the derivative to zero.	$\dfrac{dP}{dr} \propto 2\left(2r - \dfrac{r^2}{r_B} \right)\left(2 - \dfrac{2r}{r_B} \right) e^{-r/r_B} - \dfrac{1}{r_B}\left(2r - \dfrac{r^2}{r_B} \right)^2 e^{-r/r_B} = 0$ $2\left(2 - \dfrac{2r}{r_B} \right) - \dfrac{1}{r_B}\left(2r - \dfrac{r^2}{r_B} \right) = 0$ $4 - \dfrac{4r}{r_B} - \dfrac{2r}{r_B} + \dfrac{r^2}{r_B^2} = 0$ $4 - \dfrac{6r}{r_B} + \dfrac{r^2}{r_B^2} = 0$
Finally, use the quadratic formula to solve for r in terms of the Bohr radius r_B. We choose the $+$ sign. The $-$ sign gives the smaller peak shown in Figure 42.16.	$r = \dfrac{6r_B \pm \sqrt{36r_B^2 - 4(4r_B^2)}}{2} = \boxed{5.24 r_B}$

:• **CHECK and THINK**
Compare the answer here to the ground-state level of hydrogen (Fig. 42.13). It makes sense that when hydrogen is in a higher energy state, the electron is most likely to be found farther from the nucleus.

42-6 Magnetic Dipole Moments and Spin

You may be wondering why m is called the *magnetic* quantum number. The reason has to do with how a material becomes magnetized. So we'll digress from our focus on hydrogen and turn our attention to atoms in general. From Section 30-6, a material becomes magnetized partly from the orbital motion of the electrons in its atoms. The magnetic moment of a current loop is given by $\vec{\mu} = (I\pi R^2)\hat{j}$ (Eq. 30.14), and in Problem 42.40 you will show that the magnetic moment of a charged particle of mass m in a circular orbit is given by

$$\vec{\mu}_{\text{orbit}} = \dfrac{q\vec{L}}{2m} \quad (42.22)$$

Because the magnitude of $\vec{L}$ is quantized, the magnitude of the magnetic moment is quantized. Furthermore, because the direction of $\vec{L}$ is quantized, the direction of the magnetic moment is quantized. For an atom in an external magnetic field pointing in the z direction, this quantization is best expressed as

$$(\mu_{\text{orbit}})_z = -m\mu_{\text{Bohr}} \quad (42.23)$$

where $\mu_{\text{Bohr}} = e\hbar/2m_e = 9.274 \times 10^{-24}$ J/T is the **Bohr magneton**, m is the magnetic quantum number, and m_e is the mass of an electron. The negative sign in Equation 42.23 indicates that the electron carries a negative charge and its magnetic moment points in the opposite direction as its orbital angular momentum. The magnetic quantum number m in Equation 42.23 determines the component of the magnetic moment that is antiparallel to the external magnetic field, and so justifies calling m the magnetic quantum number.

Recall that the magnetization of a material also depends on the spin of the electrons in its atoms (Section 30-6). Again, the term *spin* gives us a mental picture of an object, such as a planet, that spins as it orbits. Although this is a useful image, it is not a good description for the actual motion of the electrons. First, classical mechanics cannot even account for the spin of an electron because particles have no spatial extent and cannot rotate. Second, quantum mechanics abandons the particle model in favor of a wave model for the electron, and it is hard even to come up with a way to visualize a "spinning" probability wave. Instead, spin is in an intrinsic property of a particle, much like charge is an intrinsic property.

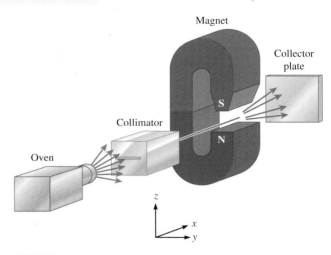

FIGURE 42.17 In the Stern–Gerlach experiment, atoms are released by an oven, and a collimator creates a narrow beam that passes through a nonuniform magnetic field. The magnetic field deflects the beam before encountering the collector plate.

Stern–Gerlach Experiment

How do we know that particles such as electrons have this intrinsic property? The answer comes from experimental evidence. In 1920, the German physicists Otto Stern (1888–1969) and Walther Gerlach (1889–1979) performed an experiment that provided that evidence. Figure 42.17 shows the basics of their experiment. An oven is used to vaporize silver. (A similar experiment was conducted by other researchers about seven years later using hydrogen.) A collimator creates a beam, which then passes through a nonuniform magnetic field. Let's model atoms in the beam as small dipole magnets with magnetic moment $\vec{\mu}$. We studied what happens when a magnetic dipole is in a *uniform* magnetic field; we found there is a net torque but no net force (Section 30-13). Now we must consider what happens when a magnetic dipole is in a *nonuniform* magnetic field. The result is a net force (and a net torque). The magnitude of the force depends both on the magnetic dipole moment and on the gradient of the magnetic field. The Stern–Gerlach experiment is set up so there is only a gradient in the z direction (Fig. 42.17). As a result, the atoms may be deflected in the positive or negative z direction based on their magnetic moment. Figure 42.18 shows that if the magnetic moment is along the x axis the atom is not deflected, but some atoms are deflected upward and others downward.

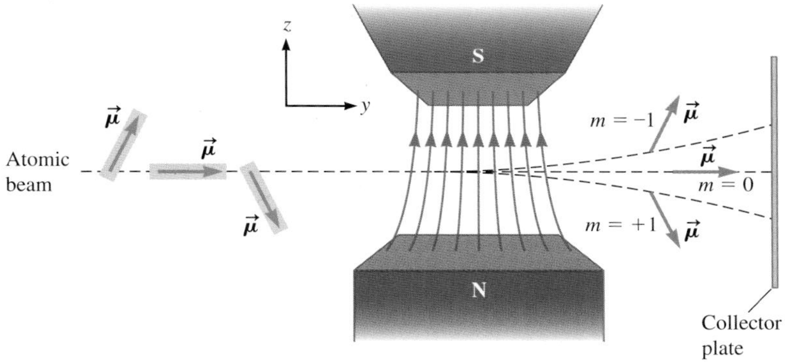

According to classical mechanics, atoms can have any magnetic moment, oriented in any direction. So when the beam passes through a magnetic field, atoms should be deflected up and down, landing on the collector plate at a wide range of positions. The data on the collector plate should completely fill the region over which the beam is deflected (Fig. 42.19A).

Compare this to the quantum-mechanical prediction. If the atom has $\ell = 1$, then there are three possible values for m: $m = 0$ and $m = \pm 1$. This means there are only three discrete orientations for the magnetic moment. So there are only three possible ways each atom may be deflected, and the data on the collector plate should show three different bands (Fig. 42.19B). Because the number of values m may take

FIGURE 42.18 Magnetic dipoles are deflected by a nonuniform magnetic field. Remember you can think of a magnetic dipole as a small bar magnet as shown on the left side of the figure with the magnetic dipole vectors. For clarity only the magnetic dipole vectors are shown on the right side of the figure.

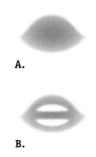

FIGURE 42.19 A. Classical prediction. **B.** Quantum prediction.

FIGURE 42.20 Two bands observed when only one band was predicted.

$(2\ell + 1)$ is an odd number, quantum mechanics predicts that an odd number of bands should be observed on the collector plate. In the Stern–Gerlach experiment using silver and in the subsequent experiment using hydrogen, the atoms where in the ground state with $\ell = 0$. Quantum mechanics predicts that one band should be observed. However, two bands were observed (Fig. 42.20). This data suggested that the atom must have a nonzero magnetic moment. So there must be another source (other than orbital angular momentum of the electron) for that magnetic moment. This other source was attributed to the intrinsic spin of the electron. Further, it was affirmed that spin must be quantized in order to form separate bands.

Spin

The spin angular momentum $\vec{S}$ of any electron (trapped or free) has a magnitude

$$S = \sqrt{s(s+1)}\hbar \qquad (42.24)$$

where s is called the **spin quantum number**. For an electron, $s = 1/2$, and so $S = \sqrt{3}\hbar/2$. The magnetic moment due to an electron's spin is $\vec{\mu}_{\text{spin}} = q\vec{S}/m$. (Notice the similarity between this and Equation 42.22, $\vec{\mu}_{\text{orbit}} = q\vec{L}/2m$.) Because the magnitude of the spin angular momentum is single-valued, the magnitude of the spin magnetic moment is also single-valued, and for an electron is given by $\mu_{\text{spin}} = \sqrt{3}\hbar e/2m_e$.

We cannot measure $\vec{S}$ or $\vec{\mu}_{\text{spin}}$, but we can measure one of their components. As in the case of $\vec{L}$ and $\vec{\mu}_{\text{orbit}}$, we choose to measure the z component. The z component of $\vec{S}$ is quantized:

$$S_z = m_s \hbar \qquad (42.25)$$

where m_s is the **spin magnetic quantum number**, having only one of two values: either $\pm 1/2$. If $m_s = +1/2$, we say the particle is *spin up*, and if $m_s = -1/2$, we say the particle is *spin down*. Table 42.4 summarizes the five quantum numbers. Because S_z is quantized, the z component of the spin magnetic moment is quantized:

$$(\mu_{\text{spin}})_z = -2m_s \mu_{\text{Bohr}} \qquad (42.26)$$

The spin magnetic moment in Equation 42.26 and the orbital magnetic moment in Equation 42.23, $(\mu_{\text{orbit}})_z = -m\mu_{\text{Bohr}}$, obey very similar relations.

TABLE 42.4 Allowed values for quantum numbers.

Name	Symbol	Allowed Values
Principal	n	1, 2, 3, …
Orbital	ℓ	0, 1, 2, … $(n-1)$
Orbital magnetic	m	0, ± 1, ± 2, … $\pm \ell$
Spin	s	½
Spin magnetic	m_s	$\pm ½$

EXAMPLE 42.7 Excited States of Hydrogen

Suppose a hydrogen atom's energy is -1.51 eV. How many separate states are available to it? Take into account the possible values of ℓ, m, and m_s.

∶• INTERPRET and ANTICIPATE

This example is much like Example 42.3, where we used the energy-level diagram in Figure 42.7 to find that the principal quantum number in this case is $n = 3$. Now we must find the possible values for ℓ, m, and m_s.

∶• SOLVE

From Table 42.4, use $\ell = 0, 1, 2, \ldots (n-1)$ to find the values for the orbital quantum number when $n = 3$.

$\ell = 0, 1, 2$

ℓ	m	Number of states
0	0	1
1	−1, 0, 1	3
2	−2, −1, 0, 1, 2	5

Consulting Table 42.4 again, use $m = 0, \pm 1, \pm 2, \ldots \pm \ell$ to find the values for the orbital magnetic quantum number for each possible value of ℓ. Count the number of states for each unique value of ℓ and m.

For each state, the electron can be either spin up ($m_s = 1/2$) or spin down ($m_s = -1/2$). Add up the number of unique states specified by ℓ and m, and then multiply by 2.

Number of states = $2(1 + 3 + 5) = 18$

:• CHECK and THINK

In all 18 of these states, hydrogen has the same energy E (−1.51 eV). So the 18 states form a single shell. This shell is broken into three subshells for the three values of l. Each subshell has the same energy and the same orbital angular momentum L, but within the subshells for $\ell = 1$ and $\ell = 2$ there is variation in the orbital magnetic moment $(\mu_{\text{orbit}})_z$. In addition, in each state the spin magnetic moment $(\mu_{\text{spin}})_z$ is either up or down.

42-7 Other Atoms

Hydrogen is the simplest atom; its atomic number $Z = 1$ means that it has one proton in the nucleus, and because it is neutral, it must also have one electron. All other atoms have multiple electron clouds surrounding a nucleus (consisting of both neutrons and protons). So in other atoms the potential well that confines any one electron is more complicated, and solving Schrödinger's equation is more difficult. Typically, it is solved numerically with a computer. Fortunately, we don't need to consider such solutions to understand the structure of atoms. We can use our understanding of hydrogen instead.

Because hydrogen has only one electron, we need only one set of four quantum numbers ($n, \ell, m,$ and m_s) to specify its state (Table 42.4). (We don't bother to list s because it is single-valued.) For other atoms, we need to specify these four quantum numbers for each of their electrons. For hydrogen, the energy is determined by n alone; so the energy is the same for each shell (Fig. 42.21). For other atoms, the energy depends on n and ℓ; so the energy is the same for each *sub*shell (Fig. 42.22).

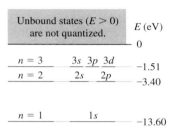

FIGURE 42.21 Compare the energy-level diagram labeled simply in terms of n to the one labeled in terms of n and ℓ for the first three energy levels in hydrogen.

The Pauli Exclusion Principle

The state with the lowest energy is the ground state. You might expect that in the ground state of a multielectron atom, all the electrons are in the 1s subshell. For example, carbon has an atomic number of 6, so it has six protons and six electrons. If all six electrons were in the 1s subshell, then on average they would be very close to the nucleus. Because the electrons are attracted to the protons, we would expect the electrons to be drawn close to the nucleus, making carbon a small atom. You might even argue that a carbon atom should be smaller than a hydrogen atom with its electron in the 1s subshell because there is a greater attractive force between six protons and six electrons than there is between one proton and one electron. However, we observe the opposite: Hydrogen atoms are smaller than carbon atoms. The answer to this puzzle is that not all the electrons in carbon can occupy the lowest subshell.

The idea that not all the electrons can be in the lowest subshell is part of a broader principle formulated by the theoretical physicist Wolfgang Pauli (1900–1958). According to the **Pauli's exclusion principle,** *no two electrons confined in the same well can have the same set of values for their quantum numbers.* We have expressed Pauli's exclusion principle in terms of electrons, but it applies equally well to any particles (such as protons and neutrons) which have $s = 1/2$. In practice, Pauli's exclusion principle means that in an atom, no two electrons can have exactly the same set of the four quantum numbers n, ℓ, m and m_s.

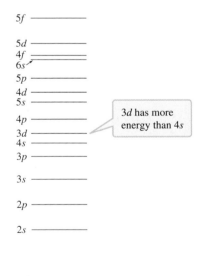

FIGURE 42.22 The energy levels in multielectron atoms depend on both n and ℓ. Notice that in some cases a state with a higher quantum number n has less energy than a state with lower n.

PAULI'S EXCLUSION PRINCIPLE
ⓘ Underlying Principle

Pauli's exclusion principle limits the number of electrons that can be in any one subshell. For example, any electron in the 1s subshell has three quantum numbers in common $(n, \ell, m) = (1, 0, 0)$. So two electrons can be in this subshell because one can be spin up with $m_s = 1/2$, and one can be spin down with $m_s = -1/2$. Put another way, each subshell is divided into *orbitals,* and each orbital may have up to two electrons of opposite spin.

When an atom is in its ground state, the electrons fill the lowest-energy subshells first while obeying Pauli's exclusion principle. Let's consider two examples. Helium has two electrons, and in the ground state both are in the 1s subshell with $(n, \ell, m) = (1, 0, 0)$: One is spin up, so the other must be spin down (Fig. 42.23A). Nitrogen has seven electrons. Two of these electrons are in the 1s state, just as they are in helium. Then two electrons are in the 2s subshell, with quantum numbers $(n, \ell, m) = (2, 0, 0)$: One is spin up and the other spin down. Finally, the last three electrons are in the 2p subshell with $(n, \ell) = (2, 1)$. Here there are three possible values for m ($-1, 0, 1$) and two values for m_s ($-1/2, 1/2$); so there are six different combinations of quantum numbers, but only three electrons. How do we find their quantum numbers? The answer is that we follow **Hund's rule**, which says that *every orbital in a subshell is singly occupied with one electron before any one orbital is doubly occupied, and all electrons in singly occupied orbital have the same spin.* This means that the maximum number of electrons have unpaired spins. Hund's rule is based on experimental evidence, which shows that the maximum number of unpaired spins corresponds to the most stable state with the lowest energy level. So, remaining electrons in nitrogen have the same m_s, say spin up, but three different values for m (Fig. 42.23B). A listing of the subshells occupied by the electrons in an atom is called the **electron configuration**. The electron configuration for helium is $1s^2$ and for nitrogen it is $1s^2 2s^2 2p^3$.

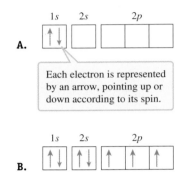

FIGURE 42.23 Each box represents one orbital; each orbital can hold up to two electrons. **A.** The two electrons in He are both in the 1s subshell. One is spin up and the other is spin down. This state is written in the compact form as $1s^2$. **B.** The seven electrons in N fill 1s and 2s, but only half of 2p. Following Hund's rule, all the electrons in 2p are spin up. This state is written as $1s^2 2s^2 2p^3$.

EXAMPLE 42.8 The Maximum Number of Electrons in Each Subshell

Argue that Pauli's exclusion principle means that the maximum number of electrons allowed in each subshell is

$$N = 2(2\ell + 1) \quad (42.27)$$

Check this for the 2p subshell.

⋮• INTERPRET and ANTICIPATE
A subshell is characterized by a particular value of n and ℓ. Table 42.4 will help us to determine the possible quantum numbers for a subshell, and Pauli's exclusion principle says there can only be one electron with each unique set of quantum numbers.

⋮• SOLVE

Because $m = 0, \pm 1, \pm 2, \ldots \pm \ell$, for each value of ℓ, there are $2\ell + 1$ possible values of m.	The range in m allows for $2\ell + 1$ electrons in a subshell.
For each value of m, the electron can be either spin up ($m_s = 1/2$) or spin down ($m_s = -1/2$). The spin magnetic quantum number doubles the number of electrons (found above) that can possibly be in the subshell.	$N = 2(2\ell + 1)$ ✓ (42.27)

⋮• CHECK and THINK

The 2p subshell has $\ell = 1$, so we expect $N = 2(2 \cdot 1 + 1) = 6$ electrons. Check this by listing all the unique combinations of quantum numbers. (Since $n = 2$ and $\ell = 1$ for the 2p subshell, we don't bother to list them separately here.)	$(m, m_s) = (-1, -1/2)$ $(m, m_s) = (-1, +1/2)$ $(m, m_s) = (0, -1/2)$ $(m, m_s) = (0, +1/2)$ $(m, m_s) = (1, -1/2)$ $(m, m_s) = (1, +1/2)$ 6 combinations

42-8 Organizing Atoms

By the mid-1800s, over 60 elements had been discovered and studied. Chemists knew their relative masses, their chemical activity, and some of their other physical properties. However, scientists didn't know why there are different types of elements, or why they have different properties. One way to answer such a question is to look for patterns or correlations among these properties. In 1869, a chemist from Siberia, Dmitri Mendeleev (1834–1907), found that if he arranged the elements by atomic mass in a grid, certain properties recurred periodically. Mendeleev's table is the basis of our periodic table; he successfully predicted the existence of undiscovered elements that filled in "blank" places in the grid. However, Mendeleev's table was not completely useful because atomic mass is not always well correlated with an element's chemical properties. The British physicist Henry Moseley (1887–1915) later showed that atomic number is better correlated with chemical properties, so it is better to arrange the elements by their atomic number, as we do in today's periodic table.

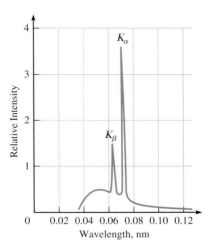

FIGURE 42.24 X-ray spectrum of molybdenum. The spectrum looks like a blackbody spectrum with two tall spikes—called lines. The K_α line comes from the transition from $n = 2$ to $n = 1$. The K_β line comes from the transition from $n = 3$ to $n = 1$. Moseley measured the wavelength of the K_α line for about 40 elements to arrive at his plot similar to the one shown in Figure 42.25.

Moseley's Experiment

Moseley devised an experiment that revealed new discoveries about the internal structure of 38 elements. He placed a sample in an evacuated tube and fired a beam of electrons at it. When an energetic electron collided with an atom, one of its low-energy electrons was freed so that a "hole" was opened in a shell near the atom's nucleus. An electron from a higher shell then dropped into this hole, releasing energy in the form of an X-ray. Moseley studied the X-ray generated by an electron that drops from the $n = 2$ shell to the $n = 1$ shell, which he labeled as K_α (Fig. 42.24). He measured the frequency and therefore the energy of this emission line for approximately 40 elements and found that the X-ray energy was correlated with each element's atomic number Z, not with its atomic mass (Fig. 42.25). Moseley's law is an empirical fit given by:

$$f(Z) = C(Z - 1)^2$$

where $C \approx 2.47 \times 10^{15}$ Hz. Today we know that arranging the periodic table by atomic number leads to a better prediction of the behavior and properties of atoms.

Shielding

In a moment we'll consider other evidence that supports the ordering of the elements in the periodic table. But now we'll digress to consider why an X-ray emission from an atom yields more information about the atom's nuclear charge Z than optical emission does. X-ray emission lines are produced by atomic transitions between shells near the nucleus, such from the $n = 2$ shell to the $n = 1$ shell. By contrast, optical lines come from transitions that involve higher-numbered shells farther from the nucleus, and the electrons that occupy such distant shells are *shielded* from the nucleus by other, closer electrons. Shielding means that the effective charge that attracts these outer electrons to the nucleus is smaller than for "core" electrons, and the Coulomb force is weaker. So when an outer electron makes a transition to another (lower) outer shell, the photon released has relatively low energy. Such low-energy photons may be observed in the optical, but not in the X-ray part of the electromagnetic spectrum. This idea of shielding also helps explain the chemical behavior of various types of elements in the periodic table.

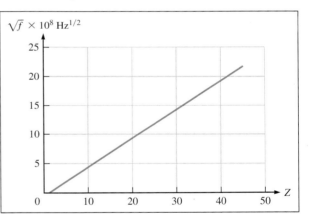

FIGURE 42.25 When trying to fit data, it is often best to choose axes that produce a straight line rather than a curve as Moseley did when he plotted his data. Similarly, this graph has been constructed to display Moseley's law as a straight line. To do this, the vertical axis is $\sqrt{f}$ on the graph here.

Atomic Size and Ionization Energy

The outermost electrons known as the *valence electrons* in an atom are primarily responsible for determining how atoms make contact or bond together. The radius of an atom is measured by how close together atoms must come to one another to form bonds or to make nonbonded contact. The **first ionization energy** is the energy required to remove the first electron from a neutral atom in its ground state. Figure 42.26A shows atomic radius as a function of atomic number, whereas Figure 42.26B shows the first ionization energy as a function of atomic number. Both graphs display a *general* trend and a *periodic* trend.

First, the radius graph (Fig. 42.26A) shows that overall a greater atomic number means a larger atom. We explain this *general* trend in radius by noting that electrons in an atom fill their energy levels, from the lowest to the highest, while obeying Pauli's exclusion principle. (Not all the electrons can be in the ground state, for example.) Atoms with large Z have a great number of electrons, some of which must occupy outer shells. Such atoms have a large radius.

Second, the *general* trend in the ionization energy graph (Fig. 42.26B) shows that overall a greater atomic number means a lower ionization energy. We can explain this trend in terms of shielding. If an atom has a great number of electrons, then a few outer electrons are greatly shielded by many core electrons. These outermost electrons are weakly attracted to the nucleus, which means the atom is large and one of the outermost electrons is relatively easily removed. Thus, the ionization energy is low.

A regular *periodic* pattern within the overall trend also occurs in both graphs. The radius graph shows that the atomic radius decreases periodically with increasing atomic number within a certain range of Z values. The ionization graph shows that ionization energy increases periodically with increasing Z within each similar range. We can also explain these periodic trends in terms of shielding. Core electrons are

FIGURE 42.26 A. Plotting the radii of atoms as a function of atomic number shows a regular periodic pattern. **B.** Plotting the first ionization energy as a function of atomic number shows a periodic repetition. In both graphs alkali metals are labeled in blue and noble gases in red.

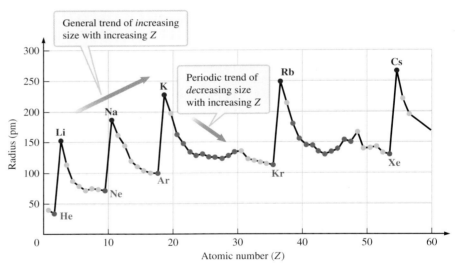

A. Atomic Radii

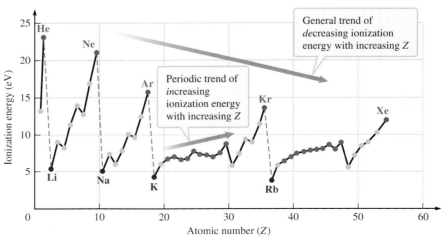

B. First Ionization Energies

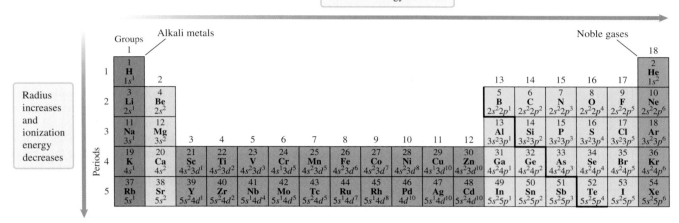

FIGURE 42.27 The first 54 elements in the periodic Table help us to see how the table is constructed. The complete periodic table is in Appendix B. Here we show 5 periods and 18 groups.

very effective at shielding outer electrons, but the valence electrons in the same subshell are not very effective at shielding one another. So, if the valence electrons fill a subshell, then there is only a little shielding due to the core electrons, and these outer electrons experience a great Coulomb attraction to the nucleus, and the atom is small. It is also difficult to remove an electron from such an atom, so the ionization energy is great. For example, neon (Ne) has 10 electrons, with electron configuration $1s^2 2s^2 2p^6$ indicating that all of its subshells are filled. Neon is relatively small (Fig. 42.26A) and it has a relatively large ionization energy (Fig. 42.26B). By contrast, if an atom has just one more electron than is required to fill a subshell, this outer electron is greatly shielded and weakly attracted to the nucleus. Such an atom is relatively large and has a relatively low ionization energy. For example, helium is the smallest atom because its $1s$ subshell is filled and it only has two electrons. Lithium (Li) has just one more electron, so its electron configuration is $1s^2 2s^1$. The core electrons in the $1s$ subshell shield the valence electron in the $2s$, so this outer electron is weakly bound. As a result, lithium has a lower ionization energy and is larger than helium.

The Periodic Table

Moseley's experiment, along with the graphs of radius and ionization energy as functions of atomic number, gives us good reasons to expect that if the elements are ordered by atomic number, we'll find trends that explain their chemical behavior. The elements in the modern periodic table are ordered by their atomic number into rows known as *periods* and columns known as *groups* (Fig. 42.27). An element's chemical behavior—how reactive it is—results from the configuration of its valence electrons. The valence electrons are important in chemical bonding because they are the most weakly attracted to the atom, and so they are the easiest to lose or share in a chemical reaction. These tend to be the outermost electrons (in s or p subshells), but for some elements the outermost d subshell electrons also act as valence electrons. The electron configuration of an element is related to its placement on the periodic table, and its placement on the table is related to its chemical and physical properties.

The structure of an element is given by its electron configuration in the ground state. To find an element's electron configuration, first find its atomic number on the periodic table. Then fill in its shells and subshells starting with the lowest energy, while following Pauli's exclusion principle and Hund's rule. The energy levels are shown in Figure 42.22. It is helpful to represent each energy level with a series of empty boxes in which arrows representing electrons are drawn. Figure 42.28 is such a blank diagram for all the energy levels in

PERIODIC TABLE OF THE ELEMENTS
✪ **Major Concept**

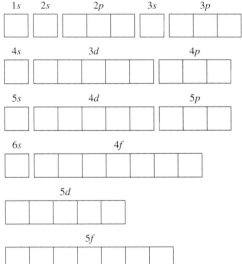

FIGURE 42.28 A blank diagram such as this can help you to find the electron configuration of an atom. Start by filling in electrons from the lowest energy on the left. Follow Pauli's exclusion principle, which says that any one box can have up to two electrons in it: one with spin up, and the other with spin down. Also, follow Hund's rule, which says that you must fill a subshell with single electrons before adding a second electron to a box.

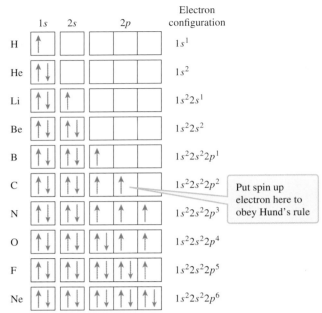

FIGURE 42.29 Use a diagram like that in Figure 42.28 to find the electron configuration of an atom. Here is the electron configuration for the first 10 elements in the periodic table.

Figure 42.22; it contains far more blocks than we are likely to use. There may be up to two electrons in each block: one with spin up represented by an upward-pointing arrow, and one with spin down represented by a downward-pointing arrow. The number of blocks corresponds to half the number of electrons allowed in each subshell, given by $(2\ell + 1)$. Figure 42.29 shows how to determine the electron configuration for the first 10 elements. By a similar process, we can find the electron configuration of all the atoms in the periodic table.

Beneath the symbol for each element in Figure 42.27, we have listed the electron configuration for the valence electrons. First, notice that the period number of an element is the same as its highest principal quantum number. According to the general trends in the graphs shown in Figure 42.26, as we consider elements down a particular group (vertical column), we find that their atomic radius increases and their ionization energy decreases. Next, notice that the first group is the alkali metals and the last group is the noble gases. So according to the periodic trends in the graphs, as we consider elements across a period from left to right, we find that their atomic radius decreases and their ionization energy increases.

We can account for the chemical behavior of the elements by means of their placement on the periodic table. For example, the alkali metals are in the first group, and their valence electrons all have configurations of the form ns^1. These metals have one valence electron that is well shielded by the core electrons in the lower subshells. This electron is weakly attracted to the nucleus, so in chemical reactions, alkali metals tend to lose their valence electron. For another example, consider the noble gases in the last group. Their valence electron configurations are all of the form ns^2np^6. These eight valence electrons fill the atom's outer subshell, making them very stable, so noble gases are generally nonreactive. In general, elements in a group tend to show similar chemical behavior.

EXAMPLE 42.9 Silicon's Electron Configuration

Find the electron configuration for silicon (Si).

: INTERPRET and ANTICIPATE
According to the periodic table, Si has 14 electrons. We expect its configuration to be longer than the configuration of neon, which has only 10 electrons.

: SOLVE
Fill in a diagram (Fig. 42.30) following these three rules: (1) start with the lowest energy subshell, (2) obey Pauli's exclusion principle, and (3) obey Hund's rule.

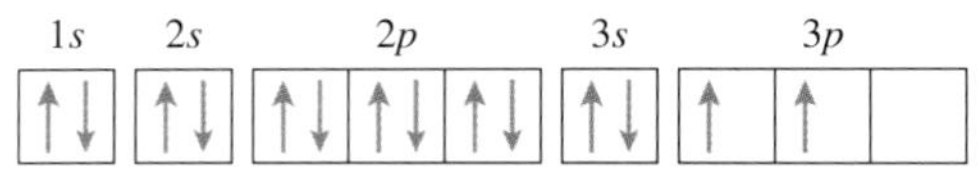

FIGURE 42.30

Write the electron configuration based on this figure.

$1s^2 2s^2 2p^6 3s^2 3p^2$

: CHECK and THINK
As expected, the electron configuration for silicon is longer than that for neon ($1s^2 2s^2 2p^6$).

42-9 The Zeeman Effect

ZEEMAN EFFECT ● Major Concept

In 1902, the Dutch physicists Pieter Zeeman (1865–1943) and Hendrik A. Lorentz (1853–1928) shared the Nobel Prize in physics for what is known as the **normal Zeeman effect**. (Lorentz is the theorist for whom the transformation equations in Chapter 39 are named.) Zeeman was an experimentalist who discovered in 1896 that the spectral lines of certain atoms split into three lines when a sample is placed in a magnetic field. The Zeeman effect confirms that angular momentum is quantized.

To see how, we'll consider the hydrogen atom, though our arguments can be applied to more complicated atoms. The energy levels E_n we have discussed so far have been for hydrogen without the presence of an external field; our system consisted of the atom's electron and nucleus. Because of the electron's orbital angular momentum $\vec{L}$, the electron has a magnetic moment given by $\vec{\mu}_{\text{orbit}} = q\vec{L}/2m$ (Eq. 42.22). Now let's consider a hydrogen atom in a uniform external magnetic field pointing in the z direction. In this case, our system consists of the electron, the nucleus, and the external field. The external magnetic field adds a potential energy term U to the energy E_n, given by $U = -\vec{\mu}_{\text{orbit}} \cdot \vec{B}$ (Eq. 30.48). Because the angular momentum is quantized in both magnitude and direction, so is the magnetic moment. And because the magnetic moment is quantized, so is this additional potential energy.

Contrast these ideas with a classical interpretation. If a current loop is placed in an external magnetic field, the potential energy of the loop-field system depends on the direction of the magnetic moment with respect to the magnetic field (Fig. 42.31A). The minimum potential energy occurs when the magnetic moment is aligned with the magnetic field: then $U_{\min} = -\mu B$. The maximum potential occurs when the magnetic moment is antiparallel to the field: then $U_{\max} = +\mu B$. As a result, the system's potential energy may be anywhere in the range from $U_{\min}$ to $U_{\max}$.

For the system consisting of an atom in an external magnetic field, not only is its potential energy quantized, but $U_{\min}$ and $U_{\max}$ are restricted by quantization of the direction of angular momentum and magnetic moment. According to Concept Exercise 42.4, for a magnetic field in the z direction, the angular momentum $\vec{L}$ can never be in the $\pm z$ direction. The electron's magnetic moment is antiparallel to its orbital angular momentum. Thus, the magnetic moment cannot be parallel (or antiparallel) to the external magnetic field. Instead, the z component of magnetic moment is given by $(\mu_{\text{orbit}})_z = -m\mu_{\text{Bohr}}$ (Eq. 42.23) where m is the magnetic quantum number. The maximum and minimum potential energy are determined by this z component, rather than by the magnitude of the magnetic moment (Fig. 42.31B).

When there is no external magnetic field, the energy of a subshell depends on the quantum numbers n and ℓ, so we label it $E_{n\ell}$. All the orbitals of that subshell have the same energy, independent of the orbital quantum number m. The energy of the subshell is said to be *degenerate*. But when an atom is in an external magnetic field pointing in the z direction, the additional potential energy is found by substituting Equation 42.23 into Equation 30.48:

$$U = -\vec{\mu}_{\text{orbit}} \cdot \vec{B} = -(\mu_{\text{orbit}})_z B_z = -(-m\mu_{\text{Bohr}})B_z = m\mu_{\text{Bohr}} B$$

where $B_z = B$. The energy of an orbital $E_{n\ell m}$ in a particular subshell depends not only on n and ℓ, but also on m:

$$E_{n\ell m} = E_{n\ell} + m\mu_{\text{Bohr}} B \quad (42.28)$$

We say that the external magnetic field removes the degeneracy, so that each orbital has its own distinct energy.

The energy of the subshells of hydrogen only depends on the principal quantum number n. In the case of hydrogen, Equation 42.28 becomes:

$$E_{nm} = E_n + m\mu_{\text{Bohr}} B$$

$$E_{nm} = -\frac{13.6 \text{ eV}}{n^2} + m\mu_{\text{Bohr}} B \quad (42.29)$$

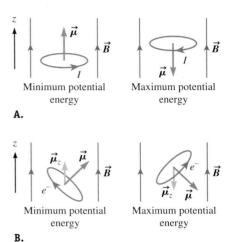

FIGURE 42.31 A. The extrema of the potential energy for a current loop–magnetic field system. **B.** The same for an atom in a magnetic field.

We can use Equation 42.29 to construct an energy-level diagram for hydrogen in an external magnetic field pointing in the z direction. For the ground state, $n = 1$ and $m = 0$, so $E_{10} = -13.6$ eV. Because there is only one orbital, the ground state in this case is exactly the same as the ground state when there is no external magnetic field. However, the $2p$ state has three orbitals, with $m = -1, 0$ and $+1$, and three corresponding energies:

$$E_{2(-1)} = -\frac{13.6 \text{ eV}}{2^2} + (-1)\mu_{\text{Bohr}}B = -3.4 \text{ eV} - \mu_{\text{Bohr}}B$$

$$E_{20} = -3.4 \text{ eV} + (0)\mu_{\text{Bohr}}B = -3.4 \text{ eV}$$

$$E_{21} = -3.4 \text{ eV} + (1)\mu_{\text{Bohr}}B = -3.4 \text{ eV} + \mu_{\text{Bohr}}B$$

The $m = 0$ orbital has the same energy as if there were no external magnetic field, whereas the $m = -1$ orbital has a lower energy and the $m = 1$ orbital has a higher energy. For easy comparison, Figure 42.32 shows the energy levels computed here for hydrogen both with and without an external magnetic field. (In Problems 42.44 and 45, you will be asked to calculate the energy of other orbitals.) Figure 42.32A shows that if a hydrogen atom undergoes a transition from either the $2s$ or $2p$ subshell to the ground state, a Lyman α photon (of energy 10.2 eV) is released if there is no external magnetic field. However, if there is an external magnetic field, the $2p$ subshell has three distinct energies, one for each orbital. So there are three distinct possible transitions from the $2p$ subshell to the ground state, and the Lyman α line is split into three closely spaced lines (Fig. 42.32B). The energies of the corresponding photons are given by:

$$E_{2(-1)} - E_1 = (-3.4 \text{ eV} - \mu_{\text{Bohr}}B) - (-13.6 \text{ eV}) = 10.2 \text{ eV} - \mu_{\text{Bohr}}B$$

$$E_{20} - E_1 = -3.4 \text{ eV} - (-13.6 \text{ eV}) = 10.2 \text{ eV}$$

$$E_{21} = (-3.4 \text{ eV} + \mu_{\text{Bohr}}B) - (-13.6 \text{ eV}) = 10.2 \text{ eV} + \mu_{\text{Bohr}}B$$

So in general, the energy of the emitted photon is given by

$$E = E_0 - \Delta m \mu_{\text{Bohr}} B \qquad (42.30)$$

where E_0 is the energy of the corresponding photon in the absence of an external magnetic field, and $\Delta m = m_f - m_i$ is the difference in the orbital magnetic quantum number between the two states. Not all transitions are equally probable, however. Although we won't show the details, applying Schrodinger's equation reveals that the most probable transitions have $\Delta \ell = \pm 1$ and $\Delta m = 0, \pm 1$; these are known as **allowed transitions**. So in the normal Zeeman effect, a single spectral line may be split into three lines, one for each allowed value of Δm. Transitions with very low probability, by contrast, are known as **forbidden transitions**. Although forbidden transitions are not important under normal conditions on the Earth, forbidden transitions are observed from astronomical sources. Observations of forbidden lines (produced by forbidden transitions) in astronomical sources tell us a great deal about the conditions in these sources such as their temperature and density.

Zeeman's observation that spectral lines divide into several closely spaced lines in the presence of an external magnetic field confirms the theory that the angular momentum vector $\vec{L}$ is quantized. Let's review the argument. Quantization of $\vec{L}$ means that the magnetic moment $\vec{\mu}_{\text{orbit}}$ is quantized, so the angle between $\vec{\mu}_{\text{orbit}}$ and the external field $\vec{B}$ is also quantized. The additional potential energy of the atom–external field system $U = -\vec{\mu}_{\text{orbit}} \cdot \vec{B}$ depends on this relative angle, so it too must be quantized. When an atom is in an external magnetic field, the energy of a particular orbital is divided into discrete energy levels. Zeeman detected these discrete energy levels when he saw one spectral line split into distinct spectral lines if an external magnetic field was applied.

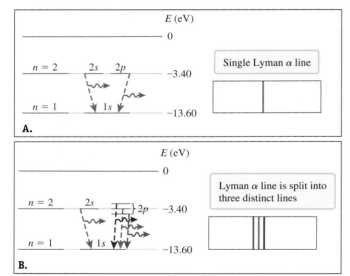

FIGURE 42.32 A. Energy-level diagram for the first two shells of hydrogen when no external magnetic field is present. A transition from $n = 2$ to $n = 1$ results in a single Lyman α line. **B.** The same applies when an external magnetic field points in the z direction. Notice that the $2p$ orbital energies are divided into three levels. Now the Lyman α line is split into three lines corresponding to $m = -1, 0,$ and 1. (Energy levels are not drawn to scale, and line colors are illustrative only.) The wavy arrows represent the photons that are emitted.

EXAMPLE 42.10 CASE STUDY The Zeeman Effect in the Sun's Hydrogen Spectrum

In 1908, George Hale was the first person to exploit the Zeeman effect to measure the magnetic field in sunspots. Although Hale observed heavy elements such as iron, we'll consider how we might observe the Zeeman effect in the Sun's hydrogen spectrum.

A Assume that the magnetic field strength in a typical sunspot is 0.1 T, and that only allowed transitions occur under the conditions in the Sun. Estimate the magnitude of the difference in wavelength $\Delta \lambda$ between either of the two Hγ lines that result from the normal Zeeman effect and the original ($B = 0$) central line. What is $\Delta \lambda / \lambda_0$, where $\lambda_0 = 434.047$ nm is the wavelength of original line?

INTERPRET and ANTICIPATE
The difference in energy between the central line and either of the other two lines comes from Equation 42.30. This energy difference corresponds to a wavelength difference, which we can derive using calculus.

SOLVE

The allowed transitions have $\Delta m = 0, \pm 1$. Start with Equation 42.30 to write an expression for the energy difference between the central line and the higher energy line generated by $\Delta m = 1$.	$E = E_0 + \Delta m \mu_{Bohr} B$ (42.30) $dE = E_0 - E = (1)\mu_{Bohr} B$ (1)
Write the energy of a photon in terms of its wavelength by using $f = c/\lambda$ (Eq. 34.20) and $E = hf$ (Eq. 40.8).	$E = hf = \dfrac{hc}{\lambda}$ (2)
Differentiate Equation (2) with respect to wavelength λ.	$\dfrac{dE}{d\lambda} = -\dfrac{hc}{\lambda^2}$ (3)
Eliminate dE from Equations (1) and (3) to arrive at an expression for $d\lambda$.	$d\lambda = -\dfrac{\mu_{Bohr} B}{hc}\lambda^2$
Use this expression to find the magnitude of the finite difference $\Delta \lambda$ and $\Delta \lambda / \lambda_0$.	$\Delta \lambda = \dfrac{\mu_{Bohr} B}{hc}\lambda^2$ $\Delta \lambda = \dfrac{(9.274 \times 10^{-24} \text{J/T})(0.1 \text{ T})}{(6.626 \times 10^{-34} \text{ J} \cdot \text{s})(3.00 \times 10^8 \text{ m/s})}(434 \times 10^{-9} \text{ m})^2$ $\Delta \lambda = 8.8 \times 10^{-13}$ m $\approx 9 \times 10^{-4}$ nm $\dfrac{\Delta \lambda}{\lambda_0} = \dfrac{8.8 \times 10^{-4} \text{ nm}}{434.047 \text{ nm}} \approx 2 \times 10^{-6}$

CHECK and THINK
The Zeeman effect is very subtle. Next we'll consider the quality of the diffraction grating required.

B Hale reported that to see the Zeeman effect with his diffraction grating, he had to use the third-order part of the spectrum. If we wish to see the splitting of Hγ, with a resolving power that is three times the minimum required in the third order, how many rulings will our grating need?

INTERPRET and ANTICIPATE
This part of the problem requires a review of Section 36-5. The minimum resolving power needed is given by $R = \lambda_{av}/\Delta \lambda$ (Eq. 36.20), where λ_{av} is the average wavelength of two lines that are just barely separated, and $\Delta \lambda$ is the difference in their wavelength. For these lines, $\lambda_{av} \approx \lambda_0$, so $R = \lambda_0/\Delta \lambda$.

SOLVE

The minimum resolving power needed is the inverse of the expression we found in part A.	$R_{min} = \dfrac{\lambda_0}{\Delta\lambda} = \dfrac{1}{2 \times 10^{-6}}$
The resolving power is related to the number of rulings according to Equation 36.21. (In this equation, m refers to the order number, not the quantum number.)	$R = Nm \qquad (36.21)$
We'd like R to be three times the minimum required resolving power, and like Hale we plan to use the third-order part of the spectrum.	$R = Nm = 3R_{min}$ $N = \dfrac{3R_{min}}{m} = \dfrac{3}{3}\left(\dfrac{1}{2 \times 10^{-6}}\right)$ $N = 5 \times 10^5 \text{ rulings}$

C Hale also reported that his grating had 567 rulings per millimeter. Assuming the same for our grating, what is the angular separation between the two lines?

INTERPRET and ANTICIPATE

This part of the problem also requires information from Section 36-5. The dispersion is given by Equation 36.19,

$$D = \dfrac{\Delta\theta}{\Delta\lambda} = \dfrac{m}{d\cos\theta}$$

which we can solve for $\Delta\theta$. First, find the angular position of the third-order spectrum, using $d\sin\theta = m\lambda$ (Eq. 36.15).

SOLVE

Find the distance between rulings, using the given 567 rulings per millimeter.	$d = \dfrac{1 \text{ mm}}{567 \text{ rulings}} = 1.76 \times 10^{-3} \text{ mm}$ $d = 1.76 \times 10^3 \text{ nm}$
Both lines have very nearly the same wavelength, so their angular position is roughly the same.	$d\sin\theta = m\lambda$ $\theta = \sin^{-1}\left(\dfrac{m\lambda}{d}\right) = \sin^{-1}\left(\dfrac{3(434.047 \text{ nm})}{1.76 \times 10^3 \text{ nm}}\right)$ $\theta = 47.7°$
Now use Equation 36.19 for the dispersion. Substitute $m = 3$ (third-order part of spectrum) and information found previously.	$D = \dfrac{\Delta\theta}{\Delta\lambda} = \dfrac{m}{d\cos\theta}$ $\Delta\theta = \dfrac{m\Delta\lambda}{d\cos\theta} = \dfrac{3(8.8 \times 10^{-4} \text{ nm})}{(1.76 \times 10^3 \text{ nm})(\cos 47.7°)}$ $\Delta\theta = 2.2 \times 10^{-6} \text{ rad}$

CHECK and THINK

To see if this angular separation is sufficient to resolve the two lines, compare it to either line's half-width. The angular separation between the two lines is about three times greater than their half-width. Lines are said to be barely resolved if the angular separation equals the half-width. In this case, the lines are better than barely resolved.	$\Delta\theta_{hw} = \dfrac{\lambda}{Nd\cos\theta} \qquad (36.16)$ $\Delta\theta_{hw} = \dfrac{434.047 \text{ nm}}{(5 \times 10^5)(1.76 \times 10^3 \text{ nm})(\cos 47.7°)}$ $\Delta\theta_{hw} \approx 7 \times 10^{-7} \text{ rad}$

Although hydrogen is the most abundant element in the Universe and is very often observed in astronomy, there are reasons to observe other elements. Hale observed the Zeeman effect in the spectra of heavier elements in order to estimate the magnetic field strength at various depths within sunspots. He found that the magnetic fields in sunspots are about three to four orders of magnitude greater than the Earth's magnetic field. The more we learn about atoms in the laboratory, the more we can learn about the Universe.

42-10 Practical Devices

In this chapter so far, we've applied the theory of quantum mechanics to understand atoms better. In Example 42.10, we saw that this deeper understanding of atoms leads us to a deeper understanding of the Universe. In this final section we will see how a deeper understanding of atoms also leads to the development of practical devices.

The Cesium Clock

Now that we are nearly at the end of this textbook. We are in a position to explain statements we made near the very beginning of the book. In Section 1-5, we learned that the time standard is determined by a Cesium clock (Fig. 1.5, page 7). Now we can describe the basic operation of such a clock. A sample of Cesium gas is heated, so that its atoms are in an excited state. The atoms then lose energy and return to their lower energy state. Of course, in losing energy the atoms must emit photons. Each photon's energy is determined by the energy difference in the two states ΔE. Their frequency is given by $f = \Delta E/h$ (Eq. 40.8). For the two energy levels exploited in a Cesium clock, this frequency is very high: 9,192,631,770 Hz. The frequency of the emitted light is the basis for the definition of a second. You can think of the frequency as 9,192,631,770 "ticks" per second. Then the second is accuracy defined as duration of 9,192,631,770 ticks. We could not have built such a precise timepiece without a quantum-mechanical model of the atom because this level of precision requires knowing that atomic energies are quantized and that an atom emits a photon with energy equal to the difference in two energy levels whenever it transitions to a lower energy level.

Fluorescent Bulbs

We can also use our deeper understanding of atoms to see why the spectrum produced by a fluorescent bulb is made up of a discrete number of emission lines (Fig. 36.17, page 1166). First, let's look at a neon bulb, such as those used in a red neon sign (Fig. 42.33A). The neon gas is confined in a tube with metal plates near its ends (Fig. 42.33B). These plates are connected to a power supply, so that an electrical current can be passed through the gas. The current ionizes the gas, producing some free electrons and ions. So now there are free electrons, ions, and neutral atoms in the tube. These numerous particles undergo many collisions, which cause many of the neutral atoms to be in an excited state. But then these excited atoms quickly transition back down to a lower energy state and release a photon. Each photon's energy is determined by the energy difference between the two states ΔE, and its frequency is given by $f = \Delta E/h$ (Eq. 40.8).

Of course, not all the neon atoms will be in the same excited state or will transition to the same lower energy state. But because only a few possible transitions are available to the neon in the tube, the emitted photons can only have a few possible frequencies. As a result, the light produced by such a neon tube is limited to just a few spectral lines. In the case of neon, this limitation makes the light reddish. A red light may be desirable in many situations, but often we would prefer a white light. In such cases, fluorescent bulbs use mercury, which produces lines in the ultraviolet. The tube of a mercury bulb is coated with a material that absorbs the ultraviolet light. The atoms in the coating are in an excited state. Then they transition to a lower energy state, emitting photons at various frequencies in the visible spectrum. These photons blend together to form nearly white light.

CASE STUDY Lasers

As we mentioned when we first introduced the case study for this chapter, lasers have many practical everyday uses. They have also played a major role in our understanding of physics. You have probably used a laser in your laboratory to re-create experiments such as Young's double-slit experiment (Fig. 35.10, page 1128). Here we consider the basic operation of lasers.

A.

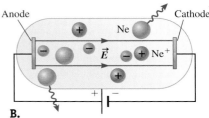

B.

FIGURE 42.33 A. A neon sign glows red. **B.** Gas in a neon bulb is excited by an electric current. Photons of a few particular frequencies (colors) are released with the atoms' transition to a lower energy state.

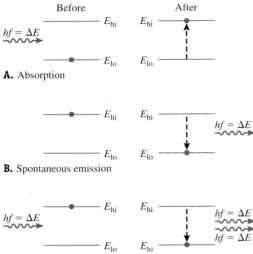

FIGURE 42.34 **A.** An atom can absorb a photon if the photon's energy equals the difference between two available energy states. **B.** An excited atom can spontaneously emit a photon and transition to a lower energy state. **C.** An excited atom can be stimulated to transition to a lower energy state by a photon with an energy equal to the energy difference of the two states.

To understand lasers, we need to think about ways that photons and atoms can interact. As we have been discussing, when a photon encounters an atom, it can be absorbed if the energy of the photon equals the energy difference between two of the atom's energy states (Fig. 42.34A). We have also discussed *spontaneous emission*, which occurs when an atom in a high-energy state transitions to a lower one spontaneously and emits a photon whose energy equals the energy difference between the two atomic states (Fig. 42.34B).

There is another way for photons and atoms to interact. If an atom is in a high state with energy E_{hi} when a photon encounters it, it is possible for the photon to *stimulate* the atom to transition to a lower state with energy E_{lo} (Fig. 42.34C). This **stimulated emission** occurs when the photon's energy equals the difference between the two energies: $E_{photon} = E_{hi} - E_{lo}$. In such a case, the atom transitions to a lower state and *releases a photon that is identical to the original stimulating photon*. So now there are two photons where before there had been one. Lasers are based on stimulated emission.

Under normal conditions, stimulated emission is rare because it requires an atom to be in an excited state when a photon of just the right energy encounters it. To justify this claim, we consider **Boltzmann's distribution law**, which relates the number density of atoms in the high-energy n_{hi} to the number density of atoms in the lower energy state n_{lo}:

$$n_{hi} = n_{lo} e^{-\Delta E / k_B T} \quad (42.31)$$

where $\Delta E = E_{hi} - E_{lo}$ and T is the temperature in kelvin. (We won't derive Boltzmann's distribution, but in Problem 42.80 you will show that Boltzmann's distribution law can be used to derive the Maxwell–Boltzmann distribution we studied in Chapter 20.) The energy difference ΔE is positive, so all the terms in the exponential function are positive. The negative sign in the exponential function means that $n_{hi} < n_{lo}$. In fact, under normal conditions there are very few atoms in the higher energy state. If such an atom were to spontaneously transition to the lower energy state and release a photon, this photon would be very unlikely to encounter another atom in the higher energy state. This photon is much more likely to encounter an atom in the lower energy state and get absorbed.

Laser is an acronym for *l*ight *a*mplification by *s*timulated *e*mission *r*adiation. Lasers depend on stimulated emission. We have just argued that under normal conditions stimulated emission is rare because there are more atoms in the lower energy state than in the higher energy state $n_{hi} < n_{lo}$. So the first step in designing a laser is to create a **population inversion**; that is, we need to have more atoms in the higher energy state than in the lower one $n_{hi} > n_{lo}$.

But a population inversion can only be created if the atoms have an available **metastable state** above their ground state. (We touched on such a state in Section 42-8 when we discussed forbidden transitions.) An atom can exist in a metastable state for a relatively long time because the transition to a lower state has a very low probability. An atom in a normal energy state may exist in that state for roughly 10^{-8} s before spontaneous emission occurs, but an atom in a metastable state can remain in that state for orders of magnitude longer making it available for stimulated emission to cause it to transition to a lower energy state.

So the laser must **pump** the atoms up to a metastable state. We'll start by describing how this is done in a pulsed laser. In a laser that uses a gas, the pumping process is similar to the one found in a neon bulb. An electric current is passed through the gas, creating some ions and free electrons (Fig. 42.35). Collisions between the atoms, ions, and electrons put the atoms into a higher energy state. So far, this is exactly like the gas in the neon bulb. The difference is that the atoms in the laser have a metastable state available to them. (In the neon gas, the atoms spontaneously emit photons.)

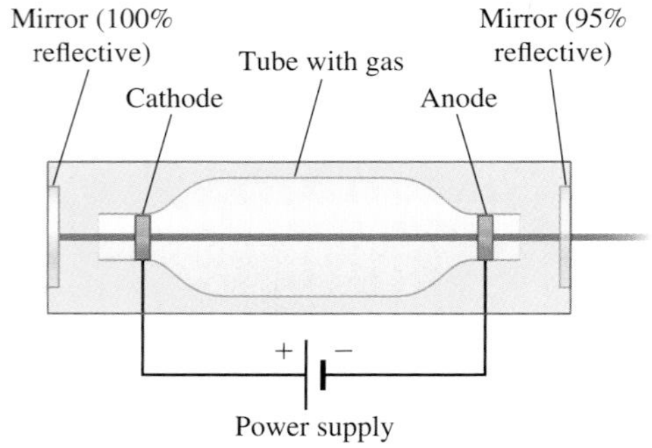

FIGURE 42.35 A schematic of a basic laser design.

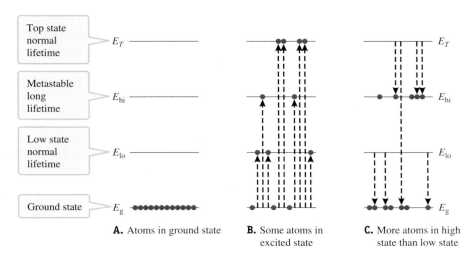

FIGURE 42.36 A. Before pumping, the atoms are in the ground state. **B.** Soon after pumping atoms occupy all four states. **C.** After a short time, spontaneous emission causes the atoms' transition to lower states. Atoms in the metastable state do not spontaneously transition in this short time. So there is a population inversion between the high and low states.

Figure 42.36 shows the energy levels of atoms in a laser. The dots represent atoms in their various energy states. These atoms have a metastable state with energy E_{hi}. Below the metastable state are two other states: One with energy E_{lo}, which we'll call the low state and another which is the ground state, with energy E_g. Above the metastable state is another energy state, which we'll call the top state, with energy E_T. Before the current is turned on, the atoms are in their ground state (Fig. 42.36A). But once pumping begins, atoms are in all four possible states (Fig. 42.36B). After a short time (about 10^{-8} s), the atoms in the top state and those in the middle state spontaneously transition to a lower energy state. However, those atoms in the metastable state do not transition. In addition, some atoms that were in the top state will transition to the metastable state. There will be very few atoms, if any, in the low state because these atoms would have transitioned to the ground state. The result is a population inversion between atoms in the higher metastable state and those in the low state (Fig. 42.36C).

In about a millisecond, a few atoms in the metastable state will spontaneously transition to the low state, releasing a photon of energy $E_{photon} = E_{hi} - E_{lo}$. Because of the population inversion, a spontaneously released photon is now *likely* to encounter an atom in the metastable state, causing it to release an identical photon. To ensure that these photons stimulate other atoms, mirrors are used to send them back and forth through the gas many times (Fig. 42.35). One of these mirrors is partially transparent, so that some of the resulting light can escape in a beam. In a *pulsed* laser, there is a burst of laser light, and then the pumping must begin again.

Because the photons produced by stimulated emission are identical, they have the same frequency, are in phase, are similarly polarized, and travel in the same direction. The result is laser light that is coherent (in phase) and monochromatic. The reflections back and forth across the tube ensure that the laser light diverges little. (Photons traveling along other paths do not stimulate many atoms.) These properties make laser light particularly useful in our laboratory experiments.

In your laboratory, you have probably used a helium–neon laser. Such a laser is a *continuous* laser rather than a pulsed laser of the kind we just described. The workings of such a laser are similar to the pulsed laser. An electric current is applied to the gas, which is a mixture of 10% He and 90% Ne. The helium is excited by this current to a $2s$ state. The excited helium then collides with neon atoms in the ground state. Because the energy of helium's $2s$ state is nearly the same as neon's $3s$ state, these collisions efficiently cause the neon to transition to its $3s$ state; this is a metastable state for neon. The neon atoms will stay in these metastable states until they undergo stimulated emission. These lasers are continuous rather than pulsed because the collision process between the helium and neon can maintain the population inversion.

CHAPTER 42 Atoms

Experiments like Young's done nearly two hundred years ago—long before lasers were invented in the 1960s—used sunlight (which is much more inconvenient than lasers). But lasers are **not** just more convenient to use. They have played a critical role in aiding scientific progress. For example, in 1997 the American physicist Steven Chu (1948–), who later became the Secretary of Energy under President Obama, shared the Nobel Prize with William Daniel Phillips (1948–) and Claude Cohen-Tannoudji (1933–) for their work involving lasers. As we know, atoms are always in motion owing to their thermal energy. But a physicist would like to slow atoms down to study them better. Chu used six lasers aimed toward the center of sodium gas to slow down and trap sodium atoms. He could cool the atoms down to a few hundred microkelvin for about half a second. So our knowledge of atoms allows us to build lasers, which allows us to trap atoms to study them better!

EXAMPLE 42.11 CASE STUDY Got to Pump

In order for a helium–neon laser to operate, the neon atoms must be in an excited state. The wavelength of light from a helium–neon laser is 633 nm. At normal room temperature, what is the ratio of the number density of neon atoms in the high-energy state to that of the atoms in the low-energy state?

• INTERPRET and ANTICIPATE
We can use the wavelength of the emitted light to find the energy difference between the two energy states that emitted that light. Then we can use Boltzmann's distribution law to find the ratio.

• SOLVE

The energy of a photon can be found by combining $f = c/\lambda$ (Eq. 34.20) and $E = hf$ (Eq. 40.8).	$E_{\text{photon}} = hf = \dfrac{hc}{\lambda}$
The energy of the photon must equal the difference in the energy of the two states.	$E_{\text{photon}} = E_{\text{hi}} - E_{\text{lo}} = \Delta E$ $\Delta E = \dfrac{hc}{\lambda} = \dfrac{(6.63 \times 10^{-34} \text{J} \cdot \text{s})(3.00 \times 10^{8} \text{ m/s})}{633 \times 10^{-9} \text{ m}}$ $\Delta E = 3.14 \times 10^{-19}$ J
To use Boltzmann's law (Eq. 42.31), we must also estimate room temperature. Typically, room temperature is about 20° C = 293 K. Rearrange to find the ratio and substitute values.	$n_{\text{hi}} = n_{\text{lo}} e^{-\Delta E / k_B T}$ (42.31) $\dfrac{n_{\text{hi}}}{n_{\text{lo}}} = e^{-\Delta E / k_B T}$ $\dfrac{n_{\text{hi}}}{n_{\text{lo}}} = \exp\left(\dfrac{-3.14 \times 10^{-19} \text{J}}{(1.38 \times 10^{-23} \text{J} \cdot \text{K})(293 \text{ K})}\right)$ $\dfrac{n_{\text{hi}}}{n_{\text{lo}}} = 1.78 \times 10^{-34}$

• CHECK and THINK
We just found a very small ratio. To make this more understandable, if there were 1 mole of neon gas, and so there were 6.0×10^{23} atoms, then no atom would likely be in the high-energy state. Now we see why pumping and the existence of a metastable state are so important. Stimulated emission requires atoms in the higher energy state. Atoms must be pumped to the higher state and remain there until they can be simulated to emit photons.

SUMMARY

❶ Underlying Principles

According to **Pauli's exclusion principle**, *no two particles with s = 1/2 confined in the same well can have the same set of values for their quantum numbers.* In practice, this means that in an atom, no two electrons can have exactly the same set of the four quantum numbers n, ℓ, m, and m_s.

✪ Major Concepts

1. **Thomson's plum pudding model** for an atom consists of two parts: a positive homogeneous sphere and electrons embedded in this positive sphere (Fig. 42.3). Although Thomson's model is consistent with Thomson's experiments on cathode rays, it cannot explain the spectral lines that we observe from atoms (Section 36-4).

2. In **Rutherford's solar system model**, the positive part of the atom is concentrated at the center and surrounded by orbiting electrons. Most of the atom is empty space (Fig. 42.1). This model accounts for Rutherford's experimental results. However, in this model the electrons must orbit the nucleus in order for the atom to be stable, a requirement contradicted by quantum mechanics.

3. As in Rutherford's model, in **Bohr's atomic model** the positive nucleus is at the center of the atom and the electrons orbit that nucleus much as the planets orbit the Sun. The two major differences with Rutherford's model are known as **Bohr's postulates**.

   a. **Bohr's first postulate** is that only certain electron orbits are allowed, with angular momentum L given by:
   $$L = n\left(\frac{h}{2\pi}\right) = n\hbar \quad (n = 1, 2, 3\ldots) \quad (42.4)$$
   where $\hbar = h/2\pi = 6.58211814 \times 10^{-16}$ eV·s. The integer n is called the **principal quantum number**. The energy levels of hydrogen are then:
   $$E_n = -\frac{13.6\,\text{eV}}{n^2} \quad (n = 1, 2, 3\ldots) \quad (42.10)$$

   b. **Bohr's second postulate** is that electrons in orbit around the nucleus do not radiate. An electron only radiates when it moves from one possible orbit to another possible orbit that has lower energy.

   For hydrogen, the wavelength of the radiated photon is given by Rydberg's formula
   $$\frac{1}{\lambda} = R\left(\frac{1}{m^2} - \frac{1}{n^2}\right) \quad (42.3)$$

4. In applying **de Broglie's model** to the atom, we model the electrons as circular standing waves with quantized wavelengths given by:
   $$\lambda_n = \frac{2\pi r}{n} \quad (42.13)$$
   Like Bohr's model, de Broglie's model successfully predicts the hydrogen spectrum.

5. When **Schrödinger's equation** is applied to an atom and the boundary conditions are applied in three dimensions, the result is three quantities (energy and the magnitude and direction of the orbital angular momentum) that are quantized and described by three quantum numbers. In addition, **spin**—an intrinsic property of a particle—is also quantized in magnitude and direction, as described by two more quantum numbers. Table 42.4 summarizes all five quantum numbers.

6. The **periodic table** (Appendix B) is the ordering of the elements by atomic number into periods and groups. The chemical and physical properties of the elements are correlated with their place in the table.

7. In the normal **Zeeman effect**, the spectral lines of certain atoms split into three lines when the sample is in an external magnetic field B. The energy of the emitted photon is given by
   $$E = E_0 - \Delta m \mu_{\text{Bohr}} B \quad (42.30)$$
   where E_0 is the energy of the corresponding photon in the absence of an external field and the difference in the orbital magnetic quantum number is $\Delta m = 0, \pm 1$ for allowed transitions.

PROBLEMS AND QUESTIONS

A = algebraic   C = conceptual   E = estimation   G = graphical   N = numerical

42-1 Early Atomic Models

Problems 1 and 4 are paired.

1. **C** One thing we know about atoms is that they are stable; that is, they don't collapse. What accounts for an atom's stability in the plum pudding model?
2. **E Review** Estimate the number of atoms that exist in a cubic centimeter of a solid material. This process will be aided by choosing an element and examining its density in the solid phase.
3. **C** Suppose you cut a wooden log or an iron bar in half repeatedly. Based on the natural models of the ancient Greeks, how would these two experiments end?

42-2 Rutherford's Model of the Atom

4. **C** One thing we know about atoms is that they are stable; that is, they don't collapse. What accounts for an atom's stability in the solar system model?
5. **C** Why was Rutherford so surprised by his experimental results?
6. **C** Explain this sentence: "You are mostly empty space."
7. **E** In the development of atomic models, it was realized that the atom is mostly empty. Consider a model for the hydrogen atom where its nucleus is a sphere with a radius of roughly 10^{-15} m, and assume the electron orbits in a circle with a radius of roughly 10^{-10} m. In order to get a better sense for the emptiness of the atom, choose an object and estimate its width. This object will be your "nucleus". How far away would the "electron" be located away from your "nucleus?"

42-3 Bohr's Model and Atomic Spectra

8. **N** Show that the constant $(2\pi^2 k^2 m_e e^4)/(h^3 c)$ is equal to the Rydberg constant $R = 1.10 \times 10^7\,\text{m}^{-1}$ (Eq. 42.2). To keep this task manageable, work with four significant figures, and report your answer to three significant figures.
9. **N** In the Bohr model of hydrogen, an electron makes a transition from the $n = 5$ excited state to the $n = 3$ excited state. What is the change in the orbital radius of the electron?
10. **N** Show that the constant $(2\pi^2 k^2 m_e e^4)/h^2$ is equal to -13.6 eV (Eq. 42.9). To keep this task manageable, work with four significant figures, and report your answer to three significant figures.
11. **N** According to Bohr's model, what is the diameter of hydrogen in its ground state?
12. **C** Suppose that an electron in the ground state of a hydrogen atom absorbs a photon with 15.0 eV of energy. Is the electron still bound to the nucleus? What is the energy of the electron after it absorbs this photon? What kind of energy does the electron then possess? Explain your answer.
13. **N** Consider photons incident on a hydrogen atom.
    a. A transition from the $n = 3$ to the $n = 7$ excited state requires the absorption of a photon of what minimum energy?
    b. A transition from the $n = 1$ ground state to the $n = 5$ excited state requires the absorption of a photon of what minimum energy?

Problems 14, 15, and 16 are grouped.

14. **A** Use a solar system model to come up with an expression for the speed of an electron orbiting in the ground state of hydrogen. Express your answer in terms of the Bohr radius, the mass of the electron, and the elementary electric charge.
15. **A** Use Bohr's model to come up with an expression for the speed of an electron orbiting in the ground state of hydrogen. Express your answer in terms of the fundamental constants. Your final expression should not involve the mass of the electron.
16. **C** Using Bohr's model for hydrogen, you can come up with a speed for the orbiting electron that does not involve the electron's mass. Explain this result, present an example of a similar situation, and explain what it means for the atom's energy.
17. **N** An atom in an excited state can, on average, exist in that state for 10^{-8} s. If a hydrogen atom is in the $n = 2$ state, about how many orbits will it undergo before the atom returns to the ground state? Assume Bohr's model for hydrogen.
18. **N** A hydrogen atom transitions from the $n = 6$ excited state to the $n = 2$ excited state, emitting a photon. **a.** What is the energy, in electron volts, of the photon emitted by the hydrogen atom? **b.** What is the wavelength of the photon emitted by the hydrogen atom? **c.** What is the frequency of the photon emitted by the hydrogen atom?
19. **N** What is the maximum photon wavelength that would free an electron in a hydrogen atom when it is in the $n = 6$ excited state?

Problems 20, 21, and 22 are grouped.

20. **N** The Balmer series consists of the spectral lines from hydrogen for an electron making a transition from an excited state to the $m = 2$ state. The Lyman series consists of the spectral lines from hydrogen for an electron making a transition from an excited state to the $m = 1$ state. Determine the wavelengths of the first four spectral lines of the Lyman series ($n = 2, 3, 4,$ and 5).
21. **N** The Balmer series consists of the spectral lines from hydrogen for an electron making a transition from an excited state to the $m = 2$ state. The Paschen series (or Bohr series) consists of the spectral lines from hydrogen for an electron making a transition from an excited state to the $m = 3$ state. Determine the wavelengths of the first four spectral lines of the Paschen series ($n = 4, 5, 6,$ and 7).
22. **N** The Balmer series consists of the spectral lines from hydrogen for an electron making a transition from an excited state to the $m = 2$ state. The Brackett series consists of the spectral lines from hydrogen for an electron making a transition from an excited state to the $m = 4$ state. Determine the wavelengths of the first four spectral lines of the Brackett series ($n = 5, 6, 7,$ and 8).

Problems 23, 24, and 25 are grouped.

23. **C** The Bohr model for the hydrogen atom can be extended to cover other atoms when they are stripped free of all but one electron. When this occurs, the energy levels for the single electron in an atom with atomic number, Z, are given by $E_n = (-13.6\,\text{eV})Z^2/n^2$ (see Example 42.4). As we examine atoms with greater atomic number, the ground-state energy becomes more negative. Explain why this would be expected and what it means regarding the requirements for freeing the electron in the ground state.
24. **N** The Bohr model for the hydrogen atom can be extended to cover other atoms when they are stripped free of all but one electron. When this occurs, the energy levels for the single electron in an atom with atomic number, Z, are given by

$E_n = (-13.6 \text{ eV})Z^2/n^2$ (see Example 42.4). Calculate the electron energy for the first five energy levels ($n = 1$ to $n = 5$) of ionized lithium (Li^{++}).

25. **N** The Bohr model for the hydrogen atom can be extended to cover other atoms when they are stripped free of all but one electron. When this occurs, the energy levels for the single electron in an atom with atomic number, Z, are given by $E_n = (-13.6 \text{ eV})Z^2/n^2$ (see Example 42.4). What wavelength of photon would be emitted if an electron transitions from the $n = 4$ excited state to the ground state in ionized lithium (Li^{++}).

42-4 De Broglie's Theory and Atoms

26. **N, C** Compare the circular standing waves of an electron in a hydrogen atom (Fig. 42.9, page 1387) to the standing waves on a guitar string (Fig. 18.25, page 535). How many nodes and antinodes exist in the third harmonic for the guitar string? How many nodes and antinodes exist for the third quantum state of an electron in hydrogen? Comment on your findings.
27. **N** The angular momentum of an electron in a hydrogen atom is $2h/\pi$. **a.** What is the radius of the electron's orbit? **b.** What is the de Broglie wavelength of the electron?
28. **C** An electron in a hydrogen atom undergoes a transition such that its de Broglie's wavelength doubles. What is the relationship between the initial and final energy levels occupied by the electron? Explain your answer.
29. **N** A hydrogen atom is in its $n = 6$ state. Find the de Broglie wavelength of its electron.

42.5 Schrödinger's Equation Applied to Hydrogen

30. **N** An electron in a hydrogen atom is in a state with a principal quantum number of 4. How many possible subshells could the electron occupy?

Problems 31 and 32 are paired.

31. **N** If the state of hydrogen is given by $n = 3$ and $\ell = 2$, find the possible values of m, L and L_z.
32. **G** If the state of hydrogen is given by $n = 3$ and $\ell = 2$, find the possible values of m, L, and L_z. Assume that the vector $\vec{L}$ lies in the x–z plane and sketch its possible values (magnitude and direction). Indicate angles and magnitudes.

Problems 33 and 35 are paired.

33. **N** If hydrogen is in its $n = 3$ state, what are the possible magnitudes of the electron's orbital angular momentum L?
34. **N** An electron in a hydrogen atom is in a state with $n = 4$. Find the orbital angular momentum and z component of the orbital angular momentum for each possible value of ℓ and m for the electron.

42.6 Magnetic Dipole Moments and Spin

35. **N** If hydrogen is in its $n = 3$ state, what are the possible values of $(\mu_{\text{orbit}})_z$ in terms of the Bohr magneton?

Problems 36 and 37 are paired.

36. **N** An electron in a hydrogen atom is in a state with a principal quantum number of 3 and an orbital quantum number of 2. What is the number of possible states for the electron? Take into account the possible values of m, and m_s.
37. **N** An electron is bound to a hydrogen atom. What is the number of possible states for the electron in each of the following cases? Take into account the possible values of m, and m_s. **a.** $n = 4$ **b.** $n = 5$ **c.** $n = 6$.

38. **A** Show that an electron's spin magnetic moment is a constant, $\mu_{\text{spin}} = \sqrt{3}\hbar e/2m_e$, regardless if it is bound or free.
39. **N** An electron in a hydrogen atom is in a state with $n = 3$. Find the orbital angular momentum and the z component of the orbital magnetic moment for each possible value of ℓ and m for the electron.
40. **A** The magnetic moment of a current loop is given by $\vec{\mu} = (I\pi R^2)\hat{j}$ (Eq. 30.14). Show that the magnetic moment of a charged particle in a circular orbit of radius R is given by $\vec{\mu}_{\text{orbit}} = (q\vec{L})/(2m)$.

42.7 Other Atoms

Problems 41 and 42 are paired.

41. **C** Is it possible for the z component of the spin magnetic moment of an electron to equal its z component of the orbital magnetic moment? If so, what value(s) of m make(s) this possible? If not, explain why not.
42. **N, C** Is it possible for the spin magnetic moment of an electron to cancel its orbital magnetic moment? If so, what value(s) of m make this possible? If not, explain why not.
43. **N** Find the maximum number of electrons in each of the following subshells: **a.** s **b.** p **c.** d **d.** f **e.** g
44. **C** Your friend tells you a story about her two cousins, Paul and Paula. She jokes about how just when one leaves the other seems to always show up, and that it doesn't seem that they can ever be in the same place twice. In light of the physics class you share, she jokes that they must be obeying Pauli's exclusion principle. Is this an accurate comparison? In what way is she using the analogy correctly? In what way is the analogy not correct? Give an example.
45. **C** When a shell is full, does it always have an even number of electrons? Explain.

42.8 Organizing Atoms

46. **C** How do you account for the fact that lithium (Li) and sodium (Na) exhibit similar chemical behavior?
47. **N** What is the ground-state electron configuration of iodine (I)?
48. **C** Do you expect carbon (C) and calcium (Ca) to exhibit similar chemical behavior? Explain.

Problems 49 and 50 are paired.

49. **N** Write the ground-state configurations for the first three noble gases on the periodic table: helium (He), neon (Ne), and argon (Ar).
50. **C** Consider the electron configurations of the noble gases in the rightmost column on the periodic table. What aspect of their electron configuration makes them so nonreactive?
51. **N** What atom's ground-state electron configuration is given by $1s^2 2s^2 2p^6 3s^2 3p^6 4s^2 3d^8$?

42.9 The Zeeman Effect

52. **G** Make an energy-level diagram for the first three shells of hydrogen when it is in an external magnetic field B pointing in the z direction. Include those same shells for the case of no external field for comparison. Draw a distinct diagram for each subshell, labeling n, ℓ, m.
53. **N** A hydrogen atom is exposed to a magnetic field of 25.00 T. An electron is in the state $n = 5$, $\ell = 1$. Find each of the possible wavelengths of the photon emitted when this electron transitions directly from its current state to the ground state of the atom.

54. **A** Calculate the energy levels for hydrogen's 3d subshell when hydrogen is in an external magnetic field B pointing in the z direction. Express your answer in terms of the Bohr magneton.

Problems 55 and 56 are paired.

55. **N** A hydrogen atom with an electron in an excited state with $n = 3$, $\ell = 2$ undergoes a single transition to the state $n = 2$, $\ell = 1$, $m = 0$ while in an external magnetic field with magnitude 4.000 T. How many possible wavelengths could be observed for the emitted photon?

56. **N** A hydrogen atom with an electron in an excited state with $n = 3$, $\ell = 2$ undergoes a single transition to the state $n = 2$, $\ell = 1$, $m = 0$ while in an external magnetic field with magnitude 4.000 T. What is the maximum possible wavelength of an observed photon from this transition?

57. **N** An electron in a hydrogen atom is in the $n = 5$ shell and $\ell = 4$ subshell. What is the difference between the highest and lowest possible energies of the electron when the atom is exposed to a 4.000 T magnetic field?

58. **C** For states of the electron in the hydrogen atom with $n > 1$, is it safe to say that there is at least one subshell that would show three distinct energies when the atom is exposed to an external magnetic field? Explain.

General Problems

59. **N** An isolated hydrogen atom is found to be in the $n = 5$ excited state. **a.** What is the radius of the Bohr orbit of the electron in this atom? **b.** What is the de Broglie wavelength for this electron?

60. **N** The Balmer series consists of the spectral lines from hydrogen for an electron making a transition from an excited state to the $m = 2$ state. The Pfund series consists of the spectral lines from hydrogen for an electron making a transition from an excited state to the $m = 5$ state. Determine the wavelengths of the first four spectral lines of the Pfund series ($n = 6, 7, 8,$ and 9).

Problems 61 and 62 are paired.

61. **N** The magnitude of the orbital angular momentum is $L = \sqrt{30}\hbar$. What is ℓ?

62. **A** The magnitude of the orbital angular momentum is $L = \sqrt{30}\hbar$. What are the possible values of m?

63. The Bohr model for the hydrogen atom can be extended to cover other atoms when they are stripped free of all but one electron, as in Example 42.4, $E_n = (-13.6 \text{ eV})Z^2/n^2$. An electron in the $n = 3$ state of a single-electron Lithium atom (Li^{++}) makes a transition to the ground state by emitting a photon.
  a. **N** What is the wavelength of the emitted photon?
  b. **C** If this emitted photon were absorbed by an electron in the ground state of a hydrogen atom, would it be enough to free the electron? Explain your answer.

64. **A** To find the most likely location for the electron, we need the *radial* probability density P, not the *volume* probability density $|\psi|^2$. The two probability densities are related by $P(r)dr = |\psi|^2 dV$. The volume element dV is the volume of a spherical shell of thickness dr, so $dV = 4\pi r^2 dr$. Show that $P(r) = (4/r_B^3)r^2 e^{-2r/r_B}$.

65. **N** An electron in a hydrogen atom makes a transition from the ground state to the $n = 4$ state after absorbing a photon. **a.** What is the minimum energy of the photon? **b.** What is the frequency of the photon with this minimum energy?

66. **C** Covalent bonds form between atoms or molecules that share electrons and often allow each atom to fill its outer shell via the sharing process. Consider hydrochloric acid HCl. Which shell is filled for each atom when they share electrons?

Problems 67 and 68 are paired.

67. **C** An electron in a hydrogen atom is in the $n = 5$ shell. Is there a subshell in which the electron could have five distinct energies when the atom is exposed to an external magnetic field? If so, which subshell? Explain. If not, why not?

68. **N** An electron in a hydrogen atom is in the $n = 5$ shell and $\ell = 1$ subshell. Find the three orbital energies of the electron, when an external magnetic field of 2.00 T is applied.

Problems 69, 70, and 71 are grouped.

69. **N** An imaginary atom has only one electron, and its energy levels are given by $E_n = (-19.5 \text{ eV})/n^2$ where $n = 1, 2, 3 \ldots$. Find the first five energy levels for this atom.

70. **N** An imaginary atom has only one electron, and its energy levels are given by $E_n = (-19.5 \text{ eV})/n^2$ where $n = 1, 2, 3 \ldots$. If the atom goes from the $n = 3$ state to the $n = 1$ state, what is the frequency of the emitted photon?

71. **N** An imaginary atom has only one electron, and its energy levels are given by $E_n = (-19.5 \text{ eV})/n^2$ where $n = 1, 2, 3 \ldots$. Initially, the atom is in its ground state. If the atom absorbs a photon with a frequency of 4.42×10^{15} Hz, in what state is the atom now?

72. **N** An electron is bound in a hydrogen atom such that it is in the $n = 6$ shell. What is the de Broglie wavelength of this electron?

73. Consider an electron in a hydrogen atom in the $n = 2$, $\ell = 1$, $m = 1$ excited state.
  a. **N** What magnitude of an applied magnetic field would cause the energy of this electron to be equal to zero?
  b. **C** Given that the energy of this electron would be zero under these conditions (if the electron does not undergo a transition to another state), what would be true about the electron?

74. **C** A common saying is that "you can be two places at the same time." How might this colloquialism be reworded to describe Pauli's exclusion principle?

Problems 75, 76, and 77 are grouped.

75. **C** Suppose an electron in a hydrogen atom transitions from its current state, n, to the state, $n/2$. What, if any, limitations can you place on the value of n? Are there any values not allowed?

76. **A** Suppose an electron in a hydrogen atom transitions from its current state, n, to the state, $n/2$. Use Eq. 42.2 to express the wavelength of the resulting emitted photon in terms of R, and n.

77. **N** Consider the result for Problem 76. Find the wavelength of the emitted photons for the cases where **a.** $n = 2$, **b.** $n = 6$, and **c.** $n = 12$.

Problems 78 and 79 are paired.

78. **C** An electron in a hydrogen atom is to make a transition from the $n = 4$ excited state to the ground state. There are three ways the electron could make this transition from the $n = 4$ state to the ground state by emitting photons. Describe each of the ways the electron could accomplish this feat.

79. **N** An electron in a hydrogen atom is to make a transition from the $n = 4$ excited state to the ground state. There are three ways the electron could make this transition from the $n = 4$ state to the ground state by emitting photons. Determine the energy of the photons that would be emitted in each way the electron can make the transition.

80. **A** Start with Boltzmann's distribution $n_{\text{hi}} = n_{\text{lo}} e^{-\Delta E/k_B T}$ (Eq. 42.31) and derive Maxwell–Boltzmann's equation $f(v) = 4\pi \left(\frac{m}{2\pi k_B T}\right)^{3/2} v^2 e^{-(mv^2/2k_B T)}$ (Eq. 20.19).

Nuclear and Particle Physics

43

❶ Underlying Principles

1. Models of nuclei
2. Strong nuclear force
3. Weak nuclear force
4. Electroweak theory
5. QED and QCD
6. The standard model

✪ Major Concepts

1. Nuclear radius
2. Nuclear decay
3. Mass deficit and binding energy
4. Fusion and fission
5. Absorbed, equivalent, and effective dose
6. Bosons and fermions
7. Leptons and hadrons

◐ Special Cases

1. Gamma rays
2. Alpha particles (and rays)
3. Beta particles (and rays)
4. Electron capture

Key Questions

How do we model the nucleus and subatomic particles?

What are nuclear reactions?

What is radioactivity, and how does it affect living organisms?

43-1	Describing the Nucleus 1414
43-2	The Strong Force 1416
43-3	Models of Nuclei 1417
43-4	Radioactive Decay 1420
43-5	The Weak Force 1424
43-6	Binding Energy 1428
43-7	Fission Reactions 1433
43-8	Fusion Reactions 1436
43-9	Human Exposure to Radiation 1441
43-10	The Standard Model 1446

Here's something odd. According to physics, there are four fundamental forces; but aside from a few brief mentions, for the past 42 chapters we discussed only two of these four forces. How did we avoid the last two forces for so long, and why do we need to discuss them now? The answer is that both of these remaining forces are *nuclear* forces. So far we have treated the nucleus as a positively charged particle, but in this chapter we explore nuclear structure and behavior. Along the way, we'll learn where we all came from, how to measure the age of fossils, and the basic principles of nuclear reactors. We'll see how all of this is organized in a well-confirmed theory known as the *standard model*.

A nuclear species is referred to as a *nuclide*; however, we will typically use the terms *nucleus* (singular) and *nuclei* (plural).

43-1 Describing the Nucleus

You already have a good idea of what a nucleus is: an object at the center of an atom, made up of protons and neutrons. You also know that the nucleus was discovered in Rutherford's scattering experiment (Section 42-2). Rutherford concluded that the nucleus is about 10^4 times smaller than the atom.

Most of an atom's mass is concentrated in its nucleus, which is composed of particles called **nucleons** (either protons or neutrons). Although neutrons are slightly more massive than protons, each nucleon's mass is about 1800 times the mass of an electron. Both kinds of nucleons have spin ½, so both obey Pauli's exclusion principle (Section 42-7). The number of neutrons in a nucleus is known as the **neutron number** N. The number of protons in the nucleus is called the **atomic number** Z. (For a neutral atom, Z is also the number of electrons.) The total number of nucleons is called the **mass number** A, and $A = N + Z$. The simplest nucleus is that of hydrogen, consisting of a single proton. All other nuclei have both protons and neutrons. The most massive naturally occurring nucleus is that of uranium, with 92 protons and 146 neutrons. Heavier nuclei have been created artificially.

Nuclei are represented by their chemical symbol X (taken from the periodic table, Appendix B) and by a preceding superscript and subscript, as in $^A_Z X$. The superscript A is the mass number, and the subscript Z is the atomic number. For example, the most massive naturally occurring uranium *isotope* (more on this term below) may be written as $^{238}_{92}U$. Often the subscript is dropped because the chemical symbol indicates the atomic mass number, making the subscript redundant. (Uranium is the 92nd element in the periodic table.) There is no need to indicate the neutron number because it can always be found from A and Z: $N = A - Z$. (Uranium has $238 - 92 = 146$ neutrons.) We often speak of an isotope using its name and mass number: for example, we call $^{238}_{92}U$ "uranium-238."

An element is characterized by its atomic number Z; that is, its chemical properties are determined by its number of electrons. An **isotope** is a species of an element (with the same Z) that differs in its neutron number, and therefore its mass number. Because all isotopes of an element have the same number of protons (and of electrons), they occupy the same place on the periodic table, have the same electron configuration, and therefore the same chemical behavior. For example, deuterium is an isotope of hydrogen with one proton and one neutron in its nucleus. The atom still has one electron, though, and like hydrogen, deuterium bonds with oxygen to form (heavy) water. As another example, uranium-235 and uranium-238 have the same chemical properties and so cannot be separated by a chemical reaction. However, because their masses are different, they can be separated mechanically with, for example, a centrifuge.

Although the chemical behavior of isotopes of a particular atom is the same, the stability of its nucleus depends on the isotope. A periodic table is not useful for showing the differences in isotopes, so we need another sort of chart. Typically, we plot Z as a function of N, creating an **isotope chart** (Fig. 43.1). There are over 3000 known isotopes, but only about 266 are stable and occur naturally. The stable isotopes are shown in blue in Figure 43.1. The other isotopes are radioactive, meaning that they **decay** or break apart by emitting a particle or radiation. Most isotopes are radioactive, and will decay (sometimes repeatedly) until the end result is a stable isotope. The stable isotopes tend to form a thin line in the middle of the chart. The *small* stable isotopes (with N less than about 20) tend to fall on the $N = Z$ line, so small stable isotopes have roughly equal numbers of protons and neutrons. However, the larger stable isotopes fall below the $N = Z$ line, meaning that they tend to have fewer protons than neutrons. Stable as well as unstable isotopes occur naturally.

The **natural abundance** of an isotope is the number of a naturally occurring isotope relative to the total number of naturally occurring isotopes of that element. For example, chlorine comes in two naturally occurring isotopes, and the natural abundance of ^{35}Cl is about 76% and that of ^{37}Cl is about 24%. The isotopes of chlorine have different numbers of nucleons, so they must have different masses. The atomic mass reported for an element, for example, on the periodic table, is the weighted average of the atomic masses of the naturally occurring isotopes.

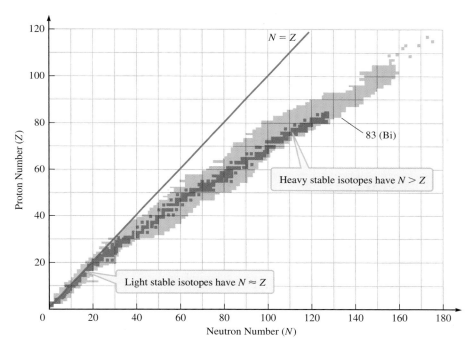

FIGURE 43.1 Isotopes are shown on a graph of Z as a function of N. The stable isotopes are in blue. The red line indicates isotopes for which the atomic number equals the neutron number. This line is well above the high-N isotopes but passes through the low-N ones. There are no stable nuclei for $Z > 83$ (bismuth). This type of graph is sometimes called a *Segrè chart*. Notice that we have plotted Z on the vertical axis and N on the horizontal, but you may encounter the graph with these axes reversed.

Atomic masses are often reported in atomic mass units (u); the mass of ^{12}C is defined as exactly 12 u. So the mass number A gives an approximate mass in atomic mass units. For example, the atomic mass of chorine is (roughly) given by: $m_{Cl} \approx 0.76(35 \text{ u}) + 0.24(37 \text{ u}) \approx 35.5$ u, which is close to the value of 35.453 u found in the periodic table. In Appendix B, conversion factors are generally given to five significant figures, but you may need more significant figures to convert from atomic mass units to kilograms:

$$1 \text{ u} = 1.66053879 \times 10^{-27} \text{ kg} \tag{43.1}$$

Often we will need to know the mass to five or so significant figures, so estimating mass from A is not sufficient.

Nuclear Radius

The radius of a nucleus is best measured in femtometers (1 fm = 10^{-15} m). A nucleus is not necessarily spherical. Nevertheless, experiments have shown that we can model some nuclei as spheres with an effective radius given by:

$$R = r_0 A^{1/3} \tag{43.2}$$

where $r_0 = 1.2$ fm and A is the mass number. The volume of a sphere is $V = 4\pi R^3/3$, so the volume of a nucleus is proportional to A. From this we can conclude that nucleons are tightly packed into the nucleus and are incompressible. Also, because the volume is proportional to the number of nucleons, the density of nuclear matter is nearly independent of the type of nucleus (Example 43.1).

A femtometer is also known as a *fermi* (both abbreviated "fm"), named for the Italian physicist Enrico Fermi (1901–1954).

Some nuclei produced in laboratories are not well modeled by Equation 43.2; these nuclei are larger than what the model predicts.

NUCLEAR RADIUS ⊗ **Major Concept**

CASE STUDY   Part 1: Where did you come from?

Your body is made of many different elements. Much of it is water, made of hydrogen and oxygen. Your blood contains iron and your thyroid contains iodine. In this case study, we explore the origins of these four elements. We'll find that some elements in your body are as old as the Universe itself (13.7 billion years), whereas other elements were made in a supernova explosion over 4.5 billion years ago. In learning where the elements in your body came from, we'll learn also about the history of the Universe.

CONCEPT EXERCISE 43.1

We concluded that because the volume of a nucleus is proportional to the number of nucleons, the nucleons must be tightly packed and incompressible. Use the analogy of packing marbles tightly into a flexible bag to explain why nucleons must be incompressible. What would happen if the marbles were replaced by highly compressible foam balls (like small Nerf® balls)?

EXAMPLE 43.1 The Density of Nuclear Material

Estimate the density of nuclear material.

INTERPRET and ANTICIPATE
Because both the volume of a nucleus and its mass are proportional to A, we expect to find that its density is a constant, independent of A.

SOLVE
Find the density from the mass and the volume.

$$\rho = \frac{m}{V} \tag{1.1}$$

The mass is approximately given by A in atomic mass units. Substitute $R = r_0 A^{1/3}$ (Eq. 43.2) for r in the volume expression.

$$\rho \approx \left(\frac{A}{\frac{4}{3}\pi R^3}\right) = \left(\frac{A}{\frac{4}{3}\pi r_0^3 A}\right)$$

$$\rho \approx \frac{3(1\ \text{u})}{4\pi r_0^3} \approx \frac{3(1.66 \times 10^{-27}\ \text{kg})}{4\pi(1.2 \times 10^{-15}\ \text{m})^3}$$

$$\rho \approx 2 \times 10^{17}\ \text{kg/m}^3$$

CHECK and THINK
As expected, our answer does not depend on A, so it is constant for all nuclei. Our result is much denser than familiar objects such as the Earth ($\rho_\oplus \approx 5500\ \text{kg/m}^3$) or water ($\rho_\text{water} \approx 1000\ \text{kg/m}^3$), but this isn't surprising because we know that atoms and therefore objects are mostly empty space.

43-2 The Strong Force

STRONG NUCLEAR FORCE
❶ **Underlying Principle**

You already know that protons in a nucleus experience a repulsive force due to Coulomb's law, and from the case study in Chapter 41 (page 1358), another force binds the protons together in the nucleus. This force, called the **strong force** or the **strong nuclear force**, is one of the four fundamental forces of physics. The strong force has a short range, acting over distances of roughly 1 fm or less. Within this short range, the strong force is about 100 times stronger than the Coulomb force. In this section, we take a closer look at the strong force, which is essential in modeling the energy levels of a nucleus.

So far, we have treated protons and neutrons as elementary particles. However, unlike electrons, protons and neutrons are actually made up of other particles known as **quarks**. Quarks come in six types, with the whimsical names *up*, *down*, *strange*, *charm*, *bottom* and *top*. A proton, for example, is made of two up quarks and one down quark; a neutron is made of two down quarks and one up quark. The strong force binds the quarks together to form these nucleons. Because protons are made of quarks, and because the strong force is an attractive force between quarks, protons are bound together in the nucleus by the strong force. The same can be said for neutrons: They are also bound together by the strong force. Electrons, which are elementary particles that are *not* made up of quarks, do not experience the strong force.

Protons experience a Coulomb repulsive force, but neutrons do not. This difference explains why light stable isotopes have nearly equal numbers of protons and neutrons, whereas heavy stable isotopes have a greater number of neutrons than protons (Fig. 43.1). Although the nucleons are always tightly packed in the nucleus, protons on the opposite side of a large nucleus may be several fermis apart. At this relatively large distance, the strong force is negligible and the net force on such protons is repulsive due to the dominance of the Coulomb force. But the neutrons in a large nucleus experience no such repulsion. They are attracted to and attract neighboring nucleons, so they keep Coulomb repulsion from splitting the nucleus. In a small nucleus, one neutron for each proton is sufficient for stability. However, larger nuclei need more neutrons than protons to ensure that the strong force dominates over the Coulomb repulsion.

CASE STUDY   Part 2: Making Protons and Neutrons

According to the **Big Bang model**, the Universe began in a very hot and very dense state about 13.7 billion years ago. Since that time, the Universe has been expanding, cooling down, and becoming less dense. Cosmologists can model the state of the early Universe and make predictions about what can be observed today. There have been some minor and major adjustments to the model, but observational evidence supports the basic idea that the Universe was once very hot and very dense. Conditions in the (relatively) early Universe have been reproduced in high-energy accelerators, with results that support the Big Bang model.

So what was the very early Universe like? In those extreme conditions, the particles, atoms, and molecules that we are familiar with could not exist. Instead the early Universe was a sea of elementary particles, including quarks. As the Universe cooled, the strong force bound quarks together to form protons and neutrons. Because a proton is the nucleus of hydrogen, hydrogen was likely the first element to form in the Universe. This kind of hydrogen is called *primordial* hydrogen.

CONCEPT EXERCISE 43.2

When we consider the attraction between the Earth and Moon, we only need to take gravity into account, and we can ignore the Coulomb force they exert on one another because they are electrically neutral. We can also ignore the strong force. Why?

43-3 Models of Nuclei

We have been using the language of particles to describe the nucleus. However, these "particles" must be modeled as probability waves governed by Schrödinger's equation. We won't solve this equation for a nucleus; instead we'll use our work on trapped particles and atoms to model the nucleus.

The Shell Model

The first step in applying Schrödinger's equation is to find the potential energy of the system. Nuclear potential energy is complicated because the system involves many nucleons exerting both attractive and repulsive forces on one another. The American physicist Maria Goeppert-Mayer (1906–1972) and the German physicist J. Hans D. Jensen (1907–1973) came up with a clever way to think about the potential energy of nucleons: the **shell nuclear model**, which is based on Bohr's model of hydrogen. According to this model, each nucleon can be treated as a particle in a system whose remainder is the rest of the nucleus. The potential energy of the system then results

MODELS OF NUCLEI
❶ **Underlying Principle**

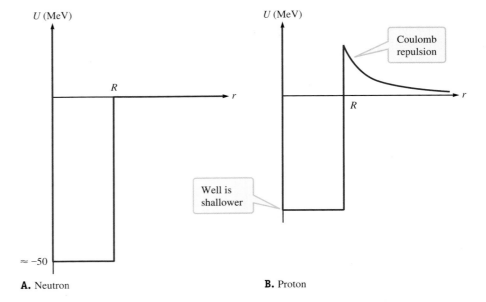

FIGURE 43.2 Schematic potential energy curves for nucleons. **A.** A neutron only experiences an attractive force due to the other nucleons in the nucleus, so the potential energy curve is a simple well. **B.** A proton experiences an attractive force due to all the other nucleons, but also a repulsive force due to the other protons; so the potential energy curve contains both a well and a barrier.

For brevity, we'll drop the full name of each system, and just refer to the proton system and the neutron system, or even just to the proton and the neutron.

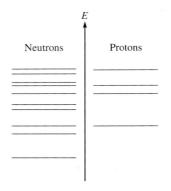

FIGURE 43.3 Energy levels for the nucleus are quantized, with the difference in energy levels on the order of several MeV. The neutron system energy levels are more closely spaced than those of the proton system. The lowest energy for protons is greater than the lowest energy for neutrons.

from the average force experienced by this nucleon due to the rest of the nucleus. For a neutron, that average force is simply attractive, but for a proton, the force is attractive for short distances and repulsive for longer distances.

Figure 43.2 shows rough potential energy curves as a function of r, where r is the distance between the nucleon and the center of the nucleus, and R is the radius of the nucleus given by $R = r_0 A^{1/3}$ (Eq. 43.2). The potential energy curve for a system consisting of a neutron and the remainder of the nucleus is a simple well (Fig. 43.2A) with a depth of roughly 50 MeV for all nuclei. By contrast, the potential energy curve for a system consisting of a proton and the remainder of the nucleus is a somewhat shallower well with a positive barrier (Fig. 43.2B). Figure 43.2B is a crude diagram; the depth and shape of the proton potential energy curve depend on the mass of the nucleus because protons in a more massive nucleus experience a greater Coulomb repulsion than those in a low-mass nucleus.

When we apply Schrödinger's equation to a trapped particle, such as a nucleon bound to a nucleus, the system's energy is quantized (Section 41-5). The spacing of energy levels depends on the depth and shape of the potential energy well. So we expect the energy levels in the proton system to be different from those in the neutron system. For high-mass nuclei, the proton energy levels are farther apart than are the neutron energy levels (Fig. 43.3). However, for low-mass nuclei ($Z \approx 8$ or less), Coulomb repulsion of the protons is weak compared to the strong force; therefore, in such nuclei the proton system and neutron system have similar potential wells and thus their energy levels are more similarly spaced.

Describing the structure and behavior of a nucleus is much like describing the structure and behavior of an atom: In both cases we do so by finding which energy levels are occupied. Because nucleons have spin ½, they obey Pauli's exclusion principle (Section 42-7). Just as when finding the electron configuration of an atom, we find the configuration of a nucleus by filling up energy levels starting with the lowest level while obeying Pauli's exclusion principle. In Chapter 42, we filled boxes in a diagram with upward- and downward-pointing arrows representing electrons with a particular set of quantum numbers (Fig. 42.23, page 1396). We can use a similar diagram here (Fig. 43.4). Each box on the left can be filled by up to two arrows (one up and the other down), representing two neutrons, while each box on the right can be filled by up to two arrows, representing protons. The vertical spacing on both sides of Figure 43.4 is equal for our convenience; remember that proton energy levels are actually farther apart than those of neutrons. Examine Figure 43.3 to see why $N > Z$ for stable heavy nuclei. Stable nuclei tend to have their energy levels filled, from the lowest level up to some highest level, for both protons and neutrons. If the highest proton energy level is roughly equal to the highest neutron energy level, then there must be more neutrons to fill the greater number of energy levels below this top level, so $N > Z$ for a heavy stable nucleus.

As in the case of atoms, the nuclear energy levels are referred to as **shells**. Filled shells produce a stable, tightly bound system. For example, noble gases are atoms with a particular number of electrons (2, 10, 18, 54 …) that allows the atomic electron shells to be completely filled (Section 42-8). Recall, for example, that helium is a noble gas and its $1s$ subshell is completely filled by its two electrons. Also recall that lithium has just one more electron, so its electron configuration is $1s^2 2s^1$; lithium's valence electron is outside of a completely filled shell and is weakly bound. As a result, lithium has a lower ionization energy than helium. Helium is more stable than lithium.

Similarly, it is easier to remove a nucleon that is outside a filled nuclear shell than it is to remove one from a completely filled shell. In nuclear physics, the particular number of protons or neutrons that completely fills a nuclear shells is known as a **magic number**. The magic numbers are 2, 8, 20, 28, 50, 82, 126…. so a nucleus that has either an atomic number Z or a neutron number N equal to a magic number will have a filled shell and be tightly bound and stable. For example, ^{18}O ($Z = 8$) and ^{92}Mo ($N = 50$) are very stable. For some nuclei, both Z and N are magic numbers; these are referred to as **doubly magic** nuclei. For example, ^{4}He ($Z = N = 2$), ^{40}Ca ($Z = N = 20$) and ^{208}Pb ($Z = 82$, $N = 126$) are doubly magic. In fact, the helium nucleus particle is so tightly bound that we often model it as a single particle called an *alpha particle*. Further, it is impossible to form a stable nucleus by adding another neutron or proton to an alpha particle.

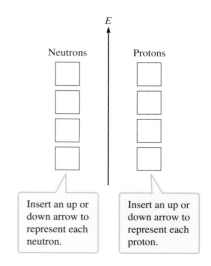

FIGURE 43.4 Use this diagram to find a nucleus's occupied energy levels. Up to two nucleons can be in each box. A proton and a neutron in the same nucleus can have the same set of quantum numbers because they are in different potential energy wells.

Other Models

Before the shell model, Niels Bohr and the American physicist John Wheeler (1911–2008) came up with the **liquid drop model** for the nucleus. According to this model, nucleons move around the nucleus much like molecules moving in a drop of liquid, and there are frequent collisions between the nucleons. The liquid drop model does a good job of accounting for **fission**, the splitting of a nucleus into smaller nuclei. Fission can be spontaneous, but it can also be triggered by the absorption of a particle such as a proton or neutron into the original nucleus (Fig. 43.5). According to the liquid drop model, when such a particle is absorbed, it quickly shares its energy with the nucleons that were already present in the nucleus. The nuclear shape is altered from nearly spherical to a two-lobed form. The nucleus then divides at the waist, forming two daughter nuclei and releasing energy (and a few nucleons).

Today's nuclear models are a combination of the shell model and the liquid drop model. In such a combined model, the *core* consists of filled shells that contain a magic number of nucleons. The core is modeled as a liquid drop. The outer nucleons fit the shell model with its quantized states. They also interact with the core nucleons, causing the core to vibrate, rotate, and distort.

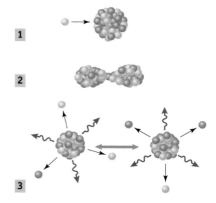

FIGURE 43.5 **1.** A particle, such as a neutron, is absorbed by a nucleus. **2.** The neutron's energy is shared by the other nucleons; the nucleus's shape is distorted as a result. **3.** The nucleus breaks into pieces; some nucleons are emitted and energy is released.

EXAMPLE 43.2 Carbon and Nitrogen

The most abundant isotope of carbon is Carbon-12, which has six protons and six neutrons, and the most abundant isotope of nitrogen is Nitrogen-14. However, for this problem consider Nitrogen-12, which has seven protons and five neutrons. Fill in energy-level diagrams (Fig. 43.4) for both nuclei. In the **CHECK and THINK** step, answer the question: Which nucleus has more energy?

INTERPRET and ANTICIPATE
Start at the lowest energy level and fill in protons and neutrons, while following Pauli's exclusion principle.

Example continues on page 1420 ▶

SOLVE

Start with carbon. The six neutrons fill three boxes (spin up and spin down in each box), and likewise for the six protons (Fig. 43.6).

Follow the same procedure for nitrogen (Fig. 43.7).

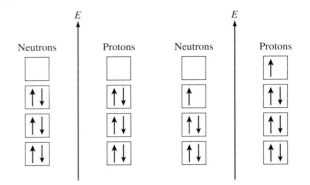

FIGURE 43.6 Nuclear energy levels in Carbon-12.

FIGURE 43.7 Nuclear energy levels in Nitrogen-12.

CHECK and THINK

Because the last proton in nitrogen is in the fourth box, nitrogen has more energy than carbon, whose highest energy nucleons are only in the third box. However, this isotope of nitrogen is unstable, so one of its protons will transform into a neutron through a process known as *inverse beta decay* (Section 43-5).

43-4 Radioactive Decay

In 1903, Antoine-Henri Becquerel (1852–1908) shared the Nobel Prize in physics with Pierre (1859–1906) and Marie Curie (1867–1934). Becquerel discovered spontaneous radioactivity, and the Curies investigated the resulting radiation. Becquerel had learned of Wilhelm Röntgen's discovery of X-rays and was investigating whether there is a connection between visible light and invisible radiation when he serendipitously discovered that uranium gives off penetrating radiation. Marie Curie named this phenomenon **radioactivity**. Ernest Rutherford then showed that radioactive substances emit more than one kind of radiation. Some radiation more readily penetrates a substance such as aluminum foil than other radiation. Rutherford named the less penetrating radiation **alpha rays** and the more penetrating radiation **beta rays**. In 1900, the French chemist Paul Villard (1860–1934) discovered an even more penetrating form of radiation known as **gamma rays**. When these three types of rays pass through a magnetic field, only the alpha and beta rays are deflected. From the direction of the deflection, we know that alpha rays are positively charged and beta rays are negatively charged. Gamma rays are not deflected, so they must be neutral.

ALPHA AND BETA RAYS
Special Case

The term *ray* is somewhat misleading. Today we know that X-rays and gamma rays are both photons (electromagnetic radiation), and that alpha rays and beta rays are also particles. **Alpha particles** are helium nuclei consisting of two protons and two neutrons. **Beta particles** are electrons. We continue to use the somewhat archaic term *ray*, but remember we mean a beam of particles. In this section we explain the origins of gamma rays and alpha particles. In Section 43.5, we'll take up beta particles.

Gamma Rays

GAMMA RAYS Special Case

The production of gamma rays is much like that of visible light or of X-rays emitted by an atom. An excited atom emits a photon when it transitions to a lower energy level. The energy of the photon equals the difference between the two energy levels. For example, the Balmer series of lines are due to photons in the visible part of the spectrum (Section 42-3). (And, as we saw in Section 42-8, atoms may generate X-rays.)

The same is true for the origin of nuclear gamma rays, except that a nucleus, instead of an atom, transitions to a lower energy level. The difference in nuclear energy levels is on the order of hundreds of keV to a few MeV. This is about 10^5 times the energy levels in the atom, so the photons released when a nucleon drops to a lower energy level are about 10^5 times more energetic. As a result, gamma rays are *typically* emitted.

Alpha Particles

Nucleons within a nucleus occasionally form themselves into a collection of two protons and two neutrons. This is an **alpha particle** or, equivalently, a helium nucleus. We already know that a helium nucleus is very stable; now we'll consider it as a particle on its own. We can model a nuclear system as having two components: an alpha particle and the rest of the nucleus. The alpha particle is positively charged, so it experiences Coulomb repulsion due to the other protons and strong force attraction due to the other nucleons. So the alpha particle is in a potential well much like the one shown in Figure 43.2B. The system's energy level is well below the peak of the potential energy curve (Fig. 43.8), so the alpha particle can only leave the nucleus by tunneling out of it. This is the reverse of the process that brings two protons together in the Sun, as discussed in the case study in Chapter 41 (Section 41-7). The probability of the alpha particle tunneling out of the nucleus depends on the height and width of the barrier.

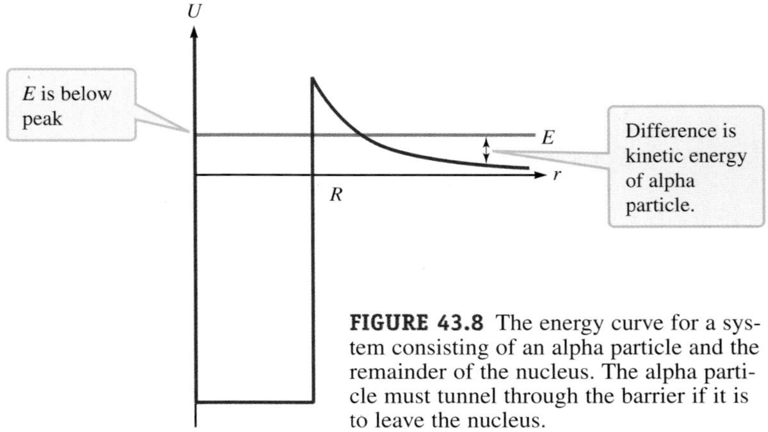

ALPHA PARTICLES ⊙ **Special Case**

FIGURE 43.8 The energy curve for a system consisting of an alpha particle and the remainder of the nucleus. The alpha particle must tunnel through the barrier if it is to leave the nucleus.

The process of emitting an alpha particle is called **alpha decay**. The original nucleus is called the *p*arent nucleus and is symbolized by $^A_Z P$, where the superscript and subscript stand for the mass number A and the atomic number Z, respectively. When the parent nucleus undergoes alpha decay, it changes into a new nucleus called the *d*aughter nucleus, symbolized by D. We write alpha decay as:

$$^A_Z P \rightarrow ^{A-4}_{Z-2} D + ^4_2 He + \text{energy} \qquad (43.3)$$

The energy released in this reaction is (mostly) in the form of kinetic energy of the escaped alpha particle (Fig. 43.8).

P stands for *p*arent and D for *d*aughter.

Decay Rate

When a parent nucleus decays by either giving off an alpha particle (this section) or beta particle (Section 43-5), it doesn't vanish; instead, it becomes a daughter nucleus. So when there is a radioactive substance made from a number of such parent nuclei, the substance itself doesn't vanish; instead, over time it turns into a sample of the daughter substance. According to quantum mechanics, the decay process is probabilistic. We cannot know precisely when a particular nucleus will decay. Instead, we can say that each nucleus has a probability $\eta \, dt$ of decaying in time dt, where the lowercase Greek letter eta (η) is the **decay constant** with dimensions $[T^{-1}]$. In a sample of N nuclei, the average number decaying in time dt is $dN = N\eta \, dt$. Thus, the number of parent nuclei in the sample changes at a rate of

$$\frac{dN}{dt} = -\eta N \qquad (43.4)$$

NUCLEAR DECAY ⊙ **Major Concept**

where the negative sign indicates that the number of parent nuclei is decreasing. The absolute value of the decay rate is called the **activity** a, so $a = -dN/dt$. In Problem 26 you will integrate Equation 43.4 to show that the number of parent nuclei decreases exponentially:

$$N = N_0 e^{-\eta t} \qquad (43.5)$$

where N_0 is the number of parent nuclei at time $t = 0$.

The activity as a function of time comes from taking the derivative of Equation 43.5:

$$a = a_0 e^{-\eta t} \qquad (43.6)$$

The letter N stands both for the neutron number and for the number of parent nuclei. You must discern the meaning from the context.

where $a_0 = \eta N_0$ is the activity at $t = 0$ (Problem 27). The SI unit of activity is the **becquerel** (Bq); however, activity is often reported in **curies** (Ci), where

$$1\,\text{Ci} = 3.7 \times 10^{10}\,\text{Bq} = 3.7 \times 10^{10}\,\text{decays/s}$$

It is often convenient to replace the decay constant η with either the half-life $T_{1/2}$ or the time constant τ. The **half-life** is the time it takes the number of nuclei in the sample to decrease to half that number, so at $t = T_{1/2}$

$$\frac{N}{N_0} = \frac{1}{2} = e^{-\eta T_{1/2}}$$

In Problem 28, you will show that the half-life is given by

$$T_{1/2} = \frac{\ln 2}{\eta} \approx \frac{0.693}{\eta} \tag{43.7}$$

The half-life of a small sample of isotopes is given in Table 43.1. The **time constant** is defined as $\tau \equiv 1/\eta$, so at $t = \tau$ the number of parent nuclei in the sample is given by:

$$N = N_0 e^{-\eta/\eta} = \frac{N_0}{e} \approx 0.368 N_0$$

This means that when $t = \tau$, the number of nuclei remaining is $0.368 N_0$. Each radioactive isotope is characterized by its own decay constant, or equivalently, by its half-life or by its time constant. The number of parent nuclei in the same isotope at time t may be written in terms of its half-life $T_{1/2}$ (Problem 29) or time constant τ as

$$N = N_0 \left(\frac{1}{2}\right)^{t/T_{1/2}} = N_0 e^{-t/\tau} \tag{43.8}$$

TABLE 43.1 Half-life for selected nuclei

Isotope	Half-life	A	Z	Decay mode
^{12}B	0.0202 s	12	5	Beta decay
^{12}N	11.00 ms	12	7	Inverse beta decay and alpha decay
^{14}C	5730 yr	14	6	Beta decay
^{16}N	7.13 s	16	7	Beta decay
^{40}K	1.25×10^9 yr	40	19	Beta decay
^{57}Co	271 d	57	27	Electron capture
^{131}I	8.04 d	131	53	Beta decay
^{222}Rn	3.82 d	222	86	Alpha decay
^{226}Ra	1600 yr	226	88	Alpha decay
^{238}U	4.46×10^9 yr	238	92	Alpha decay

Radioactive Dating

When a sample contains parent nuclei that decay with some known half-life, it is possible to measure the relative amount of parent nuclei to find the sample's age. Different isotopes are used, depending on the approximate age of the sample. For example, potassium is found in rocks and soil, and the isotope ^{40}K has a half-life of 1.25×10^9 yr. It decays to a stable isotope of argon, ^{40}Ar. By measuring the ratio of ^{40}K/^{40}Ar, it is possible to estimate the ages of rock and soil samples that formed between roughly 10^5 and 4.5×10^9 (age of the Earth) years ago.

Carbon dating is another example of radioactive dating. A naturally occurring stable isotope of carbon is ^{12}C. Radioactive ^{14}C with a half-life of 5730 ± 40 yr is produced in the Earth's atmosphere when cosmic rays collide with nitrogen. A constant ratio of ^{14}C/^{12}C is maintained in the atmosphere due to the balance between radioactive decay and the production of new ^{14}C. Both isotopes of carbon have the same chemical behavior, so both form the carbon dioxide molecule CO_2. All living organisms absorb CO_2, so all living organisms maintain ^{14}C and ^{12}C in the constant ratio of $^{14}C/^{12}C = 1.3 \times 10^{-12}$, found in the atmosphere. When these organisms die, they stop absorbing CO_2, and any absorbed ^{14}C decays without being resupplied. The ratio $^{14}C/^{12}C$ decreases over time in the organism's remains, owing to the decay of ^{14}C. Carbon dating is used to measure ages between about 500 and 50,000 yr.

CONCEPT EXERCISE 43.3

A group of college students discusses a homework problem.

Shannon: It says that iodine-131 has a half-life of 8 days. We have to find what fraction of the sample is still iodine-131 after 16 days.

Cameron: That's one I can answer. None. In the first 8 days half the sample decays, and in the second 8 days the other half decays.

Shannon: No. The answer depends on how much iodine-131 was in the sample to begin with. You assumed the whole sample was iodine-131 at the beginning of the 16 days. I don't think they gave us enough information to do the problem numerically. I think it is a conceptual problem, and we are just meant to explain that not enough information is given.

Avi: I don't think this is a conceptual problem that we do in our heads. I think we can come up with a numerical answer. I agree that the answer isn't zero because every eight days the amount of iodine-131 is halved; but that doesn't mean it is all gone in 16 days.

Shannon: But that means I'm right about needing to know how much iodine-131 there was at the beginning of the 16 days.

With whom do you agree, and why?

EXAMPLE 43.3 An Old Rock

Another isotope used in radioactive dating is ^{238}U, which decays into ^{206}Pb and has a half-life of 4.46×10^9 yr. Suppose a rock contains a ratio of lead-206 to uranium-238 nuclei of $N_{Pb}/N_U = 0.65$. How old is the rock? (Assume the rock was originally all uranium-238.)

:• **INTERPRET and ANTICIPATE**
Because a substantial amount of the uranium has decayed into lead, we expect that the age of the rock will be close to the uranium half-life of several billion years.

:• **SOLVE**

At $t = 0$, the sample contained N_0 nuclei of uranium. Today's number N_U is given by Equation 43.8.	$N_U = N_0 \left(\dfrac{1}{2}\right)^{t/T_{1/2}}$	(43.8)
The number of lead nuclei N_{Pb} in the rock today is the number of original uranium nuclei minus the number of uranium nuclei today.	$N_{Pb} = N_0 - N_U$	(1)
Substitute Equation 43.8 into Equation (1).	$N_{Pb} = N_0 - N_0 \left(\dfrac{1}{2}\right)^{t/T_{1/2}}$ $N_{Pb} = N_0 \left[1 - \left(\dfrac{1}{2}\right)^{t/T_{1/2}}\right]$	(2)

Example continues on page 1424 ▶

Use Equations 43.8 and (2) to write a ratio for N_{Pb}/N_U.	$\dfrac{N_{Pb}}{N_U} = \dfrac{N_0[1 - (\frac{1}{2})^{t/T_{1/2}}]}{N_0(\frac{1}{2})^{t/T_{1/2}}}$ $\dfrac{N_{Pb}}{N_U} = \left(\dfrac{1}{2}\right)^{-t/T_{1/2}} - 1$
Substitute values and solve for t.	$0.65 = \left(\dfrac{1}{2}\right)^{-t/T_{1/2}} - 1$ $\left(\dfrac{1}{2}\right)^{-t/T_{1/2}} = 1.65$ $t = \dfrac{T_{1/2} \ln 1.65}{\ln 2} = \dfrac{(4.46 \times 10^9 \text{ yr}) \ln 1.65}{\ln 2}$ $t = 3.2 \times 10^9 \text{ yr}$

CHECK and THINK
As expected, our answer is several billion years. When working such a problem, you shouldn't expect to find rocks that are older than the age of the Earth (about 4.5 billion years).

43-5 The Weak Force

Beta decay is another way for an unstable isotope to decay. In **beta decay**, a neutron becomes a proton, and in the process an electron is emitted. The electron is called a **beta particle** in this context. Consider ^{12}B, which has seven neutrons and five protons (Fig. 43.9). If ^{12}B undergoes beta decay, its neutron number is reduced by 1, its atomic number increases by 1, and its mass number remains unchanged. Because the result has a new atomic number, it is a different element: in this case ^{12}C. Carbon-12 is more tightly bound than boron-12, so it is stable. To explain beta decay, we must discuss an elementary particle known as a **neutrino**, and the fourth fundamental force: the **weak nuclear force**.

Neutrinos and the Weak Force

Rutherford named beta rays in the early 1900s, so in a sense beta decay was discovered then. However, later studies found problems with the theory of beta decay. It seemed in particular that energy was not conserved by beta decay; instead, it appeared that energy was lost. The supporting data was so convincing that Bohr proposed that perhaps energy was really not conserved in this situation, violating one of the most fundamental principles of physics.

Pauli and Fermi responded with a solution to the problem, proposing the existence of another particle emitted during beta decay that carries the missing energy. Fermi called this particle a *neutrino*. He reasoned that the particle must be neutral, or the reaction would have a charge imbalance, and that the particle must have low mass, or it would have been easy to discover. So he gave it a name that means "little neutral one" in Italian. The neutrino also has an antiparticle, known as an **antineutrino**. The antineutrino was discovered in the mid-1950s by a team of American physicists led by Frederick Reines (1918–1998) and Clyde Cowan (1919–1974), both shown in Figure 43.10. (Reines shared the Nobel Prize with Martin Perl (1927–2014) for this work in 1995; the Nobel Prize is not awarded posthumously, so Cowan was not eligible. Perl was awarded the prize for his discovery of the tau lepton.) The symbol

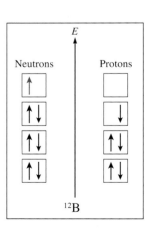
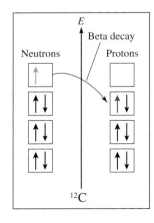

FIGURE 43.9 In the beta decay of ^{12}B, one of boron's neutrons becomes a proton. The result is carbon. Both nuclei have the same mass number A.

for the neutrino is the lowercase Greek letter nu (ν), whereas the antineutrino is designated by a bar over the symbol ($\bar{\nu}$). So we write beta decay as

$$n \rightarrow p + e^- + \bar{\nu} \quad (43.9)$$

The number of particles and antiparticles is balanced. There is one particle on the left. On the right there are two particles and one antiparticle; one particle is balanced by one antiparticle. The net result is a matter particle on each side of the reaction.

The beta decay reaction (Eq. 43.9) says that a neutron is transmuted into a proton. This almost sounds like alchemy, a pseudoscience based on the medieval belief that cheap metal could be transmuted into gold. But beta decay has been observed, and alchemy has not. So we need an explanation for how neutrons can be transformed into protons. The answer involves another fundamental force of nature—the **weak force** (or the **weak nuclear force**), which is short-ranged (about 10^{-3} fm) and is exerted on quarks and neutrinos. It transforms a neutron into a proton at the level of quarks, so in particular it turns a down quark into an up quark. (It can also turn an up quark into a down quark, which transforms a proton into a neutron.) The weak force is so named because it is about 10^7 times weaker than the strong force. Any process that converts one type of quark into another involves the weak force.

The neutrino is a neutral particle, so the electromagnetic force does not act on it. In addition, because it is not made of quarks, the strong force does not act on neutrinos. Neutrinos also have a very low mass, so they are only weakly affected by gravity. The only force that really affects neutrinos is the weak force. For this reason, neutrinos are extremely difficult to detect: They hardly interact with matter. Their mean free path through water is thousands of light-years. So although they are the most numerous particles in the Universe (see the **CASE STUDY** at the end of this section), billions of them pass through your body every second without leaving a trace.

WEAK NUCLEAR FORCE
❶ Underlying Principle

FIGURE 43.10 Frederick Reines and Clyde Cowan worked on the Poltergeist project—an experiment to confirm the existence of the neutrino. Here Reines (left) and Cowan (in hat) demonstrate their methods in searching for the existence of the neutrino, using one of their colleagues.

Beta Decay, Inverse Beta Decay, and Electron Capture

Now let's examine three reactions that involve the weak force and neutrinos. In beta decay a neutron is transmuted into a proton, so the beta decay reaction for a parent nucleus with N neutrons and Z protons is:

$$^A_Z P \rightarrow\ ^A_{Z+1} D + e^- + \bar{\nu} \quad (43.10)$$

The daughter nucleus has one more proton and one fewer neutron than its parent, but both nuclei have the same mass number A. Beta decay from a given type of isotope always yields the same amount of energy, but that energy can be split in an infinite number of ways between the emitted electron and antineutrino.

The weak force can also transmute a proton into a neutron, as given by:

$$p \rightarrow n + e^+ + \nu \text{ (inside nucleus)} \quad (43.11)$$

Such a reaction is called **inverse beta decay**. Free protons do not decay, but a proton inside a nucleus can decay. For a parent nucleus with N neutrons and Z protons, the inverse beta decay reaction is:

$$^A_Z P \rightarrow\ ^A_{Z-1} D + e^+ + \nu \quad (43.12)$$

The daughter nucleus has one fewer protons, one additional neutron, and the same mass number as its parent. In the case of inverse beta decay, a **positron** e^+ and a neutrino ν are released. A positron is the **antiparticle** of an electron; it's like the electron except that it has a positive charge $+e$. In Equation 43.12 there is one particle on the left. On the right there are two particles and one antiparticle; one particle is balanced by one antiparticle. So the net result is that there is one matter particle on each side of the reaction.

Nuclei that are rich in neutrons are likely to undergo beta decay, and atoms that are rich in protons are likely to undergo inverse beta decay. So beta decay and inverse

The beta decay reaction is also referred to as β^- decay because a negative beta ray (electron) is released.

The inverse beta decay reaction is also referred to as β^+ decay because a positive beta particle (positron) is released.

ELECTRON CAPTURE ▶ Special Case

Electron capture is also called *K* capture because the innermost electrons are in the *K* shell.

beta decay act to stabilize nuclei and to even out the numbers of protons and neutrons. The half-lives of beta decay and inverse beta decay for nuclei vary greatly, ranging from a few minutes to tens of millions of years. A free neutron (outside the nucleus) will undergo beta decay in a short time ($T_{1/2} \approx 10.2$ min). However, free protons do not spontaneously undergo inverse beta decay into neutrons.

Finally, inner electrons have a nonzero probability of being in the nucleus (Section 42-5). So it is possible that one of these electrons can be *captured* by a proton in the process called **electron capture**. This process is more likely to happen in high Z atoms because their inner electron shells are more tightly bound to the nucleus. The reaction for electron capture is:

$$^{A}_{Z}\text{P} + e^- \rightarrow\, ^{A}_{Z-1}\text{D} + \nu \tag{43.13}$$

The daughter nucleus has one more neutron and one fewer proton than its parent. After the inner electron is captured, an outer electron will fall into the resulting gap, releasing an X-ray photon. In principle, we can observe electron capture by observing the released neutrino in Equation 43.13. However, because of the difficulty observing neutrinos, we infer that electron capture has taken place when we observe the released X-ray photon.

Uranium: A Naturally Occurring Radioactive Element

The most abundant naturally occurring radioactive element on the Earth is uranium, $^{238}_{92}\text{U}$. Uranium decays in a series of reactions into lead, $^{206}_{82}\text{Pb}$. Figure 43.11 illustrates this series of reactions on a portion of a **Segrè chart**. The horizontal axis shows the neutron number and the vertical axis the atomic number. Alpha decay decreases both the neutron number and the atomic number by two, so on the chart it is represented by a diagonal line: two boxes down and two boxes to the left. For example, $^{238}_{92}\text{U}$ alpha decays into $^{234}_{90}\text{Th}$ according to

$$^{238}_{92}\text{U} \rightarrow\, ^{234}_{90}\text{Th} + \,^{4}_{2}\text{He}$$

as represented by the blue arrow between $^{238}_{92}\text{U}$ and $^{234}_{90}\text{Th}$. Beta decay lowers the neutron number by one and increases the atomic number by one, so it is represented on the chart by a diagonal line: one box to the left and one box up. For example, $^{234}_{90}\text{Th}$ beta decays into $^{234}_{91}\text{Pa}$ according to:

$$^{234}_{90}\text{Th} \rightarrow\, ^{234}_{91}\text{Pa}^* + e^- + \overline{\nu}$$

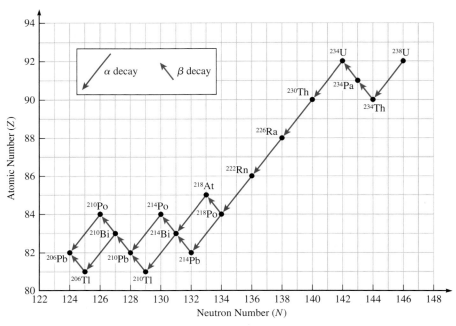

FIGURE 43.11 Uranium-238 is unstable and decays into lead-206, which is stable. Radon is part of this series of reactions. Radon gas is found in some homes, and is a hazard to human health.

as represented by the red arrow between $^{234}_{90}$Th and $^{234}_{91}$Pa. After this beta decay, $^{234}_{91}$Pa* is an in excited state, indicated by the asterisk * next to its abbreviation. By emitting a gamma ray, $^{234}_{91}$Pa* drops down to its ground state:

$$^{234}_{91}\text{Pa}^* \rightarrow {}^{234}_{91}\text{Pa} + \gamma$$

The energy released due to the decay of uranium in the Earth heats up the Earth's interior. This process is partially responsible for plate tectonics, which in turn causes volcanic activity and earthquakes.

CONCEPT EXERCISE 43.4

Positron emission tomography (PET) is a technique used to see inside a living body. When a positron encounters an electron, the two annihilate and give off two gamma photons that are used to create the image. Fluorine-18 is used as the source of positrons. The fluorine decays to oxygen-18. Which type of decay process (beta, inverse-beta, or electron capture) produces the necessary positrons? Write down the equation for decay of fluorine-18 to oxygen-18.

CONCEPT EXERCISE 43.5

If ^{12}N (Fig. 43.7) undergoes inverse beta decay, what element is the daughter nucleus?

CASE STUDY   Part 3: Making Hydrogen and Helium

Quarks in the early Universe were bound together to form protons and neutrons. Other high-energy particles were present too, such as electrons, positrons, neutrinos, and antineutrinos. These particles were kept in thermal equilibrium by the weak force through a few reactions:

$$p + e^- \rightleftarrows n + \nu$$
$$p + \bar{\nu} \rightleftarrows n + e^+$$
(43.14)

where the double arrow $\rightleftarrows$ means that the reactions happened in both directions. So protons were converted into neutrons, and neutrons into protons. When the Universe was just tens of seconds old, the positrons were annihilated by reactions with electrons. The number of electrons was greatly reduced, but not to zero. Likewise, the antineutrinos were annihilated by reactions with neutrinos; the number of neutrinos decreased, but not to zero. The remaining neutrinos still exist today and are believed to be the most numerous particles in the Universe. Without a great number of electrons, positrons, neutrinos, and antineutrinos, the reactions in Equation 43.14 could not maintain equilibrium. Instead, free neutrons began to undergo beta decay into protons (Eq. 43.9), with no compensating reaction to restore protons into neutrons.

The half-life of a free neutron via beta decay is about 10 minutes. However, neutrons in a nucleus have much longer half-lives. So many of the neutrons survived by combining with a proton, forming an isotope of hydrogen known as deuterium, $^2_1$H. In the first few minutes of the early Universe, much of the deuterium that formed was used up in making helium nuclei, $^4_2$He. A few other low-mass elements formed in trace amounts, but only hydrogen and helium nuclei were produced in great abundance as the Universe cooled. Once the temperatures were low enough, electrons began to bind to these nuclei to form atoms. Because no other nuclei were available, only hydrogen and helium atoms were present. So where did the iron in your blood and the iodine in your thyroid come from? We will explain that in Section 43-8.

> **EXAMPLE 43.4** **CASE STUDY** Evidence in Favor of the Big Bang Model
>
> One convincing way to confirm the Big Bang model is by calculating the expected amount of primordial helium as a ratio $N_{He}/(N_H + N_{He})$. Helium began to form after sufficient time had passed for many neutrons to undergo beta decay. Roughly speaking, the theoretically calculated ratio of neutrons to protons was $N_n/N_p \approx 1/7$ when helium began to form. Assume that all the available neutrons went into helium and the remaining protons went into hydrogen, and estimate the ratio $N_{He}/(N_H + N_{He})$. Express your answer as a percentage.
>
> **INTERPRET and ANTICIPATE**
> A single-helium nucleus contains two protons and two neutrons. If we assume that all the available neutrons go into helium, then an equal number of protons also go into helium. The remainder becomes hydrogen. So if the number of neutrons equaled the number of protons, there would be no hydrogen at all.
>
> **SOLVE**
> Every helium nucleus contains 2 neutrons, so the number of helium nuclei that can form is half the number of neutrons. Since one proton must go into helium for every neutron that does so, the number of hydrogen nuclei is the number of protons minus the number of neutrons.
>
> $$\frac{N_{He}}{N_{He} + N_H} = \frac{\frac{1}{2}N_n}{\frac{1}{2}N_n + (N_p - N_n)}$$
>
> $$\frac{N_{He}}{N_{He} + N_H} = \frac{\frac{1}{2}N_n}{N_p - \frac{1}{2}N_n}$$
>
> Divide the numerator and denominator by N_p.
>
> $$\frac{N_{He}}{N_{He} + N_H} = \frac{\frac{1}{2}(N_n/N_p)}{1 - \frac{1}{2}(N_n/N_p)}$$
>
> Substitute $\frac{N_n}{N_p} \approx \frac{1}{7}$.
>
> $$\frac{N_{He}}{N_{He} + N_H} = \frac{\frac{1}{2}(\frac{1}{7})}{1 - \frac{1}{2}(\frac{1}{7})} = \frac{1}{13}$$
>
> $$\frac{N_{He}}{N_{He} + N_H} \approx 8\%$$
>
> **CHECK and THINK**
> This result is very close to the observation that about 90% of the ordinary matter in the Universe is primordial hydrogen and about 10% is primordial helium. More careful calculations have been done that include other elements created in the early Universe, and careful observations have confirmed these calculations. Although there have been great changes to our understanding of the evolution of the Universe, these fundamental ideas about how elements formed, taken from the original Big Bang theory, have changed very little and are well supported by observations.

43-6 Binding Energy

FUSION AND FISSION
★ Major Concept

Nuclear reactors on the Earth supply us with some of the electrical power we need to run useful devices, such as our refrigerators, TVs, and computers. The energy produced by these reactors comes from nuclear *fission* reactions that break up large nuclei into smaller ones. The Sun is the ultimate source of energy in the solar system, and its core is essentially a nuclear reactor. However, the energy generated by the Sun's core is the result of *fusion* reactions that combine small nuclei into larger ones. In the following two sections, we take a closer look at fission and fusion reactions. In this section, we'll explain why both fission reactions and fusion reactions can release energy. First, we need to consider the energy stored in a nucleus.

Mass Deficit and Binding Energy

MASS DEFICIT AND BINDING ENERGY
★ Major Concept

Imagine assembling a bound system, such as a nucleus, from free particles such as nucleons. If you could do this, you would find that the mass of the bound system is *less* than the mass of the individual free particles. The resulting system has a **mass deficit** Δm, given by:

$$\Delta m = \sum m_i - M \qquad (43.15)$$

where $\sum m_i$ is the sum of the free particles' individual (rest) masses and M is the (rest) mass of the bound system. According to Einstein's theory of relativity, mass is a form of energy, as given by $E = mc^2$ (Eq. 39.44). The mass deficit that results when free particles are combined into a bound system is a form of energy known as the **binding energy** E_B. Mathematically, the binding energy is given by

$$E_B = \Delta mc^2 \qquad (43.16)$$

Some authors define binding energy with a negative sign: $E_B = -\Delta mc^2$, but in this book there is no negative sign in Equation 43.16.

Energy is required to break apart the bound system of the nucleus. The binding energy is the amount of energy, supplied by an outside source, required to pull the particles out of the bound system and separate them into free particles at rest. If more energy than the binding energy is supplied, the separated free particles would have some nonzero kinetic energy.

In some *fusion* reactions, the binding energy may be released. The released energy may go into some combination of particles, photons, and their kinetic energy. To understand this process, let's use an analogy. Suppose an asteroid is initially at rest and far (but not infinity far) from the Earth and then falls to the Earth. The asteroid–Earth system loses gravitational potential energy, and gains kinetic energy. Figure 43.12 shows the familiar energy bar chart for such a process. Let's think about this familiar process in a slightly different way. We can say that, initially, the system had energy stored as gravitational potential energy. Think of the asteroid's fall to Earth in terms of it "fusing" with the Earth to form a new complex body. (Keep in mind that this is an analogy. The asteroid was initially bound to the Earth.) Some of the stored energy is released in the process of "fusing" the asteroid and the Earth; this process is represented by the kinetic energy bar in Figure 43.12. Often we think of this released energy as going into increasing the asteroid's kinetic energy, but we can improve our analogy somewhat if think about other places where this released energy may go. For example, the released energy goes into increasing the atmosphere's thermal energy, light is emitted, and particles of debris fly into the air.

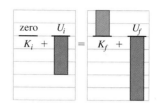

FIGURE 43.12 Energy bar chart for an asteroid-Earth system.

Our language here is somewhat loose. Remember: energy describes the motion, configuration and mass of a system. Energy is not a substance.

A similar process happens during the nuclear fusion of small nuclei. When such small reactants (free particles) are fused to form a larger product, the nucleons are bound together by the strong force, and an amount of energy equal to the binding energy must be added to separate the nucleus back into its original reactants. Such fusion reactions release energy. Consider two free particles having mass m_1 and m_2. Each particle stores energy in the form of mass, and according to Einstein's famous equation, these stored energies are $m_1 c^2$ and $m_2 c^2$. So the total energy stored in their mass is given by $(m_1 + m_2)c^2$. Figure 43.13A illustrates these ideas about the energy stored in their mass. Now suppose the two particles fuse together. In some fusion reactions, the reactants (the free particles) store more energy in their individual masses than the product stores in its mass M. This difference in stored mass energy is the binding energy (Fig. 43.13B), and the binding energy is released in the fusion reaction. Analogous to the energy released when an asteroid falls to Earth, the energy released by a fusion reaction can go into kinetic energy, but it can also go into new particles and photons.

But the analogy with gravity only goes so far. Gravity is an attractive force, and gravitational "fusion" always increases a system's binding energy. You would have to put energy into the bound asteroid-Earth system to separate them; separating the two is analogous to fission. So for gravity, "fusion" always releases stored energy and "fission" always requires a supply of energy. However, in the nucleus there are two competing forces—the attractive strong nuclear force and the repulsive Coulomb force. So some nuclear *fusion* reactions release energy and others require a supply of energy. Similarly, some nuclear *fission* reactions release energy and other require it.

The repulsive force tends to be more important in larger nuclei. If a large nucleus undergoes a *fission* reaction that breaks it into two smaller nuclei, the products may have a greater binding energy than the original nucleus. In this case, it would take

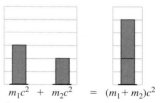

A. Before fusion

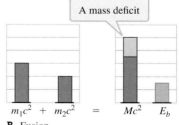

B. Fusion

FIGURE 43.13 A. The total rest mass energy of two particles. **B.** In this fusion reaction, the sum of the rest mass energy of the reactants is greater than the rest mass energy of the product: $(m_1 + m_2)c^2 > Mc^2$.

energy from an outside source to fuse these smaller nuclei back together to form a larger nucleus.

Putting this together, in either a fusion or fission reaction, energy may be released or be required. If there is a positive mass deficit such that the product has less mass than the sum of the reactant's masses, then energy is released. In other words, energy is released when the reaction's products have a greater binding energy than its reactants. The following may help you to remember: *A reaction that increases binding energy releases energy.* Similarly, energy must be supplied by an external source when the reaction's products have less binding energy (more mass) than its reactants. *A reaction that decreases binding energy requires energy.*

Convenient Units of Mass and Binding Energy

Because of the intimate connection between mass and energy, it is often convenient to express rest mass in the *pseudo-units* of MeV/c^2. It is also common to express the rest mass of a nucleon in atomic mass units. The translation between these two sets of units can be found as follows. The conversion factor between atomic mass units and kilograms comes from Equation 43.1 (1 u = $1.66053879 \times 10^{-27}$ kg), and the rest mass energy of 1 u is found by applying $E = mc^2$ (Eq. 39.44):

> We use the term *pseudo-unit* here because MeV/c^2 is a combination of unit for energy and the speed of light constant.

$$E = (1\,\text{u})c^2 = (1.66053879 \times 10^{-27}\,\text{kg})(2.99792458 \times 10^8\,\text{m/s})^2$$
$$= 1.492417837 \times 10^{-10}\,\text{J}$$

This is more conveniently expressed in MeV:

$$E_B = \frac{1.492417837 \times 10^{-10}\,\text{J}}{1.602176565 \times 10^{-13}\,\text{J/MeV}} = 931.494\,\text{MeV}$$

So 1 u of mass is equivalent to 931.494 MeV (rounded to six significant figures for convenience) of energy, and the translation can be written as:

> It is common to drop the "/c^2" and report rest mass in units of MeV.

$$1\,\text{u} = 931.494\,\text{MeV}/c^2 \qquad (43.17)$$

For convenience, the rest mass energies of the proton, the neutron, and a few nuclei are provided in Table 43.2.

The Binding Energy Curve

To compare the properties of various nuclei, it is helpful to calculate the binding energy per nucleon ε by dividing the binding energy of a nucleus by its mass

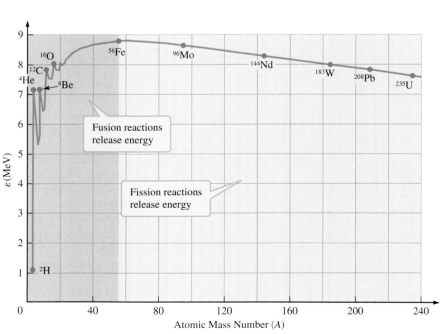

FIGURE 43.14 Binding energy per nucleon ε as a function of mass number A. The curve peaks at iron ^{56}Fe.

number: $\varepsilon \equiv E_B/A$. A graph of ε as a function of A for the known nuclei forms the **binding energy curve** (Fig. 43.14). This curve rises steeply from deuterium up to around oxygen. Then it peaks at $A \approx 56$, corresponding to iron, after which it becomes relatively flat for heavier nuclei. The flat shape of the binding energy curve is evidence that the strong force is short-ranged, only extending far enough that a nucleon is bound by the few nucleons that are relatively close to it and not by those farther away.

A nucleus with high binding energy per nucleon is tightly bound; it will take considerable energy per nucleon to separate it into free particles. Iron is the most tightly bound nucleus, but there are other local peaks in the binding energy curve, most notably ^4_2He, ^8_4Be, $^{12}_6\text{C}$, and $^{16}_8\text{O}$. These nuclei are more tightly bound than are their neighbors on the binding energy curve. We have already seen that such nuclei are particularly stable, having the same number of protons and neutrons. Each can be modeled as one or more alpha particles.

The binding energy curve explains why some fusion and fission reactions release energy and why others require energy. *When nuclear reactions result in a product that has a greater binding energy and so is more tightly bound than its reactants, energy is released.* (Recall our analogy: Energy is released when an object such as an asteroid falls to the Earth.) Otherwise, energy would be required to make the reaction possible.

Because iron is at the peak of the binding energy curve, both the fusion of nuclei that are lighter (to the left of the peak) and the fission of nuclei that are heavier (to the right of the peak) release energy. So the fission of nuclei lighter than iron into even lighter nuclei requires the input of energy. Likewise, the fusion of nuclei heavier than iron into even heaver nuclei also requires the input of energy.

Because the curve is very flat, we should more precisely say that iron is *roughly* at the peak. We won't worry about such fine distinctions in our language here.

TABLE 43.2 Rest mass of selected particles and nuclei

Nucleon or Nucleus	m (in MeV/c^2)
Proton	938.27
Neutron	939.57
Deuteron ^2_1H	1875.612859
Helium ^4_2He	3727.379
Carbon $^{12}_6\text{C}$	11177.9
Oxygen $^{16}_8\text{O}$	14899.2
Iron $^{56}_{26}\text{Fe}$	52103.06
Copper $^{63}_{29}\text{Cu}$	58603.84
Gold $^{197}_{79}\text{Au}$	183433.33
Lead $^{208}_{82}\text{Pb}$	193687.68
Uranium $^{238}_{92}\text{U}$	221696.64

In this book we ignore the mass of neutrinos, but strictly speaking they should be taken into account.

Energy Released by Reactions

The energy released in a nuclear reaction, including the decay processes we have discussed, depends on the change in mass $\Delta M = M_f - M_i$, where M_i is the total mass of the reactants before the reaction (such as nuclei, electrons, and positrons), and M_f is the total mass of the products that result from the reaction. The energy released is then,

$$E_{\text{released}} = -\Delta M c^2 \qquad (43.18)$$

This quantity is positive as long as the mass of the reactants before the reaction is greater than the mass of the products that result. If this isn't the case, then energy must be provided to make the reaction possible. So a spontaneous reaction will occur only if $\Delta M > 0$. For example, alpha decay is possible when the mass of the original parent nucleus is greater than the sum of the masses of the daughter nucleus and helium-4 (the alpha particle). When using Equation 43.18, you may need to know the mass of the electron or the positron in atomic mass units; each of them has mass $m_e = 5.48579909 \times 10^{-4}$ u.

CONCEPT EXERCISE 43.6

Why are the units MeV/c^2 referred to as *pseudo*-units? Why call Equation 43.17 a translation instead of a conversion factor? *Hint*: Why did we consider 1 kg = 2.2 lb to be a translation instead of a conversion factor (Section 5-7, page 131)?

CONCEPT EXERCISE 43.7

The ground-state energy of hydrogen is -13.6 eV. What is its binding energy? What is another name for binding energy in this case?

EXAMPLE 43.5 Binding Energy of a Deuteron

A deuteron is the nucleus of deuterium (an isotope of hydrogen, ^2_1H), consisting of one proton and one neutron. Use the information in Table 43.2 to find the binding energy of a deuteron. Use no more than six significant figures in your work.

INTERPRET and ANTICIPATE
The mass of a deuteron is less than the mass of a proton plus a neutron. The mass deficit is proportional to the binding energy.

SOLVE
First, find the sum of the masses of the proton and neutron.

$$m_p + m_n = (938.27 + 939.57) \text{ MeV}/c^2$$
$$m_p + m_n = 1877.84 \text{ MeV}/c^2$$

Next, find the mass deficit (Eq. 43.15).

$$\Delta m = \sum m_i - M \qquad (43.15)$$
$$\Delta m = (m_p + m_n) - m_{\text{deuteron}} = 1877.84 \text{ MeV}/c^2 - 1875.61 \text{ MeV}/c^2$$
$$\Delta m = 2.23 \text{ MeV}/c^2$$

Multiply by c^2 to find the binding energy. Now we see why these pseudo-units are so convenient—the c^2 cancels out.

$$E_B = \Delta mc^2 = (2.23 \text{ MeV}/c^2)c^2$$
$$E_B = 2.23 \text{ MeV}$$

CHECK and THINK
This may not seem like much energy at first, but compare it to the energy required to ionize hydrogen (13.6 eV). The binding energy of a nucleus is about 10^5 times greater than the binding energy of an atom.

EXAMPLE 43.6 What Happens to Cobalt-57?

The rest mass of $^{57}_{27}Co$ is 56.936296 u, and the rest mass of $^{57}_{26}Fe$ is 56.935399 u. Can $^{57}_{27}Co$ undergo spontaneous electron capture to become $^{57}_{26}Fe$? If so, how much energy is released? Give your answer in MeV to three significant figures. To keep this simple ignore the neutrino.

INTERPRET and ANTICIPATE
We need to know if the mass of the reactants is greater than or less than the mass of the products. It is helpful to write the electron capture reaction: $^{57}_{27}Co + e^- \rightarrow ^{57}_{26}Fe + \nu$.

SOLVE
First, find the sum of the masses of the electron and $^{57}_{27}Co$. Use $m_e = 5.48579909 \times 10^{-4}$ u for the electron's mass.

$$m_e + m_{Co} = 5.48579909 \times 10^{-4} \text{ u} + 56.936296 \text{ u}$$
$$m_e + m_{Co} = 56.936845 \text{ u}$$

Compare the mass of the reactants to that of the products.

$$m_e + m_{Co} > m_{Fe}$$

Electron capture is possible.

Find the difference in mass and multiply by c^2 to find the energy released.

$$\Delta M = 56.936845 \text{ u} - 56.935399 \text{ u} = 1.4460 \times 10^{-3} \text{ u}$$
$$E_{\text{released}} = \Delta M c^2 = 1.4460 \times 10^{-3} \text{ u} \left(\frac{931.494 \text{ MeV}/c^2}{1 \text{ u}} \right) c^2$$
$$E_{\text{released}} = 1.35 \text{ MeV}$$

:• **CHECK and THINK**
This released energy goes into the total energy of the neutrino and the kinetic energy of the iron. The neutrino travels outward without interacting further, but the iron can interact with other particles, sharing this energy.

43-7 Fission Reactions

FISSION ✪ Major Concept

Perhaps you remember the Great Sendai Earthquake on March 11, 2011. It began off the northern coast of Japan's main island. Later it caused large tsunamis, triggering the second worst nuclear accident at a nuclear power plant. The power plant—Fukushima Daiichi—melted down and released radiation because the tsunami damaged the backup generators and the loss of power meant the cooling systems could not operate. You might think that such tragedies mean that nuclear power plants are too dangerous. In fact, some nations (such as Germany) shut down many of their power plants after the meltdown at Fukushima Daiichi. However, when you consider how carbon emission causes climate change (CASE STUDY in Chapter 20), you might consider nuclear power plants a "greener" source of energy than power plants that use fossil fuels. So other nations (such as France and the United States) are keeping their existing nuclear plants running, and some nations (such as China) are building many new ones. In this section, we consider the fission reactions and some of the practical concerns about generating energy with these reactions.

Germany's carbon emission per capita is actually pretty low.

In a fission reaction, a heavy nucleus splits into smaller daughter nuclei. From the previous section, we know that energy is released if the combined mass of the daughter nuclei is less than that of the parent nuclei. In other words, there must be a positive mass deficit (an increase in binding energy) in order to release energy in the fission reaction.

Although other parent nuclei are possible, we'll consider uranium-235 to make our discussion concrete. The fission reaction is triggered by the absorption of a neutron (Fig. 43.15). Because neutrons are neutral, they don't experience a Coulomb repulsion and can easily penetrate a nuclei. The addition of the neutron means that uranium-235 becomes uranium-236:

$$^{1}_{0}n + ^{235}_{92}U \rightarrow ^{236}_{92}U^{*}$$

Although the half-life of uranium-235 is many millions of years, the half-life of uranium-236 is very short and it only lasts for about a picosecond before it splits into two smaller daughter nuclei and releases more neutrons. There are many possible daughter nuclei. For example,

$$^{236}_{92}U^{*} \rightarrow ^{141}_{56}Ba + ^{92}_{36}Kr + 3(^{1}_{0}n)$$

Or, as another example,

$$^{236}_{92}U^{*} \rightarrow ^{147}_{57}La + ^{87}_{35}Br + 2(^{1}_{0}n)$$

In these two reactions we see that 2 or 3 neutrons are released. In fact, on average 2.5 neutrons are released. It make sense that a net number of neutrons are released in a fission reaction because, in general, larger nuclei require a higher neutron-to-proton ratio for stability (Fig. 43.1), so the smaller daughter nuclei don't require as many neutrons (per proton) and these extra neutrons are released. These released neutrons can trigger more fission reactions, leading to a chain reaction (Fig. 43.15).

The daughter nuclei typically have atomic mass numbers between 70 and 160 (Fig. 43.16). So, they are about half the size of the uranium nucleus. The most probable values are 96 and 135. The binding

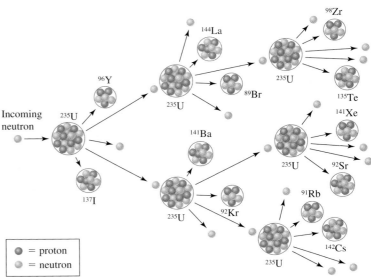

FIGURE 43.15 A single neutron triggers a chain of fission reactions. Each fission reaction releases more neutrons that trigger more fission reactions.

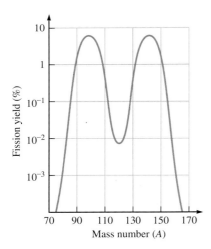

FIGURE 43.16 When uranium-235 undergoes neutron-induced fission, many possible daughter nuclei are produced. This figure roughly shows the distribution of these daughter nuclei as a function of mass number A.

energy per nucleon for uranium-235 is $\varepsilon_U \approx 7.6$ MeV. The daughter nuclei are more tightly bound, with a binding energy per nucleon that is nearly 1 MeV greater; so about 1 MeV per nucleon is released by each fission reaction. Uranium-235 has 235 nucleons, so each fission reaction releases nearly 235 MeV. A better estimate is about 200 MeV released per fission reaction. This released energy goes into producing gamma rays and into the kinetic energy of the reactants.

Furthermore, the daughter nuclei are generally not stable because they have a high neutron-to-proton ratio. So either they undergo beta decay (a neutron is transformed into a proton), or a neutron is emitted. This beta decay releases energy but not as much as the fission reaction itself. So, on the one hand, these daughter nuclei can contribute to energy production, but on the other hand they can slow the chain reaction by capturing neutrons, which would otherwise be contributing to it.

Nuclear Power Plants

Nuclear fission reactions have been put to practical use in generating energy that is used to power ships, submarines, spacecraft, cities and towns. Let's take a look at some of the practical aspects of a nuclear power plant used to produce electricity. The power plant must be able to sustain and control fission chain reactions such as those shown in Figure 43.15. The first challenge is finding the right fuel source. For example, the natural abundance of uranium-235 is 0.7%, and the rest (99.3%) is uranium-238. Uranium-238 tends to absorb neutrons without a subsequent fission reaction. So naturally occurring uranium is difficult to use as a fuel source in a power plant. One solution to this difficulty is to *enrich* the uranium fuel, such that uranium-235 achieves an artificial abundance of a few percent. Although many power plants use other fuel sources, we'll discuss enriched uranium to make our discussion specific. But the general ideas apply to other power plant designs as well.

Figure 43.17 shows the basic design of a nuclear power plant. Figure 43.17A shows the reactor core. The reactor contains a *moderator*—often liquid water. One of the moderator's jobs is to slow down the free neutrons. Slower neutrons are more likely to interact with nuclei and cause a fission reaction. *Fuel rods* (comprised of enriched uranium) are placed in the moderator, and these rods may be replaced when the fuel is exhausted. In addition to the fuel rods, there are *control rods* made of a material such as cadmium or boron, which is good at absorbing free neutrons. Both the control rod and the fuel rods may be inserted or removed to control the rate of fission reactions. For example, if the reaction rate is too high, inserting control rods to absorb neutrons can slow down the reaction rate.

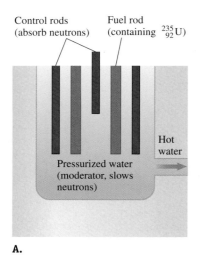

A.

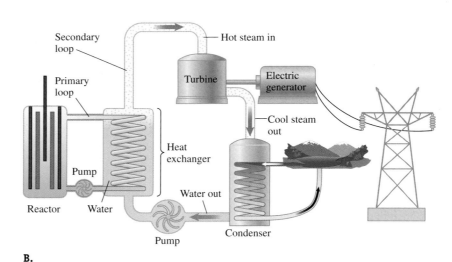

B.

FIGURE 43.17 A. Inside the reactor core are fuel rods and control rods in a water bath. These rods are removable. **B.** The water is heated and pumped into a heat exchanger. The hot water heats water in a secondary loop. The water in the secondary loop becomes steam, which turns the turbines that generate electricity. The steam loses energy and cools. The cool steam is condensed by cold water.

Let's take a close look at the moderator's other job. For the purposes of our discussion, we'll assume the moderator is water. The fission reactions cause this water to heat up, and the circulation of this water results in the production of electricity. The water in the primary loop is kept at high pressure so that it doesn't become steam. This hot liquid water from the reactor circulates through its own closed primary loop so that it doesn't contaminate water in a secondary loop, which contains water as shown in Figure 43.17B. The hot water from the reactor heats the water in the secondary loop in the heat exchanger. The result of our discussion does not depend on the fact that the power plant has a nuclear reactor; it could also be a fossil fuel power plant. The water in the secondary loop boils into steam. The steam drives the blades of the turbine. In Section 32-6, we learned that the basis of an AC electrical generator involves turning coils of wire in a magnetic field (Fig. 32.21, page 1027). So the rotational motion of the turbine is used to turn wire coils and produce electricity. Next, cool water is pumped through a third loop in the condenser. This cool water is used to condense the steam in the secondary loop back into liquid water so that it can circulate back through the heat exchanger. As a result, the water in the third loop is heated. Typically, power plants are built near a natural body of water, and the water in the third loop comes from that natural source. The result is that power plants heat the natural body of water, which can have harmful consequences for organisms that live in that water. So, many newer power plants do not return hot water directly to the natural body of water.

Nuclear fission power plants have other environmental and public health risks. When the fuel rods are no longer effective, they are replaced. However, the "spent" rods are still radioactive and considered toxic for centuries. Safely storing these spent rods is a major challenge. In a disaster, the chain reaction in the reactor could run away leading to a meltdown, and even an improperly functioning reactor can leak radioactive material into the atmosphere or water. In addition, transporting both nuclear fuel and spent fuel rods has the potential for harming the environment if an accident occurs en route.

Nuclear Fission Bombs

In principle the physics of a fission bomb and of a fission power plant are the same. However, in a nuclear power plant, the goal is to achieve a sustainable and controlled fission chain reaction so that the energy is released slowly over a long period of time. In a fission bomb, the goal is to have a runaway chain reaction so that the energy is released very quickly. As before, we consider uranium-235 to be the fuel. In a power plant, uranium must be enriched so that the abundance of uranium-235 is about 3%, but for a bomb the abundance of uranium-235 must be about 90%.

In a power plant, the enriched uranium is shaped into pellets that are about the size and shape of an eraser at the end of your pencil (Figure 43.18). These pellets are stacked to form fuel rods that are several meters tall. The shape and size of the fuel matters because nuclei near the surface are likely to release neutrons that escape rather than trigger further fission reactions. So a small sample with a large surface area-to-volume ratio may not be able to sustain the chain reaction for very long. The shape with the smallest surface area-to-volume ratio is the sphere. A bomb requires a low surface area-to-volume ratio; this means a relatively large sample of enriched uranium shaped into a sphere. The minimum mass of nuclear fuel required to sustain a chain reaction is called the *critical mass*. For uranium, the critical mass is about 50 kg. In a bomb, the uranium is kept in two or more separate pieces, each below the critical mass. (This is known as a subcritical mass.) When the nuclear bomb is detonated, those pieces are brought together to form a single object whose mass is greater than the critical mass. (This is called a supercritical mass.) These pieces must be brought together quickly, and a conventional bomb or explosion is used. Figure 43.19 shows the schematic of the bomb (known as *Little Boy*) dropped over Hiroshima near the end of World War II. The uranium was shaped into two hemispheres and kept separate until detonation. A conventional explosion inside Little Boy forced the two hemispheres together to form a single sphere of roughly 64 kg (about 14 kg above the critical mass).

FIGURE 43.18 Nuclear fuel pellets.

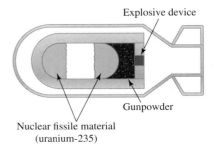

FIGURE 43.19 In Little Boy—the bomb dropped over Hiroshima—a conventional explosion brought the two subcritical hemispheres together to form one supercritical sphere of nuclear fissile material.

EXAMPLE 43.7 Uranium versus Fossil Fuel

Figure 43.18 shows pellets of enriched uranium used in a nuclear power plant. Estimate the total energy released by just one of the pellets. In the **CHECK and THINK** step, compare your answer to energy released by same mass of fossil fuel such as coal or gasoline. *Hint*: See Table 9.1 (page 265).

INTERPRET and ANTICIPATE
We can use the photo to estimate the mass of single pellet. Once we know the mass of a whole pellet, we'll estimate the mass of uranium-235. From this we can estimate the number of uranium-235 nuclei. We know that each fission reaction releases about 200 MeV per nuclei. We'll assume all the nuclei undergo fission to find the total energy released.

SOLVE

The pellets look fairly light. Perhaps each one is the mass of a few quarters. From Appendix B, we find that the mass of a single quarter is about 6 g. So let's assume a single pellet is about 20 g. We'll also assume that it has been enriched so that the abundance of uranium-235 is about 3%. We'll assume the pellet is comprised of uranium isotopes of roughly the same mass. (That is not really true. The uranium isotopes have slightly different masses and the pellets are not pure uranium, but this is an estimate and such details don't make a huge difference.)	$m_{235} = 0.03 m_{\text{pellet}} = 0.03(20\,\text{g}) = 0.6\,\text{g}$
Next we need the number of uranium-235 nuclei. The molar mass of uranium-235 is 235 g, which means 1 mole of uranium-235 has a mass of 235 g. We calculate the number of nuclei using Avogadro's number N_A found on the back inside cover.	$N = m_{235} \dfrac{N_A}{235\,\text{g}}$ $N = (0.6\,\text{g}) \dfrac{6.02 \times 10^{23}\,\text{nuclei}}{235\,\text{g}}$ $N = 1.5 \times 10^{21}\,\text{nuclei}$
Assume every uranium-235 nuclei undergoes a fission reaction and the each reaction releases 200 MeV of energy.	$E_{\text{released}} = (200\,\text{MeV/nuclei})N$ $\phantom{E_{\text{released}}} = (200\,\text{MeV/nuclei})(1.5 \times 10^{21})$ $E_{\text{released}} = 3 \times 10^{23}\,\text{MeV} = 5 \times 10^{10}\,\text{J}$

CHECK and THINK
First, to check our answer we can use the fact that Table 9.1 says that 1 lb of uranium-235 contains 3.7×10^{13} J. 1 lb is the weight of 453 g. So, according to Table 9.1, our 0.6 g of uranium-235 contains $0.6\,\text{g}\left(\frac{3.7 \times 10^{13}\,\text{J}}{453\,\text{g}}\right) = 5 \times 10^{10}$ J exactly as we estimated. Next, Table 9.1 says that 1 lb of coal contains 1.6×10^7 J. So 20 g of coal contains $20\,\text{g}\left(\frac{1.6 \times 10^7\,\text{J}}{453\,\text{g}}\right) = 7 \times 10^5$ J. Our results are amazing; a pellet of uranium releases tens of thousands of times more energy than an equal amount of coal. If you would like, you can repeat this comparison for gasoline, which you will also find in Table 9.1. The results are similar; fission releases millions of times more energy than burning fossil fuel.

FUSION ○ Major Concept

43-8 Fusion Reactions

We've just looked at the practical application of *fission* reactions. As described in Section 43-7, the *fusion* of nuclei that are lighter than iron (to the left of the peak) releases energy. The practical application of fusion reactors on Earth could meet our energy needs without many of the complications of fission reactors. For example, fission reactors require the mining and enrichment of fuel. Fusion begins with light elements such as hydrogen or deuterium (an isotope of hydrogen with one extra neutron). Such light elements can easily be found in water. Fusion produces very few radioactive byproducts. So there are fewer environmental concerns.

FIGURE 43.20 ITER construction (May 2015).

However, many obstacles must be overcome before fusion reactors power our towns and cities. Several experimental fusion reactors have been built on Earth, which have overcome many of these obstacles. One in particular, known as ITER (International Thermonuclear Experimental Reactor) (pronounced "eater"), is currently being built in the south of France (Figure 43.20). The name, ITER, comes from the Latin word for "the way," and the goal of ITER is to generate 10 times more energy by fusion than is required to run the reactor. Perhaps, ITER will make its goal by 2019.

Fusion in the Sun's Core

Many great technological advancements often come from observing nature. So we turn our attention to the fusion reactor in the core of our Sun. The reactions in the Sun's core that fuse hydrogen into helium (**CASE STUDY** Chapter 41) are:

$$\begin{aligned} {}_1^1\mathrm{H} + {}_1^1\mathrm{H} &\rightarrow {}_1^2\mathrm{H} + e^+ + \nu \\ {}_1^2\mathrm{H} + {}_1^1\mathrm{H} &\rightarrow {}_2^3\mathrm{He} + \gamma \\ {}_2^3\mathrm{He} + {}_2^3\mathrm{He} &\rightarrow {}_2^4\mathrm{He} + {}_1^1\mathrm{H} + {}_1^1\mathrm{H} \end{aligned} \quad (43.19)$$

According to these reactions, two protons (hydrogen nuclei) fuse together to form deuterium, and then deuterium and a proton fuse to make an isotope of helium. Because the reaction chain begins with the fusion of two protons, it is called the **proton-proton chain**. Finally, two isotopes of helium fuse to form helium-4. Helium-4 is at a local peak on the binding energy curve (Fig. 43.14); it has more binding energy per nucleon than the hydrogen and helium isotopes from which it is formed.

Notice that the third reaction requires two helium-3 nuclei, which means the first two reactions must occur twice as often as the third reaction:

$$\begin{aligned} 2({}_1^1\mathrm{H} + {}_1^1\mathrm{H} &\rightarrow {}_1^2\mathrm{H} + e^+ + \nu) \\ 2({}_1^2\mathrm{H} + {}_1^1\mathrm{H} &\rightarrow {}_2^3\mathrm{He} + \gamma) \\ {}_2^3\mathrm{He} + {}_2^3\mathrm{He} &\rightarrow {}_2^4\mathrm{He} + {}_1^1\mathrm{H} + {}_1^1\mathrm{H} \end{aligned}$$

Also notice that the third reaction releases two protons that are free to undergo fusion. The net result is that four protons fuse to form helium-4, gamma rays, positrons, and neutrinos:

$$4({}_1^1\mathrm{H}) \rightarrow {}_2^4\mathrm{He} + 2e^+ + 2\nu + 2\gamma \quad (43.20)$$

Compare Equation 43.20 to the one given in Example 39.10 (page 1295) and you will see we've included more detail here, but the essence of that example still applies.

Astronomers often refer to the proton–proton chain as the *pp* chain, without even cracking a smile.

Each complete *pp* chain releases 4.28×10^{-12} J (26.7 MeV) and there are about 8.97×10^{37} reactions per second occurring in the Sun to maintain its power output.

In the case study of Chapter 41 (page 1358), we learned that the first reaction cannot be explained by classical physics. The two protons must fuse and be held together by the strong force, but according to classical physics, Coulomb repulsion would prevent them from coming close enough for the strong force to bind them. According to quantum mechanics, however, there is a nonzero probability that they will tunnel through the repulsive barrier and fuse (Example 41.7, page 1358). The probability of tunneling increases with temperature and density; the Sun's core is very hot (1.5×10^7 K) and very dense (1.5×10^5 kg/m^3). Such a high temperature and density requirements make the *pp* chain impractical on Earth. These conditions are met in the Sun because of the great gravitational force that crushes the core into a very dense and very hot sphere.

Fusion on Earth

Using fusion reactors to meet our energy needs is an important goal. Here we look at some of the practical concerns of achieving that goal with a focus on ITER because it is currently under construction.

First, we need to think about the possible fusion reactions. As described above, the *pp* chain is not a great option because of the difficulty of fusing two protons. It is much easier to fuse isotopes of hydrogen. The most efficient fusion reaction possible on Earth begins with deuterium and tritium:

$$^{2}_{1}\text{H} + ^{3}_{1}\text{H} \rightarrow ^{4}_{2}\text{He} + ^{1}_{0}\text{n} \qquad (43.21)$$

This reaction releases an energy of 17.59 MeV. Deuterium can be distilled from water and so is easily available. Tritium is radioactive with a short half-life (12.3 yr), and it only occurs in a trace amount in nature. So tritium must be manufactured; this is known as *breeding*. There are about 20 kg of tritium available worldwide, some of which ITER will use in its early stages. However, tritium is bred from lithium, which is plentiful in the Earth's crust. So in later stages, ITER will breed tritium using lithium. So the fuels for fusion are deuterium and lithium, both of which are abundant, easy to obtain, and safe to handle. (Compare this to the fuel required for fission—enriched uranium.)

Second, the deuterium and tritium are both positively charged. So like the first reaction in the *pp* chain, there is a repulsive barrier and the particles must tunnel through this barrier. Fortunately tunneling can be achieved on Earth without requiring the enormous gravitational force exerted on the Sun's core. All that is required is a temperature that is about 10 times higher than the Sun's core temperature, and this is achievable on Earth. However, achieving such a high temperature is not easy. ITER will use three techniques to achieve this high temperature.

Before discussing ITER's heating techniques, we must look at a third concern—confinement. As the fuel temperature is increased, it becomes a very hot plasma. Like a gas, a plasma fills its container, but there are no walls that can be built that would withstand the high temperature of the plasma. The plasma must not come in contact with the walls of its container. Instead, it is confined by a magnetic field. ITER will use a tokamak chamber (see Fig. P31.48, page 1007). It is a donut-shaped container with a strong magnetic field produced by superconductors.

Now we can continue our discussion of ITER's three techniques for heating the plasma. The first involves the tokamak's magnetic field. The tokamak uses a changing magnetic field, and from Faraday's law, we know that a changing magnetic field produces an electric field. This electric field creates a current in the plasma. As we've seen before, a current heats a conductor through collisions (Chapter 25). Similarly the plasma is heated, however, this will not achieve the required temperature. So ITER will use two external heating methods; these are the second and third techniques. The second technique involves injecting the plasma with high-energy neutral deuterium. Outside the tokamak, charged deuterium is accelerated and then neutralized before being injected into the plasma. The deuterium transfers energy to the plasma through collisions. This increases the temperature but still not high enough for fusion. The third technique is similar to the way a microwave oven operates.

Energy is transferred by high-frequency electromagnetic waves to the plasma by matching the resonance frequency of the ions and of the electrons. Once fusion begins, ITER's goal is to achieve a *burning plasma*. In such a burning plasma, the temperature is maintained by the fusion reaction itself, and then the external heating methods can be phased out.

If ITER meets its goals, then we may be on our way to achieving fusion-powered cities, towns, cars and trains. Such an energy source may be important to preventing further climate change while sustaining our way of life. Fusion emits no pollution or greenhouse gas. Its end product—helium—is an inert, nontoxic gas. There is no possibility of runaway reaction because the conditions for fusion are too difficult to maintain: if the conditions are altered, the plasma cools and fusion stops. So keep your fingers crossed, and check the ITER website for further developments.

CASE STUDY   Part 4: Where Did Oxygen and Iron Come From?

We've seen that small nuclei—primarily those of hydrogen and helium—formed in the early universe. As the Universe continued to expand and cool, these nuclei bound with electrons to form atoms. This doesn't explain where the larger elements, such as oxygen and iron, came from. The water (H_2O) in your body and the iron in your blood cannot be primordial, so how were their elements made?

As the Universe continued to expand and cool, matter clumped together due to gravitational attraction, forming clouds of hydrogen (and helium) gas. (One of the contemporary problems physicists face is to explain how that clumping took place on timescales that match our observations.) The clouds of gas then collapsed due to gravity, forming galaxies full of stars. The core of every living star is a nuclear fusion reactor, and the primary fuel in such a reactor is hydrogen (Eq. 43.19). This hydrogen must be hot enough for quantum tunneling to occur (case study, Chapter 41, page 1358), so only the hydrogen in the core of a star undergoes fusion. Other elements are not able to undergo fusion at all because they require even higher temperatures in order to achieve quantum tunneling.

But eventually, there isn't enough hydrogen to fuel the reactions in a star's core. At this point, the star begins to collapse under its own weight. The temperature near the core increases, making the fusion of heavier elements possible. In stars several times more massive than the Sun, this process continues until the fusion reactions produce iron. Because iron is at the top of the binding energy curve (Fig. 43.14), more fusion reactions would require the addition of energy instead of releasing energy. Fission reactions could release energy, but they require elements heavier than iron—and none existed in the earliest stars. (Even later generations of stars do not have enough large nuclei to make fission possible.) So the early heavy stars ran out of fuel and died in explosions known as supernova events. However, before they died, they produced elements that fall between helium and iron on the binding energy curve—elements including oxygen. Before our Sun was born, one such heavy star exploded, and material from that star went to form everything in our solar system. This answers our question about oxygen, but we still need to find out where even heavier elements, such as the iodine in your thyroid, came from.

CASE STUDY   Part 5: Where did Iodine Come From?

In the core of even the most massive stars, fusion can only produce elements up to iron. Where did the heavier elements, such as the iodine in your thyroid, come from? (Insufficient iodine can lead to goiters; see Fig. 43.21.) It may seem impossible that such heavy elements exist because they have less binding energy than iron, and so making them through fusion requires energy. Where does the energy come from, and what reactants fuse together?

The answer is that in a supernova event, much gravitational potential energy is released and a great number of neutrons are produced. These neutrons penetrate nearby nuclei, creating products that are heavier than iron. Let's consider this process in a little more detail.

First, we need a source of free neutrons. These are produced in nuclear reactions between relatively small nuclei, such as:

$$^4_2\text{He} + ^{13}_6\text{C} \rightarrow ^{16}_8\text{O} + \text{n}$$

$$^{16}_8\text{O} + ^{16}_8\text{O} \rightarrow ^{31}_{16}\text{S} + \text{n}$$

Next, the free neutrons made in such reactions penetrate nearby nuclei. Because the neutron is electrically neutral, there is no Coulomb barrier to overcome, and no quantum tunneling is necessary. The daughter nucleus that results has a higher mass number than the parent, but no change in atomic number:

$$^A_Z\text{P} + \text{n} \rightarrow ^{A+1}_Z\text{D} + \gamma$$

Thus, the daughter in these reactions is an isotope of the parent. Such neutron penetration can happen in the relatively normal conditions inside a star or during a supernova event.

Under the normal conditions inside a star, neutrons are generated and absorbed by nuclei at a slow rate. When a nucleus captures a neutron under these slow conditions, it undergoes beta decay before another neutron can be captured. Of course, beta decay decreases the neutron number and increases the atomic number. So the nucleus is no longer an isotope of the original parent; instead it becomes a new element. This is called the **s-process**, where *s* stands for slow.

Under the extraordinary conditions during a supernova event, neutrons are produced and absorbed rapidly. Under these conditions, a nucleus will likely absorb many neutrons before undergoing beta decay. So the daughter nucleus may be a higher mass number isotope of the parent. After the period of rapid neutron penetration, these isotopes are generally unstable and undergo either alpha or beta decay. This is called the **r-process**, where *r* stands for rapid.

Today's isotopes beyond iron were generally formed in either the *s*-process or the *r*-process, though some isotopes can be formed in both. How was iodine formed? To answer such a question, astronomers measure the abundance of various elements in stars, while laboratory scientists measure the properties of various nuclei and theorists model the conditions in stars and in supernova events. These models must match the astronomers' observations and the experimentalists' data. Some elements, such as copper, silver, and gold, were most likely produced in the *s*-process. Other elements, such as radium, uranium, plutonium, and thorium, were almost certainly produced in the *r*-process. In general, elements as massive as bismuth ($^{209}_{83}$Bi) can be produced in the *s*-process, and heavier elements in the *r*-process. So in principle, iodine ($^{127}_{53}$I) can be produced in either process, but theoretical models and experimental data suggest that it was most likely produced in the *r*-process.

Perhaps one day a child will ask you, "Where did I come from?" You might answer, "You were made by the cosmos. You began as hot particles soon after time began. You grew in a star that exploded and gave birth to our Sun and to our planet, and to everyone you know, including you." One of the main purposes of these case studies is to connect physics to the human experience. With this final case study we see that physics can even explain how the human experience came to be.

FIGURE 43.21 An iodine deficiency in the thyroid gland can cause goiters. Iodine added to our salt helps to prevent this condition.

EXAMPLE 43.8 Deuterium versus Fossil Fuel

In seawater there is about 1 deuterium atom for every 6400 hydrogen atoms. If the deuterium from 1 gal of such water were used in the fusion reaction given in Equation 43.21, estimate how much energy would be released? Don't worry about details such as obtaining the tritium. In the **CHECK and THINK** step compare your answer to energy released by the same volume of fossil fuel such as coal or gasoline. *Hints*: See Table 9.1 (page 265) and Table 15.1 (page 420); also 1 m^3 = 264 gal. Report your final answer to two significant figures.

INTERPRET and ANTICIPATE
We need to find the number of deuterium atoms in a gallon of seawater. We don't need to worry about the details such as ionizing and heating the deuterium. We'll assume that the number of deuterium nuclei available for the fusion reaction equals the number of deuterium atoms. Finally, we'll use the fact that each fusion reaction releases energy equal to 17.59 MeV.

SOLVE

First, we need the mass of 1 gallon of seawater. The density of seawater is 1025.18 kg/m³ (Table 15.1).	$m_{water} = 1 \text{ gal}\left(\dfrac{1 \text{ m}^3}{264 \text{ gal}}\right)\left(\dfrac{1025 \text{ kg}}{\text{m}^3}\right) = 3.88 \text{ kg}$
Next we need the number of water molecules in the gallon. The molar mass of H_2O is 18 g, which means 1 mole of water has a mass of 18 g. (Of course, heavy water is slightly heavier, but we are making an estimate to two significant figures, and we don't need to be concerned about such details.)	$N_{molecules} = m_{water}\dfrac{N_A}{18 \text{ g}}$ $N_{molecules} = m_{water}\dfrac{6.02 \times 10^{23} \text{ molecules}}{18 \text{ g}}$ $N_{molecules} = (3.88 \times 10^3 \text{ g})\dfrac{6.02 \times 10^{23} \text{ nuclei}}{18 \text{ g}}$ $N_{molecules} = 1.3 \times 10^{26}$ molecules
There are two hydrogen atoms in each molecule of water. So we'll multiply the number of molecules by 2 to find the number of hydrogen atoms.	$N_H = 2N_{molecules} = 2.6 \times 10^{26}$
We need the number of deuterium atoms. Because there is 1 deuterium atom for every 6400 hydrogen atoms, we divide by 6400. As described, this is also the number of nuclei that undergo fusion.	$N_D = \dfrac{N_H}{6400} = \dfrac{2.6 \times 10^{26}}{6400} = 4.1 \times 10^{22}$ nuclei
Assume deuterium nuclei undergo a fusion reaction and that each reaction releases 17.59 MeV of energy.	$E_{released} = (17.59 \text{ MeV/nuclei})N_D$ $E_{released} = (17.59 \text{ MeV/nuclei})(4.1 \times 10^{22} \text{ nuclei})$ $E_{released} = 7.1 \times 10^{23} \text{ MeV} = 1.1 \times 10^{11}$ J

CHECK and THINK
According to Table 9.1, 1 gal of gasoline contains 1.3×10^8 J. Fusing the deuterium in 1 gal of seawater releases about a thousand times more energy than burning 1 gal of gasoline, and fusion doesn't produce greenhouse gases.

43-9 Human Exposure to Radiation

Throughout your life, you are exposed to both artificially-generated and naturally-occurring radiation. This radiation includes the X-rays and gamma rays that are part of the electromagnetic spectrum, as well as alpha rays, beta rays, and other emitted particles. Artificial sources of radiation include nuclear power plants, medical examinations and procedures, fallout from past nuclear tests of weapons, and accidents such as the 1986 disaster at the Chernobyl nuclear power plant in Ukraine. Natural sources of radiation include cosmic rays from space and radioactive elements in the Earth. Radiation may break apart molecules and atoms, which can then damage living cells. The good news is that life on Earth has evolved in an environment with a low level of radiation so that living tissue can repair itself. However, living organisms cannot repair all the damaged cells, and damaged cells can produce tumors, cause cancer, and destroy bone marrow—to name just a few health problems. In this section, we explore the effects of nuclear radiation on living organisms, with a focus on how this radiation affects humans. We'll also learn how the exposure to nuclear radiation is quantified.

DNA (deoxyribonucleic acid) contains the genetic information that is transmitted during cell reproduction.

DNA is arranged into chromosomes in the cell. During cell reproduction, each chromosome is disassembled.

First, how does radiation damage tissue? The radiation transfers energy to the atoms and molecules that make up the tissue. (To make our discussion simpler, we'll refer to the molecules, but the same description could be applied to the atoms.) This energy either excites or ionizes the molecules. If a molecule is particularly important for life such as DNA, the damage may be detrimental to the tissue. The ionization of other molecules may create *free radicals.* Free radicals are neutral atoms with an unpaired electron. Such free radicals are highly reactive, and they can diffuse into the organism, setting off chemical reactions in critical sites such as in organs. These chemical reactions may involve important biological systems such as chromosomes. In such cases, the reactions can kill cells or cause genetic mutations. So some of the damage done by radiation exposure may be apparent only a few hours after exposure as critical biological systems are destroyed. Other consequences such as cancer or transferring genetic defects may take years or generations to appear.

The extent of the damage that occurs depends on the type of radiation. It is helpful to consider four types: alpha rays, beta rays, neutrons, and photons (such as gamma rays). Alpha particles transfer energy directly to living tissues by colliding with electrons in the tissue. These collisions excite the molecules in the tissue and can ionize them. The alpha particles are easily stopped by the skin and don't usually penetrate past the dead layer of skin cells. So normally, alpha particles do little damage.

If, however, the source of alpha particles is ingested or inhaled, they can do damage. The series of reactions in Figure 43.11 includes radon $^{222}_{86}$Rn. Radon is a colorless, odorless, radioactive noble gas, with a half-life of 3.82 days. Radon gas is found in some homes. Although its half-life is less than four days, it is continually replenished by the decay of $^{226}_{88}$Ra(radium), which is found in the rock and soil on which some homes are built. Radon is a particularly dangerous radioactive element because, as a gas, it can be inhaled into the lungs of a home's inhabitants. Once inside the lungs, radon undergoes alpha decay. The reactions in Figure 43.11 may take place inside the lungs, releasing more alpha (and beta) particles into the body.

According to the U.S. Surgeon General's office, as many as 20,000 deaths due to lung cancer are caused by radon annually. Smokers are at a higher risk than nonsmokers because radon decay particles attach to tobacco leaves and lodge in smoke particles in the lungs and bronchi. According to the U.S. Environmental Protection Agency, 1 in 15 homes in the United States has such a high radon level that the family living there should take action. Recommended actions include ventilation to remove radon gas from the home, and sealing the floors and walls to prevent more radon from leaking in.

Next, let's consider how beta particles affect living tissue. Alpha particles are heavy, and though they collide, they travel in nearly straight lines through tissue. However, beta particles are electrons. Electrons are much lighter, and so when they collide, they are easily scattered and travel in a zigzag path through tissue. Beta particles don't lose much energy with each collision, so they penetrate farther than alpha particles. It would take a few millimeters of metal to stop beta particles with an energy of 1 MeV, whereas alpha particles of that energy are stopped by a few 10^{-2} millimeters of skin tissue. Beta particles also lose energy by radiating photons, and these photons may be absorbed elsewhere in the organism. So the energy from beta particles is deposited over a greater volume than the energy of the heavier alpha particles.

Now we consider how neutrons deposit energy in tissues. Of course, neutrons are neutral, and so they penetrate and interact with nuclei. The most common element in a living organism is hydrogen because living organisms contain a lot of water. So, a neutron is most likely to interact with a proton in hydrogen. When a high-energy neutron (more than 1 keV) collides with a proton (in hydrogen), the neutron loses a large fraction of its energy to the proton. The proton then travels through tissue, ionizing atoms and molecules as it loses energy.

Finally, we consider the effect of photons on tissue. Photons transfer energy to electrons through the Compton effect (Section 40-4). This transfer is most dramatic for photons with energies between 40 keV and a few tens of MeV. The energy may be deposited very deep in the tissue. Gamma rays with energies of a few MeV are very penetrating.

Dosimetry

Now that we've discussed how *too much* radiation can damage living tissue, we want to know: How much is *too much*? To answer this question, we first need to know how radiation exposure is measured.

Radiation dosimetry is the branch of medical physics that involves the calculation and measurement of ionizing radiation absorbed by tissue. In this subsection we learn about the many quantities that are defined and measured in this branch of physics. The first quantity is the **absorbed dose** D; it is the energy delivered to tissue per unit mass of the tissue. The SI unit of dose is called the **gray** and abbreviated Gy: 1 Gy = 1 J/kg. However, an earlier system of dosimetry units is still in popular use today. In that system, the absorbed dose D is measured in rad, which stands for *r*adiation *a*bsorbed *d*ose. The conversion factors between the two systems are:

$$1 \text{ rad} = 0.01 \text{ J/kg} = 0.01 \text{ Gy}$$

ABSORBED DOSE ● Major Concept

It is also helpful to use the mass **average absorbed dose** D_{av} for an organ or a particular tissue such as the lungs or bone marrow. The average absorbed dose is the total energy E transferred to the organ (or tissue) divided by its mass m:

$$D_{av} = \frac{E}{m} \quad (43.22)$$

Living tissue is equipped to repair itself. So if the radiation dose is delivered slowly, it is possible that the tissue will repair itself before permanent damage is done. This means that it is important to quantify not just the dose of radiation absorbed, but also the rate at which radiation is absorbed. This is particularly important when radiation is used as a therapy to fight disease such as cancer. The large dose of radiation required to kill the cancer cells is given in several smaller doses spread over time so that the healthy cells have time to repair themselves between treatments.

The degree of permanent injury also depends on the degree of ionization. Because the four types of radiation (alpha rays, beta rays, neutrons, and photons) differ in their ability to ionize molecules, the same dose of each type of radiation will injure a particular tissue differently. To quantify the impact different types of radiation have on living tissue, a dimensionless quantity called the **radiation weighting factor** Q is introduced. For convenience, the quality factor scale is set such that X-ray radiation has $Q = 1$. So, for example, radiation with $Q = 5$ is equivalent to X-ray radiation with a 5 times higher dose. The values of Q come from experimentation, and some values are provided in Table 43.3. The effect of ionizing radiation on living tissue depends on both the dose and the type of radiation. This is quantified by the **equivalent dose H**, which is given by

The radiation weighting factor was formerly called the quality factor; hence, we use Q in this textbook. Q is based on the relative biological effectiveness (RBE)—the empirically determined ratio of the biological effectiveness of one type of radiation to X-ray radiation. Q is then an average arrived at by the consensus of regulators, industries, and governments.

$$H = QD \quad (43.23)$$

EQUIVALENT DOSE ● Major Concept

The SI unit for H is the sievert (Sv), which is a joule per kilogram. Although Q is dimensionless, it is reported in sieverts per gray. When $Q = 1$, the equivalent dose in sieverts equals the absorbed dose in gray. There is an older system of units still in use today. The equivalent dose H in that system is in rem, which stands for *r*oentgen *e*quivalent for *m*an and 1 rem = 0.01 Sv. (The roentgen is named for Wilhelm Röntgen, 1845–1923, who discovered the X-ray.) One roentgen is 2.58×10^{-4} C/kg.

TABLE 43.3 Some radiation weighting factors

Type of radiation	Q (Sv/Gy)
α particles	20
Neutrons (in range 100 keV – 2 MeV)	20
Neutrons (in range 10-100 keV or 2 MeV – 20 MeV)	10
Neutrons (of less than 10 keV or greater than 20 MeV)	5.0
Photons (of 30 keV or more)	1.0
β particles (of 30 keV or more)	1.0
β particles (of less than 30 keV)	1.7

EFFECTIVE DOSE ⊗ Major Concept

Another factor that must be taken into account is the part of the organism that is exposed. Some parts are more sensitive to radiation than others. For example, the lungs are 12 times more sensitive than the skin. The **tissue weighting factor** w is a dimensionless quantity that takes the sensitivity of tissues into account. The human body is divided into 15 parts. Each part has a tissue weighting factor. The sum of all 15 tissue weighting factors is 1.00 (Table 43.4). The **effective dose** D_{eff} takes the tissue weighting factor into account and is given by

$$D_{\text{eff}} = wH \qquad (43.24)$$

Now that we know how exposure to radiation is quantified, we can consider our exposure to radiation. Table 43.5 gives the average annual exposure to both natural and artificial sources of radiation both in the United States and worldwide. The typical natural radiation dose equivalent in the United States is roughly 3 mSv annually, and the average annual dose equivalent is roughly another 3 mSv. So Americans are exposed to about 6 mSv annually, about double the average worldwide. Perhaps this is due to American's access to medical procedures. For example, a full mammogram exam with a total of six images has a dose equivalent of about 8 mSv, and the average annual dose due to medical procedures in the United States is 3 mSv. (This average includes other procedures including diagnostic and treatment procedures.) The American Medical Association recommends that women over 40 have an annual mammogram, but in the United Kingdom the National Health Service recommends that woman between the ages of 40 and 70 get a mammogram every three years. To help make sense of how all this exposure to radiation affects our health, we end with this statistic: On average 5% of the people exposed to an effective dose of 1000 mSv will contract a fatal cancer as a result.

TABLE 43.4 Tissue weighting factors

Organ or tissue	w
Bladder	0.04
Bone surface	0.01
Brain	0.01
Breast	0.12
Colon	0.12
Esophagus	0.04
Gonads	0.08
Liver	0.04
Lung	0.12
Red bone marrow	0.12
Salivary glands	0.01
Skin	0.01
Stomach	0.12
Thyroid	0.04
Remainder of the body	0.12
Total	**1.00**

The data is from ICRP Publication 103 37 (2007).

TABLE 43.5 Average annual human exposure to radiation in mSv

Source	U.S.	Worldwide
Inhalation of air (primarily radon)	2.28	1.26
Ingestion of food and water	0.28	0.29
Ground	0.21	0.48
Cosmic rays	0.33	0.39
Total natural sources	**3.10**	**2.40**
Medical	3.00	0.60
Other	0.14	0.012
Total artificial sources	**3.14**	**0.612**

EXAMPLE 43.9 Are bananas safe to eat?

Potassium K is an important mineral in the human body. Your muscles need potassium to operate, potassium keeps the sodium levels in your body under control, and all of your cells need potassium to function normally. The average adult needs 4700 mg of potassium per day. Many people try to meet this need by eating bananas. A single typical banana has a mass of 125 g, about 450 mg of which is potassium, nearly 10% of your daily requirement. However, potassium-40 is a radioactive isotope of potassium with a natural abundance of 0.012%. Potassium-40 has a half-life of 1.25×10^9 years, and it radiates gamma and beta rays. The average energy released with each decay is about 0.5 MeV. The body takes around 50 hours to expel digested food after it has been consumed. Assume the radiation from the potassium is mostly absorbed by the tissue and organs along the digestive tract: the esophagus, liver, salivary glands and stomach. Further assume the combined mass of all of these tissues is about 2 kg. Estimate the absorbed dose, the equivalent dose, and the effective dose delivered to these tissues. Report your estimates to one significant figure.

INTERPRET and ANTICIPATE
The absorbed dose D is the energy delivered to tissue per unit mass of the tissue. We know the mass of the tissue, so we just need to find the energy delivered. We also know the energy released by each decay. So we must find the number of decays in 50 hours to find the total energy delivered to the tissues. Once we know the absorbed dose, the effective dose D_{eff} is found from the equivalent dose and the tissue weighting factor.

SOLVE

To find the number of decays, we must first find the initial number N_0 of potassium-40 nuclei when the banana first entered the body. The banana contains 0.45 g of potassium. The molar mass of potassium is 39 g, and the natural abundance of potassium-40 is 0.012%.	$N_0 = (\text{fraction of }^{40}\text{K}) \times \dfrac{(\text{mass of K ingested})}{(\text{molar mass of K})} \times N_A$ $N_0 = 0.00012 \dfrac{0.45 \text{ g}}{39 \text{ g/mol}} 6.02 \times 10^{23}$ nuclei/mol $N_0 = 8.3 \times 10^{17}$ nuclei
From the half-life, we find the number of decays that occur in the 50 hours in which the potassium is in the body. The number of parent nuclei as a function of time is given by Equation 43.8. The number of decays is the initial number of parent nuclei minus the number of parent nuclei remaining after 50 hours. The half-life must be converted from years to hours (1.09×10^{13} h); the units have been left off the exponent for clarity.	$N = N_0 \left(\dfrac{1}{2}\right)^{t/T_{1/2}}$ (43.8) $N_{\text{decays}} = N_0 - N$ $N_{\text{decays}} = N_0 \left[1 - \left(\dfrac{1}{2}\right)^{t/T_{1/2}}\right]$ $N_{\text{decays}} = 8.3 \times 10^{17} \left[1 - \left(\dfrac{1}{2}\right)^{50/1.09 \times 10^{13}}\right]$ $N_{\text{decays}} = 2.7 \times 10^6$
Each decay releases 0.5 MeV of energy. So we find the total energy delivered by multiplying the number of decays by 0.5 MeV.	$E = (0.5 \text{ MeV}) N_{\text{decays}} = (0.5 \text{ MeV})(2.7 \times 10^6)$ $E = 1.3 \times 10^6 \text{ MeV} = 2 \times 10^{-7}$ J
The absorbed dose D is found by dividing the delivered energy by the mass of the tissue.	$D = \dfrac{E}{m} = \dfrac{2 \times 10^{-7} \text{ J}}{2 \text{ kg}} = 1 \times 10^{-7}$ Gy
The radiation weighting factor Q is 1 Sv/Gy for both photons and beta particles with energy above 30 keV (Table 43.3). So the equivalent dose (Eq. 43.23) is numerically equal to the absorbed dose.	$H = QD = (1 \text{ Sv/Gy})(1 \times 10^{-7} \text{ Gy})$ $H = 1 \times 10^{-7}$ Sv
The effective dose takes into account the sensitivity of the particular organs and tissues that absorb the radiation. We add the tissue weighting factors for these organs and tissues (Table 43.4). Then find D_{eff} using Equation 43.24. (The subscripts stand for *esoph*agus, *liver*, *sal*ivary *gl*an*d*s and *stm*a*ch*.)	$w = w_{\text{esoph}} + w_{\text{liver}} + w_{\text{sal gld}} + w_{\text{stmch}}$ $w = 0.04 + 0.04 + 0.01 + 0.12 = 0.21$ $D_{\text{eff}} = wH$ (43.24) $D_{\text{eff}} = (0.21)(1 \times 10^{-7} \text{ Sv}) = 2 \times 10^{-8}$ Sv

CHECK and THINK
If you search the Internet, you are likely to find the vague statement that the radiation dose of a banana is 0.1 μSv. Perhaps the statement is referring to the equivalent dose H, in which case the statement exactly matches what we found. Your search will also reveal an informal unit for measuring a radiation dose—the banana equivalent dose (BED). The idea of this informal unit is to help make radiation doses more understandable. For example, the maximum permitted radiation leakage for a nuclear power plant is 2500 BED. We all feel safe eating a banana, but how about 2500 bananas?

THE STANDARD MODEL

⚠ **Underlying Principle**

43-10 The Standard Model

We've reached the last section of this textbook. Here we introduce one final theory—known as the *standard model*. The **standard model** is a theory that lists the particles that make up that the Universe, explains their properties, and describes the forces by which they interact. So one of the wonderful things about saving the standard model for the end of this book is we get to review many forces and particles we have studied so far, and order them in a logical scheme.

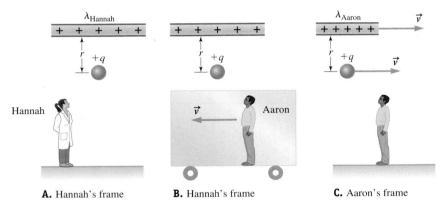

FIGURE 43.22 A. Hannah in the laboratory frame observes a particle with charge $+q$ near a charged rod with linear charge density λ_{Hannah}. **B.** *According to Hannah,* Aaron observes the same particle and rod from a frame moving at constant velocity to the left. **C.** To Aaron the rod and the particle seem to move to the right at constant velocity. To Aaron the rod is shorter, and the linear charge density is greater $\lambda_{\text{Aaron}} > \lambda_{\text{Hannah}}$.

Electricity, Magnetism, and Relativity

We begin by taking another look at the electric force, the magnetic force, and special relativity. Hannah and Aaron both observe the force between a positively charged particle and a nearby positively charged rod (Fig. 43.22). Hannah is in the laboratory frame, which is at rest with respect to the particle and the rod. Aaron is in a frame moving to the left at constant velocity. So he sees the rod and the particle moving to the right.

Hannah observes that force on the particle is repulsive and so points straight down (Fig. 43.23A). To Hannah, the magnitude of the electric field produced by the rod is

$$E_{\text{Hannah}} = \frac{1}{2\pi\varepsilon_0} \frac{\lambda_{\text{Hannah}}}{r} \qquad (25.13)$$

and the electric force on the charged particle is $F_{\text{Hannah}} = qE_{\text{Hannah}}$.

Aaron sees the rod and the particle moving to the right. Because of this relative motion, there is a length contraction (Section 39-5), and to Aaron the rod appears shorter. Because for him the rod is shorter, Aaron sees the same amount of charge spread over a shorter rod, and therefore, a greater linear charge density. So Aaron claims that the electric field points downward but has a greater magnitude than that observed by Hannah.

However, Einstein postulated that all laws of physics are true in all inertial reference frames. Put simply, according to Einstein, Aaron and Hannah must agree on the force exerted on the charged particle; if they disagree, something else must be going on. But there is something else going on. Aaron sees the charged rod is moving; to him the charged rod is producing a current. According to the simple right-hand rule for currents (Fig. 30.11, page 939) the magnetic field is directed into the page at the location of charged particle. According to Equation 30.17 $\vec{F} = q(\vec{v} \times \vec{B})$, the magnetic force on the charged particle is upward toward the rod. The downward electric force and the upward magnetic force observed by Aaron give a net force on the particle that exactly matches the electric force observed by Hannah (Fig. 43.23B).

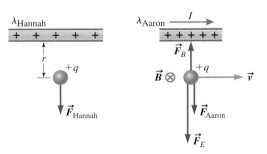

FIGURE 43.23 A. Hannah observes only a downward-pointing electric force. **B.** Aaron observes a downward-pointing electric force and an upward-pointing magnetic force. The net force he observes is exactly equal to the net force Hannah observes.

This thought experiment illustrates the idea behind unified forces. Two forces are unified when they are understood as two aspects of the same underlying force. Here, a net force on a charged particle can be understood in terms of the electric force alone, or in terms of a combination of the electric and magnetic forces, depending on the relative motion between the particle and the observer. So we say the two forces are aspects of a single underlying *electromagnetic force*.

Quantum Electrodynamics (QED)

Several times throughout this textbook, we have mentioned that Maxwell's equations give a unified theory of electricity and magnetism. Furthermore, this unified theory explains a phenomenon—light—that neither theory alone can explain. The American physicists Richard Feynman (1918–1988) and Julian Schwinger (1918–1994), and the Japanese physicist Sin-Itiro Tomonaga (1906–1979), shared the Nobel Prize in 1965 for developing a theory known as **quantum electrodynamics (QED)**. This theory combines electromagnetism with quantum mechanics. According to QED, photons are more than just particles that make up the phenomenon of light. They play a role in all phenomena that involve the electromagnetic force.

To understand the role played by photons, we consider a crude analogy. Figure 43.24A shows Sallie and Pete playing a game of catch. Pete has just thrown the ball. He pushes on the ball, and by Newton's third law the ball pushes on him: he experiences a recoil. When Sallie catches the ball she exerts a force on the ball, and the ball exerts a force on her. We can think of the ball as a particle that carries a force between Pete and Sallie.

According to QED, when two charged particles exert forces on one another, a photon is exchanged (Fig. 43.24B). The photons carry the force from one particle to the other, much as the ball carries a force between Pete and Sallie. Both Pete and Sallie feel pushed away from each other, so the ball carries a repulsive force. This analogy is crude because it cannot account for an attractive force, but according to QED, both attractive and repulsive forces are attributed to the exchange of photons. We'll soon learn that other forces are accounted for by the exchange of other particles. The particles, such as photons, that are exchanged when forces are exerted are referred to as "messenger particles," "force-carrying particles," or "exchange particles." The standard model includes a number of other exchange particles, and we say that a particular force is "mediated" by its particular exchange particles. But next, we must digress to learn about the particles that make up the standard model.

QED ❶ **Underlying Principle**

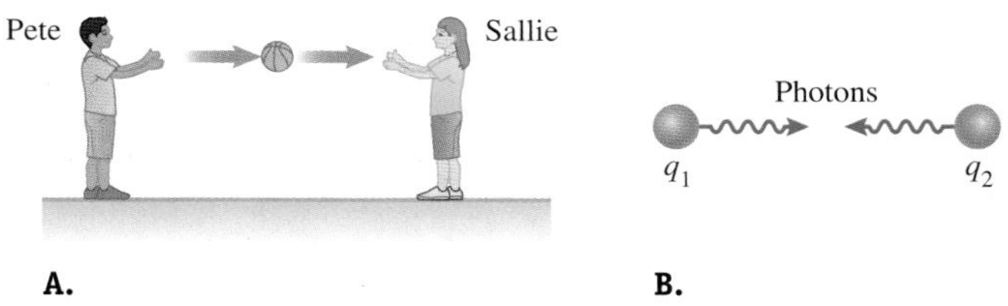

FIGURE 43.24 A. Pete throws a ball to Sallie. **B.** Two charged particles interact. In this analogy, the ball is like a photon carrying a force from Pete to Sallie.

A Zoo of Particles

The standard model classifies particles into several different categories. Some categories overlap, and you can find particles in more than one category. In this section we'll discuss these categories and some of the particles in the standard model. Many of the particles will be familiar to you, but there are a few new ones as well.

Like the electromagnetic force, each fundamental force is thought to be propagated by exchange particles. The exchange particles are **bosons**. Bosons do not obey Pauli's exclusion principle (Section 42-6), so they can have the same quantum numbers. Electrons are an example of another type of particle, called **fermions**. Fermions do obey the Pauli exclusion principle and cannot have the same quantum numbers. We saw that, in the case of electrons, Pauli's exclusion principle explains why atoms take up space (not all the electrons can be in the atom's ground state.) Put simply, solid objects, such as the chair you are sitting on, must be made of fermions because they obey Pauli's exclusion principle and take up space. Because bosons do not obey Pauli's exclusion principle, however, they can be packed together without taking up space. Bosons also differ from fermions in their spins. All known fermions, such as electrons, are spin 1/2 particles, whereas bosons have integer spin. For example, the photon has a spin of 1.

Just as the photon is a boson that mediates the electromagnetic force, there are eight **gluons** that are bosons that mediate the strong force. There are also three bosons that mediate the weak force. All of these bosons have been observed in a laboratory and are part of the standard model. In addition, there is a theory that gravity should also be mediated by a boson. This boson is called the graviton, but its

BOSON AND FERMIONS
✪ **Major Concept**

existence hasn't been confirmed by experimentation. The standard model does not account for gravity, so the graviton isn't properly part of the standard model either. Table 43.6 summarizes all of these bosons.

TABLE 43.6 Bosons

Force	Name or symbol	Mass (GeV/c^2)	Charge	Spin
Electromagnetic	Photon	0	0	1
Strong nuclear	(8) Gluons	0	0	1
Weak nuclear	W^+	80.4	+1	1
	W^-	80.4	−1	1
	Z	91.2	0	1
Gravity	Graviton	0	0	2

LEPTONS AND HADRONS
✪ Major Concepts

The standard model also divides particles into **leptons** and **hadrons**. Leptons are fundamental particles that cannot be subdivided. We are already familiar with two of them: the electron and the neutrino. In fact, there are six leptons and six antiparticles, called **antileptons**. The neutrino we are familiar with from inverse beta decay (Eq. 43.12) is actually called an **electron neutrino**. The right part of Table 43.7 summarizes the six leptons. Notice that they fall into three pairs, such that each particle is paired with a neutrino. Leptons are subject to the weak force but not the strong force. Charged leptons are further subject to the electromagnetic force. To help understand reactions involving leptons, it is convenient to assign each lepton a lepton number of +1 and each antilepton a lepton number of −1.

Hadrons, in contrast, are made of quarks. They are further subdivided into **baryons** and **mesons**. Baryons are made of three quarks. Familiar examples of baryons are protons and neutrons. Mesons are made of two quarks. Both quarks and leptons are fermions, so they obey Pauli's exclusion principle. Quarks are charged and have a charge of either $\pm e/3$ or $\pm 2e/3$. Table 43.7 summarizes the fermions. All hadrons are composed of quarks, so they are subject to the strong force. Also, because quarks are charged, they are subject to the electromagnetic force. Lastly, they are subject to the weak force. Table 43.8 provides examples of baryons and mesons. Each meson is made of a quark and an antiquark. A bar over the symbol for a quark denotes an antiquark; for example, $\bar{s}$ is a strange antiquark. A meson or baryon's charge is the sum of the charge of its quarks. For example, a neutron is made of two down quarks and one up quark, and its charge is $2(-e/3) + 2e/3 = 0$ as expected. Each quark has a baryon number of +1/3, and each antiquark has a baryon number of −1/3. So the baryon number for all the mesons is 0. However, the baryon number for a proton or a neutron is +1, and the baryon number for an antiproton or an antineutron is −1.

TABLE 43.7 Fermions

Quark	Symbol	Charge	Mass (GeV/c^2)
Up	u	$+2e/3$	0.002
Down	d	$-e/3$	0.005
Charm	c	$+2e/3$	1.5
Strange	s	$-e/3$	0.15
Top	t	$+2e/3$	172
Bottom	b	$-e/3$	4.7

Lepton	Symbol	Charge	Mass
Electron	e	$-e$	0.511 MeV/c^2
Electron neutrino	ν_e	0	Between 0.05 and 2 eV/c^2
Muon	μ	$-e$	106 MeV/c^2
Muon neutrino	ν_μ	0	Less than 0.19 MeV/c^2
Tau	τ	$-e$	1.78 GeV/c^2
Tau neutrino	ν_τ	0	Less than 18 MeV/c^2

TABLE 43.8 Some baryons and mesons

Baryon	Symbol	Quarks	Meson	Symbol	Quarks
Proton	p	uud	Pion	π^+	$u\bar{d}$
Neutron	n	udd	Phi	ϕ	$s\bar{s}$
Omega	Ω^-	sss	Kaon (plus)	K^+	$u\bar{s}$
Lambda	Λ	sud	Kaon (minus)	K^-	$\bar{u}s$

Forces in the Standard Model

Now let's see what the standard model says about how all these particles interact. First, the standard model is a combination of two theories. One theory is the **electroweak theory**—the unification of the electromagnetic force and the weak force (Fig. 23.2, page 684). Because we have studied the electromagnetic force and the weak force, we can surmise that the electroweak force acts on charged particles, leptons, and quarks. The other theory describes how the strong force is mediated by gluons and acts on quarks. This part of the standard model is called **quantum chromodynamics** (QCD). So the standard model accounts for the behaviors involving all the forces except gravity. Here we briefly look at some of these behaviors.

No one has found a free particle with a fractional charge. This means that no one has found a free quark. According to the standard model, quarks are confined in hadrons because the energy required to separate quarks *grows* as their separation increases. At a large enough separation, this energy is large enough to produce a new quark–antiquark pair, which binds to form a hadron.

Quarks are fermions, and they obey Pauli's exclusion principle. So you might be wondering how two up quarks can exist in a single proton. The answer is that quarks have another intrinsic quantum property, called *color*. Quarks come in three colors: red, green, and blue. Of course, these are not actual colors. You can think of color much like spin. Spin is an intrinsic quantum property, but quantum particles are not really spinning like a top. Just as two electrons can be in the same subshell because one can be spin up with $m_s = 1/2$, and one can be spin down with $m_s = -1/2$, two up quarks can be in the same proton, as long as they are different colors.

The standard model makes a prediction that helps us understand some of the reactions we have studied. According to QCD, baryon number is conserved during a reaction. The standard model also requires that the lepton number be conserved. As an illustration, consider this reaction (inside a nucleus): $p \rightarrow n + e^+ + \nu$ (Eq. 43.11). We start with a baryon number of +1 for the proton. After the reaction, the baryon number is +1 for the neutron. Before the reaction, the lepton number is 0. After the reaction, the lepton numbers are −1 for the positron and +1 for the neutrino, which sum to 0. Both the baryon number and the lepton number are conserved in this reaction. The standard model also requires that the total charge of quarks confined in a hadron be either $+e$, $-e$ or 0. Check Table 43.8, and you will find that those hadrons meet the charge requirement.

ELECTROWEAK THEORY AND QCD
❶ **Underlying Principles**

The Higgs Boson

As we have mentioned several times, the electroweak theory unifies the electromagnetic and the weak forces. The Nobel Prize was awarded in 1979 to the American physicists Sheldon Glashow (1932–) and Steven Weinberg (1933–), and the Pakistani physicist Abdus Salam (1926–1996), for developing the electroweak theory. Their electroweak theory explains phenomena that cannot be accounted for by either the electromagnetic theory or the weak theory alone. Perhaps more amazingly, the electroweak theory predicted the existence of three bosons: W^+, W^-, and Z (Table 43.6). The existence of these bosons was later confirmed by experimentation.

The standard model also predicts the existence of another boson, known as the **Higgs boson**. The Higgs boson is named for the British physicist Peter Higgs (1929–), one of the theorists who predicted its existence. Take a look at Table 43.6, and you might be struck by the fact that for 3 out of the 4 fundamental forces, the force-carrying bosons are massless, whereas the force-carrying bosons for the weak force have mass. In the 1960s, particle theorists were disturbed by this too, and they wanted the standard model to account for this discrepancy. According to the particle physics theories at that time, all force-carrying particles should be massless. So the question they asked was "what could cause these force-carrying bosons to acquire mass?" For that matter, they wanted to account for why other massive particles have mass. Their answer predicted the Higgs boson. Unlike the other bosons we've discussed, the Higgs boson has a spin of 0.

According to the standard model, the Higgs boson is a very massive (found to be 125 GeV/c^2) neutral particle that decays very rapidly. Because the Higgs is so massive, it requires a lot of energy to create it. Because the Higgs is neutral and decays rapidly, once created, it is hard to detect.

The Large Hadron Collider (LHC) is an experiment at CERN, the European Organization for Nuclear Research, located on the French-Swiss border. CERN operates the world's largest and most powerful particle physics laboratory. The LHC was designed to find the Higgs boson by smashing together very high-energy protons. After the protons collide, a plethora of particles emerge. By studying these particles, physicists look for evidence of the Higgs boson. On July, 4, 2012, two research groups who independently analyzed the data from the LHC announced that they had found evidence for the Higgs boson. Soon after the announcement, champagne corks popped as physicists celebrated the power of theorists to make predictions and the power of experimentalists to discover evidence testing those predictions.

CONCEPT EXERCISE 43.8

On an exam, you've been asked to write down the expression for beta decay. You remember the daughter nucleus has an extra proton and there are two products. You know one is either an electron or a positron and the other is either a neutrino or an antineutrino: $^A_Z P \rightarrow\ ^A_{Z+1} D\ +\ \underline{e^-\ \text{or}\ e^+}\ +\ \underline{\nu\ \text{or}\ \bar{\nu}}$. Use conservation principles to figure out which product belongs on each line. Explain.

A Final Word

Congratulations! You have just finished reading this entire, enormous book. But studying physics does not end when you close this book, not even after you ace your final exam. Physics takes a long time to digest fully. It is the sort of subject you need to live with. So my advice is: Keep thinking about physics. Keep your book. Keep your notes. Keep your old homework and tests. You will find times when you will look at all of these. Next semester, you may be asked to solve a homework problem similar to something you saw in this class. Perhaps one day, you will need to decide whether to buy a diesel car. Before seeing the dealer, it will help to reread Chapter 22 on engines. Or you may read that there are more neural connections in your brain than atoms in the universe, and you may want to look up a few facts in Appendix B to see if that makes sense. Maybe you will just be sentimental one day, and take a look at your old college work. Rutherford (the inventor of the solar system model of the atom) once took a look at his old notes. When he did he said,

> I've just finished reading some of my early papers, and you know, when I'd finished I said to myself, "Rutherford, my boy, you used to be a damned clever fellow."

You are clever too. Don't throw your work away, or sell your used book for the price of a pizza. Keep connecting physics with your life. Thank you for reading this book.

Debora M. Katz

SUMMARY

❶ Underlying Principles

1. **Models of nuclei**
   a. In the **shell nuclear model**, each nucleon can be treated as a particle in the system, while the other part of the system is the rest of the nucleus. In this model the system's energy is quantized. In general, proton energy levels are farther apart than are neutron energy levels.
   b. In the **liquid drop model**, nucleons move around the nucleus much like molecules moving in a drop of liquid, and collide frequently.
   Today's nuclear models are a combination of the shell model and the liquid drop model, in which the core is modeled as a liquid drop and the outer nucleons fit the shell model with its quantized states.
2. The **strong nuclear force** is one of the four fundamental forces of physics; it binds quarks together to form nucleons and holds the nucleus together. The strong force has a short range, acting over distances of roughly 1 fm or less. At this short range, the strong force is about 100 times stronger than the Coulomb force.
3. The **weak nuclear force** is another of the four fundamental forces of physics; at the level of quarks it transforms a neutron into a proton or a proton into a neutron. The weak force has a short range and is about 10^7 times weaker than the strong force.
4. **Electroweak theory** unifies the electromagnetic and the weak forces. The electroweak theory explains phenomena that cannot be accounted for by either the electromagnetic theory or the weak theory alone. Perhaps more amazingly, the electroweak theory predicted the existence of three bosons: W^+, W^- and Z. The existence of these bosons was later confirmed by experimentation.
5. **QED and QCD**
   a. **Quantum electrodynamics** (QED) is a confirmed theory combining electromagnetism with quantum mechanics. According to QED, photons are force carrying particles that play a role in all phenomena that involve the electromagnetic force.
   b. **Quantum chromodynamics** (QCD) is another confirmed theory that describes how the strong force is carried by gluons and acts on quarks.
6. **The standard model** is a combination of the electroweak theory and QCD. So the standard model accounts for the behaviors involving all the forces except gravity. Put simply the standard model lists the particles that make up that the universe, explains their properties and describes the forces by which they interact.

✪ Major Concepts

1. The effective **nuclear radius** is based on a spherical model and is given by:
$$R = r_0 A^{1/3} \quad (43.2)$$
where $r_0 = 1.2$ fm and A is the mass number.
2. **Nuclear decay**: The parent nucleus of a radioactive substance may **decay** into a daughter nucleus, so over time a sample of radioactive material is transformed into of the daughter nuclei. The number of parent nuclei decreases exponentially:
$$N = N_0 e^{-\eta t} \quad (43.5)$$
where N_0 is the number of parent nuclei at time $t = 0$. The decay constant η is more conveniently expressed in terms of the half-life $T_{1/2}$:
$$T_{1/2} = \frac{\ln 2}{\eta} \quad (43.7)$$
or in terms of the **time constant**, defined as $\tau \equiv 1/\eta$.
3. A bound system has a **mass deficit** Δm given by:
$$\Delta m = \sum m_i - M \quad (43.15)$$
where $\sum m_i$ is the sum of the free particles' individual masses, and M is the mass of the bound system. The **binding energy** is the amount of energy supplied by an outside source required to pull the particles out of the bound system and separate them into free particles at rest. Mathematically, the binding energy is
$$E_B = \Delta m c^2 \quad (43.16)$$
Putting this together, if there is a positive mass deficit such that the product has less mass than the sum of the reactant's masses, then binding energy *increases* and energy is released.
4. In a **fusion** reaction lighter nuclei combine to form a heavier product. In a **fission** reaction a heavy nucleus divides in to two (or more) fragments.
5. a. The **absorbed dose** D is the energy delivered to tissue per unit mass of the tissue. The SI unit of dose is called the gray and abbreviated Gy: 1 Gy = 1 J/kg.
   b. The **equivalent dose** is given by
$$H = QD \quad (43.23)$$

Major Concepts (Continued)

where Q is a dimensionless quantity called the **radiation weighting factor** Q, which quantifies the impact different types of radiation have on living tissue

c. The **effective dose** D_{eff} takes the tissue weighting factor into account and is given by

$$D_{eff} = wH \quad (43.24)$$

6. **Bosons** (Table 43.6) do not obey Pauli's exclusion principle, but **fermions** (Table 32.7) do. So bosons can have the same quantum numbers, but fermions cannot have the same quantum numbers. Further, all known fermions, such as electrons, are spin ½ particles, whereas bosons have integer spin.

7. **Leptons** are fundamental particles that cannot be subdivided. There are six leptons and six antileptons (Table 43.7). **Hadrons**, on the other hand, are made of quarks. Hadrons are further subdivided into **baryons** and **mesons**. Baryons are made of three quarks, and mesons are made of two quarks (Table 43.8). Both quarks and leptons are fermions.

Special Cases

1. **Gamma rays** are photons that are emitted when a nucleon drops from a high energy level to a lower energy level.
2. **Alpha rays** are also called **alpha particles**, and are equivalent to helium nuclei. The process of emitting an alpha particle is called **alpha decay**:

$$^A_Z P \to ^{A-4}_{Z-2} D + ^4_2 He + \text{energy} \quad (43.3)$$

3. **Beta rays, or beta particles**, is another name for electrons. The process of emitting a beta particle is called **beta decay**, in which a neutron in the parent nucleus becomes a proton:

$$^A_Z P \to ^A_{Z+1} D + e^- + \bar{\nu} \quad (43.10)$$

In **inverse beta decay, a proton in** the parent nucleus becomes neutron:

$$^A_Z P \to ^A_{Z-1} D + e^+ + \nu \quad (43.12)$$

4. In **electron capture**, an inner atomic electron is captured by a nuclear proton. So the nucleus gains a neutron and loses a proton:

$$^A_Z P + e^- \to ^A_{Z-1} D + \nu \quad (43.13)$$

PROBLEMS AND QUESTIONS

A = algebraic   C = conceptual   E = estimation   G = graphical   N = numerical

43-1 Describing the Nucleus

1. **C** A nucleus is symbolized by $^{202}_{87}$Fr.
   a. What is the name of the element? b. How many protons are in the nucleus? c. How many neutrons are in the nucleus?
2. **E** One way to probe the nucleus is to bombard a sample with high-energy electrons. To learn about the nuclear structures in a sample, the de Broglie wavelengths of these electrons would need to be a little smaller than a nuclear radius. Estimate the energy of such electrons. Give your answer in electron-volts.
3. **C** An isotope of lead has 132 neutrons. Write the symbol for this isotope in the form $^A_Z X$.
4. **N** The element nitrogen (N) has two stable isotopes—one with an atomic mass number of 14 and the other with an atomic mass number of 15. a. How many neutrons are in each of these nuclei? b. If the natural abundance of ^{15}N is 0.37%, what is the natural abundance of ^{14}N?

Problems 5 and 6 are paired.

5. **N** The lightest naturally occurring element on Earth is hydrogen (H), and the heaviest is uranium (U). a. What is the nuclear radius of hydrogen? b. What is the nuclear radius of uranium?
6. **E, N** In order to get a sense of scale of the size difference between hydrogen and uranium nuclei, assume that a baseball represents the size of the hydrogen nucleus. In this scaled system, what is the diameter of the uranium nucleus?
7. **N** Find the nuclear radius of ^{40}Ca in fermis.

43-2 The Strong Force

Problems 8 and 9 are paired.

8. **N, C** Consider two protons that are separated by 1.5 fm. What is the magnitude of the Coulomb repulsive force between them? Assume that these protons are one another's nearest neighbors and that the strong force between them is about 2000 N. Will these protons hold together or fly apart?
9. **N, C** Consider two protons that are separated by 7.0 fm. What is the magnitude of the Coulomb repulsive force between them? Assume that these protons are on opposite sides of a nucleus and that the strong force on their nearest neighbors is about 2000 N, but nearly zero between nucleons on opposite sides of the nucleus. Comment on the stability of this nucleus.
10. **C** Could an isotope of helium with two protons and no neutrons exist? Why or why not?
11. **C** Protons in a nucleus experience four fundamental forces: Coulomb repulsion, the weak nuclear force, the strong force attraction, and gravitational attraction. Why did we ignore the gravitational force in our discussions?

43-3 Models of Nuclei

Problems 12, 13, and 52 are grouped.

12. **G, C** Fill in an energy-level diagram (Fig. P43.12) for $^{16}_{8}$O. Based on your diagram, do you expect $^{16}_{8}$O to be stable? Explain.

13. **G, C** Fill in an energy-level diagram (Fig. P43.12) for $^{19}_{8}$O. Based on your diagram, do you expect $^{19}_{8}$O to be stable? Explain.

14. **C** Express the symbol for each of the first five doubly magic isotopes in the form $^{A}_{Z}$X, beginning with helium.

15. **C** Consider the energy-level diagram for a particular nucleus shown in Figure P43.15. Express the symbol for this isotope in the form $^{A}_{Z}$X.

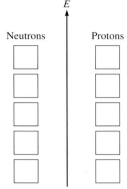

FIGURE P43.12 Problems 12 and 13.

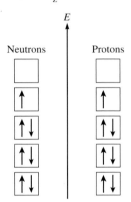

FIGURE P43.15

16. **C** There are only two stable nuclei ($^{1}_{1}$H and $^{3}_{2}$He) that have more protons than neutrons. Why does $Z > N$ in general cause a nucleus to be unstable? Why do you suppose that these two nuclei are exceptions to the general rule?

17. **C** Consider the energy-level diagram for a particular nucleus shown in Figure P43.17. Is this nucleus stable? Why or why not?

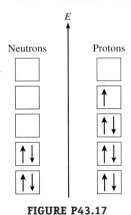

FIGURE P43.17

43-4 Radioactive Decay

18. **N** The half-life of ^{14}C is 5730 yr, and a constant ratio of ^{14}C/^{12}C = 1.3×10^{-12} is maintained in all living tissues. A fossil is found to have ^{14}C/^{12}C = 2.65×10^{-13}. How old is the fossil?

19. **N** At $t = 0$, a radioactive sample has exactly 20,000 radioactive nuclei. How many radioactive nuclei does it have after two half-lives have passed?

Problems 20 and 21 are paired.

20. **N** The number of radioactive isotopes in a sample is found to drop to 22.50% of its original value in 924.5 s. What is the decay time constant of the isotope?

21. **N** The number of radioactive isotopes in a sample is found to drop to 22.50% of its original value in 924.5 s. What is the half-life of the isotope?

Problems 22 and 23 are paired.

22. **N** In the Fukushima Daiichi nuclear disaster (March 2011), one of the major radioactive contaminants released was iodine-131. The half-life of iodine-131 is 8.04 days. **a.** What is the decay constant in s^{-1} for iodine-131? **b.** What is the decay time constant in seconds for iodine-131?

23. **N** During the disaster at Fukushima, an estimate for the activity of iodine-131 released in the event was 355 PBq (Petabecquerels, 10^{15} Bq). The half-life of iodine-131 is 8.04 days. **a.** Assuming this was the initial activity, what will be the activity 2 days later? **b.** How many days must pass for the activity to drop to 1.00% of this initial value?

24. **N** Nitrogen-16 has a half-life of 7.13 s and decays to oxygen-16. If half of the ^{16}N nuclei in a sample have decayed, how much time has passed since the sample was created?

25. **C** Why don't we use carbon dating to measure the age of fossils that are roughly a hundred million years old?

26. **A** Integrate $dN/dt = -\eta N$ (Eq. 43.4) to show that the number of parent nuclei decreases exponentially according to $N = N_0 e^{-\eta t}$ (Eq. 43.5).

27. **A** Show that $a = a_0 e^{-\eta t}$ (Eq. 43.6), where $a_0 = \eta N_0$ is the activity at $t = 0$. Start with $N = N_0 e^{-\eta t}$ (Eq. 43.5).

28. **A** Show that the half-life is given by Equation 43.7, $T_{1/2} = (\ln 2)/\eta$.

29. **A** Show that $N = N_0(1/2)^{t/T_{1/2}}$ (Eq. 43.8).

43-5 The Weak Force

30. **E CASE STUDY** Before free neutrons in the early Universe had much time to beta decay, the ratio of neutrons to protons was $N_n/N_p \approx 1/5$. At this point, what was the abundance of helium $N_{He}/(N_H + N_{He})$? Express your answer as a percentage.

31. **C** Carbon-14 beta decays into another isotope as follows: $^{14}_{6}C \rightarrow ^{?}_{?}D + e^{-} + \bar{\nu}$. What is the daughter nucleus? Be sure to include the missing numbers.

32. **C** Use Equation 43.10 to express the beta decay process of each of the following parent isotopes: **a.** $^{60}_{26}$Fe, **b.** $^{10}_{4}$Be, **c.** $^{129}_{52}$Te.

33. **C** Use Equation 43.13 to express the electron capture process of each of the following parent isotopes: **a.** $^{188}_{77}$Ir, **b.** $^{127}_{54}$Xe, **c.** $^{57}_{27}$Co.

34. **C** Iodine undergoes a decay process as follows: $^{124}_{?}I \rightarrow ^{?}_{?}Te + e^{+} + \nu$. Rewrite this reaction, filling in the missing numbers. What sort of decay is this?

35. **C** Oxygen undergoes the following reaction: $^{15}_{?}O + e^{-} \rightarrow ^{15}_{?}? + \nu$. Rewrite this reaction, filling in all the missing information. What sort of reaction is this?

36. **C** Use Equation 43.12 to express the inverse beta decay process of each of the following parent isotopes: **a.** $^{23}_{12}$Mg, **b.** $^{40}_{19}$K, **c.** $^{15}_{8}$O.

37. C Oxygen undergoes the following reaction: $^{19}_{?}O \rightarrow ^{19}_{?}F + ? + ?$. Rewrite this reaction, filling in all the missing information. What sort of reaction is this?

43-6 Binding Energy

38. N In one example of nuclear fusion, deuterium (^{2}H) fuses with tritium (^{3}H) to form an alpha particle and a neutron. The rest mass energies of the deuterium and the tritium are 1875.62 MeV and 2808.92 MeV, respectively, whereas the rest mass energies for the alpha particle and the neutron are 3727.38 MeV and 939.57 MeV, respectively. **a.** What is the energy released in this fusion reaction? **b.** What is the mass deficit in this reaction?

39. N According to Table 43.2, the rest mass of iron $^{56}_{26}$Fe is 52103.06 MeV/c^2. Find its binding energy and its binding energy per nucleon. Check your answer with Figure 43.14.

40. The mass of a proton and that of a neutron are not exactly the same. The mass of a proton is 1.6726×10^{-27} kg, whereas the mass of a neutron is 1.6749×10^{-27} kg.
  a. N Calculate the rest mass energy of each nucleon.
  b. N A neutron can decay into a proton. The decay also results in the creation of other particles. What must be the total rest mass energy of these particles?
  c. C Compare the rest mass you found in part (b) to the rest mass energy of an electron (or positron).

41. N Uranium $^{238}_{92}$U has a binding energy of 1802 MeV. What is its mass deficit in atomic mass units?

42. N According to Figure 43.11, uranium-238 alpha decays into thorium-234. Calculate the energy released in this alpha decay. The rest masses of $^{238}_{92}$U, $^{234}_{90}$Th, and $^{4}_{2}$He are 238.05079 u, 234.04363 u, and 4.00260 u, respectively.

43. N Use the binding energy curve (Fig. 43.14) to find the binding energy per nucleon for $^{4}_{2}$He. Then find the binding energy for the whole nucleus.

44. N Consider the masses of ^{12}C (12.0000 u), ^{13}C (13.0034 u), and ^{11}C (11.0114 u). What is the binding energy per nucleon for each of these isotopes?

45. N Consider the masses of ^{6}Li (6.015122 u) and ^{7}Li (7.016004 u). What is the binding energy per nucleon for each of these isotopes?

43-7 Fission Reactions

46. C In each of the following cases, state whether or not the fission depicted is possible and explain why or why not.
  a. $^{236}_{92}U \rightarrow ^{141}_{56}Ba + ^{92}_{36}Kr + ^{1}_{0}n$
  b. $^{239}_{94}Pu + ^{1}_{0}n \rightarrow ^{148}_{58}Ce + ^{89}_{36}Kr + 3(^{1}_{0}n)$
  c. $^{235}_{92}U + ^{1}_{0}n \rightarrow ^{141}_{56}Ba + ^{92}_{36}Kr + 2(^{1}_{0}n)$

Problems 47, 48, and 49 are grouped

47. C Complete the following reaction by replacing ^A_ZX with the appropriate isotope: $^{239}_{94}Pu + ^{1}_{0}n \rightarrow ^{148}_{58}Ce + 3(^{1}_{0}n) + ^A_Z X$.

48. N Consider the fission reaction from the previous problem, where $^{239}_{94}$Pu (239.05216 u) undergoes fission into $^{148}_{58}$Ce (147.9242 u) and ^A_ZX (88.91764 u). Compute the binding energy of **a.** $^{239}_{94}$Pu, **b.** $^{148}_{58}$Ce, and **c.** ^A_ZX.

49. N Consider the fission reaction in Problem 47 and the atomic masses given in Problem 48. **a.** Determine the energy released when the reaction occurs once. **b.** How many of these reactions must occur to equal the total output energy of the Sun each day, about 2.1×10^{44} MeV.

50. C Suppose that an operator at a nuclear power plant loses the ability to control the movement of both the fuel and control rods in a fission reactor. What are the possible, disastrous consequences, and explain how they would occur?

43-8 Fusion Reactions

51. C In each of the following cases, state whether or not the fusion depicted is possible and explain why or why not.
  a. $^1_1H + ^3_1H \rightarrow ^3_2He + \gamma$
  b. $^4_2He + ^4_2He + ^4_2He \rightarrow ^{12}_6C$
  c. $^2_1H + ^3_1H \rightarrow ^4_2He + ^1_0n$

Problems 52 and 53 are paired

52. N The mass of tritium is 2808.9261 MeV/c^2. Use this and Table 43.2 to compute the mass deficit in the fusion reaction depicted in Equation 43.21, $^2_1H + ^3_1H \rightarrow ^4_2He + ^1_0n$.

53. N The mass of tritium is 2808.9261 MeV/c^2. Use this and Table 43.2 to verify that the energy released in the fusion reaction depicted in Equation 43.21, $^2_1H + ^3_1H \rightarrow ^4_2He + ^1_0n$ is 17.59 MeV.

Problems 54 and 55 are paired

54. N A fusion process called the *triple-alpha process* involves the fusing of three alpha particles, resulting in $^{12}_{6}$C. Although there is actually an intermediate step in the process, determine the mass deficit for the combination of three alpha particles resulting in $^{12}_{6}$C. Express your answer using the units MeV/c^2.

55. E, N A fusion process called the *triple-alpha process* involves the fusing of three alpha particles, resulting in $^{12}_{6}$C. Although there is actually an intermediate step in the process, determine an estimate for the energy released in this process.

43-9 Human Exposure to Radiation

56. C The quality factor for alpha radiation is much greater than that of X-rays, gamma radiation, or beta radiation. What is the meaning behind the difference in these values? In other words, why is the quality factor so much higher for alpha radiation?

57. N During a chest X-ray a patient is exposed to a dose equivalent of 0.3 mSv. The X-ray has an energy of 25 keV. If the portion of the chest that absorbed the radiation has a mass of 2.5 kg, how many X-ray photons were absorbed?

58. C How does the activity, a, affect the average absorbed dose, D_{av}? Suppose that you have a substance that emits alpha radiation of a particular energy. How will the activity affect the average absorbed dose? For example, will the effect be linear?

Problems 59 and 60 are paired.

59. N While walking outside of New Vegas, you pick up a piece of metal and are exposed to alpha radiation with a quality factor of 20. The dose absorbed by your body is 30.0 rad. **a.** What was the absorbed dose expressed in the units of grays (Gy)? **b.** What was the dose equivalent expressed in units of sieverts (Sv)? **c.** What was the dose equivalent expressed in units of rem?

60. C While walking outside of New Vegas, you pick up a piece of metal and are exposed to alpha radiation with a quality factor of 20. The dose absorbed by your body is 30.0 rad. Should you be worried about your health? Explain your answer.

61. N Beginning in the early 1920s, people sometimes exposed themselves to X-rays at the shoe store in search of a better-fitting shoe. The X-ray fluoroscope (Figure P43.61) was installed in shoe stores and offered customers the opportunity to see their feet in their shoes to observe the fit. Though primarily a gimmick, the machines remained in use in some stores until some time in the 1970s, at which point many states had banned their use. On average, during a single viewing, a user would be exposed to an equivalent dose of about 13 rem. The quality factor of X-rays is 1.0. What was the dose to which people were exposed when using the X-ray fluoroscope. Express you answer in both Gy and rad.

FIGURE P43.61

43-10 The Standard Model

62. **C** Given that the electron is a fermion, should the positron (e^+) be a fermion or a baryon? Justify your answer by considering a reaction that involves a positron.

63. **C** Is the following reaction possible: $^1_1p + ^1_0n \rightarrow ^1_1p + \mu^- + \mu^+$? Consider the baryon number before and after the reaction number. Explain your answer.

64. **C a.** Prior to the discovery of the electron antineutrino, its existence was predicted because of the varied amount of kinetic energy of the electron resulting from neutron decay. Given what you have learned about lepton number in this chapter, why is the following reaction dissatisfying: $^1_0n \rightarrow ^1_1p + e$? **b.** Include an electron antineutrino in this reaction such that the issue you identified in part (a) is resolved. Write the resulting reaction.

65. **N** The neutron is composed of an up quark and two down quarks. How many color combinations are possible for the two down quarks?

General Problems

66. **N** A bone found in a crypt is being dated. If 25% of the original amount of carbon-14 has decayed, what is the age of the bone? The half-life of ^{14}C is 5730 yr.

67. **C** A Bose–Einstein condensate is a form of matter consisting of bosons, all occupying the same ground state. Could a form of matter consisting of fermions occupying the same ground state be created? Explain your answer.

68. **C** The isotope $^{19}_8O$ is unstable. What do you expect will happen to make a more stable nucleus?

69. **N** The carbon cycle is a series of reactions beginning with the fusion of $^{12}_6C$ and ^1_1p, whereby the end result is $^{12}_6C$ and ^4_2He. If the addition of a proton is involved at three more points during the cycle, how many of these protons must also decay into positron and neutrino pairs?

70. **N** The element neon (Ne) has three stable isotopes with atomic mass numbers of 20, 21, and 22. The ratios of the natural abundances are $^{20}Ne/^{21}Ne = 335.11$ and $^{20}Ne/^{22}Ne = 9.78$. Find the natural abundance of each isotope.

71. **N** What atomic nucleus would have a volume that is three times greater than the volume occupied by a carbon-12 nucleus? Assume that the nuclei are spheres. Make your selection by comparing to atomic masses as listed on the periodic table of elements.

72. **N** Consider the masses of ^{16}O (15.994915 u), ^{17}O (16.999132 u), and ^{18}O (17.999160 u). What is the binding energy per nucleon for each of these isotopes?

Problems 73 and 74 are paired.

73. **N** A sample material is observed to have an activity of 1.45×10^9 decays/s. Express the activity in **a.** Bq, and **b.** Ci. **c.** If the material has a half-life of 30.07 yr, how long will it take for the activity to decrease by 10.0%? **d.** How long will it take for the activity to decrease by 20.0%?

74. **N** Consider the sample material in Problem 73. **a.** What is the number of radioactive nuclei when the activity is 1.45×10^9 decays/s? **b.** How long will it take for 10% of the nuclei to decay?

75. **N** The mass of ^{107}Ag is 106.9051 u. What is the binding energy of this silver nucleus?

76. One of the few stable isotopes with more protons than neutrons is ^3He (3.016 u).
    a. C Why is this such a rarity? Why are there not more stable isotopes of elements with more protons than neutrons?
    b. N What is the binding energy per nucleon of ^3He?

Problems 77 and 78 are paired

77. **N** The Chernobyl power station was capable of producing 4.00×10^3 MW of electrical power, prior to the disaster at one of its four reactors in 1986. Assuming that each fission reaction produces about 200 MeV and that the power plant is about 30.0% efficient, determine the number of fission events that must occur each second in the operation of the power plant. Retain three significant figures in your answer.

78. **N** The Chernobyl power station was capable of producing 4.00×10^3 MW of electrical power, prior to the disaster at one of its four reactors in 1986. Assuming that each fission reaction produces about 200 MeV and that the power plant is about 30.0% efficient, determine the mass of $^{235}_{92}U$ that must be used in operating the plant for 1.00 hours. Retain three significant figures in your answer.

Problems 79 and 80 are paired.

79. In nuclear fission, ^{235}U interacts with a neutron and could split into ^{90}Rb, ^{143}Cs, and three neutrons, releasing energy.
    a. C Express this nuclear reaction as an equation with ^{235}U and the neutron it absorbs on the left side of the equation.
    b. N What is the binding energy of ^{235}U (235.0439 u)?
    c. N ^{90}Rb (89.9148 u)?
    d. N ^{143}Cs (142.9278 u)?

80. **N** In nuclear fission, ^{235}U (235.0439 u) interacts with a neutron and could split into ^{90}Rb (89.9148 u), ^{143}Cs (142.9278 u), and three neutrons, releasing energy. What is the energy released in this reaction?

Mathematics

APPENDIX A

This appendix is not meant to serve as a tutorial or review. It is a short list of useful formulas. A more comprehensive list may be found in various handbooks such as the *Handbook of Chemistry and Physics* (Boca Raton, FL: CRC Press, published annually). These handbooks are available in hardcover and as e-books, and they make great birthday presents for science and engineering students.

A-1 Algebra and geometry App-1
A-2 Trigonometry App-2
A-3 Calculus App-3
A-4 Propagation of uncertainty App-5

A-1 Algebra and Geometry

Quadratic formula: If $ax^2 + bx + c = 0$, then $x = \dfrac{-b \pm \sqrt{b^2 - 4ac}}{2a}$

Factorial notation: $n! = n(n-1)\cdots 2 \cdot 1$

Binomial theorem: $(1 + x)^n = 1 + \dfrac{nx}{1!} + \dfrac{n(n-1)x^2}{2!} + \cdots \quad (x < 1)$

Commonly used approximation: $(1 + x)^n \approx 1 + nx \ (x \ll 1)$

Exponential expansion: $e^x = 1 + x + \dfrac{x^2}{2!} + \dfrac{x^3}{3!} + \cdots$

Logarithms (any base):
$$\log(x)^n = n \log x$$
$$\log(AB) = \log A + \log B$$
$$\log(A/B) = \log A - \log B$$

Logarithms (base 10):
$$10^{\log x} = x$$
$$\log 10^x = x$$

Natural logarithms (base e):
$$e^{\ln x} = x$$
$$\ln e^x = x$$

App-1

Area of common shapes:

Rectangle
$A = \ell w$

Circle
$A = \pi r^2$

Parallelogram
$A = bh$

Ellipse
$A = \pi ab$

Triangle
$A = \frac{1}{2}bh$

Equation of a straight line of slope m and y intercept b: $y = mx + b$

Equation of a parabola: $y = ax^2 + bx + c$

Equation of a circle: $x^2 + y^2 = r^2$

Equation of an ellipse: $\left(\dfrac{x}{a}\right)^2 + \left(\dfrac{y}{b}\right)^2 = 1$

Circumference of a circle: $c = 2\pi r$

Volume and surface area of common solids:

Rectangular box $V = \ell wh$ $A = 2(\ell w + \ell h + wh)$

Right circular cylinder $V = \pi r^2 h$ $A = 2\pi r^2 + 2\pi rh$

Sphere $V = \frac{4}{3}\pi r^3$ $A = 4\pi r^2$

A-2 Trigonometry

Pythagorean theorem (applied to a right triangle): $x^2 + y^2 = r^2$

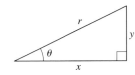

Trigonometric functions (applied to a right triangle):

$\sin \theta = \dfrac{y}{r}$ $\csc \theta = \dfrac{1}{\sin \theta} = \dfrac{r}{y}$

$\cos \theta = \dfrac{x}{r}$ $\sec \theta = \dfrac{1}{\cos \theta} = \dfrac{r}{x}$

$\tan \theta = \dfrac{y}{x}$ $\cot \theta = \dfrac{1}{\tan \theta} = \dfrac{x}{y}$

Commonly used approximations:

$$\sin\theta \approx \tan\theta \approx \theta \text{ for small } \theta \text{ in radians}$$
$$\cos\theta \approx 1 \text{ for small } \theta$$

Trigonometric identities:

$\sin(-\theta) = -\sin\theta$	$\sin 2\theta = 2\sin\theta\cos\theta$
$\cos(-\theta) = \cos\theta$	$\cos 2\theta = \cos^2\theta - \sin^2\theta$
$\tan(-\theta) = -\tan\theta$	
$\sin(90° - \theta) = \sin\left(\frac{\pi}{2} - \theta\right) = \cos\theta$	$\sin(\alpha \pm \beta) = \sin\alpha\cos\beta \pm \cos\alpha\sin\beta$
	$\cos(\alpha \pm \beta) = \cos\alpha\cos\beta \mp \sin\alpha\sin\beta$
$\cos(90° - \theta) = \cos\left(\frac{\pi}{2} - \theta\right) = \sin\theta$	
$\sin(90° + \theta) = \sin\left(\frac{\pi}{2} + \theta\right) = \cos\theta$	$\cos\theta = \sin(\theta + 90°)$
$\cos(90° + \theta) = \cos\left(\frac{\pi}{2} + \theta\right) = -\sin\theta$	
$\tan\theta = \dfrac{\sin\theta}{\cos\theta}$	$\sin\alpha \pm \sin\beta = 2\sin\tfrac{1}{2}(\alpha \pm \beta)\cos\tfrac{1}{2}(\alpha \mp \beta)$
	$\cos\alpha + \cos\beta = 2\cos\tfrac{1}{2}(\alpha + \beta)\cos\tfrac{1}{2}(\alpha - \beta)$
$\sin^2\theta + \cos^2\theta = 1$	$\cos\alpha - \cos\beta = -2\sin\tfrac{1}{2}(\alpha + \beta)\sin\tfrac{1}{2}(\alpha - \beta)$

A-3 Calculus

Derivatives

In this section, f, g, u, and v are functions of x; a, b, C, and n are constants.

Derivative of $f(x)$: $\dfrac{df}{dx} = \lim\limits_{\delta x \to 0} \dfrac{f(x + \delta x) + f(x)}{\delta x}$

Derivative of a constant: $\dfrac{dC}{dx} = 0$

The power rule: $\dfrac{dx^n}{dx} = nx^{n-1}$

The derivative of a sum: $\dfrac{d}{dx}[f(x) + g(x)] = \dfrac{df}{dx} + \dfrac{dg}{dx}$

The product rule: $\dfrac{d}{dx}[f(x)g(x)] = g\dfrac{df}{dx} + f\dfrac{dg}{dx}$

Special case of the product rule: $\dfrac{d}{dx}[Cf(x)] = C\dfrac{df}{dx}$

The chain rule: $\dfrac{df}{dx} = \dfrac{df}{du}\dfrac{du}{dx}$

Second derivative: $\dfrac{d^2f}{dx^2} = \dfrac{d}{dx}\left(\dfrac{df}{dx}\right)$

Derivatives of special functions:

$$\frac{d}{dx}\ln x = \frac{1}{x} \qquad \frac{d}{dx}\sin x = \cos x$$

$$\frac{d}{dx}e^x = e^x \qquad \frac{d}{dx}\cos x = -\sin x$$

$$\frac{d}{dx}e^u = e^u \frac{du}{dx} \qquad \frac{d}{dx}\tan x = \sec^2 x$$

L'Hôpital's rule: If the limit of the numerator $f(x)$ and of the denominator $g(x)$ of a fraction both approach zero or both approach infinity, the limit of the fraction is indeterminate in the form of type $0/0$ or ∞/∞. In such cases, it may be possible to find the limit using L'Hôpital's rule:

$$\lim \frac{f(x)}{g(x)} = \lim \frac{df/dx}{dg/dx}$$

Essentially, L'Hôpital's rule replaces the limit of a fraction with the limit of a new fraction, where the numerator and denominator are the derivatives of their original counterparts.

Integrals

Indefinite integral or antiderivative: $\int df(x) = f(x) + C$

Indefinite integral of a constant: $\int a\,dx = ax + C$

The power rule: $\int x^n\,dx = \frac{x^{n+1}}{n+1} + C\ (n \neq -1)$

Indefinite integral of a sum: $\int [f(x) + g(x)]\,dx = \int f(x)\,dx + \int g(x)\,dx$

Indefinite integrals of particular functions:

$$\int \frac{1}{x}\,dx = \ln|x| + C$$

$$\int e^x\,dx = e^x + C$$

$$\int e^{-ax}\,dx = -\frac{1}{a}e^{-ax} + C$$

$$\int (\sin ax)\,dx = -\frac{1}{a}\cos ax + C$$

$$\int (\cos ax)\,dx = \frac{1}{a}\sin ax + C$$

$$\int (\tan ax)\,dx = -\frac{1}{a}\ln(\cos ax) + C$$

$$\int \sin^2 ax\,dx = \frac{x}{2} - \frac{\sin 2ax}{4a} + C$$

$$\int \cos^2 ax\,dx = \frac{x}{2} + \frac{\sin 2ax}{4a} + C$$

$$\int \frac{dx}{\sqrt{x^2+a^2}} = \ln|x + \sqrt{x^2+a^2}| + C$$

$$\int \frac{dx}{\sqrt{a^2-x^2}} = \sin^{-1}\frac{x}{|a|} + C$$

$$\int \frac{dx}{x^2+a^2} = \frac{1}{a}\tan^{-1}\frac{x}{a} + C$$

$$\int \frac{dx}{(x^2+a^2)^{3/2}} = \frac{x}{a^2\sqrt{x^2+a^2}} + C$$

$$\int \frac{x\,dx}{(x^2+a^2)^{3/2}} = -\frac{1}{\sqrt{x^2+a^2}} + C$$

U substitution integration method: $\int f(x)\frac{df}{dx}\,dx = \int u\,du$, where $u \equiv f(x)$ and $du = \frac{df}{dx}\,dx$

Integration by parts: $\int u\,dv = uv - \int v\,du$

Definite integrals and the Fundamental Theorem of Calculus:
If $F(x)$ is continuous on the interval from $x = a$ to b, then

$$\int_a^b F(x)\,dx = f(x)\big|_a^b = f(b) - f(a)$$

where $f(x)$ is the antiderivative of $F(x)$ and a and b are known as the **limits**.

Average value of a function in the interval from $x = a$ to $x = b$: $f_{av} = \dfrac{1}{b-a}\displaystyle\int_a^b f(x)\,dx$

Definite integrals of particular functions:

$$\int_0^\infty x^n e^{-ax}\,dx = \frac{n!}{a^{n+1}}$$

$$\int_0^\pi (\sin^2 ax)\,dx = \frac{\pi}{2}$$

$$\int_0^\infty e^{-ax^2}\,dx = \frac{1}{2}\sqrt{\frac{\pi}{a}}$$

$$\int_0^\pi (\cos^2 ax)\,dx = \frac{\pi}{2}$$

$$\int_0^\infty x e^{-ax^2}\,dx = \frac{2}{a}$$

$$\int_0^\infty e^{-ax}(\sin nx)\,dx = \frac{n}{a^2+n^2}\quad (a>0)$$

$$\int_0^\infty x^2 e^{-ax^2}\,dx = \frac{1}{4}\sqrt{\frac{\pi}{a^3}}$$

$$\int_0^\infty e^{-ax}(\cos nx)\,dx = \frac{a}{a^2+n^2}\quad (a>0)$$

A-4 Propagation of Uncertainty

Physics is an experimental science that relies on measurements. All measurements have uncertainty or error. We often seek quantities that result from the combination of uncertain measurements, so the resulting quantity is also uncertain. One way to estimate the uncertainty δQ in a quantity q calculated from uncertain measurements is to find the best estimate Q and then the extreme possible values of q.

For example, if $q = 2(a+b)/c$, $a = 5.2 \pm 0.2$, $b = 7.5 \pm 0.3$, and $c = 57.6 \pm 0.5$, the best estimate of q is

$$Q = \frac{2(5.2 + 7.5)}{57.6} = 0.44$$

The maximum value of q is

$$Q_{max} = \frac{2(5.4 + 7.8)}{57.1} = 0.46$$

where we chose the maximum possible values in the numerator and the minimum possible value in the denominator. Similarly, we find that the minimum value of q is

$$Q_{min} = \frac{2(5.0 + 7.2)}{58.1} = 0.42$$

The quantity must fall between its minimum and maximum values: $0.42 \le q \le 0.46$. So,

$$q = Q \pm \delta Q = 0.44 \pm 0.02$$

where $\delta Q = Q_{max} - Q = Q - Q_{min}$.

We apply this technique to come up with three rules for propagating uncertainty.

Sums and Differences

If $q = a + b$, $a = A \pm \delta A$, and $b = B \pm \delta B$, then

$$q = Q \pm \delta Q = (A + B) \pm (\delta A + \delta B)$$

If $q = a - b$, $a = A \pm \delta A$, and $b = B \pm \delta B$, then

$$q = Q \pm \delta Q = (A - B) \pm (\delta A + \delta B)$$

When measured quantities are added or subtracted, their errors add.[1]

---

[1] If the original uncertainties δA and δB are random and independent, then $\delta A + \delta B$ is an overestimate of the propagated error.

Products, Quotients, and Powers

If $a = A \pm \delta A$, the *fractional uncertainty* in a is $\delta A/|A|$. If $q = ab$, $a = A \pm \delta A$, and $b = B \pm \delta B$, then $Q = AB$, and the fractional uncertainty in q is approximately

$$\frac{\delta Q}{|Q|} \approx \frac{\delta A}{|A|} + \frac{\delta B}{|B|} \qquad (1)$$

and

$$q = Q \pm \delta Q \approx AB \pm |AB|\left(\frac{\delta A}{|A|} + \frac{\delta B}{|B|}\right)$$

If $q = a/b$, $a = A \pm \delta A$, and $b = B \pm \delta B$, then $Q = A/B$. The fractional uncertainty in q is given by Equation (1), and

$$q = Q \pm \delta Q \approx \frac{A}{B} \pm \left|\frac{A}{B}\right|\left(\frac{\delta A}{|A|} + \frac{\delta B}{|B|}\right)$$

If $q = a^n$ and $a = A \pm \delta A$, then $Q = A^n$, and the fractional uncertainty in q is

$$\frac{\delta Q}{|Q|} \approx n\frac{\delta A}{|A|}$$

and

$$q = Q \pm \delta Q \approx A^n \pm |A^n|\left(n\frac{\delta A}{|A|}\right)$$

When measured quantities are multiplied or divided, their fractional errors add.

Multiplication by an Exact Number

If $q = ab$, a is exact, and $b = B \pm \delta B$, then

$$q = Q \pm \delta Q = aB \pm |a|\delta B$$

Reference Tables

APPENDIX B

- B-1 Symbols and units App-7
- B-2 Conversion factors App-9
- B-3 Some astronomical data App-10
- B-4 Rough magnitudes and scales App-11
- Periodic table of the elements App-14

B-1 Symbols and Units

Prefixes for powers of 10

Name	Abbreviation	Value
yocto	y	10^{-24}
zepto	z	10^{-21}
atto	a	10^{-18}
femto	f	10^{-15}
pico	p	10^{-12}
nano	n	10^{-9}
micro	μ (Greek letter "mu")	10^{-6}
milli	m	10^{-3}
centi	c	10^{-2}
deci	d	10^{-1}
deka	da	10^{1}
hecto	h	10^{2}
kilo	k	10^{3}
mega	M	10^{6}
giga	G	10^{9}
tera	T	10^{12}
peta	P	10^{15}
exa	E	10^{18}
zetta	Z	10^{21}
yotta	Y	10^{24}

Greek alphabet

Name	Uppercase	Lowercase	Name	Uppercase	Lowercase	Name	Uppercase	Lowercase
Alpha	A	α	Iota	I	ι	Rho	P	ρ
Beta	B	β	Kappa	K	κ	Sigma	Σ	σ
Gamma	Γ	γ	Lambda	Λ	λ	Tau	T	τ
Delta	Δ	δ	Mu	M	μ	Upsilon	Y	υ
Epsilon	E	ε	Nu	N	ν	Phi	Φ	φ
Zeta	Z	ζ	Xi	Ξ	ξ	Chi	X	χ
Eta	H	η	Omicron	O	o	Psi	Ψ	ψ
Theta	Θ	θ	Pi	Π	π	Omega	Ω	ω

SI base units

Dimension	SI unit	Symbol	Definition
Time	second	s	1 second is the duration of 9,192,631,770 periods of the radiation (corresponding to the transition between hyperfine levels of the ground state) of the cesium-133 atom.
Length	meter	m	1 meter is the distance light travels through empty space in 1/299,729,458 second.
Mass	kilogram	kg	1 kilogram is the mass of a prototype (a particular platinum-iridium cylinder).
Thermodynamic temperature	kelvin	K	1 kelvin is the fraction 1/273.16 of the thermodynamic temperature of the triple point of water.
Amount of substance	mole	mol	1 mole is the amount of a substance of a system that contains as many elementary entities as there are atoms in 0.012 kg of carbon-12.
Electrical current	ampere	A	1 ampere is the constant current that, if maintained in two straight parallel conductors of infinite length, of negligible circular cross section, and placed 1 m apart in a vacuum, would produce between these conductors a force per unit length equal to 2×10^{-7} N/m.
Luminous intensity	candela	cd	1 candela is the luminous intensity, in a given direction, of a source that emits monochromatic radiation of frequency 540×10^{12} Hz and that has a radiant intensity in that direction of 1/683 watt per steradian.

Source: Adapted from "Definitions of the SI base units," National Institute of Standards and Technology. See http://physics.nist.gov/cuu/Units/current.html.

Symbols and abbreviations for units

Unit	Symbol	Unit	Symbol
ampere	A	light-year	ly
atmosphere	atm	liter	L
atomic mass unit	U	meter	m
British thermal unit	Btu	mile	mi
calorie	cal	miles per hour	mph
coulomb	C	millimeter of mercury (torricelli)	mm Hg (torr)
day	d	minute	min
degree Celsius	°C	mole	mol
degree Fahrenheit	°F	newton	N
electron volt	eV	ohm	Ω
farad	F	pascal	Pa
foot	ft	pound	lb
gallon	gal	pounds per square inch	psi
gauss	G	radian	rad
gram	g	revolution	rev
henry	H	revolutions per minute	rpm
hertz	Hz	second	s
horsepower	hp	tesla	T
inch	in.	volt	V
joule	J	watt	W
kelvin	K	weber	Wb
kilocalorie	Cal	yard	yd
kilogram	kg	year	yr
kilowatt-hour	kWh		

B-2 Conversion Factors

Length

	meter	cm	km	in.	ft	mi
1 meter	1	10^2	10^{-3}	39.37	3.281	6.214×10^{-4}
1 centimeter	10^{-2}	1	10^{-5}	0.3937	3.281×10^{-2}	6.214×10^{-6}
1 kilometer	10^3	10^5	1	3.937×10^4	3281	0.6214
1 inch	2.540×10^{-2}	2.540	2.540×10^{-5}	1	8.333×10^{-2}	1.578×10^{-5}
1 foot	0.3048	30.48	3.048×10^{-4}	12	1	1.894×10^{-4}
1 mile	1609	1.609×10^5	1.609	6.336×10^4	5280	1
1 angstrom	10^{-10}	10^{-8}	10^{-13}	3.937×10^{-9}	3.281×10^{-10}	6.214×10^{-14}
1 AU	1.496×10^{11}	1.496×10^{13}	1.496×10^8	5.890×10^{12}	4.908×10^{11}	9.296×10^7
1 nautical mile	1852	1.852×10^5	1.852	7.291×10^4	6076	1.151
1 light-year	9.461×10^{15}	9.461×10^{17}	9.461×10^{12}	3.725×10^{17}	3.104×10^{16}	5.878×10^{12}
1 parsec	3.086×10^{16}	3.086×10^{18}	3.086×10^{13}	1.215×10^{18}	1.012×10^{17}	1.917×10^{13}
1 yard	0.9144	91.44	9.144×10^{-4}	36	3	5.682×10^{-4}

Mass

	kilogram	g	slug	u
1 kilogram	1	10^3	6.852×10^{-2}	6.022×10^{26}
1 gram	10^{-3}	1	6.852×10^{-5}	6.022×10^{23}
1 slug	14.59	1.459×10^4	1	8.786×10^{27}
1 atomic mass unit	$1.6605402 \times 10^{-27}$	1.661×10^{-24}	1.138×10^{-28}	1

Force

	newton	dyne	lb	oz	ton
1 newton	1	10^5	0.2248	3.597	1.124×10^{-4}
1 dyne	10^{-5}	1	2.248×10^{-6}	3.597×10^{-5}	1.124×10^{-9}
1 pound	4.448	4.448×10^5	1	16	5×10^{-4}
1 ounce	0.2780	2.780×10^4	6.250×10^{-2}	1	3.125×10^{-5}
1 ton	8.896×10^3	8.896×10^8	2000	3.2×10^4	1

Pressure

	pascal	atm	Torr (mm Hg)	psi	dyne/cm²
1 pascal	1	9.869×10^{-6}	7.501×10^{-3}	1.450×10^{-4}	10
1 atm	1.013×10^5	1	760	14.70	1.013×10^6
1 Torr	1333	1.316×10^{-2}	1	0.1934	1.333×10^4
1 psi	6.895×10^3	6.805×10^{-2}	51.71	1	6.895×10^4
1 dyne/cm²	0.1	9.869×10^{-7}	7.501×10^{-4}	1.405×10^{-5}	1

Energy

	joule	erg	ft·lb	cal	eV
1 joule	1	10^7	0.7376	0.2389	6.242×10^{18}
1 erg	10^{-7}	1	7.376×10^{-8}	2.389×10^{-8}	6.242×10^{11}
1 ft·lb	1.356	1.356×10^7	1	0.3238	8.464×10^{18}
1 cal	4.184	4.184×10^7	3.088	1	2.612×10^{19}
1 eV	1.602×10^{-19}	1.602×10^{-19}	1.182×10^{-19}	3.827×10^{-20}	1

B-3 Some Astronomical Data

Object	Symbol	Rotation period (hh:mm:ss.s or days)	Mass ($\times 10^{24}$ kg)	Equatorial radius ($\times 10^6$ m)	Free-fall acceleration near surface (m/s^2)	Escape speed (km/s)	Blackbody temperature (K)
Sun	☉	≈ 25 to 36 days[1]	1.9891×10^6	695.51	274	618	5777
Mercury	☿	58.65 days	0.3302	2.4397	3.7	4.3	440.1
Venus	♀	243 days	4.87	6.052	8.9	10.36	184.2
Earth	⊕	23:56:4.1	5.9736	6.378136	9.81	11.186	254.3
Moon	☾	27.3 days	0.07	1.738	1.6	2.38	270.7
Mars	♂	24:37:22.6	0.64	3.397	3.7	5.03	210.1
Ceres	⚳	09:04:19	9.6×10^{-4}	0.48			239
Jupiter	♃	9:50:30	1900	71.493	24.8	59.5	110.0
Saturn	♄	10:14:00	569	60.268	10.4	35.5	81.1
Uranus	♅	17:14:00	87	25.559	8.87	21.3	58.2
Neptune	♆	16:03:00	103	24.764	11.2	23.5	46.6
Pluto	♇	6.387 days	0.01	1.135	0.58	1.2	37.5
Eris			≈ 10^{-2}	1.2			30

[1]The Sun is gaseous and does not rotate as a solid body; its period near the equator is shorter than at the poles.

Orbital parameters for objects that orbit the Sun

Object	Orbital period (days or years)	Semimajor axis (AU)	Eccentricity
Mercury	87.969 days	0.387	0.2056
Venus	224.701 days	0.723	0.0067
Earth	365.26 days	1.000	0.0167
Mars	1.8808 years	1.524	0.0935
Ceres	4.603 years	2.767	0.097
Jupiter	11.8618 years	5.204	0.0489
Saturn	29.4567 years	9.5482	0.0565
Uranus	84.0107 years	19.201	0.0457
Neptune	164.79 years	30.047	0.0113
Pluto	247.68 years	39.482	0.2488
Eris	559 years	67.89	0.4378

Some natural satellites

Satellite	Planet	Orbital period (days)	Semimajor axis (10^6 m)	Mass (10^{22} kg)	Radius (10^6 m)
Moon	Earth	27.322	384.4	7.349	1.7371
Io	Jupiter	1.769	421.6	8.932	1.8216
Europa	Jupiter	3.551	670.9	4.800	1.5608
Ganymede	Jupiter	7.155	1070.4	14.819	2.6312
Callisto	Jupiter	16.689	1882.7	10.759	2.4103
Titan	Saturn	15.945	1221.8	13.455	2.575
Triton	Neptune	5.877	354.8	2.14	1.3534

B-4 Rough Magnitudes and Scales

Numbers

Quantity	Approximate value or order of magnitude
Number of atoms in the Earth	10^{50}
Number of atoms in a 70-kg person	7×10^{27}
Number of cells in a person	5×10^{13}
Number of mobile phones in U.S.	330 million
Number of mobile phones worldwide	5 billion
Number of dogs in U.S.	78 million
Number of dogs in Italy	8 million
Population of the Earth	7 billion
Population of students at University of CA	220,000
Population of U.S.	300 million
Population of China	1.3 billion
Population of New York City	8 million
Population of Annapolis, MD	38,000
Population of Chesterton, IN	13,000
Population of Morris, MN	5,000
Veterans in U.S.	23 million
Percentage of people in U.S. under age of 18	24%
Money spent in film investments in U.S.	$15 billion
Money spend in film investments in India	$200 million

Sizes: Lengths, diameters, areas, and volumes

Quantity	SI or metric units	U.S. customary units
Area of a $1 bill	100 cm^2	17 in.^{2}
Area of a typical college campus	15 km^2	6.5 mi^2
Area of a cell phone	30 cm^2	5 in.^{2}
Area of continents	1.5×10^{14} m^2	6×10^7 mi^2
Area of oceans	3.6×10^{14} m^2	1.4×10^8 mi^2
Area of palm	40 cm^2	6 in.^{2}
Area of U.S. land	9×10^{12} m^2	3.5×10^6 mi^2
Average human stride	1 m	1 yd
Diameter of a hydrogen atom	10^{-10} m	
Diameter of a pollen grain	10–100 μm	
Diameter of a proton	10^{-15} m	
Diameter of a U.S. nickel	2.121 cm	0.835 in.
Diameter of the Milky Way galaxy	10^{21} m	10^5 ly
Height of a typical adult human	2 m	5–6 ft
Height of a typical story	3 m	10 ft
Length of a house fly	0.5 cm	0.2 in.
Length of a human thumb	5 cm	2 in.
Length of a match stick	5 cm	2 in.
Size of a living cell	10 μm	
Size of the smallest visible dust particle	0.1 μm	
Thickness of a human hair	50 μm	
Thickness of a U.S. nickel	1.95 mm	
Width of human finger	1–3 cm	

Speeds

	SI or metric units	U.S. customary units
Top speed of a car	200 km/h	120 mph
Top speed of a *typical* bicycle	50 km/h	30 mph
Walking	1.3 m/s	3 mph
Running or jogging	15 km/h	6-min. mile (10 mph)
Commercial airplane cruising speed	250 m/s	550 mph
Speed of a snail	1 mm/s	2–3 inch/min
Speed of a cheetah	28 m/s	62 mph
Speed of a rifle bullet	700 m/s	1600 mph

Weights and masses

	SI or metric units	U.S. customary units
Mass of a U.S. nickel	5.000 g	3×10^{-4} slug (approx)
Weight of a car	10000–20000 N	1–2 tons
Mass of a car	1000–2000 kg	70–140 slug
Weight of a physics book	50 N	10 lb
Mass of a physics book	5 kg	0.4 slug
Weight of a U.S. quarter	6×10^{-2} N	0.2 oz
Mass of a U.S. quarter	6 g	4×10^{-4} slug
Mass of the Milky Way galaxy	10^{42} kg	10^{41} slug
Mass of an elephant	5×10^3 kg	340 slug
Mass of a frog	100 g	7×10^{-3} slug
Mass of a house fly	8–20 mg	$(5–14) \times 10^{-4}$ slug
Weight of an adult human	500–1000 N	110–200 lbs
Mass of an adult human	50–100 kg	4–7 slug

Times, ages, periods, frequencies, and angular momentum

Quantity	Convenient units
Resting heart beat	60–80 per min
Age of the Universe	14 billion years
Age of human written history	10^4 years
Age of the Earth	4.5 billion years
Age of oldest fossil	2.7 billion years
Time for light to travel from the Sun to the Earth	10 min
Time for light to cross the diameter of a proton	3.3×10^{-24} s
Period of Halley's comet	2.4×10^9 s
Period of a typical x-ray	10^{-19} s
Time for light to travel from nearest star	4.3 years
Angular speed of record turntable	33 rpm
Angular momentum of record (33 rpm)	6 mJ · s
Angular momentum of electric fan	1 J · s
Angular momentum of Frisbee	0.1 J · s
Angular momentum of helicopter rotor (320 rpm)	5×10^4 J · s

Periodic Table of the Elements

Legend:
- Symbol: Ca
- Atomic number: 20
- Atomic mass†: 40.078
- Electron configuration: $4s^2$

Group I	Group II				Transition elements				
H 1 1.0079 $1s$									
Li 3 6.941 $2s^1$	**Be** 4 9.0122 $2s^2$								
Na 11 22.990 $3s^1$	**Mg** 12 24.305 $3s^2$								
K 19 39.098 $4s^1$	**Ca** 20 40.078 $4s^2$	**Sc** 21 44.956 $3d^14s^2$	**Ti** 22 47.867 $3d^24s^2$	**V** 23 50.942 $3d^34s^2$	**Cr** 24 51.996 $3d^54s^1$	**Mn** 25 54.938 $3d^54s^2$	**Fe** 26 55.845 $3d^64s^2$	**Co** 27 58.933 $3d^74s^2$	
Rb 37 85.468 $5s^1$	**Sr** 38 87.62 $5s^2$	**Y** 39 88.906 $4d^15s^2$	**Zr** 40 91.224 $4d^25s^2$	**Nb** 41 92.906 $4d^45s^1$	**Mo** 42 95.94 $4d^55s^1$	**Tc** 43 (98) $4d^55s^2$	**Ru** 44 101.07 $4d^75s^1$	**Rh** 45 102.91 $4d^85s^1$	
Cs 55 132.91 $6s^1$	**Ba** 56 137.33 $6s^2$	57–71*	**Hf** 72 178.49 $5d^26s^2$	**Ta** 73 180.95 $5d^36s^2$	**W** 74 183.84 $5d^46s^2$	**Re** 75 186.21 $5d^56s^2$	**Os** 76 190.23 $5d^66s^2$	**Ir** 77 192.2 $5d^76s^2$	
Fr 87 (223) $7s^1$	**Ra** 88 (226) $7s^2$	89–103**	**Rf** 104 (261) $6d^27s^2$	**Db** 105 (262) $6d^37s^2$	**Sg** 106 (266)	**Bh** 107 (264)	**Hs** 108 (277)	**Mt** 109 (268)	

*Lanthanide series

La 57 138.91 $5d^16s^2$	**Ce** 58 140.12 $5d^14f^16s^2$	**Pr** 59 140.91 $4f^36s^2$	**Nd** 60 144.24 $4f^46s^2$	**Pm** 61 (145) $4f^56s^2$	**Sm** 62 150.36 $4f^66s^2$

**Actinide series

Ac 89 (227) $6d^17s^2$	**Th** 90 232.04 $6d^27s^2$	**Pa** 91 231.04 $5f^26d^17s^2$	**U** 92 238.03 $5f^36d^17s^2$	**Np** 93 (237) $5f^46d^17s^2$	**Pu** 94 (244) $5f^67s^2$

Note: Atomic mass values given are averaged over isotopes in the percentages in which they exist in nature.
† For an unstable element, mass number of the most stable known isotope is given in parentheses.

Periodic Table of the Elements

	Group III	Group IV	Group V	Group VI	Group VII	Group 0
					H 1 1.007 9 $1s^1$	**He** 2 4.002 6 $1s^2$
	B 5 10.811 $2p^1$	**C** 6 12.011 $2p^2$	**N** 7 14.007 $2p^3$	**O** 8 15.999 $2p^4$	**F** 9 18.998 $2p^5$	**Ne** 10 20.180 $2p^6$
	Al 13 26.982 $3p^1$	**Si** 14 28.086 $3p^2$	**P** 15 30.974 $3p^3$	**S** 16 32.066 $3p^4$	**Cl** 17 35.453 $3p^5$	**Ar** 18 39.948 $3p^6$

Ni 28 58.693 $3d^84s^2$	**Cu** 29 63.546 $3d^{10}4s^1$	**Zn** 30 65.41 $3d^{10}4s^2$	**Ga** 31 69.723 $4p^1$	**Ge** 32 72.64 $4p^2$	**As** 33 74.922 $4p^3$	**Se** 34 78.96 $4p^4$	**Br** 35 79.904 $4p^5$	**Kr** 36 83.80 $4p^6$
Pd 46 106.42 $4d^{10}$	**Ag** 47 107.87 $4d^{10}5s^1$	**Cd** 48 112.41 $4d^{10}5s^2$	**In** 49 114.82 $5p^1$	**Sn** 50 118.71 $5p^2$	**Sb** 51 121.76 $5p^3$	**Te** 52 127.60 $5p^4$	**I** 53 126.90 $5p^5$	**Xe** 54 131.29 $5p^6$
Pt 78 195.08 $5d^96s^1$	**Au** 79 196.97 $5d^{10}6s^1$	**Hg** 80 200.59 $5d^{10}6s^2$	**Tl** 81 204.38 $6p^1$	**Pb** 82 207.2 $6p^2$	**Bi** 83 208.98 $6p^3$	**Po** 84 (209) $6p^4$	**At** 85 (210) $6p^5$	**Rn** 86 (222) $6p^6$
Ds 110 (271)	**Rg** 111 (272)	**Cn** 112 (285)	113†† (284)	**Fl** 114 (289)	115†† (288)	**Lv** 116 (293)	117†† (294)	118†† (294)

Eu 63 151.96 $4f^76s^2$	**Gd** 64 157.25 $4f^75d^16s^2$	**Tb** 65 158.93 $4f^85d^16s^2$	**Dy** 66 162.50 $4f^{10}6s^2$	**Ho** 67 164.93 $4f^{11}6s^2$	**Er** 68 167.26 $4f^{12}6s^2$	**Tm** 69 168.93 $4f^{13}6s^2$	**Yb** 70 173.04 $4f^{14}6s^2$	**Lu** 71 174.97 $4f^{14}5d^16s^2$
Am 95 (243) $5f^77s^2$	**Cm** 96 (247) $5f^76d^17s^2$	**Bk** 97 (247) $5f^86d^17s^2$	**Cf** 98 (251) $5f^{10}7s^2$	**Es** 99 (252) $5f^{11}7s^2$	**Fm** 100 (257) $5f^{12}7s^2$	**Md** 101 (258) $5f^{13}7s^2$	**No** 102 (259) $5f^{14}7s^2$	**Lr** 103 (262) $5f^{14}6d^17s^2$

††Elements 113, 115, 117, and 118 have not yet been officially named. Only small numbers of atoms of these elements have been observed.

Note: For a description of the atomic data, visit *physics.nist.gov/PhysRefData/Elements/per_text.html*.

Answers to Concept Exercises and Odd-Numbered Problems

CHAPTER 23: Concept Exercises

23.1 Because electrons are transferred from the glass to the silk, the silk has the same number of excess electrons as the number of excess protons in the glass. So the charge of the silk is negative as given by Equation 23.2:

$$q = -Ne = -(3.33 \times 10^{11})(1.60 \times 10^{-19} \text{ C})$$
$$q = 5.33 \times 10^{-8} \text{ C} = 53.3 \text{ nC}$$

23.2 a. The red rod is rubbed with the red cloth, so they have charges of equal magnitude and opposite sign. We represent this fact by drawing the same number of signs on the red rod and the red cloth, three plus signs on the rod and three minus signs on the cloth. **b.** The green rod has the greatest positive charge because it is shown with the greatest number of plus signs—five in this figure.

23.3 To determine whether two charged objects are attracted to or repelled by each other, remember that *opposites attract*. The two rods are attracted to each other because the glass rod is positively charged and the amber rod is negatively charged. The two cloths are attracted to each other because the silk is negatively charged and the wool is positively charged. The silk is repelled by the amber rod because both are negatively charged. The wool and the glass rod are mutually repelled because both are positively charged.

23.4 a. Yes. Because A is repelled by B, we know that A and B must have charges of the same sign. Suppose both are positive. Because A is attracted to C, we know that A and C have charges of the opposite sign; if A is positive, then C is negative. So B must also be attracted to C.
b. No. If A is attracted to B, then A and B must have charges of opposite sign; if A is positive, then B is negative. In addition, if A is attracted to C, then A and C must also have charges of opposite sign; if A is positive, then C is negative. That means B is repelled by C because B and C are both the same sign—negative.

If you ever discover three charged objects such that A is attracted to B and both are attracted to C, you will be dealing with charge that doesn't come in only two types—positive and negative. It must come in three types.

23.5 a. No. The metal ball may be either positive or neutral. If the metal ball is positive, it is attracted to the negative rod because *opposites attract*. If the ball is neutral, it is attracted to the charged rod because the rod causes the ball to be polarized, and it is attracted to the rod just like the can in Figure 23.16B (page 693).
b. Yes. The metal ball must have a negative charge. If the ball were positive or neutral, it would be attracted to the rod as explained in part (a). The only way for the ball to be repelled is if the ball and the rod have charges of the same sign—in this case, negative.

23.6 a. Silk is an insulator. The silk rope is used so that the boy is not grounded. If he were grounded, he would discharge and the fun would be over. A metal chain would ground the boy by providing a pathway for electrons to go from the boy through the chain, through the building, and into the Earth.
b. The boy is charged, so he can induce dipoles in the paper. Like the comb in Figure 23.5C (page 686), the boy is able to attract paper and turn the pages of a book without touching it.
c. The woman is about to provide a pathway from the boy to the Earth; in other words, she is about to ground him. If the boy has built up a large charge, when the woman gets very close to him, the air will act as a conductor, momentarily completing the pathway to ground even before she makes contact. When that happens, there will be a spark as in the case of a lightning strike (but smaller!).

CHAPTER 23: Problems and Questions

1. A contact force is exerted when the source and subject touch; a field force does not require contact. The only field force presented in the first 22 chapters is gravity. All other forces (normal, friction, drag, tension, spring, buoyant) are contact forces and are manifestations of the electromagnetic field force. So, in a sense, all these "contact" forces are field forces.

3. 2.2×10^{11}

5. 6.25×10^{18}

7. a. The rod has lost mass. Electrons have been stripped from the rod, leaving behind a net positive charge. **b.** 2.60×10^{-16} kg

9. a. 2.63×10^{13} **b.** 1.81×10^{-12} **c.** 2.39×10^{-17} kg

11. Yes. The glass and the plastic are attracted to each other because they have opposite charges. The silk and the wool must also have opposite charges and be attracted because each was electrically neutral, like the glass and the plastic, but became charged when they exchanged electrons with those objects.

13. The technician wants to be protected and prevent electric charge from transferring from the lines to his body. Because rubber is an insulator, charges that are transferred to the rubber will stay at or near the point where they contact the rubber and not move through the technician and to the ground.

15. 9.54×10^{12} electrons/m²

Ans-1

Ans-2 CHAPTER 24 Answers

17. The charged insulator's net charge remains unchanged, $+30.0$ μC, while the other maintains a net charge of 0. It does matter how the contact is made. If either of the insulators is moved while they are in contact such that a friction force acts on each insulator over a short distance, charges can be transferred from one to the other.

19. The charge is free to move once it is on the conductor. The charge spreads out uniformly across the surface of the conductor.

21. a. 4.92×10^{-8} N **b.** Attractive

23. 2.4×10^{-5} C

25. $r_i/2$

27. a. $6.66 \times 10^{-6} \hat{i}$ N
b. $-6.66 \times 10^{-6} \hat{i}$ N

29. a. $d\sqrt{2}$ **b.** $d/\sqrt{2}$

31. a. 9.21×10^{-8} N **b.** 4.16×10^{42}

33. $\pm 4.3 \times 10^{-13}$ C

35. 2.28×10^{-9}

37. $-9.66 \times 10^{-4} \hat{i}$ N

39. $(-758\hat{i} - 2.35 \times 10^3 \hat{j})$ N

41. a. $-1.02 \times 10^{-7} \hat{j}$ N **b.** $1.05 \times 10^{-7} \hat{j}$ N
c. $-2.74 \times 10^{-9} \hat{j}$ N

43. 1.45 N, 46.7° below the $-x$ axis
45. $(\sqrt{2} - 1)d$
47. 4.55×10^{-7} C
49. -1.5×10^{-9} C
51. $-q((1 + 2\sqrt{2})/4)$
53. 1.33 m from the 8.00-nC sphere
55. $\sqrt{(4L^2 mg \tan\theta \sin^2\theta)/5k}$
57. 2.80×10^{23}
59. 1.90×10^{-7} C
61. $(1.38\hat{i} - 1.93\hat{j})$ N
63. a. $8.79\hat{i}$ N **b.** $21.6\hat{i}$ N
65. $kq^2/(4L^2 g \tan\theta \sin^2\theta)$
67. 1.12×10^{-7} C
69. 2.28×10^{-5} C
71. 0.21 m
73. a. $(2kQqh)/(h^2 + a^2)^{3/2} \hat{j}$
75. $(-0.273\hat{i} - 0.273\hat{j} - 0.273\hat{k})$ N
77. $kq^2/(mgr^2)$

CHAPTER 24: Concept Exercises

24.1 According to $\vec{E}(r) = (kQ_S/r^2)\hat{r}$ (Eq. 24.3), the electric field vectors point radially. If the source is positive, the electric field vectors point outward, and if the source is negative, the electric field vectors point inward. Also according to Equation 24.3, the electric field is stronger near the source; the field drops off as $1/r^2$ as r increases. These statements are consistent with Figure 24.4C, which shows outward-pointing electric field vectors that are longer near the source.

24.2 To find the magnitude of the electric field at the position of a charged particle, take the limit $\lim_{r \to 0} E(r) = \lim_{r \to 0}(kQ_S/r^2) \to \infty$. The electric field due to a charged particle at its exact location approaches infinity.

24.3 The lines in cases 2 and 4 cannot represent electric field lines because they do not originate or terminate on charged particles.

24.4 a. The particle on the left is positively charged because the electric field lines originate from it. The particle on the right is negatively charged because the electric field lines terminate on it. **b.** The electric field is weakest in region C because the electric field lines in that region have the lowest density. **c.** The electric field is strongest in region A because the electric field lines in that region have the highest density.

24.5 The positive charge is due to 10 protons and the negative charge is due to 10 electrons, so $Q = 10e$. The dipole moment is $p = Qd = 10ed = 6.2 \times 10^{-30}$ C · m.

24.6 a. The surface charge density is

$$\sigma = \frac{Q}{A} = \frac{Q}{2\pi R L}$$

The small amount of charge dq in a length dx of the rod is

$$dq = \sigma dA = \sigma(2\pi R)dx$$

$$dq = \frac{Q}{2\pi R L}(2\pi R)dx = \frac{Q}{L}dx$$

b. The linear charge density is

$$\lambda = \frac{Q}{L}$$

The small amount of charge dq in a length dx of the rod is

$$dq = \lambda dx = \frac{Q}{L}dx$$

This is exactly what we found in part (a). We conclude that we can ignore the radius of a real rod and model the rod as an infinitely thin line.

CHAPTER 24: Problems and Questions

1. The symbol for electrostatic force is F_e and it has units of N. The symbol for electrostatic field is E and it has units of N/C. The force requires a source and a subject, whereas the field requires only a source.

3. 1.32×10^5 N/C

5. a. $-1.03 \times 10^5 \hat{j}$ N/C
b. $-5.54 \times 10^{-4} \hat{j}$ N

7. 5.92×10^5 C

9.

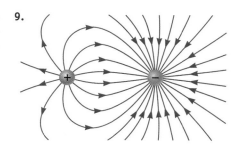

11. $(1.3 \times 10^5 \hat{\imath} + 4.0 \times 10^4 \hat{\jmath})$ N/C
13. $x = 12.8$ m
15. $x = -0.0986$ m
17. $-6.14 \times 10^{13} \hat{\imath}$ N/C
19. $-1.03 \times 10^5 \hat{\imath}$ N/C
21. $n = 1: E_P \approx 2.81 \times 10^3$ N/C,
    $n = 2: E_P \approx 3.40 \times 10^3$ N/C,
    $n = 4: E_P \approx 3.64 \times 10^3$ N/C,
    $n = 8: E_P \approx 3.72 \times 10^3$ N/C
23. $(k\lambda L)/(x^2 - xL)$
27. $y = R/\sqrt{2}$ and $\vec{E} = (2kQ)/(R^2 3\sqrt{3})\hat{\jmath}$
29. $E_{\text{disk}} = E_{\text{particle}} = 6.0 \times 10^3$ N/C
31. $3.95 \times 10^2 \hat{\imath}$ N/C
33. 428 N/C
35. $5.31 \times 10^8 \hat{\jmath}$ N/C
37. $-4.73 \times 10^5 \hat{\jmath}$ N/C
39. 3×10^{-7} C. This is about 10 times less than the charge in the case of Wilson's rod.
41. 0.77 m
43. 3.27×10^3 N/C, upward
45. 1.81×10^{-18} C/m^2
47. 2.1×10^5 m/s
49. 4.21×10^{-7} C/m
51. **a.** $(-347\hat{\imath} - 635\hat{\jmath})$ N/C
    b. $(-6.94 \times 10^{-7}\hat{\imath} - 1.27 \times 10^{-6}\hat{\jmath})$ N
53. **a.** -1.10×10^{-8} C **b.** 7.46×10^{-3} N
55. The water molecules reorient themselves as they fall due to the electric field created by the excess charge on the comb. The part of the water molecule that is attracted toward the comb rotates to be closer, while the other side is rotated away from the comb. The attracted portion of the molecule is closer to the comb, so the magnitude of the electrostatic force on that portion is greater than that on the other side of the molecule that is repelled. The water then is attracted to the comb due to the net force on the molecules.
57. **a.** 4.4×10^{-29} J **b.** 5.0×10^{-28} J
    c. -4.4×10^{-29} J
59. $x = 0.60$ m
61. **a.** 0   **b.** $kQ/(2\sqrt{2}R^2)$
63. $-2kQ/\pi R^2 \hat{\jmath}$
65. $-4kq/5\hat{\imath}$
67. **a.** 5.86×10^6 N/C **b.** 1.10×10^7 N/C
    c. 5.56×10^6 N/C **d.** 1.21×10^5 N/C
69. 9.78×10^4 N/C
71. $(3.71\hat{\imath} + 111\hat{\jmath})$ N/C
73. 4.20×10^6 N/C, to the left
75. 0.743 N
77. **a.** $\sqrt{2}kQ/(4\ell^2)\hat{\jmath}$ **b.** $kQ/(\sqrt{2}\ell^2)\hat{\jmath}$
    c. The ratio is 1/2.
79. **a.** $0.628 kQ/\ell^2 \hat{\jmath}$ **b.** $kQ/(\sqrt{2}\ell^2)\hat{\jmath}$
    c. 0.888

CHAPTER 25: Concept Exercises

25.1 a. Other uppercase letters that have the same symmetry as **A** are **M, T, U, V, W,** and **Y. b.** Other uppercase letters that have the same symmetry as **Z** are **N** and **S. c.** No other uppercase letters have the same symmetry as **O. B, C, D, E, K,** and **R** are all symmetrical with respect to a 180° rotation around the x axis. **H, I,** and **X** are symmetrical with respect to 180° rotations around all three axes. **F, G, J, L, P,** and **Q** are not symmetrical.

25.2 *Electric force* is the field force exerted between charged objects. It may be repulsive or attractive. The SI units are newtons. *Electric field* is the electric force per unit charge. It can be found by imagining the electric force on a positive test charge. The SI units are newtons per coulomb. *Electric flux* is analogous to the flow of a fluid through a loop. The electric flux is represented by the number of electric field lines that penetrate an area. The SI units are newtons · meter squared per coulomb.

25.3 Take the dot product to find the electric flux $\Phi_E = \vec{E} \cdot \vec{A}$ (Eq. 25.3):

$\Phi_E = (15\hat{\imath} + 25\hat{\jmath}) \cdot (0.65\hat{\imath} + 0.35\hat{\jmath})$ N·m^2/C $= (15)(0.65)$
$+ (25)(0.35)$ N·m^2/C $= 19$ N·m^2/C

25.4 Choices b and c are correct ways to express Gauss's law. Choice a is not correct because it is only an expression for flux.

25.5 In cases 1 and 2, the lumpy surface and the sphere both have $+10$ mC inside, so according to $\Phi_E = q_{\text{in}}/\varepsilon_0$ (Eq. 25.7), the net electric flux is

$$\Phi_E = \frac{10 \times 10^{-3}\,\text{C}}{8.85 \times 10^{-12}\,\text{C}^2/\text{N}\cdot\text{m}^2} = 1.1 \times 10^9\,\text{N}\cdot\text{m}^2/\text{C}$$

In case 3, there is no charge in the box, so according to Gauss's law, the electric flux through the box is zero: $\Phi_E = q_{\text{in}}/\varepsilon_0 = 0$.

25.6 To show that Equation 25.13 is consistent with Equation 24.15, take the limit of Equation 24.15 in the case that the line becomes very long, $\ell \gg y$:

$$\lim_{\ell \gg y}\vec{E} = \lim_{\ell \gg y} \frac{kQ}{y}\frac{1}{\sqrt{\ell^2 + y^2}}\hat{\jmath} = \frac{kQ}{y}\frac{1}{\ell}\hat{\jmath} = \frac{k2\lambda\ell}{y}\frac{1}{\ell}\hat{\jmath}$$

where $\lambda = Q/2\ell$ for a rod of length 2ℓ.

$$\lim_{\ell \gg y}\vec{E} = \frac{k2\lambda}{y}\hat{\jmath} = \frac{1}{4\pi\varepsilon_0}\frac{2\lambda}{y}\hat{\jmath} = \frac{1}{2\pi\varepsilon_0}\frac{\lambda}{y}\hat{\jmath}$$

This is for a point on the y axis. In general, we can write $y \to r$ and $\hat{\jmath} \to \hat{r}$ for a point at some distance r from the rod. Then Equation 24.15 is the same as Equation 25.13 in the limit that the rod becomes infinitely long:

$$\lim_{\ell \to \infty}\vec{E} = \frac{1}{2\pi\varepsilon_0}\frac{\lambda}{r}\hat{r} \checkmark$$

25.7 No. The sphere in Figure 25.23 cannot be a conductor in electrostatic equilibrium. If it were a conductor, the excess charge would quickly move to the surface. The sphere must consist of either an insulator or many individual charges, such as the protons in a nucleus.

CHAPTER 25: Problems and Questions

1. We proceed letter by letter. **Z** has the same symmetry as **N** and **S**, but not **C** or **B**. Thus, only **NUT** and **SUE** are possible answers. **A** has the same symmetry as **U** in each of those words, but the symmetry of **K** matches only the symmetry of **E**, not **T**. Thus, the answer is (b), **SUE**.
3. The source inside must be negative because it attracts the positively charged ball.
5. The box contains both positively and negatively charged objects. The negatively charged object is near the attractive face and the positively charged object is near the repulsive face. The net charge in the box may be zero, positive, or negative depending on the magnitudes of the charged objects near each face.
7. **a.** The net electric flux is zero. The number of electric field lines that enter and leave the box is the same. **b.** The net electric flux is positive. More field lines leave the box than enter it.
9. 1.39×10^7 N/C
11. **a.** 0   **b.** 4.18×10^4 N·m²/C
13. 2.62×10^3 N·m²/C
15. 655 N·m²/C
17. $\Phi_E = \pi r^2 E \sin \omega t$
19. $\Phi_{E,1} = 0, \Phi_{E,2} = -3.05 \times 10^6$ N·m²/C,
    $\Phi_{E,3} = -4.58 \times 10^6$ N·m²/C,
    $\Phi_{E,4} = -3.05 \times 10^6$ N·m²/C
21. $\Phi_{E,A} = 904$ N·m²/C, $\Phi_{E,B} = 0$,
    $\Phi_{E,C} = 0, \Phi_{E,D} = 1.47 \times 10^3$ N·m²/C
23. $q/8\varepsilon_0$
25. $\Phi_{E,A} = Q/\varepsilon_0, \Phi_{E,B} = -Q/\varepsilon_0$,
    $\Phi_{E,C} = 0, \Phi_{E,D} = Q/\varepsilon_0$
27. **a.** 8.81×10^3 N·m²/C   **b.** 4.41×10^3 N·m²/C
29. $\rho(r^2 - a^2)/(2\varepsilon_0 r)\hat{r}$
31. **a.** 0   **b.** 4.50×10^4 N/C   **c.** 1.80×10^3 N/C
33. $(\lambda r)/(2\pi R^2 \varepsilon_0)\hat{r}$
35. $\vec{E}_A = -2.1 \times 10^5 \hat{\imath}$ N/C,
    $\vec{E}_B = 4.6 \times 10^5 \hat{\imath}$ N/C,
    $\vec{E}_C = -4.8 \times 10^5 \hat{\imath}$ N/C
37. -3.04×10^{-4} C/m²
39. 0, 8.81×10^9 N/C, 1.76×10^{10} N/C, 7.83×10^9 N/C
41. 0
43. **a.** $cR^5/(5\varepsilon_0 r^2)$ **b.** $cr^3/(5\varepsilon_0)$
45. 5.54×10^3 N/C
47. **a.** $\vec{E}_A = -4.07 \times 10^6 \hat{\imath}$ N/C,
    $\vec{E}_B = 1.36 \times 10^6 \hat{\imath}$ N/C,
    $\vec{E}_C = 4.07 \times 10^6 \hat{\imath}$ N/C
    b. $\vec{F}_A = 6.51 \times 10^{-13} \hat{\imath}$ N,
    $\vec{F}_B = -2.17 \times 10^{-13} \hat{\imath}$ N,
    $\vec{F}_C = -6.51 \times 10^{-13} \hat{\imath}$ N
49. **a.** $\sigma_{upper} = 3.98 \times 10^{-7}$ C/m² and
    $\sigma_{lower} = -3.98 \times 10^{-7}$ C/m²
    b. $Q_{upper} = 4.34 \times 10^{-8}$ C and
    $Q_{lower} = -4.34 \times 10^{-8}$ C
51. **a.** $\rho x/\varepsilon_0$ **b.** 6.78×10^3 N/C
53. 8.3×10^9
55. 1.12×10^4 N/C, 1.80×10^{-15} N
57. $Q_{inner} = -38.3 \mu$C, $Q_{outer} = 38.3 \mu$C
59. **a.** -24.6 mC   **b.** 0
    c. -4.89×10^{-2} C/m²   **d.** 0
61. 0.72 N·m²/C
63. **a.** 0   **b.** 3.62×10^6 N/C
65. -0.120 N·m²/C
67. **a.** -1.46×10^{-7} C   **b.** 1.68×10^{-7} C
    c. $Q_{inner} = 1.46 \times 10^{-7}$ C,
    $Q_{outer} = 2.2 \times 10^{-8}$ C
69. **a.** 0   **b.** 1.65×10^6 N/C
    c. 3.30×10^6 N/C   **d.** 3.66×10^5 N/C
71. 0
73. 5.0×10^{-11} C
75. **a.** $2\pi AR^2$ **b.** $AR^2/(2\varepsilon_0 r^2)$ **c.** $A/(2\varepsilon_0)$
77. $E_{outside} = 5.6 \times 10^5$ N/C
79. 5.45×10^4 N·m²/C

CHAPTER 26: Concept Exercises

26.1 **a.** Gravitational potential. Gravitational force depends on the mass of *both* the source and the subject, whereas gravitational field depends *only* on the mass of the source. Similarly, gravitational potential energy depends on the mass of *both* the source and the subject, whereas gravitational potential depends *only* on the mass of the source. **b.** Gravitational potential. Gravitational force is a *vector* that depends on the mass of both the source and the subject, whereas gravitational potential energy is a *scalar* that depends on the mass of both the source and the subject. Similarly, gravitational field is a *vector* that depends only on the mass of the source, whereas gravitational potential is a *scalar* that depends only on the mass of the source.

26.2 **a.** The sphere must be positively charged. Because the electric potential energy is positive, the source must have charge of the same sign as the subject (a proton).
b. $U_E = -15.0 \times 10^{-20}$ J

26.3 Start with $U_E = kQq/r$ (Eq. 26.3):

$U_E = k\dfrac{(10e)(-10e)}{r} = -100k\dfrac{e^2}{r} = -100(8.99 \times 10^9$ N·m²/C²$)$
$\times \left[\dfrac{(1.60 \times 10^{-19} \text{ C})^2}{3.9 \times 10^{-12} \text{ m}}\right]$
$U_E = -5.90 \times 10^{-15}$ J

The negative sign means that energy is transferred from the system to the environment in bringing the particles together from infinity, and it would take positive work to separate the charged particles of a dipole.

26.4 Map A → Landscape 4, a simple symmetrical hill
  Map B → Landscape 3, a hill that is steep on one side
  Map C → Landscape 2, a sharp drop-off or cliff
  Map D → Landscape 1, two hills

26.5 Only *voltage* is a synonym. It is a common term used for electric potential.

26.6 The landscape would look like two ramps meeting at a thin, long peak, something like the peaked roof line shown in the next column.

26.7 No. You cannot use an EKG to plot E versus position because an EKG is a graph of V versus time. You need a graph of V versus *position* because E comes from the spatial derivative of V. Remember that E is the slope only on a V-versus-*position* graph.

CHAPTER 26: Problems and Questions

1. When a vector, like force, is negative, its direction is opposite the direction that is defined to be positive. When the potential energy is negative, it means the potential energy is less than that of some reference configuration where the potential energy is zero.

3. While all of the quantities change as the planet moves, the gravitational potential and the field at the location of the planet do not depend on the mass of the planet. They are properties of the space affected by the star. The other quantities depend on the mass of the planet, its distance from the star, and its speed at any moment in time.

5.

Term	Mathematical symbol	Vector or scalar?	Source and subject or source only?	SI units
Electrostatic force	$\vec{F}_E$	vector	both	N
Electric field	$\vec{E}$	vector	source only	N/C or V/m
Electric potential energy	U_E	scalar	both	J
Electric potential	V	scalar	source only	V

7. -5.6×10^{-6} J
9. a. $U_E = kq^2/r$ **b.** $U_E = -kq^2/r$ **c.** $U_E = kq^2/r$
11. 8.81×10^{-8} J
13. 3.8×10^{-11} m
15. a. 58.2 J **b.** The particles will fly apart and their kinetic energies will increase. The kinetic energy of the system will eventually equal 58.2 J.
17. 2.05×10^{-5} J
19. a. 84.4 V **b.** The potential is higher at $y = 2.00$ cm.
21. a. 2.91 m **b.** 8.27×10^{-7} C
23. -1.5 eV, -2.4×10^{-19} J
25. 13.8 V
27. a. -8.21×10^5 V **b.** -6.65×10^5 V
29. 90.0 V
31. a. 2.2×10^4 V **b.** 4.3×10^4 V **c.** 4.3×10^4 V
33. 1.68×10^3 V
35. 2.22 J
37. -1.07×10^{-6} C
39. -1.07×10^7 V
41. a. $k\lambda \ln\left[(a + \sqrt{a^2 + y^2})/(-a + \sqrt{a^2 + y^2})\right]$
  b. $kq\lambda \ln\left[(a + \sqrt{a^2 + y^2})/(-a + \sqrt{a^2 + y^2})\right]$
45. 111 V
47. $(Aa^2/2) - (Bb^3/3)$
49. a. $V = RE$ **b.** 4.8×10^5 V
51. $(V_0 e^{-ar}/r)(a + 1/r)$
53. a. -5.50 V, 12.5 V, 30.5 V **b.** 4.50 V/m in the $-y$ direction
55. a. $\vec{E} = -(8xz + 2y^2)\hat{\imath} - (4xy - 8z^2)\hat{\jmath} - (4x^2 - 16yz)\hat{k}$
  b. 58.8 V/m

57. a. Constant V **b.** 0

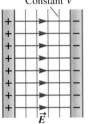

59. a. 0, 3.45×10^4 V **b.** 1.57×10^5 V/m, 3.45×10^4 V **c.** 3.04×10^4 V/m, 1.52×10^4 V
61. a. -0.766 J **b.** -1.29×10^5 V
63. 23.4 m/s and 6.63 m/s
65. a. -125 V **b.** -125 V **c.** No. The electric force is a conservative force; thus, only the initial and final positions of a particle's movement determine the change in electric potential.
67. Yes. The sign of the charged particle does matter. In either case, the change in electric potential is the same (positive), but the change in electric potential energy is not. If the particle has a net negative charge, the change in electric potential energy is negative according to $\Delta U_E = q\Delta V$. Likewise, if the particle has a net positive charge, the change in electric potential energy is positive.
69. 46.9 J
71. -2.69×10^4 V
73. $2\pi k\sigma\left[1 - x/(\sqrt{R^2 + x^2})\right]\hat{\imath}$
75. a. $V = Er \ln(R/r)$ **b.** -6.3×10^3 V
77. $(2\pi ka/3)\left\{\sqrt{R^2 + x^2}[R^2 - 2x^2] + 2x^3\right\}$
79. $2kq/3$

CHAPTER 27: Concept Exercises

27.1 Consider the spring, the object, and the Earth to be the system. Assume the object undergoes purely translational motion, and ignore air resistance. Then the potential energy stored by the spring at maximum compression equals the gravitational potential energy when the object is at its maximum height:

$$\frac{1}{2}ky^2 = mgy_{max}$$

$$y = \sqrt{\frac{2mgy_{max}}{k}} = \sqrt{\frac{2(0.09\,\text{kg})(9.81\,\text{m/s}^2)(120\,\text{m})}{1000\,\text{N/m}}}$$

$$y = 0.46\,\text{m} \approx 0.5\,\text{m}$$

This means that even a stiff spring has to be longer than 0.5 m (or about the length of your forearm) in order to compress it far enough.

27.2 With $Q = C\Delta V$ (Eq. 27.1), the slope of each graph (Fig. 27.8, page 832) is $1/C$. Find the slope for capacitor A:

$$\frac{1}{C_A} = \frac{30\,\text{V}}{3\,\text{nC}} = \frac{30\,\text{V}}{3 \times 10^{-9}\,\text{C}} = 1 \times 10^{10}\,\text{V/C}$$

Take the reciprocal of the slope to find the capacitance:

$$C_A = \frac{1}{1 \times 10^{10}\,\text{V/C}} = 1 \times 10^{-10}\,\text{F} = 100\,\text{pF}$$

Repeat this process for capacitor B:

$$\frac{1}{C_B} = \frac{30\,\text{V}}{1.5\,\text{nC}} = \frac{30\,\text{V}}{1.5 \times 10^{-9}\,\text{C}} = 2 \times 10^{10}\,\text{V/C}$$

$$C_B = 5 \times 10^{-11}\,\text{F} = 50\,\text{pF}$$

27.3 a. From the graph for capacitor B, when $Q = 0.5$ nC, the potential difference between the plates is $\Delta V = 10$ V. Combine $Q = C\Delta V$ with $U_E = Q^2/2C$ to find $U_E = \frac{1}{2}Q\Delta V$. Substitute values:

$$U_E = \frac{1}{2}(0.5\,\text{nC})(10\,\text{V}) = 2.5 \times 10^{-9}\,\text{J} = 2.5\,\text{nJ}$$

b. From Concept Exercise 27.2, $C_A = 100$ pF. Find the charge from $U_E = Q_A^2/2C_A$:

$$Q_A = \sqrt{2C_A U_E} = \sqrt{2(100\,\text{pF})(2.5\,\text{nJ})}$$

$$Q_A = 7.1 \times 10^{-10}\,\text{C} = 0.7\,\text{nC}$$

Find the potential difference from $U_E = \frac{1}{2}C_A \Delta V_A^2$:

$$\Delta V_A = \sqrt{\frac{2U_E}{C_A}} = \sqrt{\frac{2(2.5\,\text{nJ})}{(100\,\text{pF})}} = 7\,\text{V}$$

Use the graph (Fig. 27.8) to confirm this.

27.4 First consider connecting the terminals of a battery with a simple wire (Fig. 27.12B, page 835). Because one terminal is positive and the other is negative, there is an electric field between the terminals, and electrons are forced to move from the negative terminal toward the positive terminal. Now consider charging a capacitor with a battery (Fig. 27.14). An electric field forces electrons to move from the negative terminal toward the positive terminal. Because of the gap in the capacitor, electrons build up on one plate and an excess of protons (lack of electrons) builds up on the other plate. When the potential difference across the capacitor equals the terminal potential, $V_C = \mathcal{E}$, the potential of the negative capacitor plate equals the potential of the battery's negative terminal. Once the negative capacitor plate and the negative battery terminal are at the same potential, there is no electric field between those locations and electrons are no longer forced to move through the wire.

27.5 The **I**-shaped conductor remains neutral. Because electrons are free to flow inside the conductor, they are repelled by the negatively charged bottom plate of the capacitor and attracted to the positively charged top plate. The net result is that the bottom plate of the **I**-shaped conductor has excess positive charge $q_1 = +Q$, and its top plate has excess negative charge $q_2 = -Q$. The capacitor plates are not changed by the insertion of the **I**-shaped conductor, so the potential difference between the top and bottom plates is still the terminal potential of the battery: $V_C = \mathcal{E}$.

27.6 The six capacitors are in parallel, so we find the capacitance of each individual capacitor from $C_{eq} = \sum_{i=1}^{N} C_i$ (Eq. 27.8):

$$C_{eq} = C + C + C + C + C + C = 6C$$

$$C = \frac{C_{eq}}{6} = \frac{94\,\mu\text{F}}{6} = 16\,\mu\text{F}$$

27.7 Find the energy density from $u_E = \frac{1}{2}\kappa\varepsilon_0 E^2$ (Eq. 27.22) with $\kappa = 1$ (because the tube is evacuated):

$$u_E = \frac{1}{2}(8.85 \times 10^{-12}\,\text{C}^2/\text{N}\cdot\text{m}^2)(5 \times 10^6\,\text{N/C})^2 = 1.1 \times 10^2\,\text{J/m}^3$$

CHAPTER 27: Problems and Questions

1. Air resistance dissipates the mechanical energy, so more potential energy would be needed initially to get the ring to reach the same final height.

3. *Leyden jar* has the advantage of sounding like a storage device, but the disadvantage of making charge seem like a liquid contained in the jar. *Capacitor* and *condenser* do not sound like storage devices, although *capacitor* might remind you of the word *capacity*, perhaps suggesting that it has the ability to store charge. Similarly, *condenser* might remind you that charges are "condensed," or brought together on the plates.

5. 2.9×10^{-4} C
7. a. $Q_1 = 8.00 \times 10^2\,\mu\text{C}$, $Q_2 = 2.00 \times 10^2\,\mu\text{C}$ **b.** 20%
9. a. 1.82×10^{-4} J **b.** 1.30×10^{-3} J
11. a. 4.69×10^{-3} C **b.** 2.93×10^{-4} F
13. 0.150 J
15. 9.002 V
17. a. 10.1 μF **b.** 2.34 μF
19. 15 V
21. $C_1 = 0.364\,\mu\text{F}$, $C_2 = 0.366\,\mu\text{F}$
23. $186C/241$
25. $1.00 \times 10^2\,\mu\text{F}$ and 9.84 μF
27. 8.22 μF

29. 22.00 pF, 44.00 pF, and 66.00 pF
31. a. 1.69×10^{-12} F **b.** 1.52×10^{-11} C
c. 3.91×10^3 V/m
33. $4\pi\varepsilon_0 r_{in}$
35. 3.56×10^{-5} m
37. a. Charge is halved. **b.** Energy is halved.
39. 1.88×10^3 V
41. 5.6×10^6 m^2
43. a. 2.71×10^5 V **b.** $r_{in} = 1.53 \times 10^{-2}$ m and $r_{out} = 2.18 \times 10^{-2}$ m
45. 2.39×10^{-9} C
47. a. 2.28×10^{-10} C **b.** 1.02×10^{-11} F
c. 22.3 V **d.** 1.40×10^{-8} J

49. 35.0
51. a. 0.333 J **b.** 851 V
53. $C = 3C_0$, $Q = 3Q_0$, $V = V_0$, $U = 3U_0$
55. a. 1.59×10^{-8} J **b.** 4.00×10^3 V/m
c. 946 V/m **d.** 3.05×10^3 V/m
e. 3.77×10^{-9} J
57. 1.25×10^4 s
59. a. 1.27 J/m³ **b.** 1.08×10^9 J/m³
61. $Q/(\kappa\varepsilon_0 \pi R^2)$
63. 2.0
65. 2.2×10^{-11} C/m
67. a. 1.20×10^{-8} J **b.** 4.00 V
69. a. 2.4 μF **b.** 9.72×10^{-5} J **c.** 2.16×10^{-5} C **d.** $U_A = 1.94 \times 10^{-5}$ J, $U_B = 7.78 \times 10^{-5}$ J
71. 2.12×10^{-8} C

73. 16.5 mJ. To achieve the highest terminal potential, connect the batteries in series. To achieve the greatest capacitance, connect the capacitors in parallel.

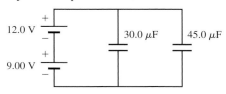

75. a. 2.91×10^{-6} F
b. $Q_A = 3.50 \times 10^{-5}$ C,
$Q_B = 2.61 \times 10^{-5}$ C,
$Q_C = 8.93 \times 10^{-6}$ C,
$Q_D = 8.93 \times 10^{-6}$ C

77. a. 3.51×10^{-12} F **b.** 4.21×10^{-11} C
c. 8.28×10^3 V/m **d.** 7.32×10^{-8} C/m²
79. 8.30 μF and 4.70 μF
81. 34.6 μF
83. a. $U_1 C_1 C_2/(C_1^2 + 2C_1 C_2 + C_2^2)$
85. a. $V_1 = V_4 = 6.67$ V,
$V_2 = V_3 = 3.33$ V, $Q_1 = Q_2 = 40.0$ μC,
and $Q_3 = Q_4 = 26.7$ μC
b. $V_1 = V_3 = 5.33$ V, $V_2 = V_4 = 4.67$ V,
$Q_1 = 32.0$ μC, $Q_2 = 56.0$ μC,
$Q_3 = 42.7$ μC, and $Q_4 = 18.7$ μC
87. a. 0.5C **b.** 0.6C **c.** 0.62C

CHAPTER 28: Concept Exercises

28.1 Imagine a single conduction electron in a swarm of other electrons, with each exerting a repulsive force on that single electron in different directions and with different magnitudes. If you sum all those random force vectors, the net force is zero (or nearly so).

28.2 If the current is constant, Equation 28.2 becomes $q = I\int_0^t dt = It$. The time in each case is 1 s, so the charge that passes through the cross section in each case is shown in the table.

Device	Charge in 1 s
a. Flashlight	$q = (0.5 \text{ A})(1 \text{ s}) = 0.5$ C
b. Starter motor	$q = (200 \text{ A})(1 \text{ s}) = 200$ C
c. Computer circuit	$q = (1 \times 10^{-12} \text{ A})(1 \text{ s}) = 1 \times 10^{-12}$ C = 1 pC

28.3 Yes. Considering the actual conduction electrons moving in the negative x direction ($\vec{v}_{\text{drift}} = -v_{\text{drift}}\hat{\imath}$) and carrying charge $-e$, we find

$$\vec{J} = n(-e)(-v_{\text{drift}})\hat{\imath} = nev_{\text{drift}}\hat{\imath}$$

This result still shows the current density pointing in the same direction as the electric field—in this case, the positive x direction. The magnitude of the current density is also the same because the drift speed of electrons is equal to the drift speed of the imaginary positive particles: $v_{\text{drift}} = v_d$.

28.4 When the filament breaks, the circuit is open. Because air is normally an insulator, there can be no current in the network. The resistance measured between the ends of the filament is infinite (or at least very high).

28.5 The potential difference across the broken filament is $\mathcal{E}$, the same as the potential across the battery because no current is in the wire or the (broken) filament.

28.6 For a broken filament, $I = 0$, $R \to \infty$, and $\Delta V = \mathcal{E}$. Any of the equations for power gives $P = 0$; for example, $P = I\Delta V = 0\mathcal{E} = 0$.

CHAPTER 28: Problems and Questions

1. Metal 1 is the better conductor because there are fewer collisions as the electrons drift through the metal.
3. The cooler penny is the better conductor because electrons undergo fewer collisions with vibrating ions as they drift through the metal.
5. Answers will vary but should be on the order of 10^5 m/s.
7. 0.48 A, to the left
9. $I = -q_0\omega \sin \omega t$
11. a. The current oscillates back and forth, changing direction. Its maximum value in either direction is $q_0\omega$. **b.** Yes. This occurs when $\sin \omega t = 0$, when $\omega t = n\pi$ ($n = 0, 1, 2, 3, \ldots$), or when $t = n\pi/\omega$ ($n = 0, 1, 2, 3, \ldots$).
13. 1.7×10^{-3} m/s
15. 2.2×10^2 C
17. a. 56 A **b.** 1.8×10^5 A/m²
19. 1.04×10^5 s

21. 6.25 mA, 4.00×10^3 A/m²
23. 4.70×10^{-5} m/s
25. 2.94×10^7 Ω^{-1}·m^{-1}
27. 1.15×10^{-2} V/m
29. a. decreases **b.** increases **c.** no change **d.** no change **e.** decreases. An increase in temperature means the atoms vibrate more and so have a larger effective cross-sectional area. An increased area decreases the time between collisions, increases the rate of collision, and so decreases the conductivity.
31. 6.8×10^8 A
33. 7.6×10^7 A
35. 0.564 A
37. 15.0 m
39. 1.50×10^4 Ω
41. a. Conductivity is a microscopic or materials property. **b.** Resistivity, the inverse of conductivity, is also microscopic. **c.** Conductance depends on both microscopic properties (through the conductivity) and macroscopic properties (the length and cross-sectional area of the conductor). **d.** Resistance is the inverse of conductance and, therefore, also depends on microscopic and macroscopic properties.
43. 30.0%
45. 1.52 m
47. a. 4.27 V/m from left to right **b.** 0.168 Ω **c.** 19.0 A from left to right **d.** 2.69×10^8 A/m²
49. No. The ratio of the potential difference to the current is not constant.
51. The word *rule* may be better because Ohm's law, like Hooke's law, is an empirical law that does not hold for all materials. It works well for modeling the behavior of many materials, but it is not universally true.
53. a. 0.4545 A **b.** 484.0 Ω
55. a. 7.083 A **b.** 16.94 Ω
57. 1.82 Ω
59. a. $2I_0$ **b.** $4V_0I_0$ **c.** $4V_0I_0$

61. 335 s
63. 2.50 W
65. 0.31 C
67. a. 24.0 Ω **b.** 2.40 × 10³ Ω
69. $v_d = E/(\rho n e)$
71. a. 14.4 W and 10.3 W **b.** the 250.0-Ω resistor
73. 0.708
75. 4.16 Ω
77. $I/[ne\pi(r_b^2 - r_a^2)]$
79. 149°C
81. 1.19 × 10⁹ s

CHAPTER 29: Concept Exercises

29.1 Volts

29.2 The voltmeter in experiment 2 will read +1.45 V, as it does in experiment 1. You might have thought the voltage would be negative because of the negative sign in the resistor rule (Eq. 29.1), but that sign holds only when the current is from the black lead to the red lead. In this circuit, the current is clockwise, so the current is from point *b* (red) to point *a* (black). Thus, the potential difference measured ($V_{red} - V_{black} = V_b + V_a$) is positive.

29.3 a. Start by solving $\mathcal{E}_1 - IR_1 - \mathcal{E}_2 - IR_2 = 0$ for current:

$$I = \frac{\mathcal{E}_1 - \mathcal{E}_2}{R_1 + R_2}$$

When battery 2's emf is nearly 0 ($\mathcal{E}_2 \approx 0$), the current is very high and the bulb is bright. As battery 2's emf increases, the current decreases and the bulb becomes fainter. In fact, if $\mathcal{E}_2 = \mathcal{E}_1$, the current is 0 and the bulb is not lit. So, when the indicator bulb goes out, the battery is recharged. **b.** The term *recharge* is misleading because it seems to imply that the function of an emf device is to supply charged particles. Of course, the emf device supplies energy, not charge. A better term may be *re-energize*.

29.4 In the pegboard analogy, a resistor is like a pegboard and the current is like rubber balls falling through and hitting the pegs. If you imagine a ball falling through two small pegboards in series, as in the accompanying figure, the resistance to the ball's motion is equal to the resistance the ball would encounter if it fell through a larger pegboard that is the sum of the two small pegboards. So, it makes sense to add resistances in series.

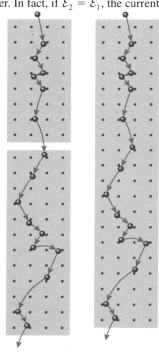

29.5 a. According to the emf rule, the potential difference is positive: $\Delta V = \mathcal{E}$. **b.** Whether we measure the potential difference across resistor 1 or resistor 2, the current is from the red lead to the black lead. According to the resistor rule, the potential difference is positive. So, we have $\Delta V = \mathcal{E} = I_1 R_1 = I_2 R_2$. It makes sense that the potential difference across all three circuit elements is the same because the voltmeter can give only one reading. When the voltmeter is connected to points *a* and *b*, it measures the voltage across all three circuit elements at once.

29.6 Applying the junction rule to either the junction at *a* or the junction at *b* gives $I_0 = I_1 + I_2$. From Concept Exercise 29.5, we have $I_1 R_1 = I_2 R_2$.

If the resistances are equal, then $R_1 = R_2$ and the current in branch 1 must equal the current in branch 2: $I_1 = I_2$. So, $I_1 = I_2 = \frac{1}{2}I_0 = \frac{1}{2}(3.0\text{ A}) = 1.5\text{ A}$.

29.7 Start with

$$\frac{1}{R_{eq}} = \frac{1}{R_1} + \frac{1}{R_2}$$

(Eq. 29.7 for two resistors in parallel). Then

$$\frac{1}{R_{eq}} = \frac{R_2}{R_1 R_2} + \frac{R_1}{R_1 R_2} = \frac{R_1 + R_2}{R_1 R_2}$$

and

$$R_{eq} = \left(\frac{R_1 + R_2}{R_1 R_2}\right)^{-1} = \frac{R_1 R_2}{R_1 + R_2} \checkmark$$

CHAPTER 29: Problems and Questions

1. a. ─── b. ─o o─ c. ─/\/\/─ d. ─|⊢─ e. ─⊙─

3. True. The magnitude may change, but the sign of each terminal is fixed for a DC emf.

5. a. 15 V **b.** 0 **c.** −15 V **d.** 0

7. a. decreases **b.** decreases **c.** decreases

9. a. 1.67 V **b.** 3.34 V **c.** 46.5 mA **d.** 93.0 mA

11. a. 12.0 Ω **b.** 2.32 Ω

13. 0

15. a. 0.339 A **b.** $\Delta V_1 = 4.54$ V, $\Delta V_2 = 6.94$ V, $\Delta V_3 = 3.32$ V **c.** $P_1 = 1.54$ W, $P_2 = 2.35$ W, $P_3 = 1.12$ W, $P_\mathcal{E} = 5.01$ W **d.** $I = 0.277$ A, $\Delta V_1 = 3.71$ V, $\Delta V_2 = 5.67$ V, $\Delta V_3 = 5.42$ V, $P_1 = 1.03$ W, $P_2 = 1.57$ W, $P_3 = 1.50$ W, $P_\mathcal{E} = 4.09$ W

17. a. The water is moving uphill, probably due to a pump of some kind. **b.** It is gradually flowing downhill, with a slight decrease in height. **c.** The water undergoes a large decrease in height as it flows downhill. **d.** The water is more or less back at the level where it started, undergoing a slight decrease in level as it finally returns to where it started.

19. 436 Ω

21. a. 9.6 Ω **b.** 11 Ω **c.** No. The same current would flow through each if they were in series, and there would be no means to select the desired resistance.

23. 67R/4

25. a. 9.00 mA **b.** 18.0 mA

27. $I_2 = 7.0$ A, $I_5 = I_6 = 10.0$ A, $I_7 = 18.0$ A

29. a. $I_1 = (6/11)I_0$, $I_2 = (3/11)I_0$, $I_3 = (2/11)I_0$ b. The same as part (a) because they are the same circuit.
31. $21R/164$
33. The current is greater in B because the equivalent resistance of the monks is higher in A. (The monks are in series in A.) The current is also greater through each monk in B because each is in parallel with the Volta pile.
35. $11R/6$
37. 9.600 A
39. a. $I_0 = 0.917$ A, $I_1 = 0.500$ A, $I_2 = 0.250$ A, $I_3 = 0.167$ A b. same as in part (a)
41. a. $I_1 = 1.51$ A, $I_2 = 0.386$ A, $I_3 = 1.89$ A b. $P_1 = 31.8$ W, $P_2 = 1.19$ W, $P_3 = 35.8$ W
43. a. $I_1 = I_2 = I_3 = 0.133$ A, $I_4 = 0$
b. $I_1 = 0.179$ A, $I_2 = I_3 = 0.124$ A, $I_4 = 0.0552$ A
45. a. $14.7\ \Omega$ b. $I_{20.0} = 0.326$ A, $I_{5.00} = 1.30$ A, $I_{9.00} = 1.63$ A, $I_{2.00} = 1.40$ A, $I_{12.0} = 0.233$ A
47. 12.38 V
49. a. $338\ \Omega$ b. 1.50×10^2 W c. $\Delta V_1 = \Delta V_2 = \Delta V_3 = 113$ V
51. a. 15.0 V b. 0.0003 A. This current is 0.06% of the current through the 30.0-Ω resistor.
53. 6.2×10^{-5} s. This is nearly the same as the 5.0×10^{-5} s firing time.
55. a. 0.949 A b. -0.500 A c. $3.90\ \mu C$
57. a. 1.26×10^{-3} s b. $72.0\ \mu C$ c. $68.4\ \mu C$
59. a. 5.25×10^{-4} s b. 3.64×10^{-4} s c. 0.429 A
61. $666\ \Omega$ and $154\ \Omega$
63. 9.00 h
65. a. 7.00 A b. 4.00 A c. 13.0 V
67. 0.0738 A
69. a. 2.00 A b. 8.0 V
71. a. $I_{50} = 7.85$ A, $I_{90} = 4.63$ A, $I_{120} = 1.31$ A, $I_{175} = 1.91$ A b. 334 V
73. $33.3\ \Omega$
75. a. 3.75×10^{-4} s b. 2.60×10^{-4} s c. 0.267 A
77. a. 2.37 A b. 1.26 V
79. $1.57 \times 10^6\ \Omega$
81. 0
85. $2\mathcal{E}/R$
87. $3\mathcal{E}/2R$
89. $\mathcal{E}/R$
91. $\mathcal{E}/R$
95. The minus sign indicates that the capacitor is losing charge as it discharges. That is, the current removes charge from the capacitor.

CHAPTER 30: Concept Exercises

30.1 An isolated north pole is a monopole, and so far there have been no confirmed monopole discoveries. However, as shown here, the magnetic field of an isolated north magnetic pole would look like the electric field due to a positive point charge.

30.2 Consider the upward current in Figure 30.11 (page 939), the result of electrons moving downward. To find the magnetic field direction due to electron motion, use a simple left-hand rule: Point your left thumb in the direction of electron motion, and then your fingers wrap counterclockwise around the wire.
30.3 The constant $\mu_0/4\pi$, the current I, and the length $d\ell$ are the same for both segments. Because $\varphi_{red} + \varphi_{blue} = 180°$, we have $\sin \varphi_{red} = \sin \varphi_{blue}$. With all else being equal, because $r_{red} > r_{blue}$, the magnetic field dB due to the blue segment is stronger than that due to the red segment at point P.

30.4 Avi is correct. The magnetic force is perpendicular to both the magnetic field $\vec{B}$ and the subject's velocity $\vec{v}$. Because these vectors are both in the plane of the page, the magnetic force $\vec{F}_B$ is either into or out of the page. Use the right-hand rule to find the direction of $\vec{v} \times \vec{B}$; your thumb points into the page. Because the particle is positively charged, this is the direction of the magnetic force.
30.5 Because the radius of the spiral is larger near the equator than it is near the poles, we know the magnetic field is weaker in the region near the equator than it is near the poles.
30.6 From $\vec{F}_B = q(\vec{v} \times \vec{B})$ and the right-hand rule for cross products, the magnetic force is in the positive x direction.
30.7 Again use $\vec{F}_B = q(\vec{v} \times \vec{B})$ and the right-hand rule. Due to the negative charge, flip over your thumb. The magnetic force is to the right in the positive x direction.
30.8 Currents in the same direction attract; opposite currents repel.

CHAPTER 30: Problems and Questions

1. Magnets either repel or attract each other depending on the orientation of their poles. In the metaphor, the hands are pulling on each other in opposite directions, but if the magnets were actually pulling each other, they would be attracting each other. The instructor might describe two magnets that are repelling each other.
3. a. Gravity. The neutron has no net electric charge but has mass. b. Gravity and the electric force. The proton has a net positive charge and mass. c. All three: gravity, the electric force, and the magnetic force. A moving charge, or current, creates a magnetic field.
5. a. Yes. b. Yes. The line through the axis has no point of origination or termination and must form a closed loop.
7. By convention, the magnetic field lines point out of the north end of a bar magnet and into the south end, which is not consistent with the drawing. The lines on the right should not cross. The lines on the left should loop back through the bar magnet, and there should be no termination points for lines, as seen on the coin.
9. a. $\hat{\jmath}$ b. $-\hat{\jmath}$ c. $-\hat{k}$ d. $-\hat{k}$
11. a. The current should flow downward along the wire from the top toward the bottom. b. The current should flow along the wire from the upper left to the lower right. c. The current should flow along the wire into the page.
13. $(\mu_0 ev)/(4\pi r^2)$
15. 5.12×10^{-21} T into the page
17. 2.00×10^{-6} T
19. $\vec{B}_A = 0$, $\vec{B}_B = [(2\mu_0 I)/(3\pi r)]\hat{\jmath}$
21. 5.28×10^{-5} T
23. 3.00 A out of the page or 9.00 A into the page
25. 4.80×10^{-4} T
27. $B_1:B_2 = 4:1$
29. a. $1.02 \times 10^{-5}\hat{k}$ (out of the page) b. 0.225 m
31. 1.59×10^{-2} A·m^2
33. 8.07×10^{-10} T
35. Both poles should be attracted. Each pole will cause the magnetic moments of nearby atoms in the refrigerator to align against the field of the bar magnet. Thus, each pole will be attracted to the anti-aligned magnetic domains that form in the refrigerator near each pole.
37. a. $\vec{F}_1 = -1.90\hat{k}$ N and $\vec{F}_2 = (1.90\hat{\imath} - 3.04\hat{\jmath})$ N b. $\vec{F}_1 = 1.90\hat{k}$ N and $\vec{F}_2 = (-1.90\hat{\imath} + 3.04\hat{\jmath})$ N

CHAPTER 31 Answers

39. $(-0.109\hat{\imath} + 0.326\hat{\jmath} - 0.716\hat{k})$ N
41. **a.** negative **b.** 3.19×10^{-7} C
43. **a.** 4.61×10^{-13} N **b.** 5.06×10^{17} m/s^2
45. **a.** 90° **b.** 2.17×10^{-4} T
47. **a.** 0.104 m **b.** 4.57×10^6 Hz
49. $r_{He} = 2r_p$
51. 2.1×10^5 m/s
53. 2.41×10^{-6} V
55. **a.** 0.535 A **b.** 6.21×10^{28} m^{-3}. This is very close to the value for tungsten.
57. $-61.6\hat{\imath}$ N
59. $(-8.86\hat{\imath} - 5.85\hat{\jmath} - 2.34\hat{k})$ N
61. **a.** 7.94 N **b.** 7.94 N **c.** 11.2 N
63. **a.**

b. $-3.60 \times 10^{-6}\hat{\jmath}$ N/m
65. **a.** $(1.04 \times 10^{-6}\hat{\imath} - 5.39 \times 10^{-6}\hat{\jmath})$ N/m
b. $(5.19 \times 10^{-6}\hat{\imath} + 1.80 \times 10^{-6}\hat{\jmath})$ N/m
67. $[(\mu_0 I_1 I_2)/(2\pi r_1) + (\mu_0 I_3 I_2)/(2\pi r_2)]\hat{\imath}$
69. 2.16 A·m^2
71. If the magnetic field were uniform, the magnitude of the magnetic torque on the loop would change as it rotated. With the radial field, the magnetic torque on the loop is constant, and the loop will rotate until the torque from the spring is equal and opposite to the magnetic torque. If the loop were in a uniform field, it would attempt to rotate until the magnetic torque was 0.
73. **a.** 33.4 N·m **b.** The loop will rotate so as to align the magnetic moment with the B field, counterclockwise as seen looking down from a position on the positive z axis.
75. The two curves appear to approach the same curve above $y = 0.08$ m and diverge fairly significantly below $y = 0.05$ m, which is equal to the diameter of the loop. Therefore, at points more than one or two loop-diameters from the current loop, we are far enough away that the approximation is quite good.

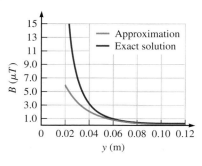

77. **a.** $3.13 \times 10^{-6}\hat{k}$ T
b. $(2.80 \times 10^{-6}\hat{\imath} - 0.800 \times 10^{-6}\hat{\jmath})$ T
79. $\left(\dfrac{N\mu_0 I}{2\pi R}\right) \sin(\pi/N)\hat{k}$
81. $2.31 \times 10^{-5}\hat{k}$ T
83. **a.**

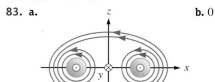

b. 0

c. $\dfrac{-(2.00 \times 10^{-6})z}{(1.21 \times 10^{-2}) + z^2}\hat{\imath}$ T, where z is in meters

85. **a.** out of the page **b.** toward the right **c.** toward the bottom of the page
87. $q(v_x B_y - v_y B_x)\hat{k}$
89. **a.** 0 **b.** 7.91×10^{-14} N
91. $f \approx -\left(\dfrac{\mu_0 I I_0}{\pi d^2}\right)x$ and $\omega = \sqrt{\dfrac{\mu_0 I I_0}{\pi \lambda d^2}}$
93. **a.** $I = \left(\dfrac{mg}{B\ell}\right)\tan\theta$ from Q to P **b.** No, $\dfrac{mg}{2g\sin\theta}$
95. 37 m

CHAPTER 31: Concept Exercises

31.1 Here is one possible experimental procedure:
1. With the bar magnet far away, find the direction of the Earth's magnetic south pole with the compass. This is the reference position.
2. Set the bar magnet perpendicular to the original position of the compass needle. If you want the compass needle to deflect toward the east, place the bar magnet to the west of the compass and point the bar magnet's north end eastward as in the accompanying figure.
3. Measure the distance between the bar magnet's pole and the compass needle.
4. Measure the deflection of the needle, and use $B = B_\oplus \tan\theta$ (Eq. 31.1) to find B.
5. Move the compass eastward or westward, and then repeat steps 3 and 4 until you have covered the region of space you are interested in.

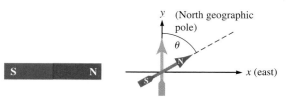

31.2 No. The integral in $\Phi_B = \int \vec{B} \cdot d\vec{A}$ (Eq. 31.2) is *not* taken over a closed (Gaussian) surface. Gauss's law for magnetism applies only to the flux through a Gaussian surface. The magnetic flux through an open surface may be nonzero.

31.3 The two Ampèrian loops are identical. Wrap the fingers of your right hand in the direction of the path; then your thumb points in the positive y direction.
Case 1: This wire penetrates the film twice: once carrying current downward in the negative y direction, and once upward in the positive y direction. The total current through the loop is $I_{thru} = I - I = 0$.
Case 2: This wire penetrates the film three times: twice carrying current downward in the negative y direction, and once upward in the positive y direction. The total current through the loop is $I_{thru} = I - 2I = -I$.

31.4 We need to find the graph of B versus r that represents Equation 30.9, which says that as $r \to 0$, $B \to \infty$, and as $r \to \infty$, $B \to 0$. Only graph 3 is consistent with the two extremes.

31.5 The deflection outside the solenoid is 0, independent of the current. Inside the solenoid, the current is found by combining $B = B_\oplus \tan\theta$ (Eq. 31.1) with $B = \mu_0 n I$ (Eq. 31.6), which yields $I = (B_\oplus/\mu_0 n)\tan\theta$. So, if we wish to measure $10° < \theta < 80°$, then 1.4×10^{-3} A $< I < 4.5 \times 10^{-2}$ A.

CHAPTER 31: Problems and Questions

1. 0.38%
3. The ball is attracted to the rod and swings toward it. The electric field is stronger near the rod, so the angle the thread makes with the vertical is larger for locations nearer the rod.
5. **a.** 0.17 A **b.** 1.4×10^2 A
7. **a.** $B_x A_x$ **b.** $B_y A_y$ **c.** 0
9. 62.6°

11. 3×10^{-5} Wb, -3×10^{-5} Wb, 9×10^{-5} Wb, -9×10^{-5} Wb, 0, and 0. The sum of the magnetic flux through all the surfaces of the cube is 0, and thus Gauss's law for magnetism holds.

13. **a.** Yes. Any Ampèrian loop (closed path) drawn will satisfy the condition. **b.** No. There is no way to find a loop through which a net current will flow.

15. **a.** Yes. Any loop in the plane of the magnetic field will work because the area vector will be perpendicular to the magnetic field. **b.** Yes. Any loop whose area vector is not perpendicular to the magnetic field will work. For example, a loop that is perpendicular to the magnetic field has an area vector that is parallel to the magnetic field and will have a nonzero magnetic flux through it.

17. The contributions for sides 1, 2, 3, and 4 are 0.0313 T·m, -0.0438 T·m, -0.0313 T·m, and 0.0438 T·m, respectively. The net current must be zero.

19. 0.2500 A

21. $(1/2)\mu_0 J$

23. $B_{in} = \dfrac{\mu_0 I r^3}{2\pi a^4}$ and $B_{out} = \dfrac{\mu_0 I}{2\pi r}$

25. 2.00 m

27. **a.** For $r < r_1$, $B = \dfrac{\mu_0 I}{2\pi r_1^2} r$; for $r_1 < r < r_2$, $B = \dfrac{\mu_0 I}{2\pi r}$; for $r_2 < r < r_3$, $B = \dfrac{\mu_0 I}{2\pi r}\left[1 - \dfrac{r^2 - r_2^2}{r_3^2 - r_2^2}\right]$; and for $r_3 < r < r_4$ (really in the entire region $r_3 < r$), $B = 0$.
b. For the long, straight wire, $B = (\mu_0 I)/(2\pi r)$ away from the wire, but for the coaxial cable, $B = 0$ outside the wire. So there is no need to worry about B interfering with outside devices in the case of the coaxial cable.

29. **a.** At $r = r_1$ **b.** 1.30×10^{-4} T

31. $\mu_0 c \left[\dfrac{r_A^2}{3} - \dfrac{R_1 r_A}{2} + \dfrac{R_1^3}{6 r_A}\right]$

33. 8.0×10^5

35. 0.612 A

37. **a.** 9.1×10^{-3} T **b.** There is no change.

39. $N = 9.78 \times 10^3$ and $\ell = 7.28 \times 10^2$ m

41. 9.47 cm

43. $n\mu_0 I_s I_w$

45. 2.0×10^1 A

47. **a.** 0.290 T **b.** 0.194 T

49. 0.230 m

51. 2.88 cm

53. 2.88×10^{12} N/(C·s)

55. The current is decreasing, so the magnitude of the magnetic field is also decreasing.

57. 1.76×10^{-7} T

59. **a.** -2.51×10^4 Wb **b.** 0

61. 3.98×10^5

63. 3.46×10^{-3} T

65. **a.** $1.60 \times 10^{-5}\hat{k}$ T
b. $-1.20 \times 10^{-5}\hat{k}$ T

67. For loop a, $-4\pi \times 10^{-7}$ T·m; for loop b, $20\pi \times 10^{-7}$ T·m; for loop c, $16\pi \times 10^{-7}$ T·m; for loop d, 0; for loop e, $4\pi \times 10^{-7}$ T·m

69. **a.** 6.44×10^{-6} Wb **b.** 3.22×10^{-6} Wb

71. $\mu_0 n I_1 I_2 s^2$

73. $\vec{B}_{inside} = -\mu_0 J z \hat{j}$ $(-L < z < L)$, $\vec{B}_{below} = \mu_0 J L \hat{j}$ $(z > L)$, $\vec{B}_{above} = -\mu_0 J L \hat{j}$ $(z < -L)$

75. $\mu_0 \varepsilon_0 \omega^2 I R^2 / 4$

CHAPTER 32: Concept Exercises

32.1 Actually, we did. We need to find the magnetic flux through the rectangular loop by breaking the integral in Equation 31.2 into two pieces. One piece is integrated over the cross-sectional area of the solenoid A_{sol}; the other piece covers the area inside the loop minus the area of the solenoid ($A_{loop} - A_{sol}$):

$$\Phi_B = \int \vec{B} \cdot d\vec{A} = \int_{sol} \vec{B} \cdot d\vec{A} + \int_{(loop - sol)} \vec{B} \cdot d\vec{A}$$

The second integral over ($A_{loop} - A_{sol}$) is 0 because the magnetic field is 0. The magnetic field is uniform over the cross-sectional area of the solenoid and $\vec{B}$ is parallel to $d\vec{A}$, so the first integral is

$$\Phi_B = \int_{sol} B dA = B \int_{sol} dA = B A_{sol}$$

So $A = A_{sol}$ in Equation 32.1.

32.2 Case 1: The angle φ changes, so an emf is induced in the loop.
Case 2: The magnetic field is strongest near the bar magnet. As the magnet swings back and forth, the magnetic flux through the loop alternately decreases and increases. An emf is induced in the loop.
Case 3: The magnet does not oscillate. The magnetic field, the loop's area, and the angle between the magnetic field and the area vector are all constant. So there is no change in the magnetic flux, and no emf is induced in the loop.

32.3 Case 1: The magnetic field $\vec{B}_{sol}$ is decreasing, so the loop's magnetic field $\vec{B}_{loop}$ must point in the same direction as $\vec{B}_{sol}$.
Case 2: The loop is tilted, so $\vec{B}_{loop}$ points either up and to the left or down and to the right. The magnetic field $\vec{B}_{sol}$ is decreasing, so the loop's magnetic field $\vec{B}_{loop}$ must have a component in same direction as $\vec{B}_{sol}$. Therefore, $\vec{B}_{loop}$ points up and to the left.

Case 3: There is no magnetic flux through the loop. When $\vec{B}_{sol}$ changes, there is no changing flux and no current induced in the loop, so there is no magnetic field $\vec{B}_{loop}$.

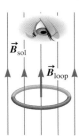
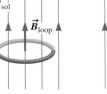

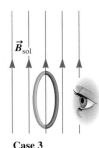

Case 1 Case 2 Case 3

32.4 The figure shows the loop's magnetic field in each case. To find the direction of the induced current, point your right thumb in the direction of $\vec{B}_{loop}$.
Case 1: The magnetic field $\vec{B}_{loop}$ points upward. Your right fingers wrap counterclockwise from the observer's perspective.
Case 2: The loop is tilted, and $\vec{B}_{loop}$ points up and to the left. Your right fingers wrap clockwise from the observer's perspective.
Case 3: There is no induced current.

32.5 The sliding magnet speeds up. Consider the sliding magnet to be the system; then both you and the fixed magnet are outside the system. Both you and the fixed magnet do positive work on the system. No friction or other forces act, so the total work done by external forces must equal the change in kinetic energy:

$$W_{tot} = \Delta K$$

Because $W_{tot} > 0$, $\Delta K > 0$, and the sliding magnet speeds up.

32.6 This device is the opposite of a magnetic brake. The relative motion of the magnet induces eddy currents in the disk. The direction of the eddy currents is given by Lenz's law; the eddy currents create their own magnetic fields that oppose the change. In this case, a region of the disk is magnetized such that it is attracted to the magnet. It might help to imagine sliding the magnet in Figure 32.18 (page 1025); how would the two loops respond in that case?

32.7 No. The current in the bulb varies sinusoidally according to $I = I_{max} \sin \omega t$ (Eq. 32.14). So, when the current is zero, the bulb is momentarily off. The fluorescent lights in your room actually flash on and off at 60 Hz, or $\omega = 2\pi(60 \text{ Hz}) = 377$ rad/s, because the coil in the power company's AC generator rotates at angular speed $\omega = 377$ rad/s. You don't see the light going on and off because your eyes cannot detect such a rapid change. AC generators have a similar effect on incandescent bulbs; they flicker slightly but don't completely flash on and off.

32.8 The peak voltage is found by multiplying the rms voltage by $\sqrt{2}$:

$$\mathcal{E}_{max} = \sqrt{2}\mathcal{E}_{rms} = \sqrt{2}(240 \text{ V})$$
$$\mathcal{E}_{max} = 340 \text{ V}$$

32.9 a. If a battery is used, the magnetic flux does not change, so there is no emf induced in the secondary coil of a transformer and no output current. **b.** If a DC generator is used, the magnetic flux increases and decreases. The result is that AC current is produced by the secondary coil; this would need to be converted back to DC for Edison's DC system.

CHAPTER 32: Problems and Questions

1. a. 5.29×10^{-4} Wb **b.** No. There is no induced current because the magnetic flux is not changing in the problem, as far as we can tell.
3. BA_{loop}
5. 8.45×10^{-2} V
7. a. 0 **b.** $-2BLW\omega/\pi$
9. 2.01 V
11. a. Increasing **b.** 2.26×10^{-4} V
13. a. 8.87×10^{-7} Wb **b.** 6.72×10^{-7} V
15. 0
17. As the magnet approaches the loop, the flux through the loop increases. The loop's magnetic field must point in the direction opposite the bar magnet's field, or to the left. Point your right thumb toward the left; then your fingers and the induced current wrap clockwise from the perspective of the observer. Now, as the magnet moves away from the loop, the magnetic flux decreases, so the loop's magnetic field must point in the same direction as the bar magnet's field, or to the right. Now point your right thumb toward the right in the direction of the loop's magnetic field: your fingers and the induced current wrap counterclockwise from the perspective of the observer.
19. 3.7×10^{-4} A, clockwise
21. a. 7.61 T/s **b.** an increase
23. $B_0 L v_y/R$, clockwise
25. 0.135 V
27. $-0.228 \sin(67.0\pi t)$
29. No. There will be no change in the magnetic flux through conductor A.
31. a. Loop B creates a magnetic field that is in the direction opposite that of loop A, so it is repelled away from loop A. **b.** Loop B creates a magnetic field in the same direction as that of loop A, so it is attracted toward loop A.
33. An electric current was created in the pipe as the magnet fell because of the changing magnetic flux as the magnet moves. Thus, there is electrical energy that accounts for the "missing energy."
35. a. 22.5 m/s **b.** No. The direction of the induced current depends on the direction of the bar's movement because the magnetic flux will either be increasing or decreasing.
37. 143 A
39. a. 0.176 V **b.** The bottom end of the bar, or the end with the lowest y value
41. 1.71×10^3 rev/min
43. 8.39×10^{-3} m^2, 9.16×10^{-2} m. The coil is about 9 cm × 9 cm, similar to the size of a cell phone.
45. a. $-114 \cos(93.2t)$ **b.** 114 V
47. a. 0.505 V **b.** 0°
49. The current in a generator switches direction and thus is an AC current. The output of the generator can be turned into a DC current through the use of a commutator, but in reality, the current in a generator is AC initially.
51. Yes. You would need a frequency of only about 0.8 Hz.
53. a. 43.8 V **b.** 0.250
55. 0.5. The average power is not 0.
57. a. 16.8 V **b.** 2.8 W
59. a. 2.18 A **b.** 28.4 W
61. a. 1.5 V **b.** −1.5 V
63. 11.1 m
65. 0.17 V
67. 21.6 A
69. 11 V
71. a. 4.73 N **b.** 5.91 W
73. 20.4 C
75. a. 2.19×10^{-6} Wb **b.** 1.09×10^{-5} V **c.** counterclockwise
77. 4×, or quadrupled
79. 1.34 A

CHAPTER 33: Concept Exercises

33.1 Yes. When the switch is moved back to position a, the magnetic flux through the circuit increases from zero. An emf is induced in the circuit (Faraday's law), and the induced emf opposes the flux change (Lenz's law). So, the induced emf is in the direction opposite the battery's emf—counterclockwise, producing a counterclockwise induced current.

33.2 Assume the initial inductance is L_i and the initial number of turns is N_i. The number of turns increases to $N_f = 1.1 N_i$. The inductance increases according to Equation 33.5, $L = \mu_0 N^2 A/\ell$:

$$L_f = \mu_0 \frac{A}{\ell} N_f^2 = \mu_0 \frac{A}{\ell} (1.1 N_i)^2 = 1.2 \left(\mu_0 \frac{A}{\ell} N_i^2 \right)$$

$$L_f = 1.2 L_i$$

So the inductance increases by 20%.

33.3 a. The current is a constant (not necessarily 0). **b.** The current may be decreasing as it goes from the black lead to the red lead. It is also possible that the current is increasing and going from the red lead to the black lead.

33.4 Mathematically, the differential equations for RL and RC circuits are identical. Compare Figures 29.39 and 33.9; the circuits look the same except the capacitor has been replaced by an inductor. Compare the charge on the capacitor in the RC circuit to the current in the RL circuit. In an RC circuit, when the switch is at A, the charge on the capacitor builds up just as the current in an RL circuit builds up. When the switch is at B in the RC circuit, the charge on the capacitor decays, similar to the decay of the current in the RL circuit.

33.5 Equation 27.22 ($u_E = \frac{1}{2}\kappa\varepsilon_0 E^2$) is analogous to Equation 33.17. In both cases, the stored energy density is proportional to the square of the field.

33.6 Mathematically, y is analogous to Q and $v = dy/dt$ is analogous to $I = dQ/dt$. On the graph (Fig. 16.5, page 453), we see that the relative phase shift between y and v is 90° just like the relative phase shift between Q and I (Fig. 33.13, page 1060). Also, the amplitude of the velocity is given by $v_{max} = \omega y_{max}$ (Eq. 16.7) and the amplitude of the current is given by $I_{max} = \omega Q_{max}$ (Eq. 33.23).

33.7 Shannon is correct. The current and voltage are out of phase, so you can't just divide the voltage at any instant by X_C to find the current at that instant. In Figure 33.20 (page 1065), initially the voltage is zero and the current is at its maximum. You cannot divide zero voltage by X_C to find the current. However, the maximum voltage divided by X_C is the maximum current.

CHAPTER 33: Problems and Questions

1. a. The short, wide inductor has the greater inductance per unit length because inductance per unit length is proportional to A, not length ℓ. **b.** They could have the same inductance if the product $A\ell$ is the same in each case.
3. Option (a) would increase the inductance. The inductance of a solenoid is given by $L = \mu_0 n^2 \ell A$, where n is the number of turns per unit length. If we compressed the inductor to a smaller length, keeping the same number of turns, $n^2\ell = N^2/\ell$ would increase and therefore the inductance would increase.
5. a. 0.750 V **b.** -3.33×10^2 A/s **c.** In both cases, the bulb would become dimmer and stop glowing as the current gets smaller. In (a), the current slowly gets smaller, so the bulb's brightness decreases slowly compared to case (b).
7. 2.53×10^{-3} H
9. a. 0.324 s **b.** 2.40 s
11. a. $4.61 L/R$ **b.** $L < 0.016$ H
13. a. 632 Ω **b.** 3.16×10^{-3} s
15. a. $\mathcal{E}\tau/eR$ **b.** $\left(\dfrac{\mathcal{E}\tau}{R}\right)\left(\dfrac{e-1}{e}\right)$ **c.** More charge passes through the resistor (in one time constant) when the current (and the energy stored in the inductor) is decreasing than when the current (and the energy stored in the inductor) is increasing. This is confirmed by comparing the areas under the curves for I versus t in each case.
17. a. $\dfrac{\mathcal{E}^2 \tau}{R}\left[\dfrac{2}{e} - \dfrac{1}{2e^2} - \dfrac{1}{2}\right]$
b. $\dfrac{\mathcal{E}^2 \tau}{2R}\left(1 - \dfrac{1}{e^2}\right)$ **c.** We find that the resistor dissipates more energy when the current is decreasing. The current is higher on average when the switch is moved to b than when the switch is at a, and the power dissipated by the resistor depends on the current squared.
19. 3.2×10^{-2} J
21. 0.975 H
23. a. 3.70×10^{-2} C **b.** 1.88 A **c.** $3.70 \times 10^{-2} \cos(50.8t)$ **d.** $-1.88 \sin(50.8t)$
25. 0.73 A
31. 8.8×10^{-8} F
33. a. 87.0 V **b.** 123 V **c.** 261 W **d.** 4.24 A
35. a. 1.4 A **b.** 1.4×10^2 V **c.** 2.0×10^2 V **d.** $(2.0 \times 10^2 \text{ V}) \sin[(120\pi \text{ rad/s})t]$
37. a. 1.16×10^3 rad/s **b.** 6.23×10^{-4} s
39. a. 0.398 Ω **b.** 171 A **c.** 242 A
41. a. 424 Ω **b.** 1.49×10^3 V **c.** 4.95 A and 2.10×10^3 V
d. $(2.10 \times 10^3 \text{ V}) \sin[(377 \text{ rad/s})t]$
43. The current through the capacitor leads the potential difference across the capacitor in an AC circuit by $\pi/2$ rad, so the current has a value of zero. Given that the circuit has only a source emf and a capacitor, both the potential difference across the capacitor and the source will be identical.
45. a. We need to use the RC filter as a low-pass filter, so we want to connect the detector (voltmeter) across the capacitor.
b. 1.27×10^{-5}. Our answer is encouraging, but really we need to compare this to the natural signal received. Because that source is much farther away, its maximum emf (at the telescope) will be much smaller. So our result doesn't tell us whether there will be interference.
47. a. 47.0 Ω **b.** 2.55 A **c.** 3.61 A
49. a. 2.4×10^2 Ω **b.** 0.063 A **c.** 0.089 A and 21.2 V
d. $(21.2 \text{ V}) \sin[(4.40 \times 10^2 \text{ rad/s})t]$
51. The current through the inductor lags the potential difference across the inductor in an AC circuit by $\pi/2$ rad, so the potential difference across the inductor is zero. Given that the circuit has only a source emf and a capacitor, both the potential difference across the capacitor and the source will be identical.
53. a. -6.91 A **b.** -4.83 A **c.** 6.91 A
55. 0.082 Hz
57. a. 0.20 mH and 31 nF **b.** 1.3×10^5 Hz
59. a. 8.41 Ω **b.** -3.66 Ω
61. a. 2.4×10^2 Ω **b.** $V_R = 2.00 \times 10^2$ V, $V_L = 4.1 \times 10^2$ V, $V_C = 849$ V
c. $-66°$ or -1.1 rad **d.** $\mathcal{E} = (4.8 \times 10^2 \text{ V}) \sin[(120\pi \text{ rad/s})t - 1.1 \text{ rad}]$ and $I = (2.00 \text{ A}) \sin[(120\pi \text{ rad/s})t]$
63. a. 0.374 A and 33.6° **b.** 92.5 V and 90.0° **c.** 99.9 V and 0° **d.** 26.1 V and $-90.0°$
65. a. 3.28×10^2 Ω **b.** 5.72×10^{-2} A **c.** 8.28°
67. 5.74×10^{-5} J
69. a. 9.5 V **b.** 3.8 A in the upward direction in Figure P33.69
71. a. 26.2 V **b.** 1.70×10^2 V **c.** 116 V
73. a. 0.148 V **b.** 0.592 V **c.** 1.00 s
75. a. 123 Ω **b.** 354 Ω **c.** 4.60×10^2 Ω **d.** 398 Ω **e.** $-30.2°$
77. a. 6.08 A/s **b.** 0.181 A/s
79. 373 mA
81. $I_{max, R} = 8.49$ A, $I_{max, C} = 12.8$ A, $I_{max, L} = 2.25$ A
83. 3.47×10^{-5} F and 1.47×10^{-6} F

CHAPTER 34: Concept Exercises

34.1 Avi makes a better argument. Little g is the acceleration due to gravity near the surface of the Earth. An alien society would find a value for g near the surface of their own planet. Unless their planet happened to be very similar to the Earth, they would not find the same value for g. Big G is the constant of proportionality in Newton's law of universal gravity. It is true that its value depends on the system of units we choose, and it even depends on how we define mass and distance. So the aliens' exact value of G would not match our value. However, an alien society would have to discover some version of G that would work in their system of units and measurements. We could find a way to convert between the two systems, and then we would discover that G is fundamentally the same on both planets.

34.2 In each case, determine the sign of the magnetic flux. Then remember Lenz's law when determining the sign of the right side of Faraday's law, Equation 34.6:

$$\oint \vec{E} \cdot d\vec{\ell} = -\frac{d}{dt}\int \vec{B} \cdot d\vec{A}$$

If the right side is positive, the electric field must be in the same direction as the path. (Negative means opposite direction.) **a.** Same direction. **b.** Same direction. **c.** Opposite direction.

34.3 a. The amplitude of the electric field is the number in front of the sine function: $E_{max} = 0.75$ V/m. The amplitude of the magnetic field is given by Equation 34.23: $B_{max} = E_{max}/c = 2.5 \times 10^{-9}$ T.

b. The angular wave number is the value in front of x: $k = 0.30$ rad/m. Find the wavelength from Equation 17.5: $\lambda = 2\pi/k = 21$ m. **c.** The propagation speed is the speed of light c. Because there is a negative sign in the middle of the sine function's argument, the wave is moving in the positive x direction: $\vec{v} = 3.00 \times 10^8 \, \hat{\imath}$ m/s. **d.** Find the angular frequency from $\omega/k = c$ (Eq. 34.20): $\omega = ck = 9.0 \times 10^7$ rad/s. Then find the frequency, $f = \omega/2\pi = 1.4 \times 10^7$ Hz, and the period, $T = 1/f = 7.0 \times 10^{-8}$ s.

34.4 Infrared radiation has a lower frequency than red light, so you might guess that *infrared* means "lower than red." (The prefix *infra* comes from the Latin word for "beneath.") Ultraviolet radiation has a higher frequency than that of violet light, so you might say that *ultraviolet* means "higher than violet." (The prefix *ultra* comes from the Latin word for "beyond.") Originally, "infrared" was called "ultrared."

34.5 The frequency can be found from Equation 34.20: $f = c/\lambda = 1.4 \times 10^9$ Hz $= 1.4$ GHz. From Table 34.2, this is in the radio or microwave part of the spectrum. Most astronomers do not distinguish between radio and microwave; they use the term *radio* for both.

34.6 Apply the right-hand rule to the cross product in $\vec{S} \equiv (1/\mu_0)(\vec{E} \times \vec{B})$ (Eq. 34.27). Point the fingers of your right hand in the direction of the electric field, and close your hand so that you push $\vec{E}$ into $\vec{B}$. Your thumb then points in the positive x direction, the direction of the Poynting vector.

34.7 *Unpolarized* light, shown in Figure 34.22, has electric fields oscillating in all possible planes (perpendicular to the direction of propagation). *All polarized* may be a better term because it might help us think of unpolarized light as *not polarized in any one particular plane*.

CHAPTER 34: Problems and Questions

1. Light is sometimes modeled as a particle, known as a photon. We think of each photon as a ball that possesses both energy and momentum. Light is modeled as a wave when it displays wave properties, such as interference.
3. $E(0.001$ s$) = 0.056$ V/m, $E(0.01$ s$) = 0.18$ V/m
5. a. $B = \mu_0 nCt$, $E = (1/2)\mu_0 nCa$
b. 4.7×10^{-6} N/C
7. $E = [(-B_0 R)/(2\tau)]e^{-t/\tau}$. Yes.
9. a. 2.99×10^{10} V·m/s **b.** 0.265 A
11. 4.59×10^{-18} T to the right
15. $(2.87 \times 10^8 \hat{\jmath} - 15.3 \times 10^8 \hat{k})$ m/s²
17. a. 1.50×10^{-3} m **b.** 3.75×10^{-9} m
19. 3.90×10^{-7} m
21. $f_{UVA} = 7.5 \times 10^{14}$ Hz $\to 9.4 \times 10^{14}$ Hz, $f_{UVB} = 9.4 \times 10^{14}$ Hz $\to 11 \times 10^{14}$ Hz
23. 7.41×10^{14} Hz
25. $1.4 \, \mu$T
27. a. 2.03×10^2 m **b.** 9.30×10^6 rad/s **c.** 3.10×10^{-2} m^{-1} **d.** 9.0×10^{-3} V/m
29. a. 20.4 V/m **b.** 4.14×10^5 Hz **c.** 725 m
31. 1.4×10^{-7} T. This magnetic field is 0.0027 times the magnetic field of the Earth.
35. No. In fact, radio waves pass through you undetected all the time. You can hear a sound wave because your ears are able to detect pressure waves in the surrounding medium.
37. Both waves travel at the speed of light in a vacuum. Only their frequency and wavelength are different.
39. A wavelength measurement of a wave that is blue-shifted appears shorter than the actual wavelength. Because the speed of the light is not affected by the measurement, the measured frequency thus appears higher than the actual frequency of the wave.
41. a. 0.0083 W/m² **b.** 1.2×10^{-6} W/m²
43. It must be about 4×10^7 times stronger.
45. 4.38×10^{-4} W/m²
47. a. 1.67×10^{-11} T **b.** 1.11×10^{-16} J/m³
49. a. $\hat{\imath}$ **b.** $\hat{\jmath}$ **c.** $\hat{k}$ **d.** $-\hat{\jmath}$
51. $VI/(2\pi RL)$
53. a. 3.33×10^{-6} Pa **b.** 1.33×10^{-7} m/s²
55. 5.66×10^{-4} Pa
57. a. 4.53×10^{-6} Pa **b.** 9.07×10^{-6} Pa
59. a. $1.72 \times 10^{-9} \hat{k}$ kg·m/s
b. $1.72 \times 10^{-9} \hat{k}$ N
61. 496 W/m²
63. a. 352 W/m² **b.** 0
65. 889 W/m²
67. a. $-\hat{\imath}$ **b.** 1.4×10^4 m^{-1}, 4.5×10^{-4} m, 6.7×10^{11} Hz
c. $E(x,t) = (12$ V/m$)\sin[(1.4 \times 10^4$ rad/m$)x + (4.2 \times 10^{12}$ rad/s$)t]$
69. 1.53×10^{-9} T
71. 224 m²
73. a. $\hat{k}$ **b.** 6.61×10^{14} Hz, 4.15×10^{15} rad/s, 1.38×10^7 m^{-1}
c. the x direction **d.** $B = (1.67 \times 10^{-11}$ T$)\sin[(1.38 \times 10^7$ m$^{-1})z - (4.15 \times 10^{15}$ rad/s$)t]$
75. 5×10^{-4} m/s²
77. $E = (1.44 \times 10^4$ V/m$)\sin[(1.01 \times 10^7$ m$^{-1})z - (3.02 \times 10^{15}$ rad/s$)t]$, $B = (4.80 \times 10^{-5}$ T$)\sin[(1.01 \times 10^7$ m$^{-1})z - (3.02 \times 10^{15}$ rad/s$)t]$
79. a. 0.00349 N **b.** 1.11×10^{-5} m/s²
81. 0
83. a. 2.61×10^3 V/m **b.** 1.08×10^3 m^{-1} **c.** 3.25×10^{11} rad/s **d.** the xy plane
e. $9.03 \times 10^3 \hat{\imath}$ W/m² **f.** 6.02×10^{-5} Pa
g. 6.98×10^{-2} m/s²

CHAPTER 35: Concept Exercises

35.1 If Newton had clearly observed a diffraction pattern, he might have abandoned his particle theory because the particle model cannot explain a diffraction pattern. Diffraction patterns are often very faint and may easily go unnoticed. We can see diffraction patterns under everyday circumstances if we deliberately look for them. Look at a lightbulb through your fingers when they are nearly touching, and you may be able to see one or two dark fringes.

35.2 As long as you are on one side of the sandbag and the source of the paint is on the other side, you do not get hit. Because the paint is made up of particles, it is absorbed by the sandbag. If you try to hide behind a rock from the water waves, you oscillate up and down because the waves are diffracted around the rock.

35.3 Moving the speaker would cause a reversal. The places where there was destructive interference would now have constructive interference, and the places where there was constructive interference would now have destructive interference. So there would be destructive interference at point C and constructive interference at D (Fig. 35.8).

35.4 The location of the bright fringes depends on the wavelength of light. In a beam of light made up of all colors, there are only certain screen positions where a particular color interferes constructively and forms a bright fringe. Because these places are different for each wavelength, the fringes are separated by color. The central maximum is white, however, because all colors experience constructive interference in the middle of the screen.

35.5 According to Figure 35.21, if the slit's width is very narrow (close to the wavelength of the light), the first dark fringes are far from the central maximum. Often we are not interested in light that falls far from the central maximum. For example, in Figure 35.9, light passes through a single slit before encountering the double slits. Only the light from the central maximum illuminates the double slits, so we are not concerned about the positions of the first dark fringes due to diffraction at the single slit.

35.6 The graph in Figure 35.25C corresponds to the light with the shortest wavelength ($\lambda = w/6$). The light with the longest wavelength produces the pattern in Figure 35.25A; its wavelength is $\lambda = w$. If light of all three wavelengths illuminated the same slit, the light with the longest wavelength (Fig. 35.25A) would produce the broadest central maximum.

CHAPTER 35: Problems and Questions

1. Model the paint as particles because the artwork has crisp, well-defined lines. There is no diffraction or bending of the paint.
3. $\lambda/4$
5. 2.08×10^{-6} m
7. 1.35×10^{-2} m
9. 31
11. 0.0222 m
13. **a.** 3.80 cm **b.** 39.3 cm **c.** The fringes are not evenly spaced. Fringes that occur at small angles appear equally spaced, but the angular separation between adjacent maxima increases as n increases.
15. 1.3×10^{-4} m
17. 8.86×10^{-5} m
19. 8.16×10^{-7} m
21. 5.33×10^{-7} m
23. 454.6 nm
25. 1.41×10^{-4} m
27. **a.** 3.58 cm **b.** 2.98 cm
29. 7.09×10^{-3} m
31. 0.5
33. Either 1.080×10^{-6} m or 2.700×10^{-7} m; no, the light is not visible in either case.
35. 5.81×10^{-7} m
37. 34
39. 1.05 rad, or 60.0°
41. **a.** 6.47 rad **b.** $I = 0.991 I_{max}$
43. $1.305 d_{green}$
45. 0.257%
47. 1.98×10^{-6} m
49. B, A, then C
51. 4.50×10^{-2}, 1.62×10^{-2}, and 8.27×10^{-3}
53. 0.0456
55. **a.** 4.50% **b.** 1.62%
57. 6.27×10^{-2}
59. three
61. The light from the two bulbs is not coherent. Even if the light were monochromatic, the phase difference would not be constant.
63. 4.73×10^{-4} m
65. ±0.32 m
67. All orders are missing except the central bright fringe ($n = 0$).
69. 5.91 m
71. 4.15×10^{-7} m
73. 5.0×10^{-4} m
75. 3.03×10^{-3} m
77. 156° or 2.72 rad
79. **a.** 29.2° **b.** 5.43×10^{-2} m
81. 2.52×10^{-6} W/m²
83. 2.80×10^{-4} m
85. 3.75 m
87. **a.** 6.03×10^{-3} m **b.** 1.21×10^{-2} m
89. $(I_0/9)\{1 + 8\cos^2[(\pi/\lambda)d \sin \theta]\}$

CHAPTER 36: Concept Exercises

36.1 According to Rayleigh's criterion, $\sin \theta_{min} = 1.22 \lambda/d$ (Eq. 36.3), a telescope with a larger diameter d has better resolution. So the minimum angular separation between two different sources that are barely resolved is smaller for the larger telescope.

36.2 The conditions for constructive and destructive interference depend on the wavelength or color of light. If the condition for a bright fringe at some location on the film is met for one particular wavelength—say, red light—then it is not met for other wavelengths. The rainbow effect is due to the condition for constructive interference being met by only one color over some small band on the film.

36.3 The pattern produced by a diffraction grating is much like the interference pattern produced by two slits. Because so many rulings (slits) are involved, it might be better to call it an *interference* grating. Of course, each slit also produces a diffraction pattern. So you might argue that the best name would be a *diffraction and interference* grating. Such a long name would probably be shortened, however.

36.4 To find the dispersion $D \equiv \Delta\theta/\Delta\lambda$ (Eq. 36.18), we need $\Delta\theta$ and $\Delta\lambda$ (Table 36.2):

$$D = \frac{(0.334 - 0.246)\,\text{rad}}{(6.56 - 4.86) \times 10^{-7}\,\text{m}} = 5.18 \times 10^5\,\text{rad/m}$$

Because these numbers apply to both gratings, both have the same dispersion.

36.5 For Young's experiment, there are two slits, so $N = 2$. For the first-order lines, $R = (2)(1) = 2$. Consider a red line of wavelength 650 nm and a violet line of wavelength 400 nm. Their average wavelength is 525 nm, and $\Delta\lambda = (650 - 400)\,\text{nm} = 250\,\text{nm}$. By Equation 36.20, the resolving power required to just barely separate them is

$$\frac{\lambda_{av}}{\Delta\lambda} = \frac{525\,\text{nm}}{250\,\text{nm}} = 2.1$$

This is about equal to the resolution of Young's double-slit experiment. So the lines are just barely resolved at the opposite ends of the spectrum. The other colors are blended together, forming nearly white light in between.

CHAPTER 36: Problems and Questions

1. As the size of the aperture decreases, the diameter of the central maximum (the Airy disk) gets larger.
3. 295 m
5. 13.8 arcmin
7. **a.** 8.13×10^{-5} rad **b.** We would choose the shortest wavelength of visible light possible—so, violet with a wavelength of about 400 nm. **c.** 6.51×10^{-5} rad
d. 4.89×10^{-5} rad
9. 1.2×10^2 m

11. 1.5×10^{-4} m
13. 1.16×10^{-7} m
15. a. $\lambda/4$ b. 9.96×10^{-8} m
17. $w_{green} = 9.35 \times 10^{-8}$ m, $w_{red} = 1.19 \times 10^{-7}$ m
19. 1.42×10^{-7} m
21. If we slowly decrease the thickness of the oil, green no longer exhibits constructive interference and is not seen. The observed color changes gradually to colors that have a shorter and shorter wavelength (from green to blue to violet).
23. a. 6.50×10^{-7} m b. 4.88×10^{-7} m
25. 1.59×10^5 rulings/m
27. 1.29×10^{-6} m
29. a. 4.36×10^3 slits/cm b. 9
31. 2.78×10^5 rulings/m
33. $9.76°$
35. a. 4 b. 9
37. $D_1 = 6.85 \times 10^5$ rad/m, $D_2 = 1.82 \times 10^6$ rad/m
39. a. 58 b. 29 c. 20
41. 3.56×10^5 rad/m
43. 6.00×10^3
45. 3162 fringes. Each fringe viewed indicates the mirror has moved 316.3 nm.
47. a. 2.5×10^{-9} m b. 8.3×10^{-18} s
49. The component in part C is more flat. Part B includes a series of curved lines, which represent regions that are not linearly increasing in gap thickness as one moves from left to right across the flat. Thus, the component in part B is more curved.
51. 2.94×10^{-7} m
53. 1.21×10^{-5} m
55. 1.29
57. 4.62 m
59. 3.57×10^{-7} m
61. 1.04×10^3 m
63. a. 1.85×10^{-10} m b. 4
67. 3.00×10^{-3} m
69. 6.19×10^{-7} m

CHAPTER 37: Concept Exercises

37.1

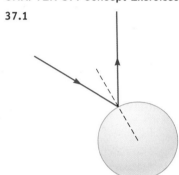

37.2 Answers will vary. Two possible sets of alternative terms for *virtual image* and *real image* are (1) *nonprojectable image* and *projectable image* and (2) *diverging image* and *converging image*. The first set of terms helps us to remember that a virtual image cannot appear on a screen, but a real one can. The second set helps us to remember that a virtual image is formed by rays that appear to diverge from each point that makes up the image, whereas each point that makes up a real image is the result of rays that converge at that point.

37.3 The spherical mirror is held in the artist's hand. It looks like the diameter of the sphere is about the same as the span of the artist's fingers, or about 9 in. or 23 cm. Then the radius of curvature is about 12 cm. The focal length is half the radius of curvature. This is a convex mirror, so according to the third sign convention, the focal length and radius of curvature are negative:

$$f = \frac{r}{2} = \frac{-12\,\text{cm}}{2} = -6\,\text{cm}$$

37.4 When traced back behind the mirror, all four rays intersect, so they produce the same results.

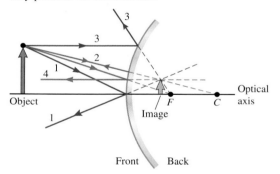

37.5

		$M > 0$ / $M < 0$	
Plane mirror	$\|d_i\| = \|d_o\|$	$M > 0$ (upright image)	$\|M\| = 1$
Convex mirror with $d_o < \|f\|$	$\|d_i\| < \|d_o\|$	$M > 0$ (upright image)	$\|M\| < 1$ In fact, $0.5 < \|M\| < 1$
Convex mirror with $d_o > \|f\|$	$\|d_i\| < \|d_o\|$	$M > 0$ (upright image)	$\|M\| < 1$ In fact, $\|M\| < 0.5$
Concave mirror with $d_o < \|f\|$	$\|d_i\| > \|d_o\|$	$M > 0$ (upright image)	$\|M\| > 1$
Concave mirror with $\|r\| > d_o > \|f\|$	$\|d_i\| > \|d_o\|$	$M < 0$ (inverted image)	$\|M\| > 1$
Concave mirror with $d_o > \|r\|$	$\|d_i\| < \|d_o\|$	$M < 0$ (inverted image)	$\|M\| < 1$

37.6 Rays from distant objects are parallel and focused at the focal point. If the secondary mirror were missing, the image of distant stars would form 57.6 m in front of the primary mirror (more than four times the length of the HST's tube). One of the roles of the secondary mirror is to "fold" the optics. Light is not focused at the secondary mirror; instead, that mirror helps to focus the light through a hole and onto a plane behind the primary mirror.

37.7 The image of the four people is upright. According to the ray diagrams in Figures 37.33 and 37.34 (page 1200–1201), if the image is upright, it must also be virtual. The people must be standing at a position closer than the focal point F, so $d_o < f$ (or $d_o < 57.6$ m for the HST).

CHAPTER 37: Problems and Questions

1. 1.5 m
3. **a.** -0.20 **b.** $d_i = (0.20)d_o$
5. 4.4 m
7. No. The rest of the electromagnetic spectrum reflects from surfaces, too. For example, radio waves reflect from a parabolic dish in a radio telescope.
9. 67.7°
11. $\theta_r = 45.0° - \theta_i$
13. 4
15. 67.5°
17. 45.0°
19. 8.5 ft
21. **a.** 4.0 ft **b.** 10.0 ft
23. 9.62 m
25. 82.5 cm
29. -1.11 cm
31. 1.20 m
33. -75.0 cm
35. -17.5 cm
37. 2.66 cm
39. **a.** $d_i = 23.7$ cm, -0.395, real, inverted **b.** $d_i = 34.0$ cm, -1.00, real, inverted
41. A convex mirror is not able to form a real image. The image must be projected onto the CCD screen. A concave mirror is capable of reflecting the light so that it passes through the location of the image and onto the CCD.
43. 1.43 cm
45. **a.** 2.10 m in front of the mirror **b.** 0.977 cm
47. 26.3 cm
49. 20.0 cm and 40.0 cm
51. **a.** 5.24 cm **b.** 2.10
53. 4.3 m
55. **a.** -3.27 cm **b.** 0.182 **c.** upright **d.** 0.636 cm
57. **FOX AND SOX**
59. **FOX AND SOX**
61. $2f$ or $4f$
63. **a.** 23° **b.** 5.7° **c.** 1.1° **d.** $(1.1 \times 10^{-5})°$
65. 10.0 cm
67. -6.0 cm
69. -8.96 cm
71. $(2n)D$
73. **a.** $\dfrac{\sqrt{x^2 + y_1^2} + \sqrt{(L-x)^2 + y_2^2}}{v}$
b. You should end up having derived the law of reflection, $\theta_i = \theta_r$. This is the condition for travel in the shortest time.

CHAPTER 38: Concept Exercises

38.1 The correct figure is sketch 1 because the angle of reflection equals the angle of incidence. Also, because the light travels from a medium with a low index of refraction to a medium with a high index of refraction, the beam is bent *toward* the normal, so the refracted angle must be smaller than the incident angle.

38.2 No. Because air has a lower index of refraction than water, when light goes from air into water, it cannot be totally reflected. Some of the light from your face is transmitted into the water.

38.3 The angle of refraction is zero in this case, which means the light does not bend in the glass and is a single white beam. The red light is still faster than the violet light, so in principle the leading edge of the beam should be red. But in practice this effect is too small to notice.

38.4 A diverging lens is like a convex mirror. Both produce virtual upright images, no matter where the real object is placed.

38.5 A converging lens is like a concave mirror. For both, whether the image produced is virtual or real, upright or inverted, depends on the placement of the real object.

38.6 Yes. Usually, the focal length of a camera lens is fairly short. In order to use the lens as a magnifier, you need to hold the feather closer to the lens than its focal point. So you should hold the lens very close to the feather, probably just a centimeter or so away.

38.7 Because we assume the near point is at 25 cm, the angular magnification is $m = 25$ cm$/f$ (Eq. 38.14). Therefore, $f = 25$ cm$/3 = 8.3$ cm.

38.8 To an astronomer, a telescope's LGP is more important than its magnification. If you know that a telescope has a large diameter, it can gather a lot of light and detect very faint objects.

CHAPTER 38: Problems and Questions

1. 36.7°
3. 41.8°
5. $\theta_t = \sin^{-1}\left(\dfrac{n_{\text{air}}}{n_{\text{water}}} \sin(90° - \theta_i)\right)$
7. 1.53
9. 32.3°
11. 1.207, 2.49×10^8 m/s, 4.29×10^{14} Hz
13. 56.7°
15. 1.41
17. 3.69 m
19. 61.3°
21. 48.6°
23. 1.34×10^{-2} m
25. All angles of incidence will cause the effect, so as long as $\theta_i < 90°$.
27. **a.** The second prism bent the monochromatic light's path, but no additional colors were seen. (See the figure.) So, Newton could conclude that prisms do not create color; prisms just spread the colors out. **b.** Using a second prism demonstrates that when all the colors come together, you get white light. Without this demonstration, you can say that white light may be separated into colors, but you cannot say that the color spectrum may be combined to make white light.

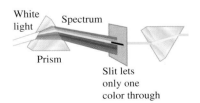

29. a. 3.7×10^{-7} m **b.** 5.5×10^{14} Hz **c.** Yes. The frequency is the same in both media.

31. a.

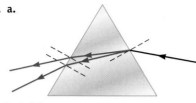

b. 2.65°
35. $d_i = -R$, $M = 1.75$
37. 2.44 m. This may be an issue when trying to catch the fish with a net if the net is not very big.
39. 7.5 cm, −0.13
41. 5.0 cm
43. −50.2 cm, diverging. The answers will not change if the object is placed on the other side of the lens.
45. 30.0 cm
47. a. −26.8 cm **b.** diverging
49. 7.72 cm
51. $d_o = 5.17$ cm, $d_i = -3.88$ cm
53. $M = 0.375$, $d_i = -3.75$ cm
55. a. 22.5 cm **b.** −4.00 cm
57. a. $1.33f$ **b.** real **c.** −0.333 **d.** inverted
59. a. −1.50 **b.** −0.075 m
61. a. 21.0 cm **b.** 25.0 cm
63. The glass and its contents act as a converging lens. The image we see is real and inverted.
65. Goggles enable you to maintain the air–eye interface so that your eye focuses as it normally does.
67. Her eyes are larger when viewed through her glasses, so their lenses act as magnifying glasses for us. She must be wearing converging lenses, with a focal length greater than the distance between her eyes and her glasses. She must be farsighted.
69. Susan's better vision is in her right eye, and she is nearsighted; $d_{\text{far, right}} = 12.9$ cm, $d_{\text{far, left}} = 12.5$ cm
71. a. $f_{\text{red}} = 7.35$ cm, $f_{\text{blue}} = 7.08$ cm
b. 3.58 cm
73. 6.00
75. $h_i = -2.14$ cm, 8.57 cm behind the second lens
77. a. 2.00 cm **b.** −0.667 cm
79. a. 17.6 cm to the right of the diverging lens, $M = -5.29$ **b.** inverted
81. a. All three students have made statements that are at least partially correct. You can *in principle* construct a microscope with these lenses, but their separation will not be the sum of their focal lengths. The resulting size may be impractical. **b.** The distance between the two lenses must be at least 112.5 cm. That is more than a meter. If we now found the position of the object, we would see that the microscope would have to be more than 2 m long. So these lenses cannot be used to construct a practical microscope given these size constraints.
83. 11.6 cm
85. 1.22
87. a. 6.74° **b.** 4.92×10^{-7} m **c.** The light wave slows down as it moves from air into water, but the sound wave speeds up by a large factor. The light wave bends toward the normal and its wavelength shortens, but the sound wave bends away from the normal and its wavelength increases.
89. 27.0°
91. a. 6.13×10^{14} Hz **b.** 3.67×10^{-7} m
c. 2.25×10^8 m/s
93. a. 34.5° **b.** 13.6°
95. a. 34.5° **b.** 15.4°
97. a. 25.4° **b.** 18.8°
99. 83.25°
101. 2.66 m
103. 48.8°
105. 53.1°
107. $\theta_i \geq 27.9°$
109. 13.1 cm
111. 0.750 m/s
113. a. 15.8 cm to the left of the diverging lens **b.** The final image is to the left of the lenses and is virtual.
115. diverging, −37.8 cm
117. −10.4 cm
119. a. 58.7 cm **b.** 25.8 cm
121. a. $d_i = -9.33$ cm, $M = 0.333$, virtual, upright **b.** $d_i = -4.20$ cm, $M = 0.700$, virtual, upright
123. a. 9.60×10^{-5} m/s **b.** toward the lens
125. b. $\alpha_{\min} = \beta(n - 1)$
127. a. 7.85° **b.** 0.007°
129. a. 11.3 cm to the right of the mirror
b. −9.00 **c.** inverted

CHAPTER 39: Concept Exercises

39.1 Your classroom and the railroad car are approximately inertial frames because neither is accelerating (except for the acceleration of the Earth itself). The car making a turn has centripetal acceleration, and the airplane taking off must increase its speed, so neither of these is an inertial frame.

39.2 The primed observer measures a shorter length because in the time it takes to make the measurement, the primed frame moves toward the second end of the rod. If we use the tower and helicopter notation from Equation 39.4, the transformation equation is

$$L = x_t - x_h = (x_t' - x_h') + v_{\text{rel}}(t_2 - t_1)$$
$$L = L' + v_{\text{rel}}(t_2 - t_1)$$

where $(t_2 - t_1)$ is the time between the two measurements. If the measurements are made simultaneously, then $t_2 = t_1$ and $L = L'$, as in Equation 39.5.

39.3 If the source is moving toward the observer, we choose the top signs in $\lambda = \lambda'\sqrt{(1 \mp \beta)/(1 \pm \beta)}$ (Eq. 39.27), and we find $\sqrt{(1 - \beta)/(1 + \beta)} < 1$. So $\lambda < \lambda'$, and the observer in the laboratory frame sees a bluer color than was emitted. If the source is moving away from the observer, we choose the bottom signs, and we find $\sqrt{(1 + \beta)/(1 - \beta)} > 1$. So $\lambda > \lambda'$, and the observer sees a redder color.

39.4 a. The ball has no energy. It isn't moving. It is cold and has almost no thermal energy. It has no potential energy because there is only one particle in the system.
b. The ball has no potential energy and essentially no thermal energy, but it has energy associated with its mass. A bowling ball's rest mass is about 6 kg; then its total energy is approximately $E = m_{\text{rest}}c^2 \approx 5 \times 10^{17}$ J (its rest mass energy). There is a considerable amount of energy stored in any massive object.

39.5 According to $R_{\text{sch}} \approx 3M/M_\odot$ km (Eq. 39.51), with $M/M_\odot = 1$, the radius would be 3 km—the distance of an easy jog.

CHAPTER 39: Problems and Questions

1. The Earth's motion causes the table of pegs to rotate under the pendulum. If the pendulum were in an inertial frame, only two pegs would be knocked over.
3. a. 2.0×10^2 s **b.** 5.6×10^2 s **c.** 1.5×10^2 s
7. a. 113.0 mph to the south **b.** 5.7 mi
9. The frame must not be accelerating.
11. a. 2.73×10^8 m/s **b.** -0.873 m **c.** 4.58×10^{-9} s
13. a. $x = 5.4 \times 10^9$ m and $y = 12.0$ m **b.** 24 s
15. a. 1.03 **b.** Although both frames could be labeled the x' frame, the convention used in this textbook is to identify the laboratory frame as the unprimed frame and the frame moving with the object as the primed frame. Realize, however, that from the electron's point of view, the laboratory is moving relative to it, and it would be fair to reverse the answers!
17. $(\sqrt{3}/2)c$
19. $(\sqrt{3}/2)c$
21. 0.724 m
23. a. 35.0 m **b.** 31.0 m **c.** $0.464c$
25. a. 1.3×10^3 yr **b.** 7.7×10^2 yr
27. $0.661c$
29. $(\sqrt{3}/2)c$
31. 4.90×10^{-9} s
33. a. 3.97 h **b.** 2.18 h
35. 1415.31 MHz and 1424.78 MHz
37. 1.83×10^6 m/s toward the observer
39. $0.496c$
41. 8.50×10^7 m/s
43. $0.96c$
45. a. $0.90c$ **b.** $0.90c$ **c.** $0.97c$
47. a. 0.145 kg **b.** 0.145 kg **c.** 0.275 kg
49. $0.877c$
51. a. 8.20×10^{-14} J or 512 keV **b.** 1.30×10^{-13} J or 812 keV **c.** $0.7762c$
53. 3.078×10^{-25} kg
55. 5.184×10^8 J whether using special relativity or classical physics
57. a. 8.20×10^{-14} J **b.** 4.12×10^{-13} J **c.** 3.30×10^{-13} J
59. 4.57×10^{14} Hz. We do not observe any gravitational shift, to three significant figures.
61. 3.25 min
63. 1.1×10^7 km $= 1700 \, R_\oplus$
65. 3.4×10^{41} kg $= 1.7 \times 10^{11} M_\odot$. This is slightly smaller than the mass of the Andromeda galaxy.
67. 1.47×10^{-3} s
69. 10.1 s
71. 16.1 μs for Joe's camera and 16.1 μs for Moe's camera
73. $0.737c$
75. $0.159c$ toward the Earth
77. Yes. The trackside observer measures the length of the train to be 70.2 m, so the train is measured to fit inside the tunnel with 9.8 m to spare.
79. $\dfrac{\rho'}{1 - (v_{\text{rel}}/c)^2}$
81. 3.33×10^{-9} J
83. a. $0.990c$ **b.** $0.989c$
85. a. 23.7 μs **b.** 37.6 μs

CHAPTER 40: Concept Exercises

40.1 a. The atom is missing one electron, so it is positively charged, with $q = +e$.
b. The pith ball's charge is positive, so the ball is missing

$$n = \frac{9 \times 10^{-9}\,\text{C}}{1.6 \times 10^{-19}\,\text{C/electron}} = 5.6 \times 10^{10}\,\text{electrons}$$

In the case of the pith ball, one electron fewer would make no difference to its total charge (to two significant figures). If you wanted the pith ball to have 1% more charge (9.09×10^{-9} C instead of 9.00×10^{-9} C), you would simply need to subtract more electrons (until 5.7×10^{10} were missing). So it seems that the pith ball can have any charge. However, in the case of a singly ionized atom, if there were one electron fewer, then its net charge would double. So you cannot increase the atom's charge by just 1%.

40.2 According to Planck's quantum theory, a system's energy comes in discrete bundles. So the idea that a digital radio can only be set to certain discrete frequencies helps us to imagine a quantity whose values are not continuous. For example, you may be able to set your digital radio to 88.1 KHz and to 88.3 KHz, but not to 88.2 KHz. Perhaps a quantum oscillator can have an energy of 88.1 eV or 88.3 eV, but not 88.2 eV. The analogy fails in that a radio is designed by a human engineer. There is no law that prevents the engineer from designing the radio so that you can tune it to 88.2 KHz. However, quantum mechanics tells us that nature sets the discrete energy levels in an oscillator.

40.3 Combine $K_{\max} = eV_0$ and $K_{\max} = E_{\text{light}} - W_0$ (Eqs. 40.9 and 40.11) with $E_{\text{light}} = hf$. Then the linear fit shown on the graph is $V_0 = \frac{h}{e}f - \frac{W_0}{e}$ and its slope is h/e. We estimate the slope $h/e = 4.1 \times 10^{-15}$ V/Hz. Solve for h to find $h = (4.1 \times 10^{-15}\,\text{V/Hz})e = 6.6 \times 10^{-34}$ J·s. Our estimate is in agreement with the accepted value. However, estimating the slope may be difficult; don't worry if your estimate is within about 3% of the accepted value.

40.4 The electron that corresponds to Figure 40.14C gains the most energy because the scattered photon has the greatest Compton shift and loses the most energy.

40.5 In classical physics, frequency, wavelength, and amplitude are properties of waves, not particles. The equation $E = hf = hc/\lambda$ links the wave properties of a photon to the amount of energy it carries as a particle, and that energy does *not* depend on amplitude. However, in the classical model the energy carried by a wave *does* depend on the wave's amplitude (Section 17-6).

CHAPTER 40: Problems and Questions

1. The pixels are very small, so when viewed on our size scale they seem continuous. Adding or subtracting a few pixels out of millions would not make much of a difference. If pixels were larger and more comparable to our size scale, we would see adjacent squares of various colors, and the detail of the image would be lost.
3. 6.443×10^{-6} m
5. a. 7.25×10^{-7} m, red **b.** 5.47×10^{-7} m, green **c.** 4.53×10^{-7} m, blue
7. 4.61×10^3 K
9. 7.00×10^{-7} m, red visible light
11. 6.29×10^{-20} J. No, it cannot. Any other frequency must be a whole-number multiple of the minimum frequency.
13. a. 5.78×10^3 K **b.** 5.01×10^{-7} m
15. a. 2.56×10^{-7} m **b.** 1.17×10^{15} Hz **c.** 2.15 eV
17. 1.2×10^{15} Hz
19. a. 5.40 eV **b.** 1.31×10^{15} Hz
21. a. 1.00 eV **b.** 2.10 eV
23. 1.09×10^{15} Hz
25. 0.61 m

27. 0.250
29. **a.** 1.274×10^{-10} m **b.** 1.560×10^{-15} J and 5.200×10^{-24} kg·m/s
31. 67.6°
33. 1.29×10^{-12} m
35. **a.** 19.84 pm **b.** 20.55 pm **c.** 24.69 pm
37. **a.** 62.54 keV **b.** 60.38 keV **c.** 50.25 keV
39. **a.** 1.72×10^{-11} m **b.** 72.0 keV **c.** 3.0 keV
41. **a.** 6.670×10^{-10} m **b.** 6.679×10^{-10} m **c.** 6.691×10^{-10} m
43. **a.** particle **b.** wave **c.** wave **d.** particle
45. **a.** 3.11×10^{-9} m **b.** 127 m/s
47. Recall that the wavelength of the wave must be comparable to the width of the opening in order to observe diffraction. A typical wavelength of a thrown baseball would be about 1×10^{-34} m. This is far less than the spacing between the slits, and so we do not expect to observe diffraction.
49. 6.66×10^{-10} m
51. The lower energy electrons will have less momentum than the higher energy electrons, which means that their wavelength is greater than that of the higher energy electrons. Because the wavelength of the lower energy electrons is greater, one might describe them as being "fatter" than the higher energy electrons.
53. 4.96×10^{-13} m
55. 0.11 m
57. 5.446×10^{-4}
59. **a.** 5.58×10^{-17} J **b.** 1.90×10^{-20} J **c.** 2.94×10^{3}
61. 0.364 V
63. **a.** 1.25×10^{-10} m **b.** 1.20×10^{-10} m
65. 1.4×10^{14}
67. 5.317×10^{6} m/s
69. 6.467×10^{-6} m
71. 4×10^{3} V
73. **a.** 0.827 V **b.** 1.65 V
75. 1.4×10^{19} s^{-1}
77. 2 eV
79. The photon's momentum would be infinite and the rest mass would be zero.
81. **a.** 4.67×10^{-2} kg·m/s **b.** 9.33×10^{-2} kg·m/s

CHAPTER 41: Concept Exercises

41.1 Cameron is correct. We relate temperature to average kinetic energy because of the great number of particles.

41.2 Answers will vary. One common occurrence in Chapter 2 is solving a quadratic equation for time and arriving at two solutions. One solution often corresponds to a time *before* the problem began, so this solution is not physically reasonable.

41.3 Figure 41.8 does not show the motion of the particle. Each panel is fixed in time, showing us the probability distribution of the trapped particle for a particular quantum state.

41.4 Imagine the particle as a cloud. The cloud is completely contained by the infinite well, but leaks out of the finite well. In both cases, the cloud's concentration in the well depends on the quantum state. For example, in the ground state the cloud is most concentrated in the center and we are most likely to find the particle there. However in the $n = 2$ state, the cloud is concentrated in two regions between the center and each boundary.

CHAPTER 41: Problems and Questions

1. $K_{av} = 1.9 \times 10^{3}$ eV $\approx$ 1 keV
3. 3.16×10^{34}
5. 1.37×10^{-50} kg·m^2
7. 2.86×10^{-12} m
9. **a.** 3 **b.** 5 **c.** It is more likely that a total of 6 would turn up on the dice because there are more possible ways to make a total of 6 with the dice.
13. $E_1 = 1.94 \times 10^{-5}$ eV, $E_2 = 7.76 \times 10^{-5}$ eV, $E_3 = 1.74 \times 10^{-4}$ eV, $E_4 = 3.10 \times 10^{-4}$ eV
15. 2.72×10^{-11} m
17. **a.** 1 **b.** 0.50
19. 9.64×10^{-21} J or 0.0602 eV
21. 0.964 eV
23. **a.** 2.34×10^{-13} m **b.** 3.00×10^{4} eV
25. 10^{-18} J or 10 eV
27. **a.** 5.68×10^{-4} m **b.** 9.09×10^{-4} m
29. **a.** 72 eV **b.** 1.7×10^{-8} m
31. **a.** 2.1×10^{-12} m **b.** 6.80×10^{-13} m **c.** 2.15×10^{-13} m
33. 5.08×10^{-12} J or 3.17×10^{7} eV
35. **a.** 1.79×10^{-12} m **b.** 4.64×10^{-13} m **c.** 1.44×10^{-13} m
37. 10^{-235}
39. 0.9591
41. 9.692×10^{-12} m
43. 89.2 eV
45. 1.79 eV
47. 1.88×10^{15} Hz
51. 7.91×10^{-36} m, so the player's claim is not legitimate.
53. **a.** 7.36×10^{-24} kg·m/s **b.** 7.16×10^{-11} m
55. 1.06×10^{-10} eV
57. **a.** $\sqrt{(15\lambda_A h)/(8m_e c)}$ **b.** $5\lambda_A/4$
59. **a.** 0.50 **b.** $(\pi - 2)/(4\pi)$
61. 2.68×10^{-22} J or 1.67×10^{-3} eV
63. $n = 3$
65. Yes. The potential barrier does not have an infinite height except as x goes to infinity, so there must be some penetration distance into the walls of the well.
67. **a.** 1.37×10^{-66} J, 5.49×10^{-66} J, 2.80×10^{-66} J **b.** 9.66×10^{-66} J
69. The energy difference will increase. As the width is decreased, the energy levels themselves increase in value, but the higher-energy levels will increase more than the lower levels, when we look at Eq. 41.16. Both are quadrupled when the width is halved, but the n level was greater than the m level, so the spread is greater.
71. 5.91×10^{-6} m
73. **a.** 71% **b.** 25%
75. Heisenberg's uncertainty principle is much more limited than Avi's summary suggests. According to Heisenberg's principle, we cannot simultaneously measure a particle's position and momentum to an unlimited precision. Avi's misconception is common. Many people think that physicists believe that the universe is completely uncertain. Understanding that quantum mechanics is probabilistic is not the same as saying that we cannot make predictions. If Avi doesn't review for the final exam, you know with great certainty what is going to happen.
77. 75%

CHAPTER 42: Concept Exercises

42.1 The word "atom" comes from the Greek word for *undivided*. So the term gives the impression that an atom's structure is fixed; but we know that an atom's electrons can move around in the atom and that an atom can even gain or lose electrons. Of course, other terms will vary, but here is a line of thought: Perhaps a better term would be based on the idea that atoms are the neutral elements that make up all matter.

42.2 Use Equation 42.1 for each value of n:

Line	n	$\frac{1}{\lambda} = R\left(\frac{1}{2^2} - \frac{1}{n^2}\right)$ (42.1)	$\frac{1}{\lambda}$ (m^{-1})	λ (nm)
Hα	3	$\frac{1}{\lambda} = R\left(\frac{1}{2^2} - \frac{1}{3^2}\right)$	1.52×10^6	656
Hβ	4	$\frac{1}{\lambda} = R\left(\frac{1}{2^2} - \frac{1}{4^2}\right)$	2.06×10^6	486
Hγ	5	$\frac{1}{\lambda} = R\left(\frac{1}{2^2} - \frac{1}{5^2}\right)$	2.30×10^6	434
Hδ	6	$\frac{1}{\lambda} = R\left(\frac{1}{2^2} - \frac{1}{6^2}\right)$	2.44×10^6	410

42.3 De Broglie's model offers some explanation for why hydrogen's energy is quantized: Quantization is a natural consequence of modeling the electron as a standing wave. By contrast, Bohr can only say that he hypothesizes that the energy is quantized because that worked well for Planck's model of a black body. Both models are ad hoc in nature. De Broglie cannot say why the electron should be modeled as a standing wave, and Bohr cannot say why his postulates are true. For a deeper understanding we must turn to Schrödinger's equation.

42.4 True. If $\vec{L} = \pm m\hbar \hat{k}$, its magnitude would be $L = \sqrt{\ell(\ell+1)}\,\hbar = m\hbar$, where m cannot be zero (or the angular momentum would be zero). This would require $m = \sqrt{\ell(\ell+1)}$, but this is not possible because both m and ℓ are nonzero integers.

CHAPTER 42: Problems and Questions

1. The plum pudding model is static. The electrons are embedded in a plum pudding of positive charge. In an odd sense, the electrons are fixed in place by something akin to a normal force and static friction, much like the chocolate chips are held in place in a cookie.

3. Based on the model of the ancient Greeks, the final result is the same in each case. Both objects would eventually be atoms, as you make the final cut between two atoms. However, the ancient Greeks also had the notion that these atoms would somehow be different from one another, as the iron and wood are different from each other.

5. The expectation was that the α particles should mostly pass through the gold foil, so the large deflections that were observed were surprising. If an α ray passed near the outside of the plum pudding sphere, you would expect a slight deflection due to the Coulomb force. However, if the α ray were to pass through the plum pudding, you would expect almost no deflection because only the charge that is in the spherical region inside the α ray's path could deflect it. But Rutherford observed large deflections, indicating that the positive charge in the pudding was concentrated and had a mass comparable to or higher than the mass of the alpha particles.

7. My dog is about 1 m long, so if he is my nucleus, the radius is 0.5 m. Then, the corresponding electron would be located at 5×10^4 m, or about 30 mi away.

9. -8.46×10^{-10} m

11. 1.06×10^{-10} m

13. a. 1.23 eV **b.** 13.1 eV

15. $2\pi ke^2/h$

17. 8×10^6

19. 3.28×10^{-6} m

21. 1.88×10^{-6} m, 1.28×10^{-6} m, 1.09×10^{-6} m, and 1.01×10^{-6} m

23. This is expected because there are more protons in the nucleus of larger atoms. This means the electron will be more attracted to the nucleus because the amount of the potential energy between an electron and the nucleus will increase. So, it will take more energy to free the electron from the atom.

25. 1.09×10^{-8} m

27. a. 8.46×10^{-10} m **b.** 1.33×10^{-9} m

29. 1.99×10^{-9} m

31. $m = 0, \pm 1, \pm 2$, $L = \sqrt{6}\hbar$, $L_z = 0, \pm\hbar, \pm 2\hbar$

33. $0, \sqrt{2}\hbar, \sqrt{6}\hbar$

35. $0, \pm\mu_{\text{Bohr}}, \pm 2\mu_{\text{Bohr}}$

37. a. 32 **b.** 50 **c.** 72

39. For $\ell = 0$, $L = 0$ and $(\mu_{\text{orbit}})_z = 0$; For $\ell = 1$, $L = \sqrt{2}\hbar$ and $(\mu_{\text{orbit}})_z = 0, \pm\mu_{\text{Bohr}}$; For $\ell = 2$, $L = \sqrt{6}\hbar$ and $(\mu_{\text{orbit}})_z = 0, \pm\mu_{\text{Bohr}}, \pm 2\mu_{\text{Bohr}}$

41. It is possible. Because the spin quantum number must be $\pm 1/2$, the orbital magnetic moment could equal the spin magnetic moment is if $m = \pm 1$. This can be seen by examining Eq. 42.23 and Eq. 42.26.

43. a. 2 **b.** 6 **c.** 10 **d.** 14 **e.** 18

45. Yes. This is because for every possible magnetic quantum number, there are two possible spin quantum numbers. Thus, when a shell is full, the total possible states, as dictated by the orbital quantum numbers and the magnetic quantum numbers, will be multiplied by 2 to account for the two possible spin states in each case.

47. $1s^2 2s^2 2p^6 3s^2 3p^6 4s^2 3d^{10} 4p^6 5s^2 4d^{10} 5p^5$

49. He: $1s^2$, Ne: $1s^2 2s^2 2p^6$, Ar: $1s^2 2s^2 2p^6 3s^2 3p^6$

51. Ni

53. 9.503×10^{-8} m, 9.504×10^{-8} m, 9.505×10^{-8} m

55. 5

57. 1.852×10^{-3} eV

59. a. 1.32×10^{-9} m **b.** 1.66×10^{-9} m

61. 5

63. a. 1.14×10^{-8} m **b.** Yes. The energy of this photon is 109 eV, which is much greater than the magnitude of the ground-state energy in hydrogen, 13.6 eV.

65. a. 12.8 eV **b.** 3.08×10^{15} Hz

67. Yes, the $\ell = 2$ subshell. The magnetic field will remove the degeneracy of each subshell. The $\ell = 2$ subshell has five degenerate states with $m = 0, \pm 1, \pm 2$.

69. -19.5 eV, -4.88 eV, -2.17 eV, -1.22 eV, -0.780 eV

71. $n = 4$

73. a. 5.87×10^4 T **b.** This would mean the electron is no longer bound to the nucleus, and would be free, if in that state.

75. Because the states must be whole numbers, n must be an even-numbered state. Thus, in this scenario, n cannot be odd, but $n/2$ could be odd.

77. a. 1.22×10^{-7} m **b.** 1.09×10^{-6} m **c.** 4.38×10^{-6} m

79. $n = 4 \to n = 1$: 12.8 eV, $n = 4 \to n = 3 \to n = 1$: 0.661 eV and 12.1 eV, $n = 4 \to n = 2 \to n = 1$: 2.55 eV and 10.2 eV, $n = 4 \to n = 3 \to n = 2 \to n = 1$: 0.661 eV, 1.89 eV, and 10.2 eV

CHAPTER 43: Concept Exercises

43.1 If you put several marbles in a flexible bag and tie it off, leaving no excess room, the bag will have a nearly spherical shape. Each marble takes up a certain fixed (constant) volume, so if you increased the number of marbles in the bag, the bag's volume would increase proportionally to the number of marbles added. If each marble were replaced by a foam ball, then the volume of these balls would not be constant. You could add more foam balls without increasing the volume of the bag, as long as you squeezed the foam balls further, reducing their individual volumes.

43.2 The strong force acts only over distances of roughly 1 fm or so. The Earth and Moon are much too far apart for the strong force to play any role.

43.3 Avi is correct. The students can use Equation 43.8, $N = N_0(\frac{1}{2})^{t/T_{1/2}}$. The problem asks for the fraction remaining, that is N/N_0, so they don't need to know N_0. The answer is not zero because every eight days *half* of the parent nuclei still remain. The number of parent nuclei goes to zero as time goes to infinity.

43.4 Inverse beta decay because it produces positrons: $^{18}_{9}\text{Fl} \rightarrow {}^{18}_{8}\text{O} + e^+ + \nu$.

43.5 The daughter nucleus becomes carbon. Compare Figures 43.6 and 43.7. If one of nitrogen's protons becomes a neutron through inverse beta decay, it will have six protons and six neutrons, exactly as carbon-12 does.

43.6 We use the term *pseudo*-units because MeV/c^2 are not true units, since they involve the constant c (the speed of light). We refer to 1 u = 931.4947 MeV/c^2 (Eq. 43.17) as a translation instead of a conversion factor because we are translating between two quantities, mass and energy, much as we did when translating between mass and weight in the expression 1 kg = 2.2 lb.

43.7 $E_B = 13.6$ eV because that is the amount of energy an external source must add to break up the hydrogen atom into free proton and a free both at rest. This is also called the ionization energy.

43.8 The correct reaction is $^{A}_{Z}\text{P} \rightarrow {}^{A}_{Z+1}\text{D} + e^- + \bar{\nu}$. You reason that charge must be conserved. Because the daughter nucleus has an extra proton, one of the products must be an electron, so there is no net gain in positive charge. You further reason that both baryon number and lepton number must be conserved. There are A baryons in the parent nucleus and A in the daughter, and the baryon number is conserved. The electron is a lepton. Because there are no leptons on the left of the reaction, there must be no net leptons on the right. The missing particle must be an antilepton to cancel out the electron's lepton number, in this case an antineutrino.

CHAPTER 43: Problems and Questions

1. a. Francium **b.** 87 **c.** 115
3. $^{214}_{82}\text{Pb}$
5. a. 1.2×10^{-15} m **b.** 7.4×10^{-15} m
7. 4.1 fm
9. 4.7 N. The nucleus should be stable because the attraction inward due to nearest neighbors is far greater than a Coulomb repulsion outward. Although there will be additional Coulomb repulsion due to nearer protons, the distance between a proton on the edge and a nearer neighbor would need to be orders of magnitude less than the distance between the two protons on the edge, which is not possible given their own dimensions.
11. We ignored it because the protons have low mass and because G, the universal gravitation constant, is small. The gravitational attraction between the protons is much weaker (about 10^{38} times weaker) than the Coulomb repulsion. It contributes very little to the stability of the nucleus. Instead, the strong nuclear force keeps the protons bound together.
13. We do not expect $^{19}_{8}\text{O}$ to be stable because there is considerably more energy in its neutron system than in its proton system.
15. $^{14}_{7}\text{N}$

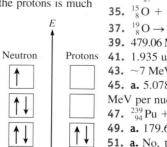

17. No. There are more protons than neutrons, and this is not one of the two exceptions to that general rule, H and He.
19. 5000
21. 429.6 s
23. a. 2.99×10^{17} Bq **b.** 53.4 days
25. The half-life of carbon-14 is roughly 6000 years. If a sample is a hundred million years old, it has lasted about 17,000 times carbon-14's half-life. So, very little carbon-14, if any, would remain in the sample. It would be very difficult to measure such a trace amount.
31. $^{14}_{7}\text{N}$
33. a. $^{188}_{77}\text{Ir} + e^- \rightarrow {}^{188}_{76}\text{Os} + \nu$
b. $^{127}_{54}\text{Xe} + e^- \rightarrow {}^{127}_{53}\text{I} + \nu$
c. $^{57}_{27}\text{Co} + e^- \rightarrow {}^{57}_{26}\text{Fe} + \nu$
35. $^{15}_{8}\text{O} + e^- \rightarrow {}^{15}_{7}\text{N} + \nu$, electron capture.
37. $^{19}_{8}\text{O} \rightarrow {}^{19}_{9}\text{F} + e^- + \bar{\nu}$, beta decay.
39. 479.06 MeV and 8.5546 MeV per nucleon
41. 1.935 u
43. ~7 MeV per nucleon and ~28 MeV
45. a. 5.0783 MeV per nucleon **b.** 5.3892 MeV per nucleon
47. $^{239}_{94}\text{Pu} + {}^{1}_{0}\text{n} \rightarrow {}^{148}_{58}\text{Ce} + 3({}^{1}_{0}\text{n}) + {}^{89}_{36}\text{Kr}$
49. a. 179.65 MeV **b.** 1.2×10^{42}
51. a. No, the total number of nucleons is not conserved or accounted for. **b.** Yes, the total number of nucleons is conserved or accounted for. **c.** Yes, the total number of nucleons is conserved or accounted for.
55. 4.237 MeV
57. 1.9×10^{11} photons
59. a. 0.300 Gy **b.** 6.00 Sv **c.** 6.00×10^2 rem
61. 0.13 Gy or 13 rad
63. No, it is not. The baryon number before the reaction is 2, and the baryon number after the reaction is 1. The baryon number should be conserved.
65. 6
67. No. Fermions cannot occupy the same state.
69. Two protons must decay. A total of four protons are added during the cycle, but the final product consists of only two extra protons. Thus, two of them must decay during the cycle.
71. Chlorine-36, Cl.
73. a. 1.45×10^9 Bq **b.** 0.0392 Ci **c.** 4.57 yr
d. 9.68 yr
75. 891.43 MeV
77. 4.17×10^{20}
79. a. $^{235}_{92}\text{U} + {}^{1}_{0}\text{n} \rightarrow {}^{143}_{55}\text{Cs} + 3({}^{1}_{0}\text{n}) + {}^{90}_{37}\text{Rb}$
b. 1737.37 MeV **c.** 758.10 MeV
d. 1150.62 MeV

Index

A

Abbreviations, for units of measure, App-8
Aberration, 1211
   chromatic, 1232, 1249
   spherical, 1211–1212
      of camera lens, 1249
      of magnifier, 1252
   stellar, 1268–1270
Absorbed dose, 1443
Absorption spectrum, 1383
AC circuits, 896, 1012–1013, 1045–1084
   with capacitance, 1064–1068, 1072–1076
   with inductance, 1046–1048, 1068–1076
   LC, 1057–1061
   oscillating, 1057–1061
   phasor diagrams for, 1063
   RC filters in, 1067–1068
   with resistance, 1062–1064, 1072–1076
   in resonance, 1075
   RL, 1052–1055
   RL filters in, 1070–1071
   RLC, 1072–1076
   vs. DC circuits, 1012–1013, 1033–1037
AC emf device, 896
AC generators, 1026–1031. *See also* AC power sources
   angular frequency of, 1064–1065
      capacitive reactance and, 1065
      impedance and, 1075–1076
      inductive reactance and, 1069–1070
      resonance and, 1076
   transformers and, 1035–1037, 1051–1052
   vs. DC generators, 1033–1035
AC motors, 1030
AC power sources, 1062–1078. *See also* AC generators
   capacitors and, 1064–1068
   inductors and, 1068–1072
   resistors and, 1062–1064
   symbol for, 1062
Accelerating reference frames, 1268, 1273, 1275, 1297
Activity, decay rate and, 1421–1422
Airy disk, 1156
Allowed transitions, 1402
Alpha decay, 1421
Alpha particles, 1419, 1421–1422
Alpha rays, 1375–1376, 1420–1421. *See also* Radiation
   tissue injury from, 1442

Ammeters, 897, 918
Amperage, 896
Ampere, 869
Ampère, André Marie, 979
Ampère's law, 979, 984–992, 1093
   for Ampèrian loop, 984–987
   displacement current and, 999–1003
   general form of, 999–1003
   for layered cylindrical wire, 990–992
   for long straight wire, 987–989
   Maxwell's addition to, 999–1003, 1093
   for solenoid, 992–996
   for toroid, 996–998
   vs. Biot-Savart law, 988
Ampèrian loop, 984–987
   current through, 985–986
   rectangular, 984–985
Angle of deviation, color spectrum and, 1224
Angle of incidence, 1188
Angle of reflection, 1188
Angle of refraction, 1219
Angular crossover frequency, 1072
Angular frequency, of AC generators. *See* AC generators, angular frequency of
Angular magnification. *See also* Magnification
   of magnifier, 1251–1252
   of microscope, 1253–1255
   of refracting telescope, 1255
Angular momentum
   of electron, 1389–1390
      Zeeman effect and, 1401–1404
   spin, 1394
   units of, App-13
Angular size, 1250
Antileptons, 1448
Antineutrinos, 1424–1425
Antiparticles, 1425
Aperture
   in camera obscura, 1185
   circular, diffraction through, 1155–1158
Aqueous humor, 1244
Arago, Dominique François, 992, 1124, 1147, 1316–1317, 1327
Area, units of, App-11
Area vectors, electric flux and, 758–759
Asteroid impacts, 1086–1087, 1111
Astronomical data, App-10

Astronomy, CCD cameras in, 1320
Atom(s), 1372–1409
   energy levels in. *See* Energy level(s)
   hydrogen. *See* Hydrogen atom
   magnetic moment of, 949–951
   models of
      Bohr's, 1377–1381, 1383, 1385
      de Broglie's, 1386–1387
      Thomson's plum pudding, 1373–1374
      Rutherford's solar-system, 1373, 1375–1379, 1389–1390, 1414
      shell. *See* Atomic shells
      standing-wave, 1386–1387
   nucleus of. *See under* Nuclear; Nucleus
   origin of, 1417
   quarks and, 687, 1309, 1416–1417, 1449
   size of, ionization energy and, 1397–1399
   strong nuclear force and, 146, 684, 1358, 1416–1417
   weak nuclear force and, 1424–1428
Atomic bombs, 1435
Atomic mass, 1414–1415
Atomic mass units, 1415, 1430
Atomic number, 1414
   in periodic table, 1397–1400
Atomic shells, 1388–1389, 1395
   Pauli's exclusion principle and, 1395–1397, 1418
   Zeeman effect and, 1401–1404

B

Back emf, 1049–1051
Balmer, John Jakob, 1378
Balmer formula, 1379
Balmer series, 1378, 1382–1384
Bar magnets. *See* Magnet(s)
Barrier
   definition of, 1356
   potential energy, 1356
Barrier tunneling, 1355–1359
   definition of, 1357
   Heisenberg's uncertainty principle and, 1364
   scanning electron microscope and, 1357–1358
   scanning tunneling microscope and, 1357
   in Sun, 1340, 1358–1359, 1364, 1438
Baryons, 1448–1449
Batteries, 833–837, 869–870
   cell phone, 865, 869–870, 889
   chemical reactions in, 834–835, 884
   circuits in, 866, 869–870. *See also* Circuit(s); Current
   defined, 834
   dry cell, 834
   in electrical network, 835–836
   functions of, 834, 884
   ideal, 835
   power from, 885–887
   thermal energy from, 884–885
   types of, 834
   wet cell, 834, 884
Beam splitter, 1176
Becquerel, 1422
Becquerel, Antoine-Henri, 1420
Beta decay, 1424–1428
   electron capture and, 1426
   inverse, 1425–1426
   neutrinos and, 1424–1426

Beta particles, 1420, 1424
Beta rays, 1375, 1420–1421, 1442. *See also* Radiation
Big Bang, 1417, 1427–1428, 1438–1439
Binding energy, 1383, 1428–1432
   definition of, 1429
   mass deficit and, 1428
   units of, 1430
Binding energy curve, 1430–1431
Binnig, Gerd, 1357
Biot, Jean-Baptiste, 940
Biot-Savart law, 940–948, 984
   for bent wire, 947–948
   for current loop, 945–946
   for moving charged particle, 940–941
   for segment of current-carrying wire, 941–942
   for straight wire, 943–945
   vs. Ampère's law, 988
Black body, 1166
   Planck's formula for, 1312–1313
   spectral intensity of, 1311–1313
Black body curves, 1310–1312
Black body radiation, 1309–1313
   ultraviolet catastrophe and, 1312–1313
Black body spectrum, 1166, 1383
Black holes
   Hawking radiation and, 1365
   primordial, 1365
Bohr, Niels Aage, 1378
Bohr, Niels Hendrick David, 1378–1379, 1386
Bohr magneton, 1393
Bohr radius, 1380
Bohr's atomic model, 1377–1381, 1383, 1385
Bohr's correspondence principle, 1347
Bohr's first postulate, 1379
Bohr's second postulate, 1379
Boltzmann's distribution law, 1406
Bombs, fission, 1435
Born, Max, 1343–1344
Boson, 1447–1448
   Higgs, 1449–1450
Boundary conditions, 1349, 1353, 1388
Brackett series, 1378, 1382, 1384
Bradley, James, 1268–1270
Bragg, William Henry, 1174
Bragg, William Lawrence, 1174
Bragg's law, 1174
Breeder reactors, 1438
Bright fringes (maxima), 1127–1312
   central, 1132
      slit width and, 1134–1136
   in double-slit interference
      intensity of, 1137–1140
      position of, 1128–1130
   secondary, 1132
   in single-slit diffraction
      intensity of, 1140–1143
      position of, 1132–1137
Buckingham Palace, lightning rods on, 736–737

C

Camera(s), 1185–1187, 1248–1250
   digital, 1320
Camera obscura. *See* Pinhole camera
Cancer, radiation-induced, 1442

Capacitance, 830, 1047, 1075
  calculation of, 843
    of cylindrical capacitor, 845–847
  defined, 830
  with dielectric, 849. *See also* Dielectric
    of parallel-plate capacitor, 844
  units of, 830
  vs. inductance, 1047
Capacitance combination rules, 912
Capacitive reactance, 1065
Capacitors, 776–777, 829–833, 1046–1047
  in AC circuits, 1064–1068
  charging, 835–836, 919–922
  cylindrical, 845–847, 851
  dielectric in, 847–856. *See also* Dielectric
  discharging, 919, 922–923
  electric fields in. *See* Dielectric
  electric potential energy in, 830–833
  energy storage in, 919–922, 1057, 1060
    derivation of, 831–832
    determinants of, 832
    with dielectric, 851–855
    energy density and, 853
  equivalent, 832, 838
  Gauss's law and, 845–846, 855–856
  in *LC* circuits, 1057–1061
  Leyden jars, 716, 828–829
  in parallel, 840–844
  parallel-plate, 844
  photoelectric effect and, 1317–1318
  plates in, 829–830
    size of, 849–850
  in *RLC* circuits, 1072–1076
  in series, 838–840, 843–844, 910
  for Thompson coil, 829, 841
  vacuum-filled, 828, 851–853
  vs. batteries, 834
  vs. inductors, 1046–1047
Carbon dating, 1422–1423
Cathode rays, 1373
CCD cameras, 1320
Cell phone batteries, 865, 869–870, 889
Center of curvature
  of lens, 1235
  of mirror, 1196
Center of mirror, 1196
Central axis, of spherical mirror, 1196
Central maxima, 1132
  slit width and, 1134–1136
Cesium clock, 1405
Changing magnetic flux. *See* Magnetic flux, changing
Charge. *See* Electric charge
Charge-coupled device, (CCD) 1320
Chemical elements. *See* Elements
Chernobyl nuclear power plant, 1441
Chromatic aberration, 1232
  of camera lens, 1249
Chromodynamics, quantum, 1449
Chu, Steven, 1408
Ciliary muscles, 1244–1245
Circuit(s), 866
  alternating current. *See* AC circuits
  in batteries, 866, 869–870
  conservation of energy and, 884–885
  current in. *See* Current
  direct current. *See* DC circuits
  filter
    RC, 1067–1068
    RL, 1070–1071
  grounded, 690–691, 899–900
  Kirchhoff's loop rule and, 903–906, 912, 1088–1089
  oscillating, 1057–1061
  phasor diagrams for, 1063
  power in, 884–889
  in resonance, 1075
  solenoids in, 992–996
Circuit analysis, 896–917
  circuit elements in series and, 899
  expected voltages in, 897–899
  for grounded circuits, 899–900
  Kirchhoff's junction rule in, 908
  Kirchhoff's loop rule in, 903–906, 912, 1088–1089
  multimeters in, 897, 917–919. *See also* Multimeters
  for real vs. ideal emf device, 898–899
  for resistors in parallel, 909–912
  for resistors in series, 906–907
  voltage rules in, 897–899
Circuit branches, 908
Circuit elements, 866. *See also* Conductors; Resistor(s)
  ohmic vs. nonohmic, 882
  power and, 887
  symbols for, 896, 897
Circuit junctions, 908
Circular aperture diffraction, 1124, 1155–1158
Circular waves, 1125
Circulation integral, 984
Classical physics
  fundamental principles of, 1086
  Galilean relativity and, 1271–1275
Classically forbidden region, 1354
Coherence, 1128
Cohen-Tannoudji, Claude, 1408
Color spectrum, 1166–1170, 1223–1226. *See also* Electromagnetic spectrum
Comets, collisions with, 1086–1087
Commutators, 968–969
  split ring, 1031
Compound microscopes, 1252–1255
Compton, Arthur Holly, 1320
Compton effect, 1320–1327
Compton shift, 1322–1326
Compton wavelength, 1322–1323
Concave mirrors, 1196. *See also* Spherical mirrors
  images formed by, 1204–1208
  sign conventions for, 1199
  spherical aberration in, 1210–1212
  vs. convex mirrors, 1205–1207
Conductance, 882
  Ohm's law and, 882
Conduction electrons, 865–866. *See also* Electron(s)
  flow of. *See* Current
Conductivity, 874–878
  defined, 875–876
  gravitational analogy for, 875–876, 884–885
  of metals, 872
  resistivity and, 876
  symbol for, 876
  temperature and, 877–878
  units of, 876

Conductors, 690–694
  capacitors and, 776–777
  charging by direct contact, 691
  charging by induction, 692
  conduction electrons in, 865–866
    Hall effect and, 935–936, 957–960. See also Current
  conductivity of. See Conductivity
  electric fields inside, 774–775, 819
    Ohm's law and, 882
  electric fields outside, 775–776
  electric potential of, 819
  in electrostatic equilibrium, 774–775
  heat from, 885–887
  metal, 872
  resistance in. See Resistance
  semiconductors and, 872, 877–878
Conservation of electric charge, 687
Conservation of energy
  circuits and, 884–885
  Lenz's law and, 1019–1020
  mass and, 1295
Conservation of momentum, 1290–1292
Constructive interference, 1127, 1138–1139
  Bragg's law and, 1174
  thin-film, 1160–1162
Contact lenses, 1245–1248. See also Lens(es)
Continuum spectrum, 1383
Contour maps, 800–801
Control rods, 1434
Convergent light rays
  divergent mirrors and, 1197
  plane mirrors and, 1191
Converging lens, 1233, 1235–1237, 1246. See also Lens(es)
  images formed by, 1240–1243, 1249–1250, 1254–1255
Conversion factors, App-9
Convex mirrors, 1196. See also Spherical mirrors
  images formed by, 1199–1204
  sign conventions for, 1199
  spherical aberration in, 1210–1212
  vs. concave mirrors, 1205–1207
Cornea, 1244
Corrective lenses, 1245–1248. See also Lens(es)
Correspondence principle, 1277, 1347
Coulomb, 687
  ampere and, 869
Coulomb force, 1358
Coulomb's constant, 697
  permittivity constant and, 761
Coulomb's law, 696–706, 1416
  applications of, 699–706, 754, 984
  brute force method for, 754, 954
  experimental basis of, 696–697
  inverse-square laws and, 697
  vector form of, 697–698
Cowan, Clyde, 1424
Critical incident angle, 1221–1222
Critical mass, in fission reactions, 1435
Crystalline lens, 1244
Curie, Marie, 1420
Curie, Pierre, 1420
Curies, 1422
Current, 864–895
  alternating. See AC circuits
  batteries and, 869–870
  conductivity and, 872, 874–878, 884–885. See also Conductivity
  defined, 868–869, 896
  direct. See DC circuits
  direction of, 869, 935–936
    Lenz's law and, 1016–1018
  displacement, 999–1003
  drift velocity and, 867, 871–874, 958–959
  eddy, 1025–1026
  electric field and, 1087–1090
  electric potential and, 878–881
  emf and, 896, 1011–1018, 1013–1015. See also Emf; Emf devices
  Faraday's law and, 1013–1015, 1087–1090
  Hall effect and, 935–936, 957–960
  in incandescent lightbulbs, 868
  macroscopic model of, 866–869
  magnitude of, 869
  measurement of, 897, 917–919
  microscopic model of, 865–866
  Ohm's law and, 881–882, 898
  resistance and, 878–881
  resistivity and, 876–878
  as scalar, 869
  temperature and, 877–878
  time-varying, RC circuits and, 919–925
  units of, 869
Current density, 870–874
  displacement, 1001
  electric fields and, 882
  of metals, 872
Current loop
  Biot-Savart law and, 945–946
  in DC motors, 968–969
  Kirchhoff's loop rule for, 903–906, 912, 1088–1089
  magnetic dipole moment for, 948–951
Cylindrical capacitor, 845–847, 851

D

Daguerre camera, 1248–1250
Dark fringes (minima), 1128–1130
  in double-slit interference
    intensity of, 1137–1140
    position of, 1128–1130, 1142–1143
  in single-slit diffraction
    intensity of, 1140–1144
    position of, 1132–1137
Daughter nucleus, 1419, 1421, 1425–1426
  in fission reactions, 1433–1434
Davisson, Clinton, 1328
Davy, Humphrey, 1373
DC circuits, 895–933, 1012–1013
  capacitors in, 776–777, 829–833. See also Capacitors
  expected voltages in, measurement of, 897–899
  grounded, 690–691, 899–900
  Kirchhoff's loop rule and, 903–906
  multimeters for, 897, 917–919
  power supply for. See Emf devices
  RC, 919–925
  resistors in, 919–925
  schematic diagram of, 896, 897
  vs. AC circuits, 1012–1013, 1033–1037
DC generators, 1030–1033
  limitations of, 1033–1034
  vs. AC generators, 1033–1037
DC motors, 968–969
DC power supply, 1049–1052

De Broglie, Louis, 1328, 1330, 1339–1341, 1386
De Broglie wavelength, 1339, 1340, 1386–1387
    Schrödinger's equation and, 1340, 1346–1347
De Broglie's atomic model, 1386–1387
Decay constant, 1421. *See also* Radioactive decay
Density
    charge, 726–727
    current, 870–874, 882
        displacement, 1001
    electric field line, 719, 758
    energy. *See* Energy density
    mass, 726
    probability, 1343–1344, 1346–1347, 1363. *See also* Schrödinger's equation
        of free particle, 1363
        radial, 1389–1392
        volume, 1390
Derivatives, App-3
    partial, 815
Descartes, René, 1124
Destructive interference, 1127, 1139
    thin-film, 1160–1162
Deuterium, in fusion reactions, 1436, 1438
Diagrams
    phasor, 1063
    ray. *See* Ray diagrams
    schematic, 835
Dielectric, 847–856
    capacitance and, 849
    defined, 847
    electric field of, 847–848, 855–856
    energy storage by, 851–855
    Gauss's law and, 855–856
    high-voltage, 850–851
    types of, 848
    vs. vacuum-filled capacitors, 851–853
Dielectric constant, 848
Dielectric strength, 848–849
Diffraction, 1124–1125
    circular aperture, 1124, 1155–1158
    defined, 1124
    double-slit, 1144–1148
    single-slit, 1132–1137, 1140–1144. *See also* Single-slit diffraction
    vs. interference, 1125, 1132, 1144–1148
    X-ray, 1173–1174
Diffraction gratings, 1166–1174
    dispersion of, 1170
    hydrogen spectrum and, 1166–1170
    maxima (lines) in
        position of, 1167
        width of, 1168–1170
    resolving power of, 1170–1172
    spectrums of, 1166–1174
    X-ray diffraction and, 1173–1177
Diffraction limited resolution, 1156
Diffraction patterns, 1124–1125
    resolution of, 1156–1158
Diffuse reflection, 1190
Digital cameras, 1320
Diodes, resistance in, 882
Dipoles. *See* Electric dipole(s); Magnetic dipole(s)
Dirac, Paul, 1339
Direct current circuits. *See* DC circuits
Dispersion, 1223–1226
    color spectrum and, 1224
    of diffraction grating, 1170
    prisms and, 1223–1226
    rainbows and, 1224
Displacement, in Galilean relativity, 1272–1273
Displacement current, 999–1003
Displacement current density, 1001
Distance
    in Galilean relativity, 1272
    near-point, 1245
    penetration, 1354–1355
Divergent light rays
    plane mirrors and, 1191
    spherical mirrors and, 1197
Diverging lens, 1233, 1235–1237. *See also* Lens(es)
    corrective, 1238–1240
    images formed by, 1238–1240, 1250
DNA, radiation damage to, 1442
Domains, magnetic, 951–952
Doppler shift, 1285–1288
    gravitational, time dilation and, 1298–1300
Dosimetry, radiation, 1443–1444
Double-slit diffraction, 1144–1148
Double-slit interference, 1127–1131, 1316, 1327, 1343–1345. *See also* Interference
    bright fringes in
        intensity of, 1137–1140
        position of, 1128–1130
    dark fringes in
        intensity of, 1137–1140
        position of, 1128–1130, 1142–1143
    probability waves and, 1343–1345
    vs. double-slit diffraction, 1144–1148
    in Young's experiment, 1127–1131, 1144, 1405, 1408
Doubly magic nucleus, 1419
Drift velocity, 867, 871–874
    conductivity and, 876
    Hall effect and, 958–959
Dry cells, 834. *See also* Batteries

E

Earth
    gravitational field of, 714. *See also* Gravitational field(s)
    as ground, 690–691
    magnetic field of, 938, 1056–1057
    speed of, Michelson-Morley experiment and, 1155, 1176–1178, 1268, 1276
Eclipses, general relativity and, 1301
Eddington luminosity, 1109–1111
Eddy currents, 1025–1026
Edison, Thomas, 1012–1013, 1033, 1043
Einstein, Albert, 1267, 1328. *See also* Relativity
    photoelectric effect and, 1313, 1316–1317
Electric charge. *See also* Electrostatic force
    capacitors and, 776–777, 829–833, 1046–1047. *See also* Capacitors
    conductors and. *See* Conductors
    conservation of, 687
    density of, 726–727
    from direct contact, 691
    electric fields and. *See* Electric field
    electric potential and, 798–799, 804–806. *See also* Electric potential
    electric potential energy and, 791. *See also* Electric potential energy

Electric charge (*continued*)
  elementary, 687
  grounds and, 690–691, 899–900
  induced dipoles and, 693
  from induction, 692
  insulators and. *See* Insulators
  Leyden jars and, 716, 828–829
  microscopic model of, 691
  negative, 686–687
  polarization and, 693
  positive, 686–687
  quantized, 687, 1309
  sketches of, 687–688
  sparks and, 693–694
  units of, 687
Electric charge flow. *See also* Current
  microscopic model of, 865–866
Electric dipole(s), 723–726
  defined, 723
  electric field of, 723–725, 741–745, 937
    potential energy of, 743
    torque on, 742
  electric potential due to, 801–802
  insulators and, 693
  polarization and, 693
  potential energy of, 743
  torque on, 742
  vs. induced dipole, 693
  water molecules as, 723, 725–726, 742
Electric dipole moment, 725
  electric potential and, 802
Electric field(s), 713–752
  calculation of
    brute force method for, 754, 954
    Gauss's law for, 753–757, 761–765, 984.
      See also Gauss's law
  in capacitors, 847–856. *See also* Dielectric
  of charged disk, 732–733
  of charged ring, 730–731
  of charged rod, 728–730
  of collection of charged particles, 720–723
  in conductors, 774–779, 819
    current density and, 882
    Ohm's law and, 882
  of continuous charge distribution, 726–728
  defined, 715
  of dipole, 723–726, 741–745, 937. *See also* Electric dipole(s)
    potential energy of, 743
    torque on, 742
  direction of, 719
  electric potential and, 807–819. *See also* Electric potential, electric fields and
  electromagnetic waves and, 1093–1100, 1103–1104, 1112–1115
  electron flow and, 867–869. *See also* Current
  electrostatic force exerted by, 737–741
  energy stored by, 853, 1055–1061, 1104
  graphing of, 799–801, 817–819
  Hall effect and, 935–936, 957–960
  in human body, 788, 814–816, 819
  induced, Faraday's law and, 1087–1090
  lightning rods and, 715–716, 733–737
  magnitude of, 719, 755
    electric flux and, 758

  near charged surface, 738
  of nonspherical objects, 715–716
  oscillating, polarization and, 1112–1115
  shell theorem and, 769
  source of, 715, 1011. *See also* Emf devices
    changing magnetic flux, 1011–1020, 1087–1090
    with linear symmetry, 766–767, 816–817
    with planar symmetry, 771–774, 813
    with spherical symmetry, 767–771, 798–799
  of spherical objects, 716–718
  subject in, 715
  units of, 813
  as vector field, 715
  visualizing, 799
  vs. gravitational fields, 717
  vs. magnetic fields, 937–938
Electric field lines, 718–720
  density of, 719, 758
  electric flux and, 758, 761
  equipotential surfaces and, 800, 809–813
  through Gaussian surfaces, 754–757, 761–765
  tips for drawing, 720–721
Electric flux, 755, 757–760
  area vectors and, 758–759
  calculation of, 758, 761–765
  electric field lines and, 758, 761
  equations for, 759
  Gauss's law and, 758, 761–765, 980–983
  in nonuniform electric field, 764
  in uniform electric field, 760
  units of, 759
Electric forces, 683–712
  in standard model, 1446
Electric guitar pickups, 1049
Electric potential, 787, 796–819
  calculation of, 797
  contour maps of, 800–801
  current and, 878–881
  differences in, 797
  due to charged disk, 805–806
  due to charged particle, 798–799, 807–808
  due to charged ring, 804–805
  due to charged sheet, 808–813, 818
  due to collection of charged particles, 801–803
  due to continuous distribution, 804–806
  due to dipole, 801–802
  electric fields and, 807–819
    equipotential surfaces and, 800, 809–813
    finding field from potential, 813–817
    finding potential from field, 807–809
    graphing of, 817–819
    in human body, 788, 814–816, 819
    of isolated charged conductor, 819
    path independence and, 811–812
    potential difference and, 807–813
  expected, in DC circuits, 897–899
  gravitational analogy for, 788–791
  in human body, 788, 797–798, 819
  measurement of, 896–903, 919. *See also* Voltmeters
  outside spherical source, 799
  reference point for, 799
  synapses and, 797–798
  terminal, 834–835
  units of, 797

visualizing, 799–801
vs. electric potential energy, 796–797
vs. gravitational potential, 796
Electric potential difference, 807–813
   current and, 878–881
   defined, 896
   measurement of, 896–903, 919. See also Voltmeters
   between terminals, 835, 896, 1011. See also Emf
Electric potential energy, 791–796
   in capacitors, 830–833
   charge and, 791
   of charged particle-sphere system, 791
   of collection of charged particles, 793–796
   defined, 791
   in Leyden jar, 828–829
   power and, 885–888
   storage of. See Batteries; Capacitors
   Thompson coils and, 829
   units of, 791
   visualizing, 799
   vs. electric potential, 796–797
   vs. gravitational potential energy, 788–792
   work and, 794–795
Electric potential energy density, stored by electric field, 853
Electrical conductivity, 874–878, 884–885. See also Conductivity
Electrical networks, 835
Electrical power transmission, 1033–1037
Electrical terminology, 896
Electrocardiography, 814
Electrodes, in batteries, 834
Electrodynamics, quantum, 1447
Electrolytes, in batteries, 834
Electromagnet, 940
Electromagnetic force, 684, 953
   Hall effect and, 935–936, 957–960
   Lorentz force and, 953, 1093
   Maxwell's equations for, 978–979, 1085, 1092–1093. See also Maxwell's equations
Electromagnetic spectrum, 1101–1103
   black-body, 1166
   diffraction gratings and, 1101–1103, 1166–1174
   visible light (color), 1166–1170, 1223–1226
Electromagnetic wave(s), 1085–1122. See also Light
   Eddington luminosity and, 1109–1111
   electric field and, 1093–1100, 1103–1104
      oscillation of, 1112–1115
   energy transferred by, 1103–1104
   frequency of, 1101
   gamma, 1102
   infrared radiation and, 1102
   intensity of, 1104–1105
      Malus' law and, 1114
      polarization and, 1113–1115
   magnetic field and, 1093–1100, 1103–1104
   Maxwell's theory and, 999–1003, 1085–1086, 1093–1100, 1155. See also Wave model
   microwaves, 1102
   momentum delivered by, 1107–1112
   polarization of, 693, 1112–1115
      linear, 1095–1098, 1112
      plane, 1112
   power flux and, 1103–1104
   pressure exerted by, 1108–1112
   radar and, 1102
   radio, 1094–1095, 1101, 1287–1288
   reflected vs. absorbed, 1108. See also Reflection
   speed of light and, 1098–1099, 1101, 1158–1159. See also Speed of light
   transmission of, 1094–1095, 1101
   transverse, 1099
   ultraviolet radiation and, 1102
   unpolarized, 1112
   visible light and, 1101–1102
   wavelength of, 1101
   X-rays, 1102
Electromagnetic wave equation, 1095–1099
Electromotive force. See Emf
Electron(s), 1373. See also Atom(s); Hydrogen atom; Nucleus
   angular momentum of, 1389–1390
   beta particle, 1424
   charge of, 686. See also Electric charge
   conduction, 865–866
   in conductors and insulators, 690–691
   magnetic moment of, 949–951
      Zeeman effect and, 1401–1404
   probability density for. See Probability density
   quantum number and, 1379, 1388–1389, 1392–1396. See also Quantum number
   radial probability density for, 1389–1392
   shells and subshells of, 1388–1389, 1395
      Pauli's exclusion principle and, 1395–1397, 1418
      Zeeman effect and, 1401–1404
   shielding of, 1397–1399
   as waves, 1328–1332
Electron capture, 1426
Electron configuration, 1396, 1399–1400
Electron flow. See also Current
   Hall effect and, 935–936, 957–960
   in magnetic field, 935–936, 957–958
Electron microscope, 1330–1332
   scanning, 1330–1332
   transmission, 577, 1330–1332
Electron neutrinos, 1448
Electron spin, 949–951. See also Spin
   in quantum physics, 1393
Electron volt, 797
Electrophorus, 708
Electroscope, 708
Electrostatic equilibrium, 774–775
Electrostatic field. See Electric field(s)
Electrostatic force. See also Electric charge
   conductors and, 690–694
   Coulomb's law and, 696–706
   defined, 688
   exerted by electric field, 737–741
   gas pump fires and, 685, 695
   insulators and, 690–694
   inverse-square laws and, 697
   models of, 685–688
   properties of, 688–689
   visualizing, 799
   vs. gravitational force, 696
Electroweak force, 684
Electroweak theory, 1449
Elementary charge, 687
Elements
   electron configuration of, 1396, 1399–1400
   origin of, 1415, 1417, 1427–1428, 1439–1440
   periodic table of, 1373, 1397, 1399–1400, App-14–App-15

Emf, 896, 1011. *See also* Electric field(s), source of
  back, 1049–1051
  changing magnetic flux and, 1011–1012
  defined, 1011
  direction of, 1016–1018
  flux linkage and, 1014
  induced, 1046–1048
    direction of, 1046, 1049–1051
    Faraday's law and, 1013–1015, 1087–1088
    Lenz's law and, 1016–1020, 1046, 1089
    magnitude of, 1050
  motional, 1020–1025. *See also* Motional emf
  root-mean-square, 1033–1034
Emf devices, 896, 1011–1012. *See also specific devices*
  AC, 896, 1026–1030
  DC, 896, 1030–1033, 1049–1051
    real vs. ideal, 899
  defined, 896
  slide generators, 1020–1025
  terminal potential of, 896
Emf rules, 899
Emission spectrum, 1383
Energy
  binding, 1383, 1428–1432
  conservation of
    circuits and, 884–885
    Lenz's law and, 1019–1020
    mass and, 1295
  of fission vs. fusion reactions, 1429–1431
  ionization, 1383, 1397–1399
  potential
    of dipole in electric field, 743
    electric. *See* Electric potential energy
    gravitational. *See* Gravitational potential energy
    magnetic, 966–967
  quantized, 1309, 1312, 1388–1392
  relativistic kinetic, 1293–1296
Energy density
  radiation intensity and, 1104
  stored by batteries, 884–885
  stored by electric field, 853, 1055–1061, 1104
  stored by magnetic field, 1046–1047, 1055–1061, 1104
Energy level(s), 1381–1384
  excited-state, 1382–1383
  ground-state, 1341–1342, 1381–1383
  for hydrogen atom, 1381–1384, 1388–1391
  for ions, 1383, 1385–1386
  for nucleus, 1418
  in Planck's quantum theory, 1313–1314, 1341, 1359–1360
  shells and subshells and, 1388–1389, 1395–1397
  transitions between
    allowed vs. forbidden, 1402
    lasers and, 1406
    Zeeman effect and, 1401–1404
Energy level diagrams, 1342, 1381–1382
Energy transfer, Poynting vector and, 1103–1104
Equilibrium, electrostatic, 774–775
Equipotential surfaces, 800
  electric field lines and, 809–813
Equivalent capacitors, 832, 838
Equivalent dose, 1443
Ether, Michelson-Morley experiment and, 1155, 1174–1178, 1268, 1276
European Organization for Nuclear Research (CERN), 1450

Event horizon, 1301
Excited atoms, 1382–1383
Expectation value, 1347
Eye
  anatomy of, 1243–1244
  far-point distance and, 1245
  focusing by, 1244–1245
  image distance in, 1244–1245
  lens of, 1243–1245
  near-point distance and, 1245
  optical model of, 1244
Eyepiece
  of microscope, 1252–1254
  of telescope, 1254–1255

F

Farad, 830
Faraday, Michael, 1010–1011, 1373
Faraday cage, 779
Faraday-Lenz's law, 1018
Faraday's generator, 1030–1031
Faraday's law, 1010–1044
  changing magnetic flux and, 1013–1015, 1087
  eddy currents and, 1025–1026
  expression of, 1013
  flux linkage and, 1014
  generalized form of, 1087–1092
  induced electric field and, 1087–1090
  induced emf and, 1013–1015, 1087
  Lenz's law and, 1018
  magnetic braking and, 1025–1026
  slide generators and, 1020–1025
Far-point distance, 1245
Femtometer, 1415
Fermi, Enrico, 1415
Fermions, 1447
Ferromagnetic materials, 951–952. *See also* Magnet(s), permanent
Feynman, Richard, 1447
Field(s), 714–716. *See also* Electric field(s); Gravitational field(s); Magnetic field(s)
  scalar, 714
  source of, 714
  subjects in, 715
  vector, 714–715
  vs. forces, 714–715
Field forces. *See also* Force(s)
  electrostatic force as, 715
  magnetic force as, 935
  subjects of, 715
Field lines, 718–720
  electric. *See* Electric field lines
  magnetic, 936, 937–938
  tips for drawing, 720–721
Filaments, 868
  resistance of, 878–881
Filter circuits
  RC, 1067–1068
  RL, 1070–1071
Fires, gas pump, 685, 695
First ionization energy, 1397–1399
Fission reactions, 1419, 1428–1432, 1433–1436
  binding energy in, 1428–1432
  in bombs, 1435

critical mass in, 1435
mass deficit in, 1428–1432
in nuclear power plants, 1428, 1434–1435
Flat mirrors, 1202
Flatland (Abbott), 1298
Fluorescent bulbs, 1405–1406
Flux
　electric. *See* Electric flux
　magnetic. *See* Magnetic flux
　power, 1103
Flux linkage, 1014, 1047
Focal length
　of camera lens, 1252
　of lens of eye, 1244–1245
　of magnifier, 1252
　of mirror, 1198
　of objective lens
　　of microscope, 1252–1253, 1255
　　of telescope, 1254–1255
　of spherical lens, 1233, 1235–1236
Focal plane, 1211
Focal point
　for lenses, 1236
　for mirrors, 1196–1197
Forbidden transitions, 1402
Force(s)
　Coulomb, 1358
　electric, 683–712
　electromagnetic. *See* Electromagnetic force
　electrostatic. *See* Electrostatic force
　electroweak, 684
　fundamental, 684, 935, 1086
　gravitational, 684, 788, 789
　　inverse-square laws and, 697
　　vs. electrostatic force, 696
　Lorentz, 953, 1093
　magnetic. *See* Magnetic force
　in standard model, 1449
　strong nuclear, 146, 684, 1358, 1416–1417
　subjects of, 715
　weak nuclear, 684, 1424–1428
Force fields. *See* Field(s)
The Forever War (Haldeman), 1281–1283, 1292
Formulas
　algebra, App-1
　geometry, App-2
　trigonometry, App-2
Franklin, Benjamin, 685–686, 687–688, 715–716, 733–737, 827, 828, 1011
Free radicals, 1442
Frequency
　angular. *See* AC generators, angular frequency of
　resonance, 1075
　units of, App-13
Frequency filter circuits, 1067–1068
Fresnel, Augustin-Jean, 1126, 1147
Fresnel's spot, 1147–1148
Fringe position. *See* Bright fringes (maxima); Dark fringes (minima)
Fuel rods, 1434–1435
Fukushima Daiichi nuclear reactor, 1433
Fundamental forces, 684, 935, 1086
Fundamental Theorem of Calculus, App-4
Fusion reactions, 1428–1432, 1436–1439
　binding energy in, 1428–1432

mass deficit in, 1428–1432
in nuclear reactors, 1437–1439
in Sun, 1295, 1358–1359, 1364, 1428, 1437–1438
　barrier tunneling in, 1340, 1358–1359, 1364, 1438

G

Galilean relativity, 1271–1275
　displacement and velocity in, 1272
　invariant quantities in, 1272
　position and distance in, 1272
Galvani, Luigi, 788, 833
Galvanometers, 917–918, 967–968
Gamma factor, 1277–1278
Gamma radiation, 1102
Gamma rays, 1420–1421, 1442. *See also* Radiation
Gas pump fires, 685, 695
Gaussian surfaces, 754–757
　box, 754–755, 762
　closed cylinder, 762–764
　in dielectric, 845–847, 855–856
　electric field lines through, 754–757, 761–765
　spherical, 755, 762, 767–771
　symmetry of, 756
Gauss's law
　for electricity, 753–786, 980–981, 1092
　　applications of, 761–765
　　conductors and, 774–779
　　dielectrics and, 855–856
　　electric flux and, 757–760
　　linear symmetry and, 766–767
　　mathematical statement of, 761
　　permittivity constant and, 761
　　planar symmetry and, 771–774
　　qualitative approach to, 754–757
　　spherical symmetry and, 767–771
　　vs. Gauss's law for magnetism, 982
　for magnetism, 980–983, 1092–1093
Geiger, Hans, 1376
General relativity. *See* Relativity, general
Generators
　AC, 1026–1030
　DC, 1030–1033
　Faraday's, 1030–1031
　slide, 1020–1025
Geometric optics, 1185–1187
　ray model and, 1185–1187
Gerlach, Walther, 1393
Germer, Lester, 1328
Glashow, Sheldon, 1449
Goeppert-Mayer, Maria, 1417
Gramme's DC generator, 1030–1033
Graphing, of electric fields, 799–801, 817–819
Gratings. *See* Diffraction gratings
Gravitational Doppler shift, time dilation and, 1298–1300
Gravitational field(s), 714, 788–789
　calculation of, 789
　of Earth, 714
　vs. accelerating reference frames, 1297
　vs. electric fields, 717
Gravitational force, 684, 788, 789
　inverse-square laws and, 697
　vs. electrostatic force, 696
Gravitational lenses, 1301
Gravitational mass-inertial mass equivalence, 1297–1300

Gravitational potential, 788–791
   calculation of, 789
   defined, 788–789
   reference point for, 788–789
   units of, 789
   vs. electric potential, 796
   vs. gravitational potential energy, 788
Gravitational potential energy, 788
   vs. electric potential energy, 788–792
   vs. gravitational potential, 788
Gravitational red shift, 1299
Gravitational time dilation, 1298–1300
Gravity
   electric potential and, 788–791
   illusory nature of, 1297–1300
   vs. electrostatic force, 696
Gray, 1443
Great Sendai Earthquake, 1433
Greek alphabet, App-7
Ground state, 11341–1342, 1381–1383
Grounds, 690–691, 899–900
Guitar pickups, 1049

H

Hadrons, 1448–1449
Hale, George, 1403–1404
Half-life, 1422
Half-width, of diffraction grating lines, 1168
Hall effect, 935–936, 957–960
Hall voltage, 958
Harmonic number, 1341
Harmonic oscillator, simple quantum, 1358–1362
Harmonic waves, 1341
   interference of, 1127
Hawking, Stephen, 1365
Hawking radiation, 1365
Heart, electric fields in, 788, 814–816, 819
Heat
   from conductors, 885–887
   defined, 870, 885
Heisenberg, Werner, 1339, 1363
Heisenberg's uncertainty principle, 1362–1365
Helium
   formation of, 1340, 1358–1359, 1364, 1427–1428, 1437–1438. *See also* Fusion reactions
   primordial, 1428
Helium-neon laser, 1407. *See also* Lasers
Henry, 1047
Henry, Joseph, 1047
Hertz, Heinrich Rudolf, 1094
Higgs, Peter, 1450
Higgs boson, 1450–1451
High-pass filters, 1068
Hooke, Robert, 1124
Hooke's law, 881–882
Hubble Space Telescope, 1190, 1209–1212
Human body
   electric fields in, 788, 814–816, 819
   electric potential in, 788, 797–798, 819
Hund's rule, 1396
Huygens, Christiaan, 1124, 1326–1327
Huygens's principle, 1126, 1132
Hydrogen
   in fusion reactions. *See* Fusion reactions
   primordial, 1417, 1428

Hydrogen atom
   energy levels in, 1381–1384, 1388–1391
      excited, 1382–1383, 1390–1391
      ground-state, 1381–1383, 1389–1390
      Zeeman effect and, 1401–1404
   magnetic quantum number for, 1388, 1392–1394
   models of. *See also* Atom(s), models of
      Bohr's, 1383, 1386
      de Broglie's, 1386–1387
   orbital magnetic quantum number for, 1388
   orbital quantum number for, 1388–1389
   principal quantum number for, 1379, 1388
   radial probability density for, 1389–1392
   Schrödinger's equation for, 1388–1392
Hydrogen fusion. *See* Fusion reactions
Hydrogen line series, 1377–1378, 1382, 1384
Hydrogen spectrum, 1166–1170, 1374–1375, 1377–1386. *See also* Electromagnetic spectrum
   absorption, 1383
   Balmer series in, 1378, 1382–1384
   black-body, 1383
   Brackett series in, 1378, 1382, 1384
   continuum, 1383
   emission, 1383
   Lyman series in, 1378, 1382, 1384, 1402
   Paschen series in, 1378, 1382, 1384
   Pfund series in, 1378, 1382, 1384
   of Sun, 1383, 1403–1404
   Zeeman effect and, 1401–1404
Hyperopia, corrective lenses for, 1245–1248

I

Ideal battery, 835
Ideal DC emf devices, 899
Ideal emf rule, 899
Ideal galvanometer, 918
Ideal inductors, 1050
Ideal solenoid, 993–995
Ideal toroid, 996–998
Ideal transformer, 1036, 1052
Image(s)
   angular size of, 1251
   depth reversal of, 1192–1193
   inverted. *See* Inverted images
   in ray diagrams, 1191
   real
      reflection and, 1191, 1199
      refraction and, 1228, 1244, 1248–1249
   reversal of, 1192–1193
   virtual, 1191
      reflection and, 1191, 1199
      refraction and, 1228, 1244
Image distance
   of camera, 1249
   in eye, 1244–1245
   reflection and, 1191–1192, 1199
   refraction and, 1228, 1230–1231
Image height
   reflection and, 1192, 1199
   refraction and, 1228
Impedance, 1074–1075
Incandescent lightbulbs
   current in, 868
   filaments in, 868
      resistance of, 878–881

Incident rays, 1219, 1222
Index of refraction, 1159, 1219, 1223
   color and, 1223
   in eye, 1244
Induced dipoles, 693
Induced emf. *See* Emf, induced
Inductance, 1047–1048, 1075
   units of, 1047
   vs. capacitance, 1047
Induction, 692
   polarization and, 693
Inductive reactance, 1069–1070
Inductor(s), 1046–1049
   defined, 1046
   energy stored by, 1046–1048, 1055–1057
   ideal, 1050
   in LC circuits, 1057–1061
   in RL circuits, 1052–1055
   in RLC circuits, 1072–1076
   solenoids, 1046–1049. *See also* Solenoids
   symbol for, 1046
   vs. capacitors, 1046–1047
Inductor rule, 1051
Inertial mass–gravitational mass equivalence, 1297–1300
Inertial reference frames, 1268–1271
   accelerating, 1268, 1273, 1275, 1297
   displacement and velocity in, 1272–1273
   invariant quantities in, 1272
   position and distance in, 1272
   speed of light in, 1276
   transformation equations for, 1272
Infrared radiation, 1102
Input voltage, 896
Insulators, 690–694
   in capacitors, 829–830, 847–856. *See also* Dielectric
   semiconductors and, 872, 877–878
Integrals, App-4
   circulation, 984
   path, 843, 848, 984
Intensity per wavelength. *See* Spectral intensity
Interference, 1125
   coherence and, 1128
   constructive, 1127, 1138–1139, 1160–1162
     Bragg's law and, 1174
     thin-film, 1160–1162
   destructive, 1127, 1139, 1160–1162
     thin-film, 1160–1162
   double-slit, 1127–1131, 1137–1140, 1316, 1327, 1343–1345. *See also* Double-slit interference
   sound wave, 1126
   thin-film, 1158–1165. *See also* Thin-film interference
   vs. diffraction, 1125, 1132, 1144–1148
   Young's experiment and, 1127–1131
Interferometer, Michelson's, 1174–1178
International Thermonuclear Experimental Reactor (ITER), 1437–1439
Invariant quantities, 1272
Inverse beta decay, 1425–1426
Inverse-square laws, 697
Inverted images
   formed by cameras, 1248–1249
   refraction and, 1228
   in vision, 1244

Iodine
   modeling of, 1361
   origin of, 1439–1440
Ionization energy, 1383, 1397–1399
Ions, energy levels of, 1383, 1385–1386
Iris, 1243
Iron, origin of, 1439
Isotope(s), 1414–1415, 1417
   decay of, 1414
   definition of, 1414
   formation of, 1439–1440
   mass of, 1414–1415
   natural abundance of, 1414
   stable, 1414
Isotope chart, 1414
ITER reactor, 1437–1439

J

Jeans, James, 1312
Jensen, J. Hans D., 1417
Joule, 797
Juice. *See* Current
Junctions, 908

K

Kamerlingh, Heike, 878
Kilowatt-hours, 887
Kinetic energy
   in barrier tunneling, 1355–1358
   of free particle, 1362–1364
   relativistic, 1293–1296
Kirchhoff's junction rule, 908, 912
Kirchhoff's loop rule, 903–906, 912, 1088–1089

L

Laboratory reference frame, 1271–1272
Land, Edwin Herbert, 1113
Large Hadron Collider (LHC), 1450
Lasers, 1374, 1406–1408
   helium-neon, 1407
   in Young's double-slit experiment, 1128, 1130–1131, 1406, 1408
Law of reflection, 1187–1190, 1219
Law of refraction, 1219–1221
LC circuit, 1057–1061
Length
   focal. *See* Focal length
   proper, 1279–1280
   units of, App-11
Length contraction, 1279–1280
Lens(es)
   camera, 1248–1250
   converging, 1233, 1235–1237, 1246
     images formed by, 1240–1243, 1249–1250, 1254–1255
   corrective, 1245–1248
   diverging, 1233, 1235–1237, 1246
     images formed by, 1238–1240, 1250
   of eye, 1243–1245
   focal length of. *See* Focal length
   gravitational, 1301
   magnification of, 1233
   magnifier, 1248, 1250–1254
   in multiple-lens systems, 1252–1257

Lens(es) (*continued*)
  Newton's rings and, 1162–1163
  objective
    of microscope, 1252–1255
    of telescope, 1254–1256
  primary rays for, 1238
  ray diagrams for, 1237–1238
  in refracting telescopes, 1254–1257
  sign conventions for, 1228, 1233, 1235–1236
  in single-lens systems, 1248, 1250–1252
  thick, 1231
  thin, 1232–1238
  thin-film interference and, 1162–1163
  thin-lens equation and, 1233
Lens maker's equation, 1233–1235
Lenz's law, 1016–1020, 1046, 1089
  conservation of energy and, 1019–1020
  Faraday's law and, 1018
Leptons, 1448–1449
Leyden jars, 716, 828–829. *See also* Capacitors
L'Hôpital's rule, App-4
Light. *See also* Electromagnetic wave(s)
  bending of, 1300–1302
  color and, 1166–1170, 1223–1226. *See also* Electromagnetic spectrum
  Doppler shift and, 1285
  Maxwell's wave theory of, 1085–1086, 1093–1100, 1155
  polarized, 1112–1115
  power of, 1086–1087
  propagation of, 1125–1126
    in ether, 1155, 1174–1178, 1268
  properties of, 1086
  ray model of, 1185–1187. *See also* Ray(s)
  reflection of. *See* Reflection
  refraction of. *See* Refraction
  speed of. *See* Speed of light
  visible, 1101, 1102
  as wave vs. particle, 1316–1320, 1326–1328. *See also* Light waves; Photons; Wave-particle duality
Light rays. *See* Ray(s)
Light waves. *See also* Electromagnetic wave(s); Wave(s); Wave model
  coherent, 1128
  diffraction of. *See* Diffraction
  Huygens's principle, 1126, 1132
  interference of. *See* Interference
  visual representation of, 1125
  vs. sound waves, 1127
Lightbulbs
  current in, 868
  filaments in, 868
    resistance of, 878–881
  fluorescent, 1405–1405
Light-gathering power, of telescope, 1255–1256
Lightning rods
  on Buckingham Palace, 736–737
  Franklin's, 715–716, 733–737
  Wilson's, 734, 736–737
Lightning strikes, 757, 779
Linear charge density, 726
Linear magnification, 1186. *See also* Magnification
  vs. angular magnification, 1251
Linear polarization, 1095–1098, 1112
Linear symmetry, 765–767, 816–817
Liquid drop model, 1419

Little Boy, 1435
Lodestones, 934–935
Lorentz, Hendrik A., 1401
Lorentz factor, 1277–1278
Lorentz force, 953, 1093
Lorentz transformations, 1276–1279
  correspondence principle and, 1277–1278
  for position, 1277–1278
  for time, 1278
Low-pass filters, 1068
Luminiferous ether, Michelson-Morley experiment and, 1155, 1174–1178, 1268
Lyman series, 1378, 1382, 1384
  Zeeman effect and, 1402

M

Magic number, 1419
Magnet(s), 934–935
  electromagnets, 940
  natural, 934–935
  permanent, 949, 951–952
Magnetic bottles, 955
Magnetic braking, 1025–1026
Magnetic dipole(s), 937–939
  magnetic field of, 948–949
  torque on, 964–969
Magnetic dipole moment, 948–951, 1392–1395
  direction of, 948–949
  of electrons and atoms, 949–951
Magnetic domains, 951–952
Magnetic field(s), 935–951
  of AC generators, 1026–1030
  of bar magnets, 940
  calculation of
    Ampère-Maxwell's law for, 999–1003, 1093
    Ampère's law for, 979, 984–1003, 1092
    for bent wire, 947–948
    Biot-Savart law for, 940–948, 984, 988. *See also* Biot-Savart law
    for current loop, 945–946, 984–986
    for layered cylindrical wire, 990–992
    for moving charged particle, 940–941
    for segment of current-carrying wire, 941–942
    for straight wire, 943–945, 987–989
  compasses and, 936–937
  current loops in, 964–969
  of dipole, 948–949
  direction of, 937, 939–941
    Lenz's law and, 1016–1018
    right-hand rule for, 939–940, 942, 948
  of Earth, 938, 1056–1057
  electromagnetic waves and, 1093–1100, 1103–1104
  of electromagnets, 940
  electron flow in, 935–936, 957–958
  energy stored by, 1046–1047, 1055–1061, 1104
  induced emf and, 1046
  magnitude of, 941, 979
  measurement of, 979–980
  motion of charged particle in, 954–957
  Orsted's demonstration of, 939
  of slide generators, 1020–1025
  sources of
    bent wire, 947–948
    current loop, 945–946, 948–951, 984–987

electrons, 949–951
layered cylindrical wire, 990–992
magnetic dipole, 948–949
moving charged particle, 940–941, 952–954
permanent magnet, 949
segment of current-carrying wire, 941–942
solenoid, 992–996, 1011–1013, 1046–1048
straight wire, 943–945, 987
toroid, 996–998
symbol for, 937
Van Allen belts and, 954
vs. electric fields, 937–938
Magnetic field lines, 936–938
Magnetic field strength, 941
Magnetic fields, Zeeman effect and, 1401–1404
Magnetic flux, 980–983
changing, 1011
as electric field source, 1011–1012, 1087–1090
Faraday's law and, 1013–1015, 1087–1088
Lenz's law and, 1016–1020, 1089
in solenoid, 1011–1012
Gauss's law and, 980–983
inductance and, 1047–1050
in solenoid, 1011–1012
switches and, 1046
Magnetic force, 684, 934–977
on current loop, 964–966
on current-carrying wire, 960–964
defined, 935
direction of, 953
in stereo speakers, 960–962
as field force, 935
magnitude of, 952
on moving charged particle, 952–954
overview of, 935–936
between parallel wires, 963–964
in standard model, 1446
Magnetic moment, 949–951
of electron, Zeeman effect and, 1401–1404
Magnetic monopoles, 938, 981–982
Magnetic poles, 936–937
Magnetic potential energy, 966–967
Magnetic quantum number, 1388, 1392–1394
Magnification
angular, 1251–1255
linear, 1186, 1251
by reflection
by plane mirrors, 1192
by spherical mirrors, 1200–1202
by refraction
by compound microscope, 1253–1254
by magnifying glass, 1248, 1250–1252
Magnifiers
magnifying glass, 1248, 1250–1252
microscope eyepiece, 1252–1254
Magnitude
of current, 869
of electric field, 719, 755, 758
of induced emf, 1050
of magnetic field, 941, 979
of magnetic force, 952
Malus's law, 1114
Marconi, Guglielmo, 1094
Marsden, Ernest, 1376

Mass
atomic, 1414–1415, 1430
critical, in fission reactions, 1435
gravitational, 1297
inertial, 1297
relativistic kinetic energy and, 1293–1296
rest, 1292, 1429, 1430
energy and, 1295
units of, 1430, App-12
Mass deficit, 1296
binding energy and, 1428
Mass density, 726
Mass number, 1414–1415
Mass spectrometers, velocity selectors for, 956–957
Mass transformation, 1290–1292
Mass-energy equivalence principle, 1294–1295
Mathematics
algebra, App-1
calculus
derivatives, App-3
integrals, App-4
geometry, App-2
propagation of uncertainty, App-5
reference tables for, App-7
trigonometry, App-2
Matter wave, 1328–1332
Schrödinger's equation for, 1338–1367.
See also Schrödinger's equation
Matthews, Thomas, 1287
Maxima. See Bright fringes (maxima)
Maxwell, James Clerk, 684, 978, 1085, 1123
Maxwell-Boltzmann speed distribution, 1405
barrier tunneling and, 1358
black-body curves and, 1311
Maxwell's equations, 978–979, 1085
Ampère-Maxwell's law, 999–1003, 1093
Faraday's law, 1010–1044, 1087–1092, 1093
Gauss's law for electricity, 753–786, 980–981, 1092
Gauss's law for magnetism, 980–983, 1092–1093
special relativity and, 1275
Maxwell's wave theory, 1085–1086, 1093–1100, 1155.
See also Wave model
Ampère's law and, 999–1003, 1093
Hertz's experiments and, 1094
Mean free time, 875
Measure of dispersion, 1224
Mendeleev, Dmitri, 1397
Mesons, 1448–1449
Metals
conductivity of, 872
current density of, 872
resistivity of, 872
Metastable state, 1406
Meteor impacts, 1086–1087
Meteor showers, 1087
Meteor, size of, 1106
Michelson, Albert Abraham, 1155, 1174–1178
Michelson-Morley experiment, 1155, 1174–1178, 1268, 1276, 1308
Michelson's interferometer, 1174–1178
Microscopes
compound, 1252–1255
scanning electron, 1330–1332
scanning tunneling, 1357–1358
transmission electron, 577, 1330–1332

Microwaves, 1102
Millikan, Robert, 737, 1318–1319
Minima. See Dark fringes (minima)
Mirror(s)
   flat, 1202
   focal length of, 1198
   magnification of, 1192, 1200–1202
   plane, 1190–1195. See also Plane mirrors
   in reflecting telescopes, 1190, 1209
   sign conventions for, 1192, 1199
   spherical, 1195–1199. See also Spherical mirrors
Mirror equation, 1200
Moderators, in nuclear reactors, 1434–1435
Modern physics, 1268. See also Relativity
Momentum
   angular, units of, App-13
   conservation of, 1290–1292
   from electromagnetic waves, 1107–1112
   of free particle, 1362–1364
   of photon, 1320–1326
   relativistic, 1293–1296, 1321
Monopoles, magnetic, 938
Morley, Edward, 1155
Moseley, Henry, 1397
Moseley's experiment, 1397
Motional emf, 1020–1025
   in slide generator, 1020–1025
Motors
   AC, 1030
   DC, 968–969
Multimeters
   as ammeters, 897, 918
   DC, 897, 912–919
   galvanometers in, 917–918, 967–968
   as ohmmeters, 897, 919
   as voltmeters, 897, 919
Multiple-lens systems, 1252–1257
   compound microscope, 1252–1255
   refracting telescope, 1254–1257
Muons, 1284–1285
Myopia, corrective lenses for, 1245–1248

N

Near-point distance, 1245
Negative charge, 686–687
Neutral objects, 686–687
Neutrinos, 1424–1427
   electron, 1448
Neutrons, 1373, 1414. See also Atom(s); Nucleus
   in alpha particle, 1421
   in beta decay, 1425
   isotopes and, 1417
   origin of, 1417
   quarks and, 1309, 1416–1417, 1449
   radiation injury and, 1442
   strong nuclear force and, 1416–1417
   from supernova events, 1439–1440
Newton, Isaac, 1124, 1223, 1268, 1326–1327
Newton's laws of motion, 1086
   second law, 1293–1296
Newton's rings, 1162–1163
Noninertial reference frames, 1268
Nonohmic circuit elements, 882
Normalization condition, 1347

Nuclear decay. See Radioactive decay
Nuclear fission. See Fission reactions
Nuclear force
   strong, 146, 684, 1358, 1416–1417
   weak, 684, 1424–1428
Nuclear fusion. See Fusion reactions
Nuclear reactors
   breeder, 1438
   fission, 1428, 1434–1435
   fusion, 1437–1439
Nucleons, 1414. See also Neutrons; Proton(s)
Nucleus
   alpha particles in, 1419, 1421–1422
   daughter, 1421, 1426
      in fission reactions, 1433–1434
   discovery of, 1376, 1414
   doubly magic, 1419
   energy levels for, 1418. See also Energy level(s)
   models of, 1417–1420
      combined, 1419
      liquid drop, 1419
      shell, 1417–1419
   origin in Big Bang, 1417
   parent, 1421, 1426
      in fission reactions, 1433
   quarks in, 687, 1309, 1416–1417, 1449
   radioactive decay and, 1420–1424
   radius of, 1415–1416
   in Rutherford's atomic model, 1376, 1414
   strong force and, 146, 684, 1358, 1416–1417
   structure of, 1414–1416
Nuclides, 1414

O

Object distance
   reflection and, 1191–1192, 1199
   refraction and, 1228
Objective lens
   of microscope, 1252–1254, 1255
   of telescope, 1254–1256
Ohm, 876
Ohmic circuit elements, 882
Ohmmeters, 897, 919
Ohm's law, 881–882, 898
Oil drop experiment, 737–738
One-lens systems
   cameras, 1185–1187, 1248–1250
   magnifiers, 1250–1252
Optic nerve, 1244
Optical axis, 1196
Optical fibers, 1222–1223
Optics
   geometric, 1185–1187
   physical, 1185
Orbital magnetic moment, 949–951
Orbital magnetic quantum number, 1388, 1394
Orbital quantum number, 1388–1389, 1394
Orbs, 1136–1137
Order numbers, 1167
Orsted, Hans Christian, 939, 979
Oscillating circuits, 1057–1061
Oscillating electric fields, polarization and, 1112–1115
Oscillator, simple harmonic quantum, 1358–1362
Oxygen, origin of, 1439

P

Parallel velocity transformation, 1288–1289
Paraxial assumption, 1226–1227
Paraxial rays, 1211
Parent nucleus, 1421, 1426
    in fission reactions, 1433
Partial derivatives, 815
Particle(s). *See also* Schrödinger's equation
    alpha, 1419–1422
    beta, 1420, 1424
    light as, 1316–1320, 1326–1328. *See also* Wave-particle duality
    in standard model, 1447–1449
Paschen series, 1378, 1382, 1384
Path integral, 843, 848, 984
Pauli's exclusion principle, 1395–1397, 1418, 1448
Penetration distance, 1354–1355
Period, units of, App-13
Periodic table of elements, 1373, 1397, 1399–1400, App-14–App-15
Perl, Martin, 1424
Permeability of free space, 941
Permittivity constant, 761
Permittivity of free space, 761
Perpendicular velocity transformation, 1288, 1290
Pfund series, 1378, 1382, 1384
Phase constant, 1064
Phasor diagrams, 1063
Phillips, William Daniel, 1408
Photoelectric effect, 1316–1320
Photons, 696, 1086, 1290, 1316–1326
    Compton effect and, 1320–1326
    gamma rays as, 1420–1421
    radiation injury and, 1442
    X-rays as, 1420
Physical optics, 1185
Physics, quantum. *See* Quantum physics
Pickups, guitar, 1049
Pinhole camera (camera obscura), 1185–1187, 1248–1249
Pixels, 1320, 1331
Planar symmetry, 771–774, 813
Planck, Max, 1309, 1312–1313
Planck's constant, 1381
Planck's quantum theory, 1313–1314, 1341, 1359–1360
Plane mirrors, 1190–1195
    depth reversal in, 1192–1193
    image distance and, 1191–1192, 1199
    image height and, 1192, 1199
    images formed by, 1191–1192. *See also* Image(s)
    left-right reversal in, 1192–1193
    magnification of, 1192
    object distance and, 1191–1192, 1199
    ray diagrams for, 1191–1192
    sign conventions for, 1199
    true, 1194
Plane of incidence
    reflection and, 1188
    refraction and, 1219
Plane parallel waves, 1125
Plane polarization, 1112
Plasma, 1438
Plates
    in capacitors, 829–830
    in Leyden jars, 829
Plum pudding model, 1373–1374
Poisson, Siméon-Denis, 1126, 1147–1148
Poisson's spot, 1147–1148
Polarization, 693, 1112–1115
    linear, 1095–1098, 1112
    plane, 1112
Polarizers, 1113–1114
Poles, magnetic, 936–937
Population inversion, 1406
Position
    in Galilean relativity, 1272
    Heisenberg's uncertainty principle and, 1362–1364
    Lorentz transformations for, 1277–1279
    Schrödinger's equation and, 1349–1357
Positive charge, 686–687
Positrons, 1425
Postulates of special relativity, 1275
Potential difference. *See* Electric potential difference
Potential energy
    electric. *See* Electric potential energy
    gravitational. *See* Gravitational potential energy
    of hydrogen atom, 1388
    magnetic, 966–967
    in quantum physics, 1340
Potential energy barrier, 1356
Potential energy well, 1349
Power
    from batteries, 885–887
    of corrective lens, 1246
    defined, 885
    light-gathering, of telescope, 1255–1256
    units of, 886, 887
    used by circuit elements, 887
Power factor, 1075–1076
Power flux, 1103
Power supply. *See* Emf devices
Power transmission, 1033–1037
Powers of 10, prefixes for, App-7
Poynting vector, 1103–1104
Pp chain, 1437
Presbyopia, corrective lenses for, 1245–1248
Pressure, radiation, 1108–1112
Primary rays, 1197–1198, 1238
    for lenses, 1238
    for mirrors, 1197–1198
Primed reference frame, 1271
Primordial black holes, 1365
Primordial hydrogen, 1417, 1428
Principal axis, of spherical mirror, 1196
Principal quantum number, 1379, 1388, 1394
Principle of equivalence, 1297–1302
    bending of light and, 1300
    black holes and, 1301–1302
    curvature of space and, 1298, 1300
    gravitational Doppler shift and, 1298–1300
    gravitational lenses and, 1301
    gravitational time dilation and, 1298–1299
    gravity as illusion and, 1297–1298
Prisms, 1223–1224
Probability density, 1343–1344, 1346–1347, 1363. *See also* Schrödinger's equation
    of free particle, 1363
    radial, 1389–1392
    volume, 1390

Probability waves, 1343–1344, 1346–1347. *See also* Schrödinger's equation
Propagation of uncertainty, App-5
Proper length, 1279–1280
Proper time, 1281–1285
Proton(s), 1373, 1414. *See also* Atom(s); Nucleus
   in alpha particle, 1421
   barrier tunneling by, 1356–1359. *See also* Barrier tunneling
   in beta decay, 1425
   charge of, 686. *See also* Electric charge
   isotopes and, 1417
   origin of, 1417
   quarks and, 687, 1309, 1416–1417, 1449
   strong nuclear force and, 1416–1417
Proton-proton chain, 1437
Pseudo-units, 1430
Pupil (eye), 1243–1244

Q

Quality factor, 1443
Quanta, 1309, 1312
Quantized charge, 687
Quantized energy, 1309, 1312, 1388–1392
Quantum chromodynamics (QCD), 1449
Quantum electrodynamics (QED), 1447
Quantum number, 1341, 1394–1395
   magnetic, 1388, 1392–1394
   orbital, 1388–1389, 1394
   orbital magnetic, 1388, 1394
   Pauli's exclusion principle and, 1395–1396
   principal, 1379, 1388, 1394
   spin, 1394
   spin magnetic, 1394
Quantum physics
   atoms and, 1372–1409. *See also* Atom(s)
   black-body radiation and, 1309–1313
   Compton effect and, 1320–1326, 1327
   energy levels in systems and, 1342
   fundamental principle of, 1309
   new era of, 1339
   origins and development of, 1308–1333
   photoelectric effect and, 1316–1320
   practical applications of, 1405–1408
      cesium clock, 1405
      fluorescent bulbs, 1405–1406
      lasers, 1374, 1406–1408
   probability density and, 1343
   probability waves and, 1343–1344
   quasi-quantum approach and, 1340–1342
   Schrödinger's equation and, 1338–1367
   theoretical basis of, 1309
   ultraviolet catastrophe and, 1312–1313
   vs. classical physics, 1341–1342, 1362
   wave-particle duality and, 1326–1328
Quantum simple harmonic oscillator, 1358–1362
Quantum stationary state, 1341
Quantum tunneling. *See* Barrier tunneling
Quarks, 687, 1309, 1416–1417, 1448–1449
Quasars, 1287–1288

R

Rad (radiation *a*bsorbed *d*ose), 1443
Radar, 1102
Radial probability density, 1389–1392
Radiation. *See also* Ray(s)
   absorbed dose of, 1443
   black-body, 1309–1313
      ultraviolet catastrophe and, 1312–1313
   electromagnetic. *See* Electromagnetic wave(s)
   equivalent dose of, 1443
   gamma ray, 1102
   Hawking, 1365
   human exposure to, 1441–1445
      limits on, 1443–1444
      tissue damage from, 1442
   infrared, 1102
   ultraviolet, 1102
   units of, 1443
   X-ray, 1102
Radiation dosimetry, 1443–1444
Radiation pressure, 1108–1112
Radiation weighting factor, 1443
Radio transmitters, 1094–1095, 1101
Radio waves, 1101
   from quasars, 1287–1288
Radioactive dating, 1422–1423
Radioactive decay, 1414, 1420–1424
   activity and, 1421–1422
   alpha, 1421
   beta, 1424–1428
   in carbon dating, 1422–1423
   half-life and, 1422
   rate of, 1421–1422
   time constant for, 1422
Radioactivity, discovery of, 1420
Radius, nuclear, 1415–1416
Radius of curvature
   reflection and, 1196, 1199
   refraction and, 1228, 1235
Radon, 1442
Rainbows, 1154, 1224
Ray(s). *See also* Radiation
   alpha, 1375–1376, 1420–1421, 1442
   beta, 1375, 1420–1421, 1442
   cathode, 1373
   convergent, 1191, 1197
   divergent, 1191, 1197
   emitted, 1191
   gamma, 1420–1421, 1442
   incident, 1219, 1222
   paraxial, 1211
   primary, 1197–1198, 1238
   reflected, 1191, 1219, 1222
   refracted, 1219
   X, 1102
      Compton effect and, 1320–1327
      discovery of, 1420
Ray diagrams, 1185–1186, 1250–1251
   for compound microscope, 1252
   elements of, 1258
   for plane mirrors, 1191–1192
   primary rays in, 1197–1198, 1238
   for spherical mirrors, 1195–1198
   for thin lenses, 1237–1238
Ray model, 1185–1187
Rayleigh, Lord, 1312
Rayleigh-Jeans's law, 1312
Rayleigh's criterion, 1156
RC circuits, 919–925

capacitor in
    charging in, 919–922
    discharging in, 919, 922–923
    time constant for, 922
RC filters, 1067–1068
Real DC emf devices, 899
Real emf rule, 899
Real focal point
    of lens, 1236
    of mirror, 1197
Real image
    reflection and, 1191, 1199
    refraction and, 1228
    in cameras, 1248–1249
    in eye, 1244
Receivers, radio, 1094–1095
Red shift, 1299
Reference frames
    accelerating, 1273, 1275, 1297
    in Galilean relativity, 1271–1273
    inertial, 1268–1271. *See also* Inertial reference frames
    laboratory, 1271–1272
    noninertial, 1268
    primed, 1271
    transformation equations for, 1272
Reflected rays, 1191, 1219, 1222
Reflecting telescopes, 1190, 1209–1212
Reflection, 1108, 1184
    angle of, 1188
    diffuse, 1190
    law of, 1187–1190
    phase changes and, 1159–1160
    by plane mirrors, 1190–1195
    specular, 1190–1191
    by spherical mirrors, 1195–1199
    total internal, 1221–1223
Refracted rays, 1219
Refracting surfaces
    concave, 1228
    convex, 1228
Refracting telescopes, 1254–1257
Refraction, 1218–1266
    defined, 1219
    image formation by, 1226–1257
    image position and, 1227–1228
    index of, 1159, 1219, 1223
        color and, 1223
        in eye, 1244
    law of, 1219–1221
    paraxial assumption and, 1226
    sign conventions for, 1228, 1233, 1235–1236
    Snell's law and, 1219
    spherical, 1226–1232. *See also* Lens(es)
        sign conventions for, 1228
    total internal reflection and, 1221–1223
Reines, Frederick, 1424
Relativistic Doppler formula, 1285
Relativistic energy, momentum and, 1321
Relativistic speed, 1277
Relativity, 1267–1307
    Galilean, 1271–1275
        postulates of, 1275
    general, 1275, 1297–1300
        bending of light and, 1300–1302
        confirmation of, 1301
        curvature of space and, 1297–1298
        eclipses and, 1301
        first postulate of, 1297
        gravitational Doppler shift and, 1298–1300
        gravitational lenses and, 1301
        second postulate of, 1297. *See also* Principle of equivalence
    inertial reference frames and, 1268–1271. *See also* Reference frames
    special
        correspondence principle and, 1277
        Doppler shift and, 1285–1288
        kinetic energy and, 1293–1296
        length contraction and, 1279–1280
        Lorentz transformations and, 1276–1279
        mass transformation and, 1290–1292
        mass-energy equivalence and, 1294–1295
        mass/momentum transformation and, 1290–1292
        momentum and, 1293
        Newton's second law of motion and, 1293–1296
        postulates of, 1275–1276
        relativity of simultaneity in, 1278
        in standard model, 1446
        time dilation and, 1281–1285
        velocity transformation and, 1288–1289
Rem (*r*oentgen *e*quivalent for *m*an), 1443
Resistance, 878–881, 1075
    defined, 879
    gravitational analogy for, 879
    measurement of, 897, 919
    Ohm's law and, 881–882
    units of, 879
    vs. resistivity, 879
Resistance combination rules, 912
Resistivity
    of metals, 872
    Ohm's law and, 882
    temperature dependence of, 877–878
    vs. resistance, 879
Resistor(s), 879–880
    in AC circuits, 1052–1055, 1062–1064, 1072–1076
    in DC circuits, 919–925
    in parallel, 909–911
    in phase, 1062
    in *RC* circuits, 919–925
    in *RL* circuits, 1052–1055
    in *RLC* circuits, 1072–1076
    in series, 906–908
    shunt, 918
Resistor rule, 898
Resistor-capacitor *(RC)* circuits, 919–925
Resistor-inductor *(RL)* circuits, 1052–1055
Resistor-inductor-capacitor *(RLC)* circuits. *See* RLC circuits
Resolution, 1156–1158
    Rayleigh's criterion and, 1156
Resolving power, of diffraction grating, 1170–1172
Resonance frequency, 1075
Rest mass, 1292, 1429–1430
Rest mass energy, 1295
Retina, 1244
Right-hand rule
    for induced current/emf, 1016–1018, 1089
    for magnetic force direction, 953
    for magnetic moment direction, 948–949, 1017
    simple, for magnetic field direction, 939–940, 942

RL circuits, 1052–1055
RL filters, 1070–1071
RLC circuits, 1072–1076
    current and voltage in, 1072–1074
    impedance in, 1074–1075
    resonance in, 1075
Rohrer, Heinrich, 1357
Röntgen, Wilhelm, 1102, 1420, 1443
Root-mean-square emf, 1033–1034
Rotation test, for symmetry, 756
Rotors, 968–969
R-process, 1440
Rutherford, Ernest, 1375, 1450
Rutherford's solar-system model, 1373, 1375–1379, 1389–1390, 1414
Rydberg constant, 1378
Rydberg formula, 1378, 1380, 1387

S

Salam, Abdus, 1449
Sandage, Allan, 1287
Savart, Felix, 940
Scalar fields, 714
Scalars, vs. vectors, 788
Scanning electron microscope, 1330–1332
Scanning tunneling microscope, 1357–1358
Schematic diagrams, 835
Schrödinger, Erwin, 1339
Schrödinger's equation, 1338–1367
    barrier tunneling and, 1355–1359, 1364
    Bohr's correspondence principle and, 1347
    boundary condition for, 1349, 1353, 1388
    classically forbidden region and, 1354
    complex conjugate and, 1347
    de Broglie wavelength and, 1340, 1346–1347
    expectation value for, 1347
    for free particle, 1362–1364
    Heisenberg's uncertainty principle, 1362–1365
    for hydrogen atom, 1388–1392
    normalization condition for, 1347
    one-dimensional, 1340
    for particle in finite square well, 1352–1355
    for particle in infinite square well, 1349–1354
    penetration distance and, 1354–1355
    plausibility of, 1346–1347
    position and, 1349–1357
    probability waves and, 1343–1345
    quantum simple harmonic oscillator and, 1358–1362
    radial solution to, 1389–1390
    simple harmonic oscillator and, 1360
    solutions to, 1347
    time-dependent version of, 1339, 1346
    time-independent version of, 1346–1348, 1362, 1389
    for trapped particle in long narrow box, 1340–1342
Schwarzschild radius, 1301–1302
Schwinger, Julian, 1447
Sclera, 1243
Secondary maxima, 1132
Segrè chart, 1415, 1426–1427
Self-inductance, 1047. *See also* Inductance
Semiconductors, 872, 877–878. *See also* Conductors
Shell(s)
    electron, 1388–1389, 1395
        Pauli's exclusion principle and, 1395–1397, 1418
    Zeeman effect and, 1401–1404
    nuclear, 1417–1419
Shell theorem, 769
Shielding, electron, 1397–1399
Shunt resistors, 918
Side lobes, 1132
Sievert, 1443
Sign conventions
    for mirrors, 1199
    for spherical refracting surfaces, 1228
    for thin lenses, 1233
Silicon, electron configuration of, 1396, 1399–1400
Simple harmonic oscillator, quantum, 1358–1362
Simple magnifier, 1248, 1250–1252
Simple right-hand rule. *See* Right-hand rule
Single-lens systems
    cameras, 1185–1187, 1248–1250
    magnifiers, 1248, 1250–1252
Single-slit diffraction, 1132–1137
    bright fringes in
        intensity of, 1140–1143
        position of, 1132–1137
    dark fringes in
        intensity of, 1140–1144
        position of, 1132–1137
    intensity and, 1140–1144
Slide generators, 1020–1025
Snell's law, 1219
Solar fusion, 1295–1296
Solar sails, 1111–1112
Solar-system model, 1373, 1375–1379, 1389–1390, 1414
Solenoids, 992–996
    ideal, 1046–1048
    as inductors, 1046–1049
    magnetic field of, 992–996, 1011–1013, 1046–1048
    magnetic flux in, 1011–1012
    in transformers, 1035–1036, 1051–1052
Sound
    Doppler shift and, 1285
    speed of, 1155
Sound waves. *See also* Wave(s)
    interference of, 1126–1127
    vs. light waves, 1127
Space, curvature of, 1297–1298
    bending of light and, 1300–1301
Sparks, 693–694
    lightning as, 715, 733
Spatial resolution, 1156
Speakers, stereo, 960–962, 1073–1074
Special relativity. *See* Relativity, special
Spectral intensity, 1311–1313
    Rayleigh-Jeans's law for, 1312
    ultraviolet catastrophe and, 1312–1313
Spectral lines, 1377–1378, 1382, 1384
    Zeeman effect and, 1401
Spectrum. *See* Electromagnetic spectrum; Hydrogen spectrum
Specular reflection, 1190–1191
Speed
    drift. *See* Drift velocity
    of Earth, Michelson-Morley experiment and, 1155, 1176–1178, 1268, 1276
    gamma factor and, 1278
    relativistic, 1277
    units of, App-12
Speed of light, 1098–1099, 1101

frequency and, 1159
index of refraction and, 1159, 1219
measurement of, 1155
reference frames and, 1276
thin-film interference and, 1158–1159
as ultimate speed limit, 1292
wavelength and, 1159
Speed of sound, 1155
Spherical aberration, 1210–1212
of camera lens, 1210–1212
of magnifier, 1252
Spherical mirrors, 1195–1199
center (vertex) of, 1196, 1198
center of curvature of, 1196
concave, 1196, 1204–1208
convex, 1196, 1199–1204
vs. concave mirrors, 1205–1207
flat mirrors as, 1202
focal length of, 1198
focal point for, 1196–1197
magnification of, 1200–1202
mirror equation for, 1200
optical (principal/central) axis of, 1196
primary rays for, 1197–1198
radius of curvature of, 1196
ray diagrams for, 1195–1198
in reflecting telescopes, 1190, 1209–1212
sign conventions for, 1199
spherical aberration and, 1210–1212
Spherical source of electric field, 716–718
electric potential of, 791–792
symmetry of, 767–771, 798–799
Spherical waves, 1125–1126
Spin, electron
magnetic moment due to, 949–951
in quantum physics, 1394
Spin angular momentum, 1394
Spin magnetic quantum number, 1394
Spin quantum number, 1394
Split ring commutator, 1031
S-process, 1440
Standard model, 1446–1450
electricity in, 1446
forces in, 1449
magnetism in, 1446
particles in, 1447–1449
quantum electrodynamics in, 1447
special relativity in, 1446
Standing wave, Schrödinger's equation and, 1346
Standing-wave atomic model, 1386–1387
Stars. *See also* Sun; Universe
apparent motion of, 1268–1270
collapse of, 1301–1302
fusion reactions in, 1439–1440
radiation pressure on, 1109–1111
State, definition of, 1341
Stellar aberration, 1268–1270
Step-up/step-down transformers, 1035–1037
Stereo speakers, 960–962, 1073–1074
Stern, Otto, 1393
Stern-Gerlach experiment, 1393–1394
Stimulated emission, 1406
Streamlines, electric flux and, 757–758
Strong nuclear force, 146, 684, 1358, 1416–1417
Subjects, of field force, 715

Subshells, electron, 1388–1389, 1395
Hund's rule for, 1396
Pauli's exclusion principle and, 1395–1397
Zeeman effect and, 1401–1404
Sun
fusion in, 1295–1296, 1358–1359, 1364, 1428, 1437–1438
barrier tunneling in, 1340, 1358–1359, 1364, 1438
hydrogen spectrum of, 1383
Zeeman effect in, 1403–1404
Superconductors, 878
Supernova reactions, 1439–1440
Superposition, of waves, 1127
Surface charge density, 726
Switch, induced emf and, 1046
Switch rule, 898–899
Symmetry
exploiting, 756–757
of Gaussian surfaces, 756–757
linear, 765–767
planar, 771–774, 813
rotation test for, 756
spherical, 767–771, 798–799
of wave functions, 1391
Synapses, electric potential and, 797–798

T

Telescopes
charge-coupled devices for, 1320
reflecting, 1190, 1209–1212
refracting, 1254–1257
Temperature
conductivity and, 877–878
of Universe, 1315
Temperature coefficient of resistivity, 877
Terminal, battery, 834
Terminal potential (voltage), 834–835
Terminology, electrical, 896
Tesla, 941
Tesla, Nikola, 1012, 1033, 1037, 1041, 1043
Theory of everything, 684
Thermal energy, from batteries, 884–885
Thick lenses, 1231
Thin lenses, 1232–1238. *See also* Lens(es)
sign conventions for, 1228, 1233, 1235–1236
thin-lens equation for, 1233
Thin-film interference, 1158–1165
constructive vs. destructive, 1160–1162
Newton's rings and, 1162–1163
speed of light and, 1158–1159
Thompson coils, 829, 832, 841
Thomson, George Paget, 1329
Thomson, Joseph John, 1329, 1373, 1375
Thomson's plum pudding model, 1373–1374
Time
Lorentz transformations for, 1278
mean free, 875
proper, 1281–1285
units of, App-13
Time constant, 922
for radioactive decay, 1422
Time dilation, 1281–1285
gravitational, 1298–1300

Tomonaga, Sin-Itiro, 1447
Toroids, 996–998
Torque
   on current loop in magnetic field, 964–969
   on electric field of dipole, 742
Total internal reflection, 1221–1223
Transformation equations, 1272
   Galilean, 1277
   Lorentz, 1276–1279
Transformers, 1035–1037, 1051–1052
Transitions
   allowed, 1402
   forbidden, 1402
Transmission axis, of polarizer, 1113–1114
Transmission electron microscope, 577, 1330–1332
Transmitters, radio, 1094–1095, 1101
Transverse electromagnetic waves, 1099
True mirrors, 1194
Tungsten, energy levels in, 1385
Tweeters, 1071–1072

U

Ultraviolet catastrophe, 1312–1313, 1379
Ultraviolet (UV) radiation, 1102
Uncertainty, propagation of, App-5
Uncertainty principle, 1362–1365
Units of measure, App-11
   conversion factors for, App-9
   pseudo-units, 1430
   SI base units, App-8
   symbols and abbreviations for, App-8
Universe
   history of, 1415, 1417, 1427–1428, 1439
   temperature of, 1315
Unpolarized waves, 1112
Uranium
   beta decay in, 1426–1427
   in fission reactions
      in bombs, 1435
      in nuclear reactors, 1434–1436

V

Van Allen belts, 954
Vector(s)
   area, electric flux and, 758–759
   Poynting, 1103–1104
   vs. scalars, 788
Vector fields, 714–715
Velocity
   drift, 867, 871–874
      conductivity and, 876
      Hall effect and, 958–959
   in Galilean relativity, 1272–1273
Velocity selectors, 955–957
Velocity transformation
   parallel, 1288–1289
   perpendicular, 1288, 1290
Venus, transit of, 1186–1187
Vertex, of mirror, 1196, 1198
Villard, Paul, 1420
Virtual focal point
   of lens, 1236
   of mirror, 1197

Virtual images
   reflection and, 1191, 1199
   refraction and, 1228
   in eye, 1244
Visible light, 1101–1102
Visible light spectrum, 1166–1170, 1223–1226. *See also* Electromagnetic spectrum
Vision, 1242–1245
Vitreous humor, 1244
Voice coil, 962
Volt, 797, 833
Volta, Allessandro, 833
Voltage, 896. *See also* Electric potential
   Hall, 958
   input, 896
   terminal, 834–835
Voltage rules, 897–899, 912
Voltmeters, 897, 919
Volume, units of, App-11
Volume charge density, 726

W

Watt, 886–887
Wave(s)
   circular, 1125
   coherent, 1128
   electromagnetic. *See* Electromagnetic wave(s)
   harmonic, 1127
   Huygens's principle, 1126, 1132
   light. *See* Light waves
   matter, 1328–1332
      Schrödinger's equation for, 1338–1367
   plane parallel, 1125
   probability, 1343–1344, 1346–1347. *See also* Schrödinger's equation
   sound, 1126–1127
   spherical, 1125–1126
   superposition of, 1127
Wave equation, 1095–1099
Wave fronts, 1125
Wave functions
   interpreting, 1350–1351
   normalizing, 1350
   of probability waves, 1343–1345
   symmetry of, 1391
Wave model, 1124–1126. *See also* Electromagnetic wave(s); Light waves
   coherence and, 1128
   diffraction and, 1124, 1166–1174. *See also* Diffraction; Diffraction gratings
   circular aperture, 1155–1158
   Hertz's experiments and, 1094
   interference and, 1125. *See also* Interference
   Maxwell's equations and, 978–979, 1085, 1092–1093. *See also* Maxwell's equations
   Michelson-Morley experiment and, 1155, 1174–1178, 1268, 1276
   physical optics and, 1185
   Poisson's spot and, 1147–1148
   single-slit diffraction and, 1132–1137, 1140–1144. *See also* Single-slit diffraction
   thin-film interference and, 1158–1165
   vs. particle model, 1316–1320, 1326–1327
   Young's experiment and, 1127–1131, 1137–1140

Wavelength
   Compton, 1322–1323
   de Broglie, 1328–1332, 1340
Wavelets, 1126
Wave-particle duality
   de Broglie wavelength and, 1328–1332
   for light, 1326–1328
   for matter, 1328–1332
   probability waves and, 1343–1345
Weak nuclear force, 684, 1424–1428
   neutrinos and, 1424–1426
Weber, 981
Weber, Wilhelm, 981
Weight, units of, App-12
Weinberg, Steven, 1449
Westinghouse, George, 1012, 1035, 1037
Wet cells, 834, 884. *See also* Batteries
Wheeler, John, 1311
Wien, Wilhelm, 1311
Wien's formula, 1311–1312
Wilson, Benjamin, 734, 736–737
Wilson, Charles, 1320

Wire rule, 898
Woofers, 1071–1072
Work, electric potential energy and, 794–795
Work function, photoelectric effect and, 1318

X

X-ray diffraction, 1173–1174
X-rays, 1102. *See also* Radiation
   Compton effect and, 1320–1327
   discovery of, 1420

Y

Young, Thomas, 1123–1124
   double-slit experiment of, 1127–1130, 1144, 1406, 1408.
      See also Double-slit interference

Z

Zeeman, Pieter, 1401–1402
Zeeman effect, 1401–1404
Zero-point energy, 1342

Some Astronomical Data

Object	Symbol	Rotation period (hh:mm:ss.s or days)	Mass (× 10^{24} kg)	Equatorial radius (× 10^6 m)	Free-fall acceleration near surface (m/s²)	Escape speed (km/s)	Blackbody temperature (K)
Sun	☉	≈ 25 to 36 days[1]	1.9891 × 10^6	695.51	274	618	5777
Mercury	☿	58.65 days	0.3302	2.4397	3.7	4.3	440.1
Venus	♀	243 days	4.87	6.052	8.9	10.36	184.2
Earth	⊕	23:56:4.1	5.9736	6.378136	9.81	11.186	254.3
Moon	☾	27.3 days	0.07	1.738	1.6	2.38	270.7
Mars	♂	24:37:22.6	0.64	3.397	3.7	5.03	210.1
Ceres	⚳	09:04:19	9.6 × 10^{-4}	0.48			239
Jupiter	♃	9:50:30	1900	71.493	24.8	59.5	110.0
Saturn	♄	10:14:00	569	60.268	10.4	35.5	81.1
Uranus	♅	17:14:00	87	25.559	8.87	21.3	58.2
Neptune	♆	16:03:00	103	24.764	11.2	23.5	46.6
Pluto	♇	6.387 days	0.01	1.135	0.58	1.2	37.5
Eris			≈ 10^{-2}	1.2			30

[1]The Sun is gaseous and does not rotate as a solid body; its period near the equator is shorter than at the poles.

Orbital parameters for objects that orbit the Sun

Object	Orbital period (days or years)	Semimajor axis (AU)	Eccentricity
Mercury	87.969 days	0.387	0.2056
Venus	224.701 days	0.723	0.0067
Earth	365.26 days	1.000	0.0167
Mars	1.8808 years	1.524	0.0935
Ceres	4.603 years	2.767	0.097
Jupiter	11.8618 years	5.204	0.0489
Saturn	29.4567 years	9.5482	0.0565
Uranus	84.0107 years	19.201	0.0457
Neptune	164.79 years	30.047	0.0113
Pluto	247.68 years	39.482	0.2488
Eris	559 years	67.89	0.4378

PEDAGOGICAL COLOR CHART

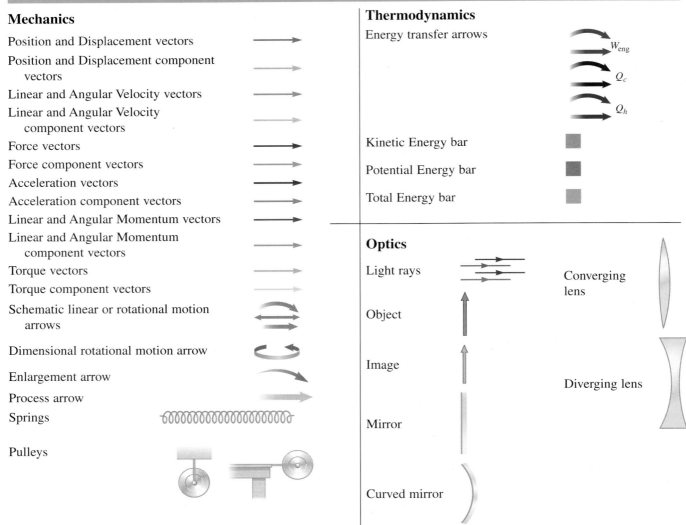

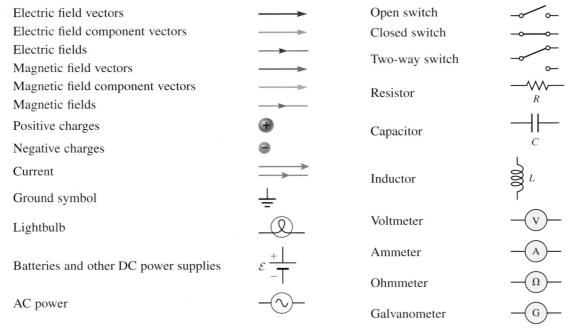

Symbols and abbreviations for units

Unit	Symbol	Unit	Symbol
ampere	A	light-year	ly
atmosphere	atm	liter	L
atomic mass unit	U	meter	m
British thermal unit	Btu	mile	mi
calorie	cal	miles per hour	mph
coulomb	C	millimeter of mercury (torricelli)	mm Hg (torr)
day	d	minute	min
degree Celsius	°C	mole	mol
degree Fahrenheit	°F	newton	N
electron volt	eV	ohm	Ω
farad	F	pascal	Pa
foot	ft	pound	lb
gallon	gal	pounds per square inch	psi
gauss	G	radian	rad
gram	g	revolution	rev
henry	H	revolutions per minute	rpm
hertz	Hz	second	s
horsepower	hp	tesla	T
inch	in.	volt	V
joule	J	watt	W
kelvin	K	weber	Wb
kilocalorie	Cal	yard	yd
kilogram	kg	year	yr
kilowatt-hour	kWh		

Greek alphabet

Name	Uppercase	Lowercase	Name	Uppercase	Lowercase	Name	Uppercase	Lowercase
Alpha	A	α	Iota	I	ι	Rho	P	ρ
Beta	B	β	Kappa	K	κ	Sigma	Σ	σ
Gamma	Γ	γ	Lambda	Λ	λ	Tau	T	τ
Delta	Δ	δ	Mu	M	μ	Upsilon	Y	υ
Epsilon	E	ε	Nu	N	ν	Phi	Φ	φ
Zeta	Z	ζ	Xi	Ξ	ξ	Chi	X	χ
Eta	H	η	Omicron	O	o	Psi	Ψ	ψ
Theta	Θ	θ	Pi	Π	π	Omega	Ω	ω

Conversion Factors

Length

	meter	cm	km	in.	ft	mi
1 meter	1	10^2	10^{-3}	39.37	3.281	6.214×10^{-4}
1 centimeter	10^{-2}	1	10^{-5}	0.3937	3.281×10^{-2}	6.214×10^{-6}
1 kilometer	10^3	10^5	1	3.937×10^4	3281	0.6214
1 inch	2.540×10^{-2}	2.540	2.540×10^{-5}	1	8.333×10^{-2}	1.578×10^{-5}
1 foot	0.3048	30.48	3.048×10^{-4}	12	1	1.894×10^{-4}
1 mile	1609	1.609×10^5	1.609	6.336×10^4	5280	1
1 angstrom	10^{-10}	10^{-8}	10^{-13}	3.937×10^{-9}	3.281×10^{-10}	6.214×10^{-14}
1 AU	1.496×10^{11}	1.496×10^{13}	1.496×10^8	5.890×10^{12}	4.908×10^{11}	9.296×10^7
1 nautical mile	1852	1.852×10^5	1.852	7.291×10^4	6076	1.151
1 light-year	9.461×10^{15}	9.461×10^{17}	9.461×10^{12}	3.725×10^{17}	3.104×10^{16}	5.878×10^{12}
1 parsec	3.086×10^{16}	3.086×10^{18}	3.086×10^{13}	1.215×10^{18}	1.012×10^{17}	1.917×10^{13}
1 yard	0.9144	91.44	9.144×10^{-4}	36	3	5.682×10^{-4}

Mass

	kilogram	g	slug	u
1 kilogram	1	10^3	6.852×10^{-2}	6.022×10^{26}
1 gram	10^{-3}	1	6.852×10^{-5}	6.022×10^{23}
1 slug	14.59	1.459×10^4	1	8.786×10^{27}
1 atomic mass unit	$1.6605402 \times 10^{-27}$	1.661×10^{-24}	1.138×10^{-28}	1

Force

	newton	dyne	lb	oz	ton
1 newton	1	10^5	0.2248	3.597	1.124×10^{-4}
1 dyne	10^{-5}	1	2.248×10^{-6}	3.597×10^{-5}	1.124×10^{-9}
1 pound	4.448	4.448×10^5	1	16	5×10^{-4}
1 ounce	0.2780	2.780×10^4	6.250×10^{-2}	1	3.125×10^{-5}
1 ton	8.896×10^3	8.896×10^8	2000	3.2×10^4	1

Pressure

	pascal	atm	Torr (mm Hg)	psi	dyne/cm²
1 pascal	1	9.869×10^{-6}	7.501×10^{-3}	1.450×10^{-4}	10
1 atm	1.013×10^5	1	760	14.70	1.013×10^6
1 Torr	1333	1.316×10^{-2}	1	0.1934	1.333×10^4
1 psi	6.895×10^3	6.805×10^{-2}	51.71	1	6.895×10^4
1 dyne/cm²	0.1	9.869×10^{-7}	7.501×10^{-4}	1.405×10^{-5}	1

Energy

	joule	erg	ft·lb	cal	eV
1 joule	1	10^7	0.7376	0.2389	6.242×10^{18}
1 erg	10^{-7}	1	7.376×10^{-8}	2.389×10^{-8}	6.242×10^{11}
1 ft·lb	1.356	1.356×10^7	1	0.3238	8.464×10^{18}
1 cal	4.184	4.184×10^7	3.088	1	2.612×10^{19}
1 eV	1.602×10^{-19}	1.602×10^{-19}	1.182×10^{-19}	3.827×10^{-20}	1